Unit Conversion Factors

Length
1 m = 100 cm =1000 mm = 10^6 μm = 10^9 nm
1 km = 1000 m = 0.6214 mi
1 m = 3.281 ft = 39.37 in
1 cm = 0.3937 in
1 in = 2.54 cm (exactly)
1 ft = 30.48 cm (exactly)
1 yd = 91.44 cm (exactly)
1 mi = 5280 ft = 1.609344 km (exactly)
1 Angstrom = 10^{-10} m =10^{-8} cm = 0.1 nm
1 nautical mile = 6080 ft = 1.152 mi
1 light-year = $9.461 \cdot 10^{15}$ m

Area
1 m^2 =10^4 cm^2 =10.76 ft^2
1 cm^2 = 0.155 in^2
1 in^2 = 6.452 cm^2
1 ft^2 =144 in^2 = 0.0929 m^2
1 hectare = 2.471 acre = 10000 m^2
1 acre = 0.4047 hectare = 43560 ft^2
1 mi^2 = 640 acre
1 yd^2 = 0.8361 m^2

Volume
1 liter =1000 cm^3 =$10^{-3}m^3$ = 0.03531 ft^3 = 61.02 in^3 =
   33.81 fluid ounce
1 ft^3 = 0.02832 m^3 = 28.32 liter = 7.477 gallon
1 gallon = 3.788 liters
1 quart = 0.9463 liter

Time
1 min = 60 s
1 h = 3,600 s
1 day = 86,400 s
1 week = 604,800 s
1 year = $3.156 \cdot 10^7$ s

Angle
1 rad = 57.30° =180°/π
1° = 0.01745 rad = (π/180) rad
1 rev = 360° = 2π rad
1 rev/min (rpm) = 0.1047 rad/s = 6°/s

Speed
1 mile per hour (mph) = 0.4470 m/s = 1.466 ft/s =
   1.609 km/h
1 m/s = 2.237 mph = 3.281 ft/s
1 km/h = 0.2778 m/s = 0.6214 mph
1 ft/s = 0.3048 m/s
1 knot = 1.151 mph = 0.5144 m/s

Acceleration
1 m/s^2 =100 cm/s^2 = 3.2
1 cm/s^2 = 0.01 m/s^2 = 0.03281 ft/s^2
1 ft/s^2 = 0.3048 m/s^2 = 30.48 cm/s^2

Mass
1 kg =1000 g = 0.0685 slug
1 slug =14.95 kg
1 kg has a weight of 2.205 lb when g = 9.807 m/s^2
1 lb has a mass of 0.4546 kg when g = 9.807 m/s^2

Force
1 N = 0.2248 lb
1 lb = 4.448 N
1 stone = 14 lb = 62.27 N

Pressure
1 Pa =1 N/m^2 =$1.450 \cdot 10^{-4}$ lb/in^2 = 0.209 lb/ft^2
1 atm =$1.013 \cdot 10^5$ Pa =101.3 kPa =14.7 lb/in^2 = 2117 lb/ft^2 =
   760 mm Hg = 29.92 in Hg
1 lb/in^2 = 6895 Pa
1 lb/ft^2 = 47.88 Pa
1 mm Hg = 1 torr = 133.3 Pa
1 bar =10^5 Pa =100 kPa

Energy
1 J = 0.239 cal
1 cal = 4.186 J
1 Btu = 1055 J = 252 cal
1 kW·h = $3.600 \cdot 10^6$ J
1 ft·lb =1.356 J
1 eV =$1.602 \cdot 10^{-19}$ J

Power
1 W = 1 J s
1 hp = 746 W = 0.746 kW = 550 ft·lb/s
1 Btu/h = 0.293 W
1 GW =1000 MW =$1.0 \cdot 10^9$ W
1 kW = 1.34 hp

Temperature
Fahrenheit to Celsius: $T_C = \frac{5}{9}(T_F - 32\ °F)$
Celsius to Fahrenheit: $T_F = \frac{9}{5} T_C + 32\ °C$
Celsius to Kelvin: $T_K = T_C + 273.15\ °C$
Kelvin to Celsius: $T_C = T_K - 273.15\ K$

University Physics

with Modern Physics

Volume Two

University
Physics
with Modern Physics

Volume Two

Wolfgang Bauer
Michigan State University

Gary D. Westfall
Michigan State University

Connect
Learn
Succeed™

UNIVERSITY PHYSICS, VOLUME 2

Published by McGraw-Hill, a business unit of The McGraw-Hill Companies, Inc., 1221 Avenue of the Americas, New York, NY 10020.

Some ancillaries, including electronic and print components, may not be available to customers outside the United States.

This book is printed on acid-free paper.

1 2 3 4 5 6 7 8 9 0 WDQ/WDQ 1 0 9 8 7 6 5 4 3 2 1 0

ISBN 978–0–07–336796–5
MHID 0–07–336796–6

Vice President & Editor-in-Chief: *Marty Lange*
Vice President, EDP: *Kimberly Meriwether-David*
Vice-President New Product Launches: *Michael Lange*
Publisher: *Ryan Blankenship*
Sponsoring Editor: *Debra B. Hash*
Senior Developmental Editor: *Mary E. Hurley*
Senior Marketing Manager: *Lisa Nicks*
Senior Project Manager: *Jayne L. Klein*
Lead Production Supervisor: *Sandy Ludovissy*
Lead Media Project Manager: *Judi David*
Senior Designer: *David W. Hash*
(USE) Cover Image: *Swirls of light around sphere, ©Ryichi Okano/Amana Images/Getty Images, Inc.; Solar panels, ©Rob Atkins/ Photographer's Choice/Getty Images, Inc.; Bose-Einstein condensate, image courtesy NIST/JILA/CU-Boulder; Scaled chrysophyte, Mallomanas lychenensis, SEM, © Dr. Peter Siver/Visuals Unlimited/Getty Images, Inc.*
Lead Photo Research Coordinator: *Carrie K. Burger*
Photo Research: *Danny Meldung/Photo Affairs, Inc.*
Compositor: *Precision Graphics*
Typeface: *10/12 Minion Pro*
Printer: *World Color Press Inc.*

Library of Congress Cataloging-in-Publication Data

Bauer, W. (Wolfgang), 1959-
  University physics with modern physics / Wolfgang Bauer, Gary D. Westfall.—1st ed.
     p. cm.
  Includes index.
  ISBN 978–0–07–285736–8 — ISBN 0–07–285736–6 (hard copy : alk. paper) 1. Physics—Textbooks. I. Westfall, Gary D. II. Title.
  QC23.2.B38 2011
  530—dc22
                                                                          2009037841

www.mhhe.com

Brief Contents

About the Authors

Wolfgang Bauer was born in Germany and obtained his Ph.D. in theoretical nuclear physics from the University of Giessen in 1987. After a post-doctoral fellowship at the California Institute of Technology, he joined the faculty at Michigan State University in 1988. He has worked on a large variety of topics in computational physics, from high-temperature superconductivity to supernova explosions, but has been especially interested in relativistic nuclear collisions. He is probably best known for his work on phase transitions of nuclear matter in heavy ion collisions. In recent years, Dr. Bauer has focused much of his research and teaching on issues concerning energy, including fossil fuel resources, ways to use energy more efficiently, and, in particular, alternative and carbon-neutral energy resources. He presently serves as chairperson of the Department of Physics and Astronomy, as well as the Director of the Insitute for Cyber-Enabled Research.

Gary D. Westfall started his career at the Center for Nuclear Studies at the University of Texas at Austin, where he completed his Ph.D. in experimental nuclear physics in 1975. From there he went to Lawrence Berkeley National Laboratory (LBNL) in Berkeley, California, to conduct his post-doctoral work in high-energy nuclear physics and then stayed on as a staff scientist. While he was at LBNL, Dr. Westfall became internationally known for his work on the nuclear fireball model and the use of fragmentation to produce nuclei far from stability. In 1981, Dr. Westfall joined the National Superconducting Cyclotron Laboratory (NSCL) at Michigan State University (MSU) as a research professor; there he conceived, constructed, and ran the MSU 4π Detector. His research using the 4π Detector produced information concerning the response of nuclear matter as it is compressed in a supernova collapse. In 1987, Dr. Westfall joined the Department of Physics and Astronomy at MSU as an associate professor, while continuing to carry out his research at NSCL. In 1994, Dr. Westfall joined the STAR Collaboration, which is carrying out experiments at the Relativistic Heavy Ion Collider (RHIC) at Brookhaven National Laboratory on Long Island, New York.

The Westfall/Bauer Partnership

Drs. Bauer and Westfall have collaborated on nuclear physics research and on physics education research for more than two decades. The partnership started in 1988, when both authors were speaking at the same conference and decided to go downhill skiing together after the session. On this occasion, Westfall recruited Bauer to join the faculty at Michigan State University (in part by threatening to push him off the ski lift if he declined). They obtained NSF funding to develop novel teaching and laboratory techniques, authored multimedia physics CDs for their students at the Lyman Briggs School, and co-authored a textbook on CD-ROM, called *cliXX Physik*. In 1992, they became early adopters of the Internet for teaching and learning by developing the first version of their online homework system. In subsequent years, they were instrumental in creating the Learning*Online* Network with CAPA, which is now used at more than 70 universities and colleges in the United States and around the world. Since 2008, Bauer and Westfall have been part of a team of instructors, engineers, and physicists, who investigate the use of peer-assisted learning in the introductory physics curriculum. This project has received funding from the NSF STEM Talent Expansion Program, and its best practices have been incorporated into this textbook.

Dedication

This book is dedicated to our families. Without their patience, encouragement, and support, we could never have completed it.

A Note from the Authors

Physics is a thriving science, alive with intellectual challenge and presenting innumerable research problems on topics ranging from the largest galaxies to the smallest subatomic particles. Physicists have managed to bring understanding, order, consistency, and predictability to our universe and will continue that endeavor into the exciting future.

However, when we open most current introductory physics textbooks, we find that a different story is being told. Physics is painted as a completed science in which the major advances happened at the time of Newton, or perhaps early in the 20th century. Only toward the end of the standard textbooks is "modern" physics covered, and even that coverage often includes only discoveries made through the 1960s.

Our main motivation to write this book is to change this perception by appropriately weaving exciting, contemporary physics throughout the text. Physics is an exciting, dynamic discipline—continuously on the verge of new discoveries and life-changing applications. In order to help students see this, we need to tell the full, exciting story of our science by appropriately integrating contemporary physics into the first-year calculus-based course. Even the very first semester offers many opportunities to do this by weaving recent results from non-linear dynamics, chaos, complexity, and high-energy physics research into the introductory curriculum. Because we are actively carrying out research in these fields, we know that many of the cutting-edge results are accessible in their essence to the first-year student.

Authors in many other fields, such as biology and chemistry, already weave contemporary research into their textbooks, recognizing the substantial changes that are affecting the foundations of their disciplines. This integration of contemporary research gives students the impression that biology and chemistry are the "hottest" research enterprises around. The foundations of physics, on the other hand, are on much firmer ground, but the new advances are just as intriguing and exciting, if not more so. We need to find a way to share the advances in physics with our students.

We believe that talking about the broad topic of energy provides a great opening gambit to capture students' interest. Concepts of energy sources (fossil, renewable, nuclear, and so forth), energy efficiency, alternative energy sources, and environmental effects of energy supply choices (global warming) are very much accessible on the introductory physics level. We find that discussions of energy spark our students' interest like no other current topic, and we have addressed different aspects of energy throughout our book.

In addition to being exposed to the exciting world of physics, students benefit greatly from gaining the ability to **problem solve and think logically about a situation.** Physics is based on a core set of ideas that is fundamental to all of science. We acknowledge this and provide a useful problem-solving method (outlined in Chapter 1) which is used throughout the entire book. This problem-solving method involves a multi-step format that both of us have developed with students in our classes.

With all of this in mind along with the desire to write a captivating textbook, we have created what we hope will be a tool to engage students' imaginations and to better prepare them for future courses in their chosen fields (admittedly, hoping that we would convert at least a few students to physics majors along the way). Having feedback from more than 300 people, including a board of advisors, several contributors, manuscript reviewers, and focus group participants, assisted greatly in this enormous undertaking, as did field testing of our ideas with approximately 4000 students in our introductory physics classes at Michigan State University. We thank you all!

—*Wolfgang Bauer and Gary D. Westfall*

Contents

Preface

University Physics is intended for use in the calculus-based introductory physics sequence at universities and colleges. It can be used in either a two-semester introductory sequence or a three-semester sequence. The course is intended for students majoring in the biological sciences, the physical sciences, mathematics, and engineering.

Problem-Solving Skills: Learning to Think Like a Scientist

Perhaps one of the greatest skills students can take from their physics course is the ability to **problem solve and think critically about a situation.** Physics is based on a core set of fundamental ideas that can be applied to various situations and problems. *University Physics* by Bauer and Westfall acknowledges this and provides a problem-solving method class tested by the authors, and used throughout the entire text. The text's problem-solving method involves a multi-step format.

> "The Problem-Solving Guidelines help students improve their problem-solving skills, by teaching them how to break a word problem down to its key components. The key steps in writing correct equations are nicely described and are very helpful for students."
>
> —*Nina Abramzon, California Polytechnic University–Pomona*

> "I often get the discouraging complaint by students, 'I don't know where to start in solving problems.' I think your systematic approach, a clearly laid-out strategy, can only help."
>
> —*Stephane Coutu, The Pennsylvania State University*

Problem-Solving Method

Solved Problem

The book's numbered **Solved Problems** are fully worked problems, each consistently following the seven-step method described in Chapter 1. Each Solved Problem begins with the Problem statement and then provides a complete Solution:

1. **THINK:** Read the problem carefully. Ask what quantities are known, what quantities might be useful but are unknown, and what quantities are asked for in the solution. Write down these quantities, representing them with commonly used symbols. Convert into SI units, if necessary.

2. **SKETCH:** Make a sketch of the physical situation to help visualize the problem. For many learning styles, a visual or graphical representation is essential, and it is often necessary for defining variables.

3. **RESEARCH:** Write down the physical principles or laws that apply to the problem. Use equations that represent these principles and connect the known and unknown quantities to each other. At times, equations may have to be derived, by combining two or more known equations, to solve for the unknown.

SOLVED PROBLEM 6.6 | **Power Produced by Niagara Falls**

PROBLEM

Niagara Falls pours an average of 5520 m^3 of water over a drop of 49.0 m every second. If all the potential energy of that water could be converted to electrical energy, how much electrical power could Niagara Falls generate?

SOLUTION

THINK

The mass of one cubic meter of water is 1000 kg. The work done by the falling water is equal to the change in its gravitational potential energy. The average power is the work per unit time.

SKETCH

A sketch of a vertical coordinate axis is superimposed on a photo of Niagara Falls in Figure 6.22.

RESEARCH

The average power is given by the work per unit time:

$$\overline{P} = \frac{W}{t}.$$

The work that is done by the water going over Niagara Falls is equal to the change in gravitational potential energy,

$$\Delta U = W.$$

The change in gravitational potential energy of a given mass m of water falling a distance h is given by

$$\Delta U = mgh.$$

SIMPLIFY

We can combine the preceding three equations to obtain

$$\overline{P} = \frac{W}{t} = \frac{mgh}{t} = \left(\frac{m}{t}\right)gh.$$

Continued—

CALCULATE

We first calculate the mass of water moving over the falls per unit time from the given volume of water per unit time, using the density of water:

$$\frac{m}{t} = \left(5520 \,\frac{m^3}{s}\right)\left(\frac{1000 \text{ kg}}{m^3}\right) = 5.52 \cdot 10^6 \text{ kg/s}.$$

The average power is then

$$\overline{P} = \left(5.52 \cdot 10^6 \text{ kg/s}\right)\left(9.81 \text{ m/s}^2\right)\left(49.0 \text{ m}\right) = 2653.4088 \text{ MW}.$$

ROUND

We round to three significant figures:

$$\overline{P} = 2.65 \text{ GW}.$$

DOUBLE-CHECK

Our result is comparable to the output of large electrical power plants, on the order of 1000 MW (1GW). The combined power generation capability of all of the hydro-electric power stations at Niagara Falls has a peak of 4.4 GW during the high water season in the spring, which is close to our answer. However, you may ask how the water produces power by simply falling over Niagara Falls. The answer is that it doesn't. Instead, a large fraction of the water of the Niagara River is diverted upstream from the falls and sent through tunnels, where it drives power generators. The water that makes it across the falls during the daytime and in the summer tourist season is only about 50% of the flow of the Niagara River. This flow is reduced even further, down to 10%, and more water is diverted for power generation during the nighttime and in the winter.

4. **SIMPLIFY:** Simplify the result algebraically as much as possible. This step is particularly helpful when more than one quantity has to be found.

5. **CALCULATE:** Substitute numbers with units into the simplified equation and calculate. Typically, a number and a physical unit are obtained as the answer.

6. **ROUND:** Consider the number of significant figures that the result should contain. A result obtained by multiplying or dividing should be rounded to the same number of significant figures as the input quantity that had the least number of significant figures. Do not round in intermediate steps, as rounding too early might give a wrong solution. Include the proper units in the answer.

7. **DOUBLE-CHECK:** Consider the result. Does the answer (both the number and the units) seem realistic? Examine the orders of magnitude. Test your solution in limiting cases.

Examples

Briefer, terser **Examples** (Problem statement and Solution only) focus on a specific point or concept. The briefer Examples also serve as a bridge between fully worked-out Solved Problems (with all seven steps) and the homework problems.

EXAMPLE 17.4 | Rise in Sea Level Due to Thermal Expansion of Water

The rise in the level of the Earth's oceans is of current concern. Oceans cover $3.6 \cdot 10^8 \text{ km}^2$, slightly more than 70% of Earth's surface area. The average ocean depth is 3700 m. The surface ocean temperature varies widely, between 35 °C in the summer in the Persian Gulf and −2 °C in the Arctic and Antarctic regions. However, even if the ocean surface temperature exceeds 20 °C, the water temperature rapidly falls off as a function of depth and reaches 4 °C at a depth of 1000 m (Figure 17.22). The global average temperature of all seawater is approximately 3 °C. Table 17.3 lists a volume expansion coefficient of zero for water at a temperature of 4 °C. Thus, it is safe to assume that the volume of ocean water changes very little at a depth greater than 1000 m. For the top 1000 m of ocean water, let's assume a global average temperature of 10.0 °C and calculate the effect of thermal expansion.

FIGURE 17.22 Average ocean water temperature as a function of depth below the surface.

PROBLEM

By how much would sea level change, solely as a result of the thermal expansion of water, if the water temperature of all the oceans increased by $\Delta T = 1.0$ °C?

SOLUTION

The volume expansion coefficient of water at 10.0 °C is $\beta = 87.5 \cdot 10^{-6} \text{ °C}^{-1}$ (from Table 17.3), and the volume change of the oceans is given by equation 17.9, $\Delta V = \beta V \Delta T$, or

$$\frac{\Delta V}{V} = \beta \Delta T. \tag{i}$$

We can express the total surface area of the oceans as $A = (0.7)4\pi R^2$, where R is the radius of Earth and the factor 0.7 reflects the fact that about 70% of the surface of the sphere is

Problem-Solving Practice

Problem-Solving Practice provides **Additional Solved Problems,** again following the full seven-step format. This section is found immediately before the end-of-chapter problems to provide a review and to emphasize the fundamental concepts of the chapter. Additional **Problem-Solving Strategies and Guidelines** are also presented here.

"They provide a useful tool for students to improve their problem-solving skills. The authors did a good job in addressing, for each chapter, the most important steps to approach the solution of the end-of-chapter problems. Students that never had physics before will find this guideline quite beneficial. I liked in particular the connection between the guideline and the solved problem. The detailed description on how to solve these problems will certainly help the students to understand the concepts better."

—*Luca Bertello, University of California–Los Angeles*

PROBLEM-SOLVING PRACTICE

Problem-Solving Guidelines: Kinetic Energy, Work, and Power

1. In all problems involving energy, the first step is to clearly identify the system and the changes in its conditions. If an object undergoes a displacement, check that the displacement is always measured from the same point on the object, such as the front edge or the center of the object. If the speed of the object changes, identify the initial and final speeds at specific points. A diagram is often helpful to show the position and the speed of the object at two different times of interest.

2. Be careful to identify the force that is doing work. Also note whether forces doing work are constant forces or variable forces, because they need to be treated differently.

3. You can calculate the sum of the work done by individual forces acting on an object or the work done by the net force acting on an object; the result should be the same. (You can use this as a way to check your calculations.)

4. Remember that the direction of the restoring force exerted by a spring is always opposite the direction of the displacement of the spring from its equilibrium point.

5. The formula for power, $P = \vec{F} \cdot \vec{v}$, is very useful, but applies only for a constant force. When using the more general definition of power, $\overline{P} = \frac{W}{\Delta t}$, and the instantaneous value of the power, $P = \frac{dW}{dt}$.

SOLVED PROBLEM 5.2 | Lifting Bricks

PROBLEM

A load of bricks at a construction site has a mass of 85.0 kg. A crane raises this load from the ground to a height of 50.0 m in 60.0 s at a low constant speed. What is the average power of the crane?

SOLUTION

THINK

Raising the bricks at a low constant speed means that the kinetic energy is negligible, so the work in this situation is done against gravity only. There is no acceleration, and friction is negligible. The average power then is just the work done against gravity divided by the time it takes to raise the load of bricks to the stated height.

SKETCH

A free-body diagram of the load of bricks is shown in Figure 5.20. Here we have defined a coordinate system in which the y-axis is vertical and positive is upward. The tension, T, exerted by the cable of the crane is a force in the upward direction, and the weight, mg, of the load of bricks is a force downward. Because the load is moving at a constant speed, the sum of the tension and the weight is zero. The load is moved vertically a distance h, as shown in Figure 5.21.

RESEARCH

The work, W, done by the crane is given by

$$W = mgh.$$

The average power, $\overline{P}$, required to lift the load in the given time Δt is

$$\overline{P} = \frac{W}{\Delta t}.$$

SIMPLIFY

Combining the above two equations gives

$$\overline{P} = \frac{mgh}{\Delta t}.$$

CALCULATE

Now we put in the numbers and get

$$\overline{P} = \frac{\left(85.0 \text{ kg}\right)\left(9.81 \text{ m/s}^2\right)\left(50.0 \text{ m}\right)}{60.0 \text{ s}} = 694.875 \text{ W}.$$

FIGURE 5.20 Free-body diagram of the load of bricks of mass m being lifted by a crane.

FIGURE 5.21 The mass m is lifted a distance h.

Continued—

End-of-Chapter Questions and Problem Sets

Along with providing problem-solving guidelines, examples, and strategies, *University Physics* also offers a **wide variety of end-of-chapter Questions and Problems.** Often professors say, "I don't need a lot of problems, just a handful of really good problems." *University Physics* has both. The end-of-chapter Questions and Problems were developed with the idea of making them interesting to the reader. The authors, along with a panel of excellent writers (who, perhaps more importantly, are also experienced physics instructors) wrote Questions and Problems for each chapter, being sure to provide wide variety in level, content, and style. Included in each chapter are a set of Multiple-Choice Questions, Questions, Problems (by section), and Additional Problems (no section "clue"). One bullet identifies slightly more challenging Problems, and two bullets identify the most challenging Problems. The problem-solving theme from the text is also carried through to the Test Bank: The same group that wrote the end-of-chapter questions and problems also wrote the Test Bank questions, providing consistency in style and coverage.

MULTIPLE-CHOICE QUESTIONS

13.1 Salt water has a greater density than freshwater. A boat floats in both freshwater and salt water. The buoyant force on the boat in salt water is _______ that in freshwater.

a) equal to b) smaller than c) larger than

13.2 You fill a tall glass with ice and then add water to the level of the glass's rim, so some fraction of the ice floats above the rim. When the ice melts, what happens to the water level? (Neglect evaporation, and assume that the ice and water remain at 0 °C during the melting process.)

a) The water overflows the rim.
b) The water level drops below the rim.
c) The water level stays at the top of the rim.
d) It depends on the difference in density between water and ice.

13.3 The figure shows four identical open-top tanks filled to the brim with water and sitting on a scale. Balls float in tanks (2) and (3), but an object sinks to the bottom in tank (4). Which of the following correctly ranks the weights shown on the scales?

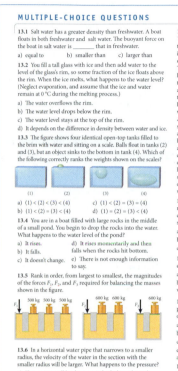

(1)
(2)
(3)
(4)

a) (1) < (2) < (3) < (4) c) (1) < (2) = (3) = (4)
b) (1) < (2) = (3) < (4) d) (1) = (2) = (3) < (4)

13.4 You are in a boat filled with large rocks in the middle of a small pond. You begin to drop the rocks into the water. What happens to the water level of the pond?

a) It rises.
b) It falls.
c) It doesn't change.
d) It rises momentarily and then falls when the rocks hit bottom.
e) There is not enough information to say.

13.5 Rank in order, from largest to smallest, the magnitudes of the forces F_1, F_2, and F_3 required for balancing the masses shown in the figure.

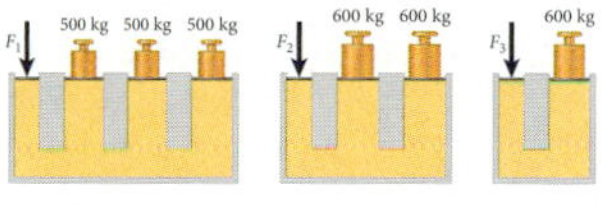

500 kg
500 kg
500 kg
600 kg
600 kg
600 kg
F_1
F_2
F_3

13.6 In a horizontal water pipe that narrows to a smaller radius, the velocity of the water in the section with the smaller radius will be larger. What happens to the pressure?

a) The pressure will be the same in both the wider and narrower sections of the pipe.
b) The pressure will be higher in the narrower section of the pipe.
c) The pressure will be higher in the wider section of the pipe.
d) It is impossible to tell.

13.7 In one of the *Star Wars*™© movies, four of the heroes are trapped in a trash compactor on the Death Star. The compactor's walls begin to close in, and the heroes need to pick an object from the trash to place between the closing walls to stop them. All the objects are the same length and have a circular cross section, but their diameters and compositions are different. Assume that each object is oriented horizontally and does not bend. They have the time and strength to hold up only one object between the walls. Which of the objects shown in the figure will work best—that is, will withstand the greatest force per unit of compression?

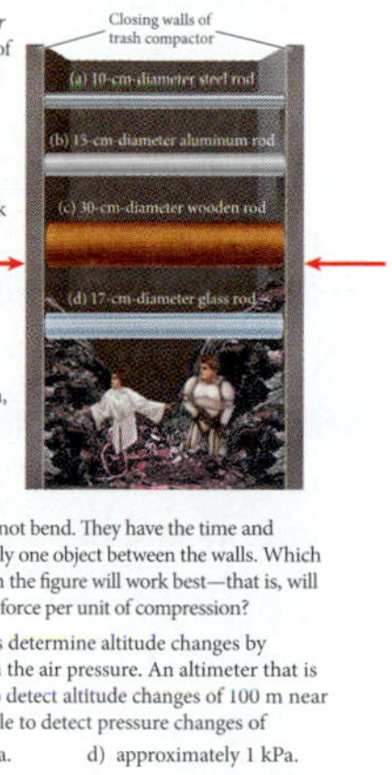

Closing walls of trash compactor
(a) 10-cm-diameter steel rod
(b) 15-cm-diameter aluminum rod
(c) 30-cm-diameter wooden rod
(d) 17-cm-diameter glass rod

13.8 Many altimeters determine altitude changes by measuring changes in the air pressure. An altimeter that is designed to be able to detect altitude changes of 100 m near sea level should be able to detect pressure changes of

a) approximately 1 Pa.
b) approximately 10 Pa.
c) approximately 100 Pa.
d) approximately 1 kPa.
e) approximately 10 kPa.

13.9 Which of the following assumptions is *not* made in the derivation of Bernoulli's Equation?

a) Streamlines do not cross.
b) There is negligible viscosity.
c) There is negligible friction.
d) There is no turbulence.
e) There is negligible gravity.

13.10 A beaker is filled with water to the rim. Gently placing a plastic toy duck in the beaker causes some of the water to spill out. The weight of the beaker with the duck floating in it is

a) greater than the weight before adding the duck.
b) less than the weight before adding the duck.
c) the same as the weight before adding the duck.
d) greater or less than the weight before the duck was added, depending on the weight of the duck.

13.11 A piece of cork (density = 0.33 g/cm^3) with a mass of 10 g is held in place under water by a string, as shown in the figure. What is the tension, T, in the string?

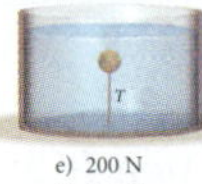

T

a) 0.10 N c) 0.30 N e) 200 N
b) 0.20 N d) 100 N f) 300 N

QUESTIONS

13.12 You know from experience that if a car you are riding in suddenly stops, heavy objects in the rear of the car move toward the front. Why does a helium-filled balloon in such a situation move, instead, toward the rear of the car?

13.13 A piece of paper is folded in half and then opened up and placed on a flat table so that it "peaks" up in the middle as shown in the figure. If you blow air between the paper and the table, will the paper move up or down? Explain.

13.14 In what direction does a force due to water flowing from a showerhead act on a shower curtain, inward toward the shower or outward? Explain.

13.15 Point out and discuss any flaws in the following statement: *The hydraulic car lift is a device that operates on the basis of Pascal's Principle. Such a device can produce large output forces with small input forces. Thus, with a small amount of work done by the input force, a much larger amount of work is produced by the output force, and the heavy weight of a car can be lifted.*

13.16 Given two springs of identical size and shape, one made of steel and the other made of aluminum, which has the higher spring constant? Why? Does the difference depend more on the shear modulus or the bulk modulus of the material?

13.17 One material has a higher density than another. Are the individual atoms or molecules of the first material necessarily more massive than those of the second?

13.18 Analytic balances are calibrated to give correct mass values for such items as steel objects of density ρ_s = 8000.00 kg/m^3. The calibration compensates for the buoyant force arising because the measurements are made in air, of density ρ_a = 1.205 kg/m^3. What compensation must be made to measure the masses of objects of a different material, of density ρ? Does the buoyant force of air matter?

13.19 If you turn on the faucet in the bathroom sink, you will observe that the stream seems to narrow from the point at which it leaves the spigot to the point at which it hits the bottom of the sink. Why does this occur?

13.20 In many problems involving application of Newton's Second Law to the motion of solid objects, friction is neglected for the sake of making the solution easier. The counterpart of friction between solids is viscosity of liquids. Do problems involving fluid flow become simpler if viscosity is neglected? Explain.

13.21 You have two identical silver spheres and two unknown fluids, A and B. You place one sphere in fluid A, and it sinks; you place the other sphere in fluid B, and it floats. What can you conclude about the buoyant force of fluid A versus that of fluid B?

13.22 Water flows from a circular faucet opening of radius r_0, directed vertically downward, at speed v_0. As the stream of water falls, it narrows. Find an expression for the radius of the stream as a function of distance fallen, $r(y)$, where y is measured downward from the opening. Neglect the eventual breakup of the stream into droplets, and any resistance due to drag or viscosity.

PROBLEMS

A blue problem number indicates a worked-out solution is available in the Student Solutions Manual. One • and two •• indicate increasing level of problem difficulty.

Sections 13.1 and 13.2

13.23 Air consists of molecules of several types, with an average molar mass of 28.95 g. An adult who inhales 0.50 L of air at sea level takes in about how many molecules?

•13.24 Ordinary table salt (NaCl) consists of sodium and chloride ions arranged in a *face-centered cubic crystal lattice*. That is, a sodium chloride crystal consists of cubic unit cells with a sodium ion on each corner and at the center of each face, and a chloride ion at the center of the cube and at the midpoint of each edge. The density of sodium chloride is $2.165 \cdot 10^3$ kg/m^3. Calculate the spacing between adjacent sodium and chloride ions in the crystal.

Section 13.3

13.25 A 20-kg chandelier is suspended from the ceiling by four vertical steel wires. Each wire has an unloaded length of 1 m and a diameter of 2 mm, and each bears an equal load. When the chandelier is hung, how far do the wires stretch?

13.26 Find the minimum diameter of a 50-m-long nylon string that will stretch no more than 1 cm when a load of 70 kg is suspended from its lower end. Assume that $Y_{nylon} = 3.51 \cdot 10^8$ N/m^2.

13.27 A 2.0-m-long steel wire in a musical instrument has a radius of 0.03 mm. When the wire is under a tension of 90 N, how much does its length change?

•13.28 A rod of length L is attached to a wall. The load on the rod increases linearly (as shown by the arrows in the figure) from zero at the left end to W newtons per unit length at the right end. Find the shear force at

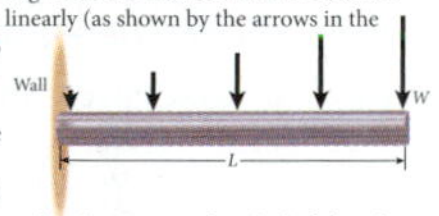

Wall
W
L

a) the right end, b) the center, and c) the left end.

•13.29 Challenger Deep in the Marianas Trench of the Pacific Ocean is the deepest known spot in the Earth's oceans, at 10.922 km below sea level. Taking density of seawater at atmospheric pressure (p_0 = 101.3 kPa) to be 1024 kg/m^3 and its bulk modulus to be $B(p) = B_0 + 6.67(p - p_0)$, with $B_0 = 2.19 \cdot 10^9$ Pa, calculate the pressure and the density of the seawater at the

"The problem-solving technique, to borrow a phrase from my students, 'doesn't suck.' I'm a skeptic when it comes to anybody else's one-size-fits-all approach to problem solving—I've seen too many that just don't work, pedagogically. The approach used by the authors, however, is one in which students are really forced to tap their intuition before they start, to reflect on the relevant first principles. . . .

Wow! There are some really nice problems at the end of the chapter. My compliments to the authors. There was a nice diversity of problems, and most of them required a lot more than simple plug-and-chug. I found many problems I'd be inclined to assign."

—Brent Corbin, University of California–Los Angeles

"The text strikes a very good balance of providing mathematical details and rigor together with a clear, intuitive presentation of physics concepts. The balance and variety of problems provided, both as worked-out examples and as end-of-chapter problems, are outstanding. Many features are found in this book that are difficult to find in other standard texts, including proper use of vector notation, explicit evaluation of multiple integrals, e.g., in moment of inertia calculations, and intriguing connections to modern physics."

—Lisa Everett, University of Wisconsin–Madison

Contemporary Topics: Capturing Students' Imaginations

University Physics incorporates a wide variety of contemporary topics as well as research-based discussions designed to help students appreciate the beauty of physics and see how physics concepts are related to the development of new technologies in the fields of engineering, medicine, astronomy and more. The "Big Picture" section at the beginning of the text is designed to introduce students to some of the amazing new frontiers of research that are being explored in various fields of physics and the results that have been obtained during the last few years. The authors return to these topics at various points within the book for more in-depth exploration.

The authors of *University Physics* also repeatedly discuss different aspects of the broad topic of energy, by addressing concepts of energy sources (fossil, nuclear, renewable, alternative, and so forth), energy efficiency, and environmental effects of energy supply choices. Alternative energy sources and renewable resources are discussed within the framework of possible solutions to the energy crisis. These discussions provide a great opportunity to capture students' interest and are accessible on the introductory physics level.

The following contemporary physics research topics and topical energy discussions (in green) are found in the text:

Chapter 1
Section 1.3 has a subsection called "Metrology" that mentions the new definition of the kilogram and the optical clock at NIST
Section 1.4 mentions the authors' research on heavy-ion collisions

Chapter 4
Section 4.2 has a subsection on the Higgs particle
Section 4.7 has a subsection on tribology

Chapter 5
Section 5.1 Energy in Our Daily Lives
Section 5.7 Power and Fuel-Efficiency of U.S. Cars

Chapter 6
Section 6.8 has a subsection titled "Preview: Atomic Physics" that discusses tunneling of particles
Solved Problem 6.6 Power Produced by Niagara Falls

Chapter 7
Example 7.5 Particle Physics
Section 7.8 discusses Sinai billiards and chaotic motion

Chapter 8
Example 8.3 mentions electromagnetic propulsion and radiation shielding in context of sending astronauts to Mars

Chapter 10
Example 10.7 Death of a Star
Example 10.8 Flybrid

Chapter 12
Section 12.1 has a subsection titled "Solar System" that mentions research on objects in Kuiper Belt
Section 12.7 Dark Matter

Chapter 13
Section 13.1 briefly discusses nanotubes and nanotechnology
Section 13.2 briefly discusses plasmas and Bose-Einstein condensates
Section 13.6 has a subsection titled "Applications of Bernoulli's Equation" that discusses lift and design of aircraft wings
Section 13.8 Turbulence and Research Frontiers in Fluid Flow

Chapter 14
Section 14.7 Chaos

Chapter 15
Section 15.5 has a subsection titled "Seismic Waves" that mentions reflection seismology
Figure 15.11b shows nanoscale guitar

Section 15.8 includes Self-Test Opportunity 15.4 on nanoscale guitar string
Section 15.9 Research on Waves

Chapter 16
Section 16.4 has a subsection titled "Mach Cone" that mentions creation of shock waves by collisions of nuclei in particle accelerators and has a paragraph on Cherenkov radiation

Chapter 17
Section 17.2 has subsections titled "Research at the Low-Temperature Frontier" and "Research at the High-Temperature Frontier"
Section 17.5 Surface Temperature of Earth
Section 17.6 Temperature of the Universe
Example 17.4 Rise in Sea Level Due to Thermal Expansion of Water

Chapter 18
Example 18.7 Roof Insulation
Solved Problem 18.2 Warming Costs for Winter
Solved Problem 18.3 Power Carried by the Gulf Stream
Example 18.8 Earth as a Blackbody
Section 18.8 Modes of Thermal Energy Transfer/Global Warming
Section 18.8 has a subsection titled "Heat in Computers"

Chapter 19
Example 19.5 The Quark-Gluon Plasma

Chapter 20
Example 20.2 Warming a House with a Heat Pump
Example 20.4 Maximum Efficiency of an Electrical Power Plant
Solved Problem 20.1 Efficiency of an Automobile Engine
Section 20.4 Real Engines and Efficiency/Hybrid Cars
Section 20.4 Real Engines and Efficiency/Efficiency and the Energy Crisis
Solved Problem 20.2 Cost to Operate an Electrical Plant
Section 20.7 Entropy Death

Chapter 21
Section 21.3 Superconductors
Section 21.5 Electrostatic Force—Coulomb's Law/Electrostatic Precipitator

Chapter 22
Example 22.4 Time Projection Chamber (STAR TPC)
Section 22.6 discusses noninvasive imaging of brain electric fields and the brain-computer interface

"I think this idea is great! It would help the instructor show the students that physics is a live and exciting subject. . . . because it shows that physics is a happening subject, relevant for discovering how the universe works, that it is necessary for developing new technologies, and how it can benefit humanity . . . . The [chapters] contain a lot of interesting modern topics and explain them very clearly."

—*Joseph Kapusta, University of Minnesota*

"Section 17.5 on the surface temperature of the Earth is excellent and is an example of what is *missing* from so many introductory textbooks: examples that are relevant and compelling for the students."

—*John William Gary, University of California–Riverside*

"I think the approach to include modern or contemporary physics throughout the text is great. Students often approach physics as a science of concepts which were discovered long ago. They view engineering as the science which has given them the advances in technology which they see today. It would be great to show students just where these advances do start, with physics."

—*Donna W. Stokes, University of Houston*

Enhanced Content: Flexibility for Your Student and Course Needs

To instructors who are looking for additional coverage of certain topics and mathematical support for those topics, *University Physics* also offers flexibility. This book includes some topics and some calculus that do not appear in many other texts. However, these topics have been presented in such a way that their exclusion will not affect the overall course. The text as a whole is written at a level appropriate for the typical introductory student. Below is a list of flex-coverage content as well as additional mathematical support:

Chapter 2

Section 2.3 The concept of the derivative is developed using an approach that is both conceptual and graphical. Examples using the derivative are provided, and students are referred to an appendix for other "refreshers." This is a more extensive approach than is taken in some other texts.

Section 2.4 Acceleration as the time derivative of velocity is introduced by analogy, and the discussion includes an example.

Section 2.6 Integration as the inverse to differentiation is introduced for finding the area under a curve. This more extensive presentation than in many texts is spread over two sections with multiple examples.

Section 2.7 Examples using differentiation are included.

Section 2.8 A derivation on minimum time arguments is shown to lead to a solution that is equivalent to Snell's Law.

End-of-chapter exercises related to this coverage include Questions 20, 22, and 23 and multiple Problems using calculus.

Chapter 3

Section 3.1 The component-wise derivative of a three-dimensional position vector into three-dimensional velocity and then into three-dimensional acceleration is presented.

Section 3.3 The tangetiality of the velocity vector to the trajectory is covered.

Section 3.4 The maximum height and range of a projectile are found by setting the derivative equal to zero.

Section 3.5 Relative motion is covered (equation 3.27).

End-of-chapter Problem 3.38 covers the derivative.

Chapter 4

Section 4.8 Example 4.10 on the best angle to pull a sled is a maximum-minimum problem.

Chapter 5

Section 5.5 Work done by a variable force is covered using definite integrals and the derivation of equation 5.20. The chain rule is also covered.

Section 5.6 Work done by spring force is discussed (equation 5.24).

Section 5.7 Power as the time derivative of work is covered (equation 5.26).

A number of end-of-chapter Problems supplement this coverage, such as Problems 5.34 through 5.37.

Chapter 6

Section 6.3 Finding the work done by a force includes use of integrals.

Section 6.4 Obtaining force from the potential includes the use of derivatives; partial derivatives and the gradient are also introduced (for example, the Lennard-Jones potential).

A number of end-of-chapter Questions and Problems supplement this coverage, such as Questions 6.24 and 6.25 and Problems 6.34, 6.35, and 6.36.

Chapter 8

The text introduces volume integrals so that the volume of a sphere and the center of mass of a half sphere can be determined in a worked example.

Chapter 9

Explicit derivatives of the radical and tangential unit vectors are provided. The text derives the equations of motion for constant angular acceleration, repeating the customary derivation of equations of motion for constant linear acceleration presented in Chapter 2.

Chapter 10

The volume integral introduced in Chapter 8 is utilized in finding the moment of inertia for different objects. The text derives the expression for angular momentum to determine the relationship between the angular momentum of a system of particles and the torque.

Chapter 11

Section 11.3 The stability condition is utilized, and the second derivative of potential energy is examined to determine the type of equilibrium through graphical interpretation of the functions.

Chapter 12

Section 12.1 Unique coverage is provided of the derivation of the gravitational force from a sphere and inside a sphere.

Chapter 14

Section 14.4 The subsection titled "Small Damping" applies the student's knowledge of simple harmonic oscillators to derive the small damping equation through differentiation. For the case of large damping, the student is again referred to the solution to the differential equation. Example 14.6 walks the student through an example of a damped harmonic oscillator. The solution to this equation is stated explicitly, but the text utilizes calculus to reach the answer. The subsection "Energy Loss in Damped Oscillations" includes a calculation of the rate of energy loss that uses the differential definition of power.

Section 14.5 A thorough discussion of forced harmonic motion takes advantage of the student's understanding of differential equations, graphically analyzes the solution, and then analyzes the outcome.

A number of end-of-chapter Problems supplement this coverage, such as Problems 14.55 and 14.73.

Chapter 15

Section 15.4 This entire section is unique among introductory physics texts as it utilizes partial differential equations to derive the wave equation.

Several of the end-of-chapter Questions and Problems related to the content of Section 15.4 require an understanding of the calculus used in this section, particularly 15.30 and 15.31.

Chapter 16

Section 16.4 Covers the Doppler effect as a function of the perpendicular distance.

Chapter 22

Section 22.9 Solved Problem 22.3 covers the electric field for a nonuniform spherical charge distribution

Chapter 25

Section 25.5 Solved Problem 25.2, "Brain Probe" covers a case of nonconstant cross-sectional area.

Chapter 32

Section 32.3 This textbook goes deeper into calculus than many others by demonstrating how calculus can be used in deducing from Newton's Laws of Motion the necessarily parabolic shape of liquid surfaces that have a net circular motion.

Some end-of-chapter Problems, such as Problem 32.43, also use calculus to solve a minimization problem.

Chapter 34

Section 34.10 This section discusses the quality of diffraction gratings using the concept of dispersion. In many similar physics textbooks, the formula for dispersion is simply given. This textbook, however, uses calculus to derive dispersion in a straightforward manner.

Chapter 35

Section 35.2 This section has a very instructive and intuitive derivation of light cone variables that goes beyond what is found in most standard textbooks.

Section 35.6 The text features calculus-based derivations for velocity transformation and energy (based on integrating the distance dependence of work). While the energy derivation follows standard integration techniques and is used in most books, the velocity transformation derivation is unique and very instructive.

Chapter 36

Section 36.2 The level of detail that characterizes the derivation of several radiation laws (Wien, Planck, Boltzmann, Raleigh-Jeans) is not found in many other textbooks.

Section 36.8 The introduction to Bose-Einstein and Fermi-Dirac statistics is important and unique to this textbook. The connection to radiation laws is especially relevant. The end-of-chapter Problems related to Section 36.8, in particular Problems 36.53 through 36.55, are challenging and utilize the relevant mathematics.

Chapter 37

Most standard textbooks teach quantum mechanics with minimal usage of calculus, using mostly a conceptual approach. This book takes a more formal approach, from Section 35.1 on wave functions. The students are exposed to derivations that are unique to modern physics, starting from the normalization condition of the wave function over operators for momentum and kinetic energy and continuing to solutions for infinite and finite potentials. Hamiltonians are introduced and applied to Schrödinger and Dirac equations. The many-particle wave function is then covered in Section 37.9. The end-of-chapter Problems that utilize calculus include Problems 37.28 through 37.39.

Chapter 38

This textbook derives the full solution of the hydrogen electron wave function and breaks it down into radial and angular parts. This complete solution allows the student to derive the degeneracy of the quantum levels rather than simply learning a simple formula for calculating the levels without understanding their physical origin.

Section 38.3 The solution of the Schrödinger equation in Section 38.3. is based on the derivations in Chapter 37, and the text further explores the full solution of the hydrogen electron wave function considered earlier in the chapter. The end-of-chapter Problems 38.35, 38.36, and 38.37 use the calculus from the chapter.

Chapter 39

The definition of the differential scattering cross section is defined in equations 39.3 and 39.4, based on classical (Rutherford) physics. (Many other texts simply show and describe graphs.) The differential cross section from quantum considerations is given in equation 39.6, and the form factor (deviation from Rutherford and deviation from point particle) is given in equations 39.7 and 39.8. Form factors are not often described in other texts. While this discussion adds to the mathematical detail, it could be easily omitted to fit the needs of the course. End-of-chapter Problem 39.32 makes use of calculus to calculate the fraction of particles scattered into a range of angles.

Chapter 40

The text presents a slightly more detailed discussion than is usually seen in deriving the Fermi energy while covering Fermi's model of the nucleus in Section 40.3. Some end-of-chapter problems could involve simple integration of the exponential function: 40.31, 40.33, 40.52, 40.53, and 40.61.

"Strongest feature . . . The use of real mathematics, especially calculus, to derive kinematic relations, the relations between quantities in circular motion, the direction of the gravitational force, the magnitude of the tidal force, the maximum extension of a set of piled blocks. Solved problems are always addressed symbolically first. Too often textbooks don't let the math do the work for them."

—*Kieran Mullen, University of Oklahoma*

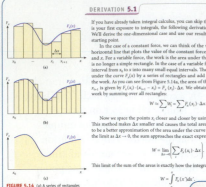

DERIVATION 5.1

If you have already taken integral calculus, you can skip this section. If equation 5.20 is your first exposure to integrals, the following derivation is a useful introduction. We'll derive the one-dimensional case and use our result for the constant force as a starting point.

In the case of a constant force, we can think of the work as the area under the horizontal line that plots the value of the constant force in the interval between x_0 and x. For a variable force, the work is the area under the curve $F_x(x)$, but that area is no longer a simple rectangle. In the case of a variable force, we need to divide the interval from x_0 to x into many small equal intervals. Then we approximate the area under the curve $F_x(x)$ by a series of rectangles and add their areas to approximate the work. As you can see from Figure 5.14a, the area of the rectangle between x_i and x_{i+1} is given by $F_x(x_i) \cdot (x_{i+1} - x_i) = F_x(x_i) \cdot \Delta x$. We obtain an approximation for the work by summing over all rectangles:

$$W \approx \sum_i W_i = \sum_i F_x(x_i) \cdot \Delta x.$$

Now we space the points x_i closer and closer by using more and more of them. This method makes Δx smaller and causes the total area of the series of rectangles to be a better approximation of the area under the curve $F_x(x)$ as in Figure 5.14b. In the limit as $\Delta x \to 0$, the sum approaches the exact expression for the work:

$$W = \lim_{\Delta x \to 0} \left(\sum_i F_x(x_i) \cdot \Delta x \right).$$

This limit of the sum of the areas is exactly how the integral is defined:

$$W = \int_{x_0}^{x} F_x(x') dx'.$$

We have derived this result for the case of one... tion of the three-dimensional case proceeds alon... in terms of algebra.

FIGURE 5.14 (a) A series of rectangles approximates the area under the curve obtained by plotting the force as a function of the displacement; (b) a better approximation using rectangles of smaller width; (c) the exact area under the curve.

Derivations

Detailed derivations are provided in the text as examples for students, who will eventually need to develop their own derivations as they review Solved Problems, work through Examples, and solve end-of-chapter Problems. The Derivations are identified in the text with numbered headings so that instructors can include these detailed features as necessary to fit the needs of their courses.

"Again the derivation resulting in equation 6.15 is outstanding. Few books that I have seen will show students even once all the math steps in derivations. This is a strength of this book. Also, in the next section, I like very much the generalization of the relation between force and potential energy to three dimensions. It is something that I always do in lecture although most books do not get close."

—*James Stone, Boston University*

Calculus Primer

A Calculus Primer can be found in the appendices. Since this course sequence is typically given in the first year of study at universities, it assumes knowledge of high school physics and mathematics. It is preferable that students have had a course in calculus before they start this course sequence, but calculus can also be taken in parallel. To facilitate this, the text contains a short calculus primer in an appendix, giving the main results of calculus without the rigorous derivations.

Building Knowledge: The Text's Learning System

Chapter Opening Outline

At the beginning of each chapter is an outline presenting the section heads within the chapter. The outline also includes the titles of the Examples and Solved Problems found in the chapter. At a quick glance, students or instructors know if a desired topic, example, or problem is in the chapter.

What We Will Learn / What We Have Learned

Each chapter of *University Physics* is organized like a good research seminar. It was once said, "Tell them what you will tell them, then tell them, and then tell them what you told them!" Each chapter starts with **What We Will Learn**—a quick summary of the main points, without any equations. And at the end of each chapter, **What We Have Learned/Exam Study Guide** contains key concepts, including major equations, symbols, and key terms. All symbols used in the chapter's formulas are also listed.

WHAT WE WILL LEARN

- A force is a vector quantity that is a measure of how an object interacts with other objects.
- Fundamental forces include gravitational attraction and electromagnetic attraction and repulsion. In daily experience, important forces include tension and normal, friction, and spring forces.
- Multiple forces acting on an object sum to a net force.
- Free-body diagrams are valuable aids in working problems.
- Newton's three laws of motion govern the motion of objects under the influence of forces.
    a) The first law deals with objects for which external forces are balanced.
    b) The second law describes those cases for which external forces are not balanced.

    c) The third law addresses equal (in magnitude) and opposite (in direction) forces that two bodies exert on each other.

- The gravitational mass and the inertial mass of an object are equivalent.
- Kinetic friction opposes the motion of moving objects; static friction opposes the impending motion of objects at rest.
- Friction is important to the unde... world motion, but its causes and... are still under investigation.
- Applications of Newton's laws of... multiple objects, multiple forces,... applying the laws to analyze a situ... the most important problem-solv... physics.

WHAT WE HAVE LEARNED | EXAM STUDY GUIDE

- The net force on an object is the vector sum of the forces acting on the object: $\vec{F}_{net} = \sum_{i=1}^{n} \vec{F}_i$.

- Mass is an intrinsic quality of an object that quantifies both the object's ability to resist acceleration and the gravitational force on the object.

- A free-body diagram is an abstraction showing all forces acting on an isolated object.

- Newton's three laws are as follows:

    Newton's First Law. In the absence of a net force on an object, the object will remain at rest, if it was at rest. If it was moving, it will remain in motion in a straight line with the same velocity.

    Newton's Second Law. If a net external force, $\vec{F}_{net}$, acts on an object with mass m, the force will cause an acceleration, $\vec{a}$, in the same direction as the force: $\vec{F}_{net} = m\vec{a}$.

    Newton's Third Law. The forces that two interacting objects exert on each other are always exactly equal in magnitude and opposite in direction: $\vec{F}_{1 \to 2} = -\vec{F}_{2 \to 1}$.

- Two types of friction occur: static and kinetic friction. Both types of friction are proportional to the normal force, N.

    Static friction describes the force of friction between an object at rest on a surface in terms of the coefficient of static friction, μ_s. The static friction force, f_s, opposes a force trying to move an object and has a maximum value, $f_{s,max}$, such that $f_s \leq \mu_s N = f_{s,max}$.

    Kinetic friction describes the force of friction between a moving object and a surface in terms of the coefficient of kinetic friction, μ_k. Kinetic friction is given by $f_k = \mu_k N$.

    In general, $\mu_s > \mu_k$.

Conceptual Introductions

Conceptual explanations are provided in the text prior to any mathematical explanations, formulas, or derivations in order to establish for students why the quantity is needed, why it is useful, and why it must be defined accurately. The authors then move from the conceptual explanation and definition to a formula and exact terms.

> "The section on thermal expansion is outstanding and the supporting example problems are very well done. This section can be put up against any text on the market and come out ahead. The authors do very well on basic concepts."
>
> —*Marllin Simon, Auburn University*

Self-Test Opportunities

Sets of questions follow the coverage of major concepts within the text to encourage students to develop an internal dialogue. These questions will help students think critically about what they have just read, decide whether they have a grasp of the concept, and develop a list of follow-up questions to ask in lecture. The answers to the Self-Tests are found at the end of each chapter.

> "The Self-Test Opportunities are effective for encouraging students to place what they have learned in this chapter in the context of the broader conceptual understanding they have been developing throughout the earlier chapters."
>
> — *Nina Abramzon, California Polytechnic University–Pomona*

6.3 Self-Test Opportunity

Why does the lighter-colored ball arrive at the bottom in Figure 6.10 before the other ball?

FIGURE 6.10 Race of two balls down different inclines of the same height.

6.3 The lighter-colored ball descends to a lower elevation earlier in its motion and thus converts more of its potential energy to kinetic energy early on. Greater kinetic energy means higher speed. Thus, the lighter-colored ball reaches higher speeds earlier and is able to move to the bottom of the track faster, even though its path length is greater.

In-Class Exercises

In-Class Exercises are designed to be used with personal response system technology. They will appear in the text so that students may begin contemplating the concepts. Answers will only be available to instructors. (Questions and answers are formatted in PowerPoint for universal use with personal response systems.)

Visual Program

Familiarity with graphic artwork on the Internet and in video games has raised the bar for the graphical presentations within textbooks, which must now be more sophisticated to excite both students and faculty. Here are some examples of techniques and ideas implemented in the *University Physics*:

- Overlays of line drawings over photographs connect sometimes very abstract physics concepts to students' realities and everyday experiences.

- A three-dimensional look for line drawings adds plasticity to the presentations. Mathematically accurate graphs and plots were created by the authors in software programs such as Mathematica and then used by the graphic artists to ensure complete accuracy along with a visually appealing style.

2.2 In-Class Exercise

Throwing a ball straight up into the air provides an example of free-fall motion. At the instant the ball reaches its maximum height, which of the following statements is true?

a) The ball's acceleration points down, and its velocity points up.

b) The ball's acceleration is zero, and its velocity points up.

c) The ball's acceleration points up, and its velocity points up.

d) The ball's acceleration points down, and its velocity is zero.

e) The ball's acceleration points up, and its velocity is zero.

f) The ball's acceleration is zero, and its velocity points down.

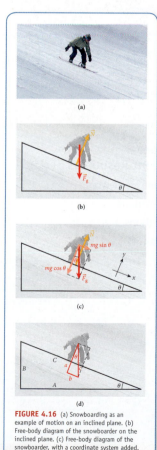

FIGURE 4.16 (a) Snowboarding as an example of motion on an inclined plane. (b) Free-body diagram of the snowboarder on the inclined plane. (c) Free-body diagram of the snowboarder, with a coordinate system added. (d) Similar triangles in the inclined-plane problem.

McGraw-Hill Higher Education.

Connect. Learn. Succeed.

McGraw-Hill Higher Education's mission is to help prepare students for the world that awaits. McGraw-Hill provides textbooks, eBooks and other digital instructional content, as well as experiential learning and assignment/assessment platforms, that connect instructors and students to valuable course content—and connect instructors and students to each other.

With the highest quality tools and content, students can engage with their coursework when, where, and however they learn best, enabling greater learning and deeper comprehension.

In turn, students can learn to their full potential and, thus, succeed academically now and in the real world.

Connect
Learn
Succeed

Mc Graw Hill

Connect:
Instructor Resources

• McGraw-Hill Connect
• Simulations
• McGraw-Hill Create
• Tegrity Campus 2.0
• Learning Solutions
• Instructor Solutions Manual
• PowerPoint Lecture Outlines
• Clicker Questions
• Electronic Images from the Text
• EzTest Test Bank

Learn:
Course Content

• Textbooks/Readers
• eBooks
• PowerPoint Presentations
• Enhanced Cartridges
• In-class Simulations
• Lecture Aids
• Custom Publishing

Succeed:
Student Resources

• Online Homework
• Simulations
• Questions
• eBook

McGraw-Hill Connect™ Physics

With Connect Physics, instructors can deliver assignments, quizzes, and tests online. The problems directly from the end-of-chapter material in Bauer/Westfall's *University Physics* are presented in an auto-gradable format. The online homework system incorporates new and exciting interactive tools and problem types: a graphing tool; a free-body diagram drawing tool; symbolic entry; a math palette; and multi-part problems. The seven-step problem-solving method from the text is also reflected in Connect Physics: Students will find the familiar seven steps outlined in the guided solutions, with additional detailed help with the math where they need it most. Over 2300 multiple-choice Test Bank questions (written by the authors and the same contributors who provided the text's end-of-chapter problems) are also included.

Instructors can edit existing questions and write entirely new problems; track individual student performance—by question, assignment, concept, or in relation to the class overall—with automatic grading; provide instant feedback to students; and secure storage of detailed grade reports online. Grade reports can be easily integrated with Learning Management Systems (LMS), such as WebCT and Blackboard.

By choosing Connect Physics, instructors are providing their students with a powerful tool for improving academic performance and truly mastering course material. Connect Physics helps reduce the time instructors spend grading homework and closes the loop on student homework and feedback.

Connect Physics allows students to practice important skills at their own pace and on their own schedule. Importantly, students' assessment results and instructors' feedback are all saved online—so students can continually review their progress and plot their course to success.

Instructors also have access to PowerPoint lecture outlines, an Instructor's Solutions Manual, electronic images from the text, clicker questions, quizzes, tutorials, simulations, video clips, and many other resources that are directly tied to specific material in *University Physics*. Students have access to self-quizzes, simulations, tutorials, and more.

Go to **www.mhhe.com/bauerwestfall** to learn more and register.

ConnectPlus™ Physics

Some instructors may also choose ConnectPlus Physics for their students. ConnectPlus offers an innovative and inexpensive electronic textbook integrated within the homework platform. Like Connect Physics, ConnectPlus Physics provides students with online assignments and assessments AND 24/7 online access to an eBook—an online edition of the *University Physics* text.

As part of the ehomework process, instructors can assign chapter and section readings from the text. With ConnectPlus, links to relevant text topics are also provided where students need them most—accessed directly from the ehomework problem!

ConnectPlus Physics:

- Provides students with a ConnectPlus eBook, allowing for anytime, anywhere access to the *University Physics* textbook to aid them in successfully completing their work, wherever and whenever they choose.

- Includes Community Notes for student-to-student or instructor-to-student note sharing to greatly enhance the user learning experience.

- Allows for insertion of lecture discussions or instructor-created additional examples using Tegrity (see page xxvi) to provide additional clarification or varied coverage on a topic.

- Merges media and assessments with the text's narrative to engage students and improve learning and retention. The eBook includes simulations, interactives, videos, and inline assessment questions.

- Pinpoints and connects key physics concepts in a snap using the powerful eBook search engine.

- Manages notes, highlights and bookmarks in one place for simple, comprehensive review.

TEGRITY Mc Graw Hill **tegrity campus**

Tegrity Campus is a service that makes class time available all the time by automatically capturing every lecture in a searchable format for students to review when they study and complete assignments. With a simple one-click start-and-stop process, you capture all computer screens and corresponding audio. Students replay any part of any class with easy-to-use browser-based viewing on a PC or Mac. Educators know that the more students can see, hear, and experience class resources, the better they learn. With Tegrity Campus, students quickly recall key moments by using Tegrity Campus's unique search feature. This search helps students efficiently find what they need, when they need it across an entire semester of class recordings. Help turn all your students' study time into learning moments immediately supported by your lecture.

To learn more about Tegrity watch a 2-minute Flash demo at http://tegritycampus.mhhe.com.

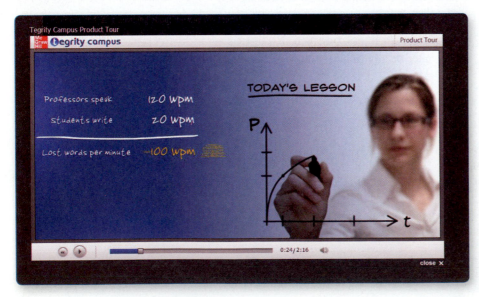

Additional Resources for Instructors and Students

Electronic Book Images and Assets for Instructors

Build instructional materials wherever, whenever, and however you want!

An online collection of Presentation Tools containing photos, artwork, and other media can be accessed from *University Physics'* Connect Physics companion site to create customized lectures, visually enhanced tests and quizzes, compelling course websites, or attractive printed support materials. Assets are copyrighted by McGraw-Hill Higher Education but can be used by instructors for classroom purposes. The visual resources in this collection are

- **Art** Full-color digital files of all illustrations in the book can be readily incorporated into lecture presentations, exams, or custom-made classroom materials. In addition, all files are pre-inserted into PowerPoint slides for ease of lecture preparation.

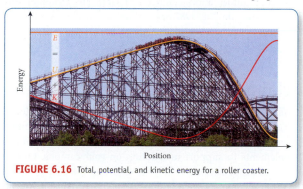

FIGURE 6.16 Total, potential, and kinetic energy for a roller coaster.

- **Photos** The photo collection contains digital files of photographs from the text, which can be reproduced for multiple classroom uses.

- **Solved Problem Library, Example Library, Table Library, and Numbered Equations Library** Access the worked examples, tables, and equations from the text in electronic format for inclusion in your classroom resources.

Also residing on the companion site are

- **PowerPoint Lecture Outlines** Ready-made presentations (created by the text authors) that combine art and lecture notes are provided for each chapter of the text. The outlines include historical information and additional examples.

- **PowerPoint Slides** For instructors who prefer to create their lectures from scratch, all illustrations and available photos are pre-inserted by chapter into PowerPoint slides.

Computerized Test Bank Online

The same group of instructors who wrote the text's end-of-chapter Questions and Problems also wrote the Test Bank questions to reinforce the important concepts from the text while ensuring consistency in style and coverage. The questions were combined with a set of questions created and tested by the authors of the text. A comprehensive bank of over 2300 test questions in multiple-choice format, organized by section of the text, at a variety of difficulty levels, is provided within a computerized Test Bank powered by McGraw-Hill's flexible electronic testing program: EZ Test Online (www.eztestonline.com). EZ Test Online allows you to easily create paper and online tests or quizzes!

Imagine being able to create and access a test or quiz anywhere, at any time, without installing the testing software. Now, with EZ Test Online, instructors can select questions from multiple McGraw-Hill test banks or create their own, and then either print the test for paper distribution or give it online. Go to www.mhhe.com/bauerwestfall for more information.

CourseSmart

CourseSmart is a new way for faculty to find and review eBooks. It's also a great option for students who are interested in accessing their course materials digitally and saving money. *CourseSmart* offers thousands of the most commonly adopted textbooks across hundreds of courses from a wide variety of higher education publishers. It is the only place for faculty to review and compare the full text of a textbook online, providing immediate access without the environmental impact of requesting a print exam copy. At *CourseSmart*, students can save up to 50% off the cost of a print book, reduce their impact on the environment, and gain access to powerful web tools for learning including full text search, notes and highlighting, and email tools for sharing notes between classmates. For further details contact your sales representative or go to www.coursesmart.com.

Create

Craft your teaching resources to match the way you teach! With McGraw-Hill Create, www.mcgrawhillcreate.com, you can easily rearrange chapters, combine material from other content sources, and quickly upload content you have written like your course syllabus or teaching notes. Find the content you need in Create by searching through thousands of leading McGraw-Hill textbooks. Arrange your book to fit your teaching style. Create even allows you to personalize your book's appearance by selecting the cover and adding your name, school, and course information. Order a Create book and you'll receive a complimentary print review copy in 3–5 business days or a complimentary electronic review copy (eComp) via email in about one hour. Go to www.mcgrawhillcreate.com today and register. Experience how McGraw-Hill Create empowers you to teach *your* students *your* way.

Personal Response Systems

Personal response systems, or "clickers," bring interactivity into the classroom or lecture hall. Wireless response systems give the instructor and students immediate feedback from the entire class. The wireless response pads are essentially remotes that are easy to use and engage students, allowing instructors to motivate student preparation, interactivity, and active learning. Instructors receive immediate feedback to gauge which concepts students understand. The In-Class Exercises from *University Physics* (formatted in PowerPoint) are available on the companion website.

Solutions Manuals

The *Instructor's Solutions Manual* includes answers to all of the text's end-of-chapter Questions and complete, worked-out solutions for the end-of-chapter Problems. Chapters 1 through 13 include worked out-solutions that follow the seven-step problem-solving method from the text for all Problems and Additional Problems. Chapters 14 through 40 continue to use the seven-step problem-solving method for all challenging (one bullet) and most challenging (two bullet) Problems and Additional Problems, while switching to a more abbreviated solution for the less challenging (no bullet) Problems and Additional Problems.

The *Student Solutions Manual* contains answers and worked-out solutions to selected end-of-chapter Questions and Problems. Again, the worked-out solutions for Chapters 1 through 13 follow the complete seven-step problem-solving method from the text for Problems and Additional Problems. Chapters 14 through 40 continue to use the seven-step method for challenging (one bullet) and most challenging (two bullet) Problems and Additional Problems, while switching to a more abbreviated solution for the less challenging (no bullet) Problems and Additional Problems.

For more information, contact a McGraw-Hill customer service representative at 800-338-3987 or by e-mail at http://www.mhhe.com/. To locate your sales representative, go to http://www.mhhe.com/ for Find My Sales Rep.

Disclaimer

As a full-service publisher of quality educational products, McGraw-Hill does much more than just sell textbooks to your students. We create and publish an extensive array of print, video, and digital supplements to support instruction on your campus. Orders of new (versus used) textbooks help us to defray the cost of developing such supplements, which is substantial. Please consult your local McGraw-Hill representative to learn about the availability of the supplements that accompany *University Physics*. If you are not sure who your representative is, you can find him or her by using the tab labeled "My Sales Rep" at www.mhhe.com.

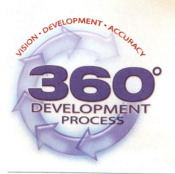

McGraw-Hill's 360° Development process is an ongoing, market-oriented approach to building accurate and innovative print and digital products. It is dedicated to continual large-scale and incremental improvement and is driven by multiple customer feedback loops and checkpoints. This process is initiated during the early planning stages of our new products, intensifies during the development and production stages, and then begins again on publication, in anticipation of the next edition.

A key principle in the development of any physics text is its ability to adapt to teaching specifications in a universal way. The only way to do so is by contacting those universal voices—and learning from their suggestions. We are confident that our book has the most current content the industry has to offer, thus pushing our desire for accuracy to the highest standard possible. In order to accomplish this, we have moved through an arduous road to production. Extensive and open-minded advice is critical in the production of a superior text.

We engaged over 200 instructors and students to provide guidance in the development of this first edition. By investing in this extensive endeavor, McGraw-Hill delivers to you a product suite that has been created, refined, tested, and validated as a successful tool for your course.

Board of Advisors

A hand-picked group of trusted instructors active in the calculus-based physics course and research groups served as the chief advisors and consultants to the author and editorial team with regard to manuscript development. The Board of Advisors reviewed the manuscript; served as a sounding board for pedagogical, media, and design concerns; helped to respond to issues raised by other reviewers; approved organizational changes; and attended a focus group to confirm the manuscript's readiness for publication.

Nina Abramzon, *California Polytechnic University–Pomona*
Rene Bellweid, *Wayne State University*
David Harrison, *University of Toronto*
John Hopkins, *The Pennsylvania State University*
David C. Ingram, *Ohio University–Athens*
Michael Lisa, *The Ohio State University*
Amy Pope, *Clemson University*
Roberto Ramos, *Drexel University*

Contributors

A panel of excellent writers created additional Questions and Problems to enhance the variety of exercises found in every chapter:

Carlos Bertulani, *Texas A&M University–Commerce*
Ken Thomas Bolland, *Ohio State University*
John Cerne, *State University of New York–Buffalo**
Ralph Chamberlain, *Arizona State University*
Eugenia Ciocan, *Clemson University**
Fivos Drymiotis, *Clemson University*
Michael Famiano, *Western Michigan University**
Yung Huh, *South Dakota State University*
Pedram Leilabady, *University of North Carolina–Charlotte**
M.A.K. Lodhi, *Texas Tech University*
Charley Myles, *Texas Tech University*
Todd Pedlar, *Luther College**
Corneliu Rablau, *Kettering University*
Roberto Ramos, *Drexel University*
Ian Redmount, *Saint Louis University*
Todd Smith, *University of Dayton**
Donna Stokes, *University of Houston**
Stephen Swingle, *City College of San Francisco*
Marshall Thomsen, *Eastern Michigan University*
Prem Vaishnava, *Kettering University**
John Vasut, *Baylor University**

*These contributors also authored questions for the Test Bank to accompany *University Physics* along with David Bannon of Oregon State University, while Richard Hallstein of Michigan State University organized and reviewed all of the contributions. Additionally, Suzanne Willis of North Illinois University helped us to compile our assets for the *University Physics* ConnectPlus eBook. Jack Cuthbert of Holmes Community College-Ridgeland composed the text's In-Class Exercises into PowerPoint files for use with clickers. Collette Marsh and Deborah Damcott of Harper College edited the PowerPoint Lectures, and finally, but certainly not least, Rob Hagood of Washtenaw Community College and Amy Pope of Clemson University spent innumerable hours reviewing and providing vital input on the quality of our Connect online homework content.

Class Testing

For five years before the production of *University Physics,* the authors tested and refined the textbook's materials with approximately 4000 of our students at Michigan State University. They collected written feedback and also conducted one-on-one interviews with a representative sample of the students, in addition to classroom testing of the in-class exercises and the PowerPoint slides. Several of the authors' colleagues (Alexandra Gade, Alex Brown, Bernard Pope, Carl Schmidt, Chong-Yu Ruan, C. P. Yuan, Dan Stump, Ed Brown, Hendrik Schatz, Kris Starosta, Lisa Lapidus, Michael Harrison, Michael Moore, Reinhard Schwienhorst, Salemeh Ahmad, S. B. Mahanti, Scott Pratt, Stan Schriber, Tibor Nagy, and Thomas Duguet), who co-taught the introductory physics sequence in parallel sections, also provided invaluable help and insight, and their contributions made this book much stronger.

1ST ROUND: AUTHORS' MANUSCRIPT

√ **Multiple rounds of review by college physics instructors**

√ **Independent accuracy check of the text, examples, and solved problems by a professional firm employing mathematicians and physicists**

√ **Second independent accuracy check of the end-of-chapter problems by a team of solution manual authors**

√ **Test bank authors' review**

√ **Clicker question author review**

2ND ROUND: TYPESET PAGES

√ **Authors**

√ **First proofreading**

√ **Third accuracy check of the text, examples, solved problems, and end-of-chapter problems by the same professional firm employing mathematicians and physicists**

3RD ROUND: REVISED TYPESET PAGES

√ **Authors**

√ **Second proofreading**

√ **Accuracy check of any unresolved issues by the professional firm employing mathematicians and physicists**

√ **A parallel fourth accuracy check of the end-of-chapter problem and solution content is made after this content is entered into the Connect Online Homework System, which allows for any further issues to be corrected in the printed text and online solutions manuals**

√ **Reviews by physics instructors**

4TH ROUND: CONFIRMING TYPESET PAGES

√ **Final check by authors**

FINAL ROUND: PRINTING

Accuracy Assurance

The authors and the publisher acknowledge the fact that inaccuracies can be a source of frustration for both instructors and students. Therefore, throughout the writing and production of this first edition, we have worked diligently to eliminate errors and inaccuracies. Ron Fitzgerald, John Klapstein, and their team at MathResources conducted an independent accuracy check and worked through all end-of-chapter questions and problems in the final draft of the manuscript. They then coordinated the resolution of discrepancies between accuracy checks, ensuring the accuracy of the text, the end-of-book answers, and the solutions manuals. Corrections were then made to the manuscript before it was typeset.

The page proofs of the text were double-proofread against the manuscript to ensure the correction of any errors introduced when the manuscript was typeset. Any issues with the textual examples, solved problems and solutions, end-of-chapter questions and problems, and problem answers were accuracy checked by MathResources again at the page proof stage after the manuscript was typeset. This last round of corrections was then cross-checked against the solutions manuals. The end-of-chapter problems from the text along with their solutions were then double-checked by two independent firms upon entry into the

Connect Physics online homework system, and again any issues were addressed in the text and solutions manuals.

Developmental Symposia

McGraw-Hill conducted four symposia and reviewer focus groups directly related to the development of *University Physics*. These events were an opportunity for editors, marketing managers, and digital producers from McGraw-Hill to gather information about the needs and challenges of instructors teaching calculus-based physics courses and to confirm the direction of the first edition of *University Physics*, its supplements, and related digital products.

Nina Abramzon, *California State Polytechnic University–Pomona*
Ed Adelson, *The Ohio State University*
Mohan Aggarwal, *Alabama A&M University*
Rene Bellweid, *Wayne State University*
Jason Brown, *Clemson University*
Ronald Brown, *California Polytechnic University San Luis Obispo*
Mike Dubson, *University of Colorado–Boulder*
David Elmore, *Purdue University*
Robert Endorf, *University of Cincinnati*
Gus Evrard, *University of Michigan*
Chris Gould, *University of Southern California*
John B. Gruber, *San Jose State University*
John Hardy, *Texas A&M University*
David Harrison, *University of Toronto*
Richard Heinz, *Indiana University*
Satoshi Hinata, *Auburn University*
John Hopkins, *The Pennsylvania State University*
T. William Houk, *Miami University–Ohio*
David C. Ingram, *Ohio University–Athens*
Elaine Kirkpatrick, *Rose-Hulman Institute of Technology*
David Lamp, *Texas Tech University*
Michael McInerney, *Rose-Hulman Institute of Technology*
Bruce Mellado, *University of Wisconsin–Madison*
C. Fred Moore, *University of Texas–Austin*
Jeffrey Morgan, *University of Northern Iowa*
Kiumars Parvin, *San Jose State University*
Amy Pope, *Clemson University*

Earl Prohofsky, *Purdue University*
Roberto Ramos, *Drexel University*
Dubravka Rupnik, *Louisiana State University*
Homeyra Sadaghiani, *California State Polytechnic University–Pomona*
Sergey Savrasov, *University of California–Davis*
Marllin Simon, *Auburn University*
Leigh Smith, *University of Cincinnati*
Donna Stokes, *University of Houston*
Michael Strauss, *University of Oklahoma*
Gregory Tarlé, *University of Michigan*

Reviewers of the Text

Numerous instructors participated in over 200 reviews of various drafts of manuscript and the proposed table of contents to provide feedback on the narrative text, content, pedagogical elements, accuracy, organization, problem sets, and general quality. This feedback was summarized by the book team and used to guide the direction of the final-draft manuscript.

Nina Abramzon, *California Polytechnic University–Pomona*
Edward Adelson, *Ohio State University*
Albert Altman, *UMASS Lowell*
Paul Avery, *University of Florida*
David T. Bannon, *Oregon State University*
Marco Battaglia, *UC Berkeley and LBNL*
Douglas R. Bergman, *Rutgers, The State University of New Jersey*
Luca Bertello, *University of California–Los Angeles*
Peter Beyersdorf, *San Jose State University*
Helmut Biritz, *Georgia Institute of Technology*
Ken Thomas Bolland, *Ohio State University*
Richard Bone, *Florida International University*
Dieter Brill, *University of Maryland–College Park*
Branton J. Campbell, *Brigham Young University*
Duncan Carlsmith, *University of Wisconsin–Madison*
Neal Cason, *University of Notre Dame*
K. Kelvin Cheng, *Texas Tech University*
Chris Church, *Miami University of Ohio–Oxford*
Eugenia Ciocan, *Clemson University*
Robert Clare, *University of California–Riverside*
Roy Clarke, *University of Michigan*

J. M. Collins, *Marquette University*
Brent A. Corbin, *University of California–Los Angeles*
Stephane Coutu, *The Pennsylvania State University*
William Dawicke, *Milwaukee School of Engineering*
Mike Dennin, *University of California–Irvine*
John Devlin, *University of Michigan–Dearborn*
John DiNardo, *Drexel University*
Fivos R. Drymiotis, *Clemson University*
Michael DuVernois, *University of Hawaii–Manoa*
David Ellis, *The University of Toledo*
Robert Endorf, *University of Cincinnati*
David Ermer, *Mississippi State University*
Harold Evensen, *University of Wisconsin–Platteville*
Lisa L. Everett, *University of Wisconsin–Madison*
Frank Ferrone, *Drexel University*
Leonard Finegold, *Drexel University*
Ray Frey, *University of Oregon*
J. William Gary, *University of California–Riverside*
Stuart Gazes, *University of Chicago*
Benjamin Grinstein, *University of California–San Diego*
John Gruber, *San Jose State University*
Kathleen A. Harper, *Denison University*
Edwin E. Hach, III, *Rochester Institute of Technology*
John Hardy, *Texas A & M University*
Laurent Hodges, *Iowa State University*
John Hopkins, *The Pennsylvania State University*
George K. Horton, *Rutgers University*
T. William Houk, *Miami University–Ohio*
Eric Hudson, *Massachusetts Institute of Technology*
A. K. Hyder, *University of Notre Dame*
David C. Ingram, *Ohio University–Athens*
Diane Jacobs, *Eastern Michigan University*
Rongying Jin, *The University of Tennessee–Knoxville*
Kate L. Jones, *University of Tennessee*
Steven E. Jones, *Brigham Young University*
Teruki Kamon, *Texas A & M University*
Lev Kaplan, *Tulane University*
Joseph Kapusta, *University of Minnesota*
Kathleen Kash, *Case Western Reserve*
Sanford Kern, *Colorado State University*
Eric Kincanon, *Gonzaga University*
Elaine Kirkpatick, *Rose-Hulman Institute of Technology*
Brian D. Koberlein, *Rochester Institute of Technology*
W. David Kulp, III, *Georgia Institute of Technology*
Fred Kuttner, *University of California–Santa Cruz*
David Lamp, *Texas Tech University*
Andre' LeClair, *Cornell University*
Patrick R. LeClair, *University of Alabama*
Luis Lehner, *Louisiana State University–Baton Rouge*
Michael Lisa, *The Ohio State University*
Samuel E. Lofland, *Rowan University*
Jerome Long, *Virginia Tech*
A. James Mallmann, *Milwaukee School of Engineering*
Pete Markowitz, *Florida International University*
Daniel Marlow, *Princeton University*
Bruce Mason, *Oklahoma University*
Martin McHugh, *Loyola University*
Michael McInerney, *Rose-Hulman Institute of Technology*
David McIntyre, *Oregon State University*
Marina Milner-Bolotin, *Ryerson University–Toronto*
Kieran Mullen, *University of Oklahoma*
Curt Nelson, *Gonzaga University*
Mark Neubauer, *University of Illinois at Urbana–Champaign*
Cindy Neyer, *Tri-State University*

Craig Ogilvie, *Iowa State University*
Bradford G. Orr, *The University of Michigan*
Karur Padmanabhan, *Wayne State University*
Jacqueline Pau, *University of California–Los Angeles*
Leo Piilonen, *Virginia Tech*
Claude Pruneau, *Wayne State University*
Johann Rafelski, *University of Arizona*
Roberto Ramos, *Drexel University*
Lawrence B. Rees, *Brigham Young University*
Andrew J. Rivers, *Northwestern University*
James W. Rohlf, *Boston University*
Philip Roos, *University of Maryland*
Dubravka Rupnik, *Louisiana State University*
Ertan Salik, *California State Polytechnic University–Pomona*
Otto Sankey, *Arizona State University*
Sergey Savrasov, *University of California–Davis*
John Schroeder, *Rensselaer Polytech*
Kunnat Sebastian, *University of Massachusetts–Lowell*
Bjoern Seipel, *Portland State University*
Jerry Shakov, *Tulane University*
Ralph Shiell, *Trent University*
Irfan Siddiqi, *University of California–Berkeley*
Marllin L. Simon, *Auburn University*
Alex Small, *California State Polytechnic University–Pomona*
Leigh Smith, *University of Cincinnati*
Xian-Ning Song, *Richland College*
Jeff Sonier, *Simon Fraser University–Surrey Central*
Chad E. Sosolik, *Clemson University*
Donna W. Stokes, *University of Houston*
James Stone, *Boston University*
Michael G. Strauss, *University of Oklahoma*
Yang Sun, *University of Notre Dame*
Maarij Syed, *Rose-Hulman Institute of Technology*
Douglas C. Tussey, *The Pennsylvania State University*
Somdev Tyagi, *Drexel University*
Erich W. Varnes, *University of Arizona*
Gautam Vemuri, *Indiana University-Purdue University–Indianapolis*
Thad Walker, *University of Wisconsin–Madison*
Fuqiang Wang, *Purdue University*
David J. Webb, *University of California–Davis*
Kurt Wiesenfeld, *Georgia Tech*
Fred Wietfeldt, *Tulane University*
Gary Williams, *University of California–Los Angeles*
Sun Yang, *University of Notre Dame*
L. You, *Georgia Tech*
Billy Younger, *College of the Albemarle*
Andrew Zangwill, *Georgia Institute of Technology*
Jens Zorn, *University of Michigan–Ann Arbor*
Michael Zudov, *University of Minnesota*

Additional International Reviewers of the Text

El Hassan El Aaoud, *University of Hail, Hail KSA*
Mohamed S. Abdelmonem, *King Fahd University of Petroleum and Minerals, Dhahran, Saudi Arabia*
Sudeb Bhattacharya, *Saha Institute of Nuclear Physics, Kolkata, India*
Shi-Jian Gu, *The Chinese University of Hong Kong, Shatin, N.T., Hong Kong*
Nasser M. Hamdan, *The American University of Sharjah*
Moustafa Hussein, *Arab Academy for Science & Engineering, Egypt*
A.K. Jain, *I.I.T. Roorkee*
Carsten Knudsen, *Technical University of Denmark*
Ajal Kumar, *The University of the South Pacific, Fiji*
Ravindra Kumar Sinha, *Delhi College of Engineering*
Nazir Mustapha, *Al-Imam University*

Acknowledgments

Reza Nejat, *McMaster University*
K. Porsezian, *Pondicherry University, Puducherry*
Wang Qing-hai, *National University of Singapore*
Kenneth J. Ragan, *McGill University*

A book like the one that you are holding in your hands is impossible to produce without tremendous work by an incredible number of dedicated individuals. First and foremost, we would like to thank the talented marketing and editorial team from McGraw-Hill: Marty Lange, Kent Peterson, Thomas Timp, Ryan Blankenship, Mary Hurley, Liz Recker, Daryl Bruflodt, Lisa Nicks, Dan Wallace, and, in particular, Deb Hash helped us in innumerable ways and managed to reignite our enthusiasm after each revision. Their team spirit, good humor, and unbending optimism kept us on track and always made it fun for us to put in the seemingly endless hours it took to produce the manuscript.

The developmental editors, Richard Heinz and David Chelton, helped us work through the near infinite number of comments and suggestions for improvements from our reviewers. They, as well as the reviewers and our board of advisors, deserve a large share of the credit for improving the quality of the final manuscript. Our colleagues on the faculty of the Department of Physics and Astronomy at Michigan State University—Alexandra Gade, Alex Brown, Bernard Pope, Carl Schmidt, Chong-Yu Ruan, C. P. Yuan, Dan Stump, Ed Brown, Hendrik Schatz, Kris Starosta, Lisa Lapidus, Michael Harrison, Michael Moore, Reinhard Schwienhorst, Salemeh Ahmad, S. B. Mahanti, Scott Pratt, Stan Schriber, Tibor Nagy, and Thomas Duguet—helped us in innumerable ways as well, teaching their classes and sections with the materials developed by us and in the process providing invaluable feedback on what worked and what needed additional refinement. We thank all of them.

We decided to involve a large number of physics instructors from around the country in the authoring of the end-of-chapter problems, in order to ensure that they are of the highest quality, relevance, and didactic value. We thank all of our problem contributors for sharing some of their best work with us, in particular, Richard Hallstein, who took on the task of organizing and processing all contributions.

At the point when we turned in the final manuscript to the publisher, a whole new army of professionals took over and added another layer of refinement, which transformed a manuscript into a book. John Klapstein and the team at MathResources worked through each and every homework problem, each exercise, and each number and equation we wrote down. The photo researchers, in particular, Danny Meldung, improved the quality of the images used in the book immensely, and they made the selection process fun for us. Pamela Crews and the team at Precision Graphics used our original drawings but improved their quality substantially, while at the same time remaining true to our original calculations that went into producing the drawings. Our copyeditor, Jane Hoover, and her team pulled it all together in the end, deciphered our scribbling, and made sure that the final product is as readable as possible. McGraw-Hill's design and production team of Jayne Klein, David Hash, Carrie Burger, Sandy Ludovissy, Judi David, and Mary Jane Lampe expertly guided the book and its ancillary materials through to publication. All of them deserve our tremendous gratitude.

Finally, we could not have made it through the last six years of effort without the support of our families, who had to put up with us working on the book through untold evenings, weekends, and even during many vacations. We hope that all of their patience and encouragement has paid off, and we thank them from the bottom of our hearts for sticking with us during the completion of this book.

—*Wolfgang Bauer*
—*Gary D. Westfall*

University Physics

with Modern Physics   Volume Two

Electrostatics

21

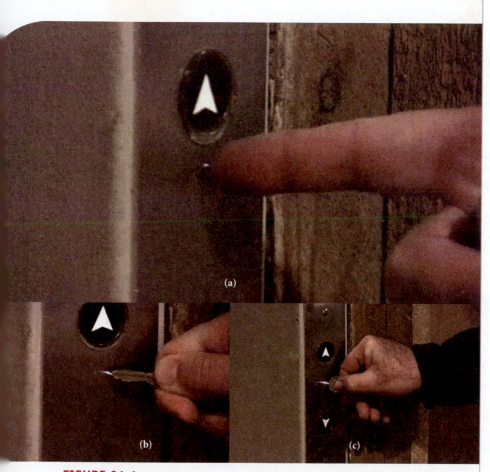

FIGURE 21.1 (a) A spark due to static electricity occurs between a person's hand and a metal surface when pushing an elevator button. (b) and (c) Similar sparks are generated when the person holds a metal object like a car key or a coin, but are painless because the spark forms between the metal surface and the metal object.

WHAT WE WILL LEARN

- Electricity and magnetism together make up electromagnetism, one of the four fundamental forces of nature.

- There are two kinds of electric charge, positive and negative. Like charges repel, and unlike charges attract.

- Electric charge is quantized, meaning that it occurs only in integral multiples of a smallest elementary quantity. Electric charge is also conserved.

- Most materials around us are electrically neutral.

- The electron is an elementary particle, and its charge is the smallest observable quantity of electric charge.

- Insulators conduct electricity poorly or not at all. Conductors conduct electricity well but not perfectly—some energy losses occur.

- Semiconductors can be made to change between a conducting state and a nonconducting state.

- Superconductors conduct electricity perfectly.

- Objects can be charged directly by contact or indirectly by induction.

- The force that two stationary electric charges exert on each other is proportional to the product of the charges and varies as the inverse square of the distance between the two charges.

Many people think of static electricity as the annoying spark that occurs when they reach for a metal object like a doorknob on a dry day, after they have been walking on a carpet (Figure 21.1). In fact, many electronics manufacturers place small metal plates on equipment so that users can discharge any spark on the plate and not damage the more sensitive parts of the equipment. However, static electricity is more than just an occasional annoyance; it is the starting point for any study of electricity and magnetism, forces that have changed human society as radically as anything since the discovery of fire or the wheel.

In this chapter, we examine the properties of electric charge. A moving electric charge gives rise to a separate phenomenon, called *magnetism,* which is covered in later chapters. Here we look at charged objects that are not moving—hence the term *electrostatics.* All objects have charge, since charged particles make up atoms and molecules. We often don't notice the effects of electrical charge because most objects are electrically neutral. The forces that hold atoms together and that keep objects separate even when they're in contact, are all electric in nature.

21.1 Electromagnetism

FIGURE 21.2 Lightning strikes over Seattle.

Perhaps no mystery puzzled ancient civilizations more than electricity, primarily in the form of lightning strikes (Figure 21.2). The destructive force inherent in lightning, which could set objects on fire and kill people and animals, seemed godlike. The ancient Greeks, for example, believed Zeus, father of the gods, had the ability to throw lightning bolts. The Germanic tribes ascribed this power to the god Thor and the Romans to the god Jupiter. Characteristically, the ability to cause lightning belonged to the god at the top (or near the top) of the hierarchy.

The ancient Greeks knew that if you rubbed a piece of amber with a piece of cloth, you could attract small, light objects with the amber. We now know that rubbing amber with a cloth transfers negatively charged particles called *electrons* from the cloth to the amber. (The words *electron* and *electricity* derive from the Greek word for amber.) Lightning also consists of a flow of electrons. The early Greeks and others also knew about naturally occurring magnetic objects called *lodestones,* which were found in deposits of magnetite, a mineral consisting of iron oxide. These objects were used to construct compasses as early as 300 BC.

The relationship between electricity and magnetism was not understood until the middle of the 19th century. The following chapters will reveal how electricity and magnetism can be unified into a common framework called *electromagnetism.* However, unification of forces does not stop there. During the early part of the 20th century, two more fundamental forces were discovered: the weak force, which operates in beta decay (in which an electron and a neutrino are spontaneously emitted from certain types of nuclei), and the strong force, which acts inside

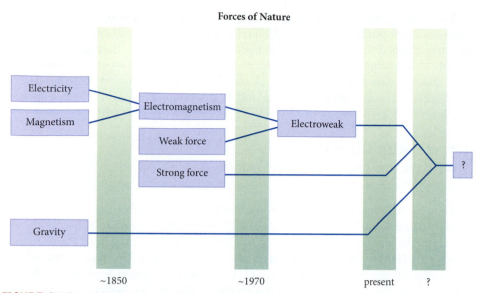

Forces of Nature

FIGURE 21.3 The history of the unification of fundamental forces.

the atomic nucleus. We'll study these forces in more detail in Chapter 39 on particle physics. Currently, the electromagnetic and weak forces are viewed as two aspects of the electroweak force (Figure 21.3). For the phenomena discussed in the following chapters, this electroweak unification has no influence; it becomes important in the highest-energy particle collisions. Because the energy scale for the electroweak unification is so high, most textbooks continue to speak of four fundamental forces: gravitational, electromagnetic, weak, and strong.

Today, a large number of physicists believe that the electroweak force and the strong force can also be unified, that is, described in a common framework. Several theories propose ways to accomplish this, but so far experimental evidence is missing. Interestingly, the force that has been known longer than any of the other fundamental forces, gravity, seems to be hardest to shoehorn into a unified framework with the other fundamental forces. Quantum gravity, supersymmetry, and string theory are current foci of cutting-edge physics research in which theorists are attempting to construct this grand unification and discover the (hubristically named) Theory of Everything. They are mainly guided by symmetry principles and the conviction that nature must be elegant and simple.

We'll return to these considerations in Chapters 39 and 40. In this chapter, we consider electric charge, how materials react to electric charge, static electricity, and the forces resulting from electric charges. **Electrostatics** covers situations where charges stay in place and do not move.

21.2 Electric Charge

Let's look a little deeper into the cause of the electric sparks that you occasionally receive on a dry winter day if you walk across a carpet and then touch a metal doorknob. (Electrostatic sparks have even ignited gas fumes while someone is filling the tank at a gas station. This is not an urban legend; a few of these cases have been caught on gas station surveillance cameras.) The process that causes this sparking is called **charging.** Charging consists of the transfer of negatively charged particles, called **electrons,** from the atoms and molecules of the material of the carpet to the soles of your shoes. This charge can move relatively easily through your body, including your hands. The built-up electric charge discharges through the metal of the doorknob, creating a spark.

The two types of electric charge found in nature are **positive charge** and **negative charge.** Normally, objects around us do not seem to be charged; instead, they are electrically neutral. Neutral objects contain roughly equal numbers of positive and negative charges that largely cancel each other. Only when positive and negative charges are not balanced do we observe the effects of electric charge.

If you rub a glass rod with a cloth, the glass rod becomes charged and the cloth acquires a charge of the opposite sign. If you rub a plastic rod with fur, the rod and fur also become oppositely charged. If you bring two charged glass rods together, they repel each other. Similarly, if you bring two charged plastic rods together, they also repel each other. However, a charged glass rod and a charged plastic rod will attract each other. This difference arises because the glass rod and the plastic rod have opposite charge. This observation leads us to the

Law of Electric Charges
Like charges repel and opposite charges attract.

The unit of electric charge is the **coulomb** (C), named after the French physicist Charles-Augustine de Coulomb (1736–1806). The coulomb is defined in terms of the SI unit for current, the ampere (A), named after another French physicist, André-Marie Ampère (1775–1836). Neither the ampere nor the coulomb can be derived in terms of the other SI units: meter, kilogram, and second. Instead, the ampere is another fundamental SI unit. For this reason, the SI system of units is sometimes called *MKSA (meter-kilogram-second-ampere) system*. The charge unit is defined as

$$1\,C = 1\,A\,s. \tag{21.1}$$

The definition of the ampere must wait until we discuss current in later chapters. However, we can define the magnitude of the coulomb by simply specifying the charge of a single electron:

$$q_e = -e \tag{21.2}$$

where q_e is the charge and e has the (currently best accepted and experimentally measured) value

$$e = 1.602176487(40) \cdot 10^{-19}\,C. \tag{21.3}$$

(Usually it is enough to carry only the first two to four significant digits of this mantissa. We will use a value of 1.602 in this chapter, but you should keep in mind that equation 21.3 gives the full accuracy to which this charge has been measured.)

The charge of the electron is an intrinsic property of the electron, just like its mass. The charge of the **proton,** another basic particle of atoms, is exactly the same magnitude as that of the electron, only the proton's charge is positive:

$$q_p = +e. \tag{21.4}$$

The choice of which charge is positive and which charge is negative is arbitrary. The conventional choice of $q_e < 0$ and $q_p > 0$ is due to the American statesman, scientist, and inventor Benjamin Franklin (1706–1790), who pioneered studies of electricity.

One coulomb is an extremely large unit of charge. We'll see later in this chapter just how big it is when we investigate the magnitude of the forces of charges on each other. Units of μC (microcoulombs, 10^{-6} C), nC (nanocoulombs, 10^{-9} C), and pC (picocoulombs, 10^{-12} C) are commonly used.

Benjamin Franklin also proposed that charge is conserved. For example, when you rub a plastic rod with fur, electrons are transferred to the plastic rod, leaving a net positive charge on the fur. (Protons are not transferred because they are usually embedded inside atomic nuclei.) The charge is not created or destroyed, simply moved from one object to another.

Law of Charge Conservation
The total electric charge of an isolated system is conserved.

This law is the fourth conservation law we have encountered so far, the first three being the conservation laws for total energy, momentum, and angular momentum. Conservation laws are a common thread that runs throughout all of physics and thus throughout this book as well.

21.1 In-Class Exercise

How many electrons does it take to make 1.00 C of charge?

a) $1.60 \cdot 10^{19}$ d) $6.24 \cdot 10^{18}$

b) $6.60 \cdot 10^{19}$ e) $6.66 \cdot 10^{17}$

c) $3.20 \cdot 10^{16}$

It is important to note that there is a conservation law for charge, but *not* for mass. We'll see later in this book that mass and energy are not independent of each other. What is sometimes described in introductory chemistry as conservation of mass is not an exact conservation law, but only an approximation used to keep track of the number of atoms in chemical reactions. (It is a good approximation to a large number of significant figures but not an exact law, like charge conservation.) Conservation of charge applies to all systems, from the macroscopic system of plastic rod and fur down to systems of subatomic particles.

Elementary Charge

Electric charge occurs only in integral multiples of a minimum size. This is expressed by saying that charge is **quantized.** The smallest observable unit of electric charge is the charge of the electron, which is $-1.602 \cdot 10^{-19}$ C (as defined in equation 21.3).

The fact that electric charge is quantized was verified in an ingenious experiment carried out in 1910 by American physicist Robert A. Millikan (1868–1953) and known as the *Millikan oil drop experiment* (Figure 21.4). In this experiment, oil drops were sprayed into a chamber where electrons were knocked out of the drops by some form of radiation, usually X-rays. The resulting positively charged drops were allowed to fall between two electrically charged plates. Adjusting the charge of the plates caused the drops to stop falling and allowed their charge to be measured. What Millikan observed was that charge was quantized rather than continuous. (A quantitative analysis of this experiment will be presented in Chapter 23 on electric potential.) That is, this experiment and its subsequent refinements established that charge comes only in integer multiples of the charge of an electron. In everyday experiences with electricity, we do not notice that charge is quantized because most electrical phenomena involve huge numbers of electrons.

In Chapter 13, we discussed the fact that matter is composed of atoms and that an atom consists of a nucleus containing charged protons and neutral neutrons. A schematic drawing of a carbon atom is shown in Figure 21.5. A carbon atom has six protons and (usually) six neutrons in its nucleus. This nucleus is surrounded by six electrons. Note that this drawing is not to scale. In the actual atom, the distance of the electrons from the nucleus is much larger (by a factor on the order of 10,000) than the size of the nucleus. In addition, the electrons are shown in circular orbits, which is also not quite correct. In Chapter 38, we'll see that the locations of electrons in the atom can be characterized only by probability distributions.

As mentioned earlier, a proton has a positive charge with a magnitude that is *exactly* equal to the magnitude of the negative charge of an electron. In a neutral atom, the number of negatively charged electrons is equal to the number of positively charged protons. The mass of the electron is much smaller than the mass of the proton or the neutron. Therefore, most of the mass of an atom resides in the nucleus. Electrons can be removed from atoms relatively easily. For this reason, electrons are typically the carriers of electricity, rather than protons or atomic nuclei.

The electron is a fundamental particle and has no substructure: It is a point particle with zero radius (at least, according to current understanding). However, high-energy probes have been used to look inside the proton. A proton is composed of charged particles called *quarks,* held together by uncharged particles called *gluons.* Quarks have a charge of $\pm\frac{1}{3}$ or $\pm\frac{2}{3}$ times the charge of the electron. These fractionally charged particles cannot exist independently and have never been observed directly, despite numerous extensive searches. Just like the charge of an electron, the charges of quarks are intrinsic properties of these elementary particles.

A proton is composed of two *up quarks* (each with charge $+\frac{2}{3}e$) and one *down quark* (with charge $-\frac{1}{3}e$), giving the proton a charge of $q_p = (2)(+\frac{2}{3}e) + (1)(-\frac{1}{3}e) = +e$ as illustrated

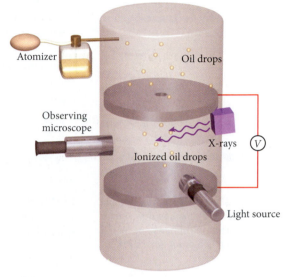

FIGURE 21.4 Schematic drawing of the Millikan oil drop experiment.

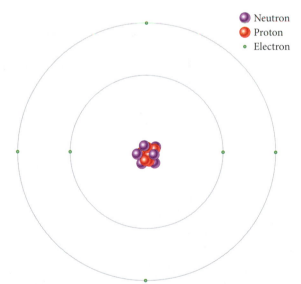

FIGURE 21.5 In a carbon atom, the nucleus contains six neutrons and six protons. The nucleus is surrounded by six electrons. Note that this drawing is schematic and not to scale.

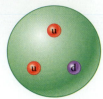

Proton

$$q_p = +\tfrac{2}{3}e + \tfrac{2}{3}e - \tfrac{1}{3}e = +e$$

(a)

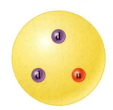

Neutron

$$q_n = +\tfrac{2}{3}e - \tfrac{1}{3}e - \tfrac{1}{3}e = 0$$

(b)

FIGURE 21.6 (a) A proton contains two up quarks (u) and one down quark (d). (b) A neutron contains one up quark (u) and two down quarks (d).

in Figure 21.6a. The electrically neutral neutron (hence the name!) is composed of an up quark and two down quarks, as shown in Figure 21.6b, so its charge is $q_n = (1)(+\tfrac{2}{3}e) + (2)(-\tfrac{1}{3}e) = 0$. In Chapter 39, we'll see that there are other, much more massive, quarks named *strange, charm, bottom,* and *top,* which have the same charges as the up and down quarks. There are also much more massive electron-like particles named *muon* and *tau.* But the basic fact remains that all of the matter in everyday experience is made up of electrons (with electrical charge –e), up and down quarks (with electrical charges $+\tfrac{2}{3}e$ and $-\tfrac{1}{3}e$, respectively), and gluons (zero charge).

It is remarkable that the charges of the quarks inside a proton add up to *exactly* the same magnitude as the charge of the electron. This fact is still a puzzle, pointing to some deep symmetry in nature that is not yet understood.

Because all macroscopic objects are made of atoms, which in turn are made of electrons and atomic nuclei consisting of protons and neutrons, the charge, q, of any object can be expressed in terms of the sum of the number of protons, N_p, minus the sum of the number of electrons, N_e, that make up the object:

$$q = e \cdot (N_p - N_e). \tag{21.5}$$

EXAMPLE 21.1 | Net Charge

PROBLEM
If we wanted a block of iron of mass 3.25 kg to acquire a positive charge of 0.100 C, what fraction of the electrons would we have to remove?

SOLUTION
Iron has mass number 56. Therefore, the number of iron atoms in the 3.25-kg block is

$$N_{atom} = \frac{(3.25 \text{ kg})(6.022 \cdot 10^{23} \text{ atoms/mole})}{0.0560 \text{ kg/mole}} = 3.495 \cdot 10^{25} = 3.50 \cdot 10^{25} \text{ atoms.}$$

Note that we have used Avogadro's number, $6.022 \cdot 10^{23}$, and the definition of the mole, which specifies that the mass of 1 mole of a substance in grams is just the mass number of the substance—in this case, 56.

Because the atomic number of iron is 26, which equals the number of protons or electrons in an iron atom, the total number of electrons in the 3.25-kg block is:

$$N_e = 26 N_{atom} = (26)(3.495 \cdot 10^{25}) = 9.09 \cdot 10^{26} \text{ electrons.}$$

We use equation 21.5 to find the number of electrons, $N_{\Delta e}$, that we would have to remove. Because the number of electrons equals the number of protons in the original uncharged object, the difference in the number of protons and electrons is the number of removed electrons, $N_{\Delta e}$:

$$q = e \cdot N_{\Delta e} \Rightarrow N_{\Delta e} = \frac{q}{e} = \frac{0.100 \text{ C}}{1.602 \cdot 10^{-19} \text{ C}} = 6.24 \cdot 10^{17}.$$

Finally, we obtain the fraction of electrons we would have to remove:

$$\frac{N_{\Delta e}}{N_e} = \frac{6.24 \cdot 10^{17}}{9.09 \cdot 10^{26}} = 6.87 \cdot 10^{-10}.$$

We would have to remove fewer than one in a billion electrons from the iron block in order to put the sizable positive charge of 0.100 C on it.

21.1 Self-Test Opportunity

Give the charge of the following elementary particles or atoms in terms of the elementary charge $e = 1.602 \cdot 10^{-19}$ C.

a) proton

b) neutron

c) helium atom (two protons, two neutron, and two electrons)

d) hydrogen atom (one proton and one electron)

e) up quark

f) down quark

g) electron

h) alpha particle (two protons and two neutrons)

21.3   Insulators, Conductors, Semiconductors, and Superconductors

Materials that conduct electricity well are called **conductors.** Materials that do not conduct electricity are called **insulators.** (Of course, there are good and poor conductors and good and poor insulators, depending on the properties of the specific materials.)

The electronic structure of a material refers to the way in which electrons are bound to nuclei, as we'll discuss in later chapters. For now, we are interested in the relative propensity of the atoms of a material to either give up or acquire electrons. For insulators, no free movement of electrons occurs because the material has no loosely bound electrons that can escape from its atoms and thereby move freely throughout the material. Even when external charge is placed on an insulator, this external charge cannot move appreciably. Typical insulators are glass, plastic, and cloth.

On the other hand, materials that are conductors have an electronic structure that allows the free movement of some electrons. The positive charges of the atoms of a conducting material do not move, since they reside in the heavy nuclei. Typical solid conductors are metals. Copper, for example, is a very good conductor used in electrical wiring.

Fluids and organic tissue can also serve as conductors. Pure distilled water is not a very good conductor. However, dissolving common table salt (NaCl), for example, in water improves its conductivity tremendously, because the positively charged sodium ions (Na^+) and negatively charged chlorine ions (Cl^-) can move within the water to conduct electricity. In liquids, unlike solids, positive as well as negative charge carriers are mobile. Organic tissue is not a very good conductor, but it conducts electricity well enough to make large currents dangerous to us. (We'll learn more about electrical current in Chapter 26, where these terms, which are in everyday use, will be defined precisely.)

Semiconductors

A class of materials called **semiconductors** can change from being an insulator to being a conductor and back to an insulator again. Semiconductors were discovered only a little more than 50 years ago but are the backbone of the entire computer and consumer electronics industries. The first widespread use of semiconductors was in transistors (Figure 21.7a); modern computer chips (Figure 21.7b) perform the functions of millions of transistors. Computers and basically all modern consumer electronics products and devices (televisions, cameras, video game players, cell phones, etc.) would be impossible without semiconductors. Gordon Moore, cofounder of Intel, famously stated that due to advancing technology, the power of the average computer's CPU (central processing unit) doubles every 18 months, which is an empirical average over the last 5 decades. This doubling phenomenon is known as *Moore's Law*. Physicists have been and will undoubtedly continue to be the driving force behind this process of scientific discovery, invention, and improvement.

Semiconductors are of two kinds: intrinsic and extrinsic. Examples of *intrinsic semiconductors* are chemically pure crystals of gallium arsenide, germanium, or, especially, silicon. Engineers produce *extrinsic semiconductors* by *doping*, which is the addition of minute amounts (typically 1 part in 10^6) of other materials that can act as electron donors or electron receptors. Semiconductors doped with electron donors are called *n-type* (n stands for "negative charge"). If the doping substance acts as an electron receptor, the hole left behind by an electron that attaches to a receptor can also travel through the semiconductor and acts as an effective positive charge carrier. These semiconductors are consequently called *p-type* (p stand for "positive charge"). Thus, unlike normal solid conductors in which only negative charges move, semiconductors have movement of negative or positive charges (which are really electron holes, that is, missing electrons).

Superconductors

Superconductors are materials that have zero resistance to the conduction of electricity, as opposed to normal conductors, which conduct electricity well but with some losses. Materials are superconducting only at very low temperatures. A typical superconductor is a niobium-titanium alloy that must be kept near the temperature of liquid helium (4.2 K) to retain its superconducting properties. During the last 20 years, new materials called *high-T_c superconductors* (T_c stands for "critical temperature," which is the maximum temperature that allows superconductivity) have been developed. These are superconducting at liquid-nitrogen temperature (77.3 K). Materials that are superconductors at room temperature (300 K) have not yet been found, but they would be extremely useful. Research directed

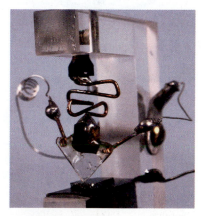

(a)

(b)

FIGURE 21.7 (a) Replica of the first transistor, invented in 1947 by John Bardeen, Walter H. Brattain, and William B. Shockley. (b) Modern computer chips made from silicon wafers contain many tens of millions of transistors.

toward developing such materials and on theoretically explaining what physical phenomena cause high-T_c superconductivity is currently in progress.

The topics of conductivity, superconductivity, and semiconductors will be discussed in more quantitative detail in the following chapters.

FIGURE 21.8 A typical electroscope used in lecture demonstrations.

21.2 In-Class Exercise

The hinged conductor moves away from the fixed conductor if a charge is applied to the electroscope, because

a) like charges repel each other.

b) like charges attract each other.

c) unlike charges attract each other.

d) unlike charges repel each other.

FIGURE 21.9 Inducing a charge: (a) An uncharged electroscope. (b) A negatively charged paddle is brought near the electroscope. (c) The negatively charged paddle is taken away.

21.4 Electrostatic Charging

Giving a static charge to an object is a process known as **electrostatic charging.** Electrostatic charging can be understood through a series of simple experiments. A power supply serves as a ready source of positive and negative charge. The battery in your car is a similar power supply; it uses chemical reactions to create a separation between positive and negative charge. Several insulating paddles can be charged with positive or negative charge from the power supply. In addition, a conducting connection is made to the Earth. The Earth is a nearly infinite reservoir of charge, capable of effectively neutralizing electrically charged objects in contact with it. This taking away of charge is called **grounding,** and an electrical connection to the Earth is called a **ground.**

An **electroscope** is a device that gives an observable response when it is charged. You can build a relatively simple electroscope by using two strips of very thin metal foil that are attached at one end and are allowed to hang straight down adjacent to each other from an isolating frame. Kitchen aluminum foil is not suitable, because it is too thick, but hobby shops sell thinner metal foils. For the isolating frame, you can use a Styrofoam coffee cup turned sideways, for example.

The lecture-demonstration-quality electroscope shown in Figure 21.8 has two conductors that in their neutral position are touching and oriented in a vertical direction. One of the conductors is hinged at its midpoint so that it will move away from the fixed conductor if a charge appears on the electroscope. These two conductors are in contact with a conducting ball on top of the electroscope, which allows charge to be applied or removed easily.

An uncharged electroscope is shown in Figure 21.9a. The power supply is used to give a negative charge to one of the insulating paddles. When the paddle is brought near the ball of the electroscope, as shown in Figure 21.9b, the electrons in the conducting ball of the electroscope are repelled, which produces a net negative charge on the conductors of the electroscope. This negative charge causes the movable conductor to rotate because the stationary conductor also has negative charge and repels it. Because the paddle did not touch the ball, the charge on the movable conductors is **induced.** If the charged paddle is then taken away, as illustrated in Figure 21.9c, the induced charge reduces to zero, and the movable conductor returns to its original position, because the total charge on the electroscope did not change in the process.

If the same process is carried out with a positively charged paddle, the electrons in the conductors are attracted to the paddle and flow into the conducting ball. This leaves a net positive charge on the conductors, causing the movable conducting arm to rotate again. Note that the net charge of the electroscope is zero in both cases and that the motion of the conductor indicates only that the paddle is charged. When the positively charged paddle

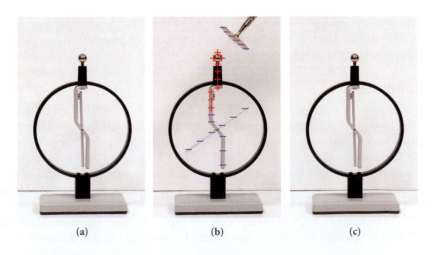

(a)   (b)   (c)

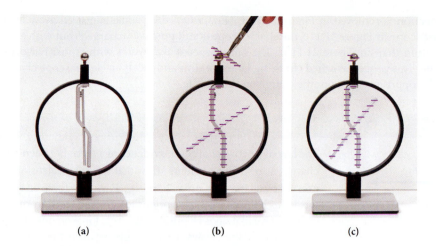

(a)   (b)   (c)

FIGURE 21.10 Charging by contact: (a) An uncharged electroscope. (b) A negatively charged paddle touches the electroscope. (c) The negatively charged paddle is removed.

is removed, the movable conductor again returns to its original position. It is important to note that we cannot determine the sign of this charge!

On the other hand, if a negatively charged insulating paddle *touches* the ball of the electroscope, as shown in Figure 21.10b, electrons will flow from the paddle to the conductor, producing a net negative charge. When the paddle is removed, the charge remains and the movable arm remains rotated, as shown in Figure 21.10c. Similarly, if a positively charged insulating paddle touches the ball of the uncharged electroscope, the electroscope transfers electrons to the positively charged paddle and becomes positively charged. Again, both a positively charged paddle and a negatively charged paddle have the same effect on the electroscope, and we have no way of determining whether the paddles are positively charged or negatively charged. This process is called **charging by contact.**

The two different kinds of charge can be demonstrated by first touching a negatively charged paddle to the electroscope, producing a rotation of the movable arm, as shown in Figure 21.10. If a positively charged paddle is then brought into contact with the electroscope, the movable arm returns to the uncharged position. The charge is neutralized (assuming both paddles originally had the same absolute value of charge). Thus, there are two kinds of charge. However, because charges are manifestations of mobile electrons, a negative charge is an excess of electrons and a positive charge is a deficit of electrons.

The electroscope can be given a charge without touching it with the charged paddle, as shown in Figure 21.11. The uncharged electroscope is shown in Figure 21.11a. A negatively charged paddle is brought close to the ball of the electroscope but not touching it, as shown in Figure 21.11b. In Figure 21.11c, the electroscope is connected to a ground. Then, while the charged paddle is still close to but not touching the ball of the electroscope, the ground

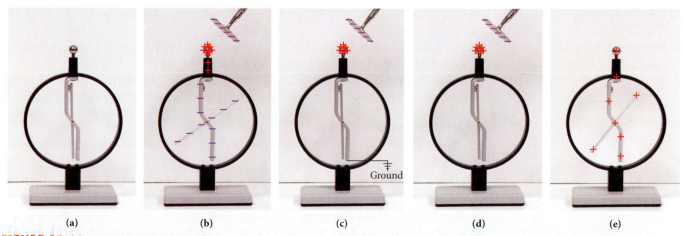

(a)   (b)   (c)   (d)   (e)

FIGURE 21.11 Charging by induction: (a) An uncharged electroscope. (b) A negatively charged paddle is brought close to the electroscope. (c) A ground is connected to the electroscope. (d) The connection to the ground is removed. (e) The negatively charged paddle is taken away, leaving the electroscope positively charged.

connection is removed in Figure 21.11d. Now, when the paddle is moved away from the electroscope in Figure 21.11e, the electroscope is still positively charged (but with a smaller deflection than in Figure 21.11b). The same process also works with a positively charged paddle. This process is called **charging by induction** and yields an electroscope charge that has the opposite sign from the charge on the paddle.

21.5   Electrostatic Force—Coulomb's Law

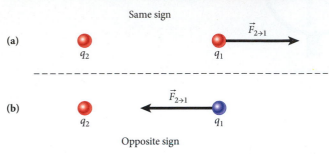

FIGURE 21.12   The force exerted by charge 2 on charge 1: (a) two charges with the same sign; (b) two charges with opposite signs.

The law of electric charges is evidence of a force between any two charges at rest. Experiments show that for the electrostatic force exerted by a charge q_2 on a charge q_1, $\vec{F}_{2\to1}$, the force on q_1 points toward q_2 if the charges have opposite signs and away from q_2 if the charges have like signs (Figure 21.12). This force on one charge due to another charge always lies on a line between the two charges. **Coulomb's Law** gives the magnitude of this force as

$$F = k\frac{\left|q_1 q_2\right|}{r^2},\qquad(21.6)$$

where q_1 and q_2 are electric charges, $r = \left|\vec{r}_1 - \vec{r}_2\right|$ is the distance between them, and

$$k = 8.99\cdot10^9\ \frac{\mathrm{N\,m}^2}{\mathrm{C}^2}\qquad(21.7)$$

is **Coulomb's constant.** You can see that one Coulomb is a *very* large charge. If two charges of 1 C each were at a distance of 1 m apart, the magnitude of the force they would exert on each other would be 8.99 billion N. For comparison, this force equals the weight of 450 fully loaded space shuttles!

The relationship between Coulomb's constant and another constant, ϵ_0, called the **electric permittivity of free space,** is

$$k = \frac{1}{4\pi\epsilon_0}.\qquad(21.8)$$

Consequently, the value of ϵ_0 is

$$\epsilon_0 = 8.85\cdot10^{-12}\ \frac{\mathrm{C}^2}{\mathrm{N\,m}^2}.\qquad(21.9)$$

An alternative way of writing equation 21.6 is then

$$F = \frac{1}{4\pi\epsilon_0}\frac{\left|q_1 q_2\right|}{r^2}.\qquad(21.10)$$

As you'll see in the next few chapters, some equations in electrostatics are more convenient to write with k, while others are more easily written in terms of $1/(4\pi\epsilon_0)$.

Note that the charges in equations 21.6 and 21.10 can be positive or negative, so the product of the charges can also be positive or negative. Since opposite charges attract and like charges repel, a negative value for the product $q_1 q_2$ signifies attraction and a positive value means repulsion.

Finally, Coulomb's Law for the force due to charge 2 on charge 1 can be written in vector form:

$$\vec{F}_{2\to1} = -k\frac{q_1 q_2}{r^3}(\vec{r}_2 - \vec{r}_1) = -k\frac{q_1 q_2}{r^2}\hat{r}_{21}.\qquad(21.11)$$

In this equation, $\hat{r}_{21}$ is a unit vector pointing from q_2 to q_1 (see Figure 21.13). The negative sign indicates that the force is repulsive if both charges are positive or both charges are negative. In that case, $\vec{F}_{2\to1}$ points away from charge 2, as depicted in Figure 21.13a. On the other hand, if one of the charges is positive and the other negative, then $\vec{F}_{2\to1}$ points toward charge 2, as shown in Figure 21.13b.

21.3   In-Class Exercise

You place two charges a distance r apart. Then you double each charge and double the distance between the charges. How does the force between the two charges change?

a) The new force is twice as large.

b) The new force is half as large.

c) The new force is four times as large.

d) The new force is four times smaller.

e) The new force is the same.

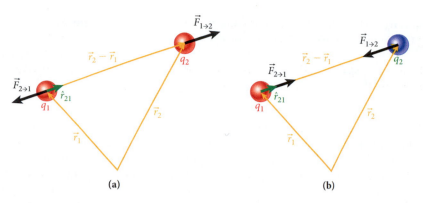

(a)   **(b)**

FIGURE 21.13 Electrostatic force vectors, which two charges exert on each other: (a) two charges of like sign; (b) two charges of opposite sign.

If charge 2 exerts the force $\vec{F}_{2\to1}$ on charge 1, then the force $\vec{F}_{1\to2}$ that charge 1 exerts on charge 2 is simply obtained from Newton's Third Law (see Chapter 4): $\vec{F}_{1\to2} = -\vec{F}_{2\to1}$.

Superposition Principle

So far in this chapter, we have been dealing with two charges. Now let's consider three point charges, q_1, q_2, and q_3, at positions x_1, x_2, and x_3, respectively, as shown in Figure 21.14. The force exerted by charge 1 on charge 3, $\vec{F}_{1\to3}$, is given by

$$\vec{F}_{1\to3} = -\frac{kq_1q_3}{\left(x_3 - x_1\right)^2}\hat{x}.$$

The force exerted by charge 2 on charge 3 is

$$\vec{F}_{2\to3} = -\frac{kq_2q_3}{\left(x_3 - x_2\right)^2}\hat{x}.$$

The force that charge 1 exerts on charge 3 is not affected by the presence of charge 2. The force that charge 2 exerts on charge 3 is not affected by the presence of charge 1. In addition, the forces exerted by charge 1 and charge 2 on charge 3 add vectorially to produce a net force on charge 3:

$$\vec{F}_{net\to3} = \vec{F}_{1\to3} + \vec{F}_{2\to3}.$$

This superposition of forces is completely analogous to that described in Chapter 4 for forces such as gravity and friction.

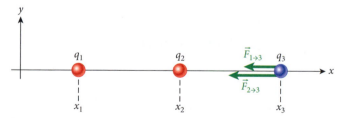

FIGURE 21.14 The forces exerted on charge 3 by charge 1 and charge 2.

21.4 In-Class Exercise

What do the forces acting on the charge q_3 in Figure 21.14 indicate about the signs of the three charges?

a) All three charges must be positive.

b) All three charges must be negative.

c) Charge q_3 must be zero.

d) Charges q_1 and q_2 must have opposite signs.

e) Charges q_1 and q_2 must have the same sign, and q_3 must have the opposite sign.

21.5 In-Class Exercise

Assuming that the lengths of the vectors in Figure 21.14 are proportional to the magnitudes of the forces they represent, what do they indicate about the magnitudes of the charges q_1 and q_2? (*Hint:* The distance between x_1 and x_2 is the same as the distance between x_2 and x_3.)

a) $|q_1| < |q_2|$

b) $|q_1| = |q_2|$

c) $|q_1| > |q_2|$

d) The answer cannot be determined from the information given in the figure.

EXAMPLE **21.2** | **Electrostatic Force inside the Atom**

PROBLEM 1

What is the magnitude of the electrostatic force that the two protons inside the nucleus of a helium atom exert on each other?

SOLUTION 1

The two protons and two neutrons in the nucleus of the helium atom are held together by the strong force; the electrostatic force is pushing the protons apart. The charge of each

Continued—

proton is $q_p = +e$. A distance of approximately $r = 2 \cdot 10^{-15}$ m separates the two protons. Using Coulomb's Law, we can find the force:

$$F = k\frac{|q_p q_p|}{r^2} = \left(8.99 \cdot 10^9 \ \frac{\text{N m}^2}{\text{C}^2}\right)\frac{\left(+1.6 \cdot 10^{-19} \ \text{C}\right)\left(+1.6 \cdot 10^{-19} \ \text{C}\right)}{\left(2 \cdot 10^{-15} \ \text{m}\right)^2} = 58 \ \text{N}.$$

Therefore, the two protons in the atomic nucleus of a helium atom are being pushed apart with a force of 58 N (approximately the weight of a small dog). Considering the size of the nucleus, this is an astonishingly large force. Why do atomic nuclei not simply explode? The answer is that an even stronger force, the aptly named strong force, keeps them together.

21.6 In-Class Exercise

Three charges are arranged on a straight line as shown in the figure. What is the direction of the electrostatic force on the *middle* charge?

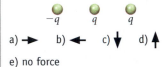

a) →   b) ←   c) ↓   d) ↑

e) no force

21.7 In-Class Exercise

Three charges are arranged on a straight line as shown in the figure. What is the direction of the electrostatic force on the *right* charge? (Note that the left charge is double what it was in In-Class Exercise 21.6.)

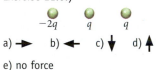

a) →   b) ←   c) ↓   d) ↑

e) no force

PROBLEM 2

What is the magnitude of the electrostatic force between a gold nucleus and an electron of the gold atom in an orbit with radius $4.88 \cdot 10^{-12}$ m?

SOLUTION 2

The negatively charged electron and the positively charged gold nucleus attract each other with a force whose magnitude is

$$F = k\frac{|q_e q_N|}{r^2},$$

where the charge of the electron is $q_e = -e$ and the charge of the gold nucleus is $q_N = +79e$. The force between the electron and the nucleus is then

$$F = k\frac{|q_e q_N|}{r^2} = \left(8.99 \cdot 10^9 \ \frac{\text{N m}^2}{\text{C}^2}\right)\frac{\left(1.60 \cdot 10^{-19} \ \text{C}\right)\left[(79)\left(1.60 \cdot 10^{-19} \ \text{C}\right)\right]}{\left(4.88 \cdot 10^{-12} \ \text{m}\right)^2} = 7.63 \cdot 10^{-4} \ \text{N}.$$

Thus, the magnitude of the electrostatic force exerted on an electron in a gold atom by the nucleus is about 100,000 times less than that between protons inside a nucleus.

Note: The gold nucleus has a mass that is approximately 400,000 times that of the electron. But the force the gold nucleus exerts on the electron has exactly the same magnitude as the force that the electron exerts on the gold nucleus. You may say that this is obvious from Newton's Third Law (see Chapter 4), which is true. But it is worth emphasizing that this basic law holds for electrostatic forces as well.

EXAMPLE 21.3 Equilibrium Position

FIGURE 21.15 Placement of three charged particles for Example 21.3. The third particle is shown with a negative charge.

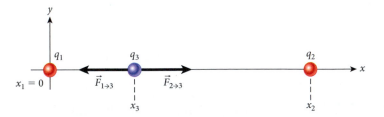

PROBLEM

Two charged particles are placed as shown in Figure 21.15: $q_1 = 0.15 \ \mu\text{C}$ is located at the origin, and $q_2 = 0.35 \ \mu\text{C}$ is located on the positive x-axis at $x_2 = 0.40$ m. Where should a third charged particle, q_3, be placed to be at an equilibrium point (the forces on it sum to zero)?

SOLUTION

Let's first determine where *not* to put the third charge. If the third charge is placed anywhere off the x-axis, there will always be a force component pointing toward or away from the x-axis. Thus, we can find an equilibrium point (a point where the forces sum to zero) only *on* the x-axis. The x-axis can be divided into three different segments: $x \le x_1 = 0$, $x_1 < x < x_2$, and $x_2 \le x$. For $x \le x_1 = 0$, the force vectors from both q_1 and q_2 acting on q_3 will point in the positive direction if the charge is negative and in the negative direction if the charge is positive. Because we are looking for a location where the two forces cancel, the segment $x \le x_1 = 0$ can be excluded. A similar argument excludes $x \ge x_2$.

In the remaining segment of the x-axis, $x_1 < x < x_2$, the forces from q_1 and q_2 on q_3 point in opposite directions. We look for the location, x_3, where the absolute magnitudes of both forces are equal and the forces thus sum to zero. We express the equality of the two forces as

$$\left| \vec{F}_{1 \to 3} \right| = \left| \vec{F}_{2 \to 3} \right|,$$

which we can rewrite as

$$k \frac{|q_1 q_3|}{(x_3 - x_1)^2} = k \frac{|q_3 q_2|}{(x_2 - x_3)^2}.$$

We now see that the magnitude and sign of the third charge do not matter because that charge cancels out, as does the constant k, giving us

$$\frac{q_1}{(x_3 - x_1)^2} = \frac{q_2}{(x_2 - x_3)^2}$$

or

$$q_1 (x_2 - x_3)^2 = q_2 (x_3 - x_1)^2. \qquad (i)$$

Taking the square root of both sides and solving for x_3, we find

$$\sqrt{q_1}(x_2 - x_3) = \sqrt{q_2}(x_3 - x_1),$$

or

$$x_3 = \frac{\sqrt{q_1} x_2 + \sqrt{q_2} x_1}{\sqrt{q_1} + \sqrt{q_2}}.$$

We can take the square root of both sides of equation (i) because $x_1 < x_3 < x_2$, and so both of the roots, $x_2 - x_3$ and $x_3 - x_1$, are assured to be positive.

Inserting the numbers given in the problem statement, we obtain

$$x_3 = \frac{\sqrt{q_1} x_2 + \sqrt{q_2} x_1}{\sqrt{q_1} + \sqrt{q_2}} = \frac{\sqrt{0.15 \ \mu C}(0.4 \ \text{m})}{\sqrt{0.15 \ \mu C} + \sqrt{0.35 \ \mu C}} = 0.16 \ \text{m}.$$

This result makes sense because we expect the equilibrium point to reside closer to the smaller charge.

SOLVED PROBLEM 21.1 | Charged Balls

PROBLEM
Two identical charged balls hang from the ceiling by insulated ropes of equal length, $\ell = 1.50$ m (Figure 21.16). A charge $q = 25.0 \ \mu C$ is applied to each ball. Then the two balls hang at rest, and each supporting rope has an angle of 25.0° with respect to the vertical (Figure 21.16a). What is the mass of each ball?

SOLUTION

THINK
Each charged ball has three forces acting on it: the force of gravity, the repulsive electrostatic force, and the tension in the supporting rope. Using the

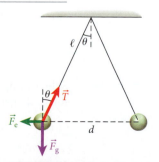
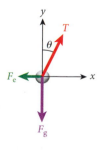

FIGURE 21.16 (a) Two charged balls hanging from the ceiling in their equilibrium position. (b) Free-body diagram for the left-hand charged ball.

Continued—

first condition for static equilibrium from Chapter 11, we know that the sum of all the forces on each ball must be zero. We can resolve the components of the three forces and set them equal to zero, allowing us to solve for the mass of the charged balls.

SKETCH
A free-body diagram for the left-hand ball is shown in Figure 21.16b.

RESEARCH
The condition for static equilibrium says that the sum of the x-components of the three forces acting on the ball must equal zero and the sum of y-components of these forces must equal zero. The sum of the x-components of the forces is

$$T \sin \theta - F_e = 0, \qquad \text{(i)}$$

where T is the magnitude of the string tension, θ is the angle of the string relative to the vertical, and F_e is the magnitude of the electrostatic force. The sum of the y-components of the forces is

$$T \cos \theta - F_g = 0. \qquad \text{(ii)}$$

The force of gravity, F_g, is just the weight of the charged ball:

$$F_g = mg, \qquad \text{(iii)}$$

where m is the mass of the charged ball. The electrostatic force the two balls exert on each other is given by

$$F_e = k \frac{q^2}{d^2}, \qquad \text{(iv)}$$

where d is the distance between the two balls. We can express the distance between the two balls in terms of the length of the string, ℓ, by looking at Figure 21.16a. We see that

$$\sin \theta = \frac{d/2}{\ell}.$$

We can then express the electrostatic force in terms of the angle with respect to the vertical, θ, and the length of the string, ℓ:

$$F_e = k \frac{q^2}{\left(2\ell \sin \theta\right)^2} = k \frac{q^2}{4\ell^2 \sin^2 \theta}. \qquad \text{(v)}$$

SIMPLIFY
We divide equation (i) by equation (ii):

$$\frac{T \sin \theta}{T \cos \theta} = \frac{F_e}{F_g},$$

eliminating the (unknown) string tension and obtaining

$$\tan \theta = \frac{F_e}{F_g}.$$

Substituting from equations (iii) and (v) for the force of gravity and the electrostatic force, we get

$$\tan \theta = \frac{k \dfrac{q^2}{4\ell^2 \sin^2 \theta}}{mg} = \frac{kq^2}{4mg\ell^2 \sin^2 \theta}.$$

Solving for the mass of the ball, we obtain

$$m = \frac{kq^2}{4g\ell^2 \sin^2 \theta \tan \theta}.$$

CALCULATE
Putting in the numerical values gives

$$m = \frac{\left(8.99 \cdot 10^9 \text{ N m}^2/\text{C}^2\right)\left(25.0 \text{ μC}\right)^2}{4\left(9.81 \text{ m/s}^2\right)\left(1.50 \text{ m}\right)^2\left(\sin^2 25.0°\right)\left(\tan 25.0°\right)} = 0.764116 \text{ kg}.$$

ROUND
We report our result to three significant figures:

$$m = 0.764 \text{ kg}.$$

DOUBLE-CHECK
To double-check, we make the small-angle approximations that $\sin\theta \approx \tan\theta \approx \theta$ and $\cos\theta \approx 1$. The tension in the string then approaches mg, and we can express the x-components of the forces as

$$T\sin\theta \approx mg\theta = F_e = k\frac{q^2}{d^2} \approx k\frac{q^2}{\left(2\ell\theta\right)^2}.$$

Solving for the mass of the charged ball, we get

$$m = \frac{kq^2}{4g\ell^2\theta^3} = \frac{\left(8.99 \cdot 10^9 \text{ N m}^2/\text{C}^2\right)\left(25.0 \text{ μC}\right)^2}{4\left(9.81 \text{ m/s}^2\right)\left(1.50 \text{ m}\right)^2\left(0.436 \text{ rad}\right)^3} = 0.768 \text{ kg},$$

which is close to our answer.

Electrostatic Precipitator

An application of electrostatic charging and electrostatic forces is the cleaning of emissions from coal-fired power plants. A device called an **electrostatic precipitator** (ESP) is used to remove ash and other particulates resulting from the burning of coal to generate electricity. Its operation is illustrated in Figure 21.17.

The ESP consists of wires and plates, with the wires held at a high negative voltage relative to a series of plates held at a positive voltage. (Here the term *voltage* is used colloquially; in Chapter 23, the concept will be defined in terms of electric potential difference.) In Figure 21.17, the exhaust from the coal-burning process enters the ESP from the left. Particulates passing near the wires pick up a negative charge. These particles are then attracted to one of the positive plates and stick there. The gas continues through the ESP, leaving the ash and other particulates behind. The accumulated material is then shaken off the plates to a hamper below. This waste can be used for many purposes, including construction materials and fertilizer. Figure 21.18 shows an example of a coal-fired power plant that incorporates an ESP.

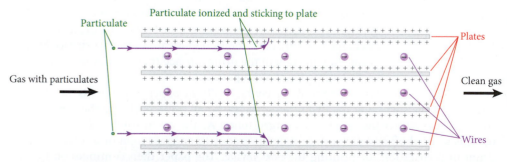

FIGURE 21.17 Operation of an electrostatic precipitator used to clean the exhaust gas of a coal-fired power plant. The view is from the top of the device.

21.2 Self-Test Opportunity

A positive point charge $+q$ is placed at point P, to the right of two charges q_1 and q_2, as shown in the figure. The net electrostatic force on the positive charge $+q$ is found to be zero. Identify each of the following statements as true or false.

a) Charge q_2 must have the opposite sign from q_1 and be smaller in magnitude.

b) The magnitude of charge q_1 must be smaller than the magnitude of charge q_2.

c) Charges q_1 and q_2 must have the same sign.

d) If q_1 is negative, then q_2 must be positive.

e) Either q_1 or q_2 must be positive.

21.8 In-Class Exercise

Consider three charges placed along the x-axis, as shown in the figure.

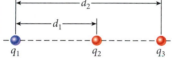

The values of the charges are $q_1 = -8.10$ μC, $q_2 = 2.16$ μC, and $q_3 = 2.16$ pC. The distance between q_1 and q_2 is $d_1 = 1.71$ m. The distance between q_1 and q_3 is $d_2 = 2.62$ m. What is the magnitude of the total electrostatic force exerted on q_3 by q_1 and q_2?

a) $2.77 \cdot 10^{-8}$ N   d) $2.22 \cdot 10^{-4}$ N

b) $7.92 \cdot 10^{-6}$ N   e) $6.71 \cdot 10^{-2}$ N

c) $1.44 \cdot 10^{-5}$ N

FIGURE 21.18 A coal-fired power plant at Michigan State University that incorporates an electrostatic precipitator to remove particulates from its emissions.

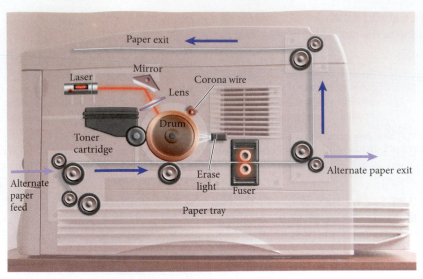

FIGURE 21.19 The operation of a typical laser printer.

Laser Printer

Another example of a device that applies electrostatic forces is the laser printer. The operation of a laser printer is illustrated in Figure 21.19. The paper path follows the blue arrows. Paper is taken from the paper tray or fed manually through the alternate paper feed. The paper passes over a drum where the toner is placed on the surface of the paper and then passes through a fuser that melts the toner and permanently affixes it to the paper.

The drum consists of a metal cylinder coated with a special photosensitive material; originally amorphous selenium was used but has been replaced with an organic material. The photosensitive surface is an insulator that retains charge in the absence of light, but discharges quickly if light is incident on the surface. The drum rotates so that its surface speed is the same as the speed of the moving paper. The basic principle of the operation of the drum is illustrated in Figure 21.20.

The drum is negatively charged with electrons using a wire held at high voltage. Then laser light is directed at the surface of the drum. Wherever the laser light strikes the surface of the drum, the surface at that point is discharged. A laser is used because its beam is narrow and remains focused. A line of the image being printed is written one pixel (picture element or dot) at a time using a laser beam directed by a moving mirror and a lens. A typical laser printer can write 300 pixels per inch, with many printers being able to write 600 or 1200 pixels per inch. The surface of the drum then passes by a roller that picks up toner from the toner cartridge. Toner consists of small, black, insulating particles composed of a plastic-like material. The toner roller is charged to the same negative voltage as the drum. Therefore, wherever the surface of the drum has been discharged, electrostatic forces deposit toner on the surface of the drum. Any portion of the drum surface that has not been exposed to the laser will not pick up toner.

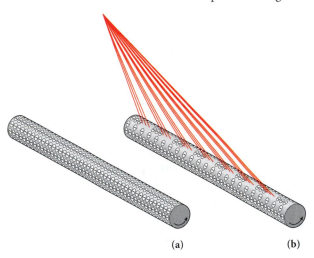

FIGURE 21.20 (a) The completely charged drum of a laser printer. This drum will produce a blank page. (b) A drum on which one line of information is being recorded by a laser. Wherever the laser strikes the charged drum, the negative charge is neutralized, and the discharged area will attract toner that will produce an image on the paper.

As the drum rotates, it next comes in contact with the paper. The toner is then transferred from the surface of the drum to the paper. Some printers charge the paper positively to help attract the negatively charged toner. As the drum rotates, any remaining toner is scraped off and the surface is neutralized with an erase light or a rotating erase drum in preparation for printing the next image. The paper then continues on to the fuser, which melts the toner, producing a permanent image on the paper. Finally the paper exits the printer.

21.6   **Coulomb's Law and Newton's Law of Gravitation**

Coulomb's Law describing the electrostatic force between two electric charges, F_e, has a similar form to Newton's Law describing the gravitational force between two masses, F_g:

$$F_g = G\frac{m_1 m_2}{r^2} \quad \text{and} \quad F_e = k\frac{|q_1 q_2|}{r^2},$$

where m_1 and m_2 are the two masses, q_1 and q_2 are the two electric charges, and r is the distance of separation. Both forces vary with the inverse square of the distance. The electric force can be attractive or repulsive because charges can have positive or negative signs. (See Figure 21.13a and b.) The gravitational force is always attractive because there is only one kind of mass. (For the gravitational force, only the case depicted in Figure 21.13b is possible.) The relative strengths of the forces are given by the proportionality constants k and G.

EXAMPLE 21.4 | Forces between Electrons

Let's evaluate the relative strengths of the two interactions by calculating the ratio of the electrostatic force and the gravitational force that two electrons exert on each other. This ratio is given by

$$\frac{F_e}{F_g} = \frac{k q_e^2}{G m_e^2}.$$

Because the dependence on distance is the same in both forces, there is no dependence on distance in the ratio of the two forces—it cancels out. The mass of an electron is $m_e = 9.109 \cdot 10^{-31}$ kg, and its charge is $q_e = -1.602 \cdot 10^{-19}$ C. Using the value of Coulomb's constant given in equation 21.7, $k = 8.99 \cdot 10^9$ N m^2/C^2, and the value of the universal gravitational constant, $G = 6.67 \cdot 10^{-11}$ N m^2/kg^2, we find numerically

$$\frac{F_e}{F_g} = \frac{(8.99 \cdot 10^9 \text{ N m}^2/\text{C}^2)(1.602 \cdot 10^{-19} \text{ C})^2}{(6.67 \cdot 10^{-11} \text{ N m}^2/\text{kg}^2)(9.109 \cdot 10^{-31} \text{ kg})^2} = 4.2 \cdot 10^{42}.$$

Therefore, the electrostatic force between electrons is stronger than the gravitational force between them by more than 42 orders of magnitude.

Despite the relative weakness of the gravitational force, on the astronomical scale, gravity is the only force that matters. The reason for this dominance is that all stars, planets, and other objects of astronomical relevance carry no net charge. Therefore, there is no net electrostatic interaction between them, and gravity dominates.

Coulomb's Law of electrostatics applies to macroscopic systems down to the atom, though subtle effects in atomic and subatomic systems require use of a more sophisticated approach called *quantum electrodynamics*. Newton's law of gravitation fails in subatomic systems and also must be modified for astronomical systems, such as the precessional motion of Mercury around the Sun. These fine details of the gravitational interaction are governed by Einstein's theory of general relativity.

The similarities between the gravitational and electrostatic interactions will be covered further in the next two chapters, which address electric fields and electric potential.

WHAT WE HAVE LEARNED | EXAM STUDY GUIDE

- There are two kinds of electric charge, positive and negative. Like charges repel, and unlike charges attract.

- The quantum (elementary quantity) of electric charge is $e = 1.602 \cdot 10^{-19}$ C.

- The electron has charge $q_e = -e$ and the proton has charge $q_p = +e$. The neutron has zero charge.

- The net charge of an object is given by e times the number of protons, N_p, minus e times the number of electrons, N_e, that make up the object: $q = e \cdot (N_p - N_e)$.

- The total charge in an isolated system is always conserved.

- Objects can be charged directly by contact or indirectly by induction.

- Coulomb's Law describes the force that two stationary charges exert on each other: $F = k\dfrac{|q_1 q_2|}{r^2} = \dfrac{1}{4\pi\epsilon_0}\dfrac{|q_1 q_2|}{r^2}$.

- The constant in Coulomb's Law is
$$k = \frac{1}{4\pi\epsilon_0} = 8.99 \cdot 10^9\ \frac{\text{N m}^2}{\text{C}^2}.$$

- The electric permittivity of free space is
$$\epsilon_0 = 8.85 \cdot 10^{-12}\ \frac{\text{C}^2}{\text{N m}^2}.$$

KEY TERMS

electrostatics, p. 685
charging, p. 685
electrons, p. 685
positive charge, p. 685
negative charge, p. 685
law of electric charges,
   p. 686

coulomb, p. 686
proton, p. 686
law of charge
   conservation, p. 686
quantized, p. 687
conductors, p. 688
insulators, p. 688

semiconductors, p. 689
superconductors, p. 689
electrostatic charging, p. 690
grounding, p. 690
ground, p. 690
electroscope, p. 690
induced, p. 690

charging by contact, p. 691
charging by induction, p. 692
Coulomb's Law, p. 692
Coulomb's constant, p. 692
electric permittivity of free
   space, p. 692
electrostatic precipitator, p. 697

NEW SYMBOLS

q, electric charge

ϵ_0, electric permittivity of free space

k, Coulomb's constant

e, the elementary quantum of charge

ANSWERS TO SELF-TEST OPPORTUNITIES

21.1 a) +1   c) 0   e) $+\frac{2}{3}$ g) −1

   b) 0   d) 0   f) $-\frac{1}{3}$ h) +2

22.2 a) true   c) false   e) true

   b) false   d) true

PROBLEM-SOLVING PRACTICE

Problem-Solving Guidelines

1. When working problems involving Coulomb's Law, drawing a free-body diagram showing the electrostatic force vectors acting on a charged particle is often helpful. Pay careful attention to signs; a negative force between two particles indicates attraction, and a positive force indicates repulsion. Be sure the directions of forces in the diagram match the signs of forces in the calculations.

2. Use symmetry to simplify your work. However, be careful to take account of charge magnitudes and signs as well as

distances. Two charges at equal distances from a third charge do not exert equal forces on that charge if they have different magnitudes or signs.

3. Units in electrostatics often have prefixes indicating powers of 10: Distances may be given in cm or mm; charges may be given in μC, nC, or pC; masses may be given in kg or g. Other units are also common. The best way to proceed is to convert all quantities to basic SI units, to be compatible with the value of k or $1/4\pi\epsilon_0$.

SOLVED PROBLEM 21.2 | Bead on a Wire

PROBLEM

A bead with charge $q_1 = +1.28\ \mu\text{C}$ is fixed in place on an insulating wire that makes an angle of $\theta = 42.3°$ with respect to the horizontal (Figure 21.21a). A second bead with charge $q_2 = -5.06\ \mu\text{C}$ slides without friction on the wire. At a distance $d = 0.380$ m between the beads, the net force on the second bead is zero. What is the mass, m_2, of the second bead?

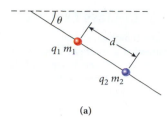

SOLUTION

THINK

The force of gravity pulling the bead of mass m_2 down the wire is compensated by the attractive electrostatic force between the positive charge on the first bead and the negative charge on the second bead. The second bead can be thought of as sliding on an inclined plane.

SKETCH

Figure 21.21b shows a free-body diagram of the forces acting on the second bead. We have defined a coordinate system in which the positive x-direction is down the wire. The force exerted on m_2 by the wire can be omitted because this force has only a y-component, and we can solve the problem by analyzing just the x-components of the forces.

RESEARCH

The attractive electrostatic force between the two beads balances the component of the force of gravity that acts on the second bead down the wire. The electrostatic force acts in the negative x-direction and its magnitude is given by

$$F_e = k\frac{|q_1 q_2|}{d^2}. \tag{i}$$

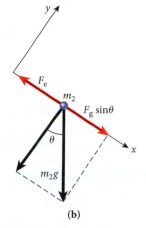

The x-component of the force of gravity acting on the second bead corresponds to the component of the weight of the second bead that is parallel to the wire. Figure 21.21b indicates that the component of the weight of the second bead down the wire is given by

$$F_g = m_2 g \sin\theta. \tag{ii}$$

FIGURE 21.21 (a) Two charged beads on a wire. (b) Free-body diagram of the forces acting on the second bead.

SIMPLIFY

For equilibrium, the electrostatic force and the gravitational force are equal: $F_e = F_g$. Substituting the expressions for these forces from equations (i) and (ii) yields

$$k\frac{|q_1 q_2|}{d^2} = m_2 g \sin\theta.$$

Solving this equation for the mass of the second bead gives us

$$m_2 = \frac{k|q_1 q_2|}{d^2 g \sin\theta}.$$

CALCULATE

We put in the numerical values and get

$$m_2 = \frac{kq_1 q_2}{d^2 g \sin\theta} = \frac{\left(8.99\cdot 10^9 \text{ N m}^2/\text{C}^2\right)\left(1.28 \text{ μC}\right)\left(5.06 \text{ μC}\right)}{\left(0.380 \text{ m}\right)^2 \left(9.81 \text{ m/s}^2\right)\left(\sin 42.3°\right)} = 0.0610746 \text{ kg}.$$

ROUND

We report our result to three significant figures:

$$m_2 = 0.0611 \text{ kg} = 61.1 \text{ g}.$$

DOUBLE-CHECK

To double-check, let's calculate the mass of the second bead assuming that the wire is vertical, that is, $\theta = 90°$. We can then set the weight of the second bead equal to the electrostatic force between the two beads:

$$k\frac{|q_1 q_2|}{d^2} = m_2 g.$$

Continued—

21.9 In-Class Exercise

Three charges are arranged at the corners of a square as shown in the figure. What is the direction of the electrostatic force on the *lower-right* charge?

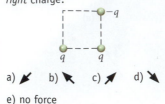

a) ✦ b) ✦ c) ✦ d) ✦

e) no force

21.10 In-Class Exercise

Four charges are arranged at the corners of a square as shown in the figure. What is the direction of the electrostatic force on the *lower-right* charge?

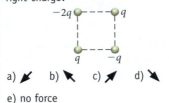

a) ✦ b) ✦ c) ✦ d) ✦

e) no force

Solving for the mass of the second bead, we obtain

$$m_2 = \frac{k q_1 q_2}{d^2 g} = \frac{\left(8.99 \cdot 10^9 \text{ N m}^2/\text{C}^2\right)\left(1.28 \text{ } \mu\text{C}\right)\left(5.06 \text{ } \mu\text{C}\right)}{\left(0.380 \text{ m}\right)^2 \left(9.81 \text{ m/s}^2\right)} = 0.0411 \text{ kg}.$$

As the angle of the wire relative to the horizontal decreases, the calculated mass of the second bead will increase. Our result of 0.0611 kg is somewhat higher than the mass that can be supported with a vertical wire, so it seems reasonable.

SOLVED PROBLEM 21.3 | Four Charged Objects

Consider four charges placed at the corners of a square with side length 1.25 m, as shown in Figure 21.22a.

PROBLEM
What are the magnitude and direction of the electrostatic force on q_4 resulting from the other three charges?

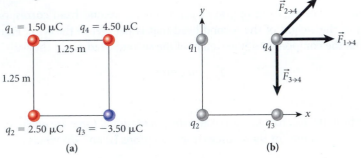

FIGURE 21.22 (a) Four charges placed at the corners of a square. (b) The forces exerted on q_4 by the other three charges.

SOLUTION

THINK
The electrostatic force on q_4 is the vector sum of the forces resulting from its interactions with the other three charges. Thus, it is important to avoid simply adding the individual force magnitudes algebraically. Instead we need to determine the individual force components in each spatial direction and add those to find the components of the net force vector. Then we need to calculate the length of this net force vector.

SKETCH
Figure 21.22b shows the four charges in an *xy*-coordinate system with its origin at the location of q_2.

RESEARCH
The net force on q_4 is the vector sum of the forces $\vec{F}_{1\to4}$, $\vec{F}_{2\to4}$, and $\vec{F}_{3\to4}$. The *x*-component of the summed forces is

$$F_x = k\frac{|q_1 q_4|}{d^2} + k\frac{|q_2 q_4|}{\left(\sqrt{2}d\right)^2}\cos 45° = \frac{k q_4}{d^2}\left(q_1 + \frac{q_2}{2}\cos 45°\right), \qquad \text{(i)}$$

where *d* is the length of a side of the square and, as Figure 21.22b indicates, the *x*-component of $\vec{F}_{3\to4}$ is zero. The *y*-component of the summed forces is

$$F_y = k\frac{|q_2 q_4|}{\left(\sqrt{2}d\right)^2}\sin 45° - k\frac{|q_3 q_4|}{d^2} = \frac{k q_4}{d^2}\left(\frac{q_2}{2}\sin 45° + q_3\right), \qquad \text{(ii)}$$

where, as Figure 21.22b indicates, the y-component of $\vec{F}_{1 \to 4}$ is zero.

The magnitude of the net force is given by

$$F = \sqrt{F_x^2 + F_y^2},\tag{iii}$$

and the angle of the net force is given by

$$\tan\theta = \frac{F_y}{F_x}.$$

SIMPLIFY

We substitute the expressions for F_x and F_y from equations (i) and (ii) into equation (iii):

$$F = \sqrt{\left[\frac{kq_4}{d^2}\left(q_1 + \frac{q_2}{2}\cos 45°\right)\right]^2 + \left[\frac{kq_4}{d^2}\left(\frac{q_2}{2}\sin 45° + q_3\right)\right]^2}.$$

We can rewrite this as

$$F = \frac{kq_4}{d^2}\sqrt{\left(q_1 + \frac{q_2}{2}\cos 45°\right)^2 + \left(\frac{q_2}{2}\sin 45° + q_3\right)^2}.$$

For the angle of the force, we get

$$\theta = \tan^{-1}\left(\frac{F_y}{F_x}\right) = \tan^{-1}\left(\frac{\dfrac{kq_4}{d^2}\left(\dfrac{q_2}{2}\sin 45° + q_3\right)}{\dfrac{kq_4}{d^2}\left(q_1 + \dfrac{q_2}{2}\cos 45°\right)}\right) = \tan^{-1}\left(\frac{\left(\dfrac{q_2}{2}\sin 45° + q_3\right)}{\left(q_1 + \dfrac{q_2}{2}\cos 45°\right)}\right).$$

CALCULATE

Putting in the numerical values, we get

$$\frac{q_2}{2}\sin 45° = \frac{q_2}{2}\cos 45° = \frac{2.50\ \mu\text{C}}{2\sqrt{2}} = 0.883883\ \mu\text{C}.$$

The magnitude of the force is then

$$F = \frac{\left(8.99 \cdot 10^9\ \text{N m}^2/\text{C}^2\right)\left(4.50\ \mu\text{C}\right)}{\left(1.25\ \text{m}\right)^2}\sqrt{\left(1.50\ \mu\text{C} + 0.883883\ \mu\text{C}\right)^2 + \left(0.883883\ \mu\text{C} - 3.50\ \mu\text{C}\right)^2}$$

$$= 0.0916379\ \text{N}.$$

For the direction of the force, we obtain

$$\theta = \tan^{-1}\left(\frac{\left(\dfrac{q_2}{2}\sin 45° + q_3\right)}{\left(q_1 + \dfrac{q_2}{2}\cos 45°\right)}\right) = \tan^{-1}\left(\frac{\left(0.883883\ \mu\text{C} - 3.50\ \mu\text{C}\right)}{\left(1.50\ \mu\text{C} + 0.883883\ \mu\text{C}\right)}\right) = -47.6593°.$$

ROUND

We report our results to three significant figures:

$$F = 0.0916\ \text{N}$$

and

$$\theta = -47.7°.$$

Continued—

DOUBLE-CHECK

To double-check our result, we calculate the magnitude of the three forces acting on q_4. For $\vec{F}_{1\rightarrow 4}$, we get

$$F_{1\rightarrow 4} = k\frac{|q_1 q_4|}{r_{14}^2} = \frac{\left(8.99\cdot 10^9\ \text{N m}^2/\text{C}^2\right)\left(1.50\ \mu\text{C}\right)\left(4.50\ \mu\text{C}\right)}{\left(1.25\ \text{m}\right)^2} = 0.0388\ \text{N}.$$

For $\vec{F}_{2\rightarrow 4}$, we get

$$F_{2\rightarrow 4} = k\frac{|q_2 q_4|}{r_{24}^2} = \frac{\left(8.99\cdot 10^9\ \text{N m}^2/\text{C}^2\right)\left(2.50\ \mu\text{C}\right)\left(4.50\ \mu\text{C}\right)}{\left[\sqrt{2}\left(1.25\ \text{m}\right)\right]^2} = 0.0324\ \text{N}.$$

For $F_{3\rightarrow 4}$, we get

$$F_{3\rightarrow 4} = k\frac{|q_3 q_4|}{r_{34}^2} = \frac{\left(8.99\cdot 10^9\ \text{N m}^2/\text{C}^2\right)\left(3.50\ \mu\text{C}\right)\left(4.50\ \mu\text{C}\right)}{\left(1.25\ \text{m}\right)^2} = 0.0906\ \text{N}.$$

All three of the magnitudes of the individual forces are of the same order as our result for the net force. This gives us confidence that our answer is not off by a large factor.

The direction we obtained also seems reasonable, because it orients the resulting force downward and to the right, as could be expected from looking at Figure 21.22b.

MULTIPLE-CHOICE QUESTIONS

21.1 When a metal plate is given a positive charge, which of the following is taking place?

a) Protons (positive charges) are transferred to the plate from another object.

b) Electrons (negative charges) are transferred from the plate to another object.

c) Electrons (negative charges) are transferred from the plate to another object, and protons (positive charges) are also transferred to the plate from another object.

d) It depends on whether the object conveying the charge is a conductor or an insulator.

21.2 The force between a charge of 25 μC and a charge of -10 μC is 8.0 N. What is the separation between the two charges?

a) 0.28 m   c) 0.45 m

b) 0.53 m   d) 0.15 m

21.3 A charge Q_1 is positioned on the x-axis at $x = a$. Where should a charge $Q_2 = -4Q_1$ be placed to produce a net electrostatic force of zero on a third charge, $Q_3 = Q_1$, located at the origin?

a) at the origin   c) at $x = -2a$

b) at $x = 2a$ d) at $x = -a$

21.4 Which one of these systems has the most negative charge?

a) 2 electrons

b) 3 electrons and 1 proton

c) 5 electrons and 5 protons

d) N electrons and $N - 3$ protons

e) 1 electron

21.5 Two point charges are fixed on the x-axis: $q_1 = 6.0$ μC is located at the origin, O, with $x_1 = 0.0$ cm, and $q_2 = -3.0$ μC is located at point A, with $x_2 = 8.0$ cm. Where should a third charge, q_3, be placed on the x-axis so that the total electrostatic force acting on it is zero?

a) 19 cm   c) 0.0 cm   e) -19 cm

b) 27 cm   d) 8.0 cm

21.6 Which of the following situations produces the largest net force on the charge Q?

a) Charge $Q = 1$ C is 1 m from a charge of -2 C.

b) Charge $Q = 1$ C is 0.5 m from a charge of -1 C.

c) Charge $Q = 1$ C is halfway between a charge of -1 C and a charge of 1 C that are 2 m apart.

d) Charge $Q = 1$ C is halfway between two charges of -2 C that are 2 m apart.

e) Charge $Q = 1$ C is a distance of 2 m from a charge of -4 C.

21.7 Two protons placed near one another with no other objects close by would

a) accelerate away from each other.

b) remain motionless.

c) accelerate toward each other.

d) be pulled together at constant speed.

e) move away from each other at constant speed.

21.8 Two lightweight metal spheres are suspended near each other from insulating threads. One sphere has a net charge; the other sphere has no net charge. The spheres will

a) attract each other.

b) exert no net electrostatic force on each other.

c) repel each other.

d) do any of these things depending on the sign of the charge on the one sphere.

21.9 A metal plate is connected by a conductor to a ground through a switch. The switch is initially closed. A charge $+Q$ is brought close to the plate without touching it, and then the switch is opened. After the switch is opened, the charge $+Q$ is removed. What is the charge on the plate then?

a) The plate is uncharged.

b) The plate is positively charged.

c) The plate is negatively charged.

d) The plate could be either positively or negatively charged, depending on the charge it had before $+Q$ was brought near.

21.10 You bring a negatively charged rubber rod close to a grounded conductor without touching it. Then you disconnect the ground. What is the sign of the charge on the conductor after you remove the charged rod?

a) negative

b) positive

c) no charge

d) cannot be determined from the given information

QUESTIONS

21.11 If two charged particles (the charge on each is Q) are separated by a distance d, there is a force F between them. What is the force if the magnitude of each charge is doubled and the distance between them changes to $2d$?

21.12 Suppose the Sun and the Earth were each given an equal amount of charge of the same sign, just sufficient to cancel their gravitational attraction. How many times the charge on an electron would that charge be? Is this number a large fraction of the number of charges of either sign in the Earth?

21.13 It is apparent that the electrostatic force is *extremely* strong, compared to gravity. In fact, the electrostatic force is the basic force governing phenomena in daily life—the tension in a string, the normal forces between surfaces, friction, chemical reactions, etc.—except weight. Why then did it take so long for scientists to understand this force? Newton came up with his gravitational law long before electricity was even crudely understood.

21.14 Occasionally, people who gain static charge by shuffling their feet on the carpet will have their hair stand on end. Why does this happen?

21.15 Two positive charges, each equal to Q, are placed a distance $2d$ apart. A third charge, $-0.2Q$, is placed exactly halfway between the two positive charges and is displaced a distance $x \ll d$ perpendicular to the line connecting the positive charges. What is the force on this charge? For $x \ll d$, how can you approximate the motion of the negative charge?

21.16 Why does a garment taken out of a clothes dryer sometimes cling to your body when you wear it?

21.17 Two charged spheres are initially a distance d apart. The magnitude of the force on each sphere is F. They are moved closer to each other such that the magnitude of the force on each of them is $9F$. By what factor has the difference between the two spheres changed?

21.18 How is it possible for one electrically neutral atom to exert an electrostatic force on another electrically neutral atom?

21.19 The scientists who first contributed to the understanding of the electrostatic force in the 18th century were well aware of Newton's law of gravitation. How could they deduce that the force they were studying was *not* a variant or some manifestation of the gravitational force?

21.20 Two charged particles move solely under the influence of the electrostatic forces between them. What shapes can their trajectories have?

21.21 Rubbing a balloon causes it to become negatively charged. The balloon then tends to cling to the wall of a room. For this to happen, must the wall be positively charged?

21.22 Two electric charges are placed on a line, as shown in the figure. Is it possible to place a charged particle (that is free to move) anywhere on the line between the two charges and have it not move?

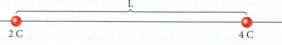

21.23 Two electric charges are placed on a line as shown in the figure. Where on the line can a third charge be placed so that the force on that charge is zero? Does the sign or the magnitude of the third charge make any difference to the answer?

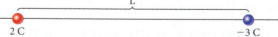

21.24 When a positively charged rod is brought close to a neutral conductor without touching it, will the rod experience an attractive force, a repulsive force, or no force at all? Explain.

21.25 When you exit a car and the humidity is low, you often experience a shock from static electricity created by sliding across the seat. How can you discharge yourself without experiencing a painful shock? Why is it dangerous to get back into your car while fueling your car?

PROBLEMS

A blue problem number indicates a worked-out solution is available in the Student Solutions Manual. One • and two •• indicate increasing level of problem difficulty.

Section 21.2

21.26 How many electrons are required to yield a total charge of 1 C?

21.27 The *faraday* is a unit of charge frequently encountered in electrochemical applications and named for the British physicist and chemist Michael Faraday. It consists of 1 mole of elementary charges. Calculate the number of coulombs in 1 faraday.

21.28 Another unit of charge is the *electrostatic unit* (esu). It is defined as follows: Two point charges, each of 1 esu and separated by 1 cm, exert a force of exactly 1 dyne on each other: 1 dyne = 1 g cm/s² = $1 \cdot 10^{-5}$ N.

a) Determine the relationship between the esu and the coulomb.

b) Determine the relationship between the esu and the elementary charge.

21.29 A current of 5 mA is enough to make your muscles twitch. Calculate how many electrons flow through your skin if you are exposed to such a current for 10 s.

•21.30 How many electrons does 1.00 kg of water contain?

•21.31 The Earth is constantly being bombarded by cosmic rays, which consist mostly of protons. These protons are incident on the Earth's atmosphere from all directions at a rate of 1245.0 protons per square meter per second. Assuming that the depth of Earth's atmosphere is 120 km, what is the total charge incident on the atmosphere in 5 min? Assume that the radius of the surface of the Earth is 6378 km.

•21.32 Performing an experiment similar to Millikan's oil drop experiment, a student measures these charge magnitudes:

$3.26 \cdot 10^{-19}$ C $5.09 \cdot 10^{-19}$ C $1.53 \cdot 10^{-19}$ C

$6.39 \cdot 10^{-19}$ C $4.66 \cdot 10^{-19}$ C

Find the charge on the electron using these measurements.

Section 21.3

•21.33 A silicon sample is doped with phosphorus at 1 part in 10^6. Phosphorus acts as an electron donor, providing one free electron per atom. The density of silicon is 2.33 g/cm³, and its atomic mass is 28.09 g/mol.

a) Calculate the number of free (conduction) electrons per unit volume of the doped silicon.

b) Compare the result from part (a) with the number of conduction electrons per unit volume of copper wire (assume each copper atom produces one free (conduction) electron). The density of copper is 8.96 g/cm³, and its atomic mass is 63.54 g/mol.

Section 21.5

21.34 Two charged spheres are 8 cm apart. They are moved closer to each other enough that the force on each of them increases four times. How far apart are they now?

21.35 Two identically charged particles separated by a distance of 1.0 m repel each other with a force of 1 N. What is the magnitude of the charges?

21.36 How far must two electrons be placed on the Earth's surface for there to be an electrostatic force between them equal to the weight of one of the electrons?

21.37 In solid sodium chloride (table salt), chloride ions have one more electron than they have protons, and sodium ions have one more proton than they have electrons. These ions are separated by about 0.28 nm. Calculate the electrostatic force between a sodium ion and a chloride ion.

21.38 In gaseous sodium chloride, chloride ions have one more electron than they have protons, and sodium ions have one more proton than they have electrons. These ions are separated by about 0.24 nm. Suppose a free electron is located 0.48 nm above the midpoint of the sodium chloride molecule. What are the magnitude and the direction of the electrostatic force the molecule exerts on it?

21.39 Calculate the magnitude of the electrostatic force the two up quarks inside a proton exert on each other if they are separated by a distance of 0.9 fm.

21.40 A −4.0-μC charge lies 20.0 cm to the right of a 2.0-μC charge on the x-axis. What is the force on the 2.0-μC charge?

•21.41 Two initially uncharged identical metal spheres, 1 and 2, are connected by an insulating spring (unstretched length $L_0 = 1.00$ m, spring constant $k = 25.0$ N/m), as shown in the figure. Charges $+q$ and $-q$ are then placed on the spheres, and the spring contracts to length $L = 0.635$ m. Recall that the force exerted by a spring is $F_s = k\Delta x$, where Δx is the change in the spring's length from its equilibrium length. Determine the charge q. If the spring is coated with metal to make it conducting, what is the new length of the spring?

Before charging   After charging

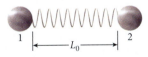   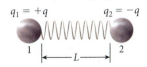

•21.42 A point charge $+3q$ is located at the origin, and a point charge $-q$ is located on the x-axis at $D = 0.500$ m. At what location on the x-axis will a third charge, q_0, experience no net force from the other two charges?

•21.43 Identical point charges Q are placed at each of the four corners of a rectangle measuring 2.0 m by 3.0 m. If $Q = 32$ μC, what is the magnitude of the electrostatic force on any one of the charges?

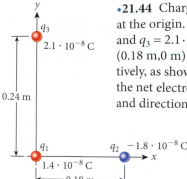

•21.44 Charge $q_1 = 1.4 \cdot 10^{-8}$ C is placed at the origin. Charges $q_2 = -1.8 \cdot 10^{-8}$ C and $q_3 = 2.1 \cdot 10^{-8}$ C are placed at points (0.18 m,0 m) and (0 m,0.24 m), respectively, as shown in the figure. Determine the net electrostatic force (magnitude and direction) on charge q_3.

•21.45 A positive charge Q is on the y-axis at a distance a from the origin, and another positive charge q is on the x-axis at a distance b from the origin.

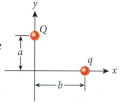

a) For what value(s) of b is the x-component of the force on q a minimum?

b) For what value(s) of b is the x-component of the force on q a maximum?

•21.46 Find the magnitude and direction of the electrostatic force acting on the electron in the figure.

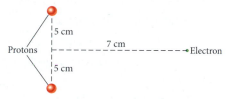

•21.47 In a region of two-dimensional space, there are three fixed charges: +1 mC at (0,0), –2 mC at (17 mm,–5 mm), and +3 mC at (–2 mm,11 mm). What is the net force on the –2-mC charge?

•21.48 Two cylindrical glass beads each of mass $m = 10$ mg are set on their flat ends on a horizontal insulating surface separated by a distance $d = 2$ cm. The coefficient of static friction between the beads and the surface is $\mu_s = 0.2$. The beads are then given identical charges (magnitude and sign). What is the minimum charge needed to start the beads moving?

•21.49 A small ball with a mass of 30 g and a charge of –0.2 μC is suspended from the ceiling by a string. The ball hangs at a distance of 5.0 cm above an insulating floor. If a second small ball with a mass of 50 g and a charge of 0.4 μC is rolled directly beneath the first ball, will the second ball leave the floor? What is the tension in the string when the second ball is directly beneath the first ball?

•21.50 A +3-mC charge and a –4-mC charge are fixed in position and separated by 5 m.

a) Where could a +7-mC charge be placed so that the net force on it is zero?

b) Where could a –7-mC charge be placed so that the net force on it is zero?

•21.51 Four point charges, q, are fixed to the four corners of a square that is 10.0 microns on a side. An electron is suspended above a point at which its weight is balanced by the electrostatic force due to the four electrons, at a distance of 15 nm above the center of the square. What is the magnitude of the fixed charges? Express both in coulombs and as a multiple of the electron's charge.

••21.52 The figure shows a uniformly charged thin rod of length L that has total charge Q. Find an expression for the magnitude of the electrostatic force acting on an electron positioned on the axis of the rod at a distance d from the midpoint of the rod.

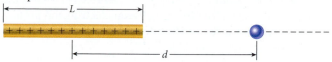

••21.53 A negative charge, $-q$, is fixed at the coordinate (0,0). It is exerting an attractive force on a positive charge, $+q$, that is initially at coordinate $(x,0)$. As a result, the positive charge accelerates toward the negative charge. Use the binomial expansion $(1 + x)^n \approx 1 + nx$, for $x \ll 1$, to show that when the positive charge moves a distance $\delta \ll x$ closer to the negative charge, the force that the negative charge exerts on it increases by $\Delta F = 2kq^2\delta/x^3$.

••21.54 Two equal magnitude negative charges $(-q$ and $-q)$ are fixed at coordinates $(-d,0)$ and $(d,0)$. A positive charge of the same magnitude, q, and with mass m is placed at coordinate (0,0), midway between the two negative charges. If the positive charge is moved a distance $\delta \ll d$ in the positive y-direction and then released, the resulting motion will be that of a harmonic oscillator—the positive charge will oscillate between coordinates $(0,\delta)$ and $(0,-\delta)$. Find the net force acting on the positive charge when it moves to $(0,\delta)$ and use the binomial expansion $(1 + x)^n \approx 1+nx$, for $x \ll 1$, to find an expression for the frequency of the resulting oscillation. (*Hint*: Keep only terms that are linear in δ.)

Section 21.6

21.55 Suppose the Earth and the Moon carried positive charges of equal magnitude. How large would the charge need to be to produce an electrostatic repulsion equal to 1% of the gravitational attraction between the two bodies?

21.56 The similarity of form of Newton's law of gravitation and Coulomb's Law caused some to speculate that the force of gravity is related to the electrostatic force. Suppose that gravitation is entirely electrical in nature—that an excess charge Q on the Earth and an equal and opposite excess charge $-Q$ on the Moon are responsible for the gravitational force that causes the observed orbital motion of the Moon about the Earth. What is the required size of Q to reproduce the observed magnitude of the gravitational force?

•21.57 In the Bohr model of the hydrogen atom, the electron moves around the one-proton nucleus on circular orbits of well-determined radii, given by $r_n = n^2 a_B$, where $n = 1, 2, 3, ...$ is an integer that defines the orbit and $a_B = 5.29 \cdot 10^{-11}$m is the radius of the first (minimum) orbit, called the *Bohr radius*. Calculate the force of electrostatic interaction between the electron and the proton in the hydrogen atom for the first four orbits. Compare the strength of this interaction to the gravitational interaction between the proton and the electron.

•**21.58** Some of the earliest atomic models held that the orbital velocity of an electron in an atom could be correlated with the radius of the atom. If the radius of the hydrogen atom is 10^{-10} m and the electrostatic force is responsible for the circular motion of the electron, what is the electron's orbital velocity? What is the kinetic energy of this orbital electron?

21.59 For the atom described in Problem 21.58, what is the ratio of the gravitational force between electron and proton to the electrostatic force? How does this ratio change if the radius of the atom is doubled?

•**21.60** In general, astronomical objects are not exactly electrically neutral. Suppose the Earth and the Moon each carry a charge of $-1.00 \cdot 10^6$ C (this is approximately correct; a more precise value is identified in Chapter 22).

a) Compare the resulting electrostatic repulsion with the gravitational attraction between the Moon and the Earth. Look up any necessary data.

b) What effects does this electrostatic force have on the size, shape, and stability of the Moon's orbit around the Earth?

Additional Problems

21.61 Eight 1-μC charges are arrayed along the y-axis located every 2 cm starting at $y = 0$ and extending to $y = 14$ cm. Find the force on the charge at $y = 4$ cm.

21.62 In a simplified Bohr model of the hydrogen atom, an electron is assumed to be traveling in a circular orbit of radius of about $5.2 \cdot 10^{-11}$ m around a proton. Calculate the speed of the electron in that orbit.

21.63 The nucleus of a carbon-14 atom (mass = 14 amu) has diameter of 3 fm. It has 6 protons and a charge of $+6e$.

a) What is the force on a proton located at 3 fm from the surface of this nucleus? Assume that the nucleus is a point charge.

b) What is the proton's acceleration?

21.64 Two charged objects experience a mutual repulsive force of 0.10 N. If the charge of one of the objects is reduced by half and the distance separating the objects is doubled, what is the new force?

21.65 A particle (charge = +19.0 μC) is located on the x-axis at $x = -10.0$ cm, and a second particle (charge = -57.0 μC) is placed on the x-axis at $x = +20.0$ cm. What is the magnitude of the total electrostatic force on a third particle (charge = -3.80 μC) placed at the origin ($x = 0$)?

21.66 Three point charges are positioned on the x-axis: $+64.0$ μC at $x = 0.00$ cm, $+80.0$ μC at $x = 25.0$ cm, and -160.0 μC at $x = 50.0$ cm. What is the magnitude of the electrostatic force acting on the $+64.0$-μC charge?

21.67 From collisions with cosmic rays and from the solar wind, the Earth has a net electric charge of approximately $-6.8 \cdot 10^5$ C. Find the charge that must be given to a 1.0-g object for it to be electrostatically levitated close to the Earth's surface.

21.68 Your sister wants to participate in the yearly science fair at her high school and asks you to suggest some exciting project. You suggest that she experiment with your recently created electron extractor to suspend her cat in the air. You tell her to buy a copper plate and bolt it to the ceiling in her room and then use your electron extractor to transfer electrons from the plate to the cat. If the cat weighs 7 kg and is suspended 2 m below the ceiling, how many electrons have to be extracted from the cat? Assume that the cat and the metal plate are point charges.

•**21.69** A 10-g mass is suspended 5 cm above a nonconducting flat plate, directly above an embedded charge of q (in coulombs). If the mass has the same charge, q, how much must q be so that the mass levitates (just floats, neither rising nor falling)? If the charge q is produced by adding electrons to the mass, by how much will the mass be changed?

•**21.70** Four point charges are placed at the following xy-coordinates:

$Q_1 = -1$ mC, at (-3 cm,0 cm)

$Q_2 = -1$ mC, at ($+3$ cm,0 cm)

$Q_3 = +1.024$ mC, at (0 cm,0 cm)

$Q_4 = +2$ mC, at (0 cm,-4 cm)

Calculate the net force on charge Q_4 due to charges Q_1, Q_2, and Q_3.

•**21.71** Three 5-g Styrofoam balls of radius 2 cm are coated with carbon black to make them conducting and then are tied to 1-m-long threads and suspended freely from a common point. Each ball is given the same charge, q. At equilibrium, the balls form an equilateral triangle with sides of length 25 cm in the horizontal plane. Determine q.

•**21.72** Two point charges lie on the x-axis. If one point charge is 6.0 μC and lies at the origin and the other is -2.0 μC and lies at 20.0 cm, at what position must a third charge be placed to be in equilibrium?

•**21.73** Two beads with charges $q_1 = q_2 = +2.67$ μC are on an insulating string that hangs straight down from the ceiling as shown in the figure. The lower bead is fixed in place on the end of the string and has a mass $m_1 = 0.280$ kg. The second bead slides without friction on the string. At a distance $d = 0.360$ m between the centers of the beads, the force of the Earth's gravity on m_2 is balanced by the electrostatic force between the two beads. What is the mass, m_2, of the second bead? (*Hint:* You can neglect the gravitational interaction between the two beads.)

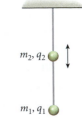

•**21.74** Find the net force on a 2.0-C charge at the origin of an xy-coordinate system if there is a +5.0-C charge at (3 m,0) and a -3.0-C charge at (0,4 m).

•**21.75** Two spheres, each of mass $M = 2.33$ g, are attached by pieces of string of length $L = 45$ cm to a common point.

The strings initially hang straight down, with the spheres just touching one another. An equal amount of charge, q, is placed on each sphere. The resulting forces on the spheres cause each string to hang at an angle of $\theta = 10.0°$ from the vertical. Determine q, the amount of charge on each sphere.

•**21.76** A point charge $q_1 = 100$ nC is at the origin of an xy-coordinate system, a point charge $q_2 = -80$ nC is on the x-axis at $x = 2.0$ m, and a point charge $q_3 = -60$ nC is on the y-axis at $y = -2.0$ m. Determine the net force (magnitude and direction) on q_1.

•**21.77** A positive charge $q_1 = 1$ μC is fixed at the origin, and a second charge $q_2 = -2$ μC is fixed at $x = 10$ cm. Where along the x-axis should a third charge be positioned so that it experiences no force?

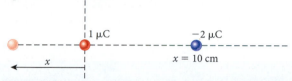

•**21.78** A bead with charge $q_1 = 1.27$ μC is fixed in place at the end of a wire that makes an angle of $\theta = 51.3°$ with the horizontal. A second bead with mass $m_2 = 3.77$ g and a charge of 6.79 μC slides without friction on the wire. What is the distance d at which the force of the Earth's gravity on m_2 is balanced by the electrostatic force between the two beads? Neglect the gravitational interaction between the two beads.

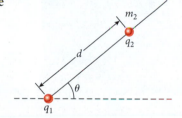

•**21.79** In the figure, the net electrostatic force on charge Q_A is zero. If $Q_A = +1$ nC, determine the magnitude of Q_0.

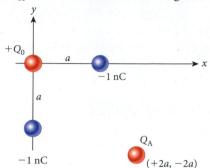

21.80 Two balls have the same mass of 0.681 kg and identical charges of 18.0 μC. They hang from the ceiling on strings of identical length as shown in the figure. If the angle with respect to the vertical of the strings is 20.0°, what is the length of the strings?

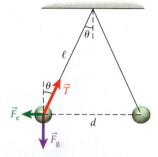

•**21.81** As shown in the figure, charge 1 is 3.94 μC and is located at $x_1 = -4.7$ m, and charge 2 is 6.14 μC and is at $x_2 = 12.2$ m. What is the x-coordinate of the point at which the net force on a point charge of 0.300 μC is zero?

22

Electric Fields and Gauss's Law

FIGURE 22.1 A great white shark can detect tiny electric fields generated by its prey.

WHAT WE WILL LEARN

- An electric field represents the electric force at different points in space.

- Electric field lines represent force vectors exerted on a unit positive electric charge. They originate on positive charges and terminate on negative charges.

- The electric field of a point charge is radial, proportional to the charge, and inversely proportional to the square of the distance from the charge.

- An electric dipole consists of a positive and a negative charge of equal magnitude.

- The electric flux is the electric field component normal to an area times the area.

- Gauss's Law states that the electric flux through a closed surface is proportional to the net electric charge enclosed within the surface.

- The electric field inside a conductor is zero.

- The magnitude of the electric field due to a uniformly charged, infinitely long wire varies as the inverse of the perpendicular distance from the wire.

- The electric field due to an infinite sheet of charge does not depend on the distance from the sheet.

- The electric field outside a spherical distribution of charge is the same as the field of a point charge at the center with the same total charge.

The great white shark is one of the most feared predators on Earth (Figure 22.1). It has several senses that have evolved for hunting prey; for example, it can smell tiny amounts of blood from as far away as 5 km (3 mi). Perhaps more amazing, it has developed special organs (called the *ampullae of Lorenzini*) that can detect the tiny electric fields generated by the movement of muscles in an organism, whether a fish, a seal, or a human. However, just what are electric fields? In addition, how are they related to electric charges?

The concept of vector fields is one of the most useful and productive ideas in all of physics. This chapter explains what an electric field is and how it is connected to electrostatic charges and forces and then examines how to determine the electric field due to some given distribution of charge. This study leads us to one of the most important laws of electricity—Gauss's Law—which provides a relationship between electric fields and electrostatic charge. However, Gauss's Law has practical application only when the charge distribution has enough geometric symmetry to simplify the calculation, and even then, some other concepts related to electric fields are necessary in order to apply the equations. We'll examine another kind of field—magnetic fields—in Chapter 27 through 29. Then, Chapter 31 will show how Gauss's Law fits into a unified description of electric and magnetic fields—one of the finest achievements in physics, from both a practical and an esthetic point of view.

22.1 Definition of an Electric Field

In Chapter 21, we discussed the force between two or more point charges. When determining the net force exerted by other charges on a particular charge at some point in space, we obtain different directions for this force, depending on the sign of the charge that is the reference point. In addition, the net force is also proportional to the magnitude of the reference charge. The techniques used in Chapter 21 require us to redo the calculation for the net force each time we consider a different charge.

Dealing with this situation requires the concept of a **field,** which can be used to describe certain forces. An **electric field,** $E(r)$, is defined at any point in space, $\vec{r}$, as the net electric force on a charge, divided by that charge:

$$\vec{E}(\vec{r}) = \frac{\vec{F}(\vec{r})}{q}. \tag{22.1}$$

The units of the electric field are newtons per coulomb (N/C). This simple definition eliminates the cumbersome dependence of the electric force on the particular charge being used to measure the force. We can quickly determine the net force on any charge by using $\vec{F}(\vec{r}) = q\vec{E}(\vec{r})$, which is a trivial rearrangement of equation 22.1.

The electric force on a charge at a point is parallel (or antiparallel, depending on the sign of the charge in question) to the electric field at that point and proportional to the charge. The magnitude of the force is given by $F = |q|E$. The direction of the force on a positive charge is along $\vec{E}(\vec{r})$; the direction of the force on a negative charge is in the direction opposite to $\vec{E}(\vec{r})$.

If several sources of electric fields are present at the same time, such as several point charges, the electric field at any given point is determined by the superposition of the electric fields from all sources. This superposition follows directly from the superposition of forces introduced in our study of mechanics and discussed in Chapter 21 for electrostatic forces. The **superposition principle** for the total electric field, $\vec{E}_t$, at any point in space with coordinate $\vec{r}$, due to n electric field sources can be stated as

$$\vec{E}_t(\vec{r}) = \vec{E}_1(\vec{r}) + \vec{E}_2(\vec{r}) + \cdots + \vec{E}_n(\vec{r}).   \tag{22.2}$$

22.2   Field Lines

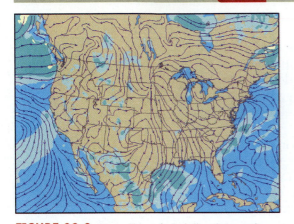

FIGURE 22.2 Streamlines of wind directions at the surface in the United States on March 23, 2008, from the National Weather Service.

An electric field can (and in most applications does) change as a function of the spatial coordinate. The changing direction and strength of the electric field can be visualized by means of **electric field lines.** These graphically represent the net vector force exerted on a unit positive test charge. The representation applies separately for each point in space where the test charge might be placed. The direction of the field line at each point is the same as the direction of the force at that point, and the density of field lines is proportional to the magnitude of the force.

Electric field lines can be compared to the streamlines of wind directions, shown in Figure 22.2. These streamlines represent the force of the wind on objects at given locations, just as the electric field lines represent the electric force at specific points. A hot-air balloon can be used as a test particle for these wind streamlines. For example, a hot-air balloon launched in Dallas, Texas, would float from north to south in the situation depicted in Figure 22.2. Where the wind streamlines are close together, the speed of the wind is higher, so the balloon would move faster.

To draw an electric field line, we imagine placing a tiny positive charge at each point in the electric field. This charge is small enough that it does not affect the surrounding field. A small charge like this is sometimes called a **test charge.** We calculate the resultant force on the charge, and the direction of the force gives the direction of the field line. For example, Figure 22.3a shows a point in an electric field. In Figure 22.3b, a charge $+q$ is placed at point P, on an electric field line. The force on the charge is in the same direction as the electric field. In Figure 22.3c, a charge $-q$ is placed at point P, and the resulting force is in the direction opposite to the electric field. In Figure 22.3d, a charge $+2q$ is placed at point P, and the resulting force on the charge is in the direction of the electric field, with twice the magnitude of the force on the charge $+q$. We will follow the convention of depicting a positive charge as red and a negative charge as blue.

In a nonuniform electric field, the electric force at a given point is tangent to the electric field lines at that point, as illustrated in Figure 22.4. The force on a positive charge is in the direction of the electric field, and the force on a negative charge is in the direction opposite to the electric field.

FIGURE 22.3 The force resulting from placing a charge in an electric field. (a) A point P on an electric field line. (b) A positive charge $+q$ placed at point P. (c) A negative charge $-q$ placed at point P. (d) A positive charge $+2q$ placed at point P.

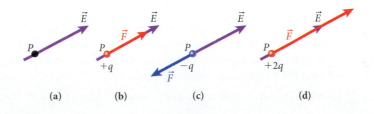

(a)   (b)   (c)   (d)

Electric field lines point away from sources of positive charge and toward sources of negative charge. Each field line starts at a charge and ends at another charge. Electric field lines always originate on positive charges and terminate on negative charges.

Electric fields exist in three dimensions (Figure 22.5); however, this chapter usually presents two-dimensional depictions of electric fields for simplicity.

Point Charge

The electric field lines arising from an isolated point charge are shown in Figure 22.6. The field lines emanate in radial directions from the point charge. If the point charge is positive (Figure 22.6a), the field lines point outward, away from the charge; if the point charge is negative, the field lines point inward, toward the charge (Figure 22.6b). For an isolated positive point charge, the electric field lines originate at the charge and terminate on negative charges at infinity, and for a negative point charge, the electric field lines originate at positive charges at infinity and terminate at the charge. Note that the electric field lines are closer together near the point charge and farther apart away from the point charge, indicating that the electric field becomes weaker with increasing distance from the charge. We'll examine the magnitude of the field quantitatively in Section 22.3.

Two Point Charges of Opposite Sign

We can use the superposition principle to determine the electric field from two point charges. Figure 22.7 shows the electric field lines for two oppositely charged point charges with the same magnitude. At each point in the plane, the electric field from the positive charge and the electric field from the negative charge add as vectors to determine the magnitude and the direction of the resulting electric field. (Figure 22.5 shows the same field lines in three dimensions.)

As noted earlier, the electric field lines originate on the positive charge and terminate on the negative charge. At a point very close to either charge, the field lines are similar to those for a single point charge, since the effect of the more distant charge is small. Near the charges, the electric field lines are close together, indicating that the field is stronger in those regions. The fact that the field lines between the two charges connect indicates that an attractive force occurs between the two charges.

Two Point Charges with the Same Sign

We can also apply the principle of superposition to two point charges with the same sign. Figure 22.8 shows the electric field lines for two point charges with the same sign and same magnitude. If both charges are positive (as in Figure 22.8), the electric field lines originate at the charges and terminate at infinity. If both charges are negative, the field lines originate at infinity and terminate at the charges. For two charges of the same sign, the field lines do not connect the two charges. Rather, the field lines terminate on opposite charges at infinity. The fact that the field lines never terminate on the other charge signifies that the charges repel each other.

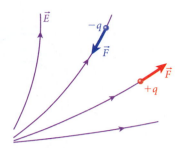

FIGURE 22.4 A nonuniform electric field. A positive charge $+q$ and a negative charge $-q$ placed in the field experience forces as shown. Each force is tangent to the electric field line.

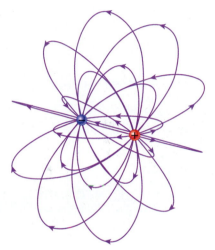

FIGURE 22.5 Three-dimensional representation of electric field lines from two point charges with opposite signs.

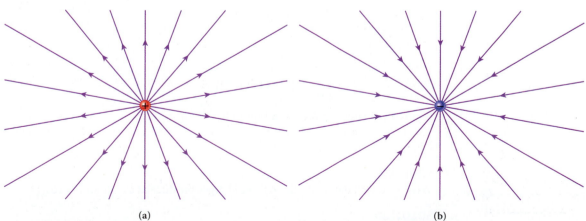

(a)   (b)

FIGURE 22.6 Electric field lines (a) from a single positive point charge and (b) to a single negative point charge.

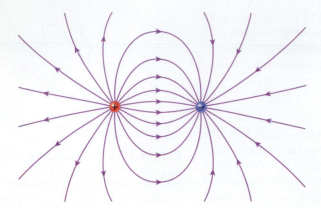

FIGURE 22.7 Electric field lines from two oppositely charged point charges. Each charge has the same magnitude.

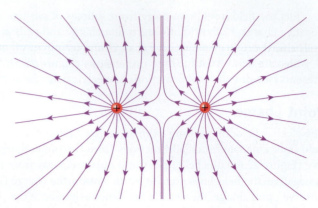

FIGURE 22.8 Electric field lines from two positive point charges with the same magnitude.

General Observations

The three simplest possible cases that we just examined lead to two general rules that apply to all field lines of all charge configurations:

1. *Field lines originate at positive charges and terminate at negative charges.*
2. *Field lines never cross.* This result is a consequence of the fact that the lines represent the electric field, which in turn is proportional to the net force that acts on a charge placed at a particular point. Field lines that crossed would imply that the net force points in two different directions at the same point, which is impossible.

22.3 Electric Field due to Point Charges

The magnitude of the electric force on a point charge q_0 due to another point charge, q, is given by

$$F = \frac{1}{4\pi\epsilon_0} \frac{|qq_0|}{r^2}.$$

(22.3)

Taking q_0 to be a small test charge, we can express the magnitude of the electric field at the point where q_0 is and due to the point charge q as

$$E = \left|\frac{F}{q_0}\right| = \frac{1}{4\pi\epsilon_0} \frac{|q|}{r^2},$$

(22.4)

where r is the distance from the test charge to the point charge. The direction of this electric field is radial. The field points outward for a positive point charge and inward for a negative point charge.

An electric field is a vector quantity and thus, the components of the field must be added separately. Example 22.1 demonstrates the addition of electric fields created by three point charges.

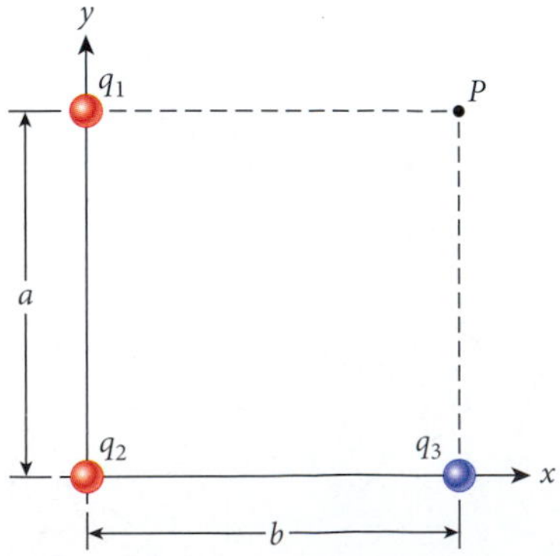

EXAMPLE 22.1   Three Charges

Figure 22.9 shows three fixed point charges: $q_1 = +1.50$ μC, $q_2 = +2.50$ μC, and $q_3 = -3.50$ μC. Charge q_1 is located at $(0,a)$, q_2 is located at $(0,0)$, and q_3 is located at $(b,0)$, where $a = 8.00$ m and $b = 6.00$ m.

PROBLEM

What electric field, $\vec{E}$, do these three charges produce at the point $P = (b,a)$?

SOLUTION

We must sum the electric fields from the three charges using equation 22.2. We proceed by summing component by component, starting with the field due to q_1:

FIGURE 22.9 Locations of three point charges.

$$\vec{E}_1 = E_{1,x}\hat{x} + E_{1,y}\hat{y}.$$

The field due to q_1 acts only in the x-direction at point (b,a), because q_1 has the same y-coordinate as P. Thus, $\vec{E}_1 = E_{1,x}\hat{x}$. We can determine $E_{1,x}$ using (equation 22.4):

$$E_{1,x} = \frac{kq_1}{b^2}.$$

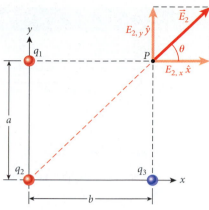

Note that the sign of $E_{1,x}$ is the same as the sign of q_1. Similarly, the field due to q_3 acts only in the y-direction at point (b,a). Thus, $\vec{E}_3 = E_{3,y}\hat{y}$, where

$$E_{3,y} = \frac{kq_3}{a^2}.$$

As shown in Figure 22.10, the electric field due to q_2 at P is given by

$$\vec{E}_2 = E_{2,x}\hat{x} + E_{2,y}\hat{y}.$$

FIGURE 22.10 Electric field due to q_2 and its x- and y-components at point P.

Note that $\vec{E}_2$, the electric field due to q_2 at point P points directly away from q_2, because $q_2 > 0$. (It would point directly toward q_2 if this charge were negative.) The magnitude of this electric field is given by

$$E_2 = \frac{k|q_2|}{a^2 + b^2}.$$

The component $E_{2,x}$ is given by $E_2 \cos\theta$, where $\theta = \tan^{-1}(a/b)$, and the component $E_{2,y}$ is given by $E_2 \sin\theta$.

Adding the components, the total electric field at point P is

$$\vec{E} = \left(E_{1,x} + E_{2,x}\right)\hat{x} + \left(E_{2,y} + E_{3,y}\right)\hat{y}$$

$$= \underbrace{\left(\frac{kq_1}{b^2} + \frac{kq_2 \cos\theta}{a^2 + b^2}\right)}_{E_x}\hat{x} + \underbrace{\left(\frac{kq_2 \sin\theta}{a^2 + b^2} + \frac{kq_3}{a^2}\right)}_{E_y}\hat{y}.$$

With the given values for a and b, we find $\theta = \tan^{-1}(8/6) = 53.1°$, and $a^2 + b^2 = (8.00 \text{ m})^2 + (6.00 \text{ m})^2 = 100 \text{ m}^2$. We can then calculate the x-component of the total electric field as

$$E_x = \left(8.99 \cdot 10^9 \text{ N m}^2/\text{C}^2\right)\left|\frac{1.50 \cdot 10^{-6} \text{ C}}{\left(6.00 \text{ m}\right)^2} + \frac{\left(2.50 \cdot 10^{-6} \text{ C}\right)\left(\cos 53.1°\right)}{100 \text{ m}^2}\right| = 509 \text{ N/C}.$$

The y-component is

$$E_y = \left(8.99 \cdot 10^9 \text{ N m}^2/\text{C}^2\right)\left|\frac{\left(2.50 \cdot 10^{-6} \text{ C}\right)\left(\sin 53.1°\right)}{100 \text{ m}^2} + \frac{-3.50 \cdot 10^{-6} \text{ C}}{\left(8.00 \text{ m}\right)^2}\right| = -312 \text{ N/C}.$$

The magnitude of the field is

$$E = \sqrt{E_x^2 + E_y^2} = \sqrt{\left(509 \text{ N/C}\right)^2 + \left(-312 \text{ N/C}\right)^2} = 597 \text{ N/C}.$$

The direction of the field at point P is

$$\varphi = \tan^{-1}\left(\frac{E_y}{E_x}\right) = \tan^{-1}\left(\frac{-312 \text{ N/C}}{509 \text{ N/C}}\right) = -31.5°,$$

which means that the electric field points to the right and downward.

Note that even though the charges in this example are in microcoulombs and the distances are in meters, the electric fields are still large, showing that a microcoulomb is a large amount of charge.

22.4 Electric Field due to a Dipole

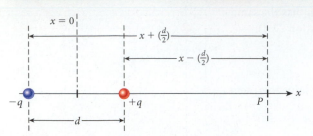

FIGURE 22.11 Calculation of the electric field from an electric dipole.

A system of two equal (in magnitude) but oppositely charged point particles is called an **electric dipole.** The electric field from an electric dipole is given by the vector sum of the electric fields from the two charges. Figure 22.7 shows the electric field lines in two dimensions for an electric dipole.

The superposition principle allows us to determine the electric field due to two point charges through vector addition of the electric fields of the two charges. Let's consider the special case of the electric field due to a dipole along the axis of the dipole, defined as the line connecting the charges. This main symmetry axis of the dipole is assumed to be oriented along the x-axis (Figure 22.11).

The electric field, $\vec{E}$, at point P on the dipole axis is the sum of the field due to $+q$, denoted as $\vec{E}_+$, and the field due to $-q$, denoted as $\vec{E}_-$:

$$\vec{E} = \vec{E}_+ + \vec{E}_-.$$

Using equation 22.4, we can express the magnitude of the dipole's electric field along the x-axis, for $x > d/2$, as

$$E = \frac{1}{4\pi\epsilon_0}\frac{q}{r_+^2} + \frac{1}{4\pi\epsilon_0}\frac{-q}{r_-^2},$$

where r_+ is the distance between P and $+q$ and r_- is the distance between P and $-q$. Absolute value bars are not needed in this equation, because the first term on the right-hand side is positive and is greater than the second (negative) term. The electric field at all points on the x-axis (except at $x = \pm d/2$, where the two charges are located) is given by

$$\vec{E} = E_x\hat{x} = \frac{1}{4\pi\epsilon_0}\frac{q(x-d/2)}{r_+^3}\hat{x} + \frac{1}{4\pi\epsilon_0}\frac{-q(x+d/2)}{r_-^3}\hat{x}. \tag{22.5}$$

Now we examine the magnitude of $\vec{E}$ and restrict the value of x to $x > d/2$, where $E = E_x > 0$. Then we have

$$E = \frac{1}{4\pi\epsilon_0}\frac{q}{\left(x-\frac{1}{2}d\right)^2} - \frac{1}{4\pi\epsilon_0}\frac{q}{\left(x+\frac{1}{2}d\right)^2}.$$

With some rearrangement and keeping in mind that we want to obtain an expression that has the same form as the electric field from a point charge, we write the preceding equation as

$$E = \frac{q}{4\pi\epsilon_0 x^2}\left[\left(1-\frac{d}{2x}\right)^{-2} - \left(1+\frac{d}{2x}\right)^{-2}\right].$$

To find an expression for the electric field at a large distance from the dipole, we can make the approximation $x \gg d$ and use the binomial expansion. (Since $x \gg d$, we can drop terms containing the square of d/x and higher powers.) We obtain

$$E \approx \frac{q}{4\pi\epsilon_0 x^2}\left[\left(1+\frac{d}{x}-\cdots\right)-\left(1-\frac{d}{x}+\cdots\right)\right] = \frac{q}{4\pi\epsilon_0 x^2}\left(\frac{2d}{x}\right),$$

which can be rewritten as

$$E \approx \frac{qd}{2\pi\epsilon_0 x^3}. \tag{22.6}$$

Equation 22.6 can be simplified by defining a vector quantity called the **electric dipole moment,** $\vec{p}$. The direction of this dipole moment is from the negative charge to the positive charge, which is opposite the direction of the electric field lines. The magnitude, p, of the electric dipole moment is given by

$$p = qd, \tag{22.7}$$

where q is the magnitude of one of the charges and d is the distance separating the two charges. With this definition, the expression for the magnitude of the electric field due to the dipole along the positive x-axis at a distance large compared with the separation between the two charges is

$$E = \frac{p}{2\pi\epsilon_0 |x|^3}. \qquad (22.8)$$

Although not shown explicitly here, equation 22.8 is also valid for $x = \ll -d$. Also, an examination of equation 22.5 for $\vec{E}$ shows that $E_x > 0$ on either side of the dipole. In contrast to the field due to a point charge, which is inversely proportional to the square of the distance, the field due to a dipole is inversely proportional to the cube of the distance, according to equation 22.8.

EXAMPLE 22.2 Water Molecule

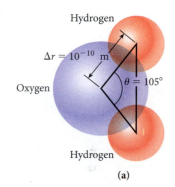

Hydrogen

$\Delta r = 10^{-10}$ m

Oxygen

$\theta = 105°$

Hydrogen

(a)

The water molecule, H_2O, is arguably the most important one for life. It has a nonzero dipole moment, which is the basic reason why many organic molecules are able to bind to water. This dipole moment also allows water to be an excellent solvent for many inorganic and organic compounds.

Each water molecule consists of two atoms of hydrogen and one atom of oxygen, as shown in Figure 22.12a. The charge distribution of each of the individual atoms is approximately spherical. The oxygen atom tends to pull the negatively charged electrons toward itself, giving the hydrogen atoms slight positive charge. The three atoms are arranged so that the lines connecting the centers of the hydrogen atoms with the center of the oxygen atom have an angle of 105° between them (see Figure 22.12a).

PROBLEM

Suppose we approximate a water molecule by two positive charges at the locations of the two hydrogen nuclei (protons) and two negative charges at the location of the oxygen nucleus, with all charges of equal magnitude. What is the resulting electric dipole moment of water?

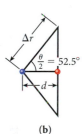

Δr

$\frac{\theta}{2} = 52.5°$

d

(b)

SOLUTION

The center of charge of the two positive charges, analogous to the center of mass of two masses, is located exactly halfway between the centers of the hydrogen atoms, as shown in Figure 22.12b. With the hydrogen-oxygen distance of $\Delta r = 10^{-10}$ m, as indicated in the figure, the distance between the positive and negative charge centers is

$$d = \Delta r \cos\left(\frac{\theta}{2}\right) = \left(10^{-10} \text{ m}\right)\left(\cos 52.5°\right) = 0.6 \cdot 10^{-10} \text{ m}.$$

This distance times the transferred charge, $q = 2e$, is the magnitude of the dipole moment of water:

$$p = 2ed = \left(3.2 \cdot 10^{-19} \text{ C}\right)\left(0.6 \cdot 10^{-10} \text{ m}\right) = 2 \cdot 10^{-29} \text{ C m}.$$

This result of an extremely oversimplified calculation actually comes close, within a factor of 3, to the measured value of $6.2 \cdot 10^{-30}$ C m. The fact that the real dipole moment of water is smaller than this calculated result is an indication that the two electrons of the hydrogen atoms are not pulled all the way to the oxygen but, on average, only one-third of the way.

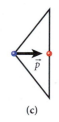

$\vec{p}$

(c)

FIGURE 22.12 (a) Schematic drawing showing the geometry of a water molecule, H_2O, with atoms as spheres. (b) Diagram showing the effective positive (red dot on the right) and negative (blue dot on the left) charge centers. (c) Dipole moment assuming pointlike charges.

22.5 General Charge Distributions

We have determined the electric fields of a single point charge and of two point charges (an electric dipole). Now let's consider the electric field due to a general charge distribution. To do this, we divide the charge into differential elements of charge, dq, and find the electric field resulting from each differential charge element as if it were a point charge. If the charge is distributed along a one-dimensional object (a line), the differential charge may be expressed in terms of a charge per unit length times a differential length, $\lambda\,dx$. If the charge is distributed over a surface (a two-dimensional object), dq is expressed in terms of a charge per unit area times a differential area, $\sigma\,dA$. And, finally, if the charge is distributed over a three-dimensional volume, then dq is written as the product of a charge per unit volume times a differential volume, $\rho\,dV$. That is,

$$\left.\begin{aligned} dq &= \lambda\,dx \\ dq &= \sigma\,dA \\ dq &= \rho\,dV \end{aligned}\right\} \text{ for a charge distribution } \left\{\begin{aligned} &\text{along a line;} \\ &\text{over a surface;} \\ &\text{throughout a volume.} \end{aligned}\right. \tag{22.9}$$

The magnitude of the electric field resulting from the charge distribution is then obtained from the differential charge:

$$dE = k\frac{dq}{r^2}. \tag{22.10}$$

In the following example, we find the electric field due to a finite line of charge.

EXAMPLE 22.3 │ Finite Line of Charge

To find the electric field along a line bisecting a finite length of wire with linear charge density λ, we integrate the contributions to the electric field from all the charge in the wire. We assume that the wire lies along the x-axis (Figure 22.13).

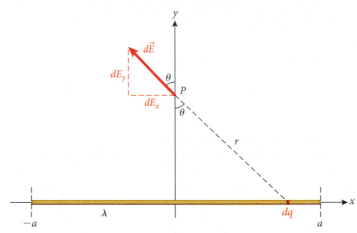

FIGURE 22.13 Calculating the electric field due to all the charge in a long wire by integrating the contributions to the electric field over the length of the wire.

We also assume that the wire is positioned with its midpoint at $x = 0$, one end at $x = a$, and the other end at $x = -a$. The symmetry of the situation then allows us to conclude that there cannot be any electric force parallel to the wire (in the x-direction) along the line bisecting the wire. Along this line, the electric field can be only in the y-direction. We can then calculate the electric field due to all the charge for $x \geq 0$ and multiply the result by 2 to get the electric field for the whole wire.

We consider a differential charge, dq, on the x-axis, as shown in Figure 22.13. The magnitude of the electric field, dE, at a point $(0,y)$ due to this charge is given by equation 22.10,

$$dE = k\frac{dq}{r^2},$$

where $r = \sqrt{x^2 + y^2}$ is the distance from dq to point P. The component of the electric field perpendicular to the wire (in the y-direction) is then given by

$$dE_y = k\frac{dq}{r^2}\cos\theta,$$

where θ is the angle between the electric field produced by dq and the y-axis (see Figure 22.13). The angle θ is related to r and y because $\cos\theta = y/r$.

We can relate the differential charge to the differential distance along the x-axis through the linear charge density, λ: $dq = \lambda\,dx$. The electric field at a distance y from the long wire is then

$$E_y = 2\int_0^a dE_y = 2\int_0^a k\frac{dq}{r^2}\cos\theta = 2k\int_0^a \frac{\lambda dx}{r^2}\frac{y}{r} = 2k\lambda y\int_0^a \frac{dx}{\left(x^2+y^2\right)^{3/2}}.$$

Evaluation of the integral on the right-hand side (with the aid of an integral table or a software package like Mathematica or Maple) gives us

$$\int_0^a \frac{dx}{\left(x^2+y^2\right)^{3/2}} = \left[\frac{1}{y^2}\frac{x}{\sqrt{x^2+y^2}}\right]_0^a = \frac{1}{y^2}\frac{a}{\sqrt{y^2+a^2}}.$$

Thus, the electric field at a distance y along a line bisecting the wire is given by

$$E_y = 2k\lambda y\frac{1}{y^2}\frac{a}{\sqrt{y^2+a^2}} = \frac{2k\lambda}{y}\frac{a}{\sqrt{y^2+a^2}}.$$

Finally, when $a \to \infty$, that is, the wire becomes infinitely long, $a/\sqrt{y^2+a^2} \to 1$, and we have for an infinitely long wire

$$E_y = \frac{2k\lambda}{y}.$$

In other words, the electric field decreases in inverse proportion to the distance from the wire.

Now let's tackle a problem with a slightly more complicated geometry, finding the electric field due to a ring of charge along the axis of the ring.

SOLVED PROBLEM 22.1   Ring of Charge

PROBLEM
Consider a charged ring with radius $R = 0.250$ m (Figure 22.14). The ring has uniform linear charge density and the total charge on the ring is $Q = +5.00$ μC. What is the electric field at a distance $d = 0.500$ m along the axis of the ring?

SOLUTION

THINK
The charge is evenly distributed around the ring. The electric field at position $x = d$ can be calculated by integrating the differential electric field due to a differential electric charge. By symmetry, the components of the electric field perpendicular to the axis of the ring integrate to zero, because the electric fields of charge elements on opposite sides of the axis cancel one another out. The resulting electric field is parallel to the axis of the circle.

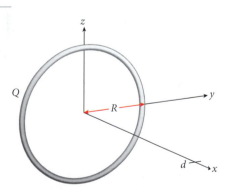

FIGURE 22.14 Charged ring with radius R and total charge Q.

SKETCH
Figure 22.15 shows the geometry for the electric field along the axis of the ring of charge.

RESEARCH
The differential electric field, dE, at $x = d$ is due to a differential charge dq located at $y = R$ (see Figure 22.15). The distance from the point $(x = d, y = 0)$ to the point $(x = 0, y = R)$ is

$$r = \sqrt{R^2 + d^2}.$$

Continued—

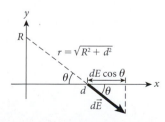

FIGURE 22.15 The geometry for the electric field along the axis of a ring of charge.

Again, the magnitude of $d\vec{E}$ is given by equation 22.10:

$$dE = k\frac{dq}{r^2}.$$

The magnitude of the component of $d\vec{E}$ parallel to the x-axis is given by

$$dE_x = dE\cos\theta = dE\frac{d}{r}.$$

SIMPLIFY

We can find the total electric field by integrating its x-components over all the charge on the ring:

$$E_x = \int\limits_{ring} dE_x = \int\limits_{ring}\frac{d}{r}k\frac{dq}{r^2}.$$

We need to integrate around the circumference of the ring of charge. We can relate the differential charge to the differential arc length, ds, as follows:

$$dq = \frac{Q}{2\pi R}ds.$$

We can then express the integral over the entire ring of charge as an integral around the arc length of a circle:

$$E_x = \int\limits_0^{2\pi R} k\left(\frac{Q}{2\pi R}ds\right)\frac{d}{r^3} = \left(\frac{kQd}{2\pi R r^3}\right)\int\limits_0^{2\pi R} ds = kQ\frac{d}{r^3} = \frac{kQd}{\left(R^2+d^2\right)^{3/2}}.$$

CALCULATE

Putting in the numerical values, we get

$$E_x = \frac{kQd}{\left(R^2+d^2\right)^{3/2}} = \frac{\left(8.99\cdot10^9\ \text{N m}^2/\text{C}^2\right)\left(5.00\cdot10^{-6}\ \text{C}\right)\left(0.500\ \text{m}\right)}{\left[\left(0.250\ \text{m}\right)^2+\left(0.500\ \text{m}\right)^2\right]^{3/2}} = 128{,}654\ \text{N/C}.$$

ROUND

We report our result to three significant figures:

$$E_x = 1.29\cdot10^5\ \text{N/C}.$$

DOUBLE-CHECK

We can check the validity of the formula we derived for the electric field by using a large distance from the ring of charge, such that $d \gg R$. In this case,

$$E_x = \frac{kQd}{\left(R^2+d^2\right)^{3/2}} \stackrel{d\gg R}{\Rightarrow} E_x = \frac{kQd}{d^3} = k\frac{Q}{d^2},$$

which is the expression for the electric field due to a point charge Q at a distance d. We can also check the formula with $d = 0$:

$$E_x = \frac{kQd}{\left(R^2+d^2\right)^{3/2}} \stackrel{d=0}{\Rightarrow} E_x = 0,$$

which is what we would expect at the center of a ring of charge. Thus, our result seems reasonable.

22.6 Force due to an Electric Field

The force $\vec{F}$ exerted by an electric field $\vec{E}$ on a point charge q is given by $\vec{F} = q\vec{E}$, a simple restatement of the definition of the electric field in equation 22.1. Thus, the force exerted by the electric field on a positive charge acts in the same direction as the electric field. The force vector is always tangent to the electric field lines and points in the direction of the electric field if $q > 0$.

22.2 In-Class Exercise

A small positively charged object could be placed in a uniform electric field at position A or position B in the figure. How do the electric forces on the object at the two positions compare?

a) The magnitude of the electric force on the object is greater at position A.

b) The magnitude of the electric force on the object is greater at position B.

c) There is no electric force on the object at either position A or position B.

d) The electric force on the object at position A has the same magnitude as the force on the object at position B but is in the opposite direction.

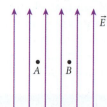

e) The electric force on the object at position A is the same nonzero electric force as that on the object at position B.

22.1 In-Class Exercise

A small positively charged object is placed at rest in a uniform electric field as shown in the figure. When the object is released, it will

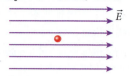

a) not move.

b) begin to move with a constant speed.

c) begin to move with a constant acceleration.

d) begin to move with an increasing acceleration.

e) move back and forth in simple harmonic motion.

The force at various locations on a positive charge due to the electric field in three dimensions is shown in Figure 22.16 for the case of two oppositely charged particles. (This is the same field as in Figure 22.5, but with some representative force vectors added.) You can see that the force on a positive charge is always tangent to the field lines and points in the same direction as the electric field. The force on a negative charge would point in the opposite direction.

22.1 Self-Test Opportunity

The figure shows a two-dimensional view of electric field lines due to two opposite charges. What is the direction of the electric field at the five points A, B, C, D, and E? At which of the five points is the magnitude of the electric field the largest?

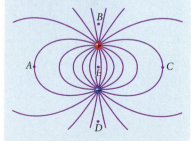

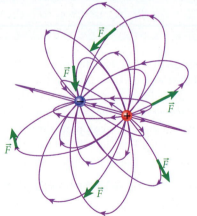

FIGURE 22.16 Direction of the force that an electric field produced by two opposite point charges exerts on a positive charge at various points in space.

EXAMPLE 22.4 | Time Projection Chamber

Nuclear physicists study new forms of matter by colliding gold nuclei at very high energies. In particle physics, new elementary particles are created and studied by colliding protons and antiprotons at the highest energies. These collisions create many particles that stream away from the interaction point at high speeds. A simple particle detector is not sufficient to identify these particles. A device that helps physicists

Continued—

Indicate whether each of the following statements about electric field lines is true or false.

a) Electric field lines point inward toward negative charges.

b) Electric field lines make circles around positive charges.

c) Electric field lines may cross.

d) Electric field lines point outward from positive charges.

e) A positive point charge released from rest will initially accelerate along the tangent to the electric field line at that point.

FIGURE 22.17 An event in the STAR TPC in which two gold nuclei have collided at very high energies at the point in the center of the image. Each colored line represents the track left behind by a subatomic particle produced in the collision.

study these collisions is a time projection chamber (TPC), found in most large particle detectors.

One example of a TPC is the STAR TPC of the Relativistic Heavy Collider at Brookhaven National Laboratory on Long Island, New York. The STAR TPC consists of a large cylinder filled with a gas (90% argon, 10% methane) that allows free electrons to move within it without recombining.

Figure 22.17 shows the results of a collision of two gold nuclei that occurred in the STAR TPC. In such a collision, thousands of charged particles are created that pass through the gas inside the TPC. As these charged particles pass through the gas, they ionize the atoms of the gas, releasing free electrons. A constant electric field of magnitude 13,500 N/C is applied between the center of the TPC and the caps on the ends of the cylinder, and the field exerts an electric force on the freed electrons. Because the electrons have a negative charge, the electric field exerts a force in the direction opposite to the electric field. The electrons attempt to accelerate in the direction of the electric force, but they interact with the electrons of the molecules of the gas and begin to drift toward the caps with a constant speed of 5 cm/μs = $5 \cdot 10^4$ m/s $\approx$ 100,000 mph.

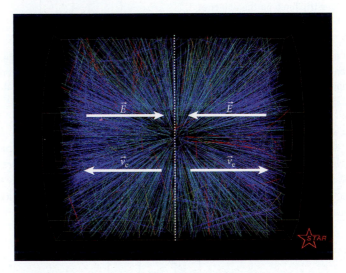

Each end cap of the cylinder has 68,304 detectors that can measure the charge as a function of the drift time of the electrons from the point where they were freed. Each detector has a specific (x,y) position. From measurements of the arrival time of the charge and the known drift speed of the electrons, the z-component of their position can be calculated. Thus, the STAR TPC can produce a complete three-dimensional representation of the ionization track of each charged particle. These tracks are shown in Figure 22.17, where the colors represent the amount of ionization produced by each track.

Dipole in an Electric Field

A point charge in an electric field experiences a force, given by equation 22.1. The electric force is always tangent to the electric field line passing through the point. The effect of an electric field on a dipole can be described in terms of the vector electric field, $\vec{E}$, and the vector electric dipole moment, $\vec{p}$, without detailed knowledge of the charges making up the electric dipole.

To examine the behavior of an electric dipole, let's consider two charges, $+q$ and $-q$, separated by a distance d in a constant uniform electric field, $\vec{E}$ (Figure 22.18). (Note that we are now considering the forces acting on a dipole placed in an external field, as opposed to considering the field caused by the dipole, which we did in Section 22.4, and we also assume the dipole field is small compared to $\vec{E}$ [so its effect on the uniform field can be ignored].) The electric field exerts an upward force on the positive charge and a downward force on the negative charge. Both forces have the magnitude qE. In Chapter 10, we saw

22.4 In-Class Exercise

A negative charge $-q$ is placed in a nonuniform electric field as shown in the figure. What is the direction of the electric force on this negative charge?

a) →

b) ↑

c) ←

d) ↓

e) The force is zero.

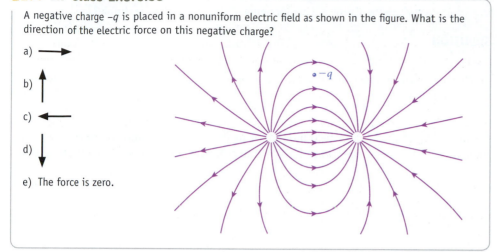

that this situation gives rise to a torque, $\vec{\tau}$, given by $\vec{\tau} = \vec{r} \times \vec{F}$, where $\vec{r}$ is the moment arm and $\vec{F}$ is the force. The magnitude of the torque is $\tau = rF \sin\theta$.

As always, we can calculate the torque about any pivot point, so we can pick the location of the negative charge. Then, only the force on the positive charge contributes to the torque, and the length of the position vector is $r = d$, that is, the length of the dipole. Since, as already stated, $F = qE$, the expression for the torque on an electric dipole in an external electric field can be written as

$$\tau = qEd \sin\theta.$$

Remembering that the electric dipole moment is defined as $p = qd$, we obtain the magnitude of the torque:

$$\tau = pE \sin\theta. \tag{22.11}$$

Because the torque is a vector and must be perpendicular to both the electric dipole moment and the electric field, the relationship in equation 22.11 can be written as a vector product:

$$\vec{\tau} = \vec{p} \times \vec{E}. \tag{22.12}$$

As with all vector products, the direction of the torque is given by a right-hand rule. As shown in Figure 22.19, the thumb indicates the direction of the first term of the vector product, in this case $\vec{p}$, and the index finger indicates the direction of the second term, $\vec{E}$. The result of the vector product, $\vec{\tau}$, is then directed along the middle finger and is perpendicular to each of the two terms.

SOLVED PROBLEM 22.2 | Electric Dipole in an Electric Field

PROBLEM

An electric dipole with dipole moment of magnitude $p = 1.40 \cdot 10^{-12}$ C m is placed in a uniform electric field of magnitude $E = 498$ N/C (Figure 22.20a).

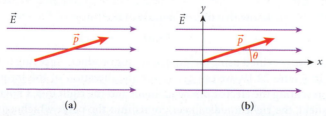

(a)   (b)

FIGURE 22.20 (a) An electric dipole in a uniform electric field. (b) The electric field oriented in the x-direction and the dipole in the xy-plane.

Continued—

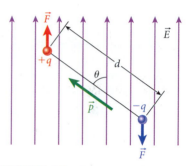

FIGURE 22.18 Electric dipole in an electric field.

22.2 Self-Test Opportunity

Use the center of mass of the dipole as the pivot point and show that you again obtain the expression $\tau = qEd \sin\theta$ for the torque.

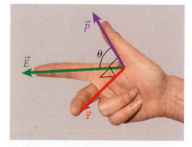

FIGURE 22.19 Right-hand rule for the vector product of the electric dipole moment and the electric field, producing the torque vector.

At some instant (in time) the angle between the electric dipole and the electric field is $\theta = 14.5°$. What are the Cartesian components of the torque on the dipole?

SOLUTION

THINK

The torque on the dipole is equal to the vector product of the electric field and the electric dipole moment.

SKETCH

We assume that the electric field lines point in the x-direction and the electric dipole moment is in the xy-plane (Figure 22.20b). The z-direction is perpendicular to the plane of the page.

RESEARCH

The torque on the electric dipole due to the electric field is given by

$$\vec{\tau} = \vec{p} \times \vec{E}.$$

Since the dipole is located in the xy-plane, Cartesian components of the electric dipole moment are

$$\vec{p} = \left(p_x, p_y, 0 \right).$$

Since the electric field is acting in the x-direction, its Cartesian components are

$$\vec{E} = \left(E_x, 0, 0 \right) = \left(E, 0, 0 \right).$$

SIMPLIFY

From the definition of the vector product, we express the Cartesian components of the torque as

$$\vec{\tau} = \left(p_y E_z - p_z E_y \right)\hat{x} + \left(p_z E_x - p_x E_z \right)\hat{y} + \left(p_x E_y - p_y E_x \right)\hat{z}.$$

In this particular case, with E_y, E_z, and p_z all equal to zero, we have

$$\vec{\tau} = - p_y E_x \hat{z}.$$

The y-component of the dipole moment is $p_y = p \sin \theta$, and the x-component of the electric field is simply $E_x = E$. The magnitude of the torque is then

$$\tau = (p \sin \theta)E = pE \sin \theta,$$

and the direction of the torque is in the negative z-direction.

CALCULATE

We insert the given numerical data and get

$$\tau = pE \sin \theta = \left(1.40 \cdot 10^{-12} \ \text{C m} \right)\left(498 \ \text{N/C} \right)\left(\sin 14.5° \right) = 1.74565 \cdot 10^{-10} \ \text{N m}.$$

ROUND

We report our result to three significant figures:

$$\tau = 1.75 \cdot 10^{-10} \ \text{N m}.$$

DOUBLE-CHECK

From equation 22.11, we know that the magnitude of the torque is

$$\tau = pE \sin \theta,$$

which is the result we obtained using the explicit vector product. Applying the right-hand rule illustrated in Figure 22.19, we can determine the direction of the torque: With the right thumb representing the electric dipole moment and the right index finger representing the electric field, the right middle finger points into the page, which agrees with the result we found via the vector product. Thus, our result is correct.

Example 22.2 looked at the dipole moment of water molecules. If water molecules are exposed to an external electric field, they experience a torque and thus begin to rotate. If the direction of the external electric field changes very rapidly, the water molecules perform rotational oscillations, which create heat. This is the principle of operation of a microwave oven (Figure 22.21). Microwave ovens use a frequency of 2.45 GHz for the oscillating electric field. (How an electric field is made to oscillate in time will be covered in Chapter 31 on electromagnetic waves.)

Electric fields also play a key role in human physiology, but these fields are time-varying and not static, like those studied in this chapter. (They will be covered in later chapters.) The evolution over time of electric fields in the human heart is measured by an electrocardiogram (ECG) (to be discussed, along with the functioning of pacemaker implants, in Chapter 26 on circuits). The human brain also generates continuously changing electrical fields through the activity of the neurons. These fields can be measured invasively by inserting electrodes through the skull and into the brain or by placing electrodes onto the surface of the exposed brain, usually during brain surgery. This method is called electrocorticography (ECoG). An intense area of current research focuses on measuring and imaging brain electric fields noninvasively by attaching electrodes to the outside of the skull. However, since the skull itself dampens the electric fields, these techniques require great instrumental sensitivity and are still in their infancy. Perhaps the most exciting (or scary, depending on your point of view) research developments are in brain-computer interfaces. In this emerging field, electrical activity in the brain is used directly to control computers, and external stimuli are used to create electric fields inside the brain. Researchers in this area are motivated by the goal of helping people overcome physical disabilities, such as blindness or paralysis.

FIGURE 22.21 Microwave oven found in most kitchens.

22.7 Electric Flux

Electric field calculations, like those in Example 22.3, can require quite a bit of work. However, in many common situations, particularly those with some geometric symmetry, a powerful technique for determining electric fields without having to explicitly calculate integrals can be used. This technique called *Gauss's Law*, is one of the fundamental relations of electric fields. However, to use Gauss's Law requires understanding of a concept called *electric flux*.

Imagine holding a ring with inside area A in a stream of water flowing with velocity $\vec{v}$, as shown in Figure 22.22. The area vector, $\vec{A}$, of the ring is defined as a vector with magnitude A pointing in a direction perpendicular to the plane of the ring. In Figure 22.22a, the area vector of the ring is parallel to the flow velocity, and the flow velocity is perpendicular to the plane of the ring. The product Av gives the amount of water passing through the ring per unit time (see Chapter 13) where v is the magnitude of the flow velocity. If the plane of the ring is tilted with respect to the direction of the flowing water (Figure 22.22b), the amount of water flowing through the ring is given by $Av\cos\theta$, where θ is the angle between the area vector of the ring and the direction of the velocity of the flowing water. The amount of water flowing through the ring is called the *flux*, $\Phi = Av\cos\theta = \vec{A}\cdot\vec{v}$. Since flux is a measure of volume per unit time, its units are cubic meters per second (m³/s).

An electric field is analogous to flowing water. Consider a uniform electric field of magnitude E passing through a given area A (Figure 22.23). Again, the area vector is $\vec{A}$, with a direction normal to the surface of the area and a magnitude A. The angle θ is the angle between the vector electric field and the area vector, as shown in Figure 22.23. The electric field passing through a given area A is called the **electric flux** and is given by

$$\Phi = EA\cos\theta. \qquad (22.13)$$

In simple terms, the electric flux is proportional to the number of electric field lines passing through the area. We'll assume that the electric field is given by $\vec{E}(\vec{r})$ and that the area is a closed surface, rather than the open surface of a simple ring in flowing water. In

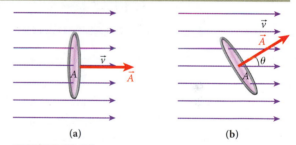

(a)   (b)

FIGURE 22.22 Water flowing with velocity of magnitude v through a ring of area A. (a) The area vector is parallel to the flow velocity. (b) The area vector is at an angle θ to the flow velocity.

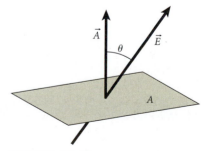

FIGURE 22.23 A uniform electric field $\vec{E}$ passing through an area $\vec{A}$.

this closed-surface case, the total, or net, electric flux is given by an integral of the electric field over the closed surface:

$$\Phi = \oiint \vec{E} \cdot d\vec{A}, \qquad (22.14)$$

where $\vec{E}$ is the electric field at each differential area element $d\vec{A}$ of the closed surface. The direction of $d\vec{A}$ is outward from the closed surface. In equation 22.14, the loop on the integrals means that the integration is over a closed surface, and the two integral signs signify an integration over two variables. (*Note:* Some books use different notation for the integral over a closed surface, $\iint_S dA$ or just $\int_S dA$, but these refer to the same integration procedure as is represented in equation 22.14.) The differential area element $d\vec{A}$ must be described by two spatial variables, such as x and y in Cartesian coordinates or θ and ϕ in spherical coordinates.

Figure 22.24 shows a nonuniform electric field, $\vec{E}$, passing through a differential area element, $d\vec{A}$. A portion of the closed surface is also shown. The angle between the electric field and the differential area element is θ.

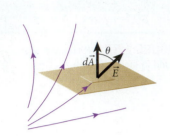

FIGURE 22.24 A nonuniform electric field, $\vec{E}$ passing through a differential area, $d\vec{A}$.

EXAMPLE 22.5   Electric Flux through a Cube

Figure 22.25 shows a cube that has faces with area A in a uniform electric field, $\vec{E}$, that is perpendicular to the plane of one face of the cube.

FIGURE 22.25 A cube with faces of area A in a uniform electric field, $\vec{E}$.

PROBLEM

What is the net electric flux passing though the cube?

SOLUTION

The electric field in Figure 22.25 is perpendicular to the plane of one of the cube's six faces and therefore is also perpendicular to the opposite face. The area vectors of these two faces, $\vec{A}_1$ and $\vec{A}_2$, are shown in Figure 22.26a. The net electric flux passing through these two faces is

$$\Phi_{12} = \Phi_1 + \Phi_2 = \vec{E} \cdot \vec{A}_1 + \vec{E} \cdot \vec{A}_2 = -EA_1 + EA_2 = 0.$$

The negative sign arises for the flux through face 1 because the electric field and the area vector, $\vec{A}_1$, are in opposite directions. The area vectors of the remaining four faces are all perpendicular to the electric field, as shown in Figure 22.26b. The net electric flux passing through these four faces is

$$\Phi_{3456} = \Phi_3 + \Phi_4 + \Phi_5 + \Phi_6 = \vec{E} \cdot \vec{A}_3 + \vec{E} \cdot \vec{A}_4 + \vec{E} \cdot \vec{A}_5 + \vec{E} \cdot \vec{A}_6 = 0.$$

All the scalar products are zero because the area vectors of these four faces are perpendicular to the electric field. Thus, the net electric flux passing through the cube is

$$\Phi = \Phi_{12} + \Phi_{3456} = 0.$$

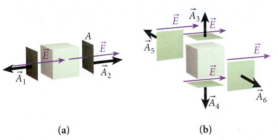

(a)   (b)

FIGURE 22.26 (a) The two faces of the cube that are perpendicular to the electric field. The area vectors are parallel and antiparallel to the electric field. (b) The four faces of the cube that are parallel to the electric field. The area vectors are perpendicular to the electric field.

22.8   Gauss's Law

To begin our discussion of Gauss's Law, let's imagine a box in the shape of a cube (Figure 22.27a), which is constructed of a material that does not affect electric fields. A positive test charge brought close to any surface of the box, will experience no force. Now suppose a positive charge is inside the box and the positive test charge is brought close to the surface of the box (Figure 22.27b). The positive test charge experiences an outward force due to the positive charge inside the box. If the test charge is close to any surface of the box, it experiences the outward force. If twice as much positive charge is inside the box, a positive test charge close to any surface of the box feels twice the outward force.

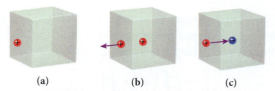

FIGURE 22.27 Three imaginary boxes constructed of material that does not affect electric fields. A positive test charge is brought up to the box from the left toward: (a) an empty box; (b) a box with a positive charge inside; (c) a box with a negative charge inside.

Now suppose there is a negative charge inside the box (Figure 22.27c). When the positive test charge is brought close to one surface of the box, the charge experiences an inward force. If the positive test charge is close to any surface of the box, it experiences an inward force. Doubling the negative charge in the box, doubles the inward force on the test charge close to any surface of the box.

In analogy with flowing water, the electric field lines seem to be flowing out of the box containing positive charge and into the box containing negative charge.

Now let's imagine an empty box in a uniform electric field (Figure 22.28). If a positive test charge is brought close to side 1, it experiences an inward force. If the charge is close to side 2, it experiences an outward force. The electric field is parallel to the other four sides, so the positive test charge does not experience any inward or outward force when brought close to those sides. Thus, in analogy with flowing water, the net amount of electric field that seems to be flowing in and out of the box is zero.

Whenever a charge is inside the box, the electric field lines seem to be flowing in or out of the box. When there is no charge in the box, the net flow of electric field lines in or out of the box is zero. These observations and the definition of electric flux, which quantifies the concept of the flow of the electric field lines, lead to **Gauss's Law:**

$$\Phi = \frac{q}{\epsilon_0}. \tag{22.15}$$

Here q is the net charge inside a closed surface, called a **Gaussian surface.** The closed surface could be a box like that we have been discussing or any arbitrarily shaped closed surface. Usually, the shape of the Gaussian surface is chosen so as to reflect the symmetries of the problem situation.

An alternative formulation of Gauss's Law incorporates the definition of the electric flux (equation 22.14):

$$\oiint \vec{E} \cdot d\vec{A} = \frac{q}{\epsilon_0}. \tag{22.16}$$

According to equation 22.16, Gauss's Law states that the surface integral of the electric field components perpendicular to the area times the area is proportional to the net charge within the closed surface. This expression may look daunting now, but it simplifies considerably in many cases and allows us to perform very quickly calculations that would otherwise be quite complicated.

Gauss's Law and Coulomb's Law

We can derive Gauss's Law from Coulomb's Law. To do this, we start with a positive point charge, q. The electric field due to this charge is radial and pointing outward, as we saw in Section 22.3. According to Coulomb's Law (Section 21.5), the magnitude of the electric field from this charge is

$$E = \frac{1}{4\pi\epsilon_0} \frac{q}{r^2}.$$

We now find the electric flux passing through a closed surface resulting from this point charge. For the Gaussian surface, we choose a spherical surface with radius r, with the charge at the center of the sphere, as shown in Figure 22.29. The electric field due to the positive point charge intersects each differential element of the surface of the Gaussian sphere perpendicularly. Therefore, at each point of this Gaussian surface, the electric field vector, $\vec{E}$, and the differential surface area vector, $d\vec{A}$, are parallel. The surface area vector will always point outward from the spherical Gaussian surface, but the electric field vector

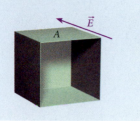

FIGURE 22.28 Imaginary box in a uniform electric field.

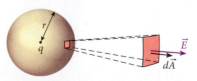

FIGURE 22.29 A spherical Gaussian surface with radius r surrounding a charge q. A closeup view of a differential surface element with area dA is shown.

can point outward or inward depending on the sign of the charge. For a positive charge, the scalar product of the electric field and the surface area element is $\vec{E} \cdot d\vec{A} = E\, dA \cos 0° = E\, dA$. The electric flux in this case according to equation 22.14, is

$$\Phi = \oiint \vec{E} \cdot d\vec{A} = \oiint E\, dA.$$

Because the electric field has the same magnitude anywhere in space at a distance r from the point charge q, we can take E outside the integral:

$$\Phi = \oiint E\, dA = E \oiint dA.$$

Now what we have left to evaluate is the integral of the differential area over a spherical surface, which is given by $\oiint dA = 4\pi r^2$. Therefore, we have found from Coulomb's Law for the case of a point charge

$$\Phi = (E)\left(\oiint dA \right) = \left(\frac{1}{4\pi\epsilon_0} \frac{q}{r^2} \right)\left(4\pi r^2 \right) = \frac{q}{\epsilon_0},$$

22.4 Self-Test Opportunity

What changes in the preceding derivation of Gauss's Law if a negative point charge is used?

which is the same as the expression for Gauss's Law in equation 22.15. We have shown that Gauss's Law can be derived from Coulomb's Law for a positive point charge, but it can also be shown that Gauss's Law holds for any distribution of charge inside a closed surface.

Shielding

Two important consequences of Gauss's Law are evident:

1. The electrostatic field inside any isolated conductor is always zero.

2. Cavities inside conductors are shielded from electric fields.

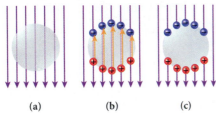

(a)   (b)   (c)

FIGURE 22.30 Shielding of an external electric field (purple vertical arrows) from the inside of a conductor.

To examine these consequences, let's suppose a net electric field exists at some moment at some point inside an isolated conductor; see Figure 22.30a. But every conductor has free electrons inside it (blue circles in Figure 22.30b), which can move rapidly in response to any net external electric field, leaving behind positively charged ions (red circles in Figure 22.30b). The charges will move to the outer surface of the conductor, leaving no net accumulation of charge inside the volume of the conductor. These charges will in turn create an electric field inside the conductor (yellow arrows in Figure 22.30b), and they will move around until the electric field produced by them exactly cancels the external electric field. The net electric field thus becomes zero everywhere inside the conductor (Figure 22.30c).

If a cavity is scooped out of a conducting body, the net charge and thus the electric field inside this cavity is always zero, no matter how strongly the conductor is charged or how strong an external electric field acts on it. To prove this, we assume a closed Gaussian surface surrounds the cavity, completely inside the conductor. From the preceding discussion (see Figure 22.30), we know that at each point of this surface, the field is zero. Therefore, the net flux over this surface is also zero. By Gauss's Law, it then follows that this surface encloses zero net charge. If there were equal amounts of positive and negative charge on the cavity surface (and thus no net charge), this charge would not be stationary, as the positive and negative charges would be attracted to each other and would be free to move around the cavity surface to cancel each other. Therefore, any cavity inside a conductor is totally shielded from any external electric field. This effect is sometimes called **electrostatic shielding.**

A convincing demonstration of this shielding is provided by placing a plastic container filled with Styrofoam peanuts on top of a Van de Graaff generator, which serves as a source of strong electric field (Figure 22.31a). Charging the generator results in a large net charge accumulation on the dome, producing a strong electric field in the vicinity. Because of this field, the charges in the Styrofoam peanuts separate slightly, and the peanuts acquire small dipole moments. If the field were uniform, there would be no force on these dipoles. However, the nonuniform electric field does exert a force, even though the peanuts are electrically neutral. The peanuts thus fly out of the container. If the same Styrofoam peanuts are placed inside an open metal can, they do not fly out when the generator is charged (Figure 22.31b). The electric field easily penetrates the walls of the plastic container and reaches the

22.5 In-Class Exercise

A hollow, conducting sphere is initially given an evenly distributed negative charge. A positive charge $+q$ is brought near the sphere and placed at rest as shown in the figure. What is the direction of the electric field inside the hollow sphere?

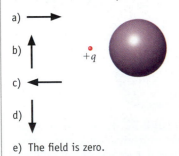

a) →

b) ↑

c) ←

d) ↓

e) The field is zero.

 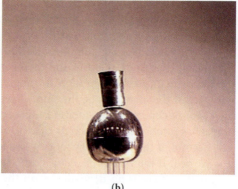

(a)   (b)

FIGURE 22.31 Styrofoam peanuts are put inside a container that is placed on top of a Van de Graaff generator, which is then charged. (a) The peanuts fly out of a nonconducting plastic container. (b) The peanuts remain within a metal can.

Styrofoam peanuts, whereas, in accord with Gauss's Law, the conducting metal can provide shielding inside and prevents the Styrofoam peanuts from acquiring dipole moments.

The conductor surrounding the cavity does not have to be a solid piece of metal; even a wire mesh is sufficient to provide shielding. This can be demonstrated most impressively by seating a person inside a cage and then hitting the cage with a lightning-like electrical discharge (Figure 22.32). The person inside the cage is unhurt, even if he or she touches the metal of the cage *from the inside*. (It is important to realize that severe injuries can result if any body parts stick out of the cage, for example, if hands are wrapped around the bars of the cage!) This cage is called a Faraday cage, after British physicist Michael Faraday (1791–1867), who invented it.

A Faraday cage has important consequences, probably the most relevant of which is the fact that your car protects you from being hit by lightning while inside it—unless you drive a convertible. The sheet metal and steel frame that surround the passenger compartment provide the necessary shielding. (But as fiberglass, plastic, and carbon fiber begin to replace sheet metal in auto bodies, this shielding is not assured any more.)

FIGURE 22.32 A person inside a Faraday cage is unharmed by a large voltage applied outside the cage, which produces a huge spark. This demonstration is performed several times daily at the Deutsches Museum in Munich, Germany.

22.6 In-Class Exercise

A hollow, conducting sphere is initially uncharged. A positive charge, $+q_1$, is placed inside the sphere, as shown in the figure. Then, a second positive charge, $+q_2$, is placed near the sphere but outside it. Which of the following statements describes the net electric force on each charge?

a) There is a net electric force on $+q_2$ but not on $+q_1$.

b) There is a net electric force on $+q_1$ but not on $+q_2$.

c) Both charges are acted on by a net electric force with the same magnitude and in the same direction.

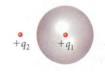

$+q_2$ $+q_1$

d) Both charges are acted on by a net electric force with the same magnitude but in opposite directions.

e) There is no net electric force on either charge.

22.9 Special Symmetries

In this section we'll determine the electric field due to charged objects of different shapes. In Section 22.5, the charge distributions for different geometries were defined; see equation 22.9. Table 22.1 lists the symbols for these charge distributions and their units.

Cylindrical Symmetry

Using Gauss's Law, we can calculate the magnitude of the electric field due to a long straight conducting wire with uniform charge per unit length $\lambda > 0$. We first imagine a Gaussian surface in the form of a right cylinder with radius r and length L surrounding the wire so that the wire is along the axis of the cylinder (Figure 22.33). We can apply

Table 22.1	Symbols for Charge Distributions	
Symbol	Name	Unit
λ	Charge per length	C/m
σ	Charge per area	C/m^2
ρ	Charge per volume	C/m^3

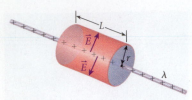

FIGURE 22.33 Long wire with charge per unit length λ surrounded by a Gaussian surface in the form of a right cylinder with radius r and length L. Representative electric field vectors are shown inside the cylinder.

22.7 In-Class Exercise

A total of $1.45 \cdot 10^6$ excess electrons are on an initially electrically neutral wire of length 1.13 m. What is the magnitude of the electric field at a point at a perpendicular distance of 0.401 m away from the center of wire? (*Hint:* Assume that 1.13 m is close enough to "infinitely long.")

a) $9.21 \cdot 10^{-3}$ N/C

b) $2.92 \cdot 10^{-1}$ N/C

c) $6.77 \cdot 10^{1}$ N/C

d) $8.12 \cdot 10^{2}$ N/C

e) $3.31 \cdot 10^{3}$ N/C

22.5 Self-Test Opportunity

By how much does the answer to In-Class Exercise 22.7 change if the assumption that the wire can be treated as being infinitely long is not made? (*Hint:* See Example 22.3.)

Gauss's Law to this Gaussian surface. From symmetry, we know that the electric field produced by the wire must be radial and perpendicular to the wire. What invoking symmetry means deserves further explanation because such arguments are very common.

First, we imagine rotating the wire about an axis along its length. This rotation would include all charges on the wire and their electric fields. However, the wire would still look the same after a rotation through any angle. The electric field created by the charge on the wire would therefore also be the same. From this argument, we conclude that the electric field cannot depend on the rotation angle around the wire. This conclusion is general: If an object has *rotational symmetry*, its electric field cannot depend on the rotation angle.

Second, if the wire is very long, it will look the same no matter where along its length it is viewed. If the wire is unchanged, its electric field is also unchanged. This observation means that there is no dependence on the coordinate along the wire. This symmetry is called *translational symmetry*. Since there is no preferred direction in space along the wire, there can be no electric field component parallel to the wire.

Returning to the Gaussian surface, we can see that the contribution to the integral in Gauss's Law (equation 22.16) from the ends of the cylinder is zero because the electric field is parallel to these surfaces and is thus perpendicular to the normal vectors from the surface. The electric field is perpendicular to the wall of the cylinder everywhere, so we have

$$\oiint \vec{E} \cdot d\vec{A} = EA = E(2\pi r L) = \frac{q}{\epsilon_0} = \frac{\lambda L}{\epsilon_0},$$

where $2\pi r L$ is the area of the wall of the cylinder. Solving this equation, we find the magnitude of the electric field due to a uniformly charged long straight wire:

$$E = \frac{\lambda}{2\pi\epsilon_0 r} = \frac{2k\lambda}{r}, \qquad (22.17)$$

where r is the perpendicular distance to the wire. For $\lambda < 0$, equation 22.17 still applies, but the electric field points inward instead of outward. Note that this is the same result we obtained in Example 22.3 for the electric field due to a wire of infinite length—but attained here in a much simpler way!

You begin to see the great computational power contained in Gauss's Law, which can be used to calculate the electric field resulting from all kinds of charge distributions, both discrete and continuous. However, it is practical to use Gauss's Law only in situations where you can exploit some symmetry; otherwise, it is too difficult to calculate the flux.

It is instructive to compare the dependence of the electric field on the distance from a point charge and from a long straight wire. For the point charge, the electric field falls off with the square of the distance, much faster than does the electric field due to the long wire, which decreases in inverse proportional to the distance.

Planar Symmetry

Assume a flat thin, infinite, nonconducting sheet of positive charge (Figure 22.34), with uniform charge per unit area of $\sigma > 0$. Let's find the electric field a distance r from the surface of this infinite plane of charge.

To do this, we choose a Gaussian surface in the form of a closed right cylinder with cross-sectional area A and length $2r$, which cuts through the plane perpendicularly, as shown in Figure 22.34. Because the plane is infinite and the charge is positive, the electric field must be perpendicular to the ends of the cylinder and parallel to the cylinder wall. Using Gauss's Law, we obtain

$$\oiint \vec{E} \cdot d\vec{A} = (EA + EA) = \frac{q}{\epsilon_0} = \frac{\sigma A}{\epsilon_0},$$

where σA is the charge enclosed in the cylinder. Thus, the magnitude of the electric field due to an infinite plane of charge is

$$E = \frac{\sigma}{2\epsilon_0}. \qquad (22.18)$$

FIGURE 22.34 Infinite flat nonconducting sheet with charge density σ. Cutting through the plane perpendicularly is a Gaussian surface in the form of a right cylinder with cross-sectional area A parallel to the plane and height r above and below the plane.

If $\sigma < 0$, then equation 22.18 still holds, but the electric field points toward the plane instead of away from it.

For an infinite conducting sheet with charge density $\sigma > 0$ on each surface, we can find the electric field by choosing a Gaussian surface in the form of a right cylinder. However, for this case, one end of the cylinder is embedded inside the conductor (Figure 22.35). The electric field inside the conductor is zero; therefore, there is no flux through the end of the cylinder enclosed in the conductor. The electric field outside the conductor must be perpendicular to the surface and therefore parallel to the wall of the cylinder and perpendicular to the end of the cylinder that is outside the conductor. Thus, the flux through the Gaussian surface is EA. The enclosed charge is given by σA, so Gauss's Law becomes

$$\oiint \vec{E} \cdot d\vec{A} = EA = \frac{\sigma A}{\epsilon_0}.$$

Thus, the magnitude of the electric field just outside the surface of a flat charged conductor is

$$E = \frac{\sigma}{\epsilon_0}. \tag{22.19}$$

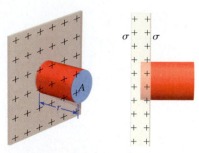

FIGURE 22.35 Infinite conducting plane with charge density σ on each surface and a Gaussian surface in the form of a right cylinder embedded in one side.

Spherical Symmetry

To find the electric field due to a spherically symmetrical distribution of charge, we consider a thin spherical shell with charge $q > 0$ and radius r_s (Figure 22.36).

Here we use a spherical Gaussian surface with $r_2 > r_s$ that is concentric with the charged sphere. Applying Gauss's Law, we get

$$\oiint \vec{E} \cdot d\vec{A} = E\left(4\pi r_2^2\right) = \frac{q}{\epsilon_0}.$$

We can solve for the magnitude of the electric field, E, which is

$$E = \frac{1}{4\pi\epsilon_0} \frac{q}{r_2^2}.$$

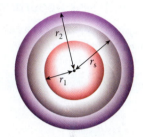

FIGURE 22.36 Spherical shell of charge with radius r_s with a Gaussian surface with radius $r_2 > r_s$ and a second Gaussian surface with $r_1 < r_s$.

If $q < 0$, the field points radially inward instead of radially outward from the spherical surfaces. For another spherical Gaussian surface, with $r_1 < r_s$, that is also concentric with the charged spherical shell, we obtain

$$\oiint \vec{E} \cdot d\vec{A} = E\left(4\pi r_1^2\right) = 0.$$

Thus, the electric field outside a spherical shell of charge behaves as if the charge were a point charge located at the center of the sphere, whereas the electric field is zero inside the spherical shell of charge.

Now let's find the electric field due to charge that is equally distributed throughout a spherical volume, with uniform charge density $\rho > 0$ (Figure 22.37). The radius of the sphere is r. We use a Gaussian surface in the form of a sphere with radius $r_1 < r$. From the symmetry of the charge distribution, we know that the electric field resulting from the charge is perpendicular to the Gaussian surface. Thus, we can write

$$\oiint \vec{E} \cdot d\vec{A} = E\left(4\pi r_1^2\right) = \frac{q}{\epsilon_0} = \frac{\rho}{\epsilon_0}\left(\frac{4}{3}\pi r_1^3\right),$$

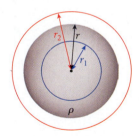

FIGURE 22.37 Spherical distribution of charge with uniform charge per unit volume ρ and radius r. Two spherical Gaussian surfaces are also shown, one with radius $r_1 < r$ and one with $r_2 > r$.

where $4\pi r_1^2$ is the area of the spherical Gaussian surface and $\frac{4}{3}\pi r_1^3$ is the volume enclosed by the Gaussian surface. From the preceding equation, we obtain the electric field at a radius r_1 inside a uniform distribution of charge:

$$E = \frac{\rho r_1}{3\epsilon_0}. \tag{22.20}$$

The total charge on the sphere can be called q_t, and it equals the total volume of the spherical charge distribution times the charge density:

$$q_t = \rho \frac{4}{3}\pi r^3.$$

The charge enclosed by the Gaussian surface then is

$$q = \frac{\text{volume inside } r_1}{\text{volume of charge distribution}} q_t = \frac{\frac{4}{3}\pi r_1^3}{\frac{4}{3}\pi r^3} q_t = \frac{r_1^3}{r^3} q_t .$$

With this the expression for the enclosed charge, we can rewrite Gauss's Law for this case as

$$\oint \vec{E} \cdot d\vec{A} = E\left(4\pi r_1^2\right) = \frac{q_t}{\epsilon_0} \frac{r_1^3}{r^3},$$

which gives us

$$E = \frac{q_t r_1}{4\pi\epsilon_0 r^3} = \frac{k q_t r_1}{r^3}. \tag{22.21}$$

If we consider a Gaussian surface with a radius larger than the radius of the charge distribution, $r_2 > r$, we can apply Gauss's Law as follows:

$$\oint \vec{E} \cdot d\vec{A} = E\left(4\pi r_2^2\right) = \frac{q_t}{\epsilon_0},$$

or

$$E = \frac{q_t}{4\pi\epsilon_0 r_2^2} = \frac{k q_t}{r_2^2}. \tag{22.22}$$

Thus, the electric field outside a uniform spherical distribution of charge is the same as the field due to a point charge of the same magnitude located at the center of the sphere.

22.6 Self-Test Opportunity

Consider a sphere of radius R with charge q uniformly distributed throughout the volume of the sphere. What is the magnitude of the electric field at a point $2R$ away from the center of the sphere?

SOLVED PROBLEM 22.3 | **Nonuniform Spherical Charge Distribution**

A spherically symmetrical but nonuniform charge distribution is given by

$$\rho(r) = \begin{cases} \rho_0\left(1 - \dfrac{r}{R}\right) & \text{for } r \le R \\[2mm] 0 & \text{for } r > R, \end{cases}$$

where $\rho_0 = 10.0 \ \mu C/m^3$ and $R = 0.250$ m.

PROBLEM
What is the electric field produced by this charge distribution at $r = 0.125$ m and at $r = 0.500$ m?

SOLUTION

THINK
We can use Gauss's Law to determine the electric field as a function of radius if we employ a spherical Gaussian surface. The radius $r = 0.125$ m is located inside the charge distribution. The charge enclosed inside the spherical surface at $r = r_1$ is given by an integral of the charge density from $r = 0$ to $r = r_1$. Outside the spherical charge distribution, the electric field is the same as that of a point charge whose magnitude is equal to the total charge of the spherical distribution.

SKETCH
The charge density, ρ, as a function of radius, r, is plotted in Figure 22.38.

RESEARCH
Gauss's Law (equation 22.16) tells us that $\oint \vec{E} \cdot d\vec{A} = q/\epsilon_0$. Inside the nonuniform spherical charge distribution at a radius $r_1 < R$, Gauss's Law becomes

$$\epsilon_0 E\left(4\pi r_1^2\right) = \int_0^{V_1} \rho(r)\,dV = \int_0^{r_1} \rho_0\left(1 - \frac{r}{R}\right)\left(4\pi r^2\right) dr. \tag{i}$$

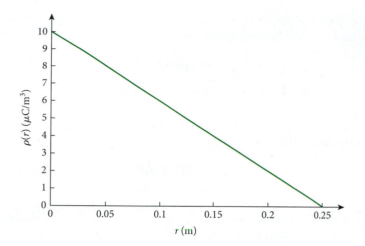

FIGURE 22.38 Charge density as a function of radius for a nonuniform spherical charge distribution.

Carrying out the integral on the right-hand side of equation (i), we obtain

$$\int_0^{r_1} \rho_0\left(1 - \frac{r}{R}\right)\left(4\pi r^2\right)dr = 4\pi\rho_0 \int_0^{r_1}\left(r^2 - \frac{r^3}{R}\right)dr = 4\pi\rho_0\left(\frac{r_1^3}{3} - \frac{r_1^4}{4R}\right). \qquad \text{(ii)}$$

SIMPLIFY

The electric field due to the charge inside $r_1 \leq R$ is then given by

$$E = \frac{4\pi\rho_0\left(\dfrac{r_1^3}{3} - \dfrac{r_1^4}{4R}\right)}{\epsilon_0\left(4\pi r_1^2\right)} = \frac{\rho_0}{\epsilon_0}\left(\frac{r_1}{3} - \frac{r_1^2}{4R}\right). \qquad \text{(iii)}$$

In order to calculate the electric field due to the charge inside $r_1 > R$, we need the total charge contained in the spherical charge distribution. We can obtain the total charge using equation (ii) with $r_1 = R$:

$$q_t = 4\pi\rho_0\left(\frac{R^3}{3} - \frac{R^4}{4R}\right) = 4\pi\rho_0\left(\frac{R^3}{3} - \frac{R^3}{4}\right) = 4\pi\rho_0\frac{R^3}{12} = \frac{\pi\rho_0 R^3}{3}.$$

The electric field outside the spherical charge distribution ($r_1 > R$) is then

$$E = \frac{1}{4\pi\epsilon_0}\frac{q_t}{r_1^2} = \frac{1}{4\pi\epsilon_0}\frac{\frac{1}{3}\pi\rho_0 R^3}{r_1^2} = \frac{\rho_0 R^3}{12\epsilon_0 r_1^2}. \qquad \text{(iv)}$$

CALCULATE

The electric field at $r_1 = 0.125$ m is

$$E = \frac{\rho_0}{\epsilon_0}\left(\frac{r_1}{3} - \frac{r_1^2}{4R}\right) = \frac{10.0\ \mu\text{C/m}^3}{8.85\cdot10^{-12}\,\text{C}^2/\text{N m}^2}\left(\frac{0.125\ \text{m}}{3} - \frac{\left(0.125\ \text{m}\right)^2}{4\left(0.250\ \text{m}\right)}\right) = 29{,}425.6\ \text{N/C}.$$

The electric field at $r_1 = 0.500$ m is

$$E = \frac{\rho_0 R^3}{12\epsilon_0 r_1^2} = \frac{\left(10.0\ \mu\text{C/m}^3\right)\left(0.250\ \text{m}\right)^3}{12\left(8.85\cdot10^{-12}\,\text{C}^2/\text{N m}^2\right)\left(0.500\ \text{m}\right)^2} = 5885.12\ \text{N/C}.$$

ROUND

We report our results to three significant figures. The electric field at $r_1 = 0.125$ m is

$$E = 2.94\cdot10^4\ \text{N/C}.$$

The electric field at $r_1 = 0.500$ m is

$$E = 5.89\cdot10^3\ \text{N/C}.$$

Continued—

DOUBLE-CHECK

The electric field at $r_1 = R$ can be calculated using equation (iii):

$$E = \frac{\rho_0}{\epsilon_0}\left(\frac{R}{3} - \frac{R^2}{4R}\right) = \frac{\rho_0 R}{12\epsilon_0} = \frac{\left(10.0 \ \mu\text{C/m}^3\right)\left(0.250 \ \text{m}\right)}{12\left(8.85\cdot10^{-12}\,\text{C}^2/\text{N m}^2\right)} = 2.35\cdot10^4 \ \text{N/C}.$$

We can also use equation (iv) to find the electric field outside the spherical charge distribution but very close to the surface, where $r_1 \approx R$:

$$E = \frac{\rho_0 R^3}{12\epsilon_0 R^2} = \frac{\rho_0 R}{12\epsilon_0},$$

which is the same result we obtained using our result for $r_1 \le R$. The calculated electric field at the surface of the charge distribution is lower than that at $r_1 = 0.125$, which may seem counterintuitive. An idea of the dependence of the magnitude of E on r_1 is provided by the plot in Figure 22.39, which was created using equations (iii) and (iv).

FIGURE 22.39 The electric field due to a nonuniform spherical distribution of charge as a function of the distance from the center of the sphere.

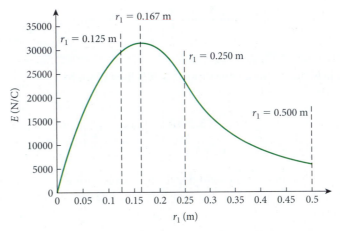

You can see that a maximum occurs in the electric field and that our result for $r_1 = 0.125$ m is less than this maximum value. We can calculate the radius at which the maximum occurs by differentiating equation (iii) with respect to r_1, setting the result equal to zero, and solving for r_1:

$$\frac{dE}{dr_1} = \frac{\rho_0}{\epsilon_0}\left(\frac{1}{3} - \frac{r_1}{2R}\right) = 0 \Rightarrow$$

$$\frac{1}{3} = \frac{r_1}{2R} \Rightarrow r_1 = \frac{2}{3}R.$$

Thus, we expect a maximum in the electric field at $r_1 = \frac{2}{3}R = 0.167$ m. The plot in Figure 22.39 does indeed show a maximum at that radius. It also shows the value of E at $r = 0.250$ to be smaller than that at $r = 0.125$ as we found in our calculation. Thus, our answers seem reasonable.

22.8 In-Class Exercise

Suppose an uncharged solid steel ball, for example, one of the steel balls used in an old-fashioned pinball machine, is resting on a perfect insulator. Some small amount of negative charge (say, a few hundred electrons) is placed at the north pole of the ball. If you could check the distribution of the charge after a few seconds, what would you detect?

a) All of the added charge has vanished, and the ball is again electrically neutral.

b) All of the added charge has moved to the center of the ball.

c) All of the added charge is distributed uniformly over the surface of the ball.

d) The added charge is still located at or very near the north pole of the ball.

e) The added charge is performing a simple harmonic oscillation on a straight line between the south and north poles of the ball.

Sharp Points and Lightning Rods

We have already seen that the electric field is perpendicular to the surface of a conductor. (To repeat, if there were a field component parallel to the conductor, then the charges inside the conductor would move until they reached equilibrium, which means no force or electric field component in the direction of motion, that is, along the surface of the conductor.) Figure 22.40a shows the distribution of charges on the surface of the end of a pointed conductor. Note that the charges are closer together at the sharp tip, where the curvature is largest.

22.9 In-Class Exercise

Suppose an uncharged hollow sphere made of a perfect insulator, for example a ping-pong ball, is resting on a perfect insulator. Some small amount of negative charge (say, a few hundred electrons) is placed at the north pole of the sphere. If you could check the distribution of the charge after a few seconds, what would you detect?

a) All of the added charge has vanished, and the sphere is again electrically neutral.

b) All of the added charge has moved to the center of the sphere.

c) All of the added charge is distributed uniformly over the surface of the sphere.

d) The added charge is still located at or very near the north pole of the sphere.

e) The added charge is performing a simple harmonic oscillation on a straight line between the south and north poles of the sphere.

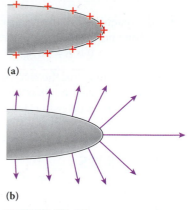

(a)

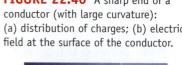

(b)

FIGURE 22.40 A sharp end of a conductor (with large curvature): (a) distribution of charges; (b) electric field at the surface of the conductor.

Near that sharp tip on the end of the conductor, the electric field looks much more like that due to a point charge, with the field lines spreading out radially (Figure 22.40b). Since the field lines are closer together near a sharp point on a conductor, the field is stronger near the sharp tip than on the flat part of the conductor.

Benjamin Franklin proposed metal rods with sharp points as lightning rods (Figure 22.41). He reasoned that the sharp points would dissipate the electric charge built up in a storm, preventing the discharge of lightning. When Franklin installed such lightning rods, they were struck by lightning instead of the buildings to which they were attached. However, recent findings indicate that lightning rods used to protect structures from lightning should have blunt, rounded ends. When charged during thunderstorm conditions, a lightning rod with a sharp point creates a strong electric field that locally ionizes the air, producing a condition that actually causes lightning. Conversely, round-ended lightning rods are just as effective in protecting structures from lightning and do not increase lightning strikes. Any lightning rod should be carefully grounded to carry charge from a lightning strike away from the structure on which the lightning rod is mounted.

FIGURE 22.41 A lightning rod with a sharp point installed on the top of a building.

WHAT WE HAVE LEARNED | EXAM STUDY GUIDE

- The electric force, $\vec{F}(\vec{r})$, on a charge, q, due to an electric field, $\vec{E}(\vec{r})$, is given by $\vec{F}(\vec{r}) = q\vec{E}(\vec{r})$.

- The electric field at any point is equal to the sum of the electric fields from all sources: $\vec{E}_t(\vec{r}) = \vec{E}_1(\vec{r}) + \vec{E}_2(\vec{r}) + \cdots + \vec{E}_n(\vec{r})$.

- The magnitude of the electric field due to a point charge q at a distance r is given by $E(r) = \dfrac{1}{4\pi\epsilon_0}\dfrac{|q|}{r^2} = \dfrac{k|q|}{r^2}$. The electric field points radially away from a positive point charge and radially toward a negative charge.

- A system of two equal (in magnitude) oppositely charged point particles is an electric dipole. The magnitude, p, of the electric dipole moment is given by $p = qd$, where q is the magnitude of either one of the charges and d is the distance separating them. The electric dipole moment is a vector pointing from the negative toward the positive charge. On the dipole

axis, the dipole produces an electric field of magnitude $E = \dfrac{p}{2\pi\epsilon_0 |x|^3}$, where $|x| \gg d$.

- Gauss's Law states that the electric flux over an entire closed surface is equal to the enclosed charge divided by ϵ_0: $\oiint \vec{E} \cdot d\vec{A} = \dfrac{q}{\epsilon_0}$.

- The differential electrical field is given by $dE = \dfrac{kdq}{r^2}$, and the differential charge is

$$
\left.\begin{array}{l} dq = \lambda\,dx \\ dq = \sigma\,dA \\ dq = \rho\,dV \end{array}\right\} \text{ for a charge distribution } \left\{\begin{array}{l} \text{along a line;} \\ \text{over a surface;} \\ \text{throughout a volume.} \end{array}\right.
$$

- The magnitude of the electric field at a distance r from a long straight wire with uniform linear charge density $\lambda > 0$ is given by $E = \dfrac{\lambda}{2\pi\epsilon_0 r} = \dfrac{2k\lambda}{r}$.

- The magnitude of the electric field produced by an infinite nonconducting plane that has uniform charge density $\sigma > 0$ is $E = \frac{1}{2}\sigma/\epsilon_0$.

- The magnitude of the electric field produced by an infinite conducting plane that has uniform charge density $\sigma > 0$ on each side is $E = \sigma/\epsilon_0$.

- The electric field inside a closed conductor is zero.

- The electric field outside a charged spherical conductor is the same as that due to a point charge of the same magnitude located at the center of the sphere.

KEY TERMS

field, p. 711
electric field, p. 711
superposition principle, p. 712

electric field lines, p. 712
test charge, p. 712
electric dipole, p. 716

electric dipole moment, p. 716
electric flux, p. 725
Gauss's Law, p. 727

Gaussian surface, p. 727
electrostatic shielding,
 p. 728

NEW SYMBOLS AND EQUATIONS

$\vec{E}$, electric field

$p = qd$, electric dipole moment

$\oiint \vec{E} \cdot d\vec{A} = \dfrac{q}{\epsilon_0}$, Gauss's Law

ϕ, electric flux

λ, charge per unit length

σ, charge per unit area

ρ, charge per unit volume

ANSWERS TO SELF-TEST OPPORTUNITIES

22.1 The direction of the electric field is downward at points A, C, and E and upward at points B and D. (There is an electric field at point E, even though there is no line drawn there; the field lines are only sample representations of the electric field, which also exists between the field lines.) The field is largest in magnitude at point E, which can be inferred from the fact that it is located where the field lines have the highest density.

22.2 The two forces acting on the two charges in the electric field create a torque on the electric dipole around its center of mass, given by

$$\tau = \left(\text{force}_+\right)\left(\text{moment arm}_+\right)\left(\sin\theta\right) + \left(\text{force}_-\right)\left(\text{moment arm}_-\right)\left(\sin\theta\right).$$

The length of the moment arm in both cases is $\frac{1}{2}d$, and the magnitude of the force is $F = qE$ for both charges. Thus, the torque on the electric dipole is

$$\tau = qE\left(\frac{d}{2}\sin\theta\right) + qE\left(\frac{d}{2}\sin\theta\right) = qEd\sin\theta.$$

22.3 The net electric flux passing though the object is EA. Remember, the object does not have a closed surface; otherwise, the result would be 0.

22.4 The sign of the scalar product changes, because the electric field points radially inward: $\vec{E} \cdot d\vec{A} = E\,dA\cos 180° = -E\,dA$. But the magnitude of the electric field due to the negative charge is $E = \dfrac{1}{4\pi\epsilon_0}\dfrac{-q}{r^2}$. The two minus signs cancel, giving the same results for Coulomb's and Gauss's Laws for a point charge, independent of the sign of the charge.

22.5 For a wire of infinite length, $E_y = \dfrac{2k\lambda}{y}$; for a wire of finite length, $E_y = \dfrac{2k\lambda}{y}\dfrac{a}{\sqrt{y^2 + a^2}}$. With the values given in the in-class exercise, $\dfrac{a}{\sqrt{y^2 + a^2}} = \dfrac{0.565}{\sqrt{0.401^2 + 0.565^2}} = 0.815$. Thus, the "infinitely long" approximation is off by $\approx 18\%$.

22.6 The charged sphere acts like a point charge, so the electric field is

$$E = k\frac{q}{\left(2R\right)^2} = k\frac{q}{4R^2}.$$

PROBLEM-SOLVING PRACTICE

Problem-Solving Guidelines

1. Be sure to distinguish between the point where an electric field is being generated and the point where the electric field is being determined.

2. Some of the same guidelines for dealing with electrostatic charges and forces also apply to electric fields: Use symmetry to simplify your calculations; remember that the field is composed of vectors and thus you have to use vector operations instead of simple addition, multiplication, and so on; convert units to meters and coulombs for consistency with the given values of constants.

3. Remember to use the correct form of the charge density for field calculations: λ for linear charge density, σ for surface charge density, and ρ for volume charge density.

4. The key to using Gauss's Law is to choose the right form of Gaussian surface for the symmetry of the problem situation. Cubical, cylindrical, and spherical Gaussian surfaces are typically useful.

5. Often, you can break a Gaussian surface into surface elements that are either perpendicular to or parallel to the electric field lines. If the field lines are perpendicular to the surface, the electric flux is simply the field strength times the area, EA, or $-EA$ if the field points inward instead of outward. If the field lines are parallel to the surface, the flux through that surface is zero. The total flux is the sum of the flux through each surface element of the Gaussian surface. Remember that zero flux through a Gaussian surface does not necessarily mean that the electric field is zero.

SOLVED PROBLEM 22.4 | Electron Moving over a Charged Plate

PROBLEM
An electron with a kinetic energy of 2000.0 eV (1 eV = $1.602 \cdot 10^{-19}$ J) is fired horizontally across a horizontally oriented charged conducting plate with surface charge density $+4.00 \cdot 10^{-6}$ C/m^2. Taking the positive direction to be upward (away from the plate), what is the vertical deflection of the electron after it has traveled a horizontal distance of 4.00 cm?

SOLUTION

THINK
The initial velocity of the electron is horizontal. During its motion, the electron experiences a constant attractive force from the positively charged plate, which causes a constant acceleration downward. We can calculate the time it takes the electron to travel 4.00 cm in the horizontal direction and use this time to calculate the vertical deflection of the electron.

SKETCH
Figure 22.42 shows the electron with initial velocity $\vec{v}_0$ in the horizontal direction. The initial position of the electron is taken to be at $x_0 = 0$ and $y = y_0$.

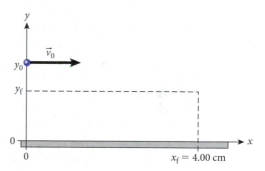

FIGURE 22.42 An electron moving to the right with initial velocity $\vec{v}_0$ over a charged conducting plate.

RESEARCH
The time the electron takes to travel the given distance is

$$t = x_f/v_0, \qquad \text{(i)}$$

where x_f is the final horizontal position and v_0 is the initial speed of the electron. While the electron is in motion, it experiences a force from the charged conducting plate. This force is directed downward (toward the plate) and has a magnitude given by

$$F = qE = e\frac{\sigma}{\epsilon_0}, \qquad \text{(ii)}$$

where σ is the charge density on the conducting plate and e is the charge of an electron. This force causes a constant acceleration in the downward direction whose magnitude is given by $a = F/m$, where m is the mass of the electron. Using the expression for the force from equation (ii), we can express the magnitude of the acceleration as

$$a = \frac{F}{m} = \frac{e\sigma}{m\epsilon_0}. \qquad \text{(iii)}$$

Note that this acceleration is constant. Thus, the vertical position of the electron as a function of time is given by

$$y_f = y_0 - \tfrac{1}{2}at^2 \Rightarrow y_f - y_0 = -\tfrac{1}{2}at^2. \qquad \text{(iv)}$$

Finally, we can relate the initial kinetic energy of the electron to the initial velocity of the electron through

$$K = \tfrac{1}{2}mv_0^2 \Rightarrow v_0^2 = \frac{2K}{m}. \qquad \text{(v)}$$

Continued—

SIMPLIFY

We substitute the expressions for the time and acceleration from equations (i) and (iii) into equation (iv) and obtain

$$y_f - y_0 = -\frac{1}{2}at^2 = -\frac{1}{2}\left(\frac{e\sigma}{m\epsilon_0}\right)\left(\frac{x_f}{v_0}\right)^2 = -\frac{e\sigma x_f^2}{2m\epsilon_0 v_0^2}. \tag{vi}$$

Now substituting the expression for the square of the initial speed from equation (v) into the expression on the right-hand side of equation (vi) gives us

$$y_f - y_0 = -\frac{e\sigma x_f^2}{2m\epsilon_0\left(\frac{2K}{m}\right)} = -\frac{e\sigma x_f^2}{4\epsilon_0 K}. \tag{vii}$$

CALCULATE

We first convert the kinetic energy of the electron from electron-volts to joules:

$$K = \left(2000.0 \text{ eV}\right)\frac{1.602 \cdot 10^{-19} \text{ J}}{1 \text{ eV}} = 3.204 \cdot 10^{-16} \text{ J}.$$

Putting the numerical values into equation (vii), we get

$$y_f - y_0 = -\frac{e\sigma x_f^2}{4\epsilon_0 K} = -\frac{\left(1.602 \cdot 10^{-19} \text{ C}\right)\left(4.00 \cdot 10^{-6} \text{ C/m}^2\right)\left(0.0400 \text{ m}\right)^2}{4\left(8.85 \cdot 10^{-12} \text{ C}^2/(\text{N m}^2)\right)\left(3.204 \cdot 10^{-16} \text{ J}\right)} = -0.0903955 \text{ m}.$$

ROUND

We report our result to three significant figures:

$$y_f - y_0 = -0.0904 \text{ m} = -9.04 \text{ cm}.$$

DOUBLE-CHECK

The vertical deflection that we calculated is about twice the distance that the electron travels in the x-direction, which seems reasonable, at least in the sense of being of the same order of magnitude. Also, equation (vii) for the deflection has several features that should be present. First, the trajectory is parabolic, which we expect for a constant force and thus constant acceleration (see Chapter 3). Second, for zero surface charge density we obtain zero deflection. Third, for very high kinetic energy there is negligible deflection, which is also intuitively what we expect.

MULTIPLE-CHOICE QUESTIONS

22.1 To be able to calculate the electric field created by a known distribution of charge using Gauss's Law, which of the following *must* be true?

a) The charge distribution must be in a nonconducting medium.

b) The charge distribution must be in a conducting medium.

c) The charge distribution must have spherical or cylindrical symmetry.

d) The charge distribution must be uniform.

e) The charge distribution must have a high degree of symmetry that allows assumptions about the symmetry of its electric field to be made.

22.2 An electric dipole consists of two equal and opposite charges situated a very small distance from each other. When the dipole is placed in a uniform electric field, which of the following statements is true?

a) The dipole will not experience any net force from the electric field; since the charges are equal and have opposite signs, the individual effects will cancel out.

b) There will be no net force and no net torque acting on the dipole.

c) There will be a net force but no net torque acting on the dipole.

d) There will be no net force, but there will (in general) be a net torque acting on dipole.

22.3 A point charge, $+Q$, is located on the x-axis at $x = a$, and a second point charge, $-Q$, is located on the x-axis at $x = -a$. A Gaussian surface with radius $r = 2a$ is centered at the origin. The flux through this Gaussian surface is

a) zero.

b) greater than zero.

c) less than zero.

d) none of the above.

22.4 A charge of $+2q$ is placed at the center of an uncharged conducting shell. What will be the charges on the inner and outer surfaces of the shell, respectively?

a) $-2q, +2q$ c) $-2q, -2q$

b) $-q, +q$ d) $-2q, +4q$

22.5 Two infinite nonconducting plates are parallel to each other, with a distance $d = 10.0$ cm between them, as shown in the figure. Each plate carries a uniform charge distribution of $\sigma = 4.5$ $\mu C/m^2$. What is the electric field, $\vec{E}$, at point P (with $x_P = 20.0$ cm)?

a) 0 N/C

b) $2.54\hat{x}$ N/C

c) $(-5.08 \cdot 10^5)\hat{x}$ N/C

d) $(5.08 \cdot 10^5)\hat{x}$ N/C

e) $(-1.02 \cdot 10^6)\hat{x}$ N/C

f) $(1.02 \cdot 10^6)\hat{x}$ N/C

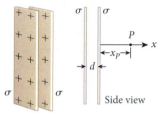

22.6 At which of the following locations is the electric field the strongest?

a) a point 1 m from a 1 C point charge

b) a point 1m (perpendicular distance) from the center of a 1-m-long wire with 1 C of charge distributed on it

c) a point 1 m (perpendicular distance) from the center of a 1-m^2 sheet of charge with 1 C of charge distributed on it

d) a point 1 m from the surface of a charged spherical shell of charge 1 C with a radius of 1m

e) a point 1 m from the surface of a charged spherical shell of charge 1 C with a radius of 0.5 m

22.7 The electric flux through a spherical Gaussian surface of radius R centered on a charge Q is 1200 N/(C m^2). What is the electric flux through a cubic Gaussian surface of side R centered on the same charge Q?

a) less than 1200 N/(C m^2)

b) more than 1200 N/(C m^2)

c) equal to 1200 N/(C m^2)

d) cannot be determined from the information given

22.8 A single positive point charge, q, is at one corner of a cube with sides of length L, as shown in the figure. The net electric flux through the three adjacent sides is zero. The net electric flux through *each* of the other three sides is

a) $q/3\epsilon_0$.

b) $q/6\epsilon_0$.

c) $q/24\epsilon_0$.

d) $q/8\epsilon_0$.

22.9 Three -9-mC point charges are located at $(0,0)$, $(3$ m,3 m$)$, and $(3$ m,-3 m$)$. What is the magnitude of the electric field at $(3$ m,$0)$?

a) $0.9 \cdot 10^7$ N/C   e) $3.6 \cdot 10^7$ N/C

b) $1.2 \cdot 10^7$ N/C   f) $5.4 \cdot 10^7$ N/C

c) $1.8 \cdot 10^7$ N/C   g) $10.8 \cdot 10^7$ N/C

d) $2.4 \cdot 10^7$ N/C

22.10 Which of the following statements is (are) true?

a) There will be *no* change in the charge on the inner surface of a hollow conducting sphere if additional charge is placed on the outer surface.

b) There will be some change in the charge on the inner surface of a hollow conducting sphere if additional charge is placed on the outer surface.

c) There will be *no* change in the charge on the inner surface of a hollow conducting sphere if additional charge is placed at the center of the sphere.

d) There will be some change in the charge on the inner surface of a hollow conducting sphere if additional charge is placed at the center of the sphere.

QUESTIONS

22.11 Many people had been sitting in a car when it was struck by lightning. Why were they able to survive such an experience?

22.12 Why is it a bad idea to stand under a tree in a thunderstorm? What should one do instead to avoid getting struck by lightning?

22.13 Why do electric field lines never cross?

22.14 How is it possible that the flux through a closed surface does not depend on where inside the surface the charge is located (that is, the charge can be moved around inside the surface with no effect whatsoever on the flux)? If the charge is moved from just inside to just outside the surface, the flux changes discontinuously to zero, according to Gauss's Law. Does this really happen? Explain.

22.15 A solid conducting sphere of radius r_1 has a total charge of $+3Q$. It is placed inside (and concentric with) a conducting spherical shell of inner radius r_2 and outer radius r_3. Find the electric field in these regions: $r < r_1$, $r_1 < r < r_2$, $r_2 < r < r_3$, and $r > r_3$.

22.16 A thin rod has end points at $x = \pm 100$ cm. There is a charge of Q uniformly distributed along the rod.

a) What is the electric field very close to the midpoint of the rod?

b) What is the electric field a few centimeters (perpendicularly) from the midpoint of the rod?

c) What is the electric field very far (perpendicularly) from the midpoint of the rod?

22.17 A dipole is completely enclosed by a spherical surface. Describe how the total electric flux through this surface varies with the strength of the dipole.

22.18 Repeat Example 22.3, assuming that the charge distribution is $-\lambda$ for $-a < x < 0$ and $+\lambda$ for $0 < x < a$.

22.19 A negative charge is placed on a solid prolate spheroidal conductor (shown in cross section in the figure). Sketch the distribution of the charge on the conductor and the electric field lines due to the charge.

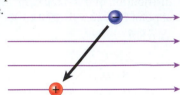

22.20 *Saint Elmo's fire* is an eerie glow that appears at the tips of masts and yardarms of sailing ships in stormy weather and at the tips and edges of the wings of aircraft in flight. St. Elmo's fire is an electrical phenomenon. Explain it, concisely.

22.21 A charge placed on a conductor of any shape forms a layer on the outer surface of the conductor. Mutual repulsion of the individual charge elements creates an outward pressure on this layer, called *electrostatic stress*. Treating the infinitesimal charge elements like tiles of a mosaic, calculate the magnitude of this electrostatic stress in terms of the surface charge density, σ. Note that σ need not be uniform over the surface.

22.22 An electric dipole is placed in a uniform electric field as shown in the figure. What motion will the dipole have in the electric field? Which way will it move? Which way will it rotate?

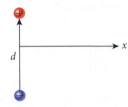

PROBLEMS

A blue problem number indicates a worked-out solution is available in the Student Solutions Manual. One • and two •• indicate increasing level of problem difficulty.

Section 22.3

22.23 A point charge, $q = 4.00 \cdot 10^{-9}$ C, is placed on the x-axis at the origin. What is the electric field produced at $x = 25.0$ cm?

22.24 A +1.6-nC point charge is placed at one corner of a square (1.0 m on a side), and a –2.4-nC charge is placed on the corner diagonally opposite. What is the magnitude of the electric field at either of the other two corners?

22.25 A +48.00-nC point charge is placed on the x-axis at $x = 4.000$ m, and a –24.00-nC point charge is placed on the y-axis at $y = -6.000$ m. What is the direction of the electric field at the origin?

•22.26 Two point charges are placed at two of the corners of a triangle as shown in the figure. Find the magnitude and the direction of the electric field at the third corner of the triangle.

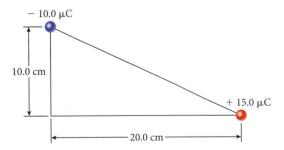

•22.27 A +5.0-C charge is located at the origin. A –3.0-C charge is placed at $x = 1.0$ m. At what finite distance(s) along the x-axis will the electric field be equal to zero?

•22.28 Three charges are on the y-axis. Two of the charges, each $-q$, are located $y = \pm d$, and the third charge, $+2q$, is located at $y = 0$. Derive an expression for the electric field at a point P on the x-axis.

Section 22.4

22.29 For the electric dipole shown in the figure, express the magnitude of the resulting electric field as a function of the perpendicular distance x from the center of the dipole axis. Comment on what the magnitude is when $x \gg d$.

•22.30 Consider an electric dipole on the x-axis and centered at the origin. At a distance h along the positive x-axis, the magnitude of electric field due to the electric dipole is given by $k(2qd)/h^3$. Find a distance perpendicular to the x-axis and measured from the origin at which the magnitude of the electric field stays the same.

Section 22.5

•22.31 A small metal ball with a mass of 4.0 g and a charge of 5.0 mC is located at a distance of 0.70 m above the ground in an electric field of 12 N/C directed to the east. The ball is then released from rest. What is the velocity of the ball after it has moved downward a vertical distance of 0.30 m?

•22.32 A charge per unit length $+\lambda$ is uniformly distributed along the positive y-axis from $y = 0$ to $y = +a$. A charge per unit length $-\lambda$ is uniformly distributed along the negative y-axis from $y = 0$ to $y = -a$. Write an expression for the electric field (magnitude and direction) at a point on the x-axis a distance x from the origin.

•22.33 A thin glass rod is bent into a semicircle of radius R. A charge $+Q$ is uniformly distributed along the upper half, and a charge $-Q$ is uniformly distributed along the lower half as shown in the figure. Find the magnitude and direction of the electric field $\vec{E}$ (in component form) at point P, the center of the semicircle.

•22.34 Two uniformly charged insulating rods are bent in a semicircular shape with radius

$r = 10$ cm. If they are positioned so they form a circle but do not touch and have opposite charges of $+1$ μC and -1 μC, find the magnitude and direction of the electric field at the center of the composite circular charge configuration.

•**22.35** A uniformly charged rod of length L with total charge Q lies along the y-axis, from $y = 0$ to $y = L$. Find an expression for the electric field at the point $(d,0)$ (that is, the point at $x = d$ on the x-axis).

••**22.36** A charge Q is distributed evenly on a wire bent into an arc of radius R, as shown in the figure. What is the electric field at the center of the arc as a function of the angle θ? Sketch a graph of the electric field as a function of θ for $0 < \theta < 180°$.

••**22.37** A thin, flat washer is a disk with an outer diameter of 10 cm and a hole in the center with a diameter of 4 cm. The washer has a uniform charge distribution and a total charge of 7 nC. What is the electric field on the axis of the washer at a distance of 30 cm from the center of the washer?

Section 22.6

22.38 Research suggests that the electric fields in some thunderstorm clouds can be on the order of 10 kN/C. Calculate the magnitude of the electric force acting on a particle with two excess electrons in the presence of a 10.0-kN/C field.

22.39 An electric dipole has opposite charges of $5.0 \cdot 10^{-15}$ C separated by a distance of 0.40 mm. It is oriented at $60°$ with respect to a uniform electric field of magnitude $2.0 \cdot 10^3$ N/C. Determine the magnitude of the torque exerted on the dipole by the electric field.

22.40 Electric dipole moments of molecules are often measured in debyes (D), where $1 D = 3.34 \cdot 10^{-30}$ C m. For instance, the dipole moment of hydrogen chloride gas molecules is 1.05 D. Calculate the maximum torque such a molecule can experience in the presence of an electric field of magnitude 160.0 N/C.

22.41 An electron is observed traveling at a speed of $27.5 \cdot 10^6$ m/s parallel to an electric field of magnitude 11,400 N/C. How far will the electron travel before coming to a stop?

22.42 Two charges, $+e$ and $-e$, are a distance of 0.68 nm apart in an electric field, E, that has a magnitude of 4.4 kN/C and is directed at an angle of $45°$ with the dipole axis. Calculate the dipole moment and thus the torque on the dipole in the electric field.

•**22.43** A body of mass M, carrying charge Q, falls from rest from a height h (above the ground) near the surface of the Earth, where the gravitational acceleration is g and there is an electric field with a constant component E in the vertical direction.

a) Find an expression for the speed, v, of the body when it reaches the ground, in terms of M, Q, h, g, and E.

b) The expression from part (a) is not meaningful for certain values of M, g, Q, and E. Explain what happens in such cases.

•**22.44** A water molecule, which is electrically neutral but has a dipole moment of magnitude $p = 6.20 \cdot 10^{-30}$ C m, is 1.00 cm away from a point charge $q = +1.00$ μC. The dipole will align with the electric field due to the charge. It will also experience a net force, since the field is not uniform.

a) Calculate the magnitude of the net force. (*Hint:* You do not need to know the precise size of the molecule, only that it is much smaller than 1 cm.)

b) Is the molecule attracted to or repelled by the point charge? Explain.

•**22.45** A total of $3.05 \cdot 10^6$ electrons are placed on an initially uncharged wire of length 1.33 m.

a) What is the magnitude of the electric field a perpendicular distance of 0.401 m away from the midpoint of the wire?

b) What is the magnitude of the acceleration of a proton placed at that point in space?

c) In which direction does the electric field force point in this case?

Sections 22.7 and 22.8

22.46 Four charges are placed in a three-dimensional space. The charges have magnitudes $+3q$, $-q$, $+2q$, and $-7q$. If a Gaussian surface encloses all the charges, what will be the electric flux through that surface?

22.47 The six faces of a cubical box each measure 20 cm by 20 cm, and the faces are numbered such that faces 1 and 6 are opposite to each other, as are faces 2 and 5, and faces 3 and 4. The flux through each face is:

Face	Flux (N m²/C)
1	−70
2	−300
3	−300
4	+300
5	−400
6	−500

Find the net charge inside the cube.

22.48 A conducting solid sphere ($R = 0.15$ m, $q = 6.1 \cdot 10^{-6}$ C) is shown in the figure. Using Gauss's Law and two different Gaussian surfaces, determine the electric field (magnitude and direction) at point A, which is 0.000001 m outside the conducting sphere. (*Hint:* One Gaussian surface is a sphere, and the other is a small right cylinder.)

22.49 Electric fields of varying magnitudes are directed either inward or outward at right angles on the faces of a cube, as shown in the figure. What is the strength and direction of the field on the face F?

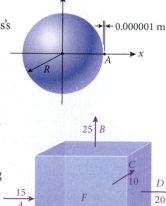

22.50 Consider a hollow spherical conductor with total charge $+5e$. The outer and inner radii are a and b, respectively. (a) Calculate the charge on the sphere's inner and outer surfaces if a charge of $-3e$ is placed at the center of the sphere. (b) What is the total net charge of the sphere?

•**22.51** A spherical aluminized Mylar balloon carries a charge Q on its surface. You are measuring the electric field at a distance R from the balloon's center. The balloon is slowly inflated, and its radius approaches but never reaches R. What happens to the electric field you measure as the balloon increases in radius. Explain.

•**22.52** A hollow conducting spherical shell has an inner radius of 8 cm and an outer radius of 10 cm. The electric field at the inner surface of the shell, E_i, has a magnitude of 80 N/C and points toward the center of the sphere, and the electric field at the outer surface, E_o, has a magnitude of 80 N/C and points away from the center of the sphere (see the figure). Determine the magnitude of the charge on the inner surface and the outer surface of the spherical shell.

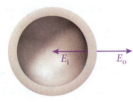

•**22.53** A -6-nC point charge is located at the center of a conducting spherical shell. The shell has an inner radius of 2 m, an outer radius of 4 m, and a charge of $+7$ nC.

a) What is the electric field at $r = 1$ m?

b) What is the electric field at $r = 3$ m?

c) What is the electric field at $r = 5$ m?

d) What is the surface charge distribution, σ, on the outside surface of the shell?

Section 22.9

22.54 A solid, nonconducting sphere of radius a has total charge Q and a uniform charge distribution. Using Gauss's Law, determine the electric field (as a vector) in the regions $r < a$ and $r > a$ in terms of Q.

22.55 There is an electric field of magnitude 150. N/C, directed downward, near the surface of the Earth. What is the net electric charge on the Earth? You can treat the Earth as a spherical conductor of radius 6371 km.

22.56 A hollow metal sphere has inner and outer radii of 20.0 cm and 30.0 cm, respectively. As shown in the figure, a solid metal sphere of radius 10.0 cm is located at the center of the hollow sphere. The electric field at a point P, a distance of 15.0 cm from the center, is found to be $E_1 = 1.00 \cdot 10^4$ N/C, directed radially inward. At point Q, a distance of 35.0 cm from the center, the electric field is found to be $E_2 = 1.00 \cdot 10^4$ N/C, directed radially outward. Determine the total charge on (a) the surface of the inner sphere, (b) the inner surface of the hollow sphere, and (c) the outer surface of the hollow sphere.

22.57 Two parallel, infinite, nonconducting plates are 10 cm apart and have charge distributions of $+1$ $\mu C/m^2$ and -1 $\mu C/m^2$. What is the force on an electron in the space between the plates? What is the force on an electron located outside the two plates near the surface of one of the two plates?

22.58 An infinitely long charged wire produces an electric field of magnitude $1.23 \cdot 10^3$ N/C at a distance of 50.0 cm perpendicular to the wire. The direction of the electric field is toward the wire.

a) What is the charge distribution?

b) How many electrons per unit length are on the wire?

•**22.59** A solid sphere of radius R has a nonuniform charge distribution $\rho = Ar^2$, where A is a constant. Determine the total charge, Q, within the volume of the sphere.

•**22.60** Two parallel, uniformly charged, infinitely long wires carry opposite charges with a linear charge density $\lambda = 1$ $\mu C/m$ and are 6 cm apart. What is the magnitude and direction of the electric field at a point midway between them and 40 cm above the plane containing the two wires?

•**22.61** A sphere centered at the origin has a volume charge distribution of 120 nC/cm^3 and a radius of 12 cm. The sphere is centered inside a conducting spherical shell with an inner radius of 30.0 cm and an outer radius of 50.0 cm. The charge on the spherical shell is -2.0 mC. What is the magnitude and direction of the electric field at each of the following distances from the origin?

a) at $r = 10.0$ cm   c) at $r = 40.0$ cm

b) at $r = 20.0$ cm   d) at $r = 80.0$ cm

•**22.62** A thin, hollow, metal cylinder of radius R has a surface charge distribution σ. A long, thin wire with a linear charge density $\lambda/2$ runs through the center of the cylinder. Find an expression for the electric fields and the direction of the field at each of the following locations:

a) $r \leq R$ b) $r \geq R$

•**22.63** Two infinite sheets of charge are separated by 10 cm as shown in the figure. Sheet 1 has a surface charge distribution of $\sigma_1 = 3$ $\mu C/m^2$ and sheet 2 has a surface charge distribution of $\sigma_2 = -5$ $\mu C/m^2$. Find the total electric field (magnitude and direction) at each of the following locations:

a) at point P, 6 cm to the left of sheet 1

b) at point P' 6 cm to the right of sheet 1

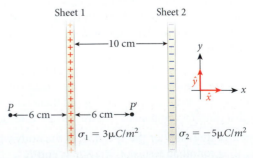

•**22.64** A conducting solid sphere of radius 20.0 cm is located with its center at the origin of a three-dimensional coordinate system. A charge of 0.271 nC is placed on the sphere.

a) What is the magnitude of the electric field at point (x,y,z) = (23.1 cm, 1.1 cm, 0 cm)?

b) What is the angle of this electric field with the x-axis at this point?

c) What is the magnitude of the electric field at point (x,y,z) = (4.1 cm, 1.1 cm, 0 cm)?

••**22.65** A solid nonconducting sphere of radius a has a total charge $+Q$ uniformly distributed throughout its volume. The surface of the sphere is coated with a very thin (negligible thickness) conducting layer of gold. A total charge of $-2Q$ is placed on this conducting layer. Use Gauss's Law to do the following.

Gold layer, Charge $-2Q$

a) Find the electric field $E(r)$ for $r < a$ (inside the sphere, up to and excluding the gold layer).

b) Find the electric field $E(r)$ for $r > a$ (outside the coated sphere, beyond the sphere and the gold layer).

c) Sketch the graph of $E(r)$ versus r. Comment on the continuity or discontinuity of the electric field, and relate this to the surface charge distribution on the gold layer.

••**22.66** A solid nonconducting sphere has a volume charge distribution given by $\rho(r) = (\beta/r) \sin(\pi r/2R)$. Find the total charge contained in the spherical volume and the electric field in the regions $r < R$ and $r > R$. Show that the two expressions for the electric field equal each other at $r = R$.

••**22.67** A very long cylindrical rod of nonconducting material with a 3-cm radius is given a uniformly distributed positive charge of 6 nC per centimeter of its length. Then a cylindrical cavity is drilled all the way through the rod, of radius 1 cm, with its axis located 1.5 cm from the axis of the rod. That is, if, at some cross section of the rod, x- and y-axes are placed so that the center of the rod is at $(x,y) = (0,0)$; then the center of the cylindrical cavity is at $(x,y) = (0,1.5)$. The creation of the cavity does not disturb the charge on the remainder of the rod that has not been drilled away; it just removes the charge from the region in the cavity. Find the electric field at the point $(x,y) = (2,1)$.

••**22.68** What is the electric field at a point P, a distance h = 20.0 cm above an infinite sheet of charge, with a charge distribution of 1.3 C/m^2 and a hole of radius 5.0 cm with P directly above the center of the hole, as shown in the figure? Plot the electric field as a function of h in units of $\sigma/(2\epsilon_0)$.

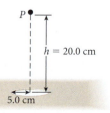

P

h = 20.0 cm

5.0 cm

Additional Problems

22.69 A cube has an edge length of 1.00 m. An electric field acting on the cube from outside has a constant magnitude of 150 N/C and its direction is also constant but unspecified (not necessarily along any edges of the cube). What is the total charge within the cube?

22.70 Consider a long horizontally oriented conducting wire with $\lambda = 4.81 \cdot 10^{-12}$ C/m. A proton (mass = $1.67 \cdot 10^{-27}$ kg) is placed 0.620 m above the wire and released. What is the magnitude of the initial acceleration of the proton?

22.71 An infinitely long, solid cylinder of radius R = 9.00 cm, with a uniform charge per unit of volume of $\rho = 6.40 \cdot 10^{-8}$ C/m^3, is centered about the y-axis. Find the magnitude of the electric field at a radius r = 4.00 cm from the center of this cylinder.

22.72 Carbon monoxide (CO) has a dipole moment of approximately $8.0 \cdot 10^{-30}$ C m. If the two atoms are separated by $1.2 \cdot 10^{-10}$ m, find the net charge on each atom and the maximum amount of torque the molecule would experience in an electric field of 500.0 N/C.

22.73 A solid metal sphere of radius 8 cm, with a total charge of 10 μC, is surrounded by a metallic shell with a radius of 15 cm carrying a –5 μC charge. The sphere and the shell are both inside a larger metallic shell of inner radius 20 cm and outer radius 24 cm. The sphere and the two shells are concentric.

a) What is the charge on the inner wall of the larger shell?

b) If the electric field outside the larger shell is zero, what is the charge on the outer wall of the shell?

22.74 Find the vector electric fields needed to counteract the weight of (a) an electron and (b) a proton at the Earth's surface.

22.75 There is an electric field of magnitude 150. N/C, directed vertically downward, near the surface of the Earth. Find the acceleration (magnitude and direction) of an electron released near the Earth's surface.

22.76 Two infinite, uniformly charged, flat surfaces are mutually perpendicular. One of the surfaces has a charge distribution of +30.0 pC/m^2, and the other has a charge distribution of –40.0 pC/m^2. What is the magnitude of the electric field at any point not on either surface?

22.77 A 30.0-cm-long uniformly charged rod is sealed in a container. The total electric flux leaving the container is $1.46 \cdot 10^6$ N m^2/C. Determine the linear charge distribution on the rod.

22.78 Suppose you have a large spherical balloon and you are able to measure the component E_n of the electric field normal to its surface. If you sum $E_n\, dA$ over the whole surface area of the balloon and obtain a magnitude of 10 N m^2/C, what is the electric charge enclosed by the balloon?

•**22.79** An object with mass m = 1.0 g and charge q is placed at point A, which is 0.05 m above an infinitely large, uniformly charged, nonconducting sheet ($\sigma = -3.5 \cdot 10^{-5}$ C/m^2), as shown in the figure. Gravity is acting downward (g = 9.81 m/s^2). Determine the number, N, of electrons that must be added to or removed from the object for the object to remain motionless above the charged plane.

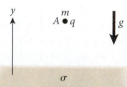

y

m

$A \bullet q$

g

σ

•**22.80** A long conducting wire with charge distribution λ and radius r produces an electric field of 2.73 N/C just outside the surface of the wire. What is the magnitude of the electric field just outside the surface of another wire with charge distribution 0.81λ and radius $6.5r$?

•**22.81** There is a uniform charge distribution of $\lambda = 8.00 \cdot 10^{-8}$ C/m along a thin wire of length $L = 6.00$ cm. The wire is then curved into a semicircle that is centered about the origin, so the radius of the semicircle is $R = L/\pi$. Find the magnitude of the electric field at the center of the semicircle.

•**22.82** A proton enters the gap between a pair of metal plates (an electrostatic separator) that produces a uniform, vertical electric field between them. Ignore the effect of gravity on the proton.

a) Assuming that the length of the plates is 15.0 cm, and that the proton will approach the plates at a speed of 15.0 km/s, what electric field strength should the plates be designed to provide, if the proton must be deflected vertically by $1.50 \cdot 10^{-3}$ rad?

b) What speed does the proton have after exiting the electric field?

c) Suppose the proton is one in a beam of protons that has been contaminated with positively charged kaons, particles whose mass is 494 MeV/c^2 ($8.81 \cdot 10^{-28}$ kg), compared to the mass of the proton, which is 938 MeV/c^2 ($1.67 \cdot 10^{-27}$ kg). The kaons have $+1e$ charge, just like the protons. If the electrostatic separator is designed to give the protons a deflection of $1.20 \cdot 10^{-3}$ rad, what deflection will kaons with the same momentum as the protons experience?

•**22.83** Consider a uniform nonconducting sphere with a charge $\rho = 3.57 \cdot 10^{-6}$ C/m^3 and a radius $R = 1.72$ m. What is the magnitude of the electric field 0.530 m from the center of the sphere?

••**22.84** A uniform sphere has a radius R and a total charge $+Q$, uniformly distributed throughout its volume. It is surrounded by a thick spherical shell carrying a total charge $-Q$, also uniformly distributed, and having an outer radius of $2R$. What is the electric field as a function of R?

••**22.85** If a charge is held in place above a large, flat, grounded, conducting slab, such as a floor, it will experience a downward force toward the floor. In fact, the electric field in the room above the floor will be exactly the same as that produced by the original charge plus a "mirror image" charge, equal in magnitude and opposite in sign, as far below the floor as the original charge is above it. Of course, there is no charge below the floor; the effect is produced by the surface charge distribution induced on the floor by the original charge.

a) Describe or sketch the electric field lines in the room above the floor.

b) If the original charge is 1.00 μC at a distance of 50.0 cm above the floor, calculate the downward force on this charge.

c) Find the electric field at (just above) the floor, as a function of the horizontal distance d from the point on the floor directly under the original charge. Assume that the original charge is a point charge, $+q$, at a distance a above the floor. Ignore any effects of walls or ceiling.

d) Find the surface charge distribution $\sigma(\rho)$ induced on the floor.

e) Calculate the total surface charge induced on the floor.

Electric Potential

23

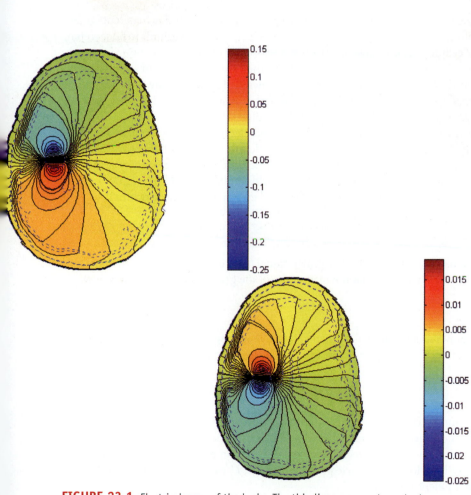

FIGURE 23.1 Electrical maps of the brain. The thin lines represent constant electric potential.

WHAT WE WILL LEARN

- Electric potential energy is analogous to gravitational potential energy.

- The change in electric potential energy is proportional to the work done by the electric field on a charge.

- The electric potential at a given point in space is a scalar.

- The electric potential, V, of a point charge, q, is inversely proportional to the distance from that point charge.

- The electric potential can be derived from the electric field by integrating the electric field over a displacement.

- The electric potential at a given point in space due to a distribution of point charges equals the algebraic sum of the electric potentials due to the individual charges.

- The electric field can be derived from the electric potential by differentiating the electric potential with respect to displacement.

The functioning of the human nervous system depends on electricity. Tiny currents travel along nerve cells to signal, for example, muscles to contract, or digestive fluids to be secreted, or white blood cells to attack an invader. The brain is a center of electrical activity; as signals come from the sense organs and the rest of the body, they are processed, and stimulate new activity, such as thought or emotion. The images shown in Figure 23.1 are electrical maps of the brain, produced in preparation for exploratory brain surgery. The thin lines indicate constant electric potential, a topic covered in this chapter.

Just as electric field strength is force per unit charge, electric potential is potential energy per unit charge. Electric potential is a property of the electric field, not of the charged object that produces the field. This distinction is important because it makes electric potential an extremely useful quantity for working with electric fields and electric circuits. However, it is necessary to be careful not to confuse electric potential energy and electric potential.

23.1 Electric Potential Energy

An electric field has many similarities to a gravitational field, including its mathematical formulation. We saw in Chapter 12 that the magnitude of the gravitational force is given by

$$F_g = G \frac{m_1 m_2}{r^2},$$

where G is the universal gravitational constant, m_1 and m_2 are two masses, and r is the distance between the two masses. In Chapter 21, we saw that the magnitude of the electrostatic force is

$$F_e = k \frac{|q_1 q_2|}{r^2} \tag{23.1}$$

where k is Coulomb's constant, q_1 and q_2 are two electric charges, and r is the distance between the two charges. Both gravitational and electrostatic forces depend only on the inverse square of the distance between the objects, and it can be shown that all such forces are conservative. Therefore, the electric potential energy, U, can be defined in analogy with the gravitational potential energy.

In Chapter 6, we saw that for any conservative force, the change in potential energy due to some spatial rearrangement of a system is equal to the negative of the work done by the conservative force during this spatial rearrangement. For a system of two or more particles, the work done by an electric force, W_e, when the system configuration changes from an initial state to a final state, is given in terms of the change in **electric potential energy, ΔU:**

$$\Delta U = U_f - U_i = -W_e, \tag{23.2}$$

where U_i is the initial electric potential energy and U_f is the final electric potential energy. Note that it does not matter *how* the system gets from the initial to the final state. The work is always the same, independent of the path taken. Chapter 6 noted that this path-independence of the work done by a force is a general feature of conservative forces.

As is the case for gravitational potential energy (see Chapter 12), a reference point for the electric potential energy must always be specified. It simplifies the equations and calculations if the zero point of the electric potential energy is assumed to be the configuration in which an infinitely large distance separates all the charges, which is exactly the same convention used for the gravitational potential energy. This assumption allows equation 23.2 for the change in electric potential energy to be rewritten as $\Delta U = U_f - 0 = U$, or

$$U = -W_{e,\infty}. \qquad (23.3)$$

Even though the convention of zero potential energy at infinity is very useful and is universally accepted for a collection of point charges, in some physical situations there is a reason to select a reference potential energy at some point in space, which will not lead to a value of zero potential energy at infinite separation. Remember, all potential energies of conservative forces are fixed only within an arbitrary additive constant. So you need to pay attention to how this constant is chosen in a particular situation. One situation in which the potential energy at infinity is not set to zero is that involving a constant electric field.

Special Case: Constant Electric Field

Let's consider a point charge, q, moving through a displacement, $\vec{d}$, in a constant electric field, $\vec{E}$ (Figure 23.2). The work done by a constant force $\vec{F}$ is $W = \vec{F} \cdot \vec{d}$. For this case, the constant force is created by a constant electric field, $\vec{F} = q\vec{E}$. Thus, the work done by the field on the charge is given by

$$W = q\vec{E} \cdot \vec{d} = qEd\cos\theta, \qquad (23.4)$$

where θ is the angle between the electric force and the displacement. When the displacement is parallel to the electric field ($\theta = 0°$), the work done by the field on the charge is $W = qEd$. When the displacement is antiparallel to the electric field ($\theta = 180°$), the work done by the field is $W = -qEd$. Because the change in electric potential energy is related to the work done on the charge by $\Delta U = -W$, if $q > 0$, the charge loses potential energy when the displacement is in the same direction as the electric field and gains potential energy when the displacement is in the direction opposite to the electric field.

Figure 23.3a shows a mass, m, near the surface of the Earth, where it can be considered to be in a constant gravitational field, which points downward. From Chapter 6, we know that when the mass moves toward the surface of the Earth a distance h, the change in the gravitational potential energy of the mass is

$$\Delta U = -W = -\vec{F}_g \cdot \vec{d} = -mgh.$$

It is intuitive that the mass has less potential energy if it is closer to the surface of the Earth. Figure 23.3b shows a positive charge, q, in a constant electric field. If the charge moves a distance, d, in the same direction as the electric field, the change in the electric potential energy is

$$\Delta U = -W = -q\vec{E} \cdot \vec{d} = -qEd. \qquad (23.5)$$

Thus, the electric potential energy of a charge in an electric field is analogous to the gravitational potential energy of a mass in Earth's gravitational field near the surface of Earth. (But, of course, the important difference between the two interactions is that masses come in only one variety, and exert gravitational attraction for one another, whereas charges can attract or repel each other. Thus, ΔU can change sign, depending on the signs of the charges.)

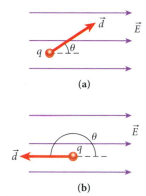

FIGURE 23.2 Work done by an electric field, $\vec{E}$, on a moving charge, q: (a) general case, (b) case where the displacement is opposite to the direction of the electric field.

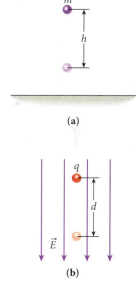

FIGURE 23.3 The analogy between gravitational potential energy and electric potential energy. (a) A mass falls in a gravitational field. (b) A positive charge moves in the same direction as an electric field.

23.2 Definition of Electric Potential

The potential energy of a charged particle, q, in an electric field depends on the magnitude of the charge as well as that of the electric field. A quantity that is independent of the charge on the particle is the **electric potential**, V, defined in terms of the electric potential energy as

$$V = \frac{U}{q}. \qquad V = \frac{J}{C} \qquad (23.6)$$

Because U is proportional to q, V is independent of q, which makes it a useful variable. The electric potential, V, characterizes an electrical property of a point in space even when no charge, q, is placed at that point. In contrast to the electric field, which is a vector, the electric potential is a scalar. It has a value everywhere in space, but has no direction.

The difference in electric potential, ΔV, between an initial point and final point, $V_f - V_i$, can be expressed in terms of the electric potential energy at each point:

$$\Delta V = V_f - V_i = \frac{U_f}{q} - \frac{U_i}{q} = \frac{\Delta U}{q}. \tag{23.7}$$

Combining equations 23.2 and 23.7 yields a relationship between the change in electric potential and the work done by an electric field on a charge:

$$\Delta V = -\frac{W_e}{q}. \tag{23.8}$$

Taking the electric potential energy to be zero at infinity as in equation (23.3) gives the electric potential at a point as

$$V = -\frac{W_{e,\infty}}{q}, \tag{23.9}$$

where $W_{e,\infty}$ is the work done by the electric field on the charge when it is brought in to the point from infinity. An electric potential can have a positive, a negative, or a zero value, but it has no direction.

The SI units for electric potential are joules/coulomb (J/C). This combination has been named the **volt** (V) for Italian physicist Alessandro Volta (1745–1827) (note the use of the roman V for the unit, whereas the italicized V is used for the physical quantity of electric potential):

$$1\,\text{V} \equiv \frac{1\,\text{J}}{1\,\text{C}}.$$

With this definition of the volt, the units for the magnitude of the electric field are

$$[E] = \frac{[F]}{[q]} = \frac{1\,\text{N}}{1\,\text{C}} = \left(\frac{1\,\text{N}}{1\,\text{C}}\right)\frac{1\,\text{V}}{\left(\frac{1\,\text{J}}{1\,\text{C}}\right)}\left(\frac{1\,\text{J}}{(1\,\text{N})(1\,\text{m})}\right) = \frac{1\,\text{V}}{1\,\text{m}}.$$

For the remainder of this book, the magnitude of an electric field will have units of V/m, which is the standard convention, instead of N/C. Note that an electric potential difference is often referred to as a "voltage," particularly in circuit analysis, because it is measured in volts.

EXAMPLE 23.1 | Energy Gain of a Proton

A proton is placed between two parallel conducting plates in a vacuum (Figure 23.4). The difference in electric potential between the two plates is 450 V. The proton is released from rest close to the positive plate.

PROBLEM

What is the kinetic energy of the proton when it reaches the negative plate?

SOLUTION

The difference in electric potential, ΔV, between the two plates is 450 V. We can relate this potential difference across the two plates to the change in electric potential energy, ΔU, of the proton using equation 23.7:

$$\Delta V = \frac{\Delta U}{q}.$$

Because of the conservation of total energy, all the electric potential energy lost by the proton in crossing between the two plates is turned into kinetic energy due to the motion

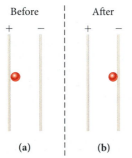

FIGURE 23.4 A proton between two charged parallel conducting plates in a vacuum. (a) The proton is released from rest. (b) The proton has moved from the positive plate to the negative plate, gaining kinetic energy.

of the proton. We apply the law of conservation of energy, $\Delta K + \Delta U = 0$, where ΔU is the change in the proton's electric potential energy:

$$\Delta K = -\Delta U = -q\Delta V.$$

Because the proton started from rest, we can express its final kinetic energy as $K = -q\Delta V$. Therefore, the kinetic energy of the proton after crossing the gap between the two plates is

$$K = -\left(1.602 \cdot 10^{-19}\ \text{C}\right)\left(-450\ \text{V}\right) = 7.21 \cdot 10^{-17}\ \text{J}.$$

23.1 In-Class Exercise

An electron is positioned and then released on the x-axis, where the electric potential has the value –20 V. Which of the following statements describes the subsequent motion of the electron?

a) The electron will move to the left (negative x-direction) because it is negatively charged.

b) The electron will move to the right (positive x-direction) because it is negatively charged.

c) The electron will move to the left (negative x-direction) because the electric potential is negative.

d) The electron will move to the right (positive x-direction) because the electric potential is negative.

e) Not enough information is given to predict the motion of the electron.

(a)

Because the acceleration of charged particles across a potential difference is often used in the measurement of physical quantities, a common unit for the kinetic energy of a singly charged particle, such as a proton or an electron, is the **electron-volt** (eV): 1 eV represents the energy gained by a proton ($q = 1.602 \cdot 10^{-19}$ C) accelerated across a potential difference of 1 V. The conversion between electron-volts and joules is

$$1\ \text{eV} = 1.602 \cdot 10^{-19}\ \text{J}.$$

The kinetic energy of the proton in Example 23.1 is then 450 eV, or 0.450 keV, which we could have obtained from the definition of the electron-volt without performing any calculations.

Batteries

A common means of creating electric potential is a battery. We'll see in Chapters 24 and 25 how a battery uses chemical reactions to provide a source of (nearly) constant potential difference between its two terminals. An assortment of batteries is shown in Figure 23.5.

At its simplest, a battery consists of two half-cells, filled with a conducting electrolyte (originally a liquid but now almost always a solid); see Figure 23.6. The electrolyte is separated into two halves by a barrier, which prevents the bulk of the electrolyte from passing through but allows charged ions to pass through. The negatively charged ions (anions) move toward the anode, and the positively charged ions move toward the cathode. This creates a potential difference between the two terminals of the battery. Thus, a battery is basically a device that converts chemical energy directly into electrical energy.

Research on battery technology is of current importance, because many mobile applications require a great deal of energy, from cell phones to laptop computers, from electrical cars to military gear. The weight of the batteries needs to be as small as possible, they need to be rapidly rechargeable for hundreds of cycles, they need to deliver as constant a potential difference as possible, and they need to be available at an affordable price. Thus, such research provides many scientific and engineering challenges.

One example of relatively recent battery technology is the lithium ion cell, which is often used in applications such as laptop computer batteries. A lithium ion battery has a much higher energy density (energy content per unit volume) than conventional batteries. A typical lithium ion cell, like the one in Figure 23.7, has a potential difference of 3.7 V. Lithium ion batteries have several other advantages over conventional batteries. They can be

(b)

FIGURE 23.5 (a) Some representative batteries (clockwise from upper left): rechargeable AA nickel metal hydride (NiMH) batteries in their charger, disposable 1.5-V AAA batteries, a 12-V lantern battery, a D-size battery, a lithium ion laptop battery, and a watch battery; (b) 330-V battery for a gas-electric hybrid SUV, filling the entire floor of the trunk.

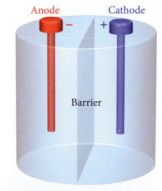

FIGURE 23.6 Schematic drawing of a battery.

FIGURE 23.7 A drawing showing the lithium ion battery pack of the Tesla electric-powered sports car. Also shown is one of the 6831 lithium ion cells that make up the battery pack.

recharged hundreds of times. They have no "memory" effect and thus do not need to be conditioned to hold their charge. They hold their charge on the shelf. They also have some disadvantages. For example, if a lithium ion battery is completely discharged, it can no longer be recharged. The battery performs best if it is not charged to more than 80% of capacity and not discharged to less than 20% of its capacity. Heat degrades lithium ion batteries. If the batteries are discharged too quickly, the constituents can catch fire or explode. To deal with these problems, most commercial lithium ion battery packs have a small built-in electronic circuit that protects the battery pack. The circuit will not allow the battery to be overcharged or overly discharged; it will not allow charge to flow out of the battery so quickly that the battery will overheat. If the battery becomes too warm, the circuit disconnects the battery.

Currently, lithium ion batteries are being used in some electric-powered cars. The following example compares the energy carried by a battery-powered car and a gasoline-powered car.

FIGURE 23.8 The Tesla electric-powered sports car.

EXAMPLE 23.2 | Battery-Powered Cars

Battery-powered cars produce no emissions and thus are an attractive alternative to gasoline-powered cars. Some of these cars, such as the Tesla sports car shown in Figure 23.8, are powered by batteries constructed of lithium ion cells.

The battery pack of the Tesla electric sports car (Figure 23.7) has the capacity to hold 53 kW h of energy. The battery pack is usually charged to 80% of its capacity and discharged to 20% of its capacity. A gasoline-powered car typically carries 50 L of gasoline, and gasoline has an energy content of 34.8 MJ/L.

PROBLEM
How does the available energy in a lithium ion battery pack of an electric-powered car compare with the energy carried by a gasoline-powered car?

SOLUTION
Because not all of the energy can be extracted from a lithium ion battery without damaging it, the total usable energy is

$$E_{\text{electric}} = \left(80\% - 20\%\right)\left(53 \text{ kW h}\right)\left(\frac{1000 \text{ W}}{1 \text{ kW}}\right)\left(\frac{3600 \text{ s}}{1 \text{ h}}\right) = 1.14 \cdot 10^8 \text{ J} = 114 \text{ MJ}.$$

A typical gasoline-powered car can carry 50 L of gasoline, which has an energy content of

$$E_{\text{gasoline}} = \left(50 \text{ L}\right)\left(34.8 \text{ MJ/L}\right) = 1740 \text{ MJ}.$$

Thus, a typical gasoline-powered car carries 15 times as much energy as the Tesla electric-powered car. However, the efficiency of a gasoline-powered car is approximately 20%, while an electric-powered car can be close to 90% efficient. Thus, the usable energy of the electric-powered car is

$$E_{\text{electric, usable}} = 0.9(114 \text{ MJ}) = 103 \text{ MJ},$$

and the usable energy of the gasoline-powered car is

$$E_{\text{gasoline, usable}} = 0.2(1740 \text{ MJ}) = 348 \text{ MJ}.$$

You can see that electric-powered cars, even with lithium ion batteries, can carry less energy than gasoline-powered cars.

Van de Graaff Generator

One means of creating large electric potentials is a **Van de Graaff generator,** a device invented by the American physicist Robert J. Van de Graaff (1901–1967). Large Van de Graaff generators can produce electric potentials of millions of volts. More modest Van de Graaff generators, such as the one shown in Figure 23.9, can produce several hundred thousand volts and are often used in physics classrooms.

A Van de Graaff generator uses a corona discharge to apply a positive charge to a non-conducting moving belt. Putting a high positive voltage on a conductor with a sharp point creates the corona discharge. The electric field on the sharp point is much stronger than on the flat surface of the conductor (see Chapter 22). The air around the sharp point is ionized. The ionized air molecules have a net positive charge, which causes the ions to be repelled away from the sharp point and deposited on the rubber belt. The moving belt, driven by an electric motor, carries the charge up into a hollow metal sphere, where the charge is taken from the belt by a pointed contact connected to the metal sphere. The charge that builds up on the metal sphere distributes itself uniformly around the outside of the sphere. On the Van de Graaff generator shown in Figure 23.9, a voltage limiter is used to keep the generator from producing sparks larger than desired.

(a)

EXAMPLE 23.3 Tandem Van de Graaff Accelerator

A Van de Graaff accelerator is a particle accelerator that uses high electric potentials for studying nuclear physics processes of astrophysical relevance. A tandem Van de Graaff accelerator with a terminal potential difference of 10.0 MV (10.0 million volts), is diagrammed in Figure 23.10. This terminal potential difference is created in the center of the accelerator by a larger, more sophisticated version of the classroom Van de Graaff generator. Negative ions are created in the ion source by attaching an electron to the atoms to be accelerated. The negative ions then accelerate toward the positively charged terminal. Inside the terminal, the ions pass through a thin foil that strips off electrons, producing positively charged ions that then are accelerated away from the terminal and out of the tandem accelerator.

FIGURE 23.10 A tandem Van de Graaff accelerator.

PROBLEM 1

What is the highest kinetic energy that carbon nuclei can attain in this tandem accelerator?

SOLUTION 1

A tandem Van de Graaff accelerator has two stages of acceleration. In the first stage, each carbon ion has a net charge of $q_1 = -e$. After the stripper foil, the maximum charge any carbon ion can have is $q_2 = +6e$. The potential difference over which the ions are accelerated is $\Delta V = 10$ MV. The kinetic energy gained by each carbon ion is

$$\Delta K = |\Delta U| = |q_1 \Delta V| + |q_2 \Delta V| = K,$$

or

$$K = e\Delta V + 6e\Delta V = 7e\Delta V,$$

assuming that the initial speed of the ions is $\approx$ zero.

Putting in the numerical values, we get

$$K = 7\left(1.602 \cdot 10^{-19} \text{ C}\right)\left(10 \cdot 10^6 \text{ V}\right) = 1.12 \cdot 10^{-11} \text{ J.}$$

Continued—

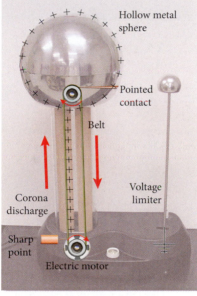

(b)

FIGURE 23.9 (a) A Van de Graaff generator used in physics classrooms. (b) The Van de Graaff generator can produce very high electric potentials by carrying charge from a corona discharge on a rubber belt up to a hollow metal sphere, where the charge is extracted from the belt by a sharp piece of metal attached to the inner surface of the hollow sphere.

23.2 In-Class Exercise

A cathode ray tube uses a potential difference of 5.0 kV to accelerate electrons and produce an electron beam that makes images on a phosphor screen. What is the speed of these electrons as a percentage of the speed of light?

a) 0.025%   d) 4.5%

b) 0.22%   e) 14%

c) 1.3%

Nuclear physicists often use electron-volts instead of joules to express the kinetic energy of accelerated nuclei:

$$K = 7e\Delta V = 7e\left(10 \cdot 10^6 \text{ V}\right) = 7 \cdot 10^7 \text{ eV} = 70 \text{ MeV}.$$

PROBLEM 2
What is the highest speed that carbon nuclei can attain in this tandem accelerator?

SOLUTION 2
To determine the speed, we use the relationship between kinetic energy and speed:

$$K = \tfrac{1}{2}mv^2,$$

where $m = 1.99 \cdot 10^{-26}$ kg is the mass of the carbon nucleus. Solving this equation for the speed, we get

$$v = \sqrt{\frac{2K}{m}} = \sqrt{\frac{2\left(1.12 \cdot 10^{-11} \text{ J}\right)}{1.99 \cdot 10^{-26} \text{ kg}}} = 3.36 \cdot 10^7 \text{ m/s},$$

which is 11% of the speed of light.

23.3 Equipotential Surfaces and Lines

Imagine you had to map out a ski resort with three peaks, like the one shown in Figure 23.11a. In Figure 23.11b, lines of equal elevation have been superimposed on the peaks. You could walk along each of these lines, without ever going uphill or downhill, and would be guaranteed to reach the point from which you started. These lines are lines of constant gravitational potential, because the gravitational potential is a function of the elevation only, and the elevation remains constant on each of the lines. Figure 23.11c shows a top view of the contour lines of equal elevation, which mark the equipotential lines for the gravitational potential. If you have understood this figure, the following discussion of electric potential lines and surfaces should be easy to follow.

When an electric field is present, the electric potential has a value everywhere in space. Points that have the same electric potential form an **equipotential surface.** Charged particles can move along an equipotential surface without having any work done on them by the electric field. According to principles of electrostatics, the surface of a conductor must be an equipotential surface; otherwise, the free electrons on the conductor surface would accelerate. The discussion in Chapter 22 established that the electric field is zero everywhere inside the body of a conductor. This means that the entire volume of the conductor must be at the same potential; that is, the entire conductor is an equipotential.

Equipotential surfaces exist in three dimensions (Figure 23.12); however, symmetries in the electric potential allow us to represent equipotential surfaces in two dimensions, as **equipotential lines** in the plane in which the charges reside. Before determining the shape and location of these equipotential surfaces, let's first look at some qualitative features of some of the simplest cases (for which the electric fields were determined in Chapter 22).

In drawing equipotential lines, we note that charges can move perpendicular to any electric field line without having any work done on them by the electric field, because

FIGURE 23.11 (a) Ski resort with three peaks; (b) the same peaks with lines of equal elevation superimposed; (c) the contour lines of equal elevation in a two-dimensional plot.

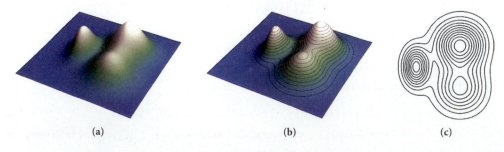

(a)   (b)   (c)

according to equation 23.4, the scalar product of the electric field and the displacement is then zero. If the work done by the electric field is zero, the potential remains the same, by equation 23.8. Thus, *equipotential lines and planes are always perpendicular to the direction of the electric field.* (In Figure 23.11b, the elevation map of the ski resort, the equivalent of electric field lines would be the lines of steepest descent, which are, of course, always perpendicular to the lines of equal elevation.)

Before examining the particular equipotential surfaces resulting from different electric field configurations, let's note the two most important general observations of this section, which hold for all of the following cases:

1. The surface of any conductor forms an equipotential surface.

2. Equipotential surfaces are always perpendicular to the electric field lines at any point in space.

FIGURE 23.12 Equipotential surface in three dimensions, resulting from eight positive point charges placed at the corners of a cube.

Constant Electric Field

A constant electric field has straight, equally spaced, and parallel field lines. Thus, such a field produces equipotential surfaces in the form of parallel planes, because of the condition that the equipotential surfaces or equipotential lines have to be perpendicular to the field lines. These planes are represented in two dimensions as equally spaced equipotential lines (Figure 23.13).

Single Point Charge

Figure 23.14 shows the electric field and corresponding equipotential lines due to a single point charge. The electric field lines extend radially from a positive point charge, as shown in Figure 23.14a. In this case, the field lines point away from the positive charge and terminate at infinity. For a negative charge as shown in Figure 23.14b, the field lines originate at infinity and terminate at the negative charge. The equipotential lines are spheres centered on the point charge. (In the two-dimensional views shown in the figure, the circles represent the lines where the plane of the page cuts through equipotential spheres.) The values of the potential difference between neighboring equipotential lines are equal, producing equipotential lines that are close together near the charge and more widely spaced away from the charge. Note again that the equipotential lines are always perpendicular to the electric field lines. Equipotential surfaces do not have arrows like the field lines, because the potential is a scalar.

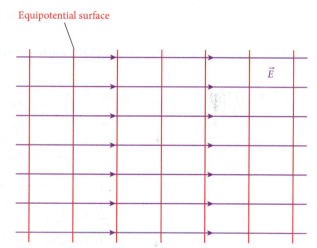

FIGURE 23.13 Equipotential surfaces (red lines) from a constant electric field. The purple lines with the arrowheads represent the electric field.

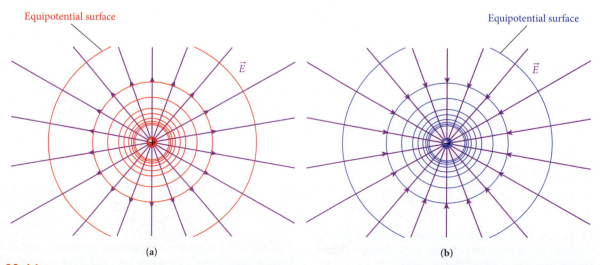

FIGURE 23.14 Equipotential surfaces and electric field lines from (a) a single positive point charge and (b) a single negative point charge.

Two Oppositely Charged Point Charges

Figure 23.15 shows the electric field lines from two oppositely charged point charges, along with equipotential surfaces depicted as equipotential lines. An electrostatic force would attract these two point charges toward each other, but this discussion assumes that the charges are fixed in space and cannot move. The electric field lines originate at the positive charge and terminate on the negative charge. Again, the equipotential lines are always perpendicular to the electric field lines. The red lines in this figure represent positive equipotential surfaces, and the blue lines represent negative equipotential surfaces. Positive charges produce positive potential, and negative charges produce negative potential (relative to the value of the potential at infinity). Close to each charge, the resultant electric field lines and the resultant equipotential lines resemble those for a single point charge. Away from the vicinity of each charge, the electric field and the electric potential are the sums of the fields and potentials due to the two charges. The electric fields add as vectors, while the electric potentials add as scalars. Thus, the electric field is defined at all points in space in terms of a magnitude and a direction, while the electric potential is defined solely by its value at a given point in space and has no direction associated with it.

Two Identical Point Charges

Figure 23.16 shows electric field lines and equipotential surfaces resulting from two identical positive point charges. These two charges experience a repulsive electrostatic force.

23.1 Self-Test Opportunity

Suppose the charges in Figure 23.15 were located at $(x,y) = (-10$ cm,$0)$ and $(x,y) = (+10$ cm,$0)$. What would the electric potential be along the y-axis $(x = 0)$?

23.2 Self-Test Opportunity

Suppose the charges in Figure 23.16 were located at $(x,y) = (-10$ cm,$0)$ and $(x,y) = (+10$ cm,$0)$. Would $(x,y) = (0,0)$ correspond to a maximum, a minimum, or a saddle point in the electric potential?

23.3 In-Class Exercise

In the figure, the lines represent equipotential lines. A charged object is moved from point P to point Q. How does the amount of work done on the object compare for these three cases?

a) All three cases involve the same work.

b) The most work is done in case 1.

c) The most work is done in case 2.

d) The most work is done in case 3.

e) Cases 1 and 3 involve the same amount of work, which is more than is involved in case 2.

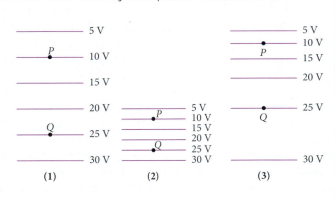

Equipotential surface positive

Equipotential surface negative

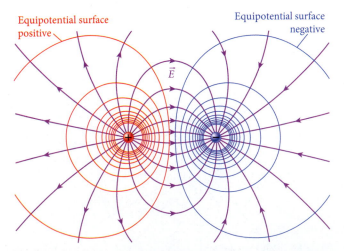

FIGURE 23.15 Equipotential surfaces created by point charges of the same magnitude but opposite sign. The red lines represent positive potential, and the blue lines represent negative potential. The purple lines with the arrowheads represent the electric field.

Equipotential surface

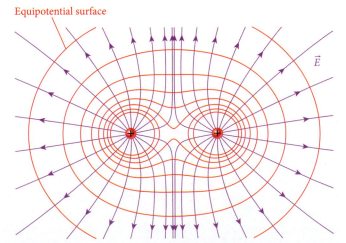

FIGURE 23.16 Equipotential surfaces (red lines) from two identical positive point charges. The purple lines with the arrowheads represent the electric field.

Because both charges are positive, the equipotential surfaces represent positive potentials. Again, the electric field and electric potential result from the sums of the fields and potentials, respectively, due to the two charges.

23.4 Electric Potential of Various Charge Distributions

The electric potential is defined as the work required to place a unit charge at a point, and work is a force acting over a distance. Also, the electric field can be defined as the force acting on a unit charge at a point. Therefore, it seems that the potential at a point should be related to the field strength at that point. In fact, electric potential and electric field are directly related; we can determine either one given an expression for the other.

To determine the electric potential from the electric field, we start with the definition of the work done on a particle with charge q by a force, $\vec{F}$, over a displacement, $d\vec{s}$:

$$dW = \vec{F} \cdot d\vec{s}.$$

In this case, the force is given by $\vec{F} = q\vec{E}$, so

$$dW = q\vec{E} \cdot d\vec{s}. \tag{23.10}$$

Integration of equation 23.10 as the particle moves in the electric field from some initial point to some final point gives

$$W = W_e = \int_i^f q\vec{E} \cdot d\vec{s} = q\int_i^f \vec{E} \cdot d\vec{s}.$$

Using equation 23.8 to relate the work done to the change in electric potential, we get

$$\Delta V = V_f - V_i = -\frac{W_e}{q} = -\int_i^f \vec{E} \cdot d\vec{s}.$$

As mentioned earlier, the usual convention is to set the electric potential to zero at infinity. With this convention, we can express the potential at some point $\vec{r}$ in space as

$$V(\vec{r}) - V(\infty) \equiv V(\vec{r}) = -\int_\infty^{\vec{r}} \vec{E} \cdot d\vec{s}. \tag{23.11}$$

Point Charge

Let's use equation 23.11 to determine the electric potential due to a point charge, q. The electric field due to a point charge, q (for now, taken as positive), at a distance r from the charge is given by

$$E = \frac{kq}{r^2}.$$

The direction of the electric field is radial from the point charge. Assume that the integration is carried out along a radial line from infinity to a point at a distance R from the point charge, such that $\vec{E} \cdot d\vec{s} = E \, dr$. Then we can use equation 23.11 to obtain

$$V(R) = -\int_\infty^R \vec{E} \cdot d\vec{s} = -\int_\infty^R \frac{kq}{r^2} dr = \left[\frac{kq}{r}\right]_\infty^R = \frac{kq}{R}.$$

Thus, the electric potential due to a point charge at a distance r from the charge is given by

$$V = \frac{kq}{r}. \tag{23.12}$$

Equation 23.12 also holds when $q < 0$. A positive charge produces a positive potential, and a negative charge produces a negative potential, as shown in Figure 23.17.

In Figure 23.17, the electric potential is calculated for all points in the xy-plane. The vertical axis represents the value of the potential at each point on the plane, $V(x, y)$, found using $r = \sqrt{x^2 + y^2}$. The potential is not calculated close to $r = 0$ because it becomes infinite there. You can see from Figure 23.17 how the circular equipotential lines shown in Figure 23.14 originate.

23.3 Self-Test Opportunity

Obtaining equation 23.12 for the electric potential from a point charge involved integrating along a radial line from infinity to a point at a distance R from the point charge. How would the result change if the integration were carried out over a different path?

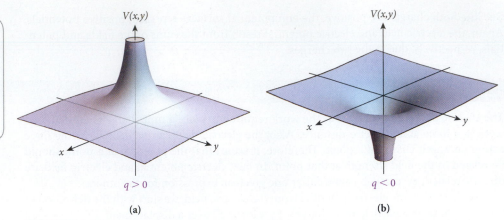

FIGURE 23.17 Electric potential due to: (a) a positive point charge and (b) a negative point charge.

SOLVED PROBLEM 23.1 | Fixed and Moving Positive Charges

PROBLEM

A positive charge of 4.50 μC is fixed in place. A particle of mass 6.00 g and charge +3.00 μC is fired with an initial speed of 66.0 m/s directly toward the fixed charge from a distance of 4.20 cm away. How close does the moving charge get to the fixed charge before it comes to rest and starts moving away from the fixed charge?

SOLUTION

THINK

The moving charge will gain electric potential energy as it nears the fixed charge. The negative of the change in potential energy of the moving charge is equal to the change in kinetic energy of the moving charge because $\Delta K + \Delta U = 0$.

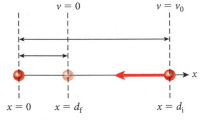

FIGURE 23.18 Two positive charges. One charge is fixed in place at $x = 0$, and the second charge begins moving with velocity $\vec{v}_0$ at $x = d_i$ and has zero velocity at $x = d_f$.

SKETCH

We set the location of the fixed charge at $x = 0$, as shown in Figure 23.18. The moving charge starts at $x = d_i$, moves with initial speed $v = v_0$, and comes to rest at $x = d_f$.

RESEARCH

The moving charge gains electric potential energy as it approaches the fixed charge and loses kinetic energy until it stops. At that point, all the original kinetic energy of the moving charge has been converted to electric potential energy. Using energy conservation, we can write this relationship as

$$\Delta K + \Delta U = 0 \Rightarrow \Delta K = -\Delta U \Rightarrow$$

$$0 - \tfrac{1}{2}mv_0^2 = -q_{moving}\Delta V \Rightarrow$$

$$\tfrac{1}{2}mv_0^2 = q_{moving}\Delta V. \qquad \text{(i)}$$

The electric potential experienced by the moving charge is due to the fixed charge, so we can write the change in potential as

$$\Delta V = V_f - V_i = k\frac{q_{fixed}}{d_f} - k\frac{q_{fixed}}{d_i} = kq_{fixed}\left(\frac{1}{d_f} - \frac{1}{d_i}\right). \qquad \text{(ii)}$$

SIMPLIFY

Substituting the expression for the potential difference from equation (ii) into equation (i), we find

$$\frac{1}{2}mv_0^2 = q_{moving}\Delta V = kq_{moving}q_{fixed}\left(\frac{1}{d_f} - \frac{1}{d_i}\right) \Rightarrow$$

$$\frac{1}{d_f} - \frac{1}{d_i} = \frac{mv_0^2}{2kq_{moving}q_{fixed}} \Rightarrow$$

$$\frac{1}{d_f} = \frac{1}{d_i} + \frac{mv_0^2}{2kq_{moving}q_{fixed}}.$$

CALCULATE

Putting in the numerical values, we get

$$\frac{1}{d_f} = \frac{1}{0.0420 \text{ m}} + \frac{(0.00600 \text{ kg})(66.0 \text{ m/s})^2}{2(8.99 \cdot 10^9 \text{ N m}^2/\text{C}^2)(3.00 \cdot 10^{-6}\text{C})(4.50 \cdot 10^{-6}\text{C})} = 131.485,$$

or

$$d_f = 0.00760545 \text{ m}.$$

ROUND

We report our result to three significant figures:

$$d_f = 0.00761 \text{ m} = 0.761 \text{ cm}.$$

DOUBLE-CHECK

The final distance of 0.761 cm is less than the initial distance of 4.20 cm. At the final distance, the electric potential energy of the moving charge is

$$U = q_{moving}V = q_{moving}\left(k\frac{q_{fixed}}{d_f}\right) = k\frac{q_{moving}q_{fixed}}{d_f}$$

$$= (8.99 \cdot 10^9 \text{ N m}^2/\text{C}^2)\frac{(3.00 \cdot 10^{-6} \text{ C})(4.50 \cdot 10^{-6} \text{ C})}{0.00761 \text{ m}} = 16.0 \text{ J}.$$

The electric potential energy at the initial distance is

$$U = q_{moving}V = q_{moving}\left(k\frac{q_{fixed}}{d_i}\right) = k\frac{q_{moving}q_{fixed}}{d_i}$$

$$= (8.99 \cdot 10^9 \text{ N m}^2/\text{C}^2)\frac{(3.00 \cdot 10^{-6} \text{ C})(4.50 \cdot 10^{-6} \text{ C})}{(0.0420 \text{ m})} = 2.9 \text{ J}.$$

The initial kinetic energy is

$$K = \frac{1}{2}mv^2 = \frac{(0.00600 \text{ kg})(66.0 \text{ m/s})^2}{2} = 13.1 \text{ J}.$$

We can see that the equation based on energy conservation, from which the solution process started, is satisfied:

$$\tfrac{1}{2}mv^2 = \Delta U$$

$$13.1 \text{ J} = 16.0 \text{ J} - 2.9 \text{ J} = 13.1 \text{ J}.$$

This gives us confidence that our result for the final distance is correct.

System of Point Charges

The electric potential due to a system of n point charges is calculated by adding the potentials due to all the charges:

$$V = \sum_{i=1}^{n} V_i = \sum_{i=1}^{n} \frac{kq_i}{r_i}. \tag{23.13}$$

Equation 23.13 can be proved by inserting the expression for the total electric field from n charges $\left(\vec{E}_t = \vec{E}_1 + \vec{E}_2 + \cdots + \vec{E}_n\right)$ into equation 23.11 and integrating term by term. The summation in equation 23.13 produces a potential at any point in space that has a value but no direction. Thus, calculating the potential due to a group of point charges is usually much simpler than calculating the electric field, which involves the addition of vectors.

EXAMPLE 23.4 | Superposition of Electric Potentials

Let's calculate the electric potential at a given point due to a system of point charges. Figure 23.19 shows three point charges: $q_1 = +1.50\ \mu C$, $q_2 = +2.50\ \mu C$, and $q_3 = -3.50\ \mu C$. Charge q_1 is located at $(0,a)$, q_2 is located at $(0,0)$, and q_3 is located at $(b,0)$, where $a = 8.00$ m and $b = 6.00$ m.

The electric potential at point P is the sum of the potentials due to the three charges:

$$V = \sum_{i=1}^{3} \frac{kq_i}{r_i} = k\left(\frac{q_1}{r_1} + \frac{q_2}{r_2} + \frac{q_3}{r_3}\right) = k\left(\frac{q_1}{b} + \frac{q_2}{\sqrt{a^2+b^2}} + \frac{q_3}{a}\right)$$

$$= \left(8.99\cdot10^9\ \text{N m}^2/\text{C}^2\right)\left|\frac{1.50\cdot10^{-6}\ \text{C}}{6.00\ \text{m}} + \frac{2.50\cdot10^{-6}\ \text{C}}{\sqrt{\left(8.00\ \text{m}\right)^2 + \left(6.00\ \text{m}\right)^2}} + \frac{-3.50\cdot10^{-6}\ \text{C}}{8.00\ \text{m}}\right|$$

$$= 562\ \text{V}.$$

Note that the potential due to q_3 is negative at point P, but the sum of the potentials is positive.

This example is similar to Example 22.1, in which we calculated the electric field at point P due to three charges. Note that this calculation of the electric potential due to three charges is much simpler than that calculation.

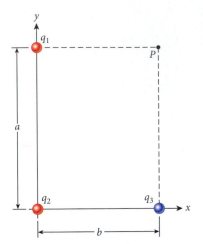

FIGURE 23.19 Electric potential at a point due to three point charges.

23.5 In-Class Exercise

Three identical positive point charges are located at fixed points in space. Then charge q_2 is moved from its initial location to a final location as shown in the figure. Four different paths, marked (a) through (d), are shown. Path (a) follows the shortest line; path (b) takes q_2 around q_3; path (c) takes q_2 around q_3 and q_1; path (d) takes q_2 out to infinity and then to the final location. Which path requires the least work?

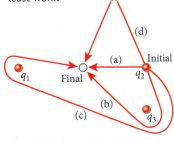

a) path (a)

b) path (b)

c) path (c)

d) path (d)

e) The work is the same for all the paths.

Continuous Charge Distribution

We can also determine the electric potential due to a continuous distribution of charge. To do this, we divide the charge into differential elements of charge, dq, and find the electric potential resulting from that differential charge as if it were a point charge. This is the way charge distributions were treated in determining electric fields in Chapter 22. The differential charge dq, can be expressed in terms of a charge per unit length times a differential length, $\lambda\,dx$; in terms of a charge per unit area times a differential area, $\sigma\,dA$; or in terms of a charge per unit volume times a differential volume, $\rho\,dV$. The electric potential resulting from the charge distribution is obtained by integrating over the contributions from the differential charges. Let's consider an example involving the electric potential due to a one-dimensional charge distribution.

EXAMPLE 23.5 | Finite Line of Charge

What is the electric potential at a distance d along the perpendicular bisector of a thin wire with length $2a$ and linear charge distribution λ (Figure 23.20)?

The differential electric potential, dV, at a distance d along the perpendicular bisector of the wire due to a differential charge, dq, is given by

$$dV = k\frac{dq}{r}.$$

The electric potential due to the whole wire is given by the integral over dV along the length of the wire:

$$V = \int_{-a}^{a} dV = \int_{-a}^{a} k\frac{dq}{r}. \qquad (i)$$

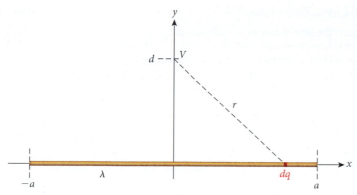

FIGURE 23.20 Calculating the electric potential due to a line of charge.

With $dq = \lambda\,dx$ and $r = \sqrt{x^2 + d^2}$, we can rewrite equation (i) as

$$V = \int_{-a}^{a} k\frac{\lambda\,dx}{\sqrt{x^2 + d^2}} = k\lambda \int_{-a}^{a} \frac{dx}{\sqrt{x^2 + d^2}}.$$

Finding this integral in a table or evaluating it with software gives

$$\int_{-a}^{a} \frac{dx}{\sqrt{x^2 + d^2}} = \left[\ln\left(x + \sqrt{x^2 + d^2} \right) \right]_{-a}^{a} = \ln\left(\frac{\sqrt{a^2 + d^2} + a}{\sqrt{a^2 + d^2} - a} \right).$$

Thus, the electric potential at a distance d along the perpendicular bisector of a finite line of charge is given by

$$V = k\lambda \ln\left(\frac{\sqrt{a^2 + d^2} + a}{\sqrt{a^2 + d^2} - a} \right).$$

23.5 Finding the Electric Field from the Electric Potential

As we mentioned earlier, we can determine the electric field starting with the electric potential. This calculation uses equations 23.8 and 23.10:

$$-q\,dV = q\vec{E}\cdot d\vec{s},$$

where $d\vec{s}$ is a vector from an initial point to a final point located a small (infinitesimal) distance away. The component of the electric field, E_s, along the direction of $d\vec{s}$ is given by the partial derivative

$$E_s = -\frac{\partial V}{\partial s}. \tag{23.14}$$

(Chapter 15 on waves applied partial derivatives, and they were treated much like conventional derivatives, which we'll continue to do here.) Thus, we can find any component of the electric field by taking the partial derivative of the potential along the direction of that component. We can then write the components of the electric field in terms of partial derivatives of the potential:

$$E_x = -\frac{\partial V}{\partial x}; \quad E_y = -\frac{\partial V}{\partial y}; \quad E_z = -\frac{\partial V}{\partial z}. \tag{23.15}$$

The equivalent vector calculus formulation is $\vec{E} = -\vec{\nabla}V \equiv -(\partial V/\partial x, \partial V/\partial y, \partial V/\partial z)$, where the operator $\vec{\nabla}$ is called the **gradient**. Thus, the electric field can be determined either graphically, by measuring the negative of the change of the potential per unit distance perpendicular to an equipotential line, or analytically, by using equation 23.15.

23.6 In-Class Exercise

Suppose an electric potential is described by $V(x, y, z) = -(5x^2 + y + z)$ in volts. Which of the following expressions describes the associated electric field, in units of volts per meter?

a) $\vec{E} = 5\hat{x} + 2\hat{y} + 2\hat{z}$

b) $\vec{E} = 10x\hat{x}$

c) $\vec{E} = 5x\hat{x} + 2\hat{y}$

d) $\vec{E} = 10x\hat{x} + \hat{y} + \hat{z}$

e) $\vec{E} = 0$

23.7 In-Class Exercise

In the figure, the lines represent equipotential lines. How does the magnitude of the electric field, E, at point P compare for the three cases?

a) $E_1 = E_2 = E_3$

b) $E_1 > E_2 > E_3$

c) $E_1 < E_2 < E_3$

d) $E_3 > E_1 > E_2$

e) $E_3 < E_1 < E_2$

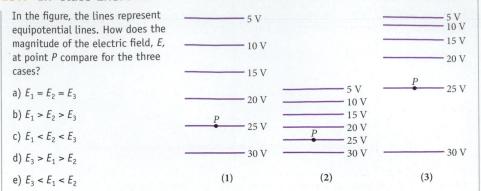

To visually reinforce the concepts of electric fields and potentials, the following example shows how a graphical technique can be used to find the field given the potential.

EXAMPLE 23.6 | Graphical Extraction of the Electric Field

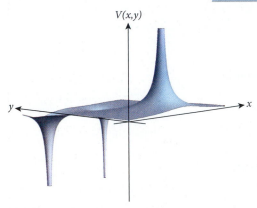

FIGURE 23.21 Electric potential due to three charges.

Let's consider a system of three point charges with values $q_1 = -6.00\ \mu C$, $q_2 = -3.00\ \mu C$, and $q_3 = +9.00\ \mu C$, located at positions $(x_1,y_1) = (1.5\ \text{cm}, 9.0\ \text{cm})$, $(x_2,y_2) = (6.0\ \text{cm}, 8.0\ \text{cm})$, and $(x_3,y_3) = (5.3\ \text{cm}, 2.0\ \text{cm})$. Figure 23.21 shows the electric potential, $V(x,y)$, resulting from these three charges, with equipotential lines calculated at potential values from -5000 V to 5000 V in 1000-V increments shown in Figure 23.22.

We can calculate the magnitude of the electric field at point P using equation 23.14 and graphical techniques. To perform this task, we use the green line in Figure 23.22, which is drawn through point P perpendicular to the equipotential line because the electric field is always perpendicular to the equipotential lines, reaching from the equipotential line of 0 V to the line of 2000 V. As you can see from Figure 23.22, the length of the green line is 1.5 cm. Therefore, the magnitude of the electric field can be approximated as

$$|E_s| = \left| -\frac{\Delta V}{\Delta s} \right| = \left| \frac{(+2000\ \text{V}) - (0\ \text{V})}{1.5\ \text{cm}} \right| = 1.3 \cdot 10^5\ \text{V/m},$$

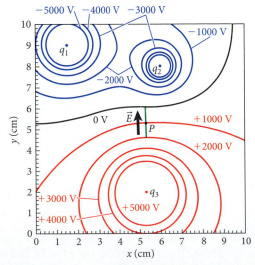

FIGURE 23.22 Equipotential lines for the electric potential due to three point charges.

where Δs is the length of the line through point P. The negative sign in equation 23.14 indicates that the direction of the electric field between neighboring equipotential lines points from the 2000-V equipotential line to the zero potential line.

In Chapter 22, we derived an expression for the electric field along the perpendicular bisector of a finite line of charge:

$$E_y = \frac{2k\lambda}{y} \frac{a}{\sqrt{y^2 + a^2}}.$$

In Example 23.5, we found an expression for the electric potential along the perpendicular bisector of a finite line of charge:

$$V = k\lambda \ln\left(\frac{\sqrt{y^2 + a^2} + a}{\sqrt{y^2 + a^2} - a}\right), \tag{23.16}$$

where the coordinate d used in Example 23.5 has been replaced with the distance in the y-direction. We can find the y-component of the electric field from the potential using equation 23.15:

$$E_y = -\frac{\partial V}{\partial y}$$

$$= -\frac{\partial\left(k\lambda \ln\left(\frac{\sqrt{y^2 + a^2} + a}{\sqrt{y^2 + a^2} - a}\right)\right)}{\partial y}$$

$$= -k\lambda\left(\frac{\partial\left(\ln\left(\sqrt{y^2 + a^2} + a\right)\right)}{\partial y} - \frac{\partial\left(\ln\left(\sqrt{y^2 + a^2} - a\right)\right)}{\partial y}\right).$$

Taking the partial derivative (remember that we can treat the partial derivate like a regular derivative), we obtain for the first term

$$\frac{\partial}{\partial y}\left(\ln\left(\sqrt{y^2 + a^2} + a\right)\right) = \underbrace{\left(\frac{1}{\sqrt{y^2 + a^2} + a}\right)}_{\text{derivative of ln}}\underbrace{\left(\frac{1}{2}\frac{1}{\sqrt{y^2 + a^2}}\right)}_{\text{derivative of }\sqrt{y^2+a^2}}\underbrace{(2y)}_{\text{derivative of }y^2} = \frac{y}{y^2 + a^2 + a\sqrt{y^2 + a^2}},$$

where the fact that the derivative of the natural log function is $d(\ln x)/dx = 1/x$ and the chain rule of differentiation have been used. (The outer and inner derivatives are indicated under the terms that they generate.) A similar expression can be found for the second term. Using the values of the derivatives, we find the component of the electric field:

$$E_y = -k\lambda\left(\frac{y}{y^2 + a^2 + a\sqrt{y^2 + a^2}} - \frac{y}{y^2 + a^2 - a\sqrt{y^2 + a^2}}\right) = \frac{2k\lambda}{y}\frac{a}{\sqrt{y^2 + a^2}}.$$

This result is the same as that for the electric field in the y-direction derived in Chapter 22 by integrating over a finite line of charge.

23.6 Electric Potential Energy of a System of Point Charges

Section 23.1 discussed the electric potential energy of a point charge in a given external electric field, and Section 23.4 described how to calculate the electric potential due to a system of point charges. This section combines these two pieces of information to find the electric

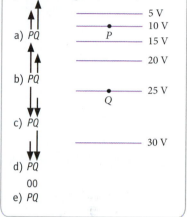

23.8 In-Class Exercise

In the figure, the lines represent equipotential lines. A positive charge is placed at point P, and then another positive charge is placed at point Q. Which set of vectors best represents the relative magnitudes and directions of the electric field forces exerted on the positive charges at P and Q?

a) PQ

b) PQ

c) PQ

d) PQ

e) PQ

23.9 In-Class Exercise

In the figure, the lines represent equipotential lines. What is the direction of the electric field at point P?

a) up

b) down

c) left

d) right

e) The electric field is zero.

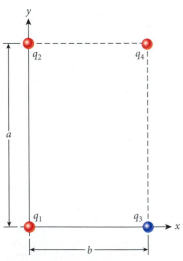

FIGURE 23.23 Two point charges separated by a distance r.

potential energy of a system of point charges. Consider a system of charges that are infinitely far apart. To bring these charges into proximity with each other, work must be done on the charges, which changes the electric potential energy of the system. The electric potential energy of a system of point charges is defined as the work required to bring the charges together from being infinitely far apart.

As an example, let's find the electric potential energy of a system of two point charges (Figure 23.23). Assume that the two charges start at an infinite separation. We then bring point charge q_1 into the system. Because the system without charges has no electric field and no corresponding electric force, this action does not require that any work be done on the charge. Keeping this charge stationary, we bring the second point charge, q_2, from infinity to a distance r from q_1. Using equation 23.6, we can write the electric potential energy of the system as

$$U = q_2 V, \qquad (23.17)$$

where

$$V = \frac{kq_1}{r}. \qquad (23.18)$$

Thus, the electric potential energy of this system of two point charges is

$$U = \frac{kq_1 q_2}{r}. \qquad (23.19)$$

From the work-energy theorem, the work, W, that must be done on the particles to bring them together and keep them stationary is equal to U. If the two charges have the same sign, $W = U > 0$, and positive work must be done to bring them together from infinity and keep them motionless. If the two charges have opposite signs, negative work must be done to bring them together from infinity and hold them motionless. To determine U for more than two point charges, we assemble them from infinity one charge at a time, in any order.

EXAMPLE 23.7 Four Point Charges

Let's calculate the electric potential energy of a system of four point charges, shown in Figure 23.24. The four point charges have the values $q_1 = +1.0\ \mu C$, $q_2 = +2.0\ \mu C$, $q_3 = -3.0\ \mu C$, and $q_4 = +4.0\ \mu C$. The charges are placed with $a = 6.0$ m and $b = 4.0$ m.

PROBLEM
What is the electric potential energy of this system of four point charges?

SOLUTION
We begin the calculation with the four charges infinitely far apart and assume that the electric potential energy is zero in that configuration. We bring in q_1 and position that charge at $(0,0)$. This action does not change the electric potential energy of the system. Now we bring in q_2 and place that charge at $(0,a)$. The electric potential energy of the system is now

$$U = \frac{kq_1 q_2}{a}.$$

FIGURE 23.24 Calculating the potential energy of a system of four point charges.

Bringing q_3 in from an infinite distance and placing it at $(b,0)$ changes the potential energy of the system through the interaction of q_3 with q_1 and the interaction of q_3 with q_2. The new potential energy is

$$U = \frac{kq_1 q_2}{a} + \frac{kq_1 q_3}{b} + \frac{kq_2 q_3}{\sqrt{a^2 + b^2}}.$$

Finally, bringing in q_4 and placing it at (b,a) changes the potential energy of the system through interactions with q_1, q_2, and q_3, bringing the total electric potential energy of the system to

$$U = \frac{kq_1 q_2}{a} + \frac{kq_1 q_3}{b} + \frac{kq_2 q_3}{\sqrt{a^2 + b^2}} + \frac{kq_1 q_4}{\sqrt{a^2 + b^2}} + \frac{kq_2 q_4}{b} + \frac{kq_3 q_4}{a}.$$

Note that the order in which the charges are brought from infinity will not change this result. (You can try a different order to verify this statement.) Putting in the numerical values, we obtain

$$U = \left(3.0 \cdot 10^{-3} \text{ J}\right) + \left(-6.7 \cdot 10^{-3} \text{ J}\right) + \left(-7.5 \cdot 10^{-3} \text{ J}\right) +$$
$$\left(5.0 \cdot 10^{-3} \text{ J}\right) + \left(1.8 \cdot 10^{-2} \text{ J}\right) + \left(-1.8 \cdot 10^{-2} \text{ J}\right) = -6.2 \cdot 10^{-3} \text{ J}.$$

From the calculation in Example 23.7, we extrapolate the result to obtain a formula for the electric potential energy of a collection of point charges:

$$U = k \sum_{ij(pairings)} \frac{q_i q_j}{r_{ij}}, \tag{23.20}$$

where i and j label each pair of charges, the summation is over each pair ij (for all $i \neq j$), and r_{ij} is the distance between the charges in each pair. An alternative way to write this double sum is

$$U = \tfrac{1}{2} k \sum_{j=1}^{n} \sum_{i=1, i \neq j}^{n} \frac{q_i q_j}{\left| \vec{r}_i - \vec{r}_j \right|},$$

which is more explicit than the equivalent formulation of equation 23.20.

WHAT WE HAVE LEARNED | EXAM STUDY GUIDE

- The change in the electric potential energy, ΔU, of a point charge moving in an electric field is equal to the negative of the work done on the point charge by the electric field W_e: $\Delta U = U_f - U_i = -W_e$.

- The change in electric potential energy, ΔU, is equal to the charge, q, times the change in electric potential, ΔV: $\Delta U = q \Delta V$.

- Equipotential surfaces and equipotential lines represent locations in space that have the same electric potential. Equipotential surfaces are always perpendicular to the electric field lines.

- A surface of a conductor is an equipotential surface.

- The change in electric potential can be determined from the electric field by integrating over the field:

  $\Delta V = -\int_i^f \vec{E} \cdot d\vec{s}$. Setting the potential equal to zero at

  infinity gives $V = \int_i^\infty \vec{E} \cdot d\vec{s}$.

- The electric potential due to a point charge, q, at a distance r from the charge is given by $V = \dfrac{kq}{r}$.

- The electric potential due to a system of n point charges can be expressed as an algebraic sum of the individual potentials: $V = \sum_{i=1}^{n} V_i$.

- The electric field can be determined from gradients of the electric potential in each component direction:

  $E_x = -\dfrac{\partial V}{\partial x}, E_y = -\dfrac{\partial V}{\partial y}, E_z = -\dfrac{\partial V}{\partial z}.$

- The electric potential energy of a system of two point charges is given by $U = \dfrac{kq_1 q_2}{r}$.

KEY TERMS

electric potential energy, p. 746
electric potential, p. 747
volt, p. 748

electron-volt, p. 749
Van de Graaff generator, p. 751
equipotential surface, p. 752

equipotential lines, p. 752
gradient, p. 759

NEW SYMBOLS

V, electric potential

ΔV, electric potential difference

eV, abbreviation for electron-volt, a unit of energy

ANSWERS TO SELF-TEST OPPORTUNITIES

23.1 The electric potential along the y-axis is zero.

23.2 $(x,y) = (0,0)$ corresponds to a saddle point.

23.3 Nothing would change. The electrostatic force is conservative, and for a conservative force, the work is path-independent.

PROBLEM-SOLVING PRACTICE

Problem-Solving Guidelines

1. A common source of error in calculations is confusing the electric field, $\vec{E}$, the electric potential energy, U, and the electric potential, V. Remember that an electric field is a vector quantity produced by a charge distribution; electric potential energy is a property of the charge distribution; and electric potential is a property of the field. Be sure you know what it is that you're calculating.

2. Be sure to identify the point with respect to which you are calculating the potential energy or potential. Like calculations involving electric fields, calculations involving potentials can use a linear charge distribution (λ), a planar charge distribution (σ), or a volume charge distribution (ρ).

3. Since potential is a scalar, the total potential due to a system of point charges is calculated by simply adding the individual potentials due to all the charges. For a continuous charge distribution, you need to calculate the potential by integrating over the differential charge. Assume that the potential produced by the differential charge is the same as the potential from a point charge!

SOLVED PROBLEM 23.2 | Beam of Oxygen Ions

PROBLEM

Fully stripped (all electrons removed) oxygen (^{16}O) ions are accelerated from rest in a particle accelerator using a total potential difference of 10.0 MV = $1.00 \cdot 10^7$ V. The ^{16}O nucleus has 8 protons and 8 neutrons. The accelerator produces a beam of $3.13 \cdot 10^{12}$ ions per second. This ion beam is completely stopped in a beam dump. What is the total power the beam dump has to absorb?

SOLUTION

THINK

Power is energy per unit time. We can calculate the energy of each ion and then the total energy in the beam per unit time to obtain the power dissipated in the beam dump.

SKETCH

Figure 23.25 illustrates a beam of fully stripped oxygen ions being stopped in a beam dump.

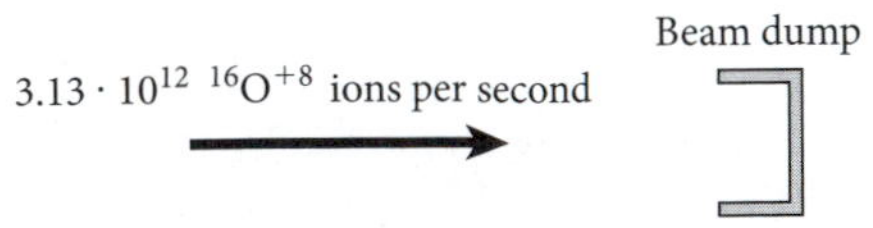

Beam dump

$3.13 \cdot 10^{12}$ ^{16}O^{+8} ions per second

FIGURE 23.25 A beam of fully stripped oxygen ions stops in a beam dump.

RESEARCH

The electric potential energy gained by each ion during the acceleration process is

$$U_{\text{ion}} = q\Delta V = ZeV,$$

where $Z = 8$ is the atomic number of oxygen, $e = 1.602 \cdot 10^{-19}$ C is the charge of a proton, and $V = 1.00 \cdot 10^7$ V is the electric potential across which the ions are accelerated.

SIMPLIFY

The power of the beam, which is dissipated in the beam dump, is then

$$P = NU_{\text{ion}} = NZeV,$$

where $N = 3.13 \cdot 10^{12}$ ions/s is the number of ions per second stopped in the beam dump.

CALCULATE

Putting in the numerical values, we get

$$P = NZeV = \left(3.13 \cdot 10^{12} \text{ s}^{-1}\right)\left(8\right)\left(1.602 \cdot 10^{-19} \text{ C}\right)\left(1.00 \cdot 10^{7} \text{ V}\right)$$

$$= 40.1141 \text{ W}.$$

ROUND

We report our result to three significant figures:

$$P = 40.1 \text{ W}.$$

DOUBLE-CHECK

We can relate the change in kinetic energy for each ion to the change in electric potential energy of each ion:

$$\Delta K = \Delta U = \tfrac{1}{2}mv^2 = U_{\text{ion}} = ZeV.$$

The mass of an oxygen nucleus is $2.66 \cdot 10^{-26}$ kg. The velocity of each ion is then

$$v = \sqrt{\frac{2ZeV}{m}} = \sqrt{\frac{2\left(8\right)\left(1.602 \cdot 10^{-19} \text{ C}\right)\left(10^{7}\right)}{2.66 \cdot 10^{-26} \text{ kg}}} = 3.10 \cdot 10^{7} \text{ m/s},$$

which is about 10% of the speed of light, which seems reasonable for the velocity of the ions. Thus, our result seems reasonable.

SOLVED PROBLEM 23.3 | Minimum Potential

PROBLEM

A charge of $q_1 = 0.829$ nC is placed at $r_1 = 0$ on the x-axis. Another charge of $q_2 = 0.275$ nC is placed at $r_2 = 11.9$ cm on the x-axis. At which point along the x-axis between the two charges does the electric potential resulting from both of them have a minimum?

SOLUTION

THINK

We can express the electric potential due to the two charges as the sum of the electric potential from each charge. To obtain the minimum potential, we take the derivative of the potential and set it equal to zero. We can then solve for the distance where the derivative is zero.

SKETCH

Figure 23.26 shows the locations of the two charges.

RESEARCH

We can express the electric potential produced along the x-axis by the two charges as

$$V = V_1 + V_2 = k\frac{q_1}{x - r_1} + k\frac{q_2}{r_2 - x} = k\frac{q_1}{x} + k\frac{q_2}{r_2 - x}.$$

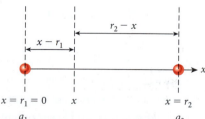

FIGURE 23.26 Two charges placed along the x-axis.

Note that the quantities x and $r_2 - x$ are always positive for $0 < x < r_2$. To find the minimum, we take the derivative of the electric potential:

$$\frac{dV}{dx} = -k\frac{q_1}{x^2} - k\frac{q_2}{\left(r_2 - x\right)^2}\left(-1\right) = k\frac{q_2}{\left(r_2 - x\right)^2} - k\frac{q_1}{x^2}.$$

Continued—

SIMPLIFY

Setting the derivative of the electric potential equal to zero and rearranging, we obtain

$$k\frac{q_2}{(r_2 - x)^2} = k\frac{q_1}{x^2}.$$

Dividing out k and rearranging, we get

$$\frac{x^2}{(r_2 - x)^2} = \frac{q_1}{q_2}.$$

Now we can take the square root and rearrange:

$$x = \pm(r_2 - x)\sqrt{\frac{q_1}{q_2}}.$$

Because $x > 0$ and $(r_2 - x) > 0$, the sign must be positive. Solving for x, we get

$$x = \frac{r_2\sqrt{\dfrac{q_1}{q_2}}}{1 + \sqrt{\dfrac{q_1}{q_2}}} = \frac{r_2}{\sqrt{\dfrac{q_2}{q_1}} + 1}.$$

CALCULATE

Putting in the numerical values results in

$$x = \frac{0.119\ \text{m}}{1 + \sqrt{\dfrac{0.275\ \text{nC}}{0.829\ \text{nC}}}} = 0.0755097\ \text{m}.$$

ROUND

We report our result to three significant figures:

$$x = 0.0755\ \text{m} = 7.55\ \text{cm}.$$

DOUBLE-CHECK

We can double-check our result by plotting (for example, with a graphing calculator) the electric potential resulting from the two charges and graphically determining the minimum (Figure 23.27).

The minimum of the electric potential is located at $x = 7.55$ cm, which confirms our calculated result.

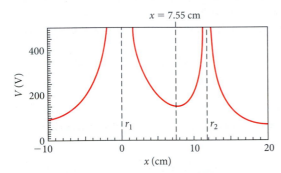

FIGURE 23.27 Graph of the electric potential resulting from two charges.

MULTIPLE-CHOICE QUESTIONS

23.1 A positive charge is released and moves along an electric field line. This charge moves to a position of

a) lower potential and lower potential energy.

b) lower potential and higher potential energy.

c) higher potential and lower potential energy.

d) higher potential and higher potential energy.

23.2 A proton is placed midway between points A and B. The potential at point A is −20 V, and the potential at point B +20 V. The potential at the midpoint is 0 V. The proton will

a) remain at rest.

b) move toward point B with constant velocity.

c) accelerate toward point A.

d) accelerate toward point B.

e) move toward point A with constant velocity.

23.3 What would be the consequence of setting the potential at +100 V at infinity, rather than taking it to be zero there?

a) Nothing; the field and the potential would have the same values at every finite point.

b) The electric potential would become infinite at every finite point, and the electric field could not be defined.

c) The electric potential everywhere would be 100 V higher, and the electric field would be the same.

d) It would depend on the situation. For example, the potential due to a positive point charge would drop off more slowly with distance, so the magnitude of the electric field would be less.

23.4 In which situation is the electric potential the highest?

a) at a point 1 m from a point charge of 1 C

b) at a point 1 m from the center of a uniformly charged spherical shell of radius 0.5 m with a total charge of 1 C

c) at a point 1 m from the center of a uniformly charged rod of length 1 m and with a total charge of 1 C

d) at a point 2 m from a point charge of 2 C

e) at a point 0.5 m from a point charge of 0.5 C

23.5 The amount of work done to move a positive point charge q on an equipotential surface of 1000 V relative to that on an equipotential surface of 10 V is

a) the same.

b) less.

c) more.

d) dependent on the distance the charge moves.

23.6 A solid conducting sphere of radius R is centered about the origin of an *xyz*-coordinate system. A total charge Q is distributed uniformly on the surface of the sphere. Assuming, as usual, that the electric potential is zero at an infinite distance, what is the electric potential at the center of the conducting sphere?

a) zero

b) $Q/\epsilon_0 R$

c) $Q/2\pi\epsilon_0 R$

d) $Q/4\pi\epsilon_0 R$

23.7 Which of the following angles between an electric dipole moment and an applied electric field will result in the most stable state?

a) 0 rad

b) $\pi/2$ rad

c) π rad

d) The electric dipole moment is not stable under any condition in an applied electric field.

23.8 A positive point charge is to be moved from point A to point B in the vicinity of an electric dipole. Which of the three paths shown in the figure will result in the most work being done by the dipole's electric field on the point charge?

a) path 1

b) path 2

c) path 3

d) The work is the same on all three paths.

23.9 Each of the following pairs of charges are separated by a distance d. Which pair has the highest potential energy?

a) +5 C and +3 C

b) +5 C and −3 C

c) −5 C and −3 C

d) −5 C and +3 C

e) All pairs have the same potential energy.

23.10 A negatively charged particle revolves in a clockwise direction around a positively charged sphere. The work done on the negatively charged particle by the electric field of the sphere is

a) positive.

b) negative.

c) zero.

QUESTIONS

23.11 High-voltage power lines are used to transport electricity cross country. These wires are favored resting places for birds. Why don't the birds die when they touch the wires?

23.12 You have heard that it is dangerous to stand under trees in electrical storms. Why?

23.13 Can two equipotential lines cross? Why or why not?

23.14 Why is it important, when soldering connectors onto a piece of electronic circuitry, to leave no pointy protrusions from the solder joints?

23.15 Using Gauss's Law and the relation between electric potential and electric field, show that the potential outside a uniformly charged sphere is identical to the potential of a point charge placed at the center of the sphere and equal to the total charge of the sphere. What is the potential at the surface of the sphere? How does the potential change if the charge distribution is not uniform but has spherical (radial) symmetry?

23.16 A metal ring has a total charge q and radius R, as shown in the figure. Without performing any calculations, predict the value

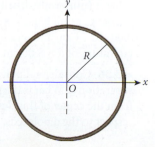

of the electric potential and electric field at the center of the circle.

23.17 Find an integral expression for the electric potential at a point on the *z*-axis a distance H from a half-disk of radius R (see the figure). The half-disk has uniformly distributed charge over its surface, with charge distribution σ.

23.18 An electron moves away from a proton. Describe how the potential it encounters changes. Describe how its potential energy is changing.

23.19 The electric potential energy of a continuous charge distribution can be found in a way similar to that used for systems of point charges in Section 23.6, by breaking the distribution up into suitable pieces. Find the electric potential energy of an *arbitrary* spherically symmetrical charge distribution, $\rho(r)$. Do **not** assume that $\rho(r)$ represents a point charge, that it is constant, that it is piecewise-constant, or that it does or does not end at any finite radius, r. Your expression must cover all possibilities. Your expression may include an integral or integrals that cannot be evaluated without knowing the specific form of $\rho(r)$. (*Hint:* A spherical pearl is built up of thin layers of nacre added one by one.)

PROBLEMS

A blue problem number indicates a worked-out solution is available in the Student Solutions Manual. One • and two •• indicate increasing level of problem difficulty.

Section 23.1

23.20 In molecules of gaseous sodium chloride, the chloride ion has one more electron than proton, and the sodium ion has one more proton than electron. These ions are separated by about 0.24 nm. How much work would be required to increase the distance between these ions to 1.0 cm?

•23.21 A metal ball with a mass of $3 \cdot 10^{-6}$ kg and a charge of +5 mC has a kinetic energy of $6 \cdot 10^8$ J. It is traveling directly at an infinite plane of charge with a charge distribution of +4 C/m^2. If it is currently 1 m away from the plane of charge, how close will it come to the plane before stopping?

Section 23.2

23.22 An electron is accelerated from rest through a potential difference of 370 V. What is its final speed?

23.23 How much work would be done by an electric field in moving a proton from a point at a potential of +180 V to a point at a potential of –60 V?

23.24 What potential difference is needed to give an alpha particle (composed of 2 protons and 2 neutrons) 200 keV of kinetic energy?

23.25 A proton, initially at rest, is accelerated through a potential difference of 500 V. What is its final velocity?

23.26 A 10-V battery is connected to two parallel metal plates placed in a vacuum. An electron is accelerated from rest from the negative plate toward the positive plate.

a) What kinetic energy does the electron have just as it reaches the positive plate?

b) What is the speed of the electron just as it reaches the positive plate?

•23.27 A proton gun fires a proton from midway between two plates, A and B, which are separated by a distance of 10.0 cm; the proton initially moves at a speed of 150.0 km/s toward plate B. Plate A is kept at zero potential, and plate B at a potential of 400.0 V.

a) Will the proton reach plate B?

b) If not, where will it turn around?

c) With what speed will it hit plate A?

•23.28 Fully stripped (all electrons removed) sulfur (^{32}S) ions are accelerated in an accelerator from rest using a total voltage of $1.00 \cdot 10^9$ V. ^{32}S has 16 protons and 16 neutrons. The accelerator produces a beam consisting of $6.61 \cdot 10^{12}$ ions per second. This beam of ions is completely stopped in a beam dump. What is the total power the beam dump has to absorb?

Section 23.4

23.29 Two point charges are located at two corners of a rectangle, as shown in the figure.

a) What is the electric potential at point A?

b) What is the potential difference between points A and B?

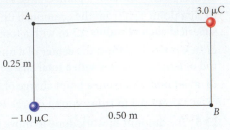

23.30 Four identical point charges (+1.61 nC) are placed at the corners of a rectangle, which measures 3.00 m by 5.00 m. If the electric potential is taken to be zero at infinity, what is the potential at the geometric center of this rectangle?

23.31 If a Van de Graff generator has an electric potential of 100,000 V and a diameter of 20 cm, find how many more protons than electrons are on its surface.

23.32 One issue encountered during the exploration of Mars has been the accumulation of static charge on land-roving vehicles, resulting in a potential of 100 V or more. Calculate how much charge must be placed on the surface of a sphere of radius 1.0 m for the electric potential just above the surface to be 100 V. Assume that the charge is uniformly distributed.

23.33 A charge $Q = +5.60$ µC is uniformly distributed on a thin cylindrical plastic shell. The radius, R, of the shell is 4.50 cm. Calculate the electric potential at the origin of the xy-coordinate system shown in the figure. Assume that the electric potential is zero at points infinitely far away from the origin.

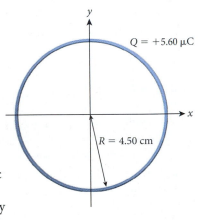

23.34 A hollow spherical conductor with a 5.0-cm radius has a surface charge of 8.0 nC.

a) What is the potential 8.0 cm from the center of the sphere?

b) What is the potential 3.0 cm from the center of the sphere?

c) What is the potential at the center of the sphere?

23.35 Find the potential at the center of curvature of the (thin) wire shown in the figure. It has a (uniformly distributed) charge per unit length of $\lambda = 3.00 \cdot 10^{-8}$ C/m and a radius of curvature of $R = 8.00$ cm.

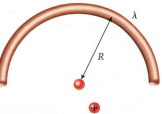

•23.36 Consider a dipole with charge q and separation d. What is the potential a distance x from the center of this dipole at an angle θ with respect to the dipole axis, as shown in the figure?

•**23.37** A spherical water drop 50 μm in diameter has a uniformly distributed charge of +20 pC. Find (a) the potential at its surface and (b) the potential at its center.

•**23.38** Consider an electron in the ground state of the hydrogen atom, separated from the proton by a distance of 0.0529 nm.

a) Viewing the electron as a satellite orbiting the proton in the electrostatic potential, calculate the speed of the electron in its orbit.

b) Calculate an effective escape speed for the electron.

c) Calculate the energy of an electron having this speed, and from it determine the energy that must be given to the electron to ionize the hydrogen atom.

•**23.39** Four point charges are arranged in a square with side length $2a$, where $a = 2.7$ cm. Three of the charges have magnitude 1.5 nC, and one of them has magnitude -1.5 nC, as shown in the figure. What is the value of the electric potential generated by these four point charges at point $P = (0,0,c)$, where $c = 4.1$ cm?

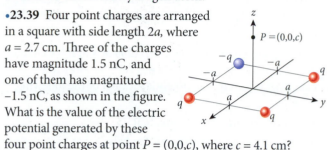

•**23.40** The plastic rod of length L shown in the figure has the nonuniform linear charge distribution $\lambda = cx$, where c is a positive constant. Find an expression for the electric potential at point P on the y-axis, a distance y from one end of the rod.

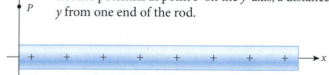

••**23.41** An electric field varies in space according to this equation: $\vec{E} = E_0 x e^{-x} \hat{x}$.

a) For what value of x does the electric field have its largest value, x_{max}?

b) What is the potential difference between the points at $x = 0$ and $x = x_{max}$?

••**23.42** Derive an expression for electric potential along the axis (the x-axis) of a disk with a hole in the center, as shown in the figure, where R_1 and R_2 are the inner and outer radii of the disk. What would the potential be if $R_1 = 0$?

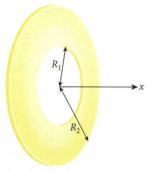

Section 23.5

23.43 An electric field is established in a nonuniform rod. A voltmeter is used to measure the potential difference between the left end of the rod and a point a distance x from the left end. The process is repeated, and it is found that the data are described by the relationship $\Delta V = 270x^2$, where ΔV has the units V/m². What is the x-component of the electric field at a point 13 cm from the left end?

23.44 Two parallel plates are held at potentials of +200.0 V and -100.0 V. The plates are separated by 1.00 cm.

a) Find the electric field between the plates.

b) An electron is initially placed halfway between the plates. Find its kinetic energy when it hits the positive plate.

23.45 A 2.50-mg dust particle with a charge of 1.00 μC falls at a point $x = 2.00$ m in a region where the electric potential varies according to $V(x) = (2.00 \text{ V/m}^2)x^2 - (3.00 \text{ V/m}^3)x^3$. With what acceleration will the particle start moving after it touches down?

23.46 The electric potential in a volume of space is given by $V(x, y, z) = x^2 + xy^2 + yz$. Determine the electric field in this region at the coordinate (3,4,5).

•**23.47** The electric potential inside a 10-m-long linear particle accelerator is given by $V = (3000 - 5x^2/\text{m}^2)$ V, where x is the distance from the left plate along the accelerator tube, as shown in the figure.

a) Determine an expression for the electric field along the accelerator tube.

b) A proton is released (from rest) at $x = 4$ m. Calculate the acceleration of the proton just after it is released.

c) What is the impact speed of the proton when (and if) it collides with the plate?

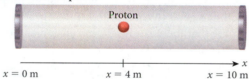

•**23.48** An infinite plane of charge has a uniform charge distribution of +4 nC/m² and is located in the yz-plane at $x = 0$. A +11 nC fixed point charge is located at $x = +2$ m.

a) Find the electric potential $V(x)$ on the x-axis from $0 < x < +2$ m.

b) At what position(s) on the x-axis between $x = 0$ and $x = +2$ m is the electric potential a minimum?

c) Where on the x-axis between $x = 0$ m and $x = +2$ m could a positive point charge be placed and not move?

•**23.49** Use $V = \dfrac{kq}{r}$, $E_x = -\dfrac{\partial V}{\partial x}$, $E_y = -\dfrac{\partial V}{\partial y}$, and $E_z = -\dfrac{\partial V}{\partial z}$ to derive the expression for the electric field of a point charge, q.

•**23.50** Show that an electron in a one-dimensional electrical potential $V(x) = Ax^2$, where the constant A is a positive real number, will execute simple harmonic motion about the origin. What is the period of that motion?

••**23.51** The electric field, $\vec{E}(\vec{r})$, and the electric potential, $V(\vec{r})$, are calculated from the charge distribution, $\rho(\vec{r})$, by integrating Coulomb's Law and then the electric field. In the other direction, the field and the charge distribution are determined from the potential by suitably differentiating. Suppose the electric potential in a large region of space is given by $V(r) = V_0 \exp(-r^2/a^2)$, where V_0 and a are constants and $r = \sqrt{x^2 + y^2 + z^2}$ is the distance from the origin.

a) Find the electric field $\vec{E}(\vec{r})$ in this region.

b) Determine the charge density $\rho(\vec{r})$ in this region, which gives rise to the potential and field.

c) Find the total charge in this region.

d) Roughly sketch the charge distribution that could give rise to such an electric field.

•• **22.52** The electron beam emitted by an electron gun is controlled (steered) with two sets of parallel conducting plates: a horizontal set to control the vertical motion of the beam, and a vertical set to control the horizontal motion of the beam. The beam is emitted with an initial velocity of $2 \cdot 10^7$ m/s. The width of the plates is $d = 5$ cm, the separation between the plates is $D = 4$ cm, and the distance between the edge of the plates and a target screen is $L = 40$ cm. In the absence of any applied voltage, the electron beam hits the origin of the xy-coordinate system on the observation screen. What voltages need to be applied to the two sets of plates for the electron beam to hit a target placed on the observation screen at coordinates $(x,y) = (0 \text{ cm}, 8 \text{ cm})$?

Section 23.6

23.53 Nuclear fusion reactions require that positively charged nuclei be brought into close proximity, against the electrostatic repulsion. As a simple example, suppose a proton is fired at a second, stationary proton from a large distance away. What kinetic energy must be given to the moving proton to get it to come within $1.00 \cdot 10^{-15}$ m of the target? Assume that there is a head-on collision and that the target is fixed in place.

23.54 Fission of a uranium nucleus (containing 92 protons) produces a barium nucleus (56 protons) and a krypton nucleus (36 protons). The fragments fly apart as a result of electrostatic repulsion; they ultimately emerge with a total of 200. MeV of kinetic energy. Use this information to estimate the size of the uranium nucleus; that is, treat the barium and krypton nuclei as point charges and calculate the separation between them at the start of the process.

23.55 A deuterium ion and a tritium ion each have charge $+e$. What work is necessary to be done on the deuterium ion in order to bring it within 10^{-14} m of the tritium ion? This is the distance within which the two ions can fuse, as a result of strong nuclear interactions that overcome electrostatic repulsion, to produce a helium-5 nucleus. Express the work in electron-volts.

• **23.56** Three charges, q_1, q_2, and q_3, are located at the corners of an equilateral triangle with side length of 1.2 m. Find the work done in each of the following cases:

a) to bring the first particle, $q_1 = 1.0$ pC, to P from infinity

b) to bring the second particle, $q_2 = 2.0$ pC, to Q from infinity

c) to bring the last particle, $q_3 = 3.0$ pC, to R from infinity

d) Find the total potential energy stored in the final configuration of q_1, q_2, and q_3.

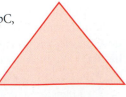

• **23.57** Two metal balls of mass $m_1 = 5$ g (diameter = 5 mm) and $m_2 = 8$ g (diameter = 8 mm) have positive charges of $q_1 = 5$ nC and $q_2 = 8$ nC, respectively. A force holds them in place so that their centers are separated by 8 mm. What will their velocities be after the force is removed and they are separated by a large distance?

Additional Problems

23.58 Two protons at rest and separated by 1.00 mm are released simultaneously. What is the speed of either at the instant when the two are 10.0 mm apart?

23.59 A 12-V battery is connected between a hollow metal sphere with a radius of 1 m and a ground, as shown in the figure. What are the electric field and the electric potential inside the hollow metal sphere?

23.60 A solid metal ball with a radius of 3 m has a charge of 4 mC. If the electric potential is zero far away from the ball, what is the electric potential at each of the following positions?

a) at $r = 0$ m, the center of the ball

b) at $r = 3$ m, on the surface of the ball

c) at $r = 5$ m

23.61 An insulating sheet in the xz-plane is uniformly charged with a charge distribution $\sigma = 3.5 \cdot 10^{-6}$ C/m^2. What is the change in potential when a charge of $Q = 1.25$ μC is moved from position A to position B in the figure?

Side view

23.62 Suppose that an electron inside a cathode ray tube starts from rest and is accelerated by the tube's voltage of 21.9 kV. What is the speed (in km/s) with which the electron (mass = $9.11 \cdot 10^{-31}$ kg) hits the screen of the tube?

23.63 A conducting solid sphere (radius of $R = 18$ cm, charge of $q = 6.1 \cdot 10^{-6}$ C) is shown in the figure. Calculate the electric potential at a point 24 cm from the center (point A), a point on the surface (point B), and at the center of the sphere (point C). Assume that the electric potential is zero at points infinitely far away from the origin of the coordinate system.

23.64 A classroom Van de Graaff generator accumulates a charge of $1.00 \cdot 10^{-6}$ C on its spherical conductor, which has a radius of 10.0 cm and stands on an insulating column.

Neglecting the effects of the generator base or any other objects or fields, find the potential at the surface of the sphere. Assume that the potential is zero at infinity.

23.65 A Van de Graaff generator has a spherical conductor with a radius of 25 cm. It can produce a maximum electric field of $2 \cdot 10^6$ V/m. What are the maximum voltage and charge that it can hold?

23.66 A proton with a speed of $1.23 \cdot 10^4$ m/s is moving from infinity directly toward a second proton. Assuming that the second proton is fixed in place, find the position where the moving proton stops momentarily before turning around.

23.67 Two metal spheres of radii $r_1 = 10$ cm and $r_2 = 20$ cm, respectively, have been positively charged so that both have a total charge of 100 μC.

a) What is the ratio of their surface charge distributions?

b) If the two spheres are connected by a copper wire, how much charge flows through the wire before the system reaches equilibrium?

23.68 The solid metal sphere of radius $a = 0.2$ m shown in the figure has a surface charge distribution of σ. The potential difference between the surface of the sphere and a point P at a distance $r_P = 0.5$ m from the center of the sphere is $\Delta V = V_{surface} - V_P = +4\pi$ V $= +12.566$ V. Determine the value of σ.

23.69 A particle with a charge of +5.0 μC is released from rest at a point on the x-axis, where $x = 0.10$ m. It begins to move as a result of the presence of a +9.0-μC charge that remains fixed at the origin. What is the kinetic energy of the particle at the instant it passes the point $x = 0.20$ m?

23.70 The sphere in the figure has a radius of 2.0 mm and carries a +2.0-μC charge uniformly distributed throughout its volume. What is the potential difference, $V_B - V_A$, if the angle between the two radii to points A and B is 60°? Is the potential difference dependent on the angle? Would the answer be the same if the charge distribution had an angular dependence, $\rho = \rho(\theta)$?

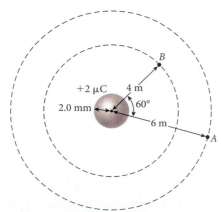

•**23.71** Two metallic spheres have radii of 10 cm and 5 cm, respectively. The magnitude of the electric field on the surface of each sphere is 3600 V/m. The two spheres are then connected by a long, thin metal wire. Determine the magnitude of the electric field on the surface of each sphere when they are connected.

•**23.72** A ring with charge Q and radius R is in the yz-plane and centered on the origin. What is the electric potential a distance x above the center of the ring? Derive the electric field from this relationship.

•**23.73** A charge of 0.681 nC is placed at $x = 0$. Another charge of 0.167 nC is placed at $x_1 = 10.9$ cm on the x-axis.

a) What is the combined electrostatic potential of these two charges at $x = 20.1$ cm, also on the x-axis?

b) At which point(s) on the x-axis does this potential have a minimum?

•**23.74** A point charge of +2.0 μC is located at (2.5 m,3.2 m). A second point charge of –3.1 μC is located at (–2.1 m,1.0 m).

a) What is the electric potential at the origin?

b) Along a line passing through both point charges, at what point(s) is (are) the electric potential(s) equal to zero?

•**23.75** A total charge of $Q = 4.2 \cdot 10^{-6}$ C is placed on a conducting sphere (sphere 1) of radius $R = 0.40$ m.

a) What is the electric potential, V_1, at the surface of sphere 1 assuming that the potential infinitely far away from it is zero? (*Hint:* What is the change in potential if a charge is brought from infinitely far away, where $V(\infty) = 0$, to the surface of the sphere?)

b) A second conducting sphere (sphere 2) of radius $r = 0.10$ m with an initial net charge of zero ($q = 0$) is connected to sphere 1 using a long thin metal wire. How much charge flows from sphere 1 to sphere 2 to bring them into equilibrium? What are the electric fields at the surfaces of the two spheres?

•**23.76** A thin line of charge is aligned along the positive y-axis from $0 \leq y \leq L$, with $L = 4.0$ cm. The charge is not uniformly distributed but has a charge per unit length of $\lambda = Ay$, with $A = 8.0 \cdot 10^{-7}$ C/m². Assuming that the electric potential is zero at infinite distance, find the electric potential at a point on the x-axis as a function of x. Give the value of the electric potential at $x = 3.0$ cm.

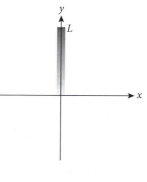

•**23.77** Two fixed point charges are on the x-axis. A charge of –3 mC is located at $x = +2$ m and a charge of +5 mC is located at $x = -4$ m.

a) Find the electric potential, $V(x)$, for an arbitrary point on the x-axis.

b) At what position(s) on the x-axis is $V(x) = 0$?

c) Find $E(x)$ for an arbitrary point on the x-axis.

•**23.78** One of the greatest physics experiments in history measured the charge-to-mass ratio of an electron, q/m. If a uniform potential difference is created between two plates, atomized particles—each with an integral amount of charge—can be suspended in space. The assumption is that the particles of unknown mass, M, contain a net number, n, of electrons of mass m and charge q. For a plate separation of d, what is the potential difference necessary to suspend a particle of mass M containing n net electrons? What is the acceleration of the particle if the voltage is cut in half? What is the acceleration of the particle if the voltage is doubled?

•**23.79** A uniform linear charge distribution of total positive charge Q has the shape of a half-circle of radius R, as shown in the figure.

a) Without performing any calculations, predict the electric potential produced by this linear charge distribution at point O.

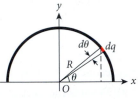

b) Confirm, through direct calculations, your prediction of part (a).

c) Make a similar prediction for the electric field.

••**23.80** A point charge Q is placed a distance R from the center of a conducting sphere of radius a, with $R > a$ (the point charge is outside the sphere). The sphere is grounded, that is, connected to a distant, unlimited source and/or sink of charge at zero potential. (Neither the distant ground nor the connection directly affects the electric field in the vicinity of the charge and sphere.) As a result, the sphere acquires a charge opposite in sign to Q, and the point charge experiences an attractive force toward the sphere.

a) Remarkably, the electric field *outside the sphere* is the same as would be produced by the point charge Q plus an imaginary *mirror-image* point charge q, with magnitude and location that make the set of points corresponding to the surface of the sphere an equipotential of potential zero. That is, the imaginary point charge produces the same field contribution outside the sphere as the actual surface charge on the sphere. Calculate the value and location of q. (*Hint:* By symmetry, q must lie somewhere on the axis that passes through the center of the sphere and the location of Q.)

b) Calculate the force exerted on point charge Q and directed toward the sphere, in terms of the original quantities Q, R, and a.

c) Determine the actual nonuniform surface charge distribution on the conducting sphere.

Capacitors

24

FIGURE 24.1 Interacting with the touch screen of an iPhone.

WHAT WE WILL LEARN

- Capacitors usually consist of two separated conductors or conducting plates.

- A capacitor can store charge on one plate, and there is typically an equal and opposite charge on the other plate.

- The capacitance of a capacitor is the charge stored on the plates divided by the resulting electric potential difference.

- A capacitor can store electric potential energy.

- A common type of capacitor is the parallel plate capacitor, consisting of two flat parallel conducting plates.

- The capacitance of a given capacitor depends on its geometry.

- In a circuit, capacitors wired in parallel or in series can be replaced by an equivalent capacitance.

- The capacitance of a given capacitor is increased when a dielectric material is placed between the plates.

- A dielectric material reduces the electric field between the plates of a capacitor as a result of the alignment of molecular dipole moments in the dielectric material.

Touch screens, such as the one shown in Figure 24.1, have become very common, found on everything from computer screens to cell phones to voting machines. They work in several ways, one of which involves using a property of conductors called *capacitance,* which we'll study in this chapter. Capacitance appears whenever two conductors—any two conductors—are separated by a small distance. The contact of a finger with a touch screen causes a change in capacitance that can be detected.

Capacitors have the very useful capability of storing electric charge and then releasing it very quickly. Thus, they are useful in camera flash attachments, cardiac defibrillators, and even experimental fusion reactors—anything that needs a large electric charge delivered quickly. Most circuits of any kind contain at least one capacitor. However, capacitance has a downside: It can also appear where it's not wanted, for example, between neighboring conductors in a tiny electronics circuit, where it can create "cross-talk"—unwanted interference between circuit components.

Because capacitors are one of the basic elements of electric circuits, this chapter examines how they function in simple circuits. The next two chapters will cover additional basic circuit elements and their uses.

24.1 Capacitance

Figure 24.2, shows that capacitors come in a variety of sizes and shapes. In general, a **capacitor** consists of two separated conductors, which are usually called *plates* even if they are not simple planes. If we take apart one of these capacitors, we might find two sheets of metal foil separated by an insulating layer of Mylar, as shown in Figure 24.3. The sandwiched layers of metal foil and Mylar can be rolled up with another insulating layer into a compact form that does not resemble two parallel conductors, as shown in Figure 24.4. This technique produces capacitors with some of the physical formats shown in Figure 24.2. The insulating layer between the two metal foils plays a crucial role in the characteristics of the capacitor.

To study the properties of capacitors, we'll assume a convenient geometry and then generalize the results. Figure 24.5 shows a **parallel plate capacitor,** which consists of two parallel conducting plates, each with area A, separated by a distance, d, and assumed to be in a vacuum. The capacitor is charged by placing a charge of $+q$ on one plate and a charge of $-q$ on the other plate. (It is not necessary to put exactly opposite charges onto the two plates of the capacitor to charge it; any difference in charge will do. But, for practical purposes, the overall device should remain neutral, and this requires charges of equal magnitude and opposite sign on the two plates.) Because the plates are conductors, they are equipotential surfaces; thus, the electrons on the plates will distribute themselves uniformly over the surfaces.

Let's apply the results obtained in Chapter 23 to determine the electric potential and electric field for the parallel plate capacitor. (In principle, we could do this by calculating the electric potential and

FIGURE 24.2 Some representative types of capacitors.

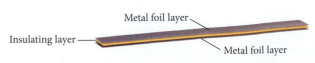

Metal foil layer

Insulating layer

Metal foil layer

FIGURE 24.3 Two sheets of metal foil separated by an insulating layer.

electric field for continuous charge distributions. However, for this physical configuration we would need to use a computer to provide the solution.) Let's place the origin of the coordinate system in the middle between the two plates, with the x-axis aligned with the two plates. Figure 24.6 shows a three-dimensional plot of the electric potential, $V(x,y)$, in the xy-plane, similar to the plots in Chapter 23.

The potential in Figure 24.6 has a very steep (and approximately linear) drop between the two plates and a more gradual drop outside the plates. This means that the electric field can be expected to be strongest between the plates and weaker outside. Figure 24.7a presents a contour plot of the electric potential shown in Figure 24.6 for the two parallel plates. Negative potential values are shaded in green, and positive values in pink. The equipotential lines, which are the lines where the three-dimensional equipotential surfaces intersect the xy-plane, displayed in Figure 24.6 are also shown in this plot, as are representations of the two plates. Note that the equipotential lines between the two plates are all parallel to each other and equally spaced.

In Figure 24.7b, the electric field lines have been added to the contour plot. The electric field is determined using $\vec{E}(\vec{r}) = -\vec{\nabla}V(\vec{r})$, introduced in Chapter 23. Far away from the two plates, the electric field looks very similar to that generated by a dipole composed of two point charges. It is easy to see that the electric field lines are perpendicular to the potential contour lines (which represent the equipotential surfaces!) everywhere in space.

But the electric field lines in Figure 24.7b do not convey adequate information about the magnitude of the electric field. Another representation of the electric field, in Figure 24.7c, displays the electric field vectors at regularly spaced grid points in the xy-plane. (The contour shading of the potential has been removed to reduce visual clutter.) In this plot, the field strength at each point of the grid is proportional to the size of the arrow at that point. You can clearly see that the electric field between the two plates is perpendicular to the plates and much larger in magnitude than the field outside the plates. The field in the space outside the plates is called the *fringe field*. If the plates are moved closer together, the electric field between the plates remains the same, while the fringe field is reduced.

The potential difference, ΔV, between the two parallel plates of the capacitor is proportional to the amount of charge on the plates. The proportionality constant is the **capacitance,** C, of the device, defined as

$$C = \left| \frac{q}{\Delta V} \right|. \qquad (24.1)$$

The capacitance of a device depends on the area of the plates and the distance between them but not on the charge or the potential difference. (This will be shown for this and other geometries in the following sections.) By definition, the capacitance is a positive number. It tells how much charge is required to produce a given potential difference between the plates. The larger the capacitance, the

FIGURE 24.4 The metal foil and Mylar sandwich shown in Figure 24.3 can be rolled up with an insulating layer to produce a capacitor with a compact geometry.

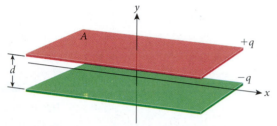

FIGURE 24.5 Parallel plate capacitor consisting of two conducting plates, each having area A, separated by a distance d.

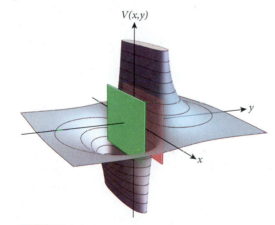

FIGURE 24.6 Electric potential in the xy-plane for the two oppositely charged parallel plates (superimposed) of Figure 24.5.

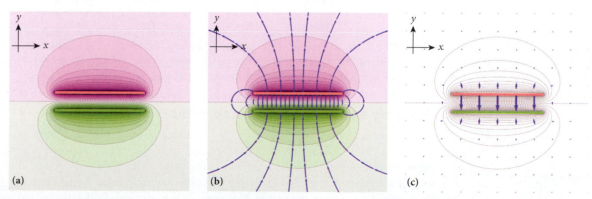

FIGURE 24.7 (a) Two-dimensional contour plot of the same potential as in Figure 24.6. (b) Contour plot with electric field lines superimposed. (c) Electric field strength at regularly spaced points in the xy-plane represented by the sizes of the arrows.

more charge is required to produce a given potential difference. (Note that it is a common practice to use V, not ΔV, to represent potential difference. Be sure you understand when V is being used for potential and when it is being used for potential difference.)

Equation 24.1, the definition of capacitance, can be rewritten in this commonly used form:

$$q = C\Delta V.$$

Equation 24.1 indicates that the units of capacitance are the units of charge divided by the units of potential, or coulombs per volt. A new unit was assigned to capacitance, named after British physicist Michael Faraday (1791–1867). This unit is called the **farad** (F):

$$1\,\text{F} = \frac{1\,\text{C}}{1\,\text{V}}. \tag{24.2}$$

One farad represents a very large capacitance. Typically, capacitors have a capacitance in the range from $1\ \mu\text{F} = 1 \cdot 10^{-6}\,\text{F}$ to $1\ \text{pF} = 1 \cdot 10^{-12}\,\text{F}$.

With the definition of the farad, we can write the electric permittivity of free space, ϵ_0 (introduced in Chapter 21), as $8.85 \cdot 10^{-12}\,\text{F/m}$.

24.2   Circuits

The next few chapters will introduce more and more complex and interesting circuits. So let's look at what a circuit is, in general.

An **electric circuit** consists of simple wires or other conducting paths that connect circuit elements. These circuit elements can be capacitors, which we'll examine in depth in this chapter. Other important circuit elements are resistors and galvanometers (introduced in Chapter 25), the voltmeter and the ammeter (introduced in Chapter 26), transistors (which can be used as on-off switches or as amplifiers; also covered in Chapter 26), and inductors (discussed in Chapter 29).

Circuits usually need some kind of power, which can be provided either by a battery or by an AC (alternating current) power source. The concept of a battery, a device that maintains a potential difference across its terminals through chemical reactions, was introduced in Chapter 23; for the purpose of a circuit, it can be viewed simply as an external source of electrostatic potential difference, something that delivers a fixed potential difference (which is commonly called *voltage*). An AC power source can produce the same result with a specially designed circuit that maintains a fixed potential difference. Chapter 29 on induction and Chapter 30 on electromagnetic oscillations and current will discuss AC power sources in more detail. Figure 24.8 lists the symbols for circuit elements, which are used throughout this and the following chapters.

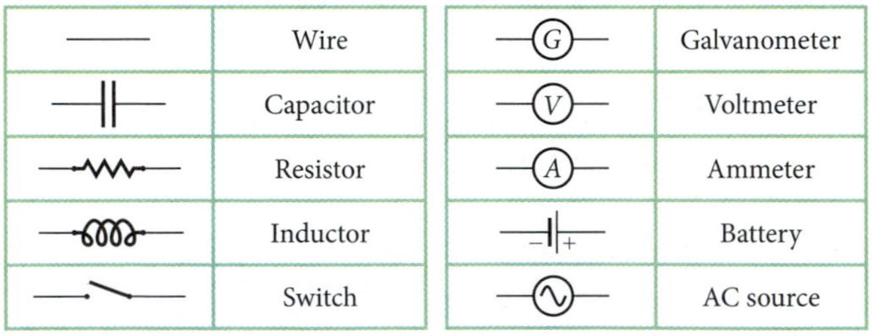

————	Wire	—Ⓖ—	Galvanometer
⊣⊢	Capacitor	—Ⓥ—	Voltmeter
—ⱽⱽⱽ—	Resistor	—Ⓐ—	Ammeter
—ⱺⱺⱺ—	Inductor	⊣⊢	Battery
—⸝—	Switch	—Ⓝ—	AC source

FIGURE 24.8 Commonly used symbols for circuit elements.

Charging and Discharging a Capacitor

A capacitor is charged by connecting it to a battery or to a constant-voltage power supply to create a circuit. Charge flows to the capacitor from the battery or power supply until the potential difference across the capacitor is the same as the supplied voltage. If the capacitor is disconnected, it retains its charge and potential difference. A real capacitor is subject to charge leaking away over time. However, in this chapter, we'll assume that an isolated capacitor retains its charge and potential difference indefinitely.

24.1 In-Class Exercise

The figure shows a charged capacitor. What is the net charge on the capacitor?

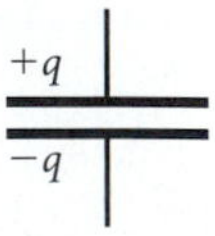

a) $(+q)+(-q) = 0$

b) $|+q| + |-q| = 0$

c) $|+q| + |-q| = 2q$

d) $(+q) + (-q) = 2q$

e) q

Figure 24.9 illustrates this charging process with a circuit diagram. In this diagram, the lines represent conducting wires. The battery (power supply) is represented by the symbol ⊤, which is labeled with plus and minus signs indicating the potential assignments of the terminals and with the potential difference, V. The capacitor is represented by the symbol ⊤, which is labeled C. This circuit also contains a switch. When the switch is between positions a and b, the battery is not connected and the circuit is open. When the switch is at position a, the circuit is closed; the battery is connected across the capacitor, and the capacitor charges. When the switch is at position b, the circuit is closed in a different manner. The battery is removed from the circuit, the two plates of the capacitor are connected to each other, and charge can flow from one plate to the other through the wire, which now forms a physical connection between the plates. When the charge has dissipated on the two plates, the potential difference between the plates drops to zero, and the capacitor is said to be discharged. (Charging and discharging of a capacitor are covered in quantitative detail in Chapter 26.)

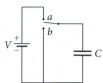

FIGURE 24.9 Simple circuit used for charging and discharging a capacitor.

24.3 Parallel Plate Capacitor

Section 24.1 discussed the general features of the electric potential and the electric field of two parallel plates of opposite charge. This section examines how to determine the electric field strength between the plates and the potential difference between the two plates. Let's consider an ideal parallel plate capacitor in the form of a pair of parallel conducting plates in a vacuum with charge $+q$ on one plate and charge $-q$ on the other plate (Figure 24.10). (This ideal parallel plate capacitor has very large plates, that are very close together, much closer than shown in Figure 24.10. This configuration allows us to neglect the fringe field, the small electric field outside the space between the plates, shown in Figure 24.7c.) When the plates are charged, the upper plate has charge $+q$ and the lower plate has charge $-q$. The electric field between the two plates points from the positively charged plate downward toward the negatively charged plate. The field near the ends of the plates, called the fringe field (compare Figure 24.7), can be neglected; that is, we can assume that the electric field is constant, with magnitude E, everywhere between the plates and zero elsewhere. The electric field is always perpendicular to the surface of the two parallel plates.

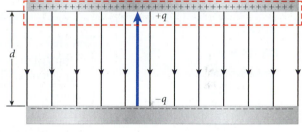

FIGURE 24.10 Side view of a parallel plate capacitor consisting of two plates of common area A separated by a small distance, d. The red dashed line is a Gaussian surface. The black arrows pointing downward represent the electric field. The blue arrow indicates an integration path.

The electric field can be found using Gauss's Law:

$$\oiint \vec{E} \cdot d\vec{A} = \frac{q}{\epsilon_0}.$$ (24.3)

How do we evaluate the integral over the Gaussian surface (whose cross section is outlined by a red dashed line in Figure 24.10)? We add the contributions from the top, the bottom, and the sides. The sides of the Gaussian surface are very small, so we can ignore the contributions from the fringe field. The top surface passes through the conductor, where the electric field is zero (remember shielding; see Chapter 22). This leaves only the bottom part of the Gaussian surface. The electric field vectors point straight down and are perpendicular to the conductor surfaces. The vector normal to the surface, $d\vec{A}$, also points in the same direction and is thus parallel to $\vec{E}$. Therefore, the scalar product is $\vec{E} \cdot d\vec{A} = E \, dA \cos 0° = E \, dA$. For the integral over the Gaussian surface, we then have

$$\oiint \vec{E} \cdot d\vec{A} = \iint_{\text{bottom}} E \, dA = E \iint_{\text{bottom}} dA = EA,$$

where A is the area of the plate. In other words, for the parallel plate capacitor, Gauss's Law yields

$$EA = \frac{q}{\epsilon_0},$$ (24.4)

where A is the surface area of the positively charged plate and q is the magnitude of the charge on the positively charged plate. The charge on each plate resides entirely on the inside surface because of the presence of opposite charge on the other plate.

The electric potential difference across the two plates in terms of the electric field is

$$\Delta V = -\int_i^f \vec{E} \cdot d\vec{s}. \tag{24.5}$$

The path of integration is chosen to be from the negatively charged plate to the positively charged plate, along the blue arrow in Figure 24.10. Since the electric field is antiparallel to this integration path (see Figure 24.10), the scalar product is $\vec{E} \cdot d\vec{s} = E\,ds\cos 180° = -E\,ds$. Thus, the integral in equation 24.5 reduces to

$$\Delta V = Ed = \frac{qd}{\epsilon_0 A},$$

where we used equation 24.4 to relate the electric field to the charge. Combining this expression for the potential difference and the definition of capacitance (equation 24.1) gives an expression for the capacitance of a parallel plate capacitor:

$$C = \left| \frac{q}{\Delta V} \right| = \frac{\epsilon_0 A}{d}. \tag{24.6}$$

Note that the capacitance of a parallel plate capacitor depends only on the area of the plates and the distance between the plates. In other words, only the geometry of a capacitor affects its capacitance. The amount of charge on the capacitor or the potential difference between its plates does not affect its capacitance.

24.2 In-Class Exercise

Suppose you charge a parallel plate capacitor using a battery and then remove the battery, isolating the capacitor and leaving it charged. You then move the plates of the capacitor farther apart. The potential difference between the plates will

a) increase.

b) decrease.

c) stay the same.

d) not be determinable.

24.1 Self-Test Opportunity

You charge a parallel plate capacitor using a battery. You then remove the battery and isolate the capacitor. If you decrease the distance between the plates of the capacitor, what will happen to the electric field between the plates?

24.3 In-Class Exercise

Suppose you have a parallel plate capacitor with area A and plate separation d, but space constraints on a circuit board force you to reduce the area of the capacitor by a factor of 2. What do you have to do to compensate and retain the same value of the capacitance?

a) reduce d by a factor of 2

b) increase d by a factor of 2

c) reduce d by a factor of 4

d) increase d by a factor of 4

EXAMPLE 24.1 | **Area of a Parallel Plate Capacitor**

A parallel plate capacitor has plates that are separated by 1.00 mm (Figure 24.11).

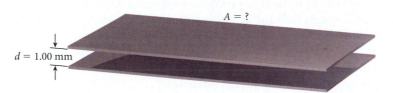

FIGURE 24.11 A parallel plate capacitor with plates separated by 1.00 mm.

PROBLEM

What is the area required to give this capacitor a capacitance of 1.00 F?

SOLUTION

The capacitance is given by

$$C = \frac{\epsilon_0 A}{d}. \tag{i}$$

Solving equation (i) for the area and putting in $d = 1.00 \cdot 10^{-3}$ m and $C = 1.00$ F, we get

$$A = \frac{dC}{\epsilon_0} = \frac{\left(1.00 \cdot 10^{-3}\text{ m}\right)\left(1.00\text{ F}\right)}{\left(8.85 \cdot 10^{-12}\text{ F/m}\right)} = 1.13 \cdot 10^8 \text{ m}^2.$$

If these plates were square, each one would be 10.6 km by 10.6 km (6.59 mi by 6.59 mi)! This result emphasizes that a farad is an extremely large amount of capacitance.

24.4 Cylindrical Capacitor

Consider a capacitor constructed of two collinear conducting cylinders with vacuum between the cylinders (Figure 24.12). The inner cylinder has radius r_1, and the outer cylinder has radius r_2. The inner cylinder has charge $-q$, and the outer cylinder has charge $+q$. The electric field between the two cylinders is then directed radially inward and perpendicular to the surfaces of both cylinders. As for a parallel plate capacitor, we assume that the cylinders are long and that there is essentially no fringe field near their ends.

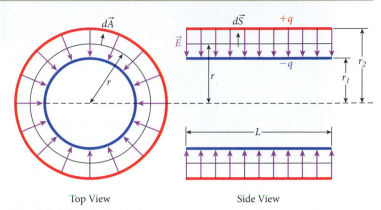

We can apply Gauss's Law to find the electric field between the two cylinders, using a Gaussian surface in the form of a cylinder with radius r and length L that is collinear with the two cylinders of the capacitor, as shown in Figure 24.12. The enclosed charge is then $-q$, because only the negatively charged surface of the capacitor is inside the Gaussian surface. The normal vector to the Gaussian surface, $d\vec{A}$, points radially outward and is thus antiparallel to the electric field. This means that $\vec{E} \cdot d\vec{A} = E\, dA \cos 180° = -E\, dA$. Applying Gauss's Law and using the fact that the surface of the cylinder has area $A = 2\pi rL$ then results in

Top View **Side View**

FIGURE 24.12 Cylindrical capacitor consisting of two long collinear conducting cylinders. The black circle represents a Gaussian surface. The purple arrows represent the electric field.

$$\oiint \vec{E} \cdot d\vec{A} = -E \oiint dA = -E2\pi rL = \frac{-q}{\epsilon_0}. \tag{24.7}$$

Equation 24.7 can be rearranged to give an expression for the magnitude of the electric field:

$$E = \frac{q}{\epsilon_0 2\pi rL}, \quad \text{for } r_1 < r < r_2.$$

The potential difference between the two cylindrical capacitor plates is obtained by integrating over the electric field, $\Delta V = -\int_i^f \vec{E} \cdot d\vec{s}$. For the integration path in the radial direction from the negatively charged cylinder at r_1 to the positively charged cylinder at r_2, the electric field is antiparallel to the path. Thus, $\vec{E} \cdot d\vec{s}$ in equation 24.5 becomes $-E\, dr$. Therefore,

$$\Delta V = -\int_i^f \vec{E} \cdot d\vec{s} = \int_{r_1}^{r_2} E\, dr = \int_{r_1}^{r_2} \frac{q}{\epsilon_0 2\pi rL}\, dr = \frac{q}{\epsilon_0 2\pi L} \ln\left(\frac{r_2}{r_1}\right).$$

This expression for the potential difference and equation 24.1 yield an expression for the capacitance:

$$C = \left|\frac{q}{\Delta V}\right| = \frac{q}{\dfrac{q}{\epsilon_0 2\pi L} \ln(r_2/r_1)} = \frac{2\pi\epsilon_0 L}{\ln(r_2/r_1)}. \tag{24.8}$$

Just as for a parallel plate capacitor, the capacitance of a cylindrical capacitor depends only on the geometry of the capacitor.

24.5 Spherical Capacitor

Now let's consider a spherical capacitor formed by two concentric conducting spheres with radii r_1 and r_2 with vacuum between the spheres (Figure 24.13). The inner sphere has charge $+q$, and the outer sphere has charge $-q$. The electric field is perpendicular to the surfaces of both spheres and points radially from the inner, positively charged sphere to the outer, negatively charged sphere, as shown by the purple arrows in Figure 24.13. (Previously, for the parallel plate and cylindrical capacitors, the integration was from the negative to the positive charge. In this section, we'll see what happens when the direction is reversed.) To determine the magnitude of the electric field, we employ Gauss's Law, using a Gaussian surface consisting of a sphere concentric with the two spherical conductors and having a

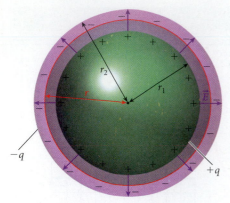

FIGURE 24.13 Spherical capacitor consisting of two concentric conducting spheres. The Gaussian surface is represented by the red circle of radius r.

radius r such that $r_1 < r < r_2$. The electric field is also perpendicular to the Gaussian surface everywhere, so we have

$$\oiint \vec{E} \cdot d\vec{A} = EA = E\left(4\pi r^2\right) = \frac{q}{\epsilon_0}. \tag{24.9}$$

Solving equation 24.9 for E gives

$$E = \frac{q}{4\pi\epsilon_0 r^2}, \qquad \text{for } r_1 < r < r_2.$$

For the potential difference, we proceed in a fashion similar to that used for the cylindrical capacitor and obtain

$$\Delta V = -\int_i^f \vec{E} \cdot d\vec{s} = -\int_{r_1}^{r_2} E\, dr = -\int_{r_1}^{r_2} \frac{q}{4\pi\epsilon_0 r^2}\, dr = -\frac{q}{4\pi\epsilon_0}\left(\frac{1}{r_1} - \frac{1}{r_2}\right).$$

In this case, $\Delta V < 0$. Why? Because we integrated from the positive charge to the negative charge! The positive charge is at a higher potential than the negative one, resulting in a negative potential difference. Equation 24.1 gives the capacitance of a spherical capacitor as the absolute value of the charge divided by the absolute value of the potential difference:

$$C = \left|\frac{q}{\Delta V}\right| = \frac{q}{\dfrac{q}{4\pi\epsilon_0}\left(\dfrac{1}{r_1} - \dfrac{1}{r_2}\right)} = \frac{4\pi\epsilon_0}{\left(\dfrac{1}{r_1} - \dfrac{1}{r_2}\right)}.$$

[handwritten: $Q = 4\pi\epsilon_0 RV$]

[handwritten: $E = \dfrac{V}{R}$]

This can be rewritten in a more convenient form:

$$C = 4\pi\epsilon_0 \frac{r_1 r_2}{r_2 - r_1}. \tag{24.10}$$

[handwritten: $u_E = \dfrac{1}{2}\epsilon_0 E^2$]

Note that again the capacitance depends only on the geometry of the device.

We can obtain the capacitance of a single conducting sphere from equation 24.10 by assuming that the outer spherical conductor is infinitely far away. With $r_2 = \infty$ and $r_1 = R$, the capacitance of an isolated spherical conductor is given by

$$C = 4\pi\epsilon_0 R. \tag{24.11}$$

24.6 Capacitors in Circuits

As stated earlier, a circuit is a set of electrical devices connected by conducting wires. Capacitors can be wired in circuits in different ways, but the two most fundamental ones are parallel connection and series connection.

Capacitors in Parallel

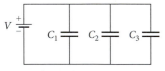

FIGURE 24.14 Simple circuit with a battery and three capacitors in parallel.

Figure 24.14 shows a circuit with three capacitors in **parallel connection.** Each of the three capacitors has one plate wired directly to the positive terminal of a battery with potential difference V and one plate wired directly to the negative terminal of that battery. The same circuit appears in the upper part of Figure 24.15, and the lower part of this figure shows the value of the potential at each part of the circuit in a three-dimensional plot. This illustrates that all capacitor plates connected to the positive terminal of the battery are at the same potential. The other plates of the capacitors are all at the potential of the negative terminal of the battery (set to zero). (The negative and positive terminals of the battery are joined with a light blue sheet to show that these two terminals are part of the same device and to provide a better visual representation of the potential difference between the two terminals. The plates of each capacitor are joined by a light gray band.)

The key insight provided by Figure 24.15 is that the potential difference across each of the three capacitors is the same, ΔV. Thus, for the three capacitors in this circuit, we have

$$q_1 = C_1 \Delta V$$
$$q_2 = C_2 \Delta V$$
$$q_3 = C_3 \Delta V.$$

In general, the charge on each capacitor can have a different value. The three capacitors can be viewed as one equivalent capacitor that holds a total charge q, given by

$$q = q_1 + q_2 + q_3 = C_1 \Delta V + C_2 \Delta V + C_3 \Delta V = (C_1 + C_2 + C_3) \Delta V.$$

Thus, the equivalent capacitance for this capacitor is

$$C_{eq} = C_1 + C_2 + C_3.$$

This result can be extended to any number, n, of capacitors connected in parallel:

$$C_{eq} = \sum_{i=1}^{n} C_i. \qquad (24.12)$$

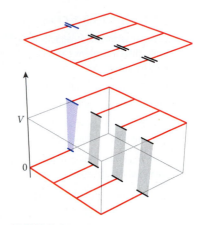

FIGURE 24.15 The potential in different parts of the circuit of Figure 24.14.

In other words, the equivalent capacitance of a system of capacitors in parallel is just the sum of the capacitances. Thus, several capacitors in parallel in a circuit can be replaced with an equivalent capacitance given by equation 24.12, as shown in Figure 24.16.

Capacitors in Series

Figure 24.17 shows a circuit with three capacitors in **series connection.** In this configuration, the battery produces an equal charge of $+q$ on the right plate of each capacitor and an equal charge of $-q$ on the left plate of each capacitor. This fact can be made clear by starting when the capacitors are uncharged. The battery is then connected to the series arrangement of the three capacitors. The positive plate of C_3 is connected to the positive terminal of the battery and begins to collect positive charge supplied by the battery. This positive charge induces a negative charge of equal magnitude onto the other plate of C_3. The negatively charged plate of C_3 is connected to the right plate of C_2, which then becomes positively charged because no net charge can accumulate on the isolated section consisting of the left plate of C_3 and the right plate of C_2. The positively charged plate of C_2 induces a negative charge of equal magnitude onto the other plate of C_2. In turn, the negatively charged plate of C_2 leaves a positive charge on the plate of C_1 connected to it, which induces a negative charge onto the left plate of C_1. The negatively charged plate of C_1 is connected to the negative terminal of the battery. Thus, charge flows from the battery, charging the positive plate of C_3 to a charge of value $+q$, and inducing a corresponding charge of $-q$ on the negatively charged plate of C_1. Therefore, each capacitor does indeed end up with the same charge.

When the three capacitors in the circuit in Figure 24.17 are charged, the sum of the potential drops across all three must equal the potential difference supplied by the battery. This is illustrated in Figure 24.18, a three-dimensional representation of the potential in the circuit with the three capacitors in series, similar to that in Figure 24.15. (Note that the potential drops at the three capacitors in series are not equal; this is true in general for a series connection.)

As you can see from Figure 24.18, the potential drops across the three capacitors must add up to the total potential difference, ΔV, supplied by the battery. Because each capacitor has the same charge, we have

$$\Delta V = \Delta V_1 + \Delta V_2 + \Delta V_3 = \frac{q}{C_1} + \frac{q}{C_2} + \frac{q}{C_3} = q \left(\frac{1}{C_1} + \frac{1}{C_2} + \frac{1}{C_3} \right).$$

The equivalent capacitance can be written as

$$\Delta V = \frac{q}{C_{eq}},$$

where

$$\frac{1}{C_{eq}} = \frac{1}{C_1} + \frac{1}{C_2} + \frac{1}{C_3}. \qquad (24.13)$$

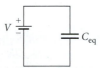

FIGURE 24.16 The three capacitors in Figure 24.14 can be replaced with an equivalent capacitance.

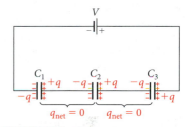

FIGURE 24.17 Simple circuit with three capacitors in series.

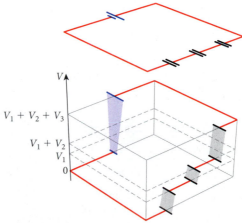

FIGURE 24.18 The potential in a circuit with three capacitors in series.

Thus, the three capacitors in series in the circuit shown in Figure 24.17 can be replaced with an equivalent capacitance given by equation 24.13, yielding the same circuit diagram as that in Figure 24.16.

24.5 In-Class Exercise

For a circuit with three capacitors in series, the equivalent capacitance must always be

a) equal to the largest of the three individual capacitances.

b) equal to the smallest of the three individual capacitances.

c) larger than the largest of the three individual capacitances.

d) smaller than the smallest of the three individual capacitances.

24.6 In-Class Exercise

The potential drop for a circuit with three capacitors of different individual capacitances in series connection is

a) the same across each capacitor and has the same value as the potential difference supplied by the battery.

b) the same across each capacitor and has $\frac{1}{3}$ of the value of the potential difference supplied by the battery.

c) largest across the capacitor with the smallest capacitance.

d) largest across the capacitor with the largest capacitance.

For a system of n capacitors, equation 24.13 generalizes to

$$\frac{1}{C_{eq}} = \sum_{i=1}^{n} \frac{1}{C_i}. \tag{24.14}$$

24.2 Self-Test Opportunity

What is the equivalent capacitance for four 10.0-μF capacitors connected in series? What is the equivalent capacitance for four 10.0-μF capacitors connected in parallel?

Thus, the capacitance of a system of capacitors in series is always less than the smallest capacitance in the system.

Finding equivalent capacitances for capacitors in series and in parallel allows problems involving complicated circuits to be solved, as the following example illustrates.

24.7 In-Class Exercise

Three capacitors, each with capacitance C, are connected as shown in the figure. What is the equivalent capacitance for this arrangement of capacitors?

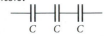

a) $C/3$ d) $9C$

b) $3C$ e) none of the above

c) $C/9$

24.8 In-Class Exercise

Three capacitors, each with capacitance C, are connected as shown in the figure. What is the equivalent capacitance for this arrangement of capacitors?

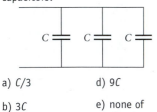

a) $C/3$ d) $9C$

b) $3C$ e) none of the above

c) $C/9$

EXAMPLE 24.2 | System of Capacitors

PROBLEM

Consider the circuit shown in Figure 24.19a, a complicated-looking arrangement of five capacitors with a battery. What is the combined capacitance of this set of five capacitors? If each capacitor has a capacitance of 5 nF, what is the equivalent capacitance of the arrangement? If the potential difference of the battery is 12 V, what is the charge on each capacitor?

SOLUTION

This problem may look complicated at first, but it can be simplified by sequential steps, using the rules for equivalent capacitances of capacitors in series and in parallel. We begin with the innermost circuit structures and work outward.

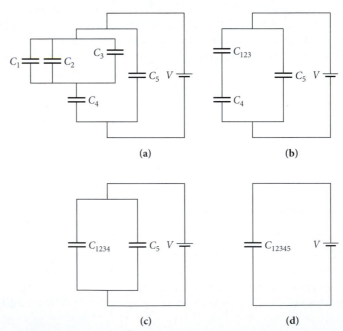

FIGURE 24.19 System of capacitors: (a) original circuit configuration; (b) reducing parallel capacitors to their equivalent; (c) reducing series capacitors to their equivalent; (d) equivalent capacitance for the entire set of capacitors.

STEP 1

Looking at capacitors 1 and 2 in Figure 24.19a, we see right away that they are in parallel. Because capacitor 3 is some distance away, it is less obvious that it is also in parallel with 1 and 2. However, the upper plates of all three of these capacitors are connected by wires and are thus at the same potential. The same goes for their lower plates, so all three are indeed in parallel. According to equation 24.12, the equivalent capacity for these three capacitors is

$$C_{123} = \sum_{i=1}^{3} C_i = C_1 + C_2 + C_3.$$

This replacement is shown in Figure 24.19b.

STEP 2

In Figure 24.19b, C_{123} and C_4 are in series. Thus, their equivalent capacitance is, according to equation 24.14,

$$\frac{1}{C_{1234}} = \frac{1}{C_{123}} + \frac{1}{C_4} \Rightarrow C_{1234} = \frac{C_{123}C_4}{C_{123}+C_4}.$$

This replacement is shown in Figure 24.19c.

STEP 3

Finally, C_{1234} and C_5 are in parallel in Figure 24.19c. Therefore, we can repeat the calculation for two capacitors in parallel and find the equivalent capacitance of all five capacitors:

$$C_{12345} = C_{1234} + C_5 = \frac{C_{123}C_4}{C_{123}+C_4} + C_5 = \frac{(C_1+C_2+C_3)C_4}{C_1+C_2+C_3+C_4} + C_5.$$

This result gives us the simple circuit shown in Figure 24.19d.

STEP 4: INSERT THE NUMBERS FOR THE CAPACITORS

We can now find the equivalent capacitance if all the capacitors have identical 5-nF capacitances:

$$\left(\frac{(5+5+5)5}{5+5+5+5} + 5 \right) \text{nF} = 8.75 \text{ nF}.$$

As you can see, more than half of the total capacitance of this arrangement is provided by capacitor 5 alone. This result shows that you need to be extremely careful about how you arrange capacitors in circuits.

STEP 5: CALCULATE THE CHARGES ON THE CAPACITORS

C_{1234} and C_5 are in parallel. Thus, they have the same potential difference across them, 12 V. The charge on C_5 is then

$$q_5 = C_5 \Delta V = (5 \text{ nF})(12 \text{ V}) = 60 \text{ nC}.$$

C_{1234} is composed of C_{123} and C_4 in series. Thus, C_{123} and C_4 must have the same charge q_4, so

$$\Delta V = \Delta V_{123} + \Delta V_4 = \frac{q_4}{C_{123}} + \frac{q_4}{C_4} = q_4 \left(\frac{1}{C_{123}} + \frac{1}{C_4} \right).$$

The charge on C_4 is then

$$q_4 = \Delta V \frac{C_{123}C_4}{C_{123}+C_4} = \Delta V \frac{(C_1+C_2+C_3)C_4}{C_1+C_2+C_3+C_4} = (12 \text{ V}) \frac{(15 \text{ nF})(5 \text{ nF})}{20 \text{ nF}} = 45 \text{ nC}.$$

C_{123} is equivalent to three capacitors in parallel, and it also has the same charge as C_4, or 45 nC. The three capacitors C_1, C_2, and C_3 have the same capacitance, the same potential difference across them as they are in parallel, and the sum of the charge on these three capacitors must equal 45 nC. Therefore, we can calculate the charge on C_1, C_2, and C_3:

$$q_1 = q_2 = q_3 = \frac{45 \text{ nC}}{3} = 15 \text{ nC}.$$

24.9 In-Class Exercise

Three capacitors are connected to a battery as shown in the figure. If $C_1 = C_2 = C_3 = 10.0$ μF and $V = 10.0$ V, what is the charge on capacitor C_3?

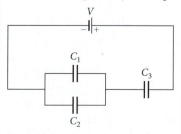

a) 66.7 μC   d) 300. μC

b) 100. μC   e) 457. μC

c) 150. μC

24.7 Energy Stored in Capacitors

Capacitors are extremely useful for storing electric potential energy. They are much more useful than batteries if the potential energy has to be converted into other energy forms very quickly. One application of capacitors for the storage and rapid release of electric potential energy is described in Example 24.4, about the use of capacitors at the National Ignition Facility. Let's examine how much energy can be stored in a capacitor.

A battery must do work to charge a capacitor. This work can be conceptualized in terms of changing the electric potential energy of the capacitor. In order to accomplish the charging process, charge has to be moved against the potential between the two capacitor plates. As noted earlier in this chapter, the larger the charge of the capacitor is, the larger the potential difference between the plates. This means that the more charge already on the capacitor, the harder it becomes to add a differential amount of charge to the capacitor. The differential work, dW, done by a battery with potential difference ΔV to put a differential charge, dq, on a capacitor with capacitance C is

$$dW = \Delta V' dq' = \frac{q'}{C} dq',$$

where $\Delta V'$ and q' are the instantaneous (increasing) potential difference and charge, respectively, on the capacitor during the charging process. The total work, W_t, required to bring the capacitor to its full charge, q, is given by

$$W_t = \int dW = \int_0^q \frac{q'}{C} dq' = \frac{1}{2} \frac{q^2}{C}.$$

This work is stored as electric potential energy:

$$U = \frac{1}{2} \frac{q^2}{C} = \frac{1}{2} C (\Delta V)^2 = \frac{1}{2} q \Delta V. \tag{24.15}$$

All three of the formulations for the stored electric potential energy in equation 24.15 are equally valid. Each can be transformed to one of the others by using $q = C\Delta V$ and eliminating one of the three quantities in favor of the other two.

The **electric energy density**, u, is defined as the electric potential energy per unit volume:

$$u = \frac{U}{\text{volume}}.$$

(*Note: V* is not used to represent the volume here, because in this context it is reserved for the potential.)

For the special case of a parallel plate capacitor that has no fringe field, it is easy to calculate the volume enclosed between two plates of area A separated by a perpendicular distance d. It is the area of each plate times the distance between the plates, or Ad. Using equation 24.15 for the electric potential energy, we obtain

$$u = \frac{U}{Ad} = \frac{\frac{1}{2} C (\Delta V)^2}{Ad} = \frac{C (\Delta V)^2}{2Ad}.$$

Using equation 24.6 for the capacitance of a parallel plate capacitor with vacuum between the plates, we get

$$u = \frac{(\epsilon_0 A/d)(\Delta V)^2}{2Ad} = \frac{1}{2} \epsilon_0 \left(\frac{\Delta V}{d} \right)^2.$$

24.10 In-Class Exercise

How much energy is stored in the 180-μF capacitor of a camera flash unit charged to 300.0 V?

a) 1.22 J   d) 115 J

b) 8.10 J   e) 300 J

c) 45.0 J

Recognizing that $\Delta V/d$ is the magnitude of the electric field, E, we obtain an expression for the electric energy density for a parallel plate capacitor:

$$u = \frac{1}{2} \epsilon_0 E^2. \tag{24.16}$$

This result, although derived for a parallel plate capacitor, is in fact much more general. The electric potential energy stored in any electric field per unit volume occupied by that field can be described using equation 24.16.

EXAMPLE 24.3 | Thundercloud

Suppose a thundercloud with a width of 2.0 km and a length of 3.0 km hovers at an altitude of 500 m over a flat area. The cloud carries a charge of 160 C, and the ground has no charge.

PROBLEM 1

What is the potential difference between the cloud and the ground?

SOLUTION 1

We can approximate the cloud-ground system as a parallel plate capacitor. Its capacitance is, according to equation 24.6,

$$C = \frac{\epsilon_0 A}{d} = \frac{(8.85 \cdot 10^{-12} \text{ F/m})(2000 \text{ m})(3000 \text{ m})}{500 \text{ m}} = 0.11 \text{ } \mu\text{F}.$$

Because we know the charge carried by the cloud, 160 C, it is tempting to insert this value into the relationship among charge, capacitance, and potential difference (equation 24.1) to find the desired answer. However, a parallel plate capacitor with a charge of $+q$ on one plate and $-q$ on the other has a charge difference of $2q$ between the plates. For the cloud-ground system, $2q = 160$ C, or $q = 80$ C. Alternatively, we can think of the cloud as a charged insulator and use the result from Section 22.9, that the field due to a plane sheet of charge is $E = \sigma/2\epsilon_0$, to justify the factor of $\frac{1}{2}$. Now we can use equation 24.1 and obtain

$$\Delta V = \frac{q}{C} = \frac{80 \text{ C}}{0.11 \text{ } \mu\text{F}} = 7.3 \cdot 10^8 \text{ V}.$$

The potential difference is more than 700 million volts!

PROBLEM 2

Lightning strikes require electric field strengths of approximately 2.5 MV/m. Are the conditions described in the problem statement sufficient for a lightning strike?

SOLUTION 2

We use the potential difference between cloud and ground and the given distance between them to calculate the electric field:

$$E = \frac{\Delta V}{d} = \frac{7.3 \cdot 10^8 \text{ V}}{500 \text{ m}} = 1.5 \text{ MV/m}.$$

From this result, we can conclude that no lightning will develop in these conditions. However, if the cloud drifted over a radio tower, the electric field strength would likely increase and lead to a lightning discharge.

PROBLEM 3

What is the total electric potential energy contained in the field between this thundercloud and the ground?

SOLUTION 3

From equation 24.15, the total electric potential energy stored in this capacitor system is

$$U = \tfrac{1}{2} q \Delta V = 0.5(80 \text{ C})(7.3 \cdot 10^8 \text{ V}) = 2.9 \cdot 10^{10} \text{ J}.$$

For comparison, this energy is sufficient to run a typical 1500-W hair dryer for more than 5000 hours.

SOLVED PROBLEM 24.1   Energy Stored in Capacitors

PROBLEM
Suppose many capacitors, each with $C = 90.0$ μF, are connected in parallel across a battery with a potential difference of $\Delta V = 160.0$ V. How many capacitors are needed to store 95.6 J of energy?

SOLUTION

THINK
The equivalent capacitance of many capacitors connected in parallel is given by the sum of the capacitances of all the capacitors. We can calculate the energy stored from the equivalent capacitance of the capacitors in parallel and the potential difference of the battery.

SKETCH
Figure 24.20 shows a circuit with n capacitors connected in parallel across a battery.

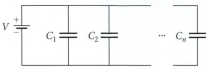

FIGURE 24.20 A circuit with n capacitors connected in parallel across a battery.

RESEARCH
The equivalent capacitance, C_{eq}, of n capacitors, each with capacitance C, connected in parallel is

$$C_{eq} = C_1 + C_2 + \cdots + C_n = nC.$$

The energy stored in the capacitors is then given by

$$U = \tfrac{1}{2} C_{eq} \left(\Delta V \right)^2 = \tfrac{1}{2} nC \left(\Delta V \right)^2. \qquad \text{(i)}$$

SIMPLIFY
Solving equation (i) for the required number of capacitors gives us

$$n = \frac{2U}{C \left(\Delta V \right)^2}.$$

CALCULATE
Putting in the numerical values, we get

$$n = \frac{2 \left(95.6 \text{ J} \right)}{\left(90.0 \cdot 10^{-6} \text{ C} \right) \left(160.0 \text{ V} \right)^2} = 82.986.$$

ROUND
We report our result as an integer number of capacitors:

$$n = 83 \text{ capacitors}.$$

DOUBLE-CHECK
The capacitance of 83 capacitors with $C = 90.0$ μF is

$$C_{eq} = 83 \left(90.0 \text{ μF} \right) = 0.00747 \text{ F}.$$

Charging this capacitor with a 160-V battery produces a stored energy of

$$U = \tfrac{1}{2} C_{eq} \left(\Delta V \right)^2 = \tfrac{1}{2} \left(0.00747 \text{ F} \right) \left(160.0 \text{ V} \right)^2 = 95.6 \text{ J}.$$

Thus, our answer for the number of capacitors is consistent.

EXAMPLE 24.4   The National Ignition Facility

The National Ignition Facility (NIF) is a high-powered laser designed to produce fusion reactions similar to those that occur in the Sun. The laser uses a short, high-energy pulse of light to heat and compress a small pellet containing isotopes of hydrogen. The laser is powered by 192 power-conditioning modules (Figure 24.21), each containing twenty 300-μF capacitors connected in parallel and charged to 24 kV. The capacitors are charged over a period of 90 s. The laser is then fired by discharging all the energy stored in the capacitors in 400 μs.

PROBLEM 1

How much energy is stored in the capacitors of NIF?

SOLUTION 1

The capacitors are connected in parallel. Thus, the equivalent capacitance of each power-conditioning module is

$$C_{eq} = 20(300 \ \mu F) = 0.006 \ F = 6 \ mF.$$

The energy stored in each power-conditioning module is

$$U = \tfrac{1}{2} C_{eq} (\Delta V)^2 = \tfrac{1}{2} (6 \cdot 10^{-3} \ F)(24 \cdot 10^3 \ V)^2 = 1.73 \ MJ.$$

Thus, the total energy stored in all the capacitors of NIF is

$$U_{total} = 192(1.73 \ MJ) = 332 \ MJ.$$

PROBLEM 2

What is the average power released by the power-conditioning modules during the laser pulse?

SOLUTION 2

The power is the energy per unit time, which is given by

$$P = \frac{\Delta U}{\Delta t} = \frac{332 \ MJ}{400 \ \mu s} = \frac{332 \cdot 10^6 \ J}{400 \cdot 10^{-6} \ s} = 8.3 \cdot 10^{11} \ W = 0.83 \ TW.$$

In comparison, the average electrical power generated in the United States in 2007 was 0.47 TW. Of course, the 0.83 TW of power delivered to the NIF laser is maintained for only a small fraction of a second.

FIGURE 24.21 Forty-eight power-conditioning modules in one of four bays at the National Ignition Facility at Lawrence Livermore National Laboratory.

Defibrillator

An important application of capacitors is the portable automatic external defibrillator (AED), a device designed to shock the heart of a person who is in ventricular fibrillation. A typical AED is shown in Figure 24.22.

Being in ventricular fibrillation means that the heart is not beating in a regular pattern. Instead, the signals that control the beating of the heart are erratic, preventing the heart from performing its function of maintaining regular blood circulation throughout the body. This condition must be treated within a few minutes to avoid permanent damage or death. Having many AED devices located in accessible public places allows quick treatment of this condition.

An AED provides a pulse of electrical current intended to stimulate the heart to beat regularly. Typically, an AED is designed to analyze a person's heartbeat automatically, determine if the person is in ventricular fibrillation, and administer the electrical pulse if required. The operator of the AED must attach the electrodes of the AED to the chest of the person experiencing the problem and push the start button. The AED will do nothing if the person is not in ventricular defibrillation. If the AED determines that the person is in ventricular fibrillation, the AED will instruct the operator to press the button to initiate the electrical pulse. Note that an AED is not designed to restart a heart that is not beating. Rather it is designed to restore a regular heartbeat when the heart is beating erratically.

Typically, an AED delivers 150 J of electric energy to the patient, administered through a pair of electrodes that are attached to the chest area (see Figure 24.22). This

(a)

(b)

FIGURE 24.22 (a) An automatic external defibrillator (AED) in its holder on a wall. (b) Schematic diagram showing where to place the hands-free electrodes.

energy is stored by charging a capacitor through a special circuit from a low-voltage battery. This capacitor typically has a capacitance of 100 μF and is charged in 10 s. The power used during charging is

$$P = \frac{E}{t} = \frac{150 \text{ J}}{10 \text{ s}} = 15 \text{ W},$$

which is within the capability of a simple battery. The energy of the capacitor is then discharged in 10 ms. The instantaneous power during the discharge is

$$P = \frac{E}{t} = \frac{150 \text{ J}}{10 \text{ ms}} = 15 \text{ kW},$$

which is beyond the capability of a small, portable battery but well within the capabilities of a well-designed capacitor.

The energy stored in the capacitor is $U = \frac{1}{2}C(\Delta V)^2$. When the capacitor is charged, its potential difference is

$$\Delta V = \sqrt{\frac{2U}{C}} = \sqrt{\frac{2(150 \text{ J})}{100 \cdot 10^{-6} \text{ F}}} = 1730 \text{ V}.$$

When the AED delivers an electrical current, the capacitor is charged from a battery contained in the AED. The capacitor is then discharged through the person to stimulate the heart to beat in a regular manner. Most AEDs can deliver the electical current many times without recharging the battery.

24.8   Capacitors with Dielectrics

The capacitors we have been discussing have air or vacuum between the plates. However, capacitors for just about any commercial application have an insulating material, called a **dielectric,** between the two plates. This dielectric serves several purposes: First, it maintains the separation between the plates. Second, the dielectric insulates the two plates from each other electrically. Third, the dielectric allows the capacitor to maintain a higher potential difference than it could with only air between the plates. Lastly, a dielectric increases the capacitance of the capacitor. We'll see that this ability to increase the capacitance is due to the molecular structure of the dielectric.

Filling the space between the plates of a capacitor completely with a dielectric increases its capacitance by a numerical factor called the **dielectric constant,** κ. We'll assume that the dielectric fills the entire volume between the capacitor plates, unless explicitly stated otherwise. Solved Problem 24.2 considers an example where the filling is only partial.

The capacitance, C, of a capacitor containing a dielectric with dielectric constant κ between its plates is given by

$$C = \kappa C_{\text{air}}, \tag{24.17}$$

where C_{air} is the capacitance of the capacitor without the dielectric.

Placing a dielectric between the plates of a capacitor has the effect of lowering the electric field between the plates (see Section 24.9 for explanation) and allowing more charge to be stored in the capacitor. For example, the electric field between the plates of a parallel plate capacitor, given by equation 24.4, is modified for a parallel plate capacitor with a dielectric to

$$E = \frac{E_{\text{air}}}{\kappa} = \frac{q}{\kappa \epsilon_0 A} = \frac{q}{\epsilon A}. \tag{24.18}$$

The constant ϵ_0 is the electric permittivity of free space, previously encountered in Coulomb's Law. The right-hand side of equation 24.18 was obtained by replacing the factor $\kappa \epsilon_0$ with ϵ, the **electric permittivity** of the dielectric. In other words, the electric permittivity of a dielectric is the product of the electric permittivity of free space (the vacuum) and the dielectric constant of the dielectric:

$$\epsilon = \kappa \epsilon_0. \tag{24.19}$$

Note that this replacement of ϵ_0 by ϵ is all that is needed to generalize expressions for the capacitance, such as equations 24.6, 24.8, and 24.10, from the values applicable for a capacitor with vacuum between its plates to those appropriate when the capacitor is completely filled with a dielectric. We can now see how the capacitance is increased by adding a dielectric between the plates. The potential difference across a parallel plate capacitor is:

$$\Delta V = Ed = \frac{qd}{\kappa \epsilon_0 A}.$$

Therefore, we can write the capacitance as

$$C = \frac{q}{\Delta V} = \frac{\kappa \epsilon_0 A}{d} = \kappa C_{air}.$$

The **dielectric strength** of a material is a measure of its ability to withstand potential difference. If the electric field strength in the dielectric exceeds the dielectric strength, the dielectric breaks down and begins to conduct charge between the plates via a spark, which usually destroys the capacitor. Thus, a useful capacitor must contain a dielectric that not only provides a given capacitance but also enables the device to hold the required potential difference without breaking down. Capacitors are normally specified by the value of their capacitance and by the maximum potential difference that they are designed to handle.

The dielectric constant of vacuum is defined to be 1, and the dielectric constant of air is close to 1.0. The dielectric constants and dielectric strengths of air and of other common materials used as dielectries are listed in Table 24.1.

24.3 Self-Test Opportunity

One way to increase the capacitance of a parallel plate capacitor, other than by adding a dielectric between the plates, is to decrease the distance between the plates. What is the minimum distance between the plates of a parallel plate capacitor in air if the maximum potential difference between the plates is to be 100.0 V? (*Hint:* Table 24.1 may be useful.)

24.11 In-Class Exercise

Suppose you charge a parallel plate capacitor with a dielectric between the plates using a battery and then remove the battery, isolating the capacitor and leaving it charged. You then remove the dielectric from between the plates. The potential difference between the plates will

a) increase.

b) decrease.

c) stay the same.

d) not be determinable.

Table 24.1	Dielectric Constants and Dielectric Strengths for Some Representative Materials	
Material	**Dielectric Constant, κ**	**Dielectric Strength (kV/mm)**
Vacuum	1	
Air (1 atm)	1.00059	2.5
Liquid nitrogen	1.454	
Teflon	2.1	60
Polyethylene	2.25	50
Benzene	2.28	
Polystyrene	2.6	24
Lexan	2.96	16
Mica	3–6	150–220
Paper	3	16
Mylar	3.1	280
Plexiglas	3.4	30
Polyvinyl chloride (PVC)	3.4	29
Glass	5	14
Neoprene	16	12
Germanium	16	
Glycerin	42.5	
Water	80.4	65
Strontium titanate	310	8

Note that these values are approximate and are for room temperature.

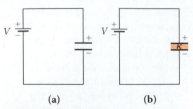

FIGURE 24.23 Parallel plate capacitor connected to a battery: (a) with no dielectric; (b) with a dielectric inserted between the plates.

EXAMPLE 24.5 Parallel Plate Capacitor with a Dielectric

PROBLEM 1

Consider a parallel plate capacitor without a dielectric and with capacitance $C = 2.00\ \mu\text{F}$ connected to a battery with potential difference $\Delta V = 12.0\ \text{V}$ (Figure 24.23a). What is the charge stored in the capacitor?

SOLUTION 1

Using the definition of capacitance (equation 24.1), we have

$$q = C\Delta V = \left(2.00 \cdot 10^{-6}\ \text{F}\right)\left(12.0\ \text{V}\right) = 2.40 \cdot 10^{-5}\ \text{C}.$$

PROBLEM 2

In Figure 24.23b, a dielectric with $\kappa = 2.5$ has been inserted between the plates of the capacitor, completely filling the space between them. Now what is the charge on the capacitor?

SOLUTION 2

The capacitance of the capacitor is increased by the dielectric:

$$C = \kappa C_{\text{air}}.$$

The charge is

$$q = \kappa C_{\text{air}} \Delta V = \left(2.50\right)\left(2.00 \cdot 10^{-6}\ \text{F}\right)\left(12.0\ \text{V}\right) = 6.00 \cdot 10^{-5}\ \text{C}.$$

The charge on the capacitor increases when the capacitance increases because the battery maintains a constant potential difference across the capacitor. The battery provides the additional charge until the capacitor is fully charged.

PROBLEM 3

Now suppose the capacitor is disconnected from the battery (Figure 24.24a). The capacitor, which is now isolated, maintains its charge of $q = 6.00 \cdot 10^{-5}$ C and its potential difference of $\Delta V = 12.0$ V. What happens to the charge and potential difference if the dielectric is removed, keeping the capacitor isolated (Figure 24.24b)?

FIGURE 24.24 Isolated capacitor: (a) with dielectric and (b) with dielectric removed.

SOLUTION 3

The charge on the isolated capacitor cannot change when the dielectric is removed because there is nowhere for the charge to flow. Thus, the potential difference on the capacitor is

$$\Delta V = \frac{q}{C} = \frac{6.00 \cdot 10^{-5}\ \text{C}}{2.00 \cdot 10^{-6}\ \text{F}} = 30.0\ \text{V}.$$

The potential difference increases because removing the dielectric increases the electric field and the resulting potential difference between the plates.

PROBLEM 4

Does removing the dielectric change the energy stored in the capacitor?

SOLUTION 4

The energy stored in a capacitor is given by equation 24.15. Before the dielectric was removed, the energy in the capacitor was

$$U = \tfrac{1}{2}C\left(\Delta V\right)^2 = \tfrac{1}{2}\kappa C_{\text{air}}\left(\Delta V\right)^2 = \tfrac{1}{2}\left(2.50\right)\left(2.00 \cdot 10^{-6}\ \text{F}\right)\left(12\ \text{V}\right)^2 = 3.60 \cdot 10^{-4}\ \text{J}.$$

After the dielectric is removed, the energy is

$$U = \tfrac{1}{2}C_{\text{air}}\left(\Delta V\right)^2 = \tfrac{1}{2}\left(2.00 \cdot 10^{-6}\ \text{F}\right)\left(30\ \text{V}\right)^2 = 9.00 \cdot 10^{-4}\ \text{J}.$$

The energy increase from $3.60 \cdot 10^{-4}$ J to $9.00 \cdot 10^{-4}$ J when the dielectric is removed is due to the work done on the dielectric in pulling it out of the electric field between the plates.

24.12 In-Class Exercise

What would happen if the dielectric in the capacitor of Example 24.5 were pulled halfway out and then released?

a) The dielectric would be pulled back into the capacitor.

b) The dielectric would heat up rapidly.

c) The dielectric would be pushed out of the capacitor.

d) The capacitor plates would heat up rapidly.

e) The dielectric would remain in the halfway position, and no heating would be observed.

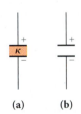

EXAMPLE 24.6 — Capacitance of a Coaxial Cable

Coaxial cables are used to transport signals, for example TV signals, between devices with minimum interference from the surroundings. A 20.0-m-long coaxial cable is composed of a conductor and a coaxial conducting shield around the conductor. The space between the conductor and the shield is filled with polystyrene. The radius of the conductor is 0.250 mm, and the radius of the shield is 2.00 mm (Figure 24.25).

PROBLEM
What is the capacitance of the coaxial cable?

SOLUTION
We can think of the conductor of the coaxial cable as a cylinder because all the charge on the conductor resides on its surface. From Table 24.1, the dielectric constant for polystyrene is 2.6. We can treat the coaxial cable as a cylindrical capacitor with $r_1 = 0.250$ mm and $r_2 = 2.00$ mm, filled with a dielectric with $\kappa = 2.6$. Then, we can use equation 24.8 to find the capacitance of the coaxial cable:

$$C = \kappa \frac{2\pi\epsilon_0 L}{\ln(r_2/r_1)} = \frac{2.6(2\pi)(8.85 \cdot 10^{-12} \text{ F/m})(20.0 \text{ m})}{\ln\left[(2.00 \cdot 10^{-3} \text{ m})/(2.5 \cdot 10^{-4} \text{ m})\right]} = 1.39 \cdot 10^{-9} \text{ F} = 1.39 \text{ nF}.$$

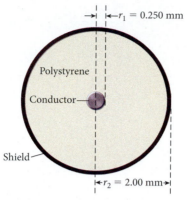

FIGURE 24.25 Cross section of a coaxial cable.

One interesting application of capacitance and the dielectric constant is in measuring liquid nitrogen levels in cryostats (containers insulated to maintain cold temperatures). It is often difficult to conduct a visual exam to determine how much liquid nitrogen is left in a cryostat. However, if one determines the capacitance, C, of the empty cryostat, then, when completely filled with liquid nitrogen, the cryostat should have a capacitance of $\kappa C = 1.454C$, because liquid nitrogen has a dielectric constant of 1.454. The capacitance varies smoothly as a function of the fullness between the maximum value $\kappa C = 1.454C$ for the completely full cryostat and the value C for the empty cryostat, giving an easy way to determine how full the cryostat is.

24.9 Microscopic Perspective on Dielectrics

Let's consider what happens at the atomic and molecular level when a dielectric is placed in an electric field. There are two types of dielectric materials: polar dielectrics and nonpolar dielectrics.

A **polar dielectric** is a material composed of molecules that have a permanent electric dipole moment due to their structure. A common example of such a molecule is water. Normally, the directions of electric dipoles are randomly distributed (Figure 24.26a). However,

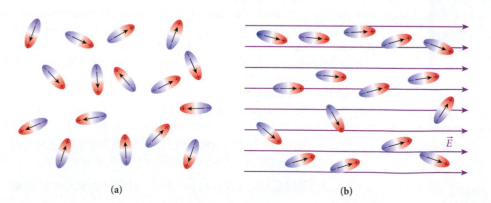

(a)   (b)

FIGURE 24.26 Polar molecules: (a) randomly distributed and (b) oriented by an external electric field.

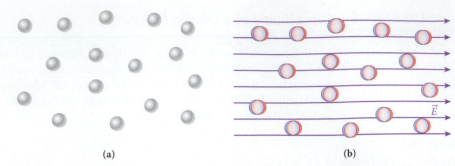

FIGURE 24.27 Nonpolar molecules: (a) with no electric dipole moment and (b) with an electric dipole moment induced by an external electric field.

when an electric field is applied to these polar molecules, they tend to align with the field (Figure 24.26b).

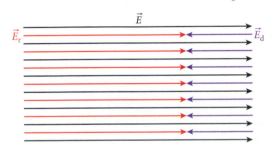

FIGURE 24.28 Partial cancellation of the applied electric field across a parallel plate capacitor by the electric dipoles of a dielectric.

A **nonpolar dielectric** is a material composed of atoms or molecules that have no inherent electric dipole moment (Figure 24.27a). These atoms or molecules can be induced to have a dipole moment under the influence of an external electric field (Figure 24.27b). The opposite directions of the electric forces acting on the negative and positive charges in the atom or molecule displace these two charge distributions and produce an induced electric dipole moment.

In both polar and nonpolar dielectrics, the fields resulting from the aligned electric dipole moments tend to partially cancel the original external electric field (Figure 24.28). For an electric field, $\vec{E}$, applied across a capacitor with a dielectric between the plates, the resulting electric field, $\vec{E}_r$, inside the capacitor is just the original field plus the electric field induced in the dielectric material, $\vec{E}_d$:

$$\vec{E}_r = \vec{E} + \vec{E}_d \ ,$$

or

$$E_r = E - E_d .$$

Note that the resulting electric field points in the same direction as the original field but is smaller in magnitude. The dielectric constant is given by $\kappa = E/E_r$.

Supercapacitors

As we've seen in this chapter, 1 F is a huge amount of capacitance. Even the National Ignition Facility (NIF), which needs the highest possible energy storage, uses only 300-μF capacitors. However, it is possible to create *supercapacitors* (also called *ultracapacitors*) with vastly greater capacitance. This is accomplished by using a material with a very large surface area between the capacitor plates. Activated charcoal is one possibility, as it has a very large surface area because of its foamlike structure at a nanoscale level. Two layers of activated charcoal are given charge of opposite polarity and are separated by an insulating material (represented by the red line in Figure 24.29b). This allows each side of the supercapacitor to store oppositely charged free ions from the electrolyte. The separation between the electrolyte ions and the charges on the activated charcoal is typically on the order of nanometers (nm), that is, millions of times smaller than in conventional capacitors. The activated charcoal provides surface areas many orders of magnitude larger than in conventional capacitors. Since, as noted in Section 24.3, the capacitance is proportional to the surface area and inversely proportional to the plate separation, this technology has resulted in commercially avail-

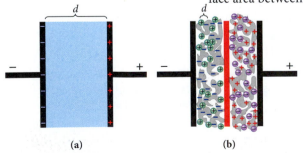

FIGURE 24.29 Comparison of (a) a conventional parallel plate capacitor and (b) a supercapacitor filled with activated charcoal.

able capacitors with capacitances on the order of kilofarads (kF), that is, millions of times larger than those used in the NIF.

Why is the NIF not using supercapacitors? The answer is that these supercapacitors can only function with potential differences of up to 2–3 V. The highest-capacity commercially available supercapacitors have capacitance values of up to 5 kF. Using $U = \frac{1}{2}C(\Delta V)^2$ and $\Delta V = 2$ V shows that a supercapacitor can hold 10 kJ. The 300-μF capacitors used at NIF, when charged to 24 kV, can hold 86.4 kJ. In addition, they can also be discharged much more rapidly, which is crucial to fulfill the high power requirement of the NIF's laser.

However, supercapacitors can reach energy storage capabilities that rival those of conventional batteries. In addition, supercapacitors can be charged and discharged millions of times, compared to perhaps thousands of times for rechargeable batteries. This, along with their very short charging time, makes them potentially suitable for many applications. For example, there is intense research into using these supercapacitors for electric vehicles. A bus based on this energy storage technology, named *capabus*, is currently in use in Shanghai, China.

A promising line of research on improving the potential difference that supercapacitors can employ is looking at the use of carbon nanotubes instead of activated charcoal. The first laboratory prototypes are very promising, and commercial products based on this approach could be in use within a few years.

WHAT WE HAVE LEARNED | EXAM STUDY GUIDE

- The capacitance of a capacitor is defined in terms of the charge, q, that can be stored on the capacitor and the potential difference, V, across the plates: $q = C\Delta V$.

- The farad is the unit of capacitance: $1 \text{ F} = \dfrac{1 \text{ C}}{1 \text{ V}}$.

- The capacitance of a parallel plate capacitor with plates of area A with a vacuum (or air) between the plates separated by distance d is given by $C = \dfrac{\epsilon_0 A}{d}$.

- The capacitance of a cylindrical capacitor of length L consisting of two collinear cylinders with a vacuum (or air) between the cylinders with inner radius r_1 and outer radius r_2 is given by $C = \dfrac{2\pi\epsilon_0 L}{\ln(r_2/r_1)}$.

- The capacitance of a spherical capacitor consisting of two concentric spheres with vacuum (or air) between the spheres with inner radius r_1 and outer radius r_2 is given by $C = 4\pi\epsilon_0 \dfrac{r_1 r_2}{r_2 - r_1}$.

- The electric energy density, u, between the plates of a parallel plate capacitor with vacuum (or air) between the plates is given by $u = \frac{1}{2}\epsilon_0 E^2$.

- A system of n capacitors connected in parallel in a circuit can be replaced by an equivalent capacitance given by the sum of the capacitances of the capacitors: $C_{eq} = \sum_{i=1}^{n} C_i$.

- A system of n capacitors connected in series in a circuit can be replaced by an equivalent capacitance given by the reciprocal of the sum of the reciprocal capacitances of the capacitors: $\dfrac{1}{C_{eq}} = \sum_{i=1}^{n} \dfrac{1}{C_i}$.

- When the space between the plates of a capacitor is filled with a dielectric of dielectric constant κ, the capacitance increases relative to the capacitance in air: $C = \kappa C_{air}$.

KEY TERMS

NEW SYMBOLS AND EQUATIONS

C, capacitance of a capacitor

$C_{eq} = \sum_{i=1}^{n} C_i$, equivalent capacitance of a number of capacitors in parallel

$\dfrac{1}{C_{eq}} = \sum_{i=1}^{n} \dfrac{1}{C_i}$, equivalent capacitance of a number of capacitors in series

F, farad, unit of capacitance

κ, dielectric constant

$u = \frac{1}{2}\epsilon_0 E^2$, electric energy density

ANSWERS TO SELF-TEST OPPORTUNITIES

24.1 The electric field remains constant.

24.2 Series:

$$\frac{1}{C_{eq}} = \frac{4}{C} \Rightarrow C_{eq} = \frac{1}{4}C = 2.50\ \mu F.$$

Parallel:

$$C_{eq} = 4C = 40.0\ \mu F.$$

24.3 $100\ V = d(2500\ V/mm) \Rightarrow d = 0.04\ mm.$

PROBLEM-SOLVING PRACTICE

Problem-Solving Guidelines

1. Remember that saying that a capacitor has charge q means that one plate has charge $+q$ and the other plate has charge $-q$. Be sure you understand how a charge applied to a capacitor is distributed between the two conducting plates; review Example 24.3 if you're uncertain about this.

2. It is always a good idea to draw a circuit diagram when solving problems involving circuits, if one is not supplied. Identifying series and parallel connections can take some practice, but is usually an important first step in reducing a complicated-looking circuit to an equivalent circuit that is straightforward to deal with. Remember that capacitors connected in series all have the same charge, and capacitors connected in parallel all have the same potential difference.

3. You can remember most of the important results for a capacitor with a dielectric if you remember that a dielectric increases the capacitance. (This is what makes a dielectric useful.) If your calculations show a reduced capacitance with a dielectric, recheck your work.

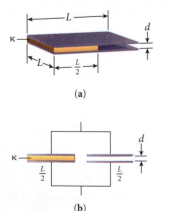

(a)

(b)

FIGURE 24.30 (a) A parallel plate capacitor with square plates of side length L separated by distance d with a dielectric that is L wide and $L/2$ long with dielectric constant κ inserted between the plates. (b) The partially filled capacitor represented as two capacitors in parallel.

SOLVED PROBLEM 24.2 | **Capacitor Partially Filled with a Dielectric**

PROBLEM

A parallel plate capacitor is constructed of two square conducting plates with side length $L = 10.0$ cm (Figure 24.30a). The distance between the plates is $d = 0.250$ cm. A dielectric with dielectric constant $\kappa = 15.0$ and thickness 0.250 cm is inserted between the plates. The dielectric is $L = 10.0$ cm wide and $L/2 = 5.00$ cm long, as shown in Figure 24.30a. What is the capacitance of this capacitor?

SOLUTION

THINK

We have a parallel plate capacitor that is partially filled with a dielectric. We can treat this capacitor as two capacitors in parallel. One capacitor is a parallel plate capacitor with plate area $A = L(L/2)$ and air between the plates; the second capacitor is a parallel plate capacitor with plate area $A = L(L/2)$ and a dielectric between the plates.

SKETCH

Figure 24.30b shows a representation of the partially filled capacitor as two capacitors in parallel: one capacitor filled with a dielectric and the other an air-filled capacitor.

RESEARCH

The capacitance, C_1, of a parallel plate capacitor is given by equation 24.6:

$$C_1 = \frac{\epsilon_0 A}{d},$$

where A is the area of the plates and d is the separation between them. If a dielectric is placed between the plates, the capacitance becomes

$$C_2 = \kappa \frac{\epsilon_0 A}{d},$$

where κ is the dielectric constant. For two capacitors, C_1 and C_2, in parallel, the effective capacitance, C_{12}, is given by

$$C_{12} = C_1 + C_2.$$

SIMPLIFY

Substituting the expressions for the two individual capacitances into the sum, we get

$$C_{12} = \frac{\epsilon_0 A}{d} + \kappa \frac{\epsilon_0 A}{d} = (\kappa + 1)\frac{\epsilon_0 A}{d}. \tag{i}$$

The area of the plates for each capacitor is

$$A = (L)(L/2) = L^2/2.$$

Inserting the expression for the area into equation (i) gives the capacitance of the partially filled capacitor to be

$$C_{12} = (\kappa + 1)\frac{\epsilon_0 (L^2/2)}{d} = \frac{(\kappa + 1)\epsilon_0 L^2}{2d}.$$

CALCULATE

Putting in the numerical values, we get

$$C_{12} = \frac{(15.0 + 1)(8.85 \cdot 10^{-12} \text{ F/m})(0.100 \text{ m})^2}{2(0.00250 \text{ m})} = 2.832 \cdot 10^{-10} \text{ F}.$$

ROUND

We report our result to three significant figures:

$$C_{12} = 2.83 \cdot 10^{-10} \text{ F} = 283 \text{ pF}.$$

DOUBLE-CHECK

To double-check our answer, we calculate the capacitance of the capacitor without any dielectric:

$$C = \frac{\epsilon_0 A}{d} = \frac{(8.85 \cdot 10^{-12} \text{ F/m})(0.100 \text{ m})^2}{0.0025 \text{ m}} = 3.54 \cdot 10^{-11} \text{ F} = 35.4 \text{ pF}.$$

We then calculate the capacitance of the capacitor if completely filled with dielectric:

$$C = \kappa \frac{\epsilon_0 A}{d} = (15.0)(35.4 \text{ pF}) = 5.31 \cdot 10^{-10} \text{ F} = 531 \text{ pF}.$$

Our result for the partially filled capacitor is half of the sum of these two results, so it seems reasonable.

SOLVED PROBLEM 24.3 | Charge on a Cylindrical Capacitor

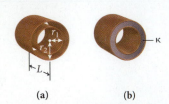

(a)   **(b)**

FIGURE 24.31 (a) A cylindrical capacitor with inner radius r_1, outer radius r_2, and length L. (b) A dielectric with dielectric constant κ is inserted between the cylinders.

FIGURE 24.32 A cylindrical capacitor connected to a battery.

PROBLEM

Consider a cylindrical capacitor with inner radius $r_1 = 10.0$ cm, outer radius $r_2 = 12.0$ cm, and length $L = 50.0$ cm (Figure 24.31a). A dielectric with dielectric constant $\kappa = 12.5$ fills the volume between the two cylinders (Figure 24.31b). The capacitor is connected to a 100.0-V battery and charged completely. What is the charge on the capacitor?

SOLUTION

THINK

We have a cylindrical capacitor filled with a dielectric. When the capacitor is connected to the battery, charge will accumulate on the capacitor until the capacitor is fully charged. We can calculate the amount of charge on the capacitor.

SKETCH

A circuit diagram with the cylindrical capacitor connected to a battery is shown in Figure 24.32.

RESEARCH

The capacitance, C, of a cylindrical capacitor is given by equation 24.8:

$$C = \frac{2\pi\epsilon_0 L}{\ln(r_2/r_1)},$$

where r_1 is the inner radius of the capacitor, r_2 is the outer radius of the capacitor, and L is the length of the capacitor. With a dielectric between the plates, the capacitance becomes

$$C = \kappa \frac{2\pi\epsilon_0 L}{\ln(r_2/r_1)}, \tag{i}$$

where κ is the dielectric constant. For a capacitor with capacitance C charged to a potential difference ΔV, the charge q is given by equation 24.1:

$$q = C\Delta V. \tag{ii}$$

SIMPLIFY

Combining equations (i) and (ii) gives

$$q = C\Delta V = \left(\kappa \frac{2\pi\epsilon_0 L}{\ln(r_2/r_1)} \right) \Delta V = \frac{2\kappa\pi\epsilon_0 L\Delta V}{\ln(r_2/r_1)}.$$

CALCULATE

Putting in the numerical values, we get

$$q = \frac{2\kappa\pi\epsilon_0 L\Delta V}{\ln(r_2/r_1)} = \frac{2(12.5)\pi(8.85\cdot10^{-12}\text{ F/m})(0.500\text{ m})(100.0\text{ V})}{\ln[(0.120\text{ m})/(0.100\text{ m})]} = 19.0618\cdot10^{-8}\text{ C}.$$

ROUND

We report our result to three significant figures:

$$q = 19.1\cdot10^{-8}\text{ C} = 191\text{ nC}.$$

DOUBLE-CHECK

Our answer is a very small fraction of a coulomb of charge, so it seems reasonable.

MULTIPLE-CHOICE QUESTIONS

24.1 In the circuit shown in the figure, the capacitance for each capacitor is C. The equivalent capacitance for these three capacitors is

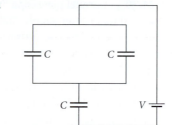

a) $\frac{1}{3}C$

b) $\frac{2}{3}C$

c) $\frac{2}{5}C$

d) $\frac{3}{5}C$

e) C

f) $\frac{5}{3}C$

24.2 A parallel plate capacitor of capacitance C has plates of area A with distance d between them. When the capacitor is connected to a battery of potential difference V, it has a charge of magnitude Q on its plates. While the capacitor is connected to the battery, the distance between the plates is decreased by a factor of 3. The magnitude of the charge on the plates and the capacitance are then

a) $\frac{1}{3}Q$ and $\frac{1}{3}C$.

b) $\frac{1}{3}Q$ and $3C$.

c) $3Q$ and $3C$.

d) $3Q$ and $\frac{1}{3}C$.

24.3 The distance between the plates of a parallel plate capacitor is reduced by half and the area of the plates is doubled. What happens to the capacitance?

a) It remains unchanged.

b) It doubles.

c) It quadruples.

d) It is reduced by half.

24.4 Which of the following capacitors has the largest charge?

a) a parallel plate capacitor with an area of 10 cm^2 and a plate separation of 2 mm connected to a 10-V battery

b) a parallel plate capacitor with an area of 5 cm^2 and a plate separation of 1 mm connected to a 10-V battery

c) a parallel plate capacitor with an area of 10 cm^2 and a plate separation of 4 mm connected to a 5-V battery

d) a parallel plate capacitor with an area of 20 cm^2 and a plate separation of 2 mm connected to a 20-V battery

e) All of the capacitors have the same charge.

24.5 Two identical parallel plate capacitors are connected in a circuit as shown in the figure. Initially the space between the plates of each capacitor is filled with air. Which of the following changes will double the total amount of charge stored on both capacitors with the same applied potential difference?

a) Fill the space between the plates of C_1 with glass (dielectric constant of 4) and leave C_2 as is.

b) Fill the space between the plates of C_1 with Teflon (dielectric constant of 2) and leave C_2 as is.

c) Fill the space between the plates of both C_1 and C_2 with Teflon (dielectric constant of 2).

d) Fill the space between the plates of both C_1 and C_2 with glass (dielectric constant of 4).

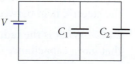

24.6 The space between the plates of an isolated parallel plate capacitor is filled with a slab of dielectric material. The magnitude of the charge Q on each plate is kept constant. If the dielectric material is removed from between the plates, the energy stored in the capacitor

a) increases.

b) stays the same.

c) decreases.

d) may increase or decrease.

24.7 Which of the following is proportional to the capacitance of a parallel plate capacitor?

a) the charge stored on each conducting plate

b) the potential difference between the two plates

c) the separation distance between the two plates

d) the area of each plate

e) all of the above

f) none of the above

24.8 A dielectric with a dielectric constant $\kappa = 4$ is inserted into a parallel plate capacitor, filling $\frac{1}{3}$ of the volume, as shown in the figure. If the capacitance of the capacitor without the dielectric is C, what is the capacitance of the capacitor with the dielectric?

a) $0.75C$

b) C

c) $2C$

d) $4C$

e) $6C$

24.9 A parallel plate capacitor is connected to a battery for charging. After some time, while the battery is still connected to the capacitor, the distance between the capacitor plates is doubled. Which of the following is (are) true?

a) The electric field between the plates is halved.

b) The potential difference of the battery is halved.

c) The capacitance doubles.

d) The potential difference across the plates does not change.

e) The charge on the plates does not change.

24.10 Referring to the figure, decide whether each of the following equations is true or false. Assume that all of the capacitors have different capacitances. The potential difference across capacitor C_1 is V_1. The potential difference across capacitor C_2 is V_2. The potential difference across capacitor C_3 is V_3. The potential difference across capacitor C_4 is V_4. The charge stored in capacitor C_1 is q_1. The charge stored in capacitor C_2 is q_2. The charge stored in capacitor C_3 is q_3. The charge stored in capacitor C_4 is q_4.

a) $q_1 = q_3$

b) $V_1 + V_2 = V$

c) $q_1 + q_2 = q_3 + q_4$

d) $V_1 + V_2 = V_3 + V_4$

e) $V_1 + V_3 = V$

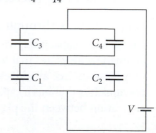

QUESTIONS

24.11 Must a capacitor's plates be made of conducting material? What would happen if two insulating plates were used instead of conducting plates?

24.12 Does it take more work to separate the plates of a charged parallel plate capacitor while it remains connected to the charging battery or after it has been disconnected from the charging battery?

24.13 When working on a piece of equipment, electricians and electronics technicians sometimes attach a grounding wire to the equipment even after turning the device off and unplugging it. Why would they do this?

24.14 Table 24.1 does not list a value of the dielectric constant for any good conductor. What value would you assign to it?

24.15 A parallel plate capacitor is charged with a battery and then disconnected from the battery, leaving a certain amount of energy stored in the capacitor. The separation between the plates is then increased. What happens to the energy stored in the capacitor? Discuss your answer in terms of energy conservation.

24.16 You have an electric device containing a 10.0-μF capacitor, but an application requires an 18.0-μF capacitor. What modification can you make to your device to increase its capacitance to 18.0-μF?

24.17 Two capacitors with capacitances C_1 and C_2 are connected in series. Show that, no matter what the values of C_1 and C_2 are, the equivalent capacitance is always less than the smaller of the two capacitances.

24.18 Two capacitors, with capacitances C_1 and C_2, are connected in series. A potential difference, V_0, is applied across the combination of capacitors. Find the potential differences V_1 and V_2 across the individual capacitors, in terms of V_0, C_1, and C_2.

24.19 An isolated solid spherical conductor of radius 5.00 cm is surrounded by dry air. It is given a charge and acquires potential V, with the potential at infinity assumed to be zero.

a) Calculate the maximum magnitude V can have.

b) Explain clearly and concisely *why* there is a maximum.

24.20 A parallel plate capacitor of capacitance C is connected to a power supply that maintains a constant potential difference, V. A close-fitting slab of dielectric, with dielectric constant κ, is then inserted and fills the previously empty space between the plates.

a) What was the energy stored on the capacitor before the insertion of the dielectric?

b) What was the energy stored after the insertion of the dielectric?

c) Was the dielectric pulled into the space between the plates, or did it have to be pushed in? Explain.

24.21 A parallel plate capacitor with square plates of edge length L separated by a distance d is given a charge Q, then disconnected from its power source. A close-fitting square slab of dielectric, with dielectric constant κ, is then inserted into the previously empty space between the plates. Calculate the force with which the slab is pulled into the capacitor during the insertion process.

24.22 Consider a cylindrical capacitor, with outer radius R and cylinder separation d. Determine what the capacitance approaches in the limit where $d \ll R$. (*Hint:* Express the capacitance in terms of the ratio d/R and then examine what happens as the ratio d/R becomes very small compared to 1.) Explain why the limit on the capacitance makes sense.

24.23 A parallel plate capacitor is constructed from two plates of different areas. If this capacitor is initially uncharged and then connected to a battery, how will the amount of charge on the big plate compare to the amount of charge on the small plate?

24.24 A parallel plate capacitor is connected to a battery. As the plates are moved farther apart, what happens to each of the following?

a) the potential difference across the plates

b) the charge on the plates

c) the electric field between the plates

PROBLEMS

A blue problem number indicates a worked-out solution is available in the Student Solutions Manual. One • and two •• indicate increasing level of problem difficulty.

Sections 24.3 through 24.5

24.25 Supercapacitors, with capacitances of 1 F or more, are made with plates that have a spongelike structure with a very large surface area. Determine the surface area of a supercapacitor that has a capacitance of 1.0 F and an effective separation between the plates of $d = 1.0$ mm.

24.26 A potential difference of 100 V is applied across the two collinear conducting cylinders shown in the figure. The radius of the outer cylinder is 15 cm, the radius of the inner cylinder is 10 cm, and the length of the two cylinders is 40.0 cm. How much charge is applied to each of the cylinders? What is the magnitude of the electric field between the two cylinders?

24.27 What is the radius of an isolated spherical conductor that has a capacitance of 1 F?

24.28 A spherical capacitor is made from two thin concentric conducting shells. The inner shell has radius r_1, and the outer shell has radius r_2. What is the fractional difference in the capacitances of this spherical capacitor and a parallel plate capacitor made from plates that have the same area as the inner sphere and the same separation $d = r_2 - r_1$ between plates?

24.29 Calculate the capacitance of the Earth. Treat the Earth as an isolated spherical conductor of radius 6371 km.

24.30 Two concentric metal spheres are found to have a potential difference of 900 V when a charge of $6.726 \cdot 10^{-8}$ C is applied to them. The radius of the outer sphere is 0.21 m. What is the radius of the inner sphere?

•**24.31** A capacitor consists of two parallel plates, but one of them can move relative to the other as shown in the figure. Air fills the space between the plates, and the capacitance is 32.0 pF when the separation between plates is $d = 0.500$ cm.

a) A battery with potential difference $V = 9.0$ V is connected to the plates. What is the charge distribution, σ, on the left plate? What are the capacitance, C', and charge distribution, σ', when d is changed to 0.250 cm?

b) With $d = 0.500$ cm, the battery is disconnected from the plates. The plates are then moved so that $d = 0.250$ cm. What is the potential difference V', between the plates?

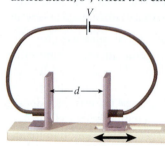

Section 24.6

24.32 Determine all the values of equivalent capacitance you can create using any combination of three identical capacitors with capacitance C.

24.33 A large parallel plate capacitor with plates that are square with side length 1 cm and are separated by a distance of 1 mm is dropped and damaged. Half of the areas of the two plates are pushed closer together to a distance of 0.5 mm. What is the capacitance of the damaged capacitor?

24.34 Three capacitors with capacitances $C_1 = 3.1$ nF, $C_2 = 1.3$ nF, and $C_3 = 3.7$ nF are wired to a battery with $V = 14.9$ V, as shown in the figure. What is the potential drop across capacitor C_2?

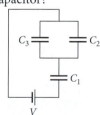

24.35 Four capacitors with capacitances $C_1 = 3.5$ nF, $C_2 = 2.1$ nF, $C_3 = 1.3$ nF, and $C_4 = 4.9$ nF are wired to a battery with $V = 10.3$ V, as shown in the figure. What is the equivalent capacitance of this set of capacitors?

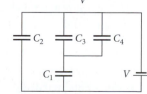

24.36 The capacitors in the circuit shown in the figure have capacitances $C_1 = 18.0$ μF, $C_2 = 11.3$ μF, $C_3 = 33.0$ μF,

and $C_4 = 44.0$ μF. The potential difference is $V = 10.0$ V. What is the total charge the power source must supply to charge this arrangement of capacitors?

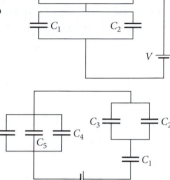

24.37 Six capacitors are connected as shown in the figure.

a) If $C_3 = 2.3$ nF, what does C_2 have to be to yield an equivalent capacitance of 5.000 nF for the combination of the two capacitors?

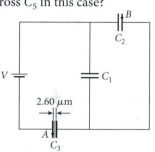

b) For the same values of C_2 and C_3 as in part (a), what is the value of C_1 that will give an equivalent capacitance of 1.914 nF for the combination of the three capacitors?

c) For the same values of C_1, C_2, and C_3 as in part (b), what is the equivalent capacitance of the whole set of capacitors if the values of the other capacitances are $C_4 = 1.3$ nF, $C_5 = 1.7$ nF, and $C_6 = 4.7$ nF?

d) If a battery with a potential difference of 11.7 V is connected to the capacitors as shown in the figure, what is the total charge on the six capacitors?

e) What is the potential drop across C_5 in this case?

•**24.38** A potential difference of $V = 80.0$ V is applied across a circuit with capacitances $C_1 = 15.0$ nF, $C_2 = 7.00$ nF, and $C_3 = 20.0$ nF, as shown in the figure. What is the magnitude and sign of q_{3l}, the charge on the left plate of C_3 (marked by point A)? What is the electric potential, V_3, across C_3? What is the magnitude and sign of the charge q_{2r}, on the right plate of C_2 (marked by point B)?

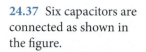

•**24.39** Fifty parallel plate capacitors are connected in series. The distance between the plates is d for the first capacitor, $2d$ for the second capacitor, $3d$ for the third capacitor, and so on. The area of the plates is the same for all the capacitors. Express the equivalent capacitance of the whole set in terms of C_1 (the capacitance of the first capacitor).

•**24.40** A 5.00-nF capacitor charged to 60.0 V and a 7.00-nF capacitor charged to 40.0 V are connected with the negative plate of each connected to the negative plate of the other. What is the final charge on the 7.00-nF capacitor?

Section 24.7

24.41 When a capacitor has a charge of magnitude 60 μC on each plate, the potential difference across the plates is 12 V. How much energy is stored in this capacitor when the potential difference across its plates is 120 V?

24.42 The capacitor in an automatic external defibrillator is charged to 7.5 kV and stores 2400 J of energy. What is its capacitance?

24.43 The Earth has an electric field of 150 N/C near its surface. Find the electrical energy contained in each cubic meter of air near the surface.

•**24.44** The potential difference across two capacitors in series is 120 V. The capacitances are $C_1 = 1000 \ \mu F$ and $C_2 = 1500 \ \mu F$.

a) What is the total capacitance of this pair of capacitors?

b) What is the charge on each capacitor?

c) What is the potential difference across each capacitor?

d) What is the total energy stored by the capacitors?

•**24.45** Neutron stars are thought to have electric dipole ($\vec{p}$) layers at their surfaces. If a neutron star with a 10-km radius has a dipole layer 1 cm thick with charge distributions of $+1 \ \mu C/cm^2$ and $-1 \ \mu C/cm^2$ on the surface, as indicated in the figure, what is the capacitance of this star? What is the electric potential energy stored in the neutron star's dipole layer?

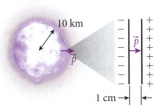

24.46 A 4000-nF parallel plate capacitor is connected to a 12-V battery and charged.

a) What is the charge Q on the positive plate of the capacitor?

b) What is the electric potential energy stored in the capacitor?

The 4000-nF capacitor is then disconnected from the 12-V battery and used to charge three uncharged capacitors, a 100-nF capacitor, a 200-nF capacitor, and a 300-nF capacitor, connected in series.

c) After charging, what is the potential difference across each of the four capacitors?

d) How much of the electrical energy stored in the 4000-nF capacitor was transferred to the other three capacitors?

•**24.47** The figure shows a circuit with $V = 12$ V, $C_1 = 500$ pF, and $C_2 = 500$ pF. The switch is closed, to A, and the capacitor C_1 is fully charged. Find (a) the energy delivered by the battery and (b) the energy stored in C_1. Then the switch is thrown to B and the circuit is allowed to reach equilibrium. Find (c) the total energy stored at C_1 and C_2. (d) Explain the energy loss, if there is any.

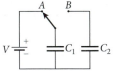

••**24.48** The Earth is held together by its own gravity. But it is also a charge-bearing conductor.

a) The Earth can be regarded as a conducting sphere of radius 6371 km, with electric field $\vec{E} = (-150. \ V/m)\hat{r}$ at its surface, where $\hat{r}$ is a unit vector directed radially outward. Calculate the total electrostatic potential energy associated with the Earth's electric charge and field.

b) The Earth has gravitational potential energy, akin to the electrostatic potential energy. Calculate this energy, treating the Earth as a uniform solid sphere. (*Hint:* $dU = -(Gm/r)dm$.)

c) Use the results of parts (a) and (b) to address this question: To what extent do electrostatic forces affect the structure of the Earth?

Section 24.8

24.49 Two parallel plate capacitors have identical plate areas and identical plate separations. The maximum energy each can store is determined by the maximum potential difference that can be applied before dielectric breakdown occurs. One capacitor has air between its plates, and the other has Mylar. Find the ratio of the maximum energy the Mylar capacitor can store to the maximum energy the air capacitor can store.

24.50 A capacitor has parallel plates, with half of the space between the plates filled with a dielectric material of constant κ and the other half filled with air as shown in the figure. Assume that the plates are square, with sides of length L, and that the separation between the plates is s. Determine the capacitance as a function of L.

24.51 Calculate the maximum surface charge distribution that can be maintained on any surface surrounded by dry air.

24.52 Thermocoax is a type of coaxial cable used for high-frequency filtering in cryogenic quantum computing experiments. Its stainless steel shield has an inner diameter of 0.35 mm, and its Nichrome conductor has a diameter of 0.17 mm. Nichrome is used because its resistance doesn't change much in going from room temperature to near absolute zero. The insulating dielectric is magnesium oxide (MgO), which has a dielectric constant of 9.7. Calculate the capacitance per meter of Thermocoax.

24.53 A parallel plate capacitor has square plates of side $L = 10$ cm and a distance $d = 1$ cm between the plates. Of the space between the plates, $\frac{1}{5}$ is filled with a dielectric with dielectric constant $\kappa_1 = 20$. The remaining $\frac{4}{5}$ of the space is filled with a different dielectric with $\kappa_2 = 5$. Find the capacitance of the capacitor.

•**24.54** A 4.0-nF parallel plate capacitor with a sheet of Mylar ($\kappa = 3.1$) filling the space between the plates is charged to a potential difference of 120 V and is then disconnected.

a) How much work is required to completely remove the sheet of Mylar from the space between the two plates?

b) What is the potential difference between the plates of the capacitor once the Mylar is completely removed?

•**24.55** The volume between the two cylinders of a cylindrical capacitor is half filled with a dielectric with dielectric constant κ and is connected to a battery with potential difference ΔV. What is the charge placed on the capacitor? What is the ratio of this charge to the charge placed on a capacitor with no dielectric connected in the same way across the same potential drop?

•**24.56** A dielectric slab with thickness d and dielectric constant $\kappa = 2.31$ is inserted in a parallel place capacitor that has been charged by a 110-V battery and having area $A = 100 \ cm^2$, and separation distance $d = 2.50$ cm.

a) Find the capacitance, C, the potential difference, V, the electric field, E, the total charge stored on the capacitor Q, and electric potential energy stored in the capacitor, U, before the dielectric material is inserted.

b) Find C, V, E, Q, and U when the dielectric slab has been inserted and the battery is still connected.

c) Find C, V, E, Q, and U when the dielectric slab is in place and the battery is disconnected.

•24.57 A parallel plate capacitor has a capacitance of 120 pF and a plate area of 100 cm^2. The space between the plates is filled with mica whose dielectric constant is 5.40. The plates of the capacitor are kept at 50 V.

a) What is the strength of the electric field in the mica?

b) What is the amount of free charge on the plates?

c) What is the amount of charge induced on the mica?

••24.58 Design a parallel plate capacitor with a capacitance of 47.0 pF and a capacity of 7.50 nC. You have available conducting plates, which can be cut to any size, and Plexiglas sheets, which can be cut to any size and machined to any thickness. Plexiglas has a dielectric constant of 3.40 and a dielectric strength of $4.00 \cdot 10^7$ V/m. You must make your capacitor as compact as possible. Specify all relevant dimensions. Ignore any fringe field at the edges of the capacitor plates.

24.59 A parallel plate capacitor consisting of a pair of rectangular plates, each measuring 1 cm by 10 cm, with a separation between the plates of 0.1 mm, is charged by a power supply at a potential difference of 1000 V. The power supply is then removed, and without being discharged, the capacitor is placed in a vertical position over a container holding de-ionized water, with the short sides of the plates in contact with the water, as shown in the figure. Using energy considerations, show that the water will rise between the plates. Neglecting other effects, determine the system of equations that can be used to calculate the height to which the water rises between the plates. You do not have to solve the system.

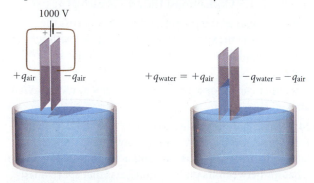

Additional Problems

24.60 Two circular metal plates of radius 0.61 m and thickness 7.1 mm are used in a parallel plate capacitor. A gap of 2.1 mm is left between the plates, and half of the space (a semicircle) between the plates is filled with a dielectric for which $\kappa = 11.1$ and the other half is filled with air. What is the capacitance of this capacitor?

24.61 Considering the dielectric strength of air, what is the maximum amount of charge that can be stored on the plates of a capacitor that are a distance of 15 mm apart and have an area of 25 cm^2?

24.62 The figure shows three capacitors in a circuit: $C_1 = 2.0$ nF and $C_2 = C_3 = 4.0$ nF. Find the charge on each capacitor when the potential difference applied is $V = 1.5$ V.

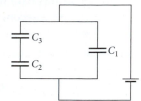

24.63 A capacitor with a vacuum between its plates is connected to a battery and then the gap is filled with Mylar. By what percentage is its energy-storing capacity increased?

24.64 A parallel plate capacitor with a plate area of 12.0 cm^2 and air in the space between the plates, which are separated by 1.5 mm, is connected to a 9-V battery. If the plates are pulled back so that the separation increases to 2.75 mm, how much work is done?

24.65 Suppose you want to make a 1.0-F capacitor using two square sheets of aluminum foil. If the sheets of foil are separated by a single piece of paper (thickness of about 0.10 mm and $\kappa \approx 5.0$), find the size of the sheets of foil (the length of each edge).

24.66 A 4-pF parallel plate capacitor has a potential difference of 10 V across it. The plates are 3 mm apart, and the space between them contains air.

a) What is the charge on the capacitor?

b) How much energy is stored in the capacitor?

c) What is the area of the plates?

d) What would the capacitance of this capacitor be if the space between the plates were filled with polystyrene?

24.67 A four-capacitor circuit is charged by a battery, as shown in the figure. The capacitances are $C_1 = 1.0$ mF, $C_2 = 2.0$ mF, $C_3 = 3.0$ mF, and $C_4 = 4.0$ mF, and the battery potential is $V_B = 1.0$ V. When the circuit is at equilibrium, point D has potential $V_D = 0$ V. What is the potential, V_A, at point A?

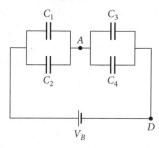

24.68 How much energy can be stored in a capacitor with two parallel plates, each with an area of 64 cm^2 and separated by a gap of 1.3 mm, filled with porcelain whose dielectric constant is 7, and holding equal and opposite charges of magnitude 420 μC?

24.69 A quantum mechanical device known as the *Josephson junction* consists of two overlapping layers of superconducting metal (for example, aluminum at 1 K) separated by 20 nm of aluminum oxide, which has a dielectric constant of 9.1. If this device has an area of 100 μm^2 and a parallel plate configuration, estimate its capacitance.

24.70 Three capacitors with capacitances $C_1 = 6$ μF, $C_2 = 3$ μF, and $C_3 = 5$ μF are connected in a circuit as shown in the figure, with an applied potential of V. After the charges on the capacitors have reached their equilibrium values, the charge Q_2 on the second capacitor is found to be 40 μC.

a) What is the charge, Q_1, on capacitor C_1?

b) What is the charge, Q_3, on capacitor C_3?

c) How much voltage, V, was applied across the capacitors?

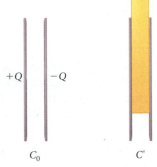

Total $Q = CV$

$C_3 Q = $

$C_1 Q = 2 \cdot C_2 Q$

$V = \dfrac{Q}{C}$

24.71 For a science project, a fourth-grader cuts the tops and bottoms off two soup cans of equal height, 7.24 cm, and with radii of 3.02 cm and 4.16 cm, puts the smaller one inside the larger, and hot-glues them both on a sheet of plastic, as shown in the figure. Then she fills the gap between the cans with a special "soup" (dielectric constant of 63). What is the capacitance of this arrangement?

24.72 The Earth can be thought of as a spherical capacitor. If the net charge on the Earth is $-7.8 \cdot 10^5$ C, find (a) the capacitance of the Earth and (b) the electric potential energy stored on the Earth's surface.

•24.73 A parallel plate capacitor with air in the gap between the plates is connected to a 6-V battery. After charging, the energy stored in the capacitor is 72 nJ. Without disconnecting the capacitor from the battery, a dielectric is inserted into the gap and an additional 317 nJ of energy flows from the battery to the capacitor.

a) What is the dielectric constant of the dielectric?

b) If each of the plates has an area of 50 cm^2, what is the charge on the positive plate of the capacitor after the dielectric has been inserted?

c) What is the magnitude of the electric field between the plates before the dielectric is inserted?

d) What is the magnitude of the electric field between the plates after the dielectric is inserted?

•24.74 An 8-μF capacitor is fully charged by a 240-V battery, which is then disconnected. Next, the capacitor is connected to an initially uncharged capacitor of capacitance C, and the potential difference across it is found to be 80 V. What is C? How much energy ends up being stored in the second capacitor?

•24.75 A parallel plate capacitor consists of square plates of edge length 2 cm separated by a distance of 1 mm. The capacitor is charged with a 15-V battery, and the battery is then removed. A 1-mm-thick sheet of nylon (dielectric constant = 3) is slid between the plates. What is the average force (magnitude and direction) on the nylon sheet as it is inserted into the capacitor?

•24.76 A proton traveling along the x-axis at a speed of $1.0 \cdot 10^6$ m/s enters the gap between the plates of a 2.0-cm-wide parallel plate capacitor. The surface charge distributions on the plates are given by $\sigma = \pm 1.0 \cdot 10^{-6}$ C/m^2. How far has the proton been deflected sideways (Δy) when it reaches the far edge of the capacitor? Assume that the electric field is uniform inside the capacitor and zero outside.

•24.77 A parallel plate capacitor has square plates of side $L = 10.0$ cm separated by a distance $d = 2.5$ mm, as shown in the figure. The capacitor is charged by a battery with potential difference $V_0 = 75.0$ V; the battery is then disconnected.

a) Determine the capacitance, C_0, and the electric potential energy, U_0, stored in the capacitor at this point.

b) A slab made of Plexiglas ($\kappa = 3.4$) is then inserted so that it fills $\frac{2}{3}$ of the volume between the plates, as shown in the figure. Determine the new capacitance, C', the new potential difference between the plates V', and the new electric potential energy, U', stored in the capacitor.

c) Neglecting gravity, did the inserter of the dielectric slab have to do work or not?

$+Q$ $-Q$

C_0 C'

Charged capacitor   Charged capacitor with
                                        Plexiglas slab inserted

•24.78 A typical AAA battery has stored energy of about 3400 J. (Battery capacity is typically listed as 625 mA h, meaning that much charge can be delivered at approximately 1.5 V.) Suppose you want to build a parallel plate capacitor to store this amount of energy, using a plate separation of 1.0 mm and with air filling the space between the plates.

a) Assuming that the potential difference across the capacitor is 1.5 V, what must the area of each plate be?

b) Assuming that the potential difference across the capacitor is the maximum that can be applied without dielectric breakdown occurring, what must the area of each plate be?

c) Is either capacitor a practical replacement for the AAA battery?

•24.79 Two parallel plate capacitors, C_1 and C_2, are connected in series to a 96-V battery. Both capacitors have plates with an area of 1.0 cm^2 and a separation of 0.1 mm; C_1 has air between its plates, and C_2 has that space filled with porcelain (dielectric constant of 7 and dielectric strength of 5.7 kV/mm).

a) After charging, what are the charges on each capacitor?

b) What is the total energy stored in the two capacitors?

c) What is the electric field between the plates of C_2?

•24.80 The plates of parallel plate capacitor A consist of two metal discs of identical radius, $R_1 = 4$ cm, separated by a distance $d = 2$ mm, as shown in the figure.

a) Calculate the capacitance of this parallel plate capacitor with the space between the plates filled with air.

b) A dielectric in the shape of a thick-walled cylinder of outer radius $R_1 = 4$ cm, inner radius $R_2 = 2$ cm, thickness $d = 2$ mm, and dielectric constant $\kappa = 2$ is placed between the plates, coaxial with the plates, as shown in the figure. Calculate the capacitance of capacitor B, with this dielectric.

c) The dielectric cylinder is removed, and instead a solid disc of radius R_1 made of the same dielectric is placed between the plates to form capacitor C, as shown in the figure. What is the new capacitance?

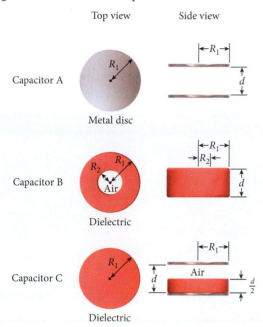

•**24.81** A 1.00-μF capacitor charged to 50.0 V and a 2.00-μF capacitor charged to 20.0 V are connected, with the positive plate of each connected to the negative plate of the other. What is the final charge on the 1.00-μF capacitor?

•**24.82** The capacitance of a spherical capacitor consisting of two concentric conducting spheres with radii r_1 and r_2 ($r_2 > r_1$) is given by $C = 4\pi\epsilon_0 r_1 r_2/(r_2 - r_1)$. Suppose that the space between the spheres, from r_1 up to a radius R ($r_1 < R < r_2$) is filled with a dielectric for which $\epsilon = 10\epsilon_0$. Find an expression for the capacitance, and check the limits when $R = r_1$ and $R = r_2$.

•**24.83** In the figure, a parallel plate capacitor is connected to a 300-V battery. With the capacitor connected, a proton is fired with a speed of $2.0 \cdot 10^5$ m/s from (through) the negative plate of the capacitor at an angle θ with the normal to the plate.

a) Show that the proton cannot reach the positive plate of the capacitor, regardless of what the angle θ is.

b) Sketch the trajectory of the proton between the plates.

c) Assuming that $V = 0$ at the negative plate, calculate the potential at the point between the plates where the proton reverses its motion in the x-direction.

d) Assuming that the plates are long enough for the proton to stay between them throughout its motion, calculate the speed (magnitude only) of the proton as it collides with the negative plate.

••**24.84** For the parallel plate capacitor with dielectric shown in the figure, prove that for a given thickness of the dielectric slab, the capacitance does not depend on the position of the slab relative to the two conducting plates (that is, it does not depend on the values of d_1 and d_3).

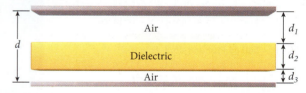

25

Current and Resistance

FIGURE 25.1 A current flowing through a wire makes this light bulb shine.

WHAT WE WILL LEARN

- Electric current at a point in a circuit is the rate at which net charge moves past that point.

- Direct current is current flowing in a direction that does not change with time. The direction of current is defined as the direction in which positive charge would be moving.

- The current density passing a given point in a conductor is the current per cross-sectional area.

- The conductivity of a material characterizes the ability of that material to conduct current. Its inverse is called resistivity.

- The resistance of a device depends on its geometry and on the material of which it is made.

- The resistivity of a conductor increases approximately linearly with temperature.

- Electromotive force (usually referred to as emf) is a potential difference in an electric circuit.

- Ohm's Law states that the potential drop across a device is equal to the current flowing through the device times the resistance of the device.

- A simple circuit consists of a source of emf and resistors connected in series or in parallel.

- In a circuit diagram, an equivalent resistance can replace resistors connected in series or in parallel.

- The power in a circuit is the product of the current and the voltage drop.

- A diode conducts current in one direction but not in the opposite direction.

Electric lighting is so commonplace that you don't even think about it. You walk into a dark room and simply flick on a switch, bringing the room to nearly daytime brightness (Figure 25.1). However, what happens when the switch is flicked depends ultimately on principles of physics and engineering devices that took decades to develop and refine.

This chapter is the first to focus on electric charges in motion. It presents some of the fundamental concepts that we'll use in Chapter 26 to analyze basic electric circuits, which are integral to all electronics applications. These chapters concentrate on the electrical effects of moving charges, but you should be aware that moving charges also give rise to other effects, which we'll start to examine in Chapter 27 on magnetism.

25.1 Electric Current

Up to this point, our study of electricity has focused on electrostatics, which deals with the properties of stationary electric charges and fields. Electric circuits were introduced in the discussion of capacitors in Chapter 24, but it covered only situations involving fully charged capacitors, where the charge is at rest. If electrostatics were all there was to electricity, it would not be nearly as important to modern society as it is. The world-changing impact of electricity is due to the properties of charges in motion, or electric current. All electrical devices rely on some kind of current for their operation.

Let's start by looking at a few very simple experiments. When you were a kid, you very likely had some toys that ran on batteries, and probably some of them contained small light bulbs. Consider a very simple circuit consisting only of a battery, a switch, and a light bulb (see Figure 25.2). If the switch is open, as in Figure 25.2a, the light bulb does not shine. If the switch is closed, as in Figure 25.2b, the light bulb turns on. We all know why this happens—because a current flows through the closed circuit. In Chapter 24, we saw that the battery provides a potential difference for the circuit. In this chapter, we'll examine what it means for the current to flow, what the physical basis of current is, and how it is related to the potential difference provided by the battery. We'll see that the light bulb acts as a resistor in the circuit, and we'll examine how resistors behave.

Let's begin by considering the simple experiment performed in Figure 25.2c, where the battery's orientation is the reverse of what it is in Figure 25.2b. The light bulb shines just the same, despite the fact that the sign of the potential difference provided by the battery is reversed. (The positive terminal of the battery is on the end with the copper color.) In Figure 25.2d, two light bulbs are in the circuit, one behind the other. (Section 25.5 will focus

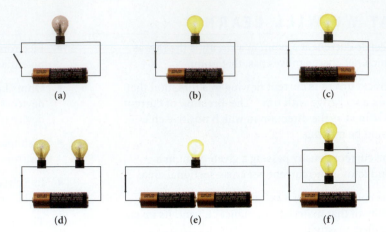

FIGURE 25.2 Experiments with batteries and light bulbs.

on this arrangement of resistors, which is a connection *in series*.) Each of the two light bulbs shines with significantly less intensity than the single bulb in Figure 25.2c, so the current in the circuit may be smaller than before. On the other hand, two batteries in series, as in Figure 25.2e, double the potential difference in the circuit, and the bulb shines significantly brighter. Finally, if separate wires are used to connect the two light bulbs to a single battery, as shown in Figure 25.2f, the light bulbs shine with about the same intensity as in Figure 25.2b or Figure 25.2c. This way of wiring resistors in a circuit is called a *parallel connection* and will be explored in Section 25.6.

Quantitatively, the **electric current,** i, is the net charge passing a given point in a given time, divided by that time. The random motion of electrons in a conductor is not a current, in spite of the fact that large amounts of charge are moving past a given point, because no *net* charge flows. If net charge dq passes a point during time dt, the current at that point is, by definition,

$$i = \frac{dq}{dt}. \tag{25.1}$$

The net amount of charge passing a given point in time t is the integral of the current with respect to time:

$$q = \int dq = \int_0^t i\,dt'. \tag{25.2}$$

Total charge is conserved, implying that charge flowing in a conductor is never lost. Therefore, the same amount of charge flows into one end of a conductor as emerges from the other end.

The unit of current, coulombs per second, was given the name **ampere** (abbreviated A, or sometimes amp), after the French physicist André Ampère (1775–1836):

$$1\,\mathrm{A} = \frac{1\,\mathrm{C}}{1\,\mathrm{s}}.$$

Some typical currents are 1 A for a light bulb, 200 A for the starter in your car, 1 mA = $1 \cdot 10^{-3}$ A to power your MP3 player, 1 nA for the currents in your brain's neurons and synaptic connections, and 10,000 A = 10^4 A in a lightning strike (for a short time). The smallest currents that can be measured are those from individual electrons tunneling in scanning tunneling microscopes and are on the order of 10 pA. The largest current in the Solar System is the solar wind, which is in the GA range. Other examples of the broad range of currents are shown in Figure 25.3.

There is a handy safety rule related to the orders of magnitudes of currents that you need to know: 1-10-100. That is, 1 mA of current flowing through a human body can be felt (as a tingle, usually), 10 mA of current makes muscles contract to the point that the person cannot let go of the wire carrying the current, and 100 mA is sufficient to stop the heart.

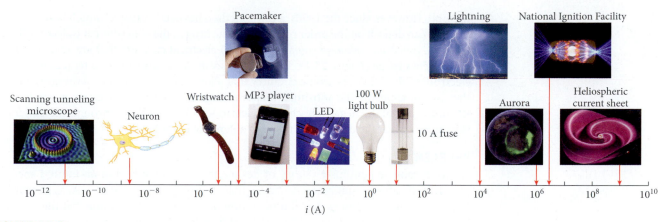

FIGURE 25.3 Examples of electrical currents ranging from 1 pA to 10 GA.

A current that flows in only one direction, which does not change with time, is called **direct current.** (Current that flows first in one direction and then in the opposite direction is called *alternating current* and will be discussed in Chapter 30.) In this chapter, the direction of the current flowing in a conductor is indicated by an arrow.

Physically, the charge carriers in a conductor are electrons, which are negatively charged. However, by convention, positive current is defined as flowing from the positive to the negative terminal. The reason for this counterintuitive definition of current direction is that the definition originated in the second half of the 19th century, when it was not known that electrons are the charge carriers responsible for current. And so the current direction was simply defined as the direction in which the positive charges would flow.

25.1 Self-Test Opportunity

A typical rechargeable AA battery is rated at 700 mAh. How long can this battery provide a current of 100 μA ?

EXAMPLE 25.1 | Iontophoresis

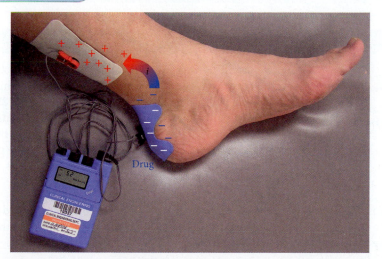

FIGURE 25.4 Iontophoresis is the application of medication under the skin with the aid of electrical current.

There are three ways to administer anti-inflammatory medication. The painless way is the oral one—simply swallowing the drug. However, this method typically leads to a small amount of the drug in the affected tissue, on the order of 1 μg. The second way is to have the drug injected locally with a needle. This hurts but can deposit on the order of 10 mg of the drug in the affected tissue—four orders of magnitude more than with the oral

Continued—

method. However, since the 1990s, a third method has been available, which is also pain-less and can deposit on the order of 100 μg of the drug in the area where it is needed. This method, called *iontophoresis*, uses (very weak) electrical currents that are sent through the patient's tissue (Figure 25.4). The iontophoresis device consists of a battery and two electrodes (plus other electronic circuitry that allows the nurse to control the strength of the current applied). The anti-inflammatory drug, usually dexamethasone, is applied to the underside of the negatively charged electrode. A current flows through the patient's skin and deposits the drug in the tissue to a depth of up to 1.7 cm.

PROBLEM

A nurse wants to administer 80 μg of dexamethasone to the heel of an injured soccer player. If she uses an iontophoresis device that applies a current of 0.14 mA, as shown in Figure 25.4, how long does the administration of the dose take? Assume that the instrument has an application rate of 650 μg/C and that the current flows at a constant rate.

SOLUTION

If the drug application rate is 650 μg/C, to apply 80 μg, requires a total charge of

$$q = \frac{80 \ \mu g}{650 \ \mu g/C} = 0.123 \ C.$$

The current flows at a constant rate, so the integral in equation 25.2 is simply

$$q = \int_0^t i \, dt' = it.$$

Solving for t and inserting the numbers, we find

$$q = it \Rightarrow t = \frac{q}{i} = \frac{0.123 \ C}{0.14 \cdot 10^{-3} \ A} = 880 \ s.$$

The iontophoresis treatment of the athlete will take approximately 15 min.

25.2 Current Density

FIGURE 25.5 Segment of a conductor (wire), with a per-pendicular plane intersecting it and forming a cross-sectional area A.

Consider a current flowing in a conductor. For a perpendicular plane through the conductor, the current per unit area flowing through the conductor at that point (the cross-sectional area A in Figure 25.5) is the **current density, $\vec{J}$**. The direction of $\vec{J}$ is defined as the direction of the velocity of the positive charges (or opposite to the direction of negative charges) crossing the plane. The current flowing through the plane is

$$i = \int \vec{J} \cdot d\vec{A}, \qquad (25.3)$$

where $d\vec{A}$ is the differential area element of the perpendicular plane, as indicated in Figure 25.5. If the current is uniform and perpendicular to the plane, then $i = JA$, and the magnitude of the current density can be expressed as

$$J = \frac{i}{A}. \qquad (25.4)$$

In a conductor that is not carrying current, the conduction electrons move randomly. When current flows through the conductor, electrons still move randomly but also have an additional **drift velocity, $\vec{v}_d$**, in the direction opposite to that of the electric field driving the current. The magnitude of the velocity of random motion is on the order of 10^6 m/s, while the magnitude of the drift velocity is on the order of 10^{-4} m/s or even less. With such a slow drift velocity, you might wonder why a light comes on almost immediately after you turn on a switch. The answer is that the switch establishes an electric field almost immediately throughout the circuit (with a speed on the order of $3 \cdot 10^8$ m/s), causing the free electrons in the entire circuit (including in the light bulb) to move almost instantly.

The current density is related to the drift velocity of the moving electrons. Consider a conductor with cross-sectional area A and electric field $\vec{E}$ applied to it. Suppose the conductor has n conduction electrons per unit volume, and assume that all the electrons have the same drift velocity and that the current density is uniform. The negatively charged electrons will drift in a direction opposite to that of the electric field. In a time interval, dt, each electron moves a net distance $v_d\,dt$. The volume of electrons passing a cross section of the conductor in time dt is then $Av_d\,dt$, and the number of electrons in this volume is $nAv_d\,dt$. Each electron has charge $-e$, so the charge dq that flows through the area in time dt is

$$dq = -nev_d A\,dt. \tag{25.5}$$

Therefore, the current is

$$i = \frac{dq}{dt} = -nev_d A. \tag{25.6}$$

The resulting current density is

$$J = \frac{i}{A} = -nev_d. \tag{25.7}$$

Equation 25.7 was derived in one spatial dimension, as is appropriate for a wire. However, it can be readily generalized to arbitrary directions in three-dimensional space:

$$\vec{J} = -(ne)\vec{v}_d.$$

You can see that the drift velocity vector is antiparallel to the current density vector, as stated before.

Figure 25.6 shows a schematic drawing of a wire carrying a current. The physical current carriers are negatively charged electrons. In Figure 25.6, these electrons are moving to the left with drift velocity $\vec{v}_d$. However, the electric field, the current density, and the current are all directed to the right because of the convention that these quantities refer to positive charges. You may find this convention somewhat confusing, but you'll need to keep it in mind.

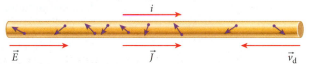

FIGURE 25.6 Electrons moving in a wire from right to left, causing a current in the direction from left to right.

SOLVED PROBLEM 25.1   **Drift Velocity of Electrons in a Copper Wire**

PROBLEM

You are playing "Galactic Destroyer" on your video game console. Your game controller operates at 12 V and is connected to the main box with an 18-gauge copper wire of length 1.5 m. As you fly your spaceship into battle, you hold the joystick in the forward position for 5.3 s, sending a current of 0.78 mA to the console. How far have the electrons in the wire moved during those few seconds, while on the screen your spaceship crossed half of a star system?

SOLUTION

THINK

To find out how far electrons in a wire move during a given time interval, we need to calculate their drift velocity. To determine the drift velocity for electrons in a copper wire carrying a current, we need to find the density of charge-carrying electrons in copper. Then, we can apply the definition of the charge density to calculate the drift velocity.

SKETCH

A copper wire with cross-sectional area A carrying a current, i, is shown in Figure 25.7, which also shows that, by convention, the electrons drift in the direction opposite to the direction of the current.

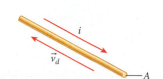

FIGURE 25.7 A copper wire with cross-sectional area A carrying a current, i.

Continued—

RESEARCH

We obtain the distance x traveled by the electrons during time t from

$$x = v_d t,$$

where v_d is the magnitude of the drift velocity of the electrons. The drift velocity is related to the current density via equation 25.7:

$$\frac{i}{A} = -nev_d \qquad (i)$$

where i is the current, A is the cross-sectional area (0.823 mm² for an 18-gauge wire), n is the density of electrons, and $-e$ is the charge of an electron. The density of electrons is defined as

$$n = \frac{\text{number of conduction electrons}}{\text{volume}}.$$

We can calculate the density of electrons by assuming there is one conduction electron per copper atom. The density of copper is

$$\rho_{Cu} = 8.96 \text{ g/cm}^3 = 8960 \text{ kg/m}^3.$$

One mole of copper has a mass of 63.5 g and contains $6.02 \cdot 10^{23}$ atoms. Thus, the density of electrons is

$$n = \left(\frac{1 \text{ electron}}{1 \text{ atom}}\right)\left(\frac{6.02 \cdot 10^{23} \text{ atoms}}{63.5 \text{ g}}\right)\left(\frac{8.96 \text{ g}}{1 \text{ cm}^3}\right)\left(\frac{10^6 \text{ cm}^3}{1 \text{ m}^3}\right) = 8.49 \cdot 10^{28} \frac{\text{electrons}}{\text{m}^3}.$$

SIMPLIFY

We solve equation (i) for the magnitude of the drift velocity:

$$v_d = \frac{i}{neA}.$$

Thus, the distance traveled by the electrons is

$$x = v_d t = \frac{i}{neA}t.$$

CALCULATE

Putting in the numerical values, we get

$$x = v_d t = \frac{it}{neA} = \frac{\left(0.78 \cdot 10^{-3} \text{ A}\right)(5.3 \text{ s})}{\left(8.49 \cdot 10^{28} \text{ m}^{-3}\right)\left(1.602 \cdot 10^{-19} \text{ C}\right)\left(0.823 \text{ mm}^2\right)}$$

$$= \left(6.96826 \cdot 10^{-8} \text{ m/s}\right)(5.3 \text{ s})$$

$$= 3.69318 \cdot 10^{-7} \text{ m}.$$

ROUND

We report our results to two significant figures:

$$v_d = 7.0 \cdot 10^{-8} \text{ m/s},$$

and

$$x = 3.7 \cdot 10^{-7} \text{ m} = 0.37 \text{ }\mu\text{m}.$$

DOUBLE-CHECK

Our result for the magnitude of the drift velocity turns out to be a stunningly small number. Earlier, it was stated that typical drift velocities are on the order of 10^{-4} m/s or smaller. Since the current is proportional to the drift velocity, a relatively small current implies a relatively small drift velocity. An 18-gauge wire can carry a current of several amperes, so the current specified in the problem statement is less than 1% of the maximum current. Therefore, the fact that our calculated drift velocity is less than 1% of 10^{-4} m/s, a typical drift velocity for high currents, is reasonable.

The distance we calculated for the movement of the electrons is less than 0.001 of the thickness of a fingernail, a very small distance compared to the length of the wire. This result provides a valuable reminder that the electromagnetic field moves with nearly the speed of light (in vacuum) inside a conductor and causes all conduction electrons to drift basically at the same time. Therefore, the signal from your game controller arrives almost instantaneously at the console, despite the incredibly slow pace of the individual electrons.

25.3 Resistivity and Resistance

Some materials conduct electricity better than others. Applying a given potential difference across a good conductor results in a relatively large current; applying the same potential difference across an insulator produces little current. The **resistivity,** ρ, is a measure of how strongly a material opposes the flow of electric current. The **resistance,** R, is a material's opposition to the flow of electric current.

If a known electric potential difference, ΔV, is applied across a conductor (some physical device or material that conducts current) and the resulting current, i, is measured, the resistance of that conductor is given by

$$R = \frac{\Delta V}{i}. \qquad (25.8)$$

The unit of resistance is the volt per ampere, which has been given the name **ohm** and symbol Ω (the capital Greek letter omega), in honor of the German physicist Georg Simon Ohm (1789–1854):

$$1\,\Omega = \frac{1\,\text{V}}{1\,\text{A}}.$$

Rearrangement of equation 25.8 results in

$$i = \frac{\Delta V}{R}, \qquad (25.9)$$

which states that for a given potential difference, ΔV, the current, i, is inversely proportional to the resistance, R. This equation is commonly referred to as **Ohm's Law.** A rearrangement of equation 25.9, $\Delta V = iR$, is sometimes also referred to as Ohm's Law.

Sometimes devices are described in terms of the **conductance,** G, defined as

$$G = \frac{i}{\Delta V} = \frac{1}{R}.$$

Conductance has the SI derived unit of siemens (S):

$$1\,\text{S} = \frac{1\,\text{A}}{1\,\text{V}} = \frac{1}{1\,\Omega}.$$

In some conductors, the resistivity depends on the direction in which the current is flowing. This chapter assumes that the resistivity of a material is uniform for all directions of the current.

The resistance of a device depends on the material of which the device is made as well as its geometry. As stated earlier, the resistivity of a material characterizes how much it opposes the flow of current. The resistivity is defined in terms of the magnitude of the applied electric field, E, and the magnitude of the resulting current density, J:

$$\rho = \frac{E}{J}. \qquad (25.10)$$

The units of resistivity are

$$[\rho] = \frac{[E]}{[J]} = \frac{\text{V/m}}{\text{A/m}^2} = \frac{\text{V m}}{\text{A}} = \Omega\,\text{m}.$$

Table 25.1 lists the resistivities of some representative conductors at 20 °C. As you can see, typical values for the resistivity of metal conductors used in wires are on the order of 10^{-8} Ω m. For example, copper has a resistivity of about $2 \cdot 10^{-8}$ Ω m. Several metal alloys listed in Table 25.1 have useful properties. For example, wire made from Nichrome (80% nickel and 20% chromium) is often used as a heating element in devices such as toasters. The next time you are toasting an English muffin, look inside the toaster. The glowing elements are probably Nichrome wires. The resistivity of Nichrome ($108 \cdot 10^{-8}$ Ω m) is about 50 times that of copper. Thus, when current is run through the Nichrome wires of the toaster, the wires dissipate power and heat up until they glow with a dull red color, while the copper wires of the electrical cord that connects the toaster to the wall remain cool.

Sometimes materials are specified in terms of their **conductivity,** σ, rather than their resistivity, ρ, which is defined as

$$\sigma = \frac{1}{\rho}.$$

The units of conductivity are $(\Omega\ m)^{-1}$.

The resistance of a conductor can be found from its resistivity and its geometry. For a homogeneous conductor of length L and constant cross-sectional area A, the equation $\Delta V = -\int \vec{E} \cdot d\vec{s}$ from Chapter 23 can be used to relate the electric field, E, and the electric potential difference, ΔV, across the conductor:

$$E = \frac{\Delta V}{L}.$$

Note that in contrast to electrostatics, where the surface of any conductor is an equipotential surface and has no electric field inside and no current flowing through it, the conductor

Table 25.1	The Resistivity and the Temperature Coefficient of the Resistivity for Some Representative Conductors	
Material	Resistivity, ρ at 20 °C (10^{-8} Ω m)	Temperature Coefficient, α (10^{-3} K^{-1})
Silver	1.62	3.8
Copper	1.72	3.9
Gold	2.44	3.4
Aluminum	2.82	3.9
Brass	3.9	2
Tungsten	5.51	4.5
Nickel	7	5.9
Iron	9.7	5
Steel	11	5
Tantalum	13	3.1
Lead	22	4.3
Constantan	49	0.01
Stainless steel	70	1
Mercury	95.8	0.89
Nichrome	108	0.4

The values for steel and stainless steel depend strongly on the type of steel.

in this situation has $\Delta V \neq 0$ and $\vec{E} \neq 0$, causing a current to flow. The magnitude of the current density is the current divided by the cross-sectional area:

$$J = \frac{i}{A}.$$

From the definition of resistivity (equation 25.10) and using $J = i/A$ and Ohm's Law (equation 25.8), we obtain

$$\rho = \frac{E}{J} = \frac{\Delta V/L}{i/A} = \frac{\Delta V}{i} \frac{A}{L} = \frac{iR}{i} \frac{A}{L} = R\frac{A}{L}.$$

Rearranging terms yields an expression for the resistance of a conductor in terms of the resistivity of its constituent material, the length, and the cross-sectional area:

$$R = \rho \frac{L}{A}. \qquad (25.11)$$

Size Convention for Wires

The American Wire Gauge (AWG) size convention for wires specifies diameters and thus cross-sectional areas on a logarithmic scale. The AWG size convention is shown in Table 25.2. The wire gauge is related to the diameter: The higher the gauge number, the thinner the wire. For large-diameter wires, gauge numbers consist of one or more zeros, as shown in Table 25.2. A 00-gauge wire is equivalent to a –1-gauge, a 000-gauge wire is equivalent to a –2-gauge, and so on. By definition, a 36-gauge wire has a diameter of exactly 0.005 in, and a 0000-gauge wire has a diameter of exactly 0.46 in. (These sizes apear in red in Table 25.2.) There are 39 gauge values between 0000-gauge and 36-gauge, and the gauge number is a logarithmic representation of the wire diameter. Therefore, the formula to convert from the AWG gauge to the wire diameter, in inches, is $d = (0.005)92^{(36-n)/39}$, where n is the gauge number. Typical residential wiring uses 12-gauge to 10-gauge wires. An important rule of

Table 25.2 Wire Diameters and Cross-Sectional Areas as Defined by the American Wire Gauge Convention

AWG	d (in)	d (mm)	A (mm²)	AWG	d (in)	d (mm)	A (mm²)	AWG	d (in)	d (mm)	A (mm²)
000000	0.5800	14.733	170.49	11	0.0907	2.3048	4.1723	26	0.0159	0.4049	0.1288
00000	0.5165	13.120	135.20	12	0.0808	2.0525	3.3088	27	0.0142	0.3606	0.1021
0000	0.46	11.684	107.22	13	0.0720	1.8278	2.6240	28	0.0126	0.3211	0.0810
000	0.4096	10.405	85.029	14	0.0641	1.6277	2.0809	29	0.0113	0.2859	0.0642
00	0.3648	9.2658	67.431	15	0.0571	1.4495	1.6502	30	0.0100	0.2546	0.0509
0	0.3249	8.2515	53.475	16	0.0508	1.2908	1.3087	31	0.0089	0.2268	0.0404
1	0.2893	7.3481	42.408	17	0.0453	1.1495	1.0378	32	0.0080	0.2019	0.0320
2	0.2576	6.5437	33.631	18	0.0403	1.0237	0.8230	33	0.0071	0.1798	0.0254
3	0.2294	5.8273	26.670	19	0.0359	0.9116	0.6527	34	0.0063	0.1601	0.0201
4	0.2043	5.1894	21.151	20	0.0320	0.8118	0.5176	35	0.0056	0.1426	0.0160
5	0.1819	4.6213	16.773	21	0.0285	0.7229	0.4105	36	0.005	0.1270	0.0127
6	0.1620	4.1154	13.302	22	0.0253	0.6438	0.3255	37	0.0045	0.1131	0.0100
7	0.1443	3.6649	10.549	23	0.0226	0.5733	0.2582	38	0.0040	0.1007	0.0080
8	0.1285	3.2636	8.3656	24	0.0201	0.5106	0.2047	39	0.0035	0.0897	0.0063
9	0.1144	2.9064	6.6342	25	0.0179	0.4547	0.1624	40	0.0031	0.0799	0.0050
10	0.1019	2.5882	5.2612								

thumb is that a reduction by 3 gauges doubles the cross-sectional area of the wire. Examining equation 25.11, you can see that to cut the resistance of a given length of wire in half, you have to reduce the gauge number by 3.

EXAMPLE 25.2 | Resistance of a Copper Wire

Standard wires that electricians put into residential housing have fairly low resistance.

PROBLEM
What is the resistance of the 100.0-m standard 12-gauge copper wire that is typically used in wiring household electrical outlets?

SOLUTION
A 12-gauge copper wire has a diameter of 2.053 mm (see Table 25.2). Its cross-sectional area is then

$$A = 3.31 \text{ mm}^2.$$

Using the value for the resistivity of copper from Table 25.1 and equation 25.11, we find

$$R = \rho \frac{L}{A} = (1.72 \cdot 10^{-8} \ \Omega \ \text{m}) \frac{100.0 \text{ m}}{3.31 \cdot 10^{-6} \text{ m}^2} = 0.520 \ \Omega.$$

25.1 In-Class Exercise

What is the resistance of a copper wire that has length $L = 70.0$ m and diameter $d = 2.60$ mm?

a) 0.119 Ω d) 0.190 Ω

b) 0.139 Ω e) 0.227 Ω

c) 0.163 Ω

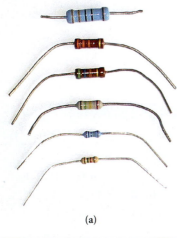

(a)

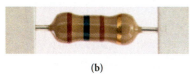

(b)

FIGURE 25.8 (a) Selection of resistors with various resistances. (b) Color-coding of a 150-Ω resistor.

Resistor Codes

In many applications, circuit design calls for a range of resistances in various parts of a circuit. Commercially available resistors, such as those shown in Figure 25.8a, have a wide range of resistances. Resistors are commonly made of carbon enclosed in a plastic cover that looks like a medicine capsule, with wires sticking out at the ends for electrical connection. The value of the resistance is indicated by three or four color bands on the plastic covering. The first two bands indicate numbers for the mantissa, the third represents a power of 10, and the fourth indicates a tolerance for the range of values. For the mantissa and power of 10, the numbers associated with the colors are black = 0, brown = 1, red = 2, orange = 3, yellow = 4, green = 5, blue = 6, purple = 7, gray = 8, and white = 9. For the tolerance, brown means 1%, red means 2%, gold means 5%, silver means 10%, and no band at all means 20%. For example, the single resistor shown in Figure 25.8b has the colors (left to right) brown, green, brown, and gold. From the code, the resistance of this resistor is $15 \cdot 10^1 \ \Omega = 150 \ \Omega$, with a tolerance of 5%.

Temperature Dependence and Superconductivity

The values of resistivity and resistance vary with temperature. For metals, this dependence on temperature is linear over a broad range of temperatures. An empirical relationship for the temperature dependence of the resistivity of a metal is

$$\rho - \rho_0 = \rho_0 \alpha (T - T_0), \tag{25.12}$$

where ρ is the resistivity at temperature T, ρ_0 is the resistivity at temperature T_0, and α is the **temperature coefficient of electric resistivity** for the particular conductor.

In everyday applications, the temperature dependence of the resistance is often important. Equation 25.11 states that the resistance of a device depends on its length and cross-sectional area. These quantities depend on temperature, as we saw in Chapter 17; however, the temperature dependence of linear expansion is much smaller than the temperature dependence of resistivity for a particular conductor. Thus, the temperature dependence of the resistance of a conductor can be approximated as

$$R - R_0 = R_0 \alpha (T - T_0). \tag{25.13}$$

Note that equations 25.12 and 25.13 deal with temperature differences, so the temperatures can be expressed in degrees Celsius or kelvins (but not in degrees Fahrenheit!).

Values of α for representative conductors are listed in Table 25.1. Note that common metal conductors such as copper have a temperature coefficient of electric resistivity on the order of $4 \cdot 10^{-3}\,K^{-1}$. However, one metal alloy, constantan (60% copper and 40% nickel), has the special characteristic that its temperature coefficient of electric resistivity is very small: $\alpha = 1 \cdot 10^{-5}\,K^{-1}$. The name of this alloy comes from shortening the phrase "constant resistance." The small temperature coefficient of constantan combined with its relatively high resistivity of $4.9 \cdot 10^{-7}\,\Omega$ m makes it useful for precision resistors whose resistances have little dependence on temperature. Note also that Nichrome has a relative small temperature coefficient, $4 \cdot 10^{-4}\,K^{-1}$, which makes it suitable for the construction of heating elements, as noted earlier.

According to equation 25.12, most materials have a resistivity that varies linearly with the temperature under ordinary circumstances. However, some materials do not follow this rule at low temperatures. At very low temperatures, the resistivity of some materials goes to exactly zero. These materials are called **superconductors.** Superconductors have applications in the construction of magnets for devices such as magnetic resonance imagers (MRI). Magnets constructed with superconductors use less power and can produce higher magnetic fields than magnets constructed with conventional resistive conductors. A more extensive discussion of superconductivity will be presented in Chapters 27 and 28.

The resistance of some semiconducting materials actually decreases as the temperature increases, which implies a negative temperature coefficient of electric resistivity. These materials are often employed in high-resolution detectors for optical measurements or in particle detectors. Such devices must be kept cold to keep their resistance high, which is accomplished with refrigerators or liquid nitrogen.

A *thermistor* is a semiconductor whose resistance depends strongly on temperature. Thermistors are used to measure temperature. The temperature dependence of the resistance of a typical thermistor is shown in Figure 25.9a. Here you can see that the resistance of a thermistor falls with increasing temperature. This drop is in contrast to the increase in resistance of a copper wire over the same temperature range, shown in Figure 25.9b.

Microscopic Basis of Conduction in Solids

Conduction of current through solids results from the motion of electrons. In a metal conductor such as copper, the atoms of the metal form a regular array called a *crystal lattice*. The outermost electrons of each atom are essentially free to move randomly in this lattice. When an electric field is applied, the electrons drift in the direction opposite to that of the electric field. Resistance to drift occurs when electrons interact with the metal atoms in the lattice. When the temperature of the metal is increased, the motion of the atoms in the lattice increases. This, in turn, increases the probability that electrons interact with the atoms, effectively increasing the resistance of the metal.

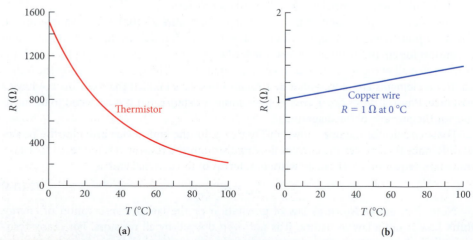

(a)   (b)

FIGURE 25.9   (a) The temperature dependence of the resistance of a thermistor. (b) The temperature dependence of the resistance of a copper wire that has a resistance of 1 Ω at $T = 0\ °C$.

The atoms of a semiconductor are also arranged in a crystal lattice. However, the outermost electrons of the atoms of the semiconductor are not free to move about within the lattice. To move about, the electrons must be given enough energy to attain an energy state where they can move freely. Thus, a typical semiconductor has a higher resistance than a metal conductor because it has many fewer conduction electrons. In addition, when a semiconductor is heated, many more electrons gain enough energy to move freely; thus, the resistance of the semiconductor goes down as its temperature increases.

25.4   Electromotive Force and Ohm's Law

For current to flow through a resistor, a potential difference must be established across the resistor. This potential difference, supplied by a battery or other device, is termed an **electromotive force,** abbreviated as **emf.** (Electromotive force is not a force at all, but rather a potential difference. The term is still in widespread use but mainly in the form of its abbreviation, pronounced "ee-em-eff.") A device that maintains a potential difference is called an *emf device* and does work on the charge carriers. The potential difference created by the emf device is represented as V_{emf}. This text assumes that emf devices have terminals to which a circuit can be connected. The emf device is assumed to maintain a constant potential difference, V_{emf}, between these terminals.

Examples of emf devices are batteries, electric generators, and solar cells. **Batteries,** discussed in Chapters 23 and 24, produce emf through chemical reactions. Electric generators create emf from mechanical motion. Solar cells convert light energy from the Sun to electric energy. If you examine a battery, you will find its potential difference (sometimes colloquially called "voltage") written on it. This "voltage" is the potential difference (emf) that the battery can provide to a circuit. (Note that a battery is a source of constant emf, it does *not* supply constant current to a circuit.) Rechargeable batteries also display a rating in mAh (milliampere-hour), which provides information on the total charge the battery can deliver when fully charged. The mAh is another unit of charge:

$$1 \text{ mAh} = (10^{-3} \text{ A})(3600 \text{ s}) = 3.6 \text{ As} = 3.6 \text{ C}.$$

Electrical components in a circuit can be sources of emf, capacitors, resistors, or other electrical devices. These components are connected with conducting wires. At least one component must be a source of emf because the potential difference created by the emf device is what drives the current through the circuit. You can think of an emf device as the pump in a water pipeline; without the pump, the water sits in the pipe and doesn't move. Once the pump is turned on, the water moves through the pipe in a continuous flow.

An electric circuit starts and ends at an emf device. Since the emf device maintains a constant potential difference, V_{emf}, between its terminals, positive current leaves the device at the higher potential of its positive terminal and enters its negative terminal at a lower potential. This lower potential is conventionally set to zero.

Consider a simple circuit of the form shown in Figure 25.10, where a source of emf provides a potential difference, V_{emf}, across a resistor with resistance R. Note an important convention for circuit diagrams: A resistor is always symbolized by a zigzag line, and it is assumed that all of the resistance, R, is concentrated there. The wires connecting the different circuit elements are represented by straight lines; it is implied that they do not have a resistance. Physical wires do, of course, have some resistance, but it is assumed to be negligible for the purpose of the diagram.

For a circuit like the one shown in Figure 25.10, the emf device provides the potential difference that creates the current flowing through the resistor. Therefore, in this case, Ohm's Law (equation 25.9) can be written in terms of the external emf as

$$V_{emf} = iR. \tag{25.14}$$

Note that, unlike Newton's law of gravitation or the law of conservation of energy, Ohm's Law is not a law of nature. It is not even obeyed by all resistors. For many resistors, called *ohmic resistors,* the current is directly proportional to the potential difference across the resistor over a wide range of temperatures and a wide range of applied potential

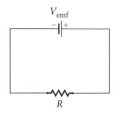

V_{emf}

R

FIGURE 25.10 Simple circuit containing a source of emf and a resistor.

differences. For other resistors, called *non-ohmic resistors,* current and potential difference are not directly proportional at all. Non-ohmic resistors include many kinds of transistors, which means that many modern electronic devices do not obey Ohm's Law. We'll take a closer look at one of these devices, the diode, in Section 25.8. Nevertheless, a large class of materials and devices (such as conventional wires, for example) do obey Ohm's Law, and thus it is worth devoting attention to its consequences. The remainder of this chapter (with the exception of Section 25.8) treats resistors as ohmic devices; that is, devices that obey Ohm's Law.

The current, i, that flows through the resistor in Figure 25.10 also flows through the source of emf and the wires connecting the components. Because the wires are assumed to have zero resistance (as noted above), the change in potential of the current must occur in the resistor, according to Ohm's Law. This change is referred to as the **potential drop** across the resistor. Thus, the circuit shown in Figure 25.10 can be represented in a different way, making it clearer where the potential drop happens and showing which parts of the circuit are at which potential. Figure 25.11a shows the circuit in Figure 25.10. Figure 25.11b shows the same circuit but with the vertical dimension representing the value of the electric potential at different points around the circuit. The potential difference is supplied by the source of emf, and the entire potential drop occurs across the single resistor. (Remember, the convention is that the lines connecting the circuit elements in a circuit diagram represent wires with no resistance. Therefore, these connecting wires are represented in Figure 25.11b by horizontal lines, signifying that the entire wire is at exactly the same potential.) Ohm's Law applies for the potential drop across the resistor, and the current in the circuit can be calculated using equation 25.9.

Figure 25.11 illustrates an important point about circuits. Sources of emf add potential difference to a circuit, and potential drops through resistors reduce potential in the circuit. However, the total potential difference on any closed path around the complete circuit must be zero. This is a straightforward consequence of the law of conservation of energy. An analogy with gravity may help: You can gain and lose potential energy by moving up and down in a gravitational field (for example, by climbing up and down hills), but if you arrive back at the same point from which you started, the net energy gained or lost is exactly zero. The same holds for current flowing in a circuit: It does not matter how many potential drops or sources of emf are encountered on any closed loop; a given point always has the same value of the electric potential. The current can flow through the loop in either direction with the same result.

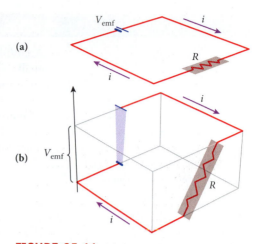

FIGURE 25.11 (a) Conventional representation of a simple circuit with a resistor and a source of emf. (b) Three-dimensional representation of the same circuit, displaying the potential at each point in the circuit. The current in the circuit is shown for both views.

25.2 Self-Test Opportunity

A resistor with $R = 10.0\ \Omega$ is connected across a source of emf with potential difference $V_{emf} = 1.50$ V. What is the current flowing though the circuit?

Resistance of the Human Body

This short introduction of resistance and Ohm's Law leads to a point about electrical safety. It was mentioned earlier that currents above 100 mA can be deadly if they flow through human heart muscle. Ohm's Law makes it clear that the resistance of the human body determines whether a given potential difference—say, from a car battery—can be dangerous. Since we usually handle tools with our hands, the most relevant measure for the human body's resistance, R_{body}, is the resistance along a path from the fingertips of one hand to the fingertips of the other hand. (Note that the heart is pretty much in the middle of this path!) For most people this resistance is in the range $500\ k\Omega < R_{body} < 2\ M\Omega$. Most of this resistance comes from the skin, in particular, the layers of dead skin on the outside. However, if the skin is wet, its conductivity is drastically increased, and consequently, the body's resistance is drastically lowered. For a given potential difference, Ohm's Law implies that the current then drastically increases. Handling electrical devices in wet environments or touching them with your tongue is thus a very bad idea.

Wires in a circuit can have sharp points where they are cut. If these points penetrate the skin at the fingertips, the resistance of the skin is eliminated, and the fingertip-to-fingertip resistance is very drastically lowered. If a wire penetrates a blood vessel, the human body's resistance decreases even further, because blood has a high salinity and is thus a good conductor. In this case, even relatively small potential differences from batteries can have a deadly effect.

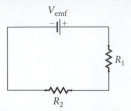

FIGURE 25.12 Circuit with two resistors in series with one source of emf.

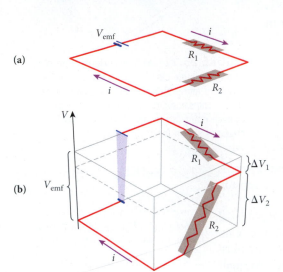

FIGURE 25.13 (a) Conventional representation of a simple circuit with two resistors in series and a source of emf. (b) Three-dimensional representation of the same circuit, displaying the potential at each point in the circuit. The current in the circuit is shown for both views.

25.2 In-Class Exercise

What are the relative values of the two resistances in Figure 25.13?

a) $R_1 < R_2$

b) $R_1 = R_2$

c) $R_1 > R_2$

d) Not enough information is given in the figure to compare the resistances.

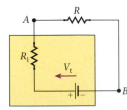

FIGURE 25.14 Battery (yellow area) with internal resistance R_i connected to an external resistor, R.

25.5 Resistors in Series

A circuit can contain more than one resistor and/or more than one source of emf. The analysis of circuits with multiple resistors requires different techniques. Let's first examine resistors connected in series.

Two resistors, R_1 and R_2, are connected in series with one source of emf with potential difference V_{emf} in the circuit shown in Figure 25.12. The potential drop across resistor R_1 is denoted by ΔV_1, and the potential drop across resistor R_2 by ΔV_2. The two potential drops must sum to the potential difference supplied by the source of emf:

$$V_{emf} = \Delta V_1 + \Delta V_2.$$

The crucial insight is that the same current must flow through all the elements of the circuit. How do we know this? Remember, at the beginning of this chapter, current was defined as the rate of change of the charge in time: $i = dq/dt$. Current has to be the same everywhere along a wire, and also in a resistor, because charge is conserved everywhere. No charge is lost or gained along the wire, and so the current is the same everywhere around the loop in Figure 25.12.

To clear up a misconception that is encountered often, note that *there is no such thing as current getting "used up" in a resistor*. No matter how many resistors are connected in series, the current that flows into the first one is the same current that flows out of the last one. An analogy to water flowing in a pipe may help: No matter how long the pipe is and how many bends it may have, all water that flows into one end has to come out the other end.

Thus, the current flowing through each resistor in Figure 25.12 is the same. For each resistor, we can apply Ohm's Law and get

$$V_{emf} = iR_1 + iR_2.$$

An equivalent resistance, R_{eq}, can replace the two individual resistances:

$$V_{emf} = iR_1 + iR_2 = iR_{eq},$$

where

$$R_{eq} = R_1 + R_2.$$

Thus, two resistors in series can be replaced with an equivalent resistance equal to the sum of the two resistances. Figure 25.13 illustrates the potential drops in the series circuit of Figure 25.12, using a three-dimensional view.

The expression for the equivalent resistance of two resistors in series can be generalized to a circuit with n resistors in series:

$$R_{eq} = \sum_{i=1}^{n} R_i \quad \text{(for resistors in series).} \tag{25.15}$$

That is, if resistors are connected in a single path so that the same current flows through all of them, their total resistance is just the sum of their individual resistances.

EXAMPLE 25.3 | Internal Resistance of a Battery

When a battery is not connected in a circuit, the potential difference across its terminals is V_t. When the battery is connected in series with a resistor with resistance R, current i flows through the circuit. When the current is flowing, the potential difference, V_{emf}, across the terminals of the battery is less than V_t. This drop occurs because the battery has an internal resistance, R_i, which can be thought of as being in series with the external resistor (Figure 25.14). That is,

$$V_t = iR_{eq} = i(R + R_i).$$

The battery is depicted by the yellow rectangle in Figure 25.14. The terminals of the battery are represented by points A and B.

PROBLEM

Consider a battery that has $V_t = 12.0$ V when it is not connected to a circuit. When a 10.0-Ω resistor is connected with the battery, the potential difference across the battery's terminals drops to 10.9 V. What is the internal resistance of the battery?

SOLUTION

The current flowing through the external resistor is given by

$$i = \frac{\Delta V}{R} = \frac{10.9 \text{ V}}{10.0 \text{ }\Omega} = 1.09 \text{ A}.$$

The current flowing in the complete circuit, including the battery, must be the same as the current flowing in the external resistor. Thus, we have

$$V_t = iR_{eq} = i(R + R_i)$$

$$(R + R_i) = \frac{V_t}{i}$$

$$R_i = \frac{V_t}{i} - R = \frac{12.0 \text{ V}}{1.09 \text{ A}} - 10.0 \text{ }\Omega = 1.0 \text{ }\Omega.$$

The internal resistance of the battery is 1.0 Ω. Batteries with internal resistance are known as non-ideal. Unless otherwise specified, batteries in circuits will be assumed to have zero internal resistance. Such batteries are known as ideal. An ideal battery maintains a constant potential difference between its terminals independent of the current flowing.

Whether a battery can still provide energy cannot be determined by simply measuring the potential difference across the terminals. Instead, you must place a resistance on the battery and then measure the potential difference. If the battery is no longer functional, it may still provide its rated potential difference when not connected, but its potential difference may drop to zero when connected to an external resistance. Some brands of batteries have built-in devices to measure the functioning potential difference simply by pressing on a particular spot on the battery and observing an indicator.

25.3 In-Class Exercise

Three identical resistors, R_1, R_2, and R_3, are wired together as shown in the figure. An electric current is flowing through the three resistors. The current through R_2

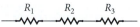

a) is the same as the current through R_1 and R_3.

b) is a third of the current through R_1 and R_3.

c) is twice the sum of the current through R_1 and R_3.

d) is three times the current through R_1 and R_3.

e) cannot be determined.

Resistor with a Non-Constant Cross Section

Up to now the discussion has assumed that a resistor has the same cross-sectional area, A, and the same resistivity, ρ, everywhere along its length (this was the implicit assumption in the derivation leading to equation 25.11). This is, of course, not always the case. How do we handle the analysis of a resistor whose cross-sectional area is a function of the position x along the resistor, $A(x)$, and/or whose resistivity can change as a function of position, $\rho(x)$? We simply divide the resistor into many very short pieces of length Δx and sum over all of them, since equation 25.15 says that the total resistance is the sum of all of the resistances of the individual short pieces; then we take the limit $\Delta x \rightarrow 0$. If this sounds like an integration to you, you are right. The general formula for computing the resistance of a resistor of length L with a nonuniform cross-sectional area, $A(x)$, is

$$R = \int_0^L \frac{\rho(x)}{A(x)} dx. \qquad (25.16)$$

A concrete example will help clarify this equation.

SOLVED PROBLEM 25.2 | **Brain Probe**

Chapter 22 mentioned the field of electrocorticography (ECoG), in which researchers measure the electric field generated by neurons in the brain. Some of these measurements can only be done by inserting very thin wires into the brain in order to probe directly into

Continued—

neurons. These wires are insulated, with only a very short tip exposed, which is pulled into a very fine conical tip. ECoG is being used to treat an epilepsy patient in Figure 25.15.

PROBLEM

If the wire used for ECoG is made of tungsten and has a diameter of 0.74 mm and the tip has a length of 2.0 mm and is sharpened to a diameter of 2.4 μm at the end, what is the resistance of the tip? (The resistivity of tungsten is listed in Table 25.1 as $5.51 \cdot 10^{-8}$ Ω m.)

SOLUTION

THINK

First, why might one want to know the resistance? To measure electrical fields or potential differences in neurons, probes with a large resistance, say, on the order of kilo-ohms, cannot be used because the fields or differences will not be detectable. However, since resistance is inversely proportional to the cross-sectional area, a very small area means a relatively large resistance. The problem statement says that the probe has a very fine tip, much pointier than any sewing needle. Hence the need to find out the resistance of the probe before inserting it into the brain!

Clearly, we are dealing with a case of nonconstant cross-sectional area, and so we will need to perform the integration of equation 25.16. However, since the tip is entirely made of tungsten, the resistivity is constant throughout its volume, which will simplify the task.

SKETCH

Figure 25.16a shows a three-dimensional view of the tip, and Figure 25.16b presents a cut through its symmetry plane and the integration path.

RESEARCH

The research part is quite simple for this problem, because we already know which equation we need to use. However, equation 25.16 needs to be altered to reflect the fact that the resistivity is constant throughout the tip:

$$R = \rho \int_0^L \frac{1}{A(x)} dx, \qquad \text{(i)}$$

where $A(x)$ is the area of a circle, $A(x) = \pi[r(x)]^2$. The radius of the circle falls linearly from r_1 to r_2 (see Figure 25.16b):

$$r(x) = r_1 + \frac{(r_2 - r_1)x}{L}. \qquad \text{(ii)}$$

SIMPLIFY

We substitute the expression for the radius from equation (ii) into the formula for the area and then substitute the resulting expression for $A(x)$ into equation (i). We arrive at

$$R = \rho \int_0^L \frac{1}{\pi \left(r_1 + (r_2 - r_1)x / L \right)^2} dx.$$

This integral may look daunting at first sight, but except for x all the other quantities are constants. We consult an integration table or software and find

$$R = -\frac{\rho L}{\pi(r_2 - r_1)\left(r_1 + (r_2 - r_1)x/L\right)}\bigg|_0^L = \frac{\rho L}{\pi r_1 r_2}.$$

CALCULATE

Putting in the numerical values, we get

$$R = \frac{\left(5.51 \cdot 10^{-8} \ \Omega \text{ m}\right)\left(2.0 \cdot 10^{-3} \text{ m}\right)}{\pi(0.37 \cdot 10^{-3} \text{ m})(1.2 \cdot 10^{-6} \text{ m})} = 7.90039 \cdot 10^{-2} \ \Omega.$$

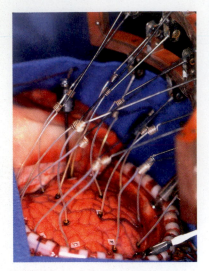

FIGURE 25.15 Electrocorticography performed with electrode grids on the cerebral cortex of an epilepsy patient.

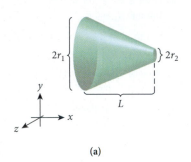

(a)

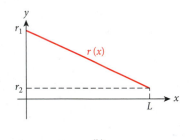

(b)

FIGURE 25.16 (a) Shape of the tip of the probe. (b) Coordinate system for the integration.

ROUND

We report our result to the two significant figures to which the geometric properties of the tip were given:

$$R = 7.9 \cdot 10^{-2} \ \Omega.$$

DOUBLE-CHECK

The value of 79 mΩ seems a very small resistance to a current that has to pass through a tip that is sharpened to a diameter of 2.4 μm. On the other hand, the tip has a very small length, which argues for a small resistance. An additional confidence builder is the fact that the units have worked out properly.

However, there are also a few tests we can perform to convince ourselves that at least the asymptotic limits of the solution to $R = \rho L/(\pi r_1 r_2)$ are reasonable. First, as the length approaches zero, so does the resistance, as expected. Second, as the radius of either end of the tip approaches zero, the formula predicts an infinite resistance, which is also expected.

25.6 Resistors in Parallel

Instead of being connected in series so that all the current must pass through both resistors, two resistors can be connected in parallel, which divides the current between them, as shown in Figure 25.17. Again, to better illustrate the potential drops, Figure 25.18 shows the same circuit in a three-dimensional view.

In this case, the potential drop across each resistor is equal to the potential difference provided by the source of emf. Using Ohm's Law (equation 25.14) for the current i_1 in R_1 and the current i_2 in R_2, we have

$$i_1 = \frac{V_{emf}}{R_1}$$

and

$$i_2 = \frac{V_{emf}}{R_2}.$$

The total current from the source of emf, i, must be

$$i = i_1 + i_2.$$

Inserting the expressions for i_1 and i_2, we obtain

$$i = i_1 + i_2 = \frac{V_{emf}}{R_1} + \frac{V_{emf}}{R_2} = V_{emf}\left(\frac{1}{R_1} + \frac{1}{R_2}\right).$$

Ohm's Law (equation 25.14) can be rewritten as

$$i = V_{emf}\left(\frac{1}{R_{eq}}\right).$$

Thus, two resistors connected in parallel can be replaced with an equivalent resistance given by

$$\frac{1}{R_{eq}} = \frac{1}{R_1} + \frac{1}{R_2}.$$

In general, the equivalent resistance for n resistors connected in parallel is given by

$$\frac{1}{R_{eq}} = \sum_{i=1}^{n} \frac{1}{R_i} \quad \text{(resistors in parallel)}. \tag{25.17}$$

Clearly, combining resistors in series and in parallel to form equivalent resistances allows circuits with various combinations of resistors to be analyzed in a way analogous to the analysis of combinations of capacitors performed in Chapter 24.

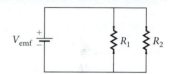

FIGURE 25.17 Circuit with two resistors connected in parallel and a single source of emf.

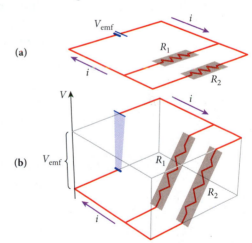

FIGURE 25.18 (a) Conventional representation of a simple circuit with two resistors in parallel and a source of emf. (b) Three-dimensional representation of the same circuit, displaying the potential at each point in the circuit.

25.4 In-Class Exercise

Three identical resistors, R_1, R_2, and R_3, are wired together as shown in the figure. An electric current is flowing from point A to point B. The current flowing through R_2

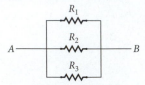

a) is the same as the current through R_1 and R_3.

b) is a third of the current through R_1 and R_3.

c) is twice the sum of the current through R_1 and R_3.

d) is three times the current through R_1 and R_3.

e) cannot be determined.

25.5 In-Class Exercise

Which combination of resistors has the highest equivalent resistance?

a) combination (a)

b) combination (b)

c) combination (c)

d) combination (d)

e) The equivalent resistance is the same for all four.

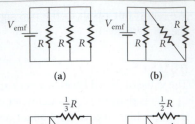

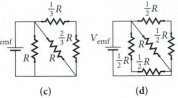

EXAMPLE 25.4 | Equivalent Resistance in a Circuit with Six Resistors

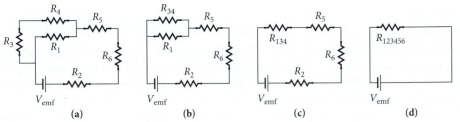

FIGURE 25.19 (a) Circuit with six resistors. (b)–(d) Steps in combining these resistors to determine the equivalent resistance.

PROBLEM

Figure 25.19a shows a circuit with six resistors, R_1 through R_6. What is the current flowing through resistors R_2 and R_3 in terms of V_{emf} and R_1 through R_6?

SOLUTION

We begin by identifying parts of the circuit that are clearly wired in parallel or in series. The current flowing through R_2 is the current flowing from the source of emf. We note that R_3 and R_4 are in series. Thus, we can write

$$R_{34} = R_3 + R_4. \tag{i}$$

This substitution is made in Figure 25.19b. This figure shows us that R_{34} and R_1 are in parallel. We can then write

$$\frac{1}{R_{134}} = \frac{1}{R_1} + \frac{1}{R_{34}},$$

or

$$R_{134} = \frac{R_1 R_{34}}{R_1 + R_{34}}. \tag{ii}$$

This substitution is depicted in Figure 25.19c. From this figure, we can see that R_2, R_5, R_6, and R_{134} are in series. Thus, we can write

$$R_{123456} = R_2 + R_5 + R_6 + R_{134}. \tag{iii}$$

This substitution is shown in Figure 25.19d. We substitute for R_{34} and R_{134} from equations (i) and (ii) into equation (iii):

$$R_{123456} = R_2 + R_5 + R_6 + \frac{R_1 R_{34}}{R_1 + R_{34}} = R_2 + R_5 + R_6 + \frac{R_1(R_3 + R_4)}{R_1 + R_3 + R_4}.$$

Thus, i_2, the current flowing through R_2, is given by

$$i_2 = \frac{V_{emf}}{R_{123456}}.$$

Now we turn to the determination of the current flowing through R_3. Current i_2 is also flowing through the equivalent resistor R_{134} that contains R_3 (see Figure 25.19c). Thus, we can write

$$V_{134} = i_2 R_{134},$$

where V_{134} is the potential drop across the equivalent resistor R_{134}. The resistor R_1 and the equivalent resistor R_{34} are in parallel. Thus, V_{34}, the potential drop across R_{34}, is the same as the potential drop across R_{134} which is V_{134}. The resistors R_3 and R_4 are in series, and thus, i_3, the current flowing through R_3, is the same as i_{34}, the current flowing through R_{34}. We can thus write

$$V_{34} = V_{134} = i_{34} R_{34} = i_3 R_{34}.$$

Now we can express i_3 in terms of V and R_1 through R_6:

$$i_3 = \frac{V_{134}}{R_{34}} = \frac{i_2 R_{134}}{R_{34}} = \frac{\left(\dfrac{V_{\text{emf}}}{R_{123456}}\right) R_{134}}{R_{34}} = \frac{V_{\text{emf}} R_{134}}{R_{34} R_{123456}} = \frac{V_{\text{emf}}\left(\dfrac{R_1 R_{34}}{R_1 + R_{34}}\right)}{R_{34} R_{123456}} = \frac{V_{\text{emf}} R_1}{R_{123456}\left(R_1 + R_{34}\right)}$$

or

$$i_3 = \frac{V_{\text{emf}} R_1}{\left(R_2 + R_5 + R_6 + \dfrac{R_1(R_3 + R_4)}{R_1 + R_3 + R_4}\right)\left(R_1 + R_3 + R_4\right)} = \frac{V_{\text{emf}} R_1}{\left(R_2 + R_5 + R_6\right)\left(R_1 + R_3 + R_4\right) + R_1\left(R_3 + R_4\right)}.$$

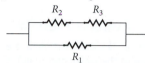
SOLVED PROBLEM 25.3 | Potential Drop across a Resistor in a Circuit

PROBLEM

The circuit shown in Figure 25.20a has four resistors and a battery with $V_{\text{emf}} = 149$ V. The values of the four resistors are $R_1 = 17.0\ \Omega$, $R_2 = 51.0\ \Omega$, $R_3 = 114.0\ \Omega$, and $R_4 = 55.0\ \Omega$. What is the magnitude of the potential drop across R_2?

SOLUTION

THINK

The resistors R_2 and R_3 are in parallel and can be replaced with an equivalent resistance, R_{23}. The resistors R_1 and R_4 are in series with R_{23}. The current flowing through R_1, R_4, and R_{23} is the same because they are in series. We can obtain the current in the circuit by calculating the equivalent resistance for R_1, R_4, and R_{23} and using Ohm's Law. The potential drop across R_{23} is equal to the current flowing in the circuit times R_{23}. The potential drop across R_2 is the same as the potential drop across R_{23} because R_2 and R_3 are in parallel.

SKETCH

The potential drop across resistor R_2 is illustrated in Figure 25.20b.

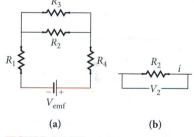

FIGURE 25.20 (a) A circuit with four resistors and a battery. (b) Potential drop across resistor R_2.

RESEARCH

The equivalent resistance for R_2 and R_3 can be calculated using equation 25.17:

$$\frac{1}{R_{23}} = \frac{1}{R_2} + \frac{1}{R_3}. \tag{i}$$

The equivalent resistance of the three resistors in series can be found using equation 25.15:

$$R_{\text{eq}} = \sum_{i=1}^{n} R_i = R_1 + R_{23} + R_4$$

Finally, we obtain the current in the circuit using Ohm's Law:

$$V_{\text{emf}} = i R_{\text{eq}} = i\left(R_1 + R_{23} + R_4\right).$$

Continued—

SIMPLIFY

The potential drop, V_2, across R_2 is equal to the potential drop, V_{23}, across the equivalent resistance R_{23}:

$$V_2 = V_{23} = iR_{23} = \frac{V_{\text{emf}}}{R_1 + R_{23} + R_4} R_{23} = \frac{R_{23} V_{\text{emf}}}{R_1 + R_{23} + R_4}. \tag{ii}$$

We can solve equation (i) for R_{23} to obtain

$$R_{23} = \frac{R_2 R_3}{R_2 + R_3}.$$

We can then use equation (ii) to determine the potential drop V_2 as

$$V_2 = \frac{\left(\dfrac{R_2 R_3}{R_2 + R_3}\right) V_{\text{emf}}}{R_1 + \left(\dfrac{R_2 R_3}{R_2 + R_3}\right) + R_4} = \frac{R_2 R_3 V_{\text{emf}}}{R_1 (R_2 + R_3) + R_2 R_3 + R_4 (R_2 + R_3)},$$

which we can rewrite as

$$V_2 = \frac{R_2 R_3 V_{\text{emf}}}{(R_1 + R_4)(R_2 + R_3) + R_2 R_3}.$$

CALCULATE

Putting in the numerical values, we get

$$V_2 = \frac{R_2 R_3 V_{\text{emf}}}{(R_1 + R_4)(R_2 + R_3) + R_2 R_3}$$

$$= \frac{(51.0\ \Omega)(114.0\ \Omega)(149\ \text{V})}{(17.0\ \Omega + 55.0\ \Omega)(51.0\ \Omega + 114.0\ \Omega) + (51.0\ \Omega)(114.0\ \Omega)}$$

$$= 48.9593\ \text{V}.$$

ROUND

We report our result to three significant figures:

$$V = 49.0\ \text{V}.$$

DOUBLE-CHECK

You may be tempted to avoid completing the analytic solution as we've done here. Instead, you may want to insert numbers earlier, for example, into the expression for R_{23}. So, to double-check our result, let's calculate the current in the circuit explicitly and then calculate the potential drop across R_{23} using that current. The equivalent resistance for R_2 and R_3 in parallel is

$$R_{23} = \frac{R_2 R_3}{R_2 + R_3} = \frac{(51.0\ \Omega)(114.0\ \Omega)}{51.0\ \Omega + 114.0\ \Omega} = 35.2\ \Omega.$$

The current in the circuit is then

$$i = \frac{V_{\text{emf}}}{R_1 + R_{23} + R_4} = \frac{149.0\ \text{V}}{17.0\ \Omega + 35.2\ \Omega + 55.0\ \Omega} = 1.39\ \text{A}.$$

The potential drop across R_2 is then

$$V_2 = iR_{23} = (1.39\ \text{A})(35.2\ \Omega) = 48.9\ \text{V},$$

which agrees with our result within rounding error. It is reassuring that both methods lead to the same answer.

We can also check that the potential drops across R_1, R_{23}, and R_4 sum to V_{emf}, as they should since R_1, R_{23}, and R_4 are in series. The potential drop across R_1 is $V_1 = iR_1 = (1.39\ \text{A})(17.0\ \Omega) = 23.6\ \text{V}$. The potential drop across R_4 is $V_4 = iR_4 = (1.39\ \text{A})(55.0\ \Omega) = 76.5\ \text{V}$. So the total potential drop is $V_{\text{total}} = V_1 + V_{23} + V_4 = (23.6\ \text{V}) + (48.9\ \text{V}) + (76.5\ \text{V}) = 149\ \text{V}$, which is equal to V_{emf}. Thus, our answer is consistent.

25.7 In-Class Exercise

As more identical resistors, R, are added to the circuit shown in the figure, the resistance between points A and B will

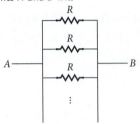

a) increase.

b) stay the same.

c) decrease.

d) change in an unpredictable manner.

25.8 In-Class Exercise

Three light bulbs are connected in series with a battery that delivers a constant potential difference, V_{emf}. When a wire is connected across light bulb 2 as shown in the figure, light bulbs 1 and 3

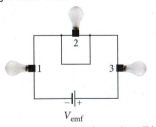

a) burn just as brightly as they did before the wire was connected.

b) burn more brightly than they did before the wire was connected.

c) burn less brightly than they did before the wire was connected.

d) go out.

25.7 Energy and Power in Electric Circuits

Consider a simple circuit in which a source of emf with potential difference ΔV causes a current, i, to flow. The work required from the emf device to move a differential amount of charge, dq, from the negative terminal to the positive terminal (within the emf device) is equal to the increase in electric potential energy of that charge, dU:

$$dU = dq\,\Delta V.$$

Remembering that current is defined as $i = dq/dt$, we can rewrite the differential electric potential energy as

$$dU = i\,dt\,\Delta V.$$

Using the definition of power, $P = dU/dt$, and substituting into it the expression for the differential potential energy, we obtain

$$P = \frac{dU}{dt} = \frac{i\,dt\,\Delta V}{dt} = i\Delta V.$$

Thus, the product of the current times the potential difference gives the power supplied by the source of emf. By conservation of energy, this power is equal to the power dissipated in a circuit containing one resistor. In a more complicated circuit, each resistor will dissipate power at the rate given by this equation, where i and ΔV refer to the current through and potential difference across that resistor. Ohm's Law (equation 25.9) leads to different formulations of the power:

$$P = i\Delta V = i^2 R = \frac{(\Delta V)^2}{R}. \tag{25.18}$$

The unit of power (as noted in Chapter 5) is the watt (W). Electrical devices, such as light bulbs, are rated in terms of how much power they consume. Your electric bill depends on how much electrical energy your appliances consume, and this energy is measured in kilowatt-hours (kW h).

Where does this energy go? This question will be addressed quantitatively in Chapter 30 when alternating currents are discussed. Qualitatively, much or most of the energy dissipated in resistors is converted into heat. This phenomenon is employed in incandescent lighting, where heating a metal filament to a very high temperature causes it to emit light. The heat dissipated in electrical circuits is a huge problem for large-scale computer systems and server farms for the biggest Internet databases. These computer systems use thousands of processors for computing applications that can be parallelized. All of these processors emit heat, and very expensive cooling has to be provided to offset it. It turns out that the cost of cooling is one of the most stringent boundary conditions limiting the maximum size of these supercomputers.

Some of the power dissipated in circuits can be converted into mechanical energy by motors. The functioning of electric motors requires an understanding of magnetism and will be covered later.

High-Voltage Direct Current Power Transmission

The transmission of electrical power from power-generating stations to users of electricity is of great practical interest. Often, electrical power-generating stations are located in remote areas, and thus the power must be transmitted long distances. This is particularly true for clean power sources, such as hydroelectric dams and large solar farms in deserts.

The power, P, transmitted to users is the product of the current, i, and the potential difference, ΔV, in the power line: $P = i\Delta V$. Thus, the current required for a given power is $i = P/\Delta V$, and a higher potential difference means a lower current in the power line. Equation 25.18 indicates that the power dissipated in an electrical power transmission line, P_{loss}, is given by $P_{loss} = i^2 R$. The resistance, R, of the power line is fixed; thus, decreasing the power lost during transmission means reducing the current carried in the transmission line. This reduction is accomplished by transmitting the power using a very high potential difference and a very low current. Looking at equation 25.18, you might

(a)

(b)

FIGURE 25.21 (a) The station that converts alternating current to direct current at the Itaipú Dam on the Paraná River in Brazil and Paraguay. (b) The station that converts the transmitted direct current back to alternating current in São Paulo, Brazil.

25.3 Self-Test Opportunity

Consider a battery with internal resistance R_i. What external resistance, R, will undergo the maximum heating when connected to this battery?

argue that we could also write $P_{loss} = (\Delta V)^2/R$ and that a high potential difference means a large power loss instead of a small power loss. However ΔV in this equation is the potential drop across the power line, not the potential difference at which the electrical power is being transmitted. The potential drop across the power line is $V_{drop} = iR$, which is much lower than the high potential difference used to transmit the electrical power. The expressions for the transmitted power and the dissipated power can be combined to give $P_{loss} = (P/\Delta V)^2 R = P^2 R/(\Delta V)^2$, which means that for a given amount of power, the dissipated power decreases as the square of the potential difference used to transmit the power.

Normally, electrical power generation and transmission use alternating currents. As we'll see in Chapter 30, alternating currents have the advantage that it is easy to raise or lower the potential difference via transformers. However, alternating currents have the inherent disadvantage of high power losses. High-voltage direct current (HVDC) transmission lines do not have this problem and suffer only power losses due to the resistance of the power line. However, HVDC transmission lines have the extra requirement that alternating current must be converted to direct current for transmission and the direct current must be converted back to alternating current at the destination.

Chapter 5 mentioned the electrical energy produced by the Itaipú Dam on the Paraná River in Brazil and Paraguay. Part of the power produced by this hydroelectric plant is transmitted via the world's largest HVDC transmission line a distance of about 800 km from the Itaipú Dam to São Paulo, Brazil, one of the ten largest metropolitan areas in the world. The transmission line carries 6300 MW of electrical power using direct current with a potential difference of ±600 kV. The station at the Itaipú Dam that converts the alternating current to direct current is shown in Figure 25.21a. The station in São Paulo that converts the transmitted direct current back to alternating current is shown in Figure 25.21b.

Future applications of HVDC power transmission include the transmission of power from solar power stations located in remote areas in the southwest United States to densely populated areas, such as large cities in California and Texas.

EXAMPLE 25.5 | Temperature Dependence of a Light Bulb's Resistance

A 100-W light bulb is connected in series to a source of emf with $V_{emf} = 100$ V. When the light bulb is lit, the temperature of its tungsten filament is 2520 °C.

PROBLEM
What is the resistance of the light bulb's tungsten filament at room temperature (20 °C)?

SOLUTION
The resistance of the filament when the light bulb is lit can be obtained using equation 25.18:

$$P = \frac{V_{emf}^2}{R}.$$

We rearrange this equation and substitute the numerical values to get the resistance of the filament:

$$R = \frac{V_{emf}^2}{P} = \frac{(100 \text{ V})^2}{100 \text{ W}} = 100 \ \Omega.$$

The temperature dependence of the filament's resistance is given by equation 25.13:

$$R - R_0 = R_0 \alpha (T - T_0).$$

We solve for the resistance at room temperature, R_0:

$$R = R_0 + R_0\alpha\left(T - T_0\right) = R_0\left[1 + \alpha\left(T - T_0\right)\right]$$

$$R_0 = \frac{R}{1 + \alpha\left(T - T_0\right)}.$$

Using the temperature coefficient of resistivity for tungsten from Table 25.1, we get

$$R_0 = \frac{R}{1 + \alpha\left(T - T_0\right)} = \frac{100\ \Omega}{1 + \left(4.5 \cdot 10^{-3}\ °C^{-1}\right)\left(2520\ °C - 20\ °C\right)} = 8.2\ \Omega.$$

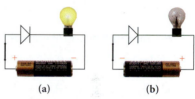

25.9 In-Class Exercise

A current of 2.00 A is maintained in a circuit with a total resistance of 5.00 Ω. How much heat is generated in 4.00 s?

a) 55.2 J   d) 168 J

b) 80.0 J   e) 244 J

c) 116 J

25.8 Diodes: One-Way Streets in Circuits

Section 25.4 stated that many resistors obey Ohm's Law. It was noted, however, that there are also non-ohmic resistors that do not obey Ohm's Law. A very common and extremely useful example is a diode. A *diode* is an electronic device that is designed to conduct current in one direction and not in the other direction. Remember that Figure 25.2c showed that a light bulb was still shining with the same intensity when the battery it was connected to was reversed. If a diode (represented by the symbol ▷|) is added to the same circuit, the diode prevents the current from flowing when the potential difference delivered by the battery is reversed; see Figure 25.22. The diode acts like a one-way street for the current.

Figure 25.23 shows current versus potential difference for a 3-Ω ohmic resistor and a silicon diode. The resistor obeys Ohm's Law, with the current flowing in the opposite direction when the potential difference is negative. The plot of current versus potential difference for the resistor is a straight line with a slope of 1/3(Ω). The silicon diode is wired so that it will not conduct any current when there is a negative potential difference. This silicon diode, like most, will conduct current if the potential difference is above 0.7 V. For potential differences above this threshold, the diode is essentially a conductor; below this threshold, the diode will not conduct current. The turn-on of the diode above the threshold potential difference increases exponentially; it can be close to instantaneous, as is visible in Figure 25.23.

Diodes are very useful for converting alternating current to direct current, as we'll see in Chapter 30. The fundamental physics principles that underlie the functioning of diodes require an understanding of quantum mechanics.

One particularly useful kind of diode is the light-emitting diode (LED), which not only regulates current in a circuit but also emits light of a single wavelength in a very controlled way. LEDs that emit light of many different wavelengths have been manufactured, and they emit light much more efficiently than conventional incandescent bulbs do. Light intensity is measured in lumens. Light sources can be compared in terms of how many lumens they produce per watt of electrical power. During the last decade, intensive research into LED technology has resulted in huge increases in the lumens per watt output for LEDs, reaching values of up to 130 to 170 lm/W. This compares very favorably with conventional incandescent lights (which are in the range from 5 to 20 lm/W), halogen lights (20 to 30 lm/W), and even fluorescent high-efficiency lights (30 to 95 lm/W). Prices for LEDs (in particular, "white" LEDs) are still comparatively high but are expected to decrease significantly. The United States uses over 100 billion kW h of electrical energy for lighting alone each year, which is approximately 10% of the total U.S. energy consumption. Universal use of LED lighting could save 70% to 90% of those 100 billion kW h, approximately the annual energy output of 10 nuclear power plants (~1 GW power each).

LEDs are also used in large display screens, where high light output is desirable. Perhaps the most impressive of these was showcased during the opening ceremony of the 2008 Beijing Olympics (Figure 25.24). It used 44,000 individual LEDs and measured an astounding 147 m by 22 m.

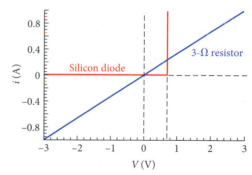

FIGURE 25.22 (a) The circuit of Figure 25.2c, but with a diode included. (b) Reversing the potential difference from the battery causes the current to stop flowing and the light bulb to stop shining.

FIGURE 25.23 Current as a function of potential difference for a resistor (blue) and a diode (red).

FIGURE 25.24 Giant LED screen used during the opening ceremony of the 2008 Olympic Games in Beijing.

WHAT WE HAVE LEARNED | EXAM STUDY GUIDE

- Current, i, is defined as the rate at which change, q, flows past a particular point: $i = \dfrac{dq}{dt}$.

- The magnitude of the average current density, J, at a given cross-sectional area, A, in a conductor is given by $J = \dfrac{i}{A}$.

- The magnitude of the current density, J, is related to the magnitude of the drift velocity, v_d, of the current-carrying charges, $-e$, by $J = \dfrac{i}{A} = -nev_d$, where n is the number of charge carriers per unit volume.

- The resistivity, ρ, of a material is defined in terms of the magnitudes of the electric field applied across the material, E, and the resulting current density, J: $\rho = \dfrac{E}{J}$.

- The resistance, R, of a specific device having resistivity ρ, length L, and constant cross-sectional area A, is $R = \rho \dfrac{L}{A}$.

- The temperature dependence of the resistivity of a material is given by $\rho - \rho_0 = \rho_0 \alpha (T - T_0)$, where ρ is the final resistivity, ρ_0 is the initial resistivity, α is the

temperature coefficient of electric resistivity, T is the final temperature, and T_0 is the initial temperature.

- The electromotive force, or emf, is a potential difference created by a device that drives current through a circuit.

- Ohm's Law states that when a potential difference, ΔV, appears across a resistor, R, the current, i, flowing through the resistor is $i = \dfrac{\Delta V}{R}$.

- Resistors connected in series can be replaced with an equivalent resistance, R_{eq}, given by the sum of the resistances of the resistors: $R_{eq} = \displaystyle\sum_{i=1}^{n} R_i$.

- Resistors connected in parallel can be replaced with an equivalent resistance, R_{eq}, given by $\dfrac{1}{R_{eq}} = \displaystyle\sum_{i=1}^{n} \dfrac{1}{R_i}$.

- The power, P, dissipated by a resistor, R, through which a current, i, flows is given by $P = i\Delta V = i^2 R = \dfrac{(\Delta V)^2}{R}$, where ΔV is the potential drop across the resistor.

KEY TERMS

electric current, p. 806
ampere, p. 806
direct current, p. 807
current density, p. 808
drift velocity, p. 808

resistivity, p. 811
resistance, p. 811
ohm, p. 811
Ohm's Law, p. 811
conductance, p. 811

conductivity, p. 812
temperature coefficient
  of electric resistivity,
  p. 814
superconductors, p. 815

electromotive force
  (emf), p. 816
batteries, p. 816
potential drop, p. 817

NEW SYMBOLS AND EQUATIONS

$i = \dfrac{dq}{dt}$, current

$\vec{v}_d$, drift velocity of current-carrying charges

$J = \dfrac{i}{A}$, current density

ρ, resistivity

$R = \rho \dfrac{L}{A}$, resistance

V_{emf}, potential difference of an emf source

α, temperature coefficient of resistivity

ANSWERS TO SELF-TEST OPPORTUNITIES

25.1 $\dfrac{700 \text{ mAh}}{0.1 \text{ mA}} = 7000 \text{ h} \approx 292 \text{ days}$.

25.2 $\Delta V = iR \Rightarrow i = \dfrac{\Delta V}{R} = \dfrac{1.50 \text{ V}}{10.0 \ \Omega} = 0.150 \text{ A}$.

25.3 The maximum heating of the external resistance occurs when the external resistance is equal to the internal resistance.

$V_t = V_{emf} + iR_i = i(R + R_i)$

$P_{heat} = i^2 R = \dfrac{V_t^2 R}{(R + R_i)^2}$

$$\frac{dP_{\text{heat}}}{dR} = -\frac{2V_t^2 R}{\left(R + R_i\right)^3} + \frac{V_t^2}{\left(R + R_i\right)^2} = 0 \text{ at extremum.}$$

From this follows $R = R_i$.

You can check that this extremum is a maximum by taking the second derivative at $R = R_i$. You find:

$$\left.\frac{d^2 P_{\text{heat}}}{dR^2}\right|_{R=R_i} = V_t^2 \left(\frac{6}{16R^3} - \frac{5}{8R^3}\right) = -\frac{V_t^2}{R^3}\left(\frac{1}{2}\right) < 0.$$

PROBLEM-SOLVING PRACTICE

Problem-Solving Guidelines

1. If a circuit diagram is not given as part of the problem statement, draw one yourself and label all the given values and unknown components. Indicate the direction of the current, starting from the emf source. (Don't worry about getting the direction of the current wrong in your diagram; if you guess wrong, your final answer for the current will be a negative number.)

2. Sources of emf supply potential to a circuit, and resistors reduce potential in the circuit. However, be careful to check

the direction of the potential of the emf source relative to that of the current; a current flowing in the direction opposite to the potential of an emf device picks up a negative potential difference.

3. The sum of the potential drops across resistors in a circuit equals the net amount of emf supplied to the circuit. (This is a consequence of the law of conservation of energy.)

4. In any given wire segment, the current is the same everywhere. (This is a consequence of the law of conservation of charge.)

SOLVED PROBLEM 25.4 | Size of Wire for a Power Line

PROBLEM

Imagine you are designing the HVDC power line from the Itaipú Dam on the Paraná River in Brazil and Paraguay to the city of São Paulo in Brazil. The power line is 800 km long and transmits 6300 MW of power at a potential difference of 1.20 MV. (Figure 25.25 shows an HVDC line.)

The electric company requires that no more than 25% of the power be lost in transmission. If the line consists of one wire made out of copper and having a circular cross section, what is the minimum diameter of the wire?

SOLUTION

THINK

Knowing the power transmitted and the potential difference with which it is transmitted, we can calculate the current carried in the line. We can then express the power lost in terms of the resistance of the transmission line. With the current and the resistance of the wire, we can write an expression for the power lost during transmission. The resistance of the wire is a function of the diameter of the wire, the length of the wire, and the resistivity of copper. We can then solve for the diameter of the wire that will keep the power loss within the specified limit.

FIGURE 25.25 HVDC power transmission line.

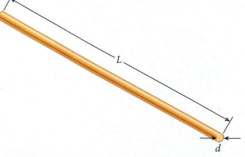

SKETCH

A sketch of a copper wire of length L and diameter d is shown in Figure 25.26.

RESEARCH

The power, P, carried in the line is related to the current, i, and the potential difference, ΔV: $P = i\Delta V$. The power lost in transmission, P_{lost}, can be related (see equation 25.18) to the current in the wire and the resistance, R, of the wire:

$$P_{\text{lost}} = i^2 R. \tag{i}$$

FIGURE 25.26 An HVDC transmission line consisting of a copper conductor (not to scale).

Continued—

The resistance of the wire is given by equation 25.11:

$$R = \rho_{Cu} \frac{L}{A}, \tag{ii}$$

where ρ_{Cu} is the resistivity of copper, L is the length of the wire, and A is the cross-sectional area of the wire.

The cross-sectional area of the wire is the area of a circle:

$$A = \pi \left(\frac{d}{2}\right)^2 = \frac{\pi d^2}{4},$$

where d is the diameter of the wire. Thus, with the area of a circle substituted for A, equation (ii) becomes

$$R = \rho_{Cu} \frac{L}{\pi d^2 / 4}. \tag{iii}$$

SIMPLIFY

We can solve $P = i\Delta V$ for the current in the wire:

$$i = \frac{P}{\Delta V}.$$

Substituting this expression for the current and that for the resistance from equation (iii) into equation (i) for the lost power gives

$$P_{lost} = \left(\frac{P}{\Delta V}\right)^2 \left(\rho_{Cu} \frac{L}{\pi d^2/4}\right) = \frac{4 P^2 \rho_{Cu} L}{\pi (\Delta V)^2 d^2}.$$

The fraction of lost power relative to total power, f, is

$$\frac{P_{lost}}{P} = \frac{\left(\dfrac{4 P^2 \rho_{Cu} L}{\pi (\Delta V)^2 d^2}\right)}{P} = \frac{4 P \rho_{Cu} L}{\pi (\Delta V)^2 d^2} = f.$$

Solving this equation for the diameter of the wire gives

$$d = \sqrt{\frac{4 P \rho_{Cu} L}{f \pi (\Delta V)^2}}.$$

CALCULATE

Putting in the numerical values gives us

$$d = \sqrt{\frac{4\left(6300 \cdot 10^6 \text{ W}\right)\left(1.72 \cdot 10^{-8} \text{ } \Omega \text{ m}\right)\left(800 \cdot 10^3 \text{ m}\right)}{(0.25)\pi\left(1.20 \cdot 10^6 \text{ V}\right)^2}} = 0.0175099 \text{ m}.$$

ROUND

Rounding to three significant figures gives us the minimum diameter of the copper wire:

$$d = 1.75 \text{ cm}.$$

DOUBLE-CHECK

To double-check our result, let's calculate the resistance of this transmission line. Using our calculated value for the diameter, we can find the cross-sectional area and then, using equation 25.11, we find

$$R = \rho_{Cu} \frac{L}{\pi d^2/4} = \frac{4\rho_{Cu}L}{\pi d^2} = \frac{4\left(1.72 \cdot 10^{-8} \ \Omega \ m\right)\left(800 \cdot 10^3 \ m\right)}{\pi \left(1.75 \cdot 10^{-2} \ m\right)^2} = 57.2 \ \Omega.$$

The current transmitted is

$$i = \frac{P}{V} = \frac{6300 \cdot 10^6 \ W}{1.20 \cdot 10^6 \ V} = 5250 \ A.$$

The power lost is then

$$P = i^2 R = \left(5250 \ A\right)^2 \left(57.2 \ \Omega\right) = 1580 \ MW,$$

which is close (within rounding error) to 25% of the total power of 6300 MW. Thus, our result seems reasonable.

MULTIPLE-CHOICE QUESTIONS

25.1 If the current through a resistor is increased by a factor of 2, how does this affect the power that is dissipated?

a) It decreases by a factor of 4.

b) It increases by a factor of 2.

c) It decreases by a factor of 8.

d) It increases by a factor of 4.

25.2 You make a parallel combination of resistors consisting of resistor A having a very large resistance and resistor B having a very small resistance. The equivalent resistance for this combination will be:

a) slightly greater than the resistance of the resistor A.

b) slightly less than the resistance of the resistor A.

c) slightly greater than the resistance of the resistor B.

d) slightly less than the resistance of the resistor B.

25.3 Two cylindrical wires, 1 and 2, made of the same material, have the same resistance. If the length of wire 2 is twice that of wire 1, what is the ratio of their cross-sectional areas, A_1 and A_2?

a) $A_1/A_2 = 2$

b) $A_1/A_2 = 4$

c) $A_1/A_2 = 0.5$

d) $A_1/A_2 = 0.25$

25.4 All three light bulbs in the circuit shown in the figure are identical. Which of the three shines the brightest?

a) A

b) B

c) C

d) A and B

e) All three are equally bright.

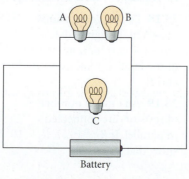

25.5 All of the six light bulbs in the circuit shown in the figure are identical. Which ordering correctly expresses the relative brightness of the bulbs? (*Hint:* The more current flowing through a light bulb, the brighter it is!)

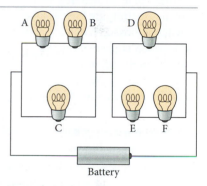

a) $A = B > C = D > E = F$

b) $A = B = E = F > C = D$

c) $C = D > A = B = E = F$

d) $A = B = C = D = E = F$

25.6 Which of the arrangements of three identical light bulbs shown in the figure draws most current from the battery?

a) A

b) B

c) C

d) All three draw equal current.

e) A and C are tied for drawing the most current.

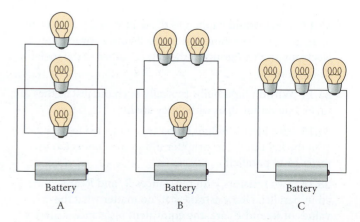

25.7 Which of the arrangements of three identical light bulbs shown in the figure has the highest resistance?

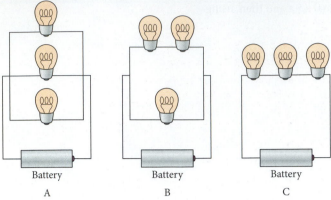

Battery   Battery   Battery

A   B   C

a) A   d) All three have equal resistance.

b) B   e) A and C are tied for having the highest

c) C   resistance.

25.8 Three identical light bulbs are connected as shown in the figure. Initially the switch is closed. When the switch is opened (as shown in the figure), bulb C goes off. What happens to bulbs A and B?

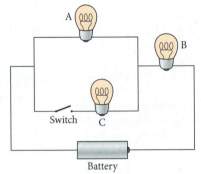

a) Bulb A gets brighter, and bulb B gets dimmer.

b) Both bulbs A and B get brighter.

c) Both bulbs A and B get dimmer.

d) Bulb A gets dimmer, and bulb B gets brighter.

25.9 Which of the following wires has the largest current flowing through it?

a) a 1-m-long copper wire of diameter 1 mm connected to a 10-V battery

b) a 0.5-m-long copper wire of diameter 0.5 mm connected to a 5-V battery

c) a 2-m-long copper wire of diameter 2 mm connected to a 20-V battery

d) a 1-m-long copper wire of diameter 0.5 mm connected to a 5-V battery

e) All of the wires have the same current flowing through them.

25.10 Ohm's Law states that the potential difference across a device is equal to

a) the current flowing through the device times the resistance of the device.

b) the current flowing through the device divided by the resistance of the device.

c) the resistance of the device divided by the current flowing through the device.

d) the current flowing through the device times the cross-sectional area of the device.

e) the current flowing through the device times the length of the device.

25.11 A constant electric field is maintained inside a semiconductor. As the temperature is lowered, the magnitude of the current density inside the semiconductor

a) increases.   c) decreases.

b) stays the same.   d) may increase or decrease.

25.12 Which of the following is an incorrect statement?

a) The currents through electronic devices connected in series are equal.

b) The potential drops across electronic devices connected in parallel are equal.

c) More current flows across the smaller resistance when two resistors are in parallel connection.

d) More current flows across the smaller resistance when two resistors are in serial connection.

QUESTIONS

25.13 What would happen to the drift velocity of electrons in a wire if the resistance due to collisions between the electrons and the atoms in the crystal lattice of the metal disappeared?

25.14 Why do light bulbs typically burn out just as they are turned on rather than while they are lit?

25.15 Two identical light bulbs are connected to a battery. Will the light bulbs be brighter if they are connected in series or in parallel?

25.16 Two resistors with resistances R_1 and R_2 are connected in parallel. Demonstrate that, no matter what the actual values of R_1 and R_2 are, the equivalent resistance is always less than the smaller of the two resistances.

25.17 Show that for resistors connected in series, it is always the highest resistance that dissipates the most power, while for resistors connected in parallel, it is always the lowest resistance that dissipates the most power.

25.18 For the connections shown in the figure, determine the current i_1 in terms of the total current, i, and R_1 and R_2.

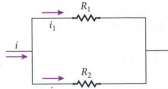

25.19 An infinite number of resistors are connected in parallel. If $R_1 = 10\ \Omega$, $R_2 = 10^2\ \Omega$, $R_3 = 10^3\ \Omega$, and so on, show that $R_{eq} = 9\ \Omega$.

25.20 You are given two identical batteries and two pieces of wire. The red wire has a higher resistance than the black wire. You place the red wire across the terminals of one battery and the black wire across the terminals of the other battery. Which wire gets hotter?

25.21 Should light bulbs (ordinary incandescent bulbs with tungsten filaments) be considered ohmic resistors? Why or why not? How would this be determined experimentally?

25.22 A charged-particle beam is used to inject a charge, Q_0, into a small, irregularly shaped region (not a cavity, just some region within the solid block) in the interior of a block of ohmic material with conductivity σ and permittivity ϵ at time $t = 0$. Eventually, all the injected charge will move to the outer surface of the block, but how quickly?

a) Derive a differential equation for the charge, $Q(t)$, in the injection region as a function of time.

b) Solve the equation from part (a) to find $Q(t)$ for all $t \geq 0$.

c) For copper, a good conductor, and for quartz (crystalline SiO_2), an insulator, calculate the time for the charge in the injection region to decrease by half. Look up the necessary values. Assume that the effective "dielectric constant" of copper is 1.00000.

25.23 Show that the drift speed of free electrons in a wire does not depend on the cross-sectional area of the wire.

25.24 Rank the brightness of the six identical light bulbs in the circuit in the figure. Each light bulb may be treated as an identical resistor with resistance R.

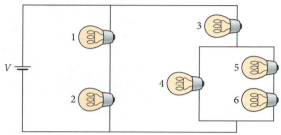

25.25 Two conductors of the same length and radius are connected to the same emf device. If the resistance of one is twice that of the other, to which conductor is more power delivered?

PROBLEMS

A blue problem number indicates a worked-out solution is available in the Student Solutions Manual. One • and two •• indicate increasing level of problem difficulty.

Sections 25.1 and 25.2

25.26 How many protons are in the beam traveling close to the speed of light in the Tevatron at Fermilab, which is carrying 11 mA of current around the 6.3-km circumference of the main Tevatron ring?

25.27 What is the current density in an aluminum wire having a radius of 1 mm and carrying a current of 1 mA? What is the drift speed of the electrons carrying this current? The density of aluminum is 2700 kg/m³, and 1 mole of aluminum has a mass of 26.98 g. There is one conduction electron per atom in aluminum.

•**25.28** A copper wire has a diameter $d_{Cu} = 0.0500$ cm, is 3.00 m long, and has a density of charge carriers of $8.50 \cdot 10^{28}$ electrons/m³. As shown in the figure, the copper wire is attached to an equal length of aluminum wire with a diameter $d_{Al} = 0.0100$ cm and density of charge carriers of $6.02 \cdot 10^{28}$ electrons/m³. A current of 0.400 A flows through the copper wire.

a) What is the ratio of the current densities in the two wires, J_{Cu}/J_{Al}?

b) What is the ratio of the drift velocities in the two wires, v_{d-Cu}/v_{d-Al}?

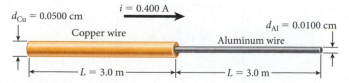

•**25.29** A current of 0.123 mA flows in a silver wire whose cross-sectional area is 0.923 mm².

a) Find the density of electrons in the wire, assuming that there is one conduction electron per silver atom.

b) Find the current density in the wire assuming that the current is uniform.

c) Find the electron's drift speed.

Section 25.3

25.30 What is the resistance of a copper wire of length $l = 10.9$ m and diameter $d = 1.3$ mm? The resistivity of copper is $1.72 \cdot 10^{-8}\ \Omega$ m.

25.31 Two conductors are made of the same material and have the same length L. Conductor A is a hollow tube with inside diameter 2.00 mm and outside diameter 3.00 mm; conductor B is a solid wire with radius R_B. What value of R_B is required for the two conductors to have the same resistance measured between their ends?

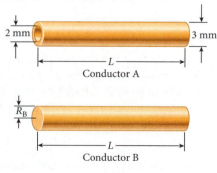

25.32 A copper coil has a resistance of 0.1 Ω at room temperature. What is its resistance when it is cooled to −100 °C?

25.33 What gauge of aluminum wire will have the same resistance per unit length as 12-gauge copper wire?

25.34 A rectangular wafer of pure silicon, with resistivity $\rho = 2300 \, \Omega$ m, measures 2.00 cm by 3.00 cm by 0.010 cm. Find the maximum resistance of this rectangular wafer between any two faces.

•**25.35** A copper wire that is 1 m long and has a radius of 0.5 mm is stretched to a length of 2 m. What is the fractional change in resistance, $\Delta R/R$, as the wire is stretched? What is $\Delta R/R$ for a wire of the same initial dimensions made out of aluminum?

•**25.36** The most common material used for sandpaper, silicon carbide, is also widely used in electrical ap-

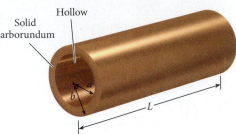
Hollow
Solid
carborundum

plications. One common device is a tubular resistor made of a special grade of silicon carbide called *carborundum*. A particular carborundum resistor (see the figure) consists of a thick-walled cylindrical shell (a pipe) of inner radius $a = 1.5$ cm, outer radius $b = 2.5$ cm, and length $L = 60$ cm. The resistance of this carborundum resistor at 20 °C is 1 Ω.

a) Calculate the resistivity of carborundum at room temperature. Compare this to the resistivities of the most commonly used conductors (copper, aluminum, and silver).

b) Carborundum has a high temperature coefficient of resistivity: $\alpha = 2.14 \cdot 10^{-3} \, K^{-1}$. If, in a particular application, the carborundum resistor heats up to 300 °C, what is the percentage change in its resistance between room temperature (20 °C) and this operating temperature?

••**25.37** As illustrated in the figure, a current, i, flows through the junction of two materials with the same cross-sectional area and with conductivities σ_1 and σ_2. Show that the total amount of charge at the junction is $\epsilon_0 i(1/\sigma_2 - 1/\sigma_1)$.

Material 1   Material 2

Section 25.4

25.38 A potential difference of 12.0 V is applied across a wire of cross-sectional area 4.5 mm^2 and length 1000 km. The current passing through the wire is $3.2 \cdot 10^{-3}$ A.

a) What is the resistance of the wire?

b) What type of wire is this?

25.39 One brand of 12.0-V automotive battery used to be advertised as providing "600 cold-cranking amps." Assuming that this is the current the battery supplies if its terminals are shorted, that is, connected to negligible resistance, determine the internal resistance of the battery. (*IMPORTANT*: **Do not** attempt such a connection as it could be lethal!)

25.40 A copper wire has radius $r = 0.0250$ cm, is 3.00 m long, has resistivity $\rho = 1.72 \cdot 10^{-8} \, \Omega$ m, and carries a current of 0.400 A. The wire has density of charge carriers of $8.50 \cdot 10^{28}$ electrons/m^3.

a) What is the resistance, R, of the wire?

b) What is the electric potential difference, ΔV, across the wire?

c) What is the electric field, E, in the wire?

•**25.41** A 34-gauge copper wire, with a constant potential difference of 0.1 V applied across its 1 m length at room temperature, is cooled to liquid nitrogen temperature (77 K = –196 °C).

a) Determine the percentage change in the wire's resistance during the drop in temperature.

b) Determine the percentage change in current flowing in the wire.

c) Compare the drift speeds of the electrons at the two temperatures.

Section 25.5

25.42 A resistor of unknown resistance and a 35-Ω resistor are connected across a 120-V emf device in such a way that an 11-A current flows. What is the value of the unknown resistance?

25.43 A battery has a potential difference of 14.50 V when it is not connected in a circuit. When a 17.91-Ω resistor is connected across the battery, the potential difference of the battery drops to 12.68 V. What is the internal resistance of the battery?

25.44 When a battery is connected to a 100-Ω resistor, the current is 4.00 A. When the same battery is connected to a 400-Ω resistor, the current is 1.01 A. Find the emf supplied by the battery and the internal resistance of the battery.

•**25.45** A light bulb is connected to a source of emf. There is a 6.20 V drop across the light bulb, and a current of 4.1 A flowing through the light bulb.

a) What is the resistance of the light bulb?

b) A second light bulb, identical to the first, is connected in series with the first bulb. The potential drop across the bulbs is now 6.29 V, and the current through the bulbs is 2.9 A. Calculate the resistance of each light bulb.

c) Why are your answers to parts (a) and (b) not the same?

Section 25.6

25.46 What is the current in the 10-Ω resistor in the circuit in the figure?

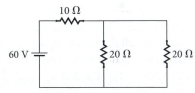

10 Ω

60 V

20 Ω

20 Ω

25.47 What is the equivalent resistance of the five resistors in the circuit in the figure?

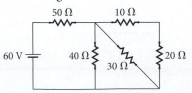

50 Ω 10 Ω

60 V

40 Ω

30 Ω

20 Ω

•**25.48** What is the current in the circuit shown in the figure when the switch is (a) open and (b) closed?

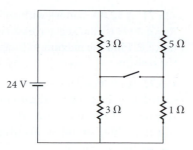

25.49 For the circuit shown in the figure, $R_1 = 6\ \Omega$, $R_2 = 6\ \Omega$, $R_3 = 2\ \Omega$, $R_4 = 4\ \Omega$, $R_5 = 3\ \Omega$, and the potential difference is 12 V.

a) What is the equivalent resistance for the circuit?

b) What is the current through R_5?

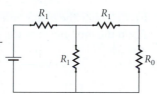

c) What is the potential drop across R_3?

25.50 Four resistors are connected in a circuit as shown in the figure. What value of R_1, expressed as a multiple of R_0, will make the equivalent resistance for the circuit equal to R_0?

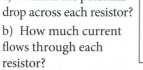

•**25.51** As shown in the figure, a circuit consists of an emf source with $V = 20$ V and six resistors. Resistors $R_1 = 5\ \Omega$ and $R_2 = 10\ \Omega$ are connected in series. Resistors $R_3 = 5\ \Omega$ and $R_4 = 5\ \Omega$ are connected in parallel and are in series with R_1 and R_2. Resistors $R_5 = 2\ \Omega$ and $R_6 = 2\ \Omega$ are connected in parallel and are also in series with R_1 and R_2.

a) What is the potential drop across each resistor?

b) How much current flows through each resistor?

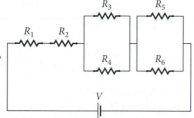

•**25.52** When a 40.0-V emf device is placed across two resistors in series, a current of 10.0 A is flowing in each of the resistors. When the same emf device is placed across the same two resistors in parallel, the current through the emf device is 50.0 A. What is the magnitude of the larger of the two resistances?

Section 25.7

25.53 A voltage spike causes the line voltage in a home to jump rapidly from 110 V to 150 V. What is the percentage increase in the power output of a 100-W tungsten-filament incandescent light bulb during this spike, assuming that the bulb's resistance remains constant?

25.54 A thundercloud similar to the one described in Example 24.3 produces a lightning bolt that strikes a radio tower. If the lightning bolt transfers 5.0 C of charge in about 0.10 ms and the potential remains constant at 70 MV, find (a) the average current, (b) the average power, (c) the total energy, and (d) the effective resistance of the air during the lightning strike.

25.55 A hair dryer consumes 1600. W of power and operates at 110. V. (Assume that the current is DC. In fact, these are root-mean-square values of AC quantities, but the calculation is not affected. Chapter 30 covers AC circuits in detail.)

a) Will the hair dryer trip a circuit breaker designed to interrupt the circuit if the current exceeds 15.0 A?

b) What is the resistance of the hair dryer when it is operating?

25.56 How much money will a homeowner owe an electric company if he turns on a 100-W incandescent light bulb and leaves it on for an entire year? (Assume that the cost of electricity is $0.12/kW h and that the light bulb lasts that long.) The same amount of light can be provided by a 26-W compact fluorescent light bulb. What would it cost the homeowner to leave one of those on for a year?

25.57 Three resistors are connected across a battery as shown in the figure.

a) How much power is dissipated across the three resistors?

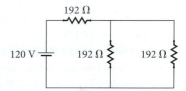

b) Determine the potential drop across each resistor.

25.58 Suppose an AAA battery is able to supply 625 mAh before its potential drops below 1.5 V. How long will it be able to supply power to a 5.0-W bulb before the potential drops below 1.5 V?

•**25.59** Show that the power supplied to the circuit in the figure by the battery with internal resistance R_i is maximum when the resistance of the resistor in the circuit, R, is equal to R_i. Determine the power supplied to R. For practice, calculate the power dissipated by a 12-V battery with an internal resistance of $2\ \Omega$ when $R = 1\ \Omega$, $R = 2\ \Omega$, and $R = 3\ \Omega$.

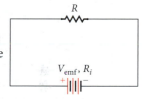

•**25.60** A water heater consisting of a metal coil that is connected across the terminals of a 15-V power supply is able to heat 250 mL of water from room temperature to boiling point in 45 s. What is the resistance of the coil?

•**25.61** A potential difference of $V = 0.500$ V is applied across a block of silicon with resistivity $8.70 \cdot 10^{-4}\ \Omega$ m. As indicated in the figure, the dimensions of the silicon block are width $a = 2.00$ mm and length $L = 15.0$ cm. The resistance of the silicon block is 50.0 Ω, and the density of charge carriers is $1.23 \cdot 10^{23}\ m^{-3}$. Assume that the current density in the block is uniform and that current flows in silicon according to Ohm's Law. The total length of 0.500-mm-diameter copper wire in the circuit is 75.0 cm, and the resistivity of copper is $1.69 \cdot 10^{-8}\ \Omega$ m.

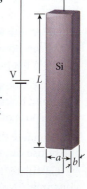

a) What is the resistance, R_w, of the copper wire?

b) What are the direction and the magnitude of the electric current, i, in the block?

c) What is the thickness, b, of the block?

d) On average, how long does it take an electron to pass from one end of the block to the other?

e) How much power, P, is dissipated by the block?

f) In what form of energy does this dissipated power appear?

Additional Problems

25.62 In an emergency, you need to run a radio that uses 30 W of power when attached to a 10-V power supply. The only power supply you have access to provides 25,000 V, but you do have a large number of 25-Ω resistors. If you want the power to the radio to be as close as possible to 30 W, how many resistors should you use, and how should they be connected (in series or in parallel)?

25.63 A certain brand of hot dog cooker applies a potential difference of 120 V to opposite ends of the hot dog and cooks it by means of the heat produced. If 48 kJ is needed to cook each hot dog, what current is needed to cook three hot dogs simultaneously in 2.0 min? Assume a parallel connection.

25.64 A circuit consists of a copper wire of length 10 m and radius 1 mm connected to a 10-V battery. An aluminum wire of length 5 m is connected to the same battery and dissipates the same amount of power. What is the radius of the aluminum wire?

25.65 The resistivity of a conductor is $\rho = 1 \cdot 10^{-5}\ \Omega$ m. If a cylindrical wire is made of this conductor, with a cross-sectional area of $1 \cdot 10^{-6}\ \text{m}^2$, what should the length of the wire be for its resistance to be 10 Ω?

25.66 Two cylindrical wires of identical length are made of copper and aluminum. If they carry the same current and have the same potential difference across their length, what is the ratio of their radii?

25.67 Two resistors with resistances 200 Ω and 400 Ω are connected (a) in series and (b) in parallel with an ideal 9-V battery. Compare the power delivered to the 200-Ω resistor.

25.68 What is (a) the conductance and (b) the radius of a 3.5-m-long iron heating element for a 110-V, 1500-W heater?

25.69 A 100-W, 240-V European light bulb is used in an American household, where the electricity is delivered at 120 V. What power will it consume?

25.70 A modern house is wired for 115 V, and the current is limited by circuit breakers to a maximum of 200 A. (For the purpose of this problem, treat these as DC quantities.)

a) Calculate the minimum total resistance the circuitry in the house can have at any time.

b) Calculate the maximum electrical power the house can consume.

•**25.71** A 12.0 V battery with an internal resistance $R_i = 4.00\ \Omega$ is attached across an external resistor of resistance R. Find the maximum power that can be delivered to the resistor.

•**25.72** A multiclad wire consists of a zinc core of radius 1 mm surrounded by a copper sheath of thickness 1 mm. The resistivity of zinc is $\rho = 5.9 \cdot 10^{-8}\ \Omega$ m. What is the resistance of a 10-m-long strand of this wire?

•**25.73** The Stanford Linear Accelerator accelerated a beam consisting of $2.0 \cdot 10^{14}$ electrons per second through a potential difference of $2.0 \cdot 10^{10}$ V.

a) Calculate the current in the beam.

b) Calculate the power of the beam.

c) Calculate the effective ohmic resistance of the accelerator.

•**25.74** In the circuit shown in the figure, $R_1 = 3\ \Omega$, $R_2 = 6\ \Omega$, $R_3 = 20\ \Omega$, and $V_{emf} = 12$ V.

a) Determine a value for the equivalent resistance.

b) Calculate the magnitude of the current flowing through R_3 on the top branch of the circuit (marked with a vertical arrow)

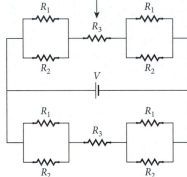

•**25.75** Three resistors are connected to a power supply with $V = 110$ V as shown in the figure.

a) Find the potential drop across R_3.

b) Find the current in R_1.

c) Find the rate at which thermal energy is dissipated from R_2.

$R_1 = 2\ \Omega$

$V = 110$ V $R_2 = 3\ \Omega$ $R_3 = 6\ \Omega$

•**25.76** A battery with $V = 1.5$ V is connected to three resistors as shown in the figure.

a) Find the potential drop across each resistor.

b) Find the current in each resistor.

$R_2 = 4\ \Omega$

$R_1 = 2\ \Omega$ $R_3 = 6\ \Omega$

$V = 1.5$ V

•**25.77** A 2.5-m-long copper cable is connected across the terminals of a 12-V car battery. Assuming that it is completely insulated from its environment, how long after the connection is made will the copper start to melt?

•**25.78** A piece of copper wire is used to form a circular loop of radius 10 cm. The wire has a cross-sectional area of 10 mm². Points A and B are 90° apart, as shown in the figure. Find the resistance between points A and B.

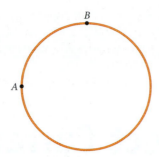

•**25.79** Two conducting wires have identical lengths $L_1 = L_2 = L = 10$ km and identical circular cross sections of radius $r_1 = r_2 = r = 1$ mm. One wire is made of steel (with resistivity $\rho_{steel} = 40 \cdot 10^{-8}$ Ω m); the other is made of copper (with resistivity $\rho_{copper} = 1.7 \cdot 10^{-8}$ Ω m).

a) Calculate the ratio of the power dissipated by the two wires, P_{copper}/P_{steel}, when they are connected in parallel; a potential difference of $V = 100$ V is applied to them.

b) Based on this result, how do you explain the fact that conductors for power transmission are made of copper and not steel?

•**25.80** Before bendable tungsten filaments were developed, Thomas Edison used carbon filaments in his light bulbs.

Though carbon has a very high melting temperature (3599 °C), its sublimation rate is high at high temperatures. So carbon-filament bulbs were kept at lower temperatures, thereby rendering them dimmer than later tungsten-based bulbs. A typical carbon-filament bulb requires an average power of 40 W, when 110 volts is applied across it, and has a filament temperature of 1800 °C. Carbon, unlike copper, has a negative temperature coefficient of resistivity: $\alpha = -0.0005$ °C^{-1}. Calculate the resistance at room temperature (20 °C) of this carbon filament.

••**25.81** A material is said to be *ohmic* if an electric field, $\vec{E}$, in the material gives rise to current density $\vec{J} = \sigma \vec{E}$, where the conductivity, σ, is a constant independent of $\vec{E}$ or $\vec{J}$. (This is the precise form of Ohm's Law.) Suppose in some material an electric field, $\vec{E}$, produces current density, $\vec{J}$, not necessarily related by Ohm's Law; that is, the material may or may not be ohmic.

a) Calculate the rate of energy dissipation (sometimes called *ohmic heating* or *joule heating*) per unit volume in this material, in terms of $\vec{E}$ and $\vec{J}$.

b) Express the result of part (a) in terms of $\vec{E}$ alone and $\vec{J}$ alone, for $\vec{E}$ and $\vec{J}$ related via Ohm's Law, that is, in an ohmic material with conductivity σ or resistivity ρ.

26

Direct Current Circuits

FIGURE 26.1 A circuit board can have hundreds of circuit components connected by metallic conducting paths.

WHAT WE WILL LEARN

- Some circuits cannot be reduced to a single loop; complex circuits can be analyzed using Kirchhoff's rules.

- Kirchhoff's Junction Rule states that the algebraic sum of the currents at any junction in a circuit must be zero.

- Kirchhoff's Loop Rule states that the algebraic sum of the potential changes around any closed loop in a circuit must be zero.

- Single-loop circuits can be analyzed using Kirchhoff's Loop Rule.

- Multiloop circuits must be analyzed using both Kirchhoff's Junction Rule and Kirchhoff's Loop Rule.

- The current in a circuit that contains a resistor and a capacitor varies exponentially with time, with a characteristic time constant given by the product of the resistance and the capacitance.

The electric circuit, such as the one shown in Figure 26.1, undoubtedly changed the world. Modern electronics continues to change human society, at a faster and faster pace. It took 38 years for radio to reach 50 million users in the United States. However, it took only 13 years for television to reach that number of users, 10 years for cable TV, 5 years for the Internet, and 3 years for cell phones.

This chapter examines the techniques used to analyze circuits that cannot be broken down into simple series and parallel connections. Modern electronics design depends on millions of different circuits, each with its own purpose and configuration. However, regardless of how complicated a circuit becomes, the basic rules for analyzing it are the ones presented in this chapter.

Some of the circuits analyzed in this chapter contain not only resistors and emf devices but also capacitors. In these circuits, the current is not steady, but changes with time. Time-varying currents will be covered more thoroughly in later chapters, which introduce additional circuit components.

26.1   Kirchhoff's Rules

In Chapter 25, we considered several kinds of direct current (DC) circuits, each containing one emf device along with resistors connected in series or in parallel. Some seemingly complicated circuits contain multiple resistors in series or in parallel that can be replaced with an equivalent resistance. However, we did not consider circuits containing multiple sources of emf. In addition, there are single-loop and multiloop circuits with emf devices and resistors that cannot be reduced to simple circuits containing parallel or series connections. Figure 26.2 shows two examples of such circuits. This chapter explains how to analyze these kinds of circuits using **Kirchhoff's Rules.**

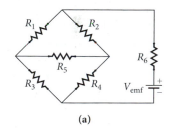

(a)

Kirchhoff's Junction Rule

A **junction** is a place in a circuit where three or more wires are connected to each other. Each connection between two junctions in a circuit is called a **branch.** A branch can contain any number of different circuit elements and the wires between them. Each branch can have a current flowing, and this current is the same everywhere in the branch. This fact leads to **Kirchhoff's Junction Rule:**

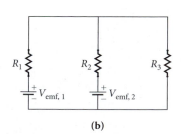

(b)

FIGURE 26.2 Two examples of circuits that cannot be reduced to simple combinations of parallel and series resistors.

The sum of the currents entering a junction must equal the sum of the currents leaving the junction.

With a positive sign assigned (arbitrarily) to currents entering the junction and a negative sign to those exiting the junction, Kirchhoff's Junction Rule is expressed mathematically as

$$\text{Junction: } \sum_{k=1}^{n} i_k = 0. \qquad (26.1)$$

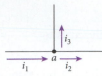

FIGURE 26.3 A single junction from a multiloop circuit.

26.1 In-Class Exercise

For the junction shown in the figure, which equation correctly expresses the sum of the currents?

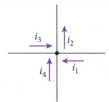

a) $i_1 + i_2 + i_3 + i_4 = 0$

b) $i_1 - i_2 + i_3 + i_4 = 0$

c) $-i_1 + i_2 + i_3 - i_4 = 0$

d) $i_1 - i_2 - i_3 - i_4 = 0$

e) $i_1 + i_2 - i_3 - i_4 = 0$

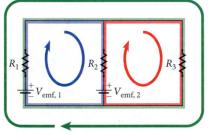

FIGURE 26.4 The three possible loops (indicated in red, green, and blue) for the circuit diagram shown in Figure 26.2b.

How do you know which currents enter a junction and which exit the junction when you make a drawing like the one shown in Figure 26.3? You don't; you simply assign a direction for each current along a given wire. If an assigned direction turns out to be wrong, you will obtain a negative number for that particular current in your final solution.

Kirchhoff's Junction Rule is a direct consequence of the conservation of electric charge. Junctions do not have the capability of storing charge. Thus, charge conservation requires that all charges streaming into a junction also leave the junction, which is exactly what Kirchhoff's Junction Rule states.

According to Kirchhoff's Junction Rule, at each junction in a multiloop circuit, the current flowing into the junction must equal the current flowing out of the junction. For example, Figure 26.3 shows a single junction, a, with a current, i_1, entering the junction and two currents, i_2 and i_3, leaving the junction. According to Kirchhoff's Junction Rule, in this case,

$$\sum_{k=1}^{3} i_k = i_1 - i_2 - i_3 = 0 \Rightarrow i_1 = i_2 + i_3.$$

Kirchhoff's Loop Rule

A **loop** in a circuit is any set of connected wires and circuit elements forming a closed path. If you follow a loop, eventually you will get to the same point from which you started. For example, in the circuit diagram shown in Figure 26.2b, three possible loops can be identified. These three loops are shown in different colors (red, green, and blue) in Figure 26.4. The blue loop includes resistors 1 and 2, emf sources 1 and 2, and their connecting wires. The red loop includes resistors 2 and 3, emf source 2, and their connecting wires. Finally, the green loop includes resistors 1 and 3, emf source 1, and their connecting wires. Note that any given wire or circuit element can be and usually is part of more than one loop.

You can move through any loop in a circuit in either a clockwise or a counterclockwise direction. Figure 26.4 shows a clockwise path through each of the loops, as indicated by the arrows. But the direction of the path taken around the loop is irrelevant as long as your choice is followed consistently all the way around the loop.

Summing the potential differences from all circuit elements encountered along any given loop yields the total potential difference of the complete path along the loop. **Kirchhoff's Loop Rule** then states:

The potential difference around a complete circuit loop must sum to zero.

Kirchhoff's Loop Rule is a direct consequence of the fact that electric potential is single-valued. This means that the electric potential energy of a conduction electron at a point in the circuit has one specific value. Suppose this rule were not valid. Then we could analyze the potential changes of a conduction electron in going around a loop and find that the electron had a different potential energy when it returned to its starting point. The potential energy of this electron would change at a point in the circuit, in obvious contradiction of energy conservation. In other words, Kirchhoff's Loop Rule is simply a consequence of the law of conservation of energy.

Application of Kirchhoff's Loop Rule requires conventions for determining the potential drop across each element of the circuit. This depends on the assumed direction of the current and the direction of the analysis. For emf sources, the rules are straightforward, since minus and plus signs (as well as short and long lines) indicate which side of the emf source is at the higher potential. The potential drop for an emf source is in the direction from minus to plus or from short line to long line. As noted earlier, the assignment of the current directions and the choice of a clockwise or counterclockwise path around a loop are arbitrary. Any direction will give the same information, as long as it is applied consistently around a loop. The conventions used to analyze circuit elements in a loop are summarized in Table 26.1 and Figure 26.5, where the magnitude of the current through the circuit element is i. (The labels in the right-most column of Table 26.1 correspond to the parts of Figure 26.5.)

If we move around a loop in a circuit in the same direction as the current, the potential changes across resistors will be negative. If we move around the loop in the opposite

Table 26.1	Conventions Used to Determine the Sign of Potential Changes Around a Single-Loop Circuit Containing Several Resistors and Sources of emf		
Element	**Direction of Analysis**	**Potential Change**	
R	Same as current	$-iR$	(a)
R	Opposite to current	$+iR$	(b)
V_{emf}	Same as emf	$+V_{emf}$	(c)
V_{emf}	Opposite to emf	$-V_{emf}$	(d)

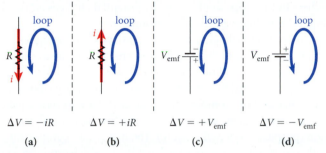

$$\Delta V = -iR \qquad \Delta V = +iR \qquad \Delta V = +V_{emf} \qquad \Delta V = -V_{emf}$$

(a)   (b)   (c)   (d)

FIGURE 26.5 Sign convention for potential changes in analyzing loops.

direction from the current, the potential changes across the resistors will be positive. If we move around a loop so that we pass through an emf source from the negative to the positive terminal, this component contributes a positive potential difference. If we pass through an emf source from the positive to the negative terminal, that component contributes a negative potential difference.

With the above conventions, Kirchhoff's Loop Rule is written in mathematical form as

$$\text{Closed loop:} \quad \sum_{j=1}^{m} V_{emf,j} - \sum_{k=1}^{n} i_k R_k = 0. \tag{26.2}$$

To clarify Kirchhoff's Loop Rule, Figure 26.6 shows a loop with two sources of emf and three resistors, using the three-dimensional display employed in Chapters 24 and 25, where the value of the electric potential, V, is represented in the vertical dimension. The most important point to be visualized from Figure 26.6 is that one complete turn around the loop always ends up at the same value of the potential as at the starting point. This is exactly what Kirchhoff's Loop Rule (equation 26.2) claims. An analogy with downhill skiing may help. When you ski, you are moving around in the gravitational potential, up and down the mountain. A ski lift corresponds to a source of emf, lifting you to a higher value of the gravitational potential. A downhill ski run corresponds to a resistor. (The annoying horizontal traverses between runs correspond to the wires in a circuit—both the wires and the horizontal traverses are at constant potential.) Thus, starting at $V_{emf,1}$ and proceeding clockwise around the loop in Figure 26.6 is analogous to a ski outing in which you take two different lifts and ski down three different runs. And the important point, which is obvious in skiing, is that you return to the same altitude (the same value of the gravitational potential) from which you started, once you have completed the round trip.

A final point about loops is illustrated by Figure 26.7. Figure 26.7a reproduces the same circuit shown in Figure 26.6 as a single isolated loop. Since this loop does not have junctions and thus has only one branch (which is the entire loop), the same current i flows everywhere in the loop. In Figure 26.7b, this loop is connected at four junctions (labeled a, b, c, and d) to other parts of a more extended circuit. Now this loop has four branches, each of which can have a different current flowing through it, as illustrated by the different-colored arrows in the figure. The point is this: In both parts of the figure, Kirchhoff's Loop Rule holds for the loop shown. The relative values of the electric potential between any two circuit elements in Figure 26.7b are the same as those shown in Figure 26.6, independent of the currents that are forced through the different branches of the loop by the rest of the

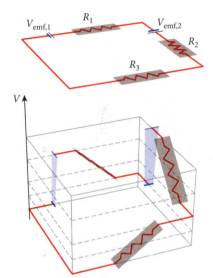

FIGURE 26.6 Loop with multiple sources of emf and multiple resistors.

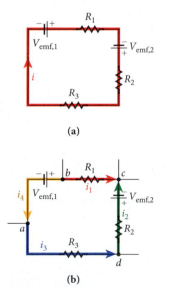

(a)

(b)

FIGURE 26.7 The same circuit loop as in Figure 26.6: (a) as an isolated single loop; (b) as a loop connected to other circuit branches.

circuit. (In our downhill skiing analogy, the currents correspond to different numbers of skiers on the lifts and the downhill runs. Obviously the number of skiers on the hill does not have any influence on the steepness of the hill.)

26.2   Single-Loop Circuits

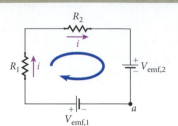

FIGURE 26.8 A single-loop circuit containing two resistors and two sources of emf in series.

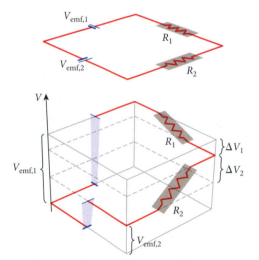

FIGURE 26.9 Three-dimensional representation of the single-loop circuit in Figure 26.8, containing two resistors and two sources of emf in series.

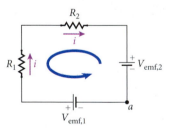

FIGURE 26.10 The same loop as in Figure 26.8, but analyzed in the counterclockwise direction.

Let's begin analyzing general circuits by considering a circuit containing two sources of emf, $V_{emf,1}$ and $V_{emf,2}$, and two resistors, R_1 and R_2, connected in series in a single loop, as shown in Figure 26.8. Note that $V_{emf,1}$ and $V_{emf,2}$ have opposite polarity. In this single-loop circuit, there are no junctions, and so the entire circuit consists of a single branch. The current is the same everywhere in the loop. To illustrate the potential changes across the components of this circuit, Figure 26.9 shows a three-dimensional view.

Although we could arbitrarily pick any point in the circuit of Figure 26.8 and assign it the value 0 V (or any other value of the potential, because we can always add a global additive constant to all potential values without changing the physical outcome), we start at point a with $V = 0$ V and proceed around the circuit in a clockwise direction (indicated by blue elliptic arrow in the figure). Because the components of the circuit are in series, the current, i, is the same in each component, and we assume that the current is flowing in the clockwise direction (purple arrows in the figure). The first circuit component along the clockwise path from point a is the source of emf, $V_{emf,1}$, which produces a positive potential gain of $V_{emf,1}$. Next is resistor R_1, which produces a potential drop given by $\Delta V_1 = iR_1$. Continuing around the loop, the next component is resistor R_2, which produces a potential drop given by $\Delta V_2 = iR_2$. Next, we encounter a second source of emf, $V_{emf,2}$. This source of emf is wired into the circuit with its polarity opposite that of $V_{emf,1}$. Thus, this component produces a potential *drop* with magnitude $V_{emf,2}$, rather than a potential *gain*. We have now completed the loop and are back at $V = 0$ V. Using equation 26.2, we sum the potential changes of this loop as follows:

$$V_{emf,1} - \Delta V_1 - \Delta V_2 - V_{emf,2} = V_{emf,1} - iR_1 - iR_2 - V_{emf,2} = 0.$$

To show that the direction in which we move through a loop, clockwise or counterclockwise, is arbitrary, let's analyze the same circuit in the counterclockwise direction, starting at point a (see Figure 26.10). The first circuit element is $V_{emf,2}$, which produces a positive potential gain. The next element is R_2. Because we have assumed that the current is in the clockwise direction and we are analyzing the loop in the counterclockwise direction, the potential change for R_2 is $+iR_2$, according to the conventions listed in Table 26.1. Proceeding to the next element in the loop, R_1, we use a similar argument to designate the potential change for this resistor as $+iR_1$. The final element in the circuit is $V_{emf,1}$, which is aligned in a direction opposite to that of our analysis, so the potential change across this element is $-V_{emf,1}$. Kirchhoff's Loop Rule then gives us

$$+V_{emf,2} + iR_2 + iR_1 - V_{emf,1} = 0.$$

You can see that the clockwise and counterclockwise loop directions give the same information, which means that the direction in which we choose to analyze the circuit does not matter.

SOLVED PROBLEM 26.1   Charging a Battery

A 12.0-V battery with internal resistance $R_i = 0.200\ \Omega$ is being charged by a battery charger that is capable of delivering a current of magnitude $i = 6.00$ A.

PROBLEM
What is the minimum emf the battery must have to be able to charge the battery?

SOLUTION

THINK
The battery charger, which is an external source of emf, must have enough potential difference to overcome the potential difference of the battery and the potential drop across

the battery's internal resistance. The battery charger must be hooked up so that its positive terminal is connected to the positive terminal of the battery to be charged. We can think of the battery's internal resistance as a resistor in a single-loop circuit that also contains two sources of emf with opposite polarities.

SKETCH

Figure 26.11 shows a diagram of the circuit, consisting of a battery with potential difference V_t and internal resistance R_i connected to an external source of emf, V_e. The yellow shaded area represents the battery's physical dimensions. Note that the positive terminal of the battery charger is connected to the positive terminal of the battery.

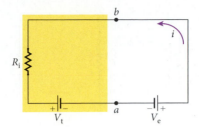

FIGURE 26.11 Circuit consisting of a battery with internal resistance connected to an external source of emf.

RESEARCH

We can apply Kirchhoff's Loop Rule to this circuit. We assume a current flowing counterclockwise around the circuit, as shown in Figure 26.11. The potential changes around the circuit must sum to zero. We sum the potential changes starting at point b and moving in a counterclockwise direction:

$$-iR_i - V_t + V_e = 0.$$

SIMPLIFY

We can solve this equation for the required potential difference of the charger:

$$V_e = iR_i + V_t,$$

where i is the current that the charger supplies.

CALCULATE

Putting in the numerical values gives us

$$V_e = iR_i + V_t = (6.00 \text{ A})(0.200 \text{ }\Omega) + 12.0 \text{ V} = 13.20 \text{ V}.$$

ROUND

We report our result to three significant figures:

$$V_e = 13.2 \text{ V}.$$

DOUBLE-CHECK

Our result indicates that the battery charger has to have a higher potential difference than the specified potential difference of the battery, which is reasonable. A typical charger for a 12-V battery has a potential difference of around 14 V.

26.3 Multiloop Circuits

Analyzing multiloop circuits requires both Kirchhoff's Loop Rule and Kirchhoff's Junction Rule. The procedure for analyzing a multiloop circuit consists of identifying complete loops and junction points in the circuit and applying Kirchhoff's rules to these parts of the circuit separately. Analyzing the single loops in a multiloop circuit with Kirchhoff's Loop Rule and the junctions with Kirchhoff's Junction Rule results in a system of coupled equations in several unknown variables. These equations can be solved for the quantities of interest using various techniques, including direct substitution. Example 26.1 illustrates the analysis of a multiloop circuit.

EXAMPLE 26.1 Multiloop Circuit

Consider the circuit shown in Figure 26.12. This circuit has three resistors, R_1, R_2, and R_3, and two sources of emf, $V_{emf,1}$ and $V_{emf,2}$. The red arrows show the direction of potential drop across the emf sources. This circuit cannot be resolved into simple series or parallel connections. To analyze this circuit, we need to assign directions to the currents flowing through the resistors. We can choose these directions arbitrarily (knowing that if we choose the wrong direction, the resulting current value will be negative). Figure 26.13 shows the circuit with assigned currents in the directions shown by the purple arrows.

Continued—

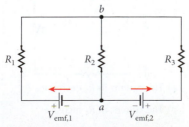

FIGURE 26.12 Multiloop circuit with three resistors and two sources of emf.

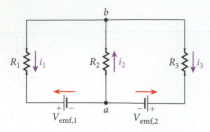

FIGURE 26.13 Multiloop circuit with the assumed direction of the current through the resistors indicated.

Let's consider junction b first. The current entering the junction must equal the current leaving it, so we can write

$$i_2 = i_1 + i_3. \tag{i}$$

Looking at junction a, we again equate the incoming current and the outgoing current to get

$$i_1 + i_3 = i_2,$$

which provides the same information obtained for junction b. Note that this is a typical result: If a circuit has n junctions, it is possible to obtain at most $n - 1$ independent equations from application of Kirchhoff's Junction Rule. (In this case, $n = 2$, so we can get only one independent equation.)

At this point we cannot determine the currents in the circuit because we have three unknown values and only one equation. Therefore, we need two more independent equations. To get these equations, we apply Kirchhoff's Loop Rule. We can identify three loops in the circuit shown in Figure 26.13:

1. the left half of the circuit, including the elements R_1, R_2, and $V_{emf,1}$;
2. the right half of the circuit, including the elements R_2, R_3, and $V_{emf,2}$; and
3. the outer loop, including the elements R_1, R_3, $V_{emf,1}$, and $V_{emf,2}$.

Applying Kirchhoff's Loop Rule to the left half of the circuit, using the assumed directions for the currents and analyzing the loop in a counterclockwise direction starting at junction b, we obtain

$$-i_1 R_1 - V_{emf,1} - i_2 R_2 = 0,$$

or

$$i_1 R_1 + V_{emf,1} + i_2 R_2 = 0. \tag{ii}$$

Applying the Loop Rule to the right half of the circuit, again starting at junction b and analyzing the loop in a clockwise direction, we get

$$-i_3 R_3 - V_{emf,2} - i_2 R_2 = 0,$$

or

$$i_3 R_3 + V_{emf,2} + i_2 R_2 = 0. \tag{iii}$$

Applying the Loop Rule to the outer loop, starting a junction b and working in a clockwise direction, gives us

$$-i_3 R_3 - V_{emf,2} + V_{emf,1} + i_1 R_1 = 0.$$

This equation provides no new information because we can also obtain it by subtracting equation (iii) from equation (ii). For all three loops, we obtain equivalent information if we analyze them in either a counterclockwise direction or a clockwise direction or if we start at any other point and move around the loops from there.

With three equations, (i), (ii), and (iii), and three unknowns, i_1, i_2, and i_3, we can solve for the unknown currents in several ways. For example, we could put the three equations in matrix format and then solve them using Kramer's method on a calculator. This is a recommended method for complicated circuits with many equations and many unknowns. However, for this example, we can proceed by substituting from equation (i) into the other two, thus eliminating i_2. We then solve one of the two resulting equations for i_1 and substitute from that into the other to obtain an expression for i_3. Substituting back then gives solutions for i_2 and i_1:

$$i_1 = -\frac{(R_2 + R_3)V_{emf,1} - R_2 V_{emf,2}}{R_1 R_2 + R_1 R_3 + R_2 R_3}$$

$$i_2 = -\frac{R_3 V_{emf,1} + R_1 V_{emf,2}}{R_1 R_2 + R_1 R_3 + R_2 R_3}$$

$$i_3 = -\frac{-R_2 V_{emf,1} + (R_1 + R_2)V_{emf,2}}{R_1 R_2 + R_1 R_3 + R_2 R_3}.$$

Note: You do not need to remember this particular solution or the linear algebra used to get there. However, the general method of applying Kirchhoff's rules for loops and junctions, and assigning currents in arbitrary directions is the central idea of circuit analysis.

26.2 In-Class Exercise

In the circuit in the figure, there are three identical resistors. The switch, S, is initially open. When the switch is closed, what happens to the current flowing in R_1?

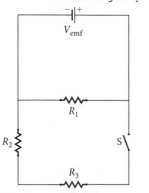

a) The current in R_1 decreases.

b) The current in R_1 increases.

c) The current in R_1 stays the same.

SOLVED PROBLEM 26.2 | The Wheatstone Bridge

The Wheatstone bridge is a particular circuit used to measure unknown resistances. The circuit diagram of a Wheatstone bridge is shown in Figure 26.14. This circuit consists of three known resistances, R_1, R_3, and a variable resistor, R_v, as well as an unknown resistance, R_u. A source of emf, V, is connected across junctions a and c. A sensitive ammeter (a device used to measure current, discussed in Section 26.4) is connected between junctions b and d. The Wheatstone bridge is used to determine R_u by varying R_v until the ammeter between b and d shows no current flowing. When the ammeter reads zero the bridge is said to be balanced.

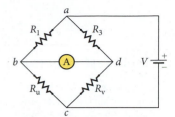

FIGURE 26.14 Circuit diagram of a Wheatstone bridge.

PROBLEM

Determine the unknown resistance, R_u, in the Wheatstone bridge shown in Figure 26.14. The known resistances are $R_1 = 100.0 \ \Omega$ and $R_3 = 110.0 \ \Omega$, and $R_v = 15.63 \ \Omega$ when the current through the ammeter is zero and thus the bridge is balanced.

SOLUTION

THINK

The circuit has four resistors and an ammeter, and each component can have a current flowing through it. However, in this case, with $R_v = 15.63 \ \Omega$, there is no current flowing through the ammeter. Setting this current to zero leaves four unknown currents through the four resistors, and we therefore need four equations. We can use Kirchhoff's rules to analyze two loops, adb and cbd, and two junctions, b and d.

SKETCH

Figure 26.15 shows the Wheatstone bridge with the assumed directions for currents i_1, i_3, i_u, i_v, and i_A.

RESEARCH

We first apply Kirchhoff's Loop Rule to loop adb, starting at a and going clockwise, to obtain

$$-i_3 R_3 + i_A R_A + i_1 R_1 = 0, \tag{i}$$

where R_A is the resistance of the ammeter. We apply Kirchhoff's Loop Rule again to loop cbd, starting at c and going clockwise, to get

$$+i_u R_u - i_A R_A - i_v R_v = 0. \tag{ii}$$

Now we can use Kirchhoff's Junction Rule at junction b to obtain

$$i_1 = i_A + i_u. \tag{iii}$$

Another application of Kirchhoff's Junction Rule, at junction d, gives

$$i_3 + i_A = i_v. \tag{iv}$$

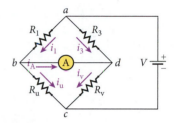

FIGURE 26.15 The Wheatstone bridge with the assumed current directions indicated.

SIMPLIFY

When the current through the ammeter is zero ($i_A = 0$), we can rewrite equations (i) through (iv) as follows:

$$i_1 R_1 = i_3 R_3 \tag{v}$$

$$i_u R_u = i_v R_v \tag{vi}$$

$$i_1 = i_u \tag{vii}$$

and

$$i_3 = i_v. \tag{viii}$$

Dividing equation (vi) by equation (v) gives us

$$\frac{i_u R_u}{i_1 R_1} = \frac{i_v R_v}{i_3 R_3},$$

which we can rewrite using equations (vii) and (viii):

$$R_u = \frac{R_1}{R_3} R_v.$$

Continued—

CALCULATE

Putting in the numerical values, we get

$$R_u = \frac{R_1}{R_3} R_v = \frac{100.0\ \Omega}{110.0\ \Omega} 15.63\ \Omega = 14.20901\ \Omega.$$

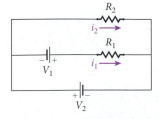
ROUND

We report our result to four significant figures:

$$R_u = 14.21\ \Omega.$$

DOUBLE-CHECK

Our result for the resistance of the unknown resistor is similar to the value for the variable resistor. Thus, our answer seems reasonable, because the other two resistors in the circuit also have resistances that are approximately equal.

General Observations on Circuit Networks

An important observation about solving circuit problems is that, in general, the complete analysis of a circuit requires knowing the current flowing in each branch of the circuit. We use Kirchhoff's Junction Rule and Loop Rule to establish equations relating the currents, and we need as many linearly independent equations as there are branches to guarantee that we can obtain a solution to the system.

Let's consider the abstract example shown in Figure 26.16, where all circuit components except the wires have been omitted. This circuit has four junctions, shown in blue in Figure 26.16a. Six branches connect these junctions, shown in Figure 26.16b. We therefore need six linearly independent equations relating the currents in these branches. It was noted earlier that not all equations obtained by applying Kirchhoff's rules to a circuit are linearly independent. This fact is worth repeating: If a circuit has n junctions, it is possible to obtain at most $n - 1$ independent equations from application of the Junction Rule. (For the circuit in Figure 26.16, $n = 4$, so we can get only three independent equations.)

The Junction Rule alone is not enough for the complete analysis of any circuit. It is generally best to write as many equations as possible for junctions and then augment them with equations obtained from loops. Figure 26.16c shows that there are six possible loops in this network, which are marked in different colors. Clearly, there are more loops than we need to analyze to obtain three equations. This is again a general observation: The system of equations that can be set up by considering all possible loops is overdetermined. Thus, you'll always have the freedom to select particular loops to augment the equations obtained from analyzing the junctions. As a general rule of thumb, it is best to choose loops with fewer circuit elements, which often makes the subsequent linear algebra considerably simpler. In particular, if you are asked to find the current in a particular branch of a network, choosing the appropriate loop may allow you to avoid setting up a lengthy set of equations and let you solve the problem with just one equation. So it pays to devote some attention to the selection of loops! Do not write down more equations than you need to solve for the unknowns in any particular problem. This will only complicate the algebra. However, once you have the solution you can use one or more of the unused loops to check your values.

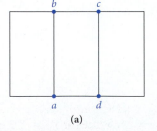
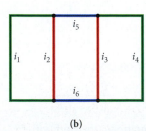
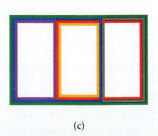

FIGURE 26.16 Circuit network consisting of (a) four junctions, (b) six branches, (c) six possible loops.

26.4 Ammeters and Voltmeters

A device used to measure current is called an **ammeter**. A device used to measure potential difference is called a **voltmeter**. To measure the current, an ammeter must be wired in a circuit in *series*. Figure 26.17 shows an ammeter connected in a circuit in a way that allows it to measure the current i. To measure the potential difference, a voltmeter must be wired in *parallel* with the component across which the potential difference is to be measured. Figure 26.17 shows a voltmeter placed in the circuit to measure the potential drop across resistor R_1.

It is important to realize that these instruments must be able to make measurements while disturbing the circuit as little as possible. Thus, ammeters are designed to have as low a resistance as possible, usually on the order of 1 Ω, so they do not have an appreciable effect on the currents they measure. Voltmeters are designed to have as high a resistance as possible, usually on the order of 10 MΩ (10^7 Ω), so they have a negligible effect on the potential differences they are measuring.

In practice, measurements of current and potential difference are made with a digital multimeter that can switch between functioning as an ammeter and functioning as a voltmeter. It displays the results with an autoranging numerical digital display, which includes the sign of the potential difference or current. Most digital multimeters can also measure the resistance of a circuit component; that is, they can function as an **ohmmeter**. The digital multimeter performs this task by applying a known potential difference and measuring the resulting current. This test is useful for determining circuit continuity and the status of fuses, as well as measuring the resistance of resistors.

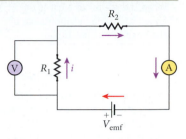

FIGURE 26.17 Placement of an ammeter and a voltmeter in a simple circuit.

EXAMPLE 26.2 Voltmeter in a Simple Circuit

Consider a simple circuit consisting of a source of emf with voltage $V_{emf} = 150.$ V and a resistor with resistance $R = 100.$ kΩ (Figure 26.18). A voltmeter with resistance $R_V = 10.0$ MΩ is connected across the resistor.

PROBLEM
What is the current in the circuit before the voltmeter is connected?

SOLUTION
Ohm's Law says that $V = iR$, so we can find the current in the circuit:

$$i = \frac{V_{emf}}{R} = \frac{150.\ \text{V}}{100.\cdot 10^3\ \Omega} = 1.50\cdot 10^{-3}\ \text{A} = 1.50\ \text{mA}.$$

PROBLEM
What is the current in the circuit when the voltmeter is connected across the resistor?

SOLUTION
The equivalent resistance of the resistor and the voltmeter connected in parallel is given by

$$\frac{1}{R_{eq}} = \frac{1}{R} + \frac{1}{R_V}.$$

Solving for the equivalent resistance and putting in the numerical values, we get

$$R_{eq} = \frac{RR_V}{R+R_V} = \frac{\left(100.\cdot 10^3\ \Omega\right)\left(10.0\cdot 10^6\ \Omega\right)}{100.\cdot 10^3\ \Omega + 10.0\cdot 10^6\ \Omega} = 9.90\cdot 10^4\ \Omega = 99.0\ \text{k}\Omega.$$

The current is then

$$i = \frac{V_{emf}}{R_{eq}} = \frac{150.\ \text{V}}{9.90\cdot 10^4\ \Omega} = 1.52\cdot 10^{-3}\ \text{A} = 1.52\ \text{mA}.$$

The current in the circuit increases by 0.02 mA when the voltmeter is connected because the parallel combination of the resistor and the voltmeter has a lower resistance than that of the resistor alone. However, the effect is small, even with this relatively large resistance ($R = 100.$ kΩ).

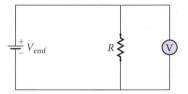

FIGURE 26.18 A simple circuit with a voltmeter connected in parallel across a resistor.

26.2 Self-Test Opportunity

When the starter of a car is engaged while the headlights are on, the headlights dim. Explain.

26.4 In-Class Exercise

Two resistors, $R_1 = 3.00\ \Omega$ and $R_2 = 5.00\ \Omega$, are connected in series with a battery with $V_{emf} = 8.00$ V and an ammeter with $R_A = 1.00\ \Omega$, as shown in the figure. What is the current measured by the ammeter?

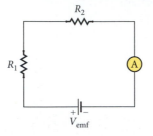

a) 0.500 A   d) 1.00 A

b) 0.750 A   e) 1.50 A

c) 0.889 A

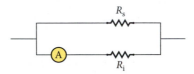

FIGURE 26.19 An ammeter with a shunt resistor connected across it in parallel.

SOLVED PROBLEM 26.3 | Increasing the Range of an Ammeter

PROBLEM

An ammeter can be used to measure different ranges of current by adding a current divider in the form of a shunt resistor connected in parallel with the ammeter. A shunt resistor is simply a resistor with a very small resistance. Its name arises from the fact that when connected in parallel with the ammeter, whose resistance is larger, most of the current is shunted through it, bypassing the meter. The sensitivity of the ammeter is therefore decreased allowing it to measure larger currents. Suppose an ammeter produces a full-scale reading when a current of $i_{int} = 5.10$ mA passes through it. The ammeter has an internal resistance of $R_i = 16.8\ \Omega$. To use this ammeter to measure a maximum current of $i_{max} = 20.2$ A, what should be the resistance of the shunt resistor, R_s, connected in parallel with the ammeter?

SOLUTION

THINK

The shunt resistor connected in parallel with the ammeter needs to have a substantially lower resistance than the internal resistance of the ammeter. Most of the current will then flow through the shunt resistor rather than the ammeter.

SKETCH

Figure 26.19 shows a shunt resistor, R_s, connected in parallel with an ammeter.

RESEARCH

The two resistors are connected in parallel, so the potential difference across each resistor is the same. The potential difference that gives a full-scale reading on the ammeter is

$$\Delta V_{fs} = i_{int} R_i. \tag{i}$$

From Chapter 25, we know that the equivalent resistance of the two resistors in parallel is given by

$$\frac{1}{R_{eq}} = \frac{1}{R_i} + \frac{1}{R_s}. \tag{ii}$$

The voltage drop across the equivalent resistance must equal the voltage drop across the ammeter that gives a full-scale reading when current i_{max} is flowing through the circuit. Therefore, we can write

$$\Delta V_{fs} = i_{max} R_{eq}. \tag{iii}$$

SIMPLIFY

Combining equations (i) and (iii) for the potential difference gives

$$\Delta V_{fs} = i_{int} R_i = i_{max} R_{eq}. \tag{iv}$$

We can rearrange equation (iv) and substitute for R_{eq} in equation (ii):

$$\frac{i_{max}}{i_{int} R_i} = \frac{1}{R_{eq}} = \frac{1}{R_i} + \frac{1}{R_s}. \tag{v}$$

Solving equation (v) for the shunt resistance gives us

$$\frac{1}{R_s} = \frac{i_{max}}{i_{int} R_i} - \frac{1}{R_i} = \frac{1}{R_i}\left(\frac{i_{max}}{i_{int}} - 1\right) = \frac{1}{R_i}\left(\frac{i_{max} - i_{int}}{i_{int}}\right),$$

or

$$R_s = R_i \frac{i_{int}}{i_{max} - i_{int}}.$$

CALCULATE

Putting in the numerical values, we get

$$R_s = R_i \frac{i_{int}}{i_{max} - i_{int}} = (16.8\ \Omega)\frac{5.10 \cdot 10^{-3}\ \text{A}}{20.2\ \text{A} - 5.10 \cdot 10^{-3}\ \text{A}}$$

$$= 0.00424266\ \Omega.$$

ROUND

We report our result to three significant figures:

$$R_s = 0.00424 \ \Omega.$$

DOUBLE-CHECK

The equivalent resistance of the ammeter and the shunt resistor connected in parallel is given by equation (ii). Solving that equation for the equivalent resistance and putting in the numbers gives

$$R_{eq} = \frac{R_i R_s}{R_i + R_s} = \frac{\left(16.8 \ \Omega\right)\left(0.00424 \ \Omega\right)}{16.8 \ \Omega + 0.00424 \ \Omega} = 0.00424 \ \Omega.$$

Thus, the equivalent resistance of the ammeter and the shunt resistor connected in parallel is approximately equal to the resistance of the shunt resistor. This low equivalent resistance is necessary for a current-measuring instrument, which must be placed in series in a circuit. If the current-measuring device has a high resistance, its presence will disturb the measurement of the current.

26.5 RC Circuits

So far in this chapter, we have dealt with circuits containing sources of emf and resistors. The currents in these circuits do not vary in time. Now we consider circuits that contain capacitors (see Chapter 24), as well as sources of emf and resistors. Called **RC circuits,** these circuits have currents that *do* vary with time. The simplest circuit operations that involve time-dependent currents are the charging and discharging of a capacitor. Understanding these time-dependent processes involves the solution of some simple differential equations. After magnetism and magnetic phenomena are introduced in Chapters 27 through 29, time-dependent currents will be discussed again in Chapter 30, which will build on the techniques introduced here.

Charging a Capacitor

Consider a circuit with a source of emf, V_{emf}, a resistor, R, and a capacitor, C (Figure 26.20). Initially, the switch is open and the capacitor is uncharged, as shown in Figure 26.20a. When the switch is closed (Figure 26.20b), current begins to flow in the circuit, building up opposite charges on the plates of the capacitor and thus creating a potential difference, ΔV, across the capacitor. Current flows because of the source of emf, which maintains a constant voltage. When the capacitor is fully charged (Figure 26.20c), no more current flows in the circuit. The potential difference across the plates is then equal to the voltage provided by the source of emf, and the size of the total charge, q_{tot}, on each plate of the capacitor is $q_{tot} = CV_{emf}$.

While the capacitor is charging, we can analyze the current, i, flowing in the circuit (assumed to flow from the negative to the positive terminal inside the voltage source) by applying Kirchhoff's Loop Rule to the loop in Figure 26.20b in the counterclockwise direction:

$$V_{emf} - V_R - V_C = V_{emf} - i(t)R - q(t)/C = 0,$$

where V_C is the potential drop across the capacitor and $q(t)$ is the charge on the capacitor at a given time t. The change of the charge on the capacitor plates due to the current is $i(t) = dq(t)/dt$, and we can rewrite the preceding equation as

$$R\frac{dq(t)}{dt} + \frac{q(t)}{C} = V_{emf},$$

or

$$\frac{dq(t)}{dt} + \frac{q(t)}{RC} = \frac{V_{emf}}{R}. \tag{26.3}$$

This differential equation relates the charge to its time derivative. The discussion of damped oscillations in Chapter 14 involved similar differential equations. It seems appropriate to

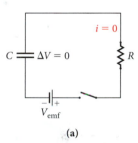

(a)

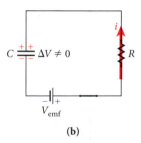

(b)

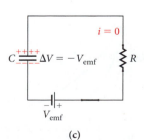

(c)

FIGURE 26.20 A basic RC circuit, containing a source of emf, a resistor, and a capacitor: (a) with the switch open; (b) a short time after the switch is closed; (c) a long time after the switch is closed.

try an exponential form for the solution of equation 26.3 because an exponential is the only function that has the property of having a derivative that is identical to itself. Because equation 26.3 also has a constant term, the trial solution needs to have a constant term. We therefore try a solution with a constant and an exponential and for which $q(0) = 0$:

$$q(t) = q_{max}\left(1 - e^{-t/\tau}\right),$$

where the constants q_{max} and τ are to be determined. Substituting this trial solution back into equation 26.3, we obtain

$$q_{max}\frac{1}{\tau}e^{-t/\tau} + \frac{1}{RC}q_{max}\left(1 - e^{-t/\tau}\right) = \frac{V_{emf}}{R}.$$

Now we collect the time-dependent terms on the left-hand side and the time-independent terms on the right-hand side.

$$q_{max}e^{-t/\tau}\left(\frac{1}{\tau} - \frac{1}{RC}\right) = \frac{V_{emf}}{R} - \frac{1}{RC}q_{max}.$$

This equation can only be true for all times if both sides are equal to zero. From the left-hand side, we then find

$$\tau = RC. \tag{26.4}$$

Thus, the constant τ (called the **time constant**) is simply the product of the capacitance and the resistance. From the right-hand side, we find an expression for the constant q_{max}:

$$q_{max} = CV_{emf}.$$

Thus, the differential equation for charging the capacitor (equation 26.3) has the solution

$$q(t) = CV_{emf}\left(1 - e^{-t/RC}\right). \tag{26.5}$$

Note that at $t = 0$, $q = 0$, which is the initial condition before the circuit components were connected. At $t = \infty$, $q = q_{max} = CV_{emf}$, which is the steady-state condition in which the capacitor is fully charged. The time dependence of the charge on the capacitor is shown in Figure 26.21a for three different values of the time constant τ.

The current flowing in the circuit is obtained by differentiating equation 26.5 with respect to time:

$$i = \frac{dq}{dt} = \left(\frac{V_{emf}}{R}\right)e^{-t/RC}. \tag{26.6}$$

From equation 26.6, at $t = 0$, the current in the circuit is V_{emf}/R, and at $t = \infty$, the current is zero, as shown in Figure 26.21b.

How do we know that the solution we have found for equation 26.3 is the only solution? This is not obvious from the preceding discussion, but the solution is unique. (A proof is generally provided in a course on differential equations.)

Discharging a Capacitor

Now let's consider a circuit containing only a resistor, R, and a fully charged capacitor, C, obtained by moving the switch in Figure 26.22 from position 1 to position 2. The charge on the capacitor before the switch is moved is q_{max}. In this case, current will flow in the circuit until the capacitor is completely discharged. While the capacitor is discharging, we can apply Kirchhoff's Loop Rule around the loop in the clockwise direction and obtain

$$-i(t)R - V_C = -i(t)R - \frac{q(t)}{C} = 0.$$

We can rewrite this equation using the definition of current:

$$\frac{Rdq(t)}{dt} + \frac{q(t)}{C} = 0. \tag{26.7}$$

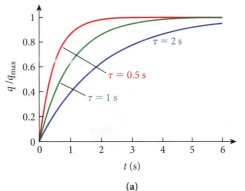

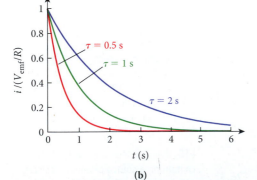

FIGURE 26.21 Charging a capacitor: (a) charge on the capacitor as a function of time; (b) current flowing through the resistor as a function of time.

The solution to equation 26.7 is obtained using the same method as for equation 26.3, except that equation 26.7 has no constant term and $q(0) > 0$. Thus, we try a solution of the form $q(t) = q_{max}e^{-t/\tau}$, which leads to

$$q(t) = q_{max}e^{-t/RC}. \tag{26.8}$$

At $t = 0$, the charge on the capacitor is q_{max}. At $t = \infty$, the charge on the capacitor is zero.

We can obtain the current by differentiating equation 26.8 as a function of time:

$$i(t) = \frac{dq}{dt} = -\left(\frac{q_{max}}{RC}\right)e^{-t/RC}. \tag{26.9}$$

At $t = 0$, the current in the circuit is $-q_{max}/RC$. At $t = \infty$, the current in the circuit is zero. Plotting the time dependence of the charge on the capacitor and the current flowing through the resistor for the discharging process would result in exponentially decreasing curves like those in Figure 26.21b.

The equations describing the time dependence of the charging and the discharging of a capacitor all involve the exponential factor $e^{-t/RC}$. Again, the product of the resistance and the capacitance is defined as the time constant of an RC circuit: $\tau = RC$. According to equation 26.5, after an amount of time equal to the time constant, the capacitor will have been charged to 63% of its maximum value. Thus, an RC circuit can be characterized by specifying the time constant. A large time constant means that it takes a long time to charge the capacitor; a small time constant means that it takes a short time to charge the capacitor.

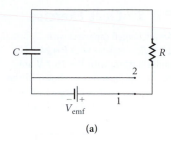

(a)

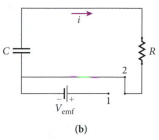

(b)

FIGURE 26.22 *RC* circuit containing an emf source, a resistor, a capacitor, and a switch. The capacitor is (a) charged with the switch in position 1 and (b) discharged with the switch in position 2.

EXAMPLE 26.3 | Time Required to Charge a Capacitor

Consider a circuit consisting of a 12.0-V battery, a 50.0-Ω resistor, and a 100.0-μF capacitor wired in series. The capacitor is initially completely discharged.

PROBLEM
How long after the circuit is closed will it take to charge the capacitor to 90% of its maximum charge?

SOLUTION
The charge on the capacitor as a function of time is given by

$$q(t) = q_{max}\left(1 - e^{-t/RC}\right),$$

where q_{max} is the maximum charge on the capacitor. We want to know the time until $q(t)/q_{max} = 0.90$, which can be obtained from

$$\left(1 - e^{-t/RC}\right) = \frac{q(t)}{q_{max}} = 0.90,$$

or

$$0.10 = e^{-t/RC}. \tag{i}$$

Taking the natural log of both sides of equation (i), we get

$$\ln 0.10 = -\frac{t}{RC},$$

or

$$t = -RC\ln 0.10 = -\left(50.0\ \Omega\right)\left(100 \cdot 10^{-6}\ \text{F}\right)\left(-2.30\right) = 0.0115\ \text{s} = 11.5\ \text{ms}.$$

26.5 In-Class Exercise

To discharge a capacitor in an RC circuit very quickly, what should the values of the resistance and the capacitance be?

a) Both should be as large as possible.

b) Resistance should be as large as possible, and capacitance as small as possible.

c) Resistance should be as small as possible, and capacitance as large as possible.

d) Both should be as small as possible.

26.3 Self-Test Opportunity

A 1.00-mF capacitor is fully charged, and a 100.0-Ω resistor is connected across the capacitor. How long will it take to remove 99.0% of the charge stored in the capacitor?

Pacemaker

A normal human heart beats at regular intervals, sending blood through the body. The heart's own electrical signals regulate its beating. These electrical signals can be measured through the skin using an electrocardiograph. This device produces a graph of potential difference

26.6 In-Class Exercise

An uncharged capacitor with $C = 14.9\ \mu\text{F}$, a resistor with $R = 24.3\ \text{k}\Omega$, and a battery with $V = 25.7\ \text{V}$ are connected in series as shown in the figure. What is the charge on the capacitor at $t = 0.3621\ \text{s}$ after the switch is closed?

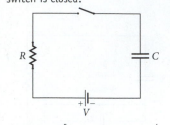

a) $5.48 \cdot 10^{-5}\ \text{C}$ d) $1.66 \cdot 10^{-4}\ \text{C}$

b) $7.94 \cdot 10^{-5}\ \text{C}$ e) $2.42 \cdot 10^{-4}\ \text{C}$

c) $1.15 \cdot 10^{-5}\ \text{C}$

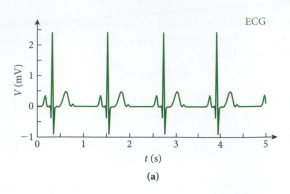

(a)

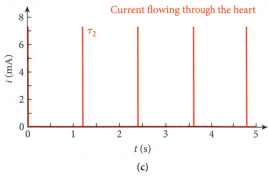

(b)

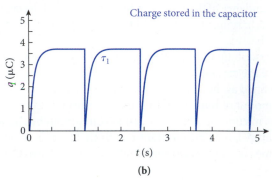

(c)

FIGURE 26.23 (a) An electrocardiogram (ECG) showing four regular heartbeats. (b) The charge stored in the pacemaker's capacitor as a function of time. (c) The current flowing through the heart due to the discharging of the pacemaker's capacitor.

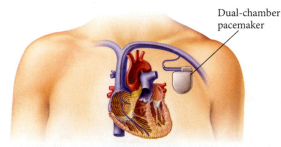

Dual-chamber pacemaker

FIGURE 26.24 A modern pacemaker implanted in a patient. The pacemaker sends electrical pulses to two chambers of the heart to help keep it beating regularly.

versus time, which is called an electrocardiogram, or ECG (sometimes EKG, from the German word *Electrokardiograph*). Figure 26.23a shows how an ECG would represent four regular heartbeats occurring at a rate of 72 beats per minute. Doctors and medical personnel can use an ECG to diagnose the health of the heart.

Sometimes the heart does not beat regularly and needs help to maintain its proper rhythm. This help can be provided by a pacemaker, an electrical circuit that sends electrical pulses to the heart at regular intervals, replacing the heart's usual electrical signals and stimulating the heart to beat at prescribed intervals. A pacemaker is implanted in the patient and connected directly to the heart, as illustrated in Figure 26.24.

EXAMPLE 26.4   Circuit Elements of a Pacemaker

Let's analyze the circuit shown in Figure 26.25, which simulates the function of a pacemaker. This pacemaker circuit operates by charging a capacitor, C, for some time using a battery with voltage V_{emf} and a resistor R_1, as illustrated in Figure 26.25a, where the switch is open. Closing the switch, as in Figure 26.25b, shorts the capacitor across the heart, and the capacitor discharges through the heart in a short time to stimulate the heart to beat. Thus, this circuit operates as a pacemaker by keeping the switch open for the time between heartbeats, closing the switch for a short time to stimulate a heartbeat, and then opening the switch again.

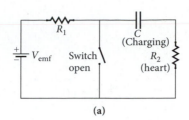

(a)

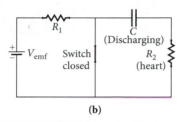

(b)

FIGURE 26.25 (a) A simplified pacemaker circuit in charging mode. (b) The pacemaker circuit in discharging mode.

PROBLEM

What values of the capacitance, C, and the resistance, R_1, should be used in a pacemaker?

SOLUTION

We assume that the heart acts as a resistor with a value $R_2 = 500 \ \Omega$ and that the source of emf is a lithium ion battery (discussed in Chapter 23). A lithium ion cell has a very high energy density and a voltage of 3.7 V. A normal heart rate is in the range from 60 to 100 beats per minute. However, the pacemaker might need to stimulate the heart to beat faster, so it should be capable of running at 180 beats per minute, which means that the capacitor may be charged up to 180 times a minute. Thus, the minimum time between discharges, t_{min}, is

$$t_{min} = \frac{1}{180 \text{ beats/min}} = \left(\frac{1 \text{ min}}{180}\right)\left(\frac{60 \text{ s}}{1 \text{ min}}\right) = 0.333 \text{ s}.$$

Equation 26.5 gives the charge, q, as a function of time, t, for the maximum charge, q_{max}, for a given time constant, $\tau_1 = R_1 C$:

$$q = q_{max}\left(1 - e^{-t/\tau_1}\right).$$

Rearranging this equation gives us

$$\frac{q}{q_{max}} = f = 1 - e^{-t/\tau_1},$$

where f is the fraction of the charge capacity of the capacitor. Let's assume the capacitor must be charged to 95% of its maximum charge in time t_{min}. Solving for the time constant gives us

$$\tau_1 = R_1 C = -\frac{t_{min}}{\ln\left(1 - f\right)} = -\frac{0.333 \text{ s}}{\ln\left(1 - 0.95\right)} = 0.111 \text{ s}.$$

Thus, the time constant for the charging should be $\tau_1 = R_1 C = 100$ ms. The time constant for discharging, τ_2, needs to be small to produce short pulses of a high current to stimulate the heart. Letting $\tau_2 = 0.500$ ms, which is on the order of the narrow electrical pulse in the ECG, we have

$$\tau_2 = R_2 C = 0.500 \text{ ms}.$$

We can solve for the required capacitance and substitute the value 500 Ω for R_2:

$$C = \frac{0.000500 \text{ s}}{500 \ \Omega} = 1.00 \ \mu\text{F}.$$

The resistance required for the charging circuit can now be related to the time constant for the charging and the value of the capacitance we just calculated:

$$\tau_1 = R_1 C = 0.100 \text{ s} = R_1\left(1.00 \ \mu\text{F}\right),$$

which gives us

$$R_1 = \frac{0.100 \text{ s}}{1.00 \cdot 10^{-6} \text{ F}} = 100 \text{ k}\Omega.$$

Figure 26.23b shows the charge in the capacitor of the simulated pacemaker as a function of time for a heart rate of 72 heartbeats per minute. You can see that the capacitor charges to nearly full capacity before it is discharged. Figure 26.23c shows the current that flows through the heart when the capacitor discharges. The current pulse is narrow, lasting less than a millisecond. This pulse stimulates the heart to beat, as illustrated by the ECG in Figure 26.23a. The rate at which the heart beats is controlled by the rate at which the pacemaker's switch is closed and opened, which is controlled by a microprocessor.

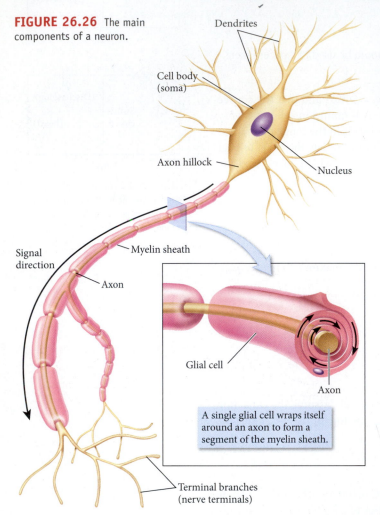

FIGURE 26.26 The main components of a neuron.

Dendrites

Cell body (soma)

Axon hillock

Nucleus

Signal direction

Myelin sheath

Axon

Glial cell

Axon

A single glial cell wraps itself around an axon to form a segment of the myelin sheath.

Terminal branches (nerve terminals)

Neuron

The type of cell responsible for transmitting and processing signals in the nervous systems and brains of humans and other animals is a neuron (Figure 26.26). The neuron conducts the necessary currents by electrochemical means, via the movement of ions (mainly Na^+, K^+, and Cl^-). Neurons receive signals from other neurons through dendrites and send signals to other neurons through an axon. The axon can be quite long (for example, in the spinal cord) and is covered with an insulating myelin sheath. The signals are received from and sent to other neurons or cells in the sense organs and other tissues. All of this, and much more, can be learned in an introductory biology or physiology course. Here we'll take a look at a neuron as a basic circuit that processes signals.

An input signal has to be strong enough to get a neuron to fire, that is, to send an output signal down the axon. It is a crude but reasonable approximation to represent the main cell body of a neuron, the soma, as a basic RC circuit that processes these signals. A diagram of this RC circuit is shown in Figure 26.27. A capacitor and a resistor are connected in parallel to an input and an output potential. The typical potential values for neurons are on the order of ±50 mV relative to the background of the surrounding tissue. If $\Delta V = V_{in} - V_{out} \neq 0$, a current flows through the circuit. Part of this current flows through the resistor, but part of it simultaneously charges the capacitor until it reaches the potential difference between input and output potentials. The potential difference between the capacitor plates rises exponentially, according to $V_C(t) = (V_{in} - V_{out})(1 - e^{-t/RC})$, just as for the process of charging a capacitor in an RC circuit. (The time constant, $\tau = RC$, is on the order of 10 ms, assuming a capacitance of 1 nF and a resistance of 10 MΩ.) If the external potential difference is then removed from the circuit, the capacitor discharges with the same time constant, and the potential across the capacitor decays exponentially, according to $V_C(t) = V_0 e^{-t/RC}$. This simple time dependence captures the basic response of a neuron. Figure 26.28 shows the potential difference across the capacitor of this model neuron, while it is charged for 30 ms and then discharged.

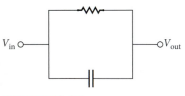

FIGURE 26.27 Simplified model of a neuron as an RC circuit.

V_{in} ○   ○ V_{out}

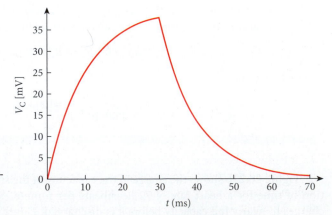

FIGURE 26.28 Potential difference on the capacitor in a model neuron.

V_C [mV]

t (ms)

WHAT WE HAVE LEARNED | EXAM STUDY GUIDE

- Kirchhoff's rules for analyzing circuits are as follows:

  - Kirchhoff's Junction Rule: The sum of the currents entering a junction must equal the sum of the currents leaving a junction.

  - Kirchhoff's Loop Rule: The potential difference around a complete loop must sum to zero.

- For applying Kirchhoff's Loop Rule, the sign of the potential change for each circuit element is determined by the direction of the current and the direction of analysis. The conventions are

  - Sources of emf in the same direction as the direction of analysis are potential gains, while sources opposite to the analysis direction are potential drops.

  - For resistors, the magnitude of the potential change is $|iR|$, where i is the assumed current and R is the resistance. The sign of the potential change depends on the (known or assumed) direction of

  the current as well as the direction of analysis. If these directions are the same, the resistor produces a potential drop. If the directions are opposite, the resistor produces a potential gain.

- An RC circuit contains a resistor of resistance R and a capacitor of capacitance C. The time constant, τ, is given by $\tau = RC$.

- In an RC circuit, the charge, q, as a function of time for a charging capacitor with capacitance C is given by $q(t) = CV_{emf}(1 - e^{-t/RC})$, where V_{emf} is the voltage supplied by the source of emf and R is the resistance of the resistor.

- In an RC circuit, the charge, q, as a function of time for a discharging capacitor with capacitance C is given by $q(t) = q_{max}e^{-t/RC}$, where q_{max} is the size of the charge on the capacitor plates at $t = 0$ and R is the resistance of the resistor.

KEY TERMS

Kirchhoff's rules, p. 839
junction, p. 839
branch, p. 839

Kirchhoff's Junction Rule, p. 839
loop, p. 840

Kirchhoff's Loop Rule, p. 840
ammeter, p. 847
voltmeter, p. 847

ohmmeter, p. 847
RC circuits, p. 849
time constant, p. 850

NEW SYMBOLS AND EQUATIONS

$\tau = RC$, time constant of an RC circuit

ANSWERS TO SELF-TEST OPPORTUNITIES

26.1 The resistors R_1 and R_u are in series and have an equivalent resistance of $R_{1u} = R_1 + R_u$. The resistors R_3 and R_v are in series and have an equivalent resistance of $R_{3v} = R_3 + R_v$. The equivalent resistances R_{1u} and R_{3v}, are in parallel. Thus, we can write

$$\frac{1}{R_{eq}} = \frac{1}{R_{1u}} + \frac{1}{R_{3v}},$$

or

$$R_{eq} = \frac{R_{1u}R_{3v}}{R_{1u}+R_{3v}} = \frac{(R_1+R_u)(R_3+R_v)}{R_1+R_u+R_3+R_v}.$$

26.2 When the headlights are on, the battery is supplying a modest amount of current to the lights, and the potential drop across the internal resistance of the battery is small. The starter motor is wired in parallel with the lights. When the starter motor engages, it draws a large current, producing a noticeable drop in potential across the internal resistance of the battery and causing less current to flow to the headlights.

26.3 $q = q_{max}e^{-t/RC}$

$$\frac{q}{q_{max}} = 0.01 = e^{-t/RC} \Rightarrow \ln 0.01 = -\frac{t}{RC}$$

$$t = -RC\ln 0.01 = -(100\ \Omega)(1.00\cdot10^{-3}\ \text{F})(\ln 0.01) = 0.461\ \text{s}.$$

PROBLEM-SOLVING PRACTICE

Problem-Solving Guidelines

1. It is always helpful to label everything in a circuit diagram, including all the given information and all the unknowns, as well as pertinent currents, branches, and junctions. Redraw the diagram at a larger scale if you need more space for clarity.

2. Remember that the directions you choose for currents and for the path around a circuit loop are arbitrary. If your choice turns out to be incorrect, a negative value for the current will result.

3. Review the signs for potential changes given in Table 26.1. Moving through a circuit loop in the same direction as the assumed current means that an emf device produces a positive potential change in the direction from negative to positive within the device and that the potential change across a resistor is negative. Sign errors are common, and it pays to stick to the conventions to avoid such errors.

4. Sources of emf or resistors may be parts of two separate loops. Count each circuit component as a part of each loop it is in, according to the sign conventions you've adopted for

that loop. A resistor may yield a potential drop in one loop and a potential gain in the other loop.

5. It is always possible to use Kirchhoff's rules to write more equations than you need to solve for unknown currents in a circuit's branches. Write as many equations as possible for junctions, and then augment them with equations representing loops. But not all loops are created equal; you need to select them carefully. As a rule of thumb, choose loops with fewer circuit elements.

SOLVED PROBLEM **26.4** | **Rate of Energy Storage in a Capacitor**

A resistor with $R = 2.50$ MΩ and a capacitor with $C = 1.25$ μF are connected in series with a battery for which $V_{emf} = 12.0$ V. At $t = 2.50$ s after the circuit is closed, what is the rate at which energy is being stored in the capacitor?

THINK
When the circuit is closed, the capacitor begins to charge. The rate at which energy is stored in the capacitor is given by the time derivative of the amount of energy stored in the capacitor, which is a function of the charge on the capacitor.

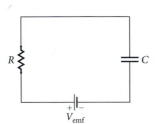

FIGURE 26.29 Series circuit containing a battery, a resistor, and a capacitor.

SKETCH
Figure 26.29 shows a diagram of the series circuit containing a battery, a resistor, and a capacitor.

RESEARCH
The charge on the capacitor as a function of time is given by equation 26.5:

$$q(t) = CV_{emf}\left(1 - e^{-t/RC}\right).$$

The energy stored in a capacitor that has charge q is given by (see Chapter 24)

$$U = \frac{1}{2}\frac{q^2}{C}. \tag{i}$$

The time derivative of the energy stored in the capacitor is then

$$\frac{dU}{dt} = \frac{d}{dt}\left(\frac{1}{2}\frac{q^2(t)}{C}\right) = \frac{q(t)}{C}\frac{dq(t)}{dt}. \tag{ii}$$

The time derivative of the charge is the current, i. Thus, we can replace dq/dt with the expression given by equation 26.6:

$$i(t) = \frac{dq(t)}{dt} = \left(\frac{V_{emf}}{R}\right)e^{-t/RC}. \tag{iii}$$

SIMPLIFY
We can express the rate of change of the energy stored in the capacitor by combining equations (i) through (iii):

$$\frac{dU}{dt} = \frac{q(t)}{C}i(t) = \frac{CV_{emf}\left(1 - e^{-t/RC}\right)}{C}\left(\frac{V_{emf}}{R}\right)e^{-t/RC} = \frac{V_{emf}^2}{R}e^{-t/RC}\left(1 - e^{-t/RC}\right).$$

CALCULATE
We first calculate the value of the time constant, $\tau = RC$:

$$RC = \left(2.50 \cdot 10^6 \ \Omega\right)\left(1.25 \cdot 10^{-6} \ F\right) = 3.125 \ s.$$

We can then calculate the rate of change of the energy stored in the capacitor:

$$\frac{dU}{dt} = \frac{(12.0 \text{ V})^2}{2.50 \cdot 10^6 \ \Omega} e^{-(2.50 \text{ s})/(3.125 \text{ s})} \left(1 - e^{-(2.50 \text{ s})/(3.125 \text{ s})}\right) = 1.42521 \cdot 10^{-5} \text{ W}.$$

ROUND

We report our result to three significant figures:

$$\frac{dU}{dt} = 1.43 \cdot 10^{-5} \text{ W}.$$

DOUBLE-CHECK

The current at $t = 2.50$ s is

$$i(2.50 \text{ s}) = \left(\frac{12.0 \text{ V}}{2.50 \text{ M}\Omega}\right) e^{-(2.50 \text{ s})/(3.125 \text{ s})} = 2.16 \cdot 10^{-6} \text{ A}.$$

The rate of energy dissipation at this time in the resistor is

$$P = \frac{dU}{dt} = i^2 R = \left(2.16 \cdot 10^{-6} \text{ A}\right)^2 \left(2.50 \cdot 10^6 \ \Omega\right) = 1.16 \cdot 10^{-5} \text{ W}.$$

The rate at which the battery delivers energy to the circuit at this time is given by

$$P = \frac{dU}{dt} = iV_{\text{emf}} = \left(2.16 \cdot 10^{-6} \text{ A}\right)\left(12.0 \text{ V}\right) = 2.59 \cdot 10^{-5} \text{ W}.$$

Energy conservation dictates that at any time the energy supplied by the battery is either dissipated as heat in the resistor or stored in the capacitor. In this case, the power supplied by the battery, $2.59 \cdot 10^{-5}$ W, is equal to the power dissipated as heat in the resistor, $1.16 \cdot 10^{-5}$ W, plus the rate at which energy is stored in the capacitor, $1.43 \cdot 10^{-5}$ W. Thus, our answer is consistent.

MULTIPLE-CHOICE QUESTIONS

26.1 A resistor and a capacitor are connected in series. If a second identical capacitor is connected in series in the same circuit, the time constant for the circuit will

a) decrease.   b) increase.   c) stay the same.

26.2 A resistor and a capacitor are connected in series. If a second identical resistor is connected in series in the same circuit, the time constant for the circuit will

a) decrease.   b) increase.   c) stay the same.

26.3 A circuit consists of a source of emf, a resistor, and a capacitor, all connected in series. The capacitor is fully charged. How much current is flowing through it?

a) $i = V/R$ b) zero   c) neither (a) nor (b)

26.4 Which of the following will reduce the time constant in an RC circuit?

a) increasing the dielectric constant of the capacitor

b) adding an additional 20 m of wire between the capacitor and the resistor

c) increasing the voltage of the battery

d) adding an additional resistor in parallel with the first resistor

e) none of the above

26.5 Kirchhoff's Junction Rule states that

a) the algebraic sum of the currents at any junction in a circuit must be zero.

b) the algebraic sum of the potential changes around any closed loop in a circuit must be zero.

c) the current in a circuit with a resistor and a capacitor varies exponentially with time.

d) the current at a junction is given by the product of the resistance and the capacitance.

e) the time for the current development at a junction is given by the product of the resistance and the capacitance.

26.6 How long would it take, in multiples of the time constant, τ, for the capacitor in an RC circuit to be 98% charged?

a) 9τ c) 90τ e) 0.98τ

b) 0.9τ d) 4τ

26.7 A capacitor C is initially uncharged. At time $t = 0$, the capacitor is attached through a resistor R to a battery. The energy stored in the capacitor increases, eventually reaching a value U as $t \rightarrow \infty$. After a time equal to the time constant $\tau = RC$, the energy stored in the capacitor is given by

a) U/e.   c) $U(1 - 1/e)^2$.

b) U/e^2.   d) $U(1 - 1/e)$.

26.8 Which of the following has the same unit as the electromotive force (emf)?

a) current

b) electric potential

c) electric field

d) electric power

e) none of the above

26.9 The capacitor in each circuit in the figure is first charged by a 10-V battery with no internal resistance. Then, the switch is flipped from position A to position B, and the capacitor is discharged through various resistors. For which circuit is the total energy dissipated by the resistor the largest?

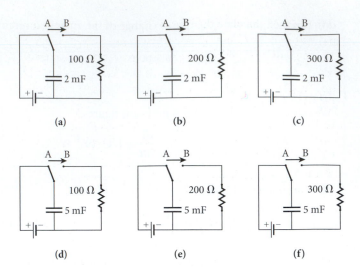

QUESTIONS

26.10 You want to measure simultaneously the potential difference across and the current through a resistor, R. As the circuit diagrams show, there are two ways to connect the two instruments—ammeter and voltmeter—in the circuit. Comment on the result of the measurement using each configuration.

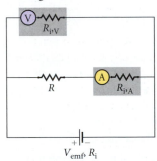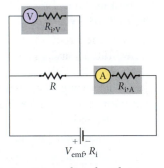

26.11 If the capacitor in an RC circuit is replaced with two identical capacitors connected in series, what happens to the time constant for the circuit?

26.12 You want to accurately measure the resistance, R_{device}, of a new device. The figure shows two ways to accomplish this task. On the left, an ohmmeter produces a current through the device and measures that current, i, and the potential difference, ΔV, across the device. This potential difference includes the potential drops across the wires leading to and from the device and across the contacts that connect the wires to the device. These extra resistances cannot always be neglected, especially if the device has low resistance. This technique is called *two-probe measurement* since two probe wires are connected to the device. The resulting current, i, is measured with an ammeter. The total resistance is then determined by dividing ΔV by i. For this configuration, what is the resistance that the ohmmeter measures? In the alternative configuration, shown on the right, a similar current source is used to produce and measure the current through the device, but the potential difference, ΔV, is measured directly across the device with a nearly ideal voltmeter

with extremely large internal resistance. This technique is called a *four-probe measurement* since four probe wires are connected to the device. What resistance is being measured in this four-probe configuration? Is it different from that being assessed by the two-probe measurement? Why or why not? (*Hint:* Four-probe measurements are used extensively by scientists and engineers and are especially useful for accurate measurements of the resistance of materials or devices with low resistance.)

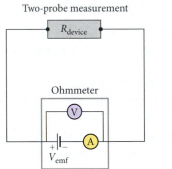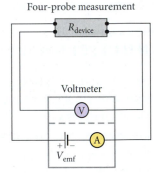

26.13 Explain why the time constant for an RC circuit increases with R and with C. (The answer "That's what the formula says" is not sufficient.)

26.14 A battery, a resistor, and a capacitor are connected in series in an RC circuit. What happens to the current through a resistor after a long time? Explain using Kirchhoff's rules.

26.15 How can you light a 1.0-W, 1.5-V bulb with your 12.0-V car battery?

26.16 A multiloop circuit contains a number of resistors and batteries. If the emf values of all the batteries are doubled, what happens to the currents in all the components of the circuit?

26.17 A multiloop circuit of resistors, capacitors, and batteries is switched on at $t = 0$, at which time all the capacitors are uncharged. The initial distribution of currents and potential differences in the circuit can be analyzed by treating the capacitors as if they were connecting wires or closed switches. The final distribution of currents and potential differences, which occurs after a long time has passed, can be analyzed by treating the capacitors as open segments or open switches. Explain why these tricks work.

26.18 Voltmeters are always connected in parallel with a circuit component, and ammeters are always connected in series. Explain why.

26.19 You wish to measure both the current through and the potential difference across some component of a circuit. It is not possible to do this simultaneously and accurately with ordinary voltmeters and ammeters. Explain why not.

26.20 Two light bulbs for use at 110 V are rated at 60 W and 100 W, respectively. Which has the filament with lower resistance?

26.21 Two capacitors in series are charged through a resistor. Identical capacitors are instead connected in parallel and charged through the same resistor. How do the times required to fully charge the two sets of capacitors compare?

26.22 The figure shows a circuit consisting of a battery connected to a resistor and a capacitor, which is fully discharged initially, in series with a switch.

a) What is the current in the circuit at any time t?

b) Calculate the total energy provided by the battery from $t = 0$ to $t = \infty$.

c) Calculate the total energy dissipated from the resistor for the same time period.

d) Is energy conserved in this circuit?

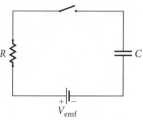

PROBLEMS

A blue problem number indicates a worked-out solution is available in the Student Solutions Manual. One • and two •• indicate increasing level of problem difficulty.

Sections 26.1 through 26.3

26.23 Two resistors, R_1 and R_2, are connected in series across a potential difference, ΔV_0. Express the potential drop across each resistor individually, in terms of these quantities. What is the significance of this arrangement?

26.24 A battery has $V_{emf} = 12.0$ V and internal resistance $r = 1.0\ \Omega$. What resistance, R, can be put across the battery to extract 10 W of power from it?

26.25 Three resistors are connected across a battery as shown in the figure. What values of R and V_{emf} will produce the indicated currents?

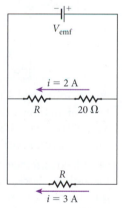

•**26.26** Find the equivalent resistance for the circuit in the figure.

•**26.27** The dead battery of your car provides a potential difference of 9.950 V and has an internal resistance of 1.100 Ω. You charge it by connecting it with jumper cables to the live battery of another car. The live battery provides a potential difference of 12.00 V and has an internal resistance of 0.0100 Ω, and the starter resistance is 0.0700 Ω.

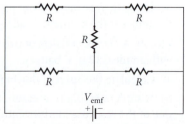

a) Draw the circuit diagram for the connected batteries.

b) Determine the current in the live battery, in the dead battery, and in the starter immediately after you closed the circuit.

•**26.28** In the circuit shown in the figure, $V_1 = 1.5$ V, $V_2 = 2.5$ V, $R_1 = 4.0\ \Omega$, and $R_2 = 5.0\ \Omega$. What is the magnitude of the current, i_1, flowing through resistor R_1?

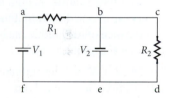

•**26.29** The circuit shown in the figure consists of two batteries with V_A and V_B and three light bulbs with resistances R_1, R_2, and R_3. Calculate the magnitudes of the currents i_1, i_2, and i_3 flowing through the bulbs. Indicate the correct directions of current flow on the diagram. Calculate the power, P_A and P_B, supplied by battery A and by battery B.

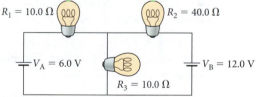

•**26.30** In the circuit shown in the figure, $R_1 = 5\ \Omega$, $R_2 = 10\ \Omega$, and $R_3 = 15\ \Omega$, $V_{emf,1} = 10$ V, and $V_{emf,2} = 15$ V. Using Kirchhoff's Loop and Junction Rules, determine the currents i_1, i_2, and i_3 flowing through R_1, R_2 and R_3, respectively, in the direction indicated in the figure.

•**26.31** For the circuit shown in the figure, find the magnitude and the direction of the current through each resistor and the power supplied by each battery, using the following values: $R_1 = 4\ \Omega$,

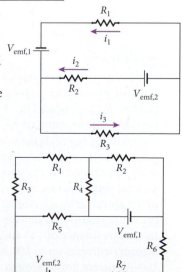

$R_2 = 6\,\Omega$, $R_3 = 8\,\Omega$, $R_4 = 6\,\Omega$, $R_5 = 5\,\Omega$, $R_6 = 10\,\Omega$, $R_7 = 3\,\Omega$, $V_{emf,1} = 6$ V, and $V_{emf,2} = 12$ V.

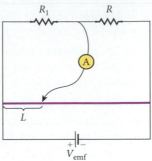

•**26.32** A Wheatstone bridge is constructed using a 1-m-long Nichrome wire (the purple line in the figure) with a conducting contact that can slide along the wire. A resistor, $R_1 = 100\,\Omega$, is placed on one side of the bridge, and another resitor, R, of unknown resistance, is placed on the other side. The contact is moved along the Nichrome wire, and it is found that the ammeter reading is zero for $L = 25$ cm. Knowing that the wire has a uniform cross section throughout its length, determine the unknown resistance.

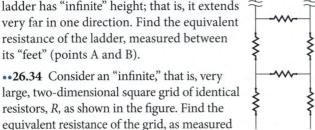

••**26.33** A "resistive ladder" is constructed with identical resistors, R, making up its legs and rungs, as shown in the figure. The ladder has "infinite" height; that is, it extends very far in one direction. Find the equivalent resistance of the ladder, measured between its "feet" (points A and B).

••**26.34** Consider an "infinite," that is, very large, two-dimensional square grid of identical resistors, R, as shown in the figure. Find the equivalent resistance of the grid, as measured across any individual resistor. (*Hint:* Symmetry and superposition are very helpful for solving this problem.)

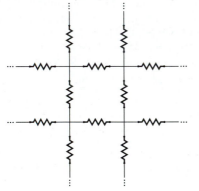

Section 26.4

26.35 To extend the useful range of an ammeter, a shunt resistor, R_{shunt}, is placed in parallel with the ammeter as shown in the figure. If the internal resistance of the ammeter is $R_{i,A}$, determine the resistance that the shunt resistor has to have to extend the useful range of the ammeter by a factor N. Then, calculate the resistance the shunt resistor has to have to allow an ammeter with an internal resistance of $1\,\Omega$ and a maximum range of 1 A to measure currents up to 100 A. What fraction of the total 100-A current flows through the ammeter, and what fraction flows through the shunt resistor?

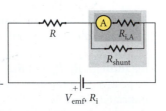

26.36 To extend the useful range of a voltmeter, an additional resistor, R_{series}, is placed in series with the voltmeter as shown in the figure. If the internal resistance of the voltmeter is $R_{i,V}$, determine the resistance that the added series resistor has to have to extend the useful range of the voltmeter by a factor N. Then, calculate the resistance the series resistor has to have to allow a voltmeter with an internal resistance of 1 MΩ ($10^6\,\Omega$) and a maximum range of 1 V to measure potential differences up to 100 V. What fraction of the total 100-V potential drop occurs across the voltmeter, and what fraction of that drop occurs across the added series resistor?

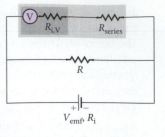

26.37 As shown in the figure, a 6.0000-V battery is used to produce a current through two identical resistors, R, each having a resistance of 100.00 kΩ. A digital multimeter (DMM) is used to measure the potential difference across the first resistor. DMMs typically have an internal resistance of 10 MΩ. Determine the potential differences V_{ab} (the potential difference between points a and b, which is the difference the DMM measures) and V_{bc} (the potential difference between points b and c, which is the difference across the second resistor). Nominally, $V_{ab} = V_{bc}$, but this may not be the case here. How can this measurement error be reduced?

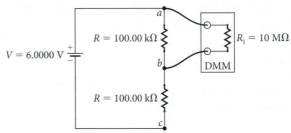

26.38 You want to make an ohmmeter to measure the resistance of unknown resistors. You have a battery with voltage $V_{emf} = 9.0$ V, a variable resistor, R, and an ammeter that measures current on a linear scale from 0 to 10 mA.

a) What resistance should the variable resistor have so that the ammeter gives its full-scale (maximum) reading when the ohmmeter is shorted?

b) Using the resistance from part (a), what is the unknown resistance if the ammeter reads $\frac{1}{4}$ of its full scale?

•**26.39** A circuit consists of two 1.00-kΩ resistors in series with an ideal 12.0-V battery.

a) Calculate the current flowing through each resistor.

b) A student trying to measure the current flowing through one of the resistors inadvertently connects an ammeter in parallel with that resistor rather than in series with it. How much current will flow through the ammeter, assuming that it has an internal resistance of $1.0\,\Omega$?

•**26.40** A circuit consists of two 100-kΩ resistors in series with an ideal 12.0-V battery.

a) Calculate the potential drop across one of the resistors.

b) A voltmeter with internal resistance 10 MΩ is connected in parallel with one of the two resistors in order to measure the potential drop across the resistor. By what percentage will the voltmeter reading deviate from the value you determined in part (a)? (*Hint:* The difference is rather small so it is helpful to solve algebraically first to avoid a rounding error.)

Section 26.5

26.41 Initially, switches S_1 and S_2 in the circuit shown in the figure are open and the capacitor has a charge of 100 mC. About how long will it take after switch S_1 is closed for the charge on the capacitor to drop to 5 mC?

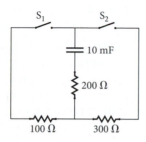

26.42 What is the time constant for the discharging of the capacitors in the circuit shown in the figure? If the 2-μF capacitor initially has a potential difference of 10 V across its plates, how much charge is left on it after the switch has been closed for a time equal to half of the time constant?

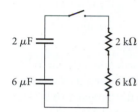

26.43 The circuit shown in the figure has a switch, S, two resistors, $R_1 = 1\ \Omega$ and $R_2 = 2\ \Omega$, a 12-V battery, and a capacitor with $C = 20\ \mu F$. After the switch is closed, what will the maximum charge on the capacitor be? How long after the switch has been closed will the capacitor have 50% of this maximum charge?

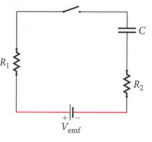

26.44 In the movie *Back to the Future*, time travel is made possible by a flux capacitor, which generates 1.21 GW of power. Assuming that a 1-F capacitor is charged to its maximum capacity with a 12-V car battery and is discharged through a resistor, what resistance is necessary to produce a peak power output of 1.21 GW in the resistor? How long would it take for a 12-V car battery to charge the capacitor to 90% of its maximum capacity through this resistor?

26.45 During a physics demonstration, a fully charged 90-μF capacitor is discharged through a 60-Ω resistor. How long will it take for the capacitor to lose 80% of its initial energy?

•26.46 Two parallel plate capacitors, C_1 and C_2, are connected in series with a 60-V battery and a 300-kΩ resistor, as shown in the figure. Both capacitors have plates with an area of 2.0 cm² and a separation of 0.1 mm. Capacitor C_1 has air between its plates, and capacitor C_2 has the gap filled with porcelain (dielectric constant of 7 and

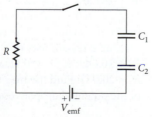

dielectric strength of 5.7 kV/mm). The switch is closed, and a long time passes.

a) What is the charge on capacitor C_1?

b) What is the charge on capacitor C_2?

c) What is the total energy stored in the two capacitors?

d) What is the electric field inside capacitor C_2?

•26.47 A parallel plate capacitor with $C = 0.050\ \mu F$ has a separation between its plates of $d = 50.0\ \mu m$. The dielectric that fills the space between the plates has dielectric constant $\kappa = 2.5$ and resistivity $\rho = 4.0 \cdot 10^{12}\ \Omega$ m. What is the time constant for this capacitor? (*Hint:* First calculate the area of the plates for the given C and κ, and then determine the resistance of the dielectric between the plates.)

•26.48 A 12-V battery is attached to a 2-mF capacitor and a 100-Ω resistor. Once the capacitor is fully charged, what is the energy stored in it? What is the energy dissipated as heat by the resistor as the capacitor is charging?

•26.49 A capacitor bank is designed to discharge 5.0 J of energy through a 10.0-kΩ resistor array in under 2.0 ms. To what potential difference must the bank be charged, and what must the capacitance of the bank be?

•26.50 The circuit in the figure has a capacitor connected to a battery, two switches, and three resistors. Initially, the capacitor is uncharged and both of the switches are open.

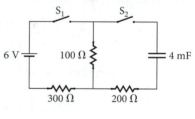

a) Switch S_1 is closed. What is the current flowing out of the battery immediately after switch S_1 is closed?

b) After about 10 min, switch S_2 is closed. What is the current flowing out of the battery immediately after switch S_2 is closed?

c) What is the current flowing out of the battery about 10 min after switch S_2 has been closed?

d) After another 10 min, switch S_1 is opened. How long will it take until the current in the 200-Ω resistor is below 1 mA?

•26.51 In the circuit shown in the figure, $R_1 = 10\ \Omega$, $R_2 = 4\ \Omega$, and $R_3 = 10\ \Omega$, and the capacitor has capacitance $C = 2\ \mu F$.

a) Determine the potential difference, ΔV_C, across the capacitor after switch S has been closed for a long time.

b) Determine the energy stored in the capacitor when switch S has been closed for a long time.

c) After switch S is opened, how much energy is dissipated through R_3?

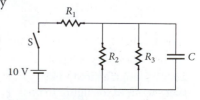

••26.52 A cube of gold that is 2.5 mm on a side is connected across the terminals of a 15-μF capacitor that initially has a potential difference of 100.0 V between its plates.

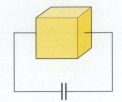

a) What time is required to fully discharge the capacitor?

b) When the capacitor is fully discharged, what is the temperature of the gold cube?

••26.53 A "capacitive ladder" is constructed with identical capacitors, C, making up its legs and rungs, as shown in the figure. The ladder has "infinite" height; that is, it extends very far in one direction. Calculate the equivalent capacitance of the ladder, measured between its "feet" (points A and B).

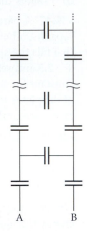

Additional Problems

26.54 In the circuit in the figure, the capacitors are completely uncharged. The switch is then closed for a long time.

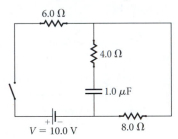

a) Calculate the current through the 4.0-Ω resistor.

b) Find the potential difference across the 4.0-Ω, 6.0-Ω, and 8.0-Ω resistors.

c) Find the potential difference across the 1.0-μF capacitor.

26.55 The ammeter your physics instructor uses for in-class demonstrations has internal resistance $R_i = 75$ Ω and measures a maximum current of 1.5 mA. The same ammeter can be used to measure currents of much greater magnitudes by wiring a shunt resistor of relatively small resistance, R_{shunt}, in parallel with the ammeter. (a) Sketch the circuit diagram, and explain why the shunt resistor connected in parallel with the ammeter allows it to measure larger currents. (b) Calculate the resistance the shunt resistor has to have to allow the ammeter to measure a maximum current of 15 A.

26.56 Many electronics devices can be dangerous even after they are shut off. Consider an RC circuit with a 150-μF capacitor and a 1-MΩ resistor connected to a 200-V power source for a long time and then disconnected and shorted, as shown in the figure. How long will it be until the potential difference across the capacitor drops to below 50 V?

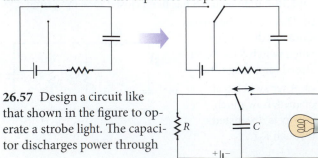

26.57 Design a circuit like that shown in the figure to operate a strobe light. The capacitor discharges power through

the light bulb filament (resistance of 2.5 kΩ) in 0.20 ms and charges through a resistor R, with a repeat cycle of 1000 Hz. What capacitor and resistor should be used?

26.58 An ammeter with an internal resistance of 53 Ω measures a current of 5.25 mA in a circuit containing a battery and a total resistance of 1130 Ω. The insertion of the ammeter alters the resistance of the circuit, and thus the measurement does not give the actual value of the current in the circuit without the ammeter. Determine the actual value of the current.

•26.59 In the circuit shown in the figure, a 10-μF capacitor is charged by a 9-V battery with the two-way switch kept in position X for a long time. Then the switch is suddenly flicked to position Y. What current flows through the 40-Ω resistor

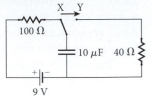

a) immediately after the switch moves to position Y?

b) 1 ms after the switch moves to position Y?

•26.60 How long will it take for the current in a circuit to drop from its initial value to 1.50 mA if the circuit contains two 3.8-μF capacitors that are initially uncharged, two 2.2-kΩ resistors, and a 12.0-V battery all connected in series?

26.61 An RC circuit has a time constant of 3.1 s. At $t = 0$, the process of charging the capacitor begins. At what time will the energy stored in the capacitor reach half of its maximum value?

•26.62 For the circuit shown in the figure, determine the charge on each capacitor when (a) switch S has been closed for a long time and (b) switch S has been open for a long time.

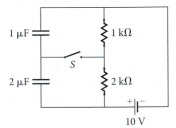

•26.63 Three resistors, $R_1 = 10$ Ω, $R_2 = 20$ Ω, and $R_3 = 30$ Ω, are connected in a multiloop circuit, as shown in the figure. Determine the amount of power dissipated in the three resistors.

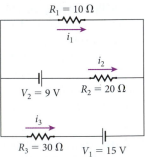

•26.64 The figure shows a circuit containing two batteries and three resistors. The batteries provide $V_{emf,1} = 12.0$ V and $V_{emf,2} = 16.0$ V and have no internal resistance. The resistors have resistances of $R_1 = 30.0$ Ω, $R_2 = 40.0$ Ω, and $R_3 = 20.0$ Ω. Find the magnitude of the potential drop across R_2.

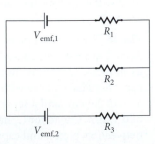

24.65 The figure shows a spherical capacitor. The inner sphere has radius $a = 1$ cm, and the outer sphere has radius $b = 1.1$ cm. The battery has $V_{emf} = 10$ V, and the resistor has a value of $R = 10$ MΩ.

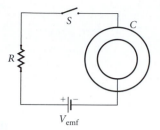

a) Determine the time constant of the RC circuit.

b) Determine how much charge has accumulated on the capacitor after switch S has been closed for 0.1 ms.

•26.66 Write the set of equations that determines the three currents in the circuit shown in the figure. (Assume that the capacitor is initially uncharged.)

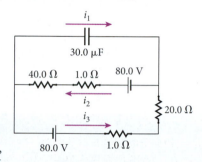

•26.67 Consider a series RC circuit with $R = 10$ Ω, $C = 10$ μF and $V = 10.0$ V.

a) How much time, expressed as a multiple of the time constant, does it take for the capacitor to be charged to half of its maximum value?

b) At this instant, what is the ratio of the energy stored in the capacitor to its maximum possible value?

c) Now suppose the capacitor is fully charged. At time $t = 0$, the original circuit is opened and the capacitor is allowed to discharge across another resistor, $R' = 1$ Ω, that is connected across the capacitor. What is the time constant for the discharging of the capacitor?

d) How many seconds does it take for the capacitor to discharge half of its maximum stored charge, Q?

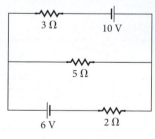

•26.68 a) What is the current in the 5-Ω resistor in the circuit shown in the figure?

b) What is the power dissipated in the 5-Ω resistor?

•26.69 In the Wheatstone bridge shown in the figure, the known resistances are $R_1 = 8.00$ Ω, $R_4 = 2.00$ Ω, and $R_5 = 6.00$ Ω, and the battery has $V_{emf} = 15.0$ V. The variable resistance R_2 is adjusted until the potential difference across R_3 is zero ($V = 0$). Find i_2 (the current through resistor R_2) at this point.

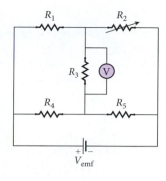

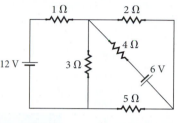

••26.70 Consider the circuit with five resistors and two batteries (with no internal resistance) shown in the figure.

a) Write a set of equations that will allow you to solve for the current in each of the resistors.

b) Solve the equations from part (a) for the current in the 4-Ω resistor.

••26.71 Consider an "infinite," that is, very large, two-dimensional square grid of identical capacitors, C, shown in the figure. Find the effective capacitance of the grid, as measured across any individual capacitor.

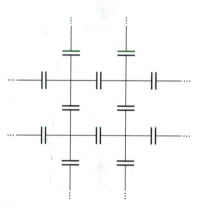

27

Magnetism

FIGURE 27.1 Hot ionized gases travel along magnetic field lines near the surface of the Sun, forming coronal loops. The superimposed image of the Earth is at the correct scale to give an idea of the size of these coronal loops.

WHAT WE WILL LEARN

- Permanent magnets exist in nature. A magnet always has a north pole and a south pole. A single magnetic north pole or south pole cannot be isolated—magnetic poles always come in pairs.

- Opposite poles attract, and like poles repel.

- Breaking a bar magnet in half results in two new magnets, each with a north and a south pole.

- A magnetic field exerts a force on a moving charged particle.

- Earth has a magnetic field.

- The force exerted on a charged particle moving in a magnetic field is perpendicular to both the magnetic field and the velocity of the particle.

- The torque on a current-carrying loop can be expressed in terms of the vector product of the magnetic dipole moment of the loop and the magnetic field.

- The Hall effect can be used to measure magnetic fields.

This chapter is the first to consider magnetism, describing magnetic fields and magnetic forces and their effects on charged particles and currents. Magnetic fields can be huge and powerful, as the image of the Sun's surface in Figure 27.1 shows. The Sun has enormous magnetic fields, and hot gases that erupt periodically from the Sun's surface tend to follow the field lines as they rise up, forming arches and loops much larger than the Earth. Astronomers believe that all stars possess powerful magnetic fields, making magnetism one of the most common and important phenomena in the universe.

We'll continue to study magnetism in the next few chapters, describing the causes of magnetic fields and their connection to electric fields. You will see that electricity and magnetism are really parts of the same universal force, called electromagnetic force; their connection is one of the most spectacular successes in physical theory.

27.1 Permanent Magnets

In the region of Magnesia (in central Greece), the ancient Greeks found several types of naturally occurring minerals that attract and repel each other and attract certain kinds of metal, such as iron. They also, if floating freely, line up with the North and South Poles of the Earth. These minerals are various forms of iron oxide and are called **permanent magnets.** Other examples of permanent magnets include refrigerator magnets and magnetic door latches, which are made of compounds of iron, nickel, or cobalt. If you touch an iron bar to a piece of the mineral lodestone (magnetic magnetite), the iron bar will be magnetized. If you float this iron bar in water, it will align with the Earth's magnetic poles. The end of the magnet that points north is called the **north magnetic pole,** and the other end is called the **south magnetic pole.**

If two permanent magnets are brought close together with the two north poles or two south poles almost touching, the magnets repel each other (Figure 27.2a). If a north pole and a south pole are brought close together, the magnets attract each other (Figure 27.2b). What is called the North Pole of Earth is actually a magnetic south pole, which is why it attracts the north pole of permanent magnets.

Breaking a permanent magnet in half does not yield one north pole and one south pole. Instead, two new magnets, each with its own north and south pole result (Figure 27.3). Unlike electric charge, which exists as separate positive (proton) and negative (electron) charges, no separate magnetic monopoles (isolated north and south poles) exist. Scientists have carried out extensive searches for magnetic monopoles, and none has been found. The discussion of the source of magnetism in this chapter will help you understand why there are no magnetic monopoles.

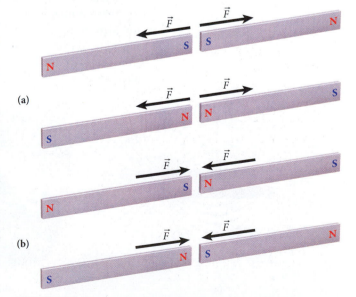

FIGURE 27.2 (a) Like magnetic poles repel; (b) unlike magnetic poles attract.

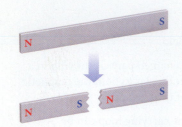

FIGURE 27.3 Breaking a bar magnet in half yields two magnets, each with its own north and south pole.

Magnetic Field Lines

Permanent magnets interact with each other at some distance, without touching. In analogy with the gravitational field and the electric field, the concept of a **magnetic field** is used to describe the magnetic force. The vector $\vec{B}(\vec{r})$ denotes the magnetic field vector at any given point in space.

Like an electric field, a magnetic field is represented using field lines. The magnetic field vector is always tangent to the **magnetic field lines.** The magnetic field lines from a permanent bar magnet are shown in Figure 27.4a. As with electric field lines, closer spacing between lines indicates higher field strength. In an electric field, the electric force on a positive test charge points in the same direction as the electric field vector. However, because no magnetic monopole exists, the magnetic force cannot be described in an analogous way.

The direction of the magnetic field is established in terms of the direction in which a compass needle points. A compass needle, with a north pole and a south pole, will orient itself so that its north pole points in the direction of the magnetic field. Thus, the direction of the field can be determined at any point by noting the direction in which a compass needle placed at that point points, as illustrated in Figure 27.5 for a bar magnet.

Externally, magnetic field lines appear to originate on north poles and terminate on south poles, but these field lines are actually closed loops that penetrate the magnet itself. This formation of loops is an important difference between electric and magnetic field lines (for static fields—this statement does not apply to time-dependent fields, as we'll see in subsequent chapters). Recall that electric field lines start at positive charges and end on negative charges. However, because no magnetic monopoles exist, magnetic field lines cannot start or stop at particular points. Instead, they form closed loops that do not start or stop anywhere. We'll see later that this difference is important in describing the interaction of electric and magnetic fields. If you see a field pattern and don't know at first if it is an electric field or a magnetic field, check for closed loops. If you find some, it is a magnetic field; if the field lines do not form loops, it is an electric field.

Earth's Magnetic Field

Earth itself is a magnet, with a magnetic field similar to the magnetic field of a bar magnet (Figure 27.4). This magnetic field is important because it protects us from high-energy radiation from space called *cosmic rays*. These cosmic rays consist mostly of charged particles that are deflected away from Earth's surface by its magnetic field. The poles of Earth's magnetic field do not coincide with the geographic poles, defined as the points where Earth's rotation axis intersects its surface.

Figure 27.6 shows a cross section of Earth's magnetic field lines. The field lines are close together, forming a surface that wraps around Earth like a doughnut. Earth's magnetic field is distorted by the solar wind, a flow of ionized particles, mainly protons, emitted by the Sun and moving outward from the Sun at approximately 400 km/s. Two bands of charged particles captured from the solar wind circle the Earth. These are called the **Van Allen radiation belts** (Figure 27.6), after James A. Van Allen (1914–2006), who discovered them in the early days of space flight by

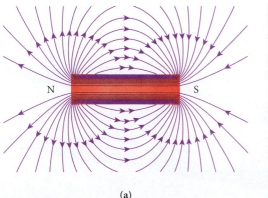

(a)   (b)

FIGURE 27.4 (a) Computer-generated magnetic field lines from a permanent bar magnet. (b) Iron filings align themselves with the magnetic field lines and make them visible.

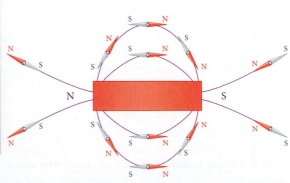

FIGURE 27.5 Using a compass needle to determine the direction of the magnetic field from a bar magnet.

putting radiation counters on satellites. The Van Allen radiation belts come closest to Earth near the north and south magnetic poles, where the charged particles trapped within the belts often collide with atoms in the atmosphere, exciting them. These excited atoms emit light of different colors as they collide and lose energy; the results are the fabulous **aurora borealis** (northern lights) in the northern latitudes (see Figure 27.7) and the **aurora australis** (southern lights) in the southern latitudes. Aurorae are not unique to Earth; they have been seen on other planets with strong magnetic fields, such as Jupiter and Saturn (shown in Figure 27.8).

Earth's magnetic poles move at a current rate of up to 40 km in a single year. Right now, the magnetic north pole is located approximately 2800 km away from the geographic South Pole, at the edge of Antarctica, and is moving toward Australia. The magnetic south pole is located in the Canadian Arctic and, if its present rate of motion continues, will reach Siberia in 2050. Earth's magnetic field has decreased steadily at a rate of about 7% per century since it was first measured accurately around 1840. At that rate, Earth's magnetic field will disappear in a few thousand years. However, some geological evidence indicates that the magnetic field of Earth has reversed itself approximately 170 times in the past 100 million years. The last reversal occurred about 770,000 years ago. Thus, rather than disappear, Earth's magnetic field may reverse its direction. What is the cause of Earth's magnetic field? Surprisingly, the answer to this question is not known exactly and is under intense current research. Most likely, it is caused by strong electrical currents inside the Earth, caused by the spinning liquid iron-nickel core. This spinning is often referred to as the *dynamo effect*. (We'll see how currents create magnetic fields in Chapter 28.)

Because the geographic North Pole and the magnetic north pole are not in the same location, a compass needle generally does not point exactly to the geographic North Pole. This difference is called the **magnetic declination.** The magnetic declination is taken to be positive when magnetic north is east of true north and negative when magnetic north is west of true north. The magnetic north pole currently lies on a line that passes through central Missouri, eastern Illinois, western Iowa, and eastern Wisconsin. Along this line, the magnetic declination is zero. West of this line, the magnetic declination is positive and reaches 18° in Seattle. East of this line, the declination is negative, up to –18° in Maine. A map showing the magnetic declinations in the United States as of 2004 is presented in Figure 27.9.

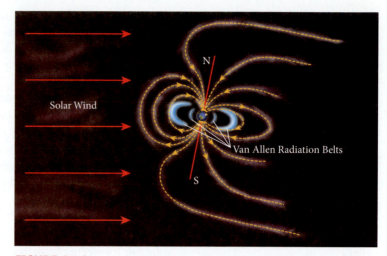

FIGURE 27.6 Cross section through Earth's magnetic field. The dashed lines represent the magnetic field lines. The axis defined by the north and south magnetic poles (red line) currently forms an angle of approximately 11° with the rotation axis.

FIGURE 27.7 Aurora borealis over Finland, photographed from the International Space Station.

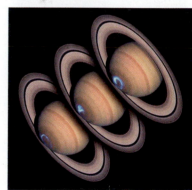

FIGURE 27.8 Aurora on Saturn, photographed by the Hubble Space Telescope.

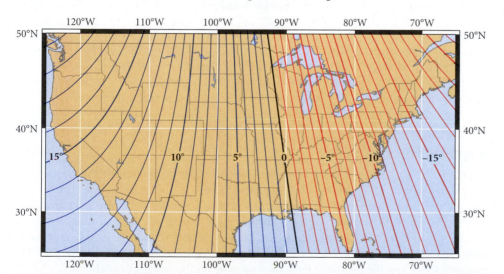

FIGURE 27.9 Magnetic declinations in the United States in 2004 measured in degrees. Red lines represent negative magnetic declinations, and blue lines signify positive magnetic declinations. Lines of magnetic declination are separated by one degree.

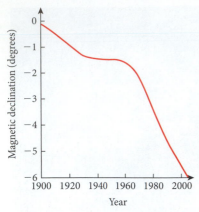

FIGURE 27.10 The magnetic declination at Lansing, Michigan, from 1900 to 2004.

Because the positions of Earth's magnetic poles move with time, the magnetic declinations for all locations on Earth's surface also change with time. For example, Figure 27.10 shows the estimated magnetic declination for Lansing, Michigan, for the period 1900–2004. A similar graph can be drawn for any location on Earth.

Superposition of Magnetic Fields

If several sources of magnetic field, such as several permanent magnets, are close together, the magnetic field at any given point in space is given by the superposition of the magnetic fields from all the sources. This superposition of fields follows directly from the superposition of forces introduced in Chapter 4. The superposition principle for the total magnetic field, $\vec{B}_{\text{total}}(\vec{r})$, due to n magnetic field sources can be stated as

$$\vec{B}_{\text{total}}(\vec{r}) = \vec{B}_1(\vec{r}) + \vec{B}_1(\vec{r}) + \cdots + \vec{B}_n(\vec{r}). \tag{27.1}$$

This superposition principle for magnetic fields is exactly analogous to the superposition principle for electric fields, presented in Chapter 22.

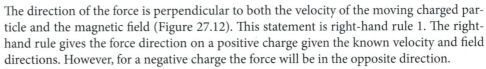

27.2 Magnetic Force

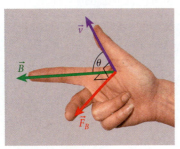

FIGURE 27.11 A beam of electrons (bluish green), made visible by a small amount of gas in an evacuated tube, is bent by a magnet (at the right edge of picture).

The qualitative discussion in the preceding section pointed out that a magnetic field has a direction, along the magnetic field lines. The magnitude of a magnetic field is determined by examining its effect on a moving charged particle. We'll start with a constant magnetic field and study its effect on a single charge. As a reminder, we saw in Chapter 22 that the electric field exerts a force on a charge given by $\vec{F}_E = q\vec{E}$. Experiments such as the one shown in Figure 27.11 show that a magnetic field does not exert a force on a charge at rest but only on a *moving* charge.

A magnetic field is defined in terms of the force exerted by the field on a moving charged particle. The magnetic force exerted by a magnetic field on a moving charged particle with charge q moving with velocity $\vec{v}$ is given by

$$\vec{F}_B = q\vec{v} \times \vec{B}. \tag{27.2}$$

The direction of the force is perpendicular to both the velocity of the moving charged particle and the magnetic field (Figure 27.12). This statement is right-hand rule 1. The right-hand rule gives the force direction on a positive charge given the known velocity and field directions. However, for a negative charge the force will be in the opposite direction.

The magnitude of the magnetic force on a moving charged particle is

$$F_B = |q|vB\sin\theta, \tag{27.3}$$

FIGURE 27.12 Right-hand rule 1 for the force exerted by a magnetic field, $\vec{B}$, on a particle with charge q moving with velocity $\vec{v}$. To find the direction of the magnetic force, point your thumb in the direction of the velocity of the moving charged particle and your index finger in the direction of the magnetic field, and your middle finger then gives you the direction of the magnetic force.

where θ is the angle between the velocity of the charged particle and the magnetic field. (The angle θ is always between 0° and 180°, and therefore, $\sin\theta \geq 0$.) You can see that no magnetic force acts on a charged particle moving parallel to a magnetic field because in that case $\theta = 0°$. If a charged particle is moving perpendicularly to the magnetic field, $\theta = 90°$ and (for fixed values of v and B) the magnitude of the magnetic force has its maximum value of

$$F_B = |q|vB \quad \left(\text{for } \vec{v} \perp \vec{B}\right). \tag{27.4}$$

Magnetic Force and Work

Equation 27.2 established that the magnetic force is the vector product of the velocity vector and magnetic field vector and thus is perpendicular to both vectors. This implies that $\vec{F}_B \cdot \vec{v} = 0$ and, since the force is the product of mass and acceleration, also that $\vec{a} \cdot \vec{v} = 0$. In Chapter 9 on circular motion, we saw that this condition means that the direction of the velocity vector can change but the magnitude of the velocity vector, the speed, remains the same. Therefore, the kinetic energy, $\frac{1}{2}mv^2$, remains constant for a particle subjected to a magnetic force, and the magnetic force *does no work* on the moving particle.

This is a profound result: A constant magnetic field cannot be used to do work on a particle. The kinetic energy of a particle moving in a constant magnetic field remains constant, even though the direction of the particle's velocity vector can change as a function of time while the particle is moving through the magnetic field. An electric field, on the other hand, can easily be used to do work on a particle.

Units of Magnetic Field Strength

To discuss the motion of charges in magnetic fields, we need to know what units are used to measure the magnetic field strength. Solving equation 27.4 for the field strength and inserting the units of the other quantities gives

$$[F_B]=[q][v][B] \Rightarrow [B]=\frac{[F_B]}{[q][v]}=\frac{N\,s}{C\,m}.$$

Because the ampere (A) is defined as 1 C/s, (N s)/(C m) = N/(A m). The unit of magnetic field strength has been named the **tesla** (T), in honor of Croatian-born American physicist and inventor Nikola Tesla (1856–1943):

$$1\,T=1\,\frac{N\,s}{C\,m}=1\,\frac{N}{A\,m}.$$

A tesla is a rather large amount of magnetic field strength. Sometimes magnetic field strength is given in gauss (G), which is not an SI unit:

$$1\,G=10^{-4}\,T.$$

For example, the strength of Earth's magnetic field at Earth's surface is on the order of 0.5 G ($5 \cdot 10^{-5}$ T). It varies with location from 0.2 G to 0.6 G, as illustrated in Figure 27.13.

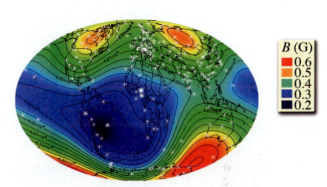

FIGURE 27.13 Global map of the strength of Earth's magnetic field.

SOLVED PROBLEM 27.1 | Cathode Ray Tube

PROBLEM

Consider a cathode ray tube similar to the one shown in Figure 27.11. In this tube, a potential difference of ΔV = 111 V accelerates electrons horizontally (starting essentially from rest) in an electron gun, as shown in Figure 27.14a. The electron gun has a specially coated filament that emits electrons when heated. A negatively charged cathode controls the number of electrons emitted. Positively charged anodes focus and accelerate the electrons into a beam. Downstream from the anodes are horizontal and vertical deflecting plates. Beyond the electron gun is a constant magnetic field with magnitude $B = 3.40 \cdot 10^{-4}$ T. The direction of the magnetic field is upward, perpendicular to the initial velocity of the electrons. What is the magnitude of the acceleration of the electrons due to the magnetic field? (The mass of an electron is $9.11 \cdot 10^{-31}$ kg.)

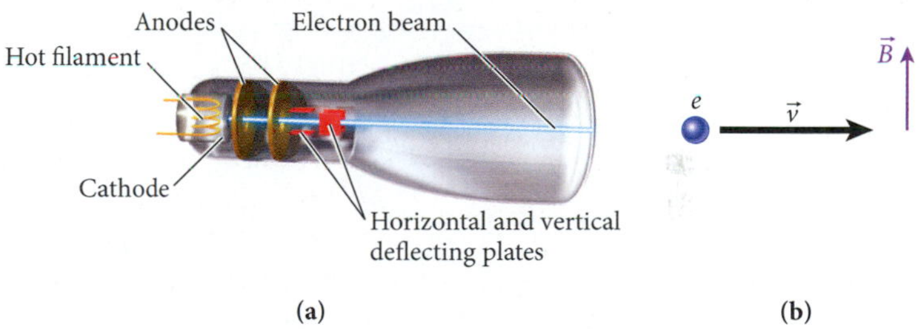

FIGURE 27.14 (a) A cathode ray tube. (b) Electrons moving with velocity $\vec{v}$ enter a constant magnetic field.

SOLUTION

THINK

The electrons gain kinetic energy in the electron gun of the cathode ray tube. The gain in kinetic energy of each electron is equal to the charge of the electron times the potential difference. The speed of the electrons can be found from the definition of kinetic energy. The magnetic force on an electron can be found from the electron charge, the electron velocity, and the strength of the magnetic field, and it is equal to the mass of the electron times its acceleration.

SKETCH

Figure 27.14b shows an electron, moving with velocity $\vec{v}$, entering a constant magnetic field that is perpendicular to the electron path.

Continued—

RESEARCH

The change in kinetic energy, ΔK, of the electrons plus the change in potential energy of the electrons is equal to zero:

$$\Delta K + \Delta U = \tfrac{1}{2}mv^2 + q\Delta V = 0.$$

Since, in this case, $q = -e$, we see that

$$e\Delta V = \tfrac{1}{2}mv^2, \tag{i}$$

where ΔV is the magnitude of the potential difference across which the electrons were accelerated and m is the mass of an electron. We can solve equation (i) for the speed of the electrons:

$$v = \sqrt{\frac{2e\Delta V}{m}}. \tag{ii}$$

The magnitude of the force exerted by the magnetic field on the electrons is given by equation 27.3:

$$F_B = evB\sin 90° = evB,$$

where $-e$ is the charge of an electron and B is the magnitude of the magnetic field. According to Newton's Second Law, $F_{net} = ma$. Since the only force present is the magnetic one, we have

$$F_B = ma = evB, \tag{iii}$$

where a is the magnitude of the acceleration of the electrons.

SIMPLIFY

We can rearrange equation (iii) and substitute the expression for the speed of the electrons from equation (ii) to obtain the acceleration of the electrons:

$$a = \frac{evB}{m} = \frac{eB\sqrt{\dfrac{2e\Delta V}{m}}}{m} = B\sqrt{2\Delta V\frac{e^3}{m^3}}.$$

CALCULATE

Putting in the numerical values gives us

$$a = \left(3.40 \cdot 10^{-4}\ \text{T}\right)\sqrt{2\left(111\ \text{V}\right)\frac{\left(1.602 \cdot 10^{-19}\ \text{C}\right)^3}{\left(9.11 \cdot 10^{-31}\ \text{kg}\right)^3}} = 3.7357 \cdot 10^{14}\ \text{m/s}^2.$$

ROUND

We report our result to three significant figures:

$$a = 3.74 \cdot 10^{14}\ \text{m/s}^2.$$

DOUBLE-CHECK

The calculated acceleration is tremendously large, almost 40 trillion times the Earth's gravitational acceleration. So we certainly want to double-check. We first calculate the speed of the electrons:

$$v = \sqrt{\frac{2e\Delta V}{m}} = \sqrt{\frac{2\left(1.602 \cdot 10^{-19}\ \text{C}\right)\left(111\ \text{V}\right)}{9.11 \cdot 10^{-31}\ \text{kg}}} = 6.25 \cdot 10^{6}\ \text{m/s}.$$

A speed of 6250 km/s may seem large, but it is reasonable for electrons because it is only 2% of the speed of light. The magnetic force on each electron is then

$$F_B = evB = \left(1.602 \cdot 10^{-19}\ \text{C}\right)\left(6.25 \cdot 10^{6}\ \text{m/s}\right)\left(3.40 \cdot 10^{-4}\ \text{T}\right) = 3.40 \cdot 10^{-16}\ \text{N}.$$

The acceleration is very large because the mass of an electron is very small.

27.1 In-Class Exercise

In what direction will the electron in Figure 27.14b be deflected as it enters the constant magnetic field?

a) into the page

b) out of the page

c) up

d) down

e) no deflection

27.1 Self-Test Opportunity

Three particles, each with charge $q = 6.15\ \mu C$ and speed $v = 465$ m/s, enter a uniform magnetic field with magnitude $B = 0.165$ T (see the figure). What is the magnitude of the magnetic force on each of the particles?

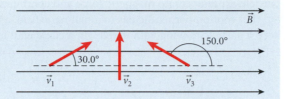

27.3 Motion of Charged Particles in a Magnetic Field

The fact that the force due to a magnetic field acting on a moving charged particle is perpendicular to both the field and the particle's velocity makes this force different from any we've considered so far. However, the tools we use to analyze this force—Newton's laws and the laws of conservation of energy, momentum, and angular momentum—are the same.

Paths of Moving Charged Particles in a Constant Magnetic Field

Suppose you drive your car at constant speed around a circular track. The friction between the tires and the road provides the centripetal force that keeps the car moving in a circle. This force always points toward the center of the circle and creates a centripetal acceleration (discussed in Chapter 9). A similar physical situation occurs when a particle with charge q and mass m moves with velocity $\vec{v}$ perpendicular to a uniform magnetic field, $\vec{B}$, as illustrated in Figure 27.15.

In this situation, the particle moves in a circle with constant speed v and the magnetic force of magnitude $F_B = |q|vB$ supplies the centripetal force that keeps the particle moving in a circle. Particles with opposite charges and the same mass will orbit in opposite directions at the same orbital radius. For example, electrons and positrons are elementary particles with the same mass; the electron has a negative charge, and the positron has a positive charge. Figure 27.16 is a bubble chamber photograph showing two electron-positron pairs. A bubble chamber is a device that can track charged particles moving in a constant magnetic field. (The pairs of electrons and positrons were created by interactions of elementary particles, which will be covered in detail in Chapter 39.) Pair 1 has an electron and a positron that have the same relatively low speed. The particles initially travel in a circle. However, as they move through the bubble chamber, they slow down. (This slowdown is not due to the magnetic force but to collisions of the particles with the molecules of the gas in the bubble chamber.) Thus, the radius of the circle gets smaller and smaller, creating a spiral. The electron and positron in pair 2 have a much higher speed. Their tracks are curved but do not form a complete circle before the particles exit the bubble chamber.

FIGURE 27.15 Electron beam bent into a circular path by the magnetic field generated by two coils.

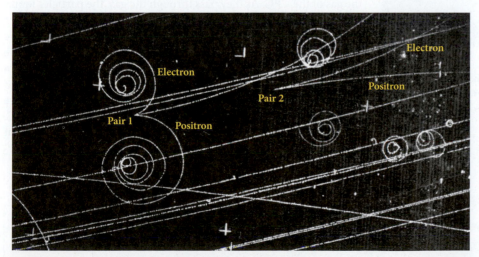

FIGURE 27.16 Bubble chamber photograph showing two electron-positron pairs. The bubble chamber is located in a constant magnetic field pointing directly out of the page.

If the velocity of a charged particle is parallel (or antiparallel) to the magnetic field, the particle experiences no magnetic force and continues to travel in a straight line.

For motion perpendicular to a magnetic field, as in Figure 27.15, the force required to keep a particle moving with speed v in a circle with radius r is the centripetal force:

$$F = \frac{mv^2}{r}.$$

Setting this expression for the centripetal force equal to that for the magnetic force, we obtain

$$vB|q| = \frac{mv^2}{r}.$$

Rearranging gives an expression for the radius of the circle in which the particle is traveling:

$$r = \frac{mv}{|q|B}. \qquad (27.5)$$

A common way to express this relationship is in terms of the magnitude of the momentum of the particle:

$$Br = \frac{p}{|q|}. \qquad (27.6)$$

If the velocity $\vec{v}$ is not perpendicular to $\vec{B}$, then the velocity component perpendicular to $\vec{B}$ causes circular motion while the parallel component of $\vec{v}$ is unaffected by $\vec{B}$ and drags this orbit into a helical shape.

Time Projection Chamber

Particle physicists create new elementary particles by colliding larger particles at the highest energies. In these collisions, many particles stream away from the interaction point at high speeds. A simple particle detector is not sufficient to identify these particles. A device that can help physicists study these collisions is a time projection chamber (TPC). The STAR TPC was described in Example 22.4 and is shown in Figure 3 of The Big Picture at the beginning of the book.

Figure 27.17 shows collisions of two protons and two gold nuclei that occurred in the center of the STAR TPC. The proton-proton collision creates dozens of particles; the gold-gold collision creates thousands of particles. Each charged particle leaves a track in the TPC. The color assigned by a computer to the track represents the ionization density of the track as particles pass through the gas of the TPC. As they pass through the gas, the particles ionize the atoms of the gas, releasing free electrons. The gas allows the free electrons to drift without recombining with positive ions. Electric fields applied between the center of the TPC and the end caps of the cylinder exert an electric force on the free electrons, making them drift toward the end caps, where they are recorded electronically. Using the drift time and the recording positions, computer software reconstructs the trajectories that the particles took through the TPC. The particles produced in the collisions have a velocity component that is perpendicular to the TPC's magnetic field and thus have circular trajectories.

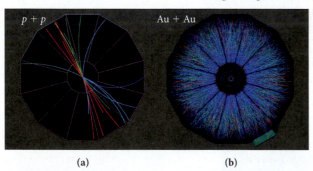

$p + p$ Au + Au

(a)   (b)

FIGURE 27.17 Curved tracks left by the motion of charged particles produced in collisions of (a) two protons, each with kinetic energy 100 GeV, and (b) two gold nuclei, each with kinetic energy of 100 GeV.

EXAMPLE **27.1**   **Transverse Momentum of a Particle in the TPC**

One track of a moving charged particle from Figure 27.17a is shown in Figure 27.18. The radius of the circular trajectory this particle is following is $r = 2.3$ m. The magnitude of the magnetic field in the TPC is $B = 0.50$ T. We can assume that the particle has charge $|q| = 1.602 \cdot 10^{-19}$ C.

PROBLEM

What is the component of the particle's momentum that is perpendicular to the magnetic field?

SOLUTION

We'll call this component the transverse momentum of the particle, p_t. We use equation 27.6, replacing p with p_t, because the magnetic force depends only on p_t, and not on the component of the momentum that is parallel to $\vec{B}$:

$$Br = \frac{p_t}{|q|}.$$

We can express the magnitude of the transverse momentum of the particle in terms of the magnitude of the TPC's magnetic field and the absolute value of the charge of the particle:

$$p_t = |q|Br = \left(1.602 \cdot 10^{-19}\ \text{C}\right)\left(0.50\ \text{T}\right)\left(2.3\ \text{m}\right) = 1.842 \cdot 10^{-19}\ \text{kg m/s}.$$

Instead of these SI units for momentum, particle physicists often use MeV/c (recall Example 7.5). Since $1\ \text{MeV} = 1.602 \cdot 10^{-13}\ \text{J}$, we have

$$p_t c = \left(1.842 \cdot 10^{-19}\ \text{kg m/s}\right)\left(3.0 \cdot 10^{8}\ \text{m/s}\right) = 5.53 \cdot 10^{-11}\ \text{J},$$

or

$$p_t = \left(5.53 \cdot 10^{-11}\ \text{J}\right)/c = 345\ \text{MeV}/c.$$

The analysis of a particle's transverse momentum carried out here can be done by automated computer algorithms on up to approximately 5000 charged particles created in a single collision of two gold nuclei. This very complex task takes about 30 seconds for a computer (3-GHz processor) to finish. In contrast, the TPC can record up to 1000 events per second, which corresponds to 1 millisecond per event.

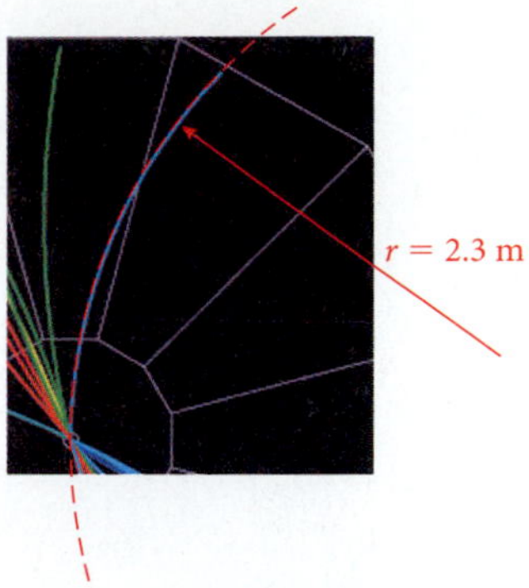

FIGURE 27.18 Circle fitted to the trajectory of one of the charged particles produced from the proton-proton collision in the STAR TPC shown in Figure 27.17a.

$r = 2.3\ \text{m}$

EXAMPLE 27.2 The Solar Wind and Earth's Magnetic Field

Section 27.1 discussed the Van Allen radiation belts that trap particles emitted from the Sun. The Sun throws approximately 1 million tons of matter into space every second. This matter is mostly protons traveling at a speed of around 400 km/s.

PROBLEM

If these protons are incident perpendicular to Earth's magnetic field (which has a magnitude of 50 μT at the Equator), what is the radius of the orbit of the protons? The mass of a proton is $1.67 \cdot 10^{-27}\ \text{kg}$.

SOLUTION

Equation 27.5 relates the magnitude of the magnetic field, B, the radius of a circular orbit, r, and speed, v, of a particle with mass m and charge q traveling perpendicular to a magnetic field:

$$r = \frac{mv}{|q|B}.$$

Putting in the numerical values, we get

$$r = \frac{\left(1.67 \cdot 10^{-27}\ \text{kg}\right)\left(400 \cdot 10^{3}\ \text{m/s}\right)}{\left(1.602 \cdot 10^{-19}\ \text{C}\right)\left(50 \cdot 10^{-6}\ \text{T}\right)} = 83.5\ \text{m}.$$

Thus, the protons of the solar wind orbit around the Earth's magnetic field lines at the Equator in circles with radius of 83.5 m. Protons that are incident on the Earth's magnetic field

Continued—

away from the Equator are not traveling perpendicular to the magnetic field so their orbital radius is larger. However, the magnetic field lines are closer together, meaning that the field is stronger toward the poles. The protons thus spiral along the field lines as they approach the poles. The shape of the Earth's magnetic field forces these protons traveling toward the poles to reverse and travel back toward the Equator, trapping the protons in the Van Allen radiation belts. Thus, the Earth's magnetic field completely blocks the solar wind from reaching the Earth's surface. This is vital, because the blocked cosmic radiation would otherwise make it impossible for higher organisms to live on Earth by ionizing (removing electrons from) atoms and destroying large molecules, for example, DNA.

Cyclotron Frequency

If a particle performs a complete circular orbit inside a uniform magnetic field—for example, like the electrons in the beam shown in Figure 27.15—then the period of revolution, T, of the particle is the circumference of the circle divided by the speed:

$$T = \frac{2\pi r}{v} = \frac{2\pi m}{|q|B}. \tag{27.7}$$

The frequency, f, of the motion of the charged particle is the inverse of the period:

$$f = \frac{1}{T} = \frac{|q|B}{2\pi m}. \tag{27.8}$$

The angular speed, ω, of the motion is

$$\omega = 2\pi f = \frac{|q|B}{m}. \tag{27.9}$$

Thus, the frequency and the angular speed of the particle's motion are independent of the particle's speed and thus independent of the particle's kinetic energy. This fact is used in cyclotrons, which is why ω as given in equation 27.9 is referred to as the **cyclotron frequency.** In a cyclotron, particles are accelerated to higher and higher kinetic energies, and the fact that the cyclotron frequency is independent of the kinetic energy makes designing a cyclotron much easier.

EXAMPLE 27.3 | Energy of a Cyclotron

A cyclotron is a particle accelerator (Figure 27.19). The golden horn-shaped pieces of metal shown in the figure (historically called *dees*) have alternating electric potentials applied to them, so a positively charged particle always has a negatively charged dee ahead when it emerges from under any dee, which is now positively charged. The resulting electric field accelerates the particle. Because the cyclotron sits in a strong magnetic field, the

FIGURE 27.19 (a) Computer-generated drawing of the central section of the K500 superconducting cyclotron at the National Superconducting Cyclotron Laboratory at Michigan State University, with the spiral trajectory of an accelerated particle superimposed. One of the three dees of the cyclotron is highlighted in green. (b) Top view of the K500, showing a proton being accelerated between two dees.

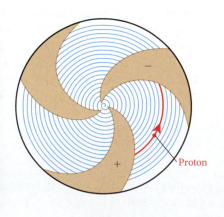

(a)   (b)

particle's trajectory is curved. The radius of the trajectory is proportional to the magnitude of the particle's momentum, according to equation 27.6, so the accelerated particle spirals outward until it reaches the edge of the magnetic field (where its path is no longer bent by the field) and is extracted. According to equation 27.9, the angular frequency is independent of the particle's momentum or energy, so the frequency with which the polarity of the dees is changed does not have to be adjusted as the particle is accelerated. (This holds true only as long as the speed of the accelerated particles does not approach a sizable fraction of the speed of light, as we'll see in Chapter 35 on relativity. To compensate for relativistic effects, the magnetic field of a cyclotron increases with the orbital radius of the accelerated particles.)

PROBLEM

What is the kinetic energy, in mega-electron-volts (MeV), of a proton extracted from a cyclotron with radius $r = 1.81$ m, if the magnetic field of the cyclotron is uniform and has magnitude $B = 0.851$ T? The mass of a proton is $1.67 \cdot 10^{-27}$ kg.

SOLUTION

We can solve equation 27.5 for the speed, v, of the proton:

$$v = \frac{r|q|B}{m}.$$

We substitute this expression for v into the equation for kinetic energy:

$$K = \frac{1}{2}mv^2 = \frac{1}{2}m\left(\frac{r|q|B}{m}\right)^2 = \frac{r^2q^2B^2}{2m}.$$

Putting in the given numbers, we get the kinetic energy in joules:

$$K = \frac{(1.81 \text{ m})^2(1.602 \cdot 10^{-19} \text{ C})^2(0.851 \text{ T})^2}{2(1.67 \cdot 10^{-27} \text{ kg})} = 1.82 \cdot 10^{-11} \text{ J}.$$

Since 1 eV $= 1.602 \cdot 10^{-19}$ J and 1 MeV $= 10^6$ eV, we have

$$K = 1.82 \cdot 10^{-11} \text{ J}\left(\frac{1 \text{ eV}}{1.602 \cdot 10^{-19} \text{ J}}\right)\left(\frac{1 \text{ MeV}}{10^6 \text{ eV}}\right) = 114 \text{ MeV}.$$

Mass Spectrometer

One application of the motion of charged particles in a magnetic field is a **mass spectrometer,** which allows precision determination of atomic and molecular masses and can be useful for carbon dating and the analysis of unknown chemical compounds. A mass spectrometer operates by ionizing the atoms or molecules to be studied and accelerating them through an electric potential. The ions are then passed through a velocity selector (described further in Solved Problem 27.2), which allows only ions with a given velocity to pass through and blocks the remaining ions. The ions then enter a region of constant magnetic field. In the magnetic field, the radius of curvature of the orbit of each ion is given by equation 27.5: $r = mv/|q|B$. Assuming that all the atoms or molecules are singly ionized (have a charge of +1 or –1), the radius of curvature is proportional to the mass of the ion. A schematic diagram of a mass spectrometer is shown in Figure 27.20.

Ions with different masses will have orbits with different radii in the constant magnetic field. For example, in Figure 27.20, ions with orbital radius r_1 have a smaller mass than ions with orbital radius r_2. The particle detector measures the distances from the entrance point, d_1 and d_2, which can be related to the orbital radii and thus the mass of the ions.

27.2 Self-Test Opportunity

A uniform magnetic field is directed out of the page (indicated by the standard "dot within a circle" notation for looking at the arrowhead representing the field lines). A charged particle is traveling in the plane of the page, as shown by the arrows in the figure.

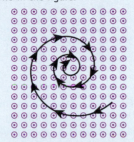

a) Is the charge of the particle positive or negative?

b) Is the particle slowing down, speeding up, or moving at constant speed?

c) Is the magnetic field doing work on the particle?

27.2 In-Class Exercise

Protons in the solar wind travel from the Sun and reach Earth's magnetic field with a speed of 400 km/s. If the magnitude of Earth's magnetic field is $5.0 \cdot 10^{-5}$ T and the velocity of the protons is perpendicular to this magnetic field, what is the cyclotron frequency of the protons in the magnetic field?

a) 122 Hz   d) 432 Hz

b) 233 Hz   e) 763 Hz

c) 321 Hz

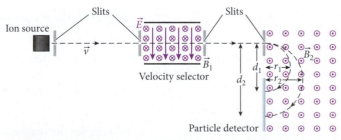

FIGURE 27.20 Schematic diagram of a mass spectrometer showing an ion source, a velocity selector consisting of crossed electric and magnetic fields (see Solved Problem 27.2), a region of constant magnetic field, and a particle detector.

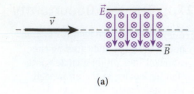

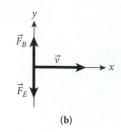

(a)

(b)

FIGURE 27.21 (a) A proton entering a velocity selector, consisting of crossed electric and magnetic fields. (b) Electric and magnetic forces on a proton passing through the combined fields.

SOLVED PROBLEM 27.2 | Velocity Selector

Protons are accelerated from rest through an electric potential difference of $\Delta V = 14.0$ kV. The protons enter a velocity selector, consisting of a parallel plate capacitor in a constant magnetic field, directed perpendicularly into the plane of the page in Figure 27.21a. The electric field between the plates of the parallel plate capacitor is $\vec{E} = 4.30 \cdot 10^5$ V/m, directed along the plane of the page and downward in Figure 27.21a. This arrangement of perpendicular electric and magnetic fields is referred to as *crossed fields*.

PROBLEM
What magnetic field is required for the protons to move through the velocity selector without being deflected?

SOLUTION

THINK
For a proton to move on a straight line without deflection requires that the net force on the proton be zero. Since the proton has a certain velocity and since the magnetic force depends on the velocity, it is plausible that this condition of zero net force cannot be realized for arbitrary speeds of the proton—hence the name *velocity selector*.

SKETCH
Figure 27.21b shows the electric and magnetic forces on the protons as they pass through the velocity selector. Note that the two forces point in opposite directions.

RESEARCH
The change in kinetic energy of the protons plus the change in electric potential energy is equal to zero, which can be expressed as

$$K = -U = \tfrac{1}{2}mv^2 = e\Delta V,$$

where m is the mass of a proton, v is the speed of the proton after acceleration, e is the charge of the proton, and ΔV is the electric potential difference across which the protons were accelerated. The speed of the proton after acceleration is

$$v = \sqrt{\frac{2e\Delta V}{m}}. \tag{i}$$

When the protons enter the velocity selector, the direction of the electric force is in the direction of the electric field, which is downward (negative y-direction). The magnitude of the electric force is

$$F_E = eE, \tag{ii}$$

where E is the magnitude of the electric field in the velocity selector. Right-hand rule 1 gives the direction of the magnetic force: With your thumb in the direction of the velocity of the protons (positive x-direction) and your index finger in the direction of the magnetic field (into the page), your middle finger points up (positive y-direction). Thus, the direction of the magnetic force on the protons is upward. The magnitude of the magnetic force is given by

$$F_B = evB, \tag{iii}$$

where B is the magnitude of the magnetic field in the velocity selector.

SIMPLIFY
The condition that allows the protons to pass though the velocity selector without being deflected is that the electric force balances the magnetic force, or $F_E = F_B$. Using equations (ii) and (iii), we can express this condition as

$$eE = evB.$$

Solving for the magnetic field, B, and substituting for v from equation (i) we obtain

$$B = \frac{E}{\sqrt{\dfrac{2e\Delta V}{m}}} = E\sqrt{\frac{m}{2e\Delta V}}.$$

CALCULATE

Putting in the numerical values gives us

$$B = \left(4.30 \cdot 10^5 \text{ V/m}\right)\sqrt{\frac{1.67 \cdot 10^{-27} \text{ kg}}{2\left(1.602 \cdot 10^{-19} \text{ C}\right)\left(14.0 \cdot 10^3 \text{ V}\right)}} = 0.262371 \text{ T}.$$

ROUND

We report our result to three significant figures:

$$B = 0.262 \text{ T}.$$

DOUBLE-CHECK

We verify that the electric force is equal to the magnetic force. The electric force is

$$F_E = eE = \left(1.602 \cdot 10^{-19} \text{ C}\right)\left(4.30 \cdot 10^5 \text{ V/m}\right) = 6.89 \cdot 10^{-14} \text{ N}.$$

To calculate the magnitude of the magnetic force, we need to find the speed of the protons:

$$v = \sqrt{\frac{2e\Delta V}{m}} = \sqrt{\frac{2\left(1.602 \cdot 10^{-19} \text{ C}\right)\left(14.0 \cdot 10^3 \text{ V}\right)}{1.67 \cdot 10^{-27} \text{ kg}}} = 1.64 \cdot 10^6 \text{ m/s}.$$

This speed is 0.55% of the speed of light, which is not totally impossible. The magnitude of the magnetic force is then

$$F_B = evB = \left(1.602 \cdot 10^{-19} \text{ C}\right)\left(1.64 \cdot 10^6 \text{ m/s}\right)\left(0.262 \text{ T}\right) = 6.88 \cdot 10^{-14} \text{ N},$$

which agrees with the value of the electric force within rounding error. Thus, our result seems reasonable.

Magnetic Levitation

An interesting application of magnetic force is **magnetic levitation,** a situation in which an upward magnetic force on an object balances the downward gravitational force, achieving static equilibrium with no need for direct contact of surfaces. But if you try to balance a magnet over another magnet by orienting the north poles (or south poles) toward each other, you'll see right away that this is not possible. Instead, one of the magnets will simply flip over, and then the opposite poles will point toward each other, and the attractive force between them will cause the two magnets to snap together. As we saw in Chapter 11, a stable equilibrium requires a local minimum of the potential energy, which does not exist for the pure repulsive interaction of two like magnetic poles.

Figure 27.22 shows a commercial toy called the Levitron demonstrating the principle of magnetic levitation. The magnetic top is spun on a plate and then lifted to the proper height and released. The top can remain suspended for several minutes. How does this toy work, considering the requirement for stable equilibrium just mentioned? The answer is that the rapid rotation of the top provides a sufficiently large angular momentum and creates a potential energy barrier that prevents the magnet from flipping over.

Of course, there are other ways to create stable magnetic levitation systems, all of them involving multiple magnets attached rigidly to each other. Magnetic levitation has real-world applications in magnetic levitation (maglev) trains. Thes trains have several advantages over normal steel rail trains: There are no moving parts to wear out, there is less vibration, and reduced friction means that high speeds are possible. Several maglev trains are already in service around the world and more are being planned. One example is the Shanghai Maglev Train (Figure 27.23a), which operates between the Shanghai Pudong Airport and downtown Shanghai and reaches speeds of up to 120 m/s (268 mph).

The Shanghai Maglev Train operates using magnets attached to the cars (Figure 27.23b). These are normal, non-superconducting magnetic coils with electronic feedback to produce

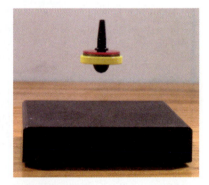

FIGURE 27.22 The Levitron, a commercial toy demonstrating the magnetic levitation of a spinning magnet above a base magnet.

FIGURE 27.23 (a) The Shanghai Maglev Train. (b) Cross section of one side of a train car. The levitation magnets lift the cars 15 cm off the guideway, and the guidance magnets keep the cars centered on the guideway. The magnets are all mounted on the moving vehicle.

(a)

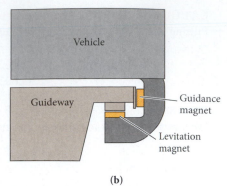

(b)

stable levitation and guidance. The train cars are held 15 cm above the guideway to allow clearance of any objects that may be on the guideway. The levitation and guidance magnets are held at a distance of 10 mm from the guideway, which is constructed of a magnetic material. The propulsion of the train is provided by magnetic fields built into the guideway. The train propulsion system operates like an electric motor (see Section 27.5) whose circular loops have been unwrapped to a linear configuration.

Maglev trains that use superconducting magnets have been tested, but some technical problems have yet to be resolved, including the maintenance of the superconducting coils and the exposure of the passengers to high magnetic fields.

27.4   Magnetic Force on a Current-Carrying Wire

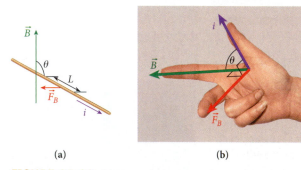

FIGURE 27.24 (a) Magnetic force on a current-carrying wire. (b) A variant of right-hand rule 1 giving the direction of the magnetic force on a current-carrying wire. To determine the direction of the force on a current-carrying wire using your right hand, point your thumb in the direction of the current and your index finger in the direction of the magnetic field; then your middle finger will point in the direction of the force.

Consider a wire carrying a current, i, in a constant magnetic field, $\vec{B}$ (Figure 27.24a). The magnetic field exerts a force on the moving charges in the wire. The charge, q, flowing past a point in the wire in a given time, t, is $q = ti$. During this time, the charge occupies a length, L, of wire given by $L = v_d t$, where v_d is the drift speed (the magnitude of the drift velocity) of the charge carriers in the wire. Thus, we obtain

$$q = ti = \frac{L}{v_d}i. \qquad (27.10)$$

The magnitude of the magnetic force is then

$$F_B = qv_d B \sin\theta = \left(\frac{L}{v_d}i\right)v_d B \sin\theta = iLB \sin\theta, \qquad (27.11)$$

where θ is the angle between the direction of the current flow and the direction of the magnetic field. The direction of the force is perpendicular to both the current and the magnetic field and is given by a variant of right-hand rule 1, with the current in the direction of the velocity of a charged particle, as illustrated in Figure 27.24b. This variant of right-hand rule 1 takes advantage of the fact that current can be thought of as charges in motion.

Equation 27.11 can be expressed as a vector product:

$$\vec{F}_B = i\vec{L} \times \vec{B}, \qquad (27.12)$$

where the notation $i\vec{L}$ represents the current in a length of wire. Equation 27.12 is simply a reformulation of equation 27.2 for the case in which the moving charges make up a current flowing in a wire. Since physical situations involving currents are far more common than those involving the motion of an isolated charged particle, equation 27.12 is the most useful form for determining the magnetic force in practical applications.

EXAMPLE 27.4 | Force on the Voice Coil of a Loudspeaker

A loudspeaker produces sound by exerting a magnetic force on a voice coil in a magnetic field, as shown in Figure 27.25. The movable voice coil is connected to a speaker cone that actually produces the sounds. The magnetic field is produced by the two permanent magnets as shown in Figure 27.25. The magnitude of the magnetic field is $B = 1.50$ T. The voice coil is composed of $n = 100$ turns of wire carrying a current, $i = 1.00$ mA. The diameter of the voice coil is $d = 2.50$ cm.

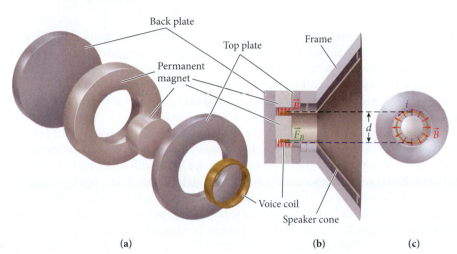

(a)   (b)   (c)

FIGURE 27.25 Schematic diagram of a loudspeaker: (a) an exploded three-dimensional view of the driver end of the loudspeaker; (b) a cross-sectional side view of the loudspeaker; (c) a front view of the driver end of the loudspeaker.

PROBLEM
What is the magnetic force exerted by the magnetic field on the loudspeaker's voice coil?

SOLUTION
The magnitude of the magnetic force on the voice coil is given by equation 27.11:

$$F = iLB\sin\theta,$$

where L is the length of wire carrying current i in the magnetic field with magnitude B. The wire makes an angle θ with the magnetic field. In this case, the wire is always perpendicular to the magnetic field, so $\theta = 90°$. The length of wire in the voice coil is given by the number of turns, n, times the circumference, πd, of each turn

$$L = n\pi d.$$

Thus, the force on the voice coil is

$$F = i(n\pi d)B(\sin 90°) = n\pi idB.$$

Putting in the numerical values, we get

$$F = n\pi idB = (100)(\pi)(1.00 \cdot 10^{-3}\ \text{A})(2.50 \cdot 10^{-2}\ \text{m})(1.50\ \text{T}) = 0.01178\ \text{N} \equiv 0.0118\ \text{N}.$$

From the right-hand rule 1 illustrated in Figure 27.24b, the direction of the force exerted by the magnetic field on the voice coil is toward the left in Figure 27.25b and perpendicularly into the page in Figure 27.25c.

If the current in the voice coil is reversed, the force will be in the opposite direction. If the current is proportional to the amplitude of a sound wave, sound waves can be reproduced in the cone of the loudspeaker. This basic idea is used in most speakers and headphones.

27.5   Torque on a Current-Carrying Loop

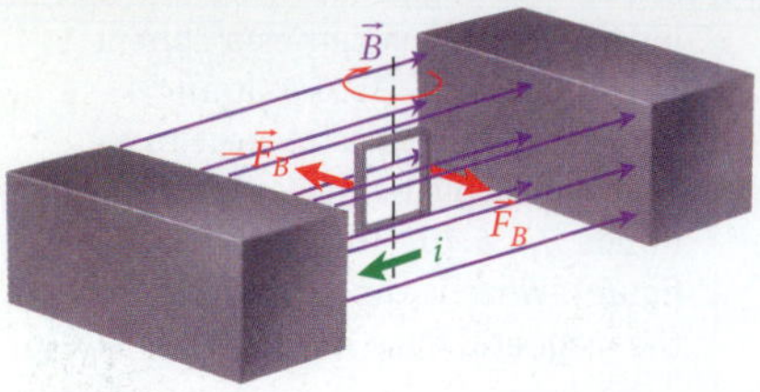

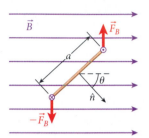

FIGURE 27.26 A primitive element of an electric motor consisting of a current-carrying loop in a magnetic field.

Electric motors rely on the magnetic force exerted on a current-carrying wire. This force is used to create a torque that turns a shaft. Let's consider a simple electric motor, consisting of a single square loop carrying a current, i, in a constant magnetic field, $\vec{B}$. The loop is oriented so that its horizontal sections are parallel to the magnetic field and its vertical sections are perpendicular to the magnetic field, as shown in Figure 27.26. The magnitude of the magnetic force on the two vertical sections of the loop is given by equation 27.11 with $\theta = 90°$:

$$F = iLB.$$

The direction of the magnetic force is given by the variant of right-hand rule 1 illustrated in Figure 27.24b. The two magnetic forces, $\vec{F}_B$ and $-\vec{F}_B$, shown in Figure 27.26 have equal magnitudes and opposite directions. These forces create a torque that tends to rotate the loop around a vertical axis of rotation. These two forces sum to zero. The two horizontal sections of the loop are parallel to the magnetic field and thus experience no magnetic force. Thus, no net force acts on the coil, even though a torque is produced.

FIGURE 27.27 Top view of a current-carrying loop in a magnetic field, showing the forces acting on the loop.

Now we consider the case where the loop rotates about its center. As the loop turns in the magnetic field, the forces on the vertical sides of the loop, perpendicular to the direction of the field, do not change. The forces on the square loop, with side length a, are illustrated in Figure 27.27, which shows a top view of the coil. In Figure 27.27, θ is the angle between a unit vector, $\hat{n}$, normal to the plane of the coil, and the magnetic field, $\vec{B}$. The unit normal vector is perpendicular to the plane of the wire loop and points in a direction given by right-hand rule 2 (Figure 27.28), based on the current flowing around the loop.

In Figure 27.27, the current is flowing out of the page in the right side of the loop, indicated by the point in a circle (representing the tip of an arrow) and flowing into the page in the left side of the loop, indicated by the cross in a circle (representing the tail of an arrow). The magnitude of the force on each of these vertical segments is

$$F = iaB.$$

The forces on the two horizontal segments of the loop are parallel or antiparallel to the axis of rotation and do not cause a torque, and these two forces sum to zero. Therefore, there is no net force on the loop.

The sum of the torques on the two vertical segments of the loop gives the net torque exerted on the loop about its center:

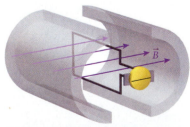

FIGURE 27.28 Right-hand rule 2 gives the direction of the unit normal vector for a current-carrying loop. According to the rule, if you curl the fingers of your right hand in the direction of the current in the loop, your thumb points in the direction of the unit normal vector.

$$\tau_1 = \left(iaB\right)\left(\frac{a}{2}\right)\sin\theta + \left(iaB\right)\left(\frac{a}{2}\right)\sin\theta = ia^2 B\sin\theta = iAB\sin\theta, \quad (27.13)$$

where the index 1 on τ_1 indicates that it is the torque on a single loop and $A = a^2$ is the area of the loop. The reason that the loop continues to rotate and doesn't stop at $\theta = 0°$ is that it is connected to a device called a **commutator,** which causes the current to change directions as the coil rotates. This commutator consists of a split ring, with one end of the loop connected to each half of the loop, as shown in Figure 27.29. The current in the loop switches direction two times for every complete rotation of the loop.

If the single loop is replaced with a coil consisting of many loops wound closely together, the torque on the coil is found by multiplying the torque on a loop, τ_1 from equation 27.13, by the number of windings (loops in the coil), N:

$$\tau = N\tau_1 = NiAB\sin\theta. \quad (27.14)$$

Does this expression for the torque hold for other shapes with area A, other than squares? The answer is yes.

27.4   In-Class Exercise

A coil is composed of circular loops of radius $r = 5.13$ cm and has $N = 47$ windings. A current, $i = 1.27$ A, flows through the coil, which is inside a homogeneous magnetic field of strength 0.911 T. What is the maximum torque on the coil due to the magnetic field?

a) 0.148 N m   b) 0.211 N m   c) 0.350 N m   d) 0.450 N m   e) 0.622 N m

FIGURE 27.29 A wire loop connected to a source of current through a commutator ring.

27.6 Magnetic Dipole Moment

A current-carrying coil can be described with one parameter, which contains information about a key characteristic of the coil in a magnetic field. The magnitude of the **magnetic dipole moment**, $\vec{\mu}$, of a current-carrying coil of wire is defined to be

$$\mu = NiA \qquad (27.15)$$

where N is the number of windings, i is the current through the wire, and A is the area of the loops. The direction of the magnetic dipole moment is given by right-hand rule 2 and is the direction of the unit normal vector, $\hat{n}$. Using equation 27.15, we can rewrite equation 27.14 as

$$\tau = \left(NiA\right)B\sin\theta = \mu B\sin\theta. \qquad (27.16)$$

The torque on a magnetic dipole is given by

$$\vec{\tau} = \vec{\mu} \times \vec{B}. \qquad (27.17)$$

That is, the torque on a current-carrying coil is the vector product of the magnetic dipole moment of the coil and the magnetic field.

A magnetic dipole has potential energy in an external magnetic field. If the dipole moment is aligned with the magnetic field, the dipole has its minimum potential energy. If the dipole moment is oriented in a direction opposite to the external field, the dipole has its maximum potential energy. From Chapter 10, the work done by a torque is

$$W = \int_{\theta_0}^{\theta} \tau\left(\theta'\right)d\theta'. \qquad (27.18)$$

Using the work-energy theorem and equation 27.16 and setting $\theta_0 = 90°$, we can express the magnetic potential energy, U, of a magnetic dipole in an external magnetic field, $\vec{B}$, as

$$W = \int_{\theta_0}^{\theta} \tau\left(\theta'\right)d\theta' = \int_{\theta_0}^{\theta} \mu B\sin\theta'\,d\theta' = -\mu B\cos\theta\Big|_{\theta_0}^{\theta} = U\left(\theta\right) - U\left(90°\right),$$

or

$$U\left(\theta\right) = -\mu B\cos\theta = -\vec{\mu}\cdot\vec{B}, \qquad (27.19)$$

where θ is the angle between the magnetic dipole moment and the external magnetic field.

The lowest value, $-\mu B$, of the potential energy of a magnetic dipole in an external magnetic field is achieved when the dipole's magnetic moment vector is parallel to the external magnetic field vector, and the highest value, $+\mu B$, results when the two vectors are antiparallel (see Figure 27.30). This dependence of potential energy on orientation occurs in diverse physical situations, for which magnetic dipoles in external magnetic fields are a simple model. So far, the only magnetic dipoles we have discussed are current-carrying loops. However, other types of magnetic dipoles exist, including bar magnets and even the Earth. In addition, elementary charged particles such as protons have intrinsic magnetic dipole moments.

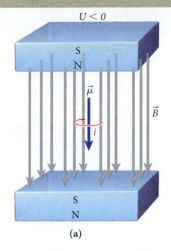

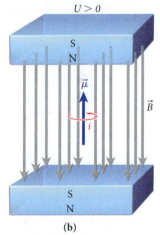

FIGURE 27.30 Magnetic dipole moment vector in an external magnetic field: (a) magnetic dipole and external magnetic field are parallel, resulting in a negative potential energy; b) magnetic dipole and external magnetic field are antiparallel, resulting in a positive potential energy.

27.3 Self-Test Opportunity

What is the maximum difference in magnetic potential energy between two orientations of a loop with area 0.100 m² carrying a current of 2.00 A in a constant magnetic field of magnitude 0.500 T?

27.7 Hall Effect

Consider a conductor carrying a current, i, flowing in a direction perpendicular to a magnetic field, $\vec{B}$ (Figure 27.31a). The electrons in the conductor are moving with velocity $\vec{v}_d$ in the direction opposite to the current. The moving electrons experience a force perpendicular to their velocity, causing them to move toward one edge of the conductor. After some time, many electrons have moved to one edge of the conductor, creating a net negative charge on that edge and leaving a net positive charge on the opposite edge of the conductor. This charge distribution creates an electric field, $\vec{E}$, which exerts a force on the electrons in

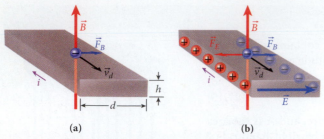

(a)   **(b)**

FIGURE 27.31 (a) A conductor carrying a current in a magnetic field. The charge carriers are electrons. (b) The electrons have drifted to one side of the conductor, leaving a net positive charge on the opposite side. This distribution of charges creates an electric field. The potential difference across the conductor is the Hall potential difference.

a direction opposite to that exerted by the magnetic field. When the magnitude of the force exerted on the electrons by the electric field is equal to the magnitude of the force exerted on them by the magnetic field, the net number of electrons on the edges of the conductor no longer changes with time. This result is called the **Hall effect.** The potential difference, ΔV_H, between the edges of the conductor when equilibrium is reached is termed the **Hall potential difference,** given by

$$\Delta V_H = Ed, \qquad (27.20)$$

where d is the width of the conductor and E is the magnitude of the created electric field. (See Chapter 23 for the relationship between the electric potential difference and the constant electric field.)

The Hall effect can be used to demonstrate that the charge carriers in metals are negatively charged. If the charge carriers in a metal were positive and moving in the direction of the current shown in Figure 27.31a, those positive charges would collect on the same edge of the conductor as the electrons in Figure 27.31b, giving an electric field with the opposite sign. Thus, the charge carriers in conductors are negatively charged and must be electrons. The Hall effect also establishes that in some semiconductors the charge carriers are electron holes (missing electrons), which appear to be positively charged carriers.

The Hall effect can also be used to determine a magnetic field by measuring the current flowing through the conductor and the resulting electric field across the conductor. To obtain the formula for the magnetic field, we start with the equilibrium condition of the Hall effect, that the magnitudes of the magnetic and electric forces are equal:

$$F_E = F_B \Rightarrow eE = v_d Be \Rightarrow B = \frac{E}{v_d} = \frac{\Delta V_H}{v_d d}, \qquad (27.21)$$

where substitution for E from equation 27.20 is used in the last step. In Chapter 25, we saw that the drift speed, v_d, of an electron in a conductor can be related to the magnitude of the current density, J, in the conductor:

$$J = \frac{i}{A} = nev_d,$$

where A is the cross-sectional area of the conductor and n is the number of electrons per unit volume in the conductor. As shown in Figure 27.31a, the cross-sectional area is given by $A = dh$, where d is the width and h is the height of the conductor. Solving $i/A = nev_d$ for the drift speed and substituting hd for A gives

$$v_d = \frac{i}{Ane} = \frac{i}{hdne}.$$

Substituting this expression for v_d into equation 27.21, we have

$$B = \frac{\Delta V_H}{v_d d} = \frac{\Delta V_H dhne}{id} = \frac{\Delta V_H hne}{i}. \qquad (27.22)$$

Thus, equation 27.22 gives the magnetic field strength (magnitude) from a measured value of the Hall potential difference, ΔV_H, and the known height, h, and density of charge carriers, n, of the conductor. Equivalently, a rearranged form of equation 27.22 can be used to find the Hall voltage if the magnetic field strength is known:

$$\Delta V_H = \frac{iB}{neh}. \qquad (27.23)$$

EXAMPLE 27.5 | Hall Effect

Suppose we use a Hall probe to measure the magnitude of a constant magnetic field. The Hall probe is a strip of copper with a height, h, of 2.00 mm. We measure a voltage of 0.250 μV across the probe when we run a current of 1.25 A through it.

PROBLEM
What is the magnitude of the magnetic field?

SOLUTION
The magnetic field is given by equation 27.22:

$$B = \frac{\Delta V_H h n e}{i}.$$

We have been given the values of V_H, h, and i, and we know e. The density of the electrons, n, is defined as the number of electrons per unit volume:

$$n = \frac{\text{number of electrons}}{\text{volume}}.$$

The density of copper is $\rho_{Cu} = 8.96 \text{ g/cm}^3 = 8960 \text{ kg/m}^3$, and 1 mole of copper has a mass of 63.5 g and $6.02 \cdot 10^{23}$ atoms. Each copper atom has one conduction electron. Thus, the density of the electrons is

$$n = \left(\frac{1 \text{ electron}}{1 \text{ atom}}\right)\left(\frac{6.02 \cdot 10^{23} \text{ atoms}}{63.5 \text{ g}}\right)\left(\frac{8.96 \text{ g}}{1 \text{ cm}^3}\right)\left(\frac{1.0 \cdot 10^6 \text{ cm}^3}{1 \text{ m}^3}\right) = 8.49 \cdot 10^{28} \frac{\text{electrons}}{\text{m}^3}.$$

We can now calculate the magnitude of the magnetic field:

$$B = \frac{\left(0.250 \cdot 10^{-6} \text{ V}\right)\left(0.002 \text{ m}\right)\left(8.49 \cdot 10^{28} \frac{\text{electrons}}{\text{m}^3}\right)\left(1.602 \cdot 10^{-19} \text{ C}\right)}{1.25 \text{ A}} = 5.44 \text{ T}.$$

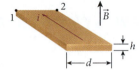
WHAT WE HAVE LEARNED | EXAM STUDY GUIDE

- Magnetic field lines indicate the direction of a magnetic field in space. Magnetic field lines do not end on magnetic poles, but form closed loops instead.

- The magnetic force on a particle with charge q moving with velocity $\vec{v}$ in a magnetic field, $\vec{B}$, is given by $\vec{F} = q\vec{v} \times \vec{B}$. Right-hand rule 1 gives the direction of the force.

- For a particle with charge q moving with speed v perpendicular to a magnetic field of magnitude B, the magnitude of the magnetic force on the moving charged particle is $F = |q|vB$.

- The unit of magnetic field is the tesla, abbreviated T.

- The average magnitude of the Earth's magnetic field at the surface is approximately $0.5 \cdot 10^{-4}$ T.

- A particle with mass m and charge q moving with speed v perpendicular to a magnetic field with magnitude B has a trajectory that is a circle with radius $r = mv / |q|B$.

- The cyclotron frequency, ω, of a particle with charge q and mass m moving in a circular orbit in a constant magnetic field of magnitude B is given by $\omega = |q|B/m$.

- The force exerted by a magnetic field, $\vec{B}$, on a length of wire, $\vec{L}$, carrying a current, i, is given by $\vec{F} = i\vec{L} \times \vec{B}$. The magnitude of this force is $F = iLB \sin \theta$, where θ is the angle between the direction of the current and the direction of the magnetic field.

- The magnitude of the torque on a loop carrying a current, i, in a magnetic field with magnitude B is $\tau = iAB \sin \theta$, where A is the area of the loop and θ is the angle between a unit vector normal to the loop and the direction of the magnetic field. Right-hand rule 2 gives the direction of the unit normal vector to the loop.

- The magnitude of the magnetic dipole moment of a coil carrying a current, i, is given by $\mu = NiA$, where N is the number of loops (windings) and A is the area of a loop. The direction of the dipole moment is given by right-hand rule 2 and is the direction in which the unit normal vector points.

- The Hall effect results when a current, i, flowing through a conductor with height h in a magnetic field of magnitude B produces a potential difference across the conductor (the Hall potential difference), given by $\Delta V_H = iB/neh$, where n is the density of electrons per unit volume and e is the magnitude of charge of an electron.

KEY TERMS

permanent magnets, p. 865
north magnetic pole, p. 865
south magnetic pole, p. 865
magnetic field, p. 866
magnetic field lines, p. 866

Van Allen radiation belts, p. 866
aurora borealis, p. 867
aurora australis, p. 867
magnetic declination, p. 867
tesla, p. 869

cyclotron frequency, p. 874
mass spectrometer, p. 875
magnetic levitation, p. 877
commutator, p. 880
magnetic dipole moment, p. 881

Hall effect, p. 882
Hall potential difference,
    p. 882

NEW SYMBOLS AND EQUATIONS

$\vec{B}$, magnetic field

$\vec{F}_B = q\vec{v} \times \vec{B}$, magnetic force on a charged particle

$\vec{F}_B = i\vec{L} \times \vec{B}$, magnetic force on a current-carrying wire

$\vec{\mu}$, magnetic dipole moment

$\Delta V_H = iB/neh$, Hall potential difference

ANSWERS TO SELF-TEST OPPORTUNITIES

27.1 Particle 1: $F_B = qvB \sin\theta =$
$(6.15 \cdot 10^{-6}\text{ C})(465\text{ m/s})(0.165\text{ T})(\sin 30.0°) = 2.36 \cdot 10^{-4}\text{ N}.$

Particle 2: $F_B = qvB \sin\theta =$
$(6.15 \cdot 10^{-6}\text{ C})(465\text{ m/s})(0.165\text{ T})(\sin 90.0°) = 4.72 \cdot 10^{-4}\text{ N}.$

Particle 3: $F_B = qvB \sin\theta =$
$(6.15 \cdot 10^{-6}\text{ C})(465\text{ m/s})(0.165\text{ T})(\sin 150.0°) = 2.36 \cdot 10^{-4}\text{ N}$

27.2 a) positive

b) slowing down

c) no (Therefore, another force must be
acting on the particle to slow it down.)

27.3 $\Delta U = U_{max} - U_{min} = 2\mu B = 2iAB =$
$2(2.00\text{ A})(0.100\text{ m}^2)(0.500\text{ T}) = 0.200\text{ J}.$

PROBLEM-SOLVING PRACTICE

Problem-Solving Guidelines

1. When working with magnetic fields and forces, you need to sketch a clear diagram of the problem situation in three dimensions. Often, a separate sketch of the velocity and magnetic field vectors (or the length and field vectors) is useful to visualize the plane in which they lie, since the magnetic force will be perpendicular to that plane.

2. Remember that the right-hand rules apply for positive charges and currents. If a charge or a current is negative, you can use the right-hand rule but the force will then be in the opposite direction.

3. A particle in both an electric and a magnetic field experiences an electric force, $\vec{F}_E = q\vec{E}$, and a magnetic force, $\vec{F}_B = q\vec{v} \times \vec{B}$. Be sure you take the vector sum of the individual forces.

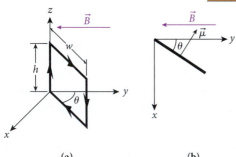

(a)   **(b)**

FIGURE 27.32 (a) A rectangular loop carrying a current in a magnetic field. (b) View of the rectangular loop looking down on the xy-plane. The magnetic dipole moment is perpendicular to the plane of the loop, with a direction determined by right-hand rule 2.

SOLVED PROBLEM **27.3** | **Torque on a Rectangular Current-Carrying Loop**

A rectangular loop with height $h = 6.50$ cm and width $w = 4.50$ cm is in a uniform magnetic field of magnitude $B = 0.250$ T, which points in the negative y-direction (Figure 27.32a). The loop makes an angle of $\theta = 33.0°$ with the y-axis, as shown in the figure. The loop carries a current of magnitude $i = 9.00$ A in the direction indicated by the arrows.

PROBLEM

What is the magnitude of the torque on the loop around the z-axis?

SOLUTION

THINK

The torque on the loop is equal to the vector cross product of the magnetic dipole moment and the magnetic field. The magnetic dipole moment is perpendicular to the plane of the loop, with the direction given by right-hand rule 2.

SKETCH

Figure 27.32b is a view of the loop looking down on the xy-plane.

RESEARCH

The magnitude of the magnetic dipole moment of the loop is

$$\mu = NiA = iwh. \qquad (i)$$

The magnitude of the torque on the loop is

$$\tau = \mu B \sin\theta_{\mu B}, \qquad (ii)$$

where $\theta_{\mu B}$ is the angle between the magnetic dipole moment and the magnetic field. From Figure 27.32b we can see that

$$\theta_{\mu B} = \theta + 90°. \qquad (iii)$$

SIMPLIFY

We can combine equations (i), (ii), and (iii) to obtain

$$\tau = iwhB \sin\left(\theta + 90°\right).$$

CALCULATE

Putting in the numerical values, we get

$$\tau = \left(9.00 \text{ A}\right)\left(4.50 \cdot 10^{-2} \text{ m}\right)\left(6.50 \cdot 10^{-2} \text{ m}\right)\left(0.250 \text{ T}\right)\left[\sin\left(33.0° + 90°\right)\right]$$

$$= 0.0055195 \text{ N m}.$$

ROUND

We report our result to three significant figures:

$$\tau = 5.52 \cdot 10^{-3} \text{ N m}.$$

DOUBLE-CHECK

The magnitude of the force on each of the vertical segments of the loop is

$$F_B = ihB = \left(9.00 \text{ A}\right)\left(6.50 \cdot 10^{-2} \text{ m}\right)\left(0.250 \text{ T}\right) = 0.146 \text{ N}.$$

The magnitude of the torque is then the magnitude of the force on the vertical segment that is not along the z-axis times the moment arm (which is w) times the sine of the angle between the force and the moment arm:

$$\tau = Fw \sin\left(33.0° + 90°\right) = 0.146 \text{ N}\left(4.5 \cdot 10^{-2} \text{ m}\right)\left[\sin\left(33.0° + 90°\right)\right] = 5.52 \cdot 10^{-3} \text{ N m}.$$

This is the same as the result calculated above.

MULTIPLE-CHOICE QUESTIONS

27.1 A magnetic field is oriented in a certain direction in a horizontal plane. An electron moves in a certain direction in the horizontal plane. For this situation, there

a) is one possible direction for the magnetic force on the electron.

b) are two possible directions for the magnetic force on the electron.

c) are infinite possible directions for the magnetic force on the electron.

27.2 A particle with charge q is at rest when a magnetic field is suddenly turned on. The field points in the z-direction. What is the direction of the net force acting on the charged particle?

a) in the x-direction

b) in the y-direction

c) The net force is zero.

d) in the z-direction

27.3 Which of the following has the largest cyclotron frequency?

a) an electron with speed v in a magnetic field with magnitude B

b) an electron with speed $2v$ in a magnetic field with magnitude B

c) an electron with speed $v/2$ in a magnetic field with magnitude B

d) an electron with speed $2v$ in a magnetic field with magnitude $B/2$

e) an electron with speed $v/2$ in a magnetic field with magnitude $2B$

27.4 In the Hall effect, a potential difference produced across a conductor of finite thickness in a magnetic field by a current flowing through the conductor is given by

a) the product of the density of electrons, the charge of an electron, and the conductor's thickness divided by the product of the magnitudes of the current and the magnetic field.

b) the reciprocal of the expression described in part (a).

c) the product of the charge on an electron and the conductor's thickness divided by the product of the density of electrons and the magnitudes of the current and the magnetic field.

d) the reciprocal of the expression described in (c).

e) none of the above.

27.5 An electron (with charge $-e$ and mass m_e) moving in the positive x-direction enters a velocity selector. The velocity selector consists of crossed electric and magnetic fields: $\vec{E}$ is directed in the positive y-direction, and $\vec{B}$ is directed in the positive z-direction. For a velocity v (in the positive x-direction), the net force on the electron is zero, and the electron moves straight through the velocity selector. With what velocity will a proton (with charge $+e$ and mass $m_p = 1836\, m_e$) move straight through the velocity selector?

a) v

b) $-v$

c) $v/1836$

d) $-v/1836$

27.6 In which direction does a magnetic force act on an electron that is moving in the positive x-direction in a magnetic field pointing in the positive z-direction?

a) the positive y-direction

b) the negative y-direction

c) the negative x-direction

d) any direction in the xy-plane

27.7 A charged particle is moving in a constant magnetic field. State whether each of the following statements concerning the magnetic force exerted on the particle is true or false? (Assume that the magnetic field is not parallel or antiparallel to the velocity.)

a) It does no work on the particle.

b) It may increase the speed of the particle.

c) It may change the velocity of the particle.

d) It can act only on the particle while the particle is in motion.

e) It does not change the kinetic energy of the particle.

27.8 An electron moves in a circular trajectory with radius r_i in a constant magnetic field. What is the final radius of the trajectory when the magnetic field is doubled?

a) $\dfrac{r_i}{4}$

b) $\dfrac{r_i}{2}$

c) r_i

d) $2r_i$

e) $4r_i$

QUESTIONS

27.9 Draw on the xyz-coordinate system and specify (in terms of the unit vectors $\hat{x}$, $\hat{y}$, and $\hat{z}$) the direction of the magnetic force on each of the moving particles shown in the figures.
Note: The positive y-axis is toward the right, the positive z-axis is toward the top of the page, and the positive x-axis is directed out of the page.

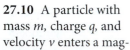

27.10 A particle with mass m, charge q, and velocity v enters a magnetic field of magnitude B and with direction perpendicular to the initial velocity of the particle. What is the work done by the magnetic field on the particle? How does this affect the particle's motion?

27.11 An electron is moving with a constant velocity. When it enters an electric field that is perpendicular to its velocity, the electron will follow a _____ trajectory. When the electron enters a magnetic field that is perpendicular to its velocity, it will follow a _____ trajectory.

27.12 A proton, moving in negative y-direction in a magnetic field, experiences a force of magnitude F, acting in the negative x-direction.

a) What is the direction of the magnetic field producing this force?

b) Does your answer change if the word "proton" in the statement is replaced by "electron"?

27.13 It would be mathematically possible, for a region with zero current density, to define a scalar magnetic potential analogous to the electrostatic potential:

$$V_B(\vec{r}) = -\int_{\vec{r}_0}^{\vec{r}} \vec{B} \cdot d\vec{s}, \text{ or } \vec{B}(\vec{r}) = -\nabla V_B(\vec{r}).$$ However, this has not been done. Explain why not.

27.14 A current-carrying wire is positioned within a large, uniform magnetic field, $\vec{B}$. However, the wire experiences no force. Explain how this might be possible.

27.15 A charged particle moves under the influence of an electric field only. Is it possible for the particle to move with a constant speed? What if the electric field is replaced with a magnetic field?

27.16 A charged particle travels with speed v, at an angle θ with respect to the z-axis. It enters at time $t = 0$ a region of space where there is a magnetic field of magnitude B in the positive z-direction. When does it emerge from this region of space?

27.17 An electron is traveling horizontally from the northwest toward the southeast in a region of space where the Earth's magnetic field is directed horizontally toward the north. What is the direction of the magnetic force on the electron?

27.18 At the Earth's surface, there is an electric field that points approximately straight down and has magnitude 150 N/m. Suppose you had a tuneable electron gun (you can release electrons with whatever kinetic energy you like) and a detector to determine the direction of motion of the electrons when they leave the gun. Explain how you could use the gun to find the direction toward the north magnetic pole. Specifically, what kinetic energy would the electrons need to have? (*Hint:* It might be easier to think about finding which direction is east or west.)

27.19 The work done by the magnetic field on a charged particle in motion in a cyclotron is zero. How, then, can a cyclotron be used as a particle accelerator, and what essential feature of the particle's motion makes it possible?

PROBLEMS

A blue problem number indicates a worked-out solution is available in the Student Solutions Manual. One • and two •• indicate increasing level of problem difficulty.

Section 27.2

27.20 A proton moving with a speed of $4.0 \cdot 10^5$ m/s in the positive y-direction enters a uniform magnetic field of 0.40 T pointing in the positive x-direction. Calculate the magnitude of the force on the proton.

27.21 The magnitude of the magnetic force on a particle with charge $-2e$ moving with speed $v = 1.0 \cdot 10^5$ m/s is $3.0 \cdot 10^{-18}$ N. What is the magnitude of the magnetic field component perpendicular to the direction of motion of the particle?

•27.22 A particle with a charge of $+10$ μC is moving at 300 m/s in the positive z-direction.

a) Find the minimum magnetic field required to keep it moving in a straight line at constant speed if there is a uniform electric field of magnitude 100 V/m pointing in the positive y-direction.

b) Find the minimum magnetic field required to keep the particle moving in a straight line at constant speed if there is a uniform electric field of magnitude 100 V/m pointing in the positive z-direction.

•27.23 A particle with a charge of 20 μC moves along the x-axis with a speed of 50 m/s. It enters a magnetic field given by $\vec{B} = 0.30\,\hat{y} + 0.70\,\hat{z}$, in teslas. Determine the magnitude and the direction of the magnetic force on the particle.

••27.24 The magnetic field in a region in space (where $x > 0$ and $y > 0$) is given by $\vec{B} = (x - az)\,\hat{y} + (xy - b)\,\hat{z}$, where a and b are positive constants. An electron moving with a constant velocity, $\vec{v} = v_0\,\hat{x}$, enters this region. What are the coordinates of the points at which the net force acting on the electron is zero?

Section 27.3

27.25 A proton is accelerated from rest by a potential difference of 400 V. The proton enters a uniform magnetic field and follows a circular path of radius 20 cm. Determine the magnitude of the magnetic field.

27.26 An electron with a speed of $4.0 \cdot 10^5$ m/s enters a uniform magnetic field of magnitude 0.040 T at an angle of 35° to the magnetic field lines. The electron will follow a helical path.

a) Determine the radius of the helical path.

b) How far forward will the electron have moved after completing one circle?

27.27 A particle with mass m and charge q is moving within both an electric field and a magnetic field, $\vec{E}$ and $\vec{B}$. The particle has velocity $\vec{v}$, momentum $\vec{p}$, and kinetic energy, K. Find general expressions for $d\vec{p}/dt$ and dK/dt, in terms of these seven quantities.

27.28 The Earth is showered with particles from space known as muons. They have a charge identical to that of an electron but are many times heavier ($m = 1.88 \cdot 10^{-28}$ kg). Suppose a strong magnetic field is established in a lab ($B = 0.50$ T) and a muon enters this field with a velocity of $3.0 \cdot 10^6$ m/s at a right angle to the field. What will be the radius of the resulting orbit of the muon?

27.29 An electron in a magnetic field moves counterclockwise on a circle in the xy-plane, with a cyclotron frequency of $\omega = 1.2 \cdot 10^{12}$ Hz. What is the magnetic field, $\vec{B}$?

27.30 An electron with energy equal to $4.00 \cdot 10^2$ eV and an electron with energy equal to $2.00 \cdot 10^2$ eV are trapped in a uniform magnetic field and move in circular paths in a plane perpendicular to the magnetic field. What is the ratio of the radii of their orbits?

•27.31 A proton with an initial velocity given by $(1.0\hat{x} + 2.0\hat{y} + 3.0\hat{z})(10^5$ m/s) enters a magnetic field given by $(0.50$ T$)\hat{z}$. Describe the motion of the proton.

•27.32 Initially at rest, a small copper sphere with a mass of $3 \cdot 10^{-6}$ kg and a charge of $5 \cdot 10^{-4}$ C is accelerated through a 7000-V potential difference before entering a magnetic field of magnitude 4 T, directed perpendicular to its velocity. What is the radius of curvature of the sphere's motion in the magnetic field?

•27.33 Two particles with masses m_1 and m_2 and charges q and $2q$ travel with the same velocity, v, and enter a magnetic field of strength B at the same point, as shown in the figure. In the magnetic field, they move in semicircles with radii R and $2R$. What is the ratio of their masses? Is it possible to apply an electric field that would cause the particles to move in a straight line in the magnetic field? If yes, what would be the magnitude and direction of the field?

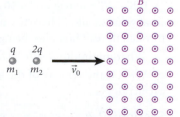

•27.34 The figure shows a schematic diagram of a simple mass spectrometer, consisting of a velocity selector and a particle detector and being used to separate singly ionized atoms ($q = +e = 1.6 \cdot 10^{-19}$ C) of gold (Au) and molybdenum (Mo). The electric field inside the velocity selector has magnitude $E = 1.789 \cdot 10^4$ V/m and points toward the top of the page, and the magnetic field has magnitude $B_1 = 1.00$ T and points out of the page.

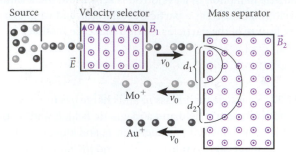

a) Draw the electric force vector, $\vec{F}_E$, and the magnetic force vector, $\vec{F}_B$, acting on the ions inside the velocity selector.

b) Calculate the velocity, v_0, of the ions that make it through the velocity selector (those that travel in a straight line). Does v_0 depend on the type of ion (gold versus molybdenum), or is it the same for both types of ions?

c) Write the equation for the radius of the semicircular path of an ion in the particle detector: $R = R(m, v_0, q, B_2)$.

d) The gold ions (represented by the black circles) exit the particle detector at a distance $d_2 = 40.00$ cm from the entrance slit, while the molybdenum ions (represented by the gray circles) exit the particle detector at a distance $d_1 = 19.81$ cm from the entrance slit. The mass of a gold ion is $m_{gold} = 3.27 \cdot 10^{-25}$ kg. Calculate the mass of a molybdenum ion.

••**27.35** A small particle accelerator for accelerating $^3\text{He}^+$ ions is shown in the figure. The $^3\text{He}^+$ ions exit the ion source with a kinetic energy of 4 keV. Regions 1 and 2 contain magnetic fields directed into the page, and region 3 contains an electric field directed from left to right. The $^3\text{He}^+$ ion beam exits the accelerator from a hole on the right that is 7 cm below the ion source, as shown in the figure.

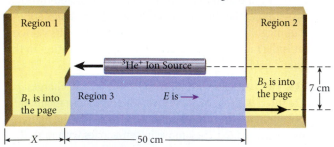

a) If $B_1 = 1$T and region 3 is 50 cm long with $E = 60$ kV/m, what value should B_2 have to cause the ions to move straight through the exit hole after being accelerated twice in region 3?

b) What minimum width X should region 1 have?

c) What is the velocity of the ions when they leave the accelerator?

Section 27.4

27.36 A straight wire of length 2.0 m carries a current of 24 A. It is placed on a horizontal tabletop in a uniform horizon-

tal magnetic field. The wire makes an angle of 30° with the magnetic field lines. If the magnitude of the force on the wire is 0.50 N, what is the magnitude of the magnetic field?

27.37 As shown in the figure, a straight conductor parallel to the x-axis can slide without friction on top of two horizontal conducting rails that are parallel to the y-axis and a distance of $L = 0.2$ m apart, in a vertical magnetic field of 1 T. A 20-A current is maintained through the conductor. If a string is connected exactly at the center of the conductor and passes over a frictionless pulley, what mass m suspended from the string allows the conductor to be at rest?

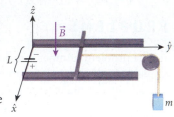

27.38 A copper wire of radius 0.5 mm is carrying a current at the Earth's Equator. Assuming that the magnetic field of the Earth has magnitude 0.5 G at the Equator and is parallel to the surface of the Earth and that the current in the wire flows toward the east, what current is required to allow the wire to levitate?

•**27.39** A copper sheet with length 1.0 m, width 0.50 m, and thickness 1.0 mm is oriented so that its largest surface area is perpendicular to a magnetic field of strength 5.0 T. The sheet carries a current of 3.0 A across its length. What is the magnitude of the force on this sheet? How does this magnitude compare to that of the force on a thin copper wire carrying the same current and oriented perpendicularly to the same magnetic field?

•**27.40** A conducting rod of length L slides freely down an inclined plane, as shown in the figure. The plane is inclined at an angle θ from the horizontal. A uniform magnetic field of strength B acts in the positive y-direction. Determine the magnitude and the direction of the current that would have to be passed through the rod to hold it in position on the inclined plane.

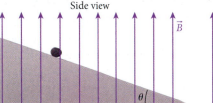

Side view   Front view

•**27.41** A square loop of wire, with side length $d = 8.0$ cm, carries a current of magnitude $i = 0.15$ A and is free to rotate. It is placed between the poles of an electromagnet that produce a uniform magnetic field of 1.0 T. 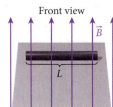 The loop is initially placed so that its normal vector, $\hat{n}$, is at a 35.0° angle relative to the direction of the magnetic field vector, with the angle θ defined as shown in the figure. The wire is copper (with a density of $\rho = 8960$ kg/m^3), and its diameter is 0.50 mm. What is the magnitude of the initial angular acceleration of the loop when it is released?

•**27.42** A rail gun accelerates a projectile from rest by using the magnetic force on a current-carrying wire. The wire has radius $r = 5.1 \cdot 10^{-4}$ m and is made of copper having a density of $\rho = 8960$ kg/m³. The gun consists of rails of length $L = 1.0$ m in a constant magnetic field of magnitude $B = 2.0$ T, oriented perpendicular to the plane defined by the rails. The wire forms an electrical connection across the rails at one end of the rails. When triggered, a current of $1.00 \cdot 10^4$ A flows through the wire, which accelerates the wire along the rails. Calculate the final speed of the wire as it leaves the rails. (Neglect friction.)

Sections 27.5 and 27.6

•**27.43** A square loop of wire of side length ℓ lies in the xy-plane, with its center at the origin and its sides parallel to the x- and y-axes. It carries a current, i, in the counterclockwise direction, as viewed looking down the z-axis from the positive direction. The loop is in a magnetic field given by $\vec{B} = (B_0/a)(z\hat{x} + x\hat{z})$, where B_0 is a constant field strength, a is a constant with the dimension of length, and $\hat{x}$ and $\hat{z}$ are unit vectors in the positive x-direction and positive z-direction. Calculate the net force on the loop.

27.44 A rectangular coil with 20 windings carries a current of 2 mA flowing in the counterclockwise direction. It has two sides that are parallel to the y-axis and have length 8 cm and two sides that are parallel to the x-axis and have length 6 cm. A uniform magnetic field of 50 μT acts in the positive x-direction. What torque must be applied to the loop to hold it steady?

27.45 A coil consists of 120 circular loops of wire of radius 4.8 cm. A current of 0.49 A runs through the coil, which is oriented vertically and is free to rotate about a vertical axis (parallel to the z-axis). It experiences a uniform horizontal magnetic field in the positive x-direction. When the coil is oriented parallel to the x-axis, a force of 1.2 N applied to the edge of the coil in the positive y-direction can keep it from rotating. Calculate the strength of the magnetic field.

27.46 Twenty loops of wire are tightly wound around a round pencil that has a diameter of 6 mm. The pencil is then placed in a uniform 5-T magnetic field, as shown in the figure. If a 3-A current is present in the coil of wire, what is the magnitude of the torque on the pencil?

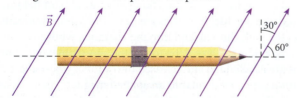

•**27.47** A copper wire with density $\rho = 8960$ kg/m³ is formed into a circular loop of radius 50.0 cm. The cross-sectional area of the wire is $1.00 \cdot 10^{-5}$ m², and a potential difference of 0.012 V is applied to the wire. What is the maximum angular acceleration of the loop when it is placed in a magnetic field of magnitude 0.25 T? The loop rotates about an axis through a diameter.

27.48 A simple galvanometer is made from a coil that consists of N loops of wire of area A. The coil is attached to a mass, M, by a light rigid rod of length L. With no current in the coil, the mass hangs straight down, and the coil lies in a horizontal plane. The coil is in a uniform magnetic field of magnitude B that is oriented horizontally. Calculate the angle from the vertical of the rigid rod as a function of the current, i, in the coil.

27.49 Show that the magnetic dipole moment of an electron orbiting in a hydrogen atom is proportional to its angular momentum, L: $\mu = -eL/2m$, where $-e$ is the charge of the electron and m is its mass.

•**27.50** The figure shows a side view of a current-carrying ring of wire having a diameter $d = 8$ cm. A 1-A current flows in the ring in the direction indicated in the figure. The ring is connected to one end of a spring with a spring constant of 100 N/m. When the ring is in the vertical position, the spring is at its equilibrium length, ℓ. Determine the extension of the spring when a magnetic field of magnitude $B = 2$ T is applied parallel to the plane of the ring as shown in the figure.

•**27.51** A coil of wire consisting of 40 rectangular loops, with width 16.0 cm and height 30.0 cm, is placed in a constant magnetic field given by $\vec{B} = 0.065T\hat{x} + 0.250T\hat{z}$. The coil is hinged to a fixed thin rod along the y-axis (along segment da in the figure) and is originally located in the xy-plane. A current of 0.200 A runs through the wire.

a) What are the magnitude and the direction of the force, $\vec{F}_{ab}$, that $\vec{B}$ exerts on segment ab of the coil?

b) What are the magnitude and the direction of force, $\vec{F}_{bc}$, that $\vec{B}$ exerts on segment bc of the coil?

c) What is the magnitude of the net force, F_{net}, that $\vec{B}$ exerts on the coil?

d) What are the magnitude and the direction of the torque, $\vec{\tau}$, that $\vec{B}$ exerts on the coil?

e) In what direction, if any, will the coil rotate about the y-axis (viewed from above and looking down that axis)?

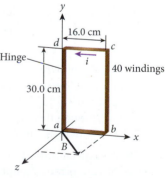

Section 27.7

27.52 A high electron mobility transistor (HEMT) controls large currents by applying a small voltage to a thin sheet of electrons. The density and mobility of the electrons in the sheet are critical for the operation of the HEMT. HEMTs consisting of AlGaN/GaN/Si are being studied because they promise better performance at higher powers, temperatures, and frequencies than conventional silicon HEMTs can achieve. In one study, the Hall effect was used to measure the density of electrons in one of these new HEMTs. When a current of 10 μA flows through the length of the electron sheet, which is 1.0 mm long, 0.30 mm wide, and 10 nm thick, a magnetic field of 1.0 T perpendicular to the sheet produces a voltage of 0.68 mV across the width of the sheet. What is the density of electrons in the sheet?

•27.53 The figure shows schematically a setup for a Hall effect measurement using a thin film of zinc oxide of thickness 1.50 μm. The current, i, across the thin film is 12.3 mA and the Hall potential, V_H, is –20.1 mV when the magnetic field of magnitude $B = 0.90$ T is applied perpendicular to the current flow.

a) What are the charge carriers in the thin film? [*Hint:* They can be either electrons with charge $-e$ or electron holes (missing electrons) with charge $+e$.]

b) Calculate the density of charge carriers in the thin film.

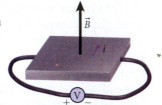

Additional Problems

27.54 A cyclotron in a magnetic field of 9 T is used to accelerate protons to 50% of the speed of light. What is the cyclotron frequency of these protons? What is the radius of their trajectory in the cyclotron? What is the cyclotron frequency and trajectory radius of the same protons in the Earth's magnetic field? Assume that the Earth's magnetic field is about 0.5 G.

27.55 A straight wire carrying a current of 3.41 A is placed at an angle of 10.0° to the horizontal between the pole tips of a magnet producing a field of 0.220 T upward. The poles' tips each have a 10.0 cm diameter. The magnetic force causes the wire to move out of the space between the poles. What is the magnitude of that force?

27.56 An electron is moving at $v = 6.00 \cdot 10^7$ m/s perpendicular to the Earth's magnetic field. If the field strength is $0.500 \cdot 10^{-4}$ T, what is the radius of the electron's circular path?

27.57 A straight wire with a constant current running through it is in Earth's magnetic field, at a location where the magnitude is 0.43 G. What is the minimum current that must flow through the wire for a 10.0-cm length of it to experience a force of 1.0 N?

27.58 A small aluminum ball with a mass of 5 g and a charge of 15 C is moving northward at 3000 m/s. You want the ball to travel in a horizontal circle with a radius of 2 m, in a clockwise sense when viewed from above. Ignoring gravity, what is the magnitude and the direction of the magnetic field that must be applied to the aluminum ball to cause it to have this motion?

27.59 The velocity selector described in Solved Problem 27.2 is used in a variety of devices to produce a beam of charged particles of uniform velocity. Suppose the fields in such a selector are given by $\vec{E} = (1.00 \cdot 10^4$ V/m$)\hat{x}$ and

$\vec{B} = (50.0$ mT$)\hat{y}$. Find the velocity in the z-direction with which a charged particle can travel through the selector without being deflected.

27.60 A circular coil with a radius of 10 cm has 100 turns of wire and carries a current, $i = 100$ mA. It is free to rotate in a region with a constant horizontal magnetic field given by $\vec{B} = (0.01$ T$)\hat{x}$. If the unit normal vector to the plane of the coil makes an angle of 30° with the horizontal, what is the magnitude of the net magnetic torque acting on the coil?

27.61 At $t = 0$ an electron crosses the positive y-axis (so $x = 0$) at 60 cm from the origin with velocity $2 \cdot 10^5$ m/s in the positive x-direction. It is in a uniform magnetic field.

a) Find the magnitude and the direction of the magnetic field that will cause the electron to cross the x-axis at $x = 60$ cm.

b) What work is done on the electron during this motion?

c) How long will the trip take from y-axis to x-axis?

•27.62 A 12-V battery is connected to a 3-Ω resistor in a rigid rectangular loop of wire measuring 3 m by 1 m. As shown in the figure, a length $\ell = 1.00$ m of wire at the end of the loop extends into a 2 m by 2 m region with a magnetic field of magnitude 5 T, directed into the page. What is the net force on the loop?

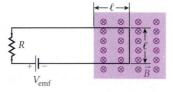

27.63 An alpha particle ($m = 6.6 \cdot 10^{-27}$ kg, $q = +2e$) is accelerated by a potential difference of 2700 V and moves in a plane perpendicular to a constant magnetic field of magnitude 0.340 T, which curves the trajectory of the alpha particle. Determine the radius of curvature and the period of revolution.

•27.64 In a certain area, the electric field near the surface of the Earth is given by $\vec{E} = (-150.$ N/C$)\hat{z}$, and the Earth's magnetic field is given by $\vec{B} = (50.0$ μT$)\hat{r}_N - (20.0$ μT$)\hat{z}$, where $\hat{z}$ is a unit vector pointing vertically upward and $\hat{r}_N$ is a horizontal unit vector pointing due north. What velocity, $\vec{v}$, will allow an electron in this region to move in a straight line at constant speed?

•27.65 A helium leak detector uses a mass spectrometer to detect tiny leaks in a vacuum chamber. The chamber is evacuated with a vacuum pump and then sprayed with helium gas on the outside. If there is any leak, the helium molecules pass through the leak and into the chamber, whose volume is sampled by the leak detector. In the spectrometer, helium ions are accelerated and released into a tube, where their motion is perpendicular to an applied magnetic field, $\vec{B}$, and they follow a circular orbit of radius r and then hit a detector. Estimate the velocity required if the orbital radius of the ions is to be no more than 5 cm, the magnetic field is 0.15 T, and the mass of a helium-4 atom is about $6.6 \cdot 10^{-27}$ kg. Assume that each ion is singly ionized (has one electron less than the neutral atom). By what factor does the required velocity change if helium-3 atoms, which have about $\frac{3}{4}$ as much mass as helium-4 atoms, are used?

•**27.66** In your laboratory, you set up an experiment with an electron gun that emits electrons with energy of 7.5 keV toward an atomic target. What deflection (magnitude and direction) would Earth's magnetic field (0.30 G) produce in the beam of electrons if the beam is initially directed due east and covers a distance of 1 m from the gun to the target? (*Hint:* First calculate the radius of curvature, and then determine how far away from a straight line the electron beam has deviated after 1 m.)

•**27.67** A proton enters the region between the two plates shown in the figure moving in the x-direction with a speed $v = 1.35 \cdot 10^6$ m/s. The potential of the top plate is 200 V, and the potential of the bottom plate is 0 V. What is the direction and the magnitude of the magnetic field, $\vec{B}$, that is required between the plates for the proton to continue traveling in a straight line along the x-direction?

y

$z \odot \longrightarrow x$

200 V

$v = 1.35 \cdot 10^6$ m/s

$d = 35.0$ mm

Proton

0 V

•**27.68** An electron moving at a constant velocity, $\vec{v} = v_0 \hat{x}$, enters a region in space where a magnetic field is present. The magnetic field, $\vec{B}$, is constant and points in the z-direction. What is the magnitude and direction of the magnetic force acting on the electron? If the width of the region where the magnetic field is present is d, what is the minimum velocity the electron must have in order to escape this region?

•**27.69** A 30-turn square coil with a mass of 0.250 kg and a side length of 0.200 m is hinged along a horizontal side and carries a 5.00-A current. It is placed in a magnetic field pointing vertically downward and having a magnitude of 0.00500 T. Determine the angle that the plane of the coil makes with the vertical when the coil is in equilibrium. Use $g = 9.81$ m/s^2.

•**27.70** A semicircular loop of wire of radius R is in the xy-plane, centered about the origin. The wire carries a current, i, counterclockwise around the semicircle, from $x = -R$ to $x = +R$ on the x-axis. A magnetic field, $\vec{B}$, is pointing out of the plane, in the positive z-direction. Calculate the net force on the semicircular loop.

•**27.71** A proton moving at speed $v = 1.0 \cdot 10^6$ m/s enters a region in space where a magnetic field given by $\vec{B} = (-0.5\ \text{T})$ $\hat{z}$ exists. The velocity vector of the proton is at an angle $\theta = 60°$ with respect to the positive z-axis.

a) Analyze the motion of the proton and describe its trajectory (in qualitative terms only).

b) Calculate the radius, r, of the trajectory projected onto a plane perpendicular to the magnetic field (in the xy-plane).

c) Calculate the period, T, and frequency, f, of the motion in that plane.

d) Calculate the pitch of the motion (the distance traveled by the proton in the direction of the magnetic field in 1 period).

28

Magnetic Fields of Moving Charges

FIGURE 28.1 Large magnets can be used to lift large metal objects.

WHAT WE WILL LEARN

- Moving charges (currents) create magnetic fields.

- The magnetic field created by a current flowing in a long, straight wire varies inversely with the distance from the wire.

- Two parallel wires carrying current in the same direction attract each other. Two parallel wires carrying current in opposite directions repel each other.

- Ampere's Law is used to calculate the magnetic field caused by certain symmetrical current distributions, just as Gauss's Law is useful in calculating electric fields in situations having spatial charge symmetry.

- The magnetic field inside a long, straight wire varies linearly with the distance from the center of the wire.

- A solenoid is an electromagnet that can be used to produce a constant magnetic field with a large volume.

- Some atoms can be thought of as small magnets created by motion of electrons in the atom.

- Materials can exhibit three kinds of intrinsic magnetism: diamagnetism, paramagnetism, and ferromagnetism.

- Superconducting magnets can be used to produce very strong magnetic fields.

Large magnets like the ones shown in Figure 28.1 are used in many industrial settings to move large metal objects. However, these magnets are not permanent magnets, but electromagnets, which can be switched on and off. But just what is an electromagnet? We saw in Chapter 27 that a magnetic field can affect the path of a charged particle or the flow of a current. In this chapter, we consider magnetic fields *caused* by electric currents. Any charged particle generates a magnetic field, if it is moving. Various magnetic fields are caused by different distributions of current. The powerful electromagnets used in industry, in physics research, in medical diagnostics, and in other applications are primarily solenoids—coils of current-carrying wire with hundreds or thousands of loops. We'll see in this chapter why solenoids generate particularly useful magnetic fields, and in the next chapter, we'll look at some other important phenomena caused by the motion of a coil in a magnetic field.

Chapter 27 introduced magnetic fields and field lines by showing how a compass needle orients itself near a permanent magnet. In a similar demonstration, a strong current is run through a (very long, straight) wire (with or without insulation). If a compass needle is then brought close to the wire, the compass needle orients itself relative to the wire in the way shown in Figure 28.2. This observation was first made by the Danish physicist Hans Oersted (1777–1851) in 1819 while conducting a demonstration for students during a lecture. We conclude that the current in the wire produces a magnetic field. Since the direction of the compass needle indicates the direction of the magnetic field, we further conclude that the magnetic field lines form circles around this current-carrying wire. Note the difference between parts (a) and (b) of Figure 28.2: When the direction of the current is reversed, the orientation of the compass needle is also reversed. Although the figure doesn't show this, if the compass needle is moved farther and farther away from the wire, eventually it again orients itself in the direction of the magnetic field of the Earth. This indicates that the magnetic field produced by the wire gets weaker as a function of increasing distance from the wire.

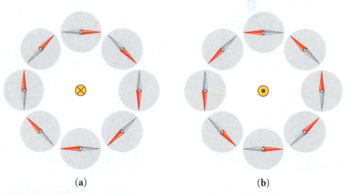

(a)   (b)

FIGURE 28.2 Wire (yellow circle) with current running through it: (a) into the page (indicated by the cross); (b) out of the page (indicated by the dot). The orientation of a compass needle placed close to the wire is shown at different locations around the wire.

28.1 Biot-Savart Law

In Chapter 27, we saw that magnetic fields could change the trajectory of moving charges. However, experiments show that this interaction works in the other direction as well: Moving charges can generate magnetic fields. How can we determine the magnetic field produced by a moving charge?

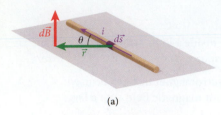

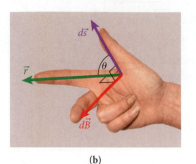

(a)

(b)

FIGURE 28.3 (a) Three-dimensional depiction of the Biot-Savart Law. The differential magnetic field is perpendicular to both the differential current element and the position vector. (b) Right-hand rule 1 applied to the quantities involved in the Biot-Savart Law.

To describe the electric field in terms of the electric charge, we showed that (see Chapter 22):

$$dE = \frac{1}{4\pi\epsilon_0}\frac{|dq|}{r^2},$$

where dq is a charge element. The electric field points in a radial direction (inward toward or outward from the electric charge, depending on the sign of the charge), so

$$d\vec{E} = \frac{1}{4\pi\epsilon_0}\frac{dq}{r^3}\vec{r} = \frac{1}{4\pi\epsilon_0}\frac{dq}{r^2}\hat{r}.$$

The situation is slightly more complicated for a magnetic field because a current element, $i\,d\vec{s}$, producing a magnetic field has a direction, as opposed to a nondirectional point charge producing an electric field. As a result of a long series of experiments involving tests similar to that depicted in Figure 28.2, conducted in the early 19th century, the French scientists Jean-Baptiste Biot (1774–1862) and Felix Savart (1791–1841) established that the magnetic field produced by a current element, $i\,d\vec{s}$, is given by

$$d\vec{B} = \frac{\mu_0}{4\pi}\frac{i\,d\vec{s}\times\vec{r}}{r^3} = \frac{\mu_0}{4\pi}\frac{i\,d\vec{s}\times\hat{r}}{r^2}. \tag{28.1}$$

Here $d\vec{s}$ is a vector of differential length ds pointing in the direction in which the current flows along the conductor and $\vec{r}$ is the position vector measured *from* the current element *to* the point at which the field is to be found. Figure 28.3 depicts the physical situation described by this formula, which is called the **Biot-Savart Law.**

The constant μ_0 in equation 28.1 is called the **magnetic permeability of free space** and has the value

$$\mu_0 = 4\pi\cdot10^{-7}\frac{\text{T m}}{\text{A}}. \tag{28.2}$$

From equation 28.1 and Figure 28.3, you can see that the direction of the magnetic field produced by the current element is perpendicular to both the position vector and the current element, $i\,d\vec{s}$. The magnitude of the magnetic field is given by

$$dB = \frac{\mu_0}{4\pi}\frac{i\,ds\sin\theta}{r^2}, \tag{28.3}$$

where θ (with possible values between 0° and 180°) is the angle between the direction of the position vector and the current element. The direction of the magnetic field is given by a variant of right-hand rule 1, introduced in Chapter 27. To determine the direction of the magnetic field using your right hand, point your thumb in the direction of the differential current element and your index finger in the direction of the position vector, and your middle finger will point in the direction of the differential magnetic field.

28.2 Magnetic Fields due to Current Distributions

Chapter 27 addressed the superposition principle for magnetic fields. Using this superposition principle, we can compute the magnetic field at any point in space as the sum of the differential magnetic fields described by the Biot-Savart Law. This section examines the magnetic fields generated by the most common configurations of current-carrying wires.

Magnetic Field from a Long, Straight Wire

Let's first examine the magnetic field from an infinitely long, straight wire carrying a current, i. We consider the magnetic field, $d\vec{B}$, at a point P at a perpendicular distance $r_\perp$ from the wire (Figure 28.4). The magnitude of the field dB at that point due to the current element $i\,ds$ is given by equation 28.3, the direction of the field is given by $d\vec{s}\times\vec{r}$, and is out of the page. We find the magnetic field from the right half of the wire and

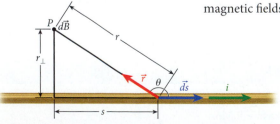

FIGURE 28.4 Magnetic field from a long, straight current-carrying wire.

multiply by 2 to get the magnetic field from the whole wire. Thus, the magnitude of the magnetic field at a perpendicular distance $r_\perp$ from the wire is given by

$$B = 2\int_0^\infty dB = 2\int_0^\infty \frac{\mu_0}{4\pi}\frac{i\,ds\sin\theta}{r^2} = \frac{\mu_0 i}{2\pi}\int_0^\infty \frac{ds\sin\theta}{r^2}.$$

We can relate r and θ to $r_\perp$ and s (r, s, and θ) by $r = \sqrt{s^2 + r_\perp^2}$ and $\sin\theta = \sin(\pi - \theta) = r_\perp/\sqrt{s^2 + r_\perp^2}$ (see Figure 28.4). Substituting for r and $\sin\theta$ in the preceding expression for B gives

$$B = \frac{\mu_0 i}{2\pi}\int_0^\infty \frac{r_\perp\,ds}{(s^2 + r_\perp^2)^{3/2}}.$$

Evaluating this definite integral, we find

$$B = \frac{\mu_0 i}{2\pi}\left[\frac{1}{r_\perp^2}\frac{r_\perp s}{(s^2 + r_\perp^2)^{1/2}}\right]_0^\infty = \frac{\mu_0 i}{2\pi r_\perp}\left[\frac{s}{(s^2 + r_\perp^2)^{1/2}}\right]_{s\to\infty} - 0.$$

For $s \gg r_\perp$, the term in the brackets approaches the value 1. Therefore, the magnitude of the magnetic field at a perpendicular distance $r_\perp$ from a long, straight wire carrying a current, i, is

$$B = \frac{\mu_0 i}{2\pi r_\perp}. \tag{28.4}$$

The direction of the magnetic field at any point is found by applying right-hand rule 1 to the current element and position vectors shown in Figure 28.4. This results in a new right-hand rule, called *right-hand rule 3*, which can be used to determine the direction of the magnetic field from a current-carrying wire. If you grab the wire with your right hand so that your thumb points in the direction of the current, your fingers will curl in the direction of the magnetic field (Figure 28.5).

Looking along a current-carrying wire would reveal that the magnetic field lines form concentric circles (Figure 28.6). Notice from the distance between the field lines that the field is strongest near the wire and drops off in proportion to $1/r_\perp$, as indicated by equation 28.4.

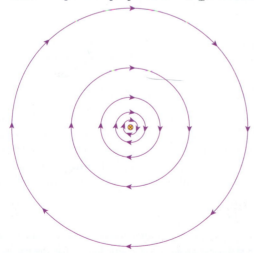

FIGURE 28.6 Magnetic field lines around a long, straight wire (yellow circle at the center) carrying a current perpendicular to the page and pointing into the page, signified by the cross.

FIGURE 28.5 Right-hand rule 3 for the magnetic field from a current-carrying wire.

28.1 In-Class Exercise

A wire is carrying a current, i_{in}, into the page as shown in the figure. In which direction does the magnetic field point at points P and Q?

• Q

⊗ i_{in} • P

a) to the right at P and upward (toward the top of the page) at Q

b) upward at P and to the right at Q

c) downward at P and to the right at Q

d) upward at P and to the left at Q

28.2 In-Class Exercise

Assume that a lightning bolt can be modeled as a long, straight line of current. If 15.0 C of charge passes by a point in $1.50\cdot10^{-3}$ s, what is the magnitude of the magnetic field at a perpendicular distance 26.0 m from the lightning bolt?

a) $7.69\cdot10^{-5}$ T   d) $1.11\cdot10^{-1}$ T

b) $9.22\cdot10^{-3}$ T   e) $2.22\cdot10^{2}$ T

c) $4.21\cdot10^{-2}$ T

28.3 In-Class Exercise

Wire 1 has a current flowing out of the page, i_{out}, as shown in the figure. Wire 2 has a current flowing into the page, i_{in}. What is the direction of the magnetic field at point P?

a) upward in the plane of the page

c) downward in the plane of the page

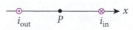

b) to the right

d) to the left

e) The magnetic field at point P is zero.

FIGURE 28.7 (a) Magnetic field line from one current-carrying wire. (b) Magnetic field created by the current in one wire exerting a force on a second current-carrying wire. (c) Magnetic field created by the current in the second wire exerting a force on the first current-carrying wire.

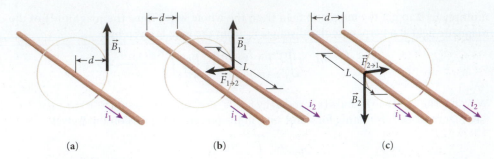

(a)   (b)   (c)

28.1 Self-Test Opportunity

The wire in the figure is carrying a current i in the positive z-direction. What is the direction of the resulting magnetic field at point P_1? What is the direction of the resulting magnetic field at point P_2?

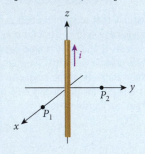

28.2 Self-Test Opportunity

Consider two parallel wires carrying the same current in the same direction. Is the force between the two wires attractive or repulsive? Now consider two parallel wires carrying current in opposite directions. What is the force between the two wires?

Two Parallel Wires

Let's examine the case in which two parallel wires are carrying current. The two wires exert magnetic forces on each other because the magnetic field of one wire exerts a force on the moving charges in the second wire. The magnitude of the magnetic field created by a current-carrying wire is given by equation 28.4. This magnetic field is always perpendicular to the wire with a direction given by right-hand rule 3 (Figure 28.5).

Let's first consider wire 1 carrying a current, i_1, toward the right, as shown in Figure 28.7a. The magnitude of the magnetic field a perpendicular distance d from wire 1 is

$$B_1 = \frac{\mu_0 i_1}{2\pi d}. \tag{28.5}$$

The direction of $\vec{B}_1$ is given by right-hand rule 3 and is shown for a particular point in Figure 28.7a.

Now consider wire 2 carrying a current, i_2, in the same direction as i_1, and placed parallel to wire 1 at a distance d from it (Figure 28.7b). The magnetic field due to wire 1 exerts a magnetic force on the moving charges in the current flowing in wire 2. In Chapter 27, we saw that the magnetic force on a current-carrying wire is given by

$$\vec{F} = i\vec{L} \times \vec{B}.$$

The magnitude of the magnetic force on a length, L, of wire 2 is then

$$F = iLB \sin\theta = i_2 LB_1, \tag{28.6}$$

because $\vec{B}_1$ is perpendicular to wire 2 and thus $\theta = 90°$. Substituting for B_1 from equation 28.5 into equation 28.6, we find the magnitude of the force exerted by wire 1 on a length L of wire 2:

$$F_{1\to 2} = i_2 L\left(\frac{\mu_0 i_1}{2\pi d}\right) = \frac{\mu_0 i_1 i_2 L}{2\pi d}. \tag{28.7}$$

According to right-hand rule 1, $\vec{F}_{1\to 2}$ points toward wire 1 and is perpendicular to both wires. An analogous calculation allows us to deduce that the force from wire 2 on a length, L, of wire 1 has the same magnitude and opposite direction: $\vec{F}_{2\to 1} = -\vec{F}_{1\to 2}$. This result is shown in Figure 28.7c and is a simple consequence of Newton's Third Law.

Definition of the Ampere

The force $F_{1\to2}$ described by equation 28.7 is used in the SI definition of the ampere: An *ampere* (A) is the constant current that, if maintained in two straight, parallel conductors of infinite length and negligible circular cross section, that are placed 1 m apart in vacuum, would produce between these conductors a force of $2 \cdot 10^{-7}$ N per meter of length. This physical situation is described by equation 28.7 for the force between two parallel current-carrying wires with $i_1 = i_2 =$ exactly 1 A, $d =$ exactly 1 m, and $F_{1\to2} =$ exactly $2 \cdot 10^{-7}$ N. We can solve that equation for μ_0:

$$\mu_0 = \frac{(2\pi d)F_{1\to2}}{i_1 i_2 L} = \frac{2\pi(1\text{ m})(2\cdot10^{-7}\text{ N})}{(1\text{ A})(1\text{ A})(1\text{ m})} = \text{exactly } 4\pi \cdot 10^{-7}\ \frac{\text{T m}}{\text{A}},$$

which indicates that the magnetic permeability of free space is *defined* to be exactly $\mu_0 = 4\pi \cdot 10^{-7}$ T m/A (see equation 28.2).

In Chapter 21, when Coulomb's Law was introduced, the value of the electric permittivity of free space, ϵ_0, was given: $\epsilon_0 = 8.85 \cdot 10^{-12}\ \text{C}^2/(\text{N m}^2)$. Since 1 A =1 C/s and 1 T = 1 (N s)/(C m) (see Chapter 27), the product of the two constants ϵ_0, and μ_0 is

$$\mu_0\epsilon_0 = \left(4\pi \cdot 10^{-7}\ \frac{\text{T m}}{\text{A}}\right)\left(8.85\cdot10^{-12}\ \frac{\text{C}^2}{\text{N m}^2}\right) = 1.11\cdot10^{-17}\ \frac{\text{s}^2}{\text{m}^2},$$

which has the units of the inverse of the square of a speed. Thus, $1/\sqrt{\mu_0\epsilon_0}$ gives the value of this speed as that of the speed of light, $c = 3.00 \cdot 10^8$ m/s. This is by no means an accident, as we'll see in later chapters. For now, it is sufficient to state the empirical finding:

$$c = \frac{1}{\sqrt{\mu_0\epsilon_0}}.$$

Since the magnetic permeability of free space is defined to be exactly $\mu_0 = 4\pi \cdot 10^{-7}$ T m/A and the speed of light is defined as exactly $c = 299{,}792{,}458$ m/s (see the discussion in Chapter 1), the expression $c = 1/\sqrt{\mu_0\epsilon_0}$ also fixes the value of the electric permittivity of free space.

EXAMPLE 28.1 Force on a Loop

A long, straight wire is carrying a current of magnitude $i_1 = 5.00$ A toward the right (Figure 28.8). A square loop with sides of length $a = 0.250$ m is placed with its sides parallel and perpendicular to the wire at a distance $d = 0.100$ m from the wire. The square loop carries a current of magnitude $i_2 = 2.20$ A in the counterclockwise direction.

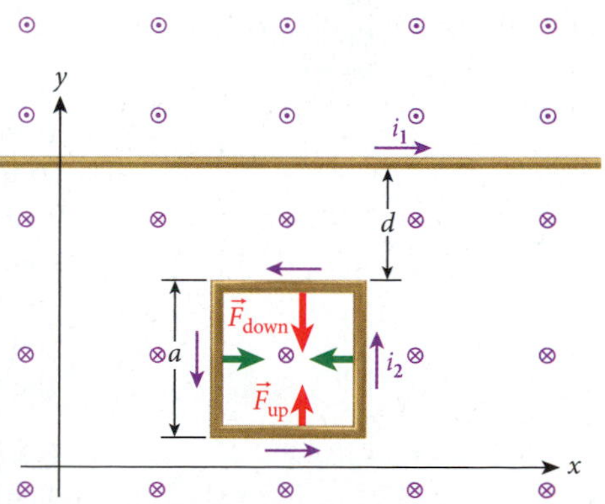

FIGURE 28.8 A current-carrying wire and a square loop.

PROBLEM

What is the net magnetic force on the square loop?

SOLUTION

The force on the square loop is due to the magnetic field created by the current flowing in the straight wire. Right-hand rule 3 tells us that the magnetic field from the current flowing in the wire is directed into the page in the region where the loop is located (see Figure 28.8). The right hand rule and equation 28.4 tell us that the resulting force on the left side of the loop is toward the right, and the force on the right side of the loop is toward the left. In Figure 28.8, these two forces are represented by green arrows. These two forces are equal in magnitude and opposite in direction, so they sum to zero. The force on the top side of the loop is downward (red arrow in Figure 28.8, pointing in the negative y-direction), and its magnitude is given by equation 28.7

$$F_{\text{down}} = \frac{\mu_0 i_1 i_2 a}{2\pi d},$$

Continued—

28.6 In-Class Exercise

A wire is carrying a current, i, in the positive y-direction, as shown in the figure. The wire is located in a uniform magnetic field, $\vec{B}$, oriented in such a way that the magnetic force on the wire is maximized. The magnetic force acting on the wire, $\vec{F}_B$, is in the negative x-direction. What is the direction of the magnetic field?

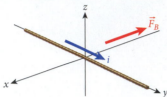

a) the positive x-direction

b) the negative x-direction

c) the negative y-direction

d) the positive z-direction

e) the negative z-direction

FIGURE 28.9 U.S. Navy rail gun.

where a is the length of the top side of the loop. The force on the bottom side of the loop is upward (the other red arrow in Figure 28.8, pointing in the positive y-direction), and its magnitude is given by

$$F_{up} = \frac{\mu_0 i_1 i_2 a}{2\pi(d+a)}.$$

Thus, we can express the net magnetic force on the loop as

$$\vec{F} = (F_{up} - F_{down})\hat{y}.$$

Putting in the numbers results in

$$\vec{F} = \frac{(4\pi \cdot 10^{-7} \text{ T m/A})(5.00 \text{ A})(2.20 \text{ A})(0.250 \text{ m})}{2\pi}\left(\frac{1}{0.350 \text{ m}} - \frac{1}{0.100 \text{ m}}\right)\hat{y} = (-3.93 \cdot 10^{-6})\hat{y} \text{ N}.$$

SOLVED PROBLEM 28.1 | Electromagnetic Rail Accelerator

Electromagnetic rail accelerators are being studied for the purposes of accelerating fuel pellets in fusion experiments and for launching spacecraft into orbit. The U.S. Navy is experimenting with electromagnetic rail accelerators that can launch projectiles at very high speeds, such as the rail gun shown in Figure 28.9. This rail gun operates by running a current through two parallel conducting rails that are connected by a movable conductor oriented perpendicular to the rails. The projectile is attached to the movable conductor. For this example, we'll assume that the rail gun consists of two parallel rails of cross-sectional radius $r = 5.00$ cm, whose centers are separated by distance $d = 25.0$ cm, and which have length $L = 5.00$ m and that the rail gun accelerates the projectile to a kinetic energy of $K = 32.0$ MJ. The projectile also functions as the movable conductor.

PROBLEM
How much current is required to accelerate the projectile?

SOLUTION

THINK
The currents in the two rails are in opposite directions. The current flowing through the movable conductor is perpendicular to the two currents in the rails. The magnetic fields from the two rails are in the same direction and exert forces on the movable conductor in the same direction. The force from the magnetic field of each rail depends on the distance from the rail. Thus, we must integrate the force along the distance between the two rails to obtain the total force. The total force on the movable conductor is twice the force from the magnetic field from one rail. The kinetic energy gained by the projectile is the total force exerted by the magnetic fields of the two rails times the distance over which the force acts.

SKETCH
Figure 28.10 shows top and cross-sectional views of the rails and the movable conductor.

RESEARCH
The current-carrying movable conductor, which completes the circuit between the two rails, is also the projectile and is accelerated by the magnetic forces produced by the two rails. The force exerted on the projectile depends on the distance, x, from the center of a rail, as illustrated in Figure 28.10b. Thus, to calculate the total force on the projectile, we must integrate over the length of the projectile. We use equation 28.4 to find the magnitude of the magnetic field, B_1, from current i flowing in rail 1 at a distance x from the center of the rail:

$$B_1 = \frac{\mu_0 i}{2\pi x}.$$

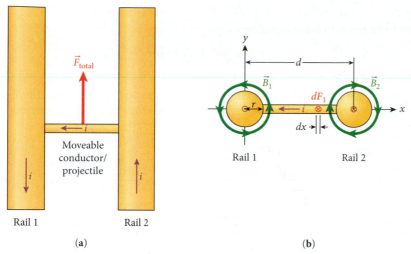

FIGURE 28.10 Schematic diagram of a rail gun: (a) top view; (b) cross-sectional view.

According to equation 28.6, the magnitude of the differential force, dF_1, exerted on a differential length, dx, of the projectile by the magnetic field from rail 1 is

$$dF_1 = i(dx)B_1 = i(dx)\left(\frac{\mu_0 i}{2\pi x}\right).$$

The direction of the force is given by right-hand rule 3, which tells us that the force is upward in the plane of the page in Figure 28.10a and into the page in Figure 28.10b. The magnitude of the force on the projectile is given by integrating dF_1 over the length of the projectile:

$$F_1 = \int_r^{d-r} dF_1 = \int_r^{d-r} \frac{\mu_0 i^2}{2\pi} \frac{dx}{x} = \frac{\mu_0 i^2}{2\pi}\Big[\ln x\Big]_r^{d-r} = \frac{\mu_0 i^2}{2\pi}\big(\ln(d-r) - \ln r\big) = \frac{\mu_0 i^2}{2\pi}\ln\left(\frac{d-r}{r}\right) \quad \text{(i)}$$

Because the magnetic field from rail 2 is in the same direction as the magnetic field from rail 1, the force exerted by the magnetic field from rail 2 on the projectile is the same as that from rail 1. Thus, the magnitude of the total force exerted on the projectile is

$$F_{\text{total}} = 2F_1. \quad \text{(ii)}$$

The kinetic energy gained by the projectile is equal (by the work-energy theorem introduced in Chapters 5 and 6) to the magnitude of the force exerted times the distance over which the force acts

$$K = F_{\text{total}}L. \quad \text{(iii)}$$

SIMPLIFY

We can combine equations (i), (ii), and (iii) to obtain

$$K = 2F_1 L = 2\left[\frac{\mu_0 i^2}{2\pi}\ln\left(\frac{d-r}{r}\right)\right]L = \frac{\mu_0 L i^2}{\pi}\ln\left(\frac{d-r}{r}\right).$$

Solving this equation for the current gives us

$$i = \sqrt{\frac{K\pi}{\mu_0 L \ln\left(\dfrac{d-r}{r}\right)}}.$$

CALCULATE

Putting in the numerical values, we get

$$i = \sqrt{\frac{K\pi}{\mu_0 L \ln\left(\dfrac{d-r}{r}\right)}} = \sqrt{\frac{\left(32.0 \cdot 10^6 \ \text{J}\right)\pi}{\left(4\pi \cdot 10^{-7} \ \dfrac{\text{T m}}{\text{A}}\right)(5.00 \ \text{m})\ln\left(\dfrac{25.0 \ \text{cm} - 5.00 \ \text{cm}}{5.00 \ \text{cm}}\right)}} = 3397287 \ \text{A}.$$

Continued—

ROUND

We report our result to three significant figures:

$$i = 3.40 \cdot 10^6 \text{ A} = 3.40 \text{ MA}.$$

DOUBLE-CHECK

Let's double-check our result by looking at the force exerted on the projectile. Putting our result for the current into equation (i) for the magnitude of the force exerted by the magnetic field of rail 1 gives

$$F_1 = \frac{\mu_0 i^2}{2\pi} \ln\left(\frac{d-r}{r}\right) = \frac{\left(4\pi \cdot 10^{-7} \, \frac{\text{T m}}{\text{A}}\right)\left(3.15 \cdot 10^6 \text{ A}\right)^2}{2\pi} \ln\left(\frac{25.0 \text{ cm} - 5.00 \text{ cm}}{5.00 \text{ cm}}\right) = 3.20 \cdot 10^6 \text{ N}.$$

We can approximate the magnitude of this force by assuming that the force is constant along the conductor and is equal to the value at $x = d/2$:

$$F_1 = i(d-2r)B = id\left|\frac{\mu_0 i(d-2r)}{2\pi\left(\frac{d}{2}\right)}\right| = \frac{\mu_0 i^2}{\pi}\frac{d-2r}{d}$$

$$= \frac{\left(4\pi \cdot 10^{-7} \, \frac{\text{T m}}{\text{A}}\right)\left(3.40 \cdot 10^6 \text{ A}\right)^2}{\pi} \frac{\left[25.0 \text{ cm} - 2(5.00 \text{ cm})\right]}{25.0 \text{ cm}} = 2.77 \cdot 10^6 \text{ N}.$$

This value is within a factor of 2 of our calculated value for the force, which seems reasonable. However, just to be sure, let's calculate the kinetic energy of the projectile using the calculated value of the force:

$$K = F_{\text{total}} L = 2F_1 L = 2\left(3.20 \cdot 10^6 \text{ N}\right)\left(5.00 \text{ m}\right) = 3.20 \cdot 10^6 \text{ J}.$$

This result agrees with the specified 3.20 MJ.

Note that if a projectile of mass $m = 5.00$ kg were given a kinetic energy of 32.0 MJ by this rail gun, its speed would be

$$v = \sqrt{\frac{2K}{m}} = \sqrt{\frac{2\left(32.0 \cdot 10^6 \text{ J}\right)}{5.00 \text{ kg}}} = 3580 \text{ m/s}.$$

The rail gun would be capable of launching a projectile at 10 times the speed of sound, which is much larger than the typical speed of a bullet, at about 3 times the speed of sound.

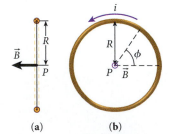

(a)   (b)

FIGURE 28.11 A circular loop of radius R carrying a current, i: (a) side view; (b) front view. The cross on the upper yellow circle in part (a) signifies that the current at the top of the loop is into the page, and the dot on the lower yellow circle in part (a) signifies that the current at the bottom of the loop is out of the page. Point P is located at the center of the loop.

Magnetic Field due to a Wire Loop

Now let's find the magnetic field at the center of a circular current-carrying loop of wire. Figure 28.11a shows a cross section of a circular loop with radius R carrying a current, i. Applying equation 28.3, $dB = \mu_0 i \, ds \sin\theta/(4\pi r^2)$, to this case, we can see that $r = R$ and $\theta = 90°$ for every current element $i \, ds$ along the loop. For the magnitude of the magnetic field at the center of the loop from each current element, we get

$$dB = \frac{\mu_0}{4\pi} \frac{i \, ds \sin 90°}{R^2} = \frac{\mu_0}{4\pi} \frac{i \, ds}{R^2}.$$

Going around the loop in Figure 28.11b, we can relate the angle ϕ to the current element by $ds = R \, d\phi$, allowing us to calculate the magnitude of the magnetic field at the center of the loop:

$$B = \int dB = \int_0^{2\pi} \frac{\mu_0}{4\pi} \frac{iR \, d\phi}{R^2} = \frac{\mu_0 i}{2R}. \tag{28.8}$$

Keep in mind that equation 28.8 only gives the magnitude of the magnetic field at the center of the loop, where the magnitude is $B(r = 0) = \frac{1}{2}\mu_0 i/R$. To determine the direction of the magnetic field, we again use a variant of right-hand rule 1. Using your right hand, point your thumb in the direction of the current element (into the page for the upper circle in Figure 28.11a marked with a cross), and your index finger in the direction of the radial vector from the current element (down); your middle finger then points to the left. Using right-hand rule 3 (Figure 28.5), we also find that the current shown in Figure 28.11 produces a magnetic field $\vec{B}$ directed toward the left.

Now let's find the magnetic field from the loop along the axis of the loop rather than at the center (Figure 28.12). We set up a coordinate system such that the axis of the loop lies along the x-axis and the center of the loop is located at $x = 0$, $y = 0$, and $z = 0$. The radial vector $\vec{r}$ is the displacement to any point along the x-axis from a current element, $i\,d\vec{s}$, along the loop. The current element shown in Figure 28.12 lies in the negative z-direction. The radial vector $\vec{r}$ lies in the xy-plane and so is perpendicular to the current element. This situation is the same for any current element around the loop. Therefore, we can employ equation 28.3 with $\theta = 90°$ and obtain an expression for the magnitude of the differential magnetic field at any point along the x-axis:

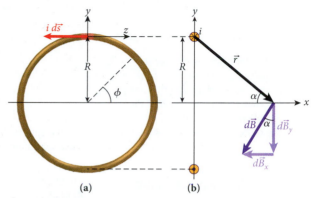

FIGURE 28.12 Geometry for calculating the magnetic field along the axis of a current-carrying loop: (a) front view, (b) side view

$$dB = \frac{\mu_0}{4\pi}\frac{i\,ds\sin 90°}{r^2} = \frac{\mu_0}{4\pi}\frac{i\,ds}{r^2}.$$

A variant of right-hand rule 1 gives the direction of the differential magnetic field: Using your right hand, point your thumb in the direction of the differential current element (negative z-direction) and your index finger in the direction of the radial vector (positive x-direction and negative y-direction); the direction of the differential magnetic field will be given by your middle finger (negative x-direction and negative y-direction). The differential magnetic field is shown in Figure 28.12. To obtain the complete magnetic field, we need to integrate over the differential current element. From the symmetry of the situation, we can see that the y-component of the differential magnetic field, dB_y, will integrate to zero. The x-component of the differential magnetic field, dB_x, is given by

$$dB_x = dB\sin\alpha = \frac{\mu_0}{4\pi}\frac{i\,ds}{r^2}\sin\alpha,$$

where α is the angle between $\vec{r}$ and the x-axis (see Figure 28.12b). We can express the magnitude of $\vec{r}$ in terms of x and R as $r = \sqrt{x^2 + R^2}$ and $\sin\alpha$ in terms of x and the radius of the loop R as $\sin\alpha = R/\sqrt{x^2 + R^2}$. We can then rewrite the expression for the x-component of the differential magnetic field as

$$dB_x = \frac{\mu_0}{4\pi}\frac{i\,ds}{x^2 + R^2}\frac{R}{\sqrt{x^2 + R^2}} = \frac{\mu_0 i\,ds}{4\pi}\frac{R}{\left(x^2 + R^2\right)^{3/2}}.$$

This expression for B_x is independent of the location of the current element, and so the integral to find the magnitude of the total field can be simplified to

$$B_x = \int dB_x = \frac{\mu_0 iR}{4\pi\left(x^2 + R^2\right)^{3/2}}\int ds.$$

Going around the loop, we can relate the angle ϕ to the current element by $ds = R\,d\phi$ (see Figure 28.12a), allowing us to calculate the magnetic field along the axis of the loop:

$$B_x = \frac{\mu_0 iR}{4\pi\left(x^2 + R^2\right)^{3/2}}\int_0^{2\pi} R\,d\phi = \frac{\mu_0 i2\pi}{4\pi}\frac{R^2}{\left(x^2 + R^2\right)^{3/2}},$$

or

$$B_x = \frac{\mu_0 i}{2}\frac{R^2}{\left(x^2 + R^2\right)^{3/2}}. \tag{28.9}$$

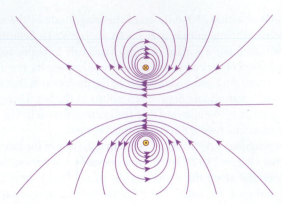

FIGURE 28.13 Magnetic field lines from a loop of wire carrying a current, looking at the loop edge-on. The upper yellow circle with a cross indicates current directed into the page, and the lower yellow circle with a dot indicates current directed out of the page.

28.3 Self-Test Opportunity

Show that equation 28.9 for the magnitude of the magnetic field along the axis of a current-carrying loop reduces to equation 28.8 for the magnitude of the magnetic field at the center of a current-carrying loop.

28.7 In-Class Exercise

Two identical wire loops carry the same current, i, as shown in the figure. What is the direction of the magnetic field at point P?

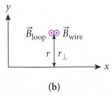

a) upward (toward the top of the page)

b) toward the right

c) downward

d) toward the left

e) The magnetic field at point P is zero.

From our earlier application of the variant of right-hand rule 1, we know that the magnetic field along the axis of the loop is in the negative x-direction, as shown in Figure 28.12. We can also apply right-hand rule 3 to obtain the direction of the magnetic field: At any point on the loop, point your thumb tangent to the loop in the direction of the current and your fingers will curl in a direction showing the field inside the loop is in the negative x-direction.

Using more advanced techniques and with the aid of a computer, we can determine the magnetic field produced by a current-carrying loop at other points in space. The magnetic field lines from a wire loop are shown in Figure 28.13. The value for the magnetic field given by equation 28.8 is valid only at the center point of Figure 28.13. The value for the magnetic field given by equation 28.9 is valid only along the axis of the loop.

SOLVED PROBLEM 28.2 | **Field from a Wire Containing a Loop**

A loop with radius $r = 8.30$ mm is formed in the middle of a long, straight insulated wire carrying a current of magnitude $i = 26.5$ mA (Figure 28.14a).

PROBLEM
What is the magnitude of the magnetic field at the center of the loop?

SOLUTION

THINK
The magnetic field at the center of the loop is equal to the vector sum of the magnetic fields from the long, straight wire and from the loop.

SKETCH
The magnetic field from the long, straight wire, $\vec{B}_{\text{wire}}$, and the magnetic field from the loop, $\vec{B}_{\text{loop}}$, are shown in Figure 28.14b.

RESEARCH
Using right-hand rule 3, we find that both magnetic fields point out of the page at the center of the loop, as illustrated in Figure 28.14b. Thus, we can add the magnitudes of the magnetic field produced by the wire and the magnetic field produced by the loop. The magnetic field produced by the wire at the center of the loop has a magnitude given by equation 28.4:

$$B_{\text{wire}} = \frac{\mu_0 i}{2\pi r_\perp},$$

where $r_\perp$ is the perpendicular distance from the wire, which is equal to r, the radius of the loop. The magnitude of the magnetic field produced by the loop at its center is given by equation 28.8:

$$B_{\text{loop}} = \frac{\mu_0 i}{2r}.$$

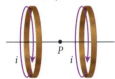

(a)

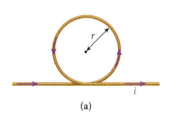

(b)

FIGURE 28.14 (a) A loop with radius r in a long, straight insulated wire carrying a current, i. (b) The magnetic field from the wire and the magnetic field from the loop, displaced slightly for clarity.

SIMPLIFY

We add the magnitudes of the two magnetic fields since the vectors are in the same direction:

$$B = B_{wire} + B_{loop} = \frac{\mu_0 i}{2\pi r} + \frac{\mu_0 i}{2r} = \frac{\mu_0 i}{2r}\left(\frac{1}{\pi} + 1\right).$$

CALCULATE

Putting in the numerical values, we get

$$B = \frac{\mu_0 i}{2r}\left(\frac{1}{\pi} + 1\right) = \frac{\left(4\pi \cdot 10^{-7} \text{ T m/A}\right)\left(26.5 \cdot 10^{-3} \text{ A}\right)}{2\left(8.30 \cdot 10^{-3} \text{ m}\right)}\left(\frac{1}{\pi} + 1\right) = 2.64463 \cdot 10^{-6} \text{ T}.$$

ROUND

We report our result to three significant figures and note the direction of the field:

$$\vec{B} = 2.64 \cdot 10^{-6} \text{ T, out of the page.}$$

DOUBLE-CHECK

To double-check our result, we calculate the magnitudes of the magnetic fields from the wire and from the loop separately. The magnitude of the magnetic field from the wire is

$$B_{wire} = \frac{\mu_0 i}{2\pi r} = \frac{\left(4\pi \cdot 10^{-7} \text{ T m/A}\right)\left(26.5 \cdot 10^{-3} \text{ A}\right)}{2\pi\left(8.30 \cdot 10^{-3} \text{ m}\right)} = 6.385 \cdot 10^{-7} \text{ T}.$$

The magnitude of the magnetic field from the loop is

$$B_{loop} = \frac{\mu_0 i}{2r} = \frac{\left(4\pi \cdot 10^{-7} \text{ T m/A}\right)\left(26.5 \cdot 10^{-3} \text{ A}\right)}{2\left(8.30 \cdot 10^{-3} \text{ m}\right)} = 2.006 \cdot 10^{-6} \text{ T}.$$

The sum of these two magnitudes matches our result:

$$6.385 \cdot 10^{-7} \text{ T} + 2.006 \cdot 10^{-6} \text{ T} = 2.64 \cdot 10^{-6} \text{ T}.$$

28.3 Ampere's Law

Recall from Chapter 22 that calculating the electric field resulting from a distribution of electric charge can require evaluating a difficult integral. However, if the charge distribution has cylindrical, spherical, or planar symmetry, we can apply Gauss's Law and obtain the electric field in an elegant manner. Similarly, calculating the magnetic field due to an arbitrary distribution of current elements using the Biot-Savart Law (equation 28.1) may involve the evaluation of a difficult integral. Alternatively, we can avoid using the Biot-Savart Law and instead apply **Ampere's Law** to calculate the magnetic field from a distribution of current elements when the distribution has cylindrical or other symmetry. Often, problems can be solved with much less effort in this way than by using a direct integration. The mathematical statement of Ampere's Law is

$$\oint \vec{B} \cdot d\vec{s} = \mu_0 i_{enc}. \tag{28.10}$$

The symbol $\oint$ means that the integrand, $\vec{B} \cdot d\vec{s}$, is integrated over a closed loop, called an **Amperian loop.** This loop is chosen so that the integral in equation 28.10 is not difficult to evaluate, a procedure similar to that used in applying Gauss's Law. The total current enclosed in this loop is i_{enc}, which is also similar to Gauss's Law, where the chosen closed surface encloses a total net charge.

As an example of how Ampere's Law is used, consider the five currents shown in Figure 28.15, which are all perpendicular to the plane. An Amperian loop, represented by the red line, encloses currents i_1, i_2, and i_3 and excludes currents i_4 and i_5. By Ampere's

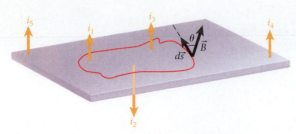

FIGURE 28.15 Five currents and an Amperian loop.

28.8 In-Class Exercise

Three wires are carrying currents of the same magnitude, i, in the directions shown in the figure. Four Amperian loops (a), (b), (c), and (d) are shown. For which Amperian loop is the magnitude of $\oint \vec{B} \cdot d\vec{s}$ the greatest?

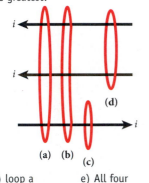

a) loop a

b) loop b

c) loop c

d) loop d

e) All four loops yield the same value of $\oint \vec{B} \cdot d\vec{s}$.

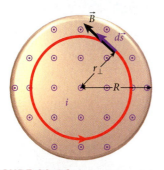

FIGURE 28.16 Using Ampere's Law to find the magnetic field produced inside a long, straight wire.

Law, the closed-loop integral over the magnetic field resulting from these three currents is given by

$$\oint \vec{B} \cdot d\vec{s} = \oint B \cos \theta \, ds = \mu_0 \left(i_1 - i_2 + i_3 \right),$$

where θ is the angle between the direction of the magnetic field, $\vec{B}$, and the direction of the element of length, $d\vec{s}$ at each point along the Amperian loop. The integration over the Amperian loop can be done in either direction. Figure 28.15 indicates a direction of integration from the direction of $d\vec{s}$, along with the resulting magnetic field. The sign of the contributing currents can be determined using a right-hand rule: Curl your fingers in the direction of integration, and then currents in the same direction as your thumb are positive. Two of the three currents in the Amperian loop are positive, and one is negative. Adding the three currents is simple, but the integral $\oint B \cos \theta \, ds$ cannot be easily evaluated. However, let's examine some special situations in which Amperian loops contain symmetrical distributions of current that can be exploited to carry out the integral.

Magnetic Field inside a Long, Straight Wire

Figure 28.16 shows a current, i, flowing out of the page in a wire with a circular cross section of radius R. This current is uniformly distributed over the cross-sectional area of the wire. To find the magnetic field due to this current, we use an Amperian loop with radius $r_\perp$, represented by the red circle. If $\vec{B}$ had an outward (or inward) component, by symmetry, it would have an outward (or inward) component at all points around the loop, and the corresponding magnetic field line could never be closed. Therefore, $\vec{B}$ must be tangential to the Amperian loop. Thus, we can rewrite the integral of Ampere's Law as

$$\oint \vec{B} \cdot d\vec{s} = B \oint d\vec{s} = B 2\pi r_\perp.$$

We can calculate the enclosed current from the ratio of the area of the Amperian loop to the cross-sectional area of the wire:

$$i_{\text{enc}} = i \frac{A_{\text{loop}}}{A_{\text{wire}}} = i \frac{\pi r_\perp^2}{\pi R^2}.$$

Thus, we obtain

$$2\pi B r_\perp = \mu_0 i \frac{\pi r_\perp^2}{\pi R^2},$$

or

$$B = \left(\frac{\mu_0 i}{2\pi R^2} \right) r_\perp. \tag{28.11}$$

Let's compare the expressions for the magnitudes of the magnetic field outside and inside the wire—equations 28.4 and 28.11. First, substituting R for $r_\perp$ in both expressions, we obtain the same result for the magnetic field magnitude at the surface of the wire in both cases: $B(R) = \mu_0 i / 2\pi R$. Both equations provide the same solution at the wire's surface. Inside the wire, we find that the magnetic field magnitude rises linearly with $r_\perp$ up to the value of $B(R) = \mu_0 i / (2\pi R)$ and from there falls off with the inverse of $r_\perp$. Figure 28.17 shows this dependence in the graph. The upper part of the figure depicts the cross section through the wire (golden area), the magnetic field lines (black circles, spaced to indicate the strength of the magnetic field), and the magnetic field vectors at selected points in space (red arrows).

28.4 Magnetic Fields of Solenoids and Toroids

We have seen that current flowing through a single loop of wire produces a magnetic field that is not uniform, as illustrated in Figure 28.13. However, real-world applications often require a uniform magnetic field. A device commonly used to produce a uniform magnetic field is the **Helmholtz coil** (Figure 28.18a). A Helmholtz coil consists of two coaxial wire loops. Each coaxial loop consists of multiple loops (windings or turns) of a single wire, which therefore acts magnetically like a single loop.

The magnetic field lines from a Helmholtz coil are shown in Figure 28.18b. You can see that a region of uniform magnetic field (characterized by horizontal parallel segments of the field lines) in the center between the loops is present, in contrast to the field from a single loop shown in Figure 28.13. Again, these field lines were calculated with the aid of a computer to provide a qualitative understanding of the geometry of magnetic fields.

Taking multiple loops a step further, Figure 28.19 shows the magnetic field lines from four coaxial wire loops. The region of uniform magnetic field in the center of the loops is expanded, but note that the field is not uniform near the wires and near the two ends.

A strong uniform magnetic field is produced by a **solenoid**, consisting of many loops of a wire wound close together. Figure 28.20 shows the magnetic field lines from a solenoid with 600 turns, or loops. You can see that the magnetic field lines are very close together on the inside of the solenoid and far apart on the outside. Like that inside the Helmholtz coil (Figure 28.18b)

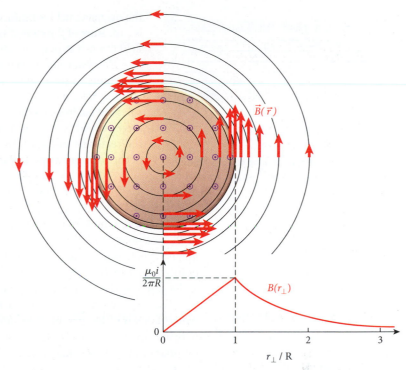

FIGURE 28.17 Radial dependence of the magnetic field for a wire with current flowing out of the page.

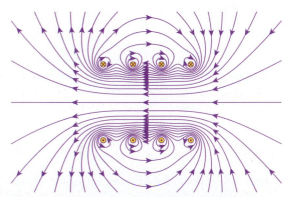

(a)

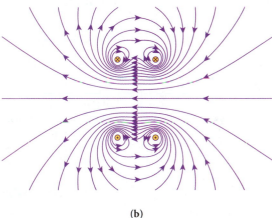

(b)

FIGURE 28.18 (a) A typical Helmholtz coil used in physics labs to generate a nearly constant magnetic field in the interior. (b) Magnetic field lines for a Helmholtz coil.

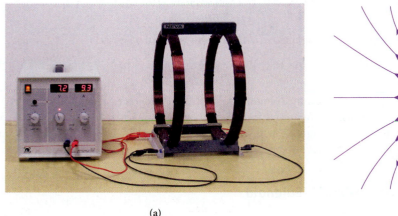

FIGURE 28.19 Magnetic field lines resulting from four coaxial wire loops with many windings.

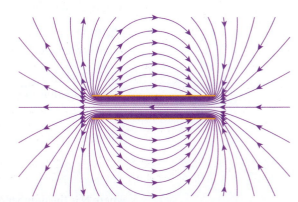

FIGURE 28.20 Magnetic field lines for a solenoid with 600 turns. The current along the top of the solenoid is directed into the page, and the current along the bottom of the solenoid is directed out of the page.

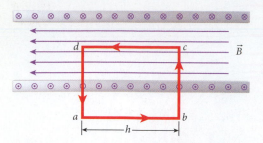

FIGURE 28.21 Amperian loop for determining the magnitude of the magnetic field of an ideal solenoid.

the magnetic field is uniform inside the solenoid coil. The spacing of the field lines is a measure of the strength of the magnetic field, and you can see that the magnetic field is much stronger inside the solenoid than outside the solenoid .

An *ideal* solenoid has a magnetic field of zero outside and of a uniform constant finite value inside. To determine the magnitude of the magnetic field inside an ideal solenoid, we can apply Ampere's Law (equation 28.10) to a section of a solenoid far from its ends (Figure 28.21). To do so, we first choose an Amperian loop over which to carry out the integration. A judicious choice, shown by the red rectangle in Figure 28.21, encloses some current and exploits the symmetry of the solenoid as well as simplifying the evaluation of the integral:

$$\oint \vec{B} \cdot d\vec{s} = \int_a^b \vec{B} \cdot d\vec{s} + \int_b^c \vec{B} \cdot d\vec{s} + \int_c^d \vec{B} \cdot d\vec{s} + \int_d^a \vec{B} \cdot d\vec{s}.$$

The value of the third integral on the right-hand side, between points c and d in the interior of the solenoid, is Bh. The values of the second and fourth integrals are zero because the magnetic field is perpendicular to the direction of integration. The first integral, between the points a and b in the exterior of the ideal solenoid, is zero because the magnetic field outside of an ideal solenoid is zero. Thus, the value of the integral over the entire Amperian loop is Bh.

The enclosed current is the current in the turns of the solenoid that are within the Amperian loop. The current is the same in each turn because the solenoid is made from one wire and the same current flows through each turn. Thus, the enclosed current is just the number of turns times the current:

$$i_{enc} = nhi,$$

where n is the number of turns per unit length. Therefore, according to Ampere's Law, we have

$$Bh = \mu_0 nhi.$$

Thus, the magnitude of the magnetic field inside an ideal solenoid is

$$B = \mu_0 ni. \tag{28.12}$$

Equation 28.12 is valid only away from the ends of the solenoid. Note that B does not depend on position inside the solenoid, so an ideal solenoid creates a constant and uniform magnetic field inside the solenoid and no field outside the solenoid. A real-world solenoid, like the one shown in Figure 28.20, has fringe fields near its ends but can still produce a high-quality uniform magnetic field.

EXAMPLE **28.2** Solenoid

The solenoid of the STAR detector at the Brookhaven National Laboratory, New York, discussed in Chapter 27 has a magnetic field of magnitude 0.50 T when carrying a current of 400 A. The solenoid is 8.0 m long.

PROBLEM

What is the number of turns in this solenoid, assuming that it is an ideal solenoid?

SOLUTION

We use equation 28.12 to calculate the magnitude of the magnetic field of an ideal solenoid:

$$B = \mu_0 ni. \tag{i}$$

The number of turns per unit length is given by

$$n = \frac{N}{L}, \tag{ii}$$

where N is the number of turns and L is the length of the solenoid. Substituting for n from equation (ii) into equation (i), we get

$$B = \mu_0 i \frac{N}{L}. \tag{iii}$$

Solving equation (iii) for the number of turns, we obtain

$$N = \frac{BL}{\mu_0 i} = \frac{(0.50\ \text{T})(8.0\ \text{m})}{\left(4\pi \cdot 10^{-7}\ \dfrac{\text{T m}}{\text{A}}\right)(400\ \text{A})} = 8000\ \text{turns}.$$

If a solenoid is bent so that the two ends meet (Figure 28.22), it acquires a doughnut shape (a torus), with the wire forming a series of loops, each with the same current flowing through it. This device is called a *toroidal magnet,* or **toroid.** Just as for an ideal solenoid, the magnetic field outside the coils of an ideal toroidal magnet is zero. The magnitude of the magnetic field inside the toroid coil can be calculated by using Ampere's Law and by assuming an Amperian loop in the form of a circle with radius r, with $r_1 < r < r_2$, where r_1 and r_2 are the inner and outer radii of the toroid. The magnetic field is always directed tangential to the Amperian loop, so we have

$$\oint \vec{B} \cdot d\vec{s} = 2\pi r B.$$

The enclosed current is the number of loops (or turns), N, in the toroid times the current, i, in the wire (in each loop); so, Ampere's law gives us

$$2\pi r B = \mu_0 N i.$$

Therefore, the magnitude of the magnetic field inside of a toroidal magnet is given by

$$B = \frac{\mu_0 N i}{2\pi r}. \qquad (28.13)$$

Note that, unlike the magnetic field inside a solenoid, the magnitude of the magnetic field inside a toroid does depend on the radius. As the radius increases, the magnitude of the magnetic field decreases. The direction of the magnetic field can be obtained using *right-hand rule 4:* If you wrap the fingers of your right hand around the toroid in the direction of the current, as shown in Figure 28.22, your thumb points in the direction of the magnetic field inside the toroid.

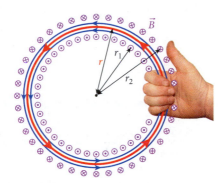

FIGURE 28.22 Toroidal magnet with Amperian loop (red) in the form of a circle with radius r. Right-hand rule 4 states that if you place the fingers of your right hand in the direction of the current flow, your thumb shows the direction of the magnetic field inside the toroid.

SOLVED PROBLEM 28.3 | Field of a Toroidal Magnet

A toroidal magnet is made from 202 m of copper wire that is capable of carrying a current of magnitude $i = 2.40$ A. The toroid has an average radius $R = 15.0$ cm and a cross-sectional diameter $d = 1.60$ cm (Figure 28.23a).

PROBLEM
What is the largest magnetic field that can be produced at the average toroidal radius, R?

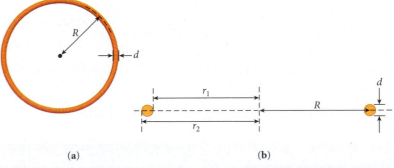

(a)   (b)

FIGURE 28.23 (a) A toroidal magnet. (b) Cross section of the toroidal magnet.

Continued—

SOLUTION

THINK

The number of turns in the toroidal magnet is given by the length of the wire divided by the circumference of the cross-sectional area of the coil. With these parameters, the magnetic field of the toroidal magnet at $r = R$ can be calculated.

SKETCH

Figure 28.23b shows a cross-sectional cut of the toroidal magnet.

RESEARCH

The magnitude of the magnetic field of a toroidal magnet is given by equation 28.13:

$$B = \frac{\mu_0 N i}{2\pi R}, \qquad \text{(i)}$$

where N is the number of turns and R is the radius at which the magnetic field is measured. The number of turns, N, is given by the length, L, of the wire divided by the circumference of the cross-sectional area:

$$N = \frac{L}{\pi d}, \qquad \text{(ii)}$$

where d is the diameter of the cross-sectional area of the toroid.

SIMPLIFY

We can combine equations (i) and (ii) to obtain an expression for B:

$$B = \frac{\mu_0 \left(L/\pi d \right) i}{2\pi R} = \frac{\mu_0 L i}{2\pi^2 R d}.$$

CALCULATE

Putting in the numerical values gives us

$$B = \frac{\mu_0 L i}{2\pi^2 R d} = \frac{\left(4\pi \cdot 10^{-7}\ \text{T m/A} \right)\left(202\ \text{m} \right)\left(2.40\ \text{A} \right)}{2\pi^2 \left(15.0 \cdot 10^{-2}\ \text{m} \right)\left(1.60 \cdot 10^{-2}\ \text{m} \right)} = 0.0128597\ \text{T}.$$

ROUND

We report our result to three significant figures:

$$B = 1.29 \cdot 10^{-2}\ \text{T}.$$

DOUBLE-CHECK

As a double-check, we calculate the magnitude of the field inside a solenoid that has the same length as the circumference of the toroidal magnet. The number of turns per unit length is

$$n = \frac{L/\pi d}{2\pi R} = \frac{L}{2\pi^2 R d} = \frac{\left(202\ \text{m} \right)}{2\pi^2 \left(15.0 \cdot 10^{-2}\ \text{m} \right)\left(1.60 \cdot 10^{-2}\ \text{m} \right)} = 4264\ \text{turns/m}.$$

The magnitude of the magnetic field of a solenoid with that number of turns per unit length is

$$B = \mu_0 n i = \left(4\pi \cdot 10^{-7}\ \text{T m/A} \right)\left(4264\ \text{m}^{-1} \right)\left(2.40\ \text{A} \right) = 1.29 \cdot 10^{-2}\ \text{T}.$$

Thus, our answer for the magnitude of the field inside the toroid seems reasonable.

28.5 Atoms as Magnets

The atoms that make up all matter contain moving electrons, which form current loops that produce magnetic fields. In most materials, these current loops are randomly oriented and produce no net magnetic field. Some materials have some fraction of these current loops aligned. These materials, called *magnetic materials* (Section 28.6), do produce a net magnetic field. Other materials can have their current loops aligned by an external magnetic field and become magnetized.

Let's consider a highly simplified model of the atom: An electron moving at constant speed v in a circular orbit with radius r (Figure 28.24). We can think of the moving charge of the electron as a current, i. Current is defined as the charge per unit time passing a particular point. For this case, the charge is the charge of the electron, with magnitude e, and the time is related to the period, T, of the electron's orbit. Thus, the magnitude of the current is given by

$$i = \frac{e}{T} = \frac{e}{2\pi r/v} = \frac{ve}{2\pi r}.$$

The magnitude of the magnetic dipole moment of the orbiting electron is given by

$$\mu_{orb} = iA = \frac{ve}{2\pi r}\left(\pi r^2\right) = \frac{ver}{2}. \tag{28.14}$$

The magnitude of the orbital angular momentum of the electron is

$$L_{orb} = rp = rmv,$$

where m is the mass of the electron. Solving equation 28.14 for v and substituting that expression into the expression for the orbital angular momentum gives us

$$L_{orb} = rm\left(\frac{2\mu_{orb}}{er}\right) = \frac{2m\mu_{orb}}{e}.$$

Because the magnetic dipole moment and the angular momentum are vector quantities, we can write

$$\vec{\mu}_{orb} = -\frac{e}{2m}\vec{L}_{orb}, \tag{28.15}$$

where the negative sign is needed because of the definition of the current as the direction of the flow of positive charge.

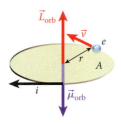

FIGURE 28.24 An electron moving with constant speed in a circular orbit in an atom.

EXAMPLE **28.3**   **Orbital Magnetic Moment of the Hydrogen Atom**

Assume that the hydrogen atom consists of an electron moving with speed v in a circular orbit with radius r around a stationary proton. Also assume that the centripetal force keeping the electron moving in a circle is the electrostatic force between the proton and the electron. The radius of the orbit of the electron is $r = 5.29 \cdot 10^{-11}$ m. (This radius is derived using concepts discussed in Chapter 38 on atomic physics.)

PROBLEM
What is the magnitude of the orbital magnetic moment of the hydrogen atom?

SOLUTION
The magnitude of the orbital magnetic moment is

$$\left|\mu_{orb}\right| = \frac{e}{2m}L_{orb} = \frac{e}{2m}(rmv) = \frac{erv}{2}. \tag{i}$$

Equating the magnitudes of the centripetal force keeping the electron moving in a circle to the electrostatic force between the electron and the proton gives us

$$\frac{mv^2}{r} = k\frac{e^2}{r^2},$$

Continued—

where k is the Coulomb constant. We can solve this equation for the speed of the electron

$$v = e\sqrt{\frac{k}{mr}}. \tag{ii}$$

Substituting for v from equation (ii) into equation (i) gives us

$$\left|\mu_{\text{orb}}\right| = \frac{er}{2}\left(e\sqrt{\frac{k}{mr}}\right) = \frac{e^2}{2}\sqrt{\frac{kr}{m}}.$$

When we put in the various numerical values, we get

$$\left|\mu_{\text{orb}}\right| = \frac{\left(1.602\cdot10^{-19}\,\text{C}\right)^2}{2}\sqrt{\frac{\left(8.99\cdot10^9\,\text{N m}^2/\text{C}^2\right)\left(5.29\cdot10^{-11}\,\text{m}\right)}{9.11\cdot10^{-31}\,\text{kg}}} = 9.27\cdot10^{-24}\,\text{A m}^2.$$

This result agrees with experimental measurements of the orbital magnetic moment of the hydrogen atom. However, other predictions about the properties of hydrogen atoms and other atoms based on the idea that electrons in atoms have circular orbits disagree with experimental observations. Thus, detailed description of the magnetic properties of atoms must incorporate phenomena described by quantum physics, which will be covered in Chapter 38.

Spin

The magnetic dipole moment from the orbital motion of electrons is not the only contribution to the magnetic moment of atoms. Electrons and other elementary particles have their own, intrinsic magnetic moments, due to their spin. The phenomenon of spin will be covered thoroughly in the discussion of quantum physics in Chapters 36 through 40, but some facts about spin and its connection to a particle's intrinsic angular momentum have been discovered experimentally and do not require an understanding of quantum mechanics. Electrons, protons, and neutrons all have a spin of magnitude $s = \frac{1}{2}$. The magnitude of the angular momentum of these particles is $S = \hbar\sqrt{s(s+1)}$, and the z-component of the angular momentum can have a value of either $S_z = -\frac{1}{2}\hbar$ or $S_z = +\frac{1}{2}\hbar$. $\hbar$ is Planck's constant divided by 2π. This spin cannot be explained by orbital motion of some substructure in the particles. Electrons, for example, are apparently true point particles. Thus, spin is an intrinsic property, similar to mass or electric charge.

The magnetic character of bulk matter is determined largely by electron spin magnetic moments. The magnetic moment of a particle with spin, $\vec{\mu}_s$, is related to its spin angular momentum, $\vec{S}$, via

$$\vec{\mu}_s = g\frac{q}{2m}\vec{S}, \tag{28.16}$$

where q is the charge of the elementary particle and m its mass. The quantity g is dimensionless and is called the *g-factor*. For the electron, its numerical value is $g = -2.0023193043622(15)$, one of the most precisely measured quantities in nature. If you compare this equation to equation 28.15 for the magnetic dipole moment due to the orbital angular momentum, you see that they are very similar.

28.6   Magnetic Properties of Matter

In Chapter 27, we saw that magnetic dipoles do not experience a net force in a homogeneous external magnetic field, but do experience a torque. This torque drives a single free dipole to an orientation in which it is antiparallel to the external field, because this is the state with the lowest magnetic potential energy. We've just seen in Section 28.5 that atoms can have magnetic dipoles. What happens when matter (which is composed of atoms) is exposed to an external magnetic field?

The dipole moments of the atoms in a material can point in different directions or in the same direction. The **magnetization, $\vec{M}$,** of a material is defined as the net dipole moment created by the dipole moments of the atoms in the material per unit volume. The magnetic field, $\vec{B}$, inside the material then depends on the external magnetic field, $\vec{B}_0$, and the magnetization, $\vec{M}$:

$$\vec{B} = \vec{B}_0 + \mu_0 \vec{M}, \qquad (28.17)$$

where μ_0 is again the magnetic permeability of free space. Instead of including the external magnetic field, $\vec{B}_0$, it is customary to use the **magnetic field strength**, $\vec{H}$:

$$\vec{H} = \frac{\vec{B}_0}{\mu_0}. \qquad (28.18)$$

With this definition of the magnetic field strength, equation 28.17 can be written

$$\vec{B} = \mu_0(\vec{H} + \vec{M}). \qquad (28.19)$$

Since the unit of magnetic field is $[B] = \text{T}$ and the unit of the magnetic permeability is $[\mu_0] = \text{T m/A}$, the units of magnetization and magnetic field strength are $[M] = [H] = \text{A/m}$.

Diamagnetism and Paramagnetism

The question not yet answered is how the magnetization depends on the external magnetic field, $\vec{B}_0$, or, equivalently, on the magnetic field strength, $\vec{H}$. For most materials (not all!) this relationship is linear:

$$\vec{M} = \chi_m \vec{H}, \qquad (28.20)$$

where the proportionality constant χ_m is called the **magnetic susceptibility** of the material (Table 28.1). But there are materials that do not obey the simple linear relationship of equation 28.20, and the most prominent among those are ferromagnets, which we'll discuss in the next subsection. Let's first examine diamagnetic and paramagnetic materials, for which equation 28.20 holds.

If $\chi_m < 0$, the dipoles inside the material tend to arrange themselves to oppose an external magnetic field, just like free dipoles. In this case, the magnetization vector points in the direction opposite to the magnetic field strength vector. Materials with $\chi_m < 0$ are said to

Table 28.1	Values of Magnetic Susceptibility for Some Common Diamagnetic and Paramagnetic Materials	
Material	**Magnetic Susceptibility ($\chi_m \cdot 10^5$)**
Aluminum	+2.2
Bismuth	−16.6
Diamond (carbon)	−2.1
Graphite (carbon)	−1.6
Hydrogen	−0.00022
Lead	−1.8
Lithium	1.4
Mercury	−2.9
Oxygen	+0.19
Platinum	+26.5
Silicon	−0.37
Sodium	+0.72
Sodium chloride (NaCl)	−1.4
Tungsten	+6.8
Uranium	+40
Water	−0.9

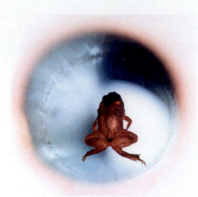

FIGURE 28.25 A live frog being levitated by a strong magnetic field at the High Field Magnet Laboratory, Radboud University Nijmegen, The Netherlands.

be *diamagnetic*. Most materials exhibit **diamagnetism.** In diamagnetic materials, a weak magnetic dipole moment is induced by an external magnetic field in a direction opposite to the direction of the external field. The induced magnetic field disappears when the external field is removed. If the external field is nonuniform, interaction of the induced dipole moment of the diamagnetic material with the external field creates a force directed from a region of greater magnetic field strength to a region of lower magnetic field strength.

An example of biological material exhibiting diamagnetism is shown in Figure 28.25. Diamagnetic forces induced by a nonuniform external magnetic field of 16 T are levitating a live frog. (This experience apparently did not bother the frog.) The normally negligible diamagnetic force is large enough in this case to overcome gravity.

If the magnetic susceptibility in equation 28.20 is greater than zero, $\chi_m > 0$, the magnetization of the material points in the same direction as the magnetic field strength. Note that χ_m for a vacuum is 0. This property is **paramagnetism,** and materials that exhibit it are said to be *paramagnetic*. Materials containing certain transition elements (including actinides and rare earths) exhibit paramagnetism. Each atom of these elements has a permanent magnetic dipole, but normally these dipole moments are randomly oriented and produce no net magnetic field. However, in the presence of an external magnetic field, some of these magnetic dipole moments align in the same direction as the external field. When the external field is removed, the induced magnetic dipole moment disappears. If the external field is nonuniform, this induced magnetic dipole moment interacts with the external field to produce a force directed from a region of lower magnetic field strength to a region of higher magnetic field strength—just the opposite of the effect of diamagnetism.

Substituting the expression for $\vec{M}$ from equation 28.20 into equation 28.19 for the magnetic field, $\vec{B}$, inside a material gives

$$\vec{B} = \mu_0(\vec{H} + \vec{M}) = \mu_0(\vec{H} + \chi_m\vec{H}) = \mu_0(1 + \chi_m)\vec{H}. \tag{28.21}$$

In analogy to the relative electric permittivity introduced in Chapter 24, the **relative magnetic permeability,** κ_m, is commonly defined as

$$\kappa_m = 1 + \chi_m. \tag{28.22}$$

Then, the magnetic permeability, μ, of a material can be expressed as

$$\mu = (1 + \chi_m)\mu_0 = \kappa_m\mu_0. \tag{28.23}$$

Replacing μ_0 with μ in the Biot-Savart Law (equation 28.1) and Ampere's Law (equation 28.10) enables us to use these laws for calculating the magnetic field in a particular material.

Finally for paramagnetic materials, there is a temperature dependence to the magnitude of the magnetization. Conventionally, this temperature dependence is expressed via *Curie's Law:*

$$M = \frac{cB}{T}, \tag{28.24}$$

where c is Curie's constant, B is the magnitude of the magnetic field, and T is the temperature in Kelvins.

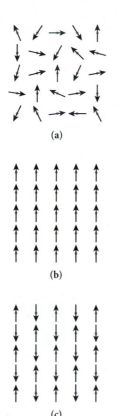

(a)

(b)

(c)

FIGURE 28.26 Magnetic domains: (a) randomly oriented; (b) perfect ferromagnetic order; (c) perfect antiferromagnetic order.

Ferromagnetism

The elements iron, nickel, cobalt, gadolinium, and dysprosium—and alloys containing these elements—exhibit **ferromagnetism.** A ferromagnetic material shows long-range ordering at the atomic level, which causes the dipole moments of atoms to line up with each other in a limited region called a **domain.** Within a domain, the magnetic field can be strong. However, in bulk samples of the material, domains are randomly oriented, leaving no net magnetic field. Figure 28.26a shows randomly oriented magnetic dipole moments in a domain, and Figure 28.26b shows perfect ferromagnetic order. Figure 28.26c illustrates the interesting case of perfect antiferromagnetic order, in which the interaction between neighboring magnetic dipole moments causes them to be oriented in opposite directions. This ordering can be realized only at very low temperatures.

An external magnetic field can align domains as shown in Figure 28.26b, as a result of the interaction between the magnetic dipole moments of the domain and the external field. As a result, a ferromagnetic material retains all or some of its induced magnetism when the external magnetic field is removed, since the domains stay aligned. In addition, the magnetic field produced by a current in a solenoid or a toroid will be larger if a ferromagnetic material is present in the device. But in contrast to diamagnetic and paramagnetic materials, ferromagnetic materials do not obey the simple linear relationship given in equation 28.20. The domains retain their orientations, and thus the material exhibits a nonzero magnetization even in the absence of an external magnetic field. (This is why permanent magnets exist.)

Figure 28.27 illustrates the dependence of the magnetization on the magnetic field strength for the three types of materials we've discussed. Figure 28.27a and Figure 28.27b show the linear dependence according to equation 28.20 for diamagnetic and paramagnetic materials, respectively. Figure 28.27c shows the typical hysteresis loop obtained for ferromagnetic materials. The arrows on the red curve show the direction in which the magnetization process develops, and the dashed lines represent the maximum magnetization (positive and negative) possible. For any point on this hysteresis loop, the magnetization can be expressed in terms of an effective value of the magnetic permeability, μ, of the ferromagnetic material, similar to what is given in equation 28.23; however, this permeability is *not* a constant but depends on the applied magnetic field strength and even on the path by which that value of the field strength was attained. Regardless, the values of the effective permeability, μ, for ferromagnetic materials can be very large compared to those measured for paramagnetic materials (greater by a factor of up to 10^4).

Ferromagnetism exhibits temperature dependence. At a certain temperature, called the *Curie temperature*, ferromagnetic materials cease to exhibit ferromagnetism. At this point, the ferromagnetic order due to the interaction of the dipole moments in these materials is overwhelmed by the thermal motion. For iron, the Curie temperature is 768 °C. Figure 28.28 shows a simple demonstration in which heating a permanent ferromagnet above its Curie temperature (Figure 28.28b) destroys the attraction between it and another permanent magnet (Figure 28.28c). As the magnet subsequently cools below its Curie temperature (Figure 28.28d), it again becomes a permanent magnet (Figure 28.28e).

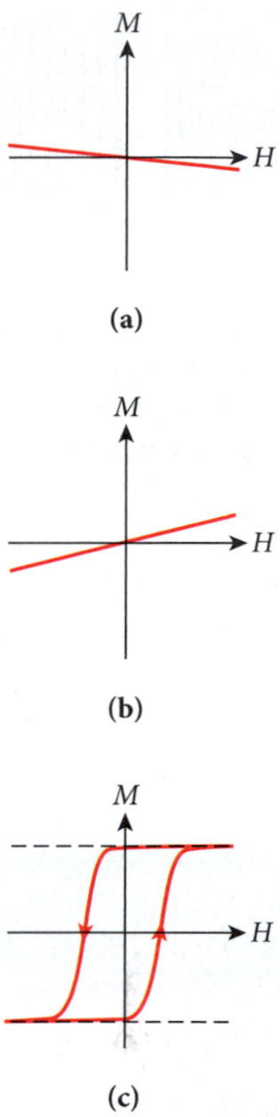

FIGURE 28.27 Magnetization as a function of the magnetic field strength: (a) for diamagnetic materials; (b) for paramagnetic materials; (c) the hysteresis loop for ferromagnetic materials.

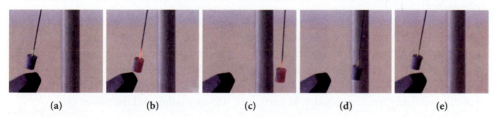

(a)   (b)   (c)   (d)   (e)

FIGURE 28.28 Demonstration of the Curie temperature: (a) A permanent magnet forms the bob of a pendulum and is deflected from the vertical and held there by another permanent magnet (lower left corner of each frame); (b) the magnet is heated and begins to glow red and approaches its Curie temperature; (c) the magnet is above its Curie temperature and hangs straight down, showing it is not magnetic any more at this temperature; (d) as the magnet cools below the Curie temperature, it begins to return to being a permanent magnet; and (e) it returns to its original equilibrium position.

28.7 Magnetism and Superconductivity

Magnets for industrial applications and scientific research can be constructed using ordinary resistive wire with current flowing through it. A typical magnet of this type is a large solenoid. The current flowing through the wire of the magnet produces resistive heating, and the heat is usually removed by low-conductivity water flowing through hollow conductors. (Low-conductivity water has been purified so that it does not conduct electricity.) These room-temperature magnets typically produce magnetic fields with strengths up to 1.5 T and are usually relatively inexpensive to construct but are expensive to operate because of the high cost of electricity.

Some applications, such as magnetic resonance imaging (MRI), require magnetic fields of the highest possible magnitude to ensure the best signal-to-noise ratio in the measurements.

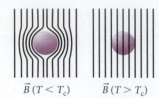

$$\vec{B}\,(T < T_c) \qquad \vec{B}\,(T > T_c)$$

FIGURE 28.29 The Meissner effect, in which a superconductor excludes external magnetic fields from its interior below the critical temperature at which the material becomes superconducting.

FIGURE 28.30 Through the Meissner effect, a superconductor expels the magnetic field of a permanent magnet, which thus hovers above it.

To achieve these fields, magnets are constructed using superconducting coils rather than resistive coils. Such a magnet can produce a stronger field than a room-temperature magnet, with a magnitude of 10 T or higher. Materials such as mercury and lead exhibit superconductivity at liquid helium temperatures, but some metals that are good conductors at room temperature, such as copper and gold, never become superconducting. The disadvantage of a superconducting magnet is that the conductor must be kept at the temperature of liquid helium, which is approximately 4 K (although recent discoveries described later in this section are easing this limitation). Thus, the magnet must be enclosed in a cryostat filled with liquid helium to keep it cold. An advantage of a superconducting magnet is the fact that once the current is established in the coil of the magnet, it will continue to flow until it is removed by external means. However, the energy saving realized by having no resistive loss in the coil is at least partially offset by the expenditure of energy required to keep the superconducting coil cold.

When current flows through superconducting mercury or lead, the material becomes almost perfectly diamagnetic ($\chi_m \approx -1$). According to equation 28.21, this means that the magnetic field inside the material becomes zero. Thus, the magnetic field is excluded from the superconducting material, and only small current densities can be achieved. The zero magnetic field inside a material cooled sufficiently so it becomes a superconductor is called the Meissner effect. Above the critical temperature, T_c, for the transition to superconductivity, the Meissner effect disappears, and the material becomes a normal conductor (Figure 28.29).

Figure 28.30 shows an impressive demonstration of the Meissner effect: A piece of a superconductor (cooled to a temperature below its critical temperature) causes a permanent magnet to float above it by expelling the magnet's intrinsic magnetic field. It achieves this because superconducting currents on its surface produce a magnetic field opposed to the applied field, which yields a net field of zero inside the superconductor and repulsion between the fields above the superconductor.

The conductor used in a superconducting magnet is specially designed to overcome the Meissner effect. Modern superconductors are constructed from filaments of niobium-titanium alloy embedded in solid copper. The niobium-titanium filaments have microscopic domains in which a magnetic field can exist without being excluded. The copper serves as a mechanical support and can take over the current load should the superconductor become normally conducting. This type of superconductor can produce magnetic fields with magnitudes as high as 15 T.

During the last two decades, physicists and engineers have discovered new materials that are superconducting at temperatures well above 4 K. Critical temperatures of up to 160 K have been reported for these *high-temperature superconductors*, which means that they can be made superconducting by cooling them with liquid nitrogen. Many researchers around the world are looking for materials that are superconducting at room temperature. These materials would revolutionize many areas of industry, in particular, transportation and the power grid.

WHAT WE HAVE LEARNED | EXAM STUDY GUIDE

- The magnetic permeability of free space, μ_0, is given by $4\pi \cdot 10^{-7}$ T m/A.

- The Biot-Savart Law, $d\vec{B} = \dfrac{\mu_0}{4\pi}\dfrac{i\,d\vec{s}\times\hat{r}}{r^2}$, describes the differential magnetic field, $d\vec{B}$, caused by a current element, $i\,d\vec{s}$, at position $\vec{r}$ relative to the current element.

- The magnitude of the magnetic field at distance $r_\perp$ from a long, straight wire carrying current i is $B = \mu_0 i/2\pi r_\perp$.

- The magnetic field magnitude at the center of a loop with radius R carrying current i is $B = \mu_0 i/2R$.

- Ampere's Law is given by $\oint \vec{B}\cdot d\vec{s} = \mu_0 i_{enc}$, where $d\vec{s}$ is the integration path and i_{enc} is the current enclosed in a chosen Amperian loop.

- The magnitude of the magnetic field inside a solenoid carrying current i and having n turns per unit length is $B = \mu_0 ni$.

- The magnitude of the magnetic field inside a toroid having N turns and carrying current i at radius r is given by $B = \mu_0 Ni/2\pi r$.

- For an electron with charge $-e$ and mass m moving in a circular orbit, the magnetic dipole moment can be related to the orbital angular momentum through $\vec{\mu}_{orb} = -\dfrac{e}{2m}\vec{L}_{orb}$.

- For diamagnetic and paramagnetic materials, magnetization is proportional to the magnetic field strength: $\vec{M} = \chi_m\vec{H}$. Ferromagnetic materials follow a hysteresis loop and thus deviate from this linear relationship.

- The magnetic field inside a diamagnetic or paramagnetic material is due to the external magnetic field strength and the magnetization:

$$\vec{B} = \mu_0(\vec{H} + \vec{M}) = \mu_0(\vec{H} + \chi_m\vec{H}) =$$

$\mu_0(1 + \chi_m)\vec{H} = \mu_0\kappa_m\vec{H} = \mu\vec{H}$, where κ_m is the relative magnetic permeability.

- The four right-hand rules related to magnetic fields are shown in Figure 28.31. Right-hand rule 1 gives the direction of the magnetic force on a charged particle moving in a magnetic field. Right-hand rule 2 gives the direction of the unit normal vector for a current-carrying loop. Right-hand rule 3 gives the direction of the magnetic field from a current-carrying wire. Right-hand rule 4 gives the direction of the magnetic field inside a toroidal magnet.

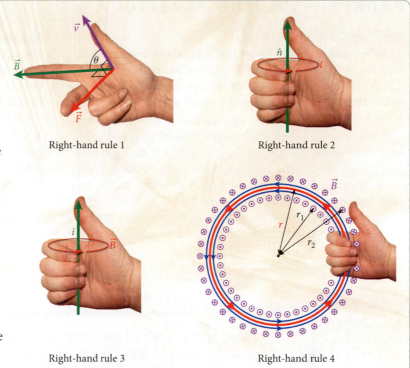

Right-hand rule 1   Right-hand rule 2

Right-hand rule 3   Right-hand rule 4

FIGURE 28.31 Four right-hand rules related to magnetic fields.

KEY TERMS

Biot-Savart Law, p. 894
magnetic permeability of
  free space, p. 894
Ampere's Law, p. 903
Amperian loop, p. 903

Helmholtz coil, p. 904
solenoid, p. 905
toroid, p. 907
magnetization, p. 910

magnetic field strength,
  p. 911
magnetic susceptibility,
  p. 911
diamagnetism, p. 912

paramagnetism, p. 912
relative magnetic
  permeability, p. 912
ferromagnetism, p. 912
domain, p. 912

NEW SYMBOLS AND EQUATIONS

$\mu_0 = 4\pi \cdot 10^{-7}$ T m/A, magnetic permeability of free space

$d\vec{s}$, vector direction of integration in Ampere's Law

i_{enc}, enclosed current inside an Amperian loop

$\oint \vec{B} \cdot d\vec{s} = \mu_0 i_{enc}$, Ampere's Law

$\vec{\mu}_{orb}$, orbital magnetic dipole moment for an electron in circular orbit

$\vec{L}_{orb}$, orbital angular momentum for an electron moving in a circular orbit in an atom

$\vec{M}$, magnetization

$\vec{H} = \vec{B}_0/\mu_0$, magnetic field strength

χ_m, magnetic susceptibility

κ_m, relative magnetic permeability

ANSWERS TO SELF-TEST OPPORTUNITIES

28.1 The magnetic field at point P_1 points in the positive y-direction. The magnetic field at point P_2 points in the negative x-direction.

28.2 Two parallel wires carrying current in the same direction attract each other. Two parallel wires carrying current in the opposite direction repel each other.

28.3 $B_x = \dfrac{\mu_0 i}{2} \dfrac{R^2}{\left(0^2 + R^2\right)^{3/2}} = \dfrac{\mu_0 i}{2} \dfrac{R^2}{R^3} = \dfrac{\mu_0 i}{2R}$.

PROBLEM-SOLVING PRACTICE

Problem-Solving Guidelines

1. When using the Biot-Savart Law, you should always draw a diagram of the situation, with the current element highlighted. Check for simplifying symmetries before proceeding with calculations; you can save yourself a significant amount of work.

2. When applying Ampere's Law, choose an Amperian loop that has some geometrical symmetry, in order to simplify the evaluation of the integral. Often, you can use right-hand rule 3 to choose the direction of integration along the loop: Point your thumb in the direction of the net current through the loop and your fingers curl in the direction of integration. This method will also remind you to sum the currents through the Amperian loop to determine the enclosed current.

3. Remember the superposition principle for magnetic fields: The net magnetic field at any point in space is the vector sum of the individual magnetic fields generated by different objects. Make sure you do not simply add the magnitudes. Instead, you generally need to add the spatial components of the different sources of magnetic field separately.

4. All of the principles governing motion of charged particles in magnetic fields and all of the problem-solving guidelines presented in Chapter 27 still apply. It does not matter if the magnetic field is due to a permanent magnet or an electromagnet.

5. In order to calculate the magnetic field in a material, you can use the formulas derived from Ampere's Law and Biot-Savart's Law, but you have to replace μ_0 with $\mu \equiv \kappa_m \mu_0 \equiv (1 + \chi_m)\mu_0$.

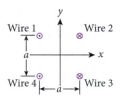

FIGURE 28.32 Four wires located at the corners of a square. Two of the wires are carrying current into the page, and the other two are carrying current out of the page.

FIGURE 28.33 The magnetic fields from the four current-carrying wires.

SOLVED PROBLEM 28.4 | Magnetic Field from Four Wires

Four wires are each carrying a current of magnitude $i = 1.00$ A. The wires are located at the four corners of a square with side $a = 3.70$ cm. Two of the wires are carrying current into the page, and the other two are carrying current out of the page (Figure 28.32).

PROBLEM
What is the y-component of the magnetic field at the center of the square?

SOLUTION

THINK
The magnetic field at the center of the square is the vector sum of the magnetic fields from the four current-carrying wires. The magnitude of the magnetic field from all four wires is the same. The direction of the magnetic field from each wire is determined using right-hand rule 3.

SKETCH
Figure 28.33 shows the magnetic fields from the four wires: $\vec{B}_1$ is the magnetic field from wire 1, $\vec{B}_2$ is the magnetic field from wire 2, $\vec{B}_3$ is the magnetic field from wire 3, and $\vec{B}_4$ is the magnetic field from wire 4. Note that $\vec{B}_2$ and $\vec{B}_4$ are equal and $\vec{B}_1$ and $\vec{B}_3$ are equal.

RESEARCH
The magnitude of the magnetic field from each of the four wires is given by

$$B = \frac{\mu_0 i}{2\pi r} = \frac{\mu_0 i}{2\pi \left(a/\sqrt{2} \right)},$$

where $a/\sqrt{2}$ is the distance from each wire to the center of the square.
Right-hand rule 3 gives us the directions of the magnetic fields, which are shown in Figure 28.33. The y-component of each of the magnetic fields is given by

$$B_y = B \sin 45°.$$

SIMPLIFY
The sum of the y-components of the four magnetic fields is

$$B_{y,\text{sum}} = 4B_y = 4B \sin 45° = 4 \frac{\mu_0 i}{2\pi \left(a/\sqrt{2} \right)} \left(\frac{1}{\sqrt{2}} \right) = \frac{2\mu_0 i}{\pi a},$$

where we have used $\sin 45° = 1/\sqrt{2}$.

CALCULATE

Putting in the numerical values gives us

$$B_{y,\text{sum}} = \frac{2\mu_0 i}{\pi a} = \frac{2\left(4\pi \cdot 10^{-7} \text{ T m/A}\right)\left(1.00 \text{ A}\right)}{\pi\left(3.70 \cdot 10^{-2} \text{ m}\right)} = 2.16216 \cdot 10^{-5} \text{ T.}$$

ROUND

We report our result to three significant figures:

$$B_{y,\text{sum}} = 2.16 \cdot 10^{-5} \text{ T.}$$

DOUBLE-CHECK

To double-check our result, we calculate the magnitude of the magnetic field from one wire at the center of the square:

$$B = \frac{\mu_0 i}{2\pi\left(a/\sqrt{2}\right)} = \frac{\left(4\pi \cdot 10^{-7} \text{ T m/A}\right)\left(1.00 \text{ A}\right)\sqrt{2}}{2\pi\left(3.70 \cdot 10^{-2} \text{ m}\right)} = 7.64 \cdot 10^{-6} \text{ T.}$$

The sum of the y-components is then

$$B_{\text{sum}} = \frac{4\left(7.64 \cdot 10^{-6} \text{ T}\right)}{\sqrt{2}} = 2.16 \cdot 10^{-5} \text{ T,}$$

which agrees with our result.

SOLVED PROBLEM 28.5 | Electron Motion in a Solenoid

An ideal solenoid has 200.0 turns/cm. An electron inside the coil of a solenoid moves in a circle with radius $r = 3.00$ cm perpendicular to the solenoid's axis. The electron moves with a speed of $v = 0.0500c$, where c is the speed of light.

PROBLEM

What is the current in the solenoid?

SOLUTION

THINK

The solenoid produces a uniform magnetic field, which is proportional to the current flowing in the solenoid. The radius of circular motion of the electron is related to the speed of the electron and the magnetic field inside the solenoid.

SKETCH

Figure 28.34 shows the circular path of the electron in the uniform magnetic field of the solenoid.

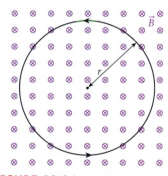

FIGURE 28.34 Electron traveling in a circular path inside a solenoid.

RESEARCH

The magnitude of the magnetic field inside the solenoid is given by

$$B = \mu_0 n i, \tag{i}$$

where i is the current in the solenoid and n is the number of turns per unit length. The magnetic force provides the centripetal force needed for the electron to move in a circle and so the radius of the electron's path can be related to B:

$$r = \frac{mv}{eB}, \tag{ii}$$

where m is the electron's mass, v is its speed, and e is the magnitude of the electron's charge.

Continued—

SIMPLIFY
Combining equations (i) and (ii), we have

$$r = \frac{mv}{e\left(\mu_0 ni\right)}.$$

Solving this equation for the current in the solenoid, we obtain

$$i = \frac{mv}{er\mu_0 n}. \tag{iii}$$

CALCULATE
The speed of the electron was specified in terms of the speed of light:

$$v = 0.0500c = 0.0500\left(3.00 \cdot 10^8 \text{ m/s}\right) = 1.50 \cdot 10^7 \text{ m/s}.$$

Putting this and the other numerical values into equation (iii), we get

$$i = \frac{mv}{er\mu_0 n} = \frac{\left(9.11 \cdot 10^{-31} \text{ kg}\right)\left(1.50 \cdot 10^7 \text{ m/s}\right)}{\left(1.602 \cdot 10^{-19} \text{ C}\right)\left(3.00 \cdot 10^{-2} \text{ m}\right)\left(4\pi \cdot 10^{-7} \text{ T m/A}\right)\left(200 \cdot 10^2 \text{ m}^{-1}\right)}$$

$$= 0.113132 \text{ A}.$$

ROUND
We report our result to three significant figures:

$$i = 0.113 \text{ A}.$$

DOUBLE-CHECK
To double-check our result, we use it to calculate the magnitude of the magnetic field inside the solenoid:

$$B = \mu_0 ni = \left(4\pi \cdot 10^{-7} \text{ T m/A}\right)\left(200 \cdot 10^2 \text{ m}^{-1}\right)\left(0.113 \text{ A}\right) = 0.00284 \text{ T}.$$

This magnitude of magnetic field seems reasonable. Thus, our calculated value for the current in the solenoid seems reasonable.

MULTIPLE-CHOICE QUESTIONS

28.1 Two long, straight wires are parallel to each other. The wires carry currents of different magnitudes. If the amount of current flowing in each wire is doubled, the magnitude of the force between the wires will be

a) twice the magnitude of the original force.

b) four times the magnitude of the original force.

c) the same as the magnitude of the original force.

d) half of the magnitude of the original force.

28.2 A current element produces a magnetic field in the region surrounding it. At any point in space, the magnetic field produced by this current element points in a direction that is

a) radial from the current element to the point in space.

b) parallel to the current element.

c) perpendicular to the current element and to the radial direction.

28.3 The number of turns in a solenoid is doubled, and its length is halved. How does its magnetic field change?

a) it doubles   c) it quadruples

b) it is halved   d) it remains unchanged

28.4 The magnetic force cannot do work on a charged particle since the force is always perpendicular to the velocity. How then can magnets pick up nails? Consider two parallel current-carrying wires. The magnetic fields cause attractive forces between the wires, so it appears that the magnetic field due to one wire is doing work on the other wire. How is this explained?

a) The magnetic force can do no work on isolated charges; this says nothing about the work it can do on charges confined in a conductor.

b) Since only an electric field can do work on charges, it is actually the electric fields doing the work here.

c) This apparent work is due to another type of force.

28.5 In a solenoid in which the wires are wound such that each loop touches the adjacent ones, which of the following will increase the magnetic field inside the magnet?

a) making the radius of the loops smaller

b) increasing the radius of the wire

c) increasing the radius of the solenoid

d) decreasing the radius of the wire

e) immersion of the solenoid in gasoline

28.6 Two insulated wires cross at a 90° angle. Currents are sent through the two wires. Which one of the figures best represents the configuration of the wires, if the current in the horizontal wire flows in the positive x-direction and the current in the vertical wire flows in the positive y-direction?

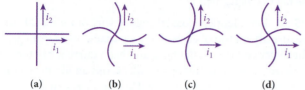

(a)   (b)   (c)   (d)

28.7 What is a good rule of thumb for designing a simple magnetic coil? Specifically, given a circular coil of radius ~1 cm, what is the approximate magnitude of the magnetic field, in gauss per amp per turn? (*Note:* 1 G = 0.0001 T.)

a) 0.0001 G/(A-turn)

b) 0.01 G/(A-turn)

c) 1 G/(A-turn)

d) 100 G/(A-turn)

28.8 A solid cylinder carries a current that is uniform over its cross section. Where is the magnitude of the magnetic field the greatest?

a) at the center of the cylinder's cross section

b) in the middle of the cylinder

c) at the surface

d) none of the above

28.9 Two long, straight wires have currents flowing in them in the same direction as shown in the figure. The force between the wires is

a) attractive.   b) repulsive.   c) zero.

28.10 In a magneto-optic experiment, a liquid sample in a 10-mL spherical vial is placed in a highly uniform magnetic field, and a laser beam is directed through the sample. Which of the following should be used to create the uniform magnetic field required by the experiment?

a) a 5-cm-diameter flat coil consisting of one turn of 4-gauge wire

b) a 10-cm-diameter, 20 turn, single layer, tightly wound coil made of 18-gauge wire

c) a 2-cm-diameter, 10-cm long, tightly wound solenoid made of 18-gauge wire

d) a set of two coaxial 10-cm-diameter coils at a distance of 5 cm apart, each consisting of one turn of 4-gauge wire

QUESTIONS

28.11 Many electrical applications use twisted-pair cables in which the ground and signal wires spiral about each other. Why?

28.12 Discuss how the accuracy of a compass needle in showing the true direction of north can be affected by the magnetic field due to currents in wires and appliances in a residential building.

28.13 Can an ideal solenoid, one with no magnetic field outside the solenoid, exist? If not, does that render the derivation of the magnetic field inside the solenoid (Section 28.4) void?

28.14 Conservative forces tend to act on objects in such a way as to minimize the system's potential energy. Use this principle to explain the direction of the force on the current-carrying loop described in Example 28.1.

28.15 Two particles, each with charge q and mass m, are traveling in a vacuum on parallel trajectories a distance d apart, both at speed v (much less than the speed of light). Calculate the ratio of the magnitude of the magnetic force that each exerts on the other to the magnitude of the electric force that each exerts on the other: F_m/F_e.

28.16 A long, straight cylindrical tube of inner radius a and outer radius b, carries a total current i uniformly across its cross section. Determine the magnitude of the magnetic field from the tube at the midpoint between the inner and outer radii.

28.17 Three identical straight wires are connected in a T, as shown in the figure. If current i flows into the junction, what is the magnetic field at point P, a distance d from the junction?

28.18 In a certain region, there is a constant and uniform magnetic field, $\vec{B}$. Any electric field in the region is also unchanging in time. Find the current density, $\vec{J}$, in this region.

28.19 The magnetic character of bulk matter is determined largely by electron spin magnetic moments, rather than by orbital dipole moments. (Nuclear contributions are negligible, as the proton's spin magnetic moment is about 658 times smaller than that of the electron.) If the atoms or molecules of a substance have unpaired electron spins, the associated magnetic moments give rise to paramagnetic behavior or to ferromagnetic behavior if the interactions between atoms or molecules are strong enough to align them in domains. If the atoms or molecules have no net unpaired spins, then magnetic perturbations of the electron orbits give rise to diamagnetic behavior.

a) Molecular hydrogen gas (H_2) is weakly diamagnetic. What does this imply about the spins of the two electrons in the hydrogen molecule?

b) What would you expect the magnetic behavior of atomic hydrogen gas (H) to be?

28.20 Exposed to sufficiently high magnetic fields, materials *saturate*, or approach a maximum magnetization. Would you expect the saturation (maximum) magnetization of paramagnetic materials to be much less than, roughly the same as, or much greater than that of ferromagnetic materials? Explain why.

28.21 A long, straight wire carries a current, as shown in the figure. A single electron is shot directly toward the wire from above. The trajectory of the electron and the wire are in the same plane. Will the electron be deflected from its initial path, and if so, in which direction?

PROBLEMS

A blue problem number indicates a worked-out solution is available in the Student Solutions Manual. One • and two •• indicate increasing level of problem difficulty.

Sections 28.1 and 28.2

28.22 Two long parallel wires are separated by 3.0 mm. The current flowing in one of the wires is twice that in the other wire. If the magnitude of the force on a 1.0-m length of one of the wires is 7.0 μN, what are the values of the two currents?

28.23 An electron is shot from an electron gun with a speed of $4.0 \cdot 10^5$ m/s and moves parallel to and a distance of 5.0 cm above a long, straight wire carrying a current of 15 A. Determine the magnitude and the direction of the acceleration of the electron the instant it leaves the electron gun.

28.24 An electron moves in a straight line at a speed of $5 \cdot 10^6$ m/s. What is the magnitude and the direction of the magnetic field created by the moving electron at a distance $d = 5$ m ahead of it on its line of motion? How does the answer change if the moving particle is a proton?

28.25 Suppose that the magnetic field of the Earth were due to a single current moving in a circle of radius 2000 km through the Earth's molten core. The strength of the Earth's magnetic field on the surface near a magnetic pole is about $6 \cdot 10^{-5}$ T. About how large a current would be required to produce such a field?

28.26 A square ammeter has sides of length 3 cm. The sides of the ammeter are capable of measuring the magnetic field they are subject to. When the ammeter is clamped around a wire carrying a DC current, as shown in the figure, the average value of the magnetic field measured in the sides is 3 G. What is the current in the wire?

•27.27 A long, straight wire carrying a 2-A current lies along the x-axis. A particle with charge $q = -3$ μC passes parallel to the z-axis through the point $(x,y,z) = (0,2,0)$. Where in the xy-plane should another long, straight wire be placed so that there is no magnetic force on the particle at the point where it crosses the plane?

•28.28 Find the magnetic field in the center of a wire semicircle like that shown in the figure, with radius $R = 10.0$ cm, if the current is $i = 12.0$ A.

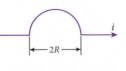

•28.29 Two very long wires run parallel to the z-axis, as shown in the figure. They each carry a current, $i_1 = i_2 = 25.0$ A, in the direction of the positive z-axis. The magnetic field of the Earth is given by $\vec{B} = (2.6 \cdot 10^{-5})\hat{y}$ T (in the xy-plane and pointing due north). A magnetic compass needle is placed

at the origin. Determine the angle θ between the compass needle and the x-axis. (*Hint:* The compass needle will align its axis along the direction of the net magnetic field.)

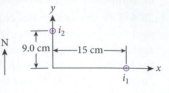

•28.30 Two identical coaxial coils of wire of radius 20 cm are directly on top of each other, separated by a 2-mm gap. The lower coil is on a flat table and has a current i in the clockwise direction; the upper coil carries an identical current and has a mass of 0.050 kg. Determine the magnitude and the direction that the current in the upper coil has to have to keep the coil levitated at its current height.

•28.31 A long, straight wire lying along the x-axis carries a current, i, flowing in the positive x-direction. A second long, straight wire lies along the y-axis and has a current i in the positive y-direction. What is the magnitude and the direction of the magnetic field at point $z = b$ on the z-axis?

•28.32 A square loop of wire with a side length of 10 cm carries a current of 0.3 A. What is the magnetic field in the center of the square loop?

•28.33 The figure shows the cross section through three long wires with a linear mass distribution of 100 g/m. They carry currents i_1, i_2, and i_3 in the directions shown. Wires 2 and 3 are 10 cm apart and are attached to a vertical surface, and each carries a current of 600 A. What current, i_1, will allow wire 1 to "float" at a perpendicular distance d from the vertical surface of 10 cm? (Neglect the thickness of the wires.)

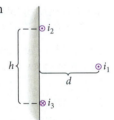

•28.34 A hairpin configuration is formed of two semi-infinite straight wires that are 2 cm apart and joined by a semicircular piece of wire (whose radius must be 1 cm and whose center is at the origin of xyz-coordinates). The top straight wire is along the line $y = 1$ cm, and the bottom straight wire is along the line $y = -1$ cm; these two wires are in the left side ($x < 0$) of the xy-plane. The current in the hairpin is 3.0 A, and it is directed toward the right in the top wire, clockwise around the semicircle, and to the left in the bottom wire. Find the magnetic field at the origin of the coordinate system.

•28.35 A long, straight wire is located along the x-axis ($y = 0$ and $z = 0$). The wire carries a current of 7 A in the positive x-direction. What is the magnitude and the direction of the force on a particle with a charge of 9 C located at $(+1 \text{ m},+2 \text{ m},0)$, when it has a velocity of 3000 m/s in each of the following directions?

a) the positive x-direction   c) the negative z-direction

b) the positive y-direction

•28.36 A long, straight wire has a 10.0-A current flowing in the positive x-direction, as shown in the figure. Close to the wire is a square loop of copper wire that carries a 2.00-A current in the direction shown. The near side of the loop is $d = 0.50$ m away from the wire. The length of each side of the square is $a = 1.00$ m.

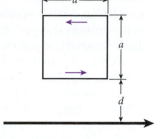

a) Find the net force between the two current-carrying objects.

b) Find the net torque on the loop.

28.37 A square box with sides of length 1 m has one corner at the origin of a coordinate system, as shown in the figure. Two coils are attached to the outside of the box. One coil is on the box face that is in the xz-plane at $y = 0$, and the second is on the box face in the yz-plane at $x = 1$ m. Each of the coils has a diameter of 1 m and contains 30 turns of wire carrying a current of 5 A in each turn. The current in each coil is clockwise when the coil is viewed from outside the box. What is the magnitude and the direction of the magnetic field at the center of the box?

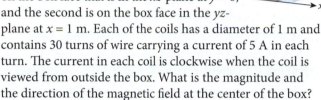

Section 28.3

28.38 A square loop, with sides of length L, carries current i. Find the magnitude of the magnetic field from the loop at the center of the loop, as a function of i and L.

28.39 The figure shows a cross section across the diameter of a long, solid, cylindrical conductor. The radius of the cylinder is $R = 10.0$ cm. A current of 1.35 A is uniformly distributed through the conductor and is flowing out of the page. Calculate the direction and the magnitude of the magnetic field at positions $r_a = 0.0$ cm, $r_b = 4.00$ cm, $r_c = 10.00$ cm, and $r_d = 16.0$ cm.

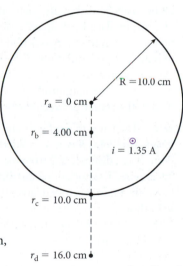

28.40 Parallel wires, a distance D apart, carry a current, i, in opposite directions as shown in the figure. A circular loop, of radius $R = D/2$, has the same current flowing in a counterclockwise direction. Determine the magnitude and the direction of the magnetic field from the loop and the parallel wires at the center of the loop as a function of i and R.

28.41 A current of constant density, J_0, flows through a very long cylindrical conducting shell with inner radius a and outer radius b. What is the magnetic field in the regions $r < a$, $a < r < b$, and $r > b$? Does $B_{a<r<b} = B_{r>b}$ for $r = b$?

28.42 The current density in a cylindrical conductor of radius R, varies as $J(r) = J_0 r/R$ (in the region from zero to R). Express the magnitude of the magnetic field in the regions $r < R$ and $r > R$. Produce a sketch of the radial dependence, $B(r)$.

•**28.43** A very large sheet of a conductor located in the xy-plane, as shown in the figure, has a uniform current flowing in the y-direction. The current density is 1.5 A/cm. Use Ampere's Law to calculate the direction and the magnitude of the magnetic field just above the center of the sheet (not close to any edges).

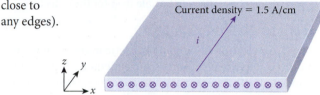

••**28.44** A coaxial wire consists of a copper core of radius 1 mm surrounded by a copper sheath of inside radius 1.5 mm and outside radius 2 mm. A current, i, flows in one direction in the core and in the opposite direction in the sheath. Graph the magnitude of the magnetic field as a function of the distance from the center of the wire.

••**28.45** The current density of a cylindrical conductor of radius R varies as $J(r) = J_0 e^{-r/R}$ (in the region from zero to R). Express the magnitude of the magnetic field in the regions $r < R$ and $r > R$. Produce a sketch of the radial dependence, $B(r)$.

Section 28.4

28.46 A current of 2 A is flowing through a 1000-turn solenoid of length $L = 40$ cm. What is the magnitude of the magnetic field inside the solenoid?

28.47 Solenoid A has twice the diameter, three times the length, and four times the number of turns of solenoid B. The two solenoids have currents of equal magnitudes flowing through them. Find the ratio of the magnitude of the magnetic field in the interior of solenoid A to that of solenoid B.

28.48 A long solenoid (diameter of 6.00 cm) is wound with 1000 turns per meter of thin wire through which a current of 0.250 A is maintained. A wire carrying a current of 10.0 A is inserted along the axis of the solenoid. What is the magnitude of the magnetic field at a point 1.00 cm from the axis?

28.49 A long, straight wire carries a current of 2.5 A.

a) What is the strength of the magnetic field at a distance of 3.9 cm from the wire?

b) If the wire still carries 2.5 A, but is used to form a long solenoid with 32 turns per centimeter and a radius of 3.9 cm, what is the strength of the magnetic field at the center of the solenoid?

28.50 Figure 28.18a shows a Helmholtz coil used to generate uniform magnetic fields. Suppose the Helmholtz coil consists of two sets of coaxial wire loops with 15 turns of radius $R = 75.0$ cm, which are separated by R, and each coil carries a current of 0.123 A flowing in the same direction. Calculate the magnitude and the direction of magnetic field in the center between the coils.

•28.51 A particle detector utilizes a solenoid that has 550 turns of wire per centimeter. The wire carries a current of 22 A. A cylindrical detector that lies within the solenoid has an inner radius of 0.80 m. Electron and positron beams are directed into the solenoid parallel to its axis. What is the minimum momentum perpendicular to the solenoid axis that a particle can have if it is to be able to enter the detector?

Sections 28.5 through 28.7

28.52 An electron has a spin magnetic moment of magnitude $\mu = 9.285 \cdot 10^{-24}$A m^2. Consequently, it has energy associated with its orientation in a magnetic field. If the difference between the energy of an electron that is "spin up" in a magnetic field of magnitude B and the energy of one that is "spin down" in the same magnetic field (where "up" and "down" refer to the direction of the magnetic field) is $9.460 \cdot 10^{-25}$ J, what is the field magnitude, B?

28.53 When a magnetic dipole is placed in a magnetic field, it has a natural tendency to minimize its potential energy by aligning itself with the field. If there is sufficient thermal energy present, however, the dipole may rotate so that it is no longer aligned with the field. Using $k_B T$ as a measure of the thermal energy, where k_B is Boltzmann's constant and T is the temperature in kelvins, determine the temperature at which there is sufficient thermal energy to rotate the magnetic dipole associated with a hydrogen atom from an orientation parallel to an applied magnetic field to one that is antiparallel to the applied field. Assume that the strength of the field is 0.15 T.

28.54 Aluminum becomes superconducting at a temperature around 1 K if exposed to a magnetic field of magnitude less than 0.0105 T. Determine the maximum current that can flow in an aluminum superconducting wire with radius $R = 1$ mm.

28.55 If you want to construct an electromagnet by running a current of 3.0 A through a solenoid with 500 windings and length 3.5 cm and you want the magnetic field inside the solenoid to have magnitude $B = 2.96$ T, you can insert a ferrite core into the solenoid. What value of the relative magnetic permeability should this ferrite core have in order to make this work?

28.56 What is the magnitude of the magnetic field inside a long, straight tungsten wire of circular cross section with diameter 2.4 mm and carrying a current of 3.5 A, at a distance of 0.6 mm from its central axis?

•28.57 You charge up a small rubber ball of mass 200 g by rubbing it over your hair. The ball acquires a charge of 2.00 μC. You then tie a 1.00-m-long string to it and swing it in a horizontal circle, providing a centripetal force of 25.0 N. What is the magnetic moment of the system?

•28.58 Consider a model of the hydrogen atom in which an electron orbits a proton in the plane perpendicular to the proton's spin angular momentum (and magnetic dipole moment) at a distance equal to the Bohr radius, $a_0 = 5.292 \cdot 10^{-11}$ m. (This is an oversimplified classical model.) The spin of the electron is allowed to be either parallel to the proton's spin or antiparallel to it; the orbit is the same in either case. But since the proton produces a magnetic field at the electron's location, and the electron has its own intrinsic magnetic dipole moment, the energy of the electron differs depending on its spin. The magnetic field produced by the proton's spin may be modeled as a dipole field, like the electric field due to an electric dipole discussed in Chapter 22. Calculate the energy difference between the two electron-spin configurations. Consider only the interaction between the magnetic dipole moment associated with the electron's spin and the field produced by the proton's spin.

••28.59 Consider an electron to be a uniformly dense sphere of charge, with a total charge of $-e = -1.60 \cdot 10^{-19}$ C, spinning at an angular frequency, ω.

a) Write an expression for its classical angular momentum of rotation, L.

b) Write an expression for its magnetic dipole moment, μ.

c) Find the ratio, $\gamma_e = \mu/L$, known as the *gyromagnetic ratio*.

Additional Problems

28.60 Two 50-turn coils, each with a diameter of 4 m, are placed 1 m apart, as shown in the figure. A current of 7 A is flowing in the wires of both coils; the direction of the current is clockwise for both coils when viewed from the left. What is the magnitude of the magnetic field in the center between the two coils?

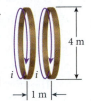

28.61 The wires in the figure are separated by a vertical distance d. Point B is at the midpoint between the two wires; point A is a distance $d/2$ from the lower wire. The horizontal distance between A and B is much larger than d. Both wires carry the same current, i. The strength of magnetic field at point A is 2 mT. What is the strength of the field at point B?

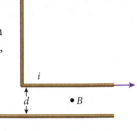

28.62 You are standing at a spot where the magnetic field of the Earth is horizontal, points due northward, and has magnitude 40.0 μT. Directly above your head, at a height of 12.0 m, a long, horizontal cable carries a steady DC current of 500.0 A due northward. Calculate the angle θ by which your magnetic compass needle is deflected from true magnetic north by the effect of the cable. Don't forget the *sign* of θ—is the deflection eastward or westward?

28.63 The magnetic dipole moment of the Earth is approximately $8.0 \cdot 10^{22}$ A m^2. The source of the Earth's magnetic field is not known; one possibility might be the circulation of ions in the Earth's molten outer core. Assume that the circulating ions move a circular loop of radius 2500 km. What "current" must they produce to yield the observed field?

28.64 A circular wire loop has radius $R = 0.12$ m and carries current $i = 0.10$ A. The loop is placed in the xy-plane in a uniform magnetic field given by $\vec{B} = -1.5\hat{z}$ T, as shown in the figure. Determine the direction and the magnitude of the loop's magnetic moment and calculate the potential energy of the loop in the position shown. If the wire loop can move freely, how will it orient itself to minimize its potential energy, and what is the value of the lowest potential energy?

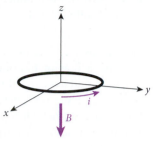

28.65 A 0.90 m-long solenoid has a radius of 5.0 mm. When the wire carries a 0.20-A current, the magnetic field in the solenoid is 5.0 mT. How many turns of wire are there in the solenoid?

28.66 In a coaxial cable, the solid core carries a current i. The sheath also carries a current i but in the opposite direction and has an inner radius a and an outer radius b. The current density is equally distributed over each conductor. Find an expression for the magnetic field at a distance $a < r < b$ from the center of the core.

•28.67 A 50-turn rectangular coil of wire of dimensions 10 cm by 20 cm lies in a horizontal plane, as shown in the figure. The axis of rotation of the coil is aligned north and south. It carries a current $i = 1$ A, and is in a magnetic field pointing from west to east. A mass of 50 g hangs from one side of the loop. Determine the strength the magnetic field has to have to keep the loop in the horizontal orientation.

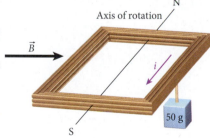

•28.68 Two long, straight parallel wires are separated by a distance of 20.0 cm. Each wire carries a current of 10.0 A in the same direction. What is the magnitude of the resulting magnetic field at a point that is 12.0 cm from each wire?

•28.69 A particle with a mass of 1 mg and a charge of q is moving at a speed of 1000 m/s along a horizontal path 10 cm below and parallel to a straight current-carrying wire. Determine q if the magnitude of the current in the wire is 10 A.

•28.70 A conducting coil consisting of n turns of wire is placed in a uniform magnetic field given by $\vec{B} = 2\hat{y}$ T, as shown in the figure. The radius of the coil is $R = 5$ cm, and the angle between the magnetic field vector and the unit normal vector to the coil is $\theta = 60°$. The current through the coil is $i = 5$ A.

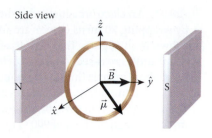

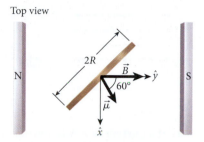

a) Specify the direction of the current in the coil, given the direction of the magnetic dipole moment, $\vec{\mu}$, in the figure.

b) Calculate the number of turns, n, the coil must have for the torque on the loop to be 3.4 N m.

c) If the radius of the loop is decreased to $R = 2.5$ cm, what should the number of turns, N, be for the torque to remain unchanged? Assume that i, B, and θ stay the same.

•28.71 A loop of wire of radius $R = 25$ cm has a smaller loop of radius $r = 0.9$ cm at its center such that the planes of the two loops are perpendicular to each other. When a current of 14.0 A is passed through both loops, the smaller loop experiences a torque due to the magnetic field produced by the larger loop. Determine this torque assuming that the smaller loop is sufficiently small so that the magnetic field due to the larger loop is same across the entire surface.

•28.72 Two wires, each 25 cm long, are connected to two separate 9-V batteries as shown in the figure. The resistance of the first wire is 5 Ω, and that of the other wire is unknown (R). If the separation between the wires is 4 mm, what value of R will produce a force of magnitude $4 \cdot 10^{-5}$ N between them? Is the force attractive or repulsive?

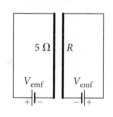

•28.73 A proton is moving under the combined influence of an electric field ($E = 1000$ V/m) and a magnetic field ($B = 1.2$ T), as shown in the figure.

a) What is the acceleration of the proton at the instant it enters the crossed fields?

b) What would the acceleration be if the direction of the proton's motion was reversed?

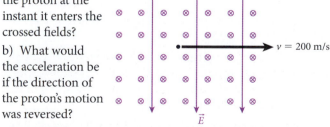

•28.74 A toy airplane of mass 0.175 kg, with a charge of 36 mC, is flying at a speed of 2.8 m/s at a height of 17.2 cm above and parallel to a wire, which is carrying a 25-A current; the airplane experiences some acceleration. Determine this acceleration.

•28.75 An electromagnetic doorbell has been constructed by wrapping 70 turns of wire around a long, thin rod, as shown in the figure. The rod has a mass of 30 g, a length of 8 cm, and a cross-sectional area of 0.2 cm^2. The rod is free to pivot about an axis through its center, which is also the center of the coil. Initially, the rod makes an angle of $\theta = 25°$ with the horizontal. When $\theta = 0°$, the rod strikes a bell. There is a uniform magnetic field of 900 G directed along $\theta = 0°$.

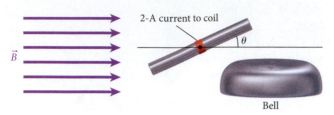

2-A current to coil

$\vec{B}$

Bell

a) If a current of 2 A is flowing in the coil, what is the torque on the rod when $\theta = 25°$?

b) What is the angular velocity of the rod when it strikes the bell?

•28.76 Two long, parallel wires separated by a distance, d, carry currents in opposite directions. If the left-hand wire carries a current $i/2$, and the right-hand wire carries a current i, determine where the magnetic field is zero.

•28.77 A horizontally oriented coil of wire of radius 5 cm and carrying a current, i, is being levitated by the south pole of a vertically oriented bar magnet suspended above the center

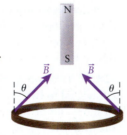

N

S

$\vec{B}$ $\vec{B}$

θ θ

of the coil. If the magnetic field on all parts of the coil makes an angle θ of 45° with the vertical, determine the magnitude and the direction of the current needed to keep the coil floating in midair. The magnitude of the magnetic field is $B = 0.010$ T, the number of turns in the coil is $N = 10$, and the total coil mass is 10 g.

•28.78 As shown in the figure, a long, hollow, conducting cylinder of inner radius a and outer radius b carries a current that is flowing out of the page. Suppose that $a = 5$ cm, $b = 7$ cm, and the current $i = 100$ mA, uniformly distributed over the cylinder wall (between a and b). Find the magnitude of the magnetic field at each of the following distances r from the center of the cylinder:

a) $r = 4$ cm

b) $r = 6.5$ cm

c) $r = 9$ cm

•28.79 A wire of radius R carries current i. The current density is given by $J = J_0(1 - r/R)$, where r is measured from the center of the wire and J_0 is a constant. Use Ampere's Law to find the magnetic field inside the wire at a distance $r < R$ from the central axis.

•28.80 A circular wire of radius 5.0 cm has a current of 3.0 A flowing in it. The wire is placed in a uniform magnetic field of 5.0 mT.

a) Determine the maximum torque on the wire.

b) Determine the range of the magnetic potential energy of the wire.

Electromagnetic Induction

29

FIGURE 29.1 The Grand Coulee Dam on the Columbia River in the state of Washington is the largest single producer of electricity in the United States. Shown here are the giant generators, which apply the physical principle of induction to produce electricity.

WHAT WE WILL LEARN

- A changing magnetic field inside a conducting loop induces a current in the loop.

- A changing current in a loop induces a current in a nearby loop.

- Faraday's Law of Induction states that a potential difference is induced in a loop when there is a change in the magnetic flux through the loop.

- Magnetic flux is the product of the magnitude of the average magnetic field and the perpendicular area that it penetrates.

- Lenz's Law states that the current induced in a loop by a changing magnetic field produces a magnetic field that opposes this change in magnetic field.

- A changing magnetic field induces an electric field.

- The inductance of a device is a measure of its opposition to changes in current flowing through it.

- Electric motors and electric generators are everyday applications of magnetic induction.

- A simple single-loop circuit with an inductor and a resistor has a characteristic time constant given by the inductance divided by the resistance.

- Energy is stored in a magnetic field.

Most of us take electric power for granted—we flip a switch and we have power for lighting, heating, and entertainment. But the vast network that supplies this power—called *the grid*—depends on large generators that convert mechanical energy into electrical energy (Figure 29.1). The physical principles that enable this conversion are the subject of this chapter.

In Chapter 27, we saw that a magnetic field can affect the path of charged particles, or electric currents, and in Chapter 28, we saw that an electric current generates a magnetic field. In this chapter, we'll see that a changing magnetic field generates an electric current, and thus an electric field. Note the word "changing" here; just as a magnetic field is generated only when electrical charges are in motion, an electric field is generated only when a magnetic field is in motion (relative to a conductor) or otherwise changes as a function of time. This symmetry will turn out to be a key part of the unified description of electricity and magnetism presented in Chapter 31.

29.1   Faraday's Experiments

Some of the great discoveries about electricity and magnetism took place in the late 18th and early 19th centuries. In 1750, the American Benjamin Franklin showed that lightning is a form of electricity, with his famous kite-flying experiment. (Perhaps the most amazing aspect of that experiment was that he was not killed by the lightning strike.) In 1799, the Italian Alessandro Volta constructed the first battery, called a *voltaic pile* at the time. In 1820, the Danish physicist Hans Christian Oersted demonstrated that an electric current could produce a magnetic field strong enough to deflect a compass needle. (He performed his experiment during a lecture to his students, making it one of the most productive lecture demonstrations in the history of science.)

However, the experiments that are most relevant to this chapter were performed in the 1830s by the British chemist and physicist Michael Faraday and independently by the American physicist Joseph Henry. Their work demonstrated that a changing magnetic field could generate a potential difference in a conductor, strong enough to produce an electric current. This discovery is of basic importance to all the electrical and magnetic devices we use every day, from computers to cell phones, from television to credit cards, from the tiniest batteries to the largest electrical power grid. Both Faraday and Henry had fundamental electrical units named after them, and justifiably so.

To understand Faraday's experiments, consider a wire loop connected to an ammeter. A bar magnet is some distance from the loop with its north pole pointing toward the loop. While the magnet is stationary, no current flows in the loop. However, if the magnet is moved toward the loop (Figure 29.2a), a counterclockwise current flows in the loop as indicated by the positive current in the ammeter. If the magnet is moved toward the loop faster, a larger current is induced in the loop. If the magnet is reversed so the south pole points toward the loop (Figure 29.2b) and moved toward the loop, current flows in the loop in the opposite direction. If the north pole of the magnet points toward the loop, and the magnet is then moved *away* from the loop (Figure 29.3a), a negative clockwise current, as indicated on the meter in Figure 29.3a, is induced in the loop. If the south pole of the magnet points toward the loop, and the magnet is moved away from the loop (Figure 29.3b), a positive current is induced.

The four results illustrated in Figures 29.2 and 29.3 can be replicated by holding the magnets stationary and moving the coils. For example, with the arrangement shown in Figure 29.2a, if the coil is moved toward the stationary magnet, a positive current flows in the loop.

Similar effects can be observed using two conducting loops (Figure 29.4). If a constant current is flowing through loop 1, no current is induced in loop 2. If the current in loop 1 is increased, a current is induced in loop 2 in the opposite direction. Thus, not only does the increasing current in the first loop induce a current in the second loop, but the induced current is in the opposite direction. Furthermore, if current is flowing in loop 1 in the same direction as before and is then *decreased* (Figure 29.5), the current induced in loop 2 flows in the *same* direction as the current in loop 1.

All of the phenomena illustrated in these four figures can be explained by Faraday's Law of Induction, discussed in Section 29.2, and by Lenz's Law, discussed in Section 29.3.

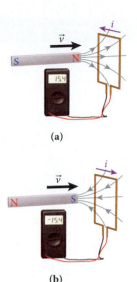

FIGURE 29.2 Moving a magnet toward a wire loop induces a current to flow in the loop. (a) With the north pole of the magnet pointing toward the loop, a positive current results. (b) With the south pole of the magnet pointing toward the loop, a negative current results.

29.1 In-Class Exercise

The four figures show a bar magnet and a low-voltage light bulb connected to the ends of a conducting loop. The plane of the loop is perpendicular to the dotted line. In case 1, the loop is stationary, and the magnet is moving away from the loop. In case 2, the magnet is stationary, and the loop is moving toward the magnet. In case 3, both the magnet and loop are stationary, but the area of the loop is increasing. In case 4, the magnet is stationary, and the loop is rotating about its center. In which of these situations will the light bulb be burning?

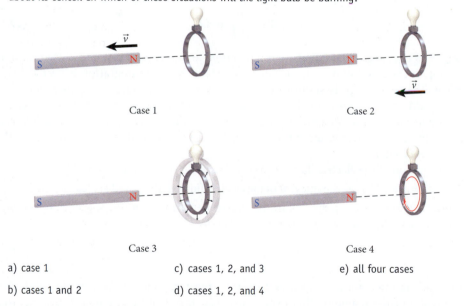

Case 1   Case 2

Case 3   Case 4

a) case 1   c) cases 1, 2, and 3   e) all four cases

b) cases 1 and 2   d) cases 1, 2, and 4

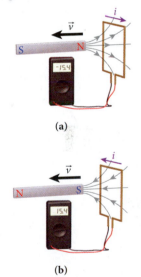

FIGURE 29.3 Moving a magnet away from a wire loop also induces a current to flow in the loop. (a) With the north pole of the magnet pointing toward the loop, a negative current results. (b) With the south pole of the magnet pointing toward the loop, a positive current results.

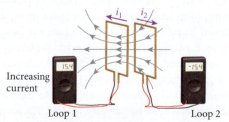

FIGURE 29.4 An increasing current in loop 1 induces a current in the opposite direction in loop 2. (The magnetic field lines shown are those produced by the current 1 flowing through loop 1.)

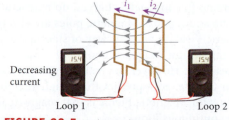

FIGURE 29.5 A decreasing current in loop 1 induces a current in the same direction in loop 2.

29.2   Faraday's Law of Induction

From the observations in the preceding section, we see that a changing magnetic field through a loop induces a current in the loop. We can visualize the change in magnetic field as a change in the number of magnetic field lines passing through the loop. **Faraday's Law of Induction** in its qualitative form states:

> A potential difference is induced in a loop when the number of magnetic field lines passing through the loop changes with time.

The rate of change of the magnetic field lines determines the induced potential difference. The existence of this potential difference means that the changing magnetic field actually creates an electric field around the loop! Thus, there are two ways of producing an electric field: from electric charges and from a changing magnetic field. If the electric field arises from a charge, the resulting electric force on a test charge is conservative. Conservative forces do no work when they act on an object whose path starts and ends at the same point in space. In contrast, electric fields generated by changing magnetic fields give rise to electric forces that are *not* conservative. Thus, a test particle that moves around a circular loop once will have work done on it by this electric field. In fact, the amount of the work done is the induced potential difference times the charge of the test particle.

The magnetic field lines are quantified by the magnetic flux, in analogy with the electric flux. Chapter 22 introduced Gauss's Law for the electric field and defined the electric flux as the surface integral of an electric field passing through a differential element of area, dA. Mathematically, $\Phi_E = \oiint \vec{E} \cdot d\vec{A}$, where $d\vec{A}$ is a vector of magnitude dA that is perpendicular to the differential area. By analogy, for a magnetic field, **magnetic flux** is defined as the surface integral of the magnetic field passing through a differential element of area:

$$\Phi_B = \oiint \vec{B} \cdot d\vec{A}, \tag{29.1}$$

where $\vec{B}$ is the magnetic field at each differential area element, $d\vec{A}$, of a closed surface. The loop in the symbol for a surface integral means that the integration is over a closed surface. The two integrals signify integration over two variables. The differential area element, $d\vec{A}$, must be described by two spatial variables, such as x and y in Cartesian coordinates or θ and ϕ in spherical coordinates. With a closed surface, the differential area vector, $d\vec{A}$, always points out of the enclosed volume and is perpendicular to the surface everywhere.

Integration of the electric flux over a closed surface (see Chapter 22) yields Gauss's Law: $\oiint \vec{E} \cdot d\vec{A} = q/\epsilon_0$. That is, the integral of the electric flux over a closed surface is equal to the enclosed electric charge, q, divided by the electric permittivity of free space, ϵ_0. Integration of the magnetic flux over a *closed* surface yields zero:

$$\oiint \vec{B} \cdot d\vec{A} = 0. \tag{29.2}$$

This result is often termed **Gauss's Law for Magnetic Fields.** You might think that the integral of the magnetic flux over a closed surface would be equal to the enclosed "magnetic charge" divided by the magnetic permeability of free space. However, there are no free magnetic charges, no magnetic monopoles, no separate north poles or separate south poles. Magnetic poles are always found in pairs. Thus, Gauss's Law for Magnetic Fields is another way of stating that magnetic monopoles do not exist. (Extensive searches for magnetic monopoles have been conducted since the 1980s, but not successfully. However, several string theories and grand unified theories—more on those in Chapter 39—predict that magnetic monopoles exist.) Another way to state Gauss's Law for Magnetic Fields is that magnetic field lines have no beginning or end but form a continuous loop.

Figure 29.6 shows a nonuniform magnetic field, $\vec{B}$, passing through a differential area element, $d\vec{A}$. A portion of the closed surface is also shown. The angle between the magnetic field and the differential area vector is θ.

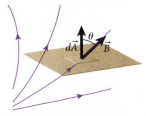

FIGURE 29.6 A nonuniform magnetic field $\vec{B}$ passing through a differential area, $d\vec{A}$.

Consider the special case of a flat loop of area A in a constant magnetic field, as illustrated in Figure 29.7. For this case, we can rewrite equation 29.1 as

$$\Phi_B = BA\cos\theta, \tag{29.3}$$

where B is the magnitude of the constant magnetic field, A is the area of the loop, and θ is the angle between the surface normal vector to the plane of the loop and the magnetic field lines. Thus, if the magnetic field is perpendicular to the plane of the loop, $\theta = 0°$ and $\Phi_B = BA$. If the magnetic field is parallel to the plane of the loop, $\theta = 90°$ and $\Phi_B = 0$.

The unit of magnetic flux is $[\Phi_B] = [B][A] = \text{T m}^2$. This unit has received a special name, the **weber** (Wb):

$$1\,\text{Wb} = 1\,\text{T m}^2. \tag{29.4}$$

Faraday's Law of Induction is stated quantitatively, in terms of the magnetic flux, as follows:

The magnitude of the potential difference, ΔV_{ind}, induced in a conducting loop is equal to the time rate of change of the magnetic flux through the loop.

Faraday's Law of Induction is thus expressed by the equation

$$\Delta V_{\text{ind}} = -\frac{d\Phi_B}{dt}. \tag{29.5}$$

The negative sign in equation 29.5 is necessary because the induced potential difference establishes an induced current whose magnetic field tends to oppose the flux change. Section 29.3 on Lenz's Law discusses this phenomenon in detail.

The magnetic flux can be changed in several ways, including changing the magnitude of the magnetic field, changing the area of the loop, or changing the angle the loop makes with respect to the magnetic field. In all situations that involve some form of motion of a conductor relative to the source of a magnetic field, the induced potential difference is called a **motional emf.**

Induction in a Flat Loop inside a Magnetic Field

Let's apply equation 29.5 to a flat wire loop inside a uniform magnetic field, where *uniform* means that the field has the same value (same magnitude and same direction) at all points in space at a given time but can vary in time. This arrangement is the simplest case we can address. According to equation 29.3, the magnetic flux in this case is given by $\Phi_B = BA\cos\theta$. According to equation 29.5, the induced potential difference is then

$$\Delta V_{\text{ind}} = -\frac{d\Phi_B}{dt} = -\frac{d}{dt}(BA\cos\theta). \tag{29.6}$$

We can use the product rule from calculus to expand this derivative:

$$\Delta V_{\text{ind}} = -A\cos\theta\frac{dB}{dt} - B\cos\theta\frac{dA}{dt} + AB\sin\theta\frac{d\theta}{dt}. \tag{29.7}$$

Because the time derivative of the angular displacement is the angular velocity, $d\theta/dt = \omega$, the induced potential difference in a flat loop inside a uniform magnetic field is

$$\Delta V_{\text{ind}} = -A\cos\theta\frac{dB}{dt} - B\cos\theta\frac{dA}{dt} + \omega AB\sin\theta. \tag{29.8}$$

Holding two of the three variables in equation 29.8 (A, B, and θ) constant results in the following *three special* cases:

1. Holding the area of the loop and its orientation relative to the magnetic field constant but varying the magnetic field in time yields

$$A\text{ and }\theta\text{ constant: }\ \Delta V_{\text{ind}} = -A\cos\theta\frac{dB}{dt}. \tag{29.9}$$

2. Holding the magnetic field as well as the orientation of the loop relative to the magnetic field constant but changing the area of the loop that is exposed to the magnetic field yields

$$B\text{ and }\theta\text{ constant: }\ \Delta V_{\text{ind}} = -B\cos\theta\frac{dA}{dt}. \tag{29.10}$$

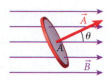

FIGURE 29.7 A flat loop of area A in a constant magnetic field, $\vec{B}$. The magnetic field makes an angle θ with respect to the surface normal vector of the loop.

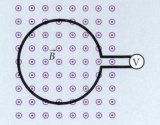

3. Holding the magnetic field constant and keeping the area of the loop fixed but allowing the angle between the two to change as a function of time yields

$$A \text{ and } B \text{ constant: } \Delta V_{ind} = \omega A B \sin\theta. \tag{29.11}$$

The following examples illustrate the first two cases. Section 29.4 addresses the third case, which has the most useful technical applications, leading directly to electric motors and generators.

EXAMPLE 29.1 Potential Difference Induced by a Changing Magnetic Field

A current of 600 mA is flowing in an ideal solenoid, resulting in a magnetic field of 0.025 T. Then the current increases with time, t, according to

$$i(t) = i_0 \left[1 + (2.4 \text{ s}^{-2})t^2 \right].$$

PROBLEM
If a circular coil of radius 3.4 cm with $N = 200$ windings is located inside the solenoid with its normal vector parallel to the magnetic field (Figure 29.8), what is the induced potential difference in the coil at $t = 2.0$ s?

SOLUTION
First, we compute the area of the coil. Since it is circular, its area is πR^2. However, it has N windings, and thus the area is that of N loops of area πR^2. The net effect is that the number of windings acts as a simple multiplier for the loop area, and the total effective area of the coil is

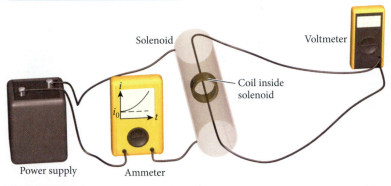

FIGURE 29.8 A time-varying current applied to a solenoid induces a potential difference in a coil.

$$A = N\pi R^2 = 200\pi (0.034 \text{ m})^2 = 0.73 \text{ m}^2. \tag{i}$$

The magnetic field inside an ideal solenoid is $B = \mu_0 n i$, where n is the number of windings per unit length, and i is the current (see Chapter 28). Because the magnetic field is proportional to the current, we immediately obtain for the time dependence of the magnetic field in this case

$$B(t) = B_0 \left[1 + (2.4 \text{ s}^{-2})t^2 \right],$$

with $B_0 = \mu_0 n i_0 = 0.025$ T, according to the problem statement.

Further, in this case, the area of the coil and the angle between each loop and the magnetic field (which is zero) are constant. Therefore, equation 29.9 applies. We then find for the induced potential difference, where the area A already accounts for the number of windings, as shown in equation (i):

$$\Delta V_{ind} = -A\cos\theta \frac{dB}{dt}$$

$$= -A\cos\theta \frac{d}{dt}(B_0(1 + \left(2.4 \text{ s}^{-2}\right)t^2))$$

$$= -AB_0 \cos\theta (2\left(2.4 \text{ s}^{-2}\right)t)$$

$$= -(0.73 \text{ m}^2)(0.025 \text{ T})(\cos 0°)\left(4.8 \text{ s}^{-2}\right)t$$

$$= \left(-0.088 \text{ V/s}\right)t.$$

At time $t = 2.0$ s, the induced potential difference in the coil is: $\Delta V_{ind} = -0.18$ V.

An important general point is made by Example 29.1: The potential difference induced in a coil with N windings and area A is simply N times the potential difference induced in a single loop of area A. Equations 29.8 through 29.11 are valid for multiloop coils, and the only way that the number of windings enters the calculations is as a multiplier in determining the effective area of the coil.

29.2 In-Class Exercise

A power supply is connected to loop 1 and an ammeter as shown in the figure. Loop 2 is close to loop 1 and is connected to a voltmeter. A graph of the current i through loop 1 as a function of time, t, is also shown in the figure. Which graph best describes the induced potential difference, ΔV_{ind}, in loop 2 as a function of time, t?

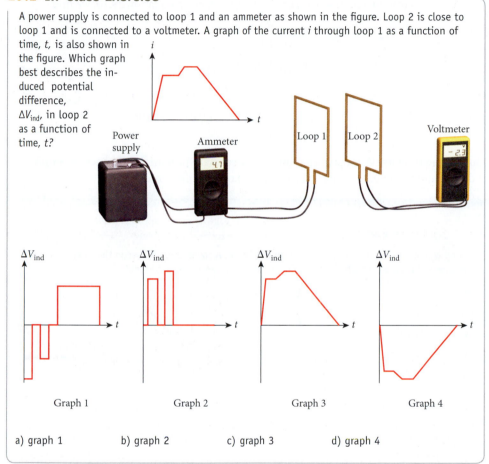

a) graph 1   b) graph 2   c) graph 3   d) graph 4

EXAMPLE 29.2 | Potential Difference Induced by a Moving Loop

A rectangular wire loop of width $w = 3.1$ cm and depth $d_0 = 4.8$ cm is pulled out of the gap between two permanent magnets. A magnetic field of magnitude $B = 0.073$ T is present throughout the gap (Figure 29.9).

PROBLEM

If the loop is removed at a constant speed of 1.6 cm/s, what is the induced voltage in the loop as a function of time?

SOLUTION

This situation corresponds to the special case of induction due to an area change, governed by equation 29.10. The magnetic field and the orientation of the loop relative to the field remain constant. We assume the angle between the magnetic field vector and the area vector is zero. What changes is the area of the loop that is exposed to the magnetic field. With such a narrow gap, shown in Figure 29.9, very little field occurs outside the gap, so the effective area of the loop exposed to the field is $A(t) = (w)(d(t))$, where $d(t) = d_0 - vt$ is the depth of the part of the loop inside the magnetic field at time t. While the

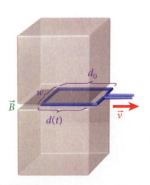

FIGURE 29.9 A wire loop (blue) is pulled out of a gap between two magnets.

Continued—

entire loop is still inside the gap, no voltage is produced. Letting the time of arrival of the right edge of the loop at the right end of the gap be $t = 0$, we have

$$A(t) = (w)\big(d(t)\big) = w(d_0 - vt).$$

This formula holds until the left edge of the loop reaches the right end of the gap, after which the area of the loop exposed to the magnetic field is zero. The left edge arrives at time $t_f = d/v = (4.8 \text{ cm})/(1.6 \text{ cm/s}) = 3.0$ s, and $A(t > t_f) = 0$. From equation 29.10, we find

$$\Delta V_{\text{ind}} = -B\cos\theta \frac{dA}{dt}$$

$$= -B\cos\theta \frac{d}{dt}\big[w(d_0 - vt)\big]$$

$$= wvB\cos\theta$$

$$= (0.031 \text{ m})(0.016 \text{ m/s})(0.073 \text{ T})\cos 0°$$

$$= 3.6 \cdot 10^{-5} \text{ V}.$$

During the time interval between 0 and 3 s, a constant potential difference of 36 μV is induced, and no potential difference is induced outside this time interval.

29.3 In-Class Exercise

A long wire carries a current, i, as shown in the figure. A square loop moves in the same plane as the wire as indicated. In which cases will the loop have an induced current?

a) cases 1 and 2

b) cases 1 and 3

c) cases 2 and 3

d) None of the loops will have an induced current.

e) All of the loops will have an induced current.

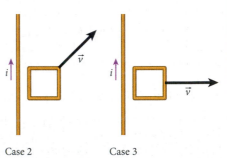

Case 1   Case 2   Case 3

29.3 Lenz's Law

Lenz's Law provides a rule for determining the direction of an induced current in a loop. An induced current will have a direction such that the magnetic field *due to* the induced current *opposes* the change in the magnetic flux that *induces* the current. The direction of the induced current can be used to determine the locations of higher and lower potential.

Let's apply Lenz's Law to the situations described in Section 29.1. The physical situation shown in Figure 29.2a involves moving a magnet toward a loop with the north pole pointed toward the loop. In this case, the magnetic field lines point away from the north pole of the magnet. As the magnet moves toward the loop, the magnitude of the magnetic field within the loop, in the direction pointing toward the loop, increases as depicted in Figure 29.10a. Lenz's Law states that the current induced in the loop tends to oppose the change in magnetic flux. The induced magnetic field, $\vec{B}_{\text{ind}}$, then points in the opposite direction from that of the field due to the magnet.

In Figure 29.2b, a magnet is moved toward a loop with the south pole pointed toward the loop. In this case, the magnetic field lines point toward the south pole of the magnet. As the magnet moves toward the loop, the magnitude of the field in the direction pointing toward the south pole increases, as depicted in Figure 29.10b. Lenz's Law states that the induced current creates a magnetic field that tends to oppose the increase in magnetic flux. This induced field points in the opposite direction from that of the field lines due to the magnet.

be induced in the metal plate in the opposite direction and will tend to oppose the increase in the current in the transmitter coil. The increasing current in the metal plate will induce a current in the receiver coil that is in the opposite direction and tends to oppose the increase in the current in the metal plate (not shown in the diagram). Thus, the metal plate induces a current in the receiver coil that flows in the same direction as the current in the transmitter coil. Without the metal plate, the increasing current in the transmitter coil induces a current in the opposite direction in the receiver coil that tends to oppose the increase in the current in the transmitter coil (as shown in the diagram). Thus, the overall effect of the metal plate in the metal detector is to decrease the observed current in the receiver coil. The metal object does not have to be a flat plate; any piece of metal, provided it is large enough, will have currents induced in it that can be detected by measuring the induced current in the receiver coil.

Metal detectors are also used to control traffic lights. In this application, a rectangular wire loop, which serves as both transmitter and receiver coil, is embedded in the road surface. A pulse of current is passed through the loop, which induces eddy currents in any metal near the loop. The current in the loop is measured after the current pulse is completed. When a car moves onto the road surface above the loop, eddy currents induced in the metal of the car cause a different current to be measured between pulses, which then triggers the traffic light to switch to green (after an appropriate delay to allow other vehicles to clear the intersection, of course!). On older road surfaces, you can often see rectangular-shaped scars in the asphalt resulting from retrofitting intersections with these induction loops.

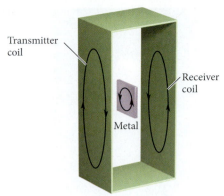

FIGURE 29.12 Schematic diagram of an airport metal detector.

Induced Potential Difference on a Wire Moving in a Magnetic Field

Consider a conducting wire of length ℓ moving with constant velocity $\vec{v}$ perpendicular to a constant magnetic field, $\vec{B}$, directed into the page (Figure 29.13). The wire is oriented so that it is perpendicular to the velocity and to the magnetic field. The magnetic field exerts a force, $\vec{F}_B$, on the conduction electrons in the wire, causing them to move downward. This motion of the electrons produces a net negative charge at the bottom end of the wire and a net positive charge at the top end of the wire. This charge separation produces an electric field, $\vec{E}$, which exerts a force, $\vec{F}_E$, on the conduction electrons that tends to cancel the magnetic force. After some time, the two forces become equal in magnitude (but opposite in direction) producing a zero net force:

$$F_B = evB = F_E = eE. \qquad (29.12)$$

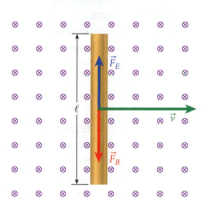

FIGURE 29.13 A moving conductor in a constant magnetic field. The magnetic and electric forces on the conduction electrons are shown.

29.4 In-Class Exercise

A metal bar is moving with constant velocity $\vec{v}$ through a uniform magnetic field pointing into the page, as shown in the figure.

Which of the following most accurately represents the charge distribution on the surface of the metal bar?

a) distribution 1

b) distribution 2

c) distribution 3

d) distribution 4

e) distribution 5

Distribution 1   Distribution 2   Distribution 3   Distribution 4   Distribution 5

FIGURE 29.14 (a) An artist's conception of the Space Shuttle *Columbia* and the tethered satellite. (b) A photograph of the tethered satellite being deployed from *Columbia*.

Thus, the induced electric field can be expressed by

$$E = vB. \qquad (29.13)$$

Because the electric field is constant in the wire, it produces a potential difference between the two ends of the wire given by

$$E = \frac{\Delta V_{ind}}{\ell} = vB. \qquad (29.14)$$

The induced potential difference between the ends of the wire is then

$$\Delta V_{ind} = v\ell B. \qquad (29.15)$$

This is one form of motional emf, mentioned in Section 29.2.

EXAMPLE 29.3 Satellite Tethered to a Space Shuttle

In 1996, the Space Shuttle *Columbia* deployed a tethered satellite on a wire out to a distance of 20 km (Figure 29.14). The wire was oriented perpendicular to the Earth's magnetic field at that point, and the magnitude of the field was $B = 5.1 \cdot 10^{-5}$ T. *Columbia* was traveling at a speed of 7.6 km/s.

PROBLEM
What was the potential difference induced between the ends of the wire?

SOLUTION
We can use equation 29.15 to determine the induced potential difference between the ends of the wire. The length of the wire is $L = 20$ km, and the speed of the wire through the magnetic field of the Earth ($B = 5.1 \cdot 10^{-5}$ T) is the same as the speed of the Space Shuttle, which is $v = 7.6$ km/s. Thus, we have

$$\Delta V_{ind} = vLB = (7.6 \cdot 10^3 \text{ m/s})(20 \cdot 10^3 \text{ m})(5.1 \cdot 10^{-5} \text{ T}) = 7800 \text{ V}.$$

The astronauts on the Space Shuttle measured a current of about 0.5 A at a voltage of 3500 V. The circuit consisted of the deployed wire and ionized atoms in space as the return path for the current. The wire broke just as the deployment length reached 20 km, but the generation of electric current from the motion of a spacecraft had been demonstrated.

Solved Problem 28.1 concerned an electromagnetic rail accelerator. The next example focuses on the phenomenon of induction in a similar system.

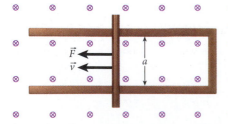

FIGURE 29.15 A conducting rod is pulled along two conducting rails with a constant velocity in a constant magnetic field directed into the page.

EXAMPLE 29.4 Pulled Conducting Rod

A conducting rod is pulled horizontally by a constant force of magnitude, $F = 5.00$ N, along a set of conducting rails separated by a distance $a = 0.500$ m (Figure 29.15). The two rails are connected, and no friction occurs between the rod and the rails. A uniform magnetic field with magnitude $B = 0.500$ T is directed into the page. The rod moves at constant speed, $v = 5.00$ m/s.

PROBLEM
What is the magnitude of the induced potential difference in the loop created by the connected rails and the moving rod?

SOLUTION
The induced potential difference is given by equation 29.10, which applies to a loop in a magnetic field when the angle and magnetic field are held constant and the area of the loop changes with time:

$$\Delta V_{ind} = -B\cos\theta \frac{dA}{dt}.$$

In this case, $\theta = 0$ and $B = 0.500$ T. The area of the loop is increasing with time. We can express the area of the loop in terms of A_0, the area before the rod started moving and an additional area given by the product of the speed of the loop and the time for which the loop has been moving times the distance, a, between the rails:

$$A = A_0 + a(vt) = A_0 + vta.$$

The change of the loop's area as a function of time is then

$$\frac{dA}{dt} = \frac{d}{dt}(A_0 + vta) = va.$$

Thus, the magnitude of the induced potential difference is

$$\Delta V_{ind} = \left| -B\cos\theta \frac{dA}{dt} \right| = vaB. \tag{i}$$

Inserting the numerical values, we obtain

$$\Delta V_{ind} = (5.00 \text{ m/s})(0.500 \text{ m})(0.500 \text{ T}) = 1.25 \text{ V}.$$

Note that equation (i), $\Delta V_{ind} = vaB$, which we derived from Faraday's Law of Induction, has the same form as equation 29.15 for the potential difference induced in a wire moving in a magnetic field, which was derived using the magnetic force on moving charges.

29.4 Generators and Motors

The third special case of the basic induction process described in Section 29.2 is by far the most interesting technologically. In this case, the angle between the conducting loop and the magnetic field is varied over time, while keeping the area of the loop as well as the magnetic field strength constant. In this situation, equation 29.11 can be used to apply Faraday's Law of Induction to the generation and application of electric current. A device that produces electric current from mechanical motion is called an **electric generator**. A device that produces mechanical motion from electric current is called an **electric motor**. Figure 29.16 shows a very simple electric motor.

A simple generator consists of a loop forced to rotate in a fixed magnetic field. The force that causes the loop to rotate can be supplied by hot steam running over a turbine, as occurs in nuclear and coal-fired power plants. (Power plants actually use multiple loops in order to increase the power output.) On the other hand, the loop can be made to rotate by flowing water or wind to generate electricity in a pollution-free way.

Figure 29.17 shows two types of simple generators. In a direct-current generator, the rotating loop is connected to an external circuit through a split commutator ring, as illustrated in Figure 29.17a. As the loop turns, the connection is reversed twice per revolution, so the induced potential difference always has the same sign. Figure 29.17b shows a similar arrangement used to produce an alternating current. An alternating current is a current that varies in time between positive and negative values, with the variation often showing a sinusoidal form. Each end of the loop is connected to the external circuit through its own solid slip ring. Thus, this generator produces an induced potential difference that varies from positive to negative and back. A generator that produces alternating voltages and the resulting alternating current is also called an **alternator**. Figure 29.18 shows the induced potential difference as a function of time for each type of generator.

The devices in Figure 29.17 could also be used as motors by supplying current to the loop and using the resulting motion of the coil to do work.

Real-world generators and motors are considerably more complex than the simple examples in Figure 29.17. For example, instead of permanent magnets, current flowing in coils creates the required magnetic field. Many closely spaced loops are employed to make more efficient use of the rotational motion. Multiple loops also resolve the problem that a simple, one-loop motor could stop in a position where current through the

FIGURE 29.16 A very simple electric motor used for lecture demonstrations. It consists of a pair of permanent magnets on the outside and two solenoids, through which current is sent, on the inside.

29.4 Self-Test Opportunity

A generator is operated by rotating a coil of N turns in a constant magnetic field of magnitude B at a frequency f. The resistance of the coil is R, and the cross-sectional area of the coil is A. Decide whether each of the following statements is true or false.

a) The average induced potential difference doubles if the frequency, f, is doubled.

b) The average induced potential difference doubles if the resistance, R, is doubled.

c) The average induced potential difference doubles if the magnetic field magnitude, B, is doubled.

d) The average induced potential difference doubles if the area, A, is doubled.

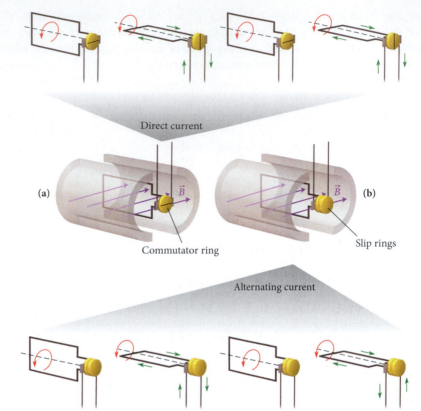

FIGURE 29.17 (a) A simple direct-current (DC) generator/motor. (b) A simple alternating-current (AC) generator/motor.

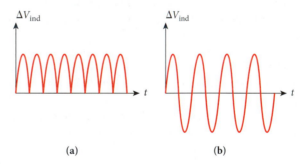

FIGURE 29.18 Induced potential difference as a function of time for (a) a simple direct-current generator; (b) a simple alternating-current generator.

FIGURE 29.19 Automobile hybrid engine, cut open to show the regenerative brake system, shown in close-up view in the inset.

loop produces no torque. The magnetic field may also change with time in phase with the rotating loop. In some generators and motors, the loops (coils) are fixed and the magnet rotates.

Regenerative Braking

Hybrid cars are propelled by a combination of gasoline power and electrical power. One attractive feature of a hybrid vehicle is that it is capable of **regenerative braking**. When the brakes are used to slow or stop a nonhybrid vehicle, the kinetic energy of the vehicle is turned into heat in the brake pads. This heat dissipates into the environment, and energy is lost. In a hybrid car, the brakes are connected to the electric motor (Figure 29.19), which functions as a generator, charging the car's battery. Thus, the kinetic energy of the car is partially recovered during braking, and this energy can later be used to propel the car, contributing to its efficiency and greatly increasing its gas mileage in stop-and-go driving.

29.5 Induced Electric Field

Faraday's Law of Induction states that a changing magnetic flux produces an induced potential difference, which can lead to an induced current. What are the consequences of this effect?

Consider a positive charge q moving in a circular path with radius r in an electric field, $\vec{E}$. The work done on the charge is equal to the integral of the scalar product of the force and the differential displacement vector. For now, let's assume that the electric field $\vec{E}$ is constant, that it has field lines that are circular, and that the charge moves along one of these lines. In one revolution of the charge, the work done on it is given by

$$\oint \vec{F} \cdot d\vec{s} = \oint q\vec{E} \cdot d\vec{s} = \oint q\cos 0° E\, ds = qE \oint ds = qE(2\pi r).$$

Since the work done by a constant electric field is $\Delta V_{\text{ind}}q$, we get

$$\Delta V_{\text{ind}} = 2\pi r E.$$

We can generalize this result by considering the work done on a charge q moving along an arbitrary closed path:

$$W = \oint \vec{F} \cdot d\vec{s} = q \oint \vec{E} \cdot d\vec{s}.$$

Again substituting $\Delta V_{\text{ind}}q$ for the work done, we obtain

$$\Delta V_{\text{ind}} = \oint \vec{E} \cdot d\vec{s}. \tag{29.16}$$

Now we can express the induced potential difference in a different way by combining equation 29.5 with equation 29.16:

$$\oint \vec{E} \cdot d\vec{s} = -\frac{d\Phi_B}{dt}. \tag{29.17}$$

Equation 27.17 states that a changing magnetic flux induces an electric field. This equation can be applied to any closed path in a changing magnetic field, even if no conductor exists in the path.

29.6 Inductance of a Solenoid

Consider a long solenoid with N turns carrying a current, i. This current creates a magnetic field in the center of the solenoid, resulting in a magnetic flux, Φ_B. The same magnetic flux goes through each of the N windings of the solenoid. It is customary to define the **flux linkage** as the product of the number of windings and the magnetic flux, or $N\Phi_B$. Equation 29.1 defined the magnetic flux as $\Phi_B = \iint \vec{B} \cdot d\vec{A}$. Inside the solenoid, the magnetic field vector and the surface normal vector, $d\vec{A}$, are parallel. And we saw in Chapter 28 that the magnitude of the magnetic field inside the solenoid is $B = \mu_0 ni$, where $\mu_0 = 4\pi \cdot 10^{-7}$ T m/A is the magnetic permeability of free space, i is the current and n is the number of windings per unit length ($n = N/\ell$). Therefore, the magnetic flux in the interior of a solenoid is proportional to the current flowing through the solenoid, which trivially means that the flux linkage is also proportional to the current. We can express this proportionality as

$$N\Phi_B = Li, \tag{29.18}$$

using a proportionality constant, L, called the **inductance**. (*Note:* Use of the letter L to represent the inductance is the convention. Although L is also used for the physical quantity of length and the physical quantity of angular momentum, inductance is not connected to either of these in any way.)

Inductance is a measure of the flux linkage produced by a solenoid per unit of current. The unit of inductance is the **henry** (H) named after American physicist Joseph Henry (1797–1878), given by

$$[L] = \frac{[\Phi_B]}{[i]} \Rightarrow 1 \text{ H} = \frac{1 \text{ T m}^2}{1 \text{ A}}. \tag{29.19}$$

The definition of the henry given in equation 29.19 means that the magnetic permeability of free space can also be given as $\mu_0 = 4\pi \cdot 10^{-7}$ H/m.

Now let's use equation 29.18 to find the inductance of a solenoid with cross-sectional area A and length ℓ. The flux linkage for this solenoid is

$$N\Phi_B = (n\ell)(BA),\tag{29.20}$$

where n is the number of turns per unit length and B is the magnitude of the magnetic field inside the solenoid. Thus, the inductance is given by

$$L = \frac{N\Phi_B}{i} = \frac{(n\ell)(\mu_0 n i)(A)}{i} = \mu_0 n^2 \ell A.\tag{29.21}$$

This expression for the inductance of a solenoid is good for long solenoids because fringe field effects at the ends of such a solenoid are small. You can see from equation 29.21 that the inductance of a solenoid depends only on the geometry (length, area, and number of turns) of the device. This dependence of inductance on geometry alone holds for all coils and solenoids, just as the capacitance of any capacitor depends only on its geometry.

Any solenoid has an inductance, and when a solenoid is used in an electric circuit, it is called an *inductor,* simply because its inductance is its most important property as far as the current flow is concerned.

29.7 Self-Inductance and Mutual Induction

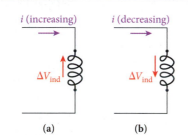

(a)   **(b)**

FIGURE 29.20 (a) Self-induced potential difference in an inductor when the current is increasing. (b) Self-induced potential difference in an inductor when the current is decreasing.

Consider the situation in which two coils, or inductors, are close to each other, and a changing current in the first coil produces a magnetic flux in the second coil. However, the changing current in the first coil also induces a potential difference in that coil, and thus the magnetic field from that coil also changes. This phenomenon is called **self-induction.** The resulting potential difference is termed the *self-induced potential difference.* Changing the current in the first coil also induces a potential difference in the second coil. This phenomenon is called **mutual induction.**

According to Faraday's Law of Induction, the self-induced potential difference for any inductor is given by

$$\Delta V_{\text{ind},L} = -\frac{d(N\Phi_B)}{dt} = -\frac{d(Li)}{dt} = -L\frac{di}{dt},\tag{29.22}$$

where equation 29.18 allows us to substitute Li for $N\Phi_B$. Thus, in any inductor, a self-induced potential difference appears when the current changes with time. This self-induced potential difference depends on the time rate of change of the current and the inductance of the device.

Lenz's Law provides the direction of the self-induced potential difference. The negative sign in equation 29.22 provides the clue that the induced potential difference always opposes any change in current. For example, Figure 29.20a shows current flowing through an inductor and increasing with time. Thus, the self-induced potential difference will oppose the increase in current. In Figure 29.20b, the current flowing through an inductor is decreasing with time. Thus, a self-induced potential difference will oppose the decrease in current. We have assumed that these inductors are ideal inductors; that is, they have no resistance. All induced potential differences manifest themselves across the connections of the inductor. Inductors with resistance are treated in the next section.

Now let's consider two adjacent coils with their central axes aligned (Figure 29.21). Coil 1 has N_1 turns, and coil 2 has N_2 turns. The current in coil 1 produces a magnetic field, $\vec{B}_1$. The flux linkage in coil 2 resulting from the

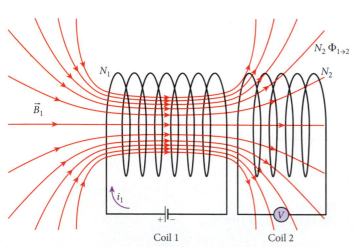

FIGURE 29.21 Coil 1 has current i_1. Coil 2 has a voltmeter capable of measuring small, induced potential differences.

magnetic field in coil 1 is $N_2 \Phi_{1 \to 2}$. The mutual inductance, $M_{1 \to 2}$, of coil 2 due to coil 1 is defined as

$$M_{1 \to 2} = \frac{N_2 \Phi_{1 \to 2}}{i_1}. \qquad (29.23)$$

Multiplying both sides of equation 29.23 by i_1 yields

$$i_1 M_{1 \to 2} = N_2 \Phi_{1 \to 2}.$$

If the current in coil 1 changes with time, we can write

$$M_{1 \to 2} \frac{di_1}{dt} = N_2 \frac{d\Phi_{1 \to 2}}{dt}.$$

The right-hand side of this equation is similar to the right-hand side of Faraday's Law of Induction (equation 29.5). Thus, we can write

$$\Delta V_{ind,2} = -M_{1 \to 2} \frac{di_1}{dt}. \qquad (29.24)$$

Now let's reverse the roles of the two coils (Figure 29.22). The current, i_2, in coil 2 produces a magnetic field, $\vec{B}_2$. The flux linkage in coil 1 resulting from the magnetic field in coil 2 is $N_1 \Phi_{2 \to 1}$. Using the same analysis applied to find the mutual inductance of coil 2 due to coil 1, we find

$$\Delta V_{ind,1} = -M_{2 \to 1} \frac{di_2}{dt}, \qquad (29.25)$$

where $M_{2 \to 1}$ is the mutual inductance of coil 1 due to coil 2.

We see that the potential difference induced in one coil is proportional to the change of current in the other coil. The proportionality constant is the mutual induction. If we switched the indexes 1 and 2 and repeated the entire analysis of the coils' effects on each other, we could show that

$$M_{1 \to 2} = M_{2 \to 1} = M.$$

We can then rewrite equations 29.24 and 29.25 as

$$\Delta V_{ind,2} = -M \frac{di_1}{dt} \qquad (29.26)$$

and

$$\Delta V_{ind,1} = -M \frac{di_2}{dt}, \qquad (29.27)$$

where M is the **mutual inductance** between the two coils. The SI unit of mutual inductance is the henry. One major application of mutual inductance is in transformers, which are discussed in Chapter 30.

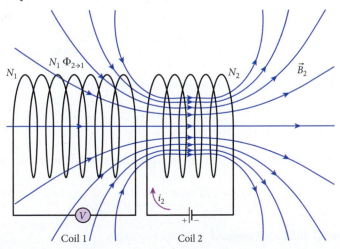

FIGURE 29.22 Coil 2 has current i_2. Coil 1 has a voltmeter capable of measuring small, induced potential differences.

SOLVED PROBLEM **29.1** | **Mutual Induction of a Solenoid and a Coil**

A long solenoid with circular cross section of radius $r_1 = 2.80$ cm and $n = 290$ turns/cm is inside and coaxial with a short coil with circular cross section of radius $r_2 = 4.90$ cm and $N = 31$ turns (Figure 29.23a). The current in the solenoid is increased at a constant rate from zero to $i = 2.20$ A over a time interval of 48.0 ms.

FIGURE 29.23 (a) A long solenoid of radius r_1 inside a short coil of radius r_2. (b) View of the two coils looking down the central axis.

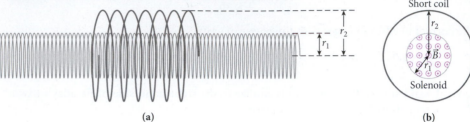

(a)   (b)

PROBLEM
What is the potential difference induced in the short coil while the current is changing?

SOLUTION

THINK
The potential difference induced in the short coil is due to the changing current flowing in the solenoid. According to equation 29.23, the mutual inductance of the short coil due to the solenoid is the number of turns in the short coil times the magnetic flux of the solenoid, divided by the current flowing in the solenoid. We can then calculate the potential difference induced in the short coil.

SKETCH
Figure 29.23b shows a view of the two coils looking down their central axis.

RESEARCH
We can formulate the mutual inductance between the coil and the solenoid as

$$M = \frac{N\Phi_{s \to c}}{i}, \qquad (i)$$

where N is the number of turns in the short coil, $N\Phi_{s \to c}$ is the flux linkage in the coil resulting from the magnetic field of the solenoid, and i is the current in the solenoid. The flux can be expressed as

$$\Phi_{s \to c} = BA, \qquad (ii)$$

where B is the magnitude of the magnetic field inside the solenoid and A is its cross-sectional area. Recall from Chapter 28 that for a solenoid, the magnitude of the magnetic field is

$$B = \mu_0 n i,$$

where n is the number of turns per unit length. The cross-sectional area of the solenoid is

$$A = \pi r_1^2. \qquad (iii)$$

The potential difference induced in the short coil is then

$$\Delta V_{ind} = -M\frac{di}{dt}.$$

SIMPLIFY
We can combine equations (i), (ii), and (iii) to obtain the mutual inductance between the two coils:

$$M = \frac{NBA}{i} = \frac{N(\mu_0 n i)(\pi r_1^2)}{i} = N\pi\mu_0 n r_1^2.$$

Then the potential difference induced in the short coil is

$$\Delta V_{ind} = -\left(N\pi\mu_0 n r_1^2\right)\frac{di}{dt}.$$

CALCULATE

The change in the current is constant, so

$$\frac{di}{dt} = \frac{2.20 \text{ A}}{48.0 \cdot 10^{-3} \text{ s}} = 45.8333 \text{ A/s}.$$

The mutual inductance between the two coils is

$$M = (31)\pi \left(4\pi \cdot 10^{-7} \text{ T m/A}\right)\left(290 \cdot 10^2 \text{ m}^{-1}\right)\left(2.80 \cdot 10^{-2} \text{ m}\right)^2 = 0.0027825 \text{ H}.$$

The potential difference induced in the short coil is then

$$\Delta V_{\text{ind}} = -(0.0027825 \text{ H})(45.8333 \text{ A/s}) = -0.127531 \text{ V}.$$

ROUND

We report our result to three significant figures:

$$\Delta V_{\text{ind}} = -0.128 \text{ V}.$$

DOUBLE-CHECK

The magnitude of the potential difference induced in the short outer coil is 128 mV, which is a magnitude that could be attained by moving a strong bar magnet in and out of a coil. Thus, our result seems reasonable.

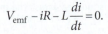

29.6 In-Class Exercise

Suppose the current in the short coil in Solved Problem 29.1 is increased steadily from zero to $i = 2.80$ A in 18.0 ms. What is the magnitude of the potential difference induced in the solenoid while the current in the short coil is changing?

a) 0.0991 V   d) 0.433 V

b) 0.128 V   e) 0.750 V

c) 0.233 V

29.8 RL Circuits

In Chapter 26, we saw that if a source of emf supplying a voltage, V_{emf}, is put into a single-loop circuit containing a resistor of resistance R and a capacitor of capacitance C, the charge, q, on the capacitor builds up over time according to

$$q = CV_{\text{emf}}\left(1 - e^{-t/\tau_{RC}}\right),$$

where the time constant of the circuit, $\tau_{RC} = RC$, is the product of the resistance and the capacitance. The same time constant governs the decrease of the initial charge, q_0, on the capacitor if the source of emf is suddenly removed and the circuit is short-circuited:

$$q = q_0 e^{-t/\tau_{RC}}.$$

If a source of emf is placed in a single-loop circuit containing a resistor with resistance R and an inductor with inductance L, called an **RL circuit,** a similar phenomenon occurs. Figure 29.24 shows a circuit in which a source of emf is connected to a resistor and an inductor in series. If the circuit included only the resistor and not the inductor, the current would increase almost instantaneously to the value given by Ohm's Law, iV_{emf}/R, as soon as the switch was closed. However, in the circuit with both the resistor and the inductor, the increasing current flowing through the inductor creates a self-induced potential difference that tends to oppose the increase in current. As time passes, the change in current decreases, and the opposing self-induced potential difference also decreases. After a long time, the current becomes steady at the value V_{emf}/R.

We can use Kirchhoff's Loop Rule to analyze this circuit, assuming that the current, i, at any given time is flowing through the circuit in a counterclockwise direction. With current flowing counterclockwise around the circuit, the source of emf represents a gain in potential, $+V_{\text{emf}}$, and the resistor represents a drop in potential, $-iR$. The self-inductance of the inductor produces a drop in potential because it is opposing the increase in current. The drop in potential due to the inductor is proportional to the time rate of change of the current, as given by equation 29.22. Thus, we can write the sum of the potential drops around the circuit as

$$V_{\text{emf}} - iR - L\frac{di}{dt} = 0.$$

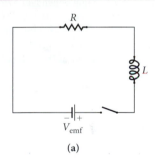

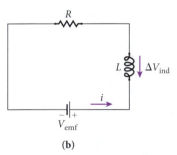

FIGURE 29.24 Single-loop circuit with a source of emf, a resistor, and an inductor: (a) switch open; (b) switch closed. When the switch is closed, the current flowing in the direction shown increases. A potential difference is induced across the inductor in the opposite direction, as shown.

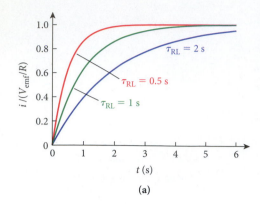

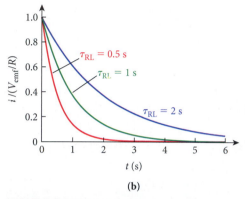

FIGURE 29.25 Time dependence of the current flowing through an RL circuit. (a) Current as a function of time when a resistor, an inductor, and a source of emf are connected in series. (b) Current as a function of time when the source of emf is suddenly removed from an RL circuit that has been connected for a long time.

29.7 In-Class Exercise

Consider the RL circuit shown in the figure. When the switch is closed, the current in the circuit increases exponentially to the value $i = V_{emf}/R$. If the inductor in this circuit is replaced with an inductor having three times the number of turns per unit length, the time required to reach a current of magnitude $0.9i$

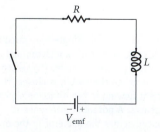

a) increases.

b) decreases.

c) stays the same.

We can rewrite this equation as

$$L\frac{di}{dt} + iR = V_{emf}. \qquad (29.28)$$

The solution to this differential equation is obtained in exactly the same way as the solution to the differential equation for the RC circuit was obtained in Chapter 26. The solution, which can be checked by substituting it into equation 29.28 is

$$i(t) = \frac{V_{emf}}{R}\left(1 - e^{-t/(L/R)}\right). \qquad (29.29)$$

The quantity L/R is the time constant of the RL circuit:

$$\tau_{RL} = \frac{L}{R}. \qquad (29.30)$$

This time dependence of the current in an RL circuit is shown in Figure 29.25a for three different values of the time constant.

Looking at equation 29.29, you can see that for $t = 0$, the current is zero. For $t \to \infty$, the current is given by $i = V_{emf}/R$, which is as expected.

Now consider the circuit depicted in Figure 29.26, in which a source of emf had been connected and is suddenly removed. We can use equation 29.28 with $V_{emf} = 0$ to describe the time dependence of this circuit:

$$L\frac{di}{dt} + iR = 0. \qquad (29.31)$$

The resistor causes a potential drop, and the inductor has a self-induced potential difference that tends to oppose the decrease in current. The solution of equation 29.31 is

$$i(t) = i_0 e^{-t/\tau_{RL}}. \qquad (29.32)$$

The initial conditions when the source of emf was connected can be used to determine the initial current: $i_0 = V_{emf}/R$. Equation 29.32 describes a single-loop circuit with a resistor and an inductor that initially has a current i_0. The current drops exponentially with time with a time constant $\tau_{RL} = L/R$, and after a long time, the current in the circuit is zero. The current in this RL circuit as a function of time for three different values of the time constant is plotted in Figure 29.25b.

RL circuits can be used as timers to turn on devices at particular intervals and can also be used to filter out noise. However, these applications are usually handled with similar RC circuits because small capacitors are available in a wider range of capacitances than are inductors. The real value of inductors becomes apparent in circuits with all three components, resistors, capacitors, and inductors, which are covered in Chapter 30.

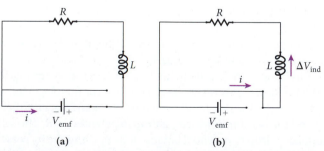

FIGURE 29.26 Single-loop circuit with a source of emf, a resistor, and an inductor. (a) The circuit with the source of emf connected. Current is flowing in the direction shown. (b) The source of emf is removed, and the resistor and the inductor are connected. Current flows in the same direction as before but is decreasing. A potential difference is induced across the inductor in the same direction as the current, as shown.

SOLVED PROBLEM 29.2 | Work Done by a Battery

A series circuit contains a battery that supplies $V_{emf} = 40.0$ V, an inductor with $L = 2.20$ H, a resistor with $R = 160.0$ Ω, and a switch, connected as shown in Figure 29.27.

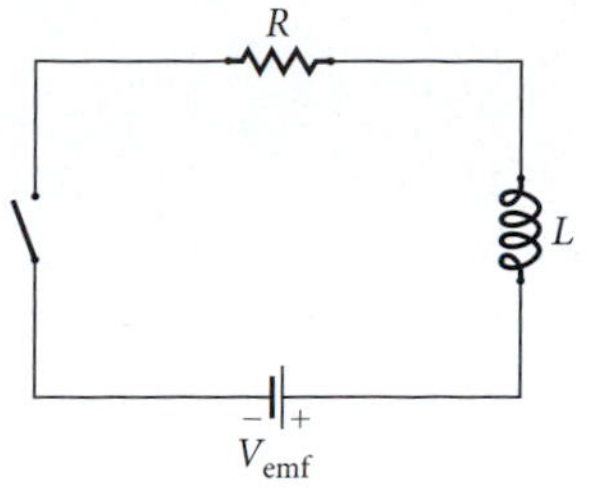

PROBLEM
The switch is closed at time $t = 0$. How much work is done by the battery between $t = 0$ and $t = 1.65 \cdot 10^{-2}$ s?

FIGURE 29.27 An RL circuit with a switch.

SOLUTION

THINK
When the switch is closed, current begins to flow and power is provided by the battery. Power is defined as the voltage times the current at any given time. Work is the integral of the power over the time during which the circuit operates.

SKETCH
Figure 29.28 shows a plot of the current in the RL circuit as a function of time.

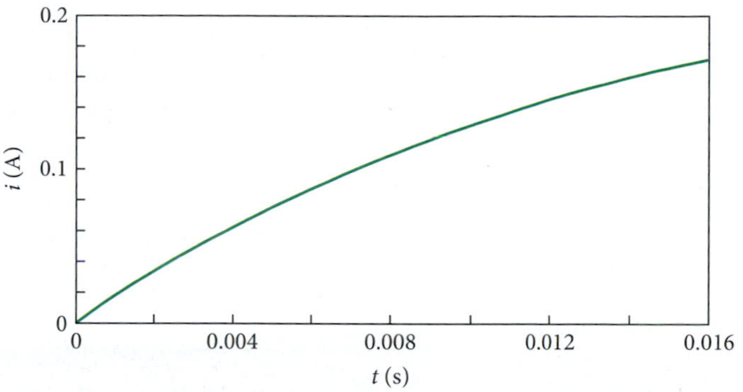

FIGURE 29.28 Current in the RL circuit as a function of time.

RESEARCH
The power in the circuit at any time t after the switch is closed is given by

$$P(t) = V_{emf} i(t), \tag{i}$$

where $i(t)$ is the current in the circuit. The current as a function of time for this circuit is given by equation 29.29,

$$i(t) = \frac{V_{emf}}{R}\left(1 - e^{-t/\tau_{RL}}\right), \tag{ii}$$

where $\tau_{RL} = L/R$. The work done by the battery is the integral of the power over the time in which the circuit has been in operation:

$$W = \int_{0}^{T} P(t)\,dt, \tag{iii}$$

where T is the time after the switch is closed.

SIMPLIFY
We can combine equations (i), (ii), and (iii) to obtain

$$W = \int_{0}^{T} \frac{V_{emf}^2}{R}\left(1 - e^{-t/\tau_{RL}}\right)dt.$$

Evaluating the definite integral gives us

$$W = \frac{V_{emf}^2}{R}\left[t + \tau_{RL}e^{-t/\tau_{RL}}\right]_{0}^{T} = \frac{V_{emf}^2}{R}\left[T + \tau_{RL}\left(e^{-T/\tau_{RL}} - 1\right)\right]. \tag{iv}$$

CALCULATE
First, we calculate the value of the time constant:

$$\tau_{RL} = \frac{L}{R} = \frac{2.20\ \text{H}}{160.0\ \Omega} = 1.375 \cdot 10^{-2}\ \text{s}.$$

Putting all the numerical values into equation (iv) then gives

$$W = \frac{(40.0\ \text{V})^2}{160.0\ \Omega}\left[\left(1.65 \cdot 10^{-2}\ \text{s}\right) + \left(1.375 \cdot 10^{-2}\ \text{s}\right)\left(e^{-\left(1.65 \cdot 10^{-2}\ \text{s}\right)/\left(1.375 \cdot 10^{-2}\ \text{s}\right)} - 1\right)\right] = 0.0689142\ \text{J}.$$

Continued—

ROUND
We report our result to three significant figures:

$$W = 6.89 \cdot 10^{-2} \text{ J}.$$

DOUBLE-CHECK
To double-check our result, we assume that the current in the circuit is constant in time and equal to half of the final current:

$$i_{ave} = \frac{i(T)}{2} = \frac{\left(V_{emf}/R\right)\left(1 - e^{-T/\tau_{RL}}\right)}{2} = \frac{\left(40.0 \text{ V}/160.0 \text{ }\Omega\right)\left(1 - e^{-\left(1.65\cdot10^{-2}\text{ s}\right)/\left(1.375\cdot10^{-2}\text{ s}\right)}\right)}{2} = 0.0874 \text{ A}.$$

This current would correspond to the average current if the current increased linearly with time. The work done would then be

$$W = PT = i_{ave}V_{emf}T = \left(0.0874 \text{ A}\right)\left(40.0 \text{ V}\right)\left(1.65 \cdot 10^{-2} \text{ s}\right) = 5.77 \cdot 10^{-2} \text{ J}$$

This value is less than, but close to, our calculated result. Thus, our result seems reasonable.

29.9 Energy and Energy Density of a Magnetic Field

We can think of an inductor as a device that can store energy in a magnetic field, similar to the way a capacitor can store energy in an electric field. The energy stored in the electric field of a capacitor is given by

$$U_E = \frac{1}{2}\frac{q^2}{C}.$$

Consider an inductor connected to a source of emf. Current begins to flow through the inductor, producing a self-induced potential difference opposing the increase in current. The instantaneous power provided by the source of emf is the product of the current and the voltage of the emf source, V_{emf}. Using equation 29.28 with $R = 0$, we can write

$$P = V_{emf}i = \left(L\frac{di}{dt}\right)i. \tag{29.33}$$

Integrating this power over the time it takes to reach a final current i in the circuit yields the energy provided by the source of emf. Since there are no resistive losses in this circuit, this amount of energy must be stored in the magnetic field of the inductor. Therefore,

$$U_B = \int_0^t P\,dt = \int_0^i Li'\,di' = \tfrac{1}{2}Li^2. \tag{29.34}$$

Equation 29.34 has a form similar to the analogous equation for a capacitor's electric field, with q replaced by i and $1/C$ replaced by L.

Now let's consider an ideal solenoid with length ℓ, cross-sectional area A, and n turns per unit length, carrying current i. The energy stored in the magnetic field of the solenoid using equation 29.21 is

$$U_B = \tfrac{1}{2}Li^2 = \tfrac{1}{2}\mu_0 n^2 \ell A i^2.$$

The magnetic field occupies the volume enclosed by the solenoid, which is given by ℓA. Thus, the energy density, u_B, of the magnetic field of the solenoid is

$$u_B = \frac{\tfrac{1}{2}\mu_0 n^2 \ell A i^2}{\ell A} = \tfrac{1}{2}\mu_0 n^2 i^2.$$

Since $B = \mu_0 ni$ for a solenoid, the energy density of the magnetic field of a solenoid can be expressed as

$$u_B = \frac{1}{2\mu_0}B^2. \tag{29.35}$$

Although we derived this expression for the special case of a solenoid, it applies to magnetic fields in general.

29.8 In-Class Exercise

Consider a long solenoid with a circular cross section of radius $r = 8.10$ cm and $n = 2.00 \cdot 10^4$ turns/m. The solenoid has length $\ell = 0.540$ m and is carrying a current of magnitude $i = 4.04 \cdot 10^{-3}$ A. How much energy is stored in the magnetic field of the solenoid?

a) $2.11 \cdot 10^{-7}$ J   d) $6.66 \cdot 10^{-3}$ J

b) $8.91 \cdot 10^{-6}$ J   e) $4.55 \cdot 10^{-1}$ J

c) $4.55 \cdot 10^{-5}$ J

Computers and many consumer electronics devices use magnetization and induction to store and retrieve information. Examples are computer hard drives, videotapes, audiotapes, and the magnetic strips on credit cards. During the last decade, the use of storage media based on other technologies, such as the optical storage of information on CDs and DVDs and the flash memory cards in digital cameras, has increased; however, magnetic storage devices are still a technological mainstay and the basis of a multibillion dollar industry.

Computer Hard Drive

One device that stores information using magnetization and induction is the computer hard drive. The hard drive stores information in the form of *bits,* the binary code consisting of zeros and ones. Eight bits make a *byte,* which can represent a number or an alphanumeric character. A modern hard drive can hold up to 2 terabyte (10^{12} bytes) of information. A hard drive consists of one or more rotating platters with a ferromagnetic coating accessed by a movable read/write head, as shown in Figure 29.29.

FIGURE 29.29 The read/write head and spinning platter inside a computer hard drive.

The read/write head can be positioned to access any one of many tracks on the rotating platter. The operation of a read/write head in a conventional hard drive is illustrated in Figure 29.30a. As the coated platter moves below the read/write head, a pulse of current in one direction magnetizes the surface of the platter to represent a binary one, or a pulse of current in the opposite direction magnetizes the surface representing a binary zero. In Figure 29.30a, a binary one is represented by a red arrow pointing to the right, and a binary zero is represented by a green arrow pointing to the left. In read mode, when the magnetized areas of the platter pass beneath the read sensor, a positive or negative current is induced, and the electronics of the hard drive can tell if the information is a zero or a one. The method used to encode and read back data shown in Figure 29.30 is called *longitudinal encoding* because the magnetic fields of the magnetized areas of the platter are parallel or antiparallel to the motion of the platter. The data storage capacity of hard drives has been increased by making the magnetized areas smaller

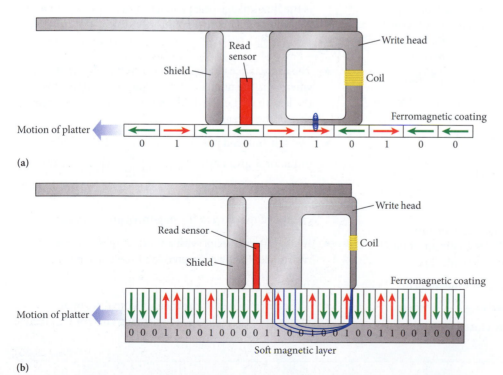

(a)

(b)

FIGURE 29.30 The read/write head of a computer hard drive. (a) Longitudinal encoding of information on the spinning platter. (b) Perpendicular encoding of information on the spinning platter.

and by adding more platters and read/write heads. However, manufacturers found it difficult to construct hard drives holding more than 250 gigabytes ($250 \cdot 10^9$ bytes) using this technique. As the manufacturers made the bits smaller, the bits interfered with each other, causing some bits to flip randomly and introduce errors in the stored data.

Recently, the technique of *perpendicular encoding* of data has been developed, illustrated in Figure 29.30b. Again, a read/write head is used above a spinning platter coated with a ferromagnetic substance. However, in this case, the magnetic fields are perpendicular to the surface of the platter, which allows a tighter packing of the bits and increases the capacity of the hard drive. The platter is constructed with a thicker ferromagnetic coating and a soft magnetic material on the bottom that acts to contain the magnetic field lines. Note that the magnetic field lines at the pointed end of the write head are very close together, whereas the magnetic field lines returning to the blunt end of the write head are widely spaced. Thus, the ferromagnetic coating of the platter is strongly magnetized in the up or down direction, depending on the direction of the current pulse through the coil of the write head, while the bits closer to the read sensor are not affected.

Hard drives using perpendicular encoding also incorporate the phenomenon called *giant magnetoresistance (GMR),* which allows the construction of a very small and sensitive read sensor. The French physicist Albert Fert and the German physicist Peter Grünberg received the Nobel Prize in Physics in 2007 for the discovery of this effect. Hard drives with information storage capacities of up to 2 terabyte (10^{12} bytes) or more that use perpendicular encoding and GMR read sensors are widely available. iPods with a storage capacity above 64 GB are an example of a device using this technology (iPod Touch and iPhones use a different storage technology that has no moving parts). The fact that you can watch full-length movies on your iPod and carry many thousands of songs in it as well is a direct result of physics research performed during the last two decades. And as research in nanoscience and nanotechnology continues to produce exciting results for technological applications, the amazing growth in the capabilities of consumer electronics devices will continue for the foreseeable future.

WHAT WE HAVE LEARNED | EXAM STUDY GUIDE

- According to Faraday's Law of Induction, the induced potential difference, ΔV_{ind}, in a conducting loop is given by the negative of the time rate of change of the magnetic flux passing through the loop: $\Delta V_{ind} = -\dfrac{d\Phi_B}{dt}$.

- The magnetic flux, Φ_B, is given by $\Phi_B = \oiint \vec{B} \cdot d\vec{A}$, where $\vec{B}$ is the magnetic field and $d\vec{A}$ is the differential area element defined by a vector normal to the surface through which the magnetic field passes.

- For a constant magnetic field, $\vec{B}$, the magnetic flux, Φ_B, passing through an area, A is given by $\Phi_B = BA \cos\theta$, where θ is the angle between the magnetic field vector and a normal vector to the area.

- Lenz's Law states that a changing magnetic flux through a conducting loop induces a current in the loop that opposes the change in magnetic flux.

- A magnetic field that is changing in time induces an electric field given by $\oint \vec{E} \cdot d\vec{s} = -\dfrac{d\Phi_B}{dt}$, where the integration is done over any closed path in the magnetic field.

- The inductance, L, of a device with conducting loops is the flux linkage (the product of the number of loops, N, and the magnetic flux, Φ_B) divided by the current, i: $L = \dfrac{N\Phi_B}{i}$.

- The inductance of a solenoid is given by $L = \mu_0 n^2 \ell A$, where n is the number of turns per unit length, ℓ is the length of the solenoid, and A is the cross-sectional area of the solenoid.

- The self-induced potential difference, $\Delta V_{ind,L}$, for any inductor is given by $\Delta V_{ind,L} = -L\dfrac{di}{dt}$, where L is the inductance of the device and $\dfrac{di}{dt}$ is the time rate of change of the current flowing through the inductor.

- A single-loop circuit with an inductance L and a resistance R has a characteristic time constant of $\tau_{RL} = \dfrac{L}{R}$.

- The energy stored in the magnetic field U_B of an inductor with inductance L carrying current i is given by $U_B = \frac{1}{2}Li^2$.

KEY TERMS

NEW SYMBOLS AND EQUATIONS

$\Phi_B = \oiint \vec{B} \bullet d\vec{A}$, magnetic flux

$\Delta V_{ind} = -\dfrac{d\Phi_B}{dt}$, Faraday's Law of Induction

$L = \dfrac{N\Phi_B}{i}$, inductance

$\tau_{RL} = \dfrac{L}{R}$, time constant for an RL circuit

ANSWERS TO SELF-TEST OPPORTUNITIES

29.1 $\dfrac{dB}{dt} = -\dfrac{0.500\ \text{T}}{0.250\ \text{s}} = -2.00\ \text{T/s}$

$\Delta V_{ind} = -\dfrac{d\Phi_B}{dt} = -\dfrac{d(BA)}{dt} = -\pi r^2 \dfrac{dB}{dt}$

$r = \sqrt{\dfrac{|\Delta V_{ind}|}{\pi |dB/dt|}} = \sqrt{\dfrac{1.24\ \text{V}}{\pi(2.00\ \text{T/s})}} = 0.444\ \text{m}.$

29.2 As the loop enters the magnetic field, the magnetic flux is increasing. The current induced in the loop will be in the counterclockwise direction to oppose the increase in the flux. As the loop exits the magnetic field, the magnetic flux

is decreasing. The current induced in the loop will be in a clockwise direction to oppose the decrease in the flux.

29.3 If the induced potential difference was equal to the change in the magnetic flux, then any increase in the flux going through a coil (perhaps from just a minute random fluctuation in the ambient magnetic field in the room) would lead to an induced potential difference, which would produce a current in the coil, which would act to increase the flux, which would lead to a larger induced potential difference, a larger current, and an even larger increase in flux. In other words, a runaway situation would result, which clearly violates energy conservation.

29.4 a) true   b) false   c) true   d) true

PROBLEM-SOLVING PRACTICE

Problem-Solving Guidelines

1. To solve a problem involving electromagnetic induction, first ask: What is making the magnetic flux change? If the magnetic field is changing, you need to use equation 29.9; if the area through which the flux passes is changing, you need to use equation 29.10; and if the orientation between the magnetic field and the area is changing, you need to use equation 29.11. You don't need to memorize these equations,

as long as you remember Faraday's Law (equation 29.5) and the definition of magnetic flux (equation 29.1).

2. Once you know which elements of the problem situation are constant and which are changing, use Lenz's Law to determine the direction of the induced current and the locations of higher and lower potential. Then, you can pick a direction for the differential area vector, $d\vec{A}$, of the flux and calculate the unknown quantities.

SOLVED PROBLEM 29.3 | Power from a Rotating Rod

A conducting rod with length $\ell = 8.17$ cm rotates around one of its ends in a uniform magnetic field that has magnitude $B = 1.53$ T and is directed parallel to the rotation axis of the rod (Figure 29.31). The other end of the rod slides on a frictionless conducting ring. The rod makes 6.00 revolutions per second. A resistor, $R = 1.63$ mΩ, is connected between the rotating rod and the conducting ring.

PROBLEM

What is the power dissipated in the resistor due to magnetic induction?

Continued—

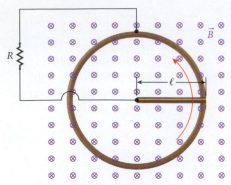

FIGURE 29.31 Conducting rod rotating in a constant magnetic field directed into the page.

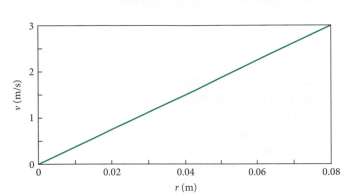

FIGURE 29.32 Speed as a function of radius for the conducting rod.

SOLUTION

THINK

We can calculate the potential difference induced in a conductor of length ℓ moving with speed v perpendicular to a magnetic field of magnitude B. However, the rotating rod has different speeds at different radii, $v(r)$. Therefore, we must calculate the potential difference induced on the rod by integrating $Bv(r)$ over the length of the rod. From the induced potential difference, we can calculate the power dissipated in the resistor.

SKETCH

Figure 29.32 shows the speed as a function of radius for the conducting rod.

RESEARCH

The potential difference, ΔV_{ind}, induced on a conductor of length ℓ moving with speed v perpendicular to a magnetic field of magnitude B is given by equation 29.15:

$$\Delta V_{ind} = v\ell B.$$

However, in this case, different parts of the conducting rod are moving at different speeds. We can express the speed of the different parts of the rod as a function of distance r from the axis of rotation:

$$v(r) = \frac{2\pi r}{T},$$

where $v(r)$ is the speed of the rod at distance r and T is the period of the rotation. We can then find the potential difference induced in the rotating rod over its length, ℓ:

$$\Delta V_{ind} = \int_0^\ell v(r)B\,dr. \tag{i}$$

The power dissipated in the resistor is given by

$$P = \frac{\Delta V_{ind}^2}{R}. \tag{ii}$$

SIMPLIFY

Evaluating the definite integral in equation (i) gives

$$\Delta V_{ind} = \int_0^\ell \left(\frac{2\pi r}{T}\right)B\,dr = \frac{2\pi B}{T}\frac{\ell^2}{2} = \frac{\pi B\ell^2}{T}. \tag{iii}$$

Substituting the expression for ΔV_{ind} from equation (iii) into equation (ii) leads to an expression for the power dissipated in the resistor:

$$P = \frac{\left(\pi B\ell^2/T\right)^2}{R} = \frac{\pi^2 B^2 \ell^4}{RT^2}.$$

CALCULATE

The period is the inverse of the frequency. The frequency is $f = 6.00$ Hz, so the period is

$$T = \frac{1}{f} = \frac{1}{6.00}\ \text{s}.$$

Putting in the numerical values gives us

$$P = \frac{\pi^2 B^2 \ell^4}{RT^2} = \frac{\pi^2 (1.53\ \text{T})^2 (0.0817\ \text{m})^4}{\left(1.63\cdot 10^{-3}\ \Omega\right)\left(\frac{1}{6.00}\ \text{s}\right)^2} = 22.7345\ \text{W}.$$

ROUND

We report our result to three significant figures:

$$P = 22.7\ \text{W}.$$

DOUBLE-CHECK

To double-check our result, we consider a conducting rod of the same length moving perpendicularly to the same magnetic field with a speed equal to the speed of the center of the rotating rod, which is

$$v(\ell/2) = \frac{2\pi(\ell/2)}{T} = \frac{2\pi L}{2T} = \frac{2\pi(0.0817 \text{ m})}{2\left(\frac{1}{6.00}\text{ s}\right)} = 1.54 \text{ m/s}.$$

The induced potential difference across the conducting rod moving perpendicularly would be

$$\Delta V_{ind} = v\ell B = (1.54 \text{ m/s})(0.0817 \text{ m})(1.53 \text{ T}) = 0.193 \text{ V}.$$

The power dissipated in the resistor would then be

$$P = \frac{\Delta V_{ind}^2}{R} = \frac{(0.193 \text{ V})^2}{1.63 \cdot 10^{-3} \text{ }\Omega} = 22.9 \text{ W},$$

which is close to our result within rounding error. Thus, our result seems reasonable.

Finally, note that there is a possible additional source of potential difference between the two ends of the rod. Our solution assumed that the potential difference between the two ends is due exclusively to the magnetic induction. However, all charge carriers inside the rod are forced on a circular path due to the rotation. This requires a centripetal force, which should in principle reduce the potential difference between the two ends of the rod. However, for the small angular velocity of the rod in this problem, this effect is negligible.

MULTIPLE-CHOICE QUESTIONS

29.1 A solenoid with 200 turns and a cross-sectional area of 60 cm² has a magnetic field of 0.60 T along its axis. If the field is confined within the solenoid and changes at a rate of 0.20 T/s, the magnitude of the induced potential difference in the solenoid will be

a) 0.0020 V.   b) 0.02 V.   c) 0.001 V.   d) 0.24 V.

29.2 The rectangular loop of wire in Figure 29.9 is pulled at constant acceleration from a region of zero magnetic field into a region of a uniform magnetic field. During this process, the current induced in the loop

a) will be zero.

b) will be some constant value that is not zero.

c) will increase linearly with time.

d) will increase exponentially with time.

e) will increase linearly with the square of the time.

29.3 Which of the following will induce a current in a loop of wire in a uniform magnetic field?

a) decreasing the strength of the field

b) rotating the loop about an axis parallel to the field

c) moving the loop within the field

d) all of the above

e) none of the above

29.4 Faraday's Law of Induction states

a) that a potential difference is induced in a loop when there is a change in the magnetic flux through the loop.

b) that the current induced in a loop by a changing magnetic field produces a magnetic field that opposes this change in magnetic field.

c) that a changing magnetic field induces an electric field.

d) that the inductance of a device is a measure of its opposition to changes in current flowing through it.

e) that magnetic flux is the product of the average magnetic field and the area perpendicular to it that it penetrates.

29.5 A conducting ring is moving from left to right through a uniform magnetic field, as shown in the figure. In which regions is there an induced current in the ring?

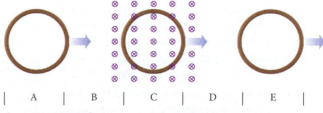

| | A | | B | | C | | D | | E | |

a) regions B and D   c) region C

b) regions B, C, and D   d) regions A through E

29.6 A circular loop of wire moving in the xy-plane with a constant velocity in the negative x-direction enters a uniform magnetic field, which covers the region in which $x < 0$, as shown in the figure. The surface normal vector of the loop points in the direction of the magnetic field. Which of the following statements is correct?

a) The induced potential difference in the loop is at a maximum as the edge of the loop just enters the region with the magnetic field.

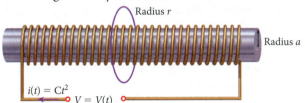

b) The induced potential difference in the loop is at a maximum when one fourth of the loop is in the region with the magnetic field.

c) The induced potential difference in the loop is at a maximum when the loop is halfway into the region with the magnetic field.

d) The induced potential difference in the loop is constant from the instant the loop starts to enter the region with the magnetic field.

29.7 Which of the following statements regarding self induction is correct?

a) Self-induction occurs only when a direct current is flowing through a circuit.

b) Self-induction occurs only when an alternating current is flowing through a circuit.

c) Self-induction occurs when either a direct current or an alternating current is flowing through a circuit.

d) Self-induction occurs when either a direct current or an alternating current is flowing through a circuit as long as the current is varying.

29.8 You have a light bulb, a bar magnet, a spool of wire that you can cut into as many pieces as you want, and nothing else. How can you get the bulb to light up?

a) You can't. The bulb needs electricity to light it, not magnetism.

b) You cut a length of wire, connect the light bulb to the two ends of the wire, and pass the magnet through the loop that is formed.

c) You cut two lengths of wire and connect the magnet and the bulb in series.

QUESTIONS

29.9 When you plug a refrigerator into a wall socket, on occasion, a spark appears between the prongs. What causes this?

29.10 People with pacemakers or other mechanical devices as implants are often warned to stay away from large machinery or motors. Why?

29.11 Chapter 14 discussed damped harmonic oscillators, in which the damping force is velocity dependent and always opposes the motion of the oscillator. One way of producing this type of force is to use a piece of metal, such as aluminum, that moves through a nonuniform magnetic field. Explain why this technique is capable of producing a damping force.

29.12 In a popular lecture demonstration, a cylindrical permanent magnet is dropped down a long aluminum tube as shown in the figure. Neglecting friction of the magnet against the inner walls of the tube and assuming that the tube is very long compared to the size of the magnet, will the magnet accelerate downward with an acceleration equal to g (free fall)? If not, describe the eventual motion of the magnet. Does it matter if the north pole or south pole of the magnet is on the lower side?

29.13 A popular demonstration of eddy currents involves dropping a magnet down a long metal tube and a long glass or plastic tube. As the magnet falls through a tube, there is changing flux as the magnet falls toward or away from each part of the tube.

a) Which tube has the larger voltage induced in it?

b) Which tube has the larger eddy currents induced in it?

29.14 The current in a very long, tightly wound solenoid with radius a and n turns per unit length varies in time according to the equation $i(t) = Ct^2$, where the current i is in amps and the time t is in seconds, and C is a constant with appropriate units. Concentric with the solenoid is a conducting ring of radius r, as shown in the figure.

a) Write an expression for the potential difference induced in the ring.

b) Write an expression for the magnitude of the electric field induced at an arbitrary point on the ring.

c) Is the ring necessary for the induced electric field to exist?

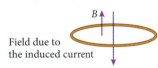

Radius r

Radius a

$i(t) = Ct^2$

$V = V(t)$

29.15 A circular wire ring experiences an increasing magnetic field in the upward direction, as shown in the figure. What is the direction of the induced current in the ring?

Field due to the induced current

B

29.16 A square conducting loop with sides of length L is rotating at a constant angular speed, ω, in a uniform magnetic field of magnitude B. At time $t = 0$, the loop is oriented so that the direction normal to the loop is aligned with the magnetic field. Find an expression for the potential difference induced in the loop as a function of time.

29.17 A solid metal disk of radius R is rotating around its center axis at a constant angular speed of ω. The disk is in a uniform magnetic field of magnitude B that is oriented normal to the surface of the disk. Calculate the magnitude of the potential difference between the center of the disk and the outside edge.

29.18 Large electric fields are certainly a hazard to the human body, as they can produce dangerous currents, but what about large magnetic fields? A man 1.80 m tall walks at 2.00 m/s perpendicular to a horizontal magnetic field of 5.0 T; that is, he walks between the pole faces of a very big magnet. (Such a magnet can, for example, be found in the National Superconducting Cyclotron Laboratory at Michigan State University.) Given that his body is full of conducting fluids, estimate the potential difference induced between his head and feet.

29.19 At Los Alamos National Laboratories, one means of producing very large magnetic fields is known as the *EPFCG* (*explosively-pumped flux compression generator*), which is used to study the effects of a high-power electromagnetic pulse (EMP) in electronic warfare. Explosives are packed and detonated in the space between a solenoid and a small copper cylinder coaxial with and inside the solenoid, as shown in the figure. The explosion occurs in a very short time and collapses the cylinder rapidly. This rapid change creates inductive currents that keep the magnetic flux constant while the cylinder's radius shrinks by a factor of r_i/r_f. Estimate the magnetic field

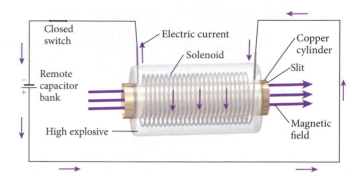

produced, assuming that the radius is compressed by a factor of 14 and the initial magnitude of the magnetic field, B_i, is 1.0 T.

29.20 A metal hoop is laid flat on the ground. A magnetic field that is directed upward, out of the ground, is increasing in magnitude. As you look down on the hoop from above, what is the direction of the induced current in the hoop?

29.21 The wire of a tightly wound solenoid is unwound and then rewound to form another solenoid with double the diameter of the first solenoid. By what factor will the inductance change?

PROBLEMS

A blue problem number indicates a worked-out solution is available in the Student Solutions Manual. One • and two •• indicate increasing level of problem difficulty.

Sections 29.1 and 29.2

29.22 A circular coil of wire with 20 turns and a radius of 40 cm is laying flat on a horizontal table as shown in the figure. There is a uniform magnetic field extending over the entire table with a magnitude of 5 T and directed to the north and downward, making an angle of 25.8° with the horizontal. What is the magnitude of the magnetic flux through the coil?

29.23 When a magnet in an MRI is abruptly shut down, the magnet is said to be *quenched*. Quenching can occur in as little as 20 s. Suppose a magnet with an initial field of 1.2 T is quenched in 20 s, and the final field is approximately zero. Under these conditions, what is the average induced potential difference around a conducting loop of radius 1.0 cm (about the size of a wedding ring) oriented perpendicular to the field?

29.24 An 8-turn coil has square loops measuring 0.200 m along a side and a resistance of 3.00 Ω. It is placed in a magnetic field that makes an angle of 40.0° with the plane of each loop. The magnitude of this field varies with time according to $B = 1.50t^3$, where t is measured in seconds and B in teslas. What is the induced current in the coil at $t = 2.00$ s?

29.25 A metal loop has an area of 0.100 m² and is placed flat on the ground. There is a uniform magnetic field pointing due west, as shown in the figure. This magnetic field initially has a magnitude of 0.123 T, which decreases steadily to 0.075 T during a period of 0.579 s. Find the potential difference induced in the loop during this time.

•**29.26** A respiration monitor has a flexible loop of copper wire, which wraps about the chest. As the wearer breathes, the radius of the loop of wire increases and decreases. When a person in the Earth's magnetic field inhales, what is the average current in the loop, assuming that it has a resistance of 30 Ω and increases in radius from 20 cm to 25 cm over 1 s? Assume that the magnetic field is perpendicular to the plane of the loop.

•**29.27** A circular conducting loop with radius a and resistance R_2 is concentric with a circular conducting loop with radius $b \gg a$ (b much greater than a) and resistance R_1. A time-dependent voltage is applied to the larger loop, having a slow sinusoidal variation in time given by $V(t) = V_0 \sin \omega t$, where V_0 and ω are constants with dimensions of voltage and inverse time, respectively. Assuming that the magnetic field throughout the inner loop is uniform (constant in space) and equal to the field at the center of the loop, derive expressions for the potential difference induced in the inner loop and the current i through that loop.

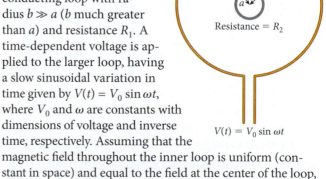

••**29.28** A long solenoid with cross-sectional area A_1 surrounds another long solenoid with cross-sectional area $A_2 < A_1$ and resistance R. Both solenoids have the same length and the same number of turns. A current given by $i = i_0 \cos \omega t$ is flowing through the outer solenoid. Find an expression for the magnetic field in the inner solenoid due to the induced current.

Section 29.3

29.29 The conducting loop in the shape of a quarter-circle shown in the figure has a radius of 10 cm and resistance of 0.2 Ω. The magnetic field strength within the dotted circle of radius 3 cm is initially 2 T. The magnetic field strength then decreases from 2 T to 1 T in 2 s. Find (a) the magnitude and (b) the direction of the induced current in the loop.

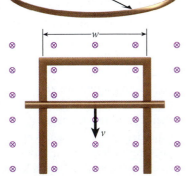

29.30 A supersonic aircraft with a wingspan of 10 m is flying over the north magnetic pole (in a magnetic field of magnitude 0.5 G perpendicular to the ground) at a speed of three times the speed of sound (Mach 3). What is the potential difference between the tips of the wings? Assume that the wings are made of aluminum.

•**29.31** A helicopter hovers above the north magnetic pole in a magnetic field of magnitude 0.5 G perpendicular to the ground. The helicopter rotors are 10 m long, are made of aluminum, and rotate about the hub with a rotational speed of 10,000 rpm. What is the potential difference from the hub of the rotor to the end?

•**29.32** An elastic circular conducting loop expands at a constant rate over time such that its radius is given by $r(t) = r_0 + vt$, where $r_0 = 0.100$ m and $v = 0.0150$ m/s. The loop has a constant resistance of $R = 12$ Ω and is placed in a uniform magnetic field of magnitude $B_0 = 0.750$ T, perpendicular to the plane of the loop, as shown in the figure. Calculate the direction and the magnitude of the induced current, i, at $t = 5$ s.

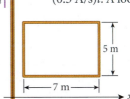

•**29.33** A rectangular frame of conducting wire has negligible resistance and width w and is held vertically in a magnetic field of magnitude B, as shown in the figure. A metal bar with mass m and resistance R is placed across the frame, maintaining contact with the frame. Derive an expression for the terminal velocity of the bar if it is allowed to fall freely along this frame starting from rest. Neglect friction between the wires and the metal bar.

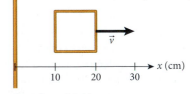

•**29.34** Two parallel conducting rails with negligible resistance are connected at one end by a resistor of resistance R, as shown in the figure. The rails are placed in a magnetic field $\vec{B}_{ext}$, which is perpendicular to the plane of the rails. This magnetic field is uniform and time independent. The distance between the rails is ℓ. A conducting rod slides without friction on top of the two rails at constant velocity $\vec{v}$.

a) Using Faraday's Law of Induction, calculate the magnitude of the potential difference induced in the moving rod.

b) Calculate the magnitude of the induced current in the rod, i_{ind}.

c) Show that for the rod to move at a constant velocity as shown, it must be pulled with an external force, $\vec{F}_{ext}$, and calculate the magnitude of this force.

d) Calculate the work done, W_{ext}, and the power generated, P_{ext}, by the external force in moving the rod.

e) Calculate the power used (dissipated) by the resistor, P_R. Explain the correlation between this result and those of part (d).

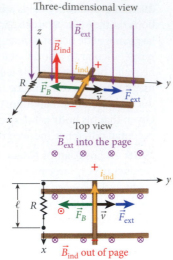

Three-dimensional view

Top view

$\vec{B}_{ext}$ into the page

$\vec{B}_{ind}$ out of page

•**29.35** A long, straight wire runs along the y-axis. The wire carries a current in the positive y-direction that is changing as a function of time according to $i = 2$ A + $(0.3 \text{ A/s})t$. A loop of wire is located in the xy-plane near the y-axis, as shown in the figure. The loop has dimensions 7 m by 5 m and is 1 m away from the wire. What is the induced potential difference in the wire loop at $t = 10$ s?

••**29.36** The long, straight wire in the figure has a current $i = 1$ A flowing in it. A square loop with 10-cm sides and a resistance of 0.02 Ω is positioned 10 cm away from the wire. The loop is then moved in the positive x-direction with speed $v = 10$ cm/s.

a) Find the direction of the induced current in the loop.

b) Identify the directions of the magnetic forces acting on all sides of the square loop.

c) Calculate the direction and the magnitude of the net force acting on the loop at the instant it starts to move.

Section 29.4

29.37 A simple generator consists of a loop rotating inside a constant magnetic field (see Figure 29.17). If the loop is rotating with frequency f, the magnetic flux is given by $\Phi(t) = BA \cos(2\pi ft)$. If $B = 1$ T and $A = 1$ m², what must the value of f be for the maximum induced potential difference to be 110 V?

•**29.38** A motor has a single loop inside a magnetic field of magnitude 0.87 T. If the area of the loop is 300. cm², find the maximum angular speed possible for this motor when connected to a source of emf providing 170 V.

•29.39 Your friend decides to produce electrical power by turning a coil of 100,000 circular loops of wire around an axis parallel to a diameter in the Earth's magnetic field, which has a local magnitude of 0.3 G. The loops have a radius of 25 cm.

a) If your friend turns the coil at a frequency of 150 Hz, what peak current will flow in a resistor, $R = 1500 \ \Omega$, connected to the coil?

b) The average current flowing in the coil will be 0.7071 times the peak current. What will be the average power obtained from this device?

Sections 29.6 and 29.7

29.40 Find the mutual inductance of the solenoid and the coil described in Example 29.1 and the potential difference at $t = 2.0$ s using the techniques described in Section 29.7. How do the results compare?

29.41 The figure shows the current through a 10-mH inductor over a time interval of 8 ms. Draw a graph showing the self-induced potential difference, $\Delta V_{ind,L}$, for the inductor over the same interval.

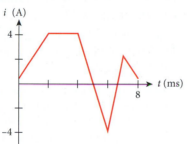

•29.42 A short coil with radius $R = 10$ cm contains $N = 30$ turns and surrounds a long solenoid with radius $r = 8.0$ cm containing $n = 60$ turns per centimeter. The current in the short coil is increased at a constant rate from zero to $i = 2.0$ A in a time of $t = 12$ s. Calculate the induced potential difference in the long solenoid while the current is increasing in the short coil.

Section 29.8

29.43 Consider an RL circuit with resistance $R = 1$ MΩ and inductance $L = 1$ H, which is powered by a 10-V battery.

a) What is the time constant of the circuit?

b) If the switch is closed at time $t = 0$, what is the current just after that time? After 2 µs? When a long time has passed?

29.44 In the circuit in the figure, $R = 120 \ \Omega$, $L = 3$ H, and $V_{emf} = 40$ V. After the switch is closed, how long will it take the current in the inductor to reach 300 mA?

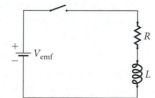

29.45 The current is increasing at a rate of 3.6 A/s in an RL circuit with $R = 3.25 \ \Omega$ and $L = 440$ mH. What is the potential difference across the circuit at the moment when the current in the circuit is 3.0 A?

•29.46 In the circuit in the figure, a battery supplies $V_{emf} = 18$ V and $R_1 = 6.0 \ \Omega$, $R_2 = 6.0 \ \Omega$, and $L = 5.0$ H.

Calculate each of the following *immediately* after the switch is closed:

a) the current flowing out of the battery

b) the current through R_1

c) the current through R_2

d) the potential difference across R_1

e) the potential difference across R_2

f) the potential difference across L

g) the rate of current change across R_1

•29.47 In the circuit in the figure, a battery supplies $V_{emf} = 18$ V and $R_1 = 6.0 \ \Omega$, $R_2 = 6.0 \ \Omega$, and $L = 5.0$ H. Calculate each of the following a *long time* after the switch is closed:

a) the current flowing out of the battery

b) the current through R_1

c) the current through R_2

d) the potential difference across R_1

e) the potential difference across R_2

f) the potential difference across L

g) the rate of current change across R_1

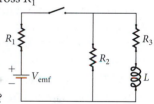

•29.48 A circuit contains a battery, three resistors, and an inductor, as shown in the figure. What will be the current through each resistor (a) immediately after the switch is closed? (b) a long time after the switch is closed? (c) Suppose the switch is reopened a long time after it has been closed. What is the current in each resistor? After a long time?

Section 29.9

29.49 Having just learned that there is energy associated with magnetic fields, an inventor sets out to tap the energy associated with the Earth's magnetic field. What volume of space near Earth's surface contains 1 J of energy, assuming the strength of the magnetic field to be $5.0 \cdot 10^{-5}$ T?

29.50 Consider a clinical MRI (magnetic resonance imaging) superconducting magnet has a diameter of 1 m, length of 1.5 m, and a uniform magnetic field of 3 T. Determine (a) the energy density of the magnetic field and (b) the total energy in the solenoid.

29.51 A *magnetar* (magnetic neutron star) has a magnetic field near its surface of magnitude $4.0 \cdot 10^{10}$ T.

a) Calculate the energy density of this magnetic field.

b) The Special Theory of Relativity associates energy with any mass m at rest according to $E_0 = mc^2$ (more on this in

Chapter 35). Find the rest mass density associated with the energy density of part (a).

•**29.52** An emf of 20 V is applied to a coil with an inductance of 40 mH and a resistance of 0.50 Ω.

a) Determine the energy stored in the magnetic field when the current reaches one fourth of its maximum value.

b) How long does it take for the current to reach this value?

•**29.53** A student wearing a 15-g gold band with radius 0.75 cm (and with a resistance of 61.9 $\mu\Omega$ and a specific heat capacity of $c = 129$ J/kg °C) on her finger moves her finger from a region having a magnetic field of 0.08 T, pointing along her finger, to a region with zero magnetic field in 40 ms. As a result of this action, thermal energy is added to the band due to the induced current, which raises the temperature of the band. Calculate the temperature rise in the band, assuming all the energy produced is used in raising the temperature.

•**29.54** A coil with N turns and area A, carrying a constant current, i, flips in an external magnetic field, $\vec{B}_{ext}$, so that its dipole moment switches from opposition to the field to alignment with the field. During this process, induction produces a potential difference that tends to reduce the current in the coil. Calculate the work done by the coil's power supply to maintain the constant current.

••**29.55** An electromagnetic wave propagating in vacuum has electric and magnetic fields given by $\vec{E}(\vec{x},t) = \vec{E}_0 \cos(\vec{k}\cdot\vec{x} - \omega t)$ and $\vec{B}(\vec{x},t) = \vec{B}_0 \cos(\vec{k}\cdot\vec{x} - \omega t)$, where $\vec{B}_0$ is given by $\vec{B}_0 = \vec{k} \times \vec{E}_0/\omega$ and the wave vector $\vec{k}$ is perpendicular to both $\vec{E}_0$ and $\vec{B}_0$. The magnitude of $\vec{k}$ and the angular frequency ω satisfy the dispersion relation, $\omega/|\vec{k}| = (\mu_0\epsilon_0)^{-1/2}$, where μ_0 and ϵ_0 are the permeability and permittivity of free space, respectively. Such a wave transports energy in both its electric and magnetic fields. Calculate the ratio of the energy densities of the magnetic and electric fields, u_B/u_E, in this wave. Simplify your final answer as much as possible.

Additional Problems

29.56 A wire of length $\ell = 10$ cm is moving with constant velocity in the xy-plane; the wire is parallel to the y-axis and moving along the x-axis. If a magnetic field of magnitude 1 T is pointing along the positive z-axis, what must the velocity of the wire be in order to induce a potential difference of 2 V across it?

29.57 The magnetic field inside the solenoid in the figure changes at the rate of 1.5 T/s. A conducting coil with 2000 turns surrounds the solenoid, as shown. The radius of the solenoid is 4 cm, and the radius of the coil is 7 cm. What is the potential difference induced in the coil?

29.58 An ideal battery (with no internal resistance) supplies V_{emf} and is connected to a superconducting (no resistance!) coil of inductance L at time $t = 0$. Find the current in the coil as a function of time, $i(t)$. Assume that all connections also have zero resistance.

29.59 A 100-turn solenoid of length 8 cm and radius 6 mm carries a current of 0.4 A from right to left. The current is then reversed so that it flows from left to right. By how much does the energy stored in the magnetic field inside the solenoid change?

29.60 What is the resistance in an RL circuit with $L = 36.94$ mH if the time taken to reach 75% of its maximum current value is 2.56 ms?

29.61 The electric field near the Earth's surface has a magnitude of 150. N/C, and the magnitude of the Earth's magnetic field near the surface is typically 50.0 μT. Calculate and compare the energy densities associated with these two fields. Assume that the electric and magnetic properties of air are essentially those of a vacuum.

29.62 A wedding ring (of diameter 2.0 cm) is tossed into the air and given a spin, resulting in an angular velocity of 17 rotations per second. The rotation axis is a diameter of the ring. Taking the magnitude of the Earth's magnetic field to be $4.0 \cdot 10^{-5}$ T, what is the maximum induced potential difference in the ring?

29.63 What is the inductance in a series RL circuit in which $R = 3.0$ kΩ if the current increases to one half of its final value in 20 μs?

29.64 A 100-V battery is connected in series with a 500-Ω resistor. According to Faraday's Law of Induction, current can never change instantaneously, so there is always some "stray" inductance. Suppose the stray inductance is 0.20 μH. How long will it take the current to build up to within 0.5% of its final value of 0.2 A after the resistor is connected to the battery?

29.65 A single loop of wire with an area of 5 m^2 is located in the plane of the page, as shown in the figure. A time-varying magnetic field in the region of the loop is directed into the page, and its magnitude is given by $B = 3$ T + (2 T/s)t. At $t = 2$ s, what is the induced potential difference in the loop and the direction of the induced current?

29.66 A 9-V battery is connected through a switch to two identical resistors and an ideal inductor, as shown in the figure. Each of the resistors has a resistance of 100 Ω, and the inductor has an inductance of 3 H. The switch is initially open.

a) Immediately after the switch is closed, what is the current in resistor R_1 and in resistor R_2?

b) At 50 ms after the switch is closed, what is the current in resistor R_1 and in resistor R_2?

c) At 500 ms after the switch is closed, what is the current in resistor R_1 and in resistor R_2?

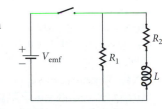

d) After a long time (> 10 s), the switch is opened again. Immediately after the switch is opened, what is the current in resistor R_1 and in resistor R_2?

e) At 50 ms after the switch is opened, what is the current in resistor R_1 and in resistor R_2?

f) At 500 ms after the switch is opened, what is the current in resistor R_1 and in resistor R_2?

•29.67 A long solenoid with length 3.0 m and $n = 290$ turns/m carries a current of 3.0 A. It stores 2.8 J of energy. What is the cross-sectional area of the solenoid?

•29.68 A rectangular conducting loop with dimensions a and b and resistance R, is placed in the xy-plane. A magnetic field of magnitude B passes through the loop. The magnetic field is in the positive z-direction and varies in time according to $B = B_0(1 + c_1 t^3)$, where c_1 is a constant with units of $1/s^3$. What is the direction of the current induced in the loop, and what is its value at $t = 1s$ (in terms of a, b, R, B_0, and c_1)?

•29.69 A circuit contains a 12-V battery, a switch, and a light bulb connected in series. When the light bulb has a current of 0.10 A flowing in it, it just starts to glow. This bulb draws 2 W when the switch has been closed for a long time. The switch is opened, and an inductor is added to the circuit, in series with the bulb. If the light bulb begins to glow 3.5 ms after the switch is closed again, what is the magnitude of the inductance? Ignore any time to heat the filament, and assume that you are able to observe a glow as soon as the current in the filament reaches the 0.10-A threshold.

•29.70 A circular loop of area A is placed perpendicular to a time-varying magnetic field of $B(t) = B_0 + at + bt^2$, where B_0, a, and b are constants.

a) What is the magnetic flux through the loop at $t = 0$?

b) Derive an equation for the induced potential difference in the loop as a function of time.

c) What is the magnitude and the direction of the induced current if the resistance of the loop is R?

•29.71 A conducting rod of length 50 cm slides over two parallel metal bars placed in a magnetic field with a magnitude of 1000 G, as shown in the figure. The ends of the rods are connected by two resistors, $R_1 = 100\ \Omega$ and $R_2 = 200\ \Omega$. The conducting rod moves with a constant speed of 8 m/s.

a) What are the currents flowing through the two resistors?

b) What power is delivered to the resistors?

c) What force is needed to keep the rod moving with constant velocity?

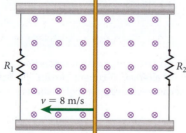

•29.72 A rectangular wire loop (dimensions of $h = 15.0$ cm and $w = 8.00$ cm) with resistance $R = 5.00\ \Omega$ is mounted on a door. The Earth's magnetic field, $B_E = 2.6 \cdot 10^{-5}$ T, is uniform and perpendicular to the surface of the closed door (the surface is in the xz-plane). At time $t = 0$, the door

is opened (right edge moves toward the y-axis) at a constant rate, with an opening angle of $\theta(t) = \omega t$, where $\omega = 3.5$ rad/s. Calculate the direction and the magnitude of the current induced in the loop, $i(t = 0.200$ s).

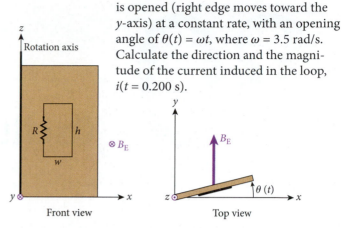

Front view   Top view

•29.73 A steel cylinder with radius 2.5 cm and length 10.0 cm rolls without slipping down a ramp that is inclined at 15° above the horizontal and has a length (along the ramp) of 3.0 m. What is the induced potential difference between the ends of the cylinder at the bottom of the ramp, if the surface of the ramp points along magnetic north?

•29.74 The figure shows a circuit in which a battery is connected to a resistor and an inductor in series.

a) What is the current in the circuit at any time t after the switch is closed?

b) Calculate the total energy provided by the battery from $t = 0$ to $t = L/R$.

c) Calculate the total energy dissipated in the resistor over the same time period.

d) Is energy conserved in this circuit?

•29.75 As shown in the figure, a rectangular (60 cm long by 15 cm wide) circuit loop with resistance 35 Ω is held parallel to the xy-plane with one half inside a uniform magnetic field. A magnetic field given by $\vec{B} = 2.0\hat{z}$ T is directed along the positive z-axis to the right of the dashed line; there is no external magnetic field to the left of the dashed line.

a) Calculate the magnitude of the force required to move the loop to the left at a constant speed of 10 cm/s while the right end of the loop is still in the magnetic field.

b) What power is expended to pull the loop out of the magnetic field at this speed?

c) What is the power dissipated by the resistor?

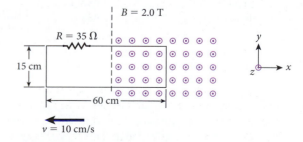

30

Electromagnetic Oscillations and Currents

(a)   (b)

FIGURE 30.1 (a) Electronic sound reproduction, as in a stereo boom box, relies on the properties of alternating-current circuits. (b) Better-quality sound is generated by having separate speakers for high-frequency and low-frequency sounds.

WHAT WE WILL LEARN

- Voltages and currents in single-loop circuits containing an inductor and a capacitor oscillate with a characteristic frequency.

- Voltages and currents in single-loop circuits containing a resistor, an inductor, and a capacitor also oscillate with a characteristic frequency, but these oscillations are damped over time.

- A single-loop circuit containing a time-varying source of emf and a resistor has currents and voltages that are in phase and time-varying.

- A single-loop circuit containing a time-varying source of emf and a capacitor has a time-varying current and voltage that are out of phase by $+\pi/2$ rad ($+90°$), with the current leading the voltage. The voltage and the current in such a circuit are related by the capacitive reactance.

- A single-loop circuit containing a time-varying source of emf and an inductor has a time-varying current and voltage that are out of phase by $-\pi/2$ rad ($-90°$), with the voltage leading the current. The voltage and the current in such a circuit are related by the inductive reactance.

- A single-loop circuit containing a time-varying source of emf and a resistor, a capacitor, and an inductor has a time-varying current and voltage. The phase difference between the current and voltage depends on the values of the resistance, the capacitance, the inductance, and on the frequency of the emf source.

- A single-loop circuit containing a time-varying source of emf and a resistor, a capacitor, and an inductor has a resonant frequency determined by the value of the inductance and the capacitance.

- The impedance of an alternating-current circuit is similar to the resistance of a direct-current circuit, but the impedance depends on the frequency of the time-varying source of emf.

- Transformers can raise (or lower) alternating voltages while lowering (or raising) alternating currents.

- Rectifiers convert alternating current to direct current.

Chapters 24 through 26 examined circuits that have a constant current or a current that increases to a constant current or decreases to a constant current. These currents did not reverse their direction of flow. This chapter introduces circuits containing a resistor, an inductor, and a capacitor. Such circuits exhibit sinusoidal oscillations in voltage and current. This chapter also covers circuits that contain a time-varying source of emf. Although these alternating-current (AC) circuits have the same circuit elements (resistors, capacitors, and inductors) as in the direct-current (DC) circuits we've considered in previous chapters, alternating-current circuits display phenomena not observed in direct-current circuits, such as resonance. Alternating currents play a big role in everyday life, for example in the operation of electronic devices such as the boom box (Figure 30.1a).

Speakers for sound reproduction (Figure 30.1b) generally have two parts: The small tweeter reproduces high-frequency sounds, and the larger woofer reproduces low-frequency sounds. But how does the electronic circuitry send one range of frequencies to one speaker and another range of frequencies to the other speaker? The answer is a filter, and we will find out how to construct one.

The DC circuits we have studied until now contain a source of emf that provides a steady potential difference to the circuit in one direction. Alternating emfs change direction in a sinusoidal pattern, usually 50 or 60 times per second, depending on location around the world. This condition results in physical behavior not possible in DC circuits and makes AC circuits the standard in most electronic devices.

30.1 LC Circuits

Previous chapters introduced three circuit elements: capacitors, resistors, and inductors. We have examined simple single-loop circuits containing resistors and capacitors (RC circuits) or resistors and inductors (RL circuits). Now we'll consider simple single-loop circuits containing inductors and capacitors: **LC circuits.** We'll see that LC circuits have currents and voltages that vary sinusoidally with time, rather than increasing or decreasing exponentially with time, as in RC and RL circuits. These variations of voltage and current in LC circuits are called **electromagnetic oscillations.**

To understand electromagnetic oscillations, consider a simple single-loop circuit consisting of an inductor and a capacitor (Figure 30.2). Recall that the energy stored in the electric field of a capacitor with capacitance C is given by (see Chapter 24)

$$U_E = \frac{1}{2}\frac{q^2}{C},$$

where q is the magnitude of the charge on the capacitor plates. The energy stored in the magnetic field of an inductor with inductance L is given by (see Chapter 29):

$$U_B = \frac{1}{2}Li^2,$$

where i is the current flowing through the inductor. Figure 30.2 shows how these energies vary with time in this LC circuit.

In Figure 30.2a, the capacitor is initially fully charged (with the positive charge on the bottom plate) and then connected to the circuit. At that time, the energy in the circuit is contained entirely in the electric field of the capacitor. The capacitor begins to discharge through the inductor in Figure 30.2b. At this point, current is flowing through the inductor, which generates a magnetic field. (A green arrow or label below each circuit diagram indicates the direction and the magnitude of the instantaneous current, i.) Now part of the energy of the circuit is stored in the electric field of the capacitor and part in the magnetic field of the inductor. The current begins to level off as the inductor's increasing magnetic

FIGURE 30.2 A single-loop circuit containing a capacitor and an inductor. (a) The capacitor is initially completely charged when it is connected to the circuit. (b)–(h) The current and the voltage in the circuit oscillate over time.

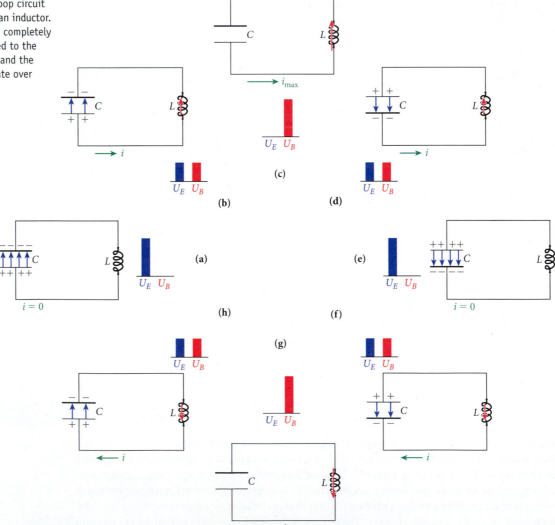

field induces an emf that opposes the current. In Figure 30.2c, the capacitor is completely discharged, and maximum current is flowing through the inductor. (When the magnitude of i has its maximum value, it is designated as i_{max} in the figure.) All the energy of the circuit is now stored in the magnetic field of the inductor. However, the current continues to flow, decreasing from its maximum value, which causes the magnetic field in the inductor to decrease. In Figure 30.2d, the capacitor begins to charge with the opposite polarity (positive charge on the top plate). Energy is again stored in the electric field of the capacitor, as well as in the magnetic field of the inductor.

In Figure 30.2e, the energy in the circuit is again entirely contained in the electric field of the capacitor. Note that the electric field now points in the opposite direction from the original field in Figure 30.2a. The current is zero, as is the magnetic field in the inductor. In Figure 30.2f, the capacitor begins to discharge again, producing a current flowing in the direction opposite to that in parts (b) through (d) of the figure; this current in turn creates a magnetic field in the opposite direction in the inductor. Again, part of the energy is stored in the electric field and part in the magnetic field. In Figure 30.2g, the energy is all stored in the magnetic field of the inductor, but with the magnetic field in the opposite direction from that in Figure 30.2c and with the maximum current in the opposite direction from that in Figure 30.2c. In Figure 30.2h, the capacitor begins to charge again, meaning there is energy in both the electric and magnetic fields. The state of the circuit then returns to that shown in Figure 30.2a. The circuit continues to oscillate indefinitely because there is no resistor in it, and the electric and magnetic fields together conserve energy. A real circuit with a capacitor and an inductor does not oscillate indefinitely; instead, the oscillations die away with time because of small resistances in the circuit (covered in Section 30.3) or electromagnetic radiation (covered in Chapter 31).

The charge on either capacitor plate and the current in the LC circuit vary sinusoidally, as shown in Figure 30.3. q_{max} refers to the maximum charge on the capacitor plate that is initially positively charged (the lower plate in Figure 30.2a). The energy in the electric field depends on the square of the charge on the capacitor, and the energy in the magnetic field depends on the square of the current in the inductor. Thus, the electric energy, U_E, and the magnetic energy, U_B, vary between zero and their respective maximum values as a function of time.

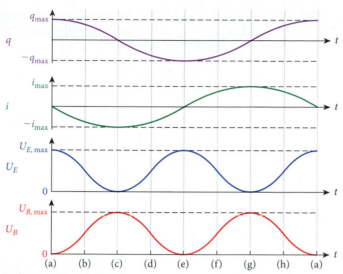

FIGURE 30.3 Variation of the charge, current, electric energy, and magnetic energy as a function of time for a simple, single-loop LC circuit. The letters along the bottom refer to the parts of Figure 30.2.

30.1 In-Class Exercise

Figure 30.2a shows that the charge on the capacitor in an LC circuit is largest when the current is zero. What about the potential difference across the capacitor?

a) The potential difference across the capacitor is largest when the current is the largest.

b) The potential difference across the capacitor is largest when the charge is the largest.

c) The potential difference across the capacitor does not change.

30.2 Analysis of LC Oscillations

This section presents a quantitative description of the phenomena described in the preceding section. Consider a single-loop circuit containing a capacitor of capacitance C and an inductor of inductance L, but no resistor and no resistive losses in the circuit wire, as illustrated in Figure 30.4. We can write the total energy in the circuit, U, as the sum of the electric energy in the capacitor and the magnetic energy in the inductor:

$$U = U_E + U_B.$$

Using the expressions for the electric energy and the magnetic energy in terms of the charge and the current, $U_E = \frac{1}{2}(q^2/C)$ and $U_B = \frac{1}{2}Li^2$, we obtain

$$U = U_E + U_B = \frac{1}{2}\frac{q^2}{C} + \frac{1}{2}Li^2.$$

FIGURE 30.4 A single-loop LC circuit containing an inductor and a capacitor.

Because we have assumed zero resistance, no energy can be lost to heat, and the energy in the circuit will remain constant, because the electric field and the magnetic field together conserve energy. Thus, the derivative with respect to time of the energy in the circuit is zero:

$$\frac{dU}{dt} = \frac{d}{dt}\left(\frac{1}{2}\frac{q^2}{C} + \frac{1}{2}Li^2\right) = \frac{q}{C}\frac{dq}{dt} + Li\frac{di}{dt} = 0.$$

By definition, the current is the time derivative of the charge, $i = dq/dt$, and therefore, the time derivative of the current is the second derivative of the charge:

$$\frac{di}{dt} = \frac{d}{dt}\left(\frac{dq}{dt}\right) = \frac{d^2q}{dt^2}.$$

With this expression for di/dt, the preceding equation for the time derivative of the total energy, dU/dt, becomes

$$\frac{q}{C}\frac{dq}{dt} + L\frac{dq}{dt}\frac{d^2q}{dt^2} = \frac{dq}{dt}\left(\frac{q}{C} + L\frac{d^2q}{dt^2}\right) = 0.$$

We can rewrite this equation as

$$\frac{d^2q}{dt^2} + \frac{q}{LC} = 0. \tag{30.1}$$

(We discard the solution $dq/dt = 0$ because it corresponds to the situations where there is initially no charge on the capacitor.) This differential equation has the same form as that for simple harmonic motion, which describes the position, x, of an object with mass m connected to a spring with spring constant k:

$$\frac{d^2x}{dt^2} + \frac{k}{m}x = 0.$$

We saw in Chapter 14 that the solution of this differential equation for the position as a function of time was a sinusoidal function: $x = x_{max} \cos(\omega_0 t + \phi)$, where ϕ is a phase constant and the angular frequency, ω_0, is given by $\omega_0 = \sqrt{k/m}$.

Simply substituting q for x and $1/LC$ for k/m in the differential equation for simple harmonic motion leads to the analogous solution for equation 30.1. Thus, the charge as a function of time in an LC circuit is given by

$$q = q_{max} \cos(\omega_0 t - \phi), \tag{30.2}$$

where q_{max} is the magnitude of the maximum charge in the circuit and ϕ is the phase constant, which is determined by the initial conditions for a given situation. (Note that the convention for electromagnetic oscillations is to use a negative sign in front of ϕ.) The angular frequency is given by

$$\omega_0 = \sqrt{\frac{1}{LC}} = \frac{1}{\sqrt{LC}}. \tag{30.3}$$

The current is given by the time derivative of equation 30.2:

$$i = \frac{dq}{dt} = \frac{d}{dt}\left(q_{max}\cos(\omega_0 t - \phi)\right) = -\omega_0 q_{max} \sin(\omega_0 t - \phi).$$

Since the maximum current in the circuit is $i_{max} = \omega_0 q_{max}$, we get

$$i = -i_{max} \sin(\omega_0 t - \phi). \tag{30.4}$$

Equations 30.2 and 30.4 correspond to the top two curves in Figure 30.3 with $\phi = 0$. We can write expressions for the electric energy and the magnetic energy as functions of time:

$$U_E = \frac{1}{2}\frac{q^2}{C} = \frac{1}{2}\frac{\left[q_{max}\cos(\omega_0 t - \phi)\right]^2}{C} = \frac{q_{max}^2}{2C}\cos^2(\omega_0 t - \phi),$$

and

$$U_B = \frac{1}{2}Li^2 = \frac{L}{2}\left[-i_{max}\sin(\omega_0 t - \phi)\right]^2 = \frac{L}{2}i_{max}^2\sin^2(\omega_0 t - \phi).$$

Since $i_{max} = \omega_0 q_{max}$ and $\omega_0 = 1/\sqrt{LC}$, we can write

$$\frac{L}{2}i_{max}^2 = \frac{L}{2}\omega_0^2 q_{max}^2 = \frac{q_{max}^2}{2C}.$$

Thus, we can express the magnetic energy as a function of time as follows:

$$U_B = \frac{q_{max}^2}{2C} \sin^2(\omega_0 t - \phi).$$

Note that both the electric energy and the magnetic energy have a maximum value equal to $q_{max}/2C$ and a minimum of zero.

We can obtain an expression for the total energy in the circuit, U, by summing the electric and magnetic energies and then using the trigonometric identity $\sin^2\theta + \cos^2\theta = 1$:

$$U = U_E + U_B = \frac{q_{max}^2}{2C} \cos^2(\omega_0 t - \phi) + \frac{q_{max}^2}{2C} \sin^2(\omega_0 t - \phi)$$

$$= \frac{q_{max}^2}{2C} \left[\sin^2(\omega_0 t - \phi) + \cos^2(\omega_0 t - \phi) \right]$$

$$= \frac{q_{max}^2}{2C} = \frac{L}{2} i_{max}^2.$$

Thus, the total energy in the circuit remains constant with time and is proportional to the square of the original charge put on the capacitor.

30.2 In-Class Exercise

In Figure 30.3, suppose $t = 0$ at point (c). What is the phase constant in this case? (Define a clockwise current as positive.)

a) 0

b) $\pi/2$

c) π

d) $3\pi/2$

e) none of the above

EXAMPLE 30.1 Characteristics of an LC Circuit

A circuit contains a capacitor, with $C = 1.50~\mu F$, and an inductor, with $L = 3.50$ mH (Figure 30.4). The capacitor is fully charged using a 12.0-V battery and is then connected to the circuit.

PROBLEMS

What is the angular frequency of the circuit? What is the total energy in the circuit? What is the charge on the capacitor after $t = 2.50$ s?

SOLUTIONS

The angular frequency of the circuit is given by

$$\omega_0 = \frac{1}{\sqrt{LC}} = \frac{1}{\sqrt{(3.50 \cdot 10^{-3}~\text{H})(1.50 \cdot 10^{-6}~\text{F})}} = 1.38 \cdot 10^4~\text{rad/s}.$$

The total energy in the circuit is

$$U = \frac{q_{max}^2}{2C}.$$

The maximum charge on the capacitor is

$$q_{max} = CV_{emf} = (1.50 \cdot 10^{-6}~\text{F})(12.0~\text{V})$$

$$= 1.80 \cdot 10^{-5}~\text{C}.$$

Thus, we can calculate the initial energy stored in the electric field of the capacitor, which is the same as the total energy in the circuit:

$$U = \frac{q_{max}^2}{2C} = \frac{(1.80 \cdot 10^{-5}~\text{C})^2}{2(1.50 \cdot 10^{-6}~\text{F})} = 1.08 \cdot 10^{-4}~\text{J}.$$

The charge on the capacitor as a function of time is given by

$$q = q_{max} \cos(\omega_0 t - \phi).$$

To determine the constant ϕ, we remember that $q = q_{max}$ at $t = 0$, so

$$q(0) = q_{max} = q_{max} \cos[(\omega_0)(0) - \phi] = q_{max} \cos(-\phi) = q_{max} \cos\phi.$$

Continued—

30.1 Self-Test Opportunity

The frequency of oscillation of an LC circuit is 200.0 kHz. At $t = 0$, the charge on the capacitor has its maximum positive charge on the lower plate. Decide whether each of the following statements is true or false.

a) At $t = 2.50~\mu s$, the charge on the lower plate has its maximum negative value.

b) At $t = 5.00~\mu s$, the current in the circuit is at its maximum value.

c) At $t = 2.50~\mu s$, the energy in the circuit is stored completely in the inductor.

d) At $t = 1.25~\mu s$, half the energy in the circuit is stored in the capacitor and half the energy is stored in the inductor.

Thus, we see that $\phi = 0$, and we can write the charge as a function of time as follows:

$$q = q_{max} \cos \omega_0 t.$$

Putting in the values $q_{max} = 1.80 \cdot 10^{-5}$ C, $\omega_0 = 1.38 \cdot 10^4$ rad/s, and $t = 2.50$ s, we get

$$q = \left(1.80 \cdot 10^{-5} \text{ C}\right) \cos\left[\left(1.38 \cdot 10^4 \text{ rad/s}\right)\left(2.50 \text{ s}\right)\right] = 1.02 \cdot 10^{-5} \text{ C}.$$

30.3   Damped Oscillations in an RLC Circuit

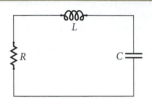

FIGURE 30.5 A single-loop RLC circuit containing a resistor, an inductor, and a capacitor.

Now let's consider a single-loop circuit that has a capacitor and an inductor but also a resistor—an **RLC circuit,** as shown in Figure 30.5. We saw in the preceding section that the energy of a circuit with a capacitor and an inductor remains constant and that the energy is transformed from electric to magnetic and back again with no losses. However, if a resistor is present in the circuit, the current flow produces ohmic losses, which show up as thermal energy. Thus, the energy of the circuit decreases because of these losses. The rate of energy loss is given by

$$\frac{dU}{dt} = -i^2 R.$$

We can rewrite the change in the energy in the circuit as a function of time:

$$\frac{dU}{dt} = \frac{q}{C}\frac{dq}{dt} + Li\frac{di}{dt} = -i^2 R.$$

Again, since $i = dq/dt$ and $di/dt = d^2q/dt^2$, we can write

$$\frac{q}{C}\frac{dq}{dt} + Li\frac{di}{dt} + i^2 R = \frac{q}{C}\frac{dq}{dt} + L\frac{dq}{dt}\frac{d^2q}{dt^2} + \left(\frac{dq}{dt}\right)^2 R = 0$$

or

$$\frac{d^2q}{dt^2} + \frac{R}{L}\frac{dq}{dt} + \frac{1}{LC}q = 0. \tag{30.5}$$

The solution of this differential equation (for small damping, meaning sufficiently small values of the resistance) is

$$q = q_{max}e^{-Rt/2L}\cos \omega t, \tag{30.6}$$

where

$$\omega = \sqrt{\omega_0^2 - \left(\frac{R}{2L}\right)^2} \tag{30.7}$$

and $\omega_0 = 1/\sqrt{LC}$.

The calculus used in arriving at the solution in equation 30.6 is not shown. You can verify that the solution satisfies equation 30.5 by straightforward substitution from equations 30.6 and 30.7 into 30.5. You can also refer back to Chapter 14, where it was shown that the equation of motion for a weakly damped (or underdamped) mechanical oscillator has a similar solution. Chapter 14 also covered overdamped and critically damped oscillations.

If the capacitor in the single-loop circuit RLC circuit of Figure 30.5 is charged and then connected in the circuit, the charge on the capacitor will vary sinusoidally with time while decreasing in amplitude (Figure 30.6). Taking the derivative of equation 30.6 shows that the current, $i = dq/dt$, has an amplitude that is damped at the same rate that the charge is damped and that this amplitude also varies sinusoidally with time. After some time, no current remains in the circuit.

We can analyze the energy in the circuit as a function of time by calculating the energy stored in the electric field of the capacitor:

$$U_E = \frac{1}{2}\frac{q^2}{C} = \frac{q_{max}^2}{2C}e^{-Rt/L}\cos^2 \omega t.$$

30.2 Self-Test Opportunity

Compare equation 30.5 for the charge on the capacitor as a function of time to the differential equation for the position of a mass on a spring presented in Chapter 14: $\frac{d^2x}{dt^2} + \frac{b}{m}\frac{dx}{dt} + \frac{k}{m}x = 0$. Which quantity in the RLC electric circuit plays the role of the mass m, which that of the spring constant k, and which that of the damping constant b?

30.3 In-Class Exercise

What is the condition for small damping that needs to be fulfilled for equation 30.6 to be a solution for equation 30.5? (*Hint:* You can find this by analogy with the damped oscillation of a mass on a spring, for which the differential equation is $\frac{d^2x}{dt^2} + \frac{b}{m}\frac{dx}{dt} + \frac{k}{m}x = 0$ and the condition for small damping is $b < 2\sqrt{mk}$. Alternatively, you can use dimensional analysis.)

a) $R < 2\sqrt{L/C}$

b) $R < \sqrt{2C/L}$

c) $R < \sqrt{2LC}$

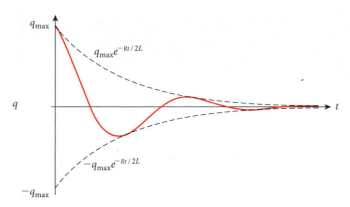

FIGURE 30.6 Graph of the charge on the capacitor as a function of time in a circuit containing a capacitor, an inductor, and a resistor.

Thus, U_E and U_B both decrease exponentially in time, and therefore so does the total energy in the circuit, $U_E + U_B$.

30.4 Driven AC Circuits

So far, we have been studying circuits that contain a source of constant emf or that start with a constant charge and contain energy that then oscillates between electric and magnetic fields. However, many interesting effects occur in a circuit in which the current oscillates continuously. This section investigates some of these effects, starting with a time-varying source of emf and then considering in turn a resistor, a capacitor, and an inductor connected to this source.

Alternating Driving emf

A source of emf can be capable of producing a time-varying voltage, as opposed to the sources of constant emf considered in previous chapters. We'll assume that the source of time-varying emf provides a sinusoidal voltage as a function of time, the *driving emf*, given by

$$V_{emf} = V_{max} \sin \omega t, \tag{30.8}$$

where ω is the angular frequency of the emf and V_{max} is the maximum amplitude or value of the emf.

The current induced in a circuit containing a source of time-varying emf will also vary sinusoidally with time. This time-varying current is called **alternating current (AC)**. However, the alternating current may not always remain in phase with the time-varying emf. The current, i, as a function of time is given by

$$i = I \sin(\omega t - \phi), \tag{30.9}$$

where I is the amplitude of the current and the angular frequency of the time-varying current is the same as that of the driving emf, but the phase constant ϕ is not zero. Note that, as is the convention, the phase constant is preceded by a negative sign.

Circuit with a Resistor

Let's begin our analysis of RLC circuits with alternating current by considering a circuit containing only a resistor and a source of time-varying emf (Figure 30.7). Applying Kirchhoff's Loop Rule to this circuit, we obtain

$$V_{emf} - v_R = 0,$$

where v_R is the voltage drop across the resistor. Substituting in v_R for V_{emf} in equation 30.8, we get

$$v_R = V_{max} \sin \omega t = V_R \sin \omega t,$$

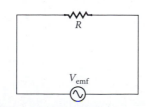

FIGURE 30.7 Single-loop circuit with a resistor and a source of time-varying emf.

FIGURE 30.8 Alternating voltage and current for a single-loop circuit containing a source of time-varying emf and a resistor: (a) voltage and current as functions of time; (b) phasors representing voltage and current, showing that they are in phase.

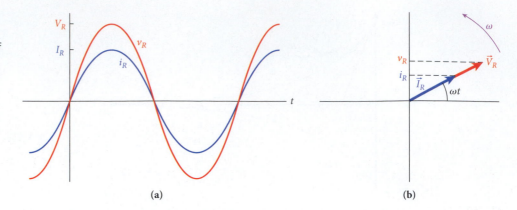

(a)   (b)

where V_R is the maximum voltage drop across the resistor. Note that the voltage as a function of time is represented by a lowercase v and the amplitude of the voltage with uppercase V. According to Ohm's Law, $V = iR$, so we can write

$$i_R = \frac{v_R}{R} = \frac{V_R}{R}\sin\omega t = I_R \sin\omega t. \tag{30.10}$$

Thus, the current amplitude and the voltage amplitude are related as follows:

$$V_R = I_R R. \tag{30.11}$$

Figure 30.8a shows the voltage across the resistor and the current through it as functions of time. The time-varying current can be represented by a phasor, $\vec{I}_R$, and the time-varying voltage by a phasor, $\vec{V}_R$ (Figure 30.8b). A **phasor** is a counterclockwise-rotating vector (with its tail fixed at the origin) whose projection on the vertical axis represents the sinusoidal variation of the particular quantity in time. The angular velocity of the phasors in Figure 30.8b is ω of equation 30.10. The current flowing through the resistor and the voltage across the resistor are in phase, which means that the phase difference between the current and the voltage is zero.

Circuit with a Capacitor

Now let's examine a circuit that contains a capacitor and a time-varying emf (Figure 30.9). The voltage across the capacitor is given by Kirchhoff's Loop Rule,

$$V_{\text{emf}} - v_C = 0,$$

where v_C is the voltage drop across the capacitor. Thus, we have

$$v_C = V_{\text{max}} \sin\omega t = V_C \sin\omega t,$$

where V_C is the maximum voltage across the capacitor. Since $q = CV$ for a capacitor, we can write

$$q = Cv_C = CV_C \sin\omega t.$$

However, we want an expression for the current (rather than the charge) as a function of time. Therefore, we take the derivative with respect to time of the preceding equation:

$$i_C = \frac{dq}{dt} = \frac{d\left(CV_C \sin\omega t\right)}{dt} = \omega CV_C \cos\omega t.$$

This equation can be written in a form comparable to that of equation 30.10 by defining a quantity that is similar to resistance, called the **capacitive reactance**, X_C:

$$X_C = \frac{1}{\omega C}. \tag{30.12}$$

This definition allows us to express the current, i_C, as

$$i_C = \frac{V_C}{X_C}\cos\omega t,$$

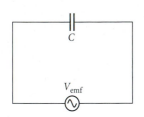

FIGURE 30.9 Single-loop circuit with a capacitor and a source of time-varying emf.

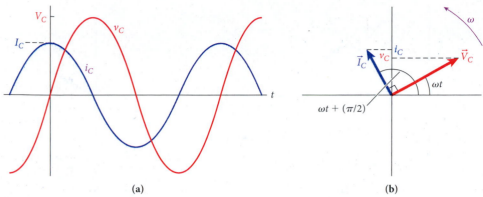

FIGURE 30.10 Alternating voltage and current for a single-loop circuit containing a source of emf and a capacitor: (a) voltage and current as functions of time; (b) phasors representing voltage and current, showing that they are out of phase by $\pi/2$ rad (90°).

or, with $I_C = V_C/X_C$, as

$$i_C = I_C \cos\omega t.$$

We can use $\cos\theta = \sin(\theta + \pi/2)$ to express this result in a form analogous to that of equation 30.10:

$$i_C = I_C \sin\left(\omega t + \pi/2\right). \tag{30.13}$$

This expression for the current flowing in a circuit with only a capacitor is similar to the expression for the current flowing in a circuit with only a resistor, except that current and voltage are out of phase by $\pi/2$ rad (90°). Figure 30.10a shows the voltage and current as functions of time.

The corresponding phasors $\vec{I}_C$ and $\vec{V}_C$, shown in Figure 30.10b, indicate that *for a circuit with only a capacitor, the current leads the voltage.* The amplitude of the voltage across the capacitor and the amplitude of the current through the capacitor are related by

$$V_C = I_C X_C. \tag{30.14}$$

This equation resembles Ohm's Law with the capacitive reactance replacing the resistance. One major difference between the capacitive reactance and the resistance is that the capacitive reactance depends on the angular frequency of the time-varying emf.

Circuit with an Inductor

Now let's consider a circuit with a source of time-varying emf and an inductor (Figure 30.11). We again apply Kirchhoff's Loop Rule to this circuit to obtain the voltage across the inductor:

$$v_L = V_{max} \sin\omega t = V_L \sin\omega t,$$

where V_L is the maximum voltage across the inductor. A changing current in an inductor induces an emf given by

$$v_L = L\frac{di_L}{dt}.$$

Note that for positive di/dt the voltage drop across the inductor is positive because the direction of current is the direction of decreasing potential. Thus, we can write

$$L\frac{di_L}{dt} = V_L \sin\omega t,$$

or

$$\frac{di_L}{dt} = \frac{V_L}{L}\sin\omega t.$$

We are interested in the current rather than its time derivative, so we integrate the preceding equation:

$$i_L = \int \frac{di_L}{dt}dt = \int \frac{V_L}{L}\sin\omega t\,dt = -\frac{V_L}{\omega L}\cos\omega t.$$

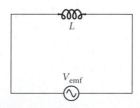

FIGURE 30.11 Single-loop circuit with an inductor and a source of time-varying emf.

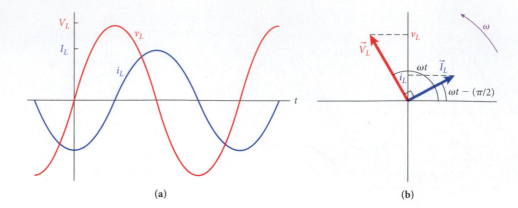

(a)   (b)

Here we set the constant of integration to zero because we are not interested in solutions that contain both an oscillating and a constant current. The **inductive reactance,** which, like the capacitive reactance, is similar to resistance, is defined as

$$X_L = \omega L. \tag{30.15}$$

Using the inductive reactance, we can express i_L as

$$i_L = -\frac{V_L}{X_L}\cos\omega t = -I_L\cos\omega t,$$

where I_L is the maximum current. Thus,

$$V_L = I_L X_L,$$

which again resembles Ohm's Law except that the inductive reactance depends on the angular frequency of the time-varying emf.

Because $-\cos\theta = \sin(\theta - \pi/2)$, we can rewrite $i_L = -I_L\cos\omega t$ as follows:

$$i_L = I_L\sin\left(\omega t - \pi/2\right). \tag{30.16}$$

Thus, the current flowing in a circuit with an inductor and a source of time-varying emf is out of phase with the voltage by $-\pi/2$ rad. Figure 30.12a shows voltage and current as functions of time. The corresponding phasors, $\vec{I}_L$ and $\vec{V}_L$, are shown in Figure 30.12b, which shows that *for a circuit with an inductor, the voltage leads the current.*

30.5 Series RLC Circuit

Now we're ready to consider a single-loop circuit that has all three circuit elements, along with a source of time-varying emf (Figure 30.13). This section will not present a full mathematical analysis of this RLC circuit but will use phasors to analyze the important aspects.

The time-varying current in the simple RLC circuit can be described by a phasor, $\vec{I}_m$ (Figure 30.14). The projection of $\vec{I}_m$ on the vertical axis represents the current i flowing in the circuit as a function of time, t, where the angle of the phasor is given by $\omega t - \phi$ such that

$$i = I_m\sin\left(\omega t - \phi\right).$$

The current i and the voltages across the circuit components have different phases with respect to the time-varying emf, as we have seen in the previous section:

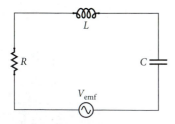

- *For the resistor,* the voltage v_R and the current i are in phase with each other, and the voltage phasor, $\vec{V}_R$, is in phase with $\vec{I}_m$.

- *For the capacitor,* the current i leads the voltage v_C by $\pi/2$ rad ($90°$), so the voltage phasor, $\vec{V}_C$, has an angle that is $\pi/2$ rad ($90°$) less than the angles of $\vec{I}_m$ and $\vec{V}_R$.

- *For the inductor,* the current i lags behind the voltage v_L by $\pi/2$ rad ($90°$), so the voltage phasor, $\vec{V}_L$, has an angle that is $\pi/2$ rad ($90°$) greater than the angles of $\vec{I}_m$ and $\vec{V}_R$.

The voltage phasors for the RLC circuit are shown in Figure 30.15. The instantaneous voltage across each component is represented by the projection of the respective phasor on the vertical axis.

The total voltage drop across all components, V, is given by

$$V = v_R + v_C + v_L. \tag{30.17}$$

The total voltage, V, can be thought of as the projection on the vertical axis of the phasor $\vec{V}_m$, representing the time-varying emf in the circuit (Figure 30.16). The phasors in Figure 30.15 rotate together, so equation 30.17 holds at any time. The voltage phasors must sum as vectors to match $\vec{V}_m$ in order to satisfy equation 30.17 at all times. This vector sum is shown in Figure 30.16. In this figure, the sum of the two phasors $\vec{V}_L$ and $\vec{V}_C$ has been replaced with $\vec{V}_L + \vec{V}_C$. The vector sum of $\vec{V}_L + \vec{V}_C$ and $\vec{V}_R$ must equal $\vec{V}_m$. Thus, we can write

$$V_m^2 = V_R^2 + \left(V_L - V_C\right)^2, \tag{30.18}$$

because the vectors $\vec{V}_L$ and $\vec{V}_C$ always point in opposite directions, and $\vec{V}_m$ is perpendicular to both. Now we can substitute our previously derived expressions for V_R, V_L, and V_C into equation 30.18, taking the amplitude of the current in all three components to be I_m because they are in series:

$$V_m^2 = \left(I_m R\right)^2 + \left(I_m X_L - I_m X_C\right)^2.$$

We can then solve for the amplitude of the current in the circuit:

$$I_m = \frac{V_m}{\sqrt{R^2 + \left(X_L - X_C\right)^2}}.$$

The denominator of the term on the right-hand side is called the **impedance,** Z:

$$Z = \sqrt{R^2 + \left(X_L - X_C\right)^2}. \tag{30.19}$$

The impedance of a circuit depends on the frequency of the time-varying emf. This time dependence is expressed explicitly when substitutions are made for the capacitive reactance, X_C, and the inductive reactance, X_L:

$$Z = \sqrt{R^2 + \left(\omega L - \frac{1}{\omega C}\right)^2}. \tag{30.20}$$

The impedance of an AC circuit has the unit ohm (Ω), just like the resistance in a DC circuit. We can then write

$$I_m = \frac{V_m}{\sqrt{R^2 + \left(\omega L - \frac{1}{\omega C}\right)^2}} = \frac{V_m}{Z}. \tag{30.21}$$

The current flowing in an AC circuit depends on the difference between the inductive reactance and the capacitive reactance and is called the total reactance. The phase constant, ϕ, can be expressed in terms of this difference. The phase constant is defined as the phase difference between the voltage phasors $\vec{V}_R$ and $\vec{V}_m$ depicted in Figure 30.16. Thus, we can express the phase constant as

$$\phi = \tan^{-1}\left(\frac{V_L - V_C}{V_R}\right).$$

Since $V_L = X_L I_m$, $V_C = X_C I_m$, and $V_R = R I_m$, this can be rewritten as follows:

$$\phi = \tan^{-1}\left(\frac{X_L - X_C}{R}\right).$$

Using $X_C = 1/\omega C$ and $X_L = \omega L$, we can then obtain the frequency dependence of the phase constant:

$$\phi = \tan^{-1}\left(\frac{\omega L - (\omega C)^{-1}}{R}\right). \tag{30.22}$$

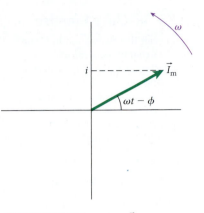

FIGURE 30.14 Phasor $\vec{I}_m$ representing the current i flowing in an RLC circuit.

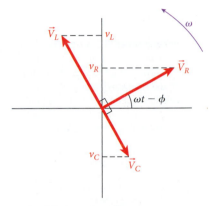

FIGURE 30.15 Voltage phasors for an RLC series circuit. The phasor $\vec{V}_R$ is in phase with the phasor $\vec{I}_m$ representing the current in the circuit.

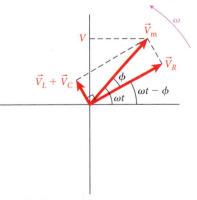

FIGURE 30.16 Sum of the voltage phasors in an RLC series circuit.

The current in the RLC circuit can now be written as

$$i = I_m \sin(\omega t - \phi), \tag{30.23}$$

where I_m is the magnitude of the phasor $\vec{I}_m$. The voltage across all the components in the circuit is given by the time-varying source of emf:

$$V = V_{emf}(t) = V_m \sin \omega t \tag{30.24}$$

where V_m is the magnitude of the phasor $\vec{V}_m$.

Thus, three conditions are possible for an AC series circuit containing a resistor, a capacitor, and an inductor:

- For $X_L > X_C$, ϕ is positive, and the current in the circuit lags behind the voltage in the circuit. This circuit is similar to a circuit with only an inductor, except that the phase constant is not necessarily $\pi/2$ rad (90°), as illustrated in Figure 30.17a.

- For $X_L < X_C$, ϕ is negative, and the current in the circuit leads the voltage in the circuit. This circuit is similar to a circuit with only a capacitor, except that the phase constant is not necessarily $-\pi/2$ rad (–90°), as illustrated in Figure 30.17b.

- For $X_L = X_C$, ϕ is zero, and the current in the circuit is in phase with the voltage in the circuit. This circuit is similar to a circuit with only a resistance, as illustrated in Figure 30.17c. When $\phi = 0$, the circuit is said to be in **resonance.**

The current amplitude, I_m, in the series RLC circuit depends on the frequency of the time-varying emf, as well as on L and C. Inspection of equation 30.21 shows that the maximum current occurs when

$$\omega L - \frac{1}{\omega C} = 0,$$

which corresponds to $\phi = 0$ and $X_L = X_C$. The angular frequency, ω_0, at which the maximum current occurs, called the **resonant angular frequency,** is

$$\omega_0 = \frac{1}{\sqrt{LC}}.$$

A Practical Example

Now let's look at a real circuit (Figure 30.18). The diagram for this circuit is shown in Figure 30.13. The circuit has a source of time-varying emf with $V_m = 7.5$ V and has $L = 8.2$ mH, $C = 100$ μF, and $R = 10$ Ω. The maximum current, I_m, was measured as a function of the ratio of the angular frequency of the time-varying emf to the resonant angular frequency, ω/ω_0 (Figure 30.19). Red circles indicate the results of the measurements. The maximum value

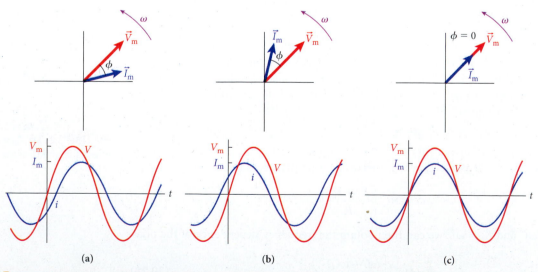

(a)   (b)   (c)

FIGURE 30.17 Current and voltage as functions of time for an RLC circuit with: (a) $X_L > X_C$; (b) $X_L < X_C$; (c) $X_L = X_C$.

of the current occurs, as expected, at the resonant angular frequency. However, with $R = 10\ \Omega$, the relationship between I_m and ω/ω_0, given by equation 30.21, results in the green curve, which does not reproduce the measured results. To better describe the current, we must remember that in a real circuit, the inductor has a resistance, even at the resonant frequency. The black curve in Figure 30.19 corresponds to equation 30.21 with $R = 15.4\ \Omega$.

The resonant behavior of an RLC circuit resembles the response of a damped mechanical oscillator (Chapter 14). Figure 30.20 shows the calculated maximum current, I_m, as a function of the ratio of the angular frequency of the time-varying emf to the resonant angular frequency, ω/ω_0, for a series RLC circuit with $V_m = 7.5$ V, $L = 8.2$ mH, $C = 100\ \mu$F, and three different resistances. You can see that as the resistance is lowered, the maximum current at the resonant angular frequency increases, producing a more pronounced peak.

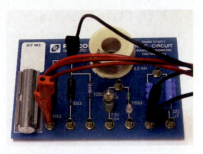

FIGURE 30.18 Real series circuit containing an 8.2-mH inductor, a 10-Ω resistor, and a 100-μF capacitor.

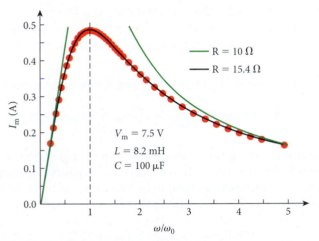

FIGURE 30.19 Graph of the maximum current, I_m, versus the ratio of the angular frequency, ω, of the time-varying emf to the resonance frequency, ω_0, for an RLC circuit. Red dots represent measurements. The text explains the green and black curves.

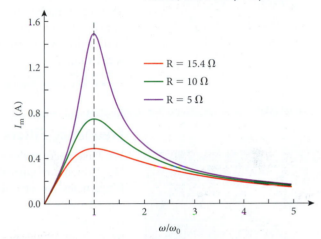

FIGURE 30.20 Graph of the maximum current, I_m, versus the ratio of the angular frequency, ω, of the time-varying emf to the resonance frequency, ω_0, for three series RLC circuits, with $L = 8.2$ mH, $C = 100\ \mu$F, and three different resistances.

EXAMPLE 30.2 | Characterizing an RLC Circuit

Suppose an RLC series circuit like the one shown in Figure 30.13 has $R = 91.0\ \Omega$, $C = 6.00\ \mu$F, and $L = 60.0$ mH. The source of time-varying emf has an angular frequency of $\omega = 64.0$ rad/s.

PROBLEM
What is the impedance of this circuit?

SOLUTION
Normally, we would solve this problem by obtaining an expression for the impedance in terms of the quantities provided. However, instead we'll calculate several intermediate numerical answers to gain insight into the characteristics of this circuit.

The impedance is given by $Z = \sqrt{R^2 + (X_L - X_C)^2}$. To see which of the quantities on the right-hand side has the greatest impact on the impedance, we calculate the quantities individually. The inductive reactance is given by

$$X_L = \omega L = (64.0\ \text{rad/s})(60.0 \cdot 10^{-3}\ \text{H}) = 3.84\ \Omega.$$

The capacitive reactance is

$$X_C = \frac{1}{\omega C} = \frac{1}{(64.0\ \text{rad/s})(6.00 \cdot 10^{-6}\ \text{F})} = 2600\ \Omega.$$

Continued—

30.4 Self-Test Opportunity

Consider a series RLC circuit like the one shown in Figure 30.13. Decide whether each of the following statements is true or false.

a) The current through the resistor is the same as the current through the inductor at all times.

b) In an ideal scenario energy is dissipated in the resistor but not in the capacitor or in the inductor.

c) The voltage drop across the resistor is the same as the voltage drop across the inductor at all times.

We see that the impedance of this circuit is dominated by the capacitive reactance at the given value of the angular frequency. This type of circuit is called *capacitive*.

Putting in our results for the capacitive and inductive reactances, we calculate the impedance:

$$Z = \sqrt{R^2 + \left(X_L - X_C\right)^2} = \sqrt{\left(91.0\ \Omega\right)^2 + \left(3.84\ \Omega - 2600\ \Omega\right)^2} = 2600\ \Omega.$$

That is, the inductive reactance and the resistance are completely negligible within rounding error. For comparison, the impedance of this circuit when it is in resonance is

$$Z = \sqrt{R^2 + \left(X_L - X_C\right)^2} = \sqrt{\left(91.0\ \Omega\right)^2 + 0} = 91.0\ \Omega.$$

This result implies that the circuit as described in the problem statement is far from resonance, which is consistent with the very different values we obtained for the capacitive and inductive reactances. (Remember that at resonance these two reactances have the same value!)

Frequency Filters

We have been analyzing circuits that have a time-varying emf with a single frequency. However, many applications involve time-varying emfs that reflect a superposition of many frequencies. In some situations, certain frequencies need to be filtered out of this kind of circuit. (Series RLC circuits can be used as frequency filters.) One example of such a circuit can be found in DSL (digital subscriber line) filters for making connections to the Internet over a household telephone line. A typical DSL filter is shown in Figure 30.21.

A DSL Internet connection operates at high frequencies and is connected to a household's regular phone line. The high operating frequency of the DSL connection causes noise on the phones in the house. Therefore, a **band-pass filter** is normally installed on all the phones in the house to filter out the high-frequency noise created by the DSL Internet connection. Frequency filters can be designed to pass low frequencies and block high frequencies (*low-pass filter*) or to pass high frequencies and block low frequencies (*high-pass filter*). A low-pass filter can be combined with a high-pass filter to allow a range of frequencies to pass (*band-pass filter*) and block the frequencies outside that range.

Figure 30.22 shows two examples of a low-pass filter, where V_{in} is a time-varying emf with many frequencies. A low-pass filter is essentially a voltage divider. Part of the original voltage passes through the circuit, while part goes to ground. For the RC version of the low-pass filter, shown in Figure 30.22a, low frequencies essentially have an open circuit, while high frequencies are preferentially sent to ground. Thus, only signals with low frequencies will pass through the filter. This behavior makes sense because current going to ground must pass through a capacitor that essentially blocks the flow of current for low frequencies because the capacitor plates are charging, while rapidly changing current does not allow charge to build up on the plates of the capacitor, allowing current to flow. For the RL version, shown in Figure 30.22b, low frequencies easily pass through the inductor while high frequencies are blocked. This effect arises because the self-induced emf in an inductor opposes rapid changes in current, effectively blocking current through the inductor at high frequencies, while a slow change in current produces a much smaller opposing emf, allowing current to flow.

To quantify the performance of the low-pass filter in Figure 30.22a, we define the input section of the circuit to be the resistor and the capacitor. The impedance of this section is $Z_{in} = \sqrt{R^2 + X_C^2}$. The impedance of the output section is just $Z_{out} = X_C$. The ratio of the emf into the filter and the emf emerging from the filter is

$$\frac{V_{out}}{V_{in}} = \frac{Z_{out}}{Z_{in}}. \tag{30.25}$$

FIGURE 30.21 A typical band-pass filter for phones connected to a household circuit that has a DSL Internet connection.

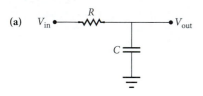

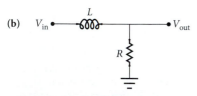

FIGURE 30.22 Two low-pass filters: (a) RC version; (b) RL version.

The ratio of emfs can then be written as

$$\frac{V_{out}}{V_{in}} = \frac{X_C}{\sqrt{R^2 + X_C^2}} = \frac{1}{\sqrt{\left(\dfrac{R}{X_C}\right)^2 + 1}} = \frac{1}{\sqrt{1 + \omega^2 R^2 C^2}}. \tag{30.26}$$

For the RL version of the low-pass filter, shown in Figure 30.22b, $Z_{in} = \sqrt{R^2 + X_L^2}$ and $Z_{out} = R$, allowing us to write

$$\frac{V_{out}}{V_{in}} = \frac{R}{\sqrt{R^2 + X_L^2}} = \frac{1}{\sqrt{1 + \left(\omega^2 L^2 / R^2\right)}}. \tag{30.27}$$

The *breakpoint frequency*, ω_B, between the response to low and high frequencies is the frequency at which the ratio V_{out}/V_{in} is $1/\sqrt{2} = 0.707$. At that frequency for the RC version, we have

$$\frac{1}{\sqrt{1 + \omega_B^2 R^2 C^2}} = \frac{1}{\sqrt{2}},$$

from which we can solve for the breakpoint frequency:

$$\omega_B = \frac{1}{RC}. \tag{30.28}$$

For the RL version of the low-pass filter, the breakpoint frequency is obtained from equation 30.27:

$$\omega_B = \frac{R}{L}. \tag{30.29}$$

Figure 30.23 shows two examples of a high-pass filter. A high-pass filter is also a voltage divider. For the RC version of the high-pass filter, shown in Figure 30.23a, signals with low frequencies cannot pass the capacitor while signals with high frequencies pass through easily. This behavior makes sense because the signal must pass through a capacitor that essentially blocks the flow of current for low frequencies because the capacitor plates are charging, while rapidly changing current does not allow charge to build up on the plates of the capacitor, allowing current to flow. For the RL version, shown in Figure 30.23b, signals with low frequency have essentially an open circuit to ground, while signals with high frequencies are blocked from reaching ground. Thus, only signals with high frequencies will be passed through the filter. This effect arises because the self-induced emf in an inductor opposes rapid changes in current, effectively blocking current through the inductor at high frequencies, while a slow change in current produces a much smaller opposing emf, allowing current to flow.

For the RC version of the high-pass filter in Figure 30.23a, the impedance of the input section is $Z_{in} = \sqrt{R^2 + X_C^2}$, while the impedance of the output section is $Z_{out} = R$. The ratio of the output emf to the input emf is then

(a)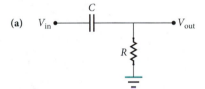

(b)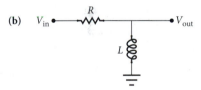

FIGURE 30.23 Two high-pass filters: (a) RC version; (b) RL version.

$$\frac{V_{out}}{V_{in}} = \frac{R}{\sqrt{R^2 + X_C^2}} = \frac{1}{\sqrt{1 + X_C^2/R^2}} = \frac{1}{\sqrt{1 + \dfrac{1}{\omega^2 R^2 C^2}}}. \tag{30.30}$$

For the RL version of the high-pass filter shown in Figure 30.23b, the ratio of the output emf to the input emf is

$$\frac{V_{out}}{V_{in}} = \frac{X_L}{\sqrt{R^2 + X_L^2}} = \frac{1}{\sqrt{\dfrac{R^2}{X_L^2} + 1}} = \frac{1}{\sqrt{1 + \dfrac{R^2}{\omega^2 L^2}}}. \tag{30.31}$$

For these high-pass filters, as the frequency increases, the ratio of output emf to input emf approaches 1, while for low frequencies, the ratio of output emf to input emf goes to zero. The breakpoint frequencies for the high-pass filters are the same as for the low-pass filters: $\omega_B = 1/(RC)$ for the RC version and $\omega_B = R/L$ for the RL version.

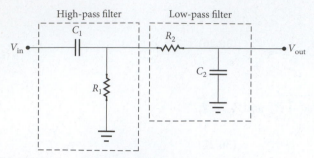

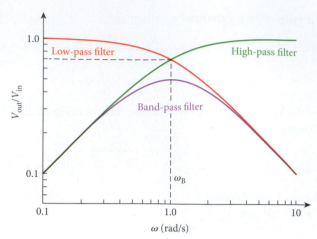

FIGURE 30.24 A band-pass filter consisting of a high-pass filter connected in series with a low-pass filter.

FIGURE 30.25 The frequency response of a low-pass filter, a high-pass filter, and a band-pass filter.

An example of a band-pass filter is shown in Figure 30.24. The band-pass filter consists of a high-pass filter in series with a low-pass filter. Thus, both high and low frequencies are suppressed and a narrow band of frequencies are allowed to pass through the filter.

Figure 30.25 shows the frequency response of a low-pass filter and a high-pass filter with $R = 50 \ \Omega$ and $C = 20$ mF. For this combination of resistance and capacitance, the breakpoint frequency is

$$\omega_B = \frac{1}{RC} = \frac{1}{\left(50 \ \Omega\right)\left(20 \cdot 10^{-3} \ \text{F}\right)} = 1 \ \text{rad/s}.$$

Also shown in Figure 30.25 is the frequency response of a band-pass filter with $R_1 = R_2 = 50 \ \Omega$ and $C_1 = C_2 = 20$ mF.

EXAMPLE **30.3** **Crossover Circuit for Audio Speakers**

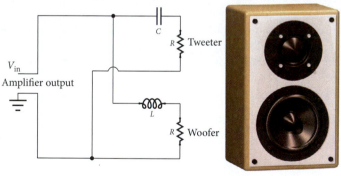

FIGURE 30.26 Crossover circuit for audio speakers.

One way to improve the performance of an audio system is to send high frequencies to a small speaker called a *tweeter* and low frequencies to a large speaker called a *woofer*. Figure 30.26 shows a simple crossover circuit that preferentially passes high frequencies to a tweeter and low frequencies to a woofer. The crossover circuit consists of an RC high-pass filter and an RL low-pass filter connected in parallel to the output of the audio amplifier. The speakers act as resistors, as shown in Figure 30.26. The capacitance and the resistance of this crossover circuit are $C = 10.0 \ \mu\text{F}$ and $L = 10.0$ mH. The speakers each have a resistance of $R = 8.00 \ \Omega$.

PROBLEM

What is the crossover frequency for this crossover circuit?

SOLUTION

We can use equation 30.27 for the response for the RL low-pass filter and equation 30.30 for the response for the RC high-pass filter and equate the two responses:

$$\frac{V_{\text{out}}}{V_{\text{in}}} = \frac{R}{\sqrt{R^2 + X_L^2}} = \frac{R}{\sqrt{R^2 + X_C^2}}.$$

Thus, the responses of the low-pass filter and the high-pass filter are the same when

$$X_L = X_C. \tag{i}$$

We can rewrite equation (i) as

$$\omega_{\text{crossover}} L = \frac{1}{\omega_{\text{crossover}} C},$$

where $\omega_{\text{crossover}}$ is the crossover angular frequency. Thus, the crossover angular frequency is

$$\omega_{\text{crossover}} = \frac{1}{\sqrt{LC}}.$$

We want to determine the crossover frequency, and since $f = \omega/2\pi$, we have

$$f_{\text{crossover}} = \frac{\omega_{\text{crossover}}}{2\pi} = \frac{1}{2\pi\sqrt{LC}}.$$

Putting in the numerical values, we get

$$f_{\text{crossover}} = \frac{1}{2\pi\sqrt{LC}} = \frac{1}{2\pi\sqrt{(10 \text{ mH})(10 \text{ }\mu\text{F})}} = 503 \text{ Hz}.$$

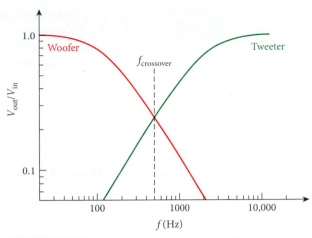

FIGURE 30.27 The response of the crossover circuit as a function of frequency.

Figure 30.27 shows the response of the crossover circuit as a function of frequency. The low-pass response and the high-pass response cross at $f_{\text{crossover}} = 503$ Hz, sending higher frequencies predominantly to the tweeter and lower frequencies predominantly to the woofer.

This simple crossover circuit would not produce ideal audio performance over a broad range of frequencies and speaker designs. More sophisticated crossover circuits that have better performance deal with mid-range frequencies as well.

30.6 Energy and Power in AC Circuits

When an RLC circuit is in operation, some of the energy in the circuit is stored in the electric field of the capacitor, some of the energy is stored in the magnetic field of the inductor, and some energy is dissipated in the form of heat in the resistor. In most applications, we are interested in the steady-state behavior of the circuit, behavior that occurs after initial (transient) effects die out. (A full mathematical analysis would also account for the transient effects, which die out exponentially in a way similar to what equation 30.6 established for a single-loop RLC circuit with no emf source.) The sum of the energy stored in the capacitor and the inductor does not change in the steady state, as we saw in Section 30.1. Therefore, the energy transferred from the source of emf to the circuit is transferred to the resistor.

The rate at which energy is dissipated in the resistor is the power, P, given by

$$P = i^2 R = \left[I \sin(\omega t - \phi) \right]^2 R = I^2 R \sin^2(\omega t - \phi), \tag{30.32}$$

where the alternating current, i, is given by equation 30.9. We can express the average power, $\langle P \rangle$, using the fact that the average value of $\sin^2(\omega t - \phi)$ over a full oscillation is $\frac{1}{2}$:

$$\langle P \rangle = \tfrac{1}{2} I^2 R = \left(\frac{I}{\sqrt{2}} \right)^2 R.$$

In calculations of power and energy, it is common to use the **root-mean-square (rms) current**, I_{rms}. (In general, *root-mean-square*, or *rms*, means the square root of the mean of the square of the specific quantity.) From equation 30.32, we have $i^2 = \left[I \sin(\omega t - \phi) \right]^2$, and the mean (or average) of i^2 is $I^2/2$. Thus, $I_{\text{rms}} = I/\sqrt{2}$. We can then write the average power as

$$\langle P \rangle = I_{\text{rms}}^2 R. \tag{30.33}$$

In a similar way, we can define the root-mean-square values of other time-varying quantities, such as the voltage:

$$V_{\text{rms}} = \frac{V_{\text{m}}}{\sqrt{2}}. \tag{30.34}$$

The current and voltage values normally quoted for alternating currents and measured by AC ammeters and voltmeters are I_{rms} and V_{rms}. For example, wall sockets in the United States provide $V_{rms} = 110$ V, which corresponds to a maximum voltage of $\sqrt{2}(110\text{ V}) \approx 156$ V.

We can rewrite equation 30.21 in terms of root-mean-square values by multiplying both sides of the equation by $1/\sqrt{2}$:

$$I_{rms} = \frac{V_{rms}}{Z} = \frac{V_{rms}}{\sqrt{R^2 + \left(\omega L - \dfrac{1}{\omega C}\right)^2}}. \tag{30.35}$$

This form is used most often to describe the characteristics of AC circuits.

We can describe the average power dissipated in an AC circuit in a different way by starting with equation 30.33:

$$\langle P \rangle = I_{rms}^2 R = \frac{V_{rms}}{Z} I_{rms} R = I_{rms} V_{rms} \frac{R}{Z}. \tag{30.36}$$

From Figure 30.16, we see that the cosine of the phase constant is equal to the ratio of the maximum value of the voltage across the resistor to the maximum value of the time-varying emf:

$$\cos\phi = \frac{V_R}{V_m} = \frac{IR}{IZ} = \frac{R}{Z}. \tag{30.37}$$

We can thus rewrite equation 30.36 as follows:

$$\langle P \rangle = I_{rms} V_{rms} \cos\phi. \tag{30.38}$$

This expression gives the average power dissipated in an AC circuit, where the term $\cos\phi$ is called the **power factor.** You can see that for $\phi = 0$, maximum power is dissipated in the circuit; that is, the maximum power is dissipated in an AC circuit when the frequency of the time-varying emf matches the resonant frequency of the circuit.

We can combine equations 30.19, 30.35, and 30.36 to obtain an expression for the average power as a function of the angular frequency, the inductance, the resistance, and the resonant frequency:

$$\langle P \rangle = I_{rms} V_{rms} \frac{R}{Z} = \frac{V_{rms}}{\sqrt{R^2 + \left(\omega L - \dfrac{1}{\omega C}\right)^2}} V_{rms} \frac{R}{\sqrt{R^2 + \left(\omega L - \dfrac{1}{\omega C}\right)^2}},$$

or simply

$$\langle P \rangle = \frac{V_{rms}^2 R}{R^2 + \left(\omega L - \dfrac{1}{\omega C}\right)^2}.$$

Since $\omega_0 = 1/\sqrt{LC}$, we can write $C = 1/(L\omega_0^2)$, and thus, we find for the average power for a series RLC circuit in terms of the angular frequency:

$$\langle P \rangle = \frac{V_{rms}^2 R}{R^2 + \left(\omega L - \dfrac{L\omega_0^2}{\omega}\right)^2} = \frac{V_{rms}^2 R\omega^2}{R^2\omega^2 + L^2\left(\omega^2 - \omega_0^2\right)^2}. \tag{30.39}$$

The **quality factor,** Q, of a series RLC circuit is defined as

$$Q = \frac{\omega_0 L}{R} = \frac{1}{R}\sqrt{\frac{L}{C}}. \tag{30.40}$$

The quality factor is the ratio of total energy stored in the system divided by the energy dissipated per cycle of the oscillation. This is the same definition used for mechanical oscillators in Chapter 14. For a series RLC circuit, the quality factor characterizes the selectivity of the circuit. The higher the value of Q, the more selective the circuit, that is, the more precisely a given frequency can be isolated (as in an AM radio receiver, discussed next). The lower the value of Q, the less selective the circuit becomes.

AM Radio Receiver

Let's look at a typical example of a selective series RLC circuit, an AM radio receiver. An AM radio receiver can be constructed using a series RLC circuit in which the time-varying emf is supplied by an antenna that picks up transmissions from a distant radio station broadcasting at a given frequency and converts those transmissions to voltage, as illustrated in Figure 30.28.

Figure 30.29 is a plot of the average power as a function of the frequency of the signal received on the antenna for the circuit shown in Figure 30.28, assuming that $R = 0.09111 \ \Omega$, $L = 5.000 \ \mu H$, $C = 6.693$ nF, and $V_m = 3.500$ mV. The resonant angular frequency for this circuit is

$$\omega_0 = \frac{1}{\sqrt{LC}} = \frac{1}{\sqrt{\left(5.000 \cdot 10^{-6} \ H\right)\left(6.693 \cdot 10^{-9} \ F\right)}} = 5.466 \cdot 10^6 \ \text{rad/s},$$

which corresponds to a resonance frequency of

$$f_0 = \frac{\omega_0}{2\pi} = \frac{5.466 \cdot 10^6 \ \text{rad/s}}{2\pi} = 870.0 \ \text{kHz}.$$

The quality factor for this circuit is

$$Q = \frac{\omega_0 L}{R} = \frac{\left(5.466 \cdot 10^6 \ \text{rad/s}\right)\left(5.000 \cdot 10^{-6} \ H\right)}{0.09111 \ \Omega} = 300.0.$$

A method for determining the approximate quality factor of a series RLC circuit uses the formula

$$Q = \frac{\omega_0}{\Delta\omega} = \frac{f_0}{\Delta f},$$

where $\Delta\omega$ and Δf are the full widths at half maximum for the angular frequency and the frequency, respectively, on the power response curve. The higher the value of Q, the narrower the power response to frequency. In Figure 30.29, $\Delta f = 2.9$ kHz, which gives a quality factor of

$$Q = \frac{f_0}{\Delta f} = \frac{870.0 \ \text{kHz}}{2.9 \ \text{kHz}} = 300.0.$$

This is the same result obtained using the formula that defines the quality factor in equation 30.40. Note that these two formulas for the quality factor have the same results only for high Q!

The alternative formula for the quality factor of a series RLC circuit is similar to the expression given in Chapter 14 for the quality of a weakly damped mechanical oscillator, $Q \cong \dfrac{\omega_0}{2\omega_\gamma}$, where ω_0 is the resonant angular frequency and ω_γ is the damping angular frequency.

In Figure 30.29, the frequencies of adjacent channels in the AM radio band are indicated by the vertical dashed lines located 10 kHz apart. The response of the series RLC circuit allows the AM receiver to tune in one station and exclude the adjacent channels.

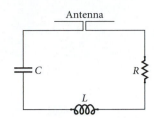

FIGURE 30.28 A series RLC circuit with the source of time-varying emf replaced by an antenna. This circuit can function as an AM radio receiver.

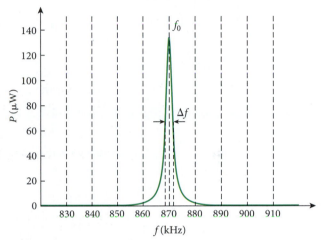

FIGURE 30.29 Power response of a series RLC circuit functioning as an AM radio receiver. The label Δf indicates the *full width at half maximum*, or the difference between the frequencies where the power has half the value that it has at its maximum value, at frequency f_0.

30.8 In-Class Exercise

Wireless WiFi networks are installed in most coffee shops and many residences to provide access to the Internet. The most common WiFi standard is known as 802.11 g, which supports communication rates of up to 54 Megabits per second. Wireless networks in the United States and Canada that follow this standard use a frequency around 2.4 GHz, in 14 different channels in the band between 2.401 GHz and 2.495 GHz. Each channel has a full width at half maximum of 22 MHz. What is the Q factor of these WiFi networks?

a) 0.1   d) 109

b) 9.9   e) 300

c) 33

SOLVED PROBLEM 30.1 | **Unknown Inductance in an RL Circuit**

Consider a series RL circuit with a time-varying source of emf. In this circuit, $V_m = 33.0$ V with $f = 7.10$ kHz and $R = 83.0 \ \Omega$. A current $I = 0.158$ A flows in the circuit.

PROBLEM

What is the magnitude of the inductance, L?

Continued—

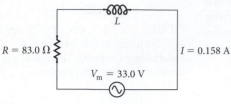

FIGURE 30.30 A series RL circuit.

SOLUTION

THINK

The specified voltage and current are implicitly root-mean-square values. We can relate the voltage and current through the impedance of the circuit. The impedance of this circuit depends on the resistance and the inductance, as well as the frequency of the source of emf.

SKETCH

A diagram of the circuit is shown in Figure 30.30.

RESEARCH

We can relate the time-varying emf, V_m, and the impedance Z in the circuit:

$$V_m = IZ. \tag{i}$$

The impedance is given by

$$Z = \sqrt{R^2 + (X_L - X_C)^2} = \sqrt{R^2 + X_L^2}, \tag{ii}$$

where R is the resistance, X_L is the inductive reactance, and X_C is the capacitive reactance, which is zero. The angular frequency, ω, of the circuit is given by

$$\omega = 2\pi f,$$

where f is the frequency. We can express the inductive reactance as

$$X_L = \omega L. \tag{iii}$$

SIMPLIFY

We can combine equations (i), (ii), and (iii) to obtain

$$Z^2 = R^2 + X_L^2 = \left(\frac{V_m}{I}\right)^2 = R^2 + (\omega L)^2. \tag{iv}$$

Rearranging equation (iv) gives

$$\omega L = \sqrt{\frac{V_m^2}{I^2} - R^2}.$$

And, finally, we find the unknown inductance:

$$L = \frac{1}{2\pi f}\sqrt{\frac{V_m^2}{I^2} - R^2}.$$

CALCULATE

Putting in the numerical values gives us

$$L = \frac{1}{2\pi\left(7.10 \cdot 10^3 \text{ s}^{-1}\right)}\sqrt{\frac{\left(33.0 \text{ V}\right)^2}{\left(0.158 \text{ A}\right)^2} - \left(83.0 \text{ }\Omega\right)^2} = 0.0042963 \text{ H}.$$

ROUND

We report our result to three significant figures:

$$L = 4.30 \cdot 10^{-3} \text{ H} = 4.30 \text{ mH}.$$

DOUBLE-CHECK

To double-check our result for the inductance, we first calculate the inductive reactance:

$$X_L = 2\pi f L = 2\pi\left(7.10 \cdot 10^3 \text{ s}^{-1}\right)\left(4.30 \cdot 10^{-3} \text{ H}\right) = 192 \text{ }\Omega.$$

The impedance of the circuit is then

$$Z = \sqrt{R^2 + X_L^2} = \sqrt{\left(83.0 \text{ }\Omega\right)^2 + \left(192 \text{ }\Omega\right)^2} = 209 \text{ }\Omega.$$

30.9 In-Class Exercise

In the series RL circuit in Solved Problem 30.1, what is the magnitude of the phase difference between the time-varying emf and the current in the circuit?

a) 30.0°   d) 75.0°

b) 45.0°   e) 90.0°

c) 66.6°

We use this value of Z to calculate a value of V_m:

$$V_m = IZ = (0.158 \text{ A})(209 \ \Omega) = 33.0 \text{ V},$$

which agrees with the value specified in the problem statement. Thus, our result is consistent.

30.7 Transformers

This section discusses the root-mean-square values of currents and voltages, rather than the maximum or instantaneous values. The result obtained for the power is always the average power when root-mean-square values are used. This practice is the convention normally followed by scientists, engineers, and electricians in dealing with AC circuits.

In an AC circuit that has only a resistor, the phase constant is zero. Thus we can express the power as

$$P = IV. \tag{30.41}$$

For a given power delivered to a circuit, the application dictates the choice of high current or high voltage. For example, to provide enough power to operate a computer or a vacuum cleaner, using a high voltage might be dangerous. The design of electric generators is complicated by the use of high voltages. Therefore, in these devices, lower voltages and higher currents are advantageous.

However, the transmission of electric power requires the opposite condition. The power dissipated in a transmission line is given by $P = I^2 R$. Thus, the power lost in a line, like those in Figure 30.31a is proportional to the square of the current in the line. As an example, consider a power plant that produces 500 MW of power. If the power is transmitted at 350 kV, the current in the power lines will be

$$I = \frac{P}{V} = \frac{500 \text{ MW}}{350 \text{ kV}} = \frac{5 \cdot 10^8 \text{ W}}{3.5 \cdot 10^5 \text{ V}} = 1430 \text{ A}.$$

If the total resistance of the power lines is 50 Ω, the power lost in the transmission lines is

$$P = I^2 R = (1430 \text{ A})^2 (50 \ \Omega) = 100 \text{ MW},$$

or 20% of the generated power. A similar calculation would show that transmitting the power at 200 kV instead of 350 kV would increase the power loss by a factor of 3.1. Thus, 60% of the power generated would be lost in transmission. This is why the transmission of electric power is always done at the highest possible voltage.

The ability to change voltage allows electric power to be generated and used at low, safe voltages but transmitted at the highest practical voltage. Alternating currents and voltages are transformed from high to low values by a device called, appropriately, a **transformer.** A transformer that takes voltages from lower to higher values is called a *step-up transformer;* a transformer that takes voltages from higher to lower values is called a *step-down transformer.* Transformers are the main components of, for example, your cell phone charger (Figure 30.32) and the power supply for your MP3 player, your laptop, and pretty much every other consumer electronics device. Most of these devices require voltages of 12 V or less, but the grid delivers 110 V to outlets in the United States, necessitating the transformers that clutter your desk drawers.

A transformer consists of two sets of coils wrapped around an iron core (Figure 30.33). The primary coil, with N_p turns, is connected to a source of emf described by

$$V_{emf} = V_{max} \sin \omega t.$$

We'll assume that the primary coil acts as an inductor. The primary circuit has the current and voltage out of phase by $\pi/2$ rad (90°), so the power factor, $\cos\phi$, is zero. Thus, the source of emf is not delivering any power to the transformer if only the primary coil is connected. In other words, if the secondary coil is not connected to a closed circuit, the transformer

30.5 Self-Test Opportunity

You might argue that power companies should simply reduce the resistance in their transmission lines to avoid the substantial power losses. Typical wires for electric power transmission lines are finger-thick. How big would they need to be to reduce the resistance by a factor of 100, assuming that all other parameters (material used, length) remain the same? (*Hint:* Consult Section 25.3.)

(a)

(b)

FIGURE 30.31 (a) High-voltage power lines; (b) transformers for residential power lines.

FIGURE 30.32 A transformer (black rectangle with yellow core) is the main component of a cell phone charger. (The charger also includes a rectifier described in Section 30.8.)

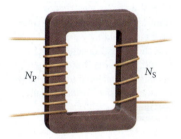

FIGURE 30.33 Transformer with N_P primary windings and N_S secondary windings.

does not draw any power. For example, your cell phone charger does not draw power if it is plugged into the wall socket but the other end is not connected to the cell phone. (This statement is not absolutely true; there is finite resistance in the wires in the primary coil, which this description is neglecting.)

The secondary coil of a transformer has N_S turns. The time-varying emf in the primary coil induces a time-varying magnetic field in the iron core. This core passes through the secondary coil. Thus, a time-varying voltage is induced in the secondary coil, as described by Faraday's Law of Induction:

$$V_{emf} = -N\frac{d\Phi_B}{dt},$$

where N is the number of turns and Φ_B is the magnetic flux. Because of the iron core, both the primary and secondary coils experience the same changing magnetic flux. Thus,

$$V_S = -N_S\frac{d\Phi_B}{dt}$$

and

$$V_P = -N_P\frac{d\Phi_B}{dt},$$

where V_S and V_P are the voltages across the secondary and primary windings, respectively. Dividing the first of these two equations by the other and rearranging gives

$$\frac{V_P}{N_P} = \frac{V_S}{N_S},$$

or

$$V_S = V_P\frac{N_S}{N_P}. \qquad (30.42)$$

The transformer changes the voltage of the primary circuit to a secondary voltage, given by the ratio of the number turns in the secondary coil divided by the number of turns in the primary coil.

If a resistor, R, is connected across the secondary windings, a current, I_S, will begin to flow through the secondary coil. The power in the secondary circuit is then $P_S = I_S V_S$. This current induces a time-varying magnetic field that induces a voltage in the primary coil, so that the emf source then produces enough current, I_P, to maintain the original voltage. This current, I_P, is in phase with the voltage because of the resistor, so power can be transmitted to the transformer. Energy conservation requires that the power delivered to the primary coil be transferred to the secondary coil, so we can write

$$P_P = I_P V_P = P_S = I_S V_S.$$

Using equation 30.42, we can express the current in the secondary circuit as

$$I_S = I_P\frac{V_P}{V_S} = I_P\frac{N_P}{N_S}. \qquad (30.43)$$

The current in the secondary circuit is equal to the current in the primary circuit multiplied by the ratio of the number of primary turns divided by the number of secondary turns.

When the secondary circuit begins to draw current, current must be supplied to the primary circuit. Since $V_S = I_S R$ in the secondary circuit, we can use equations 30.42 and 30.43 to write

$$I_P = \frac{N_S}{N_P}I_S = \frac{N_S}{N_P}\frac{V_S}{R} = \frac{N_S}{N_P}\left(V_P\frac{N_S}{N_P}\right)\frac{1}{R} = \left(\frac{N_S}{N_P}\right)^2\frac{V_P}{R}. \qquad (30.44)$$

The effective resistance of the primary circuit can be expressed in terms of $V_P = I_P R_P$, so that the effective resistance is

$$R_P = \frac{V_P}{I_P} = V_P\left(\frac{N_P}{N_S}\right)^2\frac{R}{V_P} = \left(\frac{N_P}{N_S}\right)^2 R. \qquad (30.45)$$

Note that we have assumed there are no losses in the transformer, that the primary coil is only an inductor, that there are no losses in magnetic flux between the primary and secondary coils, and that the secondary circuit has the only resistance. Real transformers do have some losses. Part of these losses result from the fact that the alternating magnetic fields from the coils induce eddy currents in the iron core of the transformer. To counter this effect, transformer cores are constructed by laminating layers of metal to inhibit the formation of eddy currents. Modern transformers can transform voltages with very little loss.

Another application of transformers is **impedance matching.** The power transfer between a source of emf and a device that uses power is at a maximum when the impedance is the same in both. Often, the source of emf and the intended device do not have the same impedance. A common example is a stereo amplifier and its speakers. Usually, the amplifier has high impedance and the speakers have low impedance. A transformer placed between the amplifier and the speakers can help match the impedances of the devices, producing a more efficient power transfer.

30.8 Rectifiers

Many electronic devices require direct current rather than alternating current. However, many common sources of electrical power provide alternating current. Therefore, this current must be converted to direct current to operate electronic equipment. A **rectifier** is a device that converts alternating current to direct current. Most rectifiers use an electronic component that was described in Section 25.8—the diode. A diode is designed to allow current to flow in one direction and not in the other direction. The symbol for a diode is ▸|—, and the direction of the arrowhead signifies the direction in which the diode will conduct current.

Let's start with a simple circuit containing a source of time-varying emf, a resistor, and a diode, as shown in Figure 30.34b. The voltage provided by the source of emf is alternately positive and negative, as shown in Figure 30.34a. Note that both ends of the source of emf are connected simultaneously so that when one end produces a positive voltage, the other end produces a negative voltage. The circuit in Figure 30.34b produces current in the resistor that indeed flows in only one direction. However, the circuit blocks half the current, as illustrated in Figure 30.34c. Thus, this type of circuit is often termed a **halfwave rectifier.**

To allow all the current to flow in one direction, the type of circuit shown in Figure 30.35 is employed. Again the voltage alternates between positive and negative, as shown in Figure 30.35a. Two equivalent circuit diagrams are shown in Figure 30.35b and Figure 30.35c. All the current in the resistor flows in one direction, as illustrated in Figure 30.35d. This type of circuit is called a **fullwave rectifier.**

To illustrate how the fullwave rectifier works, Figure 30.36 shows instantaneous views of the circuit with positive and negative voltage. In Figure 30.36a, the voltage from the

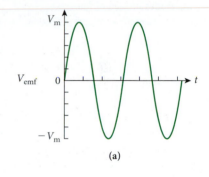

(a)

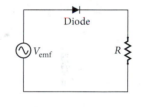

(b)

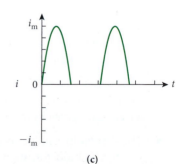

(c)

FIGURE 30.34 Circuit containing a source of time-varying emf, a resistor, and a diode, forming a halfwave rectifier: (a) the emf as a function of time; (b) the circuit diagram; (c) the current flowing through the circuit as a function of time.

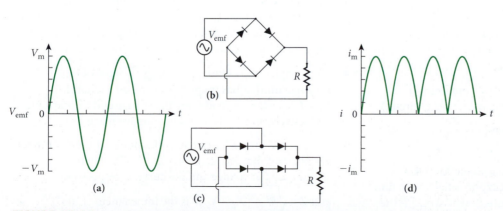

FIGURE 30.35 Circuit containing a source of time-varying emf, a resistor, and four diodes, forming a fullwave rectifier: (a) the emf as a function of time; (b) the circuit diagram; (c) alternative way of drawing the circuit diagram; (d) the current flowing through the circuit as a function of time.

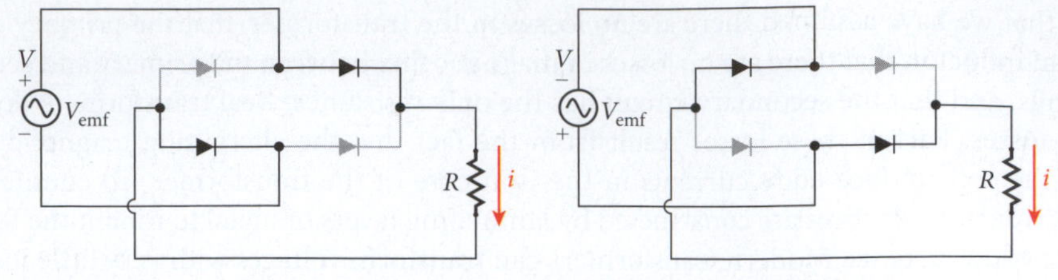

FIGURE 30.36 A fullwave rectifier with the diodes that are not conducting current at the given instant in gray. The current in the resistor always flows in the same direction. (a) Positive voltage. (b) Negative voltage.

(a)   (b)

30.6 Self-Test Opportunity

A typical alternator found in automobiles produces three-phase alternating current. Each phase is shifted by 120° from the next phase. Draw the circuit diagram for the fullwave rectifier for this alternator based on the circuit diagram in Figure 30.35c but incorporating six diodes instead of four.

source of emf is positive. The black diodes are conducting current, while the gray diodes are not. The voltage is reversed in Figure 30.36b, and the current flows through the other pair of diodes; the current in the resistor is still in the same direction.

Although the fullwave rectifier does indeed convert alternating current to direct current, the resulting direct current varies with time. This variance, often called *ripple*, can be smoothed out by adding a capacitor to the output of the rectifier, creating an RC circuit with a time constant governed by the choice of R and C, as shown in Figure 30.37. In Figure 30.37a, the time-varying emf is shown, and the circuit diagram is presented in Figure 30.37b. The direct current is filtered by the added capacitor. The resulting current as a function of time is shown in Figure 30.37c. The current still varies with time but much less so than the current flowing out of the full wave rectifier without a capacitor.

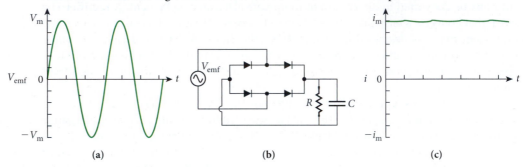

(a)   (b)   (c)

FIGURE 30.37 Circuit containing a source of time-varying emf, a resistor, a capacitor, and four diodes, forming a filtered fullwave rectifier: (a) the emf as a function of time; (b) the circuit diagram; (c) the current flowing through the circuit as a function of time.

WHAT WE HAVE LEARNED | EXAM STUDY GUIDE

- The energy stored in the electric field of a capacitor with capacitance C and charge q is given by $U_E = \frac{1}{2}(q^2/C)$; the energy stored in the magnetic field of an inductor with inductance L that is carrying current i is given by $U_B = \frac{1}{2}Li^2$.

- The current in a single-loop circuit containing an inductor and a capacitor (an LC circuit) oscillates with a frequency given by $\omega_0 = 1/\sqrt{LC}$.

- The current in a single-loop circuit containing a resistor, an inductor, and a capacitor (an RLC circuit) oscillates with a frequency given by $\omega = \sqrt{\omega_0^2 - (R/2L)^2}$, where $\omega_0 = 1/\sqrt{LC}$.

- The charge, q, on a capacitor in a single-loop RLC circuit oscillates and decreases exponentially with time according to $q = q_{max}\, e^{-Rt/2L} \cos(\omega t)$, where q_{max} is the original charge on the capacitor.

- For a single-loop circuit containing a source of time-varying emf and a resistor, R, $V_R = I_R R$, where V_R and I_R are the voltage and the current, respectively.

- For a single-loop circuit containing a source of time-varying emf that has frequency ω and a capacitor, $V_C = I_C X_C$, where V_C and I_C are the voltage and the current, respectively, and $X_C = 1/\omega C$ is the capacitive reactance.

- For a single-loop circuit containing a source of time-varying emf that has frequency ω and an inductor, $V_L = I_L X_L$, where V_L and I_L are the voltage and the current, respectively, and $X_L = \omega L$ is the inductive reactance.

- For a single-loop RLC circuit containing a source of time-varying emf that has frequency ω, $V = IZ$, where V and I are the voltage and the current, respectively, and $Z = \sqrt{R^2 + (X_L - X_C)^2}$ is the impedance.

- The phase constant, ϕ, between the current and voltage in a single-loop RLC circuit containing a source of time-varying emf that has frequency ω is given by $\phi = \tan^{-1}\left(\dfrac{X_L - X_C}{R}\right)$.

- The average power in a single-loop RLC circuit containing a source of time-varying emf that has frequency ω is given by $\langle P \rangle = I_{rms}V_{rms}\cos\phi$, where $I_{rms} = I_m/\sqrt{2}$ and $V_{rms} = V_m/\sqrt{2}$.

- All currents, voltages, and powers quoted for alternating-current (AC) circuits are typically root-mean-square values.

- A transformer with N_P windings in the primary coil and N_S windings in the secondary coil can convert a primary alternating voltage, V_P, to a secondary alternating voltage, V_S, given by $V_S = V_P \dfrac{N_S}{N_P}$, and a primary alternating current, I_P, to a secondary alternating current, I_S, given by $I_S = I_P \dfrac{N_P}{N_S}$.

KEY TERMS

LC circuit, p. 959
electromagnetic
   oscillations, p. 959
RLC circuit, p. 964
alternating current (AC),
   p. 965

phasor, p. 966
capacitive reactance,
   p. 966
inductive reactance, p. 968
impedance, p. 969
resonance, p. 970

resonant angular
   frequency, p. 970
band-pass filter, p. 972
root-mean-square (rms)
   current, p. 975
power factor, p. 976

quality factor, p. 976
transformer, p. 979
impedance matching, p. 981
rectifier, p. 981
halfwave rectifier, p. 981
fullwave rectifier, p. 981

NEW SYMBOLS AND EQUATIONS

$\omega_0 = \dfrac{1}{\sqrt{LC}}$, resonant frequency of LC and RLC circuits

$X_C = \dfrac{1}{\omega C}$, capacitive reactance

$X_L = \omega L$, inductive reactance

$Z = \sqrt{R^2 + (X_L - X_C)^2}$, impedance of an alternating-current (AC) circuit

$\langle P \rangle = I_{rms}V_{rms}\cos\phi$, average power dissipated in an AC circuit

$\phi = \tan^{-1}\left(\dfrac{X_L - X_C}{R}\right)$, phase constant between the voltage and the current in an AC circuit

$Q = \dfrac{\omega_0 L}{R}$, quality factor of a series RLC circuit

N_P and N_S, the numbers of primary and secondary windings in a transformer

ANSWERS TO SELF-TEST OPPORTUNITIES

30.1 $\omega_0 = 2\pi f = 2\pi(200 \text{ kHz}) \text{ rad/s} = 4\pi \cdot 10^5 \text{ rad/s}$; $\phi = 0$ since $q(0) = q_{max}$

a) true ($\cos\omega_0 t = -1$)

b) false ($\sin\omega_0 t = 0$)

c) false ($\cos\omega_0 t = -1$ and $\sin\omega_0 t = 0$)

d) false ($\cos\omega_0 t = 0$ and $\sin\omega_0 t = 1$)

30.2 k/m corresponds to $1/LC$, and $b/2m$ corresponds to $R/2L$. From this follows that the inductance, L, plays the role of the mass, m, the capacitance, C, corresponds to the inverse spring constant, $1/k$, and the resistance, R, has the function of the damping constant, b.

30.3 a) true   b) true   c) false

30.4 a) true   b) true   c) false

30.5 The resistance is inversely proportional to the inverse of the area of the wire and thus inversely proportional to the square of the radius. Therefore a wire, which is 10 times thicker, has a 100 times lower resistance.

30.6

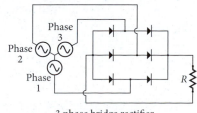

3 phase bridge rectifier

PROBLEM-SOLVING PRACTICE

Problem-Solving Guidelines

1. Most problems concerning AC circuits require you to calculate resistance, capacitive reactance, inductive reactance, or impedance. Be sure that you understand what each of these quantities is and how to use them in calculating currents and voltages.

2. You will often have to distinguish between the instantaneous current or voltage in a circuit and the root-mean-square or maximum value of current or voltage. The common convention is to use lowercase i and v for instantaneous values and uppercase I and V for constant values (with subscripts as necessary). Be sure you use notation that is clear so that you won't become confused during the calculations.

3. Remember the phase relations for AC circuits: For a resistor, current and voltage are in phase; for a capacitor, current leads voltage; for an inductor, current lags voltage.

4. Phasors add by vector operations, not by simple scalar arithmetic. Whenever you use phasors to determine current or voltage, check the results by checking the phase relationships given in the preceding guideline.

5. It is usually easier to work with angular frequency (ω) than with frequency (f) in analyzing AC circuits. Most often, you will be given an angular frequency in the problem statement, but if you are given a frequency, convert it to an angular frequency by multiplying it by 2π.

SOLVED PROBLEM 30.2 | Voltage Drop across an Inductor

A series RLC circuit has a time-varying source of emf providing $V_{rms} = 170.0$ V, a resistance $R = 820.0$ Ω, an inductance $L = 30.0$ mH, and a capacitance $C = 0.290$ mF. The circuit is operating at its resonant frequency.

PROBLEM
What is the root-mean-square voltage drop across the inductor?

SOLUTION

THINK
At the resonant frequency, the impedance of the circuit is equal to the resistance. We can calculate the root-mean-square current in the circuit. The voltage drop across the inductor is then the product of the root-mean-square current in the circuit and the inductive reactance.

SKETCH
A diagram of the series RLC circuit is shown in Figure 30.38.

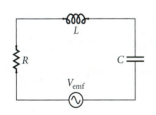

FIGURE 30.38 A series RLC circuit.

RESEARCH
At resonance, the impedance of the circuit is

$$Z = \sqrt{R^2 + \left(X_L - X_C\right)^2} = R.$$

At resonance, the root-mean-square current, I_{rms}, in the circuit is given by

$$V_{rms} = I_{rms}R.$$

The root-mean-square voltage drop across the inductor, V_L, at resonance is

$$V_L = I_{rms}X_L,$$

where the inductive reactance X_L is defined as

$$X_L = \omega L$$

and ω is the angular frequency at which the circuit is operating. The resonant frequency ω_0 of the circuit is

$$\omega_0 = \frac{1}{\sqrt{LC}}.$$

SIMPLIFY
Combining all these equations gives us the voltage drop across the inductor at resonance:

$$V_L = \left(\frac{V_{rms}}{R}\right)\left(\omega_0 L\right) = \frac{LV_{rms}}{R}\frac{1}{\sqrt{LC}} = \frac{V_{rms}}{R}\sqrt{\frac{L}{C}}.$$

CALCULATE

Putting in the numerical values gives us

$$V_L = \frac{170.0 \text{ V}}{820.0 \text{ } \Omega} \sqrt{\frac{30.0 \cdot 10^{-3} \text{ H}}{0.290 \cdot 10^{-3} \text{ F}}} = 2.10861 \text{ V}.$$

ROUND

We report our result to three significant figures:

$$V_L = 2.11 \text{ V}.$$

DOUBLE-CHECK

The root-mean-square voltage drop across the capacitor is

$$V_C = \left(\frac{V_{rms}}{R}\right)\left(\frac{1}{\omega_0 C}\right) = \frac{V_{rms}}{RC}\sqrt{LC} = \left(\frac{V_{rms}}{R}\right)\sqrt{\frac{L}{C}},$$

which is the same as the root-mean-square voltage drop across the inductor. At resonance, the instantaneous voltage drop across the inductor is the negative of the voltage drop across the capacitor. Thus, the rms voltage across the capacitor should be the same as the rms voltage across the inductor. Thus, our result seems reasonable.

SOLVED PROBLEM 30.3 | Power Dissipated in an RLC Circuit

A series RLC circuit has a source of emf providing $V_{rms} = 120.0$ V and operating at frequency $f = 50.0$ Hz, an inductor, $L = 0.500$ H, a capacitor, $C = 3.30$ μF, and a resistor, $R = 276$ Ω.

PROBLEM

What is the average power dissipated in the circuit?

SOLUTION

THINK

The average power dissipated in the circuit is the root-mean-square current times the root-mean-square voltage, but it depends on the angular frequency of the source of emf. The current in the circuit can be found using the impedance.

SKETCH

A diagram of a series RLC circuit is shown in Figure 30.38.

RESEARCH

The angular frequency, ω, of the source of emf is

$$\omega = 2\pi f.$$

The impedance, Z, of the circuit is

$$Z = \sqrt{R^2 + \left(X_L - X_C\right)^2},$$

where the inductive reactance is given by

$$X_L = \omega L$$

and the capacitive reactance is given by

$$X_C = \frac{1}{\omega C}.$$

We can find the root-mean-square current, I_{rms}, in the circuit using the relationship

$$V_{rms} = I_{rms} Z.$$

The average power dissipated in the circuit, $\langle P \rangle$, is given by

$$\langle P \rangle = I_{rms} V_{rms} \cos\phi,$$

Continued—

where ϕ is the phase constant between the voltage and the current in the circuit:

$$\phi = \tan^{-1}\left(\frac{X_L - X_C}{R}\right).$$

SIMPLIFY

We can combine all these equations to obtain an expression for the average power dissipated in the circuit:

$$\langle P \rangle = \frac{V_{rms}}{Z} V_{rms} \cos\phi = \frac{V_{rms}^2}{\sqrt{R^2 + (X_L - X_C)^2}} \cos\phi.$$

CALCULATE

First, we calculate the inductive reactance:

$$X_L = \omega L = 2\pi f L = 2\pi (50.0 \text{ Hz})(0.500 \text{ H}) = 157.1 \text{ }\Omega.$$

Next, we calculate the capacitive reactance:

$$X_C = \frac{1}{\omega C} = \frac{1}{2\pi f C} = \frac{1}{2\pi (50.0 \text{ Hz})(3.30 \cdot 10^{-6} \text{ F})} = 964.6 \text{ }\Omega.$$

The phase constant is then

$$\phi = \tan^{-1}\left(\frac{X_L - X_C}{R}\right) = \tan^{-1}\left(\frac{157.1 \text{ }\Omega - 964.6 \text{ }\Omega}{276 \text{ }\Omega}\right) = -1.241 \text{ rad} = -71.13°.$$

We now calculate the average power dissipated in the circuit:

$$\langle P \rangle = \frac{(120.0 \text{ V})^2}{\sqrt{(276 \text{ }\Omega)^2 + (157.1 \text{ }\Omega - 964.6 \text{ }\Omega)^2}} \cos(-1.241 \text{ rad}) = 5.46477 \text{ W}.$$

ROUND

We report our result to three significant figures:

$$\langle P \rangle = 5.46 \text{ W}.$$

DOUBLE-CHECK

To double-check our result, we can calculate the power that would be dissipated in the circuit if it were operating at the resonant frequency. At the resonant frequency, the maximum power is dissipated in the circuit and the impedance of the circuit is equal to the resistance of the resistor. Thus, the maximum average power is

$$\langle P \rangle_{max} = \frac{V_{rms}^2}{R} = \frac{(120.0 \text{ V})^2}{276 \text{ }\Omega} = 52.2 \text{ W}.$$

Our result for the power dissipated at $f = 50.0$ Hz is lower than the maximum average power, so it seems plausible.

MULTIPLE-CHOICE QUESTIONS

30.1 A 200-Ω resistor, a 40.0-mH inductor and a 3.0-μF capacitor are connected in series with a time-varying source of emf that provides 10.0 V at a frequency of 1000 Hz. What is the impedance of the circuit?

a) 200 Ω c) 342 Ω

b) 228 Ω d) 282 Ω

30.2 For which values of f is $X_L > X_C$?

a) $f > 2\pi(LC)^{1/2}$ c) $f > (2\pi(LC)^{1/2})^{-1}$

b) $f > (2\pi LC)^{-1}$ d) $f > 2\pi LC$

30.3 Which statement about the phase relation between the electric and magnetic fields in an LC circuit is correct?

a) When one field is at its maximum, the other is also, and the same for the minimum values.

b) When one field is at maximum strength, the other is at minimum (zero) strength.

c) The phase relation, in general, depends on the values of L and C.

30.4 For the band-pass filter shown in Figure 30.24, how can the width of the frequency response be increased?

a) increase R_1

d) increase C_2

b) decrease C_1

e) do any of the above

c) increase R_2

30.5 The phase constant, ϕ, between the voltage and the current in an AC circuit depends on the _____.

a) inductive reactance

c) resistance

b) capacitive reactance

d) all of the above

30.6 The AM radio band covers the frequency range from 520 kHz to 1610 kHz. Assuming a fixed inductance in a simple LC circuit, what ratio of capacitance is necessary to cover this frequency range? That is, what is the value of C_h/C_l,

where C_h is the capacitance for the highest frequency and C_l is the capacitance for the lowest frequency?

a) 9.59

c) 0.568

b) 0.104

d) 1.76

30.7 In the RLC circuit in the figure, $R = 60\ \Omega$, $L = 3$ mH, $C = 4$ mF, and the source of time-varying emf has a peak voltage of 120 V. What should the angular frequency, ω, be to produce the largest current in the resistor?

a) 4.2 rad/s

d) 289 rad/s

b) 8.3 rad/s

e) 5000 rad/s

c) 204 rad/s

f) 20,000 rad/s

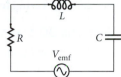

30.8 A standard North American wall socket plug is labeled 110 V. This label indicates the _____ value of the voltage.

a) average

c) root-mean-square (rms)

b) maximum

d) instantaneous

QUESTIONS

30.9 What is the impedance of a series RLC circuit when the frequency of time-varying emf is set to the resonant frequency of the circuit?

30.10 Estimate the total energy stored in the 5 km of space above Earth's surface if the average magnitude of the magnetic field at Earth's surface is about $0.50 \cdot 10^{-4}$ T.

30.11 In a DC circuit containing a capacitor, a current will flow through the circuit for only a very short time, while the capacitor is being charged or discharged. On the other hand, a steady alternating current will flow in a circuit containing the same capacitor but powered by a source of time-varying emf. Does it mean that charges are crossing the gap (dielectric) of the capacitor?

30.12 In an RL circuit with alternating current, the current lags behind the voltage. What does this mean, and how can it be explained qualitatively, based on the phenomenon of electromagnetic induction?

30.13 In Solved Problem 30.1, the voltage supplied by the source of time-varying emf is 33.0 V, the voltage across the resistor is $V_R = IR = 13.1$ V, and the voltage across the inductor is $V_L = IX_L = 30.3$ V. Does this circuit obey Kirchhoff's rules?

30.14 Why is RMS power specified for an AC circuit, not average power?

30.15 Why can't we use a universal charger that plugs into a household electrical outlet to charge all our electrical appliances— cell phone, toy dog, can opener, and so on—rather than using a separate charger with its own transformer for each device?

30.16 If you use a parallel plate capacitor with air in the gap between the plates as part of a series RLC circuit in a generator, you can measure current flowing through the generator. Why

is it that the air gap in the capacitor does not act like an open switch, blocking all current flow in the circuit?

30.17 A common configuration of wires has twisted pairs as opposed to straight, parallel wires. What is the technical advantage of using twisted pairs of wires versus straight, parallel pairs?

30.18 In a classroom demonstration, an iron core is inserted into a large solenoid connected to an AC power source. The effect of the core is to magnify the magnetic field in the solenoid by the relative magnetic permeability, κ_m, of the core (where κ_m is a dimensionless constant, substantially greater than unity for a ferromagnetic material, introduced in Chapter 28) or, equivalently, to replace the magnetic permeability of free space, μ_0, with the magnetic permeability of the core, $\mu = \kappa_m \mu_0$.

a) The measured root-mean-square current drops from approximately 10 A to less than 1 A and remains at the lower value. Explain why.

b) What would happen if the power source were DC?

30.19 Along Capitol Drive in Milwaukee, Wisconsin, there are a large number of radio broadcasting towers. Contrary to expectation, radio reception there is terrible; unwanted stations often interfere with the one tuned in. Given that a car radio tuner is a resonant oscillator—its resonant frequency is adjusted to that of the desired station—explain this crosstalk phenomenon.

30.20 A series RLC circuit is in resonance when driven by a sinusoidal voltage at its resonant frequency, $\omega_0 = (LC)^{-1/2}$. But if the same circuit is driven by a *square-wave* voltage (which is alternately on and off for equal time intervals), it will exhibit resonance at its resonant frequency and at $\frac{1}{3}$, $\frac{1}{5}$, $\frac{1}{7}$, ..., of this frequency. Explain why.

30.21 Is it possible for the voltage amplitude across the inductor in a series RLC circuit to exceed the voltage amplitude of the voltage supply? Why or why not?

30.22 Why can't a transformer be used to step up or step down the voltage in a DC circuit?

PROBLEMS

A blue problem number indicates a worked-out solution is available in the Student Solutions Manual. One • and two •• indicate increasing level of problem difficulty.

Sections 30.1 and 30.2

30.23 For the LC circuit in the figure, $L = 32.0$ mH and $C = 45.0$ μF. The capacitor is charged to $q_0 = 10.0$ μC, and at $t = 0$ s, the switch is closed. At what time is the energy stored in the capacitor first equal to the energy stored in the inductor?

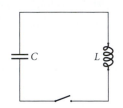

30.24 A 2.00-μF capacitor is fully charged by being connected to a 12.0-V battery. The fully charged capacitor is then connected to a 0.250-H inductor. Calculate (a) the maximum current in the inductor and (b) the frequency of oscillation of the LC circuit.

30.25 An LC circuit consists of a 1-mH inductor and a fully charged capacitor. After 2.1 ms, the energy stored in the capacitor is half of its original value. What is the capacitance?

30.26 The time-varying current in an LC circuit where $C = 10$ μF is given by $i(t) = (1A)\sin(1200t)$, where t is in seconds.

a) At what time after $t = 0$ does the current reach its maximum value?

b) What is the total energy of the circuit?

c) What is the inductance, L?

•30.27 A 10-μF capacitor is fully charged by a 12-V battery and is then disconnected from the battery and allowed to discharge through a 0.2-H inductor. Find the first three times when the charge on the capacitor is 80-μC, taking $t = 0$ as the instant when the capacitor is connected to the inductor.

•30.28 A 4-mF capacitor is connected in series with a 7-mH inductor. The peak current in the wires between the capacitor and the inductor is 3 A.

a) What is the total electric energy in this circuit?

b) Write an expression for the charge on the capacitor as a function of time, assuming the capacitor is fully charged at $t = 0$ s.

Section 30.3

30.29 A series RLC circuit has resistance R, inductance L, and capacitance C. At what time does the energy in the circuit reach half of its initial value?

•30.30 An RLC oscillator circuit contains a 50-Ω resistor and a 1-mH inductor. What capacitance is necessary for the time constant of the circuit (the 1/e value) to be equal to the oscilla-

tion period? Plot the voltage across the resistor as a function of time.

•30.31 A 2.00-μF capacitor was fully charged by being connected to a 12.0-V battery. The fully charged capacitor is then connected in series with a resistor and an inductor: $R = 50$ Ω and $L = 0.200$ H. Calculate the damped frequency of the resulting circuit.

•30.32 An LC circuit consists of a capacitor, $C = 2.50$ μF, and an inductor, $L = 4.0$ mH. The capacitor is fully charged using a battery and then connected to the inductor. An oscilloscope is used to measure the frequency of the oscillations in the circuit. Next, the circuit is opened, and a resistor, R, is inserted in series with the inductor and the capacitor. The capacitor is again fully charged using the same battery and then connected to the circuit. The angular frequency of the damped oscillations in the RLC circuit is found to be 20% less than the angular frequency of the oscillations in the LC circuit.

a) Determine the resistance of the resistor.

b) How long after the capacitor is reconnected in the circuit will the amplitude of the damped current through the circuit be 50% of the initial amplitude?

c) How many complete damped oscillations will have occurred in that time?

Section 30.4

30.33 At what frequency will a 10-μF capacitor have reactance $X_C = 200$ Ω?

30.34 A capacitor with capacitance $C = 5 \cdot 10^{-6}$ F is connected to an AC power source having a peak value of 10 V and $f = 100$ Hz. Find the reactance of the capacitor and the maximum current in the circuit.

30.35 An inductor with inductance $L = 47$ mH is connected to an AC power source having a peak value of 12 V and $f = 1000$ Hz. Find the reactance of the inductor and the maximum current in the circuit.

Section 30.5

30.36 The figure shows a circuit with a source of constant emf connected in series to a resistor, an inductor, and a capacitor. What is the steady-state current flow through the circuit?

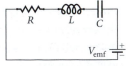

30.37 A series circuit contains a 100.0-Ω resistor, a 0.500-H inductor, a 0.400-μF capacitor, and a time-varying source of emf providing 40.0 V.

a) What is the resonant angular frequency of the circuit?

b) What current will flow through the circuit at the resonant frequency?

30.38 A variable capacitor used in an RLC circuit produces a resonant frequency of 5.0 MHz when its capacitance is set to 15 pF. What will the resonant frequency be when the capacitance is increased to 380 pF?

30.39 Determine the phase constant and the impedance of the RLC circuit shown in the figure when the frequency of the time-varying emf is 1 kHz, $C = 100 \ \mu F$, $L = 10$ mH, and $R = 100 \ \Omega$.

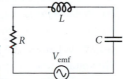

30.40 What is the resonant frequency of the series RLC circuit of Problem 30.39 if $C = 4 \ \mu F$, $L = 5$ mH, and $R = 1$ kΩ? What is the maximum current in the circuit if $V_m = 10$ V at the resonant frequency?

•30.41 In a series RLC circuit, $V = (12 \ V)(\sin \omega t)$, $R = 10 \ \Omega$, $L = 2$ H, and $C = 10 \ \mu F$. At resonance, determine the voltage amplitude across the inductor. Is the result reasonable, considering that the voltage supplied to the entire circuit has an amplitude of 12 V?

•30.42 An AC power source with $V_m = 220$ V and $f = 60.0$ Hz is connected in a series RLC circuit. The resistance, R, inductance, L, and capacitance, C, of this circuit are, respectively, 50.0 Ω, 0.200 H, and 0.040 mF. Find each of the following quantities:

a) the inductive reactance

b) the capacitive reactance

c) the impedance of the circuit

d) the maximum current through the circuit

e) the maximum potential difference across each circuit element

•30.43 The series RLC circuit shown in the figure has $R = 2.20 \ \Omega$, $L = 9.30$ mH, $C = 2.27$ mF, $V_m = 110$ V, and $\omega = 377$ rad/s.

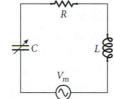

a) What is the maximum current, I_m, in this circuit?

b) What is the phase constant, 0, between the voltage and the current?

c) The capacitance, C, can be varied. What value of C will allow the largest current amplitude oscillations to occur, and what are the magnitudes of this current, I'_m, and the phase angle, ϕ', between the current and the voltage?

•30.44 Design an RC band-pass filter that passes a signal with frequency 5 kHz, has a ratio $V_{out}/V_{in} = 0.5$, and has an impedance of 1.0 kΩ at very high frequencies.

a) What components will you use?

b) What is the phase of V_{out} relative to V_{in} at the frequency of 5 kHz?

••30.45 Design an RC high-pass filter that rejects 60 Hz line noise from a circuit used in a detector. Your criteria are reduction of the amplitude of the line noise by a factor of 1000 and total impedance at high frequencies of 2.0 kΩ.

a) What components will you use?

b) What is the frequency range of the signals that will be passed with at least 90% of their amplitude?

Section 30.6

30.46 What is the maximum value of the AC voltage whose root-mean-square value is (a) 110 V or (b) 220 V?

30.47 The quality factor, Q, of a circuit can be defined by $Q = \omega(U_E + U_B)/P$. Express the quality factor of a series RLC circuit in terms of its resistance R, inductance L, and capacitance C.

30.48 A label on a hair dryer reads "110V 1250W." What is the peak current in the hair dryer, assuming that it behaves like a resistor?

30.49 A radio tuner has a resistance of 1.00 $\mu\Omega$, a capacitance of 25.0 nF, and an inductance of 3.00 mH.

a) Find the resonant frequency of this tuner.

b) Calculate the power in the circuit if a signal at the resonant frequency produces a voltage across the antenna of 1.50 mV.

•30.50 A circuit contains a 100-Ω resistor, a 0.0500-H inductor, a 0.400-μF capacitor, and a source of time-varying emf connected in series. The time-varying emf is 50.0 V at a frequency of 2000 Hz.

a) Determine the current in the circuit.

b) Determine the voltage drop across each component of the circuit.

c) How much power is drawn from the source of emf?

•30.51 The figure shows a simple FM antenna circuit in which $L = 8.22 \ \mu H$ and C is variable (the capacitor can be tuned to receive a specific station). The radio signal from your favorite FM station produces a sinusoidal time-varying emf with an amplitude of 12.9 μV and a frequency of 88.7 MHz in the antenna.

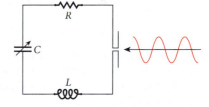

a) To what value, C_0, should you tune the capacitor in order to best receive this station?

b) Another radio station's signal produces a sinusoidal time-varying emf with the same amplitude, 12.9 μV, but with a frequency of 88.5 MHz in the antenna. With the circuit tuned to optimize reception at 88.7 MHz, what should the value, R_0, of the resistance be in order to reduce by a factor of 2 (compared to the current if the circuit were optimized for 88.5 MHz) the current produced by the signal from this station?

Section 30.7

30.52 The transmission of electric power occurs at the highest possible voltage to reduce losses. By how much could the power loss be reduced by raising the voltage by a factor of 10?

30.53 Treat the solenoid and coil of Solved Problem 29.1 as a transformer.

a) Find the root-mean-square voltage in the coil if the solenoid has a root-mean-square voltage of 120 V and a frequency of 60 Hz. The length of the solenoid is 12.0 cm.

b) What is the voltage in the coil if the frequency is 0 Hz (DC current)?

30.54 A transformer has 800 turns in the primary coil and 40 turns in the secondary coil.

a) What happens if an AC voltage of 100 V is across the primary coil?

b) If the initial AC current is 5 A, what is the output current?

c) What happens if a DC current at 100 V flows into the primary coil?

d) If the initial DC current is 5 A, what is the output current?

30.55 A transformer contains a primary coil with 200 turns and a secondary coil with 120 turns. The secondary coil drives a current I through a 1-kΩ resistor. If an input voltage $V_{rms} = 75$ V is applied across the primary coil, what is the power dissipated in the resistor?

Section 30.8

30.56 Consider the filtered fullwave rectifier shown in the figure. If the frequency of the source of time-varying emf is 60. Hz, what is the frequency of the resulting current?

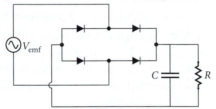

•30.57 A voltage $V_{rms} = 110$ V at a frequency of 60. Hz is applied to the primary coil of a transformer. The transformer has a ratio $N_P/N_S = 11$. The secondary coil is used as the source of V_{emf} for the filtered fullwave rectifier of Problem 30.56.

a) What is the maximum voltage in the secondary coil of the transformer?

b) What is the DC voltage provided to the resistor?

Additional Problems

30.58 A vacuum cleaner motor can be viewed as an inductor with an inductance of 100 mH. For a 60-Hz AC voltage, $V_{rms} = 115$ V, what capacitance must be in series with the motor to maximize the power output of the vacuum cleaner?

30.59 When you turn the dial on a radio to tune it, you are adjusting a variable capacitor in an LC circuit. Suppose you tune to an AM station broadcasting at a frequency of 1000 kHz, and there is a 10-mH inductor in the tuning circuit. When you have tuned in the station, what is the capacitance of the capacitor?

30.60 A series RLC circuit has a source of time-varying emf providing 12 V at a frequency f_0, with $L = 7$ mH, $R = 100$ Ω, and $C = 0.05$ mF.

a) What is the resonant frequency of this circuit?

b) What is the average power dissipated in the resistor at this resonant frequency?

30.61 What are the maximum values of (a) current and (b) voltage when an incandescent 60-W light bulb (at 110 V) is connected to a wall plug labeled 110 V?

30.62 A 360-Hz source of emf is connected in a circuit consisting of a capacitor, a 25-mH inductor, and an 0.80-Ω resistor. For the current and voltage to be in phase what should the value of C be?

30.63 What is the resistance in an RLC circuit with $L = 65$ mH and $C = 1$ μF if the circuit loses 3.5% of its total energy as thermal energy in each cycle?

30.64 A transformer with 400 turns in its primary coil and 20 turns in its secondary coil is designed to deliver an average power of 1200 W with a maximum voltage of 60 V. What is the maximum current in the primary coil?

30.65 A 5-μF capacitor in series with a 4-Ω resistor is charged with a 9-V battery for a long time by closing the switch (position a in the figure). The capacitor is then discharged through an inductor ($L = 40$ mH) by closing the switch (position b) at $t = 0$.

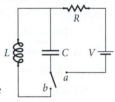

a) Determine the maximum current through the inductor.

b) What is the first time at which the current is at its maximum?

•30.66 In the RC high-pass filter shown in the figure, $R = 10$ kΩ and $C = 0.047$ μF. What is the 3-dB frequency of this circuit (where dB means basically the same for electric current as it did for sound in Chapter 16)? That is, at what frequency does the ratio of output voltage to input voltage satisfy $20 \log (V_{out}/V_{in}) = -3$?

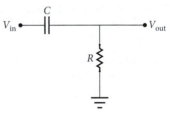

•30.67 The discussion of RL, RC, and RLC circuits in this chapter has assumed a purely resistive resistor, one whose inductance and capacitance are exactly zero. While the capacitance of a resistor can generally be neglected, inductance is an intrinsic part of the resistor. Indeed, one of the most widely used resistors, the wire-wound resistor, is nothing but a solenoid made of highly resistive wire. Suppose a wire-wound resistor of unknown resistance is connected to a DC power supply. At a voltage of $V = 10$ V across the resistor, the current through the resistor is 1 A. Next, the same resistor is connected to an AC power source providing $V_{rms} = 10$ V at a variable frequency. When the frequency is 20 kHz, a current, $I_{rms} = 0.8$ A, is measured through the resistor.

a) Calculate the resistance of the resistor.

b) Calculate the inductive reactance of the resistor.

c) Calculate the inductance of the resistor.

d) Calculate the frequency of the AC power source at which the inductive reactance of the resistor exceeds its resistance.

•**30.68** An RLC circuit has a capacitor, a resistor, and an inductor connected in parallel, as shown in the figure, and connected to a time-varying source of emf providing V_{rms} at frequency f. Find an expression for I_{rms} in terms of V_{rms}, f, L, C, and R.

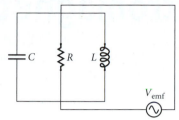

•**30.69** a) A loop of wire 5 cm in diameter is carrying a current of 2 A. What is the energy density of the magnetic field at its center?

b) What current has to flow in a straight wire to produce the same energy density at a point 4 cm from the wire?

•**30.70** A 75,000-W light bulb (yes, there are such things!) operates at $I_{rms} = 200$. A and $V_{rms} = 440$. V in a 60.0-Hz AC circuit. Find the resistance, R, and self-inductance, L, of this bulb. Its capacitive reactance is negligible.

•**30.71** Show that the power dissipated in a resistor connected to an AC power source with frequency ω oscillates with frequency 2ω.

•**30.72** A 300-Ω resistor is connected in series with a 4.0-μF capacitor and a source of time-varying emf providing 40.0 V.

a) At what frequency will the potential drop across the capacitor equal that across the resistor?

b) What is the current through the circuit when this occurs?

•**30.73** An electromagnet consists of 200 loops and has a length of 10 cm and a cross-sectional area of 5.0 cm^2. Find the resonant frequency of this electromagnet when it is attached to the Earth (treat the Earth as a spherical capacitor).

•**30.74** Laboratory experiments with series RLC circuits require some care, as these circuits can produce large voltages at resonance. Suppose you have a 1.00-H inductor (not difficult to obtain) and a variety of resistors and capacitors. Design a series RLC circuit that will resonate at a frequency (not an angular frequency) of 60.0 Hz and will produce at resonance a magnification of the voltage across the capacitor or the inductor by a factor of 20.0 times the input voltage or the voltage across the resistor.

•**30.75** A particular RC low-pass filter has a breakpoint frequency of 200 Hz. At what frequency will the output voltage divided by the input voltage be 0.10?

•**30.76** In a certain RLC circuit, a 20-Ω resistor, a 10-mH inductor, and a 5-μF capacitor are connected in series with an AC power source for which $V_{rms} = 10$ V and $f = 100$ Hz. Calculate

a) the amplitude of the current,

b) the phase between the current and the voltage, and

c) the maximum voltage across each component.

31

Electromagnetic Waves

FIGURE 31.1 Comet Hale-Bopp showing its striking two tails.

WHAT WE WILL LEARN

- Changing electric fields induce magnetic fields, and changing magnetic fields induce electric fields.

- Maxwell's equations describe electromagnetic phenomena.

- Electromagnetic waves have both electric and magnetic fields.

- Solutions of Maxwell's equations can be expressed in terms of sinusoidally varying traveling waves.

- For an electromagnetic wave, the electric field is perpendicular to the magnetic field and both fields are perpendicular to the direction in which the wave is traveling.

- The speed of light can be expressed in terms of constants related to electric and magnetic fields.

- Light is an electromagnetic wave.

- Electromagnetic waves can transport energy and momentum.

- The intensity of an electromagnetic wave is proportional to the square of the root-mean-square magnitude of the electric field of the wave.

- The direction of the electric field of a traveling electromagnetic wave is called the polarization direction.

Comet Hale-Bopp made a spectacular appearance in 1997 (Figure 31.1). Particularly striking in the photo are the comet's two tails. The white tail follows the comet's trajectory. It consists of dust from the comet's body, evaporated by the Sun's heat and illuminated by sunlight. The blue tail points directly away from the Sun and consists of gas ionized by the solar wind—the stream of high-energy particles emitted by the Sun (mentioned in Section 27.1 in connection with Earth's magnetic field).

In this chapter, we'll examine the nature of light, including how it can transmit energy and pressure to other objects. We'll see that light is one type of wave, called an *electromagnetic wave*, consisting of interacting electric and magnetic fields. Other types of electromagnetic waves that are also familiar to you range from TV and radio waves to microwaves to X-rays. We'll see how the various types are alike and how they are different. The next chapter is the first to focus on optics—the properties and behavior of light—and many of those properties apply to other electromagnetic waves as well.

Note that most of the results in this chapter apply only to electromagnetic waves propagating through vacuum. For all practical purposes, there is no difference for electromagnetic waves propagating through media like Earth's atmosphere. There are some significant differences for electromagnetic waves propagating through other media, but these are not addressed in this chapter.

31.1 Induced Magnetic Fields

In Chapter 29, we saw that a changing magnetic field induces an electric field. According to Faraday's Law of Induction,

$$\oint \vec{E} \cdot d\vec{s} = -\frac{d\Phi_B}{dt},$$
(31.1)

where $\vec{E}$ is the electric field induced around a closed loop by the changing magnetic flux, Φ_B, through that loop. In a similar way, a changing electric field induces a magnetic field. **Maxwell's Law of Induction** (named for British physicist James Clerk Maxwell, 1831–1879) describes this phenomenon as follows:

$$\oint \vec{B} \cdot d\vec{s} = \mu_0 \epsilon_0 \frac{d\Phi_E}{dt},$$
(31.2)

where $\vec{B}$ is the magnetic field induced around a closed loop by a changing electric flux, Φ_E, through that loop. This equation is similar to equation 31.1 except for the constant $\mu_0 \epsilon_0$ and the lack of a negative sign. The constant is a consequence of the SI units used for magnetic fields. The fact that the right-hand side of equation 31.2 does not have a negative sign implies that the induced magnetic field has the opposite sign from that of

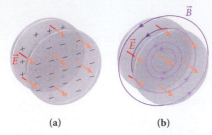

FIGURE 31.2 (a) A charged circular capacitor. The red arrows represent the electric field between the plates. (b) A capacitor with charge increasing with time. The red arrows represent the electric field, and the purple loops represent the induced magnetic field.

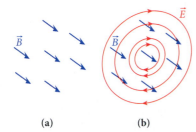

FIGURE 31.3 (a) A constant uniform magnetic field. (b) A uniform magnetic field increasing with time, which induces an electric field represented by the red loops.

an induced electric field when both are induced under similar conditions, as we'll see shortly.

First, note that it is *not at all obvious* that equation 31.2 can be written in analogy to equation 31.1. When Maxwell first wrote this equation, it represented a major step forward in the unification of electricity and magnetism. Faraday discovered his law in 1831, but it took a quarter of a century for Maxwell to come up with the counterpart. What is stated here as simple fact is in reality the first great conceptual leap toward the unification of all physical forces of nature. This unification began with Maxwell's work one and a half centuries ago and continues in modern physics research today.

A circular capacitor can be used to illustrate an induced magnetic field (Figure 31.2). For the capacitor shown in Figure 31.2a, the charge is constant, and a constant electric field appears between the plates. There is no magnetic field. For the capacitor shown in Figure 31.2b, the charge is increasing with time. Thus, the electric flux between the plates is increasing with time. A magnetic field, $\vec{B}$, is induced, represented by the purple loops, which also indicates the direction of $\vec{B}$. Along each loop, the magnetic field vector has the same magnitude and is directed tangentially to the loop. When the charge stops increasing, the electric flux remains constant and the magnetic field disappears.

Next, consider a uniform magnetic field that is also constant in time as in Figure 31.3a. In Figure 31.3b, the magnetic field is still uniform in space but is increasing with time, which induces an electric field, shown by the red loops. The electric field vector has constant magnitude along each loop and is directed tangentially to the loops as shown. Note that this induced electric field points in the opposite direction from the induced magnetic field caused by an increasing electric field (Figure 31.2b).

Now recall Ampere's Law:

$$\oint \vec{B} \cdot d\vec{s} = \mu_0 i_{\text{enc}}, \tag{31.3}$$

which relates the integral around a loop of the dot product of the magnetic field and the differential displacement along the loop, $\vec{B} \cdot d\vec{s}$, to the current flowing through the loop.

However, Maxwell realized that this equation is incomplete because it does not account for contributions to the magnetic field caused by changing electric fields. Equations 31.2 and 31.3 can be combined to produce a description of magnetic fields created by moving charges and by changing electric fields:

$$\oint \vec{B} \cdot d\vec{s} = \mu_0 \epsilon_0 \frac{d\Phi_E}{dt} + \mu_0 i_{\text{enc}}. \tag{31.4}$$

Equation 3.14 is called the **Maxwell-Ampere Law.** You can see that for the case of constant current, such as current flowing in a conductor, this equation reduces to Ampere's Law. For the case of a changing electric field without current flowing, such as the electric field between the plates of a capacitor, this equation reduces to Maxwell's Law of Induction. It is important to realize that the Maxwell-Ampere Law describes two different sources of magnetic field: the conventional current (as discussed in Chapter 28) and the time-varying electric flux (examined in more detail in the following section).

31.2 Displacement Current

Looking at the Maxwell-Ampere Law (equation 31.4), you can see that the term $\epsilon_0 \, d\Phi_E/dt$ on the right-hand side of the equation must have the units of current. Although no actual "current" is "displaced," this term is called the **displacement current,** i_d:

$$i_{\text{d}} = \epsilon_0 \frac{d\Phi_E}{dt}. \tag{31.5}$$

With this definition, we can rewrite equation 31.4 as

$$\oint \vec{B} \cdot d\vec{s} = \mu_0 \left(i_{\text{d}} + i_{\text{enc}} \right).$$

Again, let's consider a parallel plate capacitor with circular plates, now placed in a circuit in which a current, i, is flowing while the capacitor is charging (Figure 31.4). For a parallel

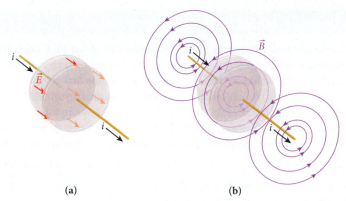

(a)   **(b)**

FIGURE 31.4 A parallel plate capacitor in a circuit being charged by a current, i: (a) the electric field between the plates at a given instant; (b) the magnetic field around the wires and between the plates of the capacitor.

plate capacitor, the charge, q, is related to the electric field between the plates, E, as follows (see Chapter 24)

$$q = \epsilon_0 A E$$

where A is the area of the plates. The current, i, in the circuit can be obtained by taking the time derivative of this equation:

$$i = \frac{dq}{dt} = \epsilon_0 A \frac{dE}{dt}. \tag{31.6}$$

Assuming that the electric field between the plates of the capacitor is uniform, we can obtain an expression for the displacement current:

$$i_d = \epsilon_0 \frac{d\Phi_E}{dt} = \epsilon_0 \frac{d(AE)}{dt} = \epsilon_0 A \frac{dE}{dt}. \tag{31.7}$$

Thus, the current in the circuit, i, given by equation 31.6, is equal to the displacement current, i_d, given by equation 31.7. Although no actual current is flowing between the plates of the capacitor, in the sense that no actual charges move across the capacitor gap from one plate to the other, the displacement current can be used to calculate the induced magnetic field.

To calculate the magnetic field between the two circular plates of the capacitor, we assume that the volume between the two plates can be replaced with a conductor of radius R carrying current i_d. In Chapter 28, we saw that the magnetic field at a perpendicular distance r from the center of the capacitor is given by

$$B = \left(\frac{\mu_0 i_d}{2\pi R^2} \right) r \quad \text{(for } r < R\text{)}.$$

The system outside the capacitor can be treated as a current-carrying wire; so the magnetic field at a perpendicular distance r from the wire is

$$B = \frac{\mu_0 i_d}{2\pi r} \quad \text{(for } r > R\text{)}.$$

31.1 In-Class Exercise

The displacement current, i_d, for the charging circular capacitor with radius R shown in the figure is equal to the conduction current, i, in the wires. Points 1 and 3 are located a perpendicular distance r from the wires, and point 2 is located the same perpendicular distance r from the center of the capacitor, such that $r > R$. Rank the magnetic fields at points 1, 2, and 3, from largest magnitude to smallest.

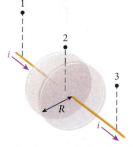

a) $B_1 > B_2 > B_3$

b) $B_3 > B_2 > B_1$

c) $B_1 = B_3 > B_2$

d) $B_2 > B_1 = B_3$

e) $B_1 = B_2 = B_3$

31.2 In-Class Exercise

The displacement current, i_d, for the charging circular capacitor with radius R shown in the figure is equal to the conduction current, i, in the wires. Points 1 and 3 are located a perpendicular distance r from the wires, and point 2 is located the same perpendicular distance r from the center of the capacitor, such that $r < R$. Rank the magnetic fields at points 1, 2, and 3, from largest magnitude to smallest.

a) $B_1 > B_2 > B_3$

b) $B_3 > B_2 > B_1$

c) $B_1 = B_3 > B_2$

d) $B_2 > B_1 = B_3$

e) $B_1 = B_2 = B_3$

31.3   Maxwell's Equations

The Maxwell-Ampere Law (equation 31.4) completes the set of four equations known as **Maxwell's equations,** which describe the interactions between electrical charges, currents, electric fields, and magnetic fields. These equations treat electricity and magnetism as two aspects of a unified force called **electromagnetism.** All of the results described previously for electricity and magnetism are still valid, but these equations show how electric and magnetic fields interact with each other, giving rise to a broad range of electromagnetic phenomena. This chapter focuses on electromagnetic waves and Chapter 34 on wave optics. A summary of Maxwell's equations is given in Table 31.1. (Again, as a reminder, the $\oiint d\vec{A}$ in the first two equations represents integration over a closed surface, and $\oint d\vec{s}$ in the last two equations indicates integration over a closed curve.)

If you scrutinize Maxwell's equations, you might notice a lack of symmetry between $\vec{E}$ and $\vec{B}$. This difference arises from the fact that electric charges exist in isolation and a corresponding current appears when charges move, but apparently no isolated, stationary magnetic charges occur in nature. Particles that hypothetically have a single magnetic charge (a north pole or a south pole, but not both) are called *magnetic monopoles,* but empirically it has been found that magnetic poles always come in pairs, a north pole together with a south pole. There is no fundamental reason for the absence of magnetic monopoles, and many experiments have searched unsuccessfully for them. The most sensitive of these experiments was MACRO, which used a massive detector that operated for many years in a laboratory located deep under the Gran Sasso mountain in Italy. MACRO searched for magnetic monopoles in cosmic rays, without success.

We'll now begin our study of **electromagnetic waves.** Electromagnetic waves consist of electric and magnetic fields, can travel through vacuum without any supporting medium, and do not involve moving charges or currents. The existence of electromagnetic waves was first demonstrated in 1888 by the German physicist Heinrich Hertz (1857–1894). Hertz used an RLC circuit that induced a current in an inductor that drove a spark gap. A spark gap consists of two electrodes that, when a potential difference is applied across them, produce a spark by exciting the gas between the electrodes. Hertz placed a loop and a small spark gap several meters apart. He observed that sparks were induced in the remote loop in a pattern that correlated with the electromagnetic oscillations in the primary RLC circuit. Thus, electromagnetic waves were able to travel through space without any medium to support them. For this contribution and others, the basic unit of oscillation, cycles per second, was named the hertz (Hz) in his honor.

Table 31.1	Maxwell's Equations Describing Electromagnetic Phenomena	
Name	Equation	Description
Gauss's Law for Electric Fields	$\oiint \vec{E} \cdot d\vec{A} = \dfrac{q_{enc}}{\epsilon_0}$	The net electric flux through a closed surface is proportional to the net enclosed electric charge.
Gauss's Law for Magnetic Fields	$\oiint \vec{B} \cdot d\vec{A} = 0$	The net magnetic flux through a closed surface is zero (no magnetic monopoles exist).
Faraday's Law of Induction	$\oint \vec{E} \cdot d\vec{s} = -\dfrac{d\Phi_B}{dt}$	An electric field is induced by a changing magnetic flux.
Maxwell-Ampere Law	$\oint \vec{B} \cdot d\vec{s} = \mu_0 \epsilon_0 \dfrac{d\Phi_E}{dt} + \mu_0 i_{enc}$	A magnetic field is induced by a changing electric flux or by a current.

31.4   Wave Solutions to Maxwell's Equations

As Section 31.11 will show, it is possible to use advanced calculus to derive a general wave equation from Maxwell's equations starting with Maxwell's equations in differential form. However, we'll first assume that electromagnetic waves propagating in vacuum (no moving charges or currents) have the form of a traveling wave and show that this form satisfies Maxwell's equations.

Proposed Solution

We assume the following equations express the electric and magnetic fields in a particular electromagnetic wave that happens to be traveling in the positive x-direction:

$$\vec{E}(\vec{r},t) = E_{max} \sin\left(\kappa x - \omega t\right)\hat{y}$$

and

$$\vec{B}(\vec{r},t) = B_{max} \sin\left(\kappa x - \omega t\right)\hat{z}, \qquad (31.8)$$

where $\kappa = 2\pi/\lambda$ is the wave number and $\omega = 2\pi f$ is the angular frequency of a wave with wavelength λ and frequency f. Note that the magnitudes of both fields have no dependence on the y- or z-coordinates, only on the x-coordinate and time. This type of wave, in which the electric and magnetic field vectors lie in a plane, is called a **plane wave.** Equation 31.8 indicates that this particular electromagnetic wave is traveling in the positive x-direction because, as the time t increases, the coordinate x has to increase to maintain the same value for the fields. The wave described by equation 31.8 is shown in Figure 31.5.

In the particular case illustrated in Figure 31.5, the electric field is completely in the y-direction and the magnetic field is completely in the z-direction; that is, both fields are perpendicular to the direction of wave propagation. It turns out that the electric field is always perpendicular to the direction the wave is traveling and is always perpendicular to the magnetic field. However, in general, for an electromagnetic wave that is propagating along the x-axis, the electric field can point anywhere in the yz-plane.

The representation of the wave in Figure 31.5 is an instantaneous abstraction. The vectors shown represent the magnitude and direction for the electric and magnetic fields; however, you should realize that these fields are not solid objects. Nothing made of matter actually moves left and right or up and down as the wave travels. The vectors pointing left and right and up and down represent the electric and magnetic fields.

Showing that the traveling wave described by equation 31.8 satisfies all of Maxwell's equations involves quite a bit of vector calculus but also uses many of the concepts that have been developed in the preceding chapters. The following subsections work through this process in detail, one equation at a time.

Gauss's Law for Electric Fields

Let's start with Gauss's Law for Electric Fields. For an electromagnetic wave in vacuum, there is no enclosed charge anywhere ($q_{enc} = 0$); thus, we must show that the proposed solution of equation 31.8 satisfies

$$\oiint \vec{E} \cdot d\vec{A} = 0. \qquad (31.9)$$

We choose a rectangular box as a Gaussian surface enclosing a portion of the vector representation of the wave (Figure 31.6). For the faces of the box in the yz-plane, $\vec{E} \cdot d\vec{A}$ is zero because the vectors $\vec{E}$ and $d\vec{A}$ are perpendicular to each other. The same is true for the faces in the xy-plane. The faces in the xz-plane contribute $+EA_1$ and $-EA_1$, where A_1 is the area of the top face and the bottom face. Thus, the integral is zero, and Gauss's Law for Electric Fields is satisfied.

If we analyzed the vector representation at different times, we would get a different electric field. However, because the electric field is always in the y-direction, the integral will always be zero.

Gauss's Law for Magnetic Fields

For Gauss's Law for Magnetic Fields, we must show that

$$\oiint \vec{B} \cdot d\vec{A} = 0. \qquad (31.10)$$

We again use the closed surface in Figure 31.6 for the integration. For the faces in the yz-plane and for the faces in the xz-plane, $\vec{B} \cdot d\vec{A}$ is zero because the vectors $\vec{B}$ and $d\vec{A}$ are perpendicular to each other. The faces in the xy-plane contribute $+BA_2$ and $-BA_2$, where

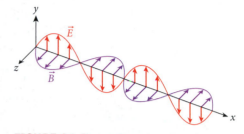

FIGURE 31.5 Representation of an electromagnetic wave traveling in the positive x-direction at a given instant.

31.3 In-Class Exercise

An electromagnetic plane wave is traveling through vacuum. The electric field of the wave is given by $\vec{E} = E_{max} \cos\left(\kappa x - \omega t\right)\hat{y}$. Which of the following equations describes the magnetic field of the wave?

a) $\vec{B} = B_{max} \cos\left(\kappa x - \omega t\right)\hat{x}$

b) $\vec{B} = B_{max} \cos\left(\kappa y - \omega t\right)\hat{y}$

c) $\vec{B} = B_{max} \cos\left(\kappa z - \omega t\right)\hat{z}$

d) $\vec{B} = B_{max} \cos\left(\kappa y - \omega t\right)\hat{z}$

e) $\vec{B} = B_{max} \cos\left(\kappa x - \omega t\right)\hat{z}$

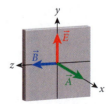

FIGURE 31.6 Gaussian surface (gray box) around a portion of the vector representation of an electromagnetic wave traveling in the positive x-direction. The area vector is shown for the front face of the Gaussian surface.

A_2 is the area of each of the two faces in the xy-plane. Thus, the integral is zero, and Gauss's Law for Magnetic Fields is satisfied.

Faraday's Law of Induction

Now let's address Faraday's Law of Induction:

$$\oint \vec{E} \cdot d\vec{s} = -\frac{d\Phi_B}{dt}. \tag{31.11}$$

To evaluate the integral on the left-hand side of this equation, we assume a closed loop in the xy-plane that has width dx and height h and goes from a to b to c to d and back to a, as depicted by the gray rectangle in Figure 31.7. The differential area $d\vec{A} = \hat{n}\,dA = \hat{n}h\,dx$ of this rectangle has its unit surface normal vector, $\hat{n}$, pointing in the positive z-direction. Note that the electric and magnetic fields change with distance along the x-axis. Thus, going from point x to point $x + dx$, the electric field changes from $\vec{E}(x)$ to $\vec{E}(x+dx) = \vec{E}(x) + d\vec{E}$.

To evaluate the integral of equation 31.11 over the closed loop, we split the loop into four pieces, integrating counterclockwise from a to b, b to c, c to d, and d to a. The contributions to the integral that are parallel to the x-axis, from integrating from b to c and from d to a, are zero because the electric field is always perpendicular to the direction of integration. For the integrations in the y-direction, a to b and c to d, the electric field is parallel or antiparallel to the direction of integration; therefore, the scalar product reduces to a conventional product. Because the electric field is independent of the y-coordinate, it can be taken out of the integral. Thus, the integral along each of the segments in the y-direction is a simple product of the integrand (the magnitude of the electric field at the corresponding x-coordinate) and the length of the integration interval (h), times -1 for the integration in the negative y-direction because $\vec{E}$ is antiparallel to the direction of integration. Thus, the integral evaluates to

$$\oint \vec{E} \cdot d\vec{s} = E\int_a^b ds - E\int_c^d ds = (E + dE)(h) - Eh = (dE)(h).$$

The right-hand side of equation 31.11 is given by

$$-\frac{d\Phi_B}{dt} = -A\frac{dB}{dt} = -(h)(dx)\frac{dB}{dt}.$$

Thus, we have

$$(h)(dE) = -(h)(dx)\frac{dB}{dt}$$

or

$$\frac{dE}{dx} = -\frac{dB}{dt}. \tag{31.12}$$

The derivatives dE/dx and dB/dt are each taken with respect to a single variable, although both E and B depend on both x and t. Thus, we can more appropriately write equation 31.12 using partial derivatives:

$$\frac{\partial E}{\partial x} = -\frac{\partial B}{\partial t}. \tag{31.13}$$

Using the assumed forms for the electric and magnetic fields (equation 31.8), we can expand the partial derivatives:

$$\frac{\partial E}{\partial x} = \frac{\partial}{\partial x}\left(E_{max}\sin(\kappa x - \omega t)\right) = \kappa E_{max}\cos(\kappa x - \omega t),$$

and

$$\frac{\partial B}{\partial t} = \frac{\partial}{\partial t}\left(B_{max}\sin(\kappa x - \omega t)\right) = -\omega B_{max}\cos(\kappa x - \omega t).$$

Substituting these expressions into equation 31.13 gives

$$\kappa E_{max}\cos(\kappa x - \omega t) = -\left[-\omega B_{max}\cos(\kappa x - \omega t)\right].$$

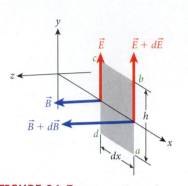

FIGURE 31.7 Two snapshots of the electric and magnetic fields in an electromagnetic wave. The gray area represents an integration loop for Faraday's Law.

The angular frequency and wave number are related via

$$\frac{\omega}{\kappa} = \frac{2\pi f}{\left(2\pi/\lambda\right)} = f\lambda = c,$$ (31.14)

where c is the speed of the wave. (In general, we could use the v for the speed of this wave. However, we choose to use c, because, as we'll see, all electromagnetic waves propagate in vacuum with a characteristic speed, the speed of light, which is conventionally represented by c.) Thus, we have

$$\frac{E_{max}}{B_{max}} = \frac{\omega}{\kappa} = c.$$

We can use equation 31.8 to rewrite this equation in terms of the ratio of the magnitudes of the fields at a fixed place and time as

$$\frac{E}{B} = \frac{\left|\vec{E}\left(\vec{r},t\right)\right|}{\left|\vec{B}\left(\vec{r},t\right)\right|} = \frac{E_{max}\left|\sin\left(\kappa x - \omega t\right)\right|}{B_{max}\left|\sin\left(\kappa x - \omega t\right)\right|} = c.$$ (31.15)

Thus, equation 31.8 satisfies Faraday's Law of Induction if the ratio of the electric and magnetic field magnitudes is c.

Maxwell-Ampere Law

Finally, we address the Maxwell-Ampere Law. For electromagnetic waves, in which no current flows, we can write

$$\oint \vec{B} \cdot d\vec{s} = \mu_0 \epsilon_0 \frac{d\Phi_E}{dt}.$$ (31.16)

To evaluate the integral on the left-hand side of this equation, we assume a closed loop in the xz-plane that has width dx and height h, represented by the gray rectangle in Figure 31.8. The differential area of this rectangle is oriented in the positive y-direction.

The integral around the loop in the counterclockwise direction (a to b to c to d to a) is given by

$$\oint \vec{B} \cdot d\vec{s} = Bh - \left(B + dB\right)\left(h\right) = -\left(dB\right)\left(h\right).$$ (31.17)

As before, the parts of the loop parallel to the x-axis do not contribute to the integral. The right-hand side of equation 31.16 can be written as

$$\mu_0\epsilon_0 \frac{d\Phi_E}{dt} = \mu_0\epsilon_0 A \frac{dE}{dt} = \mu_0\epsilon_0 \left(h\right)\left(dx\right)\frac{dE}{dt}.$$ (31.18)

Substituting from equations 31.17 and 31.18 into equation 31.16, we get

$$-\left(dB\right)\left(h\right) = \mu_0\epsilon_0 \left(h\right)\left(dx\right)\frac{dE}{dt}.$$

Expressing this equation in terms of partial derivatives, as we did for equation 31.12, we get

$$-\frac{\partial B}{\partial x} = \mu_0\epsilon_0 \frac{\partial E}{\partial t}.$$

Now, using equation 31.8, we have

$$-\left[\kappa B_{max}\cos\left(\kappa x - \omega t\right)\right] = -\mu_0\epsilon_0\omega E_{max}\cos\left(\kappa x - \omega t\right),$$

or

$$\frac{E_{max}}{B_{max}} = \frac{\kappa}{\mu_0\epsilon_0\omega} = \frac{1}{\mu_0\epsilon_0 c}.$$

We can express this equation in terms of the electric and magnetic field magnitudes as before:

$$\frac{E}{B} = \frac{1}{\mu_0\epsilon_0 c}.$$ (31.19)

Equation 31.8 satisfies the Maxwell-Ampere Law if the ratio of the electric and magnetic field magnitudes is given by $1/\mu_0\epsilon_0 c$.

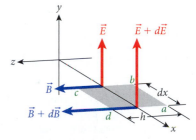

FIGURE 31.8 Representations of the electric and magnetic fields in an electromagnetic wave at a given instant. The gray area represents an integration loop for the Maxwell-Ampere Law.

31.5   The Speed of Light

From equations 31.15 and 31.19, we can conclude that

$$\frac{E}{B} = \frac{1}{\mu_0 \epsilon_0 c} = c,$$

which leads to

$$c = \frac{1}{\sqrt{\mu_0 \epsilon_0}}. \tag{31.20}$$

Thus, the speed of an electromagnetic wave can be expressed in terms of the two fundamental constants related to electric and magnetic fields: the magnetic permeability and the electric permittivity of free space (vacuum). Putting the accepted values of these constants into equation 31.20 gives

$$c = \frac{1}{\sqrt{\left(4\pi \cdot 10^{-7}\ \text{H/m}\right)\left(8.85 \cdot 10^{-12}\ \text{F/m}\right)}} = 3.00 \cdot 10^8\ \text{m/s}.$$

This calculated speed is equal to the measured speed of light. This equality means that all electromagnetic waves travel (in vacuum) at the speed of light and suggests that light is an electromagnetic wave.

Equation 31.15 states $E/B = c$. Even though c is a very large number, equation 31.15 does not mean that the electric field magnitude is much larger than the magnetic field magnitude. In fact, electric and magnetic fields are measured in different units, so a direct comparison is not possible.

The speed of light plays an important role in the theory of special relativity, which we'll examine in Chapter 35. The speed of light is always the same in any reference frame. Thus, if you send an electromagnetic wave out in a specific direction, any observer, regardless of whether that observer is moving toward you or away from you or in another direction, will see that wave moving at the speed of light. This amazing result, along with the plausible postulate that the laws of physics are the same for all inertial observers, leads to the theory of special relativity.

The speed of light can be measured extremely precisely, much more precisely than the meter could be determined from the original reference standard. Therefore, the speed of light is now defined to be precisely

$$c = 299{,}792{,}458\ \text{m/s}. \tag{31.21}$$

The definition of the meter is now simply the distance that light can traverse in vacuum in a time interval of $1/299{,}792{,}458$ s.

31.4 In-Class Exercise

What is the time required for laser light to travel from the Earth to the Moon and back again? The distance between the Earth and the Moon is $3.84 \cdot 10^8$ m.

a) 0.640 s   d) 15.2 s

b) 1.28 s   e) 85.0 s

c) 2.56 s

31.1 Self-Test Opportunity

The brightest star in the night sky is Sirius, which is at a distance of $8.30 \cdot 10^{16}$ m from Earth. When we see the light from this star, how far back in time (in years) are we looking?

31.6   The Electromagnetic Spectrum

All electromagnetic waves travel at the speed of light. However, the wavelength and the frequency of electromagnetic waves vary dramatically. The speed of light, c, the wavelength, λ, and the frequency, f, are related by

$$c = \lambda f. \tag{31.22}$$

Examples of electromagnetic waves are light, radio waves, microwaves, X-rays, and gamma rays. Three applications of electromagnetic waves are shown in Figure 31.9.

FIGURE 31.9 (a) Very Large Array radio telescope. (b) False color radar image of the surface of Venus. (c) X-ray image of a hand.

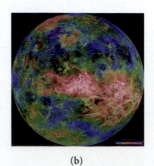
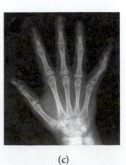

(a)   (b)   (c)

The **electromagnetic spectrum** is illustrated in Figure 31.10, which includes electromagnetic waves with wavelengths ranging from 1000 m and longer to less than 10^{-12} m, with corresponding frequencies ranging from 10^5 to 10^{20} Hz. Electromagnetic waves with wavelengths (and frequencies) in certain ranges are identified by characteristic names:

- **Visible light** refers to electromagnetic waves that we can see with our eyes, with wavelengths from 400 nm (blue) to 700 nm (red). The response of the human eye peaks at around 550 nm (green) and drops off quickly away from that wavelength. Other wavelengths of electromagnetic waves are invisible to the human eye. However, we can detect them by other means.

- **Infrared waves** (with wavelengths just longer than visible up to around 10^{-4} m) are felt as warmth. Detectors of infrared waves can be used to measure heat leaks in homes and offices, as well as locate brewing volcanoes. Many animals have developed the ability to see infrared waves, so they can see in the dark. Infrared beams are also used in automatic faucets in public restrooms and in remote control units for TV and DVD players.

- **Ultraviolet rays** with wavelengths just shorter than visible down to a few nanometers (10^{-9} m) can damage skin and cause sunburn. Fortunately, Earth's atmosphere, particularly its ozone layer, prevents most of the Sun's ultraviolet rays from reaching Earth's surface. Ultraviolet rays are used in hospitals to sterilize equipment and also produce optical properties such as fluorescence.

- **Radio waves** have frequencies ranging from several hundred kHz (AM radio) to 100 MHz (FM radio). They are also widely used in astronomy because they can pass through clouds of dust and gas that block visible light; the Very Large Array shown in Figure 31.9a is a collection of telescopes that utilize radio waves.

- **Microwaves,** used to pop popcorn in microwave ovens and transmit phone messages through relay towers or satellites, have frequencies around 10 GHz. Radar uses waves with wavelengths between those of radio waves and microwaves, which enable them to travel easily through the atmosphere and reflect off objects from the

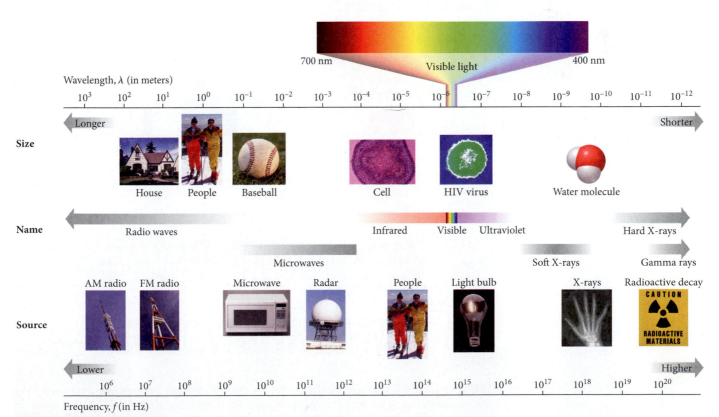

FIGURE 31.10 The electromagnetic spectrum.

31.2 Self-Test Opportunity

An FM radio station broadcasts at 90.5 MHz, and an AM radio station broadcasts at 870 kHz. What are the wavelengths of these electromagnetic waves?

size of a baseball to the size of a storm cloud. Figure 31.9b shows a radar image of the surface of Venus, which is always obscured by clouds that block visible light.

- **X-rays** used to produce medical images, such the one shown in Figure 31.9c, have wavelengths on the order of 10^{-10} m. This length is about the same as the distance between atoms in a solid crystal, so X-rays are used to determine the detailed molecular structure of any material that can be crystallized.

- **Gamma rays** emitted in the decay of radioactive nuclei have very short wavelengths, on the order of 10^{-12} m, and can cause damage to human cells. They are often used in medicine to destroy cancer cells or other malignant tissues that are hard to reach.

Communication Frequency Bands

The frequency ranges assigned to radio and television broadcasts are shown in Figure 31.11. The range of frequencies assigned to AM (amplitude modulation) radio is from 535 kHz to 1705 kHz. FM (frequency modulation) radio use the frequencies between 88.0 MHz and 108.0 MHz. VHF (very high frequency) television operates in two ranges: 54.0 MHz to 88.0 MHz for channels 2 through 6, and 174.0 MHz to 216.0 MHz for channels 7 through 13. UHF (ultra-high frequency) television channels 14 through 83 broadcast in the range from 512.0 MHz to 698.0 MHz. Most high-definition television (HDTV) broadcasts use the UHF band and channels 14 through 83.

A radio or television station transmits a carrier signal on a given frequency. The carrier signal is a sine wave with a frequency equal to the frequency of the broadcasting station. In the case of AM broadcasts, the amplitude of the carrier wave is modified by the information being transmitted, as illustrated in Figure 31.12a. The modulation of the amplitude of the carrier signal carries the transmitted message. Figure 31.12a shows a simple sine wave, indicating that a simple tone is being transmitted. The signal is received by a tuned RLC circuit whose resonant frequency is equal to the frequency of the carrier signal. The current induced in the circuit is proportional to the message being transmitted. AM transmission is vulnerable to noise and signal loss because the message is proportional to the amplitude of the signal, which can change if conditions vary.

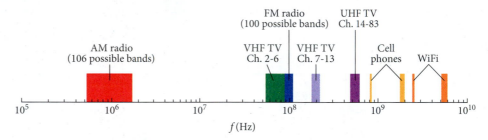

FIGURE 31.11 The frequency bands assigned to radio broadcasts, television broadcasts, cell phones, and WiFi computer connections in the United States.

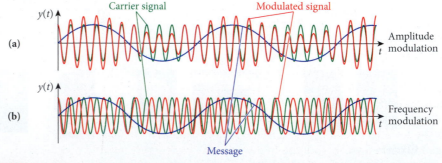

FIGURE 31.12 (a) Amplitude modulation. (b) Frequency modulation. For both cases, the green curve represents the carrier signal, the red curve represents the modulated signal, and the blue curve signifies the information being transmitted.

For FM transmission, the frequency of the carrier signal is modified by the message to produce a modulated signal, as shown in Figure 31.12b. This type of transmission is much less affected by noise and signal loss because the message is extracted from frequency shifts of the carrier signal, rather than from changes in the amplitude of the carrier signal. FM radio receivers commonly use a Foster-Seeley discriminator to demodulate the FM signal. A Foster-Seeley discriminator uses an RLC circuit tuned to the frequency of the carrier signal and connected to two diodes, resembling the fullwave rectifier discussed in Chapter 30. If the input equals the carrier frequency, the two halves of the tuned circuit produce the same rectified voltage and the output is zero. As the frequency of the carrier signal changes, the balance between the two halves of the rectified circuit changes, resulting in a voltage proportional to the frequency deviation of the carrier signal. A Foster-Seeley discriminator is sensitive to amplitude variations and is usually coupled with a limiter amplifier stage to desensitize it to variations in the strength of the carrier wave by allowing lower power signals to pass through it unaffected while removing the peaks of the signals that exceed a certain power level.

HDTV transmitters broadcast information digitally in the form of zeros and ones. One byte of information contains eight bits, where a bit is a zero or a one. The screen is subdivided into picture elements (pixels) with digital representations of the red, green, and blue color of each pixel. Currently, the highest resolution for HDTV is 1080i, which has 1920 pixels in the horizontal direction and 1080 pixels in the vertical direction. Half the picture (every other horizontal line) is updated 60 times every second, and the two halves of the image are interlaced to form the complete image. (See Section 31.10 for more information on video formats.) HDTV is broadcast using a compression-decompression (codec) technique, typically the standard known as MPEG-2, to reduce the amount of data that must be transmitted. A typical HDTV station broadcasts about 17 megabytes of information per second.

Cell phone transmissions occur in the frequency bands from 824.04 to 848.97 MHz and 1.85 to 1.99 GHz. WiFi wireless data connections for computers operate in the ranges from 2.401 to 2.484 GHz (for the international standard; the U.S. standard band has an upper limit of 2.473 GHz) and from 5.15 to 5.85 GHz. These frequencies are in the microwave range, and some people worry about prolonged exposure to electromagnetic waves emitted by cell phones and WiFi. However, the relatively low power of these devices combined with the fact that the energy of these waves is much lower than that of other waves that are commonly encountered, such as visible light, argue that there is little danger from cell phones and WiFi. Chapter 37 on quantum mechanics will discuss the energy of the photons associated with electromagnetic waves.

31.7 Traveling Electromagnetic Waves

Subatomic processes can produce electromagnetic waves such as gamma rays, X-rays, and light. Electromagnetic waves can also be produced by an RLC circuit connected to an antenna (Figure 31.13). The connection between the RLC circuit and the antenna occurs through a transformer. A dipole antenna is used to approximate an electric dipole. The voltage and current in the antenna vary sinusoidally with time and cause the flow of charge in the antenna to oscillate with the frequency, ω_0, of the RLC circuit. The accelerating charges create **traveling electromagnetic waves**. These waves travel from the antenna at speed c and frequency $f = \omega_0/(2\pi)$.

The traveling electromagnetic waves propagate as wave fronts spreading out spherically from the antenna. However, at a large distance from the antenna, the wave fronts appear to be almost flat, or planar. Thus, such a traveling wave is described by equation 31.8. If a second RLC circuit tuned to the same frequency, ω_0, as the emitting circuit is placed in the path of these electromagnetic waves, voltage and current will be induced in this second circuit. These induced oscillations are the basic idea of radio transmission and reception. If the second circuit has $\omega = 1/\sqrt{LC}$, different from ω_0, much smaller voltages and currents will be induced. Only if the resonant frequency of the receiving circuit is the same as or

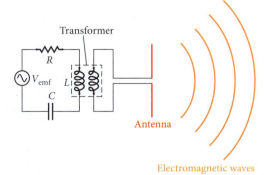

FIGURE 31.13 An RLC circuit coupled to an antenna that emits traveling electromagnetic waves.

very close to the transmitted frequency will any signal be induced in the receiving circuit. Thus, the receiver can select a transmission with a given frequency and reject all others.

The principle of transmission of electromagnetic waves was discovered by Heinrich Hertz in 1888, as described in Section 31.3, and was used by the Italian physicist Guglielmo Marconi (1874–1937) to transmit wireless signals.

31.8 Poynting Vector and Energy Transport

When you walk out into the sunlight, you feel warmth. If you stay out too long in the bright sunshine, you become sunburned. These phenomena are caused by electromagnetic waves emitted from the Sun. These electromagnetic waves carry energy that was generated in the nuclear reactions in the core of the Sun.

The rate at which energy is transported by an electromagnetic wave is usually defined in terms of a vector, $\vec{S}$, given by

$$\vec{S} = \frac{1}{\mu_0} \vec{E} \times \vec{B}. \qquad (31.23)$$

This quantity is called the **Poynting vector,** after British physicist John Poynting (1852–1914), who first discussed its properties. The magnitude of $\vec{S}$ is related to the instantaneous rate at which energy is transported by an electromagnetic wave over a given area, or more simply, the instantaneous power per unit area:

$$S = \left|\vec{S}\right| = \left(\frac{\text{power}}{\text{area}}\right)_{\text{instantaneous}}. \qquad (31.24)$$

The units of the Poynting vector are thus watts per square meter (W/m^2).

For an electromagnetic wave, where $\vec{E}$ is perpendicular to $\vec{B}$, equation 31.23 yields

$$S = \frac{1}{\mu_0} EB.$$

According to equation 31.15, the magnitudes of the electric field and the magnetic field are directly related via $E/B = c$. Thus, we can express the instantaneous power per unit area of an electromagnetic wave in terms of the magnitude of the electric field or that of the magnetic field. It is usually easier to measure an electric field than a magnetic field, the instantaneous power per unit area is given by

$$S = \frac{1}{c\mu_0} E^2. \qquad (31.25)$$

We can now substitute a sinusoidal form for the electric field, $E = E_{\text{max}} \sin(\kappa x - \omega t)$, and obtain an expression for the transmitted power per unit area. However, the usual means of describing the power per unit area in an electromagnetic wave is the intensity, I, of the wave, given by

$$I = S_{\text{ave}} = \left(\frac{\text{power}}{\text{area}}\right)_{\text{ave}} = \frac{1}{c\mu_0} \left[E_{\text{max}}^2 \sin^2\left(\kappa x - \omega t\right)\right]_{\text{ave}}.$$

The units of intensity are the same as the units of the Poynting vector, W/m^2. The time-averaged value of $\sin^2(\kappa x - \omega t)$ is $\frac{1}{2}$, so we can express the intensity as

$$I = \frac{1}{c\mu_0} E_{\text{rms}}^2, \qquad (31.26)$$

where $E_{\text{rms}} = E_{\text{max}}/\sqrt{2}$.

Because the magnitudes of the electric and magnetic fields of an electromagnetic wave are related by $E = cB$ and c is such a large number, you might conclude that the energy transported by the electric field is much larger than the energy transported by the magnetic field. Actually these energies are *the same*. To see this, recall from Chapters 24 and 29 that the energy density of an electric field is given by

$$u_E = \frac{1}{2}\epsilon_0 E^2,$$

and the energy density of a magnetic field is given by

$$u_B = \frac{1}{2\mu_0}B^2.$$

If we substitute $E = cB$ and $c = 1/\sqrt{\mu_0 \epsilon_0}$ into the expression for the energy density of the electric field, we get

$$u_E = \frac{1}{2}\epsilon_0(cB)^2 = \frac{1}{2}\epsilon_0\left(\frac{B}{\sqrt{\mu_0\epsilon_0}}\right)^2 = \frac{1}{2\mu_0}B^2 = u_B. \qquad (31.27)$$

Thus, the energy density of the electric field is the same as the energy density of the magnetic field everywhere in the electromagnetic wave.

EXAMPLE 31.1 / Using Solar Panels to Charge an Electric Car

Photovoltaic (solar power to electric power) solar panels (Figure 31.14a) can be mounted on the roof of your house at a cost per area of $\eta = \$1430/m^2$. You have an electric car (Figure 31.14b) that requires a charge corresponding to an energy of $U = 8.0$ kW h for a day of local driving. The solar panels convert solar power to electricity with an efficiency $\epsilon = 10.7\%$ and have an area A. Suppose that sunlight is incident on your solar panels for $\Delta t = 8.0$ h with an average intensity of $S_{ave} = 600$ W/m^2.

PROBLEM
How much do you need to spend on solar panels to give your electric car its daily charge?

SOLUTION
We equate the total amount of energy produced by the solar panels to the energy required to charge the car:

$$U_{produced} = P\Delta t = U.$$

The amount of power incident on the solar panels is the average intensity of the sunlight times the area of the solar panels times the efficiency of the solar panels:

$$P = \epsilon A S_{ave}.$$

Thus, the total area required is

$$A = \frac{P}{\epsilon S_{ave}} = \frac{(U/\Delta t)}{\epsilon S_{ave}} = \frac{U}{\epsilon S_{ave}\Delta t}.$$

The total cost will then be

$$\text{Cost} = \eta A = \frac{\eta U}{\epsilon S_{ave}\Delta t}.$$

Putting in the numerical values gives us

$$\text{Cost} = \frac{\eta U}{\epsilon S_{ave}\Delta t} = \frac{(\$1430/m^2)(8.00 \text{ kW h})}{(0.107)(0.6 \text{ kW/m}^2)(8.0 \text{ h})} = \$22,000.$$

(a)

(b)

FIGURE 31.14 (a) Photovoltaic solar panels mounted on the roof of a house. (b) A plug-in electric car capable of driving 40 miles on electric power.

If you drove your electric car 40 miles per day each day for 10 years, that would work out to 15 cents per mile. In contrast, the cost would be 20 cents per mile for a gasoline-powered car with a 20 mpg rating and gas costing $4.00 per gallon.

Obviously, this is not a large cost savings. However, your solar-powered electric car would be a completely carbon-neutral, zero emission mode of transportation (the same as riding your bicycle, without the exercise benefits). Material scientists are working intensely to increase the efficiency of commercially available solar cells, and mass production is expected to lower the costs dramatically. So solar-powered electric cars should soon be a viable and attractive alternative to gasoline-powered cars.

EXAMPLE 31.2 | **The Root-Mean-Square Electric and Magnetic Fields from Sunlight**

The average intensity of sunlight at the Earth's surface is approximately 1400 W/m², if the Sun is directly overhead.

PROBLEM

What are the root-mean-square electric and magnetic fields of these electromagnetic waves?

SOLUTION

The intensity of sunlight can be related to the root-mean-square electric field using equation 31.26:

$$I = \frac{1}{c\mu_0} E_{rms}^2.$$

Solving for the root-mean-square electric field gives us

$$E_{rms} = \sqrt{Ic\mu_0} = \sqrt{\left(1400 \text{ W/m}^2\right)\left(3.00 \cdot 10^8 \text{ m/s}\right)\left(4\pi \cdot 10^{-7} \text{ T m/A}\right)}$$

$$= 730 \text{ V/m}.$$

In comparison, the root-mean-square electric field in a typical home is 5–10 V/m. Standing directly under an electric power transmission line, one would experience a root-mean-square electric field of 200–10,000 V/m depending on the conditions.

The root-mean-square magnetic field is

$$B_{rms} = \frac{E_{rms}}{c} = \frac{730 \text{ V/m}}{3.00 \cdot 10^8 \text{ m/s}} = 2.4 \text{ μT}.$$

In comparison, the root-mean-square value of the Earth's magnetic field is 50 μT, the root-mean-square magnetic field found in a typical home is 0.5 μT, and the root-mean-square magnetic field under a power transmission line is 2 μT.

31.5 In-Class Exercise

The average intensity of sunlight at the Earth's surface is approximately 1400 W/m², if the Sun is directly overhead. The average distance between the Earth and the Sun is $1.50 \cdot 10^{11}$ m. What is the average power emitted by the Sun?

a) $99.9 \cdot 10^{25}$ W   d) $4.3 \cdot 10^{28}$ W

b) $4.0 \cdot 10^{26}$ W   e) $5.9 \cdot 10^{29}$ W

c) $6.3 \cdot 10^{27}$ W

31.9   Radiation Pressure

When you walk out into the sunlight, you feel warmth, but you do not feel any force from the sunlight. Sunlight is exerting a pressure on you, but that pressure is so small that you cannot notice it. Because the electromagnetic waves making up sunlight are radiated from the Sun and travel to the Earth, they are referred to as **radiation.** Chapter 18 discussed radiation as one way to transfer heat. As we'll see in Chapter 40 on nuclear physics, this type of radiation is not necessarily the same as radioactive radiation resulting from the decay of unstable nuclei. However, radio waves, infrared waves, visible light, and X-rays are all fundamentally the same electromagnetic radiation. (Which is not to say, however, that all kinds of electromagnetic radiation have the same effect on the human body. For example, UV light can give you sunburn and even trigger skin cancer, whereas there is no credible evidence that radiation emitted from cell phones can cause cancer.)

Let's calculate the magnitude of the pressure exerted by these radiated electromagnetic waves. Electromagnetic waves carry energy, U, as shown in Section 31.8. Electromagnetic waves also have linear momentum, $\vec{p}$. This concept is subtle because electromagnetic waves have no mass, and we saw in Chapter 7 that momentum is equal to mass multiplied by velocity. Maxwell showed that if a plane wave of radiation is totally absorbed on a surface (perpendicular to the direction of the plane wave) for a time interval, Δt, and an amount of energy, ΔU, is absorbed by the surface in that process, then the magnitude of the momentum transferred to that surface by the wave in that time interval is

$$\Delta p = \frac{\Delta U}{c}.$$

Chapter 35 on relativity will show that this relationship between energy and momentum holds for massless objects; for now, it is stated as a fact, without proof.

The magnitude of the force on the surface is then $F = \Delta p/\Delta t$ (Newton's Second Law). The total energy, ΔU, absorbed by area A of the surface during the time interval Δt is equal to the product of the area, the time interval, and the radiation intensity, I (introduced in Section 31.8): $\Delta U = IA\Delta t$. Therefore, the magnitude of the force exerted by the electromagnetic wave on this area is

$$F = \frac{\Delta p}{\Delta t} = \frac{\Delta U}{c\Delta t} = \frac{IA\Delta t}{c\Delta t} = \frac{IA}{c}.$$

Since pressure is defined as force (magnitude) per unit area, the radiation pressure, p_r, is

$$p_r = \frac{F}{A},$$

and, consequently,

$$p_r = \frac{I}{c} \quad \text{(for total absorption).} \tag{31.28}$$

Equation 31.28 states that the radiation pressure due to electromagnetic waves is simply the intensity divided by the speed of light, but only for the case of total absorption of the radiation on the surface.

The other limiting case is total reflection of the electromagnetic waves. In that case, the momentum transfer is *twice* as big as for total absorption, just like the momentum transfer from a ball to a wall is twice as big in a perfectly elastic collision as in a perfectly inelastic collision. In the perfectly elastic collision, the ball's initial momentum is reversed and $\Delta p = p_i - (-p_i) = 2p_i$, whereas for the totally inelastic collision, $\Delta p = p_i - 0 = p_i$, as explained in Chapter 7. So, the radiation pressure for the case of perfect reflection of the electromagnetic waves off a surface is

$$p_r = \frac{2I}{c} \quad \text{(for perfect reflection).} \tag{31.29}$$

The radiation pressure from sunlight is comparatively small. The intensity of sunlight at the Earth's surface is at most 1400 W/m^2, when the Sun is directly overhead and there are no clouds in the sky. (This can happen only between the Tropics of Cancer and Capricorn, located at ±23° latitude relative to the Equator.) Thus, the maximum radiation pressure for sunlight that is totally absorbed is

$$p_r = \frac{I}{c} = \frac{1400 \text{ W/m}^2}{3 \cdot 10^8 \text{ m/s}} = 4.67 \cdot 10^{-6} \text{ N/m}^2 = 4.67 \text{ }\mu\text{Pa}.$$

For comparison, atmospheric pressure is 101 kPa (see Chapter 13), which is greater than the sunlight's radiation pressure on the surface of Earth by more than a factor of 20 billion. Another useful comparison is the lowest pressure difference that human hearing can detect, which is generally quoted as approximately 20 μPa for sounds in the 1-kHz frequency range, where the human ear is most sensitive (see Chapter 16).

31.6 In-Class Exercise

What is the radiation pressure due to sunlight incident on a perfectly absorbing surface, whose surface normal vector is at an angle of 70° relative to the incident light?

a) $(4.67 \text{ }\mu\text{Pa})(\cos 70°)$

b) $(4.67 \text{ }\mu\text{Pa})(\sin 70°)$

c) $(4.67 \text{ }\mu\text{Pa})(\tan 70°)$

d) $(4.67 \text{ }\mu\text{Pa})(\cot 70°)$

31.7 In-Class Exercise

What is the maximum radiation pressure due to sunlight incident on a perfectly reflecting surface?

a) 0   c) 4.67 μPa

b) 2.34 μPa   d) 9.34 μPa

EXAMPLE 31.3 | Radiation Pressure from a Laser Pointer

A green laser pointer has a power of 1.00 mW. You shine the laser pointer perpendicularly on a white sheet of paper, which reflects the light. The spot of light on the paper is 2.00 mm in diameter.

PROBLEM
What force does the light from the laser pointer exert on the paper?

SOLUTION
The intensity of the light is given by

$$I = \frac{\text{power}}{\text{area}} = \frac{1.00 \cdot 10^{-3} \text{ W}}{\pi \left(1.00 \cdot 10^{-3} \text{ m}\right)^2} = 318 \text{ W/m}^2.$$

Continued—

The radiation pressure for a perfectly reflecting surface is given by equation 31.29 and also is equal to the force exerted by the light divided by the area over which it acts:

$$p_r = \frac{\text{force}}{\text{area}} = \frac{2I}{c}.$$

Thus, the force exerted on the paper is

$$\text{Force} = (\text{area})\left(\frac{2I}{c}\right) = \pi\left(1.0\cdot10^{-3}\text{ m}\right)^2 \frac{2\left(318\text{ W/m}^2\right)}{3.00\cdot10^8\text{ m/s}} = 6.66\cdot10^{-12}\text{ N}.$$

31.3 Self-Test Opportunity

Suppose you have a satellite in orbit around the Sun, as shown in the figure. The orbit is in the counterclockwise direction looking down on the north pole of the Sun. You want to deploy a solar sail consisting of a large, totally reflecting mirror, which can be oriented so that it is perpendicular to the light coming from the Sun or at an angle with respect to the light coming from the Sun. Describe the effect on the orbit of your satellite for the three deployment angles shown in the figure.

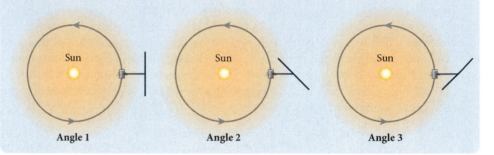

Angle 1   Angle 2   Angle 3

SOLVED PROBLEM **31.1** | **Solar Stationary Satellite**

Suppose researchers want to place a satellite above the north pole of the Sun and stationary with respect to the Sun in order to study its long-term rotational characteristics. The satellite will have a totally reflecting solar sail and be located at a distance of $1.50\cdot10^{11}$ m from the center of the Sun. The intensity of sunlight at that distance is 1400 W/m². The plane of the solar sail is perpendicular to a line connecting the satellite and the center of the Sun. The mass of the satellite and sail is 100.0 kg.

PROBLEM
What is the required area of the solar sail?

SOLUTION

THINK
In an equilibrium position for the satellite, the area of the solar sail times the radiation pressure from the Sun produces a force that is balanced by the gravitational force between the satellite and the Sun. We can equate these two forces and solve for the area of the solar sail.

SKETCH
Figure 31.15 is a diagram of the satellite with a solar sail near the Sun.

FIGURE 31.15 A satellite with a solar sail near the Sun.

RESEARCH
The satellite will be stationary if the force of gravity, F_g, is balanced by the force from the radiation pressure of sunlight, F_{rp}:

$$F_g = F_{rp}.$$

The force corresponding to the radiation pressure from sunlight is equal to the radiation pressure, p_r, times the area of the solar sail, A:

$$F_{rp} = p_r A.$$

The radiation pressure can be expressed in terms of the intensity of the sunlight, I, incident on the totally reflecting solar sail:

$$p_r = \frac{2I}{c}.$$

The force of gravity between the satellite and the Sun is given by

$$F_g = G \frac{m m_{Sun}}{d^2},$$

where G is the universal gravitational constant, m is the mass of the satellite and sail, m_{Sun} is the mass of the Sun, and d is the distance between the satellite and the Sun.

SIMPLIFY

We can combine all these equations to obtain

$$\left(\frac{2I}{c} \right) A = G \frac{m m_{Sun}}{d^2}.$$

Solving for the area of the solar sail gives us

$$A = G \frac{c m m_{Sun}}{2 I d^2}.$$

CALCULATE

Putting in the numerical values, we get

$$A = \left(6.67 \cdot 10^{-11} \text{ m}^3 \text{ kg}^{-1} \text{ s}^{-2} \right) \frac{\left(3.00 \cdot 10^8 \text{ m/s} \right) \left(100.0 \text{ kg} \right) \left(1.99 \cdot 10^{30} \text{ kg} \right)}{2 \left(1400 \text{ W/m}^2 \right) \left(1.50 \cdot 10^{11} \text{ m} \right)^2} = 63,206.2 \text{ m}^2.$$

ROUND

We report our result to three significant figures:

$$A = 6.32 \cdot 10^4 \text{ m}^2.$$

DOUBLE-CHECK

If the solar sail were circular, the radius of the sail would be

$$R = \sqrt{\frac{A}{\pi}} = \sqrt{\frac{6.32 \cdot 10^4 \text{ m}^2}{\pi}} = 142 \text{ m},$$

which is an achievable size. We can relate the thickness of the sail, t, times the density, ρ, of the material from which the sail is constructed to the mass per unit area of the sail:

$$t \rho = \frac{m}{A}.$$

If the sail were composed of a sturdy material such as kapton ($\rho = 1420 \text{ kg/m}^3$) and had a mass of 75 kg, the thickness of the sail would be

$$t = \frac{75 \text{ kg}}{\left(1420 \text{ kg/m}^3 \right) \left(6.32 \cdot 10^4 \text{ m}^2 \right)} = 8.36 \cdot 10^{-7} \text{ m} = 0.836 \text{ } \mu\text{m}.$$

Kapton is a polyimide film developed to remain stable in the wide range of temperatures found in space; from near absolute zero to over 600 K. Current production techniques cannot produce kapton this thin. However, the required areal mass density may be realizable using other materials in the future.

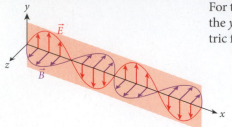

31.10 Polarization

FIGURE 31.16 An electromagnetic wave with the plane of oscillation of the electric field shown in pink.

For the electromagnetic wave represented in Figure 31.5, the electric field always points along the y-axis. The direction in which the wave is traveling is the positive x-direction, so the electric field of the electromagnetic wave lies within a plane of oscillation (Figure 31.16).

We can visualize the polarization of an electromagnetic wave by looking at the electric field vector of the wave in the yz-plane, which is perpendicular to the direction in which the wave is traveling (Figure 31.17a). The electric field vector changes from the positive y-direction to the negative y-direction and back again as the wave travels. The electric field of the wave oscillates in the y-direction only, never changing its orientation. This type of wave is called a **plane-polarized wave** in the y-direction.

The electromagnetic waves making up the light emitted by most common light sources, such as the Sun or an incandescent light bulb, have random polarizations. Each wave has its electric field vector oscillating in a different plane. Such light is called **unpolarized light.** Light from an unpolarized source can be represented by many vectors like the one shown in Figure 31.17a, but with random orientations (Figure 31.17b). Unpolarized light can also be represented by summing the y-components and the z-components separately to produce the net y- and z-components. Unpolarized light has equal components in the y- and z-directions (Figure 31.18a). If there is less net polarization in the y-direction than in the z-direction, then the light is said to be partially polarized in the z-direction (Figure 31.18b).

Unpolarized light can be transformed to polarized light by passing the unpolarized light through a polarizer. A **polarizer** allows only one component of the electric field vectors of the light waves to pass through. One way to make a polarizer is to produce a material that consists of long parallel chains of molecules, which effectively let components of the light with polarization parallel to the chains pass through and block components of the light with polarization perpendicular to that direction.

This discussion will not go into the details of the molecular structure but will simply characterize each polarizer by a polarizing direction. Unpolarized light passing through a polarizer emerges polarized in the polarizing direction (Figure 31.19). The components of the light that have the same direction as the polarizer are transmitted, but the components of the light that are perpendicular to the polarizer are absorbed.

Now let's consider the intensity of the light that passes through a polarizer. Unpolarized light with intensity I_0 has equal components in the y- and z-directions. After passing through a vertical

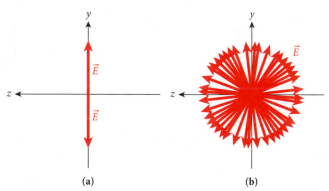

FIGURE 31.17 (a) Electric field vectors in the yz-plane, defining the plane of polarization to be the xy-plane. (b) Electric field vectors oriented at random angles.

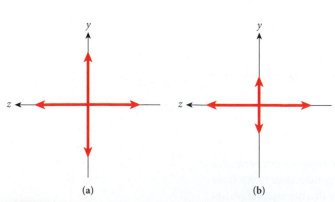

FIGURE 31.18 (a) Net components of the electric field for unpolarized light. (b) Net components of the electric field for partially polarized light.

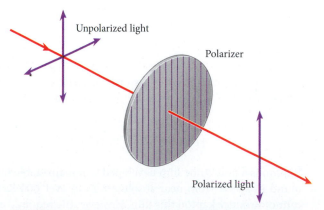

FIGURE 31.19 Unpolarized light passing through a vertical polarizer. After the light passes through the polarizer, it is vertically polarized.

polarizer, only the y-component (or vertical component) remains. The intensity, I, of the light that passes through the polarizer is given by

$$I = \tfrac{1}{2} I_0, \qquad (31.30)$$

because the unpolarized light had equal contributions from the y- and z-components and only the y-components are transmitted by the polarizer. The factor $\tfrac{1}{2}$ applies only to the case of unpolarized light passing through a polarizer.

Let's consider polarized light passing through a polarizer (Figure 31.20). If the polarizer axis is parallel to the polarization direction of the incident polarized light, all of the light will be transmitted with the original polarization (Figure 31.20a). If the polarizing angle of the polarizer is perpendicular to the polarization of polarized light, no light will be transmitted (Figure 31.20b).

What happens when polarized light is incident on a polarizer and the polarization of the light is neither parallel nor perpendicular to the polarizing angle of the polarizer (Figure 31.21)? Let's assume that the angle between the incident polarized light and the polarizing angle is θ. The magnitude of the transmitted electric field, E, is given by

$$E = E_0 \cos\theta,$$

where E_0 is the magnitude of the electric field of the incident polarized light. From equation 31.26, we can see that the intensity of the light before passing through the polarizer, I_0, is given by

$$I_0 = \frac{1}{c\mu_0} E_{\text{rms}}^2 = \frac{1}{2c\mu_0} E_0^2.$$

After the light passes through the polarizer, the intensity, I, is given by

$$I = \frac{1}{2c\mu_0} E^2.$$

We can express the transmitted intensity in terms of the initial intensity as follows:

$$I = \frac{1}{2c\mu_0} E^2 = \frac{1}{2c\mu_0}\left(E_0 \cos\theta\right)^2 = I_0 \cos^2\theta. \qquad (31.31)$$

This equation is called the **Law of Malus.** It applies only to polarized light incident on a polarizer.

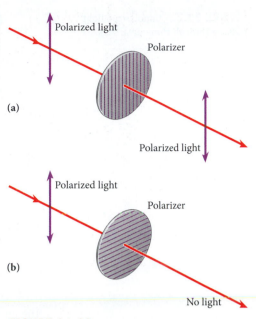

FIGURE 31.20 (a) Vertically polarized light incident on a vertical polarizer. (b) Vertically polarized light incident on a horizontal polarizer.

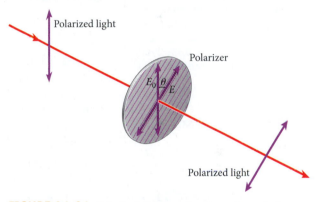

FIGURE 31.21 Polarized light passing through a polarizer whose polarizing angle is neither parallel nor perpendicular to the polarization of the incident light.

EXAMPLE 31.4 | Three Polarizers

Suppose that unpolarized light with intensity I_0 is initially incident on the first of three polarizers in a line. The first polarizer has a polarizing direction that is vertical. The second polarizer has a polarizing angle of 45° with respect to the vertical. The third polarizer has a polarizing angle of 90° with respect to the vertical.

PROBLEM
What is the intensity of the light after passing through all three polarizers, in terms of the initial intensity?

SOLUTION
Figure 31.22 illustrates the light passing through the three polarizers. The intensity of the unpolarized light is I_0. The intensity of the light after passing through the first polarizer is

$$I_1 = \tfrac{1}{2} I_0.$$

Continued—

FIGURE 31.22 Unpolarized light passing through three polarizers.

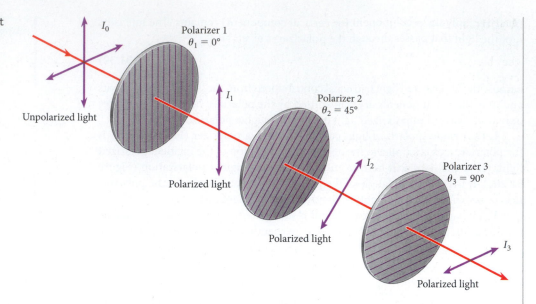

The intensity of the light after passing through the second polarizer is

$$I_2 = I_1 \cos^2\left(45° - 0°\right) = I_1 \cos^2 45° = \tfrac{1}{2} I_0 \cos^2 45°.$$

The intensity of the light after passing through the third polarizer is

$$I_3 = I_2 \cos^2\left(90° - 45°\right) = I_2 \cos^2 45° = \tfrac{1}{2} I_0 \cos^4 45°,$$

or $I_3 = I_0/8$.

The fact that $\tfrac{1}{8}$ of the light's initial intensity is transmitted is somewhat surprising because polarizers 1 and 3 have polarizing angles that are perpendicular to each other. Acting by themselves, polarizers 1 and 3 would block all of the light. Yet by adding an additional obstacle (polarizer 2) between these two polarizers, $\tfrac{1}{8}$ of the original intensity gets through. A series of polarizers with small differences in their polarizing angles can thus be used to rotate the polarization direction of light with only modest losses in intensity.

31.8 In-Class Exercise

Unpolarized light with intensity $I_{in} = 1.87$ W/m^2 passes through two polarizers. The emerging polarized light has intensity $I_{out} = 0.383$ W/m^2. What is the angle between the two polarizers?

a) 23.9°   d) 72.7°

b) 34.6°   e) 88.9°

c) 50.2°

31.9 In-Class Exercise

The figure shows unpolarized light incident on polarizer 1 with polarizer angle $\theta_1 = 0°$ and then on polarizer 2 with polarizer angle $\theta_2 = 90°$, which results in no light passing through. If polarizer 3 with polarizer angle $\theta_3 = 50°$ is placed between polarizers 1 and 2, which of the following statements is true?

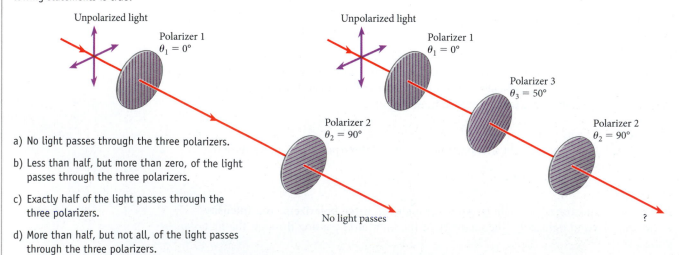

a) No light passes through the three polarizers.

b) Less than half, but more than zero, of the light passes through the three polarizers.

c) Exactly half of the light passes through the three polarizers.

d) More than half, but not all, of the light passes through the three polarizers.

e) All of the light passes through the three polarizers.

Applications of Polarization

Polarization has many practical applications. Sunglasses often have a polarized coating that blocks reflected light, which is usually polarized. A computer's or television's liquid crystal display (LCD) has an array of liquid crystals sandwiched between two polarizers whose polarizing angles are rotated 90° with respect to each other. Normally, the liquid crystal rotates the polarization of the light between the two polarizers so that light passes through. An array of addressable electrodes applies a varying voltage across each of the liquid crystals, causing the liquid crystals to rotate the polarization less, darkening the area covered by the electrode. The television or computer monitor screen (Figure 31.23) can then display a large number of picture elements, or pixels, that produce a high-resolution image.

FIGURE 31.23   An LCD computer monitor.

Figure 31.24a shows a top view of the layers of an LCD screen. Unpolarized light is emitted by a backlight. This light passes through a vertical polarizer. The polarized light then passes through a transparent layer of conducting pixel pads. These pads are designed to put varying amounts of voltage across the next layer, which is composed of liquid crystals, with respect to the transparent common electrode. If no voltage is applied across the liquid crystals, they rotate the polarization of the incident light by 90°. This light with rotated polarization can then pass through the transparent common electrode, the color filter, the horizontal polarizer, and the screen cover. When voltage of varying magnitude is applied to the pixel pad, the liquid crystals rotate the polarization of the incident light by a varying amount. When the full voltage is applied to the pixel pad, the polarization of the incident light is not rotated, and the horizontal polarizer blocks any light transmitted through the transparent common electrode and the color filter.

Figure 31.24b shows a front view of a small segment of the LCD screen, illustrating how the screen produces an image. The image is created by an array of pixels. Each pixel is subdivided into three subpixels: one red, one green, and one blue. By varying the voltage across each subpixel, a superposition of red, green, and blue light is created, producing a color on that pixel. It is difficult to connect a single wire to each subpixel, however. A high-definition 1080p LCD screen has 1080 times 1920 times 3 subpixels, or 6,220,800 subpixels. These subpixels are connected in columns and rows, as shown in Figure 31.24b. To turn on a subpixel, a voltage from both a column and a row must be applied. Thus, the subpixels are turned on one row at a time. With the voltage for one row on, the voltages for the subpixels in the desired columns are turned on. A small capacitor holds the voltage until the row is turned on again.

A high-definition 1080p LCD screen is scanned 60 times a second, producing a complete image in each scan. On a high-definition 1080i screen, every other row of the image is scanned 60 times a second and then the two images are interlaced. Another high-definition standard is 720p, which scans 720 rows 60 times a second with a horizontal resolution of 1280 pixels. The 720p and 1080i standards are in common use in television broadcasting. The standard resolution for a television image is 480p, with 480 rows being updated 60 times a second and producing 640 columns of pixels.

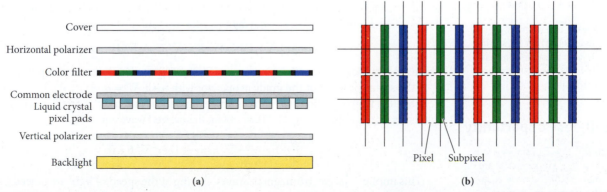

FIGURE 31.24   (a) Top view of the layers making up an LCD screen. (b) Front view of a subset of pixels and subpixels on an LCD screen.

31.11 Derivation of the Wave Equation

Table 31.1 lists the four equations known as Maxwell's equations in integral form. There is also an equivalent differential version of these equations, which is the way they are usually printed on T-shirts and posters:

$$\vec{\nabla}\bullet\vec{E}=\frac{\rho}{\epsilon_0},$$

$$\vec{\nabla}\times\vec{E}=-\frac{\partial}{\partial t}\vec{B},$$

$$\vec{\nabla}\bullet\vec{B}=0,$$

and

$$\vec{\nabla}\times\vec{B}=\epsilon_0\mu_0\frac{\partial}{\partial t}\vec{E}+\mu_0\vec{j},$$

where ρ is the charge density (charge q per unit volume) and $\vec{j}$ is the current density. In vacuum and in the absence of charges, both are zero; $\rho=0$ and $\vec{j}=0$. The symbol $\vec{\nabla}$ was introduced in Chapter 6 and represents the vector with the partial derivatives in each spatial direction. In Cartesian coordinates, it is $\vec{\nabla}=(\partial/\partial x,\partial/\partial y,\partial/\partial z)$.

In Section 31.4, we saw that electromagnetic waves as described by equation 31.8 are valid solutions to all the Maxwell equations in vacuum. However, strictly speaking, we have not yet written the wave equation that the electric field and the magnetic field obey. Now, with the aid of the differential form of Maxwell's equations, we can derive the wave equation for the electric field, which is

$$\frac{\partial^2}{\partial t^2}\vec{E}-c^2\nabla^2\vec{E}=0. \tag{31.32}$$

The wave equation for the magnetic field is

$$\frac{\partial^2}{\partial t^2}\vec{B}-c^2\nabla^2\vec{B}=0. \tag{31.33}$$

DERIVATION 31.1 | **Wave Equation for the Electric Field in Vacuum**

To derive the wave equation for the electric field in vacuum, we take the vector product of the second Maxwell equation and the gradient operator, $\vec{\nabla}$:

$$\vec{\nabla}\times\vec{\nabla}\times\vec{E}=-\vec{\nabla}\times\frac{\partial}{\partial t}\vec{B}. \tag{i}$$

On the right-hand side of equation (i), we can interchange the order of the time derivative and the spatial derivative:

$$-\vec{\nabla}\times\frac{\partial}{\partial t}\vec{B}=-\frac{\partial}{\partial t}(\vec{\nabla}\times\vec{B})=-\frac{\partial}{\partial t}\left(\epsilon_0\mu_0\frac{\partial\vec{E}}{\partial t}\right)=-\epsilon_0\mu_0\frac{\partial^2}{\partial t^2}\vec{E}. \tag{ii}$$

The second step of this transformation makes use of the third Maxwell equation with $\vec{j}=0$, which is appropriate in vacuum. The left-hand side of equation (i) is a double vector product. Chapter 10 introduced the BAC-CAB rule for double vector products: $\vec{A}\times(\vec{B}\times\vec{C})=\vec{B}(\vec{A}\bullet\vec{C})-\vec{C}(\vec{A}\bullet\vec{B})$. Applying this rule, we find

$$\vec{\nabla}\times\vec{\nabla}\times\vec{E}=\vec{\nabla}(\vec{\nabla}\bullet\vec{E})-\nabla^2\vec{E}=-\nabla^2\vec{E}, \tag{iii}$$

where the first Maxwell equation $\left(\text{in vacuum: }\vec{\nabla}\bullet\vec{E}=0\right)$ is used in the second step. The symbol ∇^2 is the scalar product of the gradient operator with itself: $\nabla^2=\partial^2/\partial x^2+\partial^2/\partial y^2+\partial^2/\partial z^2$. If we substitute from equations (ii) and (iii) into equation (i) and use the fact that the speed of light is $c=1/\sqrt{\mu_0\epsilon_0}$ (equation 31.20) we obtain the desired wave equation:

$$\frac{\partial^2}{\partial t^2}\vec{E}-\frac{1}{\epsilon_0\mu_0}\nabla^2\vec{E}=\frac{\partial^2}{\partial t^2}\vec{E}-c^2\nabla^2\vec{E}=0.$$

This implies that electromagnetic waves moving at the speed of light are indeed a solution of Maxwell's equations, as discussed (but not exactly proven) in Section 31.4.

31.4 Self-Test Opportunity

Derive the wave equation for the magnetic field (equation 31.33) in the same way as presented in Derivation 31.1 for the wave equation for the electric field.

31.5 Self-Test Opportunity

Show that $\vec{E}(\vec{r},t)=E_{max}\sin(\kappa x-\omega t)\hat{y}$ and $\vec{B}(\vec{r},t)=B_{max}\sin(\kappa x-\omega t)\hat{z}$ are indeed solutions of the wave equation for the electric and magnetic fields.

WHAT WE HAVE LEARNED | EXAM STUDY GUIDE

- When a capacitor is being charged, a displacement current can be visualized between the plates, given by $i_d = \epsilon_0\, d\Phi_E/dt$, where Φ_E is the electric flux.

- Maxwell's equations describe how electrical charges, currents, electric fields, and magnetic fields affect each other, forming a unified theory of electromagnetism.

  - Gauss's Law for Electric Fields, $\oiint \vec{E} \cdot d\vec{A} = q_{enc}/\epsilon_0$, relates the net electric flux through a closed surface to the net enclosed electric charge.

  - Gauss's Law for Magnetic Fields, $\oiint \vec{B} \cdot d\vec{A} = 0$, states that the net magnetic flux through any closed surface is zero.

  - Faraday's Law of Induction, $\oint \vec{E} \cdot d\vec{s} = -d\Phi_B/dt$, relates the induced electric field to the changing magnetic flux.

  - The Maxwell-Ampere Law, $\oint \vec{B} \cdot d\vec{s} = \mu_0\epsilon_0\, d\Phi_E/dt + \mu_0 i_{enc}$, relates the induced magnetic field to the changing electric flux and to the current.

- For an electromagnetic wave traveling in the positive x-direction, the electric and magnetic fields can be described by $\vec{E}(\vec{r},t) = E_{max} \sin(\kappa x - \omega t)\hat{y}$ and $\vec{B}(\vec{r},t) = B_{max} \sin(\kappa x - \omega t)\hat{z}$, where $\kappa = 2\pi/\lambda$ is the wave number and $\omega = 2\pi f$ is the angular frequency.

- The magnitudes of the electric and magnetic fields of an electromagnetic wave at any fixed time and place are related by the speed of light, $E = cB$.

- The speed of light can be related to the two basic electromagnetic constants: $c = 1/\sqrt{\mu_0 \epsilon_0}$.

- The instantaneous power per unit area carried by an electromagnetic wave is the magnitude of the Poynting vector $S = [1/(c\mu_0)]E^2$, where E is the magnitude of the electric field.

- The intensity of an electromagnetic wave is defined as the average power per unit area carried by the wave, $I = S_{ave} = [1/(c\mu_0)]E_{rms}^2$, where E_{rms} is the root-mean-square magnitude of the electric field.

- For an electromagnetic wave, the energy density carried by the electric field is $u_E = \frac{1}{2}\epsilon_0 E^2$, and the energy density carried by the magnetic field is $u_B = [1/(2\mu_0)]B^2$. For any such wave, $u_E = u_B$.

- The radiation pressure exerted by electromagnetic waves of intensity I is given by $p_r = I/c$ if the electromagnetic waves are totally absorbed or $p_r = 2I/c$ if the waves are perfectly reflected.

- The polarization of an electromagnetic wave is given by the direction of the electric field vector.

- The intensity of unpolarized light that has passed through a polarizer is $I = I_0/2$, where I_0 is the intensity of unpolarized light incident on the polarizer.

- The intensity of polarized light that has passed through a polarizer is $I = I_0 \cos^2\theta$, where I_0 is the intensity of polarized light incident on the polarizer and θ is the angle between the polarization of the incident polarized light and the polarizing angle of the polarizer.

KEY TERMS

Maxwell's Law of Induction, p. 993
Maxwell-Ampere Law, p. 994
displacement current, p. 994
Maxwell's equations, p. 996

electromagnetism, p. 996
electromagnetic waves, p. 996
plane wave, p. 997
electromagnetic spectrum, p. 1001
visible light, p. 1001
infrared waves, p. 1001

ultraviolet rays, p. 1001
radio waves, p. 1001
microwaves, p. 1001
X-rays, p. 1002
gamma rays, p. 1002
traveling electromagnetic waves, p. 1003

Poynting vector, p. 1004
radiation, p. 1006
plane-polarized wave, p. 1010
unpolarized light, p. 1010
polarizer, p. 1010
Law of Malus, p. 1011

NEW SYMBOLS AND EQUATIONS

$i_d = \dfrac{\epsilon_0 d\Phi_E}{dt}$, displacement current

$\kappa = \dfrac{2\pi}{\lambda}$, wave number

$S = \dfrac{E^2}{c\mu_0}$, magnitude of the Poynting vector, representing the instantaneous power per unit area carried by an electromagnetic wave

$I = S_{ave} = \dfrac{E_{rms}^2}{c\mu_0}$, average power per unit area carried by an electromagnetic wave

ANSWERS TO SELF-TEST OPPORTUNITIES

31.1 $t = \dfrac{d}{c} = \dfrac{8.30 \cdot 10^{16} \text{ m}}{3.00 \cdot 10^8 \text{ m/s}} = 2.77 \cdot 10^8 \text{ s} = 8.77 \text{ yr.}$

31.2 $c = \lambda f \Rightarrow \lambda = \dfrac{c}{f}$

$\lambda_{\text{FM}} = \dfrac{3.00 \cdot 10^8 \text{ m}}{90.5 \cdot 10^6 \text{ Hz}} = 3.31 \text{ m}$

$\lambda_{\text{AM}} = \dfrac{3.00 \cdot 10^8 \text{ m}}{870 \cdot 10^3 \text{ Hz}} = 345 \text{ m.}$

31.3 Deployment angle 1 will produce an elliptical orbit with the Sun at one focus. The force from radiation pressure depends on the inverse square of the distance, just as the force of gravity does. Thus, the orbit will become an ellipse just as if the mass of the Sun or the mass of the object were suddenly reduced slightly. Because the force is perpendicular to the velocity of the satellite, the energy of the satellite is not affected.

Deployment angle 2 will result in a growing orbit. The resulting force from the reflected light produces a component of force that is in the same direction as the velocity of the spacecraft. Thus, the spacecraft gains energy, and the radius of the orbit increases. Note that the speed of the spacecraft decreases, but its total energy increases.

Deployment angle 3 will result in a shrinking orbit. The resulting force from the reflected light produces a component of force that is in the opposite direction from the

velocity of the spacecraft. Thus, the spacecraft loses energy and the radius of the orbit decreases. Note that the speed of spacecraft increases but its total energy decreases.

31.4 Take the vector product of the gradient operator $\vec{\nabla}$ and the fourth Maxwell equation: $\vec{\nabla} \times \vec{\nabla} \times \vec{B} = \epsilon_0 \mu_0 \vec{\nabla} \times \dfrac{\partial}{\partial t} \vec{E}.$

The right-hand side of this equation is

$$\epsilon_0 \mu_0 \vec{\nabla} \times \dfrac{\partial}{\partial t} \vec{E} = \epsilon_0 \mu_0 \dfrac{\partial}{\partial t} (\vec{\nabla} \times \vec{E}) = \epsilon_0 \mu_0 \dfrac{\partial}{\partial t} \left(-\dfrac{\partial \vec{B}}{\partial t} \right) = -\epsilon_0 \mu_0 \dfrac{\partial^2}{\partial t^2} \vec{B}.$$

The left-hand side is $\vec{\nabla} \times \vec{\nabla} \times \vec{B} = \vec{\nabla}(\vec{\nabla} \cdot \vec{B}) - \nabla^2 \vec{B} = -\nabla^2 \vec{B}.$

31.5 $\dfrac{\partial^2}{\partial t^2} \sin(\kappa x - \omega t) = -\omega^2 \sin(\kappa x - \omega t)$

and

$\dfrac{\partial^2}{\partial x^2} \sin(\kappa x - \omega t) = -\kappa^2 \sin(\kappa x - \omega t).$

Thus, this function is a solution for $c = \omega/\kappa.$

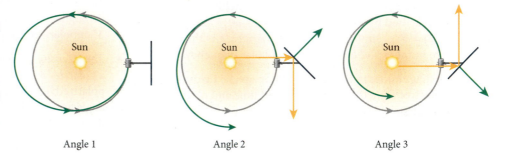

Angle 1   Angle 2   Angle 3

PROBLEM-SOLVING PRACTICE

Problem-Solving Guidelines

1. The same basic relationships that characterize any waves apply to electromagnetic waves. Remember that $c = \lambda f$ and $\omega = c\kappa$, where c is the speed of an electromagnetic wave. Review Chapter 15 if necessary.

2. It is often helpful to draw a diagram showing the direction of the wave motion and the orientation of both electric and magnetic fields. Remember the relationships between $\vec{E}$ and $\vec{B}$ for both magnitude and directions, including $E/B = (\mu_0 \epsilon_0)^{-1/2} = c$ for electromagnetic waves.

SOLVED PROBLEM 31.2 / Multiple Polarizers

Suppose you have light polarized in the vertical direction and want to rotate the polarization to the horizontal direction ($\theta = 90.0°$). If you pass the vertically polarized light through a polarizer whose polarization angle is horizontal, all the light will be blocked. If, instead, you use a series of ten polarizers, each of which has a polarizing angle θ that is 9.00° more than that of the preceding one, with the first polarizer having $\theta = 9.00°$, you can rotate the polarization by 90.0° and still have light passing through.

PROBLEM

What fraction of the intensity of the incident light is transmitted through the ten polarizers?

SOLUTION

THINK

Each polarizer is rotated 9.00° from the preceding polarizer. Thus, each polarizer transmits a fraction of the intensity equal to $f = \cos^2 9°$. The fraction transmitted is then f^{10}.

SKETCH

Figure 31.25 shows the direction of the initial polarization and the orientations of the ten polarizers.

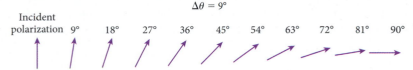

FIGURE 31.25 The direction of the polarization of the incident light and the direction of the polarizing angles of ten polarizers.

RESEARCH

The intensity, I, of polarized light passing through a polarizer whose polarizing direction makes an angle θ with the polarization of the incident light is given by

$$I = I_0 \cos^2\theta,$$

where I_0 is the intensity of the incident polarized light. In this case, the polarizing direction of each successive polarizer is $\Delta\theta = 9°$ larger than the polarizing direction of the preceding polarizer. Thus, each polarizer reduces the intensity by the factor

$$\frac{I}{I_0} = \cos^2\Delta\theta.$$

SIMPLIFY

The reduction in intensity after the light has passed through ten polarizers, each with its polarization direction differing from that of the preceding polarizer by $\Delta\theta$ is

$$\frac{I_{10}}{I_0} = \left(\cos^2\Delta\theta\right)^{10}.$$

CALCULATE

Putting in the numerical values, we get

$$\frac{I_{10}}{I_0} = \left(\cos^2 9°\right)^{10} = 0.780546.$$

ROUND

We report our result to three significant figures:

$$\frac{I_{10}}{I_0} = 0.781 = 78.1\%.$$

DOUBLE-CHECK

Using ten polarizers, each rotated by 9° more than the preceding one, the polarization of the incident polarized light was rotated by 90° and 78.1% of the light was transmitted, whereas using one polarizer rotated by 90° would have blocked all of the incident light. To see if our answer is reasonable, let's assume that instead of ten polarizers, we use n polarizers, each rotated by an angle $\Delta\theta = \theta_{max}/n$, where $\theta_{max} = 90°$. For each polarizer, the angle between that polarizer and the preceding polarizer is small, so we can use the small-angle approximation for $\cos\Delta\theta$ to write

$$\frac{I_1}{I_0} \approx \left(1 - \frac{(\Delta\theta)^2}{2}\right)^2.$$

Continued—

The intensity of the light that passes through the n polarizers is then

$$\frac{I_n}{I_0} \approx \left(1 - \frac{\left(\theta_{max}/n\right)^2}{2}\right)^{2n} = \left(1 - \frac{\theta_{max}^2}{2n^2}\right)^{2n}.$$

For large n,

$$\frac{I_n}{I_0} \approx 1 = 100\%.$$

Using ten polarizers to rotate the polarization of the incident polarized light allowed 78.1 % of the light to pass. Using more polarizers with smaller changes in the polarization direction would allow the transmission to approach 100%. Thus, our result seems reasonable.

SOLVED PROBLEM 31.3 Laser-Powered Sailing

One idea for propelling long-range spacecraft involves using a high-powered laser beam, rather than sunlight, focused on a large totally reflecting sail. The spacecraft could then be propelled from Earth. Suppose a 10.0-GW laser could be focused at long distances. The spacecraft has a mass of 200.0 kg, and its reflecting sail is large enough to intercept all of the light emitted by the laser.

PROBLEM
Neglecting gravity, how long would it take the spacecraft to reach a speed of 30.0% of the speed of light, starting from rest?

SOLUTION

THINK
The radiation pressure from the laser produces a constant force on the spacecraft, resulting in a constant acceleration. Using the constant acceleration, we can calculate the time to reach the final speed starting from rest.

SKETCH
Figure 31.26 is a diagram of a laser focusing light on the spacecraft with a totally reflecting sail.

FIGURE 31.26 A laser focusing its light on a spacecraft with a totally reflecting sail.

RESEARCH
The radiation pressure, p_r, from the light with intensity I produced by the laser is

$$p_r = \frac{2I}{c}.$$

Pressure is defined as force, F, per unit area, A, of the spot the beam produces on the sail, so we can write

$$\frac{2I}{c} = \frac{F}{A}.$$

The intensity of the laser is given by the power, P, of the laser divided by the spot's area, A. Assuming that the sail of the spacecraft can intercept the entire laser beam, we can write

$$\frac{F}{A} = \frac{2\left(P/A\right)}{c}.$$

Solving for the force exerted by the laser beam on the sail and using Newton's Second Law, we can write

$$F = \frac{2P}{c} = ma. \tag{i}$$

SIMPLIFY

We can solve equation (i) for the acceleration:

$$a = \frac{2P}{mc}.$$

Assuming that all the power of the laser remains focused on the sail of the spacecraft, the spacecraft will experience a constant acceleration. Then, the final speed, v, of the spacecraft can be related to the time it takes to reach that speed through

$$v = at = 0.300c.$$

Solving for the time gives us

$$t = \frac{0.300c}{2P/mc} = \frac{0.300mc^2}{2P}.$$

CALCULATE

Putting in the numerical values gives us

$$t = \frac{0.300mc^2}{2P} = \frac{0.300(200.0 \text{ kg})(3.00 \cdot 10^8 \text{ m/s})^2}{2(10.0 \cdot 10^9 \text{ W})} = 270,000,000 \text{ s}.$$

ROUND

We report our result to three significant figures:

$$t = 270,000,000 \text{ s} = 8.56 \text{ yr}.$$

DOUBLE-CHECK

To double-check our result, we calculate the acceleration of the spacecraft:

$$a = \frac{2P}{mc} = \frac{2(10.0 \cdot 10^9 \text{ W})}{(200 \text{ kg})(3.00 \cdot 10^8 \text{ m/s})} = 0.333 \text{ m/s}^2.$$

This acceleration is 3% of the acceleration due to gravity at the surface of the Earth. This acceleration is produced by a laser with 10 times the power of a typical power station, which must run continuously for 8.56 yr. The distance the spacecraft will travel during that time is

$$x = \tfrac{1}{2}at^2 = \tfrac{1}{2}(0.333 \text{ m/s}^2)(2.70 \cdot 10^8 \text{ s})^2 = 1.21 \cdot 10^{16} \text{ m} = 1.28 \text{ light year},$$

which is slightly more than $1\tfrac{1}{4}$ times the distance light travels in a year. The laser must remain focused on the spacecraft at this distance. Thus, although our calculations seem reasonable, the requirements for a laser-driven spacecraft with sail seem difficult to achieve. In Chapter 35, we'll see that we must modify this calculation because the speed involved is a significant fraction of the speed of light.

MULTIPLE-CHOICE QUESTIONS

31.1 Which of the following phenomena can be observed for electromagnetic waves but not for sound waves?

a) interference

b) diffraction

c) polarization

d) absorption

e) scattering

c) Only the electric field vector is perpendicular to the direction of the wave's propagation.

d) Both the electric field vector and the magnetic field vector are perpendicular to the direction of propagation.

e) An electromagnetic wave carries energy only when $E = B$.

31.2 Which of the following statements concerning electromagnetic waves are incorrect? (Select all that apply.)

a) Electromagnetic waves in vacuum travel at the speed of light.

b) The magnitudes of the electric field and the magnetic field are equal.

31.3 The international radio station Voice of Slobbovia announces that it is "transmitting to North America on the 49-meter band." Which frequency is the station transmitting on?

a) 820 kHz

b) 6.12 MHz

c) 91.7 MHz

d) The information given tells nothing about the frequency.

31.4 Which of the following exerts the largest amount of radiation pressure?

a) a 1-mW laser pointer on a 2-mm-diameter spot 1 m away

b) a 200-W light bulb on a 4-mm-diameter spot 10 m away

c) a 100-W light bulb on a 2-mm-diameter spot 4 m away

d) a 200-W light bulb on a 2-mm-diameter spot 5 m away

e) All of the above exert the same pressure.

31.5 What is the direction of the net force on the moving positive charge in the figure?

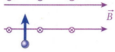

a) into the page   c) out of the page

b) toward the right   d) toward the left

31.6 A proton moves perpendicularly to crossed electric and magnetic fields as shown in the figure. What is the direction of the net force on the proton?

a) toward the left

b) toward the right

c) into the page

d) out of the page

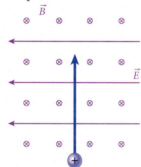

31.7 It is speculated that isolated magnetic "charges" (magnetic monopoles) may exist somewhere in the universe. Which of Maxwell's equations, (1) Gauss's Law for Electric Fields, (2) Gauss's Law for Magnetic Fields, (3) Faraday's Law of Induction, and/or (4) the Maxwell-Ampere Law, would be altered by the existence of magnetic monopoles?

a) only (2)   c) (2) and (3)

b) (1) and (2)   d) only (3)

31.8 According to Gauss's Law for Magnetic Fields, all magnetic field lines form a complete loop. Therefore, the direction of the magnetic field $\vec{B}$ points from _____ pole to _____ pole outside of an ordinary bar magnet and from _____ pole to _____ pole inside the magnet.

a) north, south, north, south

b) north, south, south, north

c) south, north, south, north

d) south, north, north, south

QUESTIONS

31.9 In a polarized light experiment, a setup similar to the one in Figure 31.22 is used. Unpolarized light with intensity I_0 is incident on polarizer 1. Polarizers 1 and 3 are crossed (at a 90° angle), and their orientations are fixed during the experiment. Initially, polarizer 2 has its polarizing angle at 45°. Then, at time $t = 0$ s, polarizer 2 starts to rotate with angular velocity ω about the direction of propagation of light in a clockwise direction as viewed by an observer looking toward the light source. A photodiode is used to monitor the intensity of the light emerging from polarizer 3.

a) Determine an expression for this intensity as a function of time.

b) How would the expression from part (a) change if polarizer 2 were rotated about an axis parallel to the direction of propagation of the light but displaced by a distance $d < R$, where R is the radius of the polarizer?

31.10 A dipole antenna is located at the origin with its axis along the z-axis. As electric current oscillates up and down the antenna, polarized electromagnetic radiation travels away from the antenna along the positive y-axis. What are the possible directions of electric and magnetic fields at point A on the y-axis? Explain.

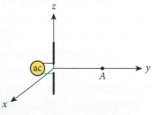

31.11 Does the information in Section 31.10 affect the answer to Example 31.2 regarding the root-mean-square magnitude of electric field at the Earth's surface from the Sun?

31.12 Maxwell's equations predict that there are no magnetic monopoles. If these monopoles existed, how would the motion of charged particles change as they approached such a monopole?

31.13 If two communication signals were sent at the same time to the Moon, one via radio waves and one via visible light, which one would arrive at the Moon first?

31.14 Show that Ampere's Law is not necessarily consistent if the surface through which the flux is to be calculated is a closed surface, but that the Maxwell-Ampere Law always is. (Hence, Maxwell's introduction of his law of induction and the displacement current are not optional; they are logically necessary.) Show also that Faraday's Law of Induction does not suffer from this consistency problem.

31.15 Maxwell's equations and Newton's laws of motion are mutually inconsistent; the great edifice of classical physics is fatally flawed. Explain why.

31.16 Practically everyone who has studied the electromagnetic spectrum has wondered how the world would appear if we could see over a range of frequencies of the ten octaves over which we can hear rather than the less than one octave over which we can see. (An octave refers to a factor of 2 in frequency.) But this is fundamentally impossible. Why?

31.17 Electromagnetic waves from a small, isotropic source are not plane waves, which have constant maximum amplitudes.

a) How does the maximum amplitude of the electric field of radiation from a small, isotropic source vary with distance from the source?

b) Compare this with the electrostatic field of a point charge.

31.18 A pair of sunglasses is held in front of a flat-panel computer monitor (which is on) so that the lenses are always parallel to the display. As the lenses are rotated, it is noticed that the intensity of light coming from the display and passing through the lenses is varying. Why?

31.19 Two polarizing filters are crossed at 90°, so when light is shined from behind the pair of filters, no light passes through. A third filter is inserted between the two, initially aligned with one of them. Describe what happens as the intermediate filter is rotated through an angle of 360°.

PROBLEMS

A blue problem number indicates a worked-out solution is available in the Student Solutions Manual. One • and two •• indicate increasing level of problem difficulty.

Section 31.1

31.20 An electric field of magnitude 200 V/m is directed perpendicular to a circular planar surface with radius 6.0 cm. If the electric field increases at a rate of 10 V/ms, determine the magnitude and the direction of the magnetic field at a radial distance 10.0 cm away from the center of the circular area.

•**31.21** A wire of radius 1.0 mm carries a current of 20.0 A. The wire is connected to a parallel plate capacitor with circular plates of radius $R = 4.0$ cm and a separation between the plates of $s = 2.0$ mm. What is the magnitude of the magnetic field due to the changing electric field at a point that is a radial distance of $r = 1.0$ cm from the center of the parallel plates? Neglect edge effects.

•**31.22** The current flowing in a solenoid that is 20 cm long and has a radius of 2 cm and 500 turns decreases from 3 A to 1 A in 0.1 s. Determine the magnitude of the induced electric field inside the solenoid 1 cm from its center.

Section 31.2

31.23 A parallel plate capacitor has air between disk-shaped plates of radius 4.0 mm that are coaxial and 1.0 mm apart. Charge is being accumulated on the plates of the capacitor. What is the displacement current between the plates at an instant when the rate of charge accumulation on the plates is 10.0 μC/s?

•**31.24** A parallel plate capacitor has circular plates of radius 10.0 cm that are separated by a distance of 5.0 mm. The potential across the capacitor is increased at a constant rate of 1200 V/s. Determine the magnitude of the magnetic field between the plates at a distance $r = 4.0$ cm from the center.

•**31.25** The voltage across a cylindrical conductor of radius r, length L, and resistance R varies with time. The time-varying voltage causes a time-varying current, i, to flow in the cylinder. Show that the displacement current equals $\epsilon_0 \rho di/dt$, where ρ is the resistivity of the conductor.

Section 31.5

31.26 The amplitude of the electric field of an electromagnetic wave is 250 V/m. What is the amplitude of the magnetic field of the electromagnetic wave?

31.27 Determine the distance in feet that light can travel in vacuum during 1.00 ns.

31.28 How long does it take light to travel from the Moon to the Earth? From the Sun to the Earth? From Jupiter to the Earth?

31.29 Alice made a telephone call from her home telephone in New York to her fiancé stationed in Baghdad, about 10,000 km away, and the signal was carried on a telephone cable. The following day, Alice called her fiancé again from work using her cell phone, and the signal was transmitted via a satellite 36,000 km above the Earth's surface, halfway between New York and Baghdad. Estimate the time taken for the signals sent by (a) the telephone cable and (b) via the satellite to reach Baghdad, assuming that the signal speed in both cases is the same as speed of light, c. Would there be a noticeable delay in either case?

•**31.30** Electric and magnetic fields in many materials can be analyzed using the same relationships as for fields in vacuum, only substituting relative values of the permittivity and the permeability, $\epsilon = \kappa \epsilon_0$ and $\mu = \kappa_m \mu_0$, for their vacuum values, where κ is the dielectric constant and κ_m the relative permeability of the material. Calculate the ratio of the speed of electromagnetic waves in vacuum to their speed in such a material.

Section 31.6

31.31 The wavelength range for visible light is 400 nm to 700 nm (see Figure 31.10) in air. What is the frequency range of visible light?

31.32 The antenna of a cell phone is a straight rod 8.0 cm long. Calculate the operating frequency of the signal from this phone, assuming that the antenna length is $\frac{1}{4}$ of the wavelength of the signal.

•**31.33** Suppose an RLC circuit in resonance is used to produce a radio wave of wavelength 150 m. If the circuit has a 2.0-pF capacitor, what size inductor is used?

•**31.34** Three FM radio stations covering the same geographical area broadcast at frequencies 91.1, 91.3, and 91.5 MHz, respectively. What is the maximum allowable wavelength width of the band-pass filter in a radio receiver so that the FM station 91.3 can be played free of interference from FM 91.1 or FM 91.5? Use $c = 3.0 \cdot 10^8$ m/s, and calculate the wavelength to an uncertainty of 1 mm.

Section 31.8

31.35 A monochromatic point source of light emits 1.5 W of electromagnetic power uniformly in all directions. Find the Poynting vector at a point situated at each of the following locations:

a) 0.30 m from the source

b) 0.32 m from the source

c) 1.00 m from the source

31.36 Consider an electron in a hydrogen atom, which is 0.050 nm from the proton in the nucleus.

a) What electric field does the electron experience?

b) In order to produce an electric field whose root-mean-square magnitude is the same as that of the field in part (a), what intensity must a laser light have?

31.37 A 3-kW carbon dioxide laser is used in laser welding. If the beam is 1 mm in diameter, what is the amplitude of the electric field in the beam?

31.38 Suppose that charges on a dipole antenna oscillate slowly at a rate of 1 cycle/s, and the antenna radiates electromagnetic waves in a region of space. If someone measured the time-varying magnetic field in the region and found its maximum to be 0.001 T, what would be the maximum electric field, E, in the region, in units of volts per meter? What is the period of the charge oscillation? What is the magnitude of the Poynting vector?

31.39 Calculate the average value of the Poynting vector, S_{ave}, for an electromagnetic wave having an electric field of amplitude 100 V/m.

a) What is the average energy density of this wave?

b) How large is the amplitude of the magnetic field?

•**31.40** The most intense beam of light that can propagate through dry air must have an electric field whose maximum amplitude is no greater than the breakdown value for air: $E_{max}^{air} = 3.0 \cdot 10^6$ V/m, assuming that this value is unaffected by the frequency of the wave.

a) Calculate the maximum amplitude the magnetic field of this wave can have.

b) Calculate the intensity of this wave.

c) What happens to a wave more intense than this?

••**31.41** A continuous-wave (cw) argon-ion laser beam has an average power of 10 W and a beam diameter of 1 mm. Assume that the intensity of the beam is the same throughout the cross section of the beam (which is not true, as the actual distribution of intensity is a Gaussian function).

a) Calculate the intensity of the laser beam. Compare this with the average intensity of sunlight at Earth's surface (1400 W/m²).

b) Find the root-mean-square electric field in the laser beam.

c) Find the average value of the Poynting vector over time.

d) If the wavelength of the laser beam is 514.5 nm in vacuum, write an expression for the instantaneous Poynting vector, where the instantaneous Poynting vector is zero at $t = 0$ and $x = 0$.

e) Calculate the root-mean-square value of the magnetic field in the laser beam.

••**31.42** A voltage, V, is applied across a cylindrical conductor of radius r, length L, and resistance R. As a result, a current, i, is flowing through the conductor, which gives rise to a magnetic field, B. The conductor is placed along the y-axis, and the current is flowing in the positive y-direction. Assume that the electric field is uniform throughout the conductor.

a) Find the magnitude and the direction of the Poynting vector at the surface of the conductor.

b) Show that $\int \vec{S} \cdot d\vec{A} = i^2 R$.

Section 31.9

31.43 Radiation from the Sun reaches the Earth at a rate of 1.4 kW/m² above the atmosphere and at a rate of 1 kW/m² on an ocean beach.

a) Calculate the maximum values of E and B above the atmosphere.

b) Find the pressure and the force exerted by the radiation on a person lying flat on the beach who has an area of 0.75 m² exposed to the Sun.

31.44 Scientists have proposed using the radiation pressure of sunlight for travel to other planets in the Solar System. If the intensity of the electromagnetic radiation produced by the Sun is about 1.4 kW/m² near the Earth, what size would a sail have to be to accelerate a spaceship with a mass of 10 metric tons at 1 m/s²?

a) Assume that the sail absorbs all the incident radiation.

b) Assume that the sail perfectly reflects all the incident radiation.

31.45 A solar sail is a giant circle (with a radius $R = 10$ km) made of a material that is perfectly reflecting on one side and totally absorbing on the other side. In deep space, away from other sources of light, the cosmic microwave background will provide the primary source of radiation incident on the sail. Assuming that this radiation is that of an ideal black body at $T = 2.725$ K, calculate the net force on the sail due to its reflection and absorption.

•**31.46** Two astronauts are at rest in outer space, one 20.0 m from the Space Shuttle and the other 40.0 m from the shuttle. Using a 100.0-W laser, the astronaut located 40.0 m away from the Shuttle decides to propel the other astronaut toward the Space Shuttle. He focuses the laser on a piece of totally reflecting fabric on her space suit. If her total mass with equipment is 100.0 kg, how long will it take her to reach the Space Shuttle?

•31.47 A laser that produces a spot of light that is 1.00 mm in diameter is shone perpendicularly on the center of a thin, perfectly reflecting circular (2 mm in diameter) aluminum plate mounted vertically on a flat piece of cork that floats on the surface of the water in a large beaker. The mass of this "sailboat" is 0.10 g, and it travels 2.0 mm in 63.0 s. Assuming that the laser power is constant in the region where the sailboat is located during its motion, what is the power of the laser? (Neglect air resistance and the viscosity of water.)

•31.48 A tiny particle of density 2000 kg/m^3 is at the same distance from the Sun as the Earth is $(1.5 \cdot 10^{11}$ m). Assume that the particle is spherical and perfectly reflecting. What would its radius have to be for the outward radiation pressure on it to be 1% of the inward gravitational attraction of the Sun? (Take the Sun's mass to be $2 \cdot 10^{30}$ kg.)

•31.49 Silica aerogel, an extremely porous, thermally insulating material made of silica, has a density of 1 mg/cm^3. A thin circular slice of aerogel has a diameter of 2 mm and a thickness of 0.1 mm.

a) What is the weight of the aerogel slice (in newtons)?

b) What is the intensity and radiation pressure of a 5-mW laser beam of diameter 2 mm on the sample?

c) How many 5-mW lasers with a beam diameter of 2 mm would be needed to make the slice float in the Earth's gravitational field? Use $g = 9.81$ m/s^2.

Section 31.10

31.50 Two polarizers are out of alignment by 30°. If light of intensity 1.00 W/m^2 and initially polarized halfway between the polarizing angles of the two filters passes through the two filters, what is the intensity of the transmitted light?

31.51 A 10-mW vertically polarized laser beam passes through a polarizer whose polarizing angle is 30° from the horizontal. What is the power of the laser beam when it emerges from the polarizer?

•31.52 Unpolarized light of intensity I_0 is incident on a series of five polarizers, each rotated 10° from the preceding one. What fraction of the incident light will pass through the series?

•31.53 A laser produces light that is polarized in the vertical direction. The light travels in the positive y-direction and passes through two polarizers, which have polarizing angles of 35° and 55° from the vertical, as shown in the figure. The laser beam is collimated (neither converging nor expand-

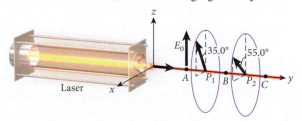

ing), has a circular cross section with a diameter of 1.00 mm, and has an average power of 15 mW at point A. At point C, what are the magnitudes of the electric and magnetic fields, and what is the intensity of the laser light?

Additional Problems

31.54 A laser beam takes 50 ms to be reflected back from a totally reflecting sail on a spacecraft. How far away is the sail?

31.55 A house with a south-facing roof has photovoltaic panels on the roof. The photovoltaic panels have an efficiency of 10% and occupy an area with dimensions 3 m by 8 m. The average solar radiation incident on the panels is 300 W/m^2, averaged over all conditions for a year. How many kilowatt hours of electricity will the solar panels generate in a 30-day month?

31.56 What is the radiation pressure due to Betelgeuse (which has a luminosity, or power output, 10,000 times that of the Sun) at a distance equal to that of Uranus's orbit from it?

31.57 A 200-W laser produces a beam with a cross-sectional area of 1 mm^2 and a wavelength of 628 m. What is the amplitude of the electric field in the beam?

31.58 What is the wavelength of the electromagnetic waves used for cell phone communications in the 850-MHz band?

31.59 As shown in the figure, sunlight is coming straight down (negative z-direction) on a solar panel (of length $L = 1.40$ m and width $W = 0.900$ m) on the Mars rover *Spirit*. The amplitude of the electric field in the solar radiation is 673 V/m and is uniform (the radiation has the same amplitude everywhere). If the solar panel has an efficiency of 18.0% in converting solar radiation into electrical power, how much average power can the panel generate?

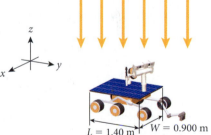

31.60 A 14.9-µF capacitor, a 24.3-kΩ resistor, a switch, and a 25.-V battery are connected in series. What is the rate of change of the electric field between the plates of the capacitor at $t = 0.3621$ s after the switch is closed? The area of the plates is 1.00 cm^2.

31.61 A focused 300-W spotlight delivers 40% of its light within a circular area with a diameter of 2 m. What is the average root-mean-square electric field in this illuminated area?

31.62 What is the electric field amplitude of an electromagnetic wave whose magnetic field amplitude is $5 \cdot 10^{-3}$ T?

31.63 What is the distance between successive heating antinodes in a microwave oven's cavity? A microwave oven typically operates at a frequency of 2.4 GHz.

31.64 The solar constant measured by Earth satellites is roughly 1400 W/m².

a) Find the maximum electric field of the electromagnetic radiation from the Sun.

b) Find the maximum magnetic field of these electromagnetic waves.

•**31.65** The peak electric field at a distance of 2.25 m from a light bulb is 21.2 V/m.

a) What is the peak magnetic field there?

b) What is the power output of the bulb?

•**31.66** If the peak electric field due to a star whose radius is twice that of the Sun is 0.015 V/m at a distance of 15 AU, what is its temperature? Treat the star as a blackbody.

•**31.67** A 5-mW laser pointer has a beam diameter of 2 mm.

a) What is the root-mean-square value of the electric field in this laser beam?

b) Calculate the total electromagnetic energy in 1 m of this laser beam.

•**31.68** At the surface of the Earth, the Sun delivers an estimated 1 kW/m² of energy. Suppose sunlight hits a 10 m by 30 m roof at an angle of 90°.

a) Estimate the total power incident on the roof.

b) Find the radiation pressure on the roof.

•**31.69** The National Ignition Facility has the most powerful laser in the world, using 192 lasers to aim 500 TW of power at a spherical pellet of diameter 2 mm. How fast would a pellet of density 2 g/cm³ move if only a single laser hits it and 1% of the light is reflected?

•**31.70** A resistor consists of a solid cylinder of radius r and length L. The resistor has resistance R and is carrying current i. Use the Poynting vector to calculate the power radiated out of the surface of the resistor.

•**31.71** What is the Poynting vector at a distance of 12.0 km from a radio tower that transmits 30,000 W of power? Assume that the radio waves that hit the Earth are reflected back into space.

a) Are the radio waves coming from this transmitter polarized or not?

b) What is the root-mean-square value of the electric force on an electron at this location?

•**31.72** Quantum theory says that electromagnetic waves actually consist of discrete packets—photons—each with energy $E = \hbar\omega$, where $\hbar = 1.054573 \cdot 10^{-34}$ J s is Planck's reduced constant and ω is the angular frequency of the wave.

a) Find the momentum of a photon.

b) Find the angular momentum of a photon. Photons are *circularly polarized;* that is, they are described by a superposition of two plane-polarized waves with equal field amplitudes, equal frequencies, and perpendicular polarizations, one-quarter of a cycle (90° or $\pi/2$ rad) out of phase, so the electric and magnetic field vectors at any fixed point rotate in a circle with the angular frequency of the waves. It can be shown that a circularly polarized wave of energy U and angular frequency ω has an angular momentum of magnitude $L = U/\omega$. (The direction of the angular momentum is given by the thumb of the right hand, when the fingers are curled in the direction in which the field vectors circulate.)

c) The ratio of the angular momentum of a particle to $\hbar$ is its spin quantum number. Determine the spin quantum number of the photon.

•**31.73** A microwave operates at 250 W. Assuming that the waves emerge from a point source emitter on one side of the oven, how long does it take to melt an ice cube 2 cm on a side that is 10 cm away from the emitter if 10% of the photons are absorbed by the cube? How many photons of wavelength 10 cm hit the ice cube per second? Assume a cube density of 0.96 g/cm³.

•**31.74** An industrial carbon dioxide laser produces a beam of radiation with average power of 6.00 kW at a wavelength of 10.6 μm. Such a laser can be used to cut steel up to 25 mm thick. The laser light is polarized in the x-direction, travels in the positive z-direction, and is collimated (neither diverging or converging) at a *constant* diameter of 100.0 μm. Write the equations for the laser light's electric and magnetic fields as a function of time and of position z along the beam. Recall that $\vec{E}$ and $\vec{B}$ are vectors. Leave the overall phase unspecified, but be sure to check the relative phase between $\vec{E}$ and $\vec{B}$.

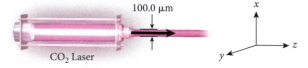

100.0 μm

x

z

y

CO₂ Laser

Geometric Optics

32

FIGURE 32.1 Primary and secondary rainbows formed by refraction and reflection of light in raindrops over Ka'anapali Beach on Maui, Hawaii.

WHAT WE WILL LEARN

- In cases where the wavelength is small compared to other length scales in a physical system, light waves can be modeled by light rays, moving on straight-line trajectories and representing the direction of a propagating light wave.

- The law of reflection states that the angle of incidence is equal to the angle of reflection.

- Mirrors can focus light and produce images governed by the mirror equation, which states that the inverse of the object distance plus the inverse of the image distance equals the inverse of the focal length of the mirror.

- Light is refracted (changes direction) when it is incident on a boundary between two optically transparent media.

- Snell's Law states that the product of the index of refraction of the medium of the incident light ray times the sine of the angle of incidence on the boundary is equal to the product of the index of refraction of the medium of the refracted light ray times the sine of the angle of the transmitted light.

- When light crosses the boundary between two media, and the index of refraction of the second medium is less than the index of refraction of the first medium, there is a critical angle of incidence above which refraction cannot take place, where instead the light is totally reflected.

The study of light is divided into three fields: geometric optics, wave optics, and quantum optics. In Chapter 31 we learned that light is an electromagnetic wave, and in Chapter 34 we will deal with the wave properties of light. This chapter discusses **geometric optics,** in which light is characterized as rays. Quantum optics makes use of the fact that light is quantized, its energy localized in point particles called photons (Chapters 36 and 37).

Rainbows (Figure 32.1) can be seen only when there are water droplets in the air and the Sun is behind the observer. Why is that? The reason has to do with how raindrops reflect and refract light—the two optical processes that are the main subjects of this chapter.

We have seen that light is a wave, but in this chapter we will examine systems in which its wavelength is small as compared to other physical dimensions of the system. Then we can ignore the wave character of light and consider only how light travels through air—or glass, or water, or any other medium. It turns out that thinking about light in this way is enough to explain the optics of mirrors, lenses, and other optical devices, including prisms and even rainbows. Later, in Chapter 34, we will examine systems in which the wavelength is not negligibly small and see how that gives rise to other optical effects.

This chapter primarily considers visible light, but keep in mind that the laws of reflection, refraction, and image formation also apply to other kinds of electromagnetic waves. For instance, many useful properties of radio waves are based on reflection and refraction.

32.1   Light Rays and Shadows

In Chapter 31 we saw that electromagnetic waves spread spherically from a point source. The concentric yellow spheres in Figure 32.2 represent the spreading of spherical **wave fronts** of the light emitted from a light bulb. (A front is a locus of points that have the same instantaneous value for the electric field.) The black arrows are the **light rays,** which are perpendicular to the wave fronts at every point in space and point in the direction of propagation of the light. The undulating red lines represent the oscillating electric field.

Light waves far away from their source can be treated as plane waves whose wave fronts are traveling in a straight line (Figure 32.3). These traveling planes can be further represented by parallel vectors or arrows perpendicular to the surface of the planes. In this chapter, we will treat light as a ray traveling in a straight line while in one homogeneous medium. Viewing light as rays will enable us to analyze and solve a broad range of practical problems, both geometrically and by means of various constructions.

Everyday experience tells us that light travels in a straight line. We cannot easily see that light has a wave structure or a quantum structure. The reason for this is simply

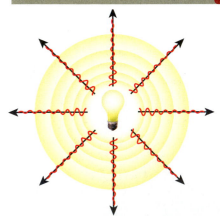

FIGURE 32.2 Light spreading from a source. Yellow: spherical wave fronts; red: oscillating electric field; black: rays.

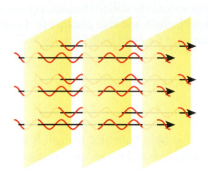

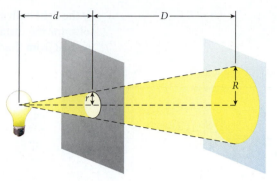

FIGURE 32.3 Planes representing wave fronts of a traveling light wave. The red sinusoidal oscillations represent the oscillating electric or magnetic field. The black arrows are the corresponding light rays that are always perpendicular to the wave front.

FIGURE 32.4 Light shining through an opening onto a screen.

that the wavelengths of visible light (400–700 nm) are small compared to the structures that form our everyday experiences. (The few notable exceptions, like soap films, will be discussed in Chapter 34.)

A person standing outside in the bright sunshine sees shadows cast by objects in the sunlight. The edges of the shadows appear reasonably sharp, so the person theorizes that light travels in straight lines and the objects block the light rays from the Sun when they strike the object. A shadow is created where light is intercepted, while bright areas are created where the unintercepted light continues in straight lines and strikes the ground or other surface. The shadow is not completely black, because light scatters from other sources and partially illuminates the shadowed area. And the edge of the shadow is also not completely sharp, because the Sun has a diameter that is not negligible. But still, the creation of shadows lends credibility to the hypothesis of straight-line motion of light.

This observation can be made in a more controlled manner by placing a piece of cardboard containing a small hole in front of a bright projector light bulb (Figure 32.4). This produces a round bright spot on the screen (the image). A smaller hole produces a smaller image on the screen. A larger hole produces a larger image.

If the size of the light source is small enough ("pointlike"), the similar triangles in Figure 32.4 can be used to find the relationship between the size of the hole (r), the size of the image (R), the distance between light source and the hole (d), and the distance between the hole and the screen (D):

$$\frac{r}{d} = \frac{R}{d+D}.$$ (32.1)

This equation is sometimes referred to as the **law of rays.**

SOLVED PROBLEM 32.1   Shadow of a Ball

The light from a small light bulb creates a shadow of a ball on a wall. The diameter of the ball is 14.3 cm and the diameter of the shadow of the ball on the wall is 27.5 cm. The ball is 1.99 m away from the wall.

PROBLEM
How far is the light bulb from the wall?

SOLUTION

THINK
The light bulb is small, which means that we can neglect its size and treat it as a point source. The triangle formed by the light bulb and the ball is similar to the triangle formed

Continued—

by the light bulb and the shadow. The distance from the ball to the wall is given, so we can solve for the distance from the light bulb to the ball, add that distance to the distance of the ball from the wall, and obtain the distance of the light bulb from the wall.

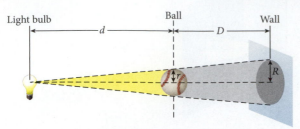

Light bulb   Ball   Wall

FIGURE 32.5   A light bulb casts a shadow of a ball on a wall.

SKETCH

Figure 32.5 shows a sketch of a light bulb casting a shadow of a ball on a wall.

RESEARCH

The triangle formed by the light bulb and the ball is similar to the triangle formed by the light bulb and the shadow. Using equation 32.1 we can write

$$\frac{r}{d} = \frac{R}{d+D},$$

where r is the radius of the ball, R is the radius of the shadow cast on the wall, d is the distance of the light bulb from the ball, and D is the distance of the ball from the wall.

SIMPLIFY

We can rearrange the law of rays to obtain

$$d+D = \frac{Rd}{r}.$$

Collecting the terms involving the distance from the light bulb to the ball on the left side gives us

$$d\left(1 - \frac{R}{r}\right) = -D.$$

Solving for the distance from the light bulb to the ball leads to

$$d = \frac{D}{\dfrac{R}{r} - 1} = \frac{rD}{R-r}.$$

The distance of the light bulb from the wall $d_{\text{lightbulb}}$ is then

$$d_{\text{lightbulb}} = d + D = \frac{rD}{R-r} + D.$$

CALCULATE

Putting in our numerical values gives us

$$d_{\text{lightbulb}} = \frac{rD}{R-r} + D = \frac{(1.99\text{ m})(0.143\text{ m})}{(0.275\text{ m}) - (0.143\text{ m})} + 1.99\text{ m} = 2.15583 + 1.99 = 4.14583\text{ m}.$$

ROUND

We report our result to three significant figures,

$$d_{\text{lightbulb}} = 4.15\text{ m}.$$

DOUBLE-CHECK

To double-check our result, we calculate the ratio of r/d and compare that result with the ratio $R/(d + D)$, which must be equal according to the law of rays, equation 32.1. We note from above, that the value of d, the distance from the bulb to the ball is 2.16 m. The first ratio is

$$\frac{r}{d} = \frac{0.143\text{ m}}{2.16\text{ m}} = 0.0662.$$

The second ratio is

$$\frac{R}{d+D} = \frac{0.275\text{ m}}{4.15\text{ m}} = 0.0663.$$

Our ratios agree to within rounding errors, so our answer seems reasonable.

As a second double-check we can examine limiting cases and see if they agree with our expectations. What happens to our result in the limiting case where the distance d from the light bulb to the ball is large compared to the distance D from the ball to the wall? This is a case similar to sunlight hitting the ball and throwing a shadow. We know that the shadow is approximately the same size as the ball in this case. And the limiting case of our formula $r/d = R/(d + D)$ bears this out, because for $d \gg D$ we see that $r/d = R/(d + D) \approx R/d$ and so $r \approx R$.

What happens in the reverse case, with $d \ll D$, where the light bulb is very close to the ball, as compared to the distance between ball and wall? The size of the shadow diverges. Our formula also shows this limiting case, because in this case $R = (d + D)r/d \approx rD/d \gg r$, that is the radius of the shadow becomes very large relative to the radius of the ball.

Let us finish this introductory discussion of light rays and ray tracing with a very important observation: The direction of the rays is reversible. In the following discussions of reflection off surfaces and refraction through boundaries between different media, light rays will be drawn with an implied direction as emerging from objects and then scattering or refracting. But all of the drawings are equally valid if the direction of the rays is reversed. Keep this in mind as we proceed through the following examples and derivations.

32.2 Reflection and Plane Mirrors

Some objects (light bulbs, fire, the Sun, …) emit light and are thus primary light sources. Objects that are not primary light sources can be seen because they reflect light. There are two different kinds of reflection, diffuse and specular. In diffuse reflection, light waves hitting the object's surface are scattered randomly. In specular reflection, they are all reflected in the same way. Most objects show diffuse reflection, where the color of the reflected light is a property not just of the wavelength of the incoming light before reflection, but also of the surface properties of the object reflecting the light. The difference between diffuse and specular reflection lies in the roughness of the surface on the scale of the wavelength of the light. For a surface showing specular reflection, the local surface normal vectors (red arrows in Figure 32.6b) are aligned, whereas they are not for a surface showing diffuse reflection (Figure 32.6a).

A **mirror** is a surface that reflects light in specular fashion. Here we will only deal with "perfect" mirrors. A perfect mirror is a mirror that does not absorb any light and that reflects 100% of incoming light, independent of the intensity of the incoming light. In Chapter 31 we saw that light is a type of electromagnetic wave. Visible light consists of electromagnetic waves with wavelengths of approximately 400 nm to 700 nm. For a mirror to be considered perfect, it must reflect 100% of the incoming light in at least this range of wavelengths. Mirror perfection is not easily achieved.

Conventional bathroom mirrors consist of a piece of glass with a metal coating on the backside. They are usually sufficient for our purposes, but they are not perfect mirrors. You can see this with the aid of a medicine cabinet with three mirrored doors. Fold out the left and right doors to the point where they are facing each other and look parallel, and then stick your head between them. You will see a huge number of your own reflections, with the light emitted from your head bouncing back and forth many times between the two mirrors until it reaches your eyes. The images formed from light that has undergone multiple reflections are noticeably dimmer (Figure 32.7). The reason is that each reflection absorbs a fraction of the incoming intensity of the light, perhaps on the order of 1%.

In 1998 Yoel Fink, then a graduate student at MIT, invented the "omnidirectional dielectric mirror," which can be considered perfect in the sense defined above, with absorption losses of less than 0.0001%. He did this by using many alternating, approximately 1 μm thick, layers of a polymer and a semiconducting glass. While this was done initially with a grant from the Defense Advanced Research Projects Agency (DARPA), this technology has now been implemented successfully to create vastly improved laser surgery tools. This

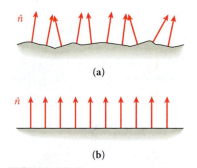

(a)

(b)

FIGURE 32.6 Orientation of surface normal vectors for a surface showing (a) diffuse reflection and one showing (b) specular reflection.

FIGURE 32.7 Light reflection between two almost exactly parallel plane mirrors.

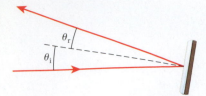

FIGURE 32.8 The angle of incidence equals the angle of reflection for reflection of light off a plane mirror. The dashed line is normal (perpendicular) to the mirror.

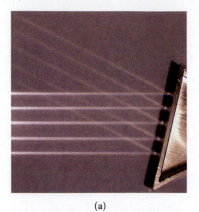

(a)

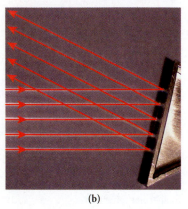

(b)

FIGURE 32.9 (a) Parallel light rays reflect off a plane mirror. (b) Arrows superimposed on the light rays.

is yet another example of how modern advanced physics research continues to lead to stunning technological breakthroughs, even in a field as long-established as geometrical optics.

We start with flat, plane mirrors. For reflection from plane mirrors, there is a simple rule for light rays incident on the surface of the mirror, known as the **law of reflection:** the angle of incidence θ_i is equal to the angle of reflection θ_r. These angles are always measured from the surface normal, which is defined to be a line perpendicular to the surface of the plane mirror. In addition, the incident ray, the normal, and the reflected ray all lie in the same plane (Figure 32.8).

Mathematically, the law of reflection is given by

$$\theta_r = \theta_i. \tag{32.2}$$

Parallel rays incident on a plane mirror are reflected such that the reflected rays are also parallel (Figure 32.9), because every normal to the surface (dashed line in Figure 32.8) is parallel to the other normals.

Image Formed by a Plane Mirror

An **image** can be formed by light reflected from a plane mirror. For example, when you stand in front of a mirror, you see an image of yourself that appears to be behind the mirror. This perception occurs because the brain assumes that light rays reaching the eyes have traveled in straight lines with no change in direction. Thus, if light rays appear to originate at a point (say behind the mirror), the eye (or a camera) sees a source of light at that point, whether or not an actual light source is there. This type of image, from which the light does not emanate, is referred to as a **virtual image.** By their nature, virtual images cannot be displayed on a screen. In contrast, **real images** are formed at a location at which you could physically put an object, like a screen or a charge coupled device (CCD) from a camera.

An example will clarify the notion of a virtual image. In Figure 32.10, every point on the surface of the candle flame emits light, with the light rays moving out in radial directions. An observer can locate the candle flame in space by tracing back the light rays to the point where they intersect. Most of these light rays are drawn in gray. The rays that hit the mirror are highlighted in black. Each of them gets reflected according to the basic law of reflection (equation 32.2), with the angle of reflection relative to the surface normal being equal to the angle of incidence. The reflected rays also have a point at which they intersect. This point is found by continuing the reflected rays behind the mirror (dashed black lines). Therefore, for the observer, the light seems to come from a point behind the mirror. This point is the location of the image formed by the mirror.

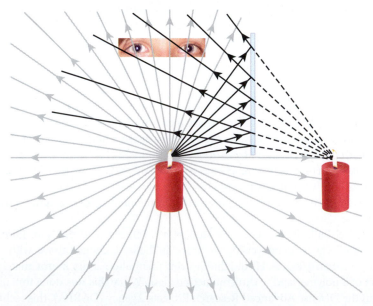

FIGURE 32.10 Image of a candle, as seen in a mirror (blue line).

Images formed by plane mirrors appear to be left-right reversed because the light rays incident on the surface of the mirror are reflected back on the other side of the normal. Thus, we have the term *mirror image*. We will deepen this point a bit more below with the aid of Figure 32.12 and Figure 32.13.

But first let's consider quantitatively the process of forming images with a plane mirror (Figure 32.11) by using the techniques of ray tracing. It turns out that a few rays are sufficient to construct the image, and we are free to select the most convenient ones.

For this image construction, we choose the case where an object with height h_o is placed a distance d_o from the mirror. Following the common convention, the object is represented by an arrow, which indicates the object's height and orientation. The object is oriented so that the tail of the arrow is on the **optic axis,** which is defined as a normal to the plane of the mirror. (For a plane mirror, the optic axis can be shown as going through any point of the mirror, but later when we examine curved mirrors we will find that there is a unique location of the optic axis.) Three light rays determine where the image is formed.

1. The first light ray emanates from the bottom of the arrow along the optic axis. This ray is reflected directly back on itself. The extrapolation of this reflected ray along the optic axis to the right of the mirror indicates that the bottom of the image is located on the optic axis.

2. The second light ray starts from the top of the arrow parallel to the optic axis and is reflected directly back on itself. The extrapolation of this ray past the mirror is shown in Figure 32.11 as a dashed line.

3. The third ray starts from the top of the arrow, strikes the mirror where the optic axis intersects the mirror, and is reflected with an angle equal to its angle of incidence. The extrapolation of the reflected ray is shown as a dashed line in Figure 32.11.

The extrapolations of these last two rays intersect at the point where the top of the image forms (Figure 32.11). It turns out that all rays from the arrow tip that hit the mirror—not just the two rays shown here—extrapolate such that they intersect at the image. Thus, the mirror produces a virtual image on the opposite side of the mirror. This image has a height h_i and is located a distance d_i to the right of the mirror. In Figure 32.11 the yellow triangle is congruent with the blue triangle. Thus,

$$h_i = h_o \tag{32.3}$$

and

$$|d_i| = |d_o|. \tag{32.4}$$

By convention, the sign of the image distance for a virtual image produced by a mirror is negative. Thus, we use the absolute value of the image distance in equation 32.4. The image produced by a plane mirror is the same distance behind the mirror as the object is in front of the mirror. In addition, the image appears upright and the same size as the object.

Why do mirror images seem to be left-right reversed but not up-down reversed? A person standing in front of a plane mirror sees an image of herself standing at the same distance behind the mirror as she is standing in front of the mirror (Figure 32.12). The image seen by the person can be constructed with two light rays as shown, but of course light rays

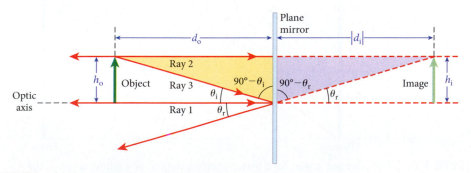

FIGURE 32.11 Ray diagram for the image formed by a plane mirror.

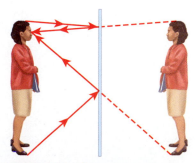

FIGURE 32.12 A person standing in front of a mirror sees a virtual image of herself.

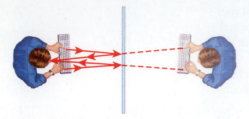

FIGURE 32.13 A person sitting in front of a mirror sees a mirror image of himself.

are coming from every visible point of the person. The image is upright (it has the same orientation as the object, not "upside-down") and virtual (the image is formed behind the surface of the mirror).

What about the apparent left-right reversal? In Figure 32.13, the virtual image is again constructed with two rays, but all other rays behave the same way. The figure shows that the mirror actually does a front-to-back reversal, not a left-right or top-bottom reversal. If you hold an arrow pointing to the right and look in the mirror, the arrow still points to the right! You can see that the real live person has his watch on his right arm. He perceives that his virtual self has his watch on his left arm only because the brain imagines that the image was formed by a 180° rotation through a vertical axis, and not by a front-back reversal. If you find this too complicated, perhaps the following visualization works for you: Suppose you paint the letters of your university on your forehead to get yourself ready for the big game. Then you take a length of clear packaging tape and tape it over the letters on your forehead. As you pull the tape off, some of the paint sticks to it. If you pull the tape straight away from you, you see the backside of the tape that contacted the paint, and the letters are on it mirror-reversed. Looking at your image in the mirror is just like looking at the backside of the tape.

If you look at an image of yourself using a webcam on your computer, you see yourself as other people see you—the image from the webcam is not reversed back-to-front. This image of yourself is confusing after your long experience with seeing yourself in a mirror. Every movement you make seems to go in the opposite direction in the image from what you expect. For this reason, some videoconferencing software now presents a reversed image that represents the back-to-front reversal inherent in a mirror image, which makes for a more natural experience.

EXAMPLE 32.1 Full-Length Mirror

PROBLEM
A person who is 184 cm (6 ft 1/2 inch) tall wants to buy a mirror in which he can see his entire body. His eyes are 8 cm from the top of his head. What is the minimum height of the mirror that is needed?

SOLUTION
For simplicity, represent the person as a pole 184 cm tall with "eyes" 8 cm from the top of the pole (this number does not matter, as the discussion below will show), as shown in Figure 32.14.

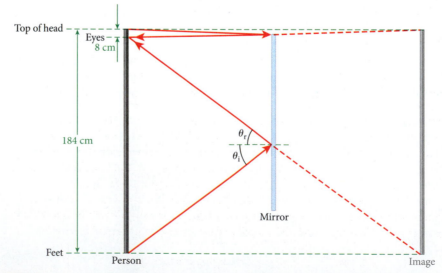

FIGURE 32.14 Distances and angles for a person standing in front of a mirror, which is the blue line.

First, consider where the light from the person's feet must travel to reach his eyes. The light leaving the feet is represented by a red arrow in Figure 32.14. The angle of incidence on the mirror, θ_i, is equal to the angle of reflection from the mirror, θ_r. We can draw two triangles that include these angles (Figure 32.15).

These two triangles are congruent because $\theta_i = \theta_r$, they each have a right angle, and they share a common side. Thus, the vertical sides of each triangle must have the same length. The sum of these two sides is equal to the height of the person minus the distance from the top of the person's head to his eyes. Therefore, the vertical side of each triangle has the length (184 cm – 8 cm)/2, as indicated in Figure 32.15.

We can now see that the bottom of the mirror only needs to extend to a height of (184 cm – 8 cm)/2 = 88 cm above the floor. A similar analysis of two congruent triangles gives us the top edge of the mirror, which needs to be (8 cm)/2 = 4 cm below the top of the person's head. Therefore, the minimum height of the mirror is 184 cm – 4 cm – 88 cm = 92 cm. This mirror is exactly half the height of the person. Thus, a mirror that is half a person's height affords the person a full-length view. This result does not depend on the distance of the eyes from the top of the person's head, or on how close to the mirror the person stands. However, it does depend on how the mirror is hung. It must be positioned so the top of the mirror is halfway between your eyes and the top of your head.

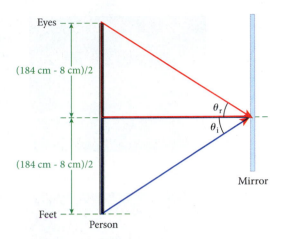

FIGURE 32.15 Two congruent triangles formed by the light from the person's feet.

32.3 Curved Mirrors

When light is reflected from the surface of a curved mirror, the light rays follow the law of reflection at each point on the surface. The light rays that are parallel before they strike the mirror are reflected in different directions, depending on the part of the mirror that they strike. Depending on whether the mirror is concave or convex, the light rays can be made to converge or diverge.

Focusing Spherical Mirrors

Consider a spherical mirror with a reflecting surface on its inside. This is a **concave mirror.** Figure 32.16 represents this spherical reflecting surface as a segment of a circle. The optic axis of the mirror, represented in this drawing by a horizontal dashed line, is a line through the center of the sphere, marked as C in Figure 32.16. Imagine that a horizontal light ray above the optic axis is incident on the surface of the mirror, parallel to the optic axis. The law of reflection applies at the point where the light ray strikes the mirror. The normal to the surface at this point, a dashed line in Figure 32.16, is a radial line that goes through the center of the sphere. On the isosceles triangle in Figure 32.16, you can see that each short side of the triangle is about half the length of the long side, provided that θ_r is small. Thus, the reflected ray crosses the optic axis approximately halfway between the mirror and point C.

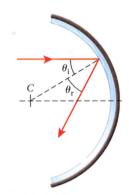

FIGURE 32.16 A horizontal light ray is reflected through the focal point of a concave mirror.

Now suppose there are many horizontal light rays incident on this spherical mirror (Figure 32.17). Each light ray obeys the law of reflection at each point. Thus, each ray will cross the optic axis halfway between the mirror and point C. This crossing point F is called the **focal point.** Note that only horizontal rays incident on the mirror close to the optic axis will be reflected through the focal point. Unless otherwise specified, all horizontal rays are assumed to be close enough to the optic axis to pass through F on reflection.

Point F is halfway between point C and the surface of the mirror. Point C is located at the center of the sphere, so the distance of C from the surface of the mirror is just the radius of the mirror, R. Therefore the **focal length** f of a converging spherical mirror is

$$f = \frac{R}{2} \quad \text{(converging mirror).} \qquad (32.5)$$

Light rays incident on an actual converging mirror are shown in Figure 32.18.

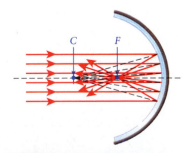

FIGURE 32.17 Many parallel light rays reflected through the focal point of a concave mirror.

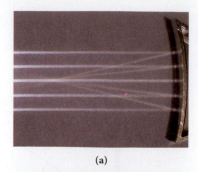

(a)

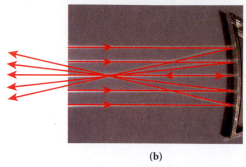

(b)

FIGURE 32.18 (a) Parallel light rays reflected to the focal point by a converging mirror. (b) Same image with arrows superimposed.

— The virtual image formed
by a convex mirror
is NOT always enlarged

A concave mirror does
NOT always form an
enlarged real image
of a real object

A convex mirror never
forms a real image of
a real object

When an object is placed
between a concave mirror
and its focal point, the
image is virtual

Now let's consider forming actual images with a converging mirror, such as the one in Figure 32.19. An object with height h_o is placed a distance d_o from the mirror, where $d_o > f$. The object is represented by an arrow, which indicates the height and orientation of the object. The object is oriented so that the tail of the arrow is on the optic axis, which, as before, is a normal to the surface of the spherical mirror along a line passing through the center C of the sphere. Four light rays determine where the image is formed.

1. The first light ray emanates from the bottom of the arrow along the optic axis and is usually not shown. This ray merely indicates that the bottom of the image is located on the optic axis.

2. The second light ray starts from the top of the arrow parallel to the optic axis and is reflected through the focal point of the mirror.

3. The third ray starts from the top of the arrow, passes through the center of the sphere, C, and is reflected back on itself.

4. The fourth ray starts from the top of the arrow, passes through the focal point, F, and is reflected back parallel to the optic axis.

The last three rays intersect at the point where the image of the top of the object is formed (Figure 32.19). It turns out that all rays from the arrow tip that strike the mirror—not just the three shown here—intersect at the image. Thus, we say that the mirror focuses the rays to form the image.

The reconstruction of the special case shown in Figure 32.19 shows a real image (on the same side of the mirror as the object, not behind the mirror) with height h_i a distance d_i from the surface of the mirror. This image has a height h_i, which is assigned a negative value to denote that the image is inverted, and is a distance d_i from the mirror. By convention, this image distance is defined as positive because the image is on the same side of the mirror as the object. The image is inverted and in this example is reduced in size relative to the object that produced the image. An image is called real when a screen placed at the image location obtains a sharp projection of the image at that point. For a virtual image, the light rays do not go through the image and thus no light reaches a screen placed at the image location.

Now let's reconstruct another case for a converging mirror, where $d_o < f$ (Figure 32.20). The object stands on the optic axis, and three light rays determine where the image is formed.

1. The first ray merely indicates that the tail of the image lies on the optic axis and is usually not shown.

2. The second ray starts from the top of the object parallel to the optic axis and is reflected through the focal point.

3. The third ray leaves the top of the object along a radius and is reflected back on itself through the center of the sphere.

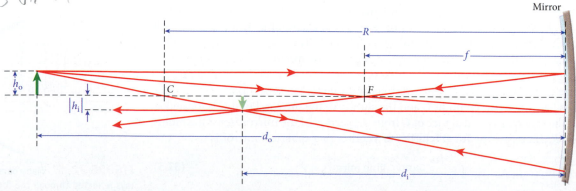

FIGURE 32.19 Image produced by a converging mirror of an object with object distance greater than the focal length of the mirror.

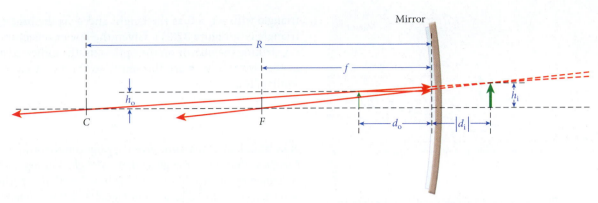

FIGURE 32.20 Image produced by a converging mirror of an object with object distance less than the focal length of the mirror.

The reflected rays are clearly diverging. To determine the location of the image, we must extrapolate the reflected rays to the other side of the mirror. These two rays intersect at a distance d_i from the surface of the mirror, producing an image with height h_i.

In this case, the image is formed on the opposite side of the mirror from the object, a virtual image. (By convention, the quantity d_i is assigned a negative value to denote that the image is virtual.) To an observer, the image appears to be behind the mirror, as was the case for the plane mirror. The image is upright and larger than the object. These results for $d_o < f$ are quite different from those for $d_o > f$. Mirrors used for shaving or applying cosmetics are usually converging mirrors; they are used with the face placed closer than the focal length, producing a large, upright image.

Before treating diverging mirrors, let's formalize the sign conventions for distances and heights.

1. All distances on the same side of the mirror as the object are defined to be positive, and all distances on the opposite side of the mirror from the object are defined to be negative. Thus f and d_o are positive for converging mirrors.

2. For real images, d_i is positive. For virtual images, d_i is negative.

3. If the image is upright, then h_i is positive, while if the image is inverted, h_i is negative.

We can derive the **mirror equation** (equation 32.6), which relates the object distance d_o, the image distance d_i, and the focal length of the mirror f:

$$\frac{1}{d_o} + \frac{1}{d_i} = \frac{1}{f}.$$

(32.6)

The signs of the terms in this equation are based on the conventions just defined.

DERIVATION 32.1 Spherical-Mirror Equation

The spherical-mirror equation can be derived by starting with a converging mirror that has a focal length f. An object h_o tall stands at a distance d_o from the surface of the mirror on the optic axis (Figure 32.21).

Trace a ray from the top of the object parallel to the optic axis and reflect it through the focal point. Trace a second ray from the top of the object through the focal point that is reflected parallel to the optic axis. The image is formed with height h_i at distance d_i from the mirror. Form a right triangle with height h_o and base d_o (green triangle 1 in Figure 32.21). Form a second right

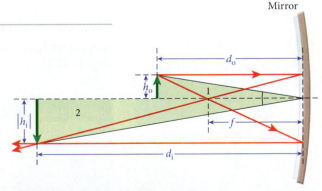

FIGURE 32.21 The two similar triangles 1 and 2 in the derivation of the spherical-mirror equation.

Continued—

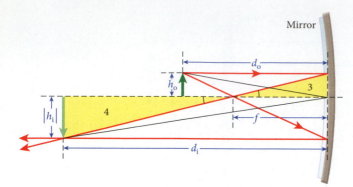

FIGURE 32.22 The two similar triangles 3 and 4 in the derivation of the spherical-mirror equation.

triangle with $-h_i > 0$ as the height and d_i as the base (green triangle 2 in Figure 32.21). Given the law of reflection for a ray striking the mirror on the optic axis, the indicated angles in the two triangles are the same, so the two triangles are similar. Thus,

$$\frac{h_o}{d_o} = \frac{-h_i}{d_i} \quad \text{or} \quad \frac{-h_i}{h_o} = \frac{d_i}{d_o}. \tag{i}$$

Now let us look at the same geometry but consider two different triangles. One right triangle (yellow triangle 4 in Figure 32.22) is defined by the height $-h_i$ and the base length $d_i - f$. The second triangle (yellow triangle 3 in Figure 32.22) is defined by the height h_o and the base f. The two triangles are similar, so

$$\frac{h_o}{f} = \frac{-h_i}{d_i - f} \quad \text{or} \quad \frac{-h_i}{h_o} = \frac{d_i - f}{f}.$$

Substituting equation (i) for the ratio of the heights derived above gives

$$\frac{d_i}{d_o} = \frac{d_i - f}{f}.$$

Multiplying both sides of this equation by f:

$$\frac{fd_i}{d_o} = d_i - f \Rightarrow \frac{fd_i}{d_o} + f = d_i.$$

Finally, dividing both sides of this equation by the product fd_i leads to the spherical-mirror equation:

$$\frac{1}{d_o} + \frac{1}{d_i} = \frac{1}{f}.$$

The **magnification** m of the mirror is defined to be

$$m = \frac{h_i}{h_o} = -\frac{d_i}{d_o}. \tag{32.7}$$

Note that the magnification m is negative for the situation used in the derivation. Algebraically, this occurs because $h_i < 0$. The significance of a negative m is that $m < 0$ tells us that the image is inverted. Table 32.1 summarizes the characteristics of images formed by a converging mirror for five different classes of object distances.

Diverging Spherical Mirrors

Suppose we have a spherical mirror with the reflecting surface on the outside of the sphere (Figure 32.23).

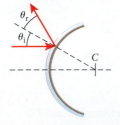

FIGURE 32.23 Reflection of a horizontal light ray from a diverging spherical mirror.

Table 32.1	Image Characteristics for Converging Mirrors		
Case	Image Type	Image Orientation	Magnification
$d_o < f$	Virtual	Upright	Enlarged
$d_o = f$	Real	Upright	Image at infinity
$f < d_o < 2f$	Real	Inverted	Enlarged
$d_o = 2f$	Real	Inverted	Same size
$d_o > 2f$	Real	Inverted	Reduced

This is a **convex mirror,** and the reflected rays diverge. In Figure 32.23 this spherical reflecting surface is indicated by a semicircle. The optic axis of the mirror is a line through the center of the sphere, represented by the horizontal dashed line. Imagine that a horizontal light ray above the optic axis is incident on the surface of the mirror. At the point where the light ray strikes the mirror, the law of reflection applies.

In contrast to the converging mirror, the normal points *away* from the center of the sphere. When we extrapolate the normal through the surface of the sphere, it intersects the optic axis of the sphere at its center, marked C in Figure 32.23. When we observe the reflected ray, it seems to be coming from inside the sphere.

Suppose many horizontal light rays are incident on this spherical mirror (Figure 32.24). Each light ray obeys the law of reflection. The rays diverge and do not seem to form any kind of image. However, if the reflected rays are extrapolated through the surface of the mirror, they all intersect the optic axis at one point. This point is called the focal point of this diverging mirror.

Figure 32.25a shows five parallel light rays incident on an actual diverging spherical mirror. The dashed lines in Figure 32.25b represent the extrapolation of the reflected rays to show the focal point.

Now let's discuss images formed by diverging mirrors (Figure 32.26). Again we use three rays.

1. The first ray establishes that the tail of the arrow lies on the optic axis and is usually not shown.

2. The second ray starts from the top of the object, traveling parallel to the optic axis, and is reflected from the surface of the mirror such that its extrapolation crosses the optic axis a distance from the surface of the mirror equal to the focal length of the mirror.

3. The third ray begins at the top of the object and is directed so that its extrapolation would intersect the center of the sphere. This ray is reflected back on itself.

The reflected rays diverge, but the extrapolated rays converge a distance d_i from the surface of the mirror on the side of the mirror opposite the object. The rays converge at a distance h_i above the optic axis, forming an upright, reduced image on the side of the mirror opposite the object. This image is virtual because the light rays do not actually go through it. These characteristics (upright, reduced, virtual image) are valid for all object distances for a diverging mirror. Such mirrors are often used in stores to give clerks at the front of the store a wide view down the aisles and for passenger side rear-view mirrors on cars.

In the case of a diverging mirror, the focal length f is negative because the focal point of the mirror is on the side opposite the object. A negative value is also assigned to the radius of a diverging mirror. Thus,

$$f = \frac{R}{2} \quad \text{(with } R < 0 \text{ for a diverging mirror).}$$

The object distance d_o is always taken to be positive. Rearranging the mirror equation (equation 32.6), which is also valid for a diverging mirror, gives

$$d_i = \frac{d_o f}{d_o - f}. \tag{32.8}$$

If d_o is always positive and f is always negative, d_i is always negative. Applying equation 32.7 for the magnification, we find that m is always positive. Looking at Figure 32.26 will also convince you that the image is always reduced in size. Thus, a diverging mirror (even if $d_o > |f|$) always produces a virtual, upright, and reduced image.

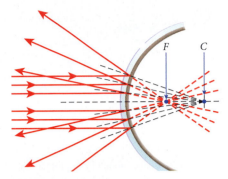

FIGURE 32.24 The reflection of parallel light rays from the surface of a diverging mirror.

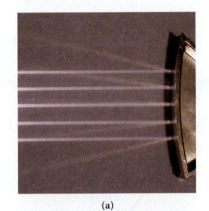

(a)

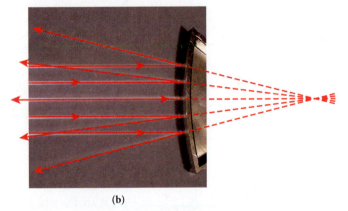

(b)

FIGURE 32.25 (a) Parallel light rays reflected from a diverging spherical mirror. (b) Arrows corresponding to the light rays, with the dashed lines representing the extrapolated light rays.

32.1 Self-Test Opportunity

You are standing 2.50 m from a diverging mirror with focal length $f = -0.500$ m. What do you see in the mirror? (The answer "Myself" is not good enough!)

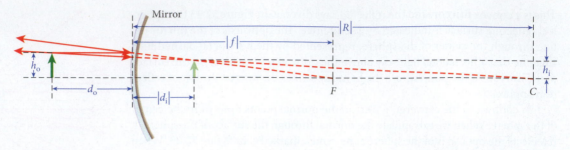

FIGURE 32.26 Image formed by an object placed in front of a diverging spherical mirror.

EXAMPLE **32.2** **Image Formed by a Converging Mirror**

Consider an object 5.00 cm tall placed 55.0 cm from a converging mirror with focal length 20.0 cm (Figure 32.27).

FIGURE 32.27 An object (green arrow) that forms an image using a converging mirror.

PROBLEM 1

Where is the image produced?

SOLUTION 1

We can use the mirror equation to find the image distance d_i in terms of the object distance d_o and the focal length of the mirror f,

$$\frac{1}{d_o} + \frac{1}{d_i} = \frac{1}{f}.$$

The mirror is specified as converging, so the focal length is positive. The image distance is

$$d_i = \frac{d_o f}{d_o - f} = \frac{(55.0 \text{ cm})(20.0 \text{ cm})}{55.0 \text{ cm} - 20.0 \text{ cm}} = 31.4 \text{ cm}.$$

PROBLEM 2

What are the size and orientation of the produced image?

SOLUTION 2

The magnification m is given by

$$m = \frac{h_i}{h_o} = -\frac{d_i}{d_o},$$

where h_o is the height of the object and h_i is the height of the produced image. The image height is then

$$h_i = -h_o \frac{d_i}{d_o} = -(5.00 \text{ cm})\frac{31.4 \text{ cm}}{55.0 \text{ cm}} = -2.85 \text{ cm}.$$

The magnification is

$$m = \frac{h_i}{h_o} = \frac{-2.85 \text{ cm}}{5.00 \text{ cm}} = -0.570.$$

Thus the produced image is inverted and reduced.

32.1 In-Class Exercise

A small object is placed in front of a converging mirror with radius $R = 7.50$ cm such that the image distance equals the object distance. How far is this small object from the mirror?

a) 2.50 cm   d) 10.0 cm

b) 5.00 cm   e) 15.0 cm

c) 7.50 cm

Spherical Aberration

The equations we have derived for spherical mirrors $\left(\dfrac{1}{d_o} + \dfrac{1}{d_i} = \dfrac{1}{f} \right.$ and $m = \dfrac{h_i}{h_o} = -\dfrac{d_i}{d_o} \left. \right)$ apply only to light rays that are close to the optic axis. If the light rays are far from the optic axis, they will not be focused through the focal point of the mirror, leading to a distorted image. Strictly speaking, there is no precise focal point in this situation. This condition is called **spherical aberration.**

Figure 32.28 shows several light rays not close to the optic axis incident on a spherical converging mirror. The rays farther from the optic axis are reflected in such a way that they cross the optic axis closer to the mirror than do rays that strike the mirror closer to the axis. As the rays approach the optic axis, they are reflected through points closer and closer to the focal point.

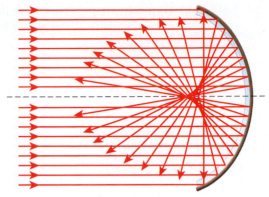

FIGURE 32.28 Parallel light rays incident on a spherical converging mirror, demonstrating spherical aberration.

DERIVATION 32.2 | Spherical Aberration for Converging Mirrors

Up until now we assumed that light rays parallel to the optic axis that are close to the axis are reflected such that they cross the optic axis at the focal point of the mirror. The focal point is at half the radius of curvature of the mirror. However, light rays that are far from the optic axis are not reflected through the focal point. To derive an expression for the point at which rays parallel to but far from the optic axis cross the axis, we draw the geometry in Figure 32.29.

Start with a light ray parallel to the optic axis and at a distance d from it. The ray reflects and crosses the axis a distance y from the mirror. The ray makes an angle θ with respect to the normal to the mirror surface, which is a radius R of the spherical mirror. The law of reflection tells us that the angle of incidence of the ray on the surface of the mirror is equal to the angle of reflection.

Define the distance from the center of the spherical mirror to the point at which the reflected ray crosses the optic axis as x. Draw a line from the point where the ray crosses the axis perpendicular to the normal. This forms two congruent right triangles with angle θ, hypotenuse x, and adjacent side $R/2$, such that

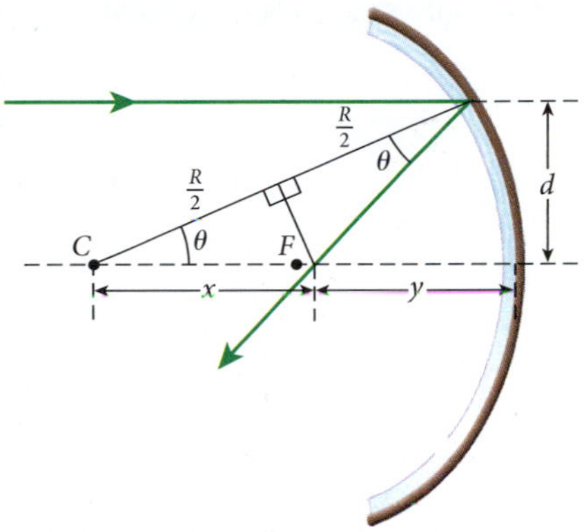

FIGURE 32.29 Geometry for the spherical aberration of a converging spherical mirror.

$$\cos\theta = \frac{R/2}{x} \Rightarrow x = \frac{R}{2\cos\theta}.$$

We can also express θ in terms of d as

$$\sin\theta = \frac{d}{R} \Rightarrow \theta = \sin^{-1}\left(\frac{d}{R}\right).$$

The distance y is given by

$$y = R - x = R - \frac{R}{2\cos\theta} = R\left(1 - \frac{1}{2\cos\theta}\right) = R\left(1 - \frac{1}{2\cos\left(\sin^{-1}\left(\dfrac{d}{R}\right)\right)}\right).$$

Remembering the trigonometric identity $\cos(\sin^{-1}(x)) = \sqrt{1-x^2}$, we can rewrite our result for y as

$$y = R\left(1 - \frac{1}{2\sqrt{1-\left(\dfrac{d}{R}\right)^2}}\right) = \frac{R}{2}\left(2 - \frac{1}{\sqrt{1-\left(\dfrac{d}{R}\right)^2}}\right) = \frac{R}{2}\left(2 - \left[1 - \left(\dfrac{d}{R}\right)^2\right]^{-1/2}\right).$$

Continued—

32.2 Self-Test Opportunity

Consider a converging spherical mirror with $R = 7.20$ cm without assuming that the incident light rays are close to the optic axis of the mirror. However, the incident light rays are parallel to the axis. Calculate the position at which the reflected light rays intersect the optic axis if the incoming ray is

a) 0.720 cm away from the optical axis.

b) 0.800 cm away from the optical axis.

c) 1.80 cm away from the optical axis.

d) 3.60 cm away from the optical axis.

We can see that for $d \ll R$, $1 - (d/R)^2 \approx 1$ and $y \approx R/2$, which agrees with equation 32.5. We can make a better approximation by writing the Maclaurin series expansion

$$\left(1 - x^2\right)^{-1/2} \approx 1 + \frac{x^2}{2} \quad (x \ll 1).$$

Taking $x = (d/R) \ll 1$, we can make the approximation

$$y = \frac{R}{2}\left(2 - \left[1 - \left(\frac{d}{R}\right)^2\right]^{-1/2}\right) \approx \frac{R}{2}\left(2 - \left[1 + \frac{1}{2}\left(\frac{d}{R}\right)^2\right]\right) = \frac{R}{2}\left(1 - \frac{d^2}{2R^2}\right).$$

Thus, the amount of spherical aberration is given approximately by

$$\frac{R}{2} - y \approx \frac{d^2}{4R} \quad \left(\frac{d}{R} \ll 1\right).$$

Parabolic Mirrors

Parabolic mirrors have a surface that reflects light from a distant source to the focal point from anywhere on the mirror. Thus, the full size of the mirror can be used to collect light and form images that do not suffer from spherical aberrations. Figure 32.30 shows vertical light rays incident on a parabolic mirror. All rays are reflected through the focal point of the mirror.

If a parabola is described by an equation $y(x) = ax^2$, then its focal point is located at the point $(x = 0, y = 1/(4a))$. Its focal length is therefore

$$f = \frac{1}{4a}. \tag{32.9}$$

Parabolic mirrors are more difficult to produce than spherical mirrors and are accordingly more expensive. Most large reflecting telescopes use parabolic mirrors in order to avoid spherical aberration. Many automobile headlights use parabolic reflectors with the same idea but send light in the opposite direction: The light source is placed at the focal point and the reflector sends the light out in a strong beam parallel to the optic axis, as illustrated in Figure 32.31. (However, in the not-so-distant future, incandescent headlights are likely going to be replaced with LED headlights, which likely will use small lenses to accomplish the same.)

Figure 32.31a shows the low-beam and high-beam headlights for a car. The parabolic reflectors for headlights are faceted. A facet is a flat area on the reflector. These facets help produce the light distribution required for the headlight. Figure 32.31b shows the light bulb for the high-beam headlight at the focus of a parabolic mirror that functions as a reflector for the light produced by the light bulb, producing a focused parallel beam of light.

The satellite TV antennas ("dishes") found on many rooftops are also parabolic in shape. These satellite dishes are not mirrors in the sense that they do not reflect visible light in specular fashion. But they are still reflecting parabolic mirrors in the wavelength range used for satellite TV transmission.

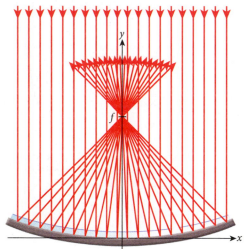

FIGURE 32.30 Light rays reflected by a parabolic mirror.

FIGURE 32.31 (a) Headlight assembly for a car. The left light is the low-beam headlight and the right light is the high-beam headlight. (b) Drawing showing the high-beam headlight with the light bulb at the focus of the parabolic reflector.

(a)

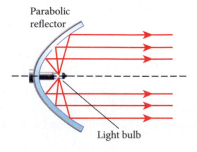

Parabolic reflector

Light bulb

(b)

Rotation Parabolas

For precise optical applications, clearly it is best to have parabolic mirrors. For example, in Chapter 33 we will see that huge parabolic mirrors are used in telescopes. One very interesting way to create parabolic mirrors is to put a liquid into rotational motion. At every point on the surface of the liquid, the surface will be perpendicular to the force from the fluid acting on that surface element. This force, $\vec{F}$, has to add with the force of gravity acting on the surface element, $-mg\hat{y}$, to provide the net centripetal force, which is needed to keep the surface element on a circular path (Figure 32.32). In the xy-coordinate system chosen here, the centripetal force is $-m\omega^2 x\hat{x}$ (compare Chapter 9 on circular motion).

The angle θ of the surface element with respect to the horizontal is given by $\tan\theta = dy/dx$ (see Figure 32.32). The same angle also can be used to express the components of the force $\vec{F}$. The vertical component of $\vec{F}$ has to balance the force of gravity, and the horizontal component has to provide the net centripetal force,

$$F\cos\theta = mg$$

$$F\sin\theta = m\omega^2 x.$$

Dividing these two equations by each other leads to $\tan\theta = (\omega^2/g)x$. Previously we showed that $\tan\theta = dy/dx$, so

$$\frac{dy}{dx} = \frac{\omega^2}{g}x.$$

Integration then results in the desired shape of the surface

$$y(x) = \frac{\omega^2}{2g}x^2,$$

which is a parabola. Because the focal length of a parabola of the form $y = ax^2$ is $f = 1/(4a)$, the focal length of this parabolic mirror made of a rotating liquid is

$$f = \frac{g}{2\omega^2}.$$

Rotation of liquid surfaces has been successfully used to construct large telescope mirrors. One such mirror is shown in Figure 32.33. There are now designs to construct a very large version of such a telescope on the Moon. While this may sound like science fiction at the present, such a telescope would be much cheaper to construct than one with a solid mirror. Since the Moon has no atmosphere, the telescope would not suffer from the atmospheric distortions that all ground-based telescopes suffer from on Earth. And it could be constructed on a much larger scale than what is possible with satellite-based telescopes like the Hubble Space Telescope. (We will cover telescopes in much greater detail in Chapter 33.)

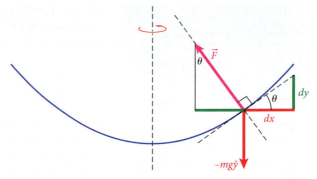

FIGURE 32.32 Geometry of a rotation parabola and free-body diagram for a fluid surface element. The force $\vec{F}$ exerted by the fluid on a surface element is shown in magenta. It needs to balance the force of gravity (red) and provide the centripetal force (green) necessary to keep the surface element moving in a circular path.

32.3 Self-Test Opportunity

Show that the derivation of $y(x) = (\omega^2/2g)x^2$ for the surface of a rotating liquid can be accomplished using energy arguments.

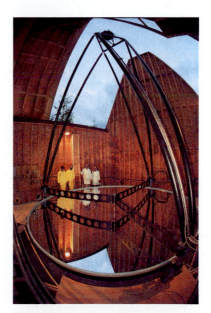

FIGURE 32.33 University of British Columbia liquid-mirror telescope.

32.2 In-Class Exercise

Suppose you have a liquid-mirror telescope of focal length f_1, and you want to double this focal length. What adjustment do you have to make to the rotational angular velocity of your mirror liquid?

a) reduce it by a factor of 0.5

b) reduce it by a factor of 0.707

c) keep it the same

d) increase it by a factor of 0.707

e) increase it by a factor of 2

32.4 Refraction and Snell's Law

Light travels at different speeds in different optically transparent materials. The ratio of the speed of light in vacuum divided by the speed of light in a material is called the **index of refraction** of the material. The index of refraction n is given by

$$n = \frac{c}{v}, \tag{32.10}$$

Table 32.2	Indices of Refraction for Some Common Materials*
Material	**Index of Refraction**
Gases	
Air	1.000271
Helium	1.000036
Carbon dioxide	1.00045
Liquids	
Water	1.333
Methyl alcohol	1.329
Ethyl alcohol	1.362
Glycerine	1.473
Benzene	1.501
Typical oil	1.5
Solids	
Ice	1.310
Calcium fluoride	1.434
Fused quartz	1.46
Salt	1.544
Polystyrene	1.49
Typical lucite	1.5
Typical glass	1.5
Quartz	1.544
Diamond	2.417

*for light with wavelength 589.3 nm

where c is the speed of light in vacuum and v is the speed of light in the medium. The speed of light in a physical medium such as glass is always less that the speed of light in vacuum. Thus, the index of refraction of a material is always greater than or equal to 1, and by definition the index of refraction of vacuum is 1. Table 32.2 lists the indices of refraction for some common materials. In general, the index of refraction is a function of the wavelength of the light, but the table gives typical average values for visible light. We will return to the wavelength dependence at the end of this section in the discussion of chromatic dispersion.

When light crosses the boundary between two transparent materials, some of it is reflected, but usually some of the light crosses the boundary into the other material and in the process is refracted. **Refraction** means that the light rays do not travel in a straight line across the boundary, but change direction. When light crosses a boundary from a medium with a lower index of refraction n_1 to a medium with a higher index of refraction n_2, the light rays change their direction and bend *toward* the normal to the boundary (Figure 32.34). Changing direction toward the normal means that the angle of refraction θ_2 is less than the angle of incidence θ_1.

Figure 32.35 shows actual light rays in air incident on the boundary between air and glass. The light rays are refracted toward the normal. (The reflected light is also observable.)

When light crosses a boundary from a medium with a higher index of refraction n_1 to a medium with a lower index of refraction n_2, the light rays are bent *away* from the normal (Figure 32.36). Bending away from the normal means $\theta_2 > \theta_1$.

Figure 32.37 shows actual light rays in glass incident on the boundary between glass and air. The light rays are refracted away from the normal.

From measurements of the angles of incident and refracted rays in media with different indices of refraction, we can construct a law of refraction based on empirical observations. This law of refraction can be expressed as

$$n_1 \sin \theta_1 = n_2 \sin \theta_2, \qquad (32.11)$$

where n_1 and θ_1 are the index of refraction and angle of incidence (relative to the surface normal) in the first medium, and n_2 and θ_2 are the index of refraction and angle of the transmitted ray (relative to the surface normal) in the second medium. This law of refraction is also called **Snell's Law.** It cannot be proven by using ray optics alone; however, in Chapter 34 this law is derived using wave optics.

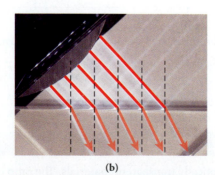

(a)

(b)

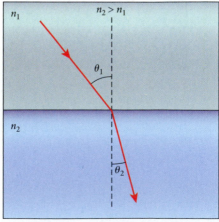

FIGURE 32.34 Light rays refracted at the boundary between two optical media with $n_1 < n_2$. (The reflected light ray is not shown.)

FIGURE 32.35 (a) Light rays refracted when crossing the boundary between air and glass. (b) Arrows superimposed on the light rays. The dashed lines are normal to the surface.

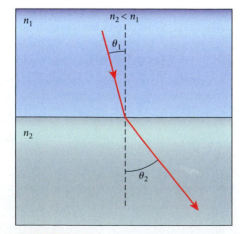

FIGURE 32.36 Light rays refracted at the boundary between two optical media with $n_1 > n_2$.

The index of refraction of air is very close to 1, as shown in Table 32.2, and in this book we will always assume that the index of refraction of air $n_{air} = 1$. It is very common for light to be incident on various media from air, so we write the formula for refraction of light incident on a surface with index of refraction n_{medium} as

$$n_{medium} = \frac{\sin \theta_{air}}{\sin \theta_{medium}}, \qquad (32.12)$$

where θ_{air} is the angle of incidence (relative to the surface normal) for light in air and θ_{medium} is the angle of the refracted light (relative to the surface normal) in the medium. Note that θ_{air} is always greater than θ_{medium}!

Also note that for light coming out of a medium into air, the formula is the same (equation 32.12)! This is a special case of the general statement advanced earlier: All ray-tracing figures remain valid if the direction of the light rays is reversed. For example, changing the direction of the arrows on the red line in Figure 32.36 still yields a physically valid path for a light ray crossing the boundary between the two media.

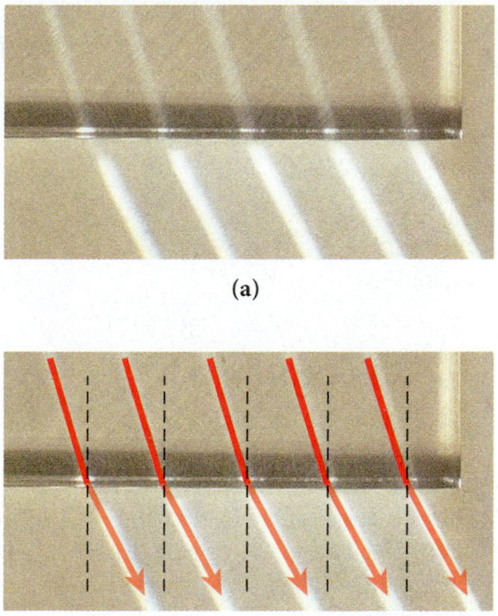

Fermat's Principle

We have just stated that Snell's Law cannot be proven by ray optics alone. But there is a qualifier to this statement, in that if you accept Fermat's Principle, then Snell's Law emerges automatically. **Fermat's Principle,** or the **principle of least time,** as expressed by Fermat in 1657, states that the path taken by a ray of light between two points in space is the path that takes the least time.

How can we prove Snell's Law this way? The key is the connection between the speed of light in a medium, v, and the index of refraction, n, which was established in equation 32.10. If light moves at different speeds in two media, then Fermat's Principle is completely equivalent to winning the aquathlon race, which we already solved in Chapter 2, Example 2.6, and the following text. In that problem, we calculated the angle at which a contestant must swim to shore in a race that consists of a swim to shore (with one given speed) and a run along the shore (with another speed). Winning the race means taking the minimum time, which is exactly what Fermat's Principle postulates.

If you accept Fermat's Principle and you want to prove Snell's Law, then all you have to do is to revisit the solution to Example 2.6. But of course you can ask where Fermat's Principle comes from. Using the minimum time to win a race seems obvious. But what makes the light rays understand that they should take the minimum time to get from point A to point B? The answer to this can be found in Huygens's Principle. However, to gain an understanding of the physics underlying that principle, we need to wait until Chapter 34, when we revisit the model of light as a wave.

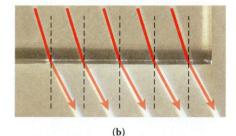

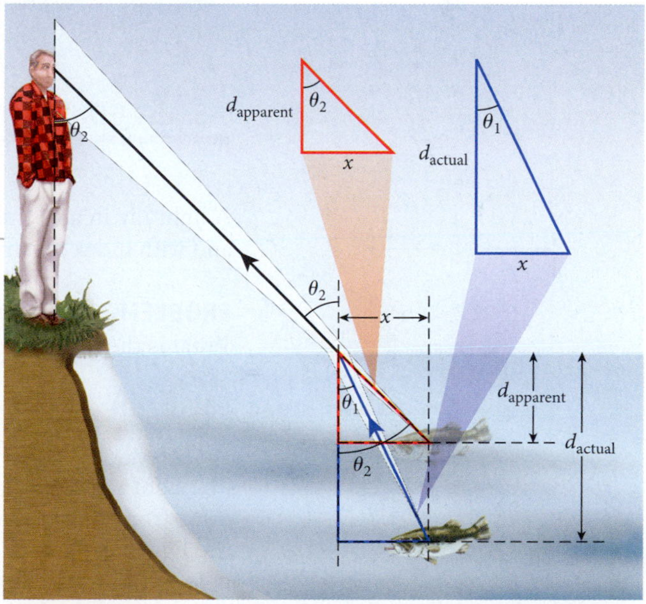

FIGURE 32.37 (a) Light rays refracted when crossing the boundary between glass and air. (b) Arrows superimposed on the light rays. The dashed lines are normal to the surface.

[handwritten annotations:] θ_{air}

$$max. \ \theta_{med} = sin^{-1}\left(\frac{1}{n}\right)$$

EXAMPLE 32.3   Apparent Depth

You are standing on the edge of a pond and looking at the water at an angle of $\theta_2 = 45.0°$ to the vertical as shown in Figure 32.38.

You see a fish in the pond. The actual depth of the fish is $d_{actual} = 1.50$ m.

PROBLEM
What is the apparent depth of fish from your vantage point?

SOLUTION
Light rays from the fish are refracted at the surface of the water into your eyes at a 45.0° angle as shown in Figure 32.38. In the figure we show not just one ray, but a collection of rays emitted from one point on the body of the fish around this central

Continued—

FIGURE 32.38 You are looking at the surface of a pond at an angle of $\theta_2 = 45.0°$, and you see a fish under the water. The rays emitted from one point on the body of the fish that hit your eye are shown as the semitransparent cone, with the center line of the cone represented by a solid line.

ray (semitransparent cone). Where the eyes see these rays intersect is where we locate the fish.

The apparent depth of the fish is located along the dotted red line shown in Figure 32.38 extrapolated from the direction of the light rays as they enter our eye. This extrapolated line makes an angle $\theta_2 = 45.0°$ with the normal to the surface of the water.

Using Snell's Law, we can calculate the angle θ_1 that the light rays from the fish must make at the surface of the water:

$$n_{air} \sin\theta_2 = n_{water} \sin\theta_1.$$

Taking $n_{air} = 1.00$ and $n_{water} = 1.333$ from Table 32.2, we can calculate θ_1:

$$\theta_1 = \sin^{-1}\left(\frac{\sin\theta_2}{n_{water}}\right) = \sin^{-1}\left(\frac{\sin 45.0°}{1.333}\right) = 32.0°.$$

We define the horizontal distance between the point where the light rays intersect the surface of the water and the fish as x, as shown in Figure 32.38. We can then define the red triangle and the blue triangle. From the red triangle we get

$$\tan\theta_2 = \frac{x}{d_{apparent}},$$

and from the blue triangle we obtain

$$\tan\theta_1 = \frac{x}{d_{actual}}.$$

Solving each of these two equations for x and setting them equal to each other gives us

$$d_{actual} \tan\theta_1 = d_{apparent} \tan\theta_2,$$

which we can solve for the apparent depth of the fish:

$$d_{apparent} = \frac{d_{actual} \tan\theta_1}{\tan\theta_2} = \frac{(1.50 \text{ m}) \tan(32.0°)}{\tan(45.0°)} = 0.937 \text{ m}.$$

Thus, the fish appears to be closer to the water surface than it actually is.

32.3 In-Class Exercise

If the same fish swimming at the same depth is seen by a different observer, but that observer sees the fish at a larger angle θ_2 than the value specified in Example 32.3, then she sees the fish at an apparent depth that is

a) larger than the one calculated in Example 32.3.

b) the same as the one calculated in Example 32.3.

c) smaller than the one calculated in Example 32.3.

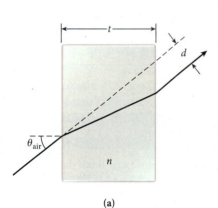

(a)

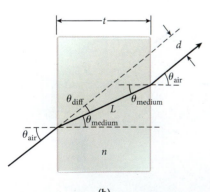

(b)

FIGURE 32.39 (a) Light ray incident on a sheet of transparent material with thickness t and index of refraction n. (b) Angles of incidence and refraction as the ray enters and exits the sheet.

SOLVED PROBLEM 32.2 | Displacement of Light Rays in Transparent Material

A light ray in air is incident on a sheet of transparent material with thickness $t = 5.90$ cm and with index of refraction $n = 1.50$. The angle of incidence is $\theta_{air} = 38.5°$ (Figure 32.39a).

PROBLEM
What is the distance d that the light ray is displaced after it passes through the sheet?

SOLUTION

THINK
The light ray refracts toward the normal as it enters the transparent sheet and refracts away from the normal as it exits the transparent sheet. After passing through the sheet, the light ray is parallel to the incident light ray, but is displaced. Using Snell's Law, we can calculate the angle of refraction in the transparent sheet. Having that angle, we can use trigonometry to calculate the displacement of the light ray passing through the sheet.

SKETCH
Figure 32.39b shows a sketch of the light ray passing through a transparent sheet.

RESEARCH

The light ray is incident on the sheet with an angle θ_{air}. Snell's Law relates the incident angle to the refracted angle θ_{medium} through the index of refraction of the transparent material n,

$$n = \frac{\sin\theta_{\text{air}}}{\sin\theta_{\text{medium}}}.$$

We can solve this equation for the refraction angle:

$$\theta_{\text{medium}} = \sin^{-1}\left(\frac{\sin\theta_{\text{air}}}{n}\right). \tag{i}$$

We call L the distance the light ray travels in the transparent sheet. Using Figure 32.39b, we can relate the thickness of the sheet t to the refraction angle:

$$\cos\theta_{\text{medium}} = \frac{t}{L}. \tag{ii}$$

If θ_{diff} is the difference between the incident angle and the refraction angle, as shown in Figure 32.39b, then

$$\theta_{\text{diff}} = \theta_{\text{air}} - \theta_{\text{medium}}. \tag{iii}$$

We can see in Figure 32.39b that the displacement of the light ray d can be related to the distance the light ray travels in the sheet and the difference between the incident angle and the refracted angle,

$$\sin\theta_{\text{diff}} = \frac{d}{L}. \tag{iv}$$

SIMPLIFY

We can combine the preceding three equations (ii), (iii), and (iv) to obtain

$$d = L\sin\theta_{\text{diff}} = \frac{t}{\cos\theta_{\text{medium}}}\sin\left(\theta_{\text{air}} - \theta_{\text{medium}}\right), \tag{v}$$

where from equation (i)

$$\theta_{\text{medium}} = \sin^{-1}\left(\frac{\sin\theta_{\text{air}}}{n}\right).$$

CALCULATE

We first calculate the angle of refraction, using equation (i)

$$\theta_{\text{medium}} = \sin^{-1}\left(\frac{\sin(38.5°)}{1.50}\right) = 24.5199°.$$

Then we calculate the displacement of the light ray, using equation (v)

$$d = \frac{5.90 \text{ cm}}{\cos(24.5199°)}\sin(38.5° - 24.5199°) = 1.56663 \text{ cm}.$$

ROUND

We report our result to three significant figures,

$$d = 1.57 \text{ cm}.$$

DOUBLE-CHECK

If the light ray were incident perpendicularly on the sheet ($\theta_{\text{air}} = 0°$), the displacement of the ray would be zero. To calculate what the displacement would be in the limit as the incident angle approaches $\theta_{\text{air}} = 90°$, we can rewrite equation (v) as

$$d = \frac{t\sin\left(\theta_{\text{air}} - \theta_{\text{medium}}\right)}{\cos\theta_{\text{medium}}} = t\,\frac{\sin\theta_{\text{air}}\cos\theta_{\text{medium}} - \cos\theta_{\text{air}}\sin\theta_{\text{medium}}}{\cos\theta_{\text{medium}}},$$

Continued—

using the trigonometric identity $\sin(\alpha - \beta) = \sin\alpha\cos\beta - \cos\alpha\sin\beta$. Taking $\theta_{\text{air}} = 90°$, we get

$$d = t\frac{1(\cos\theta_{\text{medium}}) - 0(\sin\theta_{\text{medium}})}{\cos\theta_{\text{medium}}} = t.$$

Thus, our result must be between zero and the thickness of the sheet. Our result is $d = 1.57$ cm, compared with $d = 5.90$ cm for $\theta_{\text{air}} = 90°$, so our result seems reasonable.

Total Internal Reflection

Now let's consider light traveling in an optical medium with index of refraction n_1 that crosses a boundary with another optical medium with a lower index of refraction n_2 such that $n_2 < n_1$. In this case, the light is bent away from the normal. As we increase the angle of incidence θ_1, the angle of the transmitted light θ_2 can approach 90°. When θ_1 exceeds the angle for which $\theta_2 = 90°$, **total internal reflection** takes place instead of refraction; all of the light is reflected internally. The critical angle θ_c at which total internal reflection takes place is given by

$$\frac{n_2}{n_1} = \frac{\sin\theta_1}{\sin\theta_2} = \frac{\sin\theta_c}{\sin 90°}$$

or

$$\sin\theta_c = \frac{n_2}{n_1} \quad (n_2 \le n_1), \tag{32.13}$$

because $\sin 90° = 1$. You can see from this equation that total internal reflection can occur only for light traveling from a medium with a higher index of refraction to a medium with a lower index of refraction, because the sine of an angle cannot be greater than 1. At angles less than θ_c, some light is reflected and some is transmitted. At angles greater than θ_c, all the light is reflected and none is transmitted.

If the second medium is air, then we can take $n_2 = 1$ and obtain an expression for the critical angle for total internal reflection for light leaving a medium with index of refraction n and entering air:

$$\sin\theta_c = \frac{1}{n}. \tag{32.14}$$

Total internal reflection is illustrated in Figure 32.40. In this figure, light rays are traveling in glass with index of refraction n from lower right to upper left. In Figure 32.40a, light rays are incident on the boundary between glass and air and are refracted according to Snell's Law. In Figure 32.40b, light rays are incident on the boundary at the critical angle of total internal reflection, θ_c. At this angle, the refracted rays would have an angle of $\theta_2 = 90°$, as depicted by the dashed arrow. Light rays in Figure 32.40c are incident on the boundary with an angle greater than the critical angle of total internal reflection, $\theta_1 > \theta_c$, so all the light is reflected.

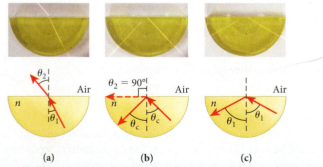

(a)   (b)   (c)

FIGURE 32.40 Light rays traveling through glass with index of refraction n from the lower right to the upper left are incident on the boundary between glass and air. The top panels are photographs and the lower panels are drawings illustrating the paths of the light rays. (a) Light is refracted at the boundary. (A small amount of light is also reflected by the boundary, but is too weak to show up in the photograph.) (b) Light is incident on the boundary at the critical angle for total internal reflection. (c) Light is totally internally reflected at the boundary.

Optical Fibers

An important application of total internal reflection is the transmission of light in **optical fibers.** Light is injected into a fiber so that the angle of incidence at the outer surface of the fiber is greater than the critical angle for total internal reflection. The light is then transported the length of the fiber as it bounces repeatedly from the fiber surface. Thus, optical fibers can be used to transport light from a source to a destination. Figure 32.41 shows a bundle of optical fibers with the end of the fibers open to the camera. The other end of the fibers is optically connected to a light source. Note that optical fibers can transport light in directions other than a straight line. The fiber may be bent, as long as the radius of curvature of the bend is not small enough to allow the light traveling in the optical fiber to have angles of incidence θ_i less than θ_c. If $\theta_i < \theta_c$, the light will be absorbed by the cladding around the surface of the fiber. These arguments are applicable to optical fibers with core diameters greater than 10 μm, that is, a core diameter large compared to the wavelength of light.

One type of optical fiber used for digital communications consists of a glass core surrounded by cladding, which is composed of glass with a lower index of refraction than the core. The cladding is then coated to prevent damage (Figure 32.42).

For a typical commercial optical fiber, the core material is SiO_2 doped with Ge to increase its index of refraction. The typical commercial fiber can transmit light 500 m with small losses. The light is generated using light-emitting diodes (LEDs) that produce light with a long wavelength. The light is generated as short pulses. Small light losses do not affect digital signals because digital signals are transmitted as binary bits rather than as analog signals. As long as the bits are registered correctly, the information is transmitted flawlessly, as opposed to analog signals, which would be directly degraded by any signal loss. Every 500 m the signals are received, amplified, and retransmitted. Thus, very high data rates that are immune to interference can be transmitted long distances. This scenario explains the physics behind the fiber-optics backbone of the modern Internet.

Optical fibers are also used for analog signal transmission, if the distance over which the signal has to be transported is not too long, up to a few meters. One application of analog signal transmission fiber optics is the endoscope, and related devices such as the borescope. An endoscope is used to look inside the human body without performing surgery, while a borescope is used to view hard-to-reach places in machinery. A typical endoscope is shown in Figure 32.43a. An endoscope uses a small lens to produce an image of the area

FIGURE 32.41 Light transported from a light source by fiber optics.

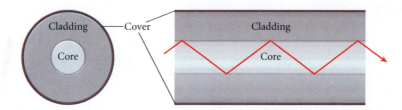

FIGURE 32.42 The structure of an optical fiber incorporating total internal reflection.

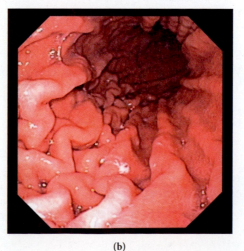

(a)   (b)

FIGURE 32.43 (a) An endoscope. (b) An image inside the stomach using an endoscope.

of interest. The image produced by the lens is focused on one end of a bundle of thousands of optical fibers. The fiber-optics bundle is small enough in diameter and flexible enough to be inserted into the human body through various orifices such as the esophagus. Each fiber transports one pixel of the image to the other end of the bundle of fibers where a second lens reproduces the image for viewing by the health professional. An endoscope may have 7,000 to 25,000 image-producing fibers. In addition, another set of optical fibers carries light to illuminate the region of interest. Often the end of the endoscope can be moved (articulated) remotely to view the desired area.

SOLVED PROBLEM 32.3 / Optical Fiber

Consider a long optical fiber with index of refraction $n = 1.265$ that is surrounded by air. (There is no cladding.) The end of the fiber is polished to be flat and is perpendicular to the length of the fiber. A light ray from a laser is incident from air onto the center of the circular face of the optical fiber.

PROBLEM
What is the maximum angle of incidence for this light ray such that it will be confined and transported by the optical fiber? (Neglect any reflection as the light ray enters the fiber.)

SOLUTION

THINK
The light ray will refract as it enters the optical fiber. Once inside the fiber, if the angle of incidence on the surface of the optical fiber is greater than the critical angle for total internal reflection, then the light is transmitted without loss.

SKETCH
Figure 32.44 shows a sketch of the light ray entering the optical fiber and reflecting off its inner surface.

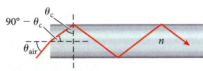

FIGURE 32.44 Light entering an optical fiber and undergoing total internal reflection.

RESEARCH
The critical angle for total internal reflection θ_c in the fiber for light entering from air is given by

$$\sin\theta_c = \frac{1}{n}, \qquad \text{(i)}$$

where n is the index of refraction of the optical fiber. For the light entering the optical fiber, Snell's Law tells us that since $n_{air} = 1$,

$$\sin\theta_{air} = n\sin\theta_{medium}, \qquad \text{(ii)}$$

where θ_{medium} is the angle of the refracted light ray in the fiber. From Figure 32.44 we can see that

$$\theta_{medium} = 90° - \theta_c. \qquad \text{(iii)}$$

SIMPLIFY
We can solve equation (ii) to obtain the maximum angle of incidence,

$$\theta_{air} = \sin^{-1}\left(n\sin\theta_{medium}\right).$$

Using equation (iii) we can write

$$\theta_{air} = \sin^{-1}\left(n\sin\left(90° - \theta_c\right)\right) = \sin^{-1}\left(n\cos\theta_c\right)$$

where we have used the trigonometric identity $\sin(90° - \alpha) = \cos\alpha$. We can then use equation (i) to obtain

$$\theta_{air} = \sin^{-1}\left(n\cos\left(\sin^{-1}\left(\frac{1}{n}\right)\right)\right).$$

CALCULATE

Putting in the numerical values, we get

$$\theta_{air} = \sin^{-1}\left((1.265)\cos\left(\sin^{-1}\left(\frac{1}{1.265}\right)\right)\right) = 50.7816°.$$

ROUND

We report our result to four significant figures, because the index of refraction was given with this accuracy:

$$\theta_{air} = 50.78°.$$

DOUBLE-CHECK

The critical angle of total internal reflection for the optical fiber is

$$\theta_c = \sin^{-1}\left(\frac{1}{1.265}\right) = 52.23°.$$

Snell's Law at the entrance of the optical fiber gives us

$$\theta_{medium} = \sin^{-1}\left(\frac{\sin(50.78°)}{1.265}\right) = 37.77°.$$

Thus $\theta_{medium} = 37.77° = 90° - \theta_c = 90° - 52.23° = 37.77°$, and our result appears reasonable.

Mirages

A mirage is often associated with traveling in the desert. You seem to see an oasis with water in the distance. As you approach the welcome sight, it disappears. However, you don't have to be traveling in the desert to observe this phenomenon. You can often see a mirage when traveling on a long, straight highway on a hot day. An example of such a mirage is shown in Figure 32.45a.

The mirage is caused by refraction in the air near the surface of the hot road. The air near the surface of the road is warmer than air farther from the surface. As shown in Figure 32.45b, the index of refraction of air goes down as its temperature goes up. Thus, the air near the road has a lower index of refraction than the cooler air above it. As light from distant objects passes through this layer, it is refracted upward, as illustrated in Figure 32.45c. The appearance of water is created by light refracted from the sky. You see the image as seeming to be light reflected off the surface of water, as you can see in Figure 32.45a. Other objects

(a)

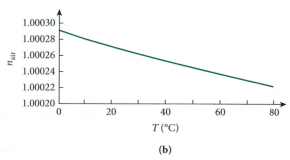

(b)

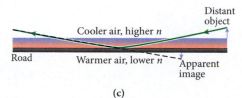

(c)

FIGURE 32.45 (a) A mirage on a hot road. There appears to be water on the road and objects appear to be reflected from the surface of that water. (b) The index of refraction of air as a function of temperature. (c) A ray diagram showing light from a distant object being refracted in the air layer near a hot road surface.

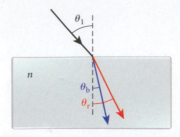

FIGURE 32.46 Chromatic dispersion for light refracted across the boundary of two optical media.

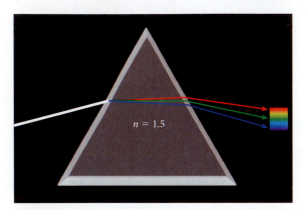

FIGURE 32.47 White light incident on a glass prism is separated into its component colors by chromatic dispersion.

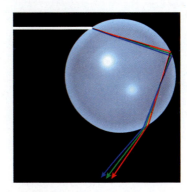

FIGURE 32.48 Chromatic dispersion in a spherical drop of water produces a rainbow.

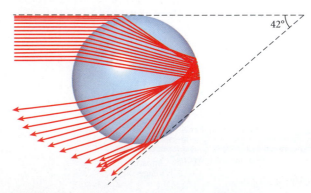

FIGURE 32.49 The paths taken by parallel light rays in a spherical drop of water.

are also visible in the mirage shown in Figure 32.45a, such as trees and car headlights. As you keep driving, you soon find that there is no water standing on the road.

Chromatic Dispersion

The index of refraction of an optical medium depends on the wavelength of the light traveling in that medium. This dependence of the index of refraction on the wavelength of the light means that light of different colors is refracted differently at the boundary between two optical media. This effect is called **chromatic dispersion.**

In general, the index of refraction for a given optical medium is greater for shorter wavelengths than for longer wavelengths. Therefore, blue light is refracted more than red light. We can see that $\theta_b < \theta_r$ in Figure 32.46.

White light consists of a superposition of all visible wavelengths. When a beam of white light is projected onto a glass prism (Figure 32.47), the incident white light separates into the different visible wavelengths, because each wavelength is refracted at a different angle. Figure 32.47 shows three refracted light rays, for red, green, and blue light, but of course white light contains a continuum of wavelengths, as seen in a rainbow.

A rainbow is a common example of chromatic dispersion (Figure 32.48). When water drops are suspended in the air and you observe these water drops with the Sun at your back, you see a rainbow. The white light from the Sun enters the water drop, refracts at the surface of the drop, and is transmitted through the drop to the far side. Here it is totally internally reflected, and is transmitted again to the surface of the drop, where it exits the drop and is refracted again. In the two refraction steps, the index of refraction is different for different wavelengths. The index of refraction for green light in water is 1.333, while the index of refraction for blue light is 1.337, and for red light it is 1.331. A continuum of indices of refraction occurs for all the colors.

A typical rainbow is shown in Figure 32.1. You can see the blue light in the inner part of the rainbow and the red light on the outside of the rainbow. The arc of the rainbow represents an average 42° angle from the direction of the sunlight. Another feature of rainbows is evident in this photograph: The region inside the arc of the rainbow appears to be brighter than the region outside the arc. You can understand this phenomenon by looking at the path light rays take as they are refracted and reflected in the raindrops (Figure 32.49).

You can see from this diagram that most of the light that is refracted and reflected back to the observer has an angle less than 42°. At angles smaller than 42°, most of the light is refracted and reflected back to the observer, although no separation of the different wavelengths is observed because dispersed colors from one ray merge with those from another ray and form white light. At larger angles, no light is sent back to an observer by this process—thus, outside the arc of the rainbow, much less light appears. However, some light still appears there because it is scattered from other sources.

Figure 32.1 also contains a secondary rainbow. The secondary rainbow appears at a larger angle than the primary rainbow, and in it the order of the colors is reversed. The secondary rainbow is created by light that reflects twice inside the raindrop (Figure 32.50). In contrast to the situation shown in Figure 32.48, when the ray in Figure 32.50 strikes the drop surface the third time (after one refraction and one reflection), not all of the light gets refracted and leaves the water drop. Instead some of it gets internally reflected again and then refracted out of the drop, forming the secondary rainbow. In Figure 32.50, you can see that the angle of the emerging blue and red light is reversed compared to the blue and red light shown in Figure 32.48 for the primary rainbow. (For the secondary rainbow there is also a

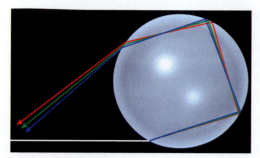

FIGURE 32.50 Chromatic dispersion and double reflection in a raindrop that produces a secondary rainbow.

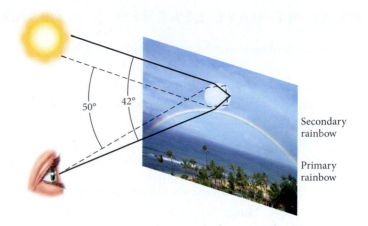

FIGURE 32.51 Geometry of the scattering of light in forming primary and secondary rainbows.

preferred angle, in this case approximately 50°, into which parallel rays entering the drop scatter for the secondary rainbow.)

Finally, combining our observations for the primary and secondary rainbows yields a quantitative understanding of the physical phenomena underlying Figure 32.1. The observed angles of the primary and secondary rainbows with respect to the sunlight are sketched in Figure 32.51.

Polarization by Reflection

The phenomenon of polarization was described in Chapter 31. Here we deal with a special way to create polarized, or at least partially polarized, light. When light is incident from air on an optical medium such as glass or water, some light is reflected and some light is refracted. The light reflected in this situation is partially polarized. When light is reflected at a certain angle, called the **Brewster angle,** θ_B, the reflected light is completely horizontally polarized, as illustrated in Figure 32.52.

The Brewster angle is named after Scottish physicist Sir David Brewster (1781–1868), who demonstrated this effect in 1815. The Brewster angle occurs when the reflected rays are perpendicular to the refracted rays. Snell's Law tells us that

$$n_{air} \sin\theta_B = n_{glass} \sin\theta_r,$$

where θ_r is the angle of the refracted ray. Figure 32.52 shows that

$$180° = \theta_B + \theta_r + 90°$$

because the reflected rays and the refracted rays are perpendicular to each other. This relationship between the angles can be rewritten as

$$\theta_r = 90° - \theta_B.$$

Substituting this result into Snell's Law gives

$$n_{air} \sin\theta_B = n_{glass} \sin\left(90° - \theta_B\right) = n_{glass} \cos\left(\theta_B\right),$$

which can be rearranged to finally obtain

$$\frac{n_{glass}}{n_{air}} = \tan\left(\theta_B\right), \tag{32.15}$$

where we have made use of the trigonometric identity $\tan\theta = \sin\theta/\cos\theta$. Equation 32.15 is called **Brewster's Law.**

The fact that light in air reflected off the surface of water is partially horizontally polarized means that the glare of sunlight off surfaces of water can be blocked by sunglasses that are covered with a polarization filter that only allows vertically polarized light to pass through.

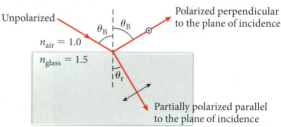

FIGURE 32.52 Unpolarized light is incident from air on glass. When the incident angle is equal to the Brewster angle, θ_B, the reflected light is 100% polarized perpendicular to the plane of incidence. The refracted light is partially polarized parallel to the plane of incidence.

WHAT WE HAVE LEARNED | EXAM STUDY GUIDE

- The angle of incidence equals the angle of reflection, $\theta_i = \theta_r$.

- The focal length of a spherical mirror is equal to half its radius of curvature, $f = \frac{1}{2}R$. The radius R is positive for converging (concave) mirrors and negative for diverging (convex) mirrors.

- For images formed with spherical mirrors, the object distance, the image distance, and the focal length of the mirror are related by the mirror equation, $\frac{1}{d_o} + \frac{1}{d_i} = \frac{1}{f}$. Here d_o is always positive, while d_i is positive if the image is on the same side of the mirror as the object and negative if the image is on the other side. The focal length f is positive for converging (concave) mirrors and negative for diverging (convex) mirrors.

- Light is refracted (changes direction) as it crosses the boundary between two media with different indices of refraction. This refraction is governed by Snell's Law: $n_1 \sin \theta_1 = n_2 \sin \theta_2$.

- The critical angle for total internal reflection at the boundary between two media with different indices of refraction is given by $\sin \theta_c = \frac{n_2}{n_1}$ $(n_2 \le n_1)$.

- The Brewster angle, θ_B, is the angle of incidence of light from air to a medium with a higher index of refraction at which the reflected light is completely horizontally polarized. The angle is given by Brewster's Law: $\tan \theta_B = n_2/n_1$, where $n_2 > n_1$.

KEY TERMS

geometric optics, p. 1026
wave fronts, p. 1026
light rays, p. 1026
law of rays, p. 1027
mirror, p. 1029
law of reflection, p. 1030
image, p. 1030

virtual image, p. 1030
real image, p. 1030
optic axis, p. 1031
concave mirror, p. 1033
focal point, p. 1033
focal length, p. 1033
mirror equation, p. 1035

magnification, p. 1036
convex mirror, p. 1037
spherical aberration, p. 1039
index of refraction, p. 1041
refraction, p. 1042
Snell's Law, p. 1042
Fermat's Principle, p. 1043

principle of least time, p. 1043
total internal reflection, p. 1046
optical fibers, p. 1047
chromatic dispersion, p. 1050
Brewster angle, p. 1051
Brewster's Law, p. 1051

NEW SYMBOLS AND EQUATIONS

θ_i, angle of incidence

θ_r, angle of reflection

$\theta_r = \theta_i$, law of reflection

θ_c, critical angle of total reflection

d_o, object distance

d_i, image distance

f, focal length

h_o, object height

h_i, image height

$n = \dfrac{c}{v}$, index of refraction

$n_1 \sin \theta_1 = n_2 \sin \theta_2$, Snell's Law

$\tan \theta_B = n_2/n_1$, $n_2 > n_1$, Brewster's Law

ANSWERS TO SELF-TEST OPPORTUNITIES

32.1 $d_i = \dfrac{d_o f}{d_o - f} = \dfrac{(2.50 \text{ m})(-0.500 \text{ m})}{(2.50 \text{ m}) - (-0.500 \text{ m})} = -0.417 \text{ m}$

$m = -\dfrac{d_i}{d_o} = -\dfrac{-0.417 \text{ m}}{2.50 \text{ m}} = 0.167$

Thus, you see an upright, smaller version of yourself.

32.2 We can use the result in Derivation 32.2

$y = R\left(1 - \dfrac{1}{2\cos\left(\sin^{-1}(d/R)\right)}\right)$

We can see that as we get closer to the optic axis, the distance of the crossing point gets closer to the focal length $f = R/2 = 3.60$ cm.

Distance from Axis (cm)	Distance from Mirror (cm)
0.72	3.58
0.8	3.58
1.8	3.48
3.6	3.04

32.3 Kinetic energy of rotation $K = \frac{1}{2}m\omega^2 x^2$; Potential energy $U = mgy$. Set them equal and find $y = x^2\omega^2/2g$.

PROBLEM-SOLVING PRACTICE

Problem-Solving Guidelines

1. Drawing a large, clear, well-labeled diagram should be the first step in solving almost any optics problem. Include all the information you know and all the information you need to find. Remember that angles are measured from the normal to a surface, not to the surface itself.

2. You normally need to draw only two principal rays to locate an image formed by mirrors, but you should draw a third ray as a check that they are drawn correctly. Even if

you need to solve a problem using the mirror equation, an accurate drawing can help you approximate the answers as a check on your calculations.

3. When you calculate distances for mirrors, remember the sign conventions and be careful to use them correctly. If a diagram indicates an inverted image, say, but your calculation does not have a negative sign, go back to the starting equation and check the signs of all distances and focal lengths.

MULTIPLE-CHOICE QUESTIONS

32.1 Legend says that Archimedes set the Roman fleet on fire as it was invading Syracuse. Archimedes created a huge _____ mirror, and he focused the Sun's rays on the Roman vessels.

a) plane

b) parabolic diverging

c) parabolic focusing

32.2 Which of the following interface combinations has the smallest critical angle?

a) light traveling from ice to diamond

b) light traveling from quartz to lucite

c) light traveling from diamond to glass

d) light traveling from lucite to diamond

e) light traveling from lucite to quartz

32.3 For specular reflection of a light ray, the angle of incidence

a) must be equal to the angle of reflection.

b) is always less than the angle of reflection.

c) is always greater than the angle of reflection.

d) is equal to 90°—the angle of reflection.

e) may be greater than, less than, or equal to the angle of reflection.

32.4 Standing by a pool filled with water, under what condition will you see a reflection of the scenery on the opposite side through total internal reflection of the light from the scenery?

a) Your eyes are level with the water.

b) You observe the pool at an angle of 41.8°.

c) Under no condition.

d) You observe the pool at an angle of 48.2°.

32.5 You are using a mirror and a camera to make a self-portrait. You focus the camera on yourself through the mirror. The mirror is a distance D away from you. To what distance should you set the range of focus on the camera?

a) D   b) 2D   c) D/2   d) 4D

32.6 What is the magnification for a plane mirror?

a) +1   c) greater than +1

b) −1   d) not defined for a plane mirror

QUESTIONS

32.7 The figure shows the difference between the refraction index profile of a so-called *step index fiber* vs. the refraction index profile of a so-called *graded index fiber*. Analyzing light propagation through the fiber from a ray-optics perspective, comment on the path followed by a light ray entering each of the two fibers.

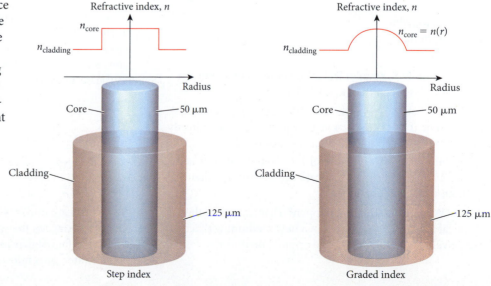

32.8 A slab of Plexiglas 2.00 cm thick with index of refraction 1.51 is placed over a physics textbook. The slab has parallel sides. The text is at height $y = 0$. Consider two rays of light that are leaving the letter "t" in the text under the Plexiglas and going toward an observer who is above the Plexiglas, looking down. ***Draw*** on the figure the apparent y-position of the text under the Plexiglas as seen by the observer. *Hint:* From where in the Plexiglas do these rays appear to originate for the observer? Using the two rays A and B after they have exited the Plexiglas, determine the apparent height (y-position) of the text under the Plexiglas as seen by the observer. You can easily do this experiment yourself. If you do not have a block of glass or Plexiglas, you can try placing a flat-bottomed drinking glass over the text.

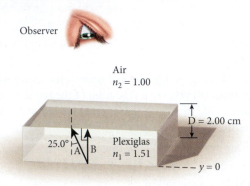

32.9 A single concave spherical mirror is used to create an image of a source 5.00 cm tall that is located at position $x = 0$ cm which is 20.0 cm to the left of Point C, the center of curvature of the mirror, as shown in the figure. The magnitude of the radius of curvature for the mirror is 10.0 cm. Without changing the mirror, how can one reduce the spherical aberration produced by this mirror? Will there be any disadvantages to your approach to reducing the spherical aberration?

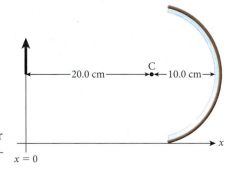

32.10 If you look at an object at the bottom of a pool, the pool looks less deep than it actually is.

a) From what you have learned, calculate how deep a pool seems to be if it is actually 4 feet deep and you look directly down on it. The refractive index of water is 1.33.

b) Would the pool look more or less deep if you look at it from an angle other than vertical? Answer this qualitatively, without using an equation.

32.11 Why does refraction happen? That is, what is the physical reason a wave moves in a new medium with a different velocity than it did in the original medium?

32.12 Many fiber-optics devices have minimum specified bending angles. Why?

32.13 A physics student is eying a steel drum, the top part of which has the approximate shape of a concave spherical surface. The surface is sufficiently polished that she can just barely make out the reflection of her finger when she places it above the drum. As she slowly moves her finger toward the surface and then away from it, you ask her what she is doing. She replies that she is estimating the radius of curvature of the drum. How can she do that?

32.14 Answer as true or false with an explanation for the following: *The wavelength of He-Ne laser light in water is less than its wavelength in the air.* (The refractive index of water is 1.33.)

32.15 Among the instruments Apollo astronauts left on the Moon were reflectors used to bounce laser beams back to Earth. These made it possible to measure the distance from the Earth to the Moon with unprecedented precision (uncertainties of a few centimeters out of 384,000 km), for the study both of celestial mechanics and of plate tectonics on Earth. The reflectors consisted not of ordinary mirrors, but of arrays of *corner cubes*, each consisting of three square plane mirrors fixed perpendicular to each other, as adjacent faces of a cube. Why? Explain the function and advantages of this design.

32.16 A 45°-45°-90° triangular prism can be used to reverse a light beam: The light enters perpendicular to the hypotenuse of the prism, reflects off each leg, and emerges perpendicular to the hypotenuse again. The surfaces of the prism are *not* silvered. If the prism is made of glass with index of refraction $n_{glass} = 1.520$ and the prism is surrounded by air, the light beam will be reflected with a minimum loss of intensity (there are reflection losses as the light enters and leaves the prism).

a) Will this work if the prism is *under water,* which has index of refraction $n_{H_2O} = 1.333$?

b) Such prisms are used, in preference to mirrors, to bend the optical path in quality binoculars. Why?

32.17 An object is imaged by a converging spherical mirror as shown in Figure 32.19, repeated below. Suppose a black cloth is put between the object and the mirror so that it covers everything above the axis of the mirror. How will the image be affected?

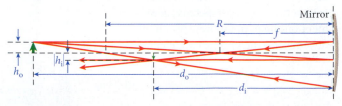

32.18 You are under water in a pond and look up at the smooth surface of the water, noticing the sun in the sky. Is the sun in fact higher in the sky than it appears to you while under water, or is it lower?

32.19 Holding a spoon in front of your face, convex side toward you, estimate the location of the image and its magnification.

32.20 A *solar furnace* uses a large parabolic mirror (mirrors several stories high have been constructed) to focus the light of the Sun to heat a target. A large solar furnace can melt metals. Is it possible to attain temperatures exceeding 6000 K (the temperature of the photosphere of the Sun) in a solar furnace? How, or why not?

PROBLEMS

A blue problem number indicates a worked-out solution is available in the Student Solutions Manual. One • and two •• indicate increasing level of problem difficulty.

Section 32.2

32.21 A person sits 1.0 m in front of a plane mirror. What is the location of the image?

32.22 A periscope consists of two flat mirrors and is used for viewing objects when an obstacle impedes direct viewing. Suppose that Curious George is looking through a periscope at the Man in the Yellow Hat, whose hat is at $d_o = 3.0$ m from the upper mirror, and suppose that the two flat mirrors are separated by a distance $L = 0.4$ m. What is the distance D of the final image of the yellow hat from the lower mirror?

32.23 A person stands at a point P relative to two plane mirrors oriented at 90°, as shown in the figure. How far away do the person's images appear from each other to the viewer?

•32.24 Even the best mirrors absorb or transmit some of the light incident on them. The highest-quality mirrors might reflect 99.997% of incident light intensity. Suppose a cubical "room," 3.00 m on an edge, were constructed with such mirrors for the walls, floor, and ceiling. How slowly would such a room get dark? Estimate the time required for the light level in such a room to fall to 1.00% of its initial value after the only light source in the room is switched off.

Section 32.3

32.25 The radius of curvature of a convex mirror is 25 cm. What is its focal length?

32.26 A single concave spherical mirror is used to create an image of a source 5.00 cm tall that is located at position $x = 0$ cm which is 20.0 cm to the left of Point C, the center of curvature of the mirror, as shown in the figure. The magnitude of the radius of curvature for the mirror is 10.0 cm. Calculate the position x_i where the image is formed. Use the coordinate system given in the drawing. What is the height h_i of the image? Is the image upright or inverted (upright = pointing up, inverted = pointing down)? Is it real or virtual?

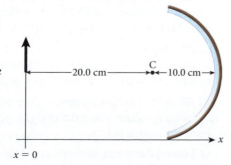

32.27 Convex mirrors are often used in sideview mirrors on cars. Many such mirrors dsiplay the warning "Objects in mirror are closer than they appear." Assume a convex mirror has a radius of curvature of 14.0 m and that there is a car that is 11.0 m behind the mirror. For a flat mirror, the image distance would be 11.0 m and the magnification would be 1. Find the image distance and magnification for this mirror.

32.28 A 5-cm object is placed 30 cm away from a convex mirror with a focal length of 10 cm. Determine the size, orientation, and position of the image.

32.29 The magnification of a convex mirror is 0.60× for an object 2.0 m from the mirror. What is the focal length of this mirror?

•32.30 An object is located at a distance of 100 cm from a concave mirror of focal length 20 cm. Another concave mirror of focal length 5 cm is located 20 cm in front of the first concave mirror. The reflecting sides of the two mirrors face each other. What is the location of the final image formed by the two mirrors and the total magnification by the combination?

••32.31 The shape of an *elliptical* mirror is described by the curve $\dfrac{x^2}{a^2} + \dfrac{y^2}{b^2} = 1$, with semimajor axis a and semiminor axis b. The foci of this ellipse are at points $(c,0)$ and $(-c,0)$, with $c = (a^2 - b^2)^{1/2}$. Show that any light ray in the xy-plane, which passes through one focus, is reflected through the other. "Whispering galleries" make use of this phenomenon with sound waves.

Section 32.4

32.32 What is the speed of light in crown glass, whose index of refraction is 1.52?

32.33 An optical fiber with an index of refraction of 1.5 is used to transport light of wavelength 400 nm. What is the critical angle for light to transport through this fiber without loss? If the fiber is immersed in water? In oil?

32.34 A helium-neon laser produces light of wavelength $\lambda_{vac} = 632.8$ nm in vacuum. If this light passes into water, with index of refraction $n = 1.333$, what then will be its

a) speed?   c) wavelength?

b) frequency?   d) color?

32.35 A light ray is incident from water of index of refraction 1.33 on a plate of glass whose index of refraction is 1.73. What is the angle of incidence, to have fully polarized reflected light?

•32.36 Suppose you are standing at the bottom of a swimming pool looking up at the surface, which we assume to be calm. Looking up, you will see a circular window to the "outer world." If your eyes are approximately 2 meters beneath the surface, what is the diameter of this circular window?

•**32.37** A ray of light is incident on an equilateral triangular prism with an index of refraction of 1.23. The ray is parallel to the base of the prism when it approaches the prism. The ray enters the prism at the midpoint of one of its sides, as shown in the figure. What is the direction of the ray when it emerges from the triangular prism?

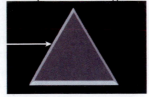

•**32.38** A collimated laser beam strikes the left side (A) of a glass block at an angle 20.0° with respect to horizontal, as shown in the figure. The block has an index of refraction of 1.55 and is surrounded by air with an index of 1.00. The left side of the glass block is vertical (90.0° from horizontal) while the right side (B) is 60.0° from horizontal. Determine the angle θ_{BT} **with respect to horizontal** at which the light exits surface B.

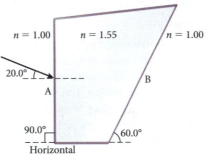

•**32.39** In a step-index fiber, the index of refraction undergoes a discontinuity (jump) at the core-cladding interface, as shown in the figure. Infrared light with wavelength 1550 nm propagates through such a step-index fiber through total internal reflection at the core-cladding interface. The index of refraction for the core at 1550 nm is $n_{core} = 1.48$. If the maximum angle α_{max} at which light can be coupled into the fiber such that no light will leak into the cladding is $\alpha_{max} = 14.033°$, calculate the percent difference between the index of refraction of the core and the index of refraction of the cladding.

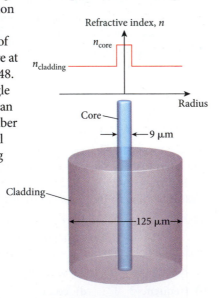

••**32.40** Refer to Figure 32.49 and prove that the arc of the primary rainbow represents the 42° angle from the direction of the sunlight.

••**32.41** Use Fermat's Principle to derive the law of reflection.

••**32.42** Fermat's Principle, from which geometric optics can be derived, states that light travels by a path that minimizes the time of travel between the points. Consider a light beam that travels a horizontal distance D and a vertical distance h, through two large flat slabs of material, with a vertical interface between the materials. One material has a thickness $D/2$ and index of refraction n_1, and the second material has a thickness $D/2$ and index of refraction n_2. Determine the equation involving the indices of refraction and angles from horizontal that the light makes at the interface (θ_1 and θ_2) which minimize the time for this travel.

Additional Problems

32.43 Suppose your height is 2.0 m and you are standing 50 cm in front of a plane mirror.

a) What is the image distance?

b) What is the image height?

c) Is the image inverted or upright?

d) Is the image real or virtual?

32.44 A light ray of wavelength 700 nm traveling in air ($n_1 = 1.00$) is incident on a boundary with a liquid ($n_2 = 1.63$).

a) What is the frequency of the refracted ray?

b) What is the speed of the refracted ray?

c) What is the wavelength of the refracted ray?

32.45 You have a spherical mirror with a radius of curvature of +20 cm (so it is concave facing you). You are looking at an object whose size you want to double in the image, so you can see it better. Where should you put the object? Where will the image be, and will it be real or virtual?

32.46 You are submerged in a swimming pool. What is the maximum angle at which you can see light coming from above the pool surface? That is, what is the angle for total internal reflection from water into air?

32.47 Light hits the surface of water at an incident angle of 30° with respect to the normal line. What is the angle between the reflected ray and the refracted ray?

32.48 A spherical metallic Christmas tree ornament has a diameter of 8.00 cm. If Saint Nicholas is by the fireplace, 1.56 m away, where will he see his reflection in the ornament? Is the image real or virtual?

32.49 One of the factors that cause a diamond to sparkle is its relatively small critical angle. Compare the critical angle of diamond in air compared to that of diamond in water.

32.50 What kinds of images, virtual or real, are formed by a converging mirror when the object is placed a distance away from the mirror that is

a) beyond the center of curvature of the mirror,

b) between the center of curvature and half the center of curvature, and

c) closer than half of the center of curvature.

32.51 At what angle θ shown in the diagram must a beam of light enter the water such that the reflected beam makes an angle of 40° with respect to the normal of the water's surface?

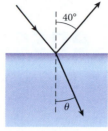

•**32.52** A concave mirror forms a real image twice as large as the object. The object is then moved such that the new real image produced is three times the size of the object. If the image was moved 75 cm from its initial position, how far was the object moved and what is the focal length of the mirror?

•**32.53** How deep does a point in the middle of a 3-m-deep pool appear to a person standing outside of it 2 meters horizontally from the point? Take the refractive index for the pool to be 1.3 and for air to be 1.

•**32.54** In the figure, what is the smallest incident angle θ_i for the beam to undergo total internal reflection at the surface of the prism having an index of refraction of 1.5?

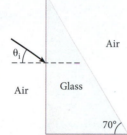

••**32.55** Reflection and refraction, like all classical features of light and other electromagnetic waves, are governed by the Maxwell equations. The Maxwell equations are *time-reversal invariant*, which means that any solution of the equations reversed in time is also a solution.

a) Suppose some configuration of electric charge density ρ, current density $\vec{j}$, electric field $\vec{E}$, and magnetic field $\vec{B}$ is a solution of the Maxwell equations. What is the corresponding time-reversed solution?

b) How, then, do "one-way mirrors" work?

••**32.56** Refer to Example 32.3 and use the numbers provided there. Further, assume that your eyes are at a height of 1.70 m above the water.

a) Calculate the time it takes for light to travel on the path from the fish to your eyes.

b) Calculate the time light would take on a straight-line path from the fish to your eyes.

c) Calculate the time light would take on a path from the fish vertically upward to the water surface and then straight to your eyes.

d) Calculate the time light would take on the straight-line path from the apparent location of the fish to your eyes.

e) What can you say about Fermat's Principle from the above numbers?

32.57 If you want to construct a liquid mirror of focal length 2.50 m, with what angular velocity do you have to rotate your liquid?

••**32.58** One proposal for a space-based telescope is to put a large rotating liquid mirror on the Moon. Suppose you want to use a liquid mirror 100.0 m in diameter, and you want it to have a focal length of 347.5 m. The gravitational acceleration on the Moon is 1.62 m/s^2.

a) What angular velocity does your mirror have?

b) What is the linear speed of a point on the perimeter of the mirror?

c) How high above the center is the perimeter of the mirror?

33

Lenses and Optical Instruments

(a)   (b)

FIGURE 33.1 (a) The Whirlpool Galaxy. (b) Marine diatoms living in Antarctica.

WHAT WE WILL LEARN

- Lenses can focus light and produce images governed by the Thin-Lens Equation, which states that the inverse of the object distance plus the inverse of the image distance equals the inverse of the focal length of the lens.

- Optical instruments are combinations of mirrors and lenses.

- A single converging lens can be used as a magnifier.

- Two-lens systems are often used in optical instruments.

- Placing a converging lens close to a diverging lens can produce a zoom lens.

- A camera is an optical instrument that records real images produced by a lens.

- The human eye is an optical instrument governed by the Thin-Lens Equation. Various types of lenses are used to correct defects in vision.

- Microscopes are systems of lenses designed to magnify the image of close but very small objects.

- Telescopes are systems of lenses or mirrors designed to magnify the image of distant but very large objects.

The two images in Figure 33.1 are some of the largest and smallest objects we can optically observe with visible light. The image in Figure 33.1a is the Whirlpool Galaxy (M51), with a diameter of about 76,000 light years ($7 \cdot 10^{20}$ m), and a distance from Earth of about 23 million light years ($2 \cdot 10^{23}$ m). The image in Figure 33.1b shows assorted diatoms found living between crystals of annual sea ice in Antarctica. Diatoms are on the order of 20 microns ($2 \cdot 10^{-5}$m) in length. Over the years, the ability to form images of these kinds of objects has completely changed our understanding of biology, astronomy, geology, engineering—in fact, just about every branch of science and technology.

All optical image-forming instruments work from a combination of lenses or mirrors. In a sense, the ideas in this chapter are simply applications of the principles discussed in Chapter 32. However, understanding how an image is formed is essential to interpreting the subject of the image. In this chapter we will examine a variety of image-forming instruments, including the human eye. The same principles of geometric optics govern image formation using other kinds of radiation; we will see some examples of this as well.

33.1 Lenses

When light is refracted while crossing a curved boundary between two different media, the light rays obey the law of refraction at each point on the boundary. The angle at which the light rays cross the boundary (with respect to the local normal to the boundary) is different along the curve, so the refracted angle is different at different points along the curve. A spherical curved boundary between two optically transparent media forms a spherical surface. If light enters a medium through one spherical surface and then returns to the original medium through another spherical surface, the device that has the spherical surfaces is called a **lens.** Light rays that are initially parallel before they strike the lens are refracted in different directions, depending on the part of the lens they strike. Depending of the shape of the lens, the light rays can be focused or can diverge.

If the front surface of a lens with index of refraction n is part of the surface of a sphere with radius of curvature R_1 and the back surface of the lens is part of the surface of a sphere with radius of curvature R_2, then we can calculate the focal length f for a thin lens using the **Lens-Maker's Formula,**

$$\frac{1}{f} = \left(n-1\right)\left(\frac{1}{R_1} - \frac{1}{R_2}\right). \tag{33.1}$$

We derive this equation, which applies to thin lenses in air, in Derivation 33.1. We will see that a *thin lens* is defined as a lens whose thickness is much smaller than any object and image distances, and so the thickness can be neglected. There we will learn a sign convention for the radii, because they can be positive or negative.

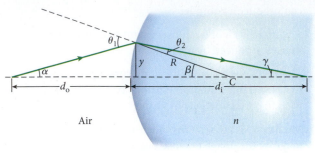

FIGURE 33.2 Light traveling in air incident on a spherical surface of an optical medium with index of refraction n.

Lens-Maker's Formula and the Lens Equation for Thin Lenses

We start the derivation of the Lens-Maker's Formula by assuming that we have light traveling in air incident on a spherical surface of an optical medium with index of refraction n and radius of curvature R (Figure 33.2). We draw the optic axis as a line perpendicular to the spherical face of the medium, through the center C of the spherical face. We assume a light ray originating from a point source at a distance d_o from the lens at an angle α with respect to the optic axis. This ray makes an angle θ_1 with respect to a normal to the surface of the optical medium. Using Snell's Law of refraction (see Chapter 32), and taking $n_1 = 1$ and $n_2 = n$, we get

$$\sin\theta_1 = n\sin\theta_2,$$

where θ_2 is the angle the refracted light ray makes with respect to the normal to the surface. If α is a small angle, then the angles θ_1 and θ_2 will be small and we can write $\theta_1 = n\theta_2$.

Looking at Figure 33.2, we can see the relationship between the angles $\theta_1 = \alpha + \beta$ and $\beta = \theta_2 + \gamma$. We can rewrite these equations as $n\theta_2 = \alpha + \beta$ and $\theta_2 = \beta - \gamma$. We insert the second into the first to eliminate θ_2, obtaining

$$\frac{\alpha+\beta}{n} = \beta - \gamma,$$

which we can rearrange to get

$$\alpha + n\gamma = \beta(n-1). \tag{i}$$

From Figure 33.2 and making the small-angle approximation, we can write

$$\alpha \approx \tan\alpha \approx \frac{y}{d_o}, \quad \beta \approx \tan\beta \approx \frac{y}{R}, \quad \gamma \approx \tan\beta \approx \frac{y}{d_i}. \tag{ii}$$

Substituting the expressions for α, β, and γ from equation (ii) into equation (i) we can write

$$\frac{y}{d_o} + \frac{ny}{d_i} = \frac{(n-1)y}{R}$$

or

$$\frac{1}{d_o} + \frac{n}{d_i} = \frac{(n-1)}{R}. \tag{iii}$$

We have now derived an expression for the image distance formed by one surface. Since d_i is independent of α, all of the light from the source goes through the same point and therefore the light is focused. Now let's put two surfaces together to make a lens (Figure 33.3).

The light ray originating from the left of the lens can be described by our result (iii) for the image distance and object distance derived above (with an added subscript "1" to denote the first surface):

$$\frac{1}{d_{o,1}} + \frac{n}{d_{i,1}} = \frac{(n-1)}{R_1}. \tag{iv}$$

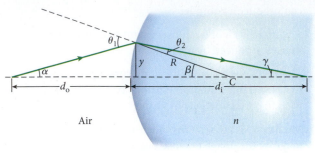

FIGURE 33.3 A lens of thickness L composed of an optical medium with index of refraction n in air. The left face of the lens has radius R_1 and the right face of the lens has radius R_2.

For the second surface, the light ray is passing from the optical medium with refractive index $n_1 = n$ into air with refractive index $n_2 = 1$. The image formed of the source by the first refracting surface acts as the "object" for this second surface. We can therefore now describe the relationship between image distance and object distance for the second surface as

$$\frac{n}{d_{o,2}} + \frac{1}{d_{i,2}} = \frac{(1-n)}{R_2}, \tag{v}$$

Where $d_{o,2}$ is the distance of the "object" from the second surface. From Figure 33.3 we can see that

$$\left|d_{i,1}\right| = L + \left|d_{o,2}\right|.$$

If we assume a thin lens, then L is much smaller than any other object or image distances and we can safely neglect the thickness of the lens. Therefore,

$$\left|d_{i,1}\right| \approx \left|d_{o,2}\right|.$$

Next, we choose a sign convention for distances. Agreeing that the light goes from left to right, object distances are positive for objects on the left of the lens and negative for objects on the right of the lens. Image distances are positive for objects on the right of the lens and negative for images on the left of the lens. Just as for mirrors, image distances are positive for images formed where the light eventually goes, that is, for real images. Because the object for the second surface of the lens is to the right of the lens, we then have

$$-d_{o,2} = d_{i,1}. \tag{vi}$$

Substituting this equation (vi) into our equation (v) for the second surface, we get

$$-\frac{n}{d_{i,1}} + \frac{1}{d_{i,2}} = \frac{(1-n)}{R_2}$$

or

$$\frac{n}{d_{i,1}} = \frac{1}{d_{i,2}} - \frac{(1-n)}{R_2}. \tag{vii}$$

Equation (vii) can now be used with our expression (iv) for the first surface to eliminate $d_{i,1}$:

$$\frac{1}{d_{o,1}} + \frac{1}{d_{i,2}} - \frac{(1-n)}{R_2} = \frac{(n-1)}{R_1}.$$

The object distance of the lens as a whole is the same as the object distance for the first surface of the lens, $d_o = d_{o,1}$. The image distance of the lens as a whole is the same as the image distance for the second surface of the lens, $d_i = d_{i,2}$. We then obtain

$$\frac{1}{d_o} + \frac{1}{d_i} = (n-1)\left(\frac{1}{R_1} - \frac{1}{R_2}\right).$$

The focal length is defined as the image distance when the object is at infinity, which gives us

$$\frac{1}{\infty} + \frac{1}{f} = \frac{1}{f} = (n-1)\left(\frac{1}{R_1} - \frac{1}{R_2}\right).$$

This result is the Lens Maker's Formula. This formula shows that a thin lens has a focus (or focal point) on both sides of the lens which are equidistant from the lens.

We can take this expression for the Lens Maker's Formula together with our expression relating the image and object distances of the lens and get the **Thin-Lens Equation,**

$$\frac{1}{d_o} + \frac{1}{d_i} = \frac{1}{f}. \tag{33.2}$$

In the above derivations, it is tacitly assumed that the light would strike a convex surface *as seen by the light*. If the surface is concave *as seen by the light,* the results are still valid provided that a negative value is used for the radius of curvature.

When we observe the second surface of a lens, we see it from the opposite perspective from that seen by the light, so care must be taken to assign the proper signs to the radii of curvature. What we would call a double convex lens would have $R_1 > 0$ and $R_2 < 0$ because the light would perceive the second surface as being concave. These complications can be

33.1 In-Class Exercise

A single lens with two convex surfaces made of sapphire with index of refraction $n = 1.77$ has surfaces with radii of curvature $R_1 = 27.0$ cm and $R_2 = -27.0$ cm. What is the focal length of this lens in air?

a) 17.5 cm   d) 54.0 cm

b) 20.0 cm   e) 60.0 cm

c) 27.0 cm

Starting Real object

-A converging lens can produce a real, inverted, reduced image

=For a converging lens an object has to be placed between the focal length and the lens in order to form a virtual image

-A converging lens can never produce a virtual, upright, reduced image

avoided by adopting the following convention: If the focal length of a lens is positive, the lens is said to be converging, while if the focal length is negative, the lens is said to be diverging.

Lenses do not necessarily have the same curvature on the entrance and exit surfaces. Lenses can be formed in several different ways, as illustrated in Figure 33.4. For example, light incident from the left on the converging meniscus lens shown in Figure 33.4 would first encounter a convex surface ($R_1 > 0$) and then a second convex surface ($R_2 > 0$). For the converging meniscus lens, $R_1 < R_2$, therefore $1/R_1 > 1/R_2$, so the focal length of the lens given by the Lens-Maker's Formula (equation 33.1) is positive, producing a converging lens. For the diverging meniscus lens shown in Figure 33.4, the first surface is convex ($R_1 > 0$) and the second surface is also convex ($R_2 > 0$). However, for the diverging meniscus lens, $R_1 > R_2$ and $1/R_1 < 1/R_2$, so the focal length is negative, producing a diverging lens. Meniscus lenses are commonly used in corrective eyeglasses.

Converging Lenses

A **converging lens,** which has $f > 0$, is shaped such that rays incident parallel to the optical axis are focused by refraction at the focal distance f from the center of the lens. Figure 33.5 shows a light ray incident on a converging glass lens. At the surface of the lens, the light ray is refracted toward the normal. When the ray leaves the lens, it is refracted away from the normal. The resulting twice-refracted ray passes through the focal point of the lens on the opposite side of the lens from the incident ray.

Consider the case of several horizontal light rays incident on a converging lens. These rays are focused to a point a distance f from the center of the lens on the opposite side of the lens from the incident rays. Figure 33.6 is a photograph of a converging lens with five parallel lines of light incident on the surface from the left.

In Figure 33.6a, the parallel lines of light are focused to one point on the right of the lens. In Figure 33.6b, red lines have been superimposed to represent the light rays. The rays enter the lens, refract at the first surface, traverse the lens in a straight line, and refract when they exit the lens. In Figure 33.6c, a black dotted line is drawn at the center of the lens. In this panel, the rays are drawn using the thin-lens approximation, whereby the incident rays refract just once at the center of the lens. Instead of following the detailed trajectory of the light rays inside a thin lens, the incident rays are drawn to the centerline and then on to the focal point. Our real-life lens (Figure 33.6a) is a thick lens, and displacement occurs between the refraction at the entrance and exit surfaces. In this book, we will only consider thin lenses and we will treat the lens as a line at which refraction takes place.

Converging lenses can be used to form images. Figure 33.7 shows the geometric construction of the formation of an image using a converging lens. An object, represented by the green arrow, stands on the optic axis. This object has a height h_o and is located a distance d_o from the center of the lens, such that $d_o > f$. Four particular rays are often useful for constructing images:

1. A ray from the bottom of the object is drawn to pass straight through the lens along the optic axis. This ray goes through the bottom of the image and is usually not shown.

2. A second ray is then drawn from the top of the object parallel to the optic axis. This ray goes through the focal point on the other side of the lens.

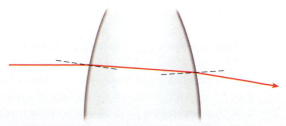

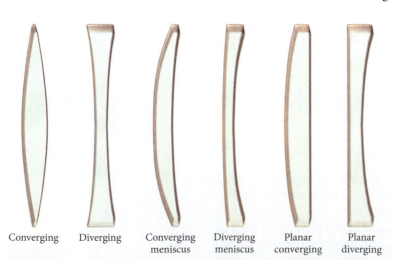

Converging   Diverging   Converging meniscus   Diverging meniscus   Planar converging   Planar diverging

FIGURE 33.4 Different types of lenses created by different radii of curvature on the entrance and exit surfaces.

FIGURE 33.5 Refraction of a horizontal light ray through a converging lens.

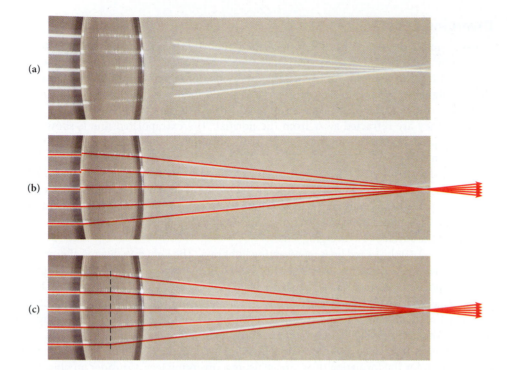

FIGURE 33.6 (a) Parallel light rays incident on a converging lens. (b) Arrows are superimposed on the light rays. (c) The optical center of the lens is indicated by a dashed line.

3. A third ray is drawn from the top of the object through the center of the lens; this ray has no net refraction in the thin-lens approximation. (It goes through surfaces that are approximately parallel, as in Solved Problem 32.2.)

4. A fourth ray is drawn from the top of the object through the focal point on the same side of the lens, which is then parallel to the optic axis after refraction.

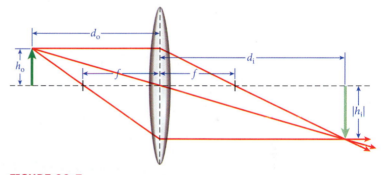

The three rays starting at the top of the object intersect, locating the top of the image formed. Any two of these three rays from the top of the object can be used to locate the top of the image.

FIGURE 33.7 Real image produced by a converging lens.

In this case, with $d_o > f$, a real, inverted, and enlarged image with height $h_i < 0$ is formed at a distance $d_i > 0$ from the center of the lens.

Now let us consider the image formed by an object with height h_o placed at a distance d_o from the center of the lens such that $d_o < f$ (Figure 33.8). Once again, four particular rays are useful for constructing the image:

1. The first ray again is drawn from the bottom of the object along the optic axis and is usually not shown.

2. The second ray is drawn from the top of the object parallel to the optic axis and is refracted through the focal point on the opposite side of the lens.

3. A third ray is drawn from the top of the object straight through the center of the lens.

4. A fourth ray is drawn from the top of the object such that it appears to have originated from the focal point on the same side of the lens and is then refracted parallel to the optic axis.

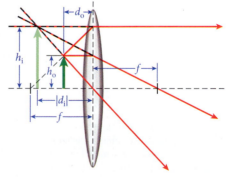

FIGURE 33.8 Virtual image produced by a converging lens.

You can see that these three rays are diverging. A virtual image is formed on the same side of the lens as the object by extrapolating the three rays back until they intersect. Red-and-black dashed lines represent the extrapolated rays. In this case, with $d_o < f$, a virtual, upright, and enlarged image is formed with $h_i > 0$ and $d_i < 0$.

FIGURE 33.9 Horizontal light ray refracted by a diverging lens.

A diverging lens cannot produce a real, inverted reduced image

Diverging Lenses

A **diverging lens,** which has $f < 0$, is shaped such that parallel rays striking the lens diverge by refraction. The extrapolation of the diverging rays would intersect at a focal distance from the center of the lens on the same side of the lens as the object. Figure 33.9 shows a light ray parallel to the optic axis incident on a diverging glass lens. At the first surface of the lens, the light rays are refracted toward the normal. When the rays leave the lens, they are refracted away from the normal. The extrapolated line is shown as a red-and-black dashed line and points to the focal point on the same side of the lens as the incident ray.

Consider the case of several horizontal light rays incident on a diverging lens. After passing through the lens, the rays diverge such that their extrapolations intersect at a point a distance f from the center of the lens, on the same side of the lens as the incident rays. Figure 33.10 is a photograph of a diverging lens with five parallel lines of light incident from the left on the surface of a diverging lens.

Figure 33.10 shows red lines representing light rays. The light rays diverge after passing through the lens. Red-and-black dashed lines show the extrapolation of the diverging rays. The extrapolated rays intersect at a distance equal to one focal length away from the center of the lens. In part (c), the diverging rays are drawn using the thin-lens approximation, whereby the incident rays refract just once at the center of the lens. The diverging rays are drawn so that their extrapolations intersect at the focal point.

Diverging lenses can also be used to form images. Figure 33.11 shows the geometric construction for the formation of an image using a diverging lens. Consider an object rep-

FIGURE 33.10 (a) Parallel light rays incident on a diverging lens. (b) Arrows are superimposed on the light rays. (c) The optical center of the lens is indicated by a vertical dashed line.

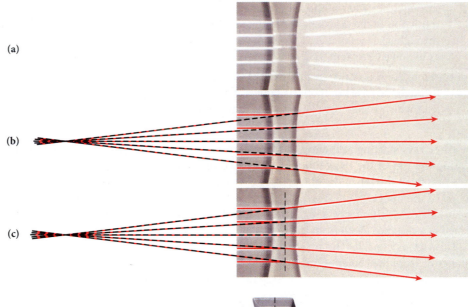

FIGURE 33.11 Image produced by a diverging lens of object placed a distance from the lens larger than the focal length of the lens.

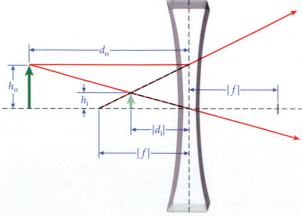

resented by the taller green arrow standing on the optic axis. This object has a height h_o and is located a distance d_o from the center of the lens, such that $d_o > |f|$.

Three rays are useful to construct the image:

1. Start with a ray along the optic axis of the lens, which passes straight through the lens and defines the bottom of the image. This ray is usually not shown.

2. A second ray is then drawn from the top of the object parallel to the optic axis. This ray is refracted such that the extrapolation of the diverging ray passes through the focal point on the left side of the lens.

3. A third ray drawn from the top of the object through the center of the lens is not refracted in the thin-lens approximation. This ray is extrapolated back along its original path.

The two extrapolated rays intersect at the top of the produced image. The image formed is virtual, upright, and reduced in size.

The images that we have formed by ray tracing can all be described algebraically with the Thin-Lens Equation (derived above),

$$\frac{1}{d_o} + \frac{1}{d_i} = \frac{1}{f}.$$

Note that this is the same relationship between focal length, image distance, and object distance that we found for mirrors.

We now introduce and review our sign conventions. We earlier defined the focal length of a converging lens to be positive and the focal length of a diverging lens to be negative. The object distance d_o and height h_o for a single lens are both positive. (With multi-lens systems, we can encounter negative object distances and heights.) If the image is on the opposite side of the lens from the object, the image distance d_i is positive and the image is real. If the image is on the same side of the lens as the object, the image distance is negative and the image is virtual. The linear magnification formula for lenses is the same as for mirrors,

$$m = \frac{h_i}{h_o} = -\frac{d_i}{d_o}.$$

If the image is upright, then h_i and the linear magnification m are positive, while if the image is inverted, h_i and m are negative.

For a converging lens, we find that for $d_o > f$ we always get a real, inverted image formed on the opposite side of the lens. If $d_o = f$, then $1/d_i = 0$ and the image is located at infinity. For a converging lens and $d_o < f$, we always get a virtual, upright, and enlarged image on the same side of the lens as the object. The results for all values of d_o are summarized in Table 33.1.

Diverging lenses always produce an image that is virtual, upright, and reduced in size.

Instrument makers often quote the power of a lens rather than its focal length. The power of a lens D, in **diopters**, a dimensionless number, is given by the equation

$$D = \frac{1 \text{ m}}{f} \qquad (33.3)$$

where f is the focal length of the lens expressed in meters. Eyeglass lenses are typically characterized in terms of diopters.

Table 33.1	Image Characteristics for Converging Lenses		
Case	Image Type	Image Orientation	Magnification
$f < d_o < 2f$	Real	Inverted	Enlarged
$d_o = 2f$	Real	Inverted	Same size
$d_o > 2f$	Real	Inverted	Reduced
$d_o = f$			Infinity
$d_o < f$	Virtual	Upright	Enlarged

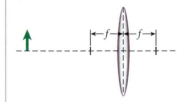

33.1 Self-Test Opportunity

In the four diagrams shown in the figure below, the solid arrow represents the object and the dashed arrow represents the image. The dashed rectangle represents a single optical element. The possible optical elements include a plane mirror, a converging mirror, a diverging mirror, a diverging lens, and a converging lens. Match each diagram with the corresponding optical element.

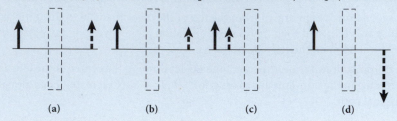

(a)   (b)   (c)   (d)

EXAMPLE 33.1 | Image Formed by a Thin Lens

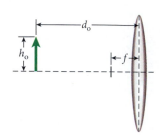

FIGURE 33.12 An object placed in front of a thin converging lens.

We place an object with height $h_o = 5.00$ cm at a distance $d_o = 16.0$ cm from a thin converging lens with focal length $f = 4.00$ cm (Figure 33.12).

PROBLEM

What is the image distance? What is the linear magnification of the image? What is the image height?

SOLUTION

The image distance can be calculated using the Thin-Lens Equation (equation 33.2),

$$\frac{1}{d_o} + \frac{1}{d_i} = \frac{1}{f}.$$

Solving for the image distance, we get

$$d_i = \frac{d_o f}{d_o - f} = \frac{(16.0 \text{ cm})(4.00 \text{ cm})}{16.0 \text{ cm} - 4.00 \text{ cm}} = 5.33 \text{ cm}.$$

The image distance is positive. Therefore, the image is real and appears on the opposite side of the lens from the object. The magnification formula for lenses is given by

$$m = \frac{h_i}{h_o} = -\frac{d_i}{d_o}.$$

Thus, we can calculate the magnification using the given object distance and the calculated image distance:

$$m = -\frac{d_i}{d_o} = -\frac{5.33 \text{ cm}}{16.0 \text{ cm}} = -0.333.$$

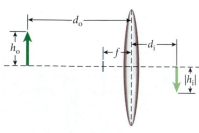

FIGURE 33.13 Image formed by a converging thin lens.

The magnification is negative, so the image is inverted. The magnitude of the magnification is less than one, so the image is reduced. We can now calculate the image height:

$$h_i = mh_o = (-0.333)(5.00 \text{ cm}) = -1.67 \text{ cm}.$$

The image height is negative, so the image is inverted, as we expected from the negative magnification.

The resulting image is illustrated in Figure 33.13.

33.2 Magnifier

One way to make an object appear larger is to bring it closer. However, if the object is brought too close to the eye, the object will appear fuzzy. The position closest to the eye that an object can be placed at and remain in focus is called the "near point," as discussed in detail below. Another way to make an object appear larger is to use a **magnifier** or magnifying glass (Figure 33.14).

A magnifier is nothing more than a converging lens that produces an enlarged virtual image of an object. This image appears at a distance that is at or beyond the near point of the eye, so an observer can clearly see the image. The **angular magnification** m_θ of the magnifier is defined as the ratio of the apparent angle subtended by the image to the angle subtended by the object when located at the near point.

Figure 33.15 shows the geometry of a magnifier. Assume an object with height h_o. Without a magnifier, the largest angle θ_1 that you can attain and still see the object clearly occurs when you place the object at the near point d_{near} (Figure 33.15a). You can get a magnified image of the object by placing the object just inside the focal length of a converging lens (Figure 33.15b). You then look through the lens at the image, which is intentionally located at least as far away as the near point. Therefore you can see the enlarged, upright, virtual image. The angle subtended by the image is θ_2.

The angular magnification of the magnifier is defined as

$$m_\theta = \frac{\theta_2}{\theta_1}. \qquad (33.4)$$

FIGURE 33.14 A typical magnifying glass.

Figure 33.15a shows that the angle subtended by the object without the magnifier is given by

$$\tan\theta_1 = \frac{h_o}{d_{near}}.$$

Figure 33.15b shows that the angle subtended by the image of the object is given by

$$\tan\theta_2 = \frac{h_o}{f},$$

where f is the focal length of the lens. We assume that the object is placed at the focal length of the lens, so the image is at minus infinity. (Recall from Section 33.1 that a ray through the center of the lens is not bent. It is this ray that forms the hypotenuse of the right triangle used for this equation.) We then make a small-angle approximation to get $\tan\theta_1 \approx \theta_1$ and $\tan\theta_2 \approx \theta_2$. Thus, the angular magnification of a magnifier can be written as

$$m_\theta = \frac{\theta_2}{\theta_1} \approx \frac{h_o/f}{h_o/d_{near}} = \frac{d_{near}}{f}.$$

Assuming a typical value for the near point of 25 cm (which is the value for a middle-aged person; for a 20-year-old the near point may be closer to 10 cm), the angular magnification can be written as

$$m_\theta \approx \frac{d_{near}}{f} \approx \frac{0.25 \text{ m}}{f}. \qquad (33.5)$$

Alternatively, the final image can be placed at the near point. Using the lens equation with $d_i = -d_{near}$ to find d_o and then using $\theta_2 = h_o/d_o$, we find $m_\theta = (0.25 \text{ m}/f) + 1$, if $d_{near} = 0.25$ m. Henceforth, unless otherwise stated, we will assume that the image is at infinity and will use equation 33.5.

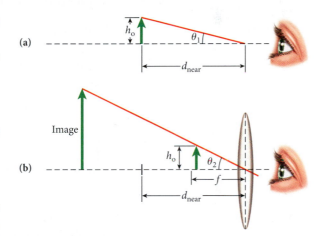

FIGURE 33.15 The geometry of a magnifier. (a) Viewing an object at the near point. (b) Viewing a magnified image of the object.

33.3 In-Class Exercise

What is the focal length (in meters) of a magnifying glass that gives an angular magnification of 6?

a) 0.010 m   d) 0.042 m

b) 0.021 m   e) 0.055 m

c) 0.035 m

33.3 Systems of Two or More Optical Elements

We have seen that a lens or mirror can produce an image of an object. This image can then, in turn, be used as the object for a second lens or mirror. The recurring theme for all two-lens systems is that the image of the first lens becomes the object of the second lens. Let's start our study of multi-lens systems by considering a two-lens system consisting of two converging lenses, placed as shown in Figure 33.16.

An object with height $h_{o,1}$ is placed a distance $d_{o,1}$ from the first lens, which has a focal length f_1. An image is produced at a distance $d_{i,1}$ given by the Thin-Lens Equation (33.2),

$$\frac{1}{d_{o,1}} + \frac{1}{d_{i,1}} = \frac{1}{f_1}.$$

Given the value of $d_{o,1}$ that we have chosen, this image is real and inverted. This image then becomes the object for the second lens. The height of this object is the same as the height of the image produced by the first lens. The object is located a distance $d_{o,2}$ from the second lens, which has a focal length f_2. An image is formed at a distance $d_{i,2}$ again governed by the Thin-Lens Equation,

$$\frac{1}{d_{o,2}} + \frac{1}{d_{i,2}} = \frac{1}{f_2}.$$

For the parameters of the system depicted in Figure 33.16, the final image of the second lens is real and inverted, compared to the second object (first image). The linear magnification of the first lens is given by $m_1 = h_{i,1}/h_{o,1}$ and the linear magnification of the second lens is given by $m_2 = h_{i,2}/h_{o,2}$. The product of the linear magnifications of the two lenses gives the linear magnification of the two-lens system:

$$m_{12} = m_1 m_2 = \left(\frac{h_{i,1}}{h_{o,1}}\right)\left(\frac{h_{i,2}}{h_{o,2}}\right) = \frac{h_{i,2}}{h_{o,1}}. \tag{33.6}$$

From equation 33.6 we can see that the image produced by this two-lens system is real and upright. Thus, this system of lenses can be used to produce real images that are not inverted. This kind of image formation cannot be done with a single converging lens.

Now let's consider a two-lens system with one converging lens and one diverging lens (Figure 33.17). An object with height $h_{o,1}$ is placed a distance $d_{o,1}$ from the first lens, which has a focal length f_1. Again, an image is produced at a distance $d_{i,1}$ given by the Thin-Lens Equation (equation 33.2). This particular image is real, inverted, and enlarged. Just like in the previous case, this image generated by the first lens then becomes the object for the second lens. This second image is virtual, upright with respect to the first image, and reduced.

FIGURE 33.16 A system of two converging lenses.

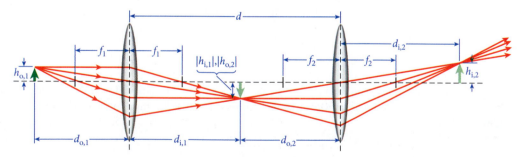

FIGURE 33.17 A two-lens system consisting of a converging lens and a diverging lens.

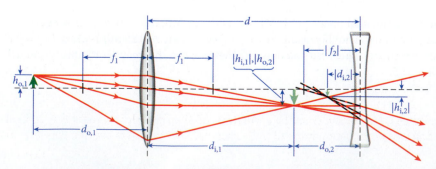

The combined linear magnification of the two lenses is again the product of the linear magnifications of the individual lenses and also given by $m_{12} = m_1 m_2 = h_{i,2}/h_{o,1}$, just like in the case discussed previously. Figure 33.17 shows that the final image produced by this two-lens system is virtual and inverted with respect to the original object.

Now let's take this same two-lens system of a converging lens followed by a diverging lens and put the two lenses close together. Place the two lenses a distance x apart (Figure 33.18). This two-lens system acts as a converging lens. By varying the distance x we can vary the effective focal length of the converging lens system. This arrangement is called a **zoom lens.**

The effective focal length of this two-lens system is defined as the distance from the center of the first lens to the position of the final image for an object originally located at infinity. The first lens has a focal length f_1, which means that objects placed at a large distance will produce an image at an image distance of $d_{i,1} = f_1$. We can understand this result using the Thin-Lens Equation,

$$\frac{1}{d_{o,1}} + \frac{1}{d_{i,1}} = \frac{1}{\infty} + \frac{1}{d_{i,1}} = 0 + \frac{1}{d_{i,1}} = \frac{1}{d_{i,1}} = \frac{1}{f_1}.$$

Assuming $f_1 > x$, the image produced by the first lens is generated on the right side of the second lens. This means that the object distance for the second lens must be negative in this case, because we have defined positive object distances to be to the left of a lens. The object distance for the second lens is

$$d_{o,2} = -\left(d_{i,1} - x\right) = x - d_{i,1} = x - f_1.$$

Using this image as the object for the second lens, we get

$$\frac{1}{d_{o,2}} + \frac{1}{d_{i,2}} = \frac{1}{x - f_1} + \frac{1}{d_{i,2}} = \frac{1}{f_2}.$$

This equation can be solved for $d_{i,2}$ to get the effective focal length of the zoom lens system,

$$f_{\text{eff}} = x + d_{i,2} = x + \frac{f_2\left(x - f_1\right)}{x - \left(f_2 + f_1\right)}. \tag{33.7}$$

Equation 33.7 and Figure 33.18 show that when the lenses are close together, the effective focal length is longer; when the lenses are farther apart, the effective focal length is shorter. Thus, by adjusting the distance between the converging lens and the diverging lens, we can produce an effective lens with varying focal length, as is done in zoom lenses of cameras. Note that for cameras, this result is useful only for $f_{\text{eff}} > x$ because objects at infinity must produce a real image on the digital image recorder (for a digital camera) or on the film (for an old-fashioned conventional camera).

SOLVED PROBLEM 33.1 | Image Produced by Two Lenses

Consider a system of two lenses. The first lens is a converging lens with a focal length $f_1 = 21.4$ cm. The second lens is a diverging lens with focal length $f_2 = -34.4$ cm. The center of the second lens is $d = 80.0$ cm to the right of the center of the first lens. An object is placed a distance $d_{o,1} = 63.5$ cm to the left of the first lens. These lenses produce an image of the object.

PROBLEM
Where is the image produced by the second lens located with respect to the center of the second lens? What are the characteristics of the image? What is the linear magnification of the final image with respect to the original object?

SOLUTION

THINK
The Thin-Lens Equation (which we will here call simply the lens equation) tells us where the first lens produces an image of the object placed to the left of the first lens. The image

Continued—

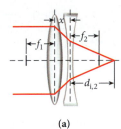

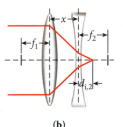

FIGURE 33.18 A zoom lens system consisting of a converging lens followed by a diverging lens. Two different distances between the two lenses are shown. (a) Long focal length. (b) Short focal length.

33.2 Self-Test Opportunity

Use equation 33.7 to show that the effective focal length f_{eff} of two thin lenses with focal lengths f_1 and f_2 close together is given by

$$\frac{1}{f_{\text{eff}}} = \frac{1}{f_1} + \frac{1}{f_2}.$$

33.3 Self-Test Opportunity

A normal 35-mm camera has a lens with a focal length of 50 mm. Suppose you replace the normal lens with a zoom lens whose focal length can be varied from 50 mm to 200 mm and use the camera to photograph an object at a very large distance. Compared to a 50-mm lens, what magnification of the image would be achieved using the 200-mm focal length?

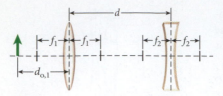

FIGURE 33.19 A system of two lenses with focal lengths and object distance marked.

produced by the first lens becomes the object for the second lens. We again use the lens equation to locate the image produced by the second lens.

SKETCH
Figure 33.19 shows a sketch of the object, the first lens, and the second lens.

RESEARCH
The lens equation applied to the first lens gives

$$\frac{1}{d_{o,1}} + \frac{1}{d_{i,1}} = \frac{1}{f_1},$$

where $d_{i,1}$ is the image distance for the first lens. Solving for the image distance of the first lens gives:

$$d_{i,1} = \frac{d_{o,1}f_1}{d_{o,1} - f_1}. \tag{i}$$

The lens equation applied to the second lens gives:

$$\frac{1}{d_{o,2}} + \frac{1}{d_{i,2}} = \frac{1}{f_2},$$

where $d_{o,2}$ is the object distance for the second lens and $d_{i,2}$ is the image distance for the second lens. We can solve for the image distance for the second lens,

$$d_{i,2} = \frac{d_{o,2}f_2}{d_{o,2} - f_2}. \tag{ii}$$

The object for the second lens is the image produced by the first lens. Thus, we can relate the object distance for the second lens to the image distance of the first lens and the distance between the two lenses:

$$d_{o,2} = d - d_{i,1}. \tag{iii}$$

SIMPLIFY
We can substitute equation (iii) into equation (ii) to obtain

$$d_{i,2} = \frac{(d - d_{i,1})f_2}{(d - d_{i,1}) - f_2}. \tag{iv}$$

We can then substitute equation (i) into equation (iv) to obtain an expression for the image distance of the second lens in terms of the quantities given in the problem:

$$d_{i,2} = \frac{\left(d - \left(\dfrac{d_{o,1}f_1}{d_{o,1} - f_1}\right)\right)f_2}{\left(d - \left(\dfrac{d_{o,1}f_1}{d_{o,1} - f_1}\right)\right) - f_2}.$$

CALCULATE
Putting in the numerical values gives us

$$d_{i,2} = \frac{\left((80.0 \text{ cm}) - \left(\dfrac{(63.5 \text{ cm})(21.4 \text{ cm})}{(63.5 \text{ cm}) - (21.4 \text{ cm})}\right)\right)(-34.4 \text{ cm})}{\left((80.0 \text{ cm}) - \left(\dfrac{(63.5 \text{ cm})(21.4 \text{ cm})}{(63.5 \text{ cm}) - (21.4 \text{ cm})}\right)\right) - (-34.4 \text{ cm})} = -19.9902 \text{ cm}.$$

ROUND
We report our result for the location of the image to three significant figures,

$$d_{i,2} = -20.0 \text{ cm}.$$

DOUBLE-CHECK

To double-check our result and calculate the quantities needed to answer the remaining parts of the problem, we calculate the position of the image produced by the first lens:

$$d_{i,1} = \frac{d_{o,1}f_1}{d_{o,1} - f_1} = \left| \frac{(63.5 \text{ cm})(21.4 \text{ cm})}{(63.5 \text{ cm}) - (21.4 \text{ cm})} \right| = 32.3 \text{ cm}.$$

The image distance for the first lens is positive, so the image is real and formed 32.3 cm to the right of the first lens. The object distance for the second lens is then

$$d_{o,2} = d - d_{i,1} = 47.7 \text{ cm}.$$

The image formed by the first lens is 47.7 cm to the left of the second lens, which seems reasonable. We can then calculate the image distance for the second lens:

$$d_{i,2} = \frac{d_{o,2}f_2}{d_{o,2} - f_2} = \left| \frac{(47.7 \text{ cm})(-34.4 \text{ cm})}{(47.7 \text{ cm}) - (-34.4 \text{ cm})} \right| = -20.0 \text{ cm},$$

which agrees with our result. Thus, our answer for the distance of the image from the center of lens 2 seems reasonable.

The final image is virtual because it is on the same side of lens 2 as the object for lens 2, which we know because the image distance for lens 2 is negative. The linear magnification for the final image compared with the original object is

$$m = m_1 m_2 = \left(-\frac{d_{i,1}}{d_{o,1}} \right) \left(-\frac{d_{i,2}}{d_{o,2}} \right) = \left(-\frac{32.3 \text{ cm}}{63.5 \text{ cm}} \right) \left(-\frac{-20.0 \text{ cm}}{47.7 \text{ cm}} \right) = -0.213.$$

The image is reduced because $|m| < 1$. The image is inverted because $m < 0$.

33.4 Human Eye

The human eye can be considered an optical instrument. The eye is nearly spherical in shape, about 2.5 cm in diameter (Figure 33.20). The front part of the eye is more sharply curved than the rest of the eye and is covered with the *cornea.* Behind the cornea is a fluid called the *aqueous humor.* Behind that is the *lens,* composed of a fibrous jelly. The lens is held in place by ligaments that connect it to the *ciliary muscles,* which allow the lens to change shape and thus change its focus. Behind the lens is the *vitreous humor.*

The index of refraction of the two fluids in the eye is 1.34, close to that of water (1.33). The index of refraction of the material making up the lens is about 1.40. Thus, most refraction of light rays occurs at the air/cornea boundary, which has the largest ratio in indices of refraction.

Refraction at the cornea and lens surfaces produces a real, inverted image on the *retina* of the eye. Cells in the retina called rods and cones convert the image from light to electrical impulses. These impulses are then sent to the brain through the *optic nerve.* The brain interprets the inverted image so that we see it upright. In front of the lens is the *iris,* the colored part of the eye, which partially opens or closes to regulate the amount of light incident on the retina.

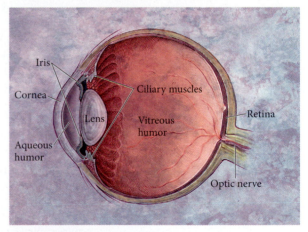

FIGURE 33.20 Drawing of the human eye, showing its major features.

For an object to be seen clearly, the image must be formed at the location of the retina (Figure 33.21a). The shape of the eye cannot be changed; so changing the shape of the lens must control the distance of the image. For a distant object, relaxing the lens focuses the image at the retina. For close objects, the ciliary muscle increases the curvature of the lens to again focus the image on the retina. This process is called **accommodation.**

The extremes over which distinct vision is possible are called the far point and near point. The **far point** of a normal eye is infinity. The **near point** of a normal eye depends on the ability of the eye

FIGURE 33.21 (a) Image produced
by the lens of a normal-sighted person;
(b) image produced by the lens of a
nearsighted person; (c) image produced
by a the lens of farsighted person.

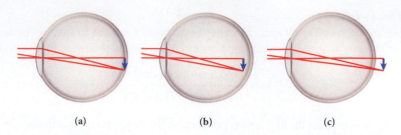

(a)   (b)   (c)

to focus. This ability changes with age. A young child can focus on objects as near as 7 cm. As a person ages, the near point increases. Typically, a 50-year-old person has a near point of 40 cm.

Several common vision defects result from incorrect focal distances. In the case of **myopia** (nearsightedness), the image is produced in front of the retina (Figure 33.21b). In the case of **hyperopia** (farsightedness), the image would be produced behind the retina (Figure 33.21c) if it could, but the retina absorbs the light before an image can be formed. Myopia can be corrected using diverging lenses, while hyperopia can be corrected using converging lenses.

EXAMPLE 33.2 Corrective Lenses

PROBLEM

Optometrists often quote the power of a corrective lens rather than its focal length. The power of a lens, D, in diopters, is given by equation 33.3 $D = 1\ m/f$, where f is the focal length of the lens in meters. What is the power of the corrective lens for a myopic (nearsighted) person whose uncorrected far point is 15 cm?

SOLUTION

The corrective lens must form a virtual, upright image of an object located at infinity, with the image located 15 cm in front of the lens (Figure 33.22).

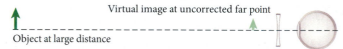

Virtual image at uncorrected far point

Object at large distance

FIGURE 33.22 Geometry of a corrective lens for nearsightedness.

Thus, the object distance d_o is ∞ and the image distance d_i is −15 cm. From the lens equation

$$\frac{1}{d_o} + \frac{1}{d_i} = \frac{1}{f}$$

we have

$$\frac{1}{\infty} + \frac{1}{-0.15\ m} = \frac{1}{f} = -6.7\ \text{diopter.}$$

The required lens is a diverging lens with a power of −6.7 diopter and a focal length of −0.15 m.

PROBLEM

A hyperopic (farsighted) person whose uncorrected near point is 75 cm wishes to read a newspaper at a distance of 25 cm. What is the power of the corrective lens needed for this person to read the paper?

SOLUTION

The corrective lens must produce a virtual, upright image of the newspaper at the uncorrected near point of the person's vision (Figure 33.23).

The object and image are on the same side of the lens, so the image distance is negative. Thus, the object distance is 25 cm and the image distance is −75 cm:

$$\frac{1}{0.25\ m} + \frac{1}{-0.75\ m} = \frac{1}{f} = +2.7\ \text{diopter.}$$

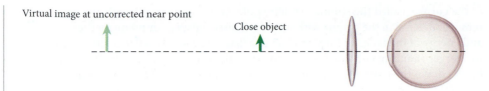

Virtual image at uncorrected near point

Close object

The required lens is a converging lens with a power of +2.7 diopter, corresponding to a focal length of +0.37 m.

Contact Lenses

The corrective lenses described in Example 33.2 consist of converging and diverging lenses. The converging lenses are convex on both surfaces and the diverging lenses are concave on both surfaces. These lenses correct vision well, but can be inconvenient.

A more convenient type of corrective lens is the **contact lens.** A contact lens is placed directly on the cornea of the eye, relieving the person from having to wear external glasses. These lenses are convex on the entrance surface and concave on the exit surface, similar to the meniscus lenses discussed in Section 33.1 (Figure 33.4). The concave exit surface is placed directly on the eye. It is possible to produce contact lenses that are converging or diverging (Figure 33.24).

We can use the Lens-Maker's Formula (equation 33.1),

$$\frac{1}{f} = (n-1)\left(\frac{1}{R_1} - \frac{1}{R_2}\right)$$

to calculate the focal length of a contact lens, where R_1 is the radius of curvature of the entrance surface and R_2 is the radius of curvature of the exit surface. Because the light entering the contact lens sees a convex surface, $R_1 > 0$. The light exiting the contact lens sees a convex surface also, so $R_2 > 0$ as well.

The contact lens shown in Figure 33.24a has $R_1 > R_2$ and thus has a negative focal length, corresponding to a diverging lens. The contact lens shown in Figure 33.24b has $R_1 < R_2$, which gives a positive focal length and a converging lens.

LASIK Surgery

An alternative to corrective lenses has been developed in which the cornea is altered to produce the desired optical response of the human eye. One such method, laser-assisted in situ keratomileusis **(LASIK) surgery,** uses a laser to modify the curvature of the cornea.

An example of LASIK surgery used to correct myopia is shown in Figure 33.25. Part (a) shows a myopic human eye, with the image produced in front of the retina. The effective focal length of this eye is too short. The LASIK procedure begins with the cutting of a flap off the surface of the cornea and folding it back (Figure 33.25b), exposing the inner part

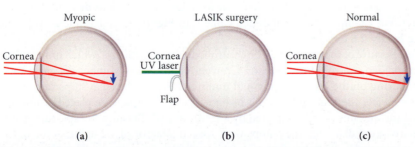

Myopic   LASIK surgery   Normal

Cornea   Cornea UV laser   Cornea

Flap

(a)   (b)   (c)

FIGURE 33.25 (a) A myopic (nearsighted) human eye, where the image is formed in front of the retina. (b) In LASIK surgery, a flap is cut from the surface of the cornea and part of the stroma is removed using a UV laser. The flap is then folded back and the cornea heals. (c) Normal vision is produced by focusing the image on the retina.

FIGURE 33.23 Geometry of a corrective lens for farsightedness.

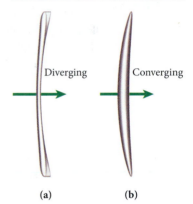

Diverging   Converging

(a)   (b)

FIGURE 33.24 (a) A diverging contact lens. (b) A converging contact lens. The green arrows represent the direction of the light traveling through the lenses.

of the cornea, called the stroma. An ultraviolet (UV) laser (wavelength 193 nm) is used to remove material in the stroma with very short laser pulses of duration on the order of 10 ns and energy of the order of 1 mJ, producing a flatter surface. This flatter surface corresponds to a larger radius of curvature for the cornea. (The laser operates with ultraviolet light that breaks down the molecular structure of the cornea without heating the surface.) The flap is then folded back and the cornea heals. The Lens-Maker's Formula (equation 33.1) tells us that increasing the front radius R_1 will increase the effective focal length f of the eye. Thus, the surgery allows the image to be produced on the retina (Figure 33.25c).

The technique described here is specific for myopic patients. The LASIK procedure does not work as well for hyperopic patients or for patients with astigmatism (irregular curvature of the cornea). To treat hyperopic patients, the laser must remove material in the stroma around the center of the cornea to increase the curvature of the surface.

33.5 Camera

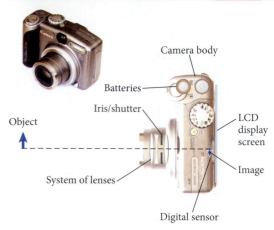

FIGURE 33.26 A typical digital camera produces an image on a digital sensor.

The **camera** is an optical instrument consisting of a body that excludes light and contains a system of lenses that focuses an image of an object on a recording medium such as photographic film or a digital sensor. A camera with a digital sensor is often called a digital camera, which is our main interest in the present section. A drawing of a digital camera is shown in Figure 33.26.

Normally we refer to the system of lenses in the camera as just the lens of the camera, neglecting the sophisticated multiple elements required to produce a high-quality image. The lens of the camera produces a real, inverted image of an object on the digital sensor. The camera lens is designed so that it can be moved to produce a sharp, focused image on the digital image recorder, depending on the image distance and the focal length of the lens. Many digital cameras have lenses that can be adjusted to have different focal lengths. A small digital camera, such as the one in Figure 33.26, with a 6× zoom lens can vary the focal length of the lens from 5.8 mm to 34.8 mm.

The lens has an opening area. An iris can be used to limit this opening area. The opening area is termed the **aperture** of the lens. The aperture is important because the amount of light that the camera can collect is proportional to the aperture. Because the aperture is usually a circle, we usually characterize the aperture by its diameter.

The size of the lens is often referred to as the **f-number** of the lens. The f-number of a lens is defined as the focal length of the lens divided by the diameter of the aperture of the lens:

$$f\text{-number} = \frac{\text{focal length}}{\text{aperture diameter}} = \frac{f}{D}. \tag{33.8}$$

A lens with a small f-number is called a fast lens, and a lens with a large f-number is called a slow lens. A small f-number implies a large aperture, which means that the lens can collect a large amount of light. A large f-number implies a small aperture, which means that the lens cannot collect a large amount of light. By convention, the f-number of a lens is usually written as f/#, where # represents the ratio of the focal length of the lens divided by the aperture diameter. For example, the small digital camera in Figure 33.26 has a lens with f-number ranging from f/2.8 to f/4.8.

The aperture of the camera can be controlled by a variable iris, which limits the amount of light incident on the digital sensor. Conventionally, the f-number of a lens was set by twisting a ring on the lens. The ring had multiple indents placed at stops to help the photographer select the required amount of light. These **f-stops** were placed at intervals that changed the amount of light accepted by a factor of two. Because the aperture depends on the square of the diameter, the f-stops were spaced a factor of $\sqrt{2}$ apart. Some of the standard f-stops are f/2, f/2.8, f/4, f/5.6, f/8, f/11, and f/16. Modern digital cameras usually set the required f-number automatically.

The digital image sensor requires a specific amount of light to form a good image. The iris can function as a shutter to control the amount of time the sensor sees and the total amount of light that is incident on the sensor. Unlike a film camera, the shutter of a digital camera is usually open to allow the photographer to see exactly what the camera sees on the LCD screen on the back of the camera. The total amount of light energy is the product of the time the light is incident on the digital sensor times the aperture times the light intensity. A digital sensor can also be programmed to accept light only over a given time interval without the use of a mechanical shutter. A typical time interval for an exposure varies from $1/60^{\text{th}}$ of a second to $1/1000^{\text{th}}$ of a second. Short exposure times allow the camera to capture a moving image without blurring. To prevent unwanted light interfering with a recorded image, the shutter closes while the image is being read out.

In addition to controlling the amount of light incident on the digital sensor, the iris can also affect the image by changing the **depth of field** of the image, which is the range in object distance where the image is in focus. If the iris is small (larger f-number), the range of angles of incident light is restricted, and a wider range of object distances will still produce a focused image on the digital sensor. If the iris is wide open (smaller f-number), a larger range of angles are admitted into the camera, meaning that only a narrow range of object distances will produce a focused image on the digital sensor. Thus, the photographer can increase the depth of field of an image by closing down the aperture and increasing the exposure time, or decrease the depth of field by opening the aperture and decreasing the exposure time.

The standard photographic film used for many years was 36 mm wide and 24 mm tall and was usually called 35-mm film. Most digital sensors are much smaller than 35-mm film. A typical small digital camera has a digital sensor that is 5.76 mm wide and 4.29 mm tall. This small size affects the magnification of the image produced by the camera, depending on the focal length of the lens. The **angle of view** of the camera, α, is more relevant to photography than the magnification. The angle of view can be defined in terms of the horizontal angle, the vertical angle, or the diagonal angle, where horizontal and vertical refer to the geometry of the film or digital sensor. We will consider the horizontal angle of view here, but the other two versions can be easily calculated from our results for the horizontal angle of view.

We can derive an expression for the horizontal angle of view, α, starting with our definition of linear magnification, m,

$$m = \frac{h_i}{h_o} = -\frac{d_i}{d_o}$$

where h_i is the image height, h_o is the object height, d_i is the image distance, and d_o is the object distance. Because photographers are not concerned about whether the image is upright or inverted, we will use the absolute value of the heights and distances in this calculation. In Figure 33.27 we show a schematic drawing containing an object, its corresponding image on film or a digital sensor, and the lens.

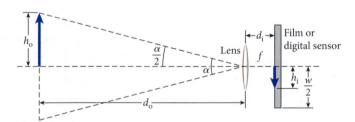

FIGURE 33.27 Schematic drawing of a camera showing the angle of view.

The horizontal width of the sensor is w. The object as shown in Figure 33.27 subtends an angle of $\alpha/2$ and the horizontal angle of view is α. The maximum image size that the film or digital sensor can detect occurs when the image height is equal to half the width of the film or digital sensor, $h_i = w/2$. If the object is not too close to the camera, the image distance is approximately equal to the focal length f of the lens, so we can use the definition of the linear magnification to write

$$\frac{w/2}{h_o} = \frac{d_i}{d_o} \approx \frac{f}{d_o} \Rightarrow \frac{h_o}{d_o} \approx \frac{w}{2f}.$$

From Figure 33.27, we can see that the horizontal angle of view can be related to the object height and object distance as

$$\frac{h_o}{d_o} = \tan(\alpha/2).$$

(a)   (b)   (c)

Combining these two equations for the ratio h_o/d_o gives us

$$\frac{h_o}{d_o} = \frac{w}{2f} = \tan(\alpha/2).$$

Thus, the horizontal angle of view is

$$\alpha = 2\tan^{-1}\left(\frac{w}{2f}\right). \tag{33.9}$$

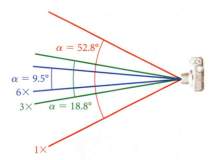

FIGURE 33.29 The horizontal angle of view for a small digital camera with 1×, 3×, and 6× zooms.

As an example of the effect of changing the focal length on a picture taken with a camera, we show three photos of a statue taken with different zoom settings in Figure 33.28.

The 1× zoom setting corresponds to a focal length of 5.8 mm, the 3× setting corresponds to a focal length of 17.4 mm, and the 6× setting corresponds to a focal length of 34.8 mm. The horizontal angle of view is shown for the each of the three zoom settings in Figure 33.29.

The horizontal angle of view for the 1× zoom setting of $\alpha = 52.8°$ is close to the horizontal angle of view of normal human vision of $\alpha \sim 50°-60°$. (Our peripheral vision extends to approximately 180°, but the inner 50°–60° dominate our perception.) The 6× zoom setting has a narrow horizontal angle of view of $\alpha = 9.5°$, providing a magnified image of the statue. For the camera to have a wider horizontal angle of view (wide-angle view), it would need a lens with a focal length less than 5.8 mm.

The digital sensor is composed of millions of electronic picture elements (pixels). For example, the small digital camera in Figure 33.26 uses a charge-coupled device (CCD) digital sensor with an array of pixels 3072 pixels wide and 2304 pixels tall for a total of 7,077,888 pixels (7.1 megapixels). Each pixel responds to light by liberating electrons. An analog-to-digital converter (ADC) reads out the pixels by digitizing the number of liberated electrons. The pixels in each row are digitized in order. Then the next row is shifted down and digitized until all the pixels are digitized. The question now remains: How does the CCD produce a color picture? The CCD responds only to the intensity of light, not the wavelength. The answer is that light from the image focused on the CCD first passes through a Bayer filter as illustrated in Figure 33.30.

The Bayer filter consists of rows of alternating pixels of red and green, followed by a row of alternating pixels of blue and green. There are more green pixels than red or blue pixels because the eye responds better to green than to red or blue. A small microprocessor in the camera calculates the best color for each pixel, based on the intensity of the red, green, and blue light passed by the Bayer filter.

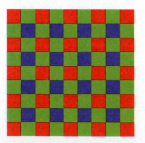

FIGURE 33.30 A portion of a Bayer filter used to produce color images with digital sensors.

A computer projector works much like a digital camera, except that the object is at the focus of the lens, which projects an image on a distant screen. The object for the projector can be a small light-transmitting LCD (liquid crystal display, see Chapter 31) screen that displays the image to be projected. A bright light shines through the LCD to produce the object for the lens. Another type of projector is the DLP (digital light processor), which uses small mirrors etched on a computer chip to produce the image. Each mirror represents a pixel. Colors are produced by alternately bathing the chip in red, green, and blue light. The image produced by the mirrors becomes the object for a lens, which projects the image on a screen. Movie theaters commonly use digital projectors based on DLP.

EXAMPLE 33.3 / Focal Length of a Simple Point-and-Shoot Camera

For many occasions, a simple point-and-shoot camera is sufficient. One example is an inexpensive commercial camera that uses 35-mm film (36 mm wide and 24 mm tall) like the one shown in Figure 33.31. You could also envision another version that uses a digital sensor that is 4.0 mm wide and 3.0 mm tall.

FIGURE 33.31 A simple point-and-shoot camera with fixed focal length.

PROBLEM
What are the required focal lengths of the lenses for these cameras if they are to have a horizontal angle of view of $\alpha = 46°$?

SOLUTION
The expression for the horizontal angle of view is given by equation 33.9:

$$\alpha = 2\tan^{-1}\left(\frac{w}{2f}\right).$$

We can rearrange this equation to obtain the focal length

$$f = \frac{w}{2\tan(\alpha/2)}.$$

The focal length for the 35-mm film version is

$$f = \frac{w}{2\tan(\alpha/2)} = \frac{36 \text{ mm}}{2\tan(46°/2)} = 42 \text{ mm}.$$

The focal length for the digital sensor version is

$$f = \frac{w}{2\tan(\alpha/2)} = \frac{4.0 \text{ mm}}{2\tan(46°/2)} = 4.7 \text{ mm}.$$

The digital sensor version would be considerably smaller than the 35-mm film version.

33.6 Microscope

The simplest **microscope** is a system of two lenses. For example, Figure 33.32 shows a microscope constructed of two thin lenses. The first lens is a converging lens of short focal length, f_o, called the **objective lens.** The second lens is another converging lens of greater focal length, f_e, called the **eyepiece.**

The object to be observed is placed just outside the focal length f_o of the objective lens, so that $d_{o,1} \approx f_o$. This arrangement allows the objective lens to form a real, inverted, and enlarged image of the object some distance from the objective lens. This image then becomes the object for the eyepiece lens. This intermediate image is placed just inside the focal length f_e of the eyepiece lens, so that the eyepiece lens produces a virtual, upright, and enlarged image of the intermediate image and $d_{o,2} \approx f_e$. The resulting magnification of the microscope is the product of the magnification from each lens.

FIGURE 33.32 Geometry of the image formed by a microscope. The objective lens, eyepiece, and object for a real microscope are also shown.

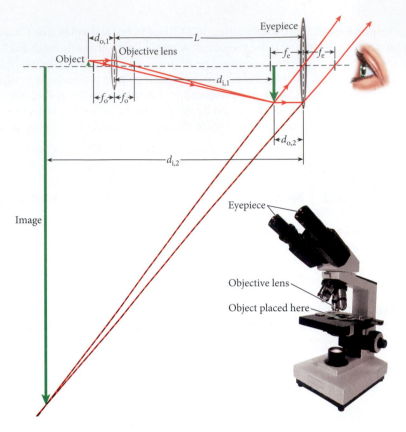

Let L be the distance between the two lenses, and assume $L \gg f_o, f_e$. Using the notation of Section 33.3, this means that $d_{i,1} \approx L$. Thus, the linear magnification of the microscope is

$$m = \frac{d_{i,1} d_{i,2}}{d_{o,1} d_{o,2}} = -\frac{(0.25 \text{ m})L}{f_o f_e}, \tag{33.10}$$

where we have used $d_{i,2} = -0.25$ m because we assume that the final image is produced at a comfortable viewing distance of 0.25 m.

EXAMPLE 33.4　Magnification of a Microscope

Consider a microscope that consists of an objective lens and an eyepiece lens separated by 15 cm. The focal length of the objective lens is 2 mm, and the focal length of the eyepiece lens is 20 mm. Assume the final image is produced at a distance of 25 cm.

PROBLEM

What is the magnitude of the linear magnification of this microscope?

SOLUTION

Taking our expression for the linear magnification of a microscope (equation 33.10), we have

$$|m| = \frac{(0.25 \text{ m})L}{f_o f_e} = \frac{(0.25 \text{ m})(0.15 \text{ m})}{(0.002 \text{ m})(0.020 \text{ m})} = 940.$$

33.6 In-Class Exercise

A microscope is intended to have a linear magnification whose magnitude is 330. If the objective lens has a power of 350 diopter and the eyepiece lens has a power of 10.0 diopter, how long must the microscope be? Assuming final image is produced at a distance of 25 cm.

a) 37.7 cm　　d) 65.0 cm

b) 40.0 cm　　e) 75.0 cm

c) 51.3 cm

33.7　Telescope

Telescopes come in many forms, including **refracting telescopes** and **reflecting telescopes.** In this chapter we study the magnification of the telescope, which is a measure of the telescope's ability to help us see large but distant objects. The resolution of a telescope, which is the capability to distinguish two nearby objects, is equally important and will be covered in Chapter 34.

Refracting Telescope

The refracting telescope consists of two converging lenses—the objective and the eyepiece. A modern commercial refracting telescope used by amateur astronomers is shown in Figure 33.33. In the following examples, we represent a telescope using two thin lenses. However, an actual refracting telescope uses more sophisticated lenses.

Because the object to be viewed is at a large distance, the incoming light rays can be considered to be parallel. Thus, the objective lens forms a real image of the distant object at distance f_o (Figure 33.34). The eyepiece is placed so that the image formed by the objective is a distance f_e from the eyepiece. Thus, the eyepiece forms a virtual, magnified image at infinity of the image formed by the objective, again producing parallel rays. In Figure 33.34, the parallel light rays from the distant object are shown incident on the objective lens. Red-and-black dashed lines depict the parallel light rays forming the virtual image.

Because the telescope deals with objects at large distances, it is not helpful to determine the magnification of the telescope using the formula for linear magnification which involves distances found from the lens equation. Therefore, we define the angular magnification of the telescope as -1 times the angle observed in the eyepiece θ_e divided by the angle subtended by the object being viewed θ_o (Figure 33.34),

$$m_\theta = -\frac{\theta_e}{\theta_o} = -\frac{f_o}{f_e}. \tag{33.11}$$

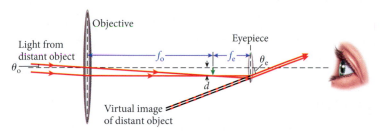

FIGURE 33.34 Geometry of an image formed by a refracting telescope.

FIGURE 33.33 A modern telescope used by amateur astronomers. This telescope has an 80-mm diameter objective lens with a 400-mm focal length.

DERIVATION 33.2 | **Angular Magnification**

The angle θ_o is the angle subtended by a distant object. The height of the image produced by the objective lens is d (Figure 33.34). This image is produced at the focal length of the objective lens, f_o. Because the focal length of the objective lens is large compared with the image size,

$$\theta_o \approx \tan\theta_o = \frac{d}{f_o}.$$

The image is placed at the focal point of the eyepiece lens. The apparent angle in the eyepiece, θ_e, can be written as

$$\theta_e \approx \tan\theta_e = \frac{d}{f_e},$$

again assuming that the image size is small compared with the focal length of the eyepiece. The ratio of the apparent eyepiece angle to the objective angle gives the angular magnification:

$$\frac{\theta_e}{\theta_o} = \frac{d/f_e}{d/f_o} = \frac{f_o}{f_e}.$$

Thus the angular magnification of a refracting telescope is

$$m_\theta = -\frac{f_o}{f_e},$$

where the minus signs means that the image is inverted.

For example, the refracting telescope shown in Figure 33.33 has an objective lens with focal length f_o = 400 mm and an eyepiece with focal length f_e = 9.70 mm, which gives the telescope a magnification of

$$m = -\frac{f_o}{f_e} = -\frac{400 \text{ mm}}{9.7 \text{ mm}} = -41.$$

EXAMPLE 33.5 | Magnification of a Refracting Telescope

The world's largest refracting telescope at that time was completed in 1897 and is still in use and housed at the Yerkes Observatory in Williams Bay, Wisconsin (between Chicago and Milwaukee). It has an objective lens of diameter 1.0 m (40 in) with a focal length of 19 m (62 ft).

PROBLEM
What should the focal length of the eyepiece be to give a magnification of magnitude 250?

SOLUTION
The focal length of the objective lens is given as f_o = 19 m. The absolute value of the magnification is to be $|m|$ = 250. The magnification is given by $|m| = f_o/f_e$. This gives us the focal length of the eyepiece lens as

$$f_e = \frac{f_o}{|m|} = \frac{19 \text{ m}}{250} = 0.076 \text{ m} = 7.6 \text{ cm}.$$

FIGURE 33.35 The SOAR telescope with a 4.1-m diameter primary mirror.

Reflecting Telescope

Most large astronomical telescopes are reflecting telescopes, with the objective lens replaced with a concave mirror. (However, the eyepiece of a reflecting telescope is still a lens.) An example of such a telescope is the Southern Astrophysical Research (SOAR) telescope shown in Figure 33.35, which has a 4.1-m diameter primary mirror. Large mirrors are necessary to gather as much light as possible to produce high-quality images from distant, faint astronomical objects. The light-gathering capability of a telescope is in practice much more important than the resulting magnification of the telescope.

Large mirrors are easier to fabricate and maintain in position than large lenses. Also, mirrors don't have chromatic aberration, so they work for a larger range of wavelengths. Finally, the size of refracting telescopes is limited because large lenses, which can be supported only around their edges, tend to sag due to their own weight. By contrast, modern reflecting telescopes use a large number (on the order of 100) individual

33.4 Self-Test Opportunity

A refracting telescope using two converging lenses is often called a Keplerian telescope. The eyepiece lens is placed a distance $d = f_o + f_e$ from the objective lens. Another type of refracting telescope uses a converging lens for the objective lens and a diverging lens for the eyepiece lens.

The eyepiece lens in this telescope is placed at a distance $d = f_o + f_e$ where $f_e < 0$. This type of telescope is often called a Galilean telescope. Discuss some advantages of a Galilean telescope versus a Keplerian telescope.

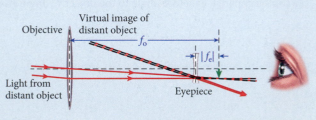

computer-controlled force actuators on the backside of the mirror to keep it as closely as possible in the ideal parabolic shape required for optimum image quality. These force actuators typically can each exert pulling or pushing forces of the order of 100 N on the mirror. They enable the mirror surface to stay within a few tens of nanometers of the ideal parabolic shape, even as the entire telescope is gradually rotated and tilted to stay locked on the astronomical object that it is tracking through the night sky during long-exposure imaging.

Various types of reflecting telescopes have been developed. Figure 33.36 shows three simple examples. The simplest form of reflecting telescope incorporates a parabolic mirror and an eyepiece (Figure 33.36a). This geometry is impractical because the observer must be placed in the direction from which the light comes. In the Newtonian solution to a reflecting telescope, a plane mirror at an angle of 45° reflects the light outside the structure of the telescope to an eyepiece (Figure 33.36b). In the Cassegrain geometry, a secondary convex hyperboloid mirror, placed perpendicular to the optic axis of the mirror, reflects the light through a hole in the center of the mirror (Figure 33.36c). In the last two cases, the secondary mirror is small enough that it absorbs a small (but not insignificant!) fraction of the incoming light. In all three cases, an eyepiece is used to magnify the image produced by the objective mirror. The SOAR telescope shown in Figure 33.35 uses the geometry shown in Figure 33.36b.

Hubble Space Telescope

The Hubble Space Telescope (HST), named for American astronomer Edwin Hubble (1889–1953), was deployed April 25, 1990, from the Space Shuttle (Figure 33.37). The HST orbits Earth 590 km above its surface, far above the atmosphere that disturbs the images gathered by ground-based telescopes. The HST is a reflecting telescope of Ritchey-Chrétian design arranged in Cassegrain geometry, but using a concave hyperbolic objective mirror, rather than a paraboloid mirror and a convex hyperbolic secondary mirror as is used in the traditional Cassegrain design. This design gives the HST a wide field of view and eliminates spherical aberration. The objective mirror is 2.40 m in diameter and has

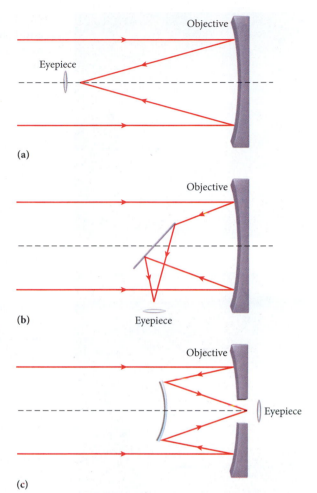

FIGURE 33.36 (a) Geometry of a standard reflecting telescope. (b) Geometry of a Newtonian reflecting telescope, with the eyepiece to the side. (c) Geometry of a Cassegrain reflecting telescope, with a secondary mirror and an eyepiece in the rear. (The relative sizes of the secondary mirror and eyepieces are not to scale with the primary mirror and are exaggerated.)

33.7 In-Class Exercise

Suppose a reflecting telescope consists of a concave spherical mirror with a radius of curvature $R = 17.0$ m and an eyepiece lens of focal length $f_e = 29.0$ cm. What is the magnitude of the magnification of this telescope?

a) 29.3   d) 66.1

b) 45.0   e) 78.9

c) 58.6

(a)   (b)

FIGURE 33.37 The Hubble Space Telescope is arguably the world's most famous optical instrument and has changed our understanding of the universe. (a) The Hubble Space Telescope in orbit. (b) Hubble photo of the gas pillars in the Eagle Nebula (M16): Pillars of Creation in a star-forming region.

(a)

(b)

FIGURE 33.38 Two images of the galaxy M100. (a) An image produced by the Hubble Space Telescope just after it was launched. (b) The same object photographed after the optics were corrected.

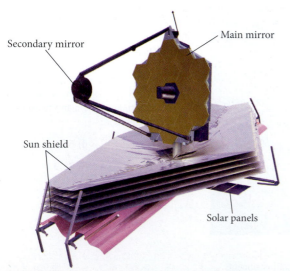

FIGURE 33.39 The planned James Webb Space Telescope. (Drawing from the Space Telescope Science Institute)

an effective focal length of 57.6 m. The secondary mirror is 0.267 m in diameter and is located 4.91 m from the objective mirror. The secondary mirror can be moved under ground control to produce the best focus. The eyepiece is replaced by a set of electronic instruments specialized for various astronomical tasks.

The original HST objective mirror was produced with a flaw caused by a defective testing instrument. The mirror had been polished very precisely, but unfortunately it was polished very precisely to a wrong shape. The maximum deviation from the perfect shape of the mirror was only 2.3 μm, but the resulting spherical aberrations were catastrophic (see Figure 33.38a). In December 1993, a Space Shuttle Servicing Mission deployed the Corrective Optics Space Telescope Axial Replacement (COSTAR) package, which corrected the flaw in the objective mirror and allowed the HST to begin revolutionizing our understanding of the universe. The two images of the galaxy M100 shown in Figure 33.38 demonstrate the image quality of the HST before and after the installation of COSTAR. In February 1997, two new instrument packages were installed in the HST on Shuttle Servicing Mission 2. These instruments contained their own optical corrections. In December 1997, the Shuttle Servicing Mission 3A was launched to correct a serious problem with the HST's stabilizing gyroscopes. In March 2002, the Shuttle Servicing Mission 3B added several new instruments. Shuttle Servicing Mission 4 in May 2009 replaced two failed instruments (spectrograph and advanced camera) and installed two new instruments (spectrograph and wide field camera) as well as new batteries and gyroscopes.

James Webb Space Telescope

The planned replacement for the HST is the James Webb Space Telescope (JWST), named for NASA's second administrator, James E. Webb (1906–1992). This project is planned for launch in the year 2014. The objective mirror for the JWST will be 6.5 m in diameter and will be composed of 18 mirror segments. An artist's conception of the JWST is shown in Figure 33.39.

The JWST will use infrared light to study the universe. Infrared light can penetrate the dust clouds that occur in our galaxy and elsewhere and block visible light. The JWST will orbit the Earth at a distance of 1.5 million km (about four times the Earth–Moon distance), in such a way that Earth will always be between the Sun and the JWST. Having the Sun always blocked by Earth will allow continuous viewing and shield the cryogenic infrared detectors aboard the JWST from changes in sunlight. The JWST is also equipped with a multiple-layer sunshield as shown in Figure 33.39.

CHANDRA X-Ray Observatory

The JWST will carry out observations using infrared light, which reflects well from mirrors. However, other types of electromagnetic waves do not reflect from the surfaces of mirrors. For example, if X-rays are incident on the surface of a mirror perpendicularly, they will penetrate the surface of the mirror rather than reflect. However, if the X-rays are incident on the surface of the mirror at a large angle, the X-rays will be reflected. The CHANDRA X-ray Observatory (CXO) produces images with X-rays using the mirrors shown in Figure 33.40a.

The mirrors of the CXO form images just as lenses and reflecting mirrors do (Figure 33.40b), although the geometry of the X-ray lenses is different. These images formed using X-rays can give us information about high-energy regions of the universe such as the very active galactic center of the Milky Way Galaxy, shown in Figure 33.40c.

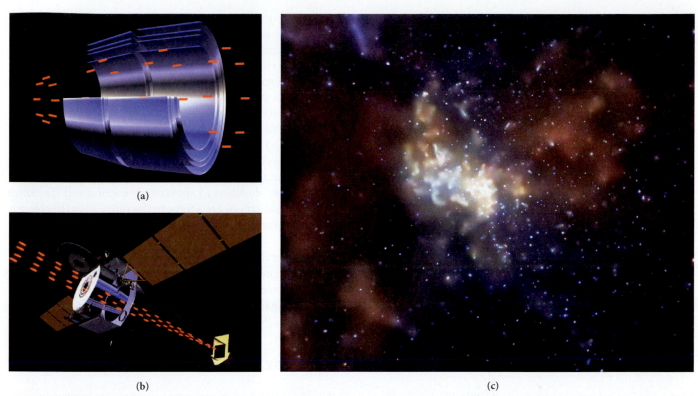

FIGURE 33.40 The CHANDRA X-ray Observatory. (a) The cylindrical mirrors used to focus X-rays. (b) The path of X-rays through CHANDRA. c) An X-ray image produced by CHANDRA of the center of the Milky Way Galaxy.

33.8 Laser Tweezers

After discussing the largest optical instruments with which we can explore the largest structures of the universe, we want to devote the last section of this chapter to an optical instrument with which we can manipulate some of the smallest structures.

As shown in Example 31.3, the force exerted on an object by light is on the order of 10^{-12} N (= 1 piconewton). This force is too small to affect macroscopic objects. However, physicists using very intense lasers focused to a small area can exert forces sufficient to manipulate objects as small as a single atom. These devices are called optical traps or **laser tweezers.**

Laser tweezers are constructed by focusing an intense laser beam to a point using the objective lens of a microscope. The force exerted by a laser pointer on a piece of paper produces a force in the direction of the original laser beam. In the physical realization of laser tweezers, the laser beam is focused in such a way that the light is more intense in the middle of the light distribution. In addition, the focusing produces light rays that converge on a point.

Let's consider the effect of the focused laser light on a spherical, optically transparent object. This object could be a small plastic sphere or a living cell that is approximately spherical. Define the original direction of the laser light as the z-direction, so the xy-plane lies perpendicular to the incident direction. Figure 33.41a shows the object in the yz-plane, displaced slightly in the negative y-direction. Light coming from

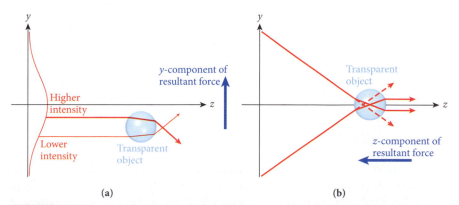

FIGURE 33.41 The effect of focused laser light on a small, spherical, optically transparent object. (a) The curved red line on the left represents the varying intensity of the wave. In addition, the restoring force in the y-direction is shown. (b) A restoring force also occurs in the z-direction.

the center of the distribution of light is more intense and also (as we will see in Chapter 34) is refracted (or bent) downward. Light coming from the edge of the distribution is less intense and is refracted upward. The resultant change in momentum of the incident light rays in the y-direction is downward. To conserve momentum, the object must recoil in the positive y-direction. Thus, the intense focused light of the laser produces a restoring force on the object if it is not positioned at $y = 0$, and the object is trapped in the y-direction.

In the z-direction, which is the direction of the incident laser light, trapping also occurs because of the converging rays produced by the focusing of the light by the lens (Figure 33.41b). In this case, the object is just to the right of the focus point. The incident rays are refracted such that they are more parallel to the incident direction than before. Thus the component of the momentum of light in the z-direction has been increased and the transparent object must recoil in the opposite direction to conserve momentum. If the object were to the left of the focus point, it would experience a force to the right. Thus the object is trapped in the direction parallel to the incident laser light as well as the direction perpendicular to the light.

This technique relies on the transmission of the incident light. If this technique is applied to an object that is not transparent, the incident light will be reflected and will produce a force pushing the object in the general direction of the incident light.

Laser tweezers have been used to trap cells, bacteria, viruses, and small polystyrene beads. They have also been used to manipulate DNA strands and to study molecular-size motors.

WHAT WE HAVE LEARNED | EXAM STUDY GUIDE

- For images formed by lenses, the object distance, the image distance, and the focal length of the lens are related by the Thin-Lens Equation, $\frac{1}{d_o} + \frac{1}{d_i} = \frac{1}{f}$. Here d_o is always positive, whereas d_i is positive if the image is on the opposite side of the lens as the object and negative if on the same side. The focal length f is positive for converging lenses and negative for diverging lenses.

- The Lens-Maker's Formula relates the curvature of both sides of a lens and its index of refraction to its focal length, $\frac{1}{f} = (n-1)\left(\frac{1}{R_1} - \frac{1}{R_2}\right)$.

- The diopter is a dimensionless number defined as the inverse of the focal length in meters.

- The angular magnification of a simple magnifier with an image at infinity is given by $m_\theta \approx \frac{0.25 \text{ m}}{f}$. (Assumed here is a typical value of 0.25 m for the near point of the average middle-aged person.)

- The linear magnification of a microscope is given by $m = -\frac{(0.25 \text{ m})L}{f_o f_e}$, where L is the distance between the two lenses, f_o is the focal length of the objective lens, and f_e is the focal length of the eyepiece lens. (Final image assumed formed at distance of 0.25 m.)

- The angular magnification of a telescope is given by $m_\theta = -\frac{\theta_e}{\theta_o} = -\frac{f_o}{f_e}$, where f_o is the focal length of the objective lens or mirror and f_e is the focal length of the eyepiece lens.

- The f-number of a camera lens is defined as the focal length of the lens f divided by the diameter of the aperture of the lens D, f-number $= \frac{\text{focal length}}{\text{aperture diameter}} = \frac{f}{D}$.

- The horizontal angle of view of a camera is $\alpha = 2 \tan^{-1}\left(\frac{w}{2f}\right)$, where w is the width of the film or digital image sensor and f is the focal length of the camera lens.

KEY TERMS

NEW SYMBOLS

$m = \dfrac{h_i}{h_o} = -\dfrac{d_i}{d_o}$, linear magnification

$D = \dfrac{1 \text{ m}}{f}$, power of a lens, in diopters

m_θ, angular magnification

f_o, focal length of the objective lens or mirror

f_e, focal length of the eyepiece

f-number, ratio of the focal length divided by the diameter of the aperture for a camera

α, horizontal angle of view of a camera

ANSWERS TO SELF-TEST OPPORTUNITIES

33.1 a) plane mirror   c) diverging lens
      b) diverging mirror   d) converging lens

33.2 Taking $x = 0$ in equation 33.7 gives us $f_{\text{eff}} = f_2 f_1 / (f_2 + f_1)$. We can invert this equation to get

$$\frac{1}{f_{\text{eff}}} = \frac{f_2 + f_1}{f_2 f_1} = \frac{1}{f_1} + \frac{1}{f_2}.$$

33.3 The magnitude of the magnification is given by $|m| = d_i / d_o$. If the object distance is infinity, then $d_i = f$. Thus the 200-mm focal length lens will produce a magnification of four times that of the normal 50-mm lens.

33.4 Comparing with Figure 33.34, we can see that a Galilean telescope produces an upright image, which is useful for terrestrial telescopes, opera glasses, and binoculars. The Keplerian telescope produces an inverted image. A Galilean telescope can be shorter than a Keplerian telescope, which is also useful for optical devices such as opera glasses or binoculars.

PROBLEM-SOLVING PRACTICE

Problem-Solving Guidelines

Basically the same guidelines as in the previous chapter also apply here, because for lenses and optical instruments we simply use the same ray-tracing laws as for mirrors.

1. Drawing a large, clear, well-labeled diagram should be the first step in solving almost any optics problem. Include all the information you know and all the information you need to find. Remember that angles are measured from the normal to a surface, not to the surface itself.

2. You need to draw only two principal rays to locate an image formed by mirrors or lenses, but you should draw a third ray as a check that they are drawn correctly. Even if you need to solve a problem using the mirror equation or Thin-Lens Equation, an accurate drawing can help you approximate the answers as a check on your calculations. But keep in mind that in some lens configurations with large magnification factors, even minute drawing inaccuracies can translate into large errors in image size or location. So it is also not a good idea to rely strongly on your drawing alone.

3. When you calculate distances for mirrors or lenses, remember the sign conventions and be careful to use them correctly. If a diagram indicates an inverted image, say, but your calculation does not have a negative sign, go back to the starting equation and check the signs of all distances and focal lengths.

SOLVED PROBLEM 33.2 | Image of the Moon

PROBLEM
An image of the Moon is focused onto a screen, using a converging lens of focal length $f = 50.0$ cm. The radius of the Moon is $R = 1.737 \cdot 10^6$ m, and the mean distance between Earth and the Moon is $d = 3.844 \cdot 10^8$ m. What is the radius of the Moon in the image on the screen?

SOLUTION

THINK
The linear magnification of the image produced by the lens can be expressed in terms of the ratio of the image size to the object size and in terms of the ratio of image distance to the object distance.

Continued—

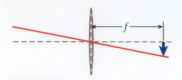

FIGURE 33.42 Image of the Moon produced by a converging lens.

SKETCH

Figure 33.42 shows a sketch of the image of the Moon produced by a converging lens. A blue arrow represents the image of the Moon.

RESEARCH

The relationship between the object distance d_o and the image distance d_i for a lens with focal length f is given by the Thin Lens Equation,

$$\frac{1}{d_o} + \frac{1}{d_i} = \frac{1}{f}.$$

We can solve this equation for the image distance,

$$d_i = \frac{d_o f}{d_o - f}.$$

In this case, the object distance is the distance from Earth to the Moon. Because the object distance is much greater than the focal length of the lens, we can write

$$d_i \approx \frac{d_o f}{d_o} = f.$$

The linear magnitude of the magnification m for a lens can be written as

$$m = -\frac{d_i}{d_o} = \frac{h_i}{h_o}$$

where h_i is the height (radius) of the image and h_o is the height of the object, which in this case is the radius of the Moon.

SIMPLIFY

We can solve the previous equation for the image height,

$$h_i = -h_o \frac{d_i}{d_o}.$$

Taking the object height to be the radius of the Moon R, the object distance to be the distance from Earth to the Moon d, and the image distance to be the focal length of the lens f, we can write

$$h_i = -R\frac{f}{d}.$$

CALCULATE

Putting in the numerical values, we have

$$h_i = -R\frac{f}{d} = \left(1.737 \cdot 10^6 \text{ m}\right)\frac{0.500 \text{ m}}{3.844 \cdot 10^8 \text{ m}} = -0.00225937 \text{ m}.$$

ROUND

We report our result for the radius of the image of the Moon on the screen to three significant figures,

$$h_i = -2.26 \cdot 10^{-3} \text{ m} = -2.26 \text{ mm}.$$

DOUBLE-CHECK

The Moon is relatively close to Earth and we can see the Moon as a disk easily with the naked eye. Thus, it seems reasonable that we could produce an image of the Moon with radius 2.26 mm on a screen with a lens of focal length 50.0 cm. The negative sign for the image height means that our image of the Moon is inverted.

SOLVED PROBLEM 33.3   **Image Produced by a Lens and a Mirror**

An object is placed a distance $d_{o,1} = 25.6$ cm to the left of a converging lens with focal length $f_1 = 20.6$ cm. A converging mirror with focal length $f_2 = 10.3$ cm is placed a distance $d = 120.77$ cm to the right of the lens.

PROBLEM

What is the magnification of the image produced by the lens-and-mirror combination?

SOLUTION

THINK

The lens will produce a real, inverted image of the object. This image becomes the object for the converging mirror. The object distance for the mirror is the distance between the lens and the mirror minus the image distance of the lens. The overall magnification is the magnification of the lens times the magnification of the mirror.

SKETCH

Figure 33.43 shows a sketch of the object, the lens, and the mirror.

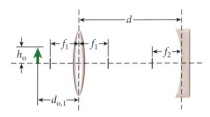

FIGURE 33.43 An object being imaged by a lens-and-mirror combination.

RESEARCH

The Thin Lens Equation tells us that the image distance $d_{i,1}$ for the lens is given by

$$d_{i,1} = \frac{d_{o,1} f_1}{d_{o,1} - f_1}. \tag{i}$$

The mirror equation tells us that the image distance for the mirror is

$$d_{i,2} = \frac{d_{o,2} f_2}{d_{o,2} - f_2}, \tag{ii}$$

where the object distance for the mirror is

$$d_{o,2} = d - d_{i,1}. \tag{iii}$$

The magnification m of the lens-and-mirror system is given by

$$m = m_1 m_2, \tag{iv}$$

where m_1 is the magnification of the lens and m_2 is the magnification of the mirror. The magnification of the lens using equation (i) is

$$m_1 = -\frac{d_{i,1}}{d_{o,1}} = -\frac{\left(\dfrac{d_{o,1} f_1}{d_{o,1} - f_1}\right)}{d_{o,1}} = -\frac{f_1}{d_{o,1} - f_1} = \frac{f_1}{f_1 - d_{o,1}}. \tag{v}$$

Similarly, the magnification of the mirror using equation (ii) is

$$m_2 = -\frac{d_{i,2}}{d_{o,2}} = -\frac{\left(\dfrac{d_{o,2} f_2}{d_{o,2} - f_2}\right)}{d_{o,2}} = -\frac{f_2}{d_{o,2} - f_2} = \frac{f_2}{f_2 - d_{o,2}}. \tag{vi}$$

SIMPLIFY

We can then write for the overall magnification, using equations (iii), (v), and (vi) in equation (iv)

$$m = \left(\frac{f_1}{f_1 - d_{o,1}}\right)\left(\frac{f_2}{f_2 - d_{o,2}}\right) = \left(\frac{f_1}{f_1 - d_{o,1}}\right)\left(\frac{f_2}{f_2 - (d - d_{i,1})}\right).$$

Finally, substituting from equation (i) for the image distance for the lens gives us

$$m = \left(\frac{f_1}{f_1 - d_{o,1}}\right)\frac{f_2}{f_2 - \left(d - \dfrac{d_{o,1} f_1}{d_{o,1} - f_1}\right)}.$$

CALCULATE

Putting in the numerical values gives us

$$m = \left(\frac{(20.6 \text{ cm})}{(20.6 \text{ cm}) - (25.6 \text{ cm})}\right)\frac{(10.3 \text{ cm})}{(10.3 \text{ cm}) - \left((120.77 \text{ cm}) - \dfrac{(25.6 \text{ cm})(20.6 \text{ cm})}{(25.6 \text{ cm}) - (20.6 \text{ cm})}\right)}.$$

$$m = 8.490596.$$

Continued—

ROUND
We report our result to three significant figures,

$$m = 8.49.$$

DOUBLE-CHECK
To double-check our result, we first calculate the image distance for the lens,

$$d_{i,1} = \frac{d_{o,1} f_1}{d_{o,1} - f_1} = \frac{(25.6 \text{ cm})(20.6 \text{ cm})}{(25.6 \text{ cm}) - (20.6 \text{ cm})} = 105.47 \text{ cm}.$$

The object distance for the mirror is then

$$d_{o,2} = d - d_{i,1} = 120.77 \text{ cm} - 105.47 \text{ cm} = 15.3 \text{ cm}.$$

The image distance for the mirror is then

$$d_{i,2} = \frac{d_{o,2} f_2}{d_{o,2} - f_2} = \frac{(15.3 \text{ cm})(10.3 \text{ cm})}{(15.3 \text{ cm}) - (10.3 \text{ cm})} = 31.52 \text{ cm}.$$

The magnification of the image is then

$$m = \frac{d_{i,1}}{d_{o,1}} \frac{d_{i,2}}{d_{o,2}} = \frac{105.5 \text{ cm}}{25.6 \text{ cm}} \frac{31.52 \text{ cm}}{15.3 \text{ cm}} = 8.49,$$

which agrees with our result.

SOLVED PROBLEM 33.4 | Two Positions of a Converging Lens

A light bulb is at a distance $d = 1.45$ m away from a screen. A converging lens with focal length $f = 15.3$ cm forms an image of the light bulb on the screen for two lens positions.

PROBLEM
What is the distance between these two positions?

SOLUTION

THINK
The sum of the object distance of the lens plus the image distance of the lens is equal to the distance of the light bulb from the screen. Using the lens equation, we can solve for the possible image distances using the quadratic equation. The distance between the two image distances is the distance between the two lens positions.

SKETCH
Figure 33.44 shows a sketch of the lens placed between the light bulb and the screen.

FIGURE 33.44 A lens is placed between a light bulb and a screen.

RESEARCH
The Thin-Lens Equation is

$$\frac{1}{d_o} + \frac{1}{d_i} = \frac{1}{f}, \qquad \text{(i)}$$

where d_o is the object distance, d_i is the image distance, and f is the focal length. We can express the object distance in terms of the image distance and the distance of the light bulb from the screen d,

$$d_o = d - d_i. \qquad \text{(ii)}$$

Substituting equation (ii) into equation (i) gives us

$$\frac{1}{d - d_i} + \frac{1}{d_i} = \frac{1}{f}.$$

We can rearrange this equation to get

$$d_i + (d - d_i) = \frac{d_i(d - d_i)}{f}.$$

Collecting terms and multiplying by f gives us

$$df = d_i d - d_i^2. \qquad \text{(iii)}$$

SIMPLIFY

We can rewrite equation (iii) so that we can recognize it as a quadratic equation, for which we can find the solutions:

$$d_i^2 - d_i d + df = 0.$$

The solutions of this equation are

$$d_i = \frac{d \pm \sqrt{d^2 - 4df}}{2},$$

where one solution corresponds to the + sign and the other solution corresponds to the − sign.

CALCULATE

The solution corresponding to the + sign is

$$d_{i+} = \frac{(1.45 \text{ m}) + \sqrt{(1.45 \text{ m})^2 - 4(1.45 \text{ m})(0.153 \text{ m})}}{2} = 1.27616 \text{ m}.$$

The solution corresponding to the − sign is

$$d_{i-} = \frac{(1.45 \text{ m}) - \sqrt{(1.45 \text{ m})^2 - 4(1.45 \text{ m})(0.153 \text{ m})}}{2} = 0.173842 \text{ m}.$$

The difference between the two positions is $\Delta d_i = 1.10232$ m.

ROUND

We report our result to three significant figures,

$$\Delta d_i = 1.10 \text{ m}.$$

(*Note*: Even though we have found two positions for the lens that allow an image to be formed, it is not correct to assume that the image has the same size in both cases.)

DOUBLE-CHECK

To double-check our answer, we substitute our solutions for the image distance into the Thin-Lens Equation and show that they work. For the first solution, $d_i = 1.276$ m, so the corresponding object distance is $d_o = d - d_i = 0.174$ m. The lens equation then tells us

$$\frac{1}{0.174} + \frac{1}{1.276} = \frac{1}{0.153 \text{ m}},$$

which agrees within round-off errors. For the second solution, we simply reverse the role of the image distance and object distance in the Thin-Lens Equation, and we get the same answer. Thus our result seems reasonable.

MULTIPLE-CHOICE QUESTIONS

33.1 For a microscope to work as intended, the separation between the objective lens and the eyepiece must be such that the intermediate image produced by the objective lens will occur at a distance (as measured from the optical center of the eyepiece)

a) slightly larger than the focal length.

b) slightly smaller than the focal length.

c) equal to the focal length.

d) The position of the intermediate image is irrelevant.

33.2 Which one of the following is not a characteristic of a simple two-lens astronomical refracting telescope?

a) The final image is virtual.

b) The objective forms a virtual image.

c) The final image is inverted.

33.3 A converging lens will be used as a magnifying glass. In order for this to work, the object must be placed at a distance

a) $d_o > f$.

b) $d_o = f$.

c) $d_o < f$.

d) None of the above.

33.4 An object is moved from a distance of 30 cm to a distance of 10 cm in front of a converging lens of focal length 20 cm. What happens to the image?

a) Image goes from real and upright to real and inverted.

b) Image goes from virtual and upright to real and inverted.

c) Image goes from virtual and inverted to real and upright.

d) Image goes from real and inverted to virtual and upright.

e) None of the above.

33.5 What type of lens is a magnifying glass?

a) converging

b) diverging

c) spherical

d) cylindrical

e) plain

33.6 LASIK surgery uses a laser to modify the

a) curvature of the retina.

b) index of refraction of the aqueous humor.

c) curvature of the lens.

d) curvature of the cornea.

33.7 What is the focal length of a flat sheet of transparent glass?

a) zero

b) infinity

c) thickness of the glass

d) undefined

33.8 Where is the image formed if an object is placed 25 cm from the eye of a nearsighted person. What kind of a corrective lens should the person wear?

a) Behind the retina. Converging lenses.

b) Behind the retina. Diverging lenses.

c) In front of the retina. Converging lenses.

d) In front of the retina. Diverging lenses.

33.9 An object is placed on the left of a converging lens at a distance that is less than the focal length of the lens. The image produced will be

a) real and inverted.

b) virtual and erect.

c) virtual and inverted.

d) real and erect.

33.10 What would you expect to happen to the magnitude of the power of a lens when it is placed in water ($n = 1.33$)?

a) It would increase.

b) It would decrease.

c) It would stay the same.

d) It would depend if the lens was converging or diverging.

33.11 An unknown lens forms an image of an object that is 24 cm away from the lens, inverted, and a factor of 4 larger in size than the object. Where is the object located?

a) 6 cm from the lens on the same side of the lens

b) 6 cm from the lens on the other side of the lens

c) 96 cm from the lens on the same side of the lens

d) 96 cm from the lens on the other side of the lens

e) No object could have formed this image.

QUESTIONS

33.12 Several small drops of paint (less than 1 mm in diameter) splatter on a painter's eyeglasses, which are approximately 2 cm in front of the painter's eyes. Do the dots appear in what the painter sees? How do the dots affect what the painter sees?

33.13 When a diver with 20/20 vision removes her mask underwater, her vision becomes blurry. Why is this the case? Does the diver become nearsighted (eye lens focuses in front of retina) or farsighted (eye lens focuses behind retina)? As the index of refraction of the medium approaches that of the lens, where does the object get imaged? Typically, the index of refraction for water is 1.33, while the index of refraction for the lens in a human eye is 1.40.

33.14 In H.G. Wells's classic story *The Invisible Man*, a man manages to change the index of refraction of his body to 1.0; thus, light would not bend as it enters his body (assuming he is in air and not swimming). If the index of refraction of his eyes were equal to one, would he be able to see? If so, how would things appear?

33.15 Astronomers sometimes place filters in the path of light as it passes through their telescopes and optical equipment. The filters allow only a single color to pass through. What are the advantages of this? What are the disadvantages?

33.16 Is it possible to start a fire by focusing the light of the Sun with ordinary eyeglasses? How, or why not?

33.17 Will the magnification produced by a simple magnifier increase, decrease, or stay the same when the object and the lens are both moved from air into water?

33.18 Is it possible to design a system that will form an image without lenses or mirrors? If so, how? and what drawbacks, if any, would it have?

33.19 A lens system widely used in machine vision for dimensional measurement is the so-called telecentric lens system. In its basic configuration, it consists of two thin lenses of focal lengths f_1 and f_2, respectively, placed a distance $d = f_1 + f_2$ apart, and a small circular aperture called a *stop* aperture placed at the common focal point between the two lenses. The purpose of such a system is to provide a magnification that is independent of the distance between the object and the lens system, within a specified range of distances that define the so-called *depth of field* of the system.

a) Draw a ray diagram through the system.

b) Determine the magnification of such a system.

c) Determine the requirements that must be met by the two lenses in order for a telecentric system to be able to image with maximum resolution an object with a diameter of 50 mm on a so-called $\frac{1}{2}$-inch CCD camera head (detector dimensions are ~ 6.5 mm × 5.0 mm).

33.20 Lenses or lens systems for photography are rated by both focal length and "speed." The "speed" of a lens is measured by its f-number or f-stop, the ratio of its focal length to its aperture diameter. For many commercially available lenses this number is approximately a power of $\sqrt{2}$, e.g., 1.4, 2.0, 2.8, 4.0, etc.; the lens is provided with a mechanical iris diaphragm that can be set to different apertures or f-stops. The smaller the f-number, the "faster" the lens. "Faster" lenses are more expensive, as a wide aperture requires a higher-quality lens.

a) Explain the connection between f-number, (i.e., aperture diameter) and "speed."

b) Look up the necessary data and calculate the f-numbers of (the primary mirrors of) the Keck 10-meter telescope, the Hubble Space Telescope, and the Arecibo radio telescope.

33.21 The f-number of a photographic system determines not just its speed but also its "depth of field," the range of distances over which objects remain in acceptable focus. Low f-numbers correspond to small depth of field, high f-numbers to large depth of field. Explain this.

33.22 Mirrors for astronomical instruments are invariably *first-surface* mirrors: The reflective coating is applied on the surface exposed to the incoming light. Household mirrors, on the other hand, are *second-surface* mirrors: The coating is applied to the back of the glass or plastic material of the mirror. (You can tell the difference by bringing the tip of an object close to the surface of the mirror. Object and image will nearly touch with a first-surface mirror; a gap will remain between them with a second-surface mirror.) Explain the reasons for these design differences.

33.23 When sharing binoculars with a friend, you notice that you have to readjust the focus when he has been using it (he wears glasses, but removes them to use the binoculars). Why?

33.24 Using ray tracing in the diagram below, find the image of the object (upright arrow) in the following system having a diverging (double-concave) lens. Is the image real or virtual? Is the image height less than or greater than the object height?

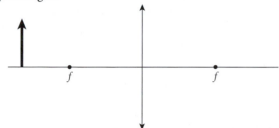

33.25 You have made a simple telescope from two convex lenses. The objective lens is the one of the two lenses that is closer to the object being observed. What kind of image is produced by the eyepiece lens if the eyepiece is closer to the objective lens than the image produced by the objective lens?

33.26 What kind of lens is used in eyeglasses to correct the vision of someone who is

a) nearsighted?

b) farsighted?

33.27 A physics student epoxies two converging lenses to the opposite ends of a $2.0 \cdot 10^1$-cm-long tube. One lens has a focal length of $f_1 = 6.0$ cm and the other has a focal length of $f_2 = 3.0$ cm. She wants to use this device as a microscope. Which end should she look through to obtain the highest magnification of an object?

PROBLEMS

A blue problem number indicates a worked-out solution is available in the Student Solutions Manual. One • and two •• indicate increasing level of problem difficulty.

Section 33.1

33.28 An object of height h is placed at a distance d_o on the left side of a converging lens of focal length f ($f < d_o$).

a) What must d_o be in order for the image to form at a distance $3f$ on the right side of the lens?

b) What will be the magnification?

33.29 An object is 6.0 cm from a converging thin lens along the axis of the lens. If the lens has a focal length of 9.0 cm, determine the image magnification.

33.30 A biconvex *ice lens* is made so that the radius of curvature of the front surface is 15 cm and that for the back is 20 cm. Determine how far you would put dry twigs if you wish to start a fire with your ice lens.

33.31 As a high-power laser engineer you need to focus a 1.06-mm diameter laser beam to a 10.0-μm diameter spot 20.0 cm behind the lens. What focal length lens would you use?

33.32 A plastic cylinder of length $3.0 \cdot 10^1$ cm has its ends ground to convex (from the rod outward) spherical surfaces, each having radius of curvature $1.0 \cdot 10^1$ cm. A small object is placed $1.0 \cdot 10^1$ cm from the left end. How far will the image of the object lie from the right end, if the index of refraction of the plastic is 1.5?

•**33.33** The object (upright arrow) in the following system has a height of 2.5 cm and is placed 5.0 cm away from a converging (convex) lens with a focal length of 3.0 cm. What is the magnification of the image? Is the image upright or inverted? Confirm your answers by ray tracing.

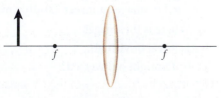

•**33.34** Demonstrate that the minimum distance possible between a real object and its real image through a thin convex lens is 4f, where f is the focal length of the lens.

•**33.35** An air-filled cavity bound by two spherical surfaces is created inside a glass block. The two spherical surfaces have radii of 30.0 cm and 20.0 cm, respectively, and the thickness of the cavity is 40.0 cm (see diagram below). A light-emitting diode (LED) is embedded inside the block a distance of 60.0 cm in front of the cavity. Given n_{glass} = 1.50 and n_{air} = 1.00, and using only paraxial light rays (i.e., in the paraxial approximation):

a) Calculate the final position of the image of the LED through the air-filled cavity.

b) Draw a ray diagram showing how the image is formed.

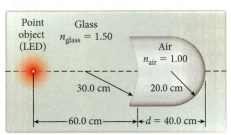

Section 33.2

33.36 To study a tissue sample better, a pathologist holds a 5-cm focal length magnifying glass 3 cm from the sample. How much magnification can he get from the lens?

33.37 Suppose a magnifying glass has a focal length of 5 cm. Find the magnifying power of this glass when the object is placed at the near point.

33.38 What is the focal length of a magnifying glass if a 1-mm object appears to be 10 mm?

•**33.39** A person with a near-point distance of 24.0 cm finds that a magnifying glass gives an angular magnification that is 1.25 times larger when the image of the magnifier is at the near point than when the image is at infinity. What is the focal length of the magnifying glass?

Section 33.3

33.40 A beam of parallel light, 1.00 mm in diameter passes through a lens with a focal length of 10.0 cm. Another lens, this one of focal length 20.0 cm, is located behind the first lens so that the light traveling out from it is again parallel.

a) What is the distance between the two lenses?

b) How wide is the outgoing beam?

33.41 How large does a 5.0-mm insect appear when viewed with a system of two identical lenses of focal length 5.0 cm separated by a distance 12 cm if the insect is 10.0 cm from the first lens? Is the image real or virtual? Inverted or upright?

•**33.42** Three converging lenses of focal length 5.0 cm are arranged with a spacing of $2.0 \cdot 10^1$ cm between them, and are used to image an insect $2.0 \cdot 10^1$ cm away.

a) Where is the image?

b) Is it real or virtual?

c) Is it upright or inverted?

•**33.43** Two identical thin convex lenses, each of focal length f, are separated by a distance $d = 2.5f$. An object is placed in front of the first lens at a distance $d_{o,1} = 2f$.

a) Calculate the position of the final image of an object through the system of lenses.

b) Calculate the total transverse magnification of the system.

c) Draw the ray diagram for this system and show the final image.

d) Describe the final image (real or virtual, erect or inverted, larger or smaller) in relation to the initial object.

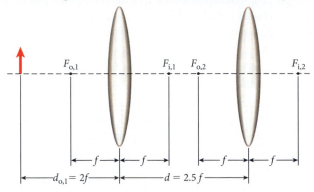

•**33.44** Two converging lenses with focal lengths 5 cm and 10 cm, respectively, are placed 30 cm apart. An object of height h = 5 cm is placed 10 cm to the left of the 5-cm lens. What will be the position and height of the final image produced by this lens system?

33.45 Two lenses are used to create an image for a source 10.0 cm tall that is located 30.0 cm to the left of the first lens, as shown in the figure below. Lens L_1 is a biconcave lens made of crown glass (index of refraction n = 1.55) and has a radius of curvature of 20.0 cm for both surface 1 and 2. Lens L_2 is 40.0 cm to the right of the first lens L_1. Lens L_2 is a converging lens with a focal length of 30.0 cm. At what distance relative to the object does the image get formed? Determine this position by sketching rays and calculating algebraically.

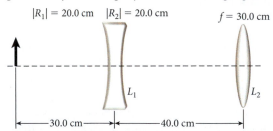

••**33.46** Sophisticated optical systems can be analyzed and designed with the aid of techniques from linear algebra. Light rays (or particle beams) are described at any point along the axis of the system by a two-component column vector containing y, the distance of the ray from the optic axis, and dy/dx, the slope of the ray. Components of the system are described by 2×2 matrices which incorporate their effects on the ray; combinations of components are described by products of these matrices.

a) A thin lens does not alter the position of a ray, but increases (diverging) or decreases (converging) its slope an amount proportional to the distance of the ray from the axis. Construct the matrix for a thin lens of focal length f.

b) A space between components does not alter the slope of a ray; the distance of the ray from the axis changes by the slope of the ray times the length of the space. Write the matrix for a space of length x.

c) Write the matrix for the two-lens "zoom lens" system described in Section 33.3.

Section 33.4

33.47 The typical length of a human eyeball is 2.50 cm.

a) What is the effective focal length of the two-lens system made from a normal person's cornea and lens when viewing objects far away?

b) What is the effective focal length for viewing objects at the near point?

33.48 Using your answers from the previous question, and given that the cornea in a typical human eye has a fixed focal length of 2.33 cm, what range of focal lengths does the lens in a typical eye have?

33.49 Jane has a near point of 125. cm and wishes to read from a computer screen 40. cm from her eye.

a) What is the object distance?

b) What is the image distance?

c) What is the focal length?

d) What is the power of the corrective lens needed?

e) Is the corrective lens diverging or converging?

33.50 Bill has a far point of 125 cm and wishes to see distant objects clearly.

a) What is the object distance?

b) What is the image distance?

c) What is the focal length?

d) What is the power of the corrective lens needed?

e) Is the corrective lens diverging or converging?

33.51 A person wearing bifocal glasses is reading a newspaper a distance of 25. cm. The lower part of the lens is converging for reading and has a focal length of 70. cm. The upper part of the lens is diverging for seeing at distances far away and has a focal length of 50. cm. What are the uncorrected near and far points for the person?

33.52 The radius of curvature for the outer part of the cornea is 8.0 mm, the inner portion is relatively flat. If the index of refraction of the cornea and the aqueous humor is 1.34:

a) Find the power of the cornea.

b) If the combination of the lens and the cornea has a power of 50. diopter, find the power of the lens (assume the two are touching).

•**33.53** As objects are moved closer to the human eye, the ciliary muscle causes the lens at the front of the eye to become more curved, thereby decreasing the focal length of the lens. The shortest focal length f_{min} of this lens is typically 2.3 cm. What is the closest one can bring an object to a normal

human eye (see left side of figure below) and still have the image of the object projected sharply on the retina, which is 2.5 cm behind the lens? Now consider a nearsighted human eye that has the same f_{min} but is stretched horizontally, with a retina that is 3.0 cm behind the lens (see right side of figure below). What is the closest one can bring an object to this nearsighted human eye and still have the image of the object projected sharply on the retina. Compare the maximum angular magnification produced by this nearsighted eye with that of the normal eye.

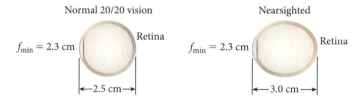

Normal 20/20 vision   Nearsighted

$f_{min} = 2.3$ cm   Retina $f_{min} = 2.3$ cm   Retina

←2.5 cm→   ←3.0 cm→

•**33.54** A person is nearsighted (myopic). The power of his eyeglass lenses is −5.75 diopters, and he wears the lenses 1.00 cm in front of his corneas. What is the prescribed power of his contact lenses?

Section 33.5

33.55 An amateur photographer attempts to build a custom zoom lens using a converging lens followed by a diverging lens. The two lenses are separated by a distance $x = 50.$ mm as presented in Figure 33.18 (reproduced below). If the focal length of the first lens is $2.0 \cdot 10^2$ mm and the focal length of the second lens is $-3.0 \cdot 10^2$ mm, what will the effective focal length of this compound lens be? What will f_{eff} be if the lens separation is changed to $1.0 \cdot 10^2$ mm?

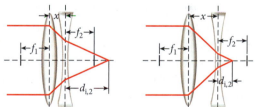

33.56 For a certain camera, the distance between the lens and the film is 10 cm. You observe that objects that are very far away appear properly focused. How far from the film would you have to move the lens in order to properly focus an object 100 cm away?

33.57 A camera has a lens with a focal length of 60. mm. Suppose you replace the normal lens with a zoom lens whose focal length can be varied from 35. mm to 250. mm and use the camera to photograph an object at infinity. Compared to a 60.-mm lens, what magnification of the image would be achieved using the 240.-mm focal length?

•**33.58** A camera lens usually consists of a combination of two or more lenses to produce a good-quality image. Suppose a camera lens has two lenses—a diverging lens of focal length 10 cm and a converging lens of focal length 5 cm. The two lenses are held 7 cm apart. A flower of length 10 cm, to be pictured, is held upright at a distance 50 cm in front of the diverging lens; the converging lens is placed behind the diverging lens. What are the location, orientation, size, and the magnification of the final image?

Section 33.6

33.59 A student finds a tube 20. cm long with a lens attached to one end. The attached lens has a focal length of 0.70 cm. The student wants to use the tube and lens to make a microscope with a magnification of $3.0 \cdot 10^2 \times$. What focal length lens should the student attach to the other end of the tube?

33.60 The objective lens in a laboratory microscope has a focal length of 3.00 cm and provides an overall magnification of $1.0 \cdot 10^2$. What is the focal length of the eyepiece if the distance between the two lenses is 30.0 cm?

33.61 You have found in the lab an old microscope, which has lost its eyepiece. It still has its objective lens, and markings indicate that its focal length is 7 mm. You can put in a new eyepiece, which goes in 20 cm from the objective. You need a magnification of about 200. Assume you want the comfortable viewing distance for the final image to be 25 cm. You find in a drawer eyepieces marked 2-, 4-, and 8-cm focal length. Which is your best choice?

33.62 A microscope has a 2.0-cm focal length eyepiece and a 0.80-cm objective lens. For a relaxed normal eye, calculate the position of the object if the distance between the lenses is 16.2 cm.

•33.63 Suppose you want to design a microscope so that for a fixed distance between the two lenses, the magnitude of the magnification will vary from 150 to 450 as you substitute eyepieces of various focal lengths.

a) If the longest focal length for an eyepiece is $6.0 \cdot 10^1$ mm, what will the shortest focal length be?

b) If you want the distance between the eyepiece and objective to be 35 cm, what should the focal length of the objective be?

Section 33.7

33.64 A refracting telescope has the objective lens of focal length 10.0 m. Assume it is used with an eyepiece of focal length 2.00 cm. What is the magnification of this telescope?

33.65 What is the magnification of a telescope with $f_o = 1.00 \cdot 10^2$ cm and $f_e = 5.00$ cm?

33.66 A simple telescope is formed using an eyepiece of focal length 25.0 mm and an objective of focal length 80.0 mm. Calculate the angle subtended by the image of the Moon when viewed through this telescope from the Earth.

33.67 Galileo discovered the moons of Jupiter in the fall of 1609. He used a telescope of his own design that had an objective lens with a focal length of $f_o = 40$ inches and an eyepiece lens with a focal length of $f_e = 2$ inches. Calculate the magnifying power of Galileo's telescope.

33.68 Two distant stars are separated by an angle of 35 arcseconds. If you have a refracting telescope whose objective lens focal length is 3.5 m, what focal length eyepiece do you need in order to observe the stars as though they were separated by 35 arcminutes?

•33.69 A 180× astronomical telescope is adjusted for a relaxed eye when the two lenses are 1.30 m apart. What is the focal length of each lens?

•33.70 Two refracting telescopes are used to look at craters on the Moon. The objective focal length of both telescopes is 95.0 cm and the eyepiece focal length of both telescopes is 3.80 cm. The telesopes are identical except for the diameter of the lenses. Telescope A has an objective diameter of 10.0 cm while the lenses of telescope B are scaled up by a factor of two, so that its objective diameter is 20.0 cm.

a) What are the angular magnifications of telescopes A and B?

b) Do the images produced by the telescopes have the same brightness? If not, which is brighter and by how much?

••33.71 Some reflecting telescope mirrors utilize a rotating tub of mercury to produce a large parabolic surface. If the tub is rotating on its axis with an angular frequency ω, show that the focal length of the resulting mirror is: $f = g/2\omega^2$.

Additional Problems

33.72 An object 4.0 cm high is projected onto a screen using a converging lens with a focal length of 35 cm. The image on the screen is 56 cm. Where are the lens and the object located relative to the screen?

33.73 A person with normal vision picks up a nearsighted friend's eyeglasses and attempts to focus on objects around her while wearing them. She is able to focus only on very distant objects while wearing the eyeglasses, and is completely unable to focus on objects that are near to her. Estimate the prescription of her friend's eyeglasses in diopters.

33.74 Suppose the near point of your eye is $2.0 \cdot 10^1$ cm and the far point is infinity. If you put on −0.20 diopter spectacles, what will be the range over which you will be able to see objects distinctly?

33.75 A fish is swimming in an aquarium at an apparent depth $d = 1.0 \cdot 10^1$ cm. At what depth should you grab in order to catch it?

33.76 A classmate claims that by using a 40-cm focal length mirror, he can project onto a screen a 10-cm tall bird located 100 m away. He claims that the image will be no less than 1 cm tall and inverted. Will he make good on his claim?

33.77 An object is 6.0 cm from a thin lens along the axis of the lens. If the lens has a focal length of 9.0 cm, determine the image distance.

33.78 A thin spherical lens is fabricated from glass so that it bulges outward in the middle on both sides. The glass lens has been ground so that the surfaces are part of a sphere with a radius of 25 cm on one side and a radius of $3.0 \cdot 10^1$ cm on the other. What is the power of this lens in diopters?

33.79 A diverging lens is fabricated from glass such that one surface of the lens is convex and the other surface is concave. The glass lens has been ground so that the convex surface is part of a sphere with a radius of 45 cm and the concave surface is part of a sphere with a radius of $2.0 \cdot 10^1$ cm. What is the power of this lens in diopters?

33.80 A person who is farsighted can see clearly an object that is at least 2.5 m away. To be able to read a book $2.0 \cdot 10^1$ cm away, what kind of corrective glasses should he purchase?

33.81 You are experimenting with a magnifying glass (consisting of a single converging lens) at a table. You discover that by holding the magnifying glass 92.0 mm above your desk, you can form a real image of a light that is directly overhead. If the distance between the light and the table is 2.35 m, what is the focal length of the lens?

33.82 A girl forgets to put on her glasses and finds that she needs to hold a book 15 cm from her eyes in order to clearly see the print.

a) If she were to hold the book 25 cm away, what type of corrective lens would she need to use to clearly see the print?

b) What is the focal length of the lens?

33.83 The focal length of the lens of a camera is 38.0 mm. How far must the lens be moved to change focus from a person 30.0 m away to one that is 5.00 m away?

33.84 A telescope is advertised as providing a magnification of magnitude 41 using an eyepiece of focal length $4.0 \cdot 10^1$ mm. What is the focal length of the objective?

•33.85 Determine the position and size of the final image formed by a system of elements consisting of an object 2.0 cm high located at $x = 0$ m, a converging lens with focal length $5.0 \cdot 10^1$ cm located at $x = 3.0 \cdot 10^1$ cm and a plane mirror located at $x = 7.0 \cdot 10^1$ cm.

33.86 The distance from the lens (actually a combination of the cornea and the crystalline lens) to the retina at the back of the eye is 2.0 cm. If light is to focus on the retina,

a) what is the focal length of the lens when viewing a distant object?

b) what is the focal length of the lens when viewing an object 25 cm away from the front of the eye?

33.87 You visit your eye doctor and discover that you require lenses having a diopter value of –8.4. Are you nearsighted or farsighted? With uncorrected vision, how far away from your eyes must you hold a book to read clearly?

33.88 Jack has a near point of 32 cm and uses a magnifier of 25 diopter.

a) What is the magnification if the final image is at infinity?

b) What is the magnification if the final image is at the near point?

33.89 Where is the image and what is the magnification if you hold a 2-inch-diameter clear glass marble a foot in front of you, and admire your image?

•33.90 A diverging lens with $f = -30.0$ cm is placed 15.0 cm behind a converging lens with $f = 20.0$ cm. Where will an object at infinity in front of the converging lens be focused?

•33.91 An instructor wants to use a lens to project a real image of a light bulb onto a screen 1.71 m from the bulb. In order to get the image to be twice as large as the bulb, what focal length lens will be needed?

•33.92 An old refracting telescope placed in a museum as an exhibit is 55 cm long and has magnification of 45. What are the focal lengths of its objective and eyepiece lenses?

•33.93 A converging lens of focal length $f = 50$ cm is placed 175 cm to the left of a metallic sphere of radius $R = 1.00 \cdot 10^2$ cm. An object of height $h = 2.0 \cdot 10^1$ cm is placed $3.0 \cdot 10^1$ cm to the left of the lens. What is the height of the image formed by the metallic sphere?

•33.94 When performing optical spectroscopy (for example, photoluminescence or Raman spectroscopy), a laser beam is focused on the sample to be investigated by means of a lens having a focal distance f. Assume that the laser beam exits a pupil D_o in diameter that is located at a distance d_o from the focusing lens. For the case when the image of the exit pupil forms on the sample, calculate

a) at what distance d_i from the lens is the sample located and

b) the diameter D_i of the laser spot (image of the exit pupil) on the sample.

c) What are the numerical results for: $f = 10.0$ cm, $D_o = 2.00$ mm, $d_o = 1.50$ m?

•33.95 For a person whose near point is 115 cm, so that he can read a computer monitor at 55 cm, what power of reading glasses should his optician prescribe, keeping the lens–eye distance of 2.0 cm for his spectacles?

•33.96 A telescope has been properly focused on the Sun. You want to observe the Sun visually, but to protect your sight you don't want to look through the eyepiece; rather, you want to project an image of the Sun on a screen 1.5 m behind (the original position of) the eyepiece, and observe that. If the focal length of the eyepiece is 8.0 cm, how must you move the eyepiece?

34

Wave Optics

FIGURE 34.1 Soap bubbles showing colors from interference phenomena.

WHAT WE WILL LEARN

- The wave nature of light leads to phenomena that cannot be explained using geometric optics.

- Superposed light waves that are in phase interfere constructively; superposed light waves that are 180° out of phase interfere destructively.

- Superposed light that has traveled different distances can interfere constructively or destructively, depending on the path length difference.

- Light waves spread out after passing through a narrow slit or after encountering an obstacle. This spreading out is called diffraction.

- Light waves that have the same phase and frequency are called coherent light waves.

- Coherent light incident on one narrow slit produces a diffraction pattern; coherent light incident on two narrow slits produces an interference pattern.

- Interference can occur in light that is partially reflected from each of the two surfaces (front and back) of a thin optical film.

- An interferometer is a device designed to measure lengths or changes in length using interference of light.

- Diffraction can limit the ability of a telescope or camera to resolve distant objects.

- A diffraction grating consists of many narrow slits or rulings that can be used to produce an intensity pattern that consists of narrow bright fringes separated by wide dark areas for a single-wavelength light source.

- X-ray diffraction can be used to study the atomic structure of materials.

Soap bubbles consist of clear water with a bit of dish soap mixed in. Why, then, do we see the colors of the rainbow when we look at soap bubbles in the air (Figure 34.1)? It turns out that the optics of soap bubbles are not the same as the reflections and refractions that give rise to rainbows (Chapter 32). Instead, the colors of bubbles appear due to interference of light of various wavelengths—a wave effect, not explained by geometric optics.

We have studied image formation by assuming that light travels in straight rays, without regard to the nature of light: for example whether light consists of particles or waves. In this chapter we look at optical effects best explained by the wave nature of light—a study sometimes called *physical optics,* as distinct from geometric optics. You may want to review material on waves in Chapter 15, especially the concept of interference of waves. We will also study a wave property called diffraction, which we did not discuss in Chapter 15 but which becomes important when dealing with very small wavelengths such as those of light. The two properties, interference and diffraction, explain not only the colors of soap bubbles, but also such issues as how well we can see distant objects as separate and not blurred together.

The material in this chapter does not completely answer the questions of what is light and how it behaves. We will see in following chapters that wondering about the nature of light led to some of the greatest developments in physics in the 20th century, ideas that are often grouped together by the name *modern physics.* The ideas in this chapter will be pivotal in our examination of the modern understanding of light, matter, time, and space.

34.1 Light Waves

We learned in Chapter 31 that light is an electromagnetic wave. However, normally we do not think of light as a wave, because its wavelength is so short that we usually do not notice its wave behavior. Thus, in Chapters 32 and 33, we discussed light as rays, a description that is appropriate for all physical situations where we can neglect the wavelength of light in comparison to all other physical dimensions. These rays travel in straight lines except when they are reflected off a mirror or refracted at the boundary between two optical media.

Now we want to address physical situations where we cannot approximate the wavelength of light as negligibly small any more. In this chapter, we discuss the wave nature of light, and we will apply to light many of the wave concepts (superposition, coherence, interference, boundary reflection, and others) we developed in Chapter 15. This will sometimes lead to rather surprising results. But all of these results can be verified experimentally.

FIGURE 34.2 Huygens construction for a plane wave traveling vertically upward.

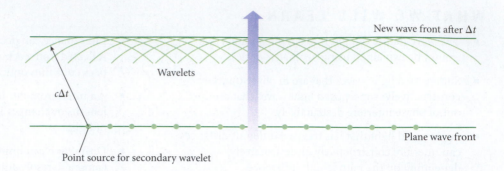

New wave front after Δt

Wavelets

$c\Delta t$

Plane wave front

Point source for secondary wavelet

One way to reconcile the wave nature of light with the geometric optical properties of light is to use **Huygens's Principle,** developed by Dutch physicist Christiaan Huygens (1629–1695). Huygens proposed a wave theory of light in 1678, long before Maxwell developed his theories of light. Huygens's Principle states that every point on a propagating wave front serves as a source of spherical secondary wavelets. Later, the envelope of these secondary wavelets becomes a new wave front. If the original wave has frequency f and speed v, the secondary wavelets have the same f and v. Diagrams of phenomena based on Huygens's Principle are called Huygens constructions.

Figure 34.2 shows a Huygens construction for a plane wave. We start with a plane wave traveling at the speed of light, c. Assume point sources of spherical wavelets along the wave front, as shown. These wavelets also travel at c, so at a time Δt the wavelets have traveled a distance of $c\Delta t$. Assuming many point sources along the wave front, Figure 34.2 shows that the envelope of these wavelets forms a new wave front parallel to the original wave front. Thus, the wave continues to travel in a straight line with the original frequency and speed.

In Chapter 32 the index of refraction n in a medium was defined as the ratio of the speed of light in vacuum, c, divided by the speed of light in that medium, v,

$$n = \frac{c}{v}. \tag{34.1}$$

Using this definition, Snell's Law can be expressed as

$$n_1 \sin\theta_1 = n_2 \sin\theta_2. \tag{34.2}$$

Chapter 32 showed that this law correctly describes refraction phenomena at the boundaries between different media, but the derivation of Snell's Law was deferred to the present chapter. With Huygens's Principle in hand, we can now perform this derivation.

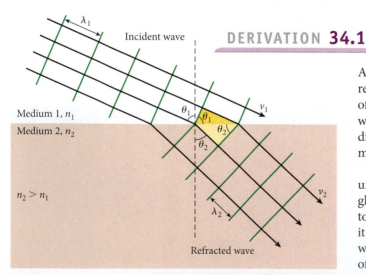

λ_1

Incident wave

Medium 1, n_1

Medium 2, n_2

θ_1 θ_1 v_1

θ_2

θ_2

$n_2 > n_1$

v_2

λ_2

Refracted wave

FIGURE 34.3 A Huygens construction of a traveling wave in an optical medium incident on the boundary with a second optical medium. The green lines represent wave fronts, while the black arrows denote rays.

DERIVATION **34.1** Snell's Law

A Huygens construction can be used to derive Snell's Law for refraction between two optical media with different indices of refraction. Assume a wave with wave fronts separated by a wavelength λ_1 traveling with speed v_1 in an optically clear medium incident on the boundary with a second optically clear medium (Figure 34.3).

The angle of the incident wave front (green line in medium 1) with respect to the boundary is θ_1, which is also the angle the light ray (black arrow) makes with respect to a normal to the boundary. When the wave enters the second medium, it travels with speed v_2. According to Huygens's Principle, the wave fronts are the result of wavelet propagation at the speed of the original wave, so the separation of the wave fronts in the second medium can be written in terms of the wavelength in the second medium λ_2. Thus, the time interval between wave fronts for the first medium is λ_1/v_1 and the time interval for the second medium is λ_2 / v_2. The essential point is that this time interval is the same for waves on either side of the boundary. (If these intervals were not equal, fronts would be appearing or disappearing mysteriously!) So:

$$\frac{\lambda_1}{v_1} = \frac{\lambda_2}{v_2} \Leftrightarrow \frac{\lambda_1}{\lambda_2} = \frac{v_1}{v_2}.$$

Thus, the wavelengths of the light in the two media are proportional to the speed of the light in those media.

We can get a relation between the angle of the incident wave fronts θ_1 with the boundary and the angle of the transmitted wave fronts θ_2 with the boundary by analyzing the yellow shaded region of Figure 34.3, shown in more detail in Figure 34.4.

From Figure 34.4, and using trigonometry, it follows that

$$\sin\theta_1 = \frac{\lambda_1}{x} \text{ and } \sin\theta_2 = \frac{\lambda_2}{x}.$$

Dividing these two equations by each other gives

$$\frac{\sin\theta_1}{\sin\theta_2} = \frac{\lambda_1}{\lambda_2} = \frac{v_1}{v_2}.$$

Given the definition of the index of refraction n of an optical medium is $n = c/v$, where c is the speed of light in a vacuum and v is the speed of light in the optical medium,

$$\frac{\sin\theta_1}{\sin\theta_2} = \frac{v_1}{v_2} = \frac{c/n_1}{c/n_2} = \frac{n_2}{n_1}$$

or

$$n_1 \sin\theta_1 = n_2 \sin\theta_2,$$

which is Snell's Law.

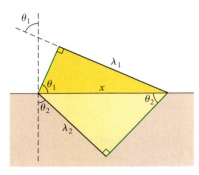

FIGURE 34.4 Expanded section of Figure 34.3, showing the incident wave front and direction as well as the transmitted wave front and direction.

We have seen that the wavelength of light changes when going from vacuum to an optical medium with index of refraction greater than one. Taking the result $\lambda_1/v_1 = \lambda_2/v_2$ from Derivation 34.1 with medium 1 being a vacuum while medium 2 has an index of refraction n, we can write

$$\lambda_n = \lambda\frac{v}{c} = \frac{\lambda}{n}.$$

Thus, the wavelength of light is shorter in a medium with index of refraction greater than one than it is in a vacuum. The frequency f of this light can be calculated from $v = \lambda f$. The frequency f_n of light traveling in the medium is then given by

$$f_n = \frac{v}{\lambda_n} = \frac{c/n}{\lambda/n} = \frac{c}{\lambda} = f. \tag{34.3}$$

Thus, the frequency of light traveling in an optical medium with $n > 1$ is the same as the frequency of that light traveling in a vacuum. (Equation 34.3 is equivalent to the statement made earlier that the time interval between wave fronts for the first medium equals the time interval between wave fronts for the second medium.)

34.2 In-Class Exercise

A light ray with wavelength $\lambda = 560.0$ nm enters a block of clear plastic from air at an incident angle of $\theta_i = 36.1°$ with respect to the normal. The angle of refraction is $\theta_r = 21.7°$. What is the speed of the light ray inside the plastic?

a) $1.16 \cdot 10^8$ m/s   c) $1.67 \cdot 10^8$ m/s   e) $3.00 \cdot 10^8$ m/s

b) $1.31 \cdot 10^8$ m/s   d) $1.88 \cdot 10^8$ m/s

34.1 In-Class Exercise

Blocks of two different transparent materials are sitting in air and have identical light rays of single wavelength incident on them at the same angle. Examining the figure, what can you say about the speed of light in these two blocks?

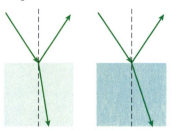

a) The speed of light is the same in both blocks.

b) The speed of light is greater in the block on the left.

c) The speed of light is greater in the block on the right.

d) The information on which speed of light is greater cannot be determined from the information given.

Definition

Coherent light is light made up of waves with the same wavelength that are in phase with each other. A great source of coherent light is a laser. Light from light bulbs or sunlight, on the other hand, is **incoherent,** meaning the waves may have different wavelengths and phase relationships with each other.

34.2   Interference

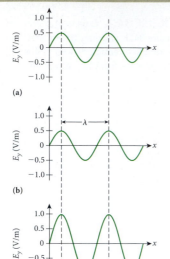

FIGURE 34.5 (a,b) Light waves in phase with the same amplitude and wavelength λ; (c) in-phase superposition of the two light waves demonstrates constructive interference, producing a wave with twice the amplitude.

Sunlight is composed of light containing a broad range of frequencies and corresponding wavelengths. We often see different colors separated out of sunlight after it refracts and reflects in raindrops, forming a rainbow. We also sometimes see various colors from sunlight due to constructive and destructive **interference** phenomena in thin transparent layers of materials, such as soap bubbles or thin films of oil floating on water. In contrast to rainbows, this thin-film effect is due to interference.

The geometric optics of Chapters 32 and 33 cannot explain interference. Interference phenomena can be understood only by taking into account the wave nature of light. This section considers light waves that have the same wavelength in vacuum. Interference takes place when such light waves are superposed. If the light waves are in phase, they interfere constructively, as illustrated in Figure 34.5.

In Figure 34.5a and Figure 34.5b, the electric field component in the y-direction is plotted for two electromagnetic waves traveling in the x-direction. The two waves are in phase. Saying that the two waves are in phase is the same as saying that the phase difference between the two waves is zero. A phase difference of 2π radians (360°), corresponding to starting with two waves in phase and then displacing one of the waves by one wavelength, will also produce two waves that are in phase. If each light wave is traveling from its own point of origin, a phase difference will occur where they meet, related to the path difference between the two waves, even though they start in phase. The criterion for **constructive interference** is characterized by a path difference Δx given by

$$\Delta x = m\lambda \quad (m = 0, \pm 1, \pm 2, \ldots). \tag{34.4}$$

Figure 34.5c shows the two waves constructively interfering. The amplitudes of the two waves add, giving a wave with the same frequency but twice the amplitude of the two original waves.

If the two light waves are out of phase by π radians (180°) (Figure 34.6a and Figure 34.6b), the amplitudes of the waves will sum to zero everywhere when they meet. This is the condition for **destructive interference** (Figure 34.6c).

Figure 34.6a and Figure 34.6b show that this situation is equivalent to starting with two waves in phase and then displacing one of the waves by half of one wavelength (λ/2). Again, if we think of the two light waves as being emitted from different sources, the phase difference can be related to the path difference. Destructive interference takes place if the path difference is a half wavelength plus an integer times the wavelength:

$$\Delta x = \left(m + \tfrac{1}{2}\right)\lambda \quad (m = 0, \pm 1, \pm 2, \ldots). \tag{34.5}$$

The following sections will discuss interference phenomena caused by light waves with the same wavelength that are initially in phase but travel different distances or travel with different speeds (in different media) to reach the same point. At this point, the light waves are superposed and can interfere. Whether the interference is constructive or destructive for coherent light depends only on the path length difference, which is a recurring theme that will dominate all of the discussions that follow.

The different interference phenomena caused by various path differences of coherent light are summarized in Figure 34.7. Most of this chapter will explore the different conditions that lead to the interference patterns shown on the screens in the three parts of this figure. Section 34.3 will explore pure double-slit interference (Figure 34.7a), which assumes that two very narrow slits of width comparable to the wavelength of the light are separated a distance

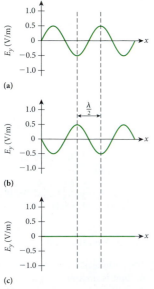

FIGURE 34.6 (a,b) Light waves out of phase by λ/2 with the same amplitude and wavelength; (c) superposition of the two light waves demonstrates destructive interference.

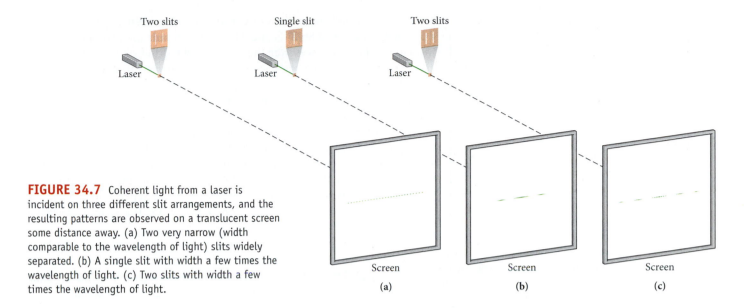

FIGURE 34.7 Coherent light from a laser is incident on three different slit arrangements, and the resulting patterns are observed on a translucent screen some distance away. (a) Two very narrow (width comparable to the wavelength of light) slits widely separated. (b) A single slit with width a few times the wavelength of light. (c) Two slits with width a few times the wavelength of light.

that is large compared to their individual widths. Section 34.7 will explore the case of coherent light hitting a single slit with width a few times the wavelength of the coherent light hitting it (Figure 34.7b). Section 34.9 combines the effects of both cases to discuss double-slit interference with diffraction (Figure 34.7c). Detailed discussion of these cases will clarify the idea that path length differences between coherent light waves are responsible for all of them.

34.3 Double-Slit Interference

Our first example of the interference of light is **Young's double-slit experiment,** named for English physicist Thomas Young (1773–1829), who carried out this investigation in 1801. Assume monochromatic coherent light such as that from a laser incident on a pair of slits of width comparable or even smaller than the wavelength of light, as shown in Figure 34.7a. (Strictly speaking, it is not necessary to use monochromatic coherent light from a laser to carry out this experiment, but doing so simplifies the following discussion and explanation of double-slit interference.) If separate sources of light are used for illuminating the two slits, then random and uncontrollable phase differences in the light from the two sources mean that these two light sources are incoherent.

For each slit we use a Huygens construction, assuming that all the light passing through the slit is due to wavelets emitted from a single point at the center of the slit (Figure 34.8a). In this figure, spherical wavelets are emitted from this point. Assume that the slit is much narrower than the wavelength of light (the wavelength of the green laser light is approximately 532 nm) so that the source of the wavelets can be represented with one point. Note that the slit opening is barely visible to the naked eye in this case. Even the two very narrow lines representing the magnified slits in the drawing of Figure 34.7a are drawn much too wide in comparison with the limiting case of extremely narrow slits that we are studying here.

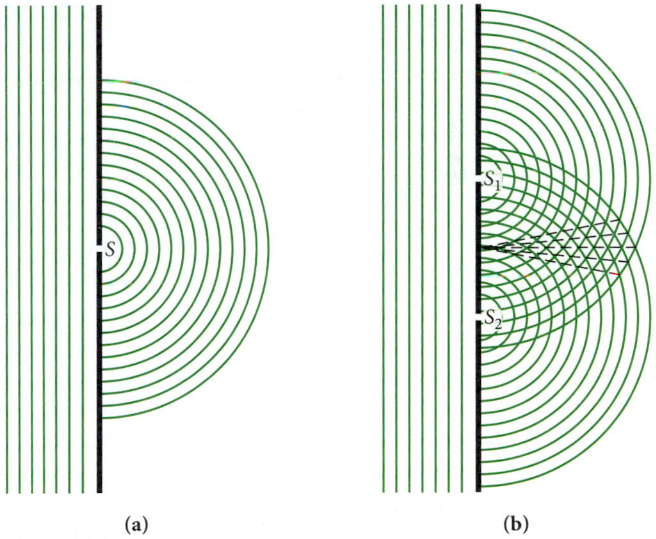

FIGURE 34.8 Huygens construction for a coherent light wave incident from the left on (a) a single slit, and (b) two slits, S_1 and S_2. The dashed lines represent lines of constructive interference.

Figure 34.8b shows two slits like the one in Figure 34.8a. A distance d separates the slits. Again, monochromatic coherent light is incident from the left and a source of spherical wavelets is at the center of each slit. The dashed lines represent lines along which constructive interference occurs. Placing a screen to the right of the slits will produce an alternating pattern of bright lines and dark lines, corresponding to constructive and destructive interference between the light waves emitted from the two slits.

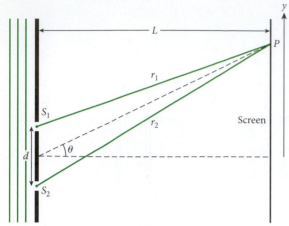

FIGURE 34.9 Expanded view of coherent light incident on two slits. The green lines to the right of the slits represent the distance light must travel from S_1 and S_2 to a point P on the screen.

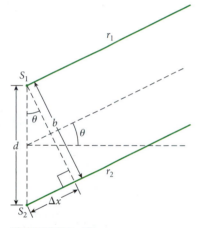

FIGURE 34.10 Expanded view of two slits, where the screen is placed far enough away that the green lines S_1P and S_2P are parallel.

To quantify these lines of constructive interference, we expand and simplify Figure 34.8b in Figure 34.9. In this figure, the two lines r_1 and r_2 represent the distances from the centers of slit S_1 and slit S_2, respectively, to a point P on a screen placed a distance L away from the slits. A line drawn from a point midway between the two slits to point P on the screen makes an angle θ with respect to a line drawn from the slits perpendicular to the screen. The point P on the screen is a distance y above this centerline.

To further quantify the two-slit geometry, we expand and simplify Figure 34.9 in Figure 34.10. In this figure, assume that the screen has been placed a large distance L away from the slits, such that the lines S_1P and S_2P are essentially parallel to each other and to the line drawn from the center of the two slits to point P. Draw a line from S_1 perpendicular to S_1P and S_2P, making a triangle with sides d, b, and Δx. The quantity Δx is the path length difference $r_2 - r_1$.

This path length difference $\Delta x = r_2 - r_1$ produces a phase difference for light originating from the two slits and illuminating the screen at point P. The path length difference can be expressed in terms of the distance between the slits d and the angle θ at which the light is observed,

$$\sin\theta = \frac{\Delta x}{d}$$

or

$$\Delta x = d\sin\theta.$$

For constructive interference, this path length difference must be an integer multiple of the wavelength of the incident light,

$$\Delta x = d\sin\theta = m\lambda \quad (m = 0, \pm 1, \pm 2, \ldots) \quad \begin{pmatrix} \text{bright fringe,} \\ \text{constructive interference} \end{pmatrix}. \quad (34.6)$$

A bright fringe on the screen signals constructive interference.

For destructive interference, the path length difference must be an integer plus one-half times the wavelength,

$$\Delta x = d\sin\theta = \left(m + \frac{1}{2}\right)\lambda \quad (m = 0, \pm 1, \pm 2, \ldots) \quad \begin{pmatrix} \text{dark fringe,} \\ \text{destructive interference} \end{pmatrix}. \quad (34.7)$$

A dark fringe on the screen signals destructive interference.

Note that for constructive interference and $m = 0$, we obtain $\theta = 0$, which means that $\Delta x = 0$ and there is a bright fringe at zero degrees. This bright fringe is called the *central maximum*. The integer m is called the **order** of the fringe. The order has a different meaning for bright fringes and for dark fringes. For example, using equation 34.6 with $m = 1$ would give the angle of the first-order bright fringe, $m = 2$ would give the second-order fringe, etc. Using equation 34.7 with $m = 0$ would give the angle of the first-order dark fringe, $m = 1$ would give the second-order fringe, etc. For both bright and dark fringes, the first-order fringe is the one closest to the central maximum.

If the screen is placed a sufficiently large distance from the slits, the angle θ is small and can be approximated as $\sin\theta \approx \tan\theta = y/L$ (see Figure 34.9). Thus, equation 34.6 can be expressed as

$$d\sin\theta = d\frac{y}{L} = m\lambda \quad (m = 0, \pm 1, \pm 2, \ldots)$$

or

$$y = \frac{m\lambda L}{d} \quad (m = 0, \pm 1, \pm 2, \ldots), \quad (34.8)$$

which gives the distances along the screen of the bright fringes from the central maximum.

Similarly, the distances along the screen of the dark fringes from the central maximum can be expressed as

$$y = \frac{\left(m + \frac{1}{2}\right)\lambda L}{d} \quad (m = 0, \pm 1, \pm 2, \ldots). \quad (34.9)$$

The positions of the centers of the bright and dark fringes are described by equations 34.8 and 34.9. But the intensity of the light at any point on the screen can also be calculated. Begin by assuming that the light emitted at each slit is in phase. The electric field of the light waves can be described by $\vec{E}_m = E_{max} \sin \omega t$, where E_{max} is the amplitude of the wave and ω is the angular frequency. When the light waves arrive at the screen from the two slits, they have traveled different distances, so can have different phases. The electric field of the light arriving at a given point on the screen from S_1 can be expressed as

$$\vec{E}_{m1} = E_{max} \sin(\omega t)$$

and the electric field of the light arriving at the same point from S_2 can be expressed as

$$\vec{E}_{m2} = E_{max} \sin(\omega t + \phi),$$

where ϕ is the phase constant of $\vec{E}_{m2}$ with respect to $\vec{E}_{m1}$. Figure 34.11a shows the two phasors $\vec{E}_{m1}$ and $\vec{E}_{m2}$.

The sum of the two phasors $\vec{E}_{m1}$ and $\vec{E}_{m2}$ is shown in Figure 34.11b, which also shows that the magnitude E of the sum of the two phasors is

$$E = 2E_{max} \cos(\phi / 2).$$

Chapter 31 showed that the intensity of an electromagnetic wave is proportional to the square of the amplitude of the electric field; so

$$\frac{I}{I_{max}} = \frac{E^2}{E_{max}^2}.$$

Using the result for the electric field just obtained, this gives

$$I = 4I_{max} \cos^2(\phi/2)$$

for the intensity of the total wave at point P as a function of the phase difference between the two light waves.

Now the phase difference must be related to the path length difference. Figure 34.10 shows that the path length difference Δx causes a phase shift given by

$$\phi = \frac{\Delta x}{\lambda}(2\pi),$$

because when $\Delta x = \lambda$, the phase shift $\phi = 2\pi$. Noting that $\Delta x = d \sin \theta$, the phase constant can be expressed as

$$\phi = \frac{2\pi d}{\lambda} \sin \theta.$$

Thus, we can write an equation for the intensity of the light produced by the interference from two slits as

$$I = 4I_{max} \cos^2\left(\frac{\pi d}{\lambda} \sin \theta\right).$$

Finally, we can obtain an expression for the intensity pattern on a screen resulting from coherent light incident on two narrow widely separated slits using the approximation discussed before that is valid when the screen is sufficiently far from the slits and θ is small that $\sin \theta \approx \tan \theta = y/L$:

$$I = 4I_{max} \cos^2\left(\frac{\pi d y}{\lambda L}\right). \qquad (34.10)$$

For example, if the screen is $L = 2.0$ m away from the slits, the slits are separated by $d = 1.0 \cdot 10^{-5}$ m, and the wavelength of the incident light is $\lambda = 550$ nm, we get the intensity pattern shown in Figure 34.12. In this figure, the intensity varies from $4I_{max}$

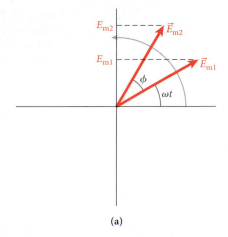

(a)

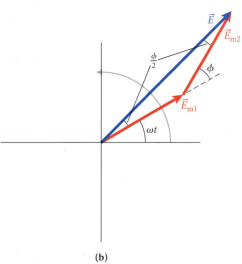

(b)

FIGURE 34.11 (a) Two phasors $\vec{E}_{m1}$ and $\vec{E}_{m2}$ separated by a phase ϕ. (b) Sum of the two phasors $\vec{E}_{m1}$ and $\vec{E}_{m2}$.

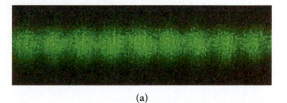

(a)

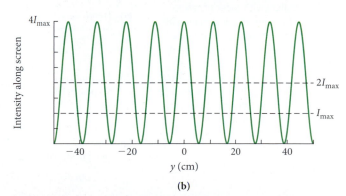

(b)

FIGURE 34.12 Intensity pattern for two-slit interference using light with wavelength 550 nm incident on two narrow slits separated by 10^{-5} m at a distance of 2 m from the screen. (a) Photograph of light intensity on the screen. (b) Calculation of the light intensity on the screen.

to zero. Covering one slit produces an intensity of I_{max} at all values of y. Illuminating both slits with light that has random phases produces an intensity of $2I_{max}$ at all values of y. Only when both slits are illuminated with coherent light do we observe the oscillatory pattern in y that is characteristic of two-slit interference.

34.3 In-Class Exercise

A pair of slits is separated by a distance $d = 1.40$ mm and is illuminated with light of wavelength $\lambda = 460.0$ nm. What is the separation of adjacent interference maxima on a screen a distance $L = 2.90$ m away?

a) 0.00332 mm   c) 0.953 mm   e) 3.23 mm

b) 0.556 mm   d) 1.45 mm

34.4 Thin-Film Interference and Newton's Rings

Another way of producing interference phenomena is by using light that is partially reflected from the front and back layers of thin films. A **thin film** is an optically clear material with thickness on the order of a few wavelengths of light. Examples of thin films include the walls of soap bubbles and thin layers of oil floating on water. When the light reflected off the front surface interferes with the light reflected off the back surface of the thin film, we see the color corresponding to the particular wavelength of light that is interfering constructively.

Consider light traveling in an optical medium with an index of refraction n_1 that strikes a second optical medium with index of refraction n_2, as illustrated in Figure 34.13.

One possibility is that the light can be transmitted through the boundary, as illustrated in Figure 34.13a and Figure 34.13b. In these cases, the phase of the light does not change independent of whether $n_1 < n_2$ or $n_1 > n_2$. A second process that can occur is that the light is reflected. In this case, the phase of the light can change, depending on the index of refraction of the two optical media. If $n_1 < n_2$, the phase of the reflected wave is changed by 180° (corresponding to half a wavelength), as shown in Figure 34.13c. If $n_2 > n_1$, then no phase change occurs, as depicted in Figure 34.13d. (The reflected waves have been displaced vertically for clarity's sake!)

The reason for this phase change upon reflection follows from the theory of electromagnetic waves, and the derivation is beyond the scope of this book. However, in Chapter 15, Section 15.4, we already studied the reflection of waves on a string from a boundary, and this mechanical analog is sufficient to understand the basic reason for a 180° phase change in one case and none in the other. Figure 34.14 reproduces the two figures showing a rigid (left) and a flexible (right) connection of the rope to the wall. The reflection of light in the optically less dense (lower value of n) boundary to the optically denser (higher value of n) medium corresponds to the situation on the left, and the reverse case to the situation on the right.

FIGURE 34.13 Light traveling in an optical medium with index of refraction n_1 is incident on a boundary with a second optical medium with index of refraction n_2. (a) Light is transmitted with no phase change for $n_1 < n_2$. (b) Light is transmitted with no phase change for $n_1 > n_2$. (c) Light is reflected with a 180° phase change for $n_1 < n_2$. (d) Light is reflected with no phase change for $n_1 > n_2$.

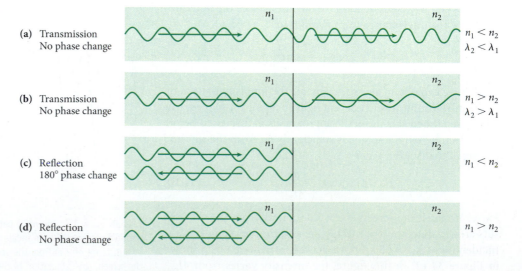

(a) Transmission No phase change — $n_1 < n_2$, $\lambda_2 < \lambda_1$

(b) Transmission No phase change — $n_1 > n_2$, $\lambda_2 > \lambda_1$

(c) Reflection 180° phase change — $n_1 < n_2$

(d) Reflection No phase change — $n_1 > n_2$

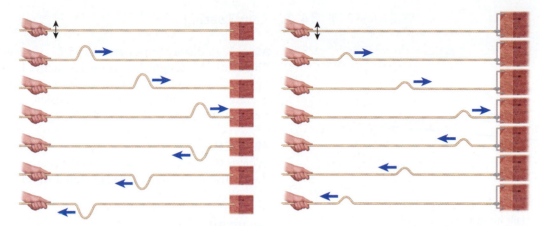

FIGURE 34.14 Sketch of the time sequence (top to bottom) of the reflection of a wave pulse on a string at a wall. Left: Rigid connection to a wall results in a 180° phase change in the reflected wave; right: A movable connection to the wall results in no phase change.

Let's begin our analysis of thin films by studying a thin film of thickness t with index of refraction n and with air on both sides of the film (Figure 34.15). Assume that monochromatic light is incident perpendicular to the surface of the film. An angle of incidence is shown for the light waves in the figure for clarity. When the light wave reaches the boundary between air and the film, part of the wave is reflected and part of the wave is transmitted. The reflected wave undergoes a phase shift of half a wavelength when it is reflected because $n_{\text{air}} < n$. The light that is transmitted has no phase shift and continues to the back surface of the film. At the back surface, again part of the wave is transmitted and part of the wave is reflected. The transmitted light passes through the film completely. (We do not show the transmitted wave in Figure 34.15, as this wave does not interest us for the present analysis.) The reflected light has no phase shift because $n > n_{\text{air}}$ and travels back to the front surface of the film. At the front surface, some of the light reflected from the back surface is transmitted and some is reflected. We don't need to consider the reflected light. The transmitted light has no phase shift and emerges from the film and interferes with the light that was reflected when the light first struck the film. The transmitted and then reflected light has traveled a longer distance than the originally reflected light and has a phase shift determined by the path length difference. This difference is twice the thickness t of the film.

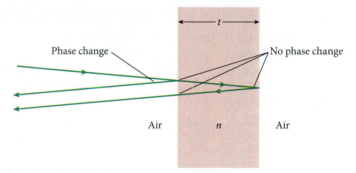

FIGURE 34.15 Light waves in air incident on a thin film with index of refraction n and thickness t.

The fact that the originally reflected light has undergone a phase shift of half a wavelength, and the transmitted and then reflected light has not, means that the criterion for *constructive* interference is given by $\Delta x = (m + \frac{1}{2})\lambda = 2t$ $(m = 0,1,2,...)$. The wavelength λ refers to the wavelength of the light traveling in the thin film, which has index of refraction n. The wavelength of the light traveling in air is related to the wavelength of the light traveling in the film by $\lambda = \lambda_{\text{air}}/n$. We can then write

$$\left(m + \frac{1}{2}\right)\frac{\lambda_{\text{air}}}{n} = 2t \quad (m = 0, \pm 1, \pm 2, ...) \quad \left(\begin{array}{l}\text{thin film,} \\ \text{constructive interference}\end{array}\right). \quad (34.11)$$

The minimum thickness t_{min} that will produce constructive interference corresponds to $m = 0$,

$$t_{\text{min}} = \frac{\lambda_{\text{air}}}{4n}. \quad (34.12)$$

An oil slick or a soap bubble has varying thicknesses and thus affects different wavelengths differently, often creating a rainbow effect.

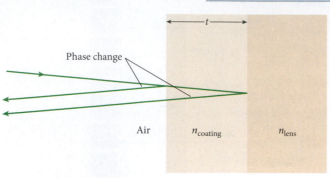

FIGURE 34.16 Light incident on a coated lens.

EXAMPLE 34.1 / Lens Coating

Many high-quality lenses are coated to prevent reflections. This coating is designed to set up destructive interference for light reflected from the surface of the lens. Assume that the coating is magnesium fluoride, which has $n_{coating} = 1.38$, and that the lens is glass with $n_{lens} = 1.51$.

PROBLEM

What is the minimum thickness of the coating that produces destructive interference for light with a wavelength in air of 550 nm?

SOLUTION

Assume that the light is incident perpendicularly (or at least close to perpendicularly) on the surface of the coated lens (Figure 34.16). Light reflected at the surface of the coating undergoes a phase change of half a wavelength, because $n_{air} < n_{coating}$. The light transmitted through the coating has no phase change. Light reflected at the boundary between the coating and the lens also undergoes a phase change of half a wavelength, because $n_{coating} < n_{lens}$. This reflected light travels back through the coating and exits with no other phase change. Thus, both the light reflected from the coating and the light reflected from the lens have undergone a phase change of half a wavelength. Therefore, the criterion for destructive interference is

$$\left(m+\frac{1}{2}\right)\frac{\lambda_{air}}{n_{coating}} = 2t \quad (m=0,\pm1,\pm2,...).$$

The minimum thickness for the coating to provide destructive interference corresponds to $m = 0$,

$$t_{min} = \frac{\lambda_{air}}{4n_{coating}} = \frac{550\cdot10^{-9}\text{ m}}{4(1.38)} = 9.96\cdot10^{-8}\text{ m} = 99.6\text{ nm}.$$

Lens coatings that fulfill this destructive interference condition are referred to as "quarter-lambda" coatings. Because they cause destructive interference for reflected light, they prevent light from reflecting off the coating and thus allow more light to enter the lens than would be the case without the coating. Expensive camera lenses have quarter-lambda coatings for wavelengths in the middle of the optical wavelength band, around 500 nm. But remember that a quarter-lambda coating can work only if the index of refraction of the coating is less than that of the lens.

34.4 In-Class Exercise

If you were to use the same coating material of the same thickness as in Example 34.1, but on a different lens with an index of refraction less than that of the coating, this would result in

a) reduced reflection and thus *more* light entering the lens than in the case without coating.

b) enhanced reflection and thus *less* light entering the lens than in the case without coating.

c) no change in the amount of light entering the lens relative to the case without coating.

d) no light at all (or close to none) entering the lens.

34.5 In-Class Exercise

A thin soap film with index of refraction $n = 1.35$ hanging in air reflects dominantly red light with $\lambda = 682$ nm. What is the minimum thickness of the film?

a) 89.5 nm   d) 302 nm

b) 126 nm   e) 432 nm

c) 198 nm

FIGURE 34.17 Newton's rings produced with white light.

A phenomenon similar to thin-film interference is **Newton's rings.** This effect is an interference pattern caused by the reflection of light between two glass surfaces, a spherical surface and an adjacent flat surface, as shown in Figure 34.17.

Newton's rings are caused by interference between light reflected from the bottom of the spherically curved glass and light reflected from the top of the flat plate of glass, as shown in Figure 34.18. (Note that the curvature of the spherical surface is exaggerated in this diagram. In real situations R would be much larger than x.)

Newton's rings are concentric circles, alternately dark and bright. The dark circles correspond to destructive interference, and the bright circles correspond to constructive interference. The photo shown in Figure 34.17 was taken in white light to show different colored rings. For the calculation of Newton's rings, assume a monochromatic, coherent light source with wavelength λ in air. The distance between the curved surface and the flat surface, t, is a function of the horizontal distance from where the spherical surface touches the flat surface, as shown in Figure 34.18. The Pythagorean Theorem applied to the yellow triangle in Figure 34.18 gives the relationship

$$t = R - \sqrt{R^2 - x^2} = R - R\sqrt{1 - \left(\frac{x}{R}\right)^2},$$

where R is the radius of curvature of the spherical glass surface and x is the horizontal distance from the point where the curved glass touches the flat glass to the point where the light enters the glass.

In Figure 34.18, the curvature of the spherical surface has been exaggerated for clarity. The actual glass surface used to produce Newton's rings has a large radius of curvature compared with the distance from the touching point, so that $x/R \ll 1$. Thus we can approximate

$$\sqrt{1 - \left(\frac{x}{R}\right)^2} \approx 1 - \frac{1}{2}\left(\frac{x}{R}\right)^2, \quad \text{for } \frac{x}{R} \ll 1.$$

Now we can write an approximate expression for the distance between the surfaces

$$t \approx R - R\left(1 - \frac{1}{2}\left(\frac{x}{R}\right)^2\right) = \frac{1}{2}\frac{x^2}{R}.$$

The path difference between the light reflecting off the bottom of the curved surface and the top of the flat surface is $2t$, which can be written

$$2t = 2\left(\frac{1}{2}\frac{x^2}{R}\right) = \frac{x^2}{R}.$$

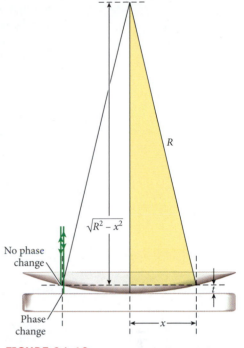

FIGURE 34.18 The geometry of Newton's rings. The bottom surface of the top glass plate is a spherical surface while the bottom plate is planar.

The light reflecting off the bottom of the curved surface has no phase change because the index of refraction of glass is higher than the index of refraction of air. The light reflecting off the top of the flat glass plate suffers a phase change of half a wavelength because the index of refraction of glass is larger than the index of refraction of air. Thus, the criterion for constructive interference is

$$\left(m + \frac{1}{2}\right)\lambda = 2t \quad (m = 0,1,2,...).$$

Combining this constructive interference criterion with the result for the thickness, and defining x_m as the radius of the mth bright circle, gives

$$\left(m + \frac{1}{2}\right)\lambda = \frac{x_m^2}{R} \quad (m = 0,1,2,...),$$

which can be solved for the radius of the bright circles

$$x_m = \sqrt{R\left(m + \frac{1}{2}\right)\lambda} \quad (m = 0,1,2,...) \quad \left(\frac{x_m}{R} \ll 1\right). \tag{34.13}$$

34.5 Interferometer

An **interferometer** is a device designed to measure lengths or small changes in length by using interference of light. An interferometer can measure changes in lengths to an accuracy of a fraction of the wavelength of the light it is using. The interferometer works by using interference fringes. The interferometer described here is similar to, but simpler than, the one constructed by Albert Michelson at the Case Institute in Cleveland, Ohio, in 1887.

A photograph and drawing of a commercial interferometer used in physics labs is shown in Figure 34.19. A more detailed drawing of this same interferometer is shown in Figure 34.20. This particular interferometer consists of a laser light source that emits coherent light with wavelength $\lambda = 632.8$ nm. The light passes through a defocusing lens to spread out the normally very narrowly focused laser beam. The light then passes through a half-silvered mirror m_1. Part of the light is reflected toward the adjustable mirror m_3 and part of the light is transmitted to the movable mirror m_2. The distance between m_1 and m_2 is x_2, and the distance between m_1 and m_3 is x_3. The transmitted light is totally reflected from m_2 back toward m_1. The reflected light is also totally reflected from m_3 back toward m_1. Part of the light reflected from m_2 is then reflected by m_1 toward the viewing screen; the rest of the light is transmitted and not considered. Part of the light from m_3 is transmitted through m_1; the rest is reflected and not considered. The actual alignment of the light is such that the two beams that strike the viewing screen are collinear; the separation shown in Figure 34.20 is simply for clarity.

The light from mirrors m_2 and m_3 that strikes the viewing screen interfere, based on their path length difference. Both paths undergo two reflections, each resulting in a phase change of half a wavelength, so the condition for constructive interference is

$$\Delta x = m\lambda \quad (m = 0, \pm 1, \pm 2, \ldots).$$

The two different paths have a path length difference of

$$\Delta x = 2x_2 - 2x_3 = 2(x_2 - x_3).$$

The viewing screen displays concentric circles or linear fringes corresponding to constructive and destructive interference, depending on the type of defocusing lens used and the tilt of the mirrors. If the movable mirror m_2 is moved a distance of $\lambda/2$, the fringes shift by one fringe. Thus, this type of interferometer can be used to measure changes in distance on the order of a fraction of a wavelength of light, depending on how well the shift of the interference fringes can be measured.

Another type of measurement can be made with this interferometer by placing a material with index of refraction n and thickness t in the path of the light traveling to the movable mirror m_2, as depicted in Figure 34.20. The path length difference in terms of the number of wavelengths will change because the wavelength of light in the material λ_n is different from λ. The wavelength in this material is related to the wavelength of light in air by

$$\lambda_n = \frac{\lambda}{n}.$$

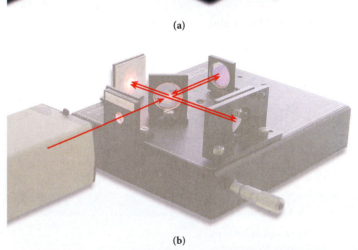

(a)

(b)

FIGURE 34.19 A Michelson interferometer used in an introductory physics laboratory. (a) Photograph; (b) schematic drawing of the light paths. The laser light is split by the central half-silvered mirror. One part of the light continues in the same direction, is reflected, returns, and is reflected by the half-silvered mirror to the screen. A second part of the light is reflected by the half-silvered mirror, and is then reflected back through the half-silvered mirror to the screen.

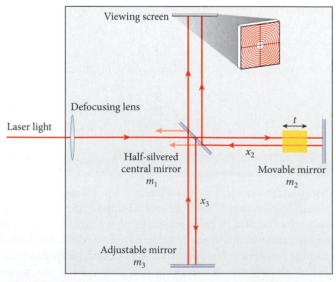

FIGURE 34.20 Top view of a Michelson-type interferometer, showing the detailed light path.

Thus the number of wavelengths in the material is

$$N_{\text{material}} = \frac{2t}{\lambda_n} = \frac{2tn}{\lambda}.$$

The number of wavelength that would have been there if the light traveled through only air is

$$N_{\text{air}} = \frac{2t}{\lambda}.$$

Thus the difference in the number of wavelengths is

$$N_{\text{material}} - N_{\text{air}} = \frac{2tn}{\lambda} - \frac{2t}{\lambda} = \frac{2t}{\lambda}(n-1). \qquad (34.14)$$

When the material is placed between m_1 and m_2, an observer will see a shift of one fringe for every wavelength shift in the path length difference. Thus, we can substitute the number of fringes shifted for $N_{\text{material}} - N_{\text{air}}$ in equation 34.14 and, given the index of refraction, we can obtain the thickness of the material. Alternatively, we can insert material with a well-known thickness and determine the index of refraction.

34.6 Diffraction

Any wave passing through an opening experiences diffraction. **Diffraction** means that the wave spreads out on the other side of the opening rather than the opening casting a sharp shadow. Diffraction is most noticeable when the opening is about the same size as the wavelength of the wave. With water surface waves, this is easily visualized, and the same effect applies to light waves. If light passes through a narrow slit, it produces a characteristic pattern of light and dark areas called a diffraction pattern. Light passing a sharp edge also exhibits a diffraction pattern.

Huygens's Principle describes this spreading out, and a Huygens construction can be used to quantify the diffraction phenomenon. For example, Figure 34.21 shows coherent light incident on an opening, which has dimensions comparable to the wavelength of the light. Rather than casting a sharp shadow, light spreads out on the other side of the opening. We can describe this spreading out by using a Huygens construction and assuming that spherical wavelets are emitted at several points inside the opening. The resulting light waves on the right side of the opening undergo interference and produce a characteristic diffraction pattern.

Light waves can also go around the edges of barriers, as shown in Figure 34.22. In this case, the light far from the edge of the barrier continues to travel like the light waves shown in Figure 34.2. The light near the edge of the barrier seems to bend around the barrier and is described by the sources of wavelets near the edge.

Figure 34.23 shows a picture of an experiment, where light from a green laser is blocked by the vertical edge of a razor blade, and the resulting diffraction pattern is viewed on a distant screen.

34.2 Self-Test Opportunity

The wavelength of a monochromatic light source is measured using a Michelson interferometer. When the movable mirror is moved a distance $d = 0.250$ mm, $N = 1200$ fringes move by the screen. What is the wavelength of the light?

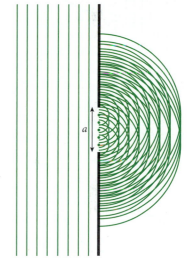

FIGURE 34.21 Coherent light incident on an opening with a width comparable to the wavelength of the light. The dots represent sources of spherical wavelets in a Huygens construction.

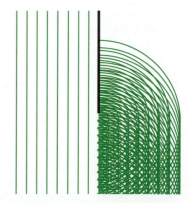

FIGURE 34.22 Coherent light incident on a barrier. The light diffracts around the barrier.

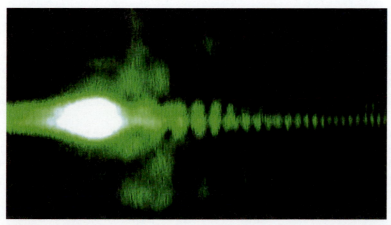

FIGURE 34.23 Light from a green laser incident on the vertical edge of a razor blade as viewed on a distant screen.

The bright spot is the central maximum and the less bright lines are higher-order diffraction maxima.

Diffraction phenomena cannot be described by geometric optics. As we will see, diffraction effects rather than geometric effects often limit the resolution of an optical instrument. The following sections will present a quantitative examination of single-slit diffraction, diffraction by a circular opening, double-slit diffraction, and diffraction by a grating. In all cases similar relationships for the location of the diffraction maxima and minima result, which are modified by the individual geometries.

34.7   Single-Slit Diffraction

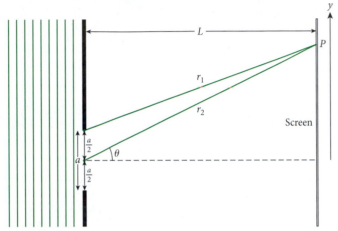

FIGURE 34.24 Geometry for determining the location of the first dark fringe from a single slit, using two rays from the slit.

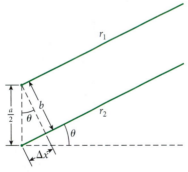

FIGURE 34.25 Expanded version of the geometry for determining the location of the first dark fringe from diffraction from a single slit.

Consider the diffraction of light through a single slit of width a, which is comparable to the wavelength of light passing through the slit, as illustrated in Figure 34.7b. The calculation is aided by a Huygens construction, as shown in Figure 34.21.

Assume that the light passing through the single slit is described by spherical wavelets emitted from a distribution of points located in the slit. The light emitted from these points will superpose and interfere, based on the path length for each wavelet at each position. At a distant screen, we observe an intensity pattern characteristic of diffraction, consisting of bright and dark fringes. For the case of two-slit interference, we were able to work out the equations for the bright fringes based on constructive interference. For diffraction, we will restrict ourselves to analyze the dark fringes of destructive interference.

To study the interference, we redraw and simplify Figure 34.21 as shown in Figure 34.24. Assume that coherent light with wavelength λ is incident on a slit with width a, producing an interference pattern on a screen a distance L away. We employ a simple method of analyzing pairs of light waves emitted from points in the slit. Start with light emitted from the top edge of the slit and from the center of the slit, as shown in Figure 34.24. To analyze the path difference, we show an expanded version of Figure 34.24 in Figure 34.25.

For this quantitative analysis, assume that the screen distance L is very large compared to the size of the slit opening a. This way the two green lines representing the paths of lengths r_1 and r_2 in Figure 34.25 are parallel and both make an angle θ with the central axis. Therefore, the path length difference for these two rays is given by

$$\sin\theta = \frac{\Delta x}{a/2} \Leftrightarrow \Delta x = \frac{1}{2}a\sin\theta.$$

The criterion for the first dark fringe is

$$\Delta x = \frac{a\sin\theta}{2} = \frac{\lambda}{2} \Rightarrow a\sin\theta = \lambda.$$

Although we chose one ray originating from the top edge of the slit and one from the middle of the slit to locate the first dark fringe, we could have used any two rays that originated $a/2$ apart inside the slit. That is, all wavelets can be paired up so that the separation within a pair is $a/2$. (Thus, we have taken into account the entire slit.) Each pair destructively interferes, so an overall dark fringe appears at the screen.

Now consider four rays instead of two (Figure 34.26). Here we choose a ray from the top edge of the slit and three more rays originating from points spaced $a/4$ apart. This drawing can be expanded to represent the case of the screen being far away, as shown in Figure 34.27. Clearly the path length difference between the pairs of rays labeled r_1 and r_2, r_2 and r_3, and r_3 and r_4 is given by

$$\sin\theta = \frac{\Delta x}{a/4} \Leftrightarrow \Delta x = \frac{1}{4}a\sin\theta.$$

The criterion for a dark fringe arising from these three pairs of rays is

$$\frac{a\sin\theta}{4}=\frac{\lambda}{2}\Rightarrow a\sin\theta=2\lambda.$$

All wavelets could be grouped into four rays similar to those shown in Figure 34.27. Each group destructively interferes, so an overall dark fringe is visible on the screen. Thus, the condition $a\sin\theta=2\lambda$ describes the second dark fringe.

At this point, we can easily generalize and take groups of six and eight rays and describe the third and fourth dark fringes, etc. The result is that the dark fringes from single-slit diffraction can be described by

$$a\sin\theta=m\lambda\quad(m=1,2,3,...).\qquad(34.15)$$

If the screen is placed a sufficiently large distance from the slits, the angle θ is small and can be approximated as $\sin\theta\approx\tan\theta=y/L$. Thus, the position of the dark fringes on the screen can be expressed as

$$\frac{ay}{L}=m\lambda\quad(m=1,2,3,...)$$

or

$$y=\frac{m\lambda L}{a}\quad(m=1,2,3,...)\begin{pmatrix}\text{positions of}\\\text{dark fringes}\end{pmatrix}.\qquad(34.16)$$

Detailed analysis of the spherical wavelets shows that the intensity I relative to I_{max} that we would get if there were no slit is

$$I=I_{max}\left(\frac{\sin\alpha}{\alpha}\right)^2,\qquad(34.17)$$

where

$$\alpha=\frac{\pi a}{\lambda}\sin\theta.\qquad(34.18)$$

Equation 34.17 shows that this expression for the intensity I is zero for $\sin\alpha=0$ (unless $\alpha=0$), which means $\alpha=m\pi$ for $m=1,2,3,...$. For the special case of $\alpha=0$ corresponding to $\theta=0$, we use that

$$\lim_{\alpha\to0}\left(\frac{\sin\alpha}{\alpha}\right)=1.$$

Therefore,

$$m\pi=\frac{\pi a}{\lambda}\sin\theta\quad(m=1,2,3,...)$$

or

$$a\sin\theta=m\lambda\quad(m=1,2,3,...),$$

which gives the same result for the diffraction minima as equation 34.15.

If the screen is placed a sufficiently large distance from the slits, the angle θ is small and can be approximated as $\sin\theta\approx\tan\theta=y/L$. Thus, equation 34.18 can be expressed as

$$\alpha=\frac{\pi ay}{\lambda L}.\qquad(34.19)$$

If the screen is $L=2.0$ m away from the slit, the slit has a width of $a=5.0\cdot10^{-6}$ m, and the wavelength of the incident light is $\lambda=550$ nm, we get the intensity pattern shown in Figure 34.28. Such an intensity distribution as shown in Figure 34.28 is called a Fraunhofer diffraction pattern.

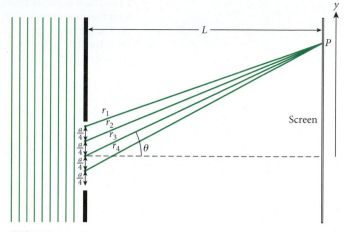

FIGURE 34.26 Geometry for determining the location of the second dark fringe from a single slit, using four rays from the slit.

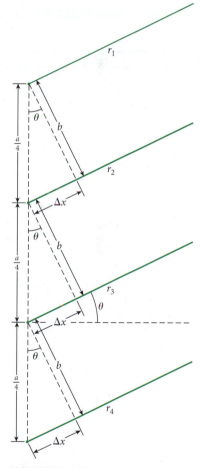

FIGURE 34.27 Expanded version of the geometry for determining the location of the second dark fringe from diffraction from a single slit.

FIGURE 34.28 Intensity pattern for diffraction through a single slit. (a) Photograph of light on the screen. (b) Calculation of the intensity along the screen.

(a)

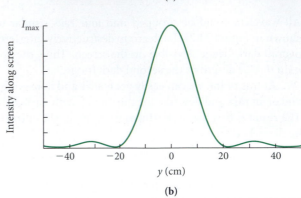

(b)

FIGURE 34.29 Single-slit diffraction pattern observed on a screen.

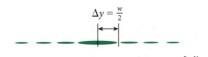

FIGURE 34.30 Single-slit diffraction pattern with the distance from the center of the central maximum to the position of the first diffraction minimum.

SOLVED PROBLEM 34.1 | **Width of Central Maximum**

The single-slit diffraction pattern shown in Figure 34.29 was produced with light of wavelength $\lambda = 510.0$ nm. The screen on which the pattern was projected was located a distance $L = 1.40$ m from the slit. The slit had a width of $a = 7.00$ mm.

PROBLEM

What is the width w of the central maximum? (The width is equal to the distance between the two first diffraction minima located on either side of the center.)

SOLUTION

THINK

The diffraction pattern produced by light incident on a single slit peaks at $y = 0$ and falls to a minimum on both sides of the peak. The first minimum corresponds to $m = 1$. Therefore, the width of the central maximum is equal to twice the y-coordinate of the first minimum.

SKETCH

Figure 34.30 shows a sketch of the single-slit diffraction pattern, marked with the distance from the center of the central maximum to the position of the first minimum.

RESEARCH

The locations of the dark fringes along a screen corresponding to diffraction minima when the screen is sufficiently far from the slit are given by equation 34.16

$$ y = \frac{m\lambda L}{a} \quad (m = 1, 2, 3, \ldots), $$

where λ is the wavelength of the light incident on the single slit, m is the order of the minimum, L is the distance from the slit to the screen, and a is the width of the slit. For single-slit diffraction, the position $y = 0$ corresponds to the position of the central maximum. The first diffraction minimum corresponds to $m = 1$. Therefore, the distance Δy from the central minimum to the position of the first minimum is given by

$$ \Delta y = \frac{\lambda L}{a}. $$

SIMPLIFY

The width of the central maximum w is then

$$ w = 2\Delta y = \frac{2\lambda L}{a}. $$

CALCULATE

Putting in the numerical values gives us

$$w = \frac{2\left(510.0 \cdot 10^{-9} \text{ m}\right)\left(1.40 \text{ m}\right)}{\left(7.00 \cdot 10^{-3} \text{ m}\right)} = 0.000204 \text{ m}.$$

ROUND

We report our result to three significant figures,

$$w = 0.000204 \text{ m} = 0.204 \text{ mm}.$$

DOUBLE-CHECK

The width of the central maximum projected on the screen is relatively small. A slit that is large compared with the wavelength of light shows little diffraction, while a slit that has a width comparable to or smaller than the wavelength of light produces a broad diffraction spectrum. The ratio of the wavelength of the light to the slit width is

$$\frac{\lambda}{a} = 510.0 \text{ nm} / 7.00 \cdot 10^{-3} \text{ m} = 7.29 \cdot 10^{-5}.$$

Thus, our answer showing a narrow central maximum appears reasonable.

34.8 Diffraction by a Circular Opening

So far, we have considered interference through two slits and diffraction through a single slit. Now we consider diffraction of light through a circular opening. Diffraction through a circular opening applies to observing objects with telescopes having circular mirrors/lenses or with cameras having a circular lens. The **resolution** of a telescope or camera (that is, its ability to distinguish two point objects as being separate) is limited by diffraction.

The first diffraction minimum from light with wavelength λ passing through a circular opening with diameter d is given by

$$\sin \theta = 1.22 \frac{\lambda}{d}, \tag{34.20}$$

where θ is the angle from the central axis through the opening to the first diffraction minimum. This result is similar to the result from a single slit except for the factor of 1.22. We will not derive this expression here. In Figure 34.31, we show a diffraction pattern for red laser light with wavelength $\lambda = 633$ nm passing through a circular opening with diameter 0.04 mm projected on a screen located 1.00 m from the opening. The diameter of the innermost dark circle is measured to be 3.9 cm. The first diffraction minimum is clearly visible in Figure 34.31.

Figure 34.32 depicts three different situations for the observation of two distant point objects using a lens. In Figure 34.32a, the angular separation is clearly large enough to resolve the objects. In Figure 34.32b, the angular separation is large enough to just barely resolve the objects, but not by much. In Figure 34.32c, the angular separation is too small to resolve the objects.

If a circular lens is used to observe two distant point objects, whose angular separation is small (such as two stars), diffraction limits the ability of the lens to distinguish these two objects. The criterion for being able to separate two point objects is based on the idea that if the central maximum of the first object is located at the first diffraction minimum of the second object, the objects are just resolved. This criterion, called **Rayleigh's Criterion,** is expressed as

$$\theta_R = \sin^{-1}\left(\frac{1.22\lambda}{d}\right), \tag{34.21}$$

where θ_R is the minimum resolvable angular separation in radians, λ is the wavelength of the light used to observe the objects, and d is the diameter of the lens or mirror.

FIGURE 34.31 The diffraction pattern for red laser light with $\lambda = 633$ nm passing through a circular opening with diameter 0.04 mm and projected on a distant screen.

FIGURE 34.32 Diffraction through a circular opening: representing the image you might see with a lens observing two distant point objects. The top row is a two-dimensional representation of the same result in a three-dimensional representation of the intensity pattern in the bottom row. (a) The angular separation is large enough to clearly resolve the two objects; (b) the angular separation is marginally large enough to resolve the two objects; (c) the angular separation of the two objects is too small to allow them to be resolved.

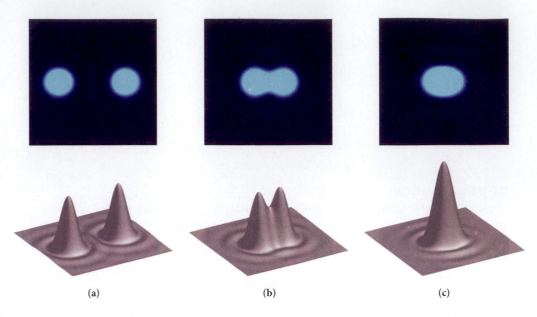

(a)   (b)   (c)

FIGURE 34.33 The main mirror of the Hubble Space Telescope has a diameter of 2.4 m. The hyperbolic shape of the mirror is accurate to 32 nm, which means that if the mirror were the size of Earth, any bumps in the glass would be 17 cm high or less.

EXAMPLE 34.2 | Rayleigh's Criterion for the Hubble Space Telescope

PROBLEM

The diameter of the main mirror in the Hubble Space Telescope (Figure 34.33) is 2.4 m. What is the minimum angular resolution of the Hubble Space Telescope for green light?

SOLUTION

Using Rayleigh's Criterion, with green light of wavelength $\lambda = 550$ nm, we get

$$\theta_R = 1.22 \frac{550 \cdot 10^{-9} \text{ m}}{2.4 \text{ m}} = 2.8 \cdot 10^{-7} \text{ rad} = 1.6 \cdot 10^{-5} \text{ degrees},$$

which corresponds to the angle subtended by a dime (diameter 17.9 mm) located 64 km away.

34.3 Self-Test Opportunity

Can individual watchtowers on the Great Wall of China be seen by astronauts on the Space Shuttle orbiting the Earth at an altitude of 190 km? Assume that the Great Wall is 10.0 m wide.

34.6 In-Class Exercise

You are driving in your car listening to music on the radio. Your car is equipped with AM radio ($f \approx 1$ MHz), FM radio ($f \approx 100$ MHz), and XM satellite radio ($f = 2.3$ GHz). You enter a tunnel with a circular opening of diameter 10 m. Which kind of radio signal will you be able to receive the longest as you continue to travel in the tunnel?

a) AM   b) FM   c) XM

34.9 Double-Slit Diffraction

Section 34.3 discussed the interference pattern produced by two slits. That analysis assumed that the slits themselves were very narrow compared with the wavelength of light, $a \ll \lambda$. For these narrow slits, the diffraction maximum is very wide, with peaks in the intensity that have the same value at all angles (see Figure 34.12).

However, with double slits for which the condition $a \ll \lambda$ is not met, as illustrated in Figure 34.7c, not all the interference fringes have the same intensity. With diffraction effects, the intensity of the interference pattern from double slits can be shown to be given by

$$I = I_{\max} \cos^2 \beta \left(\frac{\sin \alpha}{\alpha} \right)^2, \tag{34.22}$$

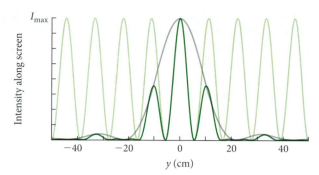

FIGURE 34.34 Intensity pattern from a double slit. The dark green line is the observed intensity pattern. The faint grey line is the intensity for a single slit with the same width as the two slits. The faint green line is the intensity distribution for two very narrow slits separated by the same distance as the double slits that produced the dark green line.

where $\alpha = \pi a \sin\theta/\lambda$, $\beta = \pi d \sin\theta/\lambda$, and d is the distance between the slits. If the screen is placed a sufficiently large distance from the slits, the angle θ is small and can be approximated as $\sin\theta \approx \tan\theta = y/L$. Then α and β can be approximated as $\alpha = \pi a y/\lambda L$ and $\beta = \pi d y/\lambda L$.

If the screen is $L = 2.0$ m away from the slits, each slit has a width of $a = 5.0 \cdot 10^{-6}$ m, the slits are $d = 1.0 \cdot 10^{-5}$ m apart, and the wavelength of the incident light is $\lambda = 550$ nm, we get the intensity pattern shown by the dark green line in Figure 34.34.

Figure 34.34 shows that the positions of the maxima on the screen are not changed from those for the double slit with very narrow slits. However, the maximum intensity is modulated by the diffraction intensity distribution, shown in faint grey. The diffraction pattern in Figure 34.34 forms an envelope for the interference intensity distribution. If one of the two slits were covered, only the diffraction pattern would be seen.

Figure 34.35 shows a photograph of an interference/diffraction pattern from a double slit projected on a screen that is 2 m away using light of wavelength $\lambda = 532$ nm from a green laser. The slits have a width of $a = 4.52 \cdot 10^{-5}$ m and are $d = 3.00 \cdot 10^{-4}$ m apart. The central maxima, which consist of all the two-slit maxima located within the single-slit central maximum envelope, are intentionally overexposed in the photograph to allow the secondary maxima to be observed. Figure 34.35 also shows the predicted intensity pattern, effectively overexposed by choosing the maximum value of the plotted intensity to be 38% of the maximum predicted intensity, which allows the secondary maxima to be visible. The first two single-slit diffraction minima on each side of the central maxima are marked with dashed lines.

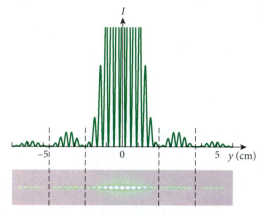

FIGURE 34.35 Photograph of the intensity pattern produced by a double slit illuminated by green light from a laser. The calculated intensity for the central lines extends above the plotted scale in the vertical direction to allow the lower intensity lines to be visible. The predicted intensity pattern is shown for $a = 0.0452$ mm, $d = 0.300$ mm, and $\lambda = 532$ nm. The dashed lines mark the diffraction minima.

34.10 Gratings

We have discussed diffraction and interference for a single slit and for two slits. How do diffraction and interference apply to a system of many slits? Putting many slits together forms a device called a **diffraction grating**. A diffraction grating has a large number of slits, or rulings, placed very close together. A diffraction grating can also be constructed using an opaque material containing grooves rather than actual slits, which is called a reflection grating. A diffraction grating produces an intensity pattern consisting of narrow bright fringes separated by wide dark areas. This characteristic pattern results because having many slits means there can be destructive interference a small distance away from the maxima.

A portion of a diffraction grating is shown in Figure 34.36. This drawing shows coherent light with wavelength λ incident on a series of narrow slits, each separated by a distance d. A diffraction pattern is produced on a screen a long distance L away.

Figure 34.36 can be expanded, as we did for the single-slit and double-slit cases (again using the limit that $L \gg d$ so that all of the rays drawn are parallel), to enable us to analyze the path length difference for the light

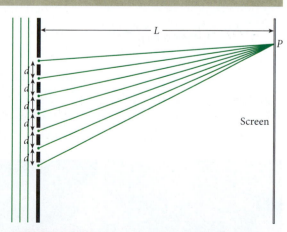

FIGURE 34.36 A portion of a diffraction grating.

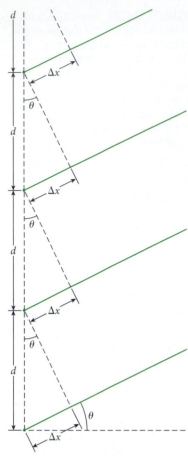

FIGURE 34.37 Expanded drawing of a diffraction grating, assuming that the screen is far away compared to the spacing between the slits of the grating.

$\#\text{lines/m} = 1/d$

$\#\text{lines/mm} = (1/d)10^{-3}$

$y = L \tan \theta$

$\Delta y = (L_{\text{grat-scrn}})(\tan \theta_1 - \tan \theta_2)$

from each of the slits to the screen (Figure 34.37). Assume that L is so large that the light rays are approximately parallel to one another.

The distance d is called the grating spacing. If the grating's width is W, the number N of slits or gratings is $N = W/d$. Diffraction gratings are often specified in terms of the number n_l of slits or rulings per unit length. We can obtain d from the specified n_l using $d = 1/n_l$.

Let's calculate the path length differences for the paths shown in Figure 34.37. Using an adjacent pair of rays, the path length difference is $\Delta x = d \sin \theta$. To produce bright lines, or constructive interference, this path length difference must be an integer multiple of the wavelength, so

$$d \sin \theta = m\lambda \quad (m = 0, 1, 2, \ldots). \tag{34.23}$$

The values of m correspond to different bright lines. For $m = 0$ we have the central maximum at $\theta = 0$. For $m = 1$ we have the first-order maximum. For $m = 2$ we have the second-order maximum, etc. Typically, diffraction gratings are designed to produce large angular separations between the maxima, so we do not make the small-angle approximation when discussing diffraction gratings.

Because diffraction gratings produce widely spaced narrow maxima, they can be used to determine the wavelength of monochromatic light by rearranging equation 34.23,

$$\lambda = \frac{d}{m} \sin \theta \quad (m = 0, 1, 2, \ldots). \tag{34.24}$$

Monochromatic light incident on a diffraction grating produces lines on a screen at widely separated angles. For example, when a focused laser beam strikes a diffraction grating, a set of widely spaced spots is created, as illustrated in Figure 34.38.

In addition, diffraction gratings can be used to separate out different wavelengths of light from a spectrum of wavelengths. For example, sunlight is dispersed into multiple sets of rainbow-like colors as a function of θ. If the light is composed of several discrete wavelengths, the light is separated into sets of lines corresponding to each of those wavelengths.

The quality of a diffraction grating can be quantified in terms of its dispersion. The **dispersion** describes the ability of a diffraction grating to spread apart the various wavelengths in a given order. Dispersion is defined by $D = \Delta\theta/\Delta\lambda$, where $\Delta\theta$ is the angular separation between two lines with wavelength difference $\Delta\lambda$. We can get an expression for the dispersion by differentiating equation 34.24 with respect to λ:

$$\frac{d\theta}{d\lambda} = \frac{d}{d\lambda}\left(\sin^{-1}\left(\frac{m\lambda}{d} \right) \right) = \frac{1}{\sqrt{1 - \left(\frac{m\lambda}{d} \right)^2}} \frac{m}{d} = \frac{m}{\sqrt{d^2 - (m\lambda)^2}}.$$

Because $m\lambda/d = \sin \theta$, we can express $d\theta/d\lambda$ as

$$\frac{d\theta}{d\lambda} = \frac{1}{\sqrt{1 - \sin^2 \theta}} \frac{m}{d} = \frac{m}{d \cos \theta},$$

where we have used the identity $\sin^2 \theta + \cos^2 \theta = 1$. Taking intervals of θ and λ that are not too large, we can thus write an expression for the dispersion of a diffraction grating as

$$D = \frac{\Delta\theta}{\Delta\lambda} = \frac{m}{d \cos \theta} \quad (m = 1, 2, 3, \ldots). \tag{34.25}$$

FIGURE 34.38 The diffraction pattern produced on a screen by a green laser light incident on a diffraction grating with line spacing $n_l = 787$ lines/cm. The spots corresponding to the central maximum, to $m = \pm 1$ and to $m = \pm 2$ are shown.

The dispersion of a diffraction grating increases as the distance d between rulings gets smaller, and as the order m gets higher. Note that the dispersion does not depend on the number of rulings N.

The **resolving power** R of a diffraction grating describes its ability to resolve closely spaced maxima. This ability depends on the width of each maximum. Consider a diffraction grating used to resolve two wavelengths λ_1 and λ_2, with $\lambda_{ave} = (\lambda_1 + \lambda_2)/2$ and $\Delta\lambda = |\lambda_2 - \lambda_1|$. Define the resolving power of the grating as

$$R = \frac{\lambda_{ave}}{\Delta\lambda_{min}}, \qquad (34.26)$$

where $\Delta\lambda_{min}$ is the minimum value of $\Delta\lambda$ such that the wavelengths are resolved.

In order to discuss the resolving power, we need an expression for the width of each maximum. The width of each maximum is determined by the position of the first minimum on one side of the central maximum. Define the angular half-width θ_{hw} of the maximum as the angle between the maximum and this first minimum (Figure 34.39).

Equation 34.25 tells us the angular spread for a given $\Delta\lambda$. The two wavelengths can barely be resolved if this spread $\Delta\theta = \theta_{hw}$. To determine the position of the first minimum, we do a single-slit diffraction analysis using the whole grating as the single slit (Figure 34.40). This approach is justified if the number of slits N is large. In any case, it provides a useful and meaningful approximation, as opposed to a precise mathematical calculation, which would tend to obscure the physics involved.

The angle of the first minimum for single-slit diffraction can be obtained from the condition $a\sin\theta = \lambda$, where Nd is substituted for the slit width a: $Nd\sin\theta_{hw} = \lambda$. Because θ_{hw} is small, we can write $\sin\theta_{hw} \approx \theta_{hw}$ or

$$\theta_{hw} = \frac{\lambda}{Nd}.$$

We can show (though we don't do so here) that the width of the maxima for other orders can be written as

$$\theta_{hw} = \frac{\lambda}{Nd\cos\theta},$$

where θ is the angle corresponding to the maximum intensity for that order. Substituting θ_{hw} for $\Delta\theta$ in equation 34.25,

$$\frac{\Delta\theta}{\Delta\lambda} = \frac{\lambda}{Nd\cos\theta\Delta\lambda} = \frac{m}{d\cos\theta}$$

or

$$R = \frac{\lambda}{\Delta\lambda} = Nm, \qquad (34.27)$$

where we have taken $\lambda \approx (\lambda + (\lambda + \Delta\lambda))/2$. Note that the resolving power of a diffraction grating depends only on the total number of rulings and the order.

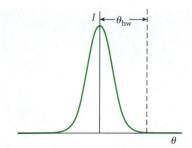

FIGURE 34.39 The angular half-width of the central maximum for a diffraction grating.

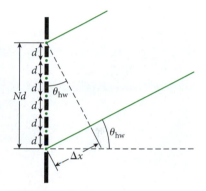

FIGURE 34.40 Calculation of the half-width of the central maximum for a diffraction grating, treating the entire grating as a single slit.

EXAMPLE 34.3 | CD or DVD as Diffraction Grating

Diffraction gratings can take the form of a series of narrow slits that light passes through or a series of closely spaced grooves that reflect light. The resulting diffraction pattern is the same. We can therefore think of the spiral grooves of a CD or a DVD as a diffraction grating. In Figure 34.41, a DVD is shown reflecting various colors from sunlight.

If we shine a green laser pointer with wavelength 532 nm perpendicular to the surface of a CD, we observe a diffraction pattern in the form of bright spots on a screen placed perpendicular to the CD located a perpendicular distance $L = 1.6$ cm away from the point where the laser beam hits the CD (Figure 34.42 and Figure 34.43). The spacing between the grooves in the CD is $d = 1.60 \cdot 10^{-6}$ m $= 1.6$ μm.

Continued—

FIGURE 34.41 Different colors resulting from constructive interference from sunlight striking a blank DVD.

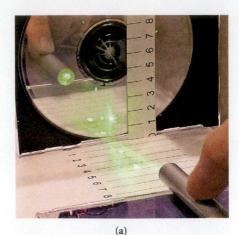

(a)

(b)

FIGURE 34.42 Using a CD as a diffraction grating. (a) A green laser pointer shining perpendicularly on the bottom surface of a CD. (b) Green lines illustrate the light from the laser pointer, and the order of each diffraction spot is labeled. The vertical and horizontal rulers have markings every centimeter.

PROBLEM

What is the horizontal distance from the surface of the CD edge to the spots observed on the screen?

SOLUTION

We start with our expression for the angle of constructive interference from a diffraction grating,

$$d \sin \phi_m = m\lambda \quad (m = 0,1,2,3,...),$$

where d is the distance between adjacent grooves, λ is the wavelength of the light, m is the order, and ϕ_m is the angle of the mth order maximum.

The wavelength that we need for this calculation is the wavelength of light in the polycarbonate plastic from which the CD is made, which has an index of refraction $n = 1.55$: $\lambda = \lambda_{air}/n$. Thus, the criteria for diffraction maxima is given by

$$d \sin \phi_m = \frac{m\lambda_{air}}{n} \Leftrightarrow n \sin \phi_m = \frac{m\lambda_{air}}{d}.$$

This light ray must now pass through the surface of the CD, where it is refracted. Applying Snell's Law at the surface and taking $n_{air} = 1$, we get

$$n \sin \phi_m = 1 \cdot \sin \theta_m = \sin \theta_m = \frac{m\lambda_{air}}{d}.$$

We can rewrite this equation as $d \sin \theta_m = m\lambda_{air}$, which is exactly what we would have gotten if we had treated the CD as a diffraction grating in air.

Starting with $m = 0$, we find that the central maximum produces a spot at $\theta_0 = 0°$, which means that the spot is produced in front of the laser pointer. You can see this diffraction maximum in the reflection of the laser pointer from the surface of the CD in Figure 34.42.

Moving to $m = 1$, we find the angle of the first-order diffraction maximum, θ_1, using

$$\theta_1 = \sin^{-1}\left(\frac{\lambda_{air}}{d}\right) = \sin^{-1}\left(\frac{532 \cdot 10^{-9} \text{ m}}{1.60 \cdot 10^{-6} \text{ m}}\right) = 19.4°.$$

Looking at Figure 34.43a, we can see that $\tan \theta_1 = L/y_1$, so we calculate the distance of the first-order spot along the screen as $y_1 = (1.6 \text{ cm})/(\tan 19.4°) = 4.54 \text{ cm}$. The photograph in Figure 34.42 shows that this calculation is in good agreement with what we observe for $m = 1$.

The angle of the second-order spot is given by

$$\theta_2 = \sin^{-1}\left(\frac{2\lambda_{air}}{d}\right) = \sin^{-1}\left(\frac{2 \cdot 532 \cdot 10^{-9} \text{ m}}{1.60 \cdot 10^{-6} \text{ m}}\right) = 41.7°.$$

FIGURE 34.43 Geometry of using a CD as a diffraction grating. (a) Side view of the geometry of the measurement shown in Figure 34.42. A horizontal white screen is used to locate the diffraction maxima. (b) Expanded drawing of the diffraction grating inside the polycarbonate structure of the CD, illustrating the diffraction and refraction that occur.

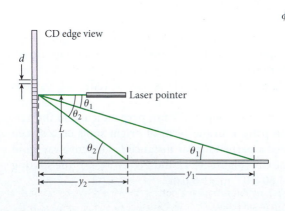

(a)

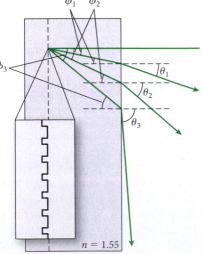

(b)

Thus, the position of the second-order spot is $y_2 = (1.6 \text{ cm})/(\tan 41.7°) = 1.80$ cm, which is also in good agreement with what we observe for $m = 2$ in Figure 34.42.

The angle of the third-order spot is

$$\theta_3 = \sin^{-1}\left(\frac{3\lambda_{\text{air}}}{d}\right) = \sin^{-1}\left(\frac{3 \cdot 532 \cdot 10^{-9} \text{ m}}{1.60 \cdot 10^{-6} \text{ m}}\right) = 85.9°.$$

The position of the third-order spot is $y_3 = (1.6 \text{ cm})/(\tan 85.9°) = 0.11$ cm. The third-order spot is not clearly visible in Figure 34.42 because the third-order maximum is dimmer than the first and second maxima and because the angle at which the spot must be observed is very close to 90°.

What about a fourth-order spot? For the angle of the fourth-order spot, we would have $\theta_4 = \sin^{-1}(4\lambda_{\text{air}}/d)$. However, for this order $4\lambda_{\text{air}}/d = 4(532 \text{ nm})/(1.6 \text{ μm}) = 1.33$, which cannot occur because $|\sin\theta| \leq 1$. Therefore, only three spots can appear on the screen, only two of which are easily visible.

If we carried out the same experiment using a common red laser with $\lambda_{\text{air}} = 633$ nm, we would get only two maxima, occurring at $\theta_1 = 23.3°$ and $\theta_2 = 52.3°$.

34.4 Self-Test Opportunity

What would happen if we repeated this experiment with a DVD and a green laser pointer? (The separation between tracks on a DVD is 740 nm compared with 1600 nm for a CD.)

Blu-Ray Discs

Earlier, in Example 9.2, we calculated the length of a CD track and also showed microscopic images of CD surfaces. Now we want to explore how Blu-ray discs and other optical discs store digital information, and how computers and consumer electronic devices read them. CDs, DVDs, and Blu-ray optical discs all operate on similar principles. Figure 34.44c and Figure 34.44d show a schematic cross section of a Blu-ray disc.

A Blu-ray disc, like a CD or DVD, stores digital information in terms of ones and zeros. These ones and zeros are encoded in the location of the edges of the high areas and low areas in the aluminum layer shown in Figure 34.44. The high and low areas rotate with the Blu-ray disc and pass over a blue solid-state laser that emits light with a wavelength of $\lambda = 405$ nm in air. A schematic drawing of the blue laser assembly is shown in Figure 34.45.

The Blu-ray player incorporates several concepts of wave optics presented in this chapter, including a diffraction grating and destructive interference, as well as polarization from Chapter 31. A solid-state laser produces light with wavelength $\lambda = 405$ nm in air. This wavelength is almost a factor of two shorter than the wavelength of the laser light used to read CDs. The shorter wavelength allows the tracks and pits to be smaller, which allows more data to be stored. This light is passed through a diffraction grating. The central maximum and the two first-order lines are shown in Figure 34.45. The higher-order lines are not used. The light in the central maximum is used to read the data from the disc and to maintain the focus of the beam.

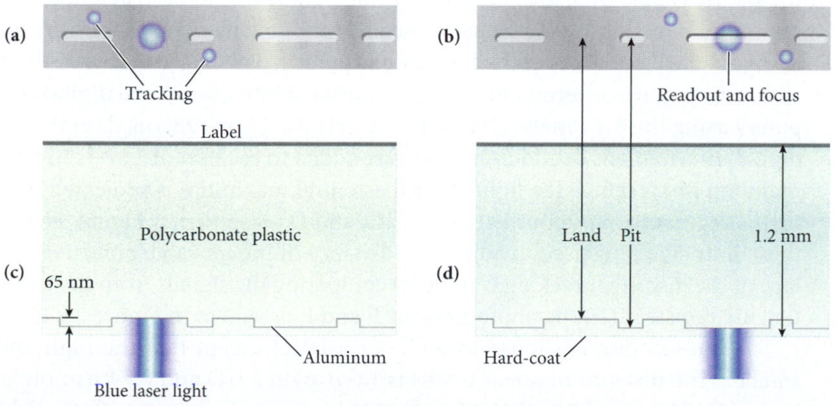

FIGURE 34.44 Cross section of a Blu-ray disc. (a) View from the bottom showing the laser data beam reflecting from only the aluminum land. (b) View from the bottom showing the laser data beam reflecting from both the aluminum land and pit simultaneously. (c) Edge view showing the laser data beam focused on the aluminum land. (d) Edge view showing the laser beam focused on the aluminum pit (and land).

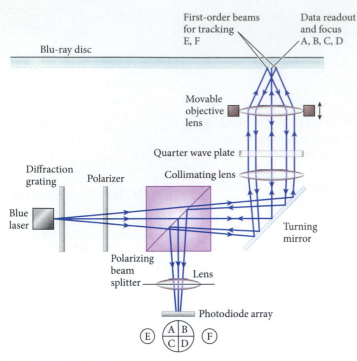

FIGURE 34.45 A schematic drawing of the blue laser assembly of a Blu-ray disc reader.

The light from the two first-order maxima is used for tracking the data on the disc.

After passing through the diffraction grating, the light passes through a polarizer and then proceeds to a polarizing beam splitter. The light is directed upward with a turning mirror. The light is then focused on the Blu-ray disc after passing through a quarter-wave plate that rotates the polarization. The effect of the polarizing elements is to separate the light reflected from the Blu-ray disc surface and direct it into the photodiode array while minimizing the direct light from the laser traveling into the photodiode array.

When the light from the central maximum is shining completely on the land, as illustrated in Figure 34.44a and Figure 34.44c, all the light is reflected into the photodiode. When the light from the laser is shining on a pit, as shown in Figure 34.44b,d, light is reflected from both the pit and the land. In this case, the light that reflects from the land area travels farther than the light reflected from the pit area. In both cases, the light undergoes a phase change when it is reflected. Thus, the analysis of destructive interference of a coating on a lens can be applied to this case. Denote the difference in height between the land and pit areas as t, so the path length difference for the light from the high and low areas is $2t$. The criterion for destructive interference is

$$\left(m+\frac{1}{2}\right)\frac{\lambda_{\text{air}}}{n_{\text{polycarbonate}}} = 2t \quad (m = 0, \pm 1, \pm 2, \ldots),$$

where $n_{\text{polycarbonate}}$ is the index of refraction of polycarbonate in the Blu-ray disc and λ_{air} is the wavelength of the light emitted by the laser. For $m = 0$,

$$t = \frac{\lambda_{\text{air}}}{4n_{\text{polycarbonate}}}.$$

For the Blu-ray laser, $\lambda_{\text{air}} = 405$ nm and the refractive index of the polycarbonate is 1.58; therefore, the thickness required for destructive interference is

$$t = \frac{\lambda_{\text{air}}}{4n_{\text{polycarbonate}}} = \frac{405 \text{ nm}}{4(1.58)} = 64.1 \text{ nm},$$

which is close to the difference of the high and low areas shown in Figure 34.44.

As the disc spins, the land and pit areas pass over the laser. When the land areas are over the laser, all of the light is reflected and the photodiode registers a given voltage. When the pit areas pass over the laser, part of the light is lost to destructive interference and the photodiode reads a lower voltage. Whenever the voltage changes—from high to low, or low to high—the electronics of the Blu-ray player records a one. Otherwise, the player records a zero. This pattern of zeros and ones get translated into the normal digital code used in computers using the 8-14 method, which converts the 14 bits encoded on the disc into 8 bits of digital information. In addition, 3 bits are added to each set of 14 bits to allow the reader to maintain its tracking. The light from the central maximum is projected onto a photodiode that is segmented into four parts, A, B, C, and D, as shown in Figure 34.45. The balance of these four signals is used to adjust the distance of the movable objective lens from the surface of the disc. The tracking is done by comparing the signals from the two first-order lines that are registered in the photodiode as E and F, as shown in Figure 34.45.

A Blu-ray disc is similar to a CD or a DVD except that the high and low areas are smaller. The distance between tracks is 1.6 μm on a CD and 0.74 μm on a DVD. In addition, a DVD uses a laser that emits light with a wavelength of 650 nm, while a CD uses a laser with a wavelength of 780 nm. A Blu-ray disc uses a track spacing of 0.30 μm and can hold 25 gigabytes of digital information. A CD can hold up to 700 megabytes of digital information, while a DVD can hold 4.7 gigabytes.

34.11 X-Ray Diffraction and Crystal Structure

Wilhelm Röntgen (1845–1923) discovered X-rays in 1895. His experiments suggested that X-rays were electromagnetic waves with a wavelength of about 10^{-10} m. At about the same time, the study of crystalline solids suggested that their atoms were arranged in a regular repeating pattern with a spacing of about 10^{-10} m between the atoms. Putting these two ideas together, Max von Laue (1879–1960) proposed in the early 1900s that a crystal could serve as a three-dimensional diffraction grating for X-rays. In 1912, von Laue, Walter Friederich (1883–1963), and Paul Knipping (1883–1935) did the first **X-ray diffraction** experiment, which showed diffraction of X-rays by a crystal. Soon afterward, Sir William Henry Bragg (1862–1942) and his son William Lawrence Bragg (1890–1971) derived Bragg's Law (given below) and carried out a series of experiments involving X-ray diffraction from crystals.

Let's assume we have a cubic crystal, with each atom in the lattice a distance a away from its neighboring atoms in all three directions (Figure 34.46). We can imagine various planes of atoms in this crystal. For example, the horizontal planes are composed of atoms spaced a distance a apart, with the planes themselves spaced a distance a from one another. If X-rays are incident on these planes, the rows of atoms in the crystalline lattice can act like a diffraction grating for the X-rays. The X-rays can be thought of as scattering from the atoms (Figure 34.47).

Interference effects are caused by path length differences. When X-rays scatter off one plane, all the waves remain in phase as long as the incident angle equals the reflected angle. However, for two adjacent planes, Figure 34.48 shows that the path length difference for the scattered X-rays from the two planes is

$$\Delta x = \Delta x_1 + \Delta x_2 = 2a\sin\theta, \qquad (34.28)$$

where θ is the angle between the incoming X-rays and the plane of atoms. (Note that unfortunately this convention in the literature is different from all of our other cases, where the angle is always measured relative to the surface normal!) Thus, the criterion for constructive interference from Bragg scattering is given by

$$2a\sin\theta = m\lambda \quad (m = 0,1,2,...). \qquad (34.29)$$

This equation is known as **Bragg's Law.**

When X-rays are incident on a crystal, several different planes can function as diffraction gratings. Some examples are illustrated in Figure 34.49. These planes do not have the spacing a between the planes.

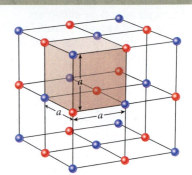

FIGURE 34.46 A cubic crystal lattice with spacing a.

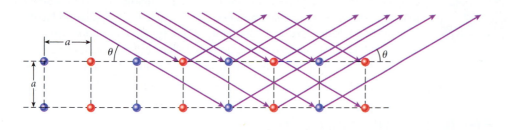

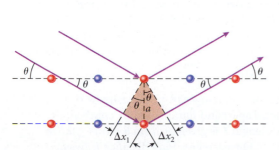

FIGURE 34.48 Path length difference for X-rays scattered from two adjacent planes.

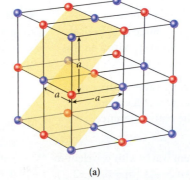

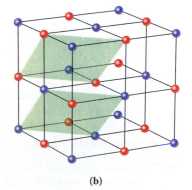

(a)　　　　(b)

FIGURE 34.47 Schematic drawing of X-rays scattering off planes of atoms in a crystal.

FIGURE 34.49 Examples of planes that could function as diffraction gratings for X-rays in a cubic crystalline lattice.

FIGURE 34.50 Two geometries for studying the atomic structure of a sample using X-ray diffraction. (a) X-rays scattered nearly parallel to the surface; (b) X-rays transmitted through the sample.

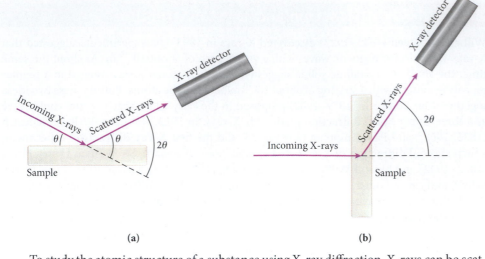

(a)   (b)

To study the atomic structure of a substance using X-ray diffraction, X-rays can be scattered nearly parallel to the surface of a sample (Figure 34.50a). Alternatively, the X-rays can be transmitted through the sample and detected on the opposite side of the sample (Figure 34.50b). For the parallel scattering method, the angle of incidence θ should equal the angle of observation. For the transmission method, the observed angle is twice the Bragg angle θ. By measuring the intensity of the X-rays as a function of θ, we can determine details of the structure of the material being studied.

Figure 34.51 shows a sample image obtained from the scattering of X-rays of a protein, in this case "3Clpro." In order for this imaging technique to work, the protein has to be fixed in a crystalline structure. From the diffraction pattern obtained, one can reconstruct the three-dimensional spatial structure with the aid of computer programs.

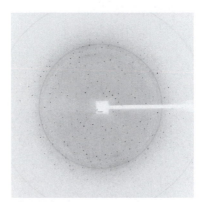

FIGURE 34.51 X-ray diffraction image of a protein.

Modern particle accelerators—such as the National Synchrotron Light Source at Brookhaven National Laboratory, the Advanced Light Source at Lawrence Berkeley National Laboratory, or the Advanced Photon Source at Argonne National Laboratory (Figure 34.52), and many others around the world—are used to produce high-quality, intense beams of X-rays to carry out research in condensed matter and materials science. In addition, we can gather similar scattering information by bombarding crystalline structures with intense neutron beams. (To understand how this works, we have to wait a few more chapters until we examine quantum mechanics.) Intense neutron beams for materials science research have now become available at the Spallation Neutron Source at Oak Ridge National Laboratory. These huge X-ray and neutron scattering facilities cost hundreds of millions of dollars, but are absolutely essential tools for investigating the nanoscale structure of materials. They are the basic research tools for modern and future nanotechnology advances.

FIGURE 34.52 Advanced Photon Source.

WHAT WE HAVE LEARNED | EXAM STUDY GUIDE

- Huygens's Principle states that every point on a propagating wave front serves as a source of spherical secondary wavelets. A geometric analysis based on this principle is called a Huygens construction.

- The requirement that two coherent waves with wavelength λ interfere constructively is $\Delta x = m\lambda$ ($m = 0,\pm1,\pm2,...$), where Δx is the path difference between the two waves.

- The requirement that two coherent waves with wavelength λ interfere destructively is $\Delta x = (m + \frac{1}{2})\lambda$ ($m = 0,\pm1,\pm2,...$), where Δx is the path difference between the two waves.

- The angle of bright fringes from two narrow slits spaced a distance d apart and illuminated by coherent light with wavelength λ is given by $d \sin\theta = m\lambda$ ($m = 0,\pm1,\pm2,...$). On a screen a long distance L away, the position y of the bright fringes from the central maximum along the screen is given by $y = \dfrac{m\lambda L}{d}$ ($m = 0,\pm1,\pm2,...$).

- The angle of dark fringes from two narrow slits spaced a distance d apart and illuminated by coherent light with wavelength λ is given by $d \sin\theta = (m + \frac{1}{2})\lambda$ ($m = 0,\pm1,\pm2,...$). On a screen a long distance L away, the position y of the dark fringes from the central maximum along the screen is given by $y = \dfrac{\left(m+\frac{1}{2}\right)\lambda L}{d}$ ($m = 0,\pm1,\pm2,...$).

- The condition for constructive interference for light with wavelength λ_{air} in a thin film of thickness t and index of refraction n in air is $\left(m+\frac{1}{2}\right)\dfrac{\lambda_{air}}{n} = 2t$ ($m = 0,\pm1,\pm2,...$).

- The radii of the bright circles in Newton's rings are given by $x_m = \sqrt{R\left(m+\frac{1}{2}\right)\lambda}$ ($m = 0,1,2,...$), where R is the radius of curvature of the upper curved glass surface and λ is the wavelength of the incident light.

- The angle of dark fringes from a single slit of width a illuminated by light with wavelength λ is given by $a \sin\theta = m\lambda$ ($m = 1,2,3,...$).

- The angle θ of the first minimum from a circular aperture with diameter d illuminated with light of wavelength λ is $\sin\theta = 1.22\dfrac{\lambda}{d}$. This expression is known as Rayleigh's Criterion. The angle in the

equation expresses the minimum resolvable angle between 2 distant objects for a telescope primary lens or mirror or camera lens with diameter d.

- The angle θ of the maxima from a diffraction grating illuminated with light of wavelength λ is given by $\theta = \sin^{-1}\left(\dfrac{m\lambda}{d}\right)$ ($m = 0,1,2,...$), where d is the distance between the rulings of the grating.

- The dispersion of a diffraction grating is given by $D = \dfrac{m}{d\cos\theta}$ ($m = 1,2,3,...$), where d is the distance between the rulings of the grating.

- The resolving power of a diffraction grating is given by $R = Nm$ ($m = 1,2,3,...$), where N is the number of rulings in the grating.

- For X-rays scattering off planes of atoms separated by a distance a, the condition for constructive interference is $2a \sin\theta = m\lambda$ ($m = 0,1,2,...$). The angle θ is the angle between incoming X-rays and the plane of atoms and the angle of observation of the X-rays.

KEY TERMS

Huygens's Principle, p. 1098

coherent light, p. 1100

incoherent, p. 1100

interference, p. 1100

constructive interference, p. 1100

destructive interference, p. 1100

Young's double-slit experiment, p. 1101

order, p. 1102

thin film, p. 1104

Newton's rings, p. 1106

interferometer, p. 1107

diffraction, p. 1109

resolution, p. 1113

Rayleigh's Criterion, p. 1113

diffraction grating, p. 1115

dispersion, p. 1116

resolving power, p. 1117

X-ray diffraction, p. 1121

Bragg's Law, p. 1121

NEW SYMBOLS AND EQUATIONS

$\Delta x = m\lambda$ ($m = 0,\pm1,\pm2,...$), path length difference for constructive interference

$\Delta x = (m + \frac{1}{2})\lambda$ ($m = 0,\pm1,\pm2,...$), path length difference for destructive interference

N, number of rulings in a diffraction grating

n_l, number of rulings per unit length for a diffraction grating

$D = \dfrac{m}{d\cos\theta}$ ($m = 1,2,3,...$), dispersion of a diffraction grating

$R = Nm$ ($m = 1,2,3,...$), resolving power of a diffraction grating

ANSWERS TO SELF-TEST OPPORTUNITIES

34.1 There is no change in colors. After all, the light has to propagate through your eyeball before it reaches your retina, and the index of refraction of your eyeball does not change when your head is underwater.

34.2 $\lambda = \dfrac{2d}{N} = \dfrac{2\left(0.25 \cdot 10^{-3}\ \text{m}\right)}{1200} = 4.17 \cdot 10^{-7}\ \text{m} = 417\ \text{nm}.$

34.3 Assume that $\lambda = 550$ nm and that the diameter of the human pupil is $d = 10.0$ mm.

$$\theta_R = \sin^{-1}\left(\frac{1.22\lambda}{d}\right) = 6.71 \cdot 10^{-5}\ \text{rad}$$

$$\theta_{wall} = \frac{y}{L} = \frac{10.0\ \text{m}}{190 \cdot 10^3\ \text{m}} = 5.26 \cdot 10^{-5}\ \text{rad}$$

Individual watchtowers on the Great Wall of China are difficult to see from the orbiting Space Shuttle.

34.4 We get for the first maximum $\theta_1 = \sin^{-1}(532\ \text{nm}/740\ \text{nm}) = 46.0°$. No other maxima are possible.

PROBLEM-SOLVING PRACTICE

Problem-Solving Guidelines

1. A sketch of the optical situation is almost always helpful. It is simplest to use rays, not wave fronts, but remember that you are dealing with wave effects. Use the diagram to clearly identify any path differences involved in the problem.

2. The basic idea of wave optics is that constructive interference occurs when the path difference is an integer number of wavelengths, while destructive interference occurs when the

path difference is an odd-integer number of half wavelengths. Always start from this concept and take into account in any additional phase changes due to reflection.

3. Remember that a phase change occurs when light reflects from a more dense medium after being incident in a less dense medium. If light reflects from a less dense medium, no phase change occurs.

SOLVED PROBLEM 34.2 | Spy Satellite

You have been assigned the job of designing a camera lens for a spy satellite. This satellite will orbit the Earth at an altitude of 201 km. The camera is sensitive to light with a wavelength of 607 nm. The camera must be able to resolve objects on the ground that are 0.490 m apart.

PROBLEM
What is the minimum diameter of the lens?

SOLUTION

THINK
This camera lens will be limited by diffraction. We can apply Rayleigh's Criterion to calculate the minimum diameter of the lens, given the angle subtended by two objects on the ground as viewed from the spy satellite orbiting above.

SKETCH
Figure 34.53 shows a sketch of the spy satellite observing two objects on the ground

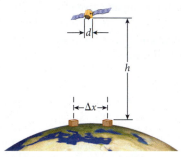

FIGURE 34.53 A spy satellite at height h observing two objects on the ground a distance Δx apart.

RESEARCH
The Rayleigh Criterion for resolving two objects separated by an angle θ_R using light with wavelength λ is given by

$$\theta_R = \sin^{-1}\left(\frac{1.22\lambda}{d}\right),$$

where d is the diameter of the circular camera lens in the spy satellite.

The angle required for the performance requirement of the spy satellite is given by

$$\theta_s = \tan^{-1}\left(\frac{\Delta x}{h}\right),$$

where Δx is the distance between the two objects on the ground and h is the height of the spy satellite above the ground.

SIMPLIFY
We can equate θ_R and θ_s to get

$$\sin^{-1}\left(\frac{1.22\lambda}{d}\right) = \tan^{-1}\left(\frac{\Delta x}{h}\right).$$

Solving for the diameter of the camera lens gives us

$$d = \frac{1.22\lambda}{\sin\left(\tan^{-1}\left(\frac{\Delta x}{h}\right)\right)}.$$

CALCULATE

Putting in the numerical values gives us

$$d = \frac{1.22\left(607\cdot10^{-9}\text{ m}\right)}{\sin\left[\tan^{-1}\left(\dfrac{0.490\text{ m}}{201\cdot10^{3}\text{ m}}\right)\right]} = 0.30377\text{ m}.$$

ROUND

We report our result to three significant figures,

$$d = 0.304\text{ m}.$$

DOUBLE-CHECK

To double-check our result, we make the small-angle approximation for the Rayleigh Criterion and the angle subtended by the two objects. For the case where the wavelength of light is much smaller than the aperture of the camera, we can write

$$\sin\left(\theta_{R}\right) = \frac{1.22\lambda}{d} \approx \theta_{R}.$$

For the angle seen by the camera in the spy satellite, we can write

$$\tan\left(\theta_{s}\right) = \frac{\Delta x}{h} \approx \theta_{s}.$$

Thus,

$$\frac{1.22\lambda}{d} = \frac{\Delta x}{h},$$

which can be solved for the minimum diameter of the camera lens

$$d = \frac{1.22\lambda h}{\Delta x} = \frac{1.22\left(607\cdot10^{-9}\text{ m}\right)\left(201\cdot10^{3}\text{ m}\right)}{0.490\text{ m}} = 0.304\text{ m},$$

which agrees with our result within round-off errors. Thus, our result seems reasonable.

SOLVED PROBLEM 34.3 | Air Wedge

Light of wavelength $\lambda = 516$ nm is incident perpendicularly on two glass plates. The glass plates are spaced at one end by a thin piece of kapton film. Due to the wedge of air created by this film, 25 bright interference fringes are observed across the top plate, with a dark fringe at the end by the film.

PROBLEM

How thick is the film?

SOLUTION

THINK

Light passes through the top plate, reflects from the top surface of the bottom plate, and then interferes with light reflected from the bottom surface of the top plate. A phase change occurs when the light is reflected from the bottom plate, so the criterion for constructive interference is that the path length is equal to an integer plus one-half times the wavelength. The criterion for destructive interference is that the path difference is an integer times the wavelength.

Continued—

FIGURE 34.54 Two glass plates with a thin film separating the plates at one end. The green arrows represent the incident light from the top. The angle of the light is exaggerated for clarity. The interfering light is reflected off the bottom of the top plate and the top of the bottom plate.

SKETCH

Figure 34.54 is a sketch showing the two glass plates with a thin piece of film separating the plates at one end. Light is incident vertically from the top.

RESEARCH

At any point along the plates, the criterion for constructive interference is given by

$$2t = \left(m + \tfrac{1}{2}\right)\lambda,$$

where t is the separation between the plates, m is an integer, and λ is the wavelength of the incident light. There are 25 bright fringes visible. The first bright fringe corresponds to $m = 0$, and the 25th bright fringe corresponds to $m = 24$.

Immediately past the 25th is a dark fringe where the piece of film is located. The criterion for destructive interference is $2t = n\lambda$, where n is an integer.

SIMPLIFY

The dark fringe located at the end of the glass plate where the film is located is given by

$$2t = \left(24 + \tfrac{1}{2} + \tfrac{1}{2}\right)\lambda = 25\lambda,$$

where the factor $24 + \tfrac{1}{2}$ describes the constructive interference and the extra $\tfrac{1}{2}$ produces the dark fringe at the end of the plate with the film. Thus we can solve for the separation of the plates, which corresponds to the thickness of the film:

$$t = \frac{25}{2}\lambda.$$

CALCULATE

Putting in the numerical values gives

$$t = \frac{25}{2}\left(516 \cdot 10^{-9}\ \text{m}\right) = 0.00000645\ \text{m}.$$

ROUND

We report our result to three significant figures,

$$t = 6.45 \cdot 10^{-6}\ \text{m}.$$

DOUBLE-CHECK

Occasionally throughout this book we need to add reminders that checking our results just for the right units and expected order of magnitude can do a lot to prevent simple errors. Here the unit m is certainly the right one for a physical length, in this case the film thickness. At first glance you may think that ~10^{-5} m, on the order of 1/100th of the thickness of a fingernail, may be impossibly thin for a solid film. However, an Internet search on *kapton film* will show that 6.5 μm is well within the range of thicknesses in which kapton film is produced. Thus, our answer seems plausible.

MULTIPLE-CHOICE QUESTIONS

34.1 Suppose the distance between the slits in a double-slit experiment is $2.00 \cdot 10^{-5}$ m. A beam of light with a wavelength of 750 nm is shone on the slits. What is the angular separation between the central maximum and the adjacent maximum?

a) $5.00 \cdot 10^{-2}$ rad   c) $3.75 \cdot 10^{-2}$ rad

b) $4.50 \cdot 10^{-2}$ rad   d) $2.50 \cdot 10^{-2}$ rad

34.2 When two light waves, both with wavelength λ and amplitude A, interfere constructively, they produce a light wave of the same wavelength but with amplitude $2A$. What will be the intensity of this light wave?

a) same intensity as before   c) quadruple the intensity

b) double the intensity   d) not enough information

34.3 A laser beam with wavelength 633 nm is split into two beams by a beam splitter. One beam goes to Mirror 1, a distance L from the beam splitter, and returns to the beam splitter, while the other beam goes to Mirror 2, a distance $L + \Delta x$ from the beam splitter, and returns to the same beam splitter. The beams then recombine and go to a detector together.

If $L = 1.00000$ m and $\Delta x = 1.00$ mm, which best describes the kind of interference at the detector? (*Hint:* To double-check your answer, you may need to use a formula that was originally intended for combining two beams in a different geometry, but which still is applicable here.)

a) purely constructive

b) purely destructive

c) mostly constructive

d) mostly destructive

e) neither constructive nor destructive

34.4 Which of the following light types on a grating with 1000 rulings with a spacing of 2.00 μm would produce the largest number of maxima on a screen 5.00 m away?

a) blue light of wavelength 450 nm

b) green light of wavelength 550 nm

c) yellow light of wavelength 575 nm

d) red light of wavelength 625 nm

e) need more information

34.5 If the wavelength of light illuminating a double slit is halved, the fringe spacing is

a) halved.

b) doubled.

c) not changed.

d) changed by a factor of $1/\sqrt{2}$.

34.6 A red laser pointer with a wavelength of 635 nm shines on a diffraction grating with 300 lines/mm. A screen is then placed a distance of 2.0 m behind the diffraction grating to observe the diffraction pattern. How far away from the central maximum will the next bright spot be on the screen?

a) 38 cm   c) 94 cm   e) 9.5 m

b) 76 cm   d) 4.2 m

34.7 Newton's rings displayed are interference patterns caused by the reflection of light between two surfaces. What color is the center of the Newton's rings when viewed with white light?

a) white   c) red

b) black   d) violet

34.8 In Young's double-slit experiment, both slits were illuminated by a laser beam and the interference pattern was observed on a screen. If the viewing screen is moved farther from the slit, what happens to the interference pattern?

a) The pattern gets brighter.

b) The pattern gets brighter and closer together.

c) The pattern gets less bright and farther apart.

d) There is no change in the pattern.

e) The pattern becomes unfocused.

f) The pattern disappears.

QUESTIONS

34.9 What would happen to a double-slit interference pattern if

a) the wavelength is increased?

b) the separation distance between the slits is increased?

c) the apparatus is placed in water?

34.10 What would be the frequency of an ultrasonic (sound) wave for which diffraction effects were as small in daily life as they are for light? (Estimate)

34.11 Why are radio telescopes so much larger than optical telescopes? Would an X-ray telescope also have to be larger than an optical telescope?

34.12 Can light pass through a single slit narrower than its wavelength? If not, why not? If so, describe the distribution of the light beyond the slit.

34.13 One type of hologram consists of bright and dark fringes produced on photographic film by interfering laser beams. If this is illuminated with white light, the image will

appear reproduced multiple times, in different pure colors at different sizes.

a) Explain why.

b) Which colors correspond to the largest and smallest images, and why?

34.14 A double slit is positioned in front of an incandescent light bulb. Will an interference pattern be produced?

34.15 Many astronomical observatories, and especially radio observatories, are coupling several telescopes together. What are the advantages of this?

34.16 In a single-slit diffraction pattern, there is a bright central maximum surrounded by successively dimmer higher-order maxima. Farther out from the central maximum, eventually no more maxima are observed. Is this because the remaining maxima are too dim? Or is there an upper limit to the number of maxima that can be observed, no matter how good the observer's eyes, for a given slit and light source?

34.17 Which close binary pair of stars will be more easily resolvable with a telescope —two red stars, or two blue ones? Assume the binary star systems are the same distance from Earth and are separated by the same angle.

34.18 A red laser pointer is shined on a diffraction grating, producing a diffraction pattern on a screen behind the diffraction grating. If the red laser pointer is replaced with a green laser pointer, will the green bright spots on the screen be closer together or farther apart than the red bright spots were?

PROBLEMS

A blue problem number indicates a worked-out solution is available in the Student Solutions Manual. One • and two •• indicate increasing level of problem difficulty.

Section 34.1

34.19 A helium-neon laser has a wavelength of 632.8 nm.

a) What is the wavelength of this light as it passes through lucite with an index of refraction $n = 1.500$?

b) What is the speed of light in the lucite?

34.20 It is common knowledge that the visible light spectrum extends approximately from 400 nm to 700 nm. Roughly, 400 nm to 500 nm corresponds to blue light, 500 nm to 550 nm corresponds to green, 550 nm to 600 nm to yellow-orange, and above 600 nm to red. In an experiment, red light with a wavelength of 632.8 nm from a HeNe laser is refracted into a fish tank filled with water with index of refraction 1.333. What is the wavelength of the same laser beam in water, and what color will the laser beam have in water?

Sections 34.2 and 34.3

34.21 What minimum path difference is needed to cause a phase shift by $\pi/4$ in light of wavelength 700 nm?

34.22 Coherent, monochromatic light of wavelength 450.0 nm is emitted from two locations and detected at another location. The path difference between the two routes taken by the light is 20.25 cm. Will the two light waves interfere destructively or constructively at the detection point?

34.23 A Young's interference experiment is performed with monochromatic green light ($\lambda = 540$ nm). The separation between the slits is 0.100 mm, and the interference pattern on a screen shows the first side maximum 5.40 mm from the center of the pattern. How far away from the slits is the screen?

34.24 For a double-slit experiment, two 1.50-mm wide slits are separated by a distance of 1.00 mm. The slits are illuminated by a laser beam with wavelength 633 nm. If a screen is placed 5.00 m away from the slits, determine the separation of the bright fringes on the screen.

•34.25 Coherent monochromatic light with wavelength $\lambda = 514$ nm is incident on two slits that are separated by a distance $d = 0.500$ mm. The intensity of the radiation at a screen 2.50 m away from each slit is 180.0 W/cm². Deter-

mine the position $y_{1/3}$ at which the intensity of the central peak (at $y = 0$) drops to $I_{max}/3$.

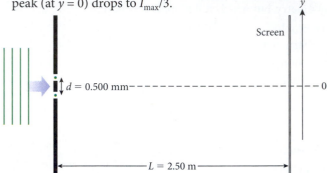

•34.26 In a double-slit experiment, He-Ne laser light of wavelength 633 nm produced an interference pattern on a screen placed at some distance from the slits. When one of the slits was covered with a thin glass slide of thickness 12.0 μm, the central fringe shifted to the point occupied earlier by the 10th dark fringe (see figure). What is the refractive index of the glass slide?

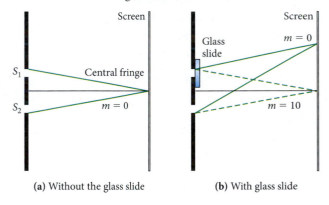

(a) Without the glass slide **(b)** With glass slide

Section 34.4

34.27 Suppose the thickness of a thin soap film ($n = 1.32$) surrounded by air is nonuniform and gradually tapers. Monochromatic light of wavelength 550 nm illuminates the film. At the thinnest end, a dark band is observed. How thick is the film at the next two dark bands closest to the first dark band?

34.28 White light (400 nm $< \lambda <$ 750 nm) shines onto a puddle of water ($n = 1.33$). There is a thin (100.0 nm thick) layer of oil ($n = 1.47$) on top of the water. What wavelengths of light would you see reflected?

34.29 Some mirrors for infrared lasers are constructed with alternating layers of hafnia and silica. Suppose you want to produce constructive interference from a thin film of hafnia ($n = 1.90$) on BK-7 glass ($n = 1.51$) when infrared radiation of wavelength 1.06 μm is used. What is the smallest film thickness that would be appropriate, assuming the laser beam is oriented at right angles to the film?

34.30 Sometimes thin films are used as filters to prevent certain colors from entering a lens. Consider an infrared filter, designed to prevent 800.0-nm light from entering a lens. Find the minimum film thickness for a layer of MgF_2 ($n = 1.38$) to prevent this light from entering the lens.

•**34.31** White light shines on a sheet of mica that has a uniform thickness of 1.30 μm. When the reflected light is viewed using a spectrometer, it is noted that light with wavelengths of 433.3 nm, 487.5 nm, 557.1 nm, 650.0 nm, and 780.0 nm is not present in the reflected light. What is the index of refraction of the mica?

•**34.32** A single beam of coherent light ($\lambda = 633 \cdot 10^{-9}$ m) is incident on two glass slides, which are touching at one end and are separated by a 0.0200-mm thick sheet of paper on the other end, as shown in the figure below. Beam 1 reflects off the bottom surface of the top slide, and Beam 2 reflects of the top surface of the bottom slide. Assume that all the beams are perfectly vertical and that they are perpendicular to both slides, i.e., the slides are nearly parallel (the angle is exaggerated in the figure); the beams are shown at angles in the figure so that they are easier to identify. Beams 1 and 2 recombine at the location of the eye in the figure below. The slides are 8.00 cm long. Starting from the left end ($x = 0$), at what positions x_{bright} do bright bands appear to the observer above the slides? How many bright bands are observed?

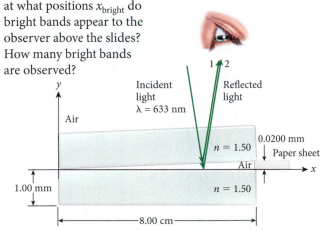

•**34.33** A common interference setup consists of a plano-convex lens placed on a plane mirror and illuminated from above at normal incidence with monochromatic light. The pattern of circular interference fringes (fringes of equal thickness)—bright and dark circles—formed due to the air wedge defined by the two glass surfaces, is known as *Newton's rings*. In an experiment using a plano-convex lens with focal length $f = 80.00$ cm and index of refraction $n_l = 1.500$, the radius of the third bright circle is found to be 0.8487 mm. Determine the wavelength of the monochromatic light.

Section 34.5

34.34 The Michelson interferometer is used in a class of commercially available optical instruments called *wavelength meters*. In a wavelength meter, the interferometer is illuminated simultaneously with the parallel beam of a reference laser of known wavelength and that of an unknown laser. The movable mirror of the interferometer is then displaced by a distance Δd, and the number of fringes produced by each laser and passing by a reference point (a photo detector) is counted. In a given wavelength meter, a red He-Ne laser ($\lambda_{Red} = 632.8$ nm) is used as a reference laser. When the movable mirror of the interferometer is displaced by a distance Δd, a number $\Delta N_{Red} = 6.000 \cdot 10^4$ red fringes and $\Delta N_{unknown} = 7.780 \cdot 10^4$ fringes pass by the reference photodiode.

a) Calculate the wavelength of the unknown laser.

b) Calculate the displacement, Δd, of the movable mirror.

34.35 Monochromatic blue light ($\lambda = 449$ nm) is beamed into a Michelson interferometer. How many fringes move by the screen when the movable mirror is moved a distance $d = 0.381$ mm?

•**34.36** At the Long-baseline Interferometer Gravitational-wave Observatory (LIGO) facilities in Hanford, Washington, and Livingston, Louisiana, laser beams of wavelength 550.0 nm travel along perpendicular paths 4.000 km long. Each beam is reflected along its path and back 100 times before the beams are combined and compared. If a gravitational wave increases the length of one path and decreases the other, each by 1.000 part in 10^{21}, what is the phase difference between the two beams as a result?

Sections 34.6 and 34.7

34.37 Light of wavelength 653 nm illuminates a slit. If the angle between the first dark fringes on either side of the central maximum is 32.0°, What is the width of the slit?

34.38 An instructor uses light of wavelength 633 nm to create a diffraction pattern with a slit of width 0.135 mm. How far away from the slit must the instructor place the screen in order for the full width of the central maximum to be 5.00 cm?

34.39 What is the largest slit width for which there are no minima when the wavelength of the incident light on the single slit is 600 nm?

34.40 Plane light waves are incident on a single slit of width 2.00 cm. The second dark fringe is observed at 43.0° from the central axis. What is the wavelength of the light?

Section 34.8

34.41 The Large Binocular Telescope (LBT), on Mount Graham near Tucson, Arizona, has two primary mirrors. The mirrors are centered a distance of 14.4 m apart, thus improving the Rayleigh limit. What is the minimum angular resolution of the LBT for green light, $\lambda = 550$ nm?

34.42 A canvas tent has a single, tiny hole in its side. On the opposite wall of the tent, 2.0 m away, you observe a dot (due to the Sun's light incident upon the hole) of width 2.0 mm, with a faint ring around it. What is the size of the hole in the tent?

34.43 Calculate and compare the angular resolutions of the Hubble Space Telescope (aperture diameter 2.40 m, wavelength 450. nm; illustrated in the text), the Keck Telescope (aperture diameter 10.0 m, wavelength 450. nm), and the Arecibo radio telescope (aperture diameter 305 m, wavelength 0.210 m). Assume that the resolution of each instrument is diffraction limited.

34.44 The Hubble Space Telescope (Figure 34.33) is capable of resolving optical images to an angular resolution of $2.8 \cdot 10^{-7}$ rad with its 2.4-m mirror. How large would a radio telescope have to be in order to image an object in the radio spectrum with the same resolution, assuming the wavelength of the waves is 10 cm?

34.45 Think of the pupil of your eye as a circular aperture 5.00 mm in diameter. Assume you are viewing light of wavelength 550 nm, to which your eyes are maximally sensitive.

a) What is the minimum angular separation at which you can distinguish two stars?

b) What is the maximum distance at which you can distinguish the two headlights of a car mounted 1.50 m apart?

Section 34.9

34.46 A red laser pointer with a wavelength of 635 nm is shined on a double slit producing a diffraction pattern on a screen that is 1.60 m behind the double slit. The central maximum of the diffraction pattern has a width of 4.20 cm, and the fourth bright spot is missing on both sides. What is the size of the individual slits, and what is the separation between them?

•34.47 A double slit is opposite the center of a 1.8-m wide screen 2.0 m from the slits. The slit separation is 24 μm and the width of each slit is 7.2 μm. How many fringes are visible on the screen if the slit is illuminated by 600-nm light?

•34.48 A two-slit apparatus is covered with a red (670 nm) filter. When white light is shone on the filter, on the screen beyond the two-slit apparatus, there are nine interference maxima within the 4.50-cm-wide central diffraction maximum. When a blue (450 nm) filter replaces the red, how many interference maxima will there be in the central diffraction maximum, and how wide will that diffraction maximum be?

34.49 The irradiance pattern observed in a two-slit interference-diffraction experiment is presented in the figure. The red line represents the actual intensity measured as a function of angle, while the green line represents the envelope of the interference patterns.

a) Determine the slit width a in terms of the wavelength λ of the light used in the experiment.

b) Determine the center-to-center slit separation d in terms of the wavelength λ.

c) Using the information in the graph, determine the ratio of slit width a to the center-to-center separation between the slits, d.

d) Can you calculate the wavelength of light, actual slit separation, and slit width?

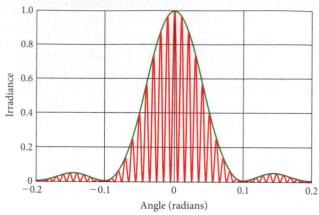

Single-Slit and Two-Slit Irradiance Distribution

Section 34.10

34.50 Two different wavelengths of light are incident on a diffraction grating. One wavelength is 600 nm and the other is unknown. If the 3rd order of the unknown wavelength appears at the same position as the 2nd order of the 600 nm light, what is the value of the unknown wavelength?

34.51 Light from an argon laser strikes a diffraction grating that has 7020 grooves per centimeter. The central and first-order principal maxima are separated by 0.332 m on a wall 1.00 m from the grating. Determine the wavelength of the laser light.

•34.52 A 5.000-cm-wide diffraction grating with 200 grooves is used to resolve two closely spaced lines (a doublet) in a spectrum. The doublet consists of two wavelengths, $\lambda_a = 629.8$ nm and $\lambda_b = 630.2$ nm. The light illuminates the entire grating at normal incidence. Calculate to *four significant digits* the angles θ_{1a} and θ_{1b} with respect to the normal at which the first-order diffracted beams for the two wavelengths, λ_a and λ_b, respectively, will be reflected from the grating. Note that this is not 0°! What order of diffraction is required to resolve these two lines using this grating?

—5.000 cm—

•34.53 A diffraction grating has $4.00 \cdot 10^3$ lines/cm and has white light (400.–700. nm) incident on it. What wavelength(s) will be visible at 45.0°?

Section 34.11

34.54 What is the wavelength of the X-rays if the first-order Bragg diffraction is observed at 23.0° related to the crystal surface, with inter atomic distance of 0.256 nm?

Additional Problems

34.55 How many lines per centimeter must a grating have if there is to be no second-order spectrum for any visible wavelength (400–750 nm)?

34.56 Many times, radio antennas occur in pairs. The effect is that they will then produce constructive interference in one direction while producing destructive interference in another direction—a directional antenna—so that their emissions don't overlap with nearby stations. How far apart at a minimum should a local radio station, operating at 88.1 MHz, place its pair of antennae operating in phase such that no emission occurs along a line 45.0° from the line joining the antennae?

34.57 A laser produces a coherent beam of light that does not spread (diffract) as much in comparison to other light sources like an incandescent bulb. Lasers therefore have been used for measuring large distances, such as the distance between the Moon and the Earth with very great accuracy. In one such experiment, a laser pulse (wavelength 633 nm) is fired at the Moon. What should be the size of the circular aperture of the laser source that would produce central maximum of 1.00-km diameter on the surface of the Moon? Distance between the Moon and the Earth is $3.84 \cdot 10^5$ km.

34.58 A diffraction grating with 1000 lines per centimeter is illuminated by a He-Ne laser of wavelength 633 nm.

a) What is the highest order of diffraction that could be observed with this grating?

b) What would be the highest order if there were 10,000 lines per centimeter?

34.59 The thermal stability of a Michelson interferometer can be improved by submerging it in water. Consider an interferometer that is submerged in water, measuring light from a monochromatic source that is in air. If the movable mirror moves a distance of $d = 0.200$ mm, $N = 800$ fringes move by the screen. What is the original wavelength (in air) of the monochromatic light?

34.60 A Blu-ray disc uses a blue laser with a free-space wavelength of 405 nm. If the disc is protected with polycarbonate ($n = 1.58$), determine the minimum thickness of the disc for destructive interference. Compare this value to that for CDs illuminated by infrared light.

34.61 An airplane is made invisible to radar by coating it with a 5.00-mm-thick layer of an antireflective polymer with the index of refraction $n = 1.50$. What is the wavelength of radar waves for which the plane is made invisible?

34.62 Coherent monochromatic light passes through parallel slits and then onto a screen that is at a distance $L = 2.40$ m from the slits. The narrow slits are a distance $d = 2.00 \cdot 10^{-5}$ m apart. If the minimum spacing between bright spots is $y = 6.00$ cm, find the wavelength of the light.

34.63 Determine the minimum thickness of a soap film ($n = 1.32$) that would produce constructive interference when illuminated by light of wavelength of 550 nm.

34.64 You are making a diffraction grating that is required to separate the two spectral lines in the sodium D doublet, at wavelengths 588.9950 and 589.5924 nm, by at least 2.0 mm on a screen that is 80 cm from the grating. The lines are to be ruled over a distance of 1.5 cm on the grating. What is the minimum number of lines you should have on the grating?

34.65 A Michelson interferometer is illuminated with a 600-nm light source. How many fringes are observed if one of the mirrors of the interferometer is moved a distance of 200 μm?

34.66 What is the smallest object separation you can resolve with your naked eye? Assume the diameter of your pupil is 3.5 mm, and that your eye has a near point of 25 cm and a far point of infinity.

34.67 When using a telescope with an objective of diameter 12.0 cm, how close can two features on the Moon be and still be resolved? Take the wavelength of light to be 550 nm, near the center of the visible spectrum.

34.68 There is air on both sides of a soap film. What is the smallest thickness of the soap film ($n = 1.420$) that would appear dark if illuminated with 500-nm light?

34.69 X-rays with a wavelength of 1.0 nm are scattered off of two small tumors in the human body. If the two tumors are a distance of 10 cm away from the X-ray detector, which has an entrance aperture of 1.0 mm, what is the minimum separation between the two tumors that will allow the X-ray detector to determine that there are two tumors instead of one?

•34.70 A glass with a refractive index of 1.50 is inserted into one arm of a Michelson interferometer that uses a 600-nm light source. This causes the fringe pattern to shift by 1000 fringes. How thick is the glass?

•34.71 White light is shone on a very thin layer of mica ($n = 1.57$), and above the mica layer, interference maxima for two wavelengths (and no other in between) are seen: one blue wavelength of 480 nm, and one yellow wavelength of 560 nm. What is the thickness of the mica layer?

•34.72 In a double-slit arrangement the slits are $1.00 \cdot 10^{-5}$ m apart. If light with wavelength 500 nm passes through the slits, what will be the distance between the $m = 1$ and $m = 3$ maxima on a screen 1.00 m away?

•34.73 A Newton's ring apparatus consists of a convex lens with a large radius of curvature R placed on a flat glass disc. (a) Show that the distance x from the center to the air, thickness d, and the radius of curvature R are given by $x^2 = 2Rd$. (b) Show that the radius of nth constructive interference is given by $x_n = [(n + \frac{1}{2}) \lambda R]^{1/2}$. (c) How many bright fringes may be seen if it is viewed by red light of wavelength 700 nm for $R = 10.0$ m, and the plane glass disc diameter is 5.00 cm?

35

Relativity

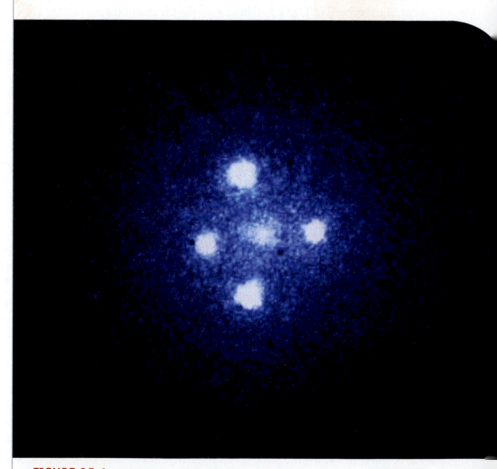

FIGURE 35.1 A photograph of Einstein's Cross, which is formed by gravitational lensing of a distant quasar by the galaxy in the center.

WHAT WE WILL LEARN

- Light always moves at the same speed in vacuum, independent of the velocity of the source or the observer.

- The two postulates of special relativity are that (1) all physical laws are the same in all inertial reference frames, and (2) the speed of light in vacuum is invariant. Inertial frames are reference frames, which move with constant velocity.

- From the postulates of special relativity, it follows that measurements of time and space intervals are different for different observers who move relative to each other.

- For high speeds, the Lorentz transformation must be used between reference frames instead of the Galilean transformation.

- Velocities do not add linearly. Velocity addition must follow the Lorentz transformation so it conforms to the postulate that the speed of light cannot be exceeded.

- Kinetic energy and momentum and their relationship need new definitions.

- All of Newtonian mechanics derives from relativistic mechanics as a limiting case for speeds small compared to the speed of light.

The picture in Figure 35.1 is not a cluster of stars, as you might assume at first glance. Instead, the central bright spot is a galaxy at a great distance from Earth, and the four spots around it are all images of an even more distant object behind it. The path traveled by the light from the more distant object (called a quasar, or quasi-stellar object) is curved by the gravitational pull of the intermediate galaxy to form four surrounding images. The fact that light can be bent by gravitation was not predicted by Newton's laws but was deduced from Einstein's theory of relativity. The image in this photograph, called Einstein's Cross, is one of many observations that confirm Einstein's theory of relativity.

The theory of relativity changes our understanding of space and time in very basic ways. And some of the consequences seem contradictory to our everyday experiences. The idea that a meter stick in a train station may not be the same length for a person standing in the station as for a person passing by on a train seems preposterous. Some people believe it is impossible, and claim that relativity only describes human perceptions or is "only a theory." However, relativity has been tested as much as any idea in science and has been confirmed every time. Time and space are not independent of the reference frame, and it is not just a wild idea or an optical illusion—it is how our universe works.

This chapter focuses primarily on the special theory of relativity, which is called *special* because it deals with the special case of motion with constant velocity; that is, zero acceleration. Brief mention will also be made of some ideas from the general theory of relativity, which does deal with accelerated motion. Although the concepts may seem strange at first, the mathematics are not particularly difficult. Newtonian mechanics can be thought of as a special case of Einstein's work, giving the same results in ordinary situations. But special relativity arrives at quite different results for motion at a significant fraction of the speed of light. These results play a major role in the physics of the very small (high-energy particle physics and quantum mechanics) and the physics of the very large (astronomy and cosmology).

35.1 Search for the Aether

Chapters 15 and 16 demonstrated that sound waves and mechanical waves need a medium in which to propagate. Chapter 31 showed that light is an electromagnetic wave, and that all electromagnetic waves can propagate through vacuum. However, this knowledge is relatively new in science, only about 100 years old. Up until 1887, scientists believed that light would also need a medium in which to propagate, and they called this medium the *luminiferous aether*, or simply the **aether**. (There is also another spelling, *ether*, for the same idea, but since this word is also used for a chemical compound, we avoid it here.) This idea of the aether brought up the question, What exactly is this medium? Light from very distant stars and galaxies reaches our eyes, so it is evidently able to propagate outside of Earth's atmosphere. This observation implies that all of space must be filled with this medium. How could this medium be detected?

If all of space were filled with aether, then Earth would have to move relative to this aether on its path around the Sun. Chapter 3 demonstrated that the motion of the medium makes a clear difference to a trajectory. For example, an airplane moving through wind has a different ground speed if it moves perpendicularly to the wind than if it moves with or against the wind. Earlier chapters showed this same basic principle at work with the propagation of sound waves though a medium.

In the 19th century, huge efforts were made to measure the speed of light. From the point of view of science history, this quest is a fascinating story in itself. However, detecting the aether does not require knowing the precise speed of light. It only requires knowing that the motion of Earth relative to the aether would imply different speeds of light measured in the lab, depending on the direction of the light's velocity with respect to the aether. This effect is exactly what Albert Michelson and Edward Morley at the Case Institute in Cleveland, Ohio, set out to measure in 1887. They used an ingenious device called an interferometer, (see Chapter 34, which also shows a picture of a Michelson-type interferometer).

What they found stunned the world of physics: A null result! No measurable difference! Light moves with exactly the same speed in every direction, and no motion relative to the aether could be detected.

Physicists struggled to explain this astounding result. Two leading theorists, Hendrik Lorentz (1853–1928) and George Fitzgerald (1851–1922), came up with the idea that objects moving through the aether become length-contracted just enough to offset the change in the speed of light with direction. It took the genius of Albert Einstein (1879–1955), however, to make the conceptual leap required for a new insight and its astounding consequences: The aether does not exist. Thinking through the fact that the speed of light is constant for all observers, independent of the observer's motion, led Einstein to the formulation of the theory of special relativity, the subject of this chapter.

35.2   Einstein's Postulates and Reference Frames

In the year 1905, a 26-year-old Swiss patent clerk fresh out of a rather undistinguished university physics career wrote three scientific articles that shook the scientific world. And what's more amazing, he did this in his spare time! These three papers were:

1. A paper explaining the so-called *photoelectric effect* as due to the quantum nature of light. This explanation earned Einstein the 1921 Nobel Prize in Physics. We will come back to this effect in Chapter 36.

2. A paper explaining the effect of Brownian motion, which is the motion of very small particles in water or other solutions, as due to collisions with molecules and atoms. This result provided compelling arguments that atoms really exist. (This fact was not at all clear before his work.)

3. Finally, the most important for the present chapter: a paper presenting the theory of special relativity.

Einstein made two postulates, from which all of special relativity followed. To understand this, we first need a definition: An **inertial reference frame** is a reference frame in which an object accelerates only when a net external force is acting on it. An inertial reference frame moves with constant velocity with respect to any other inertial reference frame. A noninertial frame is a frame where the point of origin experiences an acceleration. For example, a diver who jumps off a diving board and is in free fall is not in an inertial reference frame, because she is experiencing a net acceleration. Unless stated otherwise, the phrase *reference frame* in this chapter refers to an *inertial* reference frame. With this definition, we can state Einstein's postulates:

Postulate 1: The laws of physics are the same in every inertial reference frame, independent of the motion of this reference frame.

Postulate 2: The speed of light c is the same in every inertial reference frame.

The value of the speed of light is

$$c = 299,792,458 \text{ m/s.} \tag{35.1}$$

As we stated in Chapter 1, this value for the speed of light is the accepted exact value, because it serves as the basis for the definition of the SI unit of the meter. Handy approximate values for the speed of light are $c \approx 186,000$ miles/s or $c \approx 1$ foot/ns in British units, or the very commonly used $c \approx 3 \cdot 10^8$ m/s.

The first of Einstein's two postulates should not raise any objections. In its motion around the Sun, the Earth moves through space around the Sun with a speed of more than 29 km/s $\approx$ 65,000 mph. Because the Earth orbits very nearly in a circle, the Earth's velocity vector relative to the Sun continually changes direction. However, we still expect that a physics measurement follows laws of nature that are independent of the season in which the measurement was made (neglecting minor effects due to the small accelerations of the Earth and of the Sun relative to the Milky Way).

The second postulate explains the null result that Michelson and Morley measured. However, this idea is not quite so easy to digest. Let's conduct a thought experiment: You fly in a rocket through space with a speed of $c/2$, directly toward Earth. Now you shine a laser in the forward direction. The light of the laser has a speed c. Naively, and with what we know about velocity addition so far, we would expect the light of the laser to have a speed of $c + (c/2) = 1.5c$ when observed on Earth. This result is what we would have predicted in the section on relative motion in Chapter 3.

This velocity addition, however, only works for speeds that are small relative to the speed of light. Einstein's second postulate says that the speed of light as seen on Earth is still c. Later in this chapter we will state a rule for velocity addition that is correct for all speeds. For now, note that this constant c is the maximum possible speed that any object can have in any reference frame. This statement is astounding and leads to all kinds of interesting and seemingly counterintuitive consequences—all of which have nonetheless been experimentally verified. Therefore, we now know that this theory is almost certainly correct. We may never think of space and time the same way again!

Beta and Gamma

Because the speed of light plays such an important role in relativity, we will introduce two commonly used dimensionless quantities, beta and gamma, that depend only on the (constant) speed of light, c, and the velocity, $\vec{v}$, of an object:

$$\vec{\beta} = \frac{\vec{v}}{c} \tag{35.2}$$

and

$$\gamma = \frac{1}{\sqrt{1-\beta^2}} = \frac{1}{\sqrt{1-(v/c)^2}}. \tag{35.3}$$

We will use the notation $\beta \equiv |\vec{\beta}|$. Note that for $v \equiv |\vec{v}| \leq c$, the equations 35.2 and 35.3 mean that $\beta \leq 1$ and $\gamma \geq 1$.

It is instructive to plot γ as a function of β (Figure 35.2). For speeds that are small compared to the speed of light, β is very small, approximately equal to zero. In that case, γ is very close to the value 1. However, as β approaches 1, γ *diverges*—that is, γ grows larger and larger and eventually becomes infinite when $\beta = 1$.

A useful approximation, valid for low speeds, is also common. In this case, $|\vec{v}|$ is small compared to c and therefore β is small compared to 1. By using the mathematical series expansion $(1-x)^{-1/2} = 1 + \frac{1}{2}x^2 + \frac{3}{8}x^4 + \ldots$, we can approximate γ as

$$\gamma \approx 1 + \frac{1}{2}\beta^2 = 1 + \frac{1}{2}\left(\frac{v}{c}\right)^2 \quad \text{(for } \beta \text{ small compared to 1).} \tag{35.4}$$

35.1 Self-Test Opportunity

A light-year is the distance light travels in a year. Calculate that distance in meters.

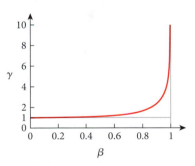

FIGURE 35.2 Dependence of γ on β.

35.2 Self-Test Opportunity

At what fraction of the speed of light would relativistic effects (deviation between the correct relativistic expression and its classical approximation) be a 5% effect? At what fraction of the speed of light would relativistic effects be a 50% effect?

EXAMPLE 35.1 | *Apollo* Spacecraft

On its way to the Moon, the *Apollo* spacecraft reached speeds of $4.0 \cdot 10^4$ km/h relative to Earth, almost an order of magnitude (10 times) faster than any jet aircraft.

PROBLEM

What are the values of the relativistic factors β and γ in this case?

SOLUTION

First convert the speed to SI units:

$$v = 4.0 \cdot 10^4 \text{ km/h} = 4.0 \cdot 10^4 \text{ km/h} \cdot (1000 \text{ m/km})/(3600 \text{ s/h}) \approx 1.1 \cdot 10^4 \text{ m/s}$$

To compute β, simply divide the spacecraft's speed by the speed of light:

$$\beta = \frac{v}{c} = \frac{1.1 \cdot 10^4 \text{ m/s}}{3.0 \cdot 10^8 \text{ m/s}} = 3.7 \cdot 10^{-5}.$$

Now substitute this result into the formula for γ and obtain

$$\gamma = \frac{1}{\sqrt{1 - \beta^2}} = \frac{1}{\sqrt{1 - (3.7 \cdot 10^{-5})^2}} = 1 + 6.9 \cdot 10^{-10} = 1.00000000069.$$

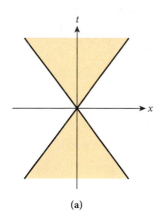

FIGURE 35.3 Light cone of a point in space.

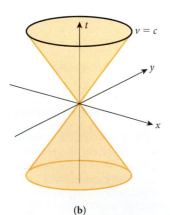

FIGURE 35.4 Conventional representation of the positive and negative light cones, with the time axis pointing vertically up. (a) 1+1 dimensional space-time, (b) 2+1 dimensional space-time.

From Example 35.1 you can see that almost all motion of macroscopic objects involves values of β that are very close to zero and values of γ that are very close to one. In all such cases, as we will see throughout this chapter, it is safe to neglect the effects of relativity and calculate the nonrelativistic approximation. However, in many interesting situations, relativity must be kept in mind. This difference is the subject of this chapter.

Light Cone

As a corollary of the two Einstein postulates, we find that nothing can propagate with a speed greater than the speed of light in vacuum. (Strictly speaking, there is the possibility that there exists particles for which the speed of light is not the *upper*, but the *lower* limit to their speed. These hypothetical particles are called tachyons. These will not be discussed any further in this book.) If nothing can propagate with speed greater than the speed of light in vacuum, then limits exist on how events can influence each other. Two people cannot exchange signals with each other at speeds exceeding the speed of light. Therefore, instantaneous effects of events originating from one point in space on another point in space are impossible. It simply takes time for a signal or cause-and-effect to propagate through space.

To investigate this result, let's consider the one-dimensional case shown in Figure 35.3. An event occurs at time $t = 0$ at location $x = x_0$ (red dot). A signal that announces this event in space and time can propagate no faster than the speed of light; that is, with a velocity in the interval between $v = -c$ and $v = +c$. The tan-colored triangle in Figure 35.3 shows the region in space and time in which the event can be observed. This region is called the positive **light cone** of the event $(0,x_0)$. The blue point located at time t_1 and position x_1 is able to receive a signal that the original event has happened; however, the green point located at time t_2 and position x_2 is not able to receive such a signal. This implies that the event represented by the red dot cannot possibly have caused the event represented by the green dot in Figure 35.3. The two events, red and green, cannot be causally connected—one cannot have caused the other.

Conversely, there is a region in the past that is able to influence the event at $(0,x_0)$. This region is the negative light cone. Figure 35.4a shows both light cones in the conventional representation of a vertical time axis. The event for which the light cone is displayed is conveniently moved to the origin of the coordinate system. Why is the light cone called a "cone," and not a "triangle"? The answer is shown in Figure 35.4b. For two space coordinates

x and y, the condition $v = \sqrt{v_x^2 + v_y^2} = c$ sweeps out a cone in the (2+1)-dimensional x,y,t space. Only events inside the negative light cone can influence events at the origin, at the apex of the light cone. Conversely, only events inside the positive light cone can be influenced by the event located at the origin.

Often the axes of light-cone diagrams are scaled so that they have the same units. One way to scale the axes is to multiply the time axis by the speed of light, so that both axes have the units of length. Another method involves dividing the x-axis by the speed of light, so that both axes have the units of time. In this way the speed of light becomes a diagonal line in the light-cone diagram, which is very useful for making quantitative arguments, as can be seen in the following example.

In a light-cone diagram we can draw world lines. A **world line** is the trajectory of an object in space and time. This kind of plot is often referred to as a graph in **space-time,** reflecting how these two dimensions are intertwined in relativity theory. A typical plot containing several world lines is shown in Figure 35.5. In this type of plot, we imagine motion in only one space dimension, x, along with time. Let's imagine we have an object initially located at $x = 0$ and $t = 0$ in this plot. If the object is not moving in the x-direction, the object traces out a vertical line in this plot (vector 1). If the object is moving in the positive x-direction with a constant speed, its trajectory is represented by a path pointing up and to the right (vector 2). If the object is moving with a constant speed in the negative x-direction, its trajectory is depicted by a path pointing up and to the left (vector 3). An object traveling with the speed of light in the positive x-direction is shown as a trajectory with a 45° angle with respect to the vertical axis (vector 4). An object traveling with the speed of light in the negative x-direction is given by a trajectory with a –45° angle with respect to the vertical axis (vector 5).

Space-Time Intervals

In classical mechanics, we can easily write down the distance between two points: $\Delta r = |\vec{r}_1 - \vec{r}_2| = \sqrt{(x_1 - x_2)^2 + (y_1 - y_2)^2 + (z_1 - z_2)^2}$. If two events take place at different times, then the time difference between them is $\Delta t = t_2 - t_1$. In view of the discussion of the light cone and causality above, we introduce the space-time interval s between two events 1 and 2. We define s through

$$s^2 = c^2(\Delta t)^2 - (\Delta r)^2. \tag{35.5}$$

Depending on the sign of s^2, we can now distinguish three types of space-time intervals—time-like intervals, light-like intervals, and space-like intervals:

$$
\left.
\begin{aligned}
s^2 > 0 &\Rightarrow c^2 \Delta t^2 > \Delta r^2 \quad \text{time-like} \\
s^2 = 0 &\Rightarrow c^2 \Delta t^2 = \Delta r^2 \quad \text{light-like} \\
s^2 < 0 &\Rightarrow c^2 \Delta t^2 < \Delta r^2 \quad \text{space-like}
\end{aligned}
\right\} \text{space-time intervals.} \tag{35.6}
$$

Light-like space-time intervals are on the surface of the light cone, time-like space-time intervals are in the interior of the light cone, and space-like intervals are on the exterior. In a time-like interval we can define a proper time interval $\Delta\tau$, which is the time between two events measured by an observer traveling with his clock in an inertial frame between these events, with the observer's path intersecting the world line of each event as that event occurs. This **proper time** is

$$\Delta\tau = \sqrt{\Delta t^2 - \Delta r^2/c^2} . \tag{35.7}$$

Note that with the time-like condition of $s^2 > 0$, the proper time is a real number (+ time unit). The existence of time-like (or light-like) intervals between two events means that these two events can be causally connected.

If two events are separated by a space-like interval, then they cannot be causally connected—that is, neither of the two events can possibly trigger the other one. We can define a proper distance $\Delta\sigma$ between these two events,

$$\Delta\sigma = \sqrt{\Delta r^2 - c^2 \Delta t^2} , \tag{35.8}$$

which is a real number (+ length unit) for space-like events.

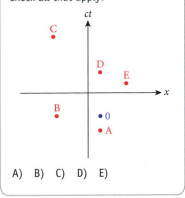
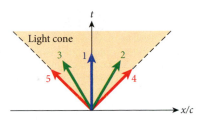

FIGURE 35.5 A space-time plot showing a positive light cone.

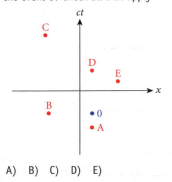

In our everyday experience, we consider time and space as absolute, without restriction or qualification. By "absolute" we mean that observers in all inertial reference frames measure the same value for the length of any object and the duration of any event. However, if we follow Einstein's postulates to their logical conclusions, this is not the case any more. The concepts of nonabsolute time and space lead to conclusions that sound like science fiction, but they have been verified experimentally.

Time Dilation

One of the most remarkable consequences of the theory of relativity is that time measurement is not independent of the reference frame. Instead, time that elapses between two events in a moving reference frame where events occur at different locations is dilated (made longer) when compared to the time interval in the rest frame (where the events occur at the same location):

$$\Delta t = \gamma \Delta t_0 = \frac{\Delta t_0}{\sqrt{1-(v/c)^2}}. \tag{35.9}$$

This **time dilation** means that if a clock advances Δt_0 while at rest, an observer sees the clock advancing by $\Delta t > \Delta t_0$ when she moves relative to the clock. That time interval Δt depends on the speed v with which she is moving relative to the clock! This result is truly revolutionary. However, it has experimentally verifiable consequences, as we will see in Example 35.2.

DERIVATION 35.1 | **Time Dilation**

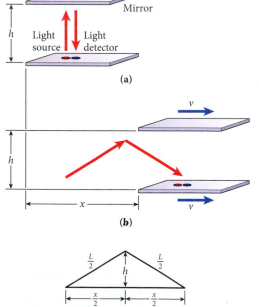

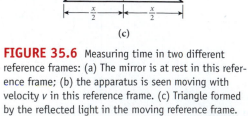

FIGURE 35.6 Measuring time in two different reference frames: (a) The mirror is at rest in this reference frame; (b) the apparatus is seen moving with velocity v in this reference frame. (c) Triangle formed by the reflected light in the moving reference frame.

How can we derive this result? We know that relativistic effects involve the speed of light, so let's construct a clock that keeps time by bouncing a vertical light beam off a mirror and detecting the reflection (Figure 35.6a).

If we know the distance h between mirror and light source, then the time for the beam to go up and down is

$$\Delta t_0 = \frac{2h}{c},$$

where the subscript "0" refers to the fact that the observer of this time interval is not moving relative to the measurement apparatus. (It takes the same time, h/c, for the light beam to go up as it does to go down.)

Now let's have an observer move to the left with a speed of v in the horizontal direction. We call this the negative x-direction, so that this observer sees the clock moving in the positive x-direction with a speed v (Figure 35.6b). You can see from Figure 35.6c that in this case the observer sees the light beam have a path length of

$$L = 2\sqrt{h^2 + (x/2)^2},$$

where x is the distance that the clock moves while the light is in the air. This result is simply a consequence of the Pythagorean Theorem. We have tacitly assumed that h is the same for observers in the two reference frames depicted in Figure 35.6. If h were not the same, then we could distinguish one reference frame from the other, in violation of Einstein's postulate 1.

This second observer would say that $x = v\Delta t$, where Δt refers now to the time interval for the light to be emitted and then detected in her or his system. We can also use the above equation relating h and Δt_0, $h = c\Delta t_0/2$, to eliminate h.

At this point the second postulate enters in its essential way. The light beams can travel only with c, the constant speed of light, which is independent of the velocity of the observer! Thus we have the relation

$$L = c\Delta t.$$

We insert this last result into the result of the Pythagorean Theorem and then substitute for h and x to get:

$$L = 2\sqrt{h^2 + (x/2)^2} = c\Delta t = 2\sqrt{(c\Delta t_0/2)^2 + (v\Delta t/2)^2} = \sqrt{(c\Delta t_0)^2 + (v\Delta t)^2}.$$

Solving this equation for Δt, we obtain

$$\Delta t = \frac{\Delta t_0}{\sqrt{1 - (v/c)^2}} = \gamma \Delta t_0.$$

In the clock's rest frame—Figure 35.6a—the time interval is Δt_0; this quantity could represent the time between two clicks of the clock. In the frame in which the clock is moving—Figure 35.6b—the time interval between these two clicks is $\Delta t = \gamma \Delta t_0 > \Delta t_0$. Thus we say that "moving clocks run slow," meaning that the time interval measured is longer when measured in any reference frame in which the clock is moving.

Because the velocity of light is independent of the speed of the observer, we had to admit that the time measured by the two observers is different. This step is remarkably bold.

Just to be clear, there is *nothing* wrong with the clocks, and there's nothing special about the moving clock. Nothing special happens to the clocks due to the high velocities. What is happening is that time itself is slowed down, according to an observer in a different reference frame. This means that everything is slowed down according to this observer, including our motion and our body clocks. We would appear to move in slow motion, and we age slower, compared to the observer in the other reference frame. It is also interesting to note that this effect is reciprocal. That is, observers in two inertial reference frames moving with respect to each other each see the other's clocks running slow! But this illustrates the central idea of special relativity—there is no absolute motion; all inertial frames are equally valid for making measurements.

Finally, a remark on notation: In the previous section we introduced the concept of proper time (see equation 35.7). Sometimes you will see the time measured in the rest frame of a clock referred to as proper time. The proper time interval used in the above Derivation 35.1 is Δt_0.

It is one thing to postulate and then mathematically prove the effect of time dilation. It is, however, altogether different to observe this effect in the laboratory or in nature. Amazingly, this has been observed, as discussed in Example 35.2.

EXAMPLE 35.2 | Muon Decay

A muon is an unstable subatomic particle with a mean lifetime τ_0 of only 2.2 μs. This lifetime can be observed easily when muons decay at rest in the lab. However, when muons are produced in flight at very high speeds, their mean lifetime becomes time dilated. In 1977, such an experiment was carried out at the European CERN particle accelerator, where the muons were produced with a speed of $v = 0.9994c \Rightarrow \beta = 0.9994$. In this case, we have for γ:

$$\gamma = \frac{1}{\sqrt{1 - \beta^2}} = \frac{1}{\sqrt{1 - 0.9994^2}} = 29.$$

Therefore, the mean lifetime, τ, of the muons is expected to be $\gamma = 29.$ longer at this speed than if they were at rest:

$$\tau = \gamma \tau_0 = 29.\,(2.2\ \mu s) = 64.\ \mu s.$$

It is quite straightforward to measure this effect by measuring the time between the production and decay of the muons. Alternatively, you could move some distance away from the production site and see if the muons still reach you. Without the effect of time dilation, during its lifetime of 2.2 μs, a muon with this speed could move a distance of only

$$x = v\tau_0 = (0.9994c)(2.2\ \mu s) = 660\ m$$

Continued—

before it decays. However, with the effect of time dilation, the travel distance becomes

$$x = v\tau = v\gamma\tau_0 = (0.9994c)\, 29.\,(2.2\ \mu s) = 19.\ km\,.$$

To do this experiment, you just need to see how far away from the muon's production site you can detect the decay products. The CERN experiment verified the relativity prediction of time dilation. This measurement shows that time dilation is indeed a real effect. Particles live longer the faster they move. (However, they live the usual time according to clocks in their rest frame.) In fact, giant particle accelerators on the drawing board right now will collide muons. For these accelerators, muons need to be transported over distances of many kilometers. The fact that these muon colliders are possible at all is due to the effect of time dilation.

35.3 In-Class Exercise

A clock on a spaceship shows that a time interval of 1.00 s has expired. If this clock is observed moving with a speed of 0.860c, what time period has expired in the frame of the stationary observer?

a) 0.860 s   d) 1.77 s

b) 1.00 s   e) 1.96 s

c) 1.25 s

Do these relativistic effects relate to subatomic particles only and have no relevance for macroscopic objects? No. In 1971, scientists flew four extremely precise atomic clocks around the Earth, once in each direction. They observed that the clocks flying eastward lost 59±10 ns, while the clocks flying west gained 275±21 ns, compared to a ground-based atomic clock. Thus, the effect of time dilation was confirmed by this experiment with macroscopic clocks. Of course, the effect was incredibly small because the speed of an airliner is small compared with the speed of light. The clocks lost 59 ns and gained 275 ns in three days (259,200 s), a few parts per trillion. The quantitative explanation of this time-dilation experiment also involves general relativity. But this is not the point we want to make here. The main point is that this experiment produced a measurable effect, and not zero, as we would have expected in the absence of time dilation. The basic fact that this experiment proves is that time can no longer be considered an absolute quantity.

Length Contraction

Relativity implies not only that time is variant and dilated as a function of speed, but also that length is not an invariant. We will see that the length L of an object moving with speed v has **length contraction** relative to its length in its own rest frame, its **proper length** L_0. We find that

$$L = \frac{L_0}{\gamma} = L_0\sqrt{1-(v/c)^2}\,. \tag{35.10}$$

DERIVATION 35.2 | Length Contraction

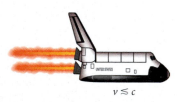

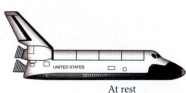

FIGURE 35.7 Illustration of length contraction (not to scale!).

Imagine that you want to measure the length of a space shuttle (Figure 35.7), that has a speed v in your reference frame and a proper length L_0. The way to measure the length without using a meter stick is simple: We can hook up a laser beam to a clock fixed in the laboratory. When the tip of the shuttle breaks the laser beam, it starts the clock, and when the end of the shuttle passes that same point and the laser beam is not blocked anymore, the clock stops. This time interval is Δt_0, or the proper time, and thus $L = v\Delta t_0$.

Inside the shuttle, using a clock fixed to the inside of the shuttle, the measured time during which the laser beam is blocked by the spaceship is $\Delta t = \gamma\Delta t_0$ because of time dilation. Thus, an astronaut inside the shuttle observes that a moving clock with speed v emitted light when passing the spacecraft tip and once again (Δt later) when passing the tail and deduces that $L_0 = v\Delta t = v\gamma\Delta t_0$. Thus we have

$$\frac{L}{L_0} = \frac{v\Delta t_0}{v\gamma\Delta t_0} \quad or \quad L = \frac{L_0}{\gamma}\,.$$

Note that for this thought experiment, it is essential that the length measured is *along the direction of motion*. All lengths perpendicular to the direction of motion remain the same (see Figure 35.7).

As you can see from Derivations 35.1 and 35.2, the phenomena of time dilation and length contraction are quite intimately related. The speed of light being constant for all observers implies time dilation, which has been experimentally confirmed many times. Time dilation, in turn, implies length contraction.

The essential fact to remember from our deliberations on length contraction is that moving objects are shorter. They don't just appear shorter—they *are* shorter as measured by an observer in the frame in which the subject is moving. This contraction is another mind-bending consequence of the postulates of special relativity.

EXAMPLE 35.3 | Length Contraction of a NASCAR Race Car

You see a NASCAR race car (Figure 35.8) go by at a constant speed of $v = 89.4$ m/s (200 mph). When stopped in the pits, the race car has a length of 5.232 m.

PROBLEM
What is the change in length of the NASCAR race car from your reference frame in the grandstands? Assume the car is moving perpendicular to your line of sight.

SOLUTION
The length of the race car will be contracted because of its motion. The proper length of the race car is $L_0 = 5.232$ m. The length in our reference frame is given by equation 35.10:

$$L = \frac{L_0}{\gamma} = L_0\sqrt{1 - (v/c)^2} \approx L_0\left(1 - \frac{1}{2}\left(\frac{v}{c}\right)^2\right) = L_0 - \Delta L,$$

where

$$\Delta L = L_0 \frac{1}{2}\left(\frac{v}{c}\right)^2$$

is the change in length of the race car. Here we have applied a series expansion $(1 - x^2)^{1/2} = 1 - \frac{1}{2}x^2 + \cdots$ as we did in equation 35.4. The car's speed is small compared with the speed of light, so $v/c \ll 1$ and our expansion is well justified. Thus, the race car appears to be shorter by

$$\Delta L = L_0 \frac{1}{2}\left(\frac{v}{c}\right)^2 = \frac{5.232 \text{ m}}{2}\left(\frac{89.4 \text{ m/s}}{3.00 \cdot 10^8 \text{ m/s}}\right)^2 = 2.32 \cdot 10^{-13} \text{ m}.$$

The car's change in length is smaller than the diameter of a typical atom. So the length contraction of objects at everyday speeds is not easy to observe.

35.4 In-Class Exercise

State whether each of the following statements is true or false.

a) For a moving object, its length along the direction of motion is shorter than when it is at rest.

b) When you are stationary, a clock moving past you at a significant fraction of the speed of light seems to run faster than the watch on your wrist.

c) When you are moving with a speed that is a significant fraction of the speed of light, and you pass by a stationary observer, you observe that your watch seems to be running faster than the watch of the stationary observer.

FIGURE 35.8 A NASCAR race car.

Twin Paradox

We have seen that a time interval (say, between clock ticks) depends on the speed of the object (say, a clock) in the frame of an observer, $\Delta t = \gamma \Delta t_0$. Let's perform a little thought experiment:

Astronaut Alice has a twin brother, Bob. At the age of 20, Alice boards a spaceship that flies to a space station 3.25 light-years away and then returns. The spaceship is a good one and can fly with a speed of 65.0% of the speed of light, resulting in a gamma factor of $\gamma = 1.32$. The total distance traveled by Alice is 2(3.25 light-years) = 6.50 light-years as seen by Bob.

In Alice's rest frame, she travels a distance of $d = 6.50$ light-years/$\gamma = 4.92$ light-years, because the distance between Earth and the space station is length-contracted in her reference frame. Thus, the time it takes Alice to complete the trip is

$$t = d/v = (4.92c \cdot \text{years})/0.650c = 7.57 \text{ years}.$$

While the entire trip back and forth takes Alice 7.6 years in Alice's reference frame, time dilation forces $7.6\gamma = 10.0$ years to pass in Bob's reference frame. Therefore, when Alice steps out of the spaceship after her trip, she will be 27.6 years old, whereas Bob is 30.0 years old.

Now we can also put ourselves into the reference frame of Alice: In Alice's frame she was at rest and Bob was moving at 65.0% of the speed of light. Therefore, Alice should have aged 1.32 times more than Bob aged. Because Alice knows that she has aged 7.57 years, she might expect her brother Bob to be only (20 + 7.57/1.32 = 25.8) years old when they meet again. Both siblings cannot each be younger than the other. This apparent inconsistency is called the twin paradox. Which of these two views is right?

The apparent paradox is resolved when we realize that although Bob remains in an inertial reference frame at rest on the Earth for the duration, astronaut Alice lives in two different inertial frames during her round trip. During the outbound leg, she is moving away from Earth and toward the distant space station. When she reaches the space station, she turns around and travels back from the space station at a constant speed to Earth. Thus, the symmetry is broken between the two twins.

We can analyze the path of the two twins in space-time by using our techniques of light cones and world lines, plotting time in an inertial rest frame versus the position of both twins in one direction, the x-direction. We analyze the problem from both the point of view of Bob and the point of view of Alice. We start by analyzing the trip in the rest frame of stay-at-home twin Bob, as shown in Figure 35.9. Here we scale the axes so that the units for both are in years.

In Figure 35.9, Bob's speed is always zero and he remains at $x = 0$. A red, vertical line represents Bob's trajectory. In contrast, Alice is moving with a speed of 65.0% of the speed of light ($v = 0.650c$) away from Earth. A blue line labeled $v = 0.650c$ depicts Alice's outbound trajectory. We define the positive x-direction as pointing from the Earth to the distant space station. Each twin wants to keep in touch with the other. Thus, each twin sends an electronic birthday card to the other twin on their birthday in their reference frame. These messages travel with the speed of light. Bob sends his electronic message directly toward the space station and Alice sends her electronic greeting back directly toward Earth. Bob's messages are shown as red arrows pointing up and to the right. Alice's messages are shown as blue arrows pointing up and to the left. When the message arrows cross the trajectory of each of the twins, the respective twin receives and enjoys the electronic birthday card.

After 5 years pass in Bob's frame and 3.79 years pass in Alice's frame, Alice reaches the space station and turns for home. Bob is getting a little worried by now because he has received only two birthday cards in five years. Alice is not feeling much better, since she has received only one message in 3.79 years. After Alice turns around, she receives eight electronic birthday cards in the next 3.79 years. Bob receives five more greetings in the remaining five years. When Alice arrives back home on Earth, she gets a firsthand 30th birthday greeting from Bob, but Alice is not ready to celebrate her 28th birthday yet. Alice's age is 27.6 years. Alice received a total of ten birthday greetings while Bob received only seven.

Now let's analyze the same trip from the inertial rest frame corresponding to Alice's outward-bound leg (Figure 35.10) to show that we can get the same answer from Alice's point of view. In this reference frame, Bob and Earth are both traveling in the negative x-direction with a velocity $v = -0.650c$. During the outbound portion of the trip, Alice's velocity is zero in this frame. The space station is traveling toward Alice with a speed of 65.0% of the speed of light, so the distance of 3.25 light-years is covered in 3.79 years due to length contraction. Note that this part of the two diagrams in Figure 35.9 and Figure 35.10 is completely symmetric.

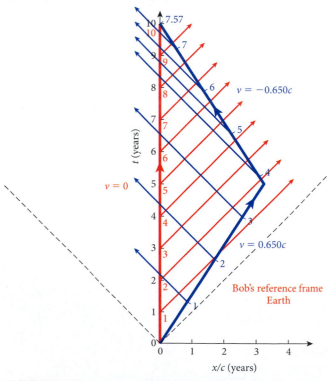

FIGURE 35.9 Plot showing the velocity of the two twins in Bob's reference frame, which is at rest on Earth. The thick, vertical red line represents Bob's trajectory. The two thick blue lines depict Alice's trajectory. Thin red lines labeled by red numbers (corresponding to the years since Alice left) represent Bob's birthday messages. Thin blue lines labeled by blue numbers (also corresponding to the years since Alice left) depict Alice's birthday messages. The dashed lines show the light cone at $t = 0$.

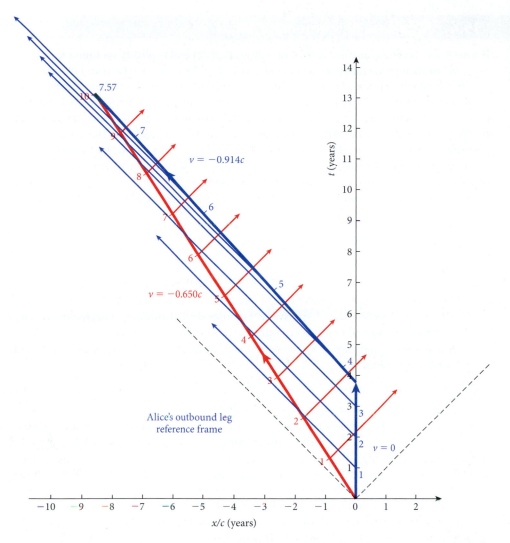

FIGURE 35.10 Plot showing the velocity of the two twins in the reference frame of Alice's outbound leg. The thick red line represents Bob's trajectory. The two thick blue lines depict Alice's trajectory. Thin red lines labeled by red numbers corresponding to the year represent Bob's birthday messages. Thin blue lines labeled by blue numbers corresponding to the year depict Alice's birthday messages. The dashed lines represent the light cone at $t = 0$.

When the space station reaches Alice, the symmetry in the two representations of Figures 35.9 and 35.10 is broken. Alice begins to travel with a speed fast enough to catch up with Earth in the negative x-direction. To establish a relative speed of 65.0% of the speed of light with respect to Earth, Alice must travel with a speed of 91.4% of the speed of light. (This relativistic addition of velocities is discussed in Section 35.6.)

Again Bob receives two birthday greetings in the first five years in his reference frame, while Alice again receives one birthday greeting before she starts moving toward Earth. As Alice streaks for Earth at a speed of 0.914c, she receives eight birthday greetings, while Bob again receives five. When the twins are reunited on Earth, Bob is 30.0 years old and Alice is 27.6 years old. This result using Alice's outbound reference frame is the same as the one we obtained using Bob's reference frame. Thus, the twin paradox is resolved.

Note that we analyzed the twin paradox in terms of special relativity only. You might worry about the parts of Alice's journey that involved acceleration. Alice had to accelerate to 65.0% of the speed of light to begin her journey to the space station, and then she had to slow down, stop, and reaccelerate back up to a speed of 65.0% of the speed of light back toward Earth. Even at a constant acceleration of three times the acceleration of gravity, it would take almost three months to reach a speed of 65.0% of the speed of light from rest and the same amount of time to stop, starting at a speed of 65.0% of the speed of light. However, we could postulate various scenarios to remove these objections or at least minimize their importance. For example, we could simply make the trip longer, making the acceleration phase negligible. Acceleration is necessary to explain the twin paradox because Alice must reverse her course to return to Earth, changing her inertial reference frame. However, the effects of general relativity are not needed to explain the twin paradox.

35.5 In-Class Exercise

The nearest star to us other than the Sun is Proxima Centauri, which is 4.22 light-years away. Suppose we had a spaceship that could travel at a speed of 90.0% of the speed of light. If you were in the spaceship, how long would it take for you to travel from the Sun to Proxima Centauri, from your point of view?

a) 2.04 years   d) 3.80 years

b) 2.29 years   e) 4.22 years

c) 3.42 years

35.4   Relativistic Frequency Shift

If time is not invariant any more, then quantities that depend explicitly on time may also be expected to change as a function of an object's velocity. The frequency f is one such quantity.

In Chapter 16 on sound waves, we discussed the Doppler effect—the change in sound's frequency due to the relative motion between observer, source, and medium. We have seen, however, that light does not need a medium in which to propagate. Therefore, the **relativistic frequency shift** is qualitatively new and only remotely connected to the classical Doppler effect. (However, it is often informally referred to as *the relativistic Doppler effect* or the *Doppler effect for electromagnetic radiation*.) If v is the relative velocity between source and observer, and the relative motion occurs in a radial direction (either directly toward each other or directly away from each other), then the formula for the observed frequency, f, of light with given frequency f_0 is

$$f = f_0 \sqrt{\frac{c \mp v}{c \pm v}}, \tag{35.11}$$

where the upper signs (– in the numerator and + in the denominator) are used for motion away from each other and the lower signs (+ in the numerator and – in the denominator) for motion toward each other. (A transverse Doppler shift also occurs, purely due to time-dilation effects, but we will not concern ourselves with this effect here.)

Because the relationship $c = \lambda f$ between speed, wavelength, and frequency is still valid, we get for the wavelength:

$$\lambda = \lambda_0 \sqrt{\frac{c \pm v}{c \mp v}}. \tag{35.12}$$

When looking out through our telescopes into the universe, we find that just about all galaxies send light toward us that is **red-shifted**, meaning $\lambda > \lambda_0$ (red is the visible color with the longest wavelength). This means that all galaxies have $v > 0$; that is, they are moving away from us. (If an object is moving toward us, then $\lambda < \lambda_0$ and the light is said to be **blue-shifted**.) In addition, astoundingly, the farther away galaxies are from us, the more they are red-shifted. This observation is clear evidence for an ever-expanding universe.

Often astronomers quote a so-called **red-shift parameter.** The red-shift parameter is often referred to as just the *red-shift*. This quantity is defined as the ratio of the wavelength shift of light divided by the wavelength of that light as observed when the source is at rest:

$$z = \frac{\Delta \lambda}{\lambda_0} = \frac{\lambda - \lambda_0}{\lambda_0} = \sqrt{\frac{c \pm v}{c \mp v}} - 1. \tag{35.13}$$

SOLVED PROBLEM 35.1   Galactic Red-Shift

During a deep-space survey, a galaxy is observed with a red-shift of $z = 0.450$.

PROBLEM
With what speed is that galaxy moving away from us?

SOLUTION

THINK
The red-shift is a function of the speed v with which the galaxy moves away from us. We can solve equation 35.13 for the speed in terms of the red-shift.

SKETCH
Figure 35.11 shows a sketch of the galaxy moving away from Earth.

FIGURE 35.11 A galaxy moving away from Earth (not to scale).

RESEARCH

We start with equation 35.13, relating the red-shift z to the speed v with which the galaxy is moving away from us,

$$z = \sqrt{\frac{c+v}{c-v}} - 1.$$

We can rearrange this equation to get

$$(z+1)^2 = \frac{c+v}{c-v}$$

$$c(z+1)^2 - v(z+1)^2 = c + v.$$

Gathering the multiplicative factors of v on one side and factors of c on the other side, we get

$$v + v(z+1)^2 = c(z+1)^2 - c.$$

SIMPLIFY

We can now write an expression for the velocity of the galaxy moving away from us in units of the speed of light,

$$\frac{v}{c} = \frac{(z+1)^2 - 1}{(z+1)^2 + 1}.$$

CALCULATE

Putting in the numerical values gives us

$$\frac{v}{c} = \frac{(0.450+1)^2 - 1}{(0.450+1)^2 + 1} = 0.355359.$$

ROUND

We report our result to three significant figures,

$$\frac{v}{c} = 0.355.$$

DOUBLE-CHECK

A galaxy moving away from us with a velocity of 35.5% of the speed of light seems reasonable, in that our answer at least does not exceed the speed of light.

35.5 Lorentz Transformation

In studying time dilation and length contraction, we have seen how time and space intervals can be transformed from one inertial reference frame to another. These transformations of time and space can be combined in going from one reference frame into another that moves with speed v relative to the first one.

To be specific, suppose we have two reference frames F (with coordinate system x, y, z and time t) and F' (with coordinate system x', y', z' and time t') that are at the same point $x = x'$, $y = y'$, $z = z'$ (this arrangement can always be accomplished by a simple shift in the location of the origin of the coordinate systems) at the same time $t = t' = 0$, moving with velocity v relative to each other along the x-axis. We want to know how to describe an event in frame F' (which is some occurrence, like a firecracker explosion) that has coordinates x, y, z and is happening at time t as observed in frame F, using coordinates x', y', z' and time t' (Figure 35.12).

Classically, this transformation is given by

$$\begin{aligned} x' &= x - vt \\ y' &= y \\ z' &= z \\ t' &= t. \end{aligned} \tag{35.14}$$

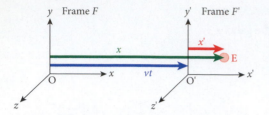

FIGURE 35.12 The Lorentz transformation connects two reference frames moving at a constant velocity v relative to one another.

This transformation is known as the **Galilean transformation,** which we encountered in Chapter 3 on motion in two and three dimensions. These equations are correct as long as v is small compared to c. Until we had talked about time dilation, the last one of these equations seemed utterly trivial. Now this is not the case any more.

The transformation that is valid for all velocities of magnitude v is called the **Lorentz transformation,** first given by the Dutch physicist Hendrik Lorentz (1853–1928):

$$x' = \gamma(x - vt)$$
$$y' = y$$
$$z' = z \tag{35.15}$$
$$t' = \gamma(t - vx/c^2).$$

We can also construct the inverse transformation, obtaining

$$x = \gamma(x' + vt')$$
$$y = y'$$
$$z = z' \tag{35.16}$$
$$t = \gamma(t' + vx'/c^2).$$

In the limit of small speeds ($\gamma = 1$, $\beta = v/c = 0$), the Galilean transformation follows from the Lorentz transformation as a special case, as can be easily seen. However, for speeds that are not small compared to the speed of light, the Lorentz transformation includes the effects of both time dilation and length contraction.

DERIVATION **35.3**   Lorentz Transformation

There is an ongoing discussion on what constitutes the "simplest" derivation of the Lorentz transformation. Here we follow the paper by J.-M. Lévy, who published this derivation in the *American Journal of Physics,* Vol. 75, No. 7 (2007), pp. 615–618. In this derivation we make use of the fact that we have already derived the effect of length contraction (Derivation 35.2).

Let's look at Figure 35.12. Assume that $y = y' = 0$, $z = z' = 0$ and describe the event E in Frame F (with coordinate system x, y, z and time t). The green arrow of length x marks the position vector of the event E relative to the origin O of frame F. We can now use simple vector addition and find that the green arrow $\overline{OE}$ is the vector sum

$$\overline{OE} = \overline{OO'} + \overline{O'E} \tag{i}$$

$\overline{OO'}$ is the displacement vector of the origin of the frame F' relative to the origin of F, and $\overline{O'E}$ is the displacement vector from the origin of F' to event E. This vector relation equation (i) is valid independent of the frame from which we are evaluating our results, and now all we need to do is to express equation (i) in both frames.

First, let's evaluate the vector addition equation (i) in frame F where the length of the vector $\overline{OE}$ is simply x and the length of the vector $\overline{OO'}$ is vt. Because x' is measured in frame F', expressing it in frame F results in length contraction; that is, in frame F the length of the vector $\overline{O'E}$ can be expressed as x'/γ. This means that in frame F we obtain for the vector addition equation (i) the expression

$$x = vt + x'/\gamma. \tag{ii}$$

Now let's perform the same task in frame F' (with coordinate system x', y', z' and time t'). In this frame, we find that the length of $\overline{O'E}$ is x' and the length of $\overline{OO'}$ is vt'. But now the frame F is moving relative to the rest frame F', and therefore in frame F' x is length-contracted, so that the length of $\overline{OE}$ is x/γ in frame F'. Consequently, in F' we evaluate the vector addition equation (i) as

$$x/\gamma = vt' + x'. \tag{iii}$$

Now we are just about done, and all that is left are some algebraic manipulations. First, we can subtract vt from both sides of equation (ii) and then multiply both sides with γ, resulting in

$$\gamma(x - vt) = x', \tag{iv}$$

which is the first equation of the Lorentz transformation (equation 35.15). Multiplying both sides of equation (iii) with γ gives us the first equation of the inverse Lorentz transformation (equation 35.16),

$$x = \gamma(x' + vt'). \tag{v}$$

If we then insert the expression for x' from equation (iv) into (v), we find

$$x = \gamma(\gamma(x - vt) + vt') = \gamma^2(x - vt) + \gamma vt' \Rightarrow$$

$$x / \gamma^2 = x - vt + vt'/\gamma \Rightarrow$$

$$-(1 - \gamma^{-2})x = -vt + vt'/\gamma.$$

Since $\gamma = 1/\sqrt{1 - v^2/c^2}$, it follows that $\gamma^2 = 1/(1 - v^2/c^2)$ and therefore $1 - \gamma^{-2} = v^2/c^2$. This leads to

$$-xv^2/c^2 = -vt + vt'/\gamma$$

$$-xv/c^2 = -t + t'/\gamma$$

$$\gamma(t - xv/c^2) = t',$$

which we recognize as the last equation in the Lorentz transformation (equation 35.15).

35.3 Self-Test Opportunity

By inserting x from equation (v) in Derivation 35.3 into equation (iv), one is able to derive the equation $t = \gamma(t' + vx'/c^2)$. Can you show this?

Invariants

Under a Lorentz transformation between inertial frames, the coordinate along the direction of relative motion of the two inertial frames (chosen to be the x-coordinate here) changes, as does the time. The two coordinates perpendicular to the velocity vector of the relative motion (chosen to be the y- and z-coordinates here) remain the same in both frames. They are **invariants** under the Lorentz transformation.

The question is: What other, if any, kinematical invariants can one construct? At this point we do not want to be exhaustive but only to focus on one, which we had introduced earlier. This invariant is the space-time interval $s^2 = c^2 \Delta t^2 - \Delta r^2$ of equation 35.5.

35.6 In-Class Exercise

Which of the following are invariants under the classical Galilean transformation of equation 35.14? (Mark all that apply!)

a) x
b) y
c) z
d) t

DERIVATION 35.4 | Lorentz-Invariance of Space-Time Intervals

We need to show that the space-time interval is the same in different inertial frames. That is, we need to show that $s^2 = s'^2$. Since we have an expression for s^2 in terms of the coordinates, and since we know how the coordinates transform under a Lorentz transformation, we can construct our proof:

$$s'^2 = c^2 \Delta t'^2 - \Delta r'^2$$

$$= c^2 \Delta t'^2 - \Delta x'^2 - \Delta y'^2 - \Delta z'^2$$

$$= c^2 \gamma^2 (\Delta t - v \Delta x/c^2)^2 - \gamma^2 (\Delta x - v \Delta t)^2 - \Delta y^2 - \Delta z^2$$

$$= c^2 \gamma^2 (\Delta t^2 - 2v \Delta x \Delta t/c^2 + v^2 \Delta x^2/c^4) - \gamma^2 (\Delta x^2 - 2v \Delta x \Delta t + v^2 \Delta t^2) - \Delta y^2 - \Delta z^2.$$

In the second step we simply replaced $\Delta r'^2$ by the sum of the squares of the components. In the third step we applied the Lorentz transformation to each coordinate interval, and in the fourth step we evaluated the squared terms. Now we can multiply out the brackets and cancel out terms:

$$s'^2 = c^2 \gamma^2 \Delta t^2 \underline{-2v\gamma^2 \Delta x \Delta t} + v^2 \gamma^2 \Delta x^2/c^2 - \gamma^2 \Delta x^2 \underline{+2v\gamma^2 \Delta x \Delta t} - v^2 \gamma^2 \Delta t^2 - \Delta y^2 - \Delta z^2$$

$$= c^2 \gamma^2 \Delta t^2 + v^2 \gamma^2 \Delta x^2/c^2 - \gamma^2 \Delta x^2 - v^2 \gamma^2 \Delta t^2 - \Delta y^2 - \Delta z^2$$

$$= c^2 \Delta t^2 (\gamma^2 - v^2 \gamma^2/c^2) - \Delta x^2 (\gamma^2 - v^2 \gamma^2/c^2) - \Delta y^2 - \Delta z^2.$$

Continued—

Here we have underlined the terms that cancel each other in the top line. Now we need to evaluate

$$\gamma^2 - v^2\gamma^2/c^2 = \frac{1}{1-v^2/c^2} - \frac{v^2/c^2}{1-v^2/c^2} = \frac{1-v^2/c^2}{1-v^2/c^2} = 1.$$

Using this identity, we find for the space-time interval:

$$s'^2 = c^2\Delta t^2 \underbrace{(\gamma^2 - v^2\gamma^2/c^2)}_{1} - \Delta x^2 \underbrace{(\gamma^2 - v^2\gamma^2/c^2)}_{1} - \Delta y^2 - \Delta z^2$$

$$= c^2\Delta t^2 - \Delta x^2 - \Delta y^2 - \Delta z^2$$

$$= c^2\Delta t^2 - \Delta r^2$$

$$= s^2.$$

We find indeed that $s^2 = s'^2$, meaning that the space-time interval is invariant under Lorentz transformation.

Other invariants will be introduced in subsequent sections.

35.6 Relativistic Velocity Transformation

Let's return to the problem we touched on in Section 35.2. Suppose you are flying in a rocket through space with a speed of $c/2$ directly toward Earth, and you shine a laser in the forward direction. Naively, we might expect that the light of the laser would have a speed of $c + (c/2) = 1.5c$ when observed on Earth. This observation would follow from the velocity-addition formula in Chapter 3, $\vec{v}_{pg} = \vec{v}_{ps} + \vec{v}_{sg}$. Here the subscripts p, g, and s refer to the projectile (light, in this case), the ground, and the spaceship. However, according to Einstein, the light of the laser has a speed c in either the rocket or the Earth reference frame.

Because the result using the classical velocity-addition formula contradicts the postulate that no speed is greater than c, we clearly need to come up with a new formula to add velocities correctly. We will restrict the use of the **relativistic velocity addition** formula to one-dimensional motion of the two frames relative to each other in the positive or negative x-directions. If the velocity of an object has a value of $\vec{u} = (u_x, u_y, u_z)$ in frame F, and if frame F' moves with velocity $\vec{v} = (v, 0, 0)$ relative to frame F, then the velocity $\vec{u}' = (u'_x, u'_y, u'_z)$ of that object in frame F' has the value

$$u'_x = \frac{u_x - v}{1 - \dfrac{vu_x}{c^2}}$$

$$u'_y = \frac{u_y/\gamma}{1 - \dfrac{vu_x}{c^2}} \tag{35.17}$$

$$u'_z = \frac{u_z/\gamma}{1 - \dfrac{vu_x}{c^2}}.$$

The inverse transformation from frame F' to frame F is given by

$$u_x = \frac{u'_x + v}{1 + \dfrac{vu'_x}{c^2}}$$

$$u_y = \frac{u'_y/\gamma}{1 + \dfrac{vu'_x}{c^2}} \tag{35.18}$$

$$u_z = \frac{u'_z/\gamma}{1 + \dfrac{vu'_x}{c^2}}.$$

We have restricted all motion to the positive and negative x-directions, just as was done to arrive at the formula for the Lorentz transformation. However, $\vec{u}$, $\vec{u}'$, and $\vec{v}$ are all vectors and can have positive and negative values, depending on the particular situation these transformation equations are applied to. In addition, these formulas can also be used to add and subtract two velocities appropriately.

DERIVATION 35.5 / Velocity Transformation

To see how a velocity transforms from one frame to another, we can start with the Lorentz transformation, which tells us how spatial and time coordinates transform. Then we can take the time derivative of the x-coordinate, which gives us the velocity component in the direction of the relative motion of the two frames with respect to each other (we call this velocity u, because we have already used the symbol v for the velocity of the frames relative to each other):

$$u'_x = \frac{dx'}{dt'} = \frac{\gamma(dx - vdt)}{\gamma(dt - vdx/c^2)}.$$

All we have done up to this point is to express the differentials dx' and dt' in the frame F, with the aid of the Lorentz transformation of equations 35.15. Now we cancel out the factor of γ in the numerator and denominator and find

$$u'_x = \frac{dx - vdt}{dt - vdx/c^2} = \frac{\dfrac{dx}{dt} - v}{1 - \dfrac{v}{c^2}\dfrac{dx}{dt}}.$$

Substituting $u_x = dx/dt$, we then get the velocity transformation for the longitudinal component (the component along the direction of the relative motion between the two frames) as

$$u'_x = \frac{u_x - v}{1 - \dfrac{vu_x}{c^2}}.$$

Now we can conduct a similar exercise for the transverse components, that is, the components of the velocity vector perpendicular to the direction of relative motion between the two frames. We show this explicitly for the y-direction, but the z-direction follows the same arguments. Again we take the derivative and Lorentz-transform the differentials, recognizing that $dy' = dy$:

$$u'_y = \frac{dy'}{dt'} = \frac{dy}{\gamma(dt - vdx/c^2)}.$$

Dividing the numerator and denominator by dt then leads to the desired result

$$u'_y = \frac{\dfrac{dy}{dt}}{\gamma\left(1 - \dfrac{v}{c^2}\dfrac{dx}{dt}\right)} = \frac{u_y/\gamma}{1 - \dfrac{vu_x}{c^2}}.$$

35.4 Self-Test Opportunity

Start with $u_x = dx/dt$, use the inverse Lorentz transformation, and show that you arrive at

$$u_x = \frac{u'_x + v}{1 + \dfrac{vu'_x}{c^2}}.$$

Note that in the limit the speeds are small compared to the speed of light, $|v| \ll c$ and $|u| \ll c$, the term vu_x/c^2 in the denominator of our velocity transformation equation becomes small compared to 1, and γ tends to 1 so that equations 35.17 and 35.18 approach the classical limit of $u'_x = u_x - v$ and $u_x = u'_x + v$, $u'_y = u_y$ and $u'_z = u_z$.

Back to our example: Now you can see that $u' = c$ "plus" $v = c/2$ does not yield $1.5c$, but instead

$$u = \frac{u' + c/2}{1 + \dfrac{u'c/2}{c^2}} = \frac{c + c/2}{1 + \dfrac{cc/2}{c^2}} = c.$$

35.7 In-Class Exercise

We have just seen that the real sum of two positive velocities consistent with relativity is less than that resulting from classical velocity addition. What about velocity differences? Assume that two velocity vectors $\vec{v}_1$ and $\vec{v}_2$ both point in the positive x-direction. Call the relative velocity between the two $\vec{v}_r$. Which of the following is correct?

a) $\left.|\vec{v}_r|\right|_{\text{classical}} < \left.|\vec{v}_r|\right|_{\text{relativistic}}$

b) $\left.|\vec{v}_r|\right|_{\text{classical}} = \left.|\vec{v}_r|\right|_{\text{relativistic}}$

c) $\left.|\vec{v}_r|\right|_{\text{classical}} > \left.|\vec{v}_r|\right|_{\text{relativistic}}$

d) Can be either (a) or (c), depending on which of the two velocities $\vec{v}_1$ and $\vec{v}_2$ is larger.

FIGURE 35.13 Electron moving past an alpha particle.

Therefore this analysis works out just right: The speed of light is the same in every reference frame! Thus, equations 35.17 and 35.18 have the proper classical limit and at the same time satisfy the postulates of special relativity.

SOLVED PROBLEM 35.2 | Particles in an Accelerator

Suppose you have an electron and an alpha particle (nucleus of a helium atom) moving through a piece of beam pipe inside a particle accelerator. The particles are traveling in the same direction. The speed of the electron is 0.830c and the speed of the alpha particle is 0.750c, both measured by a stationary observer in the lab.

PROBLEM

What is the speed of the alpha particle as observed from the electron, in units of the speed of light?

SOLUTION

THINK

The electron is overtaking the alpha particle, because it is faster, as measured from the stationary observer in the lab. One could be tempted to make constructions of the kind $\vec{v}_1 - \vec{v}_2$. However, since we now have the relativistic velocity transformation, it is far easier to cast this problem in this way. Assume that all given velocities occur in the positive x-direction. All we need to do is to be careful which velocities we identify with v, u_x, and u'_x. We can then use equation 35.17 or equation 35.18 to find our answer.

SKETCH

Figure 35.13 is a sketch of the electron passing the alpha particle.

RESEARCH

We have two reference frames, our original frame F with origin on Earth, and our frame F' with origin on the electron, moving with constant velocity v relative to the lab.

This problem of finding the velocity u'_x of the alpha particle in the frame of the electron involves the relativistic transformation of velocities, where the velocity u_x of the alpha particle in lab's frame is already known. We can then use equation 35.17 to accomplish our goal of finding

$$u'_x = \frac{u_x - v}{1 - \frac{vu_x}{c^2}}.$$

SIMPLIFY

There is not much to simplify in this case. Most of the work here was to define which quantity is measured in which frame, and what is the correct relative velocity of the frames.

CALCULATE

Putting in the numerical values gives us

$$u'_x = \frac{u_x - v}{1 - \frac{vu_x}{c^2}} = \frac{0.750c - 0.830c}{1 - \frac{(0.830c)(0.750c)}{c^2}} = -0.2119205c.$$

ROUND

We report the result to three significant figures,

$$u'_x = -0.212c.$$

DOUBLE-CHECK

The negative sign for the velocity of the alpha particle as seen by the electron makes sense because the alpha particle would be seen to be approaching the electron in the negative x-direction. If we calculate the relative velocity between the electron and the alpha particle nonrelativistically, we obtain

$$v_{\text{rel}} = v_{\text{alpha}} - v_{\text{electron}} = 0.750c - 0.830c = -0.080c.$$

Looking at our solution, we can see that the nonrelativistic velocity difference is modified by a factor of $|1/(1 - u_x v/c^2)|$. As the velocities of the electron and the alpha particle become a significant fraction of the speed of light, this factor will become larger than 1, and the magnitude of the relativistic velocity difference will be larger than the magnitude of the nonrelativistic difference. Thus, our result seems reasonable.

35.7 Relativistic Momentum and Energy

Length, time, and velocity, are not the only concepts that need revision within the theory of special relativity. Energy and momentum need revising as well. Again we are guided by the principle that physical laws should be invariant under the transformation from one frame to another, and that for speeds small compared to the speed of light we should again recover the classical relationships derived earlier in the mechanics chapters.

Momentum

Classically, momentum is defined as the product of velocity and mass: $\vec{p} = m\vec{v}$. Because we have now seen that an object's speed cannot exceed c, this result implies that either the definition of momentum has to be changed or a maximum momentum is possible for a given particle. It turns out that the first possibility is the correct one.

The definition of momentum that is consistent with the theory of special relativity is

$$\vec{p} = \gamma \, m\vec{u}, \tag{35.19}$$

where m is the **rest mass** of the particle (mass as measured in a frame where the particle is at rest), $\vec{u}$ is the velocity of the particle in some frame, $\gamma = \dfrac{1}{\sqrt{1 - u^2/c^2}}$, and $\vec{p}$ is the momentum of the particle in the frame. Equation 35.19 is valid only for particles having mass $m > 0$. We discuss massless particles later in this section. (Unfortunately, many older textbooks use the notion of *relativistic mass*, γm. This notion is now almost universally rejected. Mass is now considered invariant—that is, independent of the speed at which an object moves in a particular reference frame.)

In Figure 35.14, the two formulas for the momentum are compared. Blue shows the correct formula, and red the classical approximation. The velocity is shown in units of c, and the momentum in units of mc. Up to speeds of approximately $c/2$, the two formulas give pretty much the same result, but then the correct relativistic momentum goes to infinity as v approaches c.

Newton's Second Law does not have to be modified in the relativistic limit. It remains

$$\vec{F}_{\text{net}} = \frac{d}{dt} \, \vec{p}, \tag{35.20}$$

where $\vec{p}$ is the relativistic momentum.

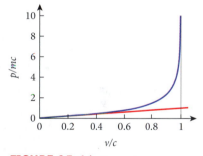

FIGURE 35.14 Momentum as a function of speed. Blue line: exact formula; red line: nonrelativistic approximation.

Energy

If momentum needs a change of definition, energy can also be expected to be in need of revision.

In our nonrelativistic considerations, we found that the total energy of a particle in the absence of an external potential is just its kinetic energy, $K = \frac{1}{2}mv^2$. In the relativistic case, we find that we have to consider the contribution of the mass to the energy of a particle. Einstein found that the energy of a particle with mass m at rest is:

$$E_0 = mc^2. \tag{35.21}$$

(Arguably the most famous formula in all of science!)

If the particle is in motion, then the energy increases by the same factor γ by which time is dilated for a moving particle. The general case for the energy is then

$$E = \gamma E_0 = \gamma mc^2. \tag{35.22}$$

The correct formula for the kinetic energy is obtained by subtracting the "rest energy" from the total energy:

$$K = E - E_0 = (\gamma - 1)E_0 = (\gamma - 1)mc^2. \tag{35.23}$$

The classical formula for kinetic energy worked for us at speeds small compared to the speed of light. So how can we recover the classical approximation $K = \frac{1}{2}mv^2$ from the general relativistic result?

This result is actually quite straightforward, because for small speeds we had already found that $\gamma = 1 + \frac{1}{2}\beta^2 = 1 + \frac{1}{2}v^2/c^2$. We can insert this formula into the kinetic energy formula and get

$$K_{\text{small } v} = (\gamma - 1)mc^2 \approx (1 + \frac{1}{2}v^2/c^2 - 1)mc^2 = \frac{1}{2}mv^2. \tag{35.24}$$

DERIVATION 35.6 | Energy

Let's go back to the work-kinetic energy theorem and calculate the consequences of using the Lorentz transformation. For simplicity we use a one-dimensional case with motion along the x-direction only. By definition, we introduced work as the integral of the force,

$$W = \int_{x_0}^{x} F dx = \int_{x_0}^{x} \frac{dp}{dt} dx.$$

Here we want to assume that x_0 is the coordinate at $t = 0$, and that the particle has 0 speed and therefore 0 kinetic energy at $t = 0$. We can calculate dp/dt by using equation 35.19 and taking the time derivative,

$$\frac{dp}{dt} = \frac{d}{dt}(\gamma mv) = \frac{d}{dt}\left(\frac{mv}{\sqrt{1-v^2/c^2}}\right) = \frac{m}{\sqrt{1-v^2/c^2}}\frac{dv}{dt} + (-\frac{1}{2})(-2v/c^2)\frac{mv}{(1-v^2/c^2)^{3/2}}\frac{dv}{dt}$$

$$= \left(\frac{m(1-v^2/c^2)}{(1-v^2/c^2)^{3/2}} + \frac{mv^2/c^2}{(1-v^2/c^2)^{3/2}}\right)\frac{dv}{dt} = \frac{m}{(1-v^2/c^2)^{3/2}}\frac{dv}{dt},$$

where we have used the product and chain rules of differentiation. We can insert this result into the above integral for the work. By using the substitution of variables $dx = vdt$, we then find

$$W = \int_0^t \frac{m}{(1-v^2/c^2)^{3/2}}\frac{dv}{dt}vdt = \int_0^v \frac{mvdv}{(1-v^2/c^2)^{3/2}}.$$

Evaluating the integral we arrive at

$$W = \frac{mc^2}{\sqrt{1-v^2/c^2}}\Bigg|_0^v = \frac{mc^2}{\sqrt{1-v^2/c^2}} - mc^2 = (\gamma - 1)mc^2.$$

Since the work-kinetic energy theorem (see Chapter 5) states that $W = \Delta K$, and since in this case we started with kinetic energy 0, this leads us to:

$$K = (\gamma - 1)mc^2,$$

which is the result of equation 35.23. We also see from this integration that the term mc^2 arises from the contribution of $v = 0$ and can thus be identified as the rest energy of equation 35.21.

Momentum-Energy Relationship

In the classical limit, we found that the kinetic energy and momentum of an object are related via $K = p^2/2m$. Therefore, it is appropriate to ask what is the correct general relationship between energy and momentum. With $E = \gamma mc^2$ and $\vec{p} = \gamma m\vec{v}$ as our starting point, we find

$$E^2 = p^2 c^2 + m^2 c^4. \qquad (35.25)$$

DERIVATION 35.7 | Energy-Momentum Relation

We start with our equations for momentum and energy, and square each of them:

$$\vec{p} = \gamma m\vec{v} \Rightarrow p^2 = \gamma^2 m^2 v^2$$
$$E = \gamma mc^2 \Rightarrow E^2 = \gamma^2 m^2 c^4.$$

The square of the relativistic gamma-factor appears in both equations. Expressed in terms of v and c, it is

$$\gamma^2 = \frac{1}{1-\beta^2} = \frac{1}{1-v^2/c^2} = \frac{c^2}{c^2-v^2}.$$

Now we can evaluate the expression for the square of the energy, obtaining

$$E^2 = \frac{c^2}{c^2-v^2} m^2 c^4 = \frac{c^2-v^2+v^2}{c^2-v^2} m^2 c^4 = m^2 c^4 + \frac{v^2}{c^2-v^2} m^2 c^4$$

$$= m^2 c^4 + \frac{c^2}{c^2-v^2} m^2 v^2 c^2 = m^2 c^4 + \gamma^2 m^2 v^2 c^2$$

$$= m^2 c^4 + p^2 c^2.$$

In the second step of this computation, we simply added and subtracted v^2 in the numerator, then realized that we could split the numerator in such a way that the factor $c^2 - v^2$ would just cancel the denominator. In the next step, we rearranged the factors in the second term of the sum so that we could factor out γ again, and finally recognized that $\gamma^2 m^2 v^2 = p^2$, which gives us our desired result.

The above energy-momentum relationship is often presented in the form of its square root,

$$E = \sqrt{p^2 c^2 + m^2 c^4}. \qquad (35.26)$$

Note also that equation 35.25 can be rewritten in the form

$$E^2 - p^2 c^2 = m^2 c^4. \qquad (35.27)$$

The mass m of a particle is a scalar invariant, and the speed of c in vacuum is also invariant, so it follows from this simple rewrite that $E^2 - p^2 c^2$ is an invariant, just like the space-time interval $s^2 = c^2 \Delta t^2 - \Delta r^2$ is an invariant.

A special case of equation 35.27 is relevant for particles with zero mass. (Photons, the subatomic particle representation for all radiation, including visible light, are examples of massless particles.) If a particle has $m = 0$, then the energy-momentum relationship simplifies considerably, and the momentum becomes proportional to the energy:

$$E = pc \quad (\text{for } m=0). \qquad (35.28)$$

Speed, Energy, and Momentum

For particles with $m > 0$, dividing the absolute value of the momentum $p = \gamma mv$ by the energy $E = \gamma mc^2$ gives

$$\frac{p}{E} = \frac{\gamma mv}{\gamma mc^2} = \frac{v}{c^2} \Rightarrow v = \frac{pc^2}{E}$$

or, equivalently,

$$\beta = \frac{v}{c} = \frac{pc}{E}.$$

(35.29)

This formula is often very useful for determining the relativistic factor β from the known energy and momentum of a particle. However, it also provides another energy-momentum relationship that is useful in practice:

$$\beta = \frac{pc}{E} \Rightarrow p = \frac{\beta E}{c} \quad \text{or} \quad E = \frac{pc}{\beta}.$$

(35.30)

EXAMPLE 35.4   Electron at 0.99c

We have seen that in some applications it is advantageous to use the energy unit electron-volt (eV), 1 eV = $1.602 \cdot 10^{-19}$ J. The rest energy of an electron is $E_0 = 5.11 \cdot 10^5$ eV = 0.511 MeV, and its rest mass is $m = 0.511$ MeV/c^2.

PROBLEM
If an electron has a speed of 99.0% that of light, what are its total energy, kinetic energy, and momentum?

SOLUTION
First we calculate γ for this case:

$$\gamma = \frac{1}{\sqrt{1-\beta^2}} = \frac{1}{\sqrt{1-(0.990)^2}} = 7.09.$$

The total energy of a particle is given by

$$E = \gamma E_0 = 7.09(0.511 \text{ MeV}) = 3.62 \text{ MeV}.$$

The kinetic energy is therefore

$$K = (\gamma - 1)E_0 = 6.09(0.511 \text{ MeV}) = 3.11 \text{ MeV}.$$

The momentum of this electron is then

$$p = \frac{\beta E}{c} = \frac{(0.990)(3.62 \text{ MeV})}{c} = 3.58 \text{ MeV}/c.$$

Interestingly, we can accelerate electrons to 99% of the speed of light with quite small accelerators, which could fit in labs of a typical physics building. However, going above 0.99c becomes very expensive. To get electrons to 99.9999999% of the speed of light, requires a giant particle accelerator. The SLAC linear accelerator at Stanford, over 3 km long, is one such machine.

The equation $E = mc^2$ implies that energy and mass are related and can be converted into each other. In chemistry class you may have heard the expression "conservation of mass," which means that the number of atoms of each kind in chemical reactions between different molecules has to remain the same. However, this is merely an accounting device, and not a strict conservation law. The conservation laws of energy and momentum still hold in relativistic kinematics, but the total mass in a reaction is *not* conserved exactly. This fact is particularly evident in decays of particles, and Example 35.5 examines one such case.

EXAMPLE 35.5   Kaon Decay

The neutral kaon is a particle that can decay into a positive pion and a negative pion:

$$K^0 \rightarrow \pi^+ + \pi^-.$$

Particles will be discussed in much greater detail in Chapter 39, but the kinematics of particle decays can be understood just from the principles of relativity.

PROBLEM

What is the kinetic energy of each of the two pions after the decay of the kaon? The mass of the neutral kaon is 497.65 MeV/c^2, and the masses of the positive and negative pions are 139.57 MeV/c^2 each. Assume that the neutral kaon is at rest before the decay occurs.

SOLUTION

Momentum and energy are conserved in this decay. The momentum before and after the decay of the kaon is zero. Thus,

$$\vec{p}_{\pi^+} + \vec{p}_{\pi^-} = 0.$$

Thus, the magnitudes of the momenta of the two pions are the same. Because the masses of the two charged pions are also the same, the kinetic energies of the two pions after the decay must be the same:

$$K_{\pi^+} = K_{\pi^-} = K,$$

where K is the kinetic energy of each pion after the decay. The total energy of the kaon before the decay is

$$E = m_{K^0} c^2,$$

because the kaon is at rest. After the decay the energy is

$$E = E_{\pi^+} + E_{\pi^-} = \left(m_{\pi^+} c^2 + K_{\pi^+} \right) + \left(m_{\pi^-} c^2 + K_{\pi^-} \right).$$

Energy conservation means that the total energy before and after the decay is the same:

$$m_{K^0} c^2 = \left(m_{\pi^+} c^2 + K_{\pi^+} \right) + \left(m_{\pi^-} c^2 + K_{\pi^-} \right).$$

We can rearrange this equation to get

$$m_{K^0} c^2 - m_{\pi^+} c^2 - m_{\pi^-} c^2 = K_{\pi^+} + K_{\pi^-} = 2K,$$

remembering that the kinetic energy of the π^+ and the π^- are the same. Solving for the kinetic energy of each pion gives us

$$K = \frac{m_{K^0} c^2 - m_{\pi^+} c^2 - m_{\pi^-} c^2}{2}.$$

Putting in the values of the masses given in the problem text leads us to the result

$$K = \tfrac{1}{2}(497.65 \text{ MeV} - 139.57 \text{ MeV} - 139.57 \text{ MeV}) = 109.26 \text{ MeV}.$$

Note that for the decay of a kaon at rest the entire kinetic energy of the two pions, which are the decay products, comes from the mass difference between the kaon and the sum of the two pions.

The possibility to convert mass into energy is used in nuclear fission reactions, where heavy atomic nuclei are split into smaller parts, liberating a large amount of kinetic energy in the process, which in turn is converted into heat and ultimately electrical energy. This is the basis for the nuclear power industry and will be covered in much greater detail in Chapter 40 on nuclear physics. There we will also discuss nuclear fusion, which ultimately is based on the same mass-energy relation, and which powers most stars and hopefully soon will also lead to commercially available fusion reactors for electrical power generation.

Lorentz Transformation

Derivation 35.3 showed that the position vector $\vec{r}$ and t in some frame F can be expressed in some other frame F' as position vector $\vec{r}'$ and time t' via the Lorentz transformation. What about momentum and energy?

In equation 35.19 the momentum vector is defined as $\vec{p} = \gamma\, m\vec{u}$. From this followed equation 35.22 for the energy, $E = \gamma m c^2$. Velocities can be transformed from one frame to another using equation 35.17 via the relationship $u'_x = (u_x - v)/(1 - v u_x/c^2)$. Therefore the Lorentz transformation of momentum and energy can be written as

$$p'_x = \gamma(p_x - vE/c^2)$$
$$p'_y = p_y$$
$$p'_z = p_z$$
$$E' = \gamma(E - v p_x). \tag{35.31}$$

We leave the proof of this relationship to an end-of-chapter exercise. With the aid of this Lorentz transformation, one can show that $E^2 - p^2 c^2 = E'^2 - p'^2 c^2$, that is, that $E^2 - p^2 c^2$ is a Lorentz invariant. This proof works in complete analogy to Derivation 35.4 and is also left as an end-of-chapter exercise.

Two-Body Collisions

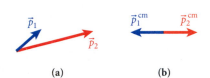

(a)   (b)

FIGURE 35.15 Momentum vectors of the two particles in two different inertial frames. (a) Arbitrary frame; (b) center-of-mass frame.

Thinking about Lorentz transformations and Lorentz invariants helps in dealing with very common problems in relativistic kinematics, those of two-body collisions. We have just stated that $E^2 - p^2 c^2$ is a Lorentz invariant. This definition can be extended to the energies and momentum vectors of two particles, as observed in a certain frame (Figure 35.15). (We use subscripts 1 and 2 to indicate the individual particle.) It is common to introduce the quantity S (simply pronounced *ess*) as

$$S \equiv (E_1 + E_2)^2 - (\vec{p}_1 + \vec{p}_2)^2 c^2, \tag{35.32}$$

where E_1, $\vec{p}_1$ are total energy and momentum of particle 1, and E_2, $\vec{p}_2$ are the total energy and momentum of particle 2. Since $E^2 - p^2 c^2$ is a Lorentz invariant, so is S. Thus we are free to transform our observables into a frame, in which they are easiest to evaluate. The section on two-body collisions in Chapter 8 showed that a convenient frame is the center-of-mass frame. We denote the quantities in the center-of-mass frame with a superscript cm. In Chapter 8 we had already found that in this frame, the two momentum vectors of the two particles are exactly equal in magnitude and opposite in direction, $\vec{p}_1^{\,cm} = -\vec{p}_2^{\,cm}$. Evaluating S in this frame, we see that

$$S = (E_1^{cm} + E_2^{cm})^2 - \underbrace{(\vec{p}_1^{\,cm} + \vec{p}_2^{\,cm})^2}_{0} c^2 = (E_1^{cm} + E_2^{cm})^2.$$

This shows the physical meaning of S: The square root of S is equal to the sum of the energies of the two particles. That is, it is the total available energy in the center-of-mass frame

$$\sqrt{S} = E_1^{cm} + E_2^{cm}. \tag{35.33}$$

Thus $\sqrt{S}$ provides information on the maximum energy that can be utilized for physical processes in two-body collisions. Because the energy of each of the two particles can also be written as the sum of kinetic energy plus mass energy, this relationship can also be written as

$$\sqrt{S} = K_1^{cm} + m_1 c^2 + K_2^{cm} + m_2 c^2.$$

For practical considerations, the other of the two most interesting inertial frames, besides the center-of-mass frame, is the laboratory frame. The following example shows useful relationships between the laboratory and the center-of-mass frames.

EXAMPLE 35.6   Colliders vs. Fixed-Target Accelerators

Suppose you want to create collisions between two protons in order to produce new particles. Two kinds of accelerators can accomplish this. In one, you can have one proton at rest in the laboratory, and shoot the other one at it with some kinetic energy. This is the fixed-target accelerator scheme. A more complicated way to generate collisions is to accelerate the two protons to the same kinetic energy and run them into each other from opposite directions. This technique requires much greater precision, but has a huge payoff if large center-of-mass energies are needed.

PROBLEM

The relativistic heavy ion collider (RHIC) at Brookhaven National Laboratory can deliver proton beams of 250 GeV kinetic energy each, traveling in opposite directions. What beam kinetic energy would be needed for a fixed-target accelerator to reach the same center-of-mass energy?

SOLUTION

The rest mass of the proton is $m_p = 0.938$ GeV/c^2. It is small compared to the kinetic energies in this problem, but we still do not want to neglect it in the following. This way we will arrive at a formula that is applicable at all beam energies.

In the center-of-mass frame, our total available energy is ($m_1 = m_2 = m_p$ and $K_1^{cm} = K_2^{cm} \equiv K^{cm}$):

$$S = (2K^{cm} + 2m_p c^2)^2. \qquad (i)$$

In the lab-frame, which is used by the fixed-target accelerator, we can start from equation 35.32, which is of course valid in all inertial frames, and evaluate

$$\begin{aligned} S &= (E_1^{lab} + E_2^{lab})^2 - (\vec{p}_1^{\,lab} + \vec{p}_2^{\,lab})^2 c^2 \\ &= (E_1^{lab})^2 + 2E_1^{lab}E_2^{lab} + (E_2^{lab})^2 - (\vec{p}_1^{\,lab})^2 c^2 - 2(\vec{p}_1^{\,lab} \cdot \vec{p}_2^{\,lab})c^2 - (\vec{p}_2^{\,lab})^2 c^2 \\ &= \left[(E_1^{lab})^2 - (\vec{p}_1^{\,lab})^2 c^2\right] + \left[(E_2^{lab})^2 - (\vec{p}_2^{\,lab})^2 c^2\right] + 2E_1^{lab}E_2^{lab} - 2(\vec{p}_1^{\,lab} \cdot \vec{p}_2^{\,lab})c^2. \end{aligned}$$

The terms in the square brackets are the squared invariant mass energies of the proton in each case; substituting for the terms in the square brackets from equation 35.27 gives

$$S = 2m_p^2 c^4 + 2E_1^{lab}E_2^{lab} - 2(\vec{p}_1^{\,lab} \cdot \vec{p}_2^{\,lab})c^2.$$

One of the protons, say proton 1, is at rest ($\vec{p}_1^{\,lab} = 0$, $E_1^{cm} = m_p c^2$). This eliminates the scalar product above. For the other proton, we can write the total energy as the sum of kinetic energy plus mass energy and then obtain

$$S = 2m_p^2 c^4 + 2m_p c^2 (K^{lab} + m_p c^2) = 4m_p^2 c^4 + 2m_p c^2 K^{lab}. \qquad (ii)$$

Because S is invariant, we can set equations (i) and (ii) equal and then find

$$\begin{aligned} 4m_p^2 c^4 + 2m_p c^2 K^{lab} &= (2K^{cm} + 2m_p c^2)^2 \\ &= 4(K^{cm})^2 + 8K^{cm}m_p c^2 + 4m_p^2 c^4. \end{aligned}$$

We solve this for the kinetic energy in the lab and find

$$K^{lab} = 4K^{cm} + \frac{2(K^{cm})^2}{m_p c^2}.$$

Clearly, for beam kinetic energies small compared to the proton mass energy, the linear term in this expression dominates. (The linear term is exactly what a nonrelativistic calculation would have found; compare Chapter 8.) But for large kinetic energies, the quadratic term dominates, as shown in Figure 35.16. Knowing that the cost of an accelerator increases with the kinetic energy, we can thus clearly see that only colliders can be used for the very high center-of-mass energies needed for modern particle and nuclear physics experiments at the energy frontier.

Finally, here is the numerical answer: If we insert 250 GeV center-of-mass kinetic energy for each of the two beams, the value for protons at RHIC, we find that we would need a fixed-target accelerator with a beam energy over 134,000 GeV (134 TeV).

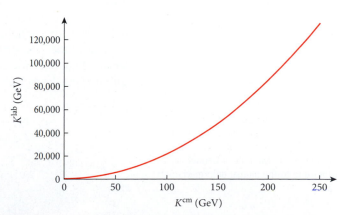

FIGURE 35.16 Equivalent fixed-target beam kinetic energy as a function of the center-of-mass kinetic energy in proton-proton collisions.

35.8 General Relativity

So far, we have talked about only special relativity. The theory of general relativity encompasses all of special relativity but, in addition, provides a theory of gravity. We have seen one theory of gravity that has been used rather successfully since the time of Newton. In the Newtonian theory of gravity, the gravitational force acting on a mass m due to another mass M is given as

$$F_g = \frac{GmM}{r^2} = ma, \tag{35.34}$$

where we have used Newton's Second Law to relate the force to m and the acceleration a. The mass m appears twice in this equation, but there is one very fine difference. The mass that appears on the right side of these equations is called the *inertial mass*. This mass is the mass that undergoes the acceleration. However, the quantity that takes a role in the force of gravity is also a mass, the *gravitational mass*. Even in our most sensitive experimental tests, we find that within experimental measurement uncertainties inertial mass and gravitational mass are exactly equal.

The Newtonian theory of gravity served us extremely well. It was in almost complete agreement with all experimental observations. However, some small problems occurred in precision observations, such as a discrepancy in the exact orbit of the planet Mercury, that could not be explained with the Newtonian theory of gravity.

Albert Einstein again had the decisive insight, in 1907: "*If a person falls freely, he will not feel his own weight*" (Figure 35.17). This observation means that you cannot distinguish if you are in an accelerating reference frame or subject to a gravitational force. This idea led to the famous **Equivalence Principle:**

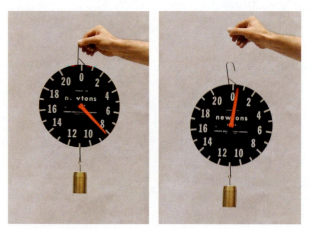

FIGURE 35.17 Einstein's observation: A free-falling object becomes weightless.

All local freely falling nonrotating laboratories are equivalent for the performance of all physical experiments.

From this principle, we can prove that space and time are locally curved due to the presence of masses, and in turn, that *curved space-time* affects the motion of masses, telling them how to move.

This concept may seem difficult to visualize, but an example might help. Consider a flat rubber sheet that symbolizes space in two dimensions. If you put a bowling ball onto this rubber sheet, it will deform the sheet in the way shown in Figure 35.18a. Any mass that now comes rolling along the surface of the rubber sheet experiences the curvature of the sheet and thus moves as if attracted to the bowling ball. From the viewpoint of general relativity, however, this scenario is a free motion along the shortest path in curved space-time. In Figure 35.18b a second mass comes along, which causes its own deformation of space, leading to a mutual attraction, but the principle remains the same.

One of the most striking predictions of general relativity concerns the motion of light. Because light does not have mass, the Newtonian law of gravity would suggest that gravity cannot affect the motion of light. The theory of general relativity, however, holds that light

FIGURE 35.18 The deformation of space due to the presence of a massive object. (a) One object deforms space around it; (b) two objects attract each other due to the mutual deformation of space.

(a)   (b)

moves through curved space-time along the shortest path and thus should be deflected by the presence of large masses.

Observations by British astronomer Sir Arthur Eddington (1882–1944) during a solar eclipse in 1919 confirmed this spectacular prediction. Figure 35.19 shows the photograph taken by Eddington's team during the solar eclipse.

To visualize what was happening, we again use our rubber sheet analogy. Light from a distant star that passes near the Sun is deflected, as in Figure 35.20. The mass of the Sun acts like a lens. Light rays that come close to the Sun on either side are deflected, as sketched in Figure 35.20. The consequence is that an observer sees two stars, or even arcs, instead of just one star. Eddington measured the undeflected directions of the stars six months later by photographing stars at night when the Sun was not near the path of the light. The observed displacement was about $\frac{1}{5}$ the diameter of the image of the distant star. General relativity predicted the angular separation of the observed image and the actual direction of the star to be 1.74", in good agreement with observed angular separation. These measurements could be carried out only during a solar eclipse; otherwise, the light from the Sun would simply overpower the light from the stars, and this effect could not be measured in this manner. The systematic accuracy of these results has been criticized, but subsequent measurements in several solar eclipses have verified Eddington's results.

Recent observations with the Hubble Space Telescope have demonstrated gravitational lensing by massive dark objects (Figure 35.21). Figure 35.21a shows a sketch of the light from a distant galaxy following a curved path around an unseen, massive dark object. This effect can produce two or more images of the distant galaxy, as illustrated, or can produce arcs of light originating from the distant galaxy. Several arcs resulting from the gravitational lensing of a massive dark object are visible in Figure 35.21b.

FIGURE 35.19 Photograph taken by Eddington's team of a solar eclipse, taken May 29, 1919, on the island of Principe, near Africa. The horizontal bars mark positions of the stars.

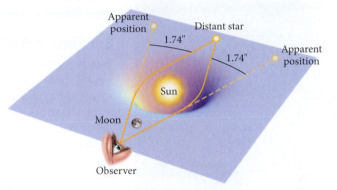

FIGURE 35.20 Light from a distant star is bent around the Sun. The angle is massively exaggerated for clarity. These stars are visible close to the Sun only during a solar eclipse.

Black Holes

When light comes close enough to an object of sufficient mass, the light cannot escape from the curvature in space-time generated by that object. What we mean by "close enough" is defined by the so-called **Schwarzschild radius**:

$$R_S = \frac{2GM}{c^2}. \tag{35.35}$$

Every mass has a Schwarzschild radius that can be easily calculated. Classically, this formula can be arrived at by setting the escape speed found in Chapter 12: $v_{esc} = \sqrt{\dfrac{2GM}{R}}$ equal to

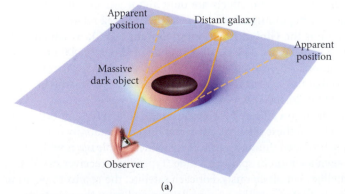

(a)

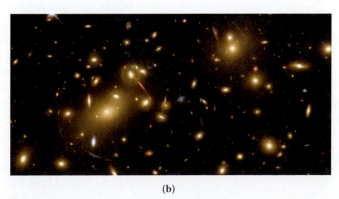

(b)

FIGURE 35.21 (a) Gravitational lensing of a distant galaxy by a massive dark object; (b) arcs due to gravitational lensing around the galaxy cluster Abell 2218.

35.8 In-Class Exercise

What is the Schwarzschild radius of a black hole with a mass of 14.6 solar masses ($m_{Sun} = 1.99 \cdot 10^{30}$ kg)?

a) 43.2 cm

b) 55.1 m

c) 1.55 km

d) 43.1 km

e) $4.55 \cdot 10^4$ km

the speed of light and rearranging. If this radius lies within the interior of the object, nothing remarkable happens. For example, the Schwarzschild radius of Earth (representing the mass of Earth with a point mass of the same magnitude) is

$$R_{S,E} = \frac{2GM_E}{c^2} = \frac{2(6.67 \cdot 10^{-11} \text{ Nm}^2/\text{kg}^2)(6.0 \cdot 10^{24} \text{ kg})}{(3.0 \cdot 10^8 \text{ m/s})^2} = 8.9 \text{ mm.}$$

This radius is less than $\frac{1}{2}$ inch and obviously much less than the radius of Earth. The Schwarzschild radius has physical meaning only if the mass M that gives rise to it is entirely contained inside the radius.

However, at the end of their lives, stars starting their existence with original masses larger than about 15 solar masses collapse to such high densities that the radius of the resulting object is less than its Schwarzschild radius. No information on this object can then escape from it to the outside. We call such an object a **black hole.** (If we could compact the entire Earth to a sphere with less than $\frac{1}{2}$ inch radius, it would turn into a black hole). Supermassive black holes of millions of solar masses are at the center of many galaxies, including our own. We have already pointed out the existence of a very massive black hole in Chapter 12 and showed in Example 12.4 that the mass of this black hole is approximately $3.6 \cdot 10^6$ times the mass of our Sun.

Gravitational Waves

One particularly intriguing prediction of general relativity is the existence of gravitational waves. These waves can be created by large masses in strongly accelerated motion. Chapter 15 already touched on the efforts to detect gravitational waves with modern gravitational wave detectors like LIGO (Laser Interferometer Gravitational-Wave Observatory). LIGO consists of two observatories, one in Hanford, Washington, and the other in Livingston, Louisiana, with a distance of 3002 km between them. An even more sensitive gravitational wave observatory is on the drawing board. Tentatively named LISA (Laser Interferometer Space Antenna), it will be space-based and will consist of three spacecraft arranged in an equilateral triangle of side length 5 million km (Figure 35.22).

Perhaps the best chance for observing a source of gravitational wave emission comes from binary pulsars, which consist of two neutron stars [extremely small (~ 20 km diameter) dense stars] orbiting each other at a very close distance, one of which is a pulsar (emitting pulses of electromagnetic radiation at very precise time intervals, usually in the range of milliseconds to seconds). These binary pulsars emit very intense gravitational waves. Since the total energy of the system is conserved, the pulsars' orbital period decreases in time. This effect was first observed in 1974 by Russell Hulse and his thesis advisor Joseph Taylor, who shared the 1993 Nobel Prize in Physics for this discovery. They were also able to show that the loss in pulsar period length is consistent with the predictions of general relativity.

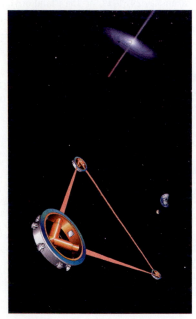

FIGURE 35.22 Artist's concept of LISA (Laser Interferometer Space Antenna), which is proposed as the next-generation gravitational wave detector.

35.9   Relativity in Our Daily Lives: GPS

While you may argue that most relativistic effects are only important in outer space, at impossibly large speeds, or at the beginning of our universe, in one area relativity touches all of our lives. This application is the **Global Positioning System (GPS),** which consists of 24 satellites that orbit the Earth (Figure 35.23), at an altitude of about 20,000 km above ground, with a period one half of a sidereal day (1 sidereal day = the time for the Earth to complete one full rotation so that the stars have the same position again in the night sky = 86,164.09074 s ≈ (1 – 1/365.2425)(86,400 s); see also Example 9.3).

GPS operates by having clocks on all satellites synchronized to a very high precision. The modern atomic clocks on board these satellites have fractional time stability to typically 1 part in 10^{13}. By sending synchronized timing signals, the satellites enable users with a GPS receiver to determine their own location in space and time. Typically, a receiver can detect signals from at least four satellites simultaneously. For each satellite, the receiver's position in space $\vec{r}_r$ and time t_r relative to that of satellite i can be determined from the equation

$$|\vec{r}_r - \vec{r}_i| = c|t_r - t_i|. \tag{35.36}$$

Because the positions in time and space of each satellite are known, equation 35.36 is an equation with four unknown quantities: the three spatial coordinates and the time on the clock of the receiver. Detecting four satellites gives us four equations for four unknown quantities. This system of equations can be solved and provides astonishing accuracy of about 1 m.

For these equations to hold, we must rely on the second postulate of special relativity. However, other relativistic effects also need to be included. The satellites move with speeds of approximately 4 km/s relative to Earth, and time-dilation effects cause frequency shifts in the clocks of 1 part in 10^{10} (satellite atomic clocks slow by about 7 μs/day compared to ground-based clocks), which is about 1000 times too large to be ignored. In addition, gravitational corrections due to general relativity are at least of the same magnitude. Therefore, a correct use of the theory of relativity is essential for the proper functioning of the Global Positioning System.

FIGURE 35.23 GPS satellite in orbit. (Image courtesy of Lockheed Martin Corporation)

WHAT WE HAVE LEARNED | EXAM STUDY GUIDE

- Light always moves at the same speed, independent of the velocity of the source and the observer: $c = 2.99792458 \cdot 10^8$ m/s.

- The two postulates of special relativity are:

  (i) The laws of physics are the same in every inertial reference frame, independent of the motion of this reference frame.

  (ii) The speed of light c is the same in every reference frame.

- The dimensionless quantities $\beta \le 1$ and $\gamma \ge 1$ are defined as:

$$\beta = \left|\vec{v}/c\right| \text{ and } \gamma = \frac{1}{\sqrt{1-\beta^2}} = \frac{1}{\sqrt{1-(v/c)^2}}.$$

- Time is dilated as a function of the speed of the observer, according to

$$\Delta t = \gamma \Delta t_0 = \frac{\Delta t_0}{\sqrt{1-(v/c)^2}}$$

where Δt_0 is the time measured in the rest frame.

- Length in the direction of motion is contracted as a function of the speed of the observer, according to

$$L = \frac{L_0}{\gamma} = L_0\sqrt{1-(v/c)^2}$$

where L_0 is the proper length, measured in its rest frame.

- The relativistic frequency shift for light from a moving source is

$$f = f_0\sqrt{\frac{c \mp v}{c \pm v}},$$

and the corresponding wavelength correction is

$$\lambda = \lambda_0\sqrt{\frac{c \pm v}{c \mp v}}.$$

The red-shift parameter is defined as

$$z = \frac{\Delta\lambda}{\lambda} = \sqrt{\frac{c \pm v}{c \mp v}} - 1.$$

- The Lorentz transformation for space and time coordinates is

$$x' = \gamma(x - vt); \ y' = y; \ z' = z; \ t' = \gamma(t - vx/c^2).$$

- The relativistic velocity transformation is

$$u' = \frac{u - v}{1 - vu/c^2},$$

and the inverse transformation is

$$u = \frac{u' + v}{1 + vu'/c^2}$$

where all velocities are in the same single direction.

- The rest energy of a particle is $E_0 = mc^2$.

- The relativistic expression for the momentum is $\vec{p} = \gamma m\vec{v}$ and for the energy is $E = \gamma E_0 = \gamma mc^2$.

- The relationship between energy and momentum is given by $E^2 = p^2 c^2 + m^2 c^4$.

- The Lorentz transformation for momentum and energy is

$p'_x = \gamma(p_x - vE/c^2)$; $p'_y = p_y$; $p'_z = p_z$; $E' = \gamma(E - vp_x)$.

- The Lorentz invariant

$$S \equiv (E_1 + E_2)^2 - (\vec{p}_1 + \vec{p}_2)^2 c^2$$

is the square of the total available energy in the center-of-mass frame.

- The equivalence principle of the theory of general relativity states that all local freely falling nonrotating laboratories are equivalent for the performance of all physical experiments.

- The Schwarzschild radius of a massive object is given by

$$R_S = \frac{2GM}{c^2}.$$

KEY TERMS

aether, p. 1133
inertial reference frame, p. 1134
light cone, p. 1136
world line, p. 1137
space-time, p. 1137
proper time, p. 1137
time dilation, p. 1138

length contraction, p. 1140
proper length, p. 1140
relativistic frequency shift, p. 1144
red-shifted, p. 1144
blue-shifted, p. 1144
red-shift parameter, p. 1144

Galilean transformation, p. 1146
Lorentz transformation, p. 1146
invariants, p. 1147
relativistic velocity addition, p. 1148
rest mass, p. 1151

Equivalence Principle, p. 1158
Schwarzschild radius, p. 1159
black hole, p. 1160
Global Positioning System (GPS), p. 1160

NEW SYMBOLS AND EQUATIONS

$\beta = |\vec{v} / c|$, speed in units of the speed of light

$\gamma = 1 / \sqrt{1 - \beta^2}$, the relativistic factor

$\Delta t = \gamma \Delta t_0$, time dilation

$L = L_0 / \gamma$, length contraction

u, u', velocity notation used in relativistic velocity transformation

$\vec{p} = \gamma m \vec{v}$, relativistic momentum

$E = \gamma E_0 = \gamma m c^2$, relativistic energy

$E_0 = mc^2$, energy of a stationary particle (rest energy) with mass m

$\sqrt{S}$, total available energy in the center-of-mass frame of two colliding particles

$R_S = \frac{2GM}{c^2}$, Schwarzschild radius

ANSWERS TO SELF-TEST OPPORTUNITIES

35.1 1 light-year $= (3.00 \cdot 10^8 \text{ m/s})(365.25 \text{ days/year})$ $(24 \text{ hours/day})(3600 \text{ s/hour}) = 9.47 \cdot 10^{15}$ m

35.2
5% effect:

Exact: $\beta = \sqrt{1 - \gamma^{-2}} = \sqrt{1 - (1.05)^{-2}} = 0.31 = 31\%$ of c

Approximation: $\beta \approx \sqrt{2(\gamma - 1)} = \sqrt{2(1.05 - 1)} = 0.32 = 33\%$ of c

50% effect:

Exact: $\beta = \sqrt{1 - \gamma^{-2}} = \sqrt{1 - (1.50)^{-2}} = 0.75 = 75\%$ of c.

35.3
$\gamma\{\gamma(x' + vt') - vt\} = \gamma^2(x' + vt') - \gamma vt = x' \Rightarrow$ (divide by γ^2)

$x' + vt' - vt / \gamma = x' / \gamma^2 \Rightarrow$ (subtract x')

$vt' - vt / \gamma = x'(\gamma^{-2} - 1) = -x'v^2 / c^2 \Rightarrow$ (divide by v)

$t' - t / \gamma = -x'v / c^2 \Rightarrow$ (rearrange)

$t' + x'v / c^2 = t / \gamma \Rightarrow$ (multiply with γ)

$t = \gamma(t' + x'v / c^2)$.

35.4 $u_x = \dfrac{dx}{dt} = \dfrac{\gamma(dx' + vdt')}{\gamma(dt' + vdx'/c^2)}$ $u' = \dfrac{\dfrac{dx'}{dt'} + v}{1 + \dfrac{v}{c^2} \dfrac{dx'}{dt'}} = \dfrac{u'_x + v}{1 + \dfrac{vu'_x}{c^2}}$.

PROBLEM-SOLVING PRACTICE

Problem-Solving Guidelines

1. The first step in attacking any relativity problem is to identify the information. What are the events involved in the problem? What are the reference frames? What is the proper time and what is the proper length? Review Section 35.3 and be sure you know how to recognize the proper time and proper length.

2. Given the subtleties of relativity theory, checking answers is very important. Check that time has dilated and length has contracted for a moving reference frame relative to a stationary frame.

3. For dealing with the Lorentz transformation equations, be sure to identify the two reference frames, as well as the direction of the velocity between them and the velocity of any motion within each frame.

SOLVED PROBLEM 35.3 / Same Velocity

For some purposes, particle physicists need to have beams of electrons that have exactly the same velocity as beams of protons. For a proton, $m_p c^2 = 938$ MeV. For an electron, $m_e c^2 = 0.511$ MeV.

PROBLEM

If you have a proton beam with kinetic energy 2.50 GeV, what is the required kinetic energy for the electron beam?

SOLUTION

THINK

Particles with the same velocity v have the same γ. We can write expressions for the electron and the proton relating γ, the total energy, and the rest mass, solve them for γ, and equate them. The total energy of a particle is equal to the rest mass plus the kinetic energy. We can then solve for the required kinetic energy of the electron.

SKETCH

Figure 35.24 is a sketch of an electron and a proton moving with the same velocity.

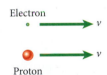

FIGURE 35.24 An electron and a proton with the same velocity.

RESEARCH

We can relate the energy E of a particle and γ through

$$E = \gamma m c^2.$$

We can therefore write the energy for an electron as $E_e = \gamma m_e c^2$ and the energy for a proton as $E_p = \gamma m_p c^2$. We can then equate γ for the electron and proton:

$$\gamma = \frac{E_e}{m_e c^2} = \frac{E_p}{m_p c^2}.$$

The energy of the electron can be written as the sum of kinetic energy and rest energy, $E_e = K_e + m_e c^2$. In the same way, we write for the proton $E_p = K_p + m_p c^2$.

SIMPLIFY

We can combine the preceding equations to get

$$\frac{K_e + m_e c^2}{m_e c^2} = \frac{K_p + m_p c^2}{m_p c^2}.$$

Solving this for the kinetic energy of the electron, we obtain

$$K_e = m_e c^2 \left(\frac{K_p + m_p c^2}{m_p c^2} \right) - m_e c^2.$$

CALCULATE

Putting in the numerical values gives us

$$K_e = (0.511 \text{ MeV}) \left(\frac{2500 \text{ MeV} + 938 \text{ MeV}}{938 \text{ MeV}} \right) - (0.511 \text{ MeV}) = 1.36194 \text{ MeV}.$$

Continued—

ROUND

We report our result with three significant figures,

$$K_e = 1.36 \text{ MeV}.$$

DOUBLE-CHECK

We can double-check our result by calculating γ for the electron and the proton. For the electron we have

$$\gamma = \frac{E_e}{m_e c^2} = \frac{1.36 \text{ MeV} + 0.511 \text{ MeV}}{0.511 \text{ MeV}} = 3.66.$$

For the proton,

$$\gamma = \frac{E_p}{m_p c^2} = \frac{2.50 \text{ GeV} + 938 \text{ MeV}}{938 \text{ MeV}} = 3.67,$$

which agrees with our calculated value for the electron within rounding errors. Thus, our result seems reasonable.

MULTIPLE-CHOICE QUESTIONS

35.1 The most important fact we learned about aether is that:

a) It was experimentally proven not to exist.

b) Its existence was proven experimentally.

c) It transmits light in all directions equally.

d) It transmits light faster in longitudinal direction.

e) It transmits light slower in longitudinal direction.

35.2 If spaceship A is traveling at 70% the speed of light relative to an observer at rest, and spaceship B is traveling at 90% the speed of light relative to an observer at rest, which of the following have the greatest velocity as measured by an observer in spaceship B?

a) A cannon shot from A to B at 50% the speed of light as measured in A's reference frame.

b) A ball thrown from B to A at 50% the speed of light as measured in B's reference frame.

c) A particle beam shot from a stationary observer to B at 70% the speed of light as measured in stationary reference frame.

d) A beam of light shot from A to B traveling at the speed of light in A's reference frame.

e) All of the above have the same velocity as measured in B's reference frame.

35.3 A particle of rest mass m_0 travels at a speed $v = 0.20c$. How fast must the particle travel in order for its momentum to increase to twice its original momentum?

a) 0.40c   c) 0.38c   e) 0.99c

b) 0.10c   d) 0.42c

35.4 Which quantity is invariant—that is, has the same value—in all reference frames?

a) time interval, Δt

b) space interval, Δx

c) velocity, v

d) space-time interval, $c^2 (\Delta t)^2 - (\Delta x)^2$

35.5 Two twins, A and B, are in deep space on similar rockets traveling in opposite directions with a relative speed of $c/4$. After a while, twin A turns around and travels back toward twin B again, so that their relative speed is $c/4$ in opposite directions. When they meet again, is one twin younger, and if so which twin is younger? Explain.

a) Twin A is younger.   d) Each twin thinks
                              the other is younger.
b) Twin B is younger.

c) The twins are the same age.

35.6 A proton with a momentum of 3.0 GeV/c is moving with what velocity relative to the observer?

a) 0.31c   c) 0.91c   e) 3.2c

b) 0.33c   d) 0.95c

35.7 A square of area 100 m^2 that is at rest in the reference frame is moving with a speed $(\sqrt{3}/2)c$. Which of the following statements is incorrect?

a) $\beta = \sqrt{3}/2$

b) $\gamma = 2$

c) To an observer at rest, it looks like another square with an area less than 100 m^2.

d) The length along the moving direction is contracted by a factor of $\frac{1}{2}$.

35.8 Consider a particle moving with a speed less than 0.5c. If the speed of the particle is doubled, by what factor will the momentum increase?

a) less than 2

b) equal to 2

c) greater than 2

QUESTIONS

35.9 In mechanics, one often uses the model of a perfectly rigid body to model and determine the motion of physical objects (see, for example, Chapter 10 on rotation). Explain how this model contradicts Einstein's special theory of relativity.

35.10 Use light cones and world lines to help solve the following problem. Eddie and Martin are throwing water balloons very fast at a target. At $t = -13$ μs, the target is at $x = 0$, Eddie is at $x = -2$ km, and Martin is at $x = 5$ km, and all three remain in these positions for all time. The target is hit at $t = 0$. Who made the successful shot? Prove this using the light cone for the target. When the target is hit, it sends out a radio signal. When does Martin know the target has been hit? When does Eddie know the target has been hit? Use the world lines to show this. Before starting to draw your diagrams, consider: If your x position is measured in km and you are plotting t versus x/c, what units must t be in, to the first significant figure?

35.11 A gravitational lens should produce a halo effect and not arcs. Given that the light travels not only to the right and left of the intervening massive object but also to the top and bottom, why do we typically see only arcs?

35.12 Suppose you are explaining the theory of relativity to a friend, and you have told him that nothing can go faster than 300,000 km/s. He says that is obviously false: Suppose a spaceship traveling past you at 200,000 km/s, which is perfectly possible according to what you are saying, fires a torpedo straight ahead whose speed is 200,000 km/s relative to the spaceship, which is also perfectly possible; then, he says, the torpedo's speed is 400,000 km/s. How would you answer him?

35.13 Consider a positively charged particle moving at constant speed parallel to a current-carrying wire, in the direction of the current. As you know (after studying Chapters 27 and 28), the particle is attracted to the wire by the magnetic force due to the current. Now suppose another observer moves along with the particle, so according to him the particle is at rest. Of course, a particle at rest feels no magnetic force. Does that observer see the particle attracted to the wire or not? How can that be? (Either answer seems to lead to a contradiction: If the particle is attracted, it must be by an electric force because there is no magnetic force, but there is no electric field from a neutral wire; if the particle is not attracted, you see that the particle is, in fact, moving toward the wire.)

35.14 At rest, a rocket has an overall length of L. A garage at rest (built for the rocket by the lowest bidder) is only $L/2$ in length. Luckily, the garage has both a front door and a back door, so that when the rocket flies at a speed of $v = 0.866c$, the rocket fits entirely into the garage. However, according to the rocket pilot, the rocket has length L and the garage has length $L/4$. How does the rocket pilot observe that the rocket does not fit into the garage?

35.15 A rod at rest on Earth makes an angle of 10° with the x-axis. If the rod is moved along the x-axis, what happens to this angle, as viewed by an observer on the ground?

35.16 An astronaut in a spaceship flying toward Earth's Equator at half the speed of light observes Earth to be an oblong solid, wider and taller than it appears deep, rotating around its long axis. A second astronaut flying toward Earth's North Pole at half the speed of light observes Earth to be a similar shape but rotating about its short axis. Why does this not present a contradiction?

35.17 Consider two clocks carried by observers in a reference frame moving at speed v in the positive x-direction relative to ours. Assume that the two reference frames have parallel axes, and that their origins coincide when clocks at that point in both frames read zero. Suppose the clocks are separated by distance l in the x'-direction in their own reference frame; for instance, $x' = 0$ for one clock and $x' = l$ for the other, with $y' = z' = 0$ for both. Determine the readings t' on both clocks as functions of the time coordinate t in our reference frame.

35.18 Prove that in all cases, two sub-light-speed velocities "added" relativistically will always yield a sub-light-speed velocity. Consider motion in one spatial dimension only.

35.19 A famous result in Newtonian dynamics is that if a particle in motion collides elastically with an identical particle at rest, the two particles emerge from the collision on perpendicular trajectories. Does the same hold in the special theory of relativity? Suppose a particle of rest mass m and total energy E collides with an identical particle at rest, the same two particles emerging from the collision with new velocities. Are those velocities necessarily perpendicular? Explain.

35.20 Suppose you are watching a spaceship orbiting Earth at 80% the speed of light. What is the length of the ship as viewed from the center of the orbit?

PROBLEMS

A blue problem number indicates a worked-out solution is available in the Student Solutions Manual. One • and two •• indicate increasing level of problem difficulty.

Sections 35.1 and 35.2

35.21 Find the speed of light in *feet per nanosecond*, to three significant figures.

35.22 Find the value of g, the gravitational acceleration at Earth's surface, in light-years per year, to three significant figures.

35.23 Michelson and Morley used an interferometer to show that the speed of light is constant, regardless of Earth's motion through any perceived luminiferous aether. An analogy can be understood from the different times it takes for a rowboat to travel two different round-trip paths in a river that flows at a constant velocity (u) downstream. Let one path be for a distance D directly across the river, then back again; and let the other path be the same distance D directly upstream, then back again. Assume that the rowboat travels at constant speed, v (with respect to the water), for both trips. Neglect the time it takes for the rowboat to turn around. Find the ratio of the cross-stream time divided by the upstream-downstream time, as a function of the given constants.

35.24 What is the value of γ for a particle moving at a speed of $0.8c$?

Section 35.3

35.25 An astronaut on a spaceship traveling at a speed of $0.50c$ is holding a meter stick parallel to the direction of motion.

a) What is the length of the meter stick as measured by another astronaut on the spaceship?

b) If an observer on Earth could observe the meter stick, what would be the length of the meter stick as measured by that observer?

35.26 A spacecraft travels along a straight line from Earth to the Moon, a distance of $3.84 \cdot 10^8$ m. Its speed measured on Earth is $0.50c$.

a) How long does the trip take, according to a clock on Earth?

b) How long does the trip take, according to a clock on the spacecraft?

c) Determine the distance between Earth and the Moon if it were measured by a person on the spacecraft.

35.27 A 30-year-old says goodbye to her 10-year-old son and leaves on an interstellar trip. When she returns to Earth, both she and her son are 40 years old. What was the speed of the spaceship?

35.28 If a muon is moving at 90% of the speed of light, how does its measured lifetime compare to when it is in the rest frame of a laboratory, where its lifetime is $2.2 \cdot 10^{-6}$ s?

35.29 A fire truck 10 meters long needs to fit into a garage 8 meters long (at least temporarily). How fast must the fire truck be going to fit entirely inside the garage, at least temporarily? How long does it take for the truck to get inside the garage, from

a) the garage's point of view?

b) the fire truck's point of view?

35.30 In Jules Verne's classic *Around the World in Eighty Days*, Phileas Fogg travels around the world in, according to his calculation, 81 days. Due to crossing the International Date Line he actually made it only 80 days. How fast would he have to go in order to have time dilation make 80 days to seem like 81? (Of course, at this speed, it would take a lot less than even 1 day to get around the world . . . .)

•**35.31** Suppose NASA discovers a planet just like Earth orbiting a star just like the Sun. This planet is 35 light-years away from our Solar System. NASA quickly plans to send astronauts to this planet, but with the condition that the astronauts would not age more than 25 years during this journey.

a) At what speed must the spaceship travel, in Earth's reference frame, so that the astronauts age 25 years during this journey?

b) According to the astronauts, what will be the distance of their trip?

•**35.32** Consider a meter stick at rest in a reference frame F. It lies in the (x,y) plane and makes an angle of 37° with the x-axis. The reference frame F now moves with a constant velocity of v parallel to the x-axis of another reference frame F'.

a) What is the velocity of the meter stick measured in F' at an angle 45° to the x-axis?

b) What is the length of the meter stick in F' under these conditions?

•**35.33** A wedge-shaped spaceship has a width of 20 m, a length of 50 m, and is shaped like an isosceles triangle. What is the angle between the base of the ship and the side of the ship as measured by a stationary observer if the ship is traveling by at a speed of $0.4c$? Plot this angle as a function of the speed of the ship.

Section 35.4

35.34 How fast must you be traveling relative to a blue light (480 nm) for it to appear red (660 nm)?

35.35 In your physics class you have just learned about the relativistic frequency shift, and you decide to amaze your friends at a party. You tell them that once you drove through a stop light and that when you were pulled over you did not get ticketed because you explained to the police officer that the relativistic Doppler shift made the red light of wavelength 650 nm appear green to you, with a wavelength of 520 nm. If your story had been true, how fast would you have been traveling?

35.36 A meteor made of pure kryptonite (Yes, we know: There really isn't such a thing as kryptonite . . .) is moving toward Earth. If the meteor eventually hits Earth, the impact will cause severe damage, threatening life as we know it. If a laser hits the meteor with wavelength 560 nm, the entire meteor will blow up. The only laser powerful enough on Earth has a 532-nm wavelength. Scientists decide to launch the laser in a spacecraft and use special relativity to get the right wavelength. The meteor is moving very slowly, so there is no correction for relative velocities. At what speed does the spaceship need to move so the laser has the right wavelength, and should it travel toward or away from the meteor?

35.37 Radar-based speed detection works by sending an electromagnetic wave out from a source and examining the Doppler shift of the reflected wave. Suppose a wave of frequency 10.6 GHz is sent toward a car moving away at a speed of 32.0 km/h. What is the difference between the frequency of the wave emitted by the source and the frequency of the wave an observer in the car would detect?

•**35.38** A HeNe laser onboard a spaceship moving toward a remote space station emits a beam of red light toward the space station. The wavelength of the beam, as measured by a wavelength meter on board the spaceship, is 632.8 nm. If the astronauts on the space station see the beam as a blue beam of light with a measured wavelength of 514.5 nm, what is the relative speed of the spaceship with respect to the space station? What is the shift parameter z in this case?

Sections 35.5 and 35.6

35.39 Sam sees two events as simultaneous:

(i) Event A occurs at the point (0,0,0) at the instant 0:00:00 universal time;

(ii) Event B occurs at the point (500 m,0,0) at the same moment.

Tim, moving past Sam with a velocity of $0.999c\hat{x}$, also observes the two events.

a) Which event occurred first in Tim's reference frame?

b) How long after the first event does the second event happen in Tim's reference frame?

35.40 Use the relativistic velocity addition to reconfirm that the speed of light with respect to any inertial reference frame is c. Assume one-dimensional motion along a common x-axis.

35.41 You are driving down a straight highway at a speed of $v = 50$ m/s relative to the ground. An oncoming car travels with the same speed in the opposite direction. With what relative speed do you observe the oncoming car?

35.42 A rocket ship approaching Earth at $0.90c$ fires a missile toward Earth with a speed of $0.50c$, relative to the rocket ship. As viewed from Earth, how fast is the missile approaching Earth?

35.43 In the twin paradox example, Alice boards a spaceship that flies to a space station 3.25 light-years away and then returns with a speed of $0.65c$.

a) Calculate the total distance Alice traveled during the trip, as measured by Alice.

b) With the aforementioned total distance, calculate the total time duration for the trip, as measured by Alice.

•**35.44** In the twin paradox example, Alice boards a spaceship that flies to a space station 3.25 light-years away and then returns with a speed of $0.650c$. This can be viewed in terms of Alice's reference frame.

a) Show that Alice must travel with a speed of $0.914c$ to establish a relative speed of $0.650c$ with respect to Earth when Alice is returning back to Earth.

b) Calculate the time duration for Alice's return flight toward Earth with the aforementioned speed.

•**35.45** Robert, standing at the rear end of a railroad car of length 100 m, shoots an arrow toward the front end of the car. He measures the velocity of the arrow as $0.30c$. Jenny,

who was standing on the platform, saw all of this as the train passed her with a velocity of $0.75c$. Determine the following as observed by Jenny:

a) the length of the car

b) the velocity of the arrow

c) the time taken by arrow to cover the length of the car

d) the distance covered by the arrow

••**35.46** Consider motion in one spatial dimension. For any velocity v, define parameter θ via the relation $v = c \tanh \theta$, where c is the vacuum speed of light. This quantity is variously called the *velocity parameter* or the *rapidity* corresponding to velocity v.

a) Prove that for two velocities, which add according to the Lorentzian rule, the corresponding velocity parameters simply add algebraically, that is, like Galilean velocities.

b) Consider two reference frames in motion at speed v in the x-direction relative to one another, with axes parallel and origins coinciding when clocks at the origin in both frames read zero. Write the Lorentz transformation between the two coordinate systems entirely in terms of the velocity parameter corresponding to v, and the coordinates.

Section 35.7

35.47 What is the speed of a particle whose momentum $p = mc$?

35.48 An electron's rest mass is 0.511 MeV/c^2.

a) How fast must an electron be moving if its energy is to be 10 times its rest energy?

b) What is the momentum of the electron at this speed?

35.49 The Relativistic Heavy Ion Collider (RHIC) can produce colliding beams of gold nuclei with beam kinetic energy of $A \cdot 100$ GeV each in the center-of-mass frame, where A is the number of nucleons in gold (197). You can approximate the mass energy of a nucleon as approximately 1 GeV. What is the equivalent fixed-target beam energy in this case? (See Example 35.6.)

35.50 How much work is required to accelerate a proton from rest up to a speed of $0.997c$?

35.51 In proton accelerators used to treat cancer patients, protons are accelerated to $0.61c$. Determine the energy of the proton, expressing your answer in MeV.

•**35.52** In some proton accelerators, proton beams are directed toward each other for head-on collisions. Suppose that in such an accelerator, protons move with a speed relative to the lab of $0.9972c$.

a) Calculate the speed of approach of one proton with respect to another one with which it is about to collide head on. Express your answer as a multiple of c, using six significant digits.

b) What is the kinetic energy of each proton beam (in units of MeV) in the laboratory reference frame?

c) What is the kinetic energy of one of the colliding protons (in units of MeV) in the rest frame of the other proton?

•**35.53** The hot filament of the electron gun in a cathode ray tube releases electrons with nearly zero kinetic energy. The electrons are next accelerated under a potential difference of 5 kV, before being steered toward the phosphor on the screen of the tube.

a) Calculate the kinetic energy acquired by the electron under this accelerating potential difference.

b) Is the electron moving at relativistic speed?

c) What is the electron's total energy and momentum? (Give both values, relativistic and nonrelativistic, for both quantities.)

35.54 Consider a one-dimensional collision at relativistic speeds between two particles with masses m_1 and m_2. Particle 1 is initially moving with a speed of $0.7c$ and collides with particle 2, which is initially at rest. After the collision, particle 1 recoils with speed $0.5c$, while particle 2 starts moving with a speed of $0.2c$. What is the ratio m_2/m_1?

•**35.55** In an elementary-particle experiment, a particle of mass m is fired, with momentum mc, at a target particle of mass $2\sqrt{2}m$. The two particles form a single new particle (completely inelastic collision). Find:

a) the speed of the projectile before the collision

b) the mass of the new particle

c) the speed of the new particle after the collision

•**35.56** Show that momentum and energy transform from one inertial frame to another as $p'_x = \gamma(p_x - vE/c^2)$; $p'_y = p_y$; $p'_z = p_z$; $E' = \gamma(E - vp_x)$. *Hint:* Look at the derivation for the space-time Lorentz transformation.

•**35.57** Show that $E^2 - p^2c^2 = E'^2 - p'^2c^2$, that is, that $E^2 - p^2c^2$ is a Lorentz invariant. *Hint:* Look at derivation showing that the space-time interval is a Lorentz invariant.

Sections 35.8 and 35.9

35.58 The deviation of the space-time geometry near the gravitating Earth from the flat space-time of the special theory of relativity can be gauged by the ratio Φ/c^2, where Φ is the Newtonian gravitational potential at the Earth's surface. Find the value of this quantity.

35.59 Calculate the Schwarzschild radius of a black hole with the mass of

a) the Sun.

b) a proton. How does this result compare with the size scale 10^{-15} m usually associated with a proton?

35.60 By assuming that the speed of GPS satellites is approximately 4 km/s relative to Earth, calculate how much slower per day the atomic clocks on the satellites run, compared to stationary atomic clocks on Earth.

35.61 What is the Schwarzschild radius of the black hole at the center of our Milky Way? *Hint:* The mass of this black hole was determined in Example 12.4.

Additional Problems

35.62 In order to fit a 50-foot-long stretch limousine into a 35-foot-long garage, how fast would the limousine driver have to be moving, in the garage's reference frame? Comment on what happens to the garage in the limousine's reference frame.

35.63 Using relativistic expressions, compare the momentum of two electrons, one moving at $2 \cdot 10^8$ m/s and the other moving at $2 \cdot 10^3$ m/s. What is the percentage difference between classical momentum values and these values?

35.64 Rocket A passes Earth at a speed of $0.75c$. At the same time, rocket B passes Earth moving $0.95c$ relative to Earth in the same direction. How fast is B moving relative to A when it passes A?

35.65 Determine the difference in kinetic energy of an electron traveling at $0.9900c$ and at $0.9999c$, first using standard Newtonian mechanics and then using special relativity.

35.66 Right before take-off, a passenger on a plane flying from town A to town B synchronizes his clock with the clock of his friend who was waiting for him in town B. The plane flies with a constant velocity of 240 m/s. The moment the plane touches the ground, the two friends check simultaneously the indication of their clocks. The clock of the passenger on the plane shows that it took exactly 3.00 h to travel from A to B. Ignoring any effects of acceleration:

a) Will the clock of the friend waiting in B show a shorter or a longer time interval?

b) What is the difference between the readings of the two clocks?

35.67 The explosive yield of the atomic bomb dropped on Hiroshima near the end of World War II was approximately 15 kilotons of TNT. One kiloton is about $4 \cdot 10^{12}$ J of energy. Find the amount of mass that was converted into energy in this bomb.

35.68 At what speed will the length of a meter stick look 90 cm?

35.69 What is the relative speed between two objects approaching each other head on, if each is traveling at speed of $0.6c$ as measured by an observer on Earth?

35.70 An old song contains the lines "While driving in my Cadillac, what to my surprise; a little Nash Rambler was following me, about one-third my size." The singer of that song assumes that the Nash Rambler is driving at a similar velocity. Suppose, though, rather than actually being one-third the Cadillac's size, the proper length of the Rambler is the same as the Cadillac. What would be the velocity of the Rambler relative to the Cadillac for the song's observation to be accurate?

35.71 You shouldn't invoke time dilation due to your relative motion with respect to the rest of the world as an excuse for being late to class. While it is true that relative to those at rest in the classroom, your time runs more slowly, the difference is likely to be negligible. Suppose over the weekend you drove from your college in the Midwest to New York City and back, a round trip of 2200 miles, driving for 20 hours each direction. By what amount, at most, would your watch differ from your professor's watch?

35.72 A spaceship is traveling at two-thirds of the speed of light directly toward a stationary asteroid. If the spaceship turns on it headlights, what will be the speed of the light traveling from the spaceship to the asteroid as observed by

a) someone on the spaceship?

b) someone on the asteroid?

35.73 Two stationary space stations are separated by a distance of 100 light-years, as measured by someone on one of the space stations. A spaceship traveling at 0.95c relative to the space stations passes by one of the space stations heading directly toward the other one. How long will it take to reach the other space station, as measured by someone on the spaceship? How much time will have passed for a traveler on the spaceship as it travels from one space station to the other, as measured by someone on one of the space stations?

35.74 An electron is accelerated from rest through a potential of $1.0 \cdot 10^6$ V. What is its final speed?

35.75 In the age of interstellar travel, an expedition is mounted to an interesting star 2000 light-years from Earth. To make it possible to get volunteers for the expedition, the planners guarantee that the round trip to the star will take no more than 10% of a normal human lifetime. (At that time the normal human lifetime is 400 years.) What is the minimum speed the ship carrying the expedition must travel?

•**35.76** What is the energy of a particle with speed of 0.8c and momentum of 10^{-20} N s?

•**35.77** In a high-speed football game, a running back traveling at 55% the speed of light relative to the field throws the ball to a receiver running at 65% the speed of light relative to the field in the same direction. The speed of the ball relative to the running back is 80% the speed of light.

a) How fast does the receiver perceive the speed of the ball to be?

b) If the running back shined a flashlight at the receiver, how fast would the photons appear to be traveling to the receiver?

•**35.78** You have been presented with a source of electrons, ^{14}C, having kinetic energy equal to 0.305 times the rest energy. Suppose you have a pair of detectors that can detect passage of the electrons without disturbing them. You wish to show that the relativistic expression for momentum is correct and the nonrelativistic expression is incorrect. If a 2.0-m-long baseline between your detectors is used, what

is the necessary timing accuracy needed to show that the relativistic momentum is correct?

•**35.79** A spacecraft travels a distance of $1.00 \cdot 10^{-3}$ light-years in 20.0 hours, as measured by an observer stationed on Earth. How long does the journey take as measured by the captain of the spacecraft?

•**35.80** More significant than the kinematic features of the special theory of relativity are the dynamical processes that it describes that Newtonian dynamics does not. Suppose a hypothetical particle with rest mass 1.000 GeV/c^2 and kinetic energy 1.000 GeV collides with an identical particle at rest. Amazingly, the two particles fuse to form a single new particle. Total energy and momentum are both conserved in the collision.

a) Find the momentum and speed of the first particle.

b) Find the rest mass and speed of the new particle.

••**35.81** Although it deals with inertial reference frames, the special theory of relativity describes accelerating objects without difficulty. Of course, uniform acceleration no longer means $dv/dt = g$, where g is a constant, since that would have v exceeding c in a finite time. Rather, it means that the acceleration *experienced* by the moving body is constant: In each increment of the body's own proper time $d\tau$, the body acquires velocity increment $dv = g d\tau$ as measured in the inertial frame in which the body is momentarily at rest. (As it accelerates, the body encounters a sequence of such frames, each moving with respect to the others.) Given this interpretation:

a) Write a differential equation for the velocity v of the body, moving in one spatial dimension, as measured in the inertial frame in which the body was initially at rest (the "ground frame"). You can simplify your equation, remembering that squares and higher powers of differentials can be neglected.

b) Solve this equation for $v(t)$, where both v and t are measured in the ground frame.

c) Verify that your solution behaves appropriately for small and large values of t.

d) Calculate the position of the body $x(t)$, as measured in the ground frame. For convenience, assume that the body is at rest at ground-frame time $t = 0$, at ground-frame position $x = c^2/g$.

e) Identify the trajectory of the body on a space-time diagram (*Minkowski diagram,* for Hermann Minkowski) with coordinates x and ct, as measured in the ground frame.

f) For $g = 9.81$ m/s^2, calculate how much time it takes the body to accelerate from rest to 70.7% of c, measured in the ground frame, and how much ground-frame distance the body covers in this time.

36

Quantum Physics

FIGURE 36.1 An image taken in near darkness with a night-vision device.

WHAT WE WILL LEARN

- On the basis of the quantum hypothesis, it is possible to derive Planck's radiation law for the power radiated in a given frequency interval. The quantum approach avoids the ultraviolet catastrophe that makes the classical explanation unphysical at short wavelengths (high frequencies), and it contains the classical radiation laws as limits.

- What we normally think of as a wave, such as light, has particle characteristics. The photoelectric effect is explained with the quantum hypothesis by saying that light consists of elementary quanta called photons. The energy of a photon is equal to its frequency times Planck's constant.

- The Compton effect is the scattering of a high-energy photon (X-ray) off an electron. The observations for the scattering of X-rays are explained by kinematics, which assume that the photon has particle characteristics.

- What we normally think of as matter also has wave characteristics. The de Broglie wavelength of a particle is defined as Planck's constant divided by the magnitude of the particle's momentum and is the fundamental wavelength associated with a matter wave.

- The Heisenberg uncertainty relation stipulates that the product of the uncertainty in momentum times the uncertainty of position, measured simultaneously, has an absolute lower bound. An energy-time uncertainty relation has the same lower bound as momentum and position.

- Elementary quantum particles have an intrinsic property called spin, which has the dimension of angular momentum. Spin is quantized. Particles are divided into two categories: fermions with spins that are half-integer multiples of Planck's constant divided by 2π, and bosons with spins of integer multiples of Planck's constant divided by 2π.

- The Pauli exclusion principle states that no two fermions can occupy the same quantum state at the same time. This means that in any given atom, no two fermions can have exactly identical quantum numbers, which are numbers that characterize the quantum state of the particle. Bosons can condense at low temperature in such a way that most of them occupy the same quantum state.

Night-vision cameras have become standard equipment for law-enforcement and military personnel. They contain lenses and produce images like most optical instruments, but their main purpose is to capture dim light and intensify it so users can see images with very little light (Figure 36.1). The way the camera works depends not on the ray or wave properties of light, but on its photon characteristics, which we will discuss in this chapter.

When light interacts with material objects, it often reveals a particle-like nature, with tiny wave packets called *photons* interacting with individual atoms or molecules or biological cells. In a night-vision camera, photons are converted to electrical signals in a process called the *photoelectric effect,* and the signals are made stronger by devices called *photomultiplier tubes* or *microchannel plates.* (We discuss the processes underlying these devices in this chapter.) Then the resulting electrons are converted back to light by striking a phosphorescent screen. The process amplifies light so we can see images in the dark, but it removes some detail and all color information, as you can see in Figure 36.1.

Does the existence of photons mean that light is not a wave after all? We will see in this chapter that the difference between a wave and a particle is not clear-cut. At very small scales, waves can act like particles, and particles can act like waves. This discovery led to revolutionary changes in our understanding of physics, as far-reaching as the changes in space and time described by relativity. The remaining chapters of this book are devoted to discussion of these changes—known as quantum physics.

36.1   The Nature of Matter, Space, and Time

By now you have probably accepted the notion that matter consists of some constituents, called atoms. Originally atoms were thought to be indivisible, hence the name *atom,* which derives from the Greek word $\alpha\tau o\mu o\varsigma$ (individual, indivisible). We will see that atoms actually do have substructure. They consist of a "cloud" of electrons surrounding a nucleus, which in turn consists of neutrons and protons. Physicists currently believe that electrons have no substructure, but protons and neutrons are known to each consist of three quarks, held together by gluons (see Section 21.2). These quarks and gluons are also thought to be elementary—that is, they are thought to lack substructure. How physicists have arrived at these deductions and conclusions is the subject of Chapters 37–40. For now, however, it suffices to point out that matter is granular—it consists of smallest indivisible pieces.

What about time and space—are they granular, too? In Chapter 35 on relativity, we found some rather surprising results on the connection between time and space. However, we have not yet considered the question of whether time or space can be subdivided into infinitesimally small quantities. Calculus assumes that time is continuous, not granular, because limits of $\Delta t \to 0$ are used to arrive at the definitions of velocity and acceleration. Energy and momentum are related to each other in ways similar to how space and time are related. This immediately raises the question whether energy and momentum are continuous quantities, or if some smallest granule of energy and some elementary quantity of momentum exist. For example, a spinning ball has rotational kinetic energy. By increasing its angular speed, we also increase its kinetic energy. But can we make the increment infinitesimally small, or is there some smallest quantum of energy that we are forced to add?

Let's start this investigation by taking another look at light. Light can be considered an electromagnetic wave, as shown in Chapter 31. In our studies of light, we first explored geometric optics in Chapters 32 and 33, examining image formation with mirrors, lenses, and other optical instruments. We considered light as rays only, assuming that light moves along straight lines. When we looked in Chapter 34 at physical effects such as interference and diffraction, we were forced to invoke the wave character of light. In Chapter 34 we found that the wave character of light comes into play only when we explore spatial dimensions on the order of the wavelength of light, and that ray optics is a very good approximation for spatial dimensions that are very large compared to the wavelength.

Does our previous description of light as an electromagnetic wave suffice to describe all the phenomena we can observe? The answer is no, as explained in the following sections.

36.2 Blackbody Radiation

When we talked about thermal radiation in Chapter 18, we introduced the idealized concept of a blackbody. This idealization can be realized to good accuracy by looking at the radiation coming from a small hole in a large cavity kept at a temperature T. If we look at the visible light that emerges from such a hole at room temperature, the hole appears black because all the light that enters the cavity through the hole is scattered and ultimately absorbed by the walls. However, at much higher temperatures this hole begins to glow in the visible part of the electromagnetic spectrum. Everyday examples of visible light from blackbody radiation include the dull red color of the cooking elements of electric stoves, the bright light from the filament of an incandescent light bulb, and the light from the Sun (Figure 36.2).

Let's first briefly review what was known about this blackbody radiation from classical wave physics and from empirical observations. The Stefan-Boltzmann radiation law for the total intensity I (energy radiated per unit time and unit area) of this blackbody radiation is

$$I = \int_0^\infty \epsilon(\lambda)d\lambda = \sigma T^4. \tag{36.1}$$

FIGURE 36.2 Volcanic lava emits light and is a very good example of a blackbody radiator.

Here $\epsilon(\lambda)$ is the **spectral emittance** (often also called the *spectral radiance*) as a function of wavelength. It is the power radiated per unit area and wavelength, and has the SI units of $[\epsilon(\lambda)] = $ W m^{-3}. The integral extends over all possible wavelengths λ from zero to infinity, and σ is the Stefan-Boltzmann constant,

$$\sigma = 5.670400(40) \cdot 10^{-8} \text{ Wm}^{-2}\text{K}^{-4}.$$

The most important feature of the Stefan-Boltzmann radiation law (equation 36.1) is that the total intensity of the radiation grows with the fourth power of the temperature.

In 1896 German physicist Wilhelm Wien (1864–1928) empirically derived **Wien's law** to describe the spectral emittance of a blackbody

$$\epsilon_{\text{Wien}}(\lambda) = \frac{a}{\lambda^5} e^{-b/\lambda T} \quad \text{(approximation for small } \lambda\text{)}, \tag{36.2}$$

where a and b are constants. Wien's law succeeded in describing the spectral emittance of blackbodies for short wavelengths but was less successful in describing the spectral emittance for long wavelengths. The **Wien displacement law** summarizes another important experimental finding about the spectral emittance. It states that the spectral emittance peaks at a certain wavelength λ_m, and that this wavelength depends on the temperature,

$$\lambda_m T = \text{constant} = 2.90 \cdot 10^{-3} \text{ K m}. \tag{36.3}$$

Using the representation of light as an electromagnetic wave, the English physicists Lord Rayleigh and Sir James Jeans managed to derive an expression for the spectral emittance of this blackbody radiation,

$$\epsilon_{\text{RJ}}(\lambda) = \frac{2\pi c k_B T}{\lambda^4} \quad \text{(approximation for large } \lambda\text{)}. \tag{36.4}$$

Here c is the speed of light and k_B is Boltzmann's constant,

$$k_B = 1.3806503(24) \cdot 10^{-23} \text{ J/K}.$$

This solution, however, had one glaring fault: as $\lambda \rightarrow 0$, this expression diverges. This problem later became known as the *ultraviolet catastrophe* (recall that ultraviolet radiation has small wavelengths). If equation 36.4 were correct for all wavelengths, then the integral in equation 36.1 would diverge and the intensity radiated by a blackbody would become infinite at any temperature. Clearly, this is impossible! However, for large wavelengths the result obtained by Rayleigh and Jeans fit the experimental observations.

In order to provide a formula for the spectral emittance that fits the observations at all wavelengths, in 1900 the German physicist Max Planck took a radical step. He proposed that the energy contained in light and all other electromagnetic radiation interacts with solid objects in discrete bundles. He hypothesized that the energy of a bundle is proportional to the light frequency,

$$E = hf, \tag{36.5}$$

where h is **Planck's constant** and has the value

$$h = 6.62606876(52) \cdot 10^{-34} \text{ J s}. \tag{36.6}$$

We have already introduced the energy unit of electron-volt, 1 eV = $1.602178 \cdot 10^{-19}$ J, so Planck's constant can also be expressed in terms of the units eV s: $h = 4.13567 \cdot 10^{-15}$ eV s.

As you will see later in this chapter, many formulas in quantum physics involve Planck's constant divided by 2π. Because this is a relatively common occurrence, it is customary to use the notation $\hbar$ to denote this ratio:

$$\hbar \equiv \frac{h}{2\pi} = 1.05457 \cdot 10^{-34} \text{ J s} = 6.5821 \cdot 10^{-16} \text{ eV s}. \tag{36.7}$$

The wavelength and frequency of light are still related to the speed via $c = \lambda f$, so we can also write, instead of equation 36.5,

$$E = hf = \frac{hc}{\lambda}. \tag{36.8}$$

Planck's radiation law, found by Planck based on the quantized-energy hypothesis, is

$$I_T(f) = \frac{2h}{c^2} \frac{f^3}{e^{hf/k_B T} - 1}. \tag{36.9}$$

Here $I_T(f)df$ is the power (amount of energy per unit time) radiated in the frequency range between f and $f + df$ by a blackbody at temperature T per unit surface area of the blackbody opening and per unit solid angle. The name for $I_T(f)$ is the *specific intensity* or **spectral brightness**, and its SI unit is W m^{-2} sr^{-1} Hz^{-1}. For now, we just write down Planck's result, but we will revisit it later at the end of this chapter in order to understand its dependence on the frequency and temperature.

Note that the spectral brightness does not depend on direction. Spectral brightness $I_T(f)$ can be integrated over the entire hemisphere of all possible directions to give the spectral emittance $\epsilon_T(f)$:

$$\epsilon_T(f) = \int_\Omega I_T(f)\cos\theta\, d\Omega = \int_0^{\pi/2}\left(\int_0^{2\pi} I_T(f)d\phi\right)\sin\theta\cos\theta\, d\theta$$

$$= I_T(f)\int_0^{2\pi} d\phi \int_0^{\pi/2}\sin\theta\cos\theta\, d\theta$$

$$= I_T(f)2\pi\frac{1}{2}$$

$$= \pi I_T(f).$$

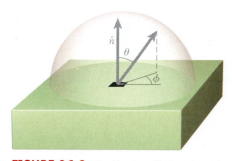

FIGURE 36.3 Blackbody radiating through the small black hole in all directions of the hemisphere.

Thus, we find that the spectral emittance is exactly a factor π bigger than the spectral brightness,

$$\epsilon_T(f) = \pi I_T(f). \tag{36.10}$$

The factor $\cos\theta$ in the first integral is the projection of the normal vector to the hole $\hat{n}$ in the direction of radiation, so $\cos\theta$ represents the effective reduction of the unit emission area as a function of the polar angle. The integral over the solid angle of the hemisphere is a double integral over the angles θ from 0 to $\pi/2$ and ϕ from 0 to 2π (Figure 36.3). These integrals are easily executed, as shown, because $I_T(f)$ does not depend on the angles. Combining equation 36.10 with equation 36.9,

$$\epsilon_T(f) = \frac{2\pi h}{c^2}\frac{f^3}{e^{hf/k_BT}-1}. \tag{36.11}$$

The spectral emittance $\epsilon_T(f)$ has the SI units of W m^{-2} Hz^{-1}.

We can also write the spectral brightness and spectral emittance as functions of the wavelength instead of the frequency. To do this, we use $c = \lambda f$, so

$$I_T(\lambda) = I_T(f)\left|\frac{df}{d\lambda}\right| = I_T(f)\left|\frac{d}{d\lambda}\left(\frac{c}{\lambda}\right)\right| = I_T(f)\frac{c}{\lambda^2}.$$

Therefore, the spectral brightness as a function of wavelength is given as

$$I_T(\lambda) = \frac{2hc^2}{\lambda^5\left(e^{hc/\lambda k_BT}-1\right)}. \tag{36.12}$$

Its SI units are W m^{-3} sr^{-1}. As done above, the spectral emittance can be obtained as a function of wavelength by integrating the spectral brightness over all emission angles, resulting in a multiplicative factor of π and giving

$$\epsilon(\lambda) = \frac{2\pi hc^2}{\lambda^5\left(e^{hc/\lambda k_BT}-1\right)}, \tag{36.13}$$

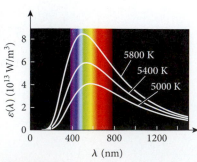

FIGURE 36.4 Planck spectral emittance as a function of wavelength for three temperatures: 5800 K, 5400 K, and 5000 K, from top to bottom.

with SI units of W m^{-3}. All four versions of Planck's law (equations 36.9, 36.11, 36.12, and 36.13) are equally valid. Care must be taken, however, when talking about the spectral brightness or the spectral emittance. They differ by a factor of π, as shown, because the spectral emittance is the integral of the spectral brightness over all emission angles in the hemisphere.

Figure 36.4 displays the shape of the Planck radiation law for the spectral emittance as a function of wavelength for three temperatures. The top curve is calculated for a temperature of 5800 K, approximately the surface temperature of the Sun. The emittance function has a maximum near a wavelength of 500 nm (green-blue), in accordance with the result of equation 36.3, the Wien displacement law. The other two curves are for 5400 K, peaking at 540 nm (yellow-green), and 5000 K, peaking at 580 nm (orange). Overlaid in this figure are the colors of the visible spectrum.

DERIVATION 36.1 | Radiation Laws

Let's show that Wien's law, the Rayleigh-Jeans law, the Stefan-Boltzmann radiation law, and the Wien displacement law can be derived from Planck's radiation law, equation 36.13.

Wien's law: For small values of λ, the argument of the exponential function in the Planck law becomes large, which allows us to write

$$\frac{1}{e^{hc/\lambda k_B T}-1}\approx e^{-hc/\lambda k_B T}.$$

We then obtain an expression for Wien's law as a limiting case of Planck's law:

$$\epsilon(\lambda)=\frac{2\pi hc^2}{\lambda^5\left(e^{hc/\lambda k_B T}-1\right)}\approx\frac{2\pi hc^2}{\lambda^5}e^{-hc/\lambda k_B T},$$

which has the same dependence on wavelength as equation 36.2, with the constants now defined as $a=2\pi hc^2$ and $b=hc/k_B$.

Rayleigh-Jeans law: For large values of λ, the argument of the exponential function in the Planck law becomes small. We can expand the exponential function for small arguments as $e^x\approx 1+x$. In this case, we then have $e^{hc/\lambda k_B T}-1\approx hc/\lambda k_B T$, and we find for large values of the wavelength,

$$\epsilon(\lambda)=\frac{2\pi hc^2}{\lambda^5\left(e^{hc/\lambda k_B T}-1\right)}\approx\frac{2\pi hc^2}{\lambda^5\left(hc/\lambda k_B T\right)}=\frac{2\pi ck_B T}{\lambda^4},$$

which is exactly what we wrote down for the Rayleigh-Jeans law in equation 36.4.

Wien displacement law: For this law, we need to find the wavelength for which $\epsilon(\lambda)$ reaches a maximum; that is, we need to take the derivative with respect to the wavelength and find the root. The derivative is

$$\frac{d\epsilon(\lambda)}{d\lambda}=\frac{d}{d\lambda}\left(\frac{2\pi hc^2}{\lambda^5\left(e^{hc/\lambda k_B T}-1\right)}\right)$$

$$=-\frac{10\pi hc^2}{\lambda^6\left(e^{hc/\lambda k_B T}-1\right)}+\frac{2\pi h^2 c^3 e^{hc/\lambda k_B T}}{\lambda^7\left(e^{hc/\lambda k_B T}-1\right)^2 k_B T}$$

$$=\frac{2\pi hc^2}{\lambda^7\left(e^{hc/\lambda k_B T}-1\right)^2 k_B T}\left(-5\lambda k_B T\left(e^{hc/\lambda k_B T}-1\right)+hce^{hc/\lambda k_B T}\right).$$

Except for the uninteresting case of $T\to\infty$, this expression can be zero only if the numerator is zero. Therefore, we need to solve

$$-5\lambda_m k_B T\left(e^{hc/\lambda_m k_B T}-1\right)+hce^{hc/\lambda_m k_B T}=0,$$

where λ_m is the value of the wavelength for which the Planck radiation law has its maximum. If we substitute $u=hc/\lambda_m k_B T$, then this equation reduces to

$$5\left(e^u-1\right)=ue^u\Rightarrow 5-5e^{-u}=u.$$

This equation can be solved by simple iteration, such as by using a spreadsheet. The trivial root is $u=0$, but we are not interested in this solution, which corresponds to an infinite wavelength. Therefore, we start our iteration at a finite value, say 1, and compute $5-5e^{-1}=3.1606$. Then we use this new value and insert again, finding $5-5e^{-3.1606}=4.7880$, and so on. You will find that you reach convergence very quickly and obtain

$$u=\frac{hc}{\lambda_m k_B T}=4.9651\Rightarrow\lambda_m T=\frac{hc}{k_B 4.9651}.$$

Continued—

Using the values of the constants h, c, k_B, we then find

$$\lambda_m T = 2.898 \cdot 10^{-3} \text{ K m},$$

which is in complete agreement with the experimentally found value of equation 36.3.

Stefan-Boltzmann law: To obtain the total intensity radiated, I, we need to integrate the Planck radiation law over the wavelength, from zero to infinity. We find:

$$I = \int_0^\infty \epsilon(\lambda) d\lambda = \int_0^\infty \frac{2\pi h c^2}{\lambda^5 \left(e^{hc/\lambda k_B T} - 1 \right)} d\lambda = \frac{2 k_B^4 \pi^5}{15 h^3 c^2} T^4.$$

Inserting the values for Planck's constant, Boltzmann's constant, and the speed of light, we can verify that the Stefan-Boltzmann constant is indeed given by

$$\sigma = \frac{2 k_B^4 \pi^5}{15 h^3 c^2} = 5.6704 \cdot 10^{-8} \text{ W m}^{-2} \text{ K}^{-4}.$$

36.1 In-Class Exercise:

The visible spectrum of light extends from approximately 380 nm (violet-blue) to 780 nm (red). What is the corresponding range of photon energies in units of electron-volts?

a) 1.59 eV to 3.26 eV

b) $2.54 \cdot 10^{-19}$ eV to $5.23 \cdot 10^{-19}$ eV

c) $0.38 \cdot 10^{15}$ eV to $0.78 \cdot 10^{15}$ eV

d) 190 eV to 390 eV

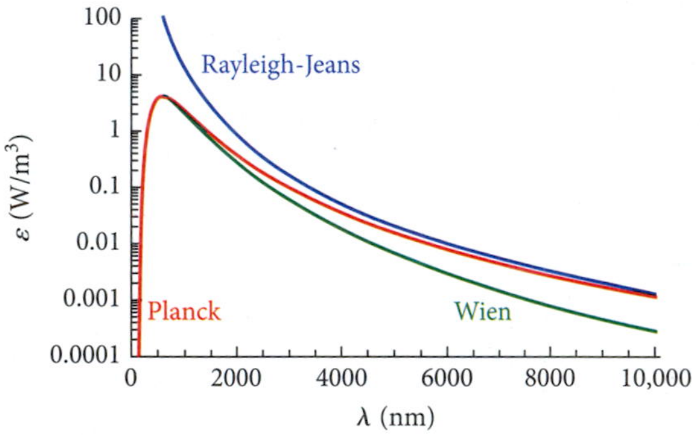

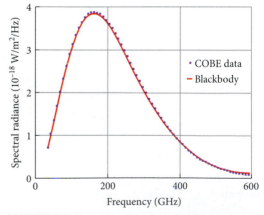

FIGURE 36.5 Comparison of Planck's radiation law, the Rayleigh-Jeans radiation law, and Wien's law (using the constants from Derivation 36.1) for $T = 5000$ K.

Thus, we see that the radiation law derived by Planck contains the previously known radiation laws as special cases, as illustrated in Figure 36.5.

Planck's law is consistent with Wien's law for short wavelengths and agrees with the Rayleigh-Jeans law at long wavelengths. This success gave immediate acceptance to the Planck radiation law for blackbodies, even though it was based on the radical assumption of quantized energy states. In particular, one can see from Derivation 36.1 how the quantum hypothesis resolves and avoids the classical ultraviolet catastrophe discussed earlier (see equation 36.4): For a given frequency f, the energy hf is needed to create a photon. As the frequency increases, it becomes less and less likely that the system can supply the energy needed for the creation of a photon. This leads to a cutoff at high frequencies and thus at low wavelengths, in agreement with observations. Thus, the observed avoidance of the ultraviolet catastrophe is a direct consequence of the quantum nature of light.

The most stunning example of a blackbody spectrum is obtained by looking at the cosmic background radiation. This radiation is a remnant of the Big Bang and is astonishingly uniform in the entire universe. The COBE satellite mission in 1990 and more recently the WMAP satellite mission have proven this in amazing detail. As indicated in Figure 36.6, the COBE mission found that the cosmic background radiation is that of a perfect blackbody at a temperature of 2.725± 0.001 K; that is, the entire universe is a perfect blackbody radiator. George Smoot and John Mather, the leaders of the COBE team, received the 2006 Nobel Prize in Physics for the achievements of this satellite mission. (A more detailed discussion of the cosmic background radiation is presented in Chapter 40.)

Just as blackbody radiation was used to measure the temperature of the universe, blackbody radiation can be used to measure the temperature of objects without physically touching them. If an object is hot enough, it will radiate photons in the visible range, as discussed at the beginning of this section. For example, the temperature of molten iron in a steel factory can be measured by analyzing the photons radiated from the red-hot molten iron. Objects close to room temperature radiate photons mainly in the infrared range. Modern infrared thermometers can be used to measure a person's temperature by observing the infrared radiation from the person's eardrum. Infrared thermometers are also used to measure the temperature of food and electrical components.

FIGURE 36.6 Data on the spectral radiance of microwave background photons as a function of frequency. The blue squares indicate data obtained by the COBE satellite, while the red curve is a best Planck spectrum fit at a temperature of 2.725 K.

36.3 Photoelectric Effect

Planck's hypothesis of some smallest possible discrete energy quanta was originally viewed as a computational construction, not as a real revolution in physics. This view changed, however, in 1905, with Einstein's explanation of the photoelectric effect. Einstein solved the puzzle of the photoelectric effect by proposing that light behaves as if it consists of localized bundles, or quanta, of energy-light. The photoelectric effect was originally discovered by Heinrich Hertz in 1886 and definitively demonstrated in 1916 by Robert A. Millikan, who quantitatively verified all of Einstein's predictions. Einstein's explanation of the photoelectric effect earned him the 1921 Nobel Prize in Physics. The quantized nature of light was conclusively demonstrated in 1923 by Compton, as we will see in Section 36.4. The American chemist Gilbert Lewis (1875–1946) coined the term **photon** in 1926 referring to these quanta of energy-light. For the remainder of this chapter, photons will be referred to as the quanta of light and all electromagnetic radiation.

In the **photoelectric effect**, light is able to knock electrons out of the surface of a suitable metal, creating an electric current. To see a practical application of the photoelectric effect, we have to look no further than the door of an elevator. How does this door sense that someone is standing in the opening? The answer is a photosensor, which usually consists of a light source and a light receptor utilizing the photoelectric effect. If an object, such as a person, is located between the light source and the receptor, the receptor no longer receives light and triggers electric switches to keep the elevator door open. The same principle applies to modern garage-door openers, which are required to use a photosensor to avoid crushing a person with the garage door on the way down.

A series of experiments can be performed to examine the photoelectric effect. Figure 36.7 shows the basic setup. On the left side is a light source, which could be a light bulb as shown or a light-emitting diode (which is utilized in many photocircuits) or just plain sunlight. On the right side is a photosensor, consisting of a piece of metal (cathode, rectangular shape) and a metal plate (anode, black line) housed in an evacuated glass container. A metal often used for the photosensor is cesium. This photosensor is part of a circuit with a voltage source and an ammeter. Between the light source and the photosensor is a filter that lets only one color of light through (blue in this case). Experiments yield the following observations:

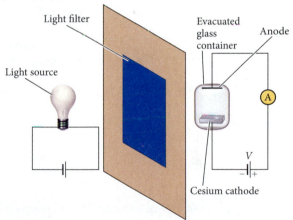

FIGURE 36.7 Schematic circuit diagram for the photoelectric effect.

- With voltage set at $V = 0$ and a blue filter, a current is detectable in the ammeter. This indicates that electrons are crossing the gap between the photo-metal and the other plate. If the intensity of the light is increased, the measured current rises as well, indicating that more electrons move across the gap.

- With a red filter, no current is detectable in the ammeter. This finding does not change as a function of the light intensity.

- With a blue filter and a positive value V, the current flowing through the ammeter rises. As voltage is changed to increasingly more negative values, the current measured in the ammeter is gradually reduced and then stops at some threshold value of the voltage.

From these experiments, one can conclude that electrons must get released from the surface of the metal by the light striking it. These electrons must have a kinetic energy, and the maximum of this kinetic energy can be measured by applying a negative voltage to the anode (top plate). The electrons need to overcome this voltage to cross the gap from the photocathode to the anode plate. If the electrons (charge $q = -e$) start with a maximum kinetic energy of K_{max} from the surface of the cathode and just reach the anode with zero kinetic energy after overcoming a potential of $V = -V_0$, then, from the work-energy theorem, we find:

$$W = \Delta K + \Delta V = (0 - K_{max}) + ((-e)(-V_0) - 0) = -K_{max} + eV_0 = 0 \Rightarrow$$

$$eV_0 = K_{max} = \tfrac{1}{2}mv_{max}^2. \qquad (36.14)$$

(The nonrelativistic approximation $\frac{1}{2}mv^2$ for the kinetic energy of the electrons can be used here, because the electrons are moving slowly in this case.) The potential V_0 is called the *stopping potential,* and for a given material it depends on the color of the light, which implies a frequency dependence of the stopping potential. Careful measurements reveal a linear dependence of V_0 on the frequency f. In addition, below a certain frequency the stopping potential becomes zero. Light with a lower frequency is not able to give the electrons in the photocathode enough energy to escape from its surface.

The key conceptual problems from the viewpoint of classical wave physics can be summarized as follows:

- Classically, a beam of light of any frequency can eject electrons from a metal, as long as the light has enough intensity. However, observations show that the incident light beam must have a frequency greater than the minimum value f_{min}, regardless of its intensity. Einstein's explanation was that energy of the energy-light quantum (now called a photon) is proportional to frequency, $E = hf$.

- Classically, the maximum kinetic energy of ejected electrons should increase with increasing intensity of the light beam. However, observations show that increasing the intensity of the light beam increases the number of electrons ejected per second, not their energy; only increasing the frequency of the light increases the energy of ejected electrons.

The physical picture of the photoelectric effect is that a photon with an energy determined by equation 36.5 hits an electron at the surface of the metal and ejects it from the material, provided the electron gains sufficient energy to overcome the electron's attraction to the material.

For a given material, a minimum energy is required to free an electron from its surface. This minimum energy is called the **work function,** ϕ, and is a constant for a given material. The maximum kinetic energy that an electron can have after the collision with the photon and being freed from the metal surface is then $K_{max} = hf - \phi$. Because K_{max} cannot be negative, this equation tells us that there is a minimum (threshold) light frequency

$$f_{min} = \phi/h \qquad (36.15)$$

necessary for the photoelectric effect to occur, consistent with the experimental results. Table 36.1 shows work functions and corresponding threshold frequencies and cutoff wave-

Table 36.1	Work Functions and Corresponding Minimum Frequencies and Maximum Wavelengths for Common Elements						
Element	ϕ (eV)	f_{min} (10^{15} Hz)	λ_{max} (nm)	Element	ϕ (eV)	f_{min} (10^{15} Hz)	λ_{max} (nm)
Aluminum	4.1	0.99	302	Magnesium	3.7	0.89	335
Beryllium	5	1.21	248	Mercury	4.5	1.09	276
Cadmium	4.1	0.99	302	Nickel	5	1.21	248
Calcium	2.9	0.70	428	Niobium	4.3	1.04	288
Carbon	4.8	1.16	258	Potassium	2.3	0.56	539
Cesium	2.1	0.51	590	Platinum	6.3	1.52	197
Cobalt	5	1.21	248	Selenium	5.1	1.23	243
Copper	4.7	1.14	264	Silver	4.7	1.14	264
Gold	5.1	1.23	243	Sodium	2.3	0.56	539
Iron	4.5	1.09	276	Uranium	3.6	0.87	344
Lead	4.1	0.99	302	Zinc	4.3	1.04	288

lengths for various materials. Using the connection between the maximum kinetic energy and the stopping potential in equation 36.14, we then find for the frequency dependence of the stopping potential in the photoelectric effect,

$$eV_0 = hf - \phi. \qquad (36.16)$$

EXAMPLE 36.1 | Work Function

Suppose you are using a circuit as depicted on the right side of Figure 36.7 and you have a photosensor with an unknown material as the photocathode. When using light of wavelength 250 nm (ultraviolet), you find that you have to apply a stopping potential of 2.86 V to eliminate the current. When using light of wavelength 400 nm (blue-violet), you measure a value of 1.00 V for the stopping potential, and using a wavelength of 630 nm (orange) you measure a stopping potential of 0.130 V.

PROBLEM

What is the value of the work function of this material?

SOLUTION

It is perhaps easiest to solve this problem in graphical form. Equation 36.16 shows that the stopping potential, the work function, and the frequency have a linear relationship, and because linear relationships can be drawn as straight lines, it is best to convert the given wavelengths into corresponding frequencies by using $f = c/\lambda$. We then plot the stopping potential V_0 as a function of frequency for the three given data points (Figure 36.8). When we fit a straight line through these data points, the line gives us a value of -2.1 V at $f = 0$. Using equation 36.16, we then find for the work function:

$$eV_0(f = 0) = -\phi \Rightarrow \phi = -eV_0(f = 0) = -e(-2.1 \text{ V}) = 2.1 \text{ eV}.$$

Looking at Table 36.1, we can see that the material used for this photosensor is probably cesium, the material with the lowest work function of all elements listed.

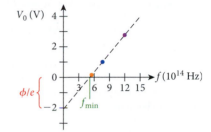

FIGURE 36.8 Applied stopping potential as a function of the frequency of light for a particular photosensor material.

36.1 Self-Test Opportunity

Assuming that the work function in Example 36.1 indeed has a value of 2.1 eV, calculate the maximum kinetic energy that emitted electrons can have in the three given cases of light of different wavelengths incident on the photosensor.

Chapter 35, on relativity, showed that the momentum and energy of any particle are related by $E^2 = p^2c^2 + m^2c^4$. A photon has zero mass, so its energy and momentum are related by $E = pc$. (When the symbol p is used with no arrow, it means the magnitude of the momentum; however, we may refer to p simply as the momentum.) For the momentum of a photon, we can then write, from equation 36.8:

$$p = \frac{E}{c} = \frac{hf}{c} = \frac{h}{\lambda}. \qquad (36.17)$$

Thus, we see that momentum and energy of a photon are both proportional to the frequency and inversely proportional to the wavelength of the corresponding electromagnetic radiation. A photon acts like a particle, even though it is described by a frequency. Light acts like a wave even though it consists of particles called photons. This **wave-particle duality** of light is conceptually hard to grasp, and it kept physicists and philosophers busy for the first part of the 20th century.

Several practical problems are associated with detecting single photons of visible light. First, as just shown, each photon has energy only in the range between 1.6 and 3.3 eV. Expressed in SI units, this is in the range between $2.6 \cdot 10^{-19}$ and $5.2 \cdot 10^{-19}$ J—a very small amount of energy. Even the best photocathodes have a quantum efficiency of only 30% or less for a photon in this energy range, meaning that at most only 30% of the photons hitting the photocathode actually manage to knock out an electron. Second, one liberated electron represents only a very tiny charge and thus only a very tiny current. To register an easily measurable current, many electrons need to be generated from each photoelectron. Many practical devices accomplish this with photomultiplier tubes.

A **photomultiplier tube** makes use of the fact that an electron that hits a metal surface with kinetic energy on the order of 100 eV usually knocks out several electrons in the process. Therefore, a photocathode and an anode are combined in an evacuated glass tube with several intermediate plates, called *dynodes*. Each dynode is kept at a potential difference of several hundred volts relative to its neighbors (Figure 36.9). Commercially available photomultiplier tubes have chains of up to $n = 14$ dynodes, and each dynode produces on average η electrons for each electron that hits it, where η can have values of up to 3.5. The total gain of such a photomultiplier tube is then η^n. For $n = 14$ and $\eta = 3.4$, for example, this effect results in a gain factor of $2.76 \cdot 10^7$—that is, almost 28 million electrons are produced at the anode for each photoelectron knocked out of the photocathode by a single photon.

Another application of the detection of photons is night-vision devices, technology mentioned in the opening of this chapter. A schematic drawing of a night-vision device is shown in Figure 36.10.

A night-vision device uses a photocathode much like a photomultiplier tube. However, in this case the entire image is focused on the photocathode by the objective lens. The incident photons cause the photocathode to release electrons. A potential difference of around 1000 V accelerates these electrons into a microchannel plate that contains an array of millions of channels that function like miniature photomultiplier tubes, multiplying the num-

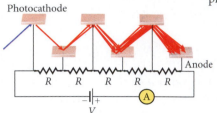

FIGURE 36.9 Schematic drawing of a photomultiplier tube. The blue arrow represents a single photon, and the red arrows represent electrons.

FIGURE 36.10 A night-vision device produces an intensified image of an object in low-light conditions.

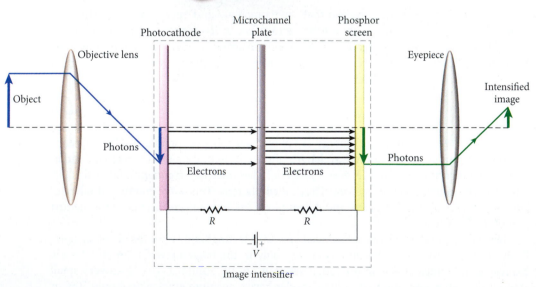

ber of electrons by a factor of around 10^4. A second potential difference of around 1000 V between the microchannel plate and a phosphor screen accelerates the multiplied electrons so that they hit the screen and cause the emission green light where they strike the screen, forming an intensified image. This light is then focused by the eyepiece to produce an image like the one shown in Figure 36.1. This night-vision device works by amplifying low light levels. There is another type of night-vision device that uses infrared light emitted by warm objects to observe those objects in the dark.

Other methods are also used to detect single photons and convert them into an electric signal. Most notable among these are charge-coupled devices (CCDs) and complementary metal oxide semiconductor (CMOS) devices, which form the basis of every digital camera and video recorder in stores now. However, an understanding of the physics underlying a CCD or a CMOS device requires study of semiconductors. Chapter 38 on atomic physics will explain how a CCD works.

EXAMPLE 36.2 | Photons from a Laser Pointer

Any object that you see emits photons that travel from this object to your retinas, where the photons trigger electric signals that are sent to your brain. Let's look at a light source to begin to understand the numbers of photons involved.

PROBLEM
What is the approximate number of photons emitted per second by a 5.00-mW green laser pointer?

SOLUTION
Green laser pointers usually operate at a wavelength of 532 nm. (Later in this chapter we will see the reason for this value.) A wavelength of 532 nm corresponds to a frequency of

$$f = \frac{c}{\lambda} = \frac{2.998 \cdot 10^8 \, \text{m/s}}{5.32 \cdot 10^{-7} \, \text{m}} = 5.635 \cdot 10^{14} \, \text{Hz}.$$

With Planck's hypothesis, $E = hf$, we can now calculate the energy contained in one single photon emitted by the green laser pointer:

$$E = hf = (6.626 \cdot 10^{-34} \, \text{J s})(5.635 \cdot 10^{14} \, \text{s}^{-1}) = 3.73 \cdot 10^{-19} \, \text{J}.$$

Because the laser pointer is rated at 5.00 mW, it emits 5.00 mJ of energy each second. The number of photons emitted in each second is therefore

$$n = \frac{5.00 \cdot 10^{-3} \, \text{J}}{3.73 \cdot 10^{-19} \, \text{J}} = 1.34 \cdot 10^{16}.$$

In other words, this little handheld laser pointer emits over thirteen million billion photons each second!

DISCUSSION
The value of Planck's constant has already been given in units of eV s, so the energy of a single photon emitted from this green laser pointer can be expressed in these units as well:

$$E = hf = (4.13567 \cdot 10^{-15} \, \text{eV s})(5.635 \cdot 10^{14} \, \text{s}^{-1}) = 2.33 \, \text{eV}.$$

Now you may begin to grasp the usefulness of the energy unit eV for dealing with atomic and quantum phenomena. The typical energy scale for processes in this realm is the electron-volt.

36.2 In-Class Exercise
You have a source of light with a given intensity and wavelength. You reduce the wavelength while leaving the intensity the same. Which of the following statements is true?

a) You will obtain more photons per second from the light source.

b) You will obtain fewer photons per second from the light source.

c) The number of photons emitted per second will remain the same, but the energy of each one is reduced.

d) The number of photons emitted per second will remain the same, but the energy of each one is increased.

e) The number of photons emitted per second will remain the same, each one will move slower.

36.2 Self-Test Opportunity
Calculate the number of photons in the visible spectrum emitted each second by our Sun. To solve this task, you need to know that the Sun's radiation has an intensity of 1370 W/m^2 on Earth, which is at a distance of 148 million km from the Sun. From this information you can calculate the total power output of the Sun. Then, from examining Figure 36.4, you can see that approximately $\frac{1}{4}$ of the photons in the Sun's radiation are in the visible spectrum.

36.4   Compton Scattering

The discussion of the electromagnetic spectrum in Chapter 31 characterized X-rays as those electromagnetic waves with frequencies approximately 100 to 100,000 times higher than visible light. Using our present picture of the electromagnetic spectrum consisting of photons, we see that X-ray photons have energies of hundreds to hundreds of thousands of electron-volts. X-rays can be produced by accelerating electrons to several thousand electron-volts (several keV) of kinetic energy and then shooting them into a metal foil. The deceleration of the electrons in the foil creates the X-rays. These X-rays are called **Bremsstrahlung,** which is the German word for deceleration radiation. (Another process also occurs in which atoms become excited and then emit X-rays of certain energies; this process is discussed in Chapter 37.) Classical electromagnetic theory can make certain predictions for electromagnetic radiation from accelerated charges, but a complete understanding really requires a theory called **quantum electrodynamics,** developed in the 1950s by American physicists Julian Schwinger (1918–1994) and Richard Feynman (1918–1988) and Japanese physicist Sin-Itiro Tomonaga (1906–1979), for which they shared the 1965 Nobel Prize in Physics.

For now, consider what happens when an X-ray scatters off an electron. First, what does the wave picture of light predict? If a wave hits a stationary small object like an electron, Huygens's Principle tells us that a spherical wave originates from the object, scattering (reflecting) the incoming wave. The scattered wave has the same frequency and wavelength as the incoming wave. However, in 1923, American physicist Arthur Holly Compton (1892–1962) discovered that X-rays scattered off electrons at rest produced X-rays with longer wavelengths than the original X-rays. A larger wavelength implies a smaller frequency, and thus a lower energy and momentum for the X-ray photons, according to equation 36.8.

If one accepts that light photons have particle-like properties of momentum and energy, then the interaction of an X-ray and an electron can be analyzed just like the scattering of one billiard ball off another. Because photons move with the speed of light, and because the X-ray energies are not negligible compared to the mass of the electron, we have to employ relativistic dynamics (Chapter 35) and cannot simply employ the formulas developed in Chapter 7 on momentum and collisions. However, what stays the same is that conservation laws of energy and momentum are used to arrive at the desired result.

Let's call the energy of the X-ray photon before the collision E and after the collision E'. Then the magnitudes of the corresponding photon momenta before and after the collision are $p = E/c$ and $p' = E'/c$. The electron has no momentum before the collision because we assume it is at rest. During the collision, it receives a momentum $\vec{p}_e$. A diagram of the scattering process is shown in Figure 36.11. The energy of the electron before the collision is simply its rest energy $m_e c^2$, and its energy after the collision is

$$E_e = \sqrt{\left(p_e c\right)^2 + \left(m_e c^2\right)^2}.$$

Energy and momentum conservation during the collision then imply

$$\vec{p}' + \vec{p}_e = \vec{p} \tag{36.18}$$

$$E' + E_e = E + m_e c^2. \tag{36.19}$$

Derivation 36.2 shows that the final wavelength of the X-ray can be expressed as

$$\lambda' = \lambda + \frac{h}{m_e c}\left(1 - \cos\theta\right), \tag{36.20}$$

where θ is the angle between the incoming and outgoing photon. This is the formula for **Compton scattering,** linking the wavelength of the photon after scattering to the wavelength of the incoming photon.

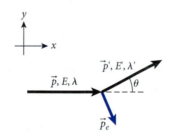

FIGURE 36.11 Momentum conservation in Compton scattering.

DERIVATION 36.2 | Compton Scattering

To derive equation 36.20, we isolate $\vec{p}_e$ in equation 36.18 and E_e in equation 36.19. For the first equation, this results in $\vec{p}_e = \vec{p} - \vec{p}'$. Square both sides, to obtain

$$p_e^2 = \left(\vec{p} - \vec{p}'\right)^2 = p^2 + p'^2 - 2pp'\cos\theta. \qquad \text{(i)}$$

Rearranging and taking the square of each side for equation 36.19 yields

$$E_e = E - E' + m_e c^2 \Rightarrow$$

$$E_e^2 = \left(E - E' + m_e c^2\right)^2 \Rightarrow$$

$$p_e^2 c^2 + m_e^2 c^4 = \left(E - E'\right)^2 + m_e^2 c^4 + 2\left(E - E'\right) m_e c^2,$$

where we have made use of the relativistic energy-momentum relation $E_e^2 = p_e^2 c^2 + m_e^2 c^4$ (see Section 35.7) on the left-hand side of the last line of this calculation. We can also use the relation between momentum and energy $E = pc$ for the photon and obtain

$$p_e^2 c^2 + m_e^2 c^4 = \left(p - p'\right)^2 c^2 + m_e^2 c^4 + 2\left(p - p'\right) m_e c^3.$$

Subtracting the term $m_e^2 c^4$ from both sides and then dividing by a common factor c^2 gives

$$p_e^2 = \left(p - p'\right)^2 + 2\left(p - p'\right) m_e c. \qquad \text{(ii)}$$

Equations (i) and (ii) have the same left-hand side, and so their right-hand sides must be equal as well:

$$p^2 + p'^2 - 2pp'\cos\theta = \left(p - p'\right)^2 + 2\left(p - p'\right) m_e c.$$

Using $(p - p')^2 = p^2 + p'^2 - 2pp'$, we then get

$$p^2 + p'^2 - 2pp'\cos\theta = p^2 + p'^2 - 2pp' + 2\left(p - p'\right) m_e c \Rightarrow$$

$$2pp'\left(1 - \cos\theta\right) = 2\left(p - p'\right) m_e c.$$

Now we use the relationship in equation 36.17 between photon momentum and wavelength, $p = h/\lambda$, and find

$$2\frac{h}{\lambda}\frac{h}{\lambda'}\left(1 - \cos\theta\right) = 2\left(\frac{h}{\lambda} - \frac{h}{\lambda'}\right) m_e c \Rightarrow \frac{h}{m_e c}\left(1 - \cos\theta\right) = \lambda' - \lambda,$$

which is the result stated in equation 36.20.

The ratio of the constants $h/m_e c$ has the dimension of a length, as can be seen from equation 36.20. This characteristic of an electron is called the **Compton wavelength** of the electron and has the value

$$\lambda_e = \frac{h}{m_e c} = \frac{6.626 \cdot 10^{-34} \text{ J s}}{(9.109 \cdot 10^{-31} \text{ kg})(2.998 \cdot 10^8 \text{ m/s})} = 2.426 \cdot 10^{-12} \text{ m}. \qquad (36.21)$$

36.3 Self-Test Opportunity

The electrons inside a metal are not quite at rest, but have kinetic energies of a few eV. Why is it permissible to assume that the electron is at rest in the derivation of the Compton scattering formula?

EXAMPLE 36.3 | Compton Scattering

An X-ray with a frequency of $3.3530 \cdot 10^{19}$ Hz strikes a metal foil and the scattered photon is detected at an angle of 32.300 degrees relative to the direction of the original X-ray.

PROBLEM

What is the energy of the incoming photon and that of the scattered photon, in units of eV?

Continued—

SOLUTION

First, let us work on the incoming photon. Converting its frequency into energy is a straightforward application of $E = hf$. Since we want to know the answer in units of eV, we should use the value of Planck's constant in units of eV s:

$$E = hf = (4.13567 \cdot 10^{-15} \text{ eV s})(3.3530 \cdot 10^{19} \text{ s}^{-1}) = 1.3867 \cdot 10^{5} \text{ keV} = 138.67 \text{ keV}.$$

In order to find the energy of the scattered photon, we need to use the Compton scattering result (see Figure 36.11). For this equation, we need the wavelength of the incoming photon, which we can obtain from the frequency via

$$\lambda = \frac{c}{f} = \frac{2.9979 \cdot 10^{8} \text{ m/s}}{3.3530 \cdot 10^{19} \text{ s}^{-1}} = 8.941 \cdot 10^{-12} \text{ m}.$$

Now we use the result we obtained for the scattering of a photon off an electron, including the value of the Compton wavelength (equation 36.23):

$$\lambda' = \lambda + \frac{h}{m_e c}\left(1 - \cos\theta\right) = \left(8.941 \cdot 10^{-12} \text{ m}\right) + \left(2.426 \cdot 10^{-12} \text{ m}\right)\left(1 - \cos\left(32.300°\right)\right)$$

$$\lambda' = 9.3164 \cdot 10^{-12} \text{ m}.$$

Converting this new wavelength back into energy results finally in

$$E' = \frac{hc}{\lambda'} = \frac{(4.13567 \cdot 10^{-15} \text{ eV s})(2.9979 \cdot 10^{8} \text{ m/s})}{9.3164 \cdot 10^{-12} \text{ m}} = 133.08 \text{ keV}.$$

PROBLEM

What is the kinetic energy of the electron after the collision? What is the magnitude of the momentum of the electron after the collision? Assume that the incoming photon traveled along the positive x-axis, that the scattering event takes place in the xy-plane, and use momentum units of keV/c.

SOLUTION

Energy is conserved in this scattering event of a photon off an electron. (This comes as no surprise, because energy conservation was one of the starting principles in the derivation of the Compton scattering formula.) Thus, the kinetic energy that the electron receives in this scattering process is simply equal to the energy loss endured by the photon:

$$K_e = E - E' = 138.67 \text{ keV} - 133.08 \text{ keV} = 5.59 \text{ keV}.$$

The total energy of the electron is its kinetic energy plus its rest energy, $m_e c^2$,

$$E_e = K_e + m_e c^2 = \sqrt{p_e^2 c^2 + m_e^2 c^4}.$$

Solving this equation for the absolute value of the momentum vector of the electron, we then find

$$p_e = \frac{1}{c}\sqrt{E_e^2 - m_e^2 c^4} = \frac{1}{c}\sqrt{\left(K_e + m_e c^2\right)^2 - m_e^2 c^4} = \frac{1}{c}\sqrt{K_e^2 + 2K_e m_e c^2} = 75.79 \text{ keV/}c.$$

ALTERNATIVE SOLUTION

The electron momentum can also be obtained using momentum conservation:

$$\vec{p}_e = \vec{p} - \vec{p}\,'.$$

Because we have calculated the energy of the photon before and after the collision, we can obtain the magnitudes of the initial and final photon momentum vectors in units of keV/c:

$$p = E/c = 138.67 \text{ keV/}c$$

$$p' = E'/c = 133.08 \text{ keV/}c.$$

The initial photon momentum, $\vec{p}$, only has an x-component, because we assumed that it traveled along the positive x-axis. It follows that

$$p_x = 138.67 \text{ keV/}c$$

$$p_y = 0.$$

The angle of the outgoing photon was specified in the initial setup of the problem as 32.300 degrees. We then obtain for the Cartesian components of the final momentum vector

$$p_x' = (133.08 \text{ keV}/c)\cos(32.300°) = 112.49 \text{ keV}/c$$
$$p_y' = (133.08 \text{ keV}/c)\sin(32.300°) = 71.11 \text{ keV}/c.$$

Thus, we have for the components of the electron momentum,

$$p_{e,x} = p_x - p_x' = 138.67 \text{ keV}/c - 112.49 \text{ keV}/c = 26.18 \text{ keV}/c$$
$$p_{e,y} = p_y - p_y' = -71.11 \text{ keV}/c.$$

Now we can obtain the absolute value of the momentum vector of the electron by our usual procedure of taking the square root of the sum of the squares of the components,

$$p_e = \sqrt{p_{e,x}^2 + p_{e,y}^2} = 75.78 \text{ keV}/c.$$

This is the same result as above, showing that our two methods of solution are consistent with each other.

36.5 Matter Waves

So far, we have established that photons are the quantum particles of light and of all other electromagnetic radiation. However, everything we have said about the wave nature of light is still true; for example, we can demonstrate interference and diffraction, which are typical wave phenomena. Looking at light as quantum particles does not invalidate the wave picture of light, just as the ray picture of light can be seen as a special limiting case of the more general wave description of light.

Given the quantum character of light and the particle character of electromagnetic waves, do things that we ordinarily think of as particles, like electrons and atoms, also have wave properties? This is exactly what Prince Louis de Broglie (1892–1987), a French graduate student at the time, proposed in 1923. His Ph.D. thesis, all of two pages long, contained this hypothesis and it won him the Nobel Prize in Physics in 1929.

If particles have wave character, then what is the appropriate wavelength? For light, we found (equation 36.17) that the momentum of a photon is $p = h/\lambda$. Thus, de Broglie tried the same for particles and proposed as the relevant wavelength for **matter waves**,

$$\lambda = \frac{h}{p} = \frac{h}{mv\gamma} = \frac{h}{mv}\sqrt{1 - \frac{v^2}{c^2}}. \tag{36.22}$$

This wavelength, the de Broglie wavelength, depends on the mass m and speed v of a particle. Equation 36.22 used the relativistic form of the momentum, $p = mv\gamma$, but the literature often presents the nonrelativistic approximation

$$\lambda = \frac{h}{mv} \quad \text{(nonrelativistic approximation)}, \tag{36.23}$$

described as the **de Broglie wavelength.** As Figure 36.12 shows for the case of an electron, up to speeds of 40% of the speed of light the nonrelativistic approximation is very close to the exact result in equation 36.22.

As you can see from Figure 36.12, the de Broglie wavelength of an electron, even one moving at 10% of the speed of light, is on the order of one-tenth of a nanometer. What are typical de Broglie wavelengths for macroscopic objects? Example 36.4 shows a calculation.

EXAMPLE 36.4 De Broglie Wavelength of a Raindrop

Raindrop sizes vary from approximately 0.50 mm diameter to 5.0 mm diameter. At the lower end of this size range, they fall with speeds of 2 m/s; at the upper end, with speeds up to 9 m/s.

Continued—

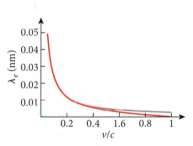

FIGURE 36.12 De Broglie wavelength of an electron as a function of its speed. Red: exact result; gray: nonrelativistic approximation.

PROBLEM

What is the range of de Broglie wavelengths for raindrops?

SOLUTION

The mass and diameter of a raindrop are related via

$$m = V\rho = \tfrac{4}{3}\pi r^3 \rho = \tfrac{1}{6}\pi d^3 \rho,$$

where ρ is density, r is radius, and d is diameter of a drop. The density of water is $\rho = 1000 \text{ kg/m}^3$, so the mass of a $d = 0.50$ mm drop is $6.5 \cdot 10^{-8}$ kg, and the mass of a $d = 5.0$-mm drop is $6.5 \cdot 10^{-5}$ kg.

For the extremely small speeds under consideration in this example, we are easily justified in using the nonrelativistic approximation $\lambda = h/mv$ for the de Broglie wavelength. We then obtain for the smallest raindrop

$$\lambda = \frac{h}{mv} = \frac{6.626 \cdot 10^{-34} \text{ J s}}{(6.5 \cdot 10^{-8} \text{ kg})(2 \text{ m/s})} = 5 \cdot 10^{-27} \text{ m}$$

and for the largest raindrops $\lambda = 1 \cdot 10^{-30}$ m.

DISCUSSION

Even the smallest and slowest-moving raindrops have de Broglie wavelengths that are many orders of magnitude smaller than diameters of individual atoms, which are approximately 10^{-10} m. Thus, we can safely ignore all implications of matter waves for macroscopic objects. Any object that is big enough for us to see it with the unaided eye, and that moves fast enough that we can discern some kind of motion at all, has a de Broglie wavelength so small that it rules out any observation of quantum wave phenomena for this object.

36.5 Self-Test Opportunity

Produce a plot of the de Broglie wavelength of an electron as a function of its kinetic energy, for kinetic energies between 1 and 1000 eV. Is there a visible difference in this range of energies if you use a nonrelativistic approximation of $p = \sqrt{2mK}$?

As we have seen in the Compton scattering example, the kinetic energy-momentum relationship is given by

$$p = \frac{1}{c}\sqrt{E^2 - m^2 c^4} = \frac{1}{c}\sqrt{(K + mc^2)^2 - m^2 c^4} = \frac{1}{c}\sqrt{K^2 + 2Kmc^2}.$$

Therefore, the de Broglie wavelength can also be written as a function of the kinetic energy of a particle:

$$\lambda = \frac{h}{p} = \frac{hc}{\sqrt{K^2 + 2Kmc^2}}.$$

So far we have discussed only the theoretical possibility of matter waves, simply by following de Broglie's postulate. Where is the experimental proof? Before we show this proof, it is helpful to go back to Chapter 34, the chapter on wave optics, and see what makes a wave a wave.

Double-Slit Experiment for Particles

To provide experimental evidence for the existence of matter waves, it is necessary to show that objects such as electrons, neutrons, protons, or whole atoms, which have mass and are normally thought of as particles, can exhibit wavelike behavior. Chapter 34 showed that the two main physical effects that characterize waves are interference and diffraction. What sort of experiment could show interference and/or diffraction of matter?

Young was able to demonstrate the wave character of light by shining light through two narrow slits, separated by a distance d. The interference pattern produced in this way is shown in Figure 36.13.

In the double-slit experiment, bright interference fringes appear on the screen a distance L away from the slits. Provided the line from the center of the two slits to a bright fringe makes a small angle with the perpendicular to the screen, the distance from this fringe to a neighboring fringe is given by

$$\Delta x = \frac{\lambda L}{d}. \tag{36.24}$$

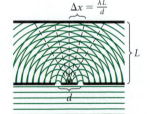

FIGURE 36.13 Double-slit interference for light.

$\Delta x = \frac{\lambda L}{d}$

However, in the derivation in Chapter 34 of this formula for the interference maxima, it was necessary to use the condition that the width of an individual slit is on the order of the wavelength of the light wave. For moderately fast-moving electrons, the de Broglie wavelength is on the order of one-tenth of a nanometer or less, and thus is more than three orders smaller than the wavelength of visible light. Producing a double slit of sufficiently small slit separation d and individual slit width a to carry out a double-slit experiment for electrons presents a sizable technical challenge. Also, if we want to use heavier particles, such as protons or neutrons, then the technical challenges become even bigger because the de Broglie wavelength is inversely proportional to the mass of a particle moving with a given speed.

For this reason, a double-slit experiment to verify the wave character of electrons was not carried out right away after de Broglie postulated his revolutionary idea. Instead, an experiment in 1927 by American physicists Clinton Davisson (1881–1958) and Lester H. Germer (1896–1971), working at the Bell Telephone Laboratories in New Jersey, provided proof for the physical existence of matter waves. Davisson and Germer continued their earlier work on Bragg scattering of X-rays off crystals and scattered a beam of electrons off a nickel crystal, observing interference patterns similar to those produced by X-rays. Today, beams of neutrons can also be scattered off crystals; American physicist Clifford G. Shull (1915–2001) and Canadian physicist Bertram Brockhouse (1918–2003) received the 1994 Nobel Prize in Physics for pioneering these techniques. X-rays, electrons, and neutrons all provide equivalent Bragg-scattering patterns, proof that matter waves are real.

Only since the early 1960s has there existed sufficiently precise technology to conduct double-slit scattering experiments with electrons. Thus, we can now think about what is going to happen in this experiment and then compare our thinking with the outcome of the experiment. The basic setup of the experiment is shown in Figure 36.14. An electron is emitted from a plate (the plate is heated in order to do this, but we do not show the heater here) and then accelerated by a voltage V. It then passes through a double slit (center) on its way to the screen above.

If electrons behave like particles, they will travel in straight lines from the electron gun through one of the slits and on to the screen. We then predict that we will see the electrons in two lines on the screen—the images of the two slits. We can allow that the electrons might get slightly deflected as they pass though a slit, so the distribution of the electrons passing through each slit will be somewhat smeared out. Because the slit separation d is very small, the two distributions of the electrons passing through the two slits will overlap on the screen. This classical particle-like expectation then leads to a distribution of the number of electrons hitting a certain region of the screen that is sketched in Figure 36.15a. Here the distribution from the electrons passing through the left slit is shown in blue, and that of the electrons passing through the right slit in red. Shown in green is the sum of the two, the total intensity distribution that should be recorded if our particle-like expectation bears out.

On the other hand, if electrons show wavelike characteristics, then the intensity should be governed by the combined effects of interference and diffraction, just as we observed for light in Chapter 34 on wave optics. In this case (compare Section 34.9), the intensity as a function of the coordinate x along the screen should be:

$$I(x) = I_{max} \cos^2\left(\frac{\pi d}{\lambda L} x\right)\left(\frac{\lambda L}{\pi a x}\sin\left(\frac{\pi a}{\lambda L}x\right)\right)^2. \qquad (36.25)$$

Here d is the slit separation, a is the width of the individual slits, and L is the distance between the double slit and the screen. This is the same terminology used in Young's interference experiment with light, but now λ is the de Broglie wavelength of the electron (equation 36.22). The function $I(x)$ of equation 36.25 is sketched in Figure 36.15b.

Figure 36.16 shows what the pattern of electrons striking the screen (yellow spots) might look like for an intensity distribution as given in equation 36.25. For this calculation, we used a de Broglie wavelength of 12.2 pm, a slit separation of 3.0 nm, a slit width of 1.0 nm, and a distance to the screen of 1.0 m. The lower part of this figure shows the counting histogram—that is, the number of electron hits in

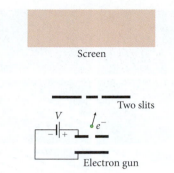

FIGURE 36.14 Experimental setup for an electron double-slit experiment.

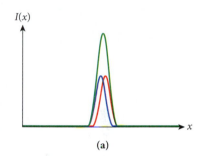

(a)

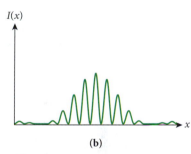

(b)

FIGURE 36.15 Intensity distribution (number of electrons hitting the screen per unit length) along the screen in the electron double-slit experiment. (a) Classical particle-like expectation; (b) expectation if electrons show matter-wave characteristics.

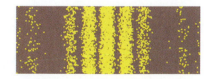

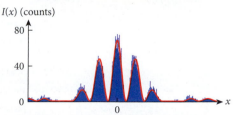

FIGURE 36.16 Connection between hits of individual electrons on the screen (yellow spots in upper part) and count histogram (blue, lower part), with predicted intensity distribution (red curve).

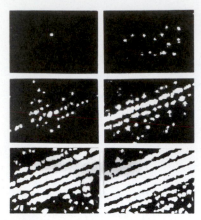

FIGURE 36.17 Experimental double-slit electron interference pattern as it develops over time.

a given interval of the coordinate x along the screen. Overlaid in red over the blue counting histogram is the intensity distribution of equation 36.25.

What is the outcome of the actual experiment? Electrons can be shot through the slits and at the screen individually, so the pattern can be observed to emerge as a function of time. This experiment was actually performed by P.G. Merli et al. in 1976; the outcome is shown in Figure 36.17. The lower right panel shows the outcome of this experiment, with the interference fringes clearly visible. The experiment clearly demonstrates that the electrons show wavelike interference phenomena.

However, the same figure shows more: You can see that each individual electron leaves a mark by hitting a certain very localized area of the screen. Therefore, the idea that each electron somehow gets distributed over the entire screen, proportional to the overall intensity distribution, is not true either.

Because photons have particle properties, it might be expected that photons striking the double slits would exhibit the granularity shown in Figure 36.17. When the double-slit experiment is performed using one photon at a time, a pattern similar to that shown for electrons in Figure 36.17 is indeed observed.

So, what makes an individual electron decide where to go? This is the central question of quantum physics. Understanding the answer will tell us most of what is essential about the atomic quantum world.

Are both slits essential for this interference pattern to emerge? In other words, does the electron somehow move through both slits at the same time? If one slit is closed, the interference pattern on the screen disappears, and only one maximum—the image of that slit—is produced. The result then corresponds to either the red or the blue curve in Figure 36.15a, depending on which slit was covered.

What if, instead, it would be possible to measure which slit the electron moves through without closing the other slit? Perhaps this could be accomplished by using the fact that the electron has a charge and thus represents a current as it moves through a slit. Perhaps this current could be measured as the electron moves through a slit.

36.4 In-Class Exercise

Before we go any further, let's see if you can guess the outcome of measuring which slit the electron passes through. If we conduct the measurement of the current representing the electron passing through the slits, which of the following will be the outcome?

a) Exactly one-half of each individual electron passes through each of the two slits.

b) Each individual electron passes through just one slit. The electrons passing through the left slit cause the left part of the interference pattern, and the electrons passing through the right slit cause the right part.

c) Each individual electron passes through just one slit. The electrons passing through the left slit cause the right part of the interference pattern, and the electrons passing through the right slit cause the left part.

d) Each individual electron passes through just one slit, but as we measure which electron passes through which slit, the interference pattern on the screen is destroyed, and we observe only the central maximum on the screen.

We find that any attempt to associate an electron with a particular slit destroys the interference pattern. This outcome should become more acceptable when we further explore the wave character of electrons and other objects that we normally think of as particles.

36.6 Uncertainty Relation

How precisely is it possible to measure physical properties such as location, momentum, energy, or time? In addition, to what precision can they be measured simultaneously? This question is never considered in classical mechanics, where it is assumed that all dynamic quantities can be measured with arbitrary precision with improved instrumentation.

However, in the subatomic quantum realm, where particles exhibit wave character (de Broglie matter waves) and where waves behave like particles (photons), the answer is not so simple. How can the precise location of a wave be specified, for example? Perhaps more important, does the process of measuring a physical property of a quantum object influence the outcome of that measurement, as well as all future measurements? For instance, when we measure the position of an object, we typically record the light waves emitted from that object. However, these emitted light waves also carry momentum, as we have seen in this chapter. Thus, we can already anticipate that the process of measuring a particle's position and its momentum cannot be done simultaneously with arbitrary precision.

Let the uncertainty in a measurement of a particle's position be Δx and the uncertainty in the measurement of its momentum be Δp_x, in the same sense as is done in statistics. In statistics, the outcome of a series of independent measurements of a quantity is quoted in terms of the mean value, which is the average of the measurements, plus/minus the standard deviation, which is a measure of the width of the distribution of the measurements. When a physical measurement is performed in the lab, the result also must be expressed in terms of the average value plus/minus the uncertainty in the measurement. This uncertainty can be of statistical or systematic origin, but for now we are not concerned with this distinction.

The astonishing statement arising from quantum physics is that the momentum and position of an object cannot be measured simultaneously with arbitrary precision. The more precisely we attempt to measure an object's momentum, the less precise the information on its position has to become, and vice versa. This physical statement is cast in mathematical terms in form of the **Heisenberg uncertainty relation,**

$$\Delta x \cdot \Delta p_x \geq \tfrac{1}{2}\hbar. \tag{36.26}$$

This relation was discovered in 1927 by the German physicist Werner Heisenberg (1901–1976) and caused a revolutionary change in our understanding of the measurement process, as well as of our fundamental ability to know the physical world. Chapter 37 will return to this uncertainty relation and use it for calculations. For now, we just want to motivate this relation, and to do this we will use the same considerations suggested by Heisenberg in his original paper. This heuristic derivation uses the so-called gamma-ray microscope.

DERIVATION 36.4 | Gamma-Ray Microscope and the Uncertainty Relation

If you want to see something in a microscope, you have to bounce light off that object and catch the reflected light in the lens of the microscope. The minimum size Δx of the object that you can still resolve with the microscope is limited by diffraction (see Chapter 34), as given by

$$\Delta x = \frac{\lambda}{2\sin\alpha} \approx \frac{\lambda\ell}{d}. \tag{i}$$

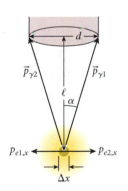

FIGURE 36.18 Geometry of a gamma-ray microscope and the momenta of photon-electron interaction.

Here λ is the wavelength of the light and α is the opening angle (Figure 36.18). In this figure, d is the size of the lens opening of the microscope, and ℓ is the distance between the object and the lens, which we assume to be large compared to the lens opening; then $2\sin\alpha \approx d/\ell$. To resolve small sizes, we need to employ light with a short wavelength, that is, gamma rays. These gamma-ray photons carry a momentum $p = h/\lambda$ (see equation 36.17).

Illuminating our object (an electron) with gamma rays (yellow circle, indicating a beam pointing into the page), the photons bounce off the object and get deflected into the lens via Compton scattering. In one extreme case, a photon can bounce off the electron and get deflected to the right edge of the lens. This photon has a momentum $\vec{p}_{\gamma 1}$, as shown in the figure. Its momentum component in the x-direction is $p_{\gamma 1,x} = p_{\gamma 1}\sin\alpha = h\sin\alpha/\lambda$. The electron receives the opposite x-momentum component $p_{e1,x}$, pointing to the left, due to recoil. At the other extreme, a photon receives momentum $\vec{p}_{\gamma 2}$ after the collision with the electron and gets scattered to the left edge of the microscope, which causes a recoil of the electron with the same magnitude as before, but in the opposite direction.

Continued—

We can detect that a photon has entered the microscope, but not where along the distance d. This means that it is undetermined which recoil the electron received. Thus, the momentum of the electron has an uncertainty of

$$\Delta p_x = 2\left|p_{\gamma 1,x}\right| = \frac{2h \sin\alpha}{\lambda}. \tag{ii}$$

Combining this equation with (i), we find

$$\Delta p_x = \frac{2h \sin\alpha}{\lambda} = \frac{h}{\Delta x}.$$

We find that the product of the minimum uncertainty in the size of the electron and the uncertainty of the momentum of the electron is Planck's constant, h.

DISCUSSION

You may feel that the above argument involves hand waving, and to a certain degree it does. The answer we find is a factor of 4π higher than the exact answer for the minimum product of momentum and coordinate uncertainties given in equation 36.26. However, an exact numerical relation is not the point of the example of the gamma-ray microscope. Instead, it shows a certain lower limit of the uncertainties in coordinate and corresponding momentum that can be observed simultaneously. This in itself is an astounding fact and a consequence of quantum physics.

Heisenberg also stated another uncertainty relation, for the uncertainties in the measurement of the energy of an object, ΔE, and the uncertainty in measuring the time, Δt. It is formally similar to the coordinate-momentum uncertainty relation of equation 36.26 and reads

$$\Delta E \cdot \Delta t \geq \tfrac{1}{2}\hbar. \tag{36.27}$$

DERIVATION 36.4 / Energy-Time Uncertainty

From the coordinate-momentum uncertainty relation, it is straightforward to derive the time-energy relation for a nonrelativistic free particle. For such a particle, the energy is all kinetic energy, and thus the uncertainty in the energy is

$$\Delta E = \Delta\left(\frac{p^2}{2m}\right) = \frac{\Delta\left(p^2\right)}{2m} = \frac{2p\Delta p}{2m} = v\Delta p.$$

The uncertainty in the time is given by

$$\Delta t = \frac{\Delta x}{v}.$$

Multiplying these two results, we find:

$$\Delta E \cdot \Delta t = (v\Delta p)\cdot(\Delta x/v) = \Delta x \cdot \Delta p \geq \tfrac{1}{2}h,$$

where we have used equation 36.26 in the last step here.

The energy-time uncertainty relation stipulates that classical energy conservation can be violated by some amount of energy for some time interval, because the quantum state may not have a sharp energy value. The larger the "violation" of energy conservation is, however, the shorter the time interval for this "violation" can be.

Do these uncertainty relations run into conflict with our everyday experience? In other words, how important are the fundamental limitations imposed by the uncertainty relation? Let us examine an example.

EXAMPLE 36.5 Trying to Get Out of a Speeding Ticket

PROBLEM

As you drive on the German autobahn, you find that some sections actually do have a speed limit. Occasionally the German police set up speed traps in these sections. They measure the speed of a vehicle and take a picture of the driver at the same time, as proof that they really have the right person committing the offense. A German physics student receives such a picture of herself and her car (BMW 318Ci, mass 1462 kg, including the driver and the gas in the tank), with a notification that she was driving 132 km/h in a 100 km/h speed limit zone. She notices that the picture the police took is very sharp and fixes her position down to an uncertainty of 1 mm. She argues that the uncertainty relation prevents the police from determining her speed precisely, and she therefore should not get a speeding ticket, or at least not get one for more than 30 km/h above the speed limit, which would cause her to lose her license. Will this strategy be successful?

SOLUTION

If the judge knows physics, the student will not be successful. Here is why: If the mass of the car is known precisely enough, then the uncertainty in the speed is given from the momentum uncertainty as

$$\Delta v = \frac{1}{m}\Delta p.$$

Using the uncertainty relation, we find the momentum uncertainty to be $\Delta p \geq \frac{1}{2}\hbar/\Delta x$, leading us to a value of the uncertainty in the speed of

$$\Delta v = \frac{1}{m}\Delta p \geq \frac{\hbar}{2m\Delta x}.$$

Using the constant (see equation 36.7) $\hbar = 1.05457 \cdot 10^{-34}$ J s and the values of $\Delta x = 10^{-3}$ m and $m = 1462$ kg given in the problem, we find numerically

$$\Delta v \geq \frac{1.05457 \cdot 10^{-34} \text{ J s}}{2(1462 \text{ kg})(10^{-3} \text{ m})} = 3.6 \cdot 10^{-35} \text{ m/s}.$$

The result is that the restriction on the minimum measurement uncertainty in the speed due to the uncertainty relation is 35 orders of magnitude too small to be useful for the defense!

36.5 In-Class Exercise

In Example 36.5, if all other parameters remain the same but the mass of the car is twice as big as the value stated, then the resulting uncertainty in the speed would have been

a) the same.

b) half as big as in the example.

c) one-quarter as big as in the example.

d) twice as big as in the example.

e) four times as big as in the example.

36.6 Self-Test Opportunity

How precisely would the position of the car in Example 36.5 have to be fixed in order for the student to be successful with her claim that she should not have to surrender her driver's license because the uncertainty in the speed had to be larger than 2.00 km/h?

The uncertainty relation is perhaps the most important result discussed in this chapter and has far-reaching consequences. The uncertainty relation sets a fundamental limit on how precisely we can measure and therefore know anything about our world. For macroscopic objects, the effect of the uncertainty relation can safely be neglected in almost all applications. This is not the point. Instead, the point is that an insurmountable limit exists to the precision of measurements of pairs of variables, such as momentum and position, or energy and time. The example of the gamma-ray microscope seems to imply that the attempt to measure the location somehow imparts a simple recoil onto the object that was to be measured. But in the quantum world, the objects normally characterized as particles have wave character, just as what we normally understand as a wave has particle character. At its core, the uncertainty relation derives from this particle-wave duality. This will be explored in detail in Chapter 37, where we will learn how to calculate properties with the tools of quantum mechanics.

Stern-Gerlach Experiment

In 1920, two German physicists, Otto Stern (1885–1969) and Walther Gerlach (1889–1979), performed one of the most influential experiments in the history of quantum physics. In this experiment, they tried to distinguish between a classical and a quantum description of atoms.

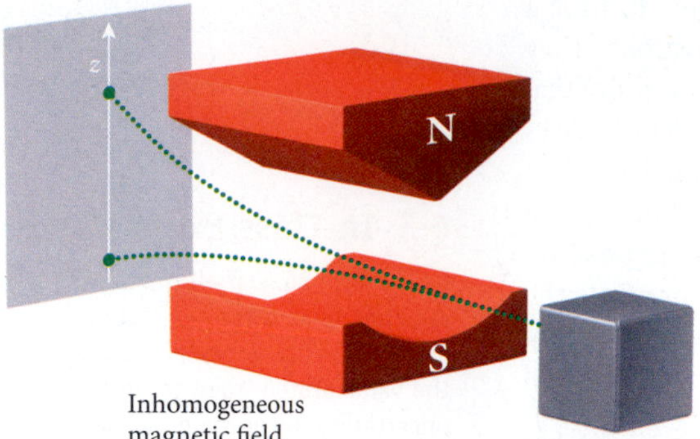

The basic setup of the Stern-Gerlach experiment is shown in Figure 36.19. An oven produces a gas of (electrically neutral) silver atoms. These atoms are allowed to escape from the oven through a hole and form a "beam" of silver atoms, which move along a straight line (green dots in the figure). The silver atoms then enter a strong inhomogeneous magnetic field, created by a magnet as shown, and continue on to a screen.

Chapter 28, Section 28.5, alluded to the fact that atoms can have a magnetic moment

$$\vec{\mu} = -\frac{e}{2m}\vec{L},$$

where $\vec{L}$ is the angular momentum. Section 28.5 considered the magnetic moment due only to a charge in a circular orbit, and the angular momentum was the orbital angular momentum. Chapter 27, Section 27.6, showed that the potential energy of a magnetic dipole in a magnetic field is $U = -\vec{\mu} \cdot \vec{B}$. The force is then given as the negative value of the gradient of the potential energy. In a magnetic field that changes only as a function of the z-coordinate, the force is

$$F_z = -\frac{\partial}{\partial z}\left(-\vec{\mu} \cdot \vec{B}\right) = \mu_z \frac{\partial B}{\partial z}.$$

Inhomogeneous magnetic field

Furnace emitting silver atoms

FIGURE 36.19 Basic setup of the Stern-Gerlach experiment.

Stern and Gerlach tried to determine whether the orbital angular momentum was quantized. Sending atoms with a magnetic moment through an inhomogeneous magnetic field causes a deflection of the beam, due to the interaction of the field gradient $\partial B/\partial z$ with the z-component of magnetic moment of the atom μ_z. Classically, the magnetic moment can have any value along the z-axis in the interval between $-|\vec{\mu}|$ and $+|\vec{\mu}|$, which should produce a line on the screen that corresponds to all possible deflections. However, in a quantum picture, only discrete values of the z-component of the angular momentum are possible, which would produce only discrete spots on the screen, separated by empty areas, as illustrated in Figure 36.19.

We know now that the orbital angular momentum of a silver atom in its ground state is zero—this will be shown in Chapter 38, where states of atoms and their orbital angular momenta are calculated. Thus, if this experiment had measured the splitting of the beam of silver atoms due only to its orbital angular momentum, the experiment would have been a failure. However, the total angular momentum of the atom is *not* just the orbital angular momentum. In addition, each elementary particle in the atom has an intrinsic angular momentum, which has no classical equivalent. The intrinsic angular momentum is called **spin.**

Spin of Elementary Particles and the Pauli Exclusion Principle

All elementary particles have characteristic intrinsic angular momentum, or spin. There are two fundamentally different groups of elementary particles: those that have integer values (in multiples of Planck's constant $\hbar$) of spin, and those that have half-integer values. Some elementary particles have spin 0, but for our purposes here, we count this as an integer spin.

Elementary particles with half-integer spins are called **fermions,** in honor of the Italian-American physicist Enrico Fermi (1901–1954). Fermions include the electron, the proton, and the neutron—in other words, all the building blocks of the matter we see around us. A fermion with spin $\frac{1}{2}\hbar$ can come in two different spin quantum states, indicated by $+\frac{1}{2}\hbar$ and $-\frac{1}{2}\hbar$, which by convention represent the projection of its spin onto the z-coordinate axis. We will encounter other quantum numbers in the following chapters, but for now all we need to know is that spin $\frac{1}{2}\hbar$ particles come in two varieties, which are usually called *spin-up* and *spin-down.*

Elementary particles with integer values of spin are called **bosons,** in honor of the Indian physicist Satyendra Nath Bose (1894–1974). Photons, which were introduced earlier in this chapter as the quanta of light, are examples of bosons and have spin $1\hbar$.

In Chapters 39 and 40 on elementary particles and nuclear physics, we will return to spin and give other examples for half-integer and integer-valued spins of elementary particles and try to find a more general organizing principle. For now, we can treat the two fundamentally different classes of particles—bosons and fermions—as working concepts.

One extremely important rule for fermions is the **Pauli exclusion principle:** No two fermions can occupy the same quantum state at the same time at the same location. In any given atom, no two fermions can have exactly identical quantum numbers. Because energy is quantized, each energy quantum state in a given system can be occupied by at most two fermions: one spin-up and the other spin-down. In the remaining chapters of this book, we will return repeatedly to the consequences of this Pauli exclusion principle. Chapter 37 introduces wave functions and then shows that two-particle wave functions for fermions and bosons have fundamentally different symmetries. Chapter 38 will show the effects of the Pauli exclusion principle in the construction of multi-electron atoms and the resulting periodic table of the elements. Chapters 39 and 40 will show that this principle gives rise to the Fermi energy inside the nucleus.

36.8 Spin and Statistics

Chapter 19 presented the probability distribution function for *classical* identical particles. For these identical particles, the distribution function is called the Maxwell-Boltzmann distribution,

$$g(E) = \frac{2}{\sqrt{\pi}} \left(\frac{1}{k_B T} \right)^{3/2} \sqrt{E} e^{-\frac{E}{k_B T}},$$

where E is the energy, T is the temperature, and k_B is Boltzmann's constant. (In deriving the Maxwell-Boltzmann distribution, we examined only the kinetic energy distribution of molecules in a gas, but this function is valid for all energy distributions of classical particles.) In this derivation, we neglected all quantum effects. Now we want to examine how quantum effects change this distribution.

The Maxwell-Boltzmann result can be rewritten in a slightly different way for particles in discrete energy states E_i. For a total of N particles in a system, the expected number of particles with energy E_i is N_i, and the expected fraction of particles in energy state E_i is $n_i = N_i/N$, where

$$n_i = \frac{N_i}{N} = \frac{g_i e^{-E_i/k_B T}}{Z} = \frac{g_i}{e^{(E_i - \mu)/k_B T}}. \tag{36.28}$$

Here g_i is the degeneracy of energy state E_i, that is, the number of different states in the system that have the same energy E_i. The quantity μ is called the *chemical potential* and has the same units as energy. The chemical potential is the amount by which the energy of the system would change if an additional particle were added, holding the remaining properties of the system fixed. The quantity Z that we have introduced is called the **partition function,**

$$Z = \sum_i g_i e^{-E_i/k_B T}. \tag{36.29}$$

The partition function encodes the thermodynamic properties of the system, and taking the appropriate derivatives of the partition function can reproduce many of the properties of the system.

The derivation of the Maxwell-Boltzmann distribution assumed that each of the particles in the ensemble is a classical particle. What we mean by *classical* is that all particles are distinguishable from one another. However, quantum particles are indistinguishable from like particles. (For example, one proton cannot be distinguished from another proton.) Thus, the distribution function has to be modified appropriately.

To see how this can work in practice, let's consider the simplest way of distributing two particles into two different states. Label these states a and b (Figure 36.20). First, consider the case of distinguishable particles, labeled 1 and 2. This is the classical **Maxwell-Boltzmann distribution.** In this case, there are four different configurations of our system. We can have both particles in state a, we can have particle 1 in state a and particle 2 in state b, we can have particle 1 in state b and particle 2 in state a, and finally we can have both particles in state b. This means that our entire two-particle system can have four different states (configurations)

FIGURE 36.20 Distributing two particles over two different states.

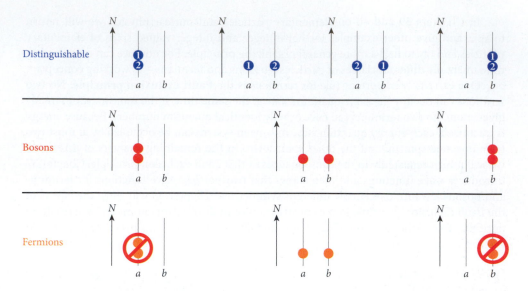

Distinguishable

Bosons

Fermions

that it can be in, if the two particles are distinguishable. Next, consider two spin-0 bosons, indistinguishable quantum particles. This is known as the **Bose-Einstein distribution.** In this case, both particles can be in state a or both in state b, just as in the case of distinguishable particles. However, if one of them is in state a and the other is in state b, then it does not matter which one is which, because they are indistinguishable. Therefore, the two-particle system has only three different states for bosons, and not four as in the case for distinguishable particles. Finally, the easiest case is obtained for two identical spin-$\frac{1}{2}$ fermions (for example, two electrons with "spin-up"). This case is called the **Fermi-Dirac distribution.** The Pauli exclusion principle states that these two identical fermions cannot occupy the same quantum state. (In the lowest panel of Figure 36.20, this fact is indicated by the *Forbidden* sign, a red circle with a diagonal bar across, over the two configurations in which both fermions would reside in the same state.) In this case, the two-particle system of identical fermions can be in only one configuration, where the quantum states a and b are each occupied by one fermion.

Now let us turn to the more complicated case of the energy distribution for the case of a system of 5 particles that share a total of 6 quanta of energy in a one-dimensional model. For classical distinguishable particles, the 6 energy quanta can be shared by one particle carrying all 6 energy quanta and the other 4 particles each with zero energy. This sharing of the total energy of 6 quanta can be written as $6 = 6+0+0+0+0$. Since the particles are distinguishable, we have to keep track of which one of the 5 carries the 6 quanta, and this energy partition then can happen in 5 different ways. The energy can also be distributed as $6 = 5+1+0+0+0$. Each of the 5 particles can be the carrier of the 5 quanta, and then each of the remaining 4 particles can carry the remaining 1. Thus, this particular energy partition can happen in $g_i = 5 \cdot 4 = 20$ different ways. In Figure 36.21, we show all possible partitions of these 6 energy quanta into 5 particles. The blue numbers above each diagram show the number of different ways the same configuration can happen by exchanging all possible

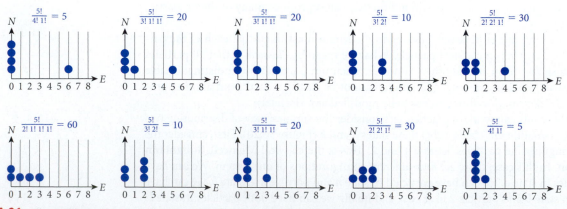

FIGURE 36.21 All possible partitions of 6 quanta of energy for 5 distinguishable particles.

particles. The total number of different energy partitions among the 5 particles is the sum of all of the numbers listed above each panel and is 210.

Now we can calculate the probability for each energy state to be occupied by counting how many partitions each energy state occurs in, multiplied by how often this partition occurs, and then dividing the result by the total number of partitions.

This is done in Figure 36.22, and the result is represented by the blue squares. For example, let's look at the probability for a particle to carry 4 quanta of energy. The only partitions for which the $E = 4$ state is populated by a particle are shown in the middle and right panels of Figure 36.21 and only one particle is in each of $E = 4$ states. The middle panel represents 20 states and the right panel 30 different states. Since the total number of states is 210, the probability of finding the $E = 4$ level occupied is $f(4) = (1 \cdot 20 + 1 \cdot 30)/210 = 0.238$. The occupation probabilities for the other values of the energy are found in the same way.

Also shown in Figure 36.22 is the result (blue line) of the analytic exponential function of equation 36.28. It is remarkable that for such a small number of particles and energy states, the exponential limit is already approximated so well. The temperature T and the chemical potential μ that appear in equation 36.28 can be extracted from the condition that the sum over all occupation probabilities has to add up to the number of particles (5 in this case), and that the sum of the products of occupation probability times the energy has to add up to the total number of energy quanta in the system (6 in this case).

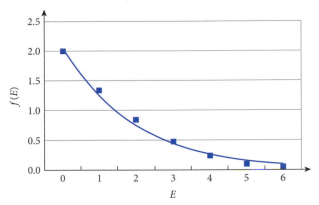

FIGURE 36.22 Average occupancy for each energy state, corresponding to the partitions of the previous figure.

Now we can ask how these considerations have to change if we are dealing with indistinguishable particles. For bosons, we have done most of the work already, because we can arrange the bosons as in Figure 36.21. However, since the bosons are indistinguishable, permutations of the different particles do not yield new states. Therefore, the weight factors that count the number of states for a given arrangement of particles among energy states are always 1 for each panel in Figure 36.21. Thus, bosons have only 10 energy partitions, instead of the 210 we found for distinguishable particles. If we now want to calculate the probability for the occupation of the $E = 4$ state, then we find it is $f(4) = (1+1)/10 = 0.2$.

The resulting distribution for identical bosons is displayed in Figure 36.23 (red symbols) and compared to the classical case (blue line) that we just discussed. The distributions are remarkably similar, but important small differences appear, which are not just numerical artifacts. For instance, the occupation of the $E = 0$ level is slightly enhanced in the boson case relative to the classical distinguishable particles. As increasingly more particles are added to the system, this effect will become increasingly pronounced. It is again a manifestation of the effect, already discussed, that bosons generally like to occupy the same states as other bosons.

For the problem of distributing the same 6 energy quanta over 5 fermions, it might appear that the fermions can be distributed into the energy states as in Figure 36.21. However, the Pauli exclusion principle does not allow this. Each energy state can be occupied by at most one spin-up fermion and one spin-down fermion. Thus, the occupancy of each level cannot exceed 2. This eliminates seven partitions from Figure 36.21, because these have three or more particles in the same energy state. The only three partitions left are those shown in Figure 36.24. To illustrate, suppose the system has three spin-up and two spin-down fermions. For the left and right partitions in Figure 36.24, only one fermion is in a singly occupied state, which therefore has to be the unpaired spin-up. The middle panel shows three fermions in

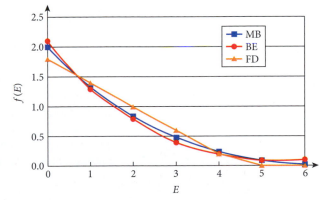

FIGURE 36.23 Comparison of the Maxwell-Boltzmann (MB), Bose-Einstein (BE), and Fermi-Dirac (FD) distributions obtained for the problem of sharing 6 energy quanta among 5 particles.

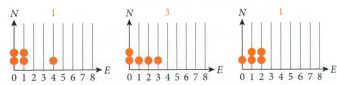

FIGURE 36.24 Possible ways to distribute 6 energy quanta among 5 fermions.

singly occupied states, and two of these have to be spin-up and one spin-down. Since each of the three energy states can hold the spin-up fermion, this middle panel represents a total of three states. This means that a total of 5 states are available to this system. Calculating the occupation of the $E = 4$ level now results in $f(4) = 1/5 = 0.2$ (which is, by chance, just the same value as for the bosons). The orange line in Figure 36.23 shows the average occupations for all energy levels. The occupation for the $E = 0$ state is suppressed in the case of fermions relative to classical particles.

36.6 In-Class Exercise

Which of the following distributions are possible ways to distribute 4 energy quanta over 5 fermions with spin $\frac{1}{2}\hbar$?

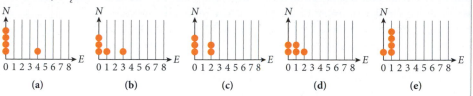

(a)　　　　　(b)　　　　　(c)　　　　　(d)　　　　　(e)

After having performed this exercise for a very small set of particles, let's finish this section by writing down the distributions for a much larger set of particles present in a physical system. For bosons, this limiting Bose-Einstein distribution of the number of particles in a given energy state E_i is

$$N_i = \frac{1}{e^{(E_i - \mu)/k_B T} - 1}. \tag{36.30}$$

For fermions the Fermi–Dirac distribution is

$$N_i = \frac{g_i}{e^{(E_i - \mu)/k_B T} + 1}. \tag{36.31}$$

where g_i is the degeneracy of the state i. (In our above fermion example, this number is $g_i = 2$ for each state, because due to the Pauli exclusion principle each state can only be occupied by one spin-up and one spin-down fermion.)

Sometimes these distributions are expressed by using the definition of the absolute activity,

$$z = e^{\mu/k_B T}.$$

The Bose-Einstein distribution is then written as

$$N_i = \frac{1}{e^{E_i/k_B T}/z - 1},$$

and the Fermi-Dirac distribution is

$$N_i = \frac{g_i}{e^{E_i/k_B T}/z + 1}.$$

Both distribution functions approach the Maxwell-Boltzmann distribution for negligibly small absolute activity, $z \ll 1$.

For energy states that are sufficiently close to one another, the discrete values of the energy states E_i can be replaced by a continuous energy variable E. Then the probability of finding a particle at energy E in the Maxwell-Boltzmann limit can be written as

$$f_{MB}(E) = \frac{1}{a e^{E/k_B T}}, \tag{36.32}$$

where a is a normalization constant. This normalization constant is a function of the chemical potential and the degeneracy. In the same way, the probability of finding a particle at energy E in the Fermi-Dirac case can be written as

$$f_{FD}(E) = \frac{1}{a e^{E/k_B T} + 1}. \tag{36.33}$$

The Fermi-Dirac distribution with $g_i = 2$ is plotted in Figure 36.25 as a function of the ratio of the energy divided by the chemical potential. The function is shown for three different temperatures. For $k_B T \ll \mu$, this function approaches a step function that falls from a value of 2 to 0 at $E = \mu$. For higher temperatures, this transition from high occupancy to low occupancy becomes increasingly smoother.

For the Bose-Einstein case,

$$f_{BE}(E) = \frac{1}{ae^{E/k_B T} - 1}. \qquad (36.34)$$

As we stated, photons are bosons and thus are subject to Bose-Einstein statistics. For the special case of photons, the chemical potential is zero, and thus the normalization constant in equation 36.34 has the value $a = 1$. If the photon energy approaches zero, then $e^{E/k_B T}$ approaches the value 1, and the denominator vanishes in equation 36.34. This means that the occupation of states with very low energy can increase without limit for photons.

Let's take another look at the Planck radiation formula (equation 36.9) for the spectral brightness as a function of the wavelength $I_T(f) = 2hc^{-2} f^3 / \left(e^{hf/k_B T} - 1 \right)$. Since $E = hf$ for photons, the denominator contains the factor $\left(e^{E/k_B T} - 1 \right)$, which we have just obtained for the Bose-Einstein distribution for photons. Thus, the Planck spectral brightness can now be rewritten as a function of photon energy,

$$I_T(E) = \frac{2E^3}{h^2 c^2 \left(e^{E/k_B T} - 1 \right)} = \frac{2E^3}{h^2 c^2} f_{BE}(E). \qquad (36.35)$$

Thus, the radiation formula obtained by Planck says that the spectral brightness as a function of photon energy is proportional to the third power of the energy times the Bose-Einstein probability of finding a photon at that energy.

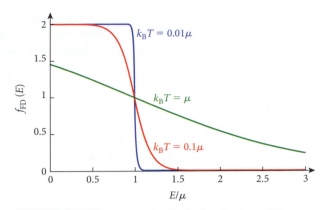

FIGURE 36.25 Fermi-Dirac distribution for three different temperatures.

Bose-Einstein Condensate

In his original paper in 1924, Bose discussed the blackbody spectrum for photons. Einstein extended this work in the same year to atoms with integer spin. Einstein noticed that at very low temperature, a large fraction of the atoms goes into the lowest energy quantum state: "One part condenses, the rest remains a 'saturated ideal gas.'" For this to happen, the atoms must be close enough to their neighbors that their de Broglie waves (see equation 36.23) overlap with one another. It can be shown that this condition implies, for the density ρ of atoms per unit volume, that $\rho \cdot \lambda^3 > 2.61$, where λ is the de Broglie wavelength.

How can we manage to obtain a Bose-Einstein condensate (BEC) in the laboratory? Carl Wieman and Eric Cornell used a magnetic trap to collect rubidium atoms, which form spin-1 bosons at low temperatures and in a strong magnetic field. They used laser cooling to reduce the temperature, but this was still not sufficient to reach the temperature for transition to the BEC. By gradually reducing the depth of the trap, they allowed the atoms with higher energies to escape from the trap, leaving only the lower-energy atoms behind. With this method of evaporative cooling, they managed to cool their atoms to temperatures down to a few nK. They then switched off their trap and allowed the trapped collection of atoms to expand. A simple ideal gas of atoms will expand due to thermal motion, but the BEC shows up as a second feature that expands only at the minimal rate required by the Heisenberg uncertainty principle. A short while later, they imaged the expanding cloud, which by this time had reached a spatial extension on the order of 0.2 mm, and found the distribution shown in Figure 36.26. This figure shows the results of three different experiments at different temperatures. Clearly, the picture at 200 nK is qualitatively different from that at 400 nK. At 200 nK the central peak indicates the presence of the BEC. The right panel shows the situation at a temperature of 50 nK, where now essentially all the rubidium atoms are part of the BEC.

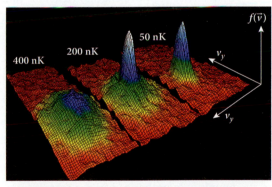

FIGURE 36.26 Bose-Einstein condensation of rubidium atoms in a magnetic trap. This picture was produced by the Wieman-Cornell group in 1995 and is included in their Nobel lecture. (The axes were added by this book's authors.) The middle panel shows the appearance of the Bose-Einstein condensate. In the right panel, almost all atoms are in the condensate.

Only six years after this discovery, this BEC work resulted in the 2001 Nobel Prize in Physics for Wieman and Cornell, which they shared with Wolfgang Ketterle, whose group had performed similar work on the BEC for other systems at the same time. The study of the BEC now flourishes in many laboratories around the world, and we continue to learn many astounding facts about atoms, condensates, and quantum mechanics.

WHAT WE HAVE LEARNED | EXAM STUDY GUIDE

- Planck's constant is $h = 6.62606876(52) \cdot 10^{-34}$ J s, and the energy of a photon is $E = hf$.

- Planck's radiation law for the power radiated in the frequency interval between f and $f + df$, the spectral brightness as a function of frequency, is

$$I_T(f) = \frac{2h}{c^2} \frac{f^3}{e^{hf/k_B T} - 1}.$$

  It avoids the ultraviolet divergence that makes its classical predecessors unphysical in the high-frequency region, and it contains the classical radiation laws as limiting cases.

- The spectral brightness as a function of wavelength is

$$I_T(\lambda) = \frac{2hc^2}{\lambda^5 \left(e^{hc/\lambda k_B T} - 1 \right)}.$$

- The spectral emittance is related to the spectral brightness by a simple multiplicative factor of π, $\epsilon_T(f) = \pi I_T(f)$.

- Integration of the spectral emittance over all frequencies (or wavelengths) yields the radiated intensity, $I = \int_0^\infty \epsilon(\lambda) d\lambda = \sigma T^4$ with the Stefan-Boltzmann constant

$$\sigma = \frac{2k_B^4 \pi^5}{15 h^3 c^2} = 5.6704 \cdot 10^{-8} \text{ W m}^{-2} \text{ K}^{-4}.$$

- What we normally think of as a wave has particle character. The photoelectric effect is explained with the quantum hypothesis by assigning particle properties to the photon, the elementary quantum of light. This quantum hypothesis gives the correct frequency dependence of the stopping potential, $eV_0 = hf - \phi$, where ϕ is the work function. The work function is a constant that depends on the material used.

- The Compton effect describes the photon wavelength λ' after a photon of wavelength λ scatters off an electron as

$$\lambda' = \lambda + \frac{h}{m_e c} (1 - \cos\theta).$$

- The Compton wavelength of the electron is

$$\lambda_e = \frac{h}{m_e c} = 2.426 \cdot 10^{-12} \text{ m}.$$

- What we normally think of as matter also has wave characteristics. The de Broglie wavelength is defined as

$$\lambda = \frac{h}{p}.$$

  This can be demonstrated by performing double-slit type interference experiments with electrons. With electrons, we find the same kind of interference patterns as for photons.

- The Heisenberg uncertainty relation stipulates that the product of the uncertainty in momentum times the uncertainty in position has an absolute lower bound, $\Delta x \cdot \Delta p_x \geq \frac{1}{2}\hbar$. The energy-time uncertainty relation is $\Delta E \cdot \Delta t \geq \frac{1}{2}\hbar$.

- Elementary quantum particles have an intrinsic property called spin, which has the dimension of an angular momentum. The spin is quantized, dividing particles into two categories: fermions with spins that are half-integer multiples of $\hbar$, and bosons with spins of integer multiples of $\hbar$.

- The Pauli exclusion principle states that no two fermions can occupy the same quantum state at the same time and at the same location. In any given atom, no two fermions can have exactly identical quantum numbers.

- Bosons can condense at low temperature in such a way that a very significant fraction of them occupy the same quantum state.

KEY TERMS

NEW SYMBOLS AND EQUATIONS

$h = 6.62606876(52) \cdot 10^{-34}$ J s, Planck's constant

ϕ, work function in photoelectric effect, $eV_0 = hf - \phi$

$\lambda' = \lambda + \dfrac{h}{m_e c}(1 - \cos\theta)$, Compton scattering formula

$\lambda_e = \dfrac{h}{m_e c} = 2.426 \cdot 10^{-12}$ m, Compton wavelength of an electron

$\Delta x \cdot \Delta p_x \geq \frac{1}{2}\hbar$, Heisenberg uncertainty relation for position and momentum

$\Delta E \cdot \Delta t \geq \frac{1}{2}\hbar$, Heisenberg uncertainty relation for energy and time

ANSWERS TO SELF-TEST OPPORTUNITIES

36.1 $K_{\max} = (f - f_{\min})h = fh - \phi$.

36.2 148 million km $= 148 \cdot 10^6 \cdot 10^3$ m $= 1.48 \cdot 10^{11}$ m

$$P = \left(4\pi r^2\right)I = 4\pi\left(1.48 \cdot 10^{11}\ \text{m}\right)^2\left(1370\ \text{W/m}^2\right) = 3.77 \cdot 10^{26}\ \text{W}$$

In visible spectrum $P_{\text{visible}} = P/4 = 9.43 \cdot 10^{25}$ W.

In 1 second, $9.43 \cdot 10^{25}$ J of energy is emitted by the Sun.

Assume the average wavelength of the visible photons is $\lambda = 550$ nm. The energy of each photon is then

$$E = hf = h\frac{c}{\lambda} = \left(6.626 \cdot 10^{-34}\ \text{J s}\right)\frac{3.00 \cdot 10^8\ \text{m/s}}{550 \cdot 10^{-9}\ \text{m}} = 3.61 \cdot 10^{-19}\ \text{J}$$

$$N = P_{\text{visible}}/E = \left(9.43 \cdot 10^{25}\ \text{W}\right)/\left(3.61 \cdot 10^{-19}\ \text{J}\right)$$

$\quad = 2.6 \cdot 10^{44}$ visible photons per second.

36.3 Momentum of 1.00 eV electron: $m_e c^2 = 511$ keV

$$E^2 = \left(m_e c^2\right)^2 + \left(pc\right)^2 = \left(m_e c^2 + K\right)^2$$

$$pc = \sqrt{\left(m_e c^2 + K\right)^2 - \left(m_e c^2\right)^2}$$

$$= \sqrt{\left(511\ \text{keV} + 1.00 \cdot 10^{-3}\ \text{keV}\right)^2 - \left(511\ \text{keV}\right)^2} = 1.01\ \text{keV}$$

Momentum of 100 keV X-ray:

$$\left(m_e c^2\right)^2 + \left(pc\right)^2 = \left(m_e c^2 + K\right)^2$$

$pc = K = 100$ keV

Momentum of electron is 1% of X-ray, negligible.

36.4 For visible light, the energy of the photon ranges from 1.59 eV to 3.27 eV.

For a 2.26 eV photon, the momentum is

$$\left(m_e c^2\right)^2 + \left(pc\right)^2 = \left(m_e c^2 + K\right)^2$$

$pc = K = 2.26$ eV

Momemtum of 1.00 eV electron: $m_e c^2 = 511$ keV

$$E^2 = \left(m_e c^2\right)^2 + \left(pc\right)^2 = \left(m_e c^2 + K\right)^2$$

$$pc = \sqrt{\left(m_e c^2 + K\right)^2 - \left(m_e c^2\right)^2}$$

$$= \sqrt{\left(511\ \text{keV} + 1.00 \cdot 10^{-3}\ \text{keV}\right)^2 - \left(511\ \text{keV}\right)^2} = 1.01\ \text{keV}$$

$(pc)_{\text{photon}}/(pc)_{\text{electron}} = (2.26\ \text{eV})/(1010\ \text{eV}) = 2.24 \cdot 10^{-3}$

The momentum of the visible photon is 0.224% of the momentum of the electron, negligible.

36.5

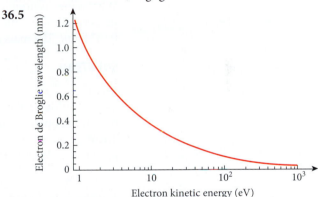

The de Broglie wavelength for an electron with a kinetic energy of 1000 eV is
$\lambda = 0.0387879$ nm.
Calculating the momentum nonrelativistically gives us
$\lambda = 0.0388068$ nm.
The difference is small.

36.6 $\Delta v = 2.00$ km/h $= 0.556$ m/s

$$\Delta x \cdot \Delta p \geq \tfrac{1}{2}\hbar$$

$$\Delta x \geq \frac{\hbar}{2\Delta p} = \frac{\hbar}{2m\Delta v} = \frac{1.05457 \cdot 10^{-34}\ \text{J s}}{2(1462\ \text{kg})(0.556\ \text{m/s})} = 6.49 \cdot 10^{-38}\ \text{m}.$$

36.7 Numbers of states are 18, 16, 495, respectively.

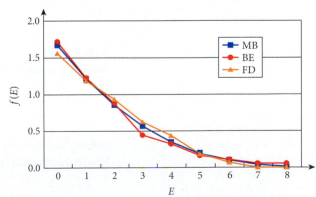

PROBLEM-SOLVING PRACTICE

Problem-Solving Guidelines

1. The starting point for most calculations involving photons or matter waves is to relate the particle properties of energy E and momentum p to the wave properties of wavelength λ and frequency f. The key relations are $E = hf$, $p = E/c = hf/c = h/\lambda$, and $\lambda = h/p$.

2. Be careful and consistent in applying units. Often, converting all units to meters and kilograms will help keep track of unit exponents. Use of electron-volts will often simplify your calculations, but be sure to use Planck's constant with the appropriate units: $h = 6.626 \cdot 10^{-34}$ J s or $h = 4.136 \cdot 10^{-15}$ eV s.

3. For checking your work, it is helpful to keep in mind some rough orders of magnitude: the size of an atom is 10^{-10} m; the mass of an electron is 10^{-30} kg; charge of a proton or electron is 10^{-19} C ; at room temperature, $k_B T = \frac{1}{40}$ eV.

SOLVED PROBLEM 36.1 / Rubidium Bose-Einstein Condensate

PROBLEM
What is the minimum density of rubidium atoms needed in a magnetic trap at a temperature of 200 nK in order to have a chance to observe the onset of Bose-Einstein condensation? (*Hint:* The mass of a rubidium-87 atom is $1.5 \cdot 10^{-25}$ kg.)

SOLUTION

THINK
In our section on the Bose-Einstein condensate, we stated that the rubidium atoms must be close enough to one another that they can overlap in a quantum mechanical sense. There we quoted this criterion as $\rho \cdot \lambda^3 > 2.61$. Thus, our task amounts to finding the de Broglie wavelength of rubidium atoms in thermal motion at a temperature of 200 nK.

SKETCH
A sketch is perhaps not really needed in this case, but in Figure 36.27 we at least try to visualize the relationship of the quantum mechanical extension of the atomic wave function, the de Broglie wavelength, to the nearest-neighbor spacing of the atoms in the trap.

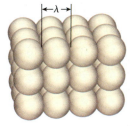

FIGURE 36.27 Sketch of overlapping rubidium atoms in the trap.

RESEARCH
We have already stated that the criterion for the density is given as

$$\rho \cdot \lambda^3 > 2.61. \tag{i}$$

The de Broglie wavelength λ was given as

$$\lambda = \frac{h}{p} \tag{ii}$$

in equation 36.22. For the momentum p of a rubidium atom we use the fact that the kinetic energy of an atom in a gas is given by the thermal energy at the given temperature,

$$E = \frac{p^2}{2m} = \frac{3}{2} k_B T \Rightarrow p = \sqrt{3mk_B T}. \tag{iii}$$

SIMPLIFY
Inserting equation (iii) into equation (ii), we find for the de Broglie wavelength in this case

$$\lambda = \frac{h}{p} = \frac{h}{\sqrt{3mk_B T}}. \tag{iv}$$

Solving equation (i) for the density and inserting our formula for the de Broglie wavelength in equation (iv) results in

$$\rho > \frac{2.61}{\lambda^3} = \frac{2.61 \left(3mk_B T \right)^{3/2}}{h^3}.$$

CALCULATE

Inserting the values of the constants ($h = 6.63 \cdot 10^{-34}$ J s, $k_B = 1.38 \cdot 10^{-23}$ J/K) and the given temperature ($T = 200$ nK $= 2 \cdot 10^{-7}$ K), we obtain for the de Broglie wavelength

$$\lambda = \frac{6.63 \cdot 10^{-34} \text{ J s}}{\sqrt{3(1.5 \cdot 10^{-25} \text{ kg})(1.38 \cdot 10^{-23} \text{ J/K})(2 \cdot 10^{-7} \text{ K})}} = 5.9491 \cdot 10^{-7} \text{ m}$$

and thus for the density

$$\rho > \frac{2.61}{\lambda^3} = 1.2396 \cdot 10^{19} \text{ m}^{-3}.$$

ROUND

Since the temperature was only given to one significant digit, our answer for the de Broglie wavelength is $\lambda = 6 \cdot 10^{-7}$ m. Note that this is approximately a factor 1000 larger than the atomic diameter of rubidium, which is approximately $5 \cdot 10^{-10}$ m.

For the minimum density needed to have a chance to observe the BEC, our appropriately rounded result is $\rho > 1 \cdot 10^{19}$ m^{-3}.

DOUBLE-CHECK

Is a density of 10^{19} atoms/m^3 a large or a small number? Let us compare it to the density of water molecules and air molecules. One cubic meter of liquid water contains $3 \cdot 10^{28}$ water molecules, and one cubic meter of air contains $3 \cdot 10^{25}$ nitrogen and oxygen molecules. Thus, the gas density of rubidium atoms in the trap is approximately a million times lower than the density of air at normal atmospheric conditions.

If we have one million atoms in a trap at a density of 10^{19} atoms/m^3 in an approximately spherical configuration, what is the radius of this sphere? The total volume occupied by the million atoms is

$$V = \frac{N}{\rho} = \frac{10^6}{10^{19} \text{ m}^{-3}} = 10^{-13} \text{ m}^3.$$

Because the volume of a sphere is $V = \frac{4}{3}\pi r^3$, we find that the radius of the sphere containing one million atoms at this density is

$$r = \left(\frac{3}{4\pi}V\right)^{1/3} = 30 \text{ } \mu\text{m}.$$

This makes it clear that the Bose-Einstein condensate could not be imaged directly in the experiment conducted by Cornell and Wieman. This is why they needed to turn off the trap and let its contents expand by at least a factor of 10 in each direction before they were able to produce the images of the kind shown in Figure 36.26.

MULTIPLE-CHOICE QUESTIONS

36.1 Ultraviolet light of wavelength 350 nm is incident on a material with a stopping potential of 0.25 volts. The work function of the material is

a) 4.0 eV.

c) 2.3 eV.

b) 3.3 eV.

d) 5.2 eV.

36.2 The existence of a cutoff frequency in the photoelectric effect

a) cannot be explained using classical physics.

b) shows that the model provided by classical physics is not correct in this case.

c) shows that a photon model of light should be used in this case.

d) shows that the energy of the photon is proportional to its frequency.

e) All of the above.

36.3 To have a larger photocurrent, which of the following should occur? (select all the correct changes)

a) brighter light

c) higher frequency

b) dimmer light

d) lower frequency

36.4 Which of the following has the smallest de Broglie wavelength?

a) an electron traveling at 80% the speed of light

b) a proton traveling at 20% the speed of light

c) a carbon nucleus traveling at 70% the speed of light

d) a helium nucleus traveling at 80% the speed of light

e) a lithium nucleus traveling at 50% the speed of light

36.5 A blackbody is an ideal system that

a) absorbs 100% of the light incident upon it, but cannot emit light of its own.

b) emits 100% of the light it generates, but cannot absorb radiation of its own.

c) either absorbs 100% of the light incident upon it, or emits 100% of the radiation it generates.

d) absorbs 50% of the light incident upon it, and emits 50% of the radiation it generates.

e) blackens completely any body that comes in contact with it

36.6 Which one of the following statements is true if the intensity of a light beam is increased while its frequency is kept the same?

a) The photons gain higher speeds.

b) The energy of the photons is increased.

c) The number of photons per unit time is increased.

d) The wavelength of the light is increased.

36.7 Which of the following has the higher temperature?

a) a white-hot object

b) a red-hot object

c) a blue-hot object

36.8 Electrons having a narrow range of kinetic energies are impinging upon a double slit, with separation D between the slits. The electrons form a pattern on a phosphorescent screen with a separation Δx between the fringes on the screen. If the spacing between the slits is reduced to $D/2$, the separation between the fringes will be:

a) Δx

b) $2\Delta x$

c) $\Delta x/2$

d) none of these

QUESTIONS

36.9 Why is a white-hot object hotter than a red-hot object?

36.10 After having read this chapter, weigh in on the discussion of whether an electron is a particle or a wave.

36.11 If I look in a mirror while wearing a blue shirt, I see a blue shirt in my reflection, not a red shirt. But according to the Compton effect, the photons that bounce back should have a lower energy and therefore a longer wavelength. Explain why my reflection shows the same color shirt as I am wearing.

36.12 Vacuum in deep space is not empty, but a boiling sea of particles and antiparticles that are constantly forming and annihilating each other. Determine the minimum lifetime for a proton-antiproton pair to form there without violating Heisenberg's uncertainty relation.

36.13 Consider a universe where Planck's constant is 5 J s. How would a game of tennis change? Consider the interactions of individual players with the ball and the interaction of the ball with the net.

36.14 In classical mechanics, for a particle with no net force on it, what information is needed in order to predict where the particle will be some later time? Why is this prediction not possible in quantum mechanics?

36.15 What would a classical physicist expect would be the result of shining a brighter UV lamp on a metal surface, in terms of the energy of emitted electrons? How does this differ from what the theory of the photoelectric effect predicts?

36.16 Which is more damaging to human tissue, a 60-W source of visible light, or a 2-mW source of X-rays? Explain your choice.

36.17 Neutrons are spin $\frac{1}{2}$ fermions. An unpolarized beam of neutrons has an equal number of spins in the +1/2 and −1/2 states. When a beam of unpolarized neutrons is passed through unpolarized ^{3}He, the neutrons can be absorbed by the ^{3}He to create ^{4}He. If the ^{3}He is then polarized, so that the spins of the neutrons in the nucleus of the ^{3}He are all aligned, will the same number of neutrons in the unpolarized neutron beam be absorbed by the polarized ^{3}He? How well is each of the two spin states of the unpolarized neutron beam absorbed by the ^{3}He?

36.18 You are performing a photoelectric effect experiment. Using a photocathode made of cesium, you first illuminate it with a green laser beam (λ = 514.5 nm) of power 100 mW. Next, you double the power of your laser beam, to 200 mW. How will the energies per electron of the electrons emitted by the cathode compare for the two cases?

PROBLEMS

A blue problem number indicates a worked-out solution is available in the Student Solutions Manual. One • and two •• indicate increasing level of problem difficulty.

Section 36.2

36.19 Calculate the peak wavelengths of

a) the solar light received by Earth, and

b) light emitted by the Earth.

Assume the surface temperatures of the Sun and the Earth are 5800 K and 300 K, respectively.

36.20 Calculate the range of temperatures for which the peak emission of the blackbody radiation from a hot filament occurs within the visible range of the electromagnetic spectrum. Take the visible spectrum as extending from 380 nm to 780 nm. What is the total intensity of the radiation from the filament at these two temperatures?

36.21 Ultra-high-energy gamma rays are found to come from the Equator of our galaxy, with energies up to $3.5 \cdot 10^{12}$ eV. What is the wavelength of this light? How does the energy of this light compare to the rest mass of a proton?

36.22 Consider an object at room temperature (20 °C) and the radiation it emits. For radiation at the peak of the spectral energy density, calculate

a) the wavelength,

b) the frequency, and

c) the energy of one photon.

•**36.23** The temperature of your skin is approximately 35 °C.

a) Assuming that it is a blackbody, what is the peak wavelength of the radiation it emits?

b) Assuming a total surface area of 2 m^2, what is the total power emitted by your skin?

c) Based on your answer in (b), why don't you glow as brightly as a light bulb?

•**36.24** A pure, defect-free semiconductor material will absorb the electromagnetic radiation incident on that material only if the energy of the individual photons in the incident beam is larger than a threshold value known as the "band-gap" of the semiconductor. The known room-temperature band-gaps for germanium, silicon, and gallium-arsenide, three widely used semiconductors, are 0.66 eV, 1.12 eV, and 1.42 eV, respectively.

a) Determine the room-temperature transparency range of these semiconductors.

b) Compare these with the transparency range of ZnSe, a semiconductor with a band-gap of 2.67 eV, and explain the yellow color observed experimentally for the ZnSe crystals.

c) Which of these materials could be used for a light detector for the 1550-nm optical communications wavelength?

•**36.25** The mass of a dime is 2.268 g, its diameter is 17.91 mm, and its thickness is 1.350 mm. Determine

a) the radiant energy coming out from a dime per second at room temperature,

b) the number of photons leaving the dime per second (Assume that all photons have the wavelength of the peak of the distribution for this estimate.), and

c) the volume of air to have energy equal to 1 second of radiation from the dime.

Section 36.3

36.26 The work function of a certain material is 5.8 eV. What is the photoelectric threshold for this material?

36.27 What is the maximum kinetic energy of the electrons ejected from a sodium surface by light of wavelength 470 nm?

36.28 The threshold wavelength for the photoelectric effect in a specific alloy is 400 nm. What is the work function in eV?

36.29 In a photoelectric effect experiment, a laser beam of unknown wavelength is shined on a cesium cathode (work function ϕ = 2.100 eV). It is found that a stopping potential of 0.310 V is needed to eliminate the current. Next, the same laser is shined on a cathode made of an unknown material, and a stopping potential of 0.110 V is found to be needed to eliminate the current.

a) What is the work function for the unknown cathode?

b) What would be a possible candidate for the material of this unknown cathode?

36.30 You illuminate a zinc surface with 550-nm light. How high do you have to turn up the stopping voltage to squelch the photoelectric current completely?

36.31 White light, λ = 400 to 750 nm, falls on barium (ϕ = 2.48 eV) .

a) What is the maximum kinetic energy of electrons ejected from it?

b) Would the longest-wavelength light eject electrons?

c) What wavelength of light would eject electrons with zero kinetic energy?

•**36.32** To determine the work function of the material of a photodiode, you measured the maximum kinetic energy of 0.5 eV corresponding to a certain wavelength. Later on you cut down the wavelength by 50% and found the maximum kinetic energy of the photoelectrons to be 3.8 eV. From this information determine

a) work function of the material, and

b) the original wavelength.

Section 36.4

36.33 X-rays of wavelength λ = 0.120 nm are scattered from carbon. What is the Compton wavelength shift for photons detected at 90° angle relative to the incident beam?

36.34 A 2.0-MeV X-ray photon is scattered from a free electron at rest into an angle of 53°. What is the wavelength of the scattered photon?

36.35 A photon with wavelength of 0.30 nm collides with an electron that is initially at rest. If the photon rebounds at an angle of 160°, how much energy did it lose in the collision?

•**36.36** X-rays having energy of 400.0 keV undergo Compton scattering from a target. The scattered rays are detected at 25.0° relative to the incident rays. Find

a) the kinetic energy of the scattered X-ray, and

b) the kinetic energy of the recoiling electron.

•**36.37** Consider the equivalent of Compton scattering, but the case in which a photon scatters off of a free proton.

a) If 140-keV X-rays bounce off of a proton at 90.0°, what is their fractional change in energy $(E_0 - E)/E_0$?

b) What energy of photon would be necessary to cause a 1% change in energy at 90.0° scattering?

•**36.38** An X-ray photon with an energy of 50.0 keV strikes an electron that is initially at rest inside a metal. The photon is scattered at an angle of 45°. What is the kinetic energy and momentum (magnitude and direction) of the electron after the collision? You may use the non-relativistic relationship connecting the kinetic energy and momentum of the electron.

Section 36.5

36.39 Calculate the wavelength of

a) a 2 eV photon, and

b) an electron with kinetic energy 2 eV.

36.40 What is the de Broglie wavelength of a $2.000 \cdot 10^3$-kg car moving at a speed of 100.0 km/h?

36.41 A nitrogen molecule of mass $m = 4.648 \cdot 10^{-26}$ kg has a speed of 300.0 m/s.

a) Determine its de Broglie wavelength.

b) How far apart are the double slits if a nitrogen molecule fringe pattern, with fringes 0.30 cm apart, is observed on a screen 70.0 cm in front of the slits?

36.42 Alpha particles are accelerated through a potential difference of 20,000 V. What is their de Broglie wavelength?

36.43 Consider an electron whose de Broglie wavelength is equal to the wavelength of green light (about 550 nm).

a) Treating the electron nonrelativistically, what is its speed?

b) Does your calculation confirm that a nonrelativistic treatment is sufficient?

c) Calculate the kinetic energy of the electron in eV.

•**36.44** After you told him about de Broglie's hypothesis that particles of momentum p have wave characteristics with wavelength $\lambda = h/p$, your 60.0-kg roommate starts thinking of his fate as a wave and asks you if he could be diffracted when passing through the 90.0-cm-wide doorway of your dorm room.

a) What is the maximum speed at which your roommate can pass through the doorway in order to be significantly diffracted?

b) If it takes one step to pass through the doorstep, how long should it take your roommate to make that step (assume the length of his step is 0.75 m) in order for him to be diffracted?

c) What is the answer to your roommate's question? *Hint:* Assume that significant diffraction occurs when the width of the diffraction aperture is less that 10.0 times the wavelength of the wave being diffracted.

••**36.45** Consider de Broglie waves for a Newtonian particle of mass m, momentum $p = mv$, and energy $E = p^2/(2m)$, that is, waves with wavelength $\lambda = h/p$ and frequency $\nu = E/h$.

a) Calculate the dispersion relation $\omega = \omega(k)$ for these waves.

b) Calculate the phase and group velocities of these waves. Which of these corresponds to the classical velocity of the particle?

••**36.46** Now consider de Broglie waves for a (relativistic) particle of mass m, momentum $p = mv\gamma$, and total energy $E = mc^2\gamma$, with $\gamma = [1 - (v/c)^2]^{-1/2}$. The waves have wavelength $\lambda = h/p$ and frequency $\nu = E/h$ as before, but with the relativistic momentum and energy.

a) Calculate the dispersion relation for these waves.

b) Calculate the phase and group velocities of these waves. Now which corresponds to the classical velocity of the particle?

Section 36.6

36.47 A 50.0-kg particle has a de Broglie wavelength of 20.0 cm.

a) How fast is the particle moving?

b) What is the smallest speed uncertainty of the particle if its position uncertainty is 20.0 cm?

36.48 During the period of time required for light to pass through a hydrogen atom ($r = 0.53 \cdot 10^{-10}$ m), what is the least uncertainty in energy for the atom? Express your answer in electron volts.

36.49 A free neutron ($m = 1.67 \cdot 10^{-27}$ kg) has a mean life of 900 s. What is the uncertainty in its mass (in kg)?

36.50 Suppose that Fuzzy, a quantum-mechanical duck, lives in a world in which Planck's constant $\hbar = 1.00$ J s. Fuzzy has a mass of 0.500 kg and initially is known to be within a 0.750-m-wide pond. What is the minimum uncertainty in Fuzzy's speed? Assuming that this uncertainty prevails for 5.00 s, how far away could Fuzzy be from the pond after 5.00 s?

36.51 An electron is confined to a box with a dimension of 20 μm. What is the minimum speed the electron can have?

•**36.52** A dust particle of mass 10^{-16} kg and diameter 5 μm is confined to a box of length 15 μm.

a) How will you know whether the particle is at rest?

b) If the particle is not at rest, what will be the range of its velocity?

c) Using the lower range of velocity, how long will it take to move a distance of 1 mm?

Section 36.8

••**36.53** Consider a quantum state of energy E, which can be occupied by any number n of some bosonic particles, including $n = 0$. At absolute temperature T, the probability of finding n particles in the state is given by $P_n = N \exp(-nE/k_BT)$, where k_B is Boltzmann's constant and the normalization factor N is determined by the requirement that all the probabilities sum to unity. Calculate the mean or expected value of n, that is, the occupancy, of this state, given this probability distribution.

•**36.54** Consider the same quantum state as the preceding problem, with a probability distribution of the same form, but with fermionic particles, so that the only possible occupation numbers are $n = 0$ and $n = 1$. Calculate the mean occupancy $\langle n \rangle$ of the state in this case.

••**36.55** Consider a system made up of N particles. The average energy per particle is given by $\langle E \rangle = (\sum E_i e^{-E_i/k_BT})/Z$ where Z is the partition function defined in equation 36.29. If this is a two-state system with $E_1 = 0$ and $E_2 = E$ and $g_1 = g_2 = 1$, calculate the heat capacity of the system, defined as $N(d\langle E \rangle/dT)$ and approximate its behavior at very high and very low temperatures (that is, $k_BT \gg 1$ and $k_BT \ll 1$).

Additional Problems

36.56 Given that the work function of tungsten is 4.55 eV, what is the stopping potential in an experiment using tungsten cathodes at 360 nm?

36.57 Find the ratios of de Broglie wavelengths of a 100-MeV proton to a 100-MeV electron.

36.58 An *Einstein (E)* is a unit of measurement equal to Avogadro's number ($6.02 \cdot 10^{23}$) of photons. How much energy is contained in 1 Einstein of violet light ($\lambda = 400$ nm)?

36.59 In baseball, a 100-g ball can travel as fast as 100 mph. What is the de Broglie wavelength of this ball? The *Voyager* spacecraft, with a mass of about 250 kg, is currently travelling at 125,000 km/h. What is its de Broglie wavelength?

36.60 What is the minimum uncertainty in the velocity of a 1.0-nanogram particle that is at rest on the head of a 1.0-mm-wide pin?

36.61 A photovoltaic device uses monochromatic light of wavelength 700 nm that is incident normally on a surface of area 10 cm^2. Calculate the photon flux rate if the light intensity is 0.30 W/cm^2.

36.62 The Solar Constant measured by Earth satellites is roughly 1400 W/m^2. Though the Sun emits light of different wavelengths, the peak of the wavelength spectrum is at 500 nm.

a) Find the corresponding photon frequency.

b) Find the corresponding photon energy.

c) Find the number flux of photons arriving at Earth, assuming that all light emitted by the Sun has the same peak wavelength.

36.63 Two silver plates in vacuum are separated by 1 cm and have a potential difference of 20 kV between them. What is the largest wavelength of light that can be shined on the cathode to produce a current through the anode?

36.64 How many photons per second must strike a surface of area 10 m^2 to produce a force of 0.1 N on the surface, if the photons are monochromatic light of wavelength 600 nm? Assume the photons are absorbed.

36.65 Suppose the wave function describing an electron predicts a statistical spread of 10^{-4} m/s in the electron's velocity. What is the corresponding statistical spread in its position?

36.66 What is the temperature of a blackbody whose peak emitted wavelength is in the X-ray portion of the spectrum?

36.67 A nocturnal bird's eye can detect monochromatic light of frequency $5.8 \cdot 10^{14}$ Hz with a power as small as $2.333 \cdot 10^{-17}$ W. What is the corresponding number of photons per second a nocturnal bird's eye can detect?

36.68 A particular ultraviolet laser produces radiation of wavelength 355 nm. Suppose this is used in a photoelectric experiment with a calcium sample. What will the stopping potential be?

36.69 What is the wavelength of an electron that is accelerated from rest through a potential difference of 10^{-5} V?

•**36.70** Compton used photons of wavelength 0.0711 nm.

a) What is the wavelength of the photons scattered at $\theta = 180°$?

b) What is energy of these photons?

c) If the target were a proton and not an electron, how would your answer in (a) change?

•**36.71** Calculate the number of photons originating at the Sun that are received in the Earth's upper atmosphere per year.

•**36.72** A free electron in a gas is struck by an 8.5-nm X-ray, which experiences an increase in wavelength of 1.5 pm. How fast is the electron moving after the interaction with the X-ray?

•**36.73** An accelerator boosts a proton's kinetic energy so that the de Broglie wavelength of the proton is $3.5 \cdot 10^{-15}$ m. What are the momentum and energy of the proton?

•**36.74** Scintillation detectors for gamma rays transfer the energy of a gamma-ray photon to an electron within a crystal, via the photoelectric effect or Compton scattering. The electron transfers its energy to atoms in the crystal, which re-emit it as a light flash detected by a photomultiplier tube. The charge pulse produced by the photomultiplier tube is proportional to the energy originally deposited in the crystal; this can be measured so an energy spectrum can be displayed. Gamma rays absorbed by the photoelectric effect are recorded as a *photopeak* in the spectrum, at the full energy of the gammas. The Compton-scattered electrons are also recorded, at a range of lower energies known as the Compton plateau. The highest-energy of these form the *Compton edge* of the plateau. Gamma-ray photons scattered 180° by the Compton effect appear as a *backscatter peak* in the spectrum. For gamma-ray photons of energy 511 KeV, calculate the energies of the Compton edge and the backscatter peak in the spectrum.

37

Quantum Mechanics

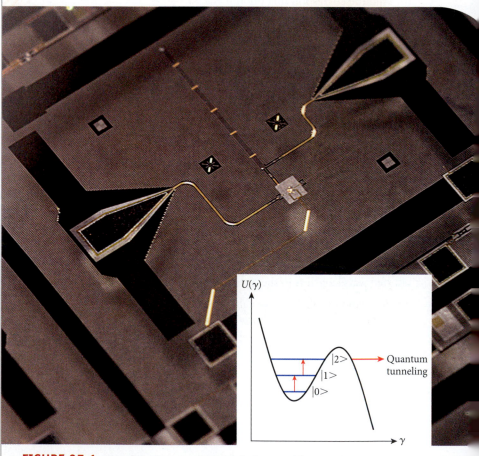

FIGURE 37.1 One of the first experimental devices used for quantum computing; inset on the lower right: schematic representation of a potential for the quantum states used in this process.

WHAT WE WILL LEARN

- A particle is described by its complex wave function. The absolute square of the complex wave function is the probability of finding a particle at some position. The integral of the square of the wave function over all of space equals 1.

- The Schrödinger equation is the nonrelativistic wave equation for a particle in a potential $U(x)$.

- Wave functions that are solutions to the problem of a particle confined to an infinite potential well are sine functions. The energy values of the solutions are proportional to the square of the quantum number of the solution. The solutions to the problem of a potential of finite height have exponential tails reaching into the classically forbidden region.

- If a particle encounters a potential barrier of finite height and width, it can tunnel through this barrier, even if it has energy less than the height of the barrier. The probability for this particle to tunnel through the barrier depends exponentially on the barrier width.

- The wave function solutions for the harmonic oscillator potential are Hermite polynomials. The corresponding energy eigenvalues are equally spaced. Oscillator wave functions with $n = 0$ have the minimum product of momentum and position uncertainty allowed by the uncertainty principle.

- The correspondence principle states that if the energy difference ΔE between neighboring energy states becomes small relative to the total energy E, the quantum solution approaches its classical limit.

- The Hamiltonian operator (or simply Hamiltonian, for short) is the operator of the total energy. It is linear. Thus a linear combination of two solutions to the Schrödinger equation, which is based on the Hamiltonian, is also a solution.

- The two-particle wave function for (non-interacting) bosons is the symmetrized product of one-particle wave functions, and the two-particle wave function for (non-interacting) fermions is the antisymmetrized product.

Chapter 36 introduced some of the basic ideas of quantum physics, including photons, matter waves, and the uncertainty principle. In this chapter, we extend these ideas to calculations of particle dynamics on the atomic scale. We emphasize physical concepts rather than mathematical details, which can become pretty involved in these calculations. However, the ideas of quantum mechanics have produced practical discoveries that could not be imagined with classical physics alone.

Many early researchers in modern physics had serious doubts about what wave functions really are and why probability distributions take a central role in predicting results. For example, the Danish physicist Niels Bohr (1885–1962) once said, "Anyone who has not been shocked by quantum physics has not understood it." And the great American physicist Richard Feynman (1918–1988) wrote, "I think I can safely say that nobody understands quantum mechanics . . . Do not keep saying to yourself, if you can possibly avoid it, 'But how can it be like that? . . .' Nobody knows how it can be like that." Yet now, in the 21st century, quantum mechanics has been established as one of the most accurate and comprehensive areas in all of physics.

When we talked about light waves and sound waves, we discussed the concepts of spherical and planar waves. These waves are specific examples of wave functions that solve a particular wave equation. In the same way, we can now ask what is the wave function of the electron, or what is the wave function of any other object that we conventionally think of as a particle. This question will lead us to look at mechanics in the quantum world and introduce us to what is conventionally called *quantum mechanics,* in which we explore the observable consequences of the wavelike behavior of particles. In the last few years there have been extensive attempts to go beyond simply understanding quantum systems, to manipulate them and to use them for purposes like quantum computing (see Figure 37.1), which carry the promise of revolutionizing nanoscience and nanotechnology.

37.1 Wave Function

Let's briefly review what we have accomplished so far in our investigation of the quantum world. Again we start with light. In Chapter 36, we saw that the photoelectric effect and Compton scattering can be explained only if light has particle properties. However, in

Chapter 31 on electromagnetic waves we also saw that light is an electromagnetic wave of the form $E = E_{\text{max}} \sin(\kappa x - \omega t)$, where κ is the wave number and ω is the angular frequency. We also saw that the intensity of the wave is proportional to the square of the electric field. We used this relationship again in Chapter 34 on wave optics to calculate the intensity of the interference pattern of two light waves. Because the oscillation of the electric field in space and time represents the wave function of light, clearly, for light, the intensity is proportional to the square of the wave function.

What is the corresponding wave function of an electron? Or more generally: What is the quantum wave function of any particle?

In common notation, $\psi(\vec{r},t)$ is written for the **wave function** to denote that it depends on the spatial coordinate $\vec{r}$ and the time t. It is common to use the lowercase Greek letter psi, ψ, for the wave function. We could have just as well used $y(\vec{r},t)$ to describe the wave function, as we did in Chapter 15 when we examined general wave phenomena. However, in the early 20th century the pioneers of quantum physics used the symbol $\psi(\vec{r},t)$, which became the standard notation. In addition, the use of y for the wave function, which is not a coordinate or a distance, could be confused with our common usage of y.

A lot of physical insight can be gained by studying mechanics problems in one spatial dimension. For the one-dimensional wave function, we use the notation $\psi(x,t)$. To start out even simpler, we first look at the wave function in coordinate space, returning to the time dependence later. If we are interested in only the spatial distribution of the wave function, we use the notation $\psi(x)$.

What is the wave function $\psi(x)$ of the electron at the screen position in the double-slit experiment of Chapter 36 (Section 36.5)? Just as for photons, the wave function is proportional to the square root of the intensity, with the intensity given by equation 36.25. The intensity is shown in Figure 37.2a. Parts (b) and (c) of the same figure show two possible wave functions that correspond to this intensity distribution, which differ from each other by a multiplicative factor of −1. Note that the wave function can have positive and negative values, whereas the intensity has a minimum value of zero (that is, it is never negative).

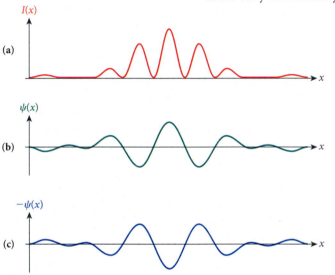

FIGURE 37.2 (a) Intensity distribution for electrons at the position of the screen in the electron double-slit experiment; (b,c) two possible wave functions corresponding to this intensity distribution.

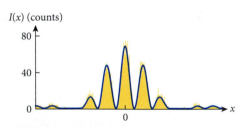

FIGURE 37.3 Intensity distribution (number of electrons hitting the screen per unit length) along the screen in the electron double-slit experiment. Yellow histogram: number of electrons hitting the screen in a coordinate interval.

Wave Function and Probability

What is the physical meaning of the quantum wave function of a particle? Figure 37.3 (reprise of Figure 36.16) shows the connection between the positions at which the electrons hit the screen and the intensity distribution resulting from double-slit interference.

Clearly, the number of counts per bin of some unit length (yellow histogram), and the intensity (blue line) are proportional to each other. The number of counts per bin, divided by the total number of electrons, is the probability that an electron hits in the interval marked by the bin. Let us denote the bin width by dx, because we want to use infinitesimally small bins. Thus, the probability of an electron striking in the interval between x and $x+dx$ is proportional to $I(x)dx$. Finally, because we have just established that the intensity is proportional to the square of the wave function, we can write for the probability $\Pi(x)$ of the electron hitting the interval between x and $x+dx$,

$$\Pi(x)dx = \left| \psi(x) \right|^2 dx. \tag{37.1}$$

This is equivalent to saying that $\left| \psi(x) \right|^2$ is the probability density of finding the particle at position x. The probability is a dimensionless number between 0 and 1, so the probability density must have the physical dimension of inverse length, because dx has the dimension of a length.

Why is the notation of the square of the absolute value of the wave function used in equation 37.1, instead of just the square of the wave function? It turns out that wave functions can have a complex amplitude. Thus, a positive number is ensured only by taking the absolute square of this amplitude.

The probability of finding a particular electron at any position in space must be 1, because it must be somewhere in space. This gives the normalization condition for the wave function,

$$\int_{-\infty}^{\infty} |\psi(x)|^2 \, dx = \int_{-\infty}^{\infty} \psi(x)^* \, \psi(x) dx = 1, \tag{37.2}$$

where $\psi(x)^*$ is the complex conjugate of $\psi(x)$. This normalization condition will allow us to determine the amplitude of the wave function. This equation says that the integral of the absolute square of the wave function over all of space has the value 1, because if we look everywhere in space, then the integrated probability of finding the electron somewhere is 1.

With the aid of complex numbers (see Appendix A for a refresher on complex numbers), we can now write down a convenient form for the coordinate space dependence of the wave function of a freely moving particle. This wave function must be a traveling wave, since the particle is traveling freely, so it can be expected that the wave function will be sinusoidal, like the waveform for an electromagnetic wave. These wave functions depend on space and time. Time dependence will be treated later. For now, we focus on the dependence on the x-coordinate alone, setting $t = 0$. The particle wave function is written as a linear combination of sine and cosine oscillations with wave number $\kappa = 2\pi/\lambda$, where λ is the de Broglie wavelength introduced in Chapter 36,

$$\psi(x) = C\cos(\kappa x) + D\sin(\kappa x).$$

Using complex number notation, the same wave function can be written as

$$\psi(x) = Ae^{i\kappa x} + Be^{-i\kappa x}. \tag{37.3}$$

The coefficients C and D, as well as A and B are, in general, complex numbers. The coefficients A and B can be chosen to suit the particular physical situation, under the condition that the overall wave function is normalized, according to equation 37.2.

Why is it more convenient to use the complex number notation? An answer to this question is given in the next subsection in relation to the momentum that corresponds to a wave function.

37.1 Self-Test Opportunity

Using Euler's formula for complex numbers, show that both of the expressions given here for $\psi(x)$ are identical, and that the amplitudes are related via $A + B = C$ and $i(A - B) = D$.

Momentum

Suppose we have a freely moving particle with wave function of the form in equation 37.3. For now, set $B = 0$. This will simplify our work, and we will soon learn the significance of the B term. Therefore, we have

$$\psi(x) = Ae^{i\kappa x} \tag{37.4}$$

as our wave function. What is the momentum associated with this wave function? According to the de Broglie relation $\lambda = h/p$, the wave number can be written

$$\kappa = \frac{2\pi}{\lambda} = \frac{2\pi}{h/p} = p\frac{2\pi}{h} = \frac{p}{\hbar},$$

again using the shorthand notation $\hbar = h/2\pi$. Consequently,

$$p = \hbar\kappa. \tag{37.5}$$

What operation applied to the wave function will result in a product of the wave function times the momentum? Let us try this *Ansatz*:

$$\mathbf{p}\psi(x) = -i\hbar\frac{d}{dx}\psi(x). \tag{37.6}$$

Here we have used the notation

$$\mathbf{p} = -i\hbar\frac{d}{dx}$$

to denote the momentum **operator;** that is, the operator that results in the product of the momentum and the wave function when applied to the wave function. This way of writing the momentum operator is motivated by noticing that taking the derivative of $\psi(x) = Ae^{i\kappa x}$ with respect to x is equivalent to multiplying the wave function by the factor $i\kappa$. However, this momentum operator turns out to be more general and works for all wave functions, not just that of equation 37.4.

We can take the derivative and convince ourselves that this *Ansatz* makes sense for our wave function:

$$-i\hbar\frac{d}{dx}\psi(x) = -i\hbar\frac{d}{dx}Ae^{i\kappa x} = -i\hbar A\frac{d}{dx}e^{i\kappa x} = -i\hbar A(i\kappa)e^{i\kappa x} = \hbar\kappa Ae^{i\kappa x} = \hbar\kappa\psi(x) = p\psi(x).$$

Thus, our momentum operator applied to the wave function $\psi(x) = Ae^{i\kappa x}$ indeed gives us the product of this wave function with its momentum $p = \hbar\kappa$. Therefore, it is appropriate to interpret our wave function (equation 37.4) as that of a free particle moving with positive momentum $p = \hbar\kappa$. Conversely, a wave function $\psi(x) = Be^{-i\kappa x}$ describes a free particle moving with negative x-component of momentum $-p$, that is, along the negative x-axis. The more general wave function (equation 37.3) therefore describes a superposition of left- and right-moving waves.

Kinetic Energy

We have found that we can introduce a momentum operator and apply it to the quantum wave function of a particle in order to find its momentum. Are there other operators that can be applied to wave functions to find the equivalent classical quantities? One such classical quantity is the kinetic energy of a particle. Classically, the kinetic energy can be written as $K = p^2/2m$, so, following the idea of equation 37.6, we introduce an operator for the kinetic energy as

$$\mathbf{K}\psi(x) = \frac{1}{2m}\mathbf{p}^2\psi(x) = \frac{1}{2m}\left(-i\hbar\frac{d}{dx}\right)^2\psi(x) = -\frac{\hbar^2}{2m}\frac{d^2}{dx^2}\psi(x). \tag{37.7}$$

This equation can be understood as the general definition of the kinetic energy operator. What happens for the special case where this operator is applied to the wave function of a freely moving particle with momentum $p = \hbar\kappa$, one that has a wave function $\psi(x) = Ae^{i\kappa x}$? This gives

$$-\frac{\hbar^2}{2m}\frac{d^2}{dx^2}\psi(x) = -\frac{\hbar^2}{2m}\frac{d^2}{dx^2}Ae^{i\kappa x} = -i\kappa\frac{\hbar^2}{2m}\frac{d}{dx}Ae^{i\kappa x} = \kappa^2\frac{\hbar^2}{2m}Ae^{i\kappa x} = \frac{p^2}{2m}Ae^{i\kappa x} = KAe^{i\kappa x}.$$

Indeed, this works as advertised. The kinetic energy operator applied to the wave function of a free particle yields its kinetic energy. Note that inserting the wave function $\psi(x) = Be^{-i\kappa x}$ would have resulted in the same value for the kinetic energy, because $(-\kappa)^2 = \kappa^2$. Thus, for the superposition of a left- and a right-moving wave with wave number κ, the kinetic energy operator applied to the wave function of equation 37.3 results in

$$\mathbf{K}(Ae^{i\kappa x} + Be^{-i\kappa x}) \equiv -\frac{\hbar^2}{2m}\frac{d^2}{dx^2}(Ae^{i\kappa x} + Be^{-i\kappa x}) = \frac{\hbar^2\kappa^2}{2m}(Ae^{i\kappa x} + Be^{-i\kappa x}). \tag{37.8}$$

37.2 Schrödinger Equation

Given that an electron can be represented by a wave function, what is the equation of motion that describes how this wave function depends on space and time? Such an equation would give solutions for the wave function that are consistent with the observations discussed so far in this chapter. In particular, the solutions should have de Broglie wavelengths of the kind found in Chapter 36, and the absolute squares of the solutions should reproduce the double-slit scattering experiment results. The Austrian physicist Erwin Schrödinger (1887–1961) found such an equation in 1925, and it now carries his name. It is the foundation for all of nonrelativistic quantum mechanics. By "nonrelativistic" we mean all physical cases where the speeds of the objects are small compared to the speed of light, so the kinetic energy can be written as $p^2/2m$.

The following discussion applies the Schrödinger equation to electrons, but the Schrödinger equation also holds for any other object conventionally thought of as a particle.

For now we look at only one-dimensional problems and investigate their static solution, that is, their solutions independent of time. The **Schrödinger equation** for this case is

$$-\frac{\hbar^2}{2m}\frac{d^2\psi(x)}{dx^2}+U(x)\psi(x)=E\psi(x). \tag{37.9}$$

In this equation, $U(x)$ represents the potential energy, which can be different for different positions, and E is the total mechanical energy of the wave. We have already introduced the first term in this equation as the operator of the kinetic energy, so we may think of the Schrödinger equation as an expression of the law of energy conservation for our wave function,

$$\big(K(x)+U(x)\big)\psi(x)=E\psi(x).$$

If the potential energy is zero everywhere, then $U(x) = 0$ and the total energy is equal to the kinetic energy. For this case we have already found the solution because the Schrödinger equation then simply reduces to equation 37.8. In the absence of a potential energy, the solution to the Schrödinger equation is thus $\psi(x) = Ae^{i\kappa x} + Be^{-i\kappa x}$.

What is the solution for a very large potential energy? Physicists like to use the limit of an *infinite* potential energy in order to obtain a simple solution. For an infinite potential energy at some point in space, the Schrödinger equation demands that either the energy E is infinite or the wave function has the value of zero at that position. We are not interested in the unphysical, infinite-energy case, so the wave function has to be zero in this region. Therefore, it is physically impossible for a particle to be in a region with infinite potential energy. We call this region a forbidden region.

Starting from the simple limiting cases of zero potential energy and infinite potential energy, more-complicated wave function solutions to potential energy distributions can be constructed. To find the solution for these wave functions, we have to keep in mind that

- the wave function must be continuous in space (that is, it must not make any "jumps," which would create positions where the derivatives of the wave function are undefined),

- the wave function must be zero in regions with infinite potential energy, as discussed just above, and

- the wave function is normalized (that is, it fulfills the normalization condition of equation 37.2).

37.3 Infinite Potential Well

Our first example of a potential energy distribution in space is that of an **infinite potential well**. This is the simplest case mathematically and it provides insight into interesting physical situations. For an infinite potential well, define the potential energy as a function of the spatial coordinate x as

$$U(x)=\begin{cases}\infty & \text{for } x<0 \\ 0 & \text{for } 0\le x\le a \\ \infty & \text{for } x>a.\end{cases} \tag{37.10}$$

For this case, the value is zero for the wave function for all values outside of the coordinate space interval between 0 and a (Figure 37.4). In particular, this also implies a value of 0 at the interval boundaries. Inside this coordinate space interval, the wave function is of the kind found in equation 37.3, $\psi(x) = Ae^{i\kappa x} + Be^{-i\kappa x}$. As stated before, this solution is mathematically equivalent to $\psi(x) = C\cos(\kappa x) + D\sin(\kappa x)$. This way of writing the wave function is advantageous in this case, because we know that $\psi(0) = 0$. Substituting $x = 0$ into the wave function gives

$$\psi(0)=C\cos(\kappa\cdot 0)+D\sin(\kappa\cdot 0)=C=0.$$

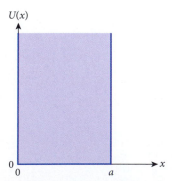

FIGURE 37.4 An infinite potential well. The allowed region with non-infinite values of the potential is shaded in blue.

Thus, the coefficient C must be zero and the cosine term cannot contribute to the solution, giving the solution of $\psi(x) = D\sin(\kappa x)$ in the interval between 0 and a. This makes physical sense, because the superposition of the left-traveling wave $Be^{-i\kappa x}$ and the right-traveling wave $Ae^{i\kappa x}$ results in a standing wave, which is the appropriate solution for a wave trapped between two infinitely high potential barriers.

Thus, our preliminary result is that our wave function can be written as

$$\psi(x) = \begin{cases} 0 & \text{for } x<0 \\ D\sin(\kappa x) & \text{for } 0 \le x \le a \\ 0 & \text{for } x>a. \end{cases}$$

Because our wave functions must be continuous, there are constraints on the possible solutions inside the interval between 0 and a, since only those solutions that vanish at the boundaries can be used. At $x = 0$, this happens automatically because the sine function is always 0 at this point. However, at $x = a$ the boundary condition is

$$\psi(x=a) = D\sin(\kappa a) = 0.$$

This implies that not all values of the wavelength $\lambda = 2\pi/\kappa$ are possible, but only those for which

$$\kappa a = \frac{2\pi a}{\lambda} = n\pi \quad \text{for all } n=1,2,3,...$$

is fulfilled, because the sine function is zero whenever its argument is an integer multiple of π. (Technically, $n = 0$ is also permitted, but for any finite value of a this implies an unphysical infinite wavelength. Negative numbers would also be permitted, but because $\sin(-x) = -\sin(x)$, these do not give us additional solutions beyond those already contained in the positive integers.) Thus, the possible wavelengths in this potential well are only those with

$$\lambda_n = \frac{2a}{n} \quad \text{for all } n=1,2,3,.... \tag{37.11}$$

Then the only possible solutions of the Schrödinger equation for the infinite potential well problem are

$$\psi(x) = \begin{cases} 0 & \text{for } x<0 \\ D\sin\left(\dfrac{n\pi x}{a}\right) \text{ with } n=1,2,3,... & \text{for } 0 \le x \le a \\ 0 & \text{for } x>a. \end{cases}$$

We are almost done, but we still have to determine the amplitude D of the wave function. In order to obtain it, we use the normalization condition (equation 37.2):

$$1 = \int_{-\infty}^{\infty} |\psi(x)|^2\, dx = \int_{-\infty}^{0} |\psi(x)|^2\, dx + \int_{0}^{a} |\psi(x)|^2\, dx + \int_{a}^{\infty} |\psi(x)|^2\, dx$$

$$= 0 + \int_{0}^{a} \left|D\sin\left(\frac{n\pi x}{a}\right)\right|^2 dx + 0$$

$$= |D|^2 \int_{0}^{a} \sin^2\left(\frac{n\pi x}{a}\right) dx$$

$$= |D|^2 \frac{a}{2}.$$

Therefore, $|D|^2 = 2/a$. Any complex number z can be written as $z = re^{i\theta}$ and $|z| = r$ for any phase angle θ. The magnitude of D is $\sqrt{2/a}$ so in general we can write $D = \sqrt{2/a}\, e^{i\theta}$. For simplicity, we select this phase angle as 0, and thus we have the complete solution for the wave function of a particle confined to a potential well with infinitely high walls:

$$\psi(x) = \begin{cases} 0 & \text{for } x < 0 \\ \sqrt{\dfrac{2}{a}} \sin\left(\dfrac{n\pi x}{a}\right) \text{ with } n = 1,2,3,\ldots & \text{for } 0 \le x \le a \\ 0 & \text{for } x > a. \end{cases} \qquad (37.12)$$

In the solution (equation 37.12), each value of n corresponds to a different possible wave function for a particle in an infinite well. To distinguish them, each wave function is labeled with an index that indicates this number n, the **principal quantum number.** Thus, $\psi_1(x)$ denotes the solution for quantum number 1, and so on. Figure 37.5a shows the wave function solutions for the lowest four quantum numbers.

Earlier, we saw that the probability of finding a particle at a given position is proportional to the absolute square of the wave function. We thus plot in Figure 37.5b the absolute square of the wave functions that correspond to the lowest four quantum numbers. This shows the relative likelihood of where in the well one could expect to find the particle when performing a position measurement.

The emergence of simple integer numbers—that is, quantum numbers—is a typical feature of quantum mechanical systems. Boundary conditions on the wave functions force the existence of solutions that are quantized. The discussions of atoms, atomic nuclei, and elementary particles in Chapters 38–39 will show that there are different quantum numbers in many situations.

Chapter 15's solution to a vibrating string has many similarities to the solution for a particle in an infinite potential well. Because the string is clamped at both ends, nodes of the wave function appear at the ends, allowing only the fundamental oscillation and its harmonics. The infinitely high potential outside the interval [0,a] accomplishes the same effect of "clamping down" the wave function of a particle at the boundaries of the potential well. Therefore, the standing wave solutions to the classical vibrating string and to the quantum particle in an infinite potential well are mathematically identical.

Finally, let's compare the quantum probability distribution to the classical one for the case we have just considered. In the classical case, the particle remains trapped between 0 and a. Between these two points, it moves with constant speed $v = \sqrt{2E/m}$.

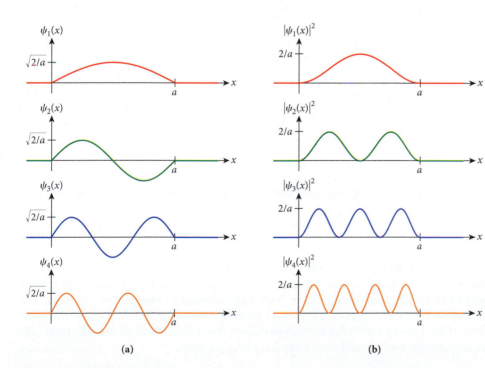

(a) (b)

FIGURE 37.5 (a) Wave functions corresponding to the lowest four quantum numbers in an infinite potential well; (b) probability of finding the particle with this wave function at any particular position in the well.

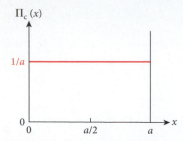

FIGURE 37.6 Classical probability distribution for finding a particle at any particular position trapped in an infinite potential well.

The probability of finding a particle at a given point in space is proportional to the fraction dt of the total time the particle spends in the vicinity dx of the point x. In turn, this time interval dt is inversely proportional to the speed $v(x)$ it has at this point, $dt = dx/v(x)$. Thus the classical probability of finding a particle in the interval between x and $x + dx$ is

$$\Pi_c(x)dx \propto \frac{1}{v(x)}. \qquad (37.13)$$

If v is independent of position, then the classical probability of finding a particle must be a constant as a function of the position. The resulting probability distribution is shown in Figure 37.6.

Comparing this result to the quantum mechanical probability distribution in Figure 37.5b, we see clear differences. The quantum mechanical probability distribution oscillates between 0 and a maximum value of $2/a$, whereas the classical value represents the average of this oscillation, $1/a$. The relationship between classical and quantum mechanical probability distributions is discussed in more detail later in this chapter.

Energy of a Particle

The Schrödinger equation 37.9 depends on the total energy of a particle. Now that we have the complete wave function solution for a particle in an infinite potential well, let's find the total energy that corresponds to this solution in this case. Outside the interval between $x = 0$ and $x = a$, the wave function is zero, so the corresponding particle has zero probability of residing there. Thus, the outside region does not contribute to the energy. Inside the interval between 0 and a there is no potential energy. Therefore, the total energy is equal to the kinetic energy of the particle. Equation 37.7 tells us how to calculate this kinetic energy:

$$\mathbf{K}\psi(x) = -\frac{\hbar^2}{2m}\frac{d^2\psi(x)}{dx^2}.$$

For each quantum number n, however, a different result should be expected for the kinetic energy and therefore the total energy. Label this total energy corresponding to the quantum number n as E_n. For the case of the particle in the infinite potential well,

$$E_n\psi_n(x) = -\frac{\hbar^2}{2m}\frac{d^2\psi_n(x)}{dx^2}$$

$$= -\frac{\hbar^2}{2m}\frac{d^2}{dx^2}\left(\sqrt{\frac{2}{a}}\sin\left(\frac{n\pi x}{a}\right)\right)$$

$$= -\frac{\hbar^2}{2m}\left(\frac{n\pi}{a}\right)\frac{d}{dx}\left(\sqrt{\frac{2}{a}}\cos\left(\frac{n\pi x}{a}\right)\right)$$

$$= \frac{\hbar^2}{2m}\left(\frac{n\pi}{a}\right)^2\left(\sqrt{\frac{2}{a}}\sin\left(\frac{n\pi x}{a}\right)\right)$$

$$= \frac{\hbar^2}{2m}\left(\frac{n\pi}{a}\right)^2\psi_n(x).$$

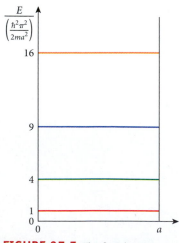

FIGURE 37.7 The four lowest energy levels in an infinite potential well. The colors of the energy levels correspond to the colors of the corresponding wave functions in Figure 37.5.

The total energy E_n corresponding to the quantum number n is thus proportional to the square of n:

$$E_n = \frac{\hbar^2\pi^2}{2ma^2}n^2, \qquad n = 1,2,3,\ldots. \qquad (37.14)$$

The energy also depends inversely on the square of the width of the potential, a, and inversely on the mass of the particle, m. These energies can be represented as horizontal lines in a plot of energy versus position. Such a plot is called an **energy-level diagram** and is shown in Figure 37.7. Because the energies of these states are lower than the (infinite) height of the potential well, the corresponding states are referred to as **bound states.** The infinite potential well has an infinite number of bound states.

EXAMPLE 37.1 | Electron in a Box

This solution for an infinite potential well is sometimes called a "particle in a rigid box" and it allows us to simply model an electron bound to an atom or a proton bound in an atomic nucleus, as the following example shows.

PROBLEM

What is the kinetic energy of the wave function with the lowest quantum number for an electron confined to a box of width $2.00 \text{ Å} \equiv 2.00 \cdot 10^{-10} \text{ m}$?

SOLUTION

The mass of the electron is $m = 9.109 \cdot 10^{-31}$ kg. In this case, the entire answer amounts to calculating

$$E_1 = \frac{\hbar^2 \pi^2}{2ma^2} = \frac{(1.0546 \cdot 10^{-34} \text{ J s})^2 \pi^2}{2(9.109 \cdot 10^{-31} \text{ kg})(2.00 \cdot 10^{-10} \text{ m})^2} = 1.51 \cdot 10^{-18} \text{ J}.$$

Alternatively, this energy value can be expressed in units of electron-volts:

$$E_1 = (1.51 \cdot 10^{-18} \text{ J})/(1.602 \cdot 10^{-19} \text{ J/eV}) = 9.43 \text{ eV}.$$

DISCUSSION

Chapter 38 will show that atoms have typical diameters of 10^{-10} m, so this example allows us to estimate that the typical energy scale for electrons in atoms has to be on the order of 10 eV. Real atoms are much more complicated than a simple model of a box with infinitely high walls, but this way of thinking allows educated guesses on the energy scales involved in a physical problem.

Performing the same exercise for a proton (with a mass approximately 2000 times that of the electron) confined to a box of width 10^{-15} m to 10^{-14} m, which is the typical dimension of an atomic nucleus, yields the answer of $5 \cdot 10^7$ eV to $5 \cdot 10^6$ eV. We conclude that typical nuclear energy scales are on the order of 5 to 50 MeV, a remarkably good first guess.

37.1 In-Class Exercise

If we reduce the width of the potential well to half its original value, then the energy of the $n = 3$ wave function will

a) stay the same.

b) be reduced by a factor of 2.

c) be reduced by a factor of 4.

d) be increased by a factor of 2.

e) be increased by a factor of 4.

37.2 Self-Test Opportunity

How do the solutions for the wave functions and energies in the problem of the infinite potential well change if we replace the potential energy function of equation 37.10 with one that is still infinite outside the interval between 0 and a, but has a constant value of $c \neq 0$ inside this interval?

Multidimensional Wells

The idea of a one-dimensional infinite potential well can be extended to that of a two-dimensional well in which a particle is confined in the xy-plane, or even to a rectangular box in three-dimensional space. We do not give a mathematical derivation of the full solution to these situations, but instead stress the general features.

First, let's think about what will be different as our calculations are extended from one to two spatial dimensions. The potential energy can now be a function of both variables, x and y, and so we write it as $U(x,y)$. This means that the wave function also must be written as a function of the same two variables, $\psi(x,y)$. In our classical considerations of kinetic energy, we saw that this can be written as

$$K = \frac{p_x^2}{2m} + \frac{p_y^2}{2m}.$$

In analogy with what we have just derived for the one-dimensional problem (see equation 37.6), the x- and y-components of the momentum operator can be written as

$$\mathbf{p}_x \psi(x,y) = -i\hbar \frac{\partial}{\partial x} \psi(x,y)$$

$$\mathbf{p}_y \psi(x,y) = -i\hbar \frac{\partial}{\partial y} \psi(x,y).$$

The only change relative to the one-dimensional problem is that partial derivatives must be used, as is appropriate for multivariable calculus. (A partial derivative with respect to one variable treats the other variables as constants.) Then the kinetic energy operator for the quantum wave function can be written as

$$\mathbf{K}\psi(x,y) = \frac{1}{2m}\mathbf{p}_x^2\psi(x,y) + \frac{1}{2m}\mathbf{p}_y^2\psi(x,y)$$

$$= -\frac{\hbar^2}{2m}\frac{\partial^2}{\partial x^2}\psi(x,y) - \frac{\hbar^2}{2m}\frac{\partial^2}{\partial y^2}\psi(x,y).$$

Finally, the two-dimensional Schrödinger equation reads

$$-\frac{\hbar^2}{2m}\frac{\partial^2\psi(x,y)}{\partial x^2} - \frac{\hbar^2}{2m}\frac{\partial^2\psi(x,y)}{\partial y^2} + U(x,y)\psi(x,y) = E\psi(x,y).$$

To proceed, we need to specify the shape of the potential energy. If the potential energy can be written as a product of a function that depends only on x and another that depends only on y—that is to say, $U(x,y) = U_1(x) \cdot U_2(y)$—then the problem becomes **separable.** This means that the wave function is also a product of two functions, and that each one depends on only one variable: $\psi(x,y) = \psi_1(x) \cdot \psi_2(y)$.

Further, if we use the simplifying case of a two-dimensional rectangular infinite potential well,

$$U(x,y) = U_1(x) \cdot U_2(y)$$

$$\text{with: } U_1(x) = \begin{cases} \infty & \text{for } x<0 \\ 0 & \text{for } 0\le x\le a, \\ \infty & \text{for } x>a \end{cases} \quad U_2(y) = \begin{cases} \infty & \text{for } y<0 \\ 0 & \text{for } 0\le y\le b \\ \infty & \text{for } y>b \end{cases}$$

then the solutions we obtain are products of the wave functions obtained in equation 37.12 in the x- and y-directions. Explicitly, these wave functions are

$$\psi(x,y) = \psi_1(x) \cdot \psi_2(y) \tag{37.15}$$

$$\psi_1(x) = \begin{cases} 0 & \text{for } x<0 \\ \sqrt{\dfrac{2}{a}}\sin\left(\dfrac{n_x\pi x}{a}\right) & \text{with } n_x = 1,2,3,\ldots \text{ for } 0\le x\le a \\ 0 & \text{for } x>a \end{cases}$$

$$\psi_2(y) = \begin{cases} 0 & \text{for } y<0 \\ \sqrt{\dfrac{2}{b}}\sin\left(\dfrac{n_y\pi y}{b}\right) & \text{with } n_y = 1,2,3,\ldots \text{ for } 0\le y\le b \\ 0 & \text{for } y>b. \end{cases}$$

These solutions are displayed in Figure 37.8 for the lowest values of the quantum numbers n_x and n_y.

In the same way that we arrived at the solution for the energy values corresponding to the one-dimensional quantum numbers n in equation 37.14, we obtain for the energies corresponding to these two-dimensional wave functions,

$$E_{n_x,n_y} = \frac{\hbar^2\pi^2}{2ma^2}n_x^2 + \frac{\hbar^2\pi^2}{2mb^2}n_y^2. \tag{37.16}$$

If, for example, $a = b$ (square potential), then the same energy value can usually be obtained in more than one way. For instance, the states with quantum numbers $(n_x = 1, n_y = 2)$ and $(n_x = 2, n_y = 1)$ have the same energy. This is referred to as "degeneracy."

If the potential wells are not rectangular, then the solution usually cannot be written in a simple form. However, in many cases the problem can still be solved numerically. Experimentally, very high (not quite infinite) two-dimensional potential wells for electrons can be generated by arranging several atoms on flat surfaces in the shapes of corrals. Figure 37.9 shows one such example, where the color coding (large image) or ripples (gray-scale inset) represent the probability distributions for the electron wave function inside corrals, consisting of iron atoms arranged in different shapes on a flat copper surface. The instrument used first to arrange the

37.3 Self-Test Opportunity

Can you use the same considerations and write down the wave functions and energies for a three-dimensional rectangular infinite potential well?

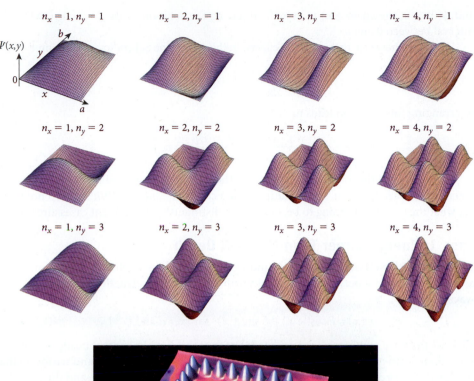

$n_x = 1, n_y = 1$ $n_x = 2, n_y = 1$ $n_x = 3, n_y = 1$ $n_x = 4, n_y = 1$

$n_x = 1, n_y = 2$ $n_x = 2, n_y = 2$ $n_x = 3, n_y = 2$ $n_x = 4, n_y = 2$

$n_x = 1, n_y = 3$ $n_x = 2, n_y = 3$ $n_x = 3, n_y = 3$ $n_x = 4, n_y = 3$

FIGURE 37.8 Wave functions with the lowest quantum numbers in a two-dimensional rectangular infinite potential well.

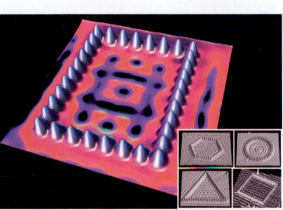

FIGURE 37.9 Two-dimensional rectangular quantum corral made of individual iron atoms, arranged on a copper surface. The color coding inside the corral shows the electron wave probability density. The gray-scale insets show several corrals of different shapes, and the ripples indicate the electron wave probability density. These arrangements were created by using a scanning tunneling microscope.

atoms in the ways shown and second to generate these experimental images is called a scanning tunneling microscope, which will be discussed in more detail later in this chapter.

37.4 Finite Potential Wells

Let's return to the one-dimensional case and solve a problem that is a bit more complicated than that of a potential well with infinitely high potential outside the well. Now we want to obtain a solution for the case where the wall of the well is not infinitely high, but instead has a finite height.

The shape of the **finite potential well** we want to study is drawn in Figure 37.10: The potential energy $U(x)$ is zero inside the interval from 0 to a, is infinite for all values of $x < 0$, and has a finite constant value of $U_1 > 0$ for $x > a$,

$$U(x) = \begin{cases} \infty & \text{for } x < 0 \\ 0 & \text{for } 0 \le x \le a \\ U_1 & \text{for } x > a. \end{cases}$$

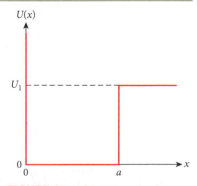

FIGURE 37.10 Finite potential energy well.

Just as in the case of the infinite potential, we construct a solution for each of the three regions separately, and then match these solutions to each other at the boundaries. The solution remains the same as before for $x < 0$, where the wave function must have a constant value of 0. In the interval from 0 to a, the wave function again must have the general form of $\psi(x) = C \cos(\kappa x) + D \sin(\kappa x)$. And again, because the wave function has to be continuous and $\psi(0) = 0$, the coefficient C must be 0, giving the solution of $\psi(x) = D\sin(\kappa x)$ in the interval between 0 and a.

For $x > a$, however, there are new effects. In this region the Schrödinger equation is

$$-\frac{\hbar^2}{2m}\frac{d^2\psi(x)}{dx^2} + U_1\psi(x) = E\psi(x).$$

Rearranging this equation leads to

$$\frac{d^2\psi(x)}{dx^2} = \frac{2m(U_1 - E)}{\hbar^2}\psi(x). \tag{37.17}$$

Let us look at the form of this equation: The second derivative of the wave function is equal to the wave function itself, multiplied by a constant, $2m(U_1 - E)/\hbar^2$. We do not know yet what the energy E is going to be, but we can distinguish two different cases already.

Case 1: Energy Larger Than the Well Depth

For $E > U_1$ we have $2m(U_1 - E)/\hbar^2 < 0$, and we obtain oscillatory solutions of the same kind as inside the interval between 0 and a, but with a different wavelength and thus a different wave number:

$$\psi(x) = E\cos(\kappa'x) + F\sin(\kappa'x) \quad \text{for } x > a \text{ and } E > U_1.$$

How are the wave numbers κ' and κ related to each other? Remember that inside the interval $[0,a]$ there is no potential energy, the total energy is all kinetic energy, and we found that $E = \hbar^2\kappa^2/2m$. For $x > a$ and $E > U_1$, the kinetic energy is then simply $E - U_1$, and thus $E - U_1 = \hbar^2\kappa'^2/2m$. Therefore

$$\kappa' = \sqrt{\kappa^2 - \frac{2mU_1}{\hbar^2}}. \tag{37.18}$$

Thus, the wave number $\kappa' < \kappa$, and so the wavelength of the spatial oscillation is larger in the region with $U(x) = U_1 > 0$ than in the region where $U(x) = 0$.

The wave function for this case of total energy larger than the height of the potential well $E > U_1$ is then

$$\psi(x) = \begin{cases} 0 & \text{for } x < 0 \\ D\sin(\kappa x) & \text{for } 0 \le x \le a \\ F\cos(\kappa'x) + G\sin(\kappa'x) & \text{for } x > a. \end{cases}$$

How do we determine the complete solution? We must find the values for the amplitude coefficients D, F, and G. These three numbers can be determined from the following three conditions:

- The wave function must be continuous at the boundary $x = a$ (as explained earlier).
- The derivative of the wave function must be continuous at $x = a$ (as explained below).
- The absolute square of the wave function must be normalized to 1.

The rationale for the second condition is that the momentum is related to the derivative of the wave function, so if the derivative of the wave function were discontinuous, then the momentum would be undefined at the point of discontinuity. Note that the wave numbers κ and κ' are not further constrained by these conditions, and neither are the possible values of the energy E. All energies larger than the value U_1 will turn out to be possible.

We do not want to work through all of the algebra at this point; this is usually done during an upper-level course on quantum mechanics. From the structure we have worked out so far, however, we can already make a sketch (Figure 37.11) of the type of wave function that we can expect. This wave function has the features that it is continuous and has a continuous derivative everywhere. Also, in general, the oscillations in the region with $U > 0$ have a larger wavelength (because of equation 37.18) and amplitude (because of the larger wavelength and the first two conditions just above) than those in the region with $U = 0$. We emphasize once more that in this case we do not have just discrete possible wavelengths, but an infinite continuum of wavelengths.

$\psi(x)$

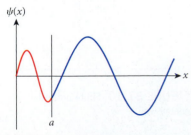

FIGURE 37.11 Wave function for the finite potential well, with energy larger than the potential step.

Case 2: Energy Smaller Than the Well Depth, Bound States

In this case, the energy of the particle is smaller than the depth of the potential well, so the wave functions that fulfill the condition $E < U_1$ form bound states. Part of our task is to find out if there are any bound states at all. Because even the lowest possible energy of a wave function in the infinite well has a finite nonzero value, we expect that for very shallow potential wells, no bound state occurs.

For $E < U_1$, the constant $2m(U_1 - E)/\hbar^2$ has a value larger than zero, so instead of oscillatory solutions, we obtain exponential solutions to the differential equation 37.17. To show this in quantitative detail, we introduce a constant γ:

$$\gamma = \sqrt{\frac{2m(U_1 - E)}{\hbar^2}}. \tag{37.19}$$

Then the differential equation 37.17 that we have to solve in the region $x > a$ becomes

$$\frac{d^2\psi(x)}{dx^2} = \gamma^2\psi(x).$$

It has the solution:

$$\psi(x) = Fe^{-\gamma x} + Ge^{\gamma x} \quad \text{for} \quad x > a \text{ and } E < U_1.$$

(That this is a solution can be verified by taking its second derivative and inserting it back into the differential equation.) Discard the exponentially rising solution $e^{\gamma x}$ because this solution becomes infinite as we approach $x \to \infty$. Then our wave function would not be normalizable; that is, the integral of the absolute square of the wave function could not have the value 1 according to equation 37.2. Thus, $G = 0$, and our solution becomes

$$\psi(x) = Fe^{-\gamma x} \quad \text{for} \quad x > a \text{ and } E < U_1.$$

Again, the wave function for all three regions can be written:

$$\psi(x) = \begin{cases} 0 & \text{for } x < 0 \\ D\sin(\kappa x) & \text{for } 0 \leq x \leq a \\ Fe^{-\gamma x} & \text{for } x > a. \end{cases} \tag{37.20}$$

The conditions to have a normalized wave function that is continuous and has a continuous derivative at $x = a$ provide three equations for the two unknown amplitudes D and F, as well as the wave number κ. The other constant that appears in this solution, γ, is not independent of κ. This can be seen by using the defining equation 37.19 for γ, and the fact that when the potential energy is zero in the region $0 \leq x \leq a$, the total energy E which is constant is simply $E = \hbar^2\kappa^2/2m$. This results in

$$\gamma^2 = \frac{2m(U_1 - E)}{\hbar^2} = \frac{2mU_1}{\hbar^2} - \frac{2mE}{\hbar^2} = \frac{2mU_1}{\hbar^2} - \kappa^2. \tag{37.21}$$

Before we calculate one representative case in greater detail, let's first think about the general features of the solution that we expect to obtain. Just as in the infinite potential well case, the wave function in the finite potential well case oscillates in sine form in the region where the potential energy vanishes. However, now the wave function is allowed to "leak" into the wall, although it has an exponentially decreasing value as x increases beyond a. This leakage allows the wavelength for $x < a$ to be somewhat longer than the wavelength of the infinite well wave function because now the function spills over into $x > a$. A longer wavelength implies a smaller wave number. Since the total energy is proportional to the square of the wave number, the energy values corresponding to the wave functions in the finite potential well can be expected to be lower than their counterparts in the infinite well. Another way of stating this result is to say that the wave functions in the infinite well are more "localized," and thus have larger kinetic energies, than their corresponding counterparts in the well of finite depth.

EXAMPLE **37.2** Finite Potential Well

PROBLEM

Suppose an electron is to have at least two bound states in a well of the shape indicated in Figure 37.10. If the width is $a = 1.30$ nm, how high does the potential step U_1 need to be for the wavelength of the $n = 2$ state to be 20% greater than it would be in an infinite potential well of the same width?

SOLUTION

This problem must be approached in steps. First, equation 37.11 gave the result for the infinite well, $\lambda_n = 2a/n$ for all $n = 1,2,3,....$ Thus, we obtain for $n = 2$ the wavelength $\lambda_2 = 2a/2 = a$. (As usual, at this point we are not going to insert the value for a, but leave the result general and insert the numbers only at the very end.)

A wavelength 20% larger than that for the infinite well in this case is $\lambda'_2 = 1.2\lambda_2 = 1.2a$. The corresponding value of the wave number is thus $\kappa'_2 = 2\pi/\lambda'_2 = 2\pi/(1.2a)$. Since the wave number is inversely proportional to the wavelength, and the wavelength needs to increase by 20%, the wave number needs to be reduced by the same factor: $\kappa'_2 = \kappa_2/1.2$. (Again, we postpone inserting numerical values for the wave number until the end.)

The general wave function for this problem is given by equation 37.20:

$$\psi_2(x) = \begin{cases} 0 & \text{for } x < 0 \\ D\sin\left(\kappa'_2 x\right) & \text{for } 0 \le x \le a \\ Fe^{-\gamma x} & \text{for } x > a. \end{cases}$$

Let's look at the boundary conditions at $x = a$. Demanding continuity of the wave function requires

$$D\sin\left(\kappa'_2 a\right) = Fe^{-\gamma a} \Rightarrow \sin\left(\kappa'_2 a\right) = \frac{F}{D}e^{-\gamma a}. \tag{i}$$

Take the derivative of the wave function, and then look at the continuity condition for the derivative of the wave function. The derivative is

$$\frac{d}{dx}\psi_2(x) = \begin{cases} 0 & \text{for } x < 0 \\ \kappa'_2 D\cos\left(\kappa'_2 x\right) & \text{for } 0 \le x \le a \\ -\gamma Fe^{-\gamma x} & \text{for } x > a. \end{cases}$$

A continuous first derivative at $x = a$ thus requires

$$\kappa'_2 D\cos\left(\kappa'_2 a\right) = -\gamma Fe^{-\gamma a} \Rightarrow -\frac{\kappa'_2}{\gamma}\cos\left(\kappa'_2 a\right) = \frac{F}{D}e^{-\gamma a}. \tag{ii}$$

We have rewritten equations (i) and (ii) so they have the same right-hand sides. Therefore their left-hand sides must also be equal, and we obtain

$$-\frac{\kappa'_2}{\gamma}\cos\left(\kappa'_2 a\right) = \sin\left(\kappa'_2 a\right) \Rightarrow$$

$$\gamma = -\kappa'_2 \cot\left(\kappa'_2 a\right). \tag{37.22}$$

Taking the square of this equation results in

$$\gamma^2 = \kappa'^2_2 \cot^2\left(\kappa'_2 a\right). \tag{iii}$$

However, we also know that we had calculated for γ^2 in equation 37.21 the relationship

$$\gamma^2 = \frac{2mU_1}{\hbar^2} - \kappa'^2_2. \tag{iv}$$

Combining equations (iii) and (iv) gives

$$\kappa_2'^2 \cot^2\left(\kappa_2' a\right) = \frac{2mU_1}{\hbar^2} - \kappa_2'^2 \Rightarrow$$

$$\frac{2mU_1}{\hbar^2} = \kappa_2'^2\left(1 + \cot^2\left(\kappa_2' a\right)\right) \Rightarrow$$

$$U_1 = \frac{\hbar^2 \kappa_2'^2}{2m}\left(1 + \cot^2\left(\kappa_2' a\right)\right).$$

Formally, this is our answer. Now we can insert our numbers. We have already noted that $\kappa_2' = 2\pi/\lambda_2' = 2\pi/(1.2a)$; therefore, $\kappa_2' a = 2\pi/1.2$. We can then calculate the factor $\cot^2(\kappa_2' a)$:

$$\cot^2\left(\kappa_2' a\right) = \cot^2\left(2\pi/1.2\right) = 0.3333 = \tfrac{1}{3}.$$

Therefore, the potential step has to have a height of $1 + \tfrac{1}{3} = \tfrac{4}{3}$ times the energy E_2 of the wave function ψ_2. This energy is

$$E_2 = \frac{\hbar^2 \kappa_2'^2}{2m} = \frac{\hbar^2 (2\pi/1.2a)^2}{2m} = \frac{h^2}{2.88ma^2}$$

$$= \frac{(6.626 \cdot 10^{-34} \text{ J s})^2}{2.88(9.109 \cdot 10^{-31} \text{ kg})(1.30 \cdot 10^{-9} \text{ m})^2}$$

$$= 9.90 \cdot 10^{-20} \text{ J} = 0.618 \text{ eV}.$$

Our final answer is $U_1 = \tfrac{4}{3}E_2 = 0.823$ eV.

Note that $\kappa_2' = \kappa_2/1.2$ as demanded by the problem. Since $E \propto \kappa^2$ for the potential well, the energy E_2 is a factor of $(1/1.2)^2 = 0.694$ lower than that of the infinite well.

DISCUSSION

This example is fairly typical. It shows how we need to use the conditions of continuity of the wave function and its derivatives at boundaries to set up equations allowing us to solve for the undetermined constants in the generic solutions. In this case, our generic solution was the same type as equation 37.20, with constants U_1, D, F, κ_2', γ, which were to be determined. The equation for the continuity of the wave function was expressed in equation (i), and for the continuity of the derivative in equation (ii). A third equation was obtained from the relationship between γ and κ_2', expressed in equation 37.21. Because κ_2' was specified in the text of the question, we would need four equations to solve for our remaining four unknown quantities. The continuity of ψ and its derivative at the boundary and the relationship between γ and κ_2' was enough to solve for U_1. To find D and F we would need to use the normalization condition (equation 37.2).

Example 37.2 presented a condition that predetermined the shape of the wave function by specifying the desired value of the wave number. A much more conventional problem is to figure out what wave functions fit into a potential of given depth. We will do this in the next example. To make our task easier, we simply use the same numbers as in the preceding example.

EXAMPLE 37.3 | Bound States

PROBLEM

If an electron is trapped in a finite potential well of the kind shown in Figure 37.10 with a well depth of $U_1 = 0.823$ eV and a width $a = 1.30$ nm, what wave numbers correspond to the possible bound states in this well?

SOLUTION

The well depth is the same as in Example 37.2, so one solution for the wave number κ should correspond to the value $2\pi/(1.2a)$, the starting point in that example. We also

Continued—

found two conditions for the exponential decay constant γ. Equation 37.22 gave $\gamma = -\kappa \cot(\kappa a)$, and equation 37.21 gave $\gamma = \sqrt{2mU_1/\hbar^2 - \kappa^2}$. Combining these results gives

$$\sqrt{\frac{2mU_1}{\hbar^2} - \kappa^2} = -\kappa \cot(\kappa a) \Rightarrow$$

$$\sqrt{\frac{2a^2 mU_1}{\hbar^2} - (\kappa a)^2} = -(\kappa a)\cot(\kappa a).$$

We have omitted the index 2 for the wave number κ, which we used in the above equations, because we want to search for all possible values of κ that satisfy this equation. This equation usually does not have an algebraic solution, but we can solve it numerically quite straightforwardly. The numerical constants in our case are $a = 1.30$ nm and $2a^2 mU_1/\hbar^2 = 36.5$. Thus, we have to solve the equation

$$\sqrt{36.5 - y^2} = -y\cot y, \text{ with } y = \kappa a.$$

In Figure 37.12, we plot the two functions $\sqrt{36.5 - y^2}$ (in blue) and $-y \cot y$ (in red) and find the positions where they intersect.

As you can see, there are only two positions where the two functions have the same value. Therefore, our potential well of depth $U_1 = 0.823$ eV and width $a = 1.30$ nm has only two bound states. Numerically, we find $y_1 = \kappa_1 a = 2.68$ and $y_2 = \kappa_2 a = 5.23$. This second value is nothing other than $2\pi/1.2$, because this is the value from which the previous example was constructed. However, the value of $\kappa_1 = 2.68/a$ is new information. Note that now $\kappa_2 \neq 2\kappa_1$, unlike the case of the infinite potential well.

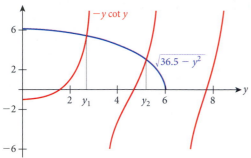

FIGURE 37.12 Finding the wave numbers for the bound states.

37.2 In-Class Exercise

If we double the width of the potential well to 2.6 nm and leave the depth the same as in Example 37.3, then the number of bound states will

a) stay the same.

b) increase.

c) decrease.

To conclude our discussion of the bound states in this example, we show in Figure 37.13 the energies that correspond to the two bound states, overlaid on the shape of the potential energy function used for the potential well.

Finally, Figure 37.14a shows the wave functions that correspond to the two values found for the wave number, as functions of the spatial coordinate x/a. The upper graph is for the wave function times $\sqrt{a}$, corresponding to wave number κ_1 and energy E_1, and the lower graph for κ_2 and E_2. The red part of the curve corresponds to the sinusoidal part of the wave function in the potential-free region. The blue part is the exponential penetration of the wave function into the region where the potential energy is greater than the energy of the electron; this is the classically forbidden region. It is apparent from the figure that the two parts of the wave functions are properly matched, since the wave function is continuous and has a continuous derivative at the boundary $x/a = 1$. In order to obtain each of these wave functions, we have to solve a system of equations of the kind used in the previous example.

The fact that an electron can penetrate the potential boundary into the classically forbidden region of $x > a$ is a phenomenon that is unique to the quantum world. Classically, the electron would simply bounce off the wall, and the electron could be found at some point between $x = 0$ and $x = a$. In the quantum picture, this is not the case any more. According to equation 37.1, the probability for the electron to be in the spatial interval dx can be calculated as $\Pi(x) = |\psi(x)|^2 dx$.

In Figure 37.14b we have calculated $a|\psi(x)|^2$ for both of our bound-state wave functions. Integrating the area under the curve gives the result that, for the bound state with wave number κ_1, the probability of finding the electron between $x = 0$ and $x = a$ is 96.9% (red area) and the probability of finding the electron in the classically forbidden region $x > a$ is 3.1% (blue area). If the electron resides in the bound state with wave number κ_2, it has a probability of 18.6% of being found in the classically forbidden region, and 81.4% in $0 < x < a$. Since the energy of the second state is higher than that of the first state and close to the depth of the potential well, the electron can penetrate farther into the classically forbidden region and consequently has a greater probability of being found there.

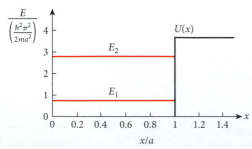

FIGURE 37.13 Energies of the lowest two possible wave functions in the finite potential well.

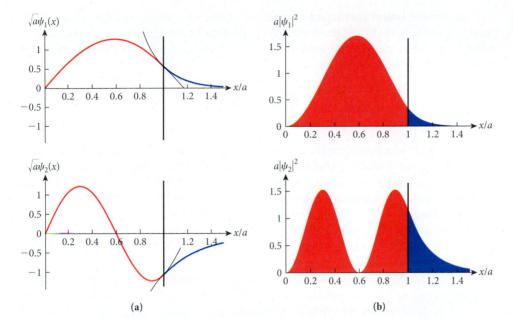

FIGURE 37.14 (a) The two wave functions with lowest values of the wave number for a finite potential step at $x = a$ as a function of x/a. The red color shows the sine function for the zero-potential region, and the blue color shows the exponential function in the classically forbidden region. (b) Corresponding probability distributions to find the electron at a given coordinate. The blue area marks the probability of finding the electron in the classically forbidden region. All four plots have scale factors such that the boundary at $x = a$ appears at the value of 1 on the abscissa and so that the areas in part (b) are the probabilities for finding the particle within the corresponding abscissa region.

Tunneling

If the wave function can reach into the classically forbidden region, then what would happen if the potential step of finite height shown in Figure 37.10 has only a finite width? This situation is shown in Figure 37.15a, which shows a plot of the potential energy function

$$U(x) = \begin{cases} \infty & \text{for } x < 0 \\ 0 & \text{for } 0 \leq x \leq a \\ U_1 & \text{for } a < x < b \\ 0 & \text{for } x \geq b. \end{cases}$$

Again, the wave function in the different regions can be written as follows:

$$\psi(x) = \begin{cases} 0 & \text{for } x < 0 \\ D\sin(\kappa x) & \text{for } 0 \leq x \leq a \\ Fe^{-\gamma x} & \text{for } a < x < b \\ G\sin(\kappa x + \phi) & \text{for } x \geq b. \end{cases}$$

Note that the same value can be used for the wave number κ on both sides of the barrier in this case because the potential is zero in both of these regions. In the preceding section we solved for the wave function in the region $0 \leq x \leq b$. The wave function has a sinusoidal part (red curve in Figure 37.15b) in the region between 0 and a and an exponentially decreasing part (blue curve in part b) between a and b. What is different now is that the wave function does not continue to decay exponentially beyond $x = b$, but it oscillates again in sinusoidal fashion for $x > b$ with the same wave number κ as between 0 and a. This process is called **tunneling**. The exact shape of the wave function (green curve in part b) for $x > b$ is again determined by matching the wave function and its first derivative at the boundary $x = b$, just as we did before at $x = a$.

Figure 37.15c shows the plot of the probability density for this wave function. Clearly there is some nonzero probability for the wave function to tunnel through the potential energy barrier and emerge on the other side. Classically, by contrast, a particle located at $0 < x < a$ with the same total energy would not have enough energy to escape and would remain trapped forever in the region $0 \leq x \leq a$.

37.4 Self-Test Opportunity

Can you use the conditions of continuity of the wave function and its derivative at the boundary $x = b$ to derive the values of the phase shift ϕ and the amplitude G in the wave function for the tunneling through a potential energy step?

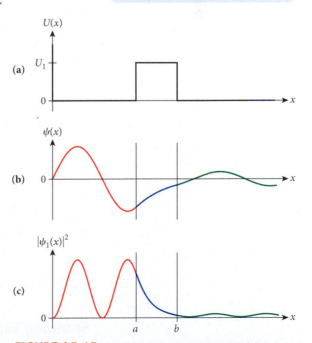

FIGURE 37.15 (a) Potential energy step of finite height and finite width; (b) wave function tunneling through the potential energy barrier; (c) probability density distribution for finding the wave function at a position x.

The **transmission coefficient** T is defined as the ratio of the absolute square of the wave amplitude at the exit of the barrier to the absolute square of the wave amplitude at the entrance to the barrier. For the wave function $Fe^{-\gamma x}$ in the region of the barrier, the transmission coefficient is found to be

$$T = \frac{\left|\psi(b)\right|^2}{\left|\psi(a)\right|^2} = \frac{\left|Fe^{-\gamma b}\right|^2}{\left|Fe^{-\gamma a}\right|^2} = \left|e^{-\gamma(b-a)}\right|^2 = e^{-2\gamma(b-a)}. \tag{37.23}$$

Thus, the transmission coefficient depends exponentially on the width of the barrier, $b-a$. Note that the probability that the particle is to the left of the barrier ($x \le a$) is proportional to $|\psi(a)|^2$. The probability it is found to the right of the barrier ($x \ge b$) is proportional to $|\psi(b)|^2$. The transmission coefficient T therefore measures the probability that a particle hitting the barrier on the left will emerge on the right of the barrier.

Fascinatingly, this process of tunneling through a potential energy barrier can be observed in nature—for example, in the alpha decay of heavy nuclei. It is a crude, but effective, model of the alpha decay process to imagine the alpha particle to be trapped inside a potential energy barrier formed by the heavy nucleus in roughly the shape shown in Figure 37.15a. Over time, the wave function of the alpha particle leaks out of the potential energy trap. When this happens, we say that the nucleus "alpha decays," meaning that it emits an alpha particle and transmutes into another nucleus. The details of this decay process will be covered in greater detail in Chapter 40.

EXAMPLE 37.4 | Neutron Tunneling

PROBLEM
If a neutron of kinetic energy 22.4 MeV encounters a rectangular potential energy barrier of height 36.2 MeV and width 8.4 fm, what is the probability that the neutron will be able to tunnel through this barrier?

SOLUTION
The tunneling probability was given in equation 37.23 as $\Pi_t = T = e^{-2\gamma(b-a)}$. Knowing that the width of the barrier is $b - a = 8.4$ fm, all we have to do to complete our task is to figure out the value of the decay constant γ. Equation 37.19 applies in the present case, and we can write

$$\gamma = \sqrt{\frac{2m(U_1 - E)}{\hbar^2}}.$$

In order to put in numbers, we can write $\hbar$ in units that are very useful in nuclear and particle physics: $\hbar c = 197.34$ MeV fm $\Leftrightarrow \hbar = 197.34$ MeV fm/c. The mass of the neutron has the value $m_n = 1.6749 \cdot 10^{-27}$ kg $= 939.57$ MeV/c^2. If we then insert our constants, we find for the decay constant

$$\gamma = \sqrt{\frac{2(939.57 \text{ MeV}/c^2)(36.2 \text{ MeV} - 22.4 \text{ MeV})}{(197.34 \text{ MeV fm}/c)^2}} = 0.816 \text{ fm}^{-1}.$$

Therefore, we obtain for the tunneling probability

$$\Pi_t = e^{-2\gamma(b-a)} = e^{-2(0.816)(8.4)} = 1.11 \cdot 10^{-6}.$$

37.3 In-Class Exercise

If we want to increase the tunneling probability in this case, we should

a) increase the barrier height and/or increase the barrier width.

b) increase the barrier height and/or decrease the barrier width.

c) decrease the barrier height and/or increase the barrier width.

d) decrease the barrier height and/or decrease the barrier width.

Scanning Tunneling Microscope

In 1981 the Swiss physicist Heinrich Rohrer (1933–) and the German physicist Gerd Binnig (1947–) discovered that the tunneling effect can be used to image surfaces of materials and, for the first time, obtain images of atoms. They were awarded half of the 1986 Nobel Prize in Physics for this discovery. Their work was a conceptual leap and a great technical achievement at the time, but the underlying basic physics is actually relatively straightforward to understand with the concepts we have developed so far.

In Figure 37.16a, the potential of an electron in an atom is shown in black. Chapter 23 showed that this potential is the Coulomb potential, $U(r) \propto 1/r$. This potential is represented with the black line as a function of one spatial coordinate, and our atom sits at x_0. We overlay a qualitative sketch of the absolute square of the electron wave function in this potential. (Chapter 38 will show exact calculations for the hydrogen atom.) In Figure 37.16b–d, a second atom, located at x_1, is moved closer and closer to the atom located at x_0. As we move the second atom, we monitor the resulting potential distribution, as well as the corresponding wave function. The potential barrier that prevents the electron from moving from the atom at x_0 to the atom at x_1 becomes narrower and at the same time lower as the atoms are moved closer to each other. Thus, the electron wave function is able to tunnel through the barrier. Quantitative calculations show that the tunnel current rises very steeply if the distance between the two atoms decreases below a certain point.

The principle of the **scanning tunneling microscope (STM)** is to very carefully move a tip the size of a single atom closer and closer to the surface of the material that we want to probe and to record the current due to tunneling electrons in this process (Figure 37.17). How is it possible to produce a tip the size of a single atom? To start with, a very thin wire is cut at an angle. This produces a very fine tip, but not the size of a single atom. However, one of the atoms at the end will usually stick out just a tiny bit farther than those surrounding it. This tiny bit is sufficient, because the tunnel current depends sensitively on the distance. Thus this tip acts like a single atom.

The tip is moved up and down in a feedback loop that attempts to keep the measured tunnel current at a constant value. Then the tip is guided over the surface in the scanning path shown by the purple line in Figure 37.17a. This process allows us to obtain images of surfaces at an atomic resolution. For example, the surface of a piece of platinum is shown in Figure 37.17b.

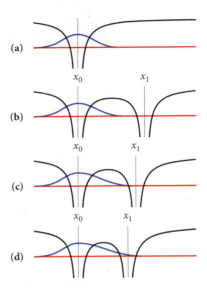

FIGURE 37.16 Black lines: potential of an electron in an atom as a function of the coordinate; blue lines: sketch of the corresponding electron probability distribution. (a) Single isolated atom; (b, c, d) the situation for two atoms at various relative distances.

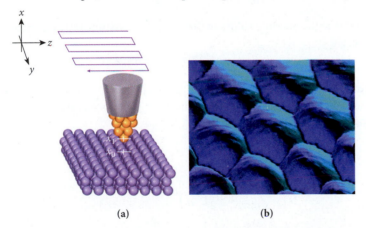

(a) (b)

FIGURE 37.17 (a) Scanning the surface with a single-atom tip; (b) actual data for the surface of a platinum sample.

37.5 Harmonic Oscillator

The three most commonly used potentials in quantum mechanical calculations are the potential well, the central $1/r$ potential, and the oscillator potential. We will address the central $1/r$ potential in Chapter 38, and Sections 37.3 and 37.4 already extensively investigated the potential well. This leaves only the oscillator potential.

Classical Harmonic Oscillator

Chapter 14, which was devoted to a study of harmonic oscillations, showed that oscillators occur in many physical situations. Now the question arises: What is the quantum representation of a harmonic oscillator? That is, we want to know the possible wave functions and energies that correspond to the potential energy function for the following harmonic oscillator:

$$U(x) = \tfrac{1}{2}kx^2 = \tfrac{1}{2}m\omega_0^2 x^2.$$

This potential energy function is plotted in Figure 37.18. Here k is the *spring constant* and is measured in units of N/m, while ω_0 is the angular frequency, $\omega_0 = \sqrt{k/m}$. (*Note:* In this

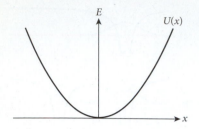

FIGURE 37.18 Potential energy as a function of position for a harmonic oscillator.

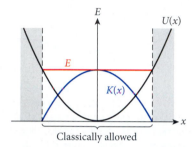

FIGURE 37.19 Classically allowed region for a particle in an oscillator potential.

chapter, we use several symbols that look like a k: k is the symbol for the spring constant, κ is used for the wave number, K is the value of the kinetic energy, and **K** is the kinetic energy operator. It is easy to confuse them, so be extra careful!)

Before we go on, let's first review the situation for a classical particle in such a harmonic oscillator potential well—for example, a mass on a spring or a pendulum. For a given total energy E, the kinetic energy of this particle can be calculated by taking the difference between the total energy and the potential energy, $K(x) = E - U(x)$. Keeping in mind that kinetic energy cannot assume negative values, we then obtain a region in coordinate space in which the particle with a given total energy is allowed to be, denoted "classically allowed" in Figure 37.19.

The points at which the kinetic energy (blue curve in Figure 37.19) reaches the value zero form the boundary of this classically allowed region. These points are called classical turning points because a classical oscillator "turns around" here. Outside the turning points lies the classically forbidden region (shaded region in Figure 37.19), into which a classical particle with total mechanical energy E is never able to penetrate.

Quantum Harmonic Oscillator

Now let's investigate the quantum oscillator. Inserting the oscillator potential energy function into the general Schrödinger equation 37.9, we find

$$-\frac{\hbar^2}{2m}\frac{d^2\psi(x)}{dx^2}+\frac{1}{2}m\omega_0^2x^2\psi(x)=E\psi(x)\Rightarrow$$

$$\frac{d^2\psi(x)}{dx^2}=-\frac{2m}{\hbar^2}\left(E-\tfrac{1}{2}m\omega_0^2x^2\right)\psi(x)\Rightarrow \tag{37.24}$$

$$\frac{d^2\psi(x)}{dx^2}=-\left(\frac{2mE}{\hbar^2}-\frac{m^2\omega_0^2}{\hbar^2}x^2\right)\psi(x).$$

Again we find that the possible values of the energy can assume only discrete values. In this case they are

$$E_n=\left(n+\tfrac{1}{2}\right)\hbar\omega_0,\quad n=0,1,2,.... \tag{37.25}$$

The wave functions have the general form of Gaussians $\left(e^{-ax^2}\right)$ multiplied by special polynomials, the so-called Hermite polynomials. The wave functions corresponding to the lowest values of the energy quantum number n are

$$\psi_0(x)=\frac{1}{\sqrt{\sigma}\pi^{1/4}}e^{-x^2/2\sigma^2}$$

$$\psi_1(x)=\frac{1}{\sqrt{\sigma}\pi^{1/4}}\frac{1}{\sqrt{2}}\left(2\frac{x}{\sigma}\right)e^{-x^2/2\sigma^2}$$

$$\psi_2(x)=\frac{1}{\sqrt{\sigma}\pi^{1/4}}\frac{1}{\sqrt{8}}\left(4\frac{x^2}{\sigma^2}-2\right)e^{-x^2/2\sigma^2} \tag{37.26}$$

$$\psi_3(x)=\frac{1}{\sqrt{\sigma}\pi^{1/4}}\frac{1}{\sqrt{48}}\left(8\frac{x^3}{\sigma^3}-12\frac{x}{\sigma}\right)e^{-x^2/2\sigma^2}$$

$$\psi_4(x)=\frac{1}{\sqrt{\sigma}\pi^{1/4}}\frac{1}{\sqrt{384}}\left(16\frac{x^4}{\sigma^4}-48\frac{x^2}{\sigma^2}+12\right)e^{-x^2/2\sigma^2}$$

where the constant σ is defined as

$$\sigma=\sqrt{\frac{\hbar}{m\omega_0}}. \tag{37.27}$$

This constant has the physical dimensions of length and is the half-width of the Gaussian. The general form of the wave function for the harmonic oscillator can be written as

$$\psi_n(x)=\frac{1}{\sqrt{\sigma}\pi^{1/4}}\frac{1}{\sqrt{n!\,2^n}}H_n(x/\sigma)e^{-x^2/2\sigma^2}. \tag{37.28}$$

The Hermite polynomial $H_n(x)$ is a polynomial of rank n; that is, the highest power of x is x^n, and it can be defined in terms of the derivatives of the Gaussian function:

$$H_n(x) = (-1)^n e^{x^2} \frac{d^n}{dx^n} e^{-x^2}.$$

DERIVATION 37.1 | Oscillator Wave Function and Energy

The entire derivation of the complete solution (equation 37.28) is somewhat lengthy and will not be shown here. Still, it is instructive to verify that one of the solutions given in equation 37.26 satisfies equation 37.24. We show as an example that $\psi_1(x)$ is indeed a solution to the Schrödinger equation 37.24 with energy $E_1 = (\frac{1}{2} + 1)\hbar\omega_0 = \frac{3}{2}\hbar\omega_0$.

Start by taking the second derivative of $\psi_1(x)$ with respect to x. To do this more efficiently, rewrite the wave function as

$$\psi_1(x) = \frac{1}{\sqrt{\sigma}\pi^{1/4}} \frac{1}{\sqrt{2}} \left(2\frac{x}{\sigma}\right) e^{-x^2/2\sigma^2} = Axe^{-x^2/2\sigma^2}.$$

Here A is just a normalization constant, so the first and second derivatives of this function are

$$\frac{d}{dx}\psi_1(x) = Ae^{-x^2/2\sigma^2} - A\frac{x^2}{\sigma^2}e^{-x^2/2\sigma^2}$$

$$\frac{d^2}{dx^2}\psi_1(x) = -A\frac{3x}{\sigma^2}e^{-x^2/2\sigma^2} + A\frac{x^3}{\sigma^4}e^{-x^2/2\sigma^2}.$$

Inserting this second derivative into the Schrödinger equation results in

$$\frac{d^2\psi(x)}{dx^2} = -\left(\frac{2mE}{\hbar^2} - \frac{m^2\omega_0^2}{\hbar^2}x^2\right)\psi(x) \Rightarrow$$

$$-A\frac{3x}{\sigma^2}e^{-x^2/2\sigma^2} + A\frac{x^3}{\sigma^4}e^{-x^2/2\sigma^2} = -\left(\frac{2mE}{\hbar^2} - \frac{m^2\omega_0^2}{\hbar^2}x^2\right)Axe^{-x^2/2\sigma^2}$$

$$= -\left(\frac{2mE}{\hbar^2}x - \frac{m^2\omega_0^2}{\hbar^2}x^3\right)Ae^{-x^2/2\sigma^2}.$$

Now factor out the common term $Ae^{-x^2/2\sigma^2}$ and sort the terms in powers of x:

$$Ae^{-x^2/2\sigma^2}\left(\left(\frac{2mE}{\hbar^2} - \frac{3}{\sigma^2}\right)x + \left(\frac{1}{\sigma^4} - \frac{m^2\omega_0^2}{\hbar^2}\right)x^3\right) = 0. \qquad \text{(i)}$$

Because $Ae^{-x^2/2\sigma^2} \neq 0$, the above equation can be fulfilled for all x only if the coefficients multiplying x and x^3 are both 0. Examining the coefficient for x^3, we find:

$$\frac{1}{\sigma^4} - \frac{m^2\omega_0^2}{\hbar^2} = 0 \Rightarrow \sigma = \sqrt{\frac{\hbar}{m\omega_0}}.$$

This confirms equation 37.27. We further find for the coefficient for x in equation (i):

$$\frac{2mE}{\hbar^2} - \frac{3}{\sigma^2} = 0 \Rightarrow E = \frac{3}{2}\frac{\hbar^2}{m\sigma^2}.$$

Using the result of equation 37.27, which we just derived, we then find:

$$E = \frac{3}{2}\frac{\hbar^2}{m\sigma^2} = \frac{3}{2}\frac{\hbar^2}{m(\hbar/m\omega_0)} = \frac{3}{2}\hbar\omega_0.$$

Thus, we have shown that $\psi_1(x)$ as defined in equation 37.26 is indeed a solution to the Schrödinger equation with energy E_1 as given in equation 37.25.

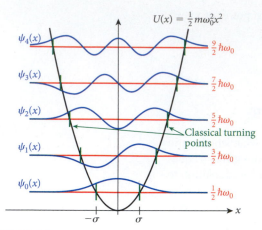

FIGURE 37.20 Oscillator wave functions (blue) corresponding to the five lowest-energy values (red) of the harmonic oscillator potential (black). The classical turning points for each wave function are indicated by short vertical green lines.

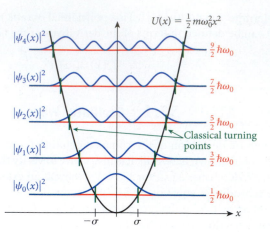

FIGURE 37.21 Probability distributions for finding an electron in an oscillator potential with a particular energy at a particular point in space.

It is important to note that the possible energy values in a one-dimensional harmonic oscillator are evenly spaced, with a constant energy difference of $\hbar\omega_0$ between neighboring energy levels. A wide variety of systems in condensed matter, atomic, and nuclear physics exhibit discrete energy spectra with constant spacing between the levels. Any time a physical system exhibits this characteristic, it can safely be assumed that it indicates some kind of quantum vibration in the system.

The oscillator wave functions of equation 37.26 are shown in Figure 37.20. To avoid possible confusion, notice that two different plots using different vertical axes are superimposed on each other here, with a common horizontal x-axis. The black parabola shows the potential energy as a function of x, and the red horizontal lines show the possible total energies that correspond to solutions of the Schrödinger equation. Both use a vertical axis in energy units. The intersections of the red total-energy lines with the potential-energy parabola mark the classical turning points and are indicated by the short vertical green lines. However, the blue lines show each wave function and do not have the units of energy, and thus correspond to a different vertical scale. For each wave function, the $\psi = 0$ line is adjusted to fall on the horizontal line that marks the energy value that corresponds to this particular wave function. This provides an elegant way of displaying the wave functions, their classical turning points, and the corresponding energy values, all in the same plot. We can quantitatively show how each wave function "leaks" into the classically forbidden region.

Figure 37.21 shows the probability distributions that correspond to each wave function. Again, they are simply the absolute squares of each wave function. Note that the higher the quantum number n, the smaller the penetration range of the corresponding wave function beyond the classical turning point turns out to be. This result can be understood by looking at the shape of the potential $U(x)$: For higher values of the energy E, the potential becomes steeper at the position of the classical turning points for that energy.

37.6 Wave Functions and Measurements

Suppose we obtain a quantum mechanical wave function for a particular situation. How can this wave function be used to gain information about the physical properties of the object with which the wave function is associated? How can the position, velocity, momentum, kinetic energy, and so on be calculated?

Equation 37.2 introduced the normalization condition for the wave function,

$$\int_{-\infty}^{\infty} |\psi(x)|^2 \, dx = \int_{-\infty}^{\infty} \psi^*(x)\psi(x)\,dx = 1.$$

To measure the average value of any observable quantity, apply an operator that corresponds to this operation to the wave function ψ (for example, $\mathbf{p}\psi = p\psi$). Then multiply the result by

ψ^* (to get the probability $\psi^*\psi$ in the integrand) and integrate over all of space. For example, to figure out the average momentum associated with a wave function, apply the momentum operator (equation 37.6) to the wave function, and integrate:

$$\langle p \rangle = \int_{-\infty}^{\infty} \psi^*(x)\mathbf{p}\psi(x)dx = -i\hbar \int_{-\infty}^{\infty} \psi^*(x)\frac{d}{dx}\psi(x)dx. \tag{37.29}$$

In a similar way, any quantity that is a function of the momentum can be calculated using integrals that involve the derivative of the wave function. For example, the average kinetic energy can be obtained as

$$\langle K \rangle = \frac{1}{2m}\langle p^2 \rangle = \frac{1}{2m}\int_{-\infty}^{\infty} \psi^*(x)\mathbf{p}^2\psi(x)dx = -\frac{\hbar^2}{2m}\int_{-\infty}^{\infty} \psi^*(x)\frac{d^2}{dx^2}\psi(x)dx. \tag{37.30}$$

It is somewhat more straightforward to find the average values for quantities that depend on the position x. To find the average position, we can calculate this integral:

$$\langle x \rangle = \int_{-\infty}^{\infty} \psi^*(x)x\psi(x)dx. \tag{37.31}$$

The average values denoted in the angular brackets $\langle \ \rangle$ are often referred to as **expectation values,** that is, the expected values of the outcomes of measurements. This explanation for the formal process of conducting a measurement is rather abstract, so let's look at an example of how to arrive at actual numbers.

EXAMPLE 37.5 | Position and Energy

PROBLEM
What is the average position for an electron (mass $m = 9.109 \cdot 10^{-31}$ kg $= 511$ keV/c^2) in the $n = 0$ wave function of the harmonic oscillator potential with an angular frequency constant of $\omega_0 = 1.00 \cdot 10^{16}$ s^{-1}? What is its average kinetic energy?

SOLUTION
According to equation 37.26, the solution to the Schrödinger equation with the harmonic oscillator potential is

$$\psi_0(x) = Ae^{-x^2/2\sigma^2} = \frac{1}{\sqrt{\sigma}\pi^{1/4}}e^{-x^2/2\sigma^2}.$$

The electron mass and the oscillator angular frequency enter into the width parameter $\sigma = \sqrt{\hbar/m\omega_0}$. Here we again use the notation of the normalization constant $A = 1/(\sqrt{\sigma}\pi^{1/4})$ to reduce the work in writing down the integrals that follow.

First, calculate the average position. Following equation 37.31, this observable can be calculated as

$$\langle x \rangle = \int_{-\infty}^{\infty} \psi^*(x)x\psi(x)dx$$

$$= \int_{-\infty}^{\infty} Ae^{-x^2/2\sigma^2}xAe^{-x^2/2\sigma^2}dx$$

$$= A^2 \int_{-\infty}^{\infty} xe^{-x^2/\sigma^2}dx$$

$$= 0.$$

These steps deserve an explanation. First, when we inserted the wave function, we used the fact that the wave function is real, with no imaginary part, so in this case $\psi^*(x) = \psi(x)$.

Continued—

In the next step we removed the constant factor of A^2 from the integral and multiplied the two Gaussians by adding their arguments. Finally, we arrived at our result of 0 for the integral by realizing that x is an odd function, and e^{-x^2/σ^2} is an even function of x; therefore their product must be an odd function of x, resulting in an integral of zero value when the integration limits are symmetrical.

In fact, we could immediately read off this result by looking at Figure 37.21 and noticing that the probability distribution $|\psi_0(x)|^2$ is symmetric with respect to 0. However, because the above integral is the first of several similar ones, it is instructive to explicitly go through all the steps and then confirm that we indeed find the expected solution.

After this warm-up, let's address the more complicated question of the average kinetic energy. According to equation 37.30, we have to solve

$$\langle K \rangle = -\frac{\hbar^2}{2m} \int_{-\infty}^{\infty} \psi^*(x) \frac{d^2}{dx^2} \psi(x) dx$$

$$= -\frac{\hbar^2}{2m} \int_{-\infty}^{\infty} Ae^{-x^2/2\sigma^2} \frac{d^2}{dx^2} Ae^{-x^2/2\sigma^2} dx.$$

Taking the second derivative of the oscillator wave function with $n = 0$ yields

$$\frac{d}{dx} Ae^{-x^2/2\sigma^2} = -A\frac{x}{\sigma^2}e^{-x^2/2\sigma^2}$$

$$\frac{d^2}{dx^2} Ae^{-x^2/2\sigma^2} = A\left(\frac{x^2}{\sigma^4} - \frac{1}{\sigma^2}\right)e^{-x^2/2\sigma^2}.$$

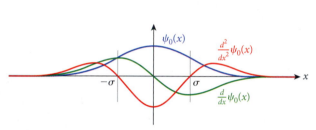

FIGURE 37.22 The $n = 0$ oscillator wave function (blue) and its first (green) and second (red) derivatives as a function of position. The gray vertical lines mark the classical turning point positions.

These first and second derivatives are plotted in Figure 37.22, where the classical turning points are also shown. For the quantum wave function, these are the points where the second derivative of the wave function has a value of 0. (We can see from equation 37.9 that if $E = U$, then $\frac{d^2}{dx^2}\psi(x) = 0$.)

Now we can proceed with the above integration and obtain

$$\langle K \rangle = -\frac{A^2\hbar^2}{2m} \int_{-\infty}^{\infty} e^{-x^2/2\sigma^2} \frac{d^2}{dx^2} e^{-x^2/2\sigma^2} dx$$

$$= -\frac{A^2\hbar^2}{2m} \int_{-\infty}^{\infty} e^{-x^2/2\sigma^2} \left(\frac{x^2}{\sigma^4} - \frac{1}{\sigma^2}\right) e^{-x^2/2\sigma^2} dx$$

$$= -\frac{A^2\hbar^2}{2m} \int_{-\infty}^{\infty} e^{-x^2/\sigma^2} \left(\frac{x^2}{\sigma^4} - \frac{1}{\sigma^2}\right) dx$$

$$= -\frac{A^2\hbar^2}{2m\sigma^4} \int_{-\infty}^{\infty} x^2 e^{-x^2/\sigma^2} dx + \frac{A^2\hbar^2}{2m\sigma^2} \int_{-\infty}^{\infty} e^{-x^2/\sigma^2} dx$$

$$= -\frac{A^2\hbar^2}{2m\sigma^4} \frac{1}{2}\sqrt{\pi}\sigma^3 + \frac{A^2\hbar^2}{2m\sigma^2} \sqrt{\pi}\sigma$$

$$= \frac{\hbar^2}{4m\sigma^2}.$$

In the last step we used $A^2 = 1/(\sigma\sqrt{\pi})$, which follows from the definition of A. We are almost done. Using $\sigma = \sqrt{\hbar/m\omega_0}$ (see equation 37.27), we find

$$\langle K \rangle = \frac{\hbar^2}{4m(\hbar/m\omega_0)} = \frac{1}{4}\hbar\omega_0.$$

Finally, inserting the given value for ω_0 gives us our numerical answer:

$$\langle K \rangle = \tfrac{1}{4}\hbar\omega_0 = \tfrac{1}{4}(1.055\cdot10^{-34}\ \text{J s})(1.00\cdot10^{16}\ \text{s}^{-1})$$

$$= 2.64\cdot10^{-19}\ \text{J}$$

$$= 1.65\ \text{eV}.$$

DISCUSSION

Our result amounts to exactly half of the total energy of the oscillator in the $n = 0$ state! Also, note that this answer depends only on the value of the angular frequency ω_0 and not on the mass m. Another particle with a different mass, trapped in the same oscillator potential, would have the same average kinetic energy in the $n = 0$ state.

37.5 Self-Test Opportunity

Calculate the average value of the momentum in this state, or simply state the result as a result of symmetry arguments.

37.6 Self-Test Opportunity

Calculate the average value of the kinetic energy in the $n = 1,2,3,...$ states of the harmonic oscillator. Is there a simple shortcut, or do you have to execute the integral for a new wave function each time?

Uncertainty Relationship for Oscillator Wave Functions

Chapter 36 explored the fundamental importance of the Heisenberg uncertainty relation, $\Delta x \cdot \Delta p \geq \tfrac{1}{2}\hbar$. At the time, we were able to motivate the relationship only by examining the "gamma-ray microscope" that Heisenberg envisioned as a *Gedankenexperiment* (thought experiment) in his original 1927 paper on the uncertainty relation. The same paper also contained a much more general mathematical proof. We will not reconstruct this proof here. However, we can calculate the momentum and position uncertainties for the oscillator wave functions that we have found and see what the result is. Let's do this for the $n = 0$ state.

By definition, the square of the uncertainty in position is $(\Delta x)^2 = \left\langle \left(x - \langle x \rangle\right)^2 \right\rangle$. We just found for the $n = 0$ state that $\langle x \rangle = 0$. In this case (and only when $\langle x \rangle = 0$), we find $(\Delta x)^2 = \langle x^2 \rangle$. Thus, to find the uncertainty in position, we have to calculate the integral

$$(\Delta x)^2 = \langle x^2 \rangle = \int_{-\infty}^{\infty} A e^{-x^2/2\sigma^2} x^2 A e^{-x^2/2\sigma^2}\, dx$$

$$= A^2 \int_{-\infty}^{\infty} x^2 e^{-x^2/\sigma^2}\, dx$$

$$= \tfrac{1}{2}A^2\sqrt{\pi}\sigma^3 = \tfrac{1}{2}\sigma^2 \Rightarrow$$

$$\Delta x = \frac{\sigma}{\sqrt{2}}.$$

For this wave function, the uncertainty in momentum is $(\Delta p)^2 = \left\langle \left(p - \langle p \rangle\right)^2 \right\rangle = \langle p^2 \rangle$, because the average momentum is also zero. Furthermore, the average kinetic energy, which is $\langle K \rangle = \langle p^2 \rangle / 2m$, for this wave function was calculated in Example 37.5. Thus we find that

$$(\Delta p)^2 = \langle p^2 \rangle = 2m\langle K^2 \rangle = 2m\frac{\hbar^2}{4m\sigma^2} = \frac{\hbar^2}{2\sigma^2} \Rightarrow$$

$$\Delta p = \frac{\hbar}{\sqrt{2}\sigma}.$$

Multiplying the two results for the uncertainties in position and momentum, the harmonic oscillator in the $n = 0$ state can be written:

$$\Delta x \cdot \Delta p = \frac{\sigma}{\sqrt{2}}\frac{\hbar}{\sqrt{2}\sigma} = \tfrac{1}{2}\hbar.$$

This means that the $n = 0$ state of the harmonic oscillator is a state with the physically minimum possible uncertainty. Also, note that this result is independent of the width σ and thus does not depend on the angular frequency or mass in the problem. It can also be shown that the uncertainty product for an oscillator wave function with quantum number n is given by $\Delta x \cdot \Delta p = (\tfrac{1}{2} + n)\hbar$.

37.7   Correspondence Principle

When we investigated relativistic mechanics, we found a smooth transition to nonrelativistic Newtonian mechanics as an object's speed becomes small compared to the speed of light. In the limit $v \ll c$, the classical physics case is recovered from the relativistic description. In a similar way, we can ask if we can recover the case of classical mechanics from quantum mechanics, and in what limit. To address this question, we examine the probability distribution for finding an electron in the oscillator potential at a particular point in coordinate space; then we compare the results for the classical and quantum oscillators.

Just as we did for the classical particle in a box, we can calculate the classical probability distribution for finding a particle with total energy E in an oscillator potential. Equation 37.13 states that the classical probability for finding a particle at some point in coordinate space is inversely proportional to the particle's speed at this point. The speed can be calculated as a function of position for the harmonic oscillator potential from energy conservation,

$$E = K + U = \tfrac{1}{2}mv^2 + \tfrac{1}{2}m\omega_0^2 x^2.$$

Solving this equation for the speed gives:

$$v = \sqrt{\frac{2E}{m} - x^2\omega_0^2}.$$

Thus, the classical probability distribution of a particle in a harmonic oscillator potential can be written:

$$\Pi_c(x) = \frac{1}{\pi\sqrt{\dfrac{2E}{m} - x^2\omega_0^2}}.$$

This function is defined only between the two classical turning points located at

$$x_t = \pm\sqrt{\frac{2E}{m\omega_0^2}}.$$

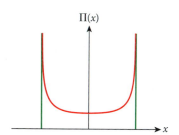

$\Pi(x)$

FIGURE 37.23 Classical probability distribution for finding a particle at a particular position in an oscillator potential (red curve); classical turning points (green vertical lines).

(The factor of π in the denominator of the probability distribution function ensures that the integral of this probability function is 1.) Figure 37.23 shows the resulting probability distribution, with the turning points marked by the two green vertical lines.

The quantum probability distributions in Figure 37.21, however, do not seem to agree at all with the classical probability distribution in Figure 37.23. In the classical case, the probability for finding the particle peaks near the classical turning points. In the quantum case shown in Figure 37.21, on the other hand, the probability distribution seems flat or even peaked in the middle. From our experience with a particle in a box, we are perhaps not surprised that the quantum wave function penetrates partially beyond the classical turning point. However, it should cause some concern that the gross features of the probability distributions in the classical and the quantum case do seem not to correspond to each other at all.

This conclusion changes when we examine a quantum oscillator wave function for a large value of the quantum number n, as done in Figure 37.24 for $n = 20$. For the higher quantum numbers, the quantum oscillator probability distribution still oscillates and still penetrates slightly beyond the classical turning points. However, now the probability starts to oscillate around an average value that corresponds to the classical limit.

Recall that the energy difference between neighboring allowed quantum oscillator energies is a constant, $\Delta E = \hbar\omega_0$. If this energy difference ΔE becomes small relative to the total energy E, $\Delta E/E \ll 1$—which is equivalent to saying that the quantum number (say n) is large—the quantum solution approaches its classical limit. This is a general feature of quantum mechanics, conventionally called the **correspondence principle**.

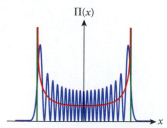

$\Pi(x)$

FIGURE 37.24 Same as Figure 37.23, but now with the probability distribution for the $n = 20$ quantum oscillator wave function superimposed.

37.8 Time-Dependent Schrödinger Equation

Up to now we have concerned ourselves with only time-independent problems—that is, problems that do not change in time. However, electrons and other particles move around. What equation can describe the time dependence of the corresponding matter wave? Besides describing the dependence on the spatial coordinate x, the wave function now also has to depend on the time t. We will still restrict ourselves to motion in one spatial direction, and we will assume that the potential energy is constant in time; that is, it is a function of the coordinate x only. The time-dependent Schrödinger equation then becomes

$$-\frac{\hbar^2}{2m}\frac{\partial^2}{\partial x^2}\Psi(x,t)+U(x)\Psi(x,t)=i\hbar\frac{\partial}{\partial t}\Psi(x,t). \tag{37.32}$$

Here we use the symbol $\Psi(x,t)$ to denote the quantum wave function and its dependence on space and time. At this point we are not interested in the general solutions to this partial differential equation, nor are we interested in deriving it from some general principles. However, we would like to see if we can find special solutions that allow us to write the general wave function as a product of two functions that depend on only one variable each

$$\Psi(x,t)=\psi(x)\chi(t).$$

This attempt to solve a partial differential equation is commonly called "separation of variables." Let us see what happens if we insert this *Ansatz* into the time-dependent Schrödinger equation 37.32:

$$-\frac{\hbar^2}{2m}\frac{\partial^2}{\partial x^2}\Big(\psi(x)\chi(t)\Big)+U(x)\Big(\psi(x)\chi(t)\Big)=i\hbar\frac{\partial}{\partial t}\Big(\psi(x)\chi(t)\Big)\Rightarrow$$

$$-\chi(t)\frac{\hbar^2}{2m}\frac{\partial^2}{\partial x^2}\psi(x)+\chi(t)U(x)\psi(x)=i\hbar\psi(x)\frac{\partial}{\partial t}\chi(t).$$

We divide both sides of this equation by the product $\psi(x)\chi(t)$ and find

$$\left(-\frac{\hbar^2}{2m}\frac{\partial^2}{\partial x^2}\psi(x)+U(x)\psi(x)\right)\frac{1}{\psi(x)}=\left(i\hbar\frac{\partial}{\partial t}\chi(t)\right)\frac{1}{\chi(t)}.$$

Now the left-hand side is a function of x only and the right-hand side is a function of t only. This equality can hold for all x and t only if each side is equal to the same constant. Motivated by the time-independent Schrödinger equation 37.9, we call this constant E, the energy that entered into equation 37.9. The left-hand side of this equation then leads to

$$\left(-\frac{\hbar^2}{2m}\frac{d^2}{dx^2}\psi(x)+U(x)\psi(x)\right)\frac{1}{\psi(x)}=E\Rightarrow$$

$$-\frac{\hbar^2}{2m}\frac{d^2}{dx^2}\psi(x)+U(x)\psi(x)=E\psi(x),$$

which we recognize as our time-independent Schrödinger equation. With the same argument, we find for the the right-hand side of our equation

$$E=\left(i\hbar\frac{d}{dt}\chi(t)\right)\frac{1}{\chi(t)}\Rightarrow$$

$$\frac{d}{dt}\chi(t)=-\frac{i}{\hbar}E\chi(t)\Rightarrow$$

$$\chi(t)=Ae^{-iEt/\hbar}=Ae^{-i\omega t}$$

where we have again introduced the angular frequency ω via $E=\hbar\omega$. We can set the normalization constant A to 1, because $\left|e^{-i\omega t}\right|^2=1$, and thus the normalization condition for the wave function is fulfilled. Our overall solution is therefore

$$\Psi(x,t)=\psi(x)e^{-i\omega t}, \tag{37.33}$$

with $\psi(x)$ being a normalized solution to the time-independent Schrödinger equation with energy $E = \hbar\omega$. A wave function of the form of equation 37.33 is called a *stationary solution* or **stationary state** of the time-dependent Schrödinger equation. The minimum condition for the existence of such stationary states is that the potential energy be time independent, that is, constant in time. A stationary state is a state with an exact energy if the potential energy is constant in time. This is just as in Newtonian mechanics, where the total mechanical energy E is conserved. Keep in mind that there can be many other solutions to the Schrödinger equation 37.32, but the special class of solutions with well-defined energy can be written in the separable form of equation 37.33.

Eigenfunctions and Eigenvalues

Just as we defined an operator **K** for the kinetic energy in equation 37.7, we can also introduce an operator **H**, which, applied to the wave function, yields a product of the energy value times the wave function,

$$\mathbf{H}\psi(x) = E\psi(x). \tag{37.34}$$

If you compare this to the time-independent Schrödinger equation, you see that this operator can be formally written as

$$\mathbf{H} = -\frac{\hbar^2}{2m}\frac{d^2}{dx^2} + U(x) = \mathbf{K} + \mathbf{U}. \tag{37.35}$$

This operator **H** is called the **Hamiltonian operator.** In linear algebra, if it is possible to apply an operator to a function and obtain a constant times that same function, then this function is called $\psi(x)$ an **eigenfunction** and the constant an **eigenvalue.** Thus, a stationary state is an eigenfunction of the Hamiltonian operator **H** with eigenvalue E.

The Hamiltonian operator is linear: For any functions $\xi_1(x)$ and $\xi_2(x)$ and constants a_1 and a_2,

$$\mathbf{H}\big(a_1\xi_1(x) + a_2\xi_2(x)\big) = a_1\mathbf{H}\xi_1(x) + a_2\mathbf{H}\xi_2(x). \tag{37.36}$$

Further, if two functions $\psi_1(x)$ and $\psi_2(x)$ are solutions with eigenvalues E_1 and E_2, then applying the Hamiltonian operator to the linear combination $\psi(x) = a_1\psi_1(x) + a_2\psi_2(x)$ yields

$$\begin{aligned}
\mathbf{H}\psi(x) &= \mathbf{H}\big(a_1\psi_1(x) + a_2\psi_2(x)\big) \\
&= a_1\mathbf{H}\psi_1(x) + a_2\mathbf{H}\psi_2(x) \\
&= a_1 E_1\psi_1(x) + a_2 E_2\psi_2(x).
\end{aligned}$$

Note that $\psi(x) = a_1\psi_1(x) + a_2\psi_2(x)$ is *not* an eigenfunction to **H** in the general case that $E_1 \neq E_2$.

37.9 Many-Particle Wave Function

So far we have discussed quantum mechanics only for the case in which a single particle is present. To proceed further, we have to discuss the general features of the wave function when two or more particles are present.

Two-Particle Wave Function

Let's start with two particles, and let us assume that we know the wave function for the case where only one particle is present. Further, let's again restrict ourselves to the static (time-independent) case. Then the Schrödinger equation for the one-particle wave function $\psi(x)$ is given by equation 37.9. If we now put two particles into the same potential, then we need to calculate how each of the particles interacts with the external potential, and how they interact with each other. The general case of the two particles interacting with each other is outside the scope of this book. However, considering the case in which the two particles do not interact with each other lets us gain important physical insight and is helpful in many physical situations.

First, let's think about the notation. We want to characterize the coordinate of particle 1 by x_1 and that of particle 2 by x_2. Then the single-particle wave function of particle 1 in state a will be called $\psi_a(x_1)$ and that of particle 2 in state b, $\psi_b(x_2)$. For the two-particle wave function as a function of both coordinates x_1 and x_2, we will use the notation $\Psi(x_1, x_2)$.

Recall the discussion of spin and statistics in Chapter 36. We now need to discuss three different cases:

Distinguishable Particles—An example is an electron and a neutron in an infinite potential well. This is the most straightforward case, because the wave function for the two-particle state is simply the product of the two single-particle states:

$$\Psi(x_1, x_2) = \psi_a(x_1) \cdot \psi_b(x_2). \tag{37.37}$$

Identical Bosons—An example is two photons inside a mirrored cavity, one of them in state a and the other in state b. Since bosons of one species (identical bosons) are indistinguishable particles, we cannot say definitely that boson 1 is in state a and the identical boson 2 is in state b. It is equally possible that boson 2 is in state a and boson 1 is in state b. The two-particle wave function has to be symmetric (stay the same) under exchange of the indices of the two particles. Instead of just writing the two-particle wave function as the product of the single-particle wave functions, we have to add an exchange term. The way to write this mathematically is

$$\Psi^B(x_1, x_2) = \frac{1}{\sqrt{2}}\left(\psi_a(x_1) \cdot \psi_b(x_2) + \psi_a(x_2) \cdot \psi_b(x_1)\right). \tag{37.38}$$

Here we use the superscript B to remind us that this is the symmetrized two-particle wave function for bosons. The factor $1/\sqrt{2}$ in front of the sum makes sure that the two-particle wave function is normalized to 1.

Identical Fermions—An example is two electrons in an atom, one of them in state a and the other in state b. Fermions of one species are also indistinguishable, but they cannot occupy the same state simultaneously. The two-particle wave function thus has to be antisymmetric (change sign) under the exchange of the indices of the two particles so that $\Psi = 0$ when $a = b$, as we discuss below. Mathematically, this is expressed as

$$\Psi^F(x_1, x_2) = \frac{1}{\sqrt{2}}\left(\psi_a(x_1) \cdot \psi_b(x_2) - \psi_a(x_2) \cdot \psi_b(x_1)\right). \tag{37.39}$$

Again we use a superscript, in this case F, to remind us that this is the antisymmetrized two-particle wave function for identical fermions.

You can see that the expressions for the two-particle wave function of bosons (equation 37.38) and fermions (equation 37.39) differ only in the sign of the exchange term, + for bosons and – for fermions. Note that the wave function in equation 37.39 fulfills the basic requirement of the Pauli exclusion principle, which states that two fermions cannot occupy the same quantum state at the same time. If we set $b = a$ in this equation, then

$$\Psi^F(x_1, x_2) = \frac{1}{\sqrt{2}}\left(\psi_a(x_1) \cdot \psi_a(x_2) - \psi_a(x_2) \cdot \psi_a(x_1)\right) = 0.$$

Further, if we exchange the position of the two fermions, then we find

$$\Psi^F(x_2, x_1) = \frac{1}{\sqrt{2}}\left(\psi_a(x_2) \cdot \psi_b(x_1) - \psi_a(x_1) \cdot \psi_b(x_2)\right)$$

$$= -\frac{1}{\sqrt{2}}\left(\psi_a(x_1) \cdot \psi_b(x_2) - \psi_a(x_2) \cdot \psi_b(x_1)\right)$$

$$= -\Psi^F(x_1, x_2).$$

Thus, if we exchange the position of the two fermions, the wave function undergoes a sign change.

This change in sign is central to several current research topics in physics: The sign change in the two-fermion wave function severely restricts the use of computer modeling for physical systems in which fermions can trade places as part of the time evolution of the system. The problem is that this sign change causes many quantities that must be calculated in

37.4 In-Class Exercise

In equation 37.39 the term $\psi_a(x_2) \cdot \psi_b(x_1)$ enters with a negative sign, and the term $\psi_a(x_1) \cdot \psi_b(x_2)$ with a positive sign. How important is this sign convention?

a) This is the only sign possible for both terms, because the second term is the "exchange term" and thus must have a negative sign.

b) It does not matter which one gets which sign; all that matters is that they have opposite signs. If you multiply the wave function by an overall factor of −1, then you still obtain a valid wave function.

c) The sign convention for the exchange term is arbitrary; it could be positive or negative. But the first term must always be positive.

d) Both terms can have both signs, and each of the four sign conventions (++,−−,−+,+−) leads to a valid wave function.

the computer to average out to zero in computer simulations; and then in many situations two quantities that must be divided by each other are each very close to zero, which results in huge uncertainties in the numerical results. This so-called "fermion sign problem" is universal in many-body theoretical physics, and physicists continue to invent many clever approaches to try to overcome the computational limitations that it imposes, so far without success.

Can we also have a symmetric wave function in coordinate space for a two-fermion system? The answer is yes! The requirement of antisymmetrization of the two-fermion wave function means that the overall wave function has to satisfy equation 37.39. However, the wave function is the product of the spin wave function and the coordinate space wave function. Thus, if the spin wave function is symmetric under exchange (two fermions with identical spin projection), then the coordinate space wave function has to be antisymmetric. Also, if the spin wave function is antisymmetric (two fermions with opposite spin projection), then the coordinate space wave function has to be symmetric under exchange, as Example 37.6 shows.

EXAMPLE 37.6 | Hydrogen Molecule

One of the most impressive examples of the importance of exchange symmetry in the wave function is covalent bonding in the hydrogen molecule. A hydrogen molecule consists of two hydrogen atoms, each consisting of one electron and one proton. In an isolated hydrogen atom, the electron is bound to the proton by the Coulomb potential, which we saw in Chapter 23 is proportional to $1/r$. Chapter 38 will show how to calculate the wave function of the electron in the hydrogen atom. For now, we mainly need to know only that this wave function has its largest value at the origin, the position of the nucleus, and falls off exponentially in the radial direction.

When two hydrogen atoms are in close proximity to each other, they can interact by sharing both of their electrons equally. This process of sharing electrons in the chemical bonding process is called *covalent bonding*, as opposed to *ionic bonding*, where one or more electrons get pulled off one atom and then predominantly orbit the other atom.

Figure 37.25 shows the antisymmetric (a) and symmetric (b) coordinate-space two-electron wave functions for the hydrogen molecule. The top row shows the wave function along the axis through both nuclei, which you can compare to the two single-electron wave functions centered at ±0.37 Å, the equilibrium separation of the two nuclei in the hydrogen molecule. The middle row shows the absolute square of the wave function, which is the probability density of finding an electron at a given value of the x-coordinate. The bottom row shows the probability density of finding the electron in the xy-plane, color-coded so that yellow corresponds to the lowest values and black to the highest.

For the symmetric coordinate-space wave function (Figure 37.25b), a significant portion of the probability of finding an electron is shifted to the region between the two nuclei. This means that both nuclei see an attractive potential in the direction toward each other, due to the Coulomb interaction of the nuclei with the electrons. The depth of this potential due to the exchange interaction is 4.52 eV at the equilibrium separation of 0.74 Å, leading to a net attractive force between the two atoms that provides the covalent bonding for the H_2 molecule. The condition for this bonding is that the two electrons must have opposite spin projections—one *spin-up* and one *spin-down*. If both have the same spin projection, then the coordinate-space wave function needs to be antisymmetric, as shown in Figure 37.25a. The resulting net depletion of the electron density in the region between the two nuclei leads to a repulsive potential at all distances of the nuclei and thus no bonding.

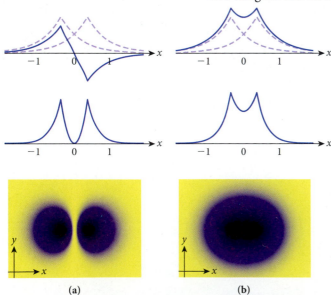

(a)   (b)

FIGURE 37.25 Coordinate-space electron wave functions in the hydrogen molecule: (a) antisymmetric wave function; (b) symmetric wave function. Shown are the two-particle wave functions (top row), probability densities (middle row) along the major axis, and the density distributions in the two-dimensional plane (bottom row).

Many-Fermion Wave Function

The structure of equation 37.39 shows that it is the same expression as we obtain for the determinant of a two-by-two matrix in the single-particle wave functions:

$$\Psi^F(x_1, x_2) = \frac{1}{\sqrt{2}} \begin{vmatrix} \psi_a(x_1) & \psi_a(x_2) \\ \psi_b(x_1) & \psi_b(x_2) \end{vmatrix}.$$

This determinant is commonly called a **Slater determinant.** The Slater determinant can be generalized to the case of many fermions in the system. If there are n fermions, then the n-fermion wave function is the determinant of an $n \times n$ matrix, where the row is a given single-particle state and the column is a particle coordinate:

$$\Psi^F(x_1, x_2, ..., x_n) = \frac{1}{\sqrt{n!}} \begin{vmatrix} \psi_1(x_1) & \psi_1(x_2) & \cdots & \psi_1(x_n) \\ \psi_2(x_1) & \psi_2(x_2) & \cdots & \psi_2(x_n) \\ \vdots & \vdots & \ddots & \vdots \\ \psi_n(x_1) & \psi_n(x_2) & \cdots & \psi_n(x_n) \end{vmatrix}. \tag{37.40}$$

This Slater determinant contains all possible permutations of the single-particle wave functions over all particle coordinates. Again, the factor $1/\sqrt{n!}$ ensures that the many-fermion wave function is normalized—that is, that the integral over all space of the absolute square of the wave function is 1 when the single-particle wave functions are properly normalized.

Quantum Computing

One of the most active current research areas in quantum mechanics is the field of quantum computing. A large number of groups of physicists around the world are using different physical systems to try to implement quantum computers. Among the many systems considered are chains of trapped atoms, quantum dots in semiconductors, electrons on the surface of liquid helium, and collections of buckyballs (C_{60} molecules in the shape of a soccer ball). New ideas regarding which quantum systems to use are generated on a regular basis, and many groups are working on refining them. What all proposed quantum computers have in common is that they work with the many-particle wave function of these systems.

To understand the potential power of quantum computing, let's first look at the workings of a classical computer. A classical computer is based on algorithms that work on digitized information, which is stored in "bits." A bit can be in two states, either on or off. Conventionally, the numbers 1 and 0 represent these two states. Note that nothing appears between 0 and 1 in today's conventional computers; it's all or nothing. A "byte" is a collection of 8 bits. Thus, a possible representation for one particular value of a byte could be 01100010, which means that the 8 bits that make up this byte are in their respective on or off positions as indicated by the numbers, one digit for each bit.

Now consider a quantum two-state system, between which transitions can be induced through some interaction with an outside force, but between which transitions do not happen spontaneously through emission or absorption of a single photon. Left alone, a particle in one of the two states should stay in that state for a long time.

Conventionally, the two quantum states are denoted as $|0\rangle$ and $|1\rangle$ to make an analogy with the classical bit that can have either value of 0 or 1. However, in quantum mechanics a system can be in a superposition of two states, as noted in the previous section. Therefore, the quantum equivalent of the bit, the **qubit,** is defined as a wave function that is a superposition of the two states $|0\rangle$ and $|1\rangle$,

$$|\psi\rangle = c_1|1\rangle + c_0|0\rangle$$

where c_1 and c_0 are complex numbers and represent the probability amplitudes that the system is in state $|1\rangle$ and $|0\rangle$, respectively (Figure 37.26). The numbers c_1 and c_0 are not completely independent, because the requirement that the sum of the probabilities for the particle to be in the state $|0\rangle$ or $|1\rangle$ equals 1 necessitates $|c_1|^2 + |c_0|^2 = 1$.

FIGURE 37.26 Simple pictorial representation of a single qubit.

From one qubit it is straightforward to generalize to n qubits, at least from a conceptual point of view. Assume a collection of individual two-state systems that are not independent of one another. Then the state of the overall system can be written as the sum of the combinations of the individual states. For a system of two qubits, the wave function can be written as

$$|\psi\rangle = c_{11}|11\rangle + c_{10}|10\rangle + c_{01}|01\rangle + c_{00}|00\rangle$$

where c_{00}, c_{01}, c_{10}, and c_{11} are again complex numbers, with the normalization condition that $|c_{11}|^2 + |c_{10}|^2 + |c_{01}|^2 + |c_{00}|^2 = 1$. For three qubits we can write

$$|\psi\rangle = c_{111}|111\rangle + c_{110}|110\rangle + c_{101}|101\rangle + c_{100}|100\rangle$$
$$+ c_{011}|011\rangle + c_{010}|010\rangle + c_{001}|001\rangle + c_{000}|000\rangle.$$

We can easily generalize this equation for larger numbers of qubits. The number of terms in the wave function for 1 qubit is 2, for 2 qubits is 4, and for 3 qubits is 8—the number of terms in the wave function for n qubits is 2^n. All of the promise of quantum computing power arises this number of 2^n. If a classical computer operates on a register of n bits, then it executes n operations simultaneously. However, if a quantum computer operates on a register of n qubits, then it executes 2^n operations simultaneously. A diagram of a particular concept for a quantum computer is shown in Figure 37.27.

Many cautionary statements are in order here. First, probing the wave function of the qubit states necessarily changes the information stored in the qubits. This requires developing special error-correction algorithms, which correct an error without destroying the quantum state. Second, an actual quantum computer that can do more than execute just a few simple demonstration operations has yet to be implemented. Furthermore, quantum computing is by its very construction probabilistic in nature. A calculation must be repeated several times to obtain a reliable result. Only for a small number of algorithms has it been proven that a quantum computer can take full advantage of the theoretically possible 2^n simultaneous operations. One of these algorithms is the factorization of large integers into prime numbers. This problem has attracted a huge amount of interest and also funding, because at present the encryption schemes most commonly used rely on the fact that it is impossible to use present-day computers to factorize large integers in a human lifetime. By comparison, a working quantum computer with approximately 100 qubits could do this job in seconds.

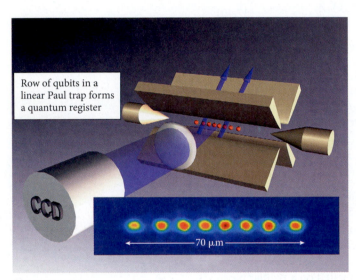

Row of qubits in a linear Paul trap forms a quantum register

CCD

← 70 μm →

FIGURE 37.27 Possible implementation of a quantum computer as a chain of 8 trapped individual Ca ions inside a Paul trap, which is a device that uses oscillating electric fields to trap ions. The figure shows a schematic drawing of the apparatus, and the inset is a picture of the induced fluorescence of the trapped ions.

37.10 Antimatter

In the previous section we discussed the Hamiltonian operator in quantum mechanics, which represents the total energy of a particle. However, we used the classical formula $K = p^2/2m$ for kinetic energy, which is appropriate only for speeds that are small compared to the speed of light. In Chapter 35 on relativity, we noted that the more general relationship between momentum and energy is given by

$$E = \pm\sqrt{p^2c^2 + m^2c^4}.$$

The question then is: Is it possible to construct a framework of quantum physics that is also correct for speeds that are *not* small compared to the speed of light? In other words, can we construct a relativistic quantum theory? The first person who figured out how to do this was the British physicist Paul Adrien Maurice Dirac (1902–1984). He published his famous **Dirac equation** in 1928 and for this accomplishment shared the 1933 Nobel Prize in Physics with Erwin Schrödinger.

We will not attempt to derive or even motivate the Dirac equation. Most university physics departments never even mention this equation in a general introductory course. However, we can write it down, explain the notation, compare it to the equivalent nonrelativistic Schrödinger equation, and explore some of its physical consequences. You can treat the following couple of pages as a resource you may want to return to in the course of your future studies. Or perhaps seeing this equation will motivate you to explore this topic further at the present time.

The Dirac equation for a "free" particle—that is, a particle that is not subjected to any external potential or force—is

$$\left(\gamma_0 mc^2 + \sum_{i=1}^{3} \gamma_i p_i c \right) \Psi(\vec{r},t) = i\hbar \frac{\partial}{\partial t} \Psi(\vec{r},t). \tag{37.41}$$

In this equation, c is the speed of light, m is the rest mass of the particle, and p_i is the momentum operator in one of the three orthogonal directions (x, y, z), which we introduced in Section 37.1. The symbols γ_0, γ_1, γ_2, γ_3 refer to the so-called Dirac matrices:

$$\gamma_0 = \begin{pmatrix} 1 & 0 & 0 & 0 \\ 0 & 1 & 0 & 0 \\ 0 & 0 & -1 & 0 \\ 0 & 0 & 0 & -1 \end{pmatrix}, \quad \gamma_1 = \begin{pmatrix} 0 & 0 & 0 & 1 \\ 0 & 0 & 1 & 0 \\ 0 & 1 & 0 & 0 \\ 1 & 0 & 0 & 0 \end{pmatrix}, \quad \gamma_2 = \begin{pmatrix} 0 & 0 & 0 & -i \\ 0 & 0 & i & 0 \\ 0 & -i & 0 & 0 \\ i & 0 & 0 & 0 \end{pmatrix}, \quad \gamma_3 = \begin{pmatrix} 0 & 0 & 1 & 0 \\ 0 & 0 & 0 & -1 \\ 1 & 0 & 0 & 0 \\ 0 & -1 & 0 & 0 \end{pmatrix}.$$

The wave function $\Psi(\vec{r},t)$ in equation 37.41 is now a four-component "spinor," instead of a one-component scalar, as is the case for the quantum mechanical wave functions that we have discussed so far:

$$\Psi(\vec{r},t) = \begin{pmatrix} \Psi_1(\vec{r},t) \\ \Psi_2(\vec{r},t) \\ \Psi_3(\vec{r},t) \\ \Psi_4(\vec{r},t) \end{pmatrix}.$$

The Dirac equation 37.41 is the relativistically correct extension of the Schrödinger equation 37.32 for the three-dimensional potential-free case:

$$\left(-\frac{\hbar^2}{2m} \frac{\partial^2}{\partial x^2} - \frac{\hbar^2}{2m} \frac{\partial^2}{\partial y^2} - \frac{\hbar^2}{2m} \frac{\partial^2}{\partial z^2} \right) \Psi(\vec{r},t) = i\hbar \frac{\partial}{\partial t} \Psi(\vec{r},t).$$

We can proceed with the solution of the Dirac equation in a manner similar to what we did for the Schrödinger equation. For example, we can write the time dependence of the solution for the free particle as

$$\Psi(\vec{r},t) = \Psi_0(\vec{r})e^{-iEt/\hbar}. \tag{37.42}$$

The static wave function $\Psi_0(\vec{r})$ satisfies the time-independent Dirac equation with the energy eigenvalue E,

$$\left(\gamma_0 mc^2 + \sum_{i=1}^{3} \gamma_i p_i c \right) \Psi_0(\vec{r}) = E\Psi_0(\vec{r}). \tag{37.43}$$

We look for plane-wave solutions of the form moving along one particular coordinate axis, conventionally chosen to be the z-axis:

$$\Psi_0(\vec{r}) = we^{ipz/\hbar}. \tag{37.44}$$

Here p is the momentum of the particle. (Because it is moving along the z-axis, its momentum component along that axis is also the absolute value of its momentum, that is, the length of its momentum vector.) Also, w is a constant four-component spinor. Inserting this *Ansatz* for $\Psi_0(\vec{r})$ into the static Dirac equation 37.43 results in the matrix equation

$$
\begin{pmatrix}
mc^2 & 0 & pc & 0 \\
0 & mc^2 & 0 & -pc \\
pc & 0 & -mc^2 & 0 \\
0 & -pc & 0 & -mc^2
\end{pmatrix}
\begin{pmatrix} w_1 \\ w_2 \\ w_3 \\ w_4 \end{pmatrix}
= E \begin{pmatrix} w_1 \\ w_2 \\ w_3 \\ w_4 \end{pmatrix}.
\tag{37.45}
$$

The solutions for the possible energies are

$$
E(p) = \pm \sqrt{m^2 c^4 + p^2 c^2}.
\tag{37.46}
$$

This is satisfying, because we recover the energy-momentum relationship that we arrived at in Chapter 35 by using relativistic mechanics. The upper two components of the spinor w, w_1 and w_2, then represent the solutions for positive energies: one for "spin-up" and one for "spin-down." The lower two components, w_3 and w_4, are the "spin-up" and "spin-down" solutions for negative energies.

The importance of the Dirac equation lies in the fact that it unifies the two extensions of classical mechanics into one consistent description. Chapter 35 showed that a relativistic description is needed for object speeds that are comparable to the speed of light, and that relativistic mechanics contains classical mechanics as a special case for speeds that are small compared to the speed of light. In Chapter 36 and the present chapter, we have found that we need a quantum description for very small systems, and that through the correspondence principle classical mechanics also emerges as a special case of quantum mechanics. Dirac's equation contains relativistic quantum mechanics and has nonrelativistic quantum mechanics as a special case; it thus provides a very satisfying and aesthetically appealing general framework.

What about the negative energy solutions in equation 37.46? How can these be interpreted? So far, in all our discussions of quantum mechanics, we have assumed that the energy of a free particle is always positive. This enables us to think of the concept of the ground state as the state with lowest energy. However, if states with negative energies are available, what prevents an electron placed into a positive energy state from decaying into a negative energy state and then on through a sequence of successively lower energy states by emitting a photon each time?

To overcome this problem of endless radiation, Dirac proposed the solution illustrated in Figure 37.28: All possible negative energy states are already occupied completely by one spin-up and one spin-down particle. This picture of the vacuum as a "sea" of occupied negative-energy electron states is commonly called the **Dirac sea.** What are these postulated particles in negative energy states? Do they have the same mass as the electron? Are they real or just a thought construction that helped a flawed model survive consistency checks? In particular, is it possible to remove one of these negative energy electrons from the Dirac sea and create a hole in its place?

The answer to this last question is yes. A hole in the Dirac sea acts just like a spin-$\frac{1}{2}$ particle with the same mass as an electron, but with an opposite charge and spin projection. To create such a hole in the Dirac sea, one of the Dirac sea electrons must be removed and lifted to a positive energy state. Lifting the electron across the forbidden gap from $-mc^2$ to mc^2 requires depositing into the vacuum an energy of at least twice the mass of the electron times the speed of light squared (Figure 37.29). Because the mass-energy of the electron is 511 keV, at least 1.022 MeV of energy is needed to lift an electron into a positive energy state and at the same time create a hole in the Dirac sea.

This positively charged hole in the Dirac sea with the same mass as the electron is not just a convenient algebraic trick, but a real particle. The American physicist Carl Anderson (1905–1991) discovered this particle, named the positron, in 1932 with photographic plates exposed to cosmic radiation. This experiment provided a triumphant confirmation for Dirac's theory. Despite this success, the idea of the Dirac sea with an infinite number of particles filling all available energy states does not seem very elegant. In particular, the idea that the vacuum has infinite energy seems counterintuitive. More-modern quantum field theories have replaced some of these conceptual ideas, but the basic difficulties of infinite quantities are still present, leading many practitioners to believe that future breakthroughs are needed.

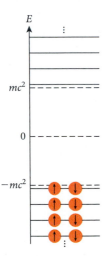

FIGURE 37.28 The Dirac vacuum, where all negative energy states are completely occupied.

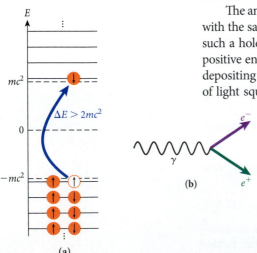

FIGURE 37.29 Creation of a hole in the Dirac sea: electron-positron pair creation. (a) Energy diagram; (b) sketch in coordinate space.

The **positron** is the **antiparticle** to the electron, an example of **antimatter.** Every fermion obeys the Dirac equation, and thus the same picture of a filled Dirac sea applies to the vacuum ground state. Thus, each fermion has an antiparticle. The Italian physicist Emilio Segrè (1905–1989) and American Owen Chamberlain (1920–2006) discovered the antiparticle to the proton, the antiproton, in 1955. In 1995, physicists detected the first atoms of antihydrogen at CERN. Since 2000, CERN has had an "antimatter factory," in which ultracold antihydrogen atoms are produced and studied. One of the many questions to be answered is whether antimatter responds to gravity in the same way that regular matter does.

An antiparticle can also annihilate if it meets its particle partner. This **annihilation** process is sketched in Figure 37.30. In the annihilation process the entire mass of both particles is converted into energy. Thus, from the combination of quantum theory and relativity emerges a new picture of the relationship between mass and energy: Mass and energy can be converted into each other, and mass is just one more form of energy.

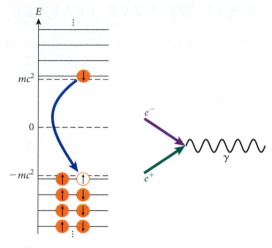

FIGURE 37.30 Electron-positron annihilation.

EXAMPLE **37.7** / **Matter Annihilation**

Matter-antimatter annihilation is a popular energy source in science fiction books and movies. This idea conveniently ignores the fact that there is no source of antimatter anywhere on Earth or in our Solar System, or, most likely, anywhere in the entire universe. (We will think more about why this is so in Chapters 39 and 40 on subatomic physics.) All antimatter that could be used as an energy source would have to be created first. Also, at least as much energy must be put into this process as could be gained later from the annihilation. Nevertheless, it is interesting to figure out how much energy would be released in the annihilation process.

PROBLEM

How much energy would be released if you could let a soft drink can full of antiwater annihilate with regular water?

SOLUTION

A soft drink can holds 0.330 liters of liquid. If this liquid is water, then its mass is 0.330 kg, because water has a density of 1000 kg/m³, so that 1 liter of water has a mass of 1 kg. Since the mass of an antiparticle (which makes up antimatter) is the same as the mass of the corresponding particle, the mass of the antiwater is also 0.330 kg. Therefore, we need to annihilate a total of 2×0.330 kg of mass (0.330 kg of antiwater and 0.330 kg of water).

The energy contained in this mass is obtained by multiplying the total mass by the square of the speed of light.

$$E = mc^2 = (0.660 \text{ kg})(3.00 \cdot 10^8 \text{ m/s})^2 = 5.94 \cdot 10^{16} \text{ J}.$$

To see how much energy this is, let's ask a follow-up question.

PROBLEM

The largest nuclear power plants produce 1200 MW. How long would such a nuclear power plant have to operate in order to produce the same amount of energy as the soft drink can of antiwater?

SOLUTION

A 1200-MW power plant produces $1.2 \cdot 10^9$ J of usable energy each second. Thus, the number of seconds it takes to produce $E = 5.94 \cdot 10^{16}$ J is

$$\frac{5.94 \cdot 10^{16} \text{ J}}{1.2 \cdot 10^9 \text{ J/s}} = 4.95 \cdot 10^7 \text{ s}.$$

This is a time of more than 1.5 years!

WHAT WE HAVE LEARNED | EXAM STUDY GUIDE

- The absolute square of the complex wave function $\psi(x)$ is the probability density of finding a particle at some position. The wave function is normalized as

$$\int_{-\infty}^{\infty} |\psi(x)|^2\, dx = \int_{-\infty}^{\infty} \psi(x)^*\psi(x)dx = 1.$$

- The operator for the momentum is $\mathbf{p}\psi(x) = -i\hbar \dfrac{d}{dx}\psi(x)$.

- The operator for the kinetic energy is

$$\mathbf{K}\psi(x) = \frac{1}{2m}\mathbf{p}^2\psi(x) = \frac{1}{2m}\left(-i\hbar\frac{d}{dx}\right)^2\psi(x) = -\frac{\hbar^2}{2m}\frac{d^2}{dx^2}\psi(x).$$

- The Schrödinger equation for a particle in a potential $U(x)$ is given by

$$-\frac{\hbar^2}{2m}\frac{d^2\psi(x)}{dx^2} + U(x)\psi(x) = E\psi(x).$$

- The wave function solution for a particle confined to an infinite well with walls located at $x = 0$ and $x = a$ is

$$\psi(x) = \begin{cases} 0 & \text{for } x < 0 \\ \sqrt{\dfrac{2}{a}}\sin\left(\dfrac{n\pi x}{a}\right) \text{ with } n = 1,2,3,\dots & \text{for } 0 \le x \le a \\ 0 & \text{for } x > a. \end{cases}$$

The corresponding energy eigenvalues are

$$E_n = \frac{\hbar^2\pi^2}{2ma^2}n^2, \quad n = 1,2,3,\dots.$$

- The solution for the finite potential

$$U(x) = \begin{cases} \infty & \text{for } x < 0 \\ 0 & \text{for } 0 \le x \le a \quad \text{for } E > U_1 \\ U_1 & \text{for } x > a \end{cases}$$

is

$$\psi(x) = \begin{cases} 0 & \text{for } x < 0 \\ D\sin(\kappa x) & \text{for } 0 \le x \le a, \\ F\cos(\kappa' x) + G\sin(\kappa' x) & \text{for } x > a \end{cases}$$

$$\text{with } \kappa' = \sqrt{\kappa^2 - \frac{2mU_1}{\hbar^2}}.$$

- The solution for the same finite potential for $E < U_1$ is

$$\psi(x) = \begin{cases} 0 & \text{for } x < 0 \\ D\sin(\kappa x) & \text{for } 0 \le x \le a, \\ Fe^{-\gamma x} & \text{for } x > a \end{cases}$$

$$\text{with } \gamma^2 = \frac{2m(U_1 - E)}{\hbar^2} = \frac{2mU_1}{\hbar^2} - \kappa^2.$$

- For a potential step of the form

$$U(x) = \begin{cases} \infty & \text{for } x < 0 \\ 0 & \text{for } 0 \le x \le a \\ U_1 & \text{for } a < x < b \\ 0 & \text{for } x \ge b \end{cases}$$

the wave function is

$$\psi(x) = \begin{cases} 0 & \text{for } x < 0 \\ D\sin(\kappa x) & \text{for } 0 \le x \le a \\ Fe^{-\gamma x} & \text{for } a < x < b \\ G\sin(\kappa x + \varphi) & \text{for } x \ge b. \end{cases}$$

The probability for a particle with energy less than the barrier height to tunnel through the barrier is

$$\Pi_t = \frac{|\psi(b)|^2}{|\psi(a)|^2} = \frac{\left|Fe^{-\gamma b}\right|^2}{\left|Fe^{-\gamma a}\right|^2} = \left|e^{-\gamma(b-a)}\right|^2 = e^{-2\gamma(b-a)}$$

and thus depends exponentially on the barrier width.

- The wave function solutions for the harmonic oscillator potential $U(x) = \frac{1}{2}kx^2 = \frac{1}{2}m\omega_0^2 x^2$ are the functions

$$\psi_n(x) = \frac{1}{\sqrt{\sigma}\pi^{1/4}}\frac{1}{\sqrt{n!\,2^n}}H_n(x/\sigma)e^{-x^2/2\sigma^2},$$

where H_n is a Hermite polynomial and

$$\sigma = \sqrt{\frac{\hbar}{m\omega_0}}.$$

The corresponding energy eigenvalues are equally spaced,

$$E_n = \left(n + \frac{1}{2}\right)\hbar\omega_0, \quad n = 0,1,2,\dots.$$

- We can find the quantum mechanical expectation value of a quantity by integrating over the entire space ψ^* times the result obtained when the corresponding operator acts on ψ. The position expectation value is

$$\langle x \rangle = \int_{-\infty}^{\infty} \psi^*(x)x\psi(x)dx,$$

and the momentum expectation value is

$$\langle p \rangle = \int_{-\infty}^{\infty} \psi^*(x)\mathbf{p}\psi(x)dx = -i\hbar\int_{-\infty}^{\infty} \psi^*(x)\frac{d}{dx}\psi(x)dx.$$

- The ground state ($n = 0$) oscillator wave function has the minimum product of momentum and coordinate uncertainty allowed by the uncertainty principle.

- The correspondence principle states that if the energy difference ΔE between neighboring quantum energy states becomes small relative to the total energy E, $\Delta E \ll E$, the quantum solution approaches its classical limit.

- The time-dependent Schrödinger equation is

$$-\frac{\hbar^2}{2m}\frac{\partial^2}{\partial x^2}\Psi(x,t)+U(x)\Psi(x,t)=i\hbar\frac{\partial}{\partial t}\Psi(x,t)$$

with the solution $\Psi(x,t)=\psi(x)e^{-i\omega t}$.

- The Hamiltonian operator is

$$\mathbf{H}=-\frac{\hbar^2}{2m}\frac{d^2}{dx^2}+U(x)=\mathbf{K}+U(x).$$

It is linear, so a linear combination of two solutions to the Schrödinger equation is also a solution.

- The two-particle wave function for (non-interacting) bosons is the symmetrized product of one-particle wave functions,

$$\Psi^B(x_1,x_2)=\frac{1}{\sqrt{2}}\big(\psi_a(x_1)\cdot\psi_b(x_2)+\psi_a(x_2)\cdot\psi_b(x_1)\big).$$

- The two-particle wave function for (non-interacting) fermions is the antisymmetrized product of one-particle wave functions

$$\Psi^F(x_1,x_2)=\frac{1}{\sqrt{2}}\big(\psi_a(x_1)\cdot\psi_b(x_2)-\psi_a(x_2)\cdot\psi_b(x_1)\big).$$

- Quantum computing defines a qubit as

$$|\psi\rangle=c_1|1\rangle+c_0|0\rangle.$$

The entanglement of many qubits gives quantum computing the (still only theoretical!) ability to evaluate many calculations simultaneously and promises a great advantage over conventional digital computing.

- The Dirac equation

$$\left(\gamma_0 mc^2+\sum_{i=1}^{3}\gamma_i p_i c\right)\Psi(\vec{r},t)=i\hbar\frac{\partial}{\partial t}\Psi(\vec{r},t)$$

is the generalization of the Schrödinger equation for the relativistically correct relationship between energy and momentum. It leads to negative energy states and antiparticles.

KEY TERMS

NEW SYMBOLS AND EQUATIONS

$\mathbf{p}\psi(x)=-i\hbar\dfrac{d}{dx}\psi(x)$, momentum operator

$\mathbf{K}\psi(x)=\dfrac{1}{2m}\mathbf{p}^2\psi(x)=\dfrac{1}{2m}\left(-i\hbar\dfrac{d}{dx}\right)^2\psi(x)=-\dfrac{\hbar^2}{2m}\dfrac{d^2}{dx^2}\psi(x)$, kinetic energy operator

$-\dfrac{\hbar^2}{2m}\dfrac{d^2\psi(x)}{dx^2}+U(x)\psi(x)=E\psi(x)$, Schrödinger equation

$E_n=\dfrac{\hbar^2\pi^2}{2ma^2}n^2$, $n=1,2,3,...$, energy eigenvalues, infinite potential well

$E_n=\left(n+\tfrac{1}{2}\right)\hbar\omega_0$, $n=0,1,2,...$, energy eigenvalues, harmonic oscillator

$\langle x\rangle=\displaystyle\int_{-\infty}^{\infty}\psi^*(x)x\psi(x)dx$, position expectation value

$\mathbf{H}=-\dfrac{\hbar^2}{2m}\dfrac{d^2}{dx^2}+U(x)=\mathbf{K}+U(x)$, Hamiltonian operator

ANSWERS TO SELF-TEST OPPORTUNITIES

37.1 Use: $e^{\pm i\theta} = \cos\theta \pm i \sin\theta$ (Euler)

$$\psi(x) = Ae^{i\kappa x} + Be^{-i\kappa x}$$

$$= A\left(\cos(\kappa x) + i\sin(\kappa x)\right) + B\left(\cos(\kappa x) - i\sin(\kappa x)\right)$$

$$= \left(A+B\right)\cos(\kappa x) + i\left(A-B\right)\sin(\kappa x)$$

$$= C\cos(\kappa x) + D\sin(\kappa x), \text{ where}$$

$$C = A + B \text{ and } D = i\left(A - B\right).$$

37.2 Adding a constant energy, c, everywhere in space leaves the wave function solutions unchanged. However, all energies are shifted up by this constant c.

37.3 The wave function is the product of three one-dimensional wave functions, $\psi(x, y, z) = \psi_1(x) \cdot \psi_2(y) \cdot \psi_3(z)$

$$\psi_1(x) = \begin{cases} 0 & \text{for } x < 0 \\ \sqrt{2/a}\,\sin\left(n_x \pi x/a\right) \text{ with } n_x = 1,2,3,\dots & \text{for } 0 \le x \le a \\ 0 & \text{for } x > a \end{cases}$$

$$\psi_2(y) = \begin{cases} 0 & \text{for } y < 0 \\ \sqrt{2/b}\,\sin\left(n_y \pi y/b\right) \text{ with } n_y = 1,2,3,\dots & \text{for } 0 \le y \le b \\ 0 & \text{for } y > b \end{cases}$$

$$\psi_3(z) = \begin{cases} 0 & \text{for } z < 0 \\ \sqrt{2/c}\,\sin\left(n_z \pi z/c\right) \text{ with } n_z = 1,2,3,\dots & \text{for } 0 \le z \le c \\ 0 & \text{for } z > c \end{cases}$$

$$E_{n_x,n_y,n_z} = \frac{\hbar^2 \pi^2}{2ma^2}n_x^2 + \frac{\hbar^2 \pi^2}{2mb^2}n_y^2 + \frac{\hbar^2 \pi^2}{2mc^2}n_z^2.$$

37.4 A continuous wave function at point b requires that

$$Fe^{-\gamma b} = G\sin(\kappa b + \phi), \qquad \text{(i)}$$

and a continuous derivative of the wave function at the same point requires that

$$-\gamma Fe^{-\gamma b} = \kappa G\cos(\kappa b + \phi). \qquad \text{(ii)}$$

Divide equation (i) by equation (ii) to obtain $\tan(\kappa b + \phi) = \dfrac{\kappa}{-\gamma}$.

Solve this for the phase shift, ϕ, to find the phase shift

$$\phi = \tan^{-1}\left(\frac{\kappa}{-\gamma}\right) - \kappa b.$$

To get the amplitude G it is easiest to divide equation (ii) by κ, then square both sides of equation (i) and equation (ii). This results in

$$F^2 e^{-2\gamma b} = G^2 \sin^2(\kappa b + \phi)$$

$$(\gamma/\kappa)^2 F^2 e^{-2\gamma b} = G^2 \cos^2(\kappa b + \phi).$$

Now add these two equations, which yields

$$(1+(\gamma/\kappa)^2)F^2 e^{-2\gamma b} = G^2.$$

Taking the square root gives us the value for G,

$$G = \sqrt{1 + (\gamma/\kappa)^2}\, Fe^{-\gamma b}.$$

37.5 Start with the definition and insert the wave function:

$$\langle p \rangle = \int_{-\infty}^{\infty} \psi^*(x)\mathbf{p}\psi(x)dx = -i\hbar \int_{-\infty}^{\infty} \psi^*(x)\frac{d}{dx}\psi(x)dx$$

$$\psi_0(x) = Ae^{-x^2/2\sigma^2}$$

$$\frac{d}{dx}\psi_0(x) = \frac{Ax}{\sigma^2}e^{-x^2/2\sigma^2}$$

$$= \int_{-\infty}^{\infty}\left(Ae^{-x^2/2\sigma^2}\right)\left(\frac{Ax}{\sigma^2}e^{-x^2/2\sigma^2}\right)dx = \frac{A}{\sigma^2}\int_{-\infty}^{\infty}xe^{-x^2/\sigma^2}dx = 0.$$

37.6 For the harmonic oscillator, we find

$$E_n = \left(n + \tfrac{1}{2}\right)\hbar\omega_0, \quad n = 0,1,2,\dots$$

$$\langle K_0 \rangle = E_0 / 2 = \frac{\hbar\omega_0}{4}$$

$$\langle K_1 \rangle = E_1 / 2 = \frac{3\hbar\omega_0}{4}$$

$$\langle K_2 \rangle = E_2 / 2 = \frac{5\hbar\omega_0}{4}$$

$$\langle K_3 \rangle = E_3 / 2 = \frac{7\hbar\omega_0}{4}.$$

PROBLEM-SOLVING PRACTICE

Problem-Solving Guidelines

1. Quantum mechanics dominates all of modern physics, and many different systems have been and continue to be investigated. However, most of these systems do not have straightforward analytic solutions and require the use of computers. On the introductory level, most problems can be mapped to (in)finite potential wells or harmonic oscillator potentials.

2. Wave functions are often determined by boundary conditions and symmetries. Be sure to exploit these in your solution attempts.

3. Quantum mechanics and classical mechanics are not totally different from each other. Classical mechanics is the limit of quantum mechanics when we can neglect that $\hbar$ is nonzero. Be sure that your quantum solutions do not violate classical mechanics in that limit.

4. Typically you will find discrete energies for your quantum solutions. Fixed energy differences between neighboring states point to oscillator potentials, whereas linearly increasing energy differences point to infinite potential wells.

SOLVED PROBLEM 37.1 / Half-Oscillator

PROBLEM

Suppose you know that an electron is confined to a potential of the form

$$U(x) = \begin{cases} a^2 x^2 \\ \infty \end{cases} \text{ for } \begin{cases} x > 0 \\ x \le 0 \end{cases}$$

with $a = $ constant. What is the energy of the first excited state, if the ground state energy of the electron in this potential has a value of 3.5 eV?

SOLUTION

THINK

The potential here is a mixture between the infinite potential well, where $U(x) = \infty$ outside certain regions in space, and the harmonic oscillator potential, $U(x) = \frac{1}{2}m\omega_0^2 x^2$, with the constant $a = \sqrt{\frac{1}{2}m\omega_0^2}$. The key to solving the present problem is that the only constraint that an infinite potential wall imposes on a wave function is that it has to be zero at the interface between the infinite and the finite part of the potential function. In the present case, this requirement can be written as $\psi(x = 0) = 0$.

SKETCH

Figure 37.31 sketches the shape of the potential function given in the problem.

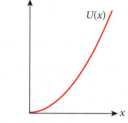

FIGURE 37.31 Half-oscillator potential.

RESEARCH

We have already established that the wave functions on the left side ($x \le 0$) have to be 0. On the right side, the wave functions need to be harmonic oscillator wave functions of the kind given in equation 37.26, where the lowest-energy wave functions in the harmonic oscillator potential are

$$\psi_0(x) = \frac{1}{\sqrt{\sigma}\pi^{1/4}} e^{-x^2/2\sigma^2}$$

$$\psi_1(x) = \frac{1}{\sqrt{\sigma}\pi^{1/4}} \frac{1}{\sqrt{2}} \left(2\frac{x}{\sigma}\right) e^{-x^2/2\sigma^2}$$

$$\psi_2(x) = \frac{1}{\sqrt{\sigma}\pi^{1/4}} \frac{1}{\sqrt{8}} \left(4\frac{x^2}{\sigma^2} - 2\right) e^{-x^2/2\sigma^2}$$

$$\psi_3(x) = \frac{1}{\sqrt{\sigma}\pi^{1/4}} \frac{1}{\sqrt{48}} \left(8\frac{x^3}{\sigma^3} - 12\frac{x}{\sigma}\right) e^{-x^2/2\sigma^2}$$

$$\psi_4(x) = \frac{1}{\sqrt{\sigma}\pi^{1/4}} \frac{1}{\sqrt{384}} \left(16\frac{x^4}{\sigma^4} - 48\frac{x^2}{\sigma^2} + 12\right) e^{-x^2/2\sigma^2}$$

with the constant σ defined as $\sigma = \sqrt{\hbar/m\omega_0}$, and the energies corresponding to these wave functions are

$$E_n = \left(n + \tfrac{1}{2}\right)\hbar\omega_0, \quad n = 0,1,2,....$$

SIMPLIFY

Clearly the functions $\psi_0(x)$, $\psi_2(x)$, $\psi_4(x)$, and all other wave functions with even n have nonzero values for $x = 0$ and thus do not qualify as solutions for the present problem of the half-oscillator. These are the even-parity wave functions with the property $\psi(-x) = \psi(x)$. However, the odd-parity wave functions $\psi_1(x)$, $\psi_3(x)$, and all other wave functions with odd n, have the required property that $\psi(x=0) = 0$. In general odd-parity wave functions have the property $\psi(-x) = -\psi(x)$ and thus must vanish at the origin in order to be continuous.

This means that the ground state for our present system must have the wave function

$$\psi_{gs}(x) = \frac{1}{\sqrt{\sigma}\pi^{1/4}} \frac{1}{\sqrt{2}} \left(2\frac{x}{\sigma}\right) e^{-x^2/2\sigma^2}$$

Continued—

with energy

$$E_{gs} = (1 + \tfrac{1}{2})\hbar\omega_0 = \tfrac{3}{2}\hbar\omega_0,$$

and the first excited state must have the wave function

$$\psi_{ex}(x) = \frac{1}{\sqrt{\sigma}\pi^{1/4}} \frac{1}{\sqrt{48}} \left(8\frac{x^3}{\sigma^3} - 12\frac{x}{\sigma} \right) e^{-x^2/2\sigma^2}$$

with energy

$$E_{ex} = (3 + \tfrac{1}{2})\hbar\omega_0 = \tfrac{7}{2}\hbar\omega_0.$$

From this it follows easily that the ratio of the energy of the first excited state to that of the ground state is

$$\frac{E_{ex}}{E_{gs}} = \frac{\tfrac{7}{2}\hbar\omega_0}{\tfrac{3}{2}\hbar\omega_0} = \frac{7}{3}.$$

CALCULATE

The energy of the ground state was given in the problem as 3.5 eV; so we find that

$$E_{ex} = \tfrac{7}{3}E_{gs} = \tfrac{7}{3} \cdot 3.5 \text{ eV} = 8.1666666 \text{ eV}.$$

ROUND

We round to the two significant digits to which the ground state energy was given in the problem, and state our final answer as

$$E_{ex} = 8.2 \text{ eV}.$$

DOUBLE-CHECK

For oscillator wave functions, it is characteristic that the energy difference between neighboring energy states is constant. This is also the case for the present setup, in which only the odd-parity oscillator wave functions are valid solutions. If one were to calculate the energy of the second excited state, then one would find a value of $\tfrac{11}{3}E_{gs} = 12.8$ eV.

MULTIPLE-CHOICE QUESTIONS

37.1 The wavelength of an electron in an infinite potential is $\alpha/2$, where α is the width of the infinite potential well. Which state is the electron in?

a) $n = 3$ c) $n = 4$

b) $n = 6$ d) $n = 2$

37.2 In an infinite square well, for which of the following states will the particle never be found in the exact center of the well?

a) the ground state   d) any of the above

b) the first excited state   e) none of the above

c) the second excited state

37.3 The probability of finding an electron in a hydrogen atom is directly proportional to

a) its energy.

b) its momentum.

c) its wave function.

d) the square of its wave function.

e) the product of the position coordinate and the square of the wave function.

f) none of the above.

37.4 Is the superposition of two wave functions, which are solutions to the Schrödinger equation for the same potential energy, also a solution to the Schrödinger equation?

a) no

b) yes

c) depends on potential energy

d) only if $\dfrac{d^2\psi(x)}{dx^2} = 0$

37.5 Let κ be the magnitude of the wave number of a particle moving in one dimension with velocity v. If the velocity of the particle is doubled, to $2v$, then the wave number is:

a) κ c) $\kappa/2$

b) 2κ d) none of these

37.6 An electron is in an infinite square well of width a: ($U(x) = \infty$ for $x < 0$ and $x > a$). If the electron is in the first excited state, $\Psi(x) = A\sin(2\pi x/a)$, at what position is the probability function a maximum?

a) 0   c) $a/2$ e) at both $a/4$

b) $a/4$ d) $3a/4$ and $3a/4$

37.7 State whether each of the following statements is true or false.

a) The energy of electrons is always discrete.

b) The energy of a bound electron is continuous.

c) The energy of a free electron is discrete.

d) The energy of an electron is discrete when it is bound to an ion.

37.8 State whether each of the following statements is true or false.

a) In a one-dimensional quantum harmonic oscillator, the energy levels are evenly spaced.

b) In an infinite one-dimensional potential well, the energy levels are evenly spaced.

c) The minimum total energy possible for a classical harmonic oscillator is zero.

d) The correspondence principle states that because the minimum possible total energy for the classical simple harmonic oscillator is zero, the expected value for the

fundamental state ($n = 0$) of the one-dimensional quantum harmonic oscillator should also be zero.

e) The $n = 0$ state of the one-dimensional quantum harmonic oscillator is the state with the minimum possible uncertainty $\Delta x \Delta p$.

37.9 Simple harmonic oscillation occurs when the potential energy function is equal to $(1/2)kx^2$, where k is a constant. What happens to the ground state energy level if k is increased?

a) It increases.

b) It remain the same.

c) It decreases.

37.10 A particle of energy $E = 5$ eV approaches an energy barrier of height $U = 8$ eV. Quantum mechanically there is a finite probability that the particle tunnels through the barrier. If the barrier height is slowly decreased, the probability that the particle will reflect from the barrier will

a) decrease.   b) increase.   c) not change.

QUESTIONS

37.11 True or False: The larger the amplitude of a Schrödinger wave function, the larger its kinetic energy. Explain your answer.

37.12 For a particle trapped in an infinite square well of length L, what happens to the probability that the particle is found between 0 and $L/2$ as the particle's energy increases?

37.13 Think about what happens to infinite square well wave functions as the quantum number n approaches infinity. Does the probability distribution in that limit obey the correspondence principle? Explain.

37.14 Show by symmetry arguments that the expectation value of the momentum for an even-n state of the one-dimensional harmonic oscillator is zero.

37.15 Is it possible for the expectation value of the position of an electron to occur at a position where the electron's probability function, $\Pi(x)$, is zero? If it is possible, give a specific example.

37.16 Sketch the two lowest energy wave functions for an electron in an infinite potential well that is 20 nm wide and a finite potential well that is 1 eV deep and is also 20 nm wide. Using your sketches, can you determine whether the energy levels in the finite potential well will be lower, the same, or higher than in the infinite potential well?

37.17 In the cores of white dwarf stars, carbon nuclei are thought to be locked into very ordered lattices because the temperature is quite cold, ~10^4 K. Consider the case of a one-dimensional lattice of carbon atoms separated by 20 fm (1 fm = $1 \cdot 10^{-15}$ m). Consider the central atom of a row of three atoms with this spacing. Approximate the Coulomb potentials of the two outside atoms to follow a quadratic relationship, as-suming small vibrations; what energy state would the central carbon atom be in at this temperature? (Use $E = 3/2k_B T$.)

37.18 For a finite square well, you have seen solutions for particle energies greater than and less than the well depth. Show that these solutions are equal outside the potential well if the particle energy is equal to the well depth. Explain your answer and the possible difficulty with it.

37.19 Consider the energies allowed for bound states of a half-harmonic oscillator, namely, a potential that is

$$U(x) = \begin{cases} \frac{1}{2}m\omega_0^2 x^2 \\ \infty \end{cases} \text{ for } \begin{cases} x > 0 \\ x \le 0 \end{cases}.$$

Using simple arguments based on the characteristics of good wave functions, what are the energies allowed for bound states in this potential?

37.20 Suppose $\psi(x)$ is a properly normalized wave function describing the state of an electron. Consider a second wave function, $\psi_{new}(x) = e^{i\phi}\psi(x)$, for some real number ϕ. How does the probability density associated with ψ_{new} compare to that associated with ψ?

37.21 The ground state energy of a particle of mass m with potential energy $U(x) = U_0 \cosh(x/a)$, where U_0 and a are constants. Show that the ground state energy of the particle can be *estimated* as:

$$E_0 \cong U_0 + \frac{1}{2}\hbar\left(\frac{U_0}{ma^2}\right)^{\frac{1}{2}}.$$

37.22 The Schrödinger equation for a nonrelativistic free particle of mass m is obtained from the energy relationship $E = p^2/(2m)$ by replacing E and p with appropriate deriva-tive operators, as suggested by the de Broglie relations. Using this procedure, derive a quantum wave equation for a *relativistic* particle of mass m, for which the energy relation is $E^2 - p^2c^2 = m^2c^4$, *without* taking any square root of this relation.

PROBLEMS

A blue problem number indicates a worked-out solution is available in the Student Solutions Manual. One • and two •• indicate increasing level of problem difficulty.

Section 37.1

37.23 A neutron has a kinetic energy of 10 MeV. What size object is necessary to observe neutron diffraction effects? Is there anything in nature of this size that could serve as a target to demonstrate the wave nature of 10-MeV neutrons?

37.24 Given the complex function $f(x) = (8 + 3i) + (7 - 2i)$ of the real variable x, what is $|f(x)|^2$?

Section 37.3

37.25 Determine the two lowest energies of a wave function of an electron in a box of width $2.0 \cdot 10^{-9}$ m.

37.26 Determine the lowest 3 energies of the wave function of a proton in a box of width $1.0 \cdot 10^{-10}$ m.

37.27 What is the ratio of energy difference between the ground state and the first excited state for an infinite square well of length L to that of length $2L$. That is, find $(E_2 - E_1)_L/(E_2 - E_1)_{2L}$.

•37.28 An electron is confined in a one-dimensional infinite potential well of 1.0 nm. Calculate the energy difference between

a) the second excited state and the ground state, and

b) the wavelength of light emitted by this radiative transition.

•37.29 Find the wave function for a particle in an infinite square well centered at the origin. The walls are at $\pm a/2$ instead of at 0 and a.

•37.30 Example 37.1 calculates the energy of the wave function with the lowest quantum number for an electron confined to a box of width 2.0 Å in the one-dimensional case. However, atoms are *three-dimensional* entities with a typical diameter of 1 Å = 10^{-10} m. It would seem then that the next, better approximation would be that of an electron trapped in a three-dimensional infinite potential well (a potential cube with sides of 1 Å).

a) Derive an expression for the electron wave function and the corresponding energies for a particle in a three-dimensional rectangular infinite potential well.

b) Calculate the lowest energy allowed for the electron in this case.

Section 37.4

37.31 An electron is confined to a potential well shaped as shown in Figure 37.10. The width of the well is $1.0 \cdot 10^{-9}$ m, and $U_1 = 2.0$ eV. Is the $n = 3$ state bound in the well?

37.32 If a proton of kinetic energy 18.0 MeV encounters a rectangular potential energy barrier of height 29.8 MeV and width $1.00 \cdot 10^{-15}$ m, what is the probability that the proton will tunnel through the barrier?

•37.33 Suppose the kinetic energy of the neutron described in Example 37.4 is increased by 15%. By what factor does the neutron's probability of tunneling through the barrier increase?

•37.34 A beam of electrons moving in the positive x-direction encounters a potential barrier that is 2.51 eV high and 1.00 nm wide. Each electron has a kinetic energy of 2.50 eV, and the electrons arrive at the barrier at a rate of 1000 electrons/s (1000 electrons every second). What is the rate I_T in electrons/s at which electrons pass through the barrier, on average? What is the rate I_R in electrons/s at which electrons reflect back from the barrier, on average? Determine and compare the wavelengths of the electrons before and after they pass through the barrier.

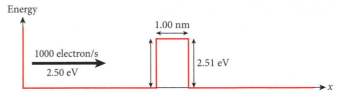

•37.35 Consider an electron approaching a potential barrier 2 nm wide and 7.00 eV high. What is the energy of the electron if it has a 10% probability of tunneling through this barrier?

•37.36 Consider an electron in a three-dimensional box—with infinite potential walls—of dimensions 1.00 nm × 2.00 nm × 3.00 nm. Find the quantum numbers n_x, n_y, n_z and energies in eV of the six lowest energy levels. Are any of these levels degenerate, that is, do any distinct quantum states have identical energies?

•37.37 In a scanning tunneling microscope, the probability that an electron from the probe will tunnel through a 0.1-nm gap is 0.1%. Calculate the work function of the probe of the scanning tunneling microscope.

••37.38 Consider an attractive square-well potential, $U(x) = 0$ for $x < -\alpha$, $U(x) = -U_0$ for $-\alpha \leq x \leq \alpha$ where U_0 is a positive constant, and $U(x) = 0$ for $x > \alpha$. For $E > 0$, the solution of the Schrödinger equation in the 3 regions will be the following:

For $x < -\alpha$, $\psi(x) = e^{i\kappa x} + Re^{-i\kappa x}$ where $\kappa^2 = 2mE/\hbar^2$ and R is the amplitude of a reflected wave.

For $-\alpha \leq x \leq \alpha$, $\psi(x) = Ae^{i\kappa' x} + Be^{-i\kappa' x}$ and $(\kappa')^2 = 2m(E + U_0)/\hbar^2$.

For $x > \alpha$, $\psi(x) = Te^{i\kappa x}$ where T is the amplitude of the transmitted wave.

Match $\psi(x)$ and $d\psi(x)/dx$ at $-\alpha$ and α and find an expression for R. What is the condition for which $R = 0$ (that is, there is no reflected wave)?

••37.39 a) Determine the wave function and the energy levels for the bound states of an electron in the symmetrical one-dimensional potential well of finite depth presented in the figure.

b) If the penetration distance η in the classically forbidden region is defined as the distance at which the wave function

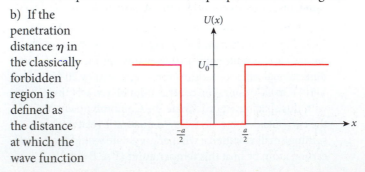

decreases to $1/e$ of its value at the edge of the well, determine an expression for this penetration distance.

c) The electrons in a typical GaAs-GaAlAs quantum-well laser diode are confined into a one-dimensional quantum well like the one in the figure, of width 1 nm and depth 0.300 eV. Numerical solutions to the Schrödinger equation show that there is only one possible bound state for the electrons in this case, with energy of 0.125 eV. Calculate the penetration depth in this case.

Section 37.5

37.40 An oxygen molecule has a vibrational mode that behaves approximately like a simple harmonic oscillator with frequency $2.99 \cdot 10^{14}$ rad/s. Calculate the energy of the ground state and the first two excited states.

37.41 An electron in a harmonic potential well emits a photon with a wavelength of 360 nm as it undergoes a $3 \rightarrow 1$ quantum jump. What wavelength photon is emitted in a $3 \rightarrow 2$ quantum jump? (*Hint:* The energy of the photon is equal to the energy difference between the initial and the final state of the electron.)

37.42 An experimental measurement of the energy levels of a hydrogen molecule, H_2, shows that the energy levels are evenly spaced and separated by about $9 \cdot 10^{-20}$ J. A reasonable model of one of the hydrogen atoms would then seem to be that of a hydrogen atom in a simple harmonic oscillator potential. Assuming that the hydrogen atom is attached by a spring with a spring constant k to the molecule, what is the spring constant k?

•37.43 Calculate the ground state energy for an electron confined to a cube with sides equal to twice the Bohr radius ($R = 0.0529$ nm). Determine the spring constant that would give this same ground state energy for a harmonic oscillator.

•37.44 A particle in the harmonic oscillator potential has the initial wave function $\Psi(x,0) = A\left[\Psi_0(x) + \Psi_1(x)\right]$. Normalize $\Psi(x,0)$.

•37.45 The ground state wave function for a harmonic oscillator is given by $\Psi_0(x) = A_2 e^{-x^2/2b^2}$.

a) Determine the normalization constant A_2.

b) Determine the probability that a quantum harmonic oscillator in the $n = 0$ state will be found in the classically forbidden region.

Section 37.6

37.46 A particle is in an infinite square well of width L and is in the $n = 3$ state. What is the probability that, when observed, the particle is found to be in the rightmost 10% of the well?

37.47 An electron is confined between $x = 0$ and $x = L$. The wave function of the electron is $\psi(x) = A\sin(2\pi x/L)$. The wave function is zero for the regions $x < 0$ and $x > L$.

a) Determine the normalization constant A.

b) What is the probability of finding the electron in the region $0 \le x \le L/3$?

37.48 Find the probability of finding an electron trapped in a one-dimensional infinite well of width 2 nm in the $n = 2$ state between 0.8 and 0.9 nm (assume that the left edge of the well is at $x = 0$ and the right edge is at $x = 2$ nm).

•37.49 An electron is trapped in a one-dimensional infinite potential well that is $L = 300$ pm wide. What is the probability that one can detect the electron in the first excited state in an interval between $x = 0.5\,L$ and $x = 0.75\,L$?

Section 37.8

•37.50 Find the uncertainty of x for a wave function $\Psi(x,t) = Ae^{-\lambda x^2}e^{-i\omega t}$.

•37.51 Write a plane-wave function $\psi(\vec{r},t)$ for a nonrelativistic free particle of mass m moving in three dimensions with momentum **p**, including correct time dependence as required by the Schrödinger equation. What is the probability density associated with this wave?

•37.52 Suppose a quantum particle is in a stationary state (energy eigenstate) with a wave function $\psi(x,t)$. The calculation of $\langle x \rangle$, the expectation value of the particle's position, is shown in the text. Calculate $d\langle x \rangle/dt$ (**not** $\langle dx/dt \rangle$).

••37.53 Although quantum systems are frequently characterized by their stationary states or energy eigenstates, a quantum particle is not required to be in such a state unless its energy has been measured. The actual state of the particle is determined by its initial conditions. Suppose a particle of mass m in a one-dimensional potential well with infinite walls (a "box") of width a is actually in a state with wave function

$$\Psi(x,t) = \frac{1}{\sqrt{2}}[\Psi_1(x,t) + \Psi_2(x,t)],$$

where Ψ_1 denotes the stationary state with quantum number $n = 1$ and Ψ_2 denotes the state with $n = 2$. Calculate the probability density distribution for the position x of the particle in this state.

Section 37.10

37.54 In Chapter 40 you will see that a nuclear-fusion reaction between two protons (creating a deuteron, an anti-electron, and a neutrino) releases 0.42 MeV of energy. Nuclear fusion is what causes the stars to shine, and we know that if we can harness it we can solve the world's energy problems completely. Compare that amount of energy with what would be released by the annihilation of a proton and an antiproton.

37.55 Particle-antiparticle pairs are occasionally created out of empty space. Looking at energy-time uncertainty, how long would such particles be expected to exist if they are:

a) an electron/positron pair?

b) a proton/antiproton pair?

37.56 A positron and an electron annihilate, producing two 2.0-MeV gamma rays moving in opposite directions. Calculate the kinetic energy of the electron when the kinetic energy of the positron is twice that of the electron.

Additional Problems

37.57 How much energy is required to promote the electron described in Example 37.1 to its first excited state?

37.58 Electrons from a scanning tunneling microscope encounter a potential barrier that has a height of $U = 4.0$ eV above their total energy. By what factor does the tunneling current change if the tip moves a net distance of 0.10 nm farther from the surface?

37.59 An electron is confined in a three-dimensional cubic space of L^3 with infinite potentials.

a) Write down the normalized solution of the wave function in the ground state.

b) How many energy states are available up to the second excited state from the ground state? (Take the electron spin into account.)

37.60 A mass-and-spring harmonic oscillator used for classroom demonstrations has angular frequency $\omega_0 = 4.45$ s^{-1}. If this oscillator is oscillating with total (kinetic plus potential) energy $E = 1.00$ J, what is its corresponding quantum excitation number n?

37.61 Calculate the energy of the first excited state of a proton in an infinite one-dimensional potential well of width $\alpha = 1$ nm.

37.62 A 5.6-MeV alpha particle inside a heavy nucleus encounters a barrier of average height 17 MeV and a width of 38 fm (1 fm = $1 \cdot 10^{-15}$ m). What is the probability that this alpha particle will tunnel through the barrier?

37.63 The neutrons in a parallel beam, each having kinetic energy 1/40 eV (which is approximately corresponding to "room temperature"), are directed through two slits 0.50 mm apart. How far apart will the interference peaks be on a screen 1.5 m away?

37.64 Find the ground state energy (in units of eV) of an electron in a one-dimensional quantum box, if the box is of length $L = 0.1$ nm.

37.65 An approximate one-dimensional quantum well can be formed by surrounding a layer of GaAs with layers of $Al_xGa_{1-x}As$. The GaAs layers can be fabricated in thicknesses that are integral multiples of the single-layer thickness, 0.28 nm. Some electrons in the GaAs layer behave as if they were trapped in a box. For simplicity, treat the box as an infinite one-dimensional well and ignore the interactions between the electrons and the Ga and As atoms (such interactions are often accounted for by replacing the actual electron mass with an effective electron mass). Calculate the energy of the ground state in this well for these cases:

a) 2 GaAs layers

b) 5 GaAs layers

37.66 Consider a water vapor molecule in a room 4.0 m × 10 m × 10 m.

a) What is the ground state energy of this molecule, treating it as a simple particle in a box?

b) Compare this energy to the average thermal energy of such a molecule, taking the temperature to be 300 K.

c) What can you conclude from the two numbers you just calculated?

37.67 A neutron moves between rigid walls 8.4 fm apart. What is the energy of its fundamental state?

•37.68 A surface is examined using a scanning tunneling microscope (STM). For the range of the working gap, L, between the tip and the sample surface, assume that the electron wave function for the atoms under investigation falls off exponentially as $|\Psi| = e^{-(10.0 \text{ nm}^{-1})a}$. The tunneling current through the STM tip is proportional to the tunneling probability. In this situation, what is the ratio of the current when the STM tip is 0.400 nm above a surface feature to the current when the tip is 0.420 nm above the surface?

•37.69 An electron is trapped in a one-dimensional infinite well of width 2 nm. It starts in the $n = 4$ state, and then goes into the $n = 2$ state, emitting radiation with energy corresponding to the energy difference in the two states. What is the wavelength of the radiation?

•37.70 Two long, straight wires that lie along the same line have a separation at their tips of 2 nm. The potential energy of an electron in the gap is about 1 eV higher than it is in the conduction band of the two wires. Conduction-band electrons have enough energy to contribute to the current flowing in the wire. What is the probability that a conduction electron in one wire will be found in the other wire after arriving at the gap?

•37.71 Consider an electron that is confined to a one-dimensional infinite potential well of width $a = 0.10$ nm, and another electron that is confined by an infinite potential well to a three-dimensional cube with sides of length $a = 0.10$ nm. Let the electron confined to the cube be in its ground state. Determine the difference in energy and the excited state of the one-dimensional electron that minimizes the difference in energy with the three-dimensional electron.

•37.72 An electron with energy of 129 KeV is trapped in a potential well defined by an infinite potential at $x < 0$ and a potential barrier of finite height U_1 extending from $x = 529.2$ fm to $x = 2116.8$ fm (1 fm = $1 \cdot 10^{-15}$ m) as shown in the figure. It is found that the electron can be detected outside the potential well beyond the barrier with a probability of 10%. Calculate the height of the potential barrier.

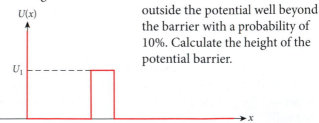

•37.73 Consider an electron that is confined to the xy-plane by a two-dimensional rectangular infinite potential well. The width of the well is w in the x-direction and $2w$ in the y-direction. What is the lowest energy that is shared by more than one distinct state, that is, where two different states have the same energy?

Atomic Physics

38

FIGURE 38.1 A sodium laser (nearly vertical yellow line) is fired into the night sky to provide a laser guide star for the adaptive optics of the telescopes of the Keck Observatory in Mauna Kea, Hawaii. The long exposure of the photograph make the stars appear as (very slightly) curved lines.

WHAT WE WILL LEARN

- The narrow lines observed in the spectra of some atoms result from the transitions of electrons from one state to another.

- The Bohr model of the atom, which is based on a quantization condition for the angular momentum, was an early success in explaining the observed line spectra in hydrogen. The orbital radii for the electrons, as well as the energy levels, obtained in the Bohr model are physically correct. However, the Bohr model does not obtain the correct values of the quantum numbers and is thus intrinsically flawed.

- To go beyond the Bohr model and solve for the correct electron wave functions of the hydrogen atom, the Schrödinger equation must be solved for the electrons in the Coulomb potential of the nucleus.

- The complete solution of the Schrödinger equation for the hydrogen atom is the product of the angular part and the radial part. The quantum numbers of these solutions are: the radial quantum number $n = 1, 2, 3, ...$; the orbital angular momentum quantum number $\ell = 0, 1, ..., n - 1$; and the magnetic quantum number $m = -\ell, ..., \ell$.

- The wave functions in other atoms can be understood in a manner similar to the hydrogen wave functions. The Pauli exclusion principle requires that each possible state be occupied by at most two electrons, one with spin down, and one with spin up.

- Lasers work on the basis of population inversion, in which electrons are lifted into a metastable state, and are subsequently stimulated to emit photons when the electrons make transitions between the metastable state and the ground state.

Lasers were first developed less than 50 years ago, but they have become common in many areas of daily life. You see them in Blu-ray, DVD, and CD players, in bar-code readers in stores, even in light shows at rock concerts and other events. They also have important applications in science, such as astronomy. Observatories fire lasers into the sky (Figure 38.1) and see how the light is distorted by atmospheric disturbance. Computers can then feed this information back to adaptive mirrors that have movable sections that correct for atmospheric motion, producing steady, sharp images.

Lasers work by producing photons of light as electrons move between energy levels in atoms, a topic we will study in this chapter. The concept of atomic energy levels was part of the Bohr model of the atom, which was one of the earliest attempts at applying quantum ideas to atomic structure. However, despite some outstanding successes, the Bohr model did not fully explain how particles function together within atoms; that explanation requires application of the quantum wave mechanics of Chapter 37. We will present an overview of this application, skipping over many of the mathematical details, and then show how the results underlie the operation of lasers and other modern devices.

Chapters 36 and 37 surveyed the phenomena that can be explained by the quantization hypothesis, and we solved some simple systems by following the laws imposed by quantum mechanics. Now that we have gained at least an introductory understanding of quantum mechanics, we have the tools that enable us to understand atoms.

38.1   Spectral Lines

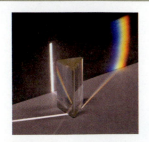

FIGURE 38.2 A prism decomposes white light into different colors (wavelengths).

Chapter 32 showed that the index of refraction of light in a prism depends on the wavelength. Figure 38.2 shows how white light (coming in from the left) is split into different wavelength components by a prism, due to the dependence of the index of refraction on wavelength.

Using a prism, a **spectrometer** can be constructed. This instrument displays the spectral components of light emitted by all kinds of objects. Figure 38.3 shows a kind of spectrometer often used in introductory physics laboratories. One arm of the spectrometer points toward a light source and provides a narrowly focused thin beam of light, which strikes the central prism and is then spectrally decomposed. The other spectrometer arm contains the ocular lenses and is used for observations. It rotates around a central pivot axis located under the center of the prism. The main idea in using the spectrometer is that we can measure the angle between the two arms and thus determine the angle of deflection of each color. Knowing the index of refraction of the prism and using Snell's law, a spectrometer allows a very precise determination of the wavelengths of light.

In the late 19th century, spectrometers were used in many laboratories to study gases. Trapping gases in electric discharge tubes allowed researchers to send a current through the tube, causing the gas to emit light. (The black box in the right part of Figure 38.3 holds the purple-glowing electric discharge tube with its power supply.) Incidentally, this way of generating light is the basis for neon lights.

A spectrometer pointed at the Sun will spectrally decompose light into a spectrum like the one shown in Figure 38.2—that is, an array of different colors. Measuring the intensity or spectral emittance of the different colors of the Sun's light makes it possible to measure the temperature of the Sun's surface, on the assumption that the Sun emits a blackbody spectrum. An expensive spectrometer is not needed to see that the Sun emits a broad spectrum of all colors; a child's toy prism can do the job.

However if you point your spectrometer at a discharge tube filled with pure hydrogen and look at the spectrum emitted by this gas, you are in for a huge surprise. Instead of a broad continuous spectrum, you will find a few thin lines of particular colors. At most you can see four lines: red (wavelength 656 nm, also historically called H-alpha), teal blue (wavelength 486 nm, H-beta), deep blue (434 nm, H-gamma), and finally violet (410 nm, H-delta). Other discrete lines appear in the hydrogen spectrum, but they lie outside the range of wavelengths accessible by the unaided human eye. Still, it is possible to determine the wavelengths of these other spectral lines, and this was done to very good precision in the 19th century. The lower part of Figure 38.4 shows where the hydrogen lines are located as a function of their wavelength λ. Shown are the four visible lines in their own colors. In addition, the invisible lines are indicated with thin white lines. These lines fall into distinct groups, named after their discoverers: the Lyman, Balmer, Paschen, and Brackett series. (Other series, not shown, were discovered at wavelengths above 2000 nm.)

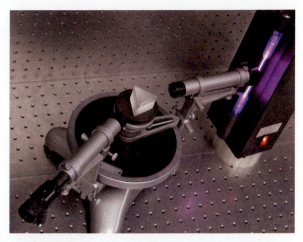

FIGURE 38.3 Spectrometer with prism.

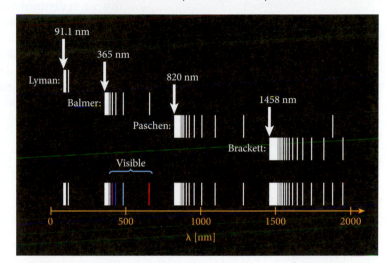

FIGURE 38.4 Simulated spectral lines of hydrogen in the spectrograph.

Purely based on studying patterns in the numbers for the wavelengths, the Swiss mathematician and hobbyist-physicist Johann Balmer (1825–1898) found in 1885 an empirical formula that predicted the wavelengths in the Balmer group of the hydrogen spectrum:

$$\lambda = (364.56 \text{ nm})\frac{n^2}{n^2 - 4}, \ n = 3, 4, 5, \dots . \tag{38.1}$$

Three years later, the Swedish physicist Johannes Rydberg (1854–1919) was able to generalize the Balmer formula in a way that also included all other line series (Lyman, Paschen, Brackett, etc.) in the hydrogen spectrum. The **Rydberg formula** for the wavelength in the hydrogen spectrum is

$$\frac{1}{\lambda} = R_H \left(\frac{1}{n_1^2} - \frac{1}{n_2^2} \right) \quad \text{with } n_1 < n_2. \tag{38.2}$$

The numbers n_1 and n_2 are simple integers; R_H is the **Rydberg constant** for hydrogen and has the dimension of an inverse length. The value of R_H is

$$R_H = 1.097373 \cdot 10^7 \ \text{m}^{-1}.$$

For $n_1 = 1$, $n_2 = 2,3,4, \dots, \infty$ we obtain the wavelengths of the Lyman series; for $n_1 = 2$, $n_2 = 3,4,5, \dots, \infty$ we obtain the Balmer series; and so on.

EXAMPLE 38.1 | Spectral Lines

PROBLEM
Figure 38.4 shows the wavelengths of the leftmost line in each of the first four series in the hydrogen spectrum. What are the corresponding values for the rightmost lines in each series?

SOLUTION
The wavelengths of the lines in the hydrogen spectrum are described by the Rydberg formula (equation 38.2),

$$\frac{1}{\lambda} = R_H \left(\frac{1}{n_1^2} - \frac{1}{n_2^2} \right).$$

Because $n_1 < n_2$, the term $1/n_1^2$ is always greater than $1/n_2^2$. For a fixed given value of n_1, an increasing value of n_2 means the subtraction of a smaller number $1/n_2^2$ from $1/n_1^2$ and thus an increase in $1/\lambda$. Therefore, the wavelength increases monotonically with increasing n_2.

Now that we understand this dependence, we can see that the leftmost and therefore smallest wavelength in each series must correspond to the largest value of n_2, which is $n_2 \sim \infty$. This is also the reason why the lines become so densely spaced at this end of the series: There is an infinite number of values of n_2 with almost exactly the same wavelength.

Conversely, the rightmost and therefore largest value of the wavelength in each band must correspond to the smallest possible value of n_2 in each series. Since we have the condition $n_1 < n_2$, this smallest possible value is $n_{2,min} = n_1 + 1$. Inserting this condition, we find for the maximum wavelength in each series:

Lyman series: $n_1 = 1$, $n_{2,min} = 1 + 1 = 2 \Rightarrow \dfrac{1}{\lambda_{max}} = R_H \left(1 - \dfrac{1}{4} \right) \Rightarrow \lambda_{max} = 121.5$ nm

Balmer series: $n_1 = 2$, $n_{2,min} = 2 + 1 = 3 \Rightarrow \dfrac{1}{\lambda_{max}} = R_H \left(\dfrac{1}{4} - \dfrac{1}{9} \right) \Rightarrow \lambda_{max} = 656$ nm

Paschen series: $n_1 = 3$, $n_{2,min} = 3 + 1 = 4 \Rightarrow \dfrac{1}{\lambda_{max}} = R_H \left(\dfrac{1}{9} - \dfrac{1}{16} \right) \Rightarrow \lambda_{max} = 1875$ nm

Brackett series: $n_1 = 4$, $n_{2,min} = 4 + 1 = 5 \Rightarrow \dfrac{1}{\lambda_{max}} = R_H \left(\dfrac{1}{16} - \dfrac{1}{25} \right) \Rightarrow \lambda_{max} = 4050$ nm

DISCUSSION
The maximum wavelength in the Balmer series corresponds to the red H-alpha line in the visible spectrum and is thus probably the most prominent line in the entire hydrogen spectrum. You can see that the highest values of the wavelengths in each of the first two series are lower than the lowest value in the next higher series. However, this changes with the Paschen series. Its longest wavelength line has a wavelength of 1875 nm, which is greater than all but a few wavelengths in the Brackett series, which has its minimum wavelength at a value of 1458 nm. The lines with the highest wavelengths in the Lyman, Balmer, and Paschen series are shown in Figure 38.4, whereas the highest wavelength of the Brackett series lies far outside the range of wavelengths displayed.

38.1 Self-Test Opportunity
Derive the values for the minimum wavelengths in each line series given in Figure 38.4.

38.1 In-Class Exercise
The number of lines in the Brackett series with wavelength greater than the maximum wavelength in the Paschen series is

a) 1.   d) 8.

b) 2.   e) ∞.

c) 4.

38.2 Self-Test Opportunity
Derive the formula originally given by Balmer for the wavelengths of the Balmer series (equation 38.1) from the more general formula given by Rydberg (equation 38.2).

Before we go on, let's emphasize the main point: Hydrogen gas can emit light when excited, but the light cannot have just any wavelength and only appears at well-defined wavelengths described by the Rydberg formula. Chapters 36 and 37, which discussed all kinds of quantized phenomena, lead us to look for an explanation for the discreteness of the hydrogen spectrum within quantum mechanics. In the next section, we describe Bohr's quantum description, which is valuable for its insight, its historical significance, and its limited success. Later in this chapter, the rigorous quantum mechanical analysis of the hydrogen atom will be presented.

Other elements in their gaseous state also show similar **line spectra** with discrete lines. In particular, some elements (such as lithium or sodium) have hydrogen-like spectra in their gaseous state. In general, however, the line spectra of other atoms are more complicated and the hydrogen atom has the simplest spectrum. The characteristic line spectra often serve as atomic "fingerprints," to detect the presence of specific chemical elements. Astronomers use this standard technique when they want to determine the elemental composition of stars, for example.

38.2 Bohr's Model of the Atom

Atoms consist of nuclei, which consist of positively charged protons and uncharged neutrons, surrounded by negatively charged electrons orbiting the nucleus and bound to it by the Coulomb interaction. In an electrically neutral atom, the number of protons is equal to the number of electrons. The ionization of an atom—that is, the removal of one or more electrons—causes the remaining ion to be positively charged. The typical size of an atom is on the order of $d \approx 10^{-10}$ m, and the typical size of the atomic nucleus is a factor of 10,000 smaller. Hydrogen—the most abundant element in the universe—is also the simplest atom in the universe, consisting of only one proton in the central nucleus with one electron orbiting it.

It is not unusual nowadays for children to learn most of these basic facts about atoms in elementary school. However, at the beginning the 20th century the atom was unknown territory, and it was not at all clear what its basic structure was. For example, one possible model of an atom was the "plum pudding" model. In this model, the entire atom was filled with positive charge, and the electrons were equally distributed over the entire atomic volume like raisins in a plum pudding. However, the 1909 scattering experiments of Ernest Rutherford, Hans Geiger, and Ernest Marsden brought clarity and led to the currently accepted model. The physics of the atomic nucleus will be discussed in Chapters 39 and 40.

After Rutherford's experiments, physicists believed that electrons orbit around the nucleus, not unlike how the Earth and the other planets orbit around the Sun. Let's write down what this means for the simplest atom, hydrogen. Chapter 9 showed that motion in a circular orbit requires a constant centripetal acceleration. Because the gravitational force acting on the electron due to the nucleus is so weak, the centripetal force is provided by the Coulomb force (see Chapter 21). This leads us to

$$k\frac{e^2}{r^2} = \mu\frac{v^2}{r} \tag{38.3}$$

where $k = 8.98755 \cdot 10^9$ N m^2/C^2 is Coulomb's constant. This equation simply restates Newton's Second Law, with the force given by the Coulomb force and the acceleration as the centripetal acceleration.

A remark on the notation of μ for the mass is in order. Up to now, we have been using the letter m for the mass of the electron. However, the electron in a hydrogen atom is not moving around the proton; instead, both are moving around a common center of mass. We can incorporate this effect by introducing the **reduced mass** μ as

$$\mu = \frac{mM}{m+M} \tag{38.4}$$

where $m = 9.10938215(45) \cdot 10^{-31}$ kg is the electron mass and $M = 1.672621637(83) \cdot 10^{-27}$ kg is the proton mass. This reduced mass has the numerical value $\mu = 9.10442 \cdot 10^{-31}$ kg. Because the proton mass is a factor of 1836 times bigger than the electron mass, the term $M/(M + m)$ is very close to 1, and thus $\mu \approx m$ to 1 part in 2000. Thus, we could still use the notation m for the mass in this case, and it would be close enough for most purposes. However, we will use the reduced mass μ from now on. An advantage in using this correct μ in our formulation is that historically, people have introduced a quantum number m in the solution to the hydrogen problem, and we want to avoid confusion. Therefore, from now on μ stands for the (reduced) mass of the electron, and we will save the letter m for later use.

This classical picture of a mini planetary system of electrons in orbit around a central nucleus, which plays the role of the Sun in our Solar System, has immense appeal and gives rise to a symmetry between the macro-cosmos and the micro-cosmos that just rings true. Right away, though, it runs into a fatal conceptual flaw: An accelerated charge radiates away energy. An electron moving on a classical circular orbit and experiencing constant centripetal acceleration would quickly lose energy and spiral into the nucleus, thus destroying the atom. Niels Bohr addressed this flaw in 1913.

Quantization of Orbital Angular Momentum

In 1913 the Danish physicist Niels Bohr (1885–1962) made an *ad hoc* assumption that the angular momentum of the electron in its orbit around the nucleus can assume only discrete values. Chapter 10 introduced the angular momentum vector of a point particle as $\vec{L} = \vec{r} \times \vec{p}$. Bohr's idea was to require that the only possible stable electron orbits should be those with angular momentum in integer multiples of $\hbar$ (Planck's constant divided by 2π):

$$L = |\vec{r} \times \vec{p}| = r\mu v = n\hbar, \quad \text{with} \quad n = 1, 2, 3, \dots. \tag{38.5}$$

Why demand this condition? Dimensionally it works out, because both angular momentum and Planck's constant have units of $\text{kg m}^2 \text{ s}^{-1} = \text{J s}$. However, of course, this fact alone is not sufficient. In the following discussion, we show that the quantization of angular momentum also leads to the quantization of energy. In addition, if the electron energy is quantized, then the electron cannot radiate arbitrarily small amounts of energy and spiral into the nucleus.

Let's first work out the consequences and predictions of the Bohr postulate and then return to the discussion of the meaning of the postulate and the validity of the model. To begin the calculation for the **Bohr model of the hydrogen atom**, start with equation 38.3 for the centripetal force and multiply both sides with a factor μr^3. This leads to:

$$k\frac{e^2}{r^2} = \mu\frac{v^2}{r} \Rightarrow r\mu ke^2 = \mu^2 v^2 r^2.$$

On the right-hand side, we can now see that $\mu^2 v^2 r^2$ is the square of the angular momentum. Using Bohr's quantization condition (equation 38.5), this is equal to $n^2\hbar^2$. Therefore,

$$r\mu ke^2 = \mu^2 v^2 r^2 = n^2\hbar^2.$$

Solving this for the orbital radius r results in

$$r = \frac{\hbar^2}{\mu ke^2}n^2 \equiv a_0 n^2. \tag{38.6}$$

This equation gives us the allowed radii in the Bohr model of the hydrogen atom. They are proportional to the square of the quantum number n, and the proportionality constant a_0 is called the **Bohr radius,**

$$a_0 = \frac{\hbar^2}{\mu ke^2} = \frac{\left(1.05457 \cdot 10^{-34} \text{ J s}\right)^2}{\left(9.10442 \cdot 10^{-31} \text{ kg}\right)\left(8.98755 \cdot 10^9 \text{ N m}^2/\text{C}^2\right)\left(1.60218 \cdot 10^{-19} \text{ C}\right)^2}$$

$$a_0 = 5.295 \cdot 10^{-11} \text{ m} = 0.05295 \text{ nm} = 0.5295 \text{ Å}.$$

Here we write the Bohr radius in units of nm as well as the unit Å, with $1 \text{ Å} = 10^{-10} \text{ m} = 0.1 \text{ nm}$. Inserting this result for the Bohr radius back into equation 38.3, we find for the speed v,

$$v = \sqrt{\frac{ke^2}{\mu a_0 n^2}} = \frac{1}{n}\sqrt{\frac{\left(8.988 \cdot 10^9 \text{ N m}^2/\text{C}^2\right)\left(1.602 \cdot 10^{-19} \text{ C}\right)^2}{\left(9.104 \cdot 10^{-31} \text{ kg}\right)\left(5.295 \cdot 10^{-11} \text{ m}\right)}} = \frac{1}{n}2.188 \cdot 10^6 \text{ m/s} = \frac{1}{n}0.007297c.$$

This speed is 0.73% of the speed of light for $n = 1$ and falls monotonically for the higher orbits. Therefore, the nonrelativistic approximation that we are using is justified. The total energy of the electron in orbit is the sum of its potential and kinetic energies,

$$E = \frac{1}{2}\mu v^2 - k\frac{e^2}{r} = -\frac{1}{2}k\frac{e^2}{r} = -\frac{1}{2}k\frac{e^2}{a_0 n^2} = -E_0\frac{1}{n^2}. \tag{38.7}$$

The second step of this equation uses the general result for satellite motion (see Chapter 12) that the kinetic energy is exactly half of the magnitude of the potential energy, so the total energy is half of the potential energy. (This relation is often referred to as the virial theorem.) The constant E_0 can be calculated by inserting the values of the constants,

$$E_0 = \frac{ke^2}{2a_0} = \frac{\left(8.988 \cdot 10^9 \text{ N m}^2/\text{C}^2\right)\left(1.602 \cdot 10^{-19} \text{ C}\right)^2}{2\left(5.295 \cdot 10^{-11} \text{ m}\right)} = 2.18 \cdot 10^{-18} \text{ J} = 13.6 \text{ eV}.$$

Thus, the allowed energies for electrons in orbit around the nucleus of the hydrogen atom are $E(n = 1) = -13.6$ eV, $E(n = 2) = -3.40$ eV, $E(n = 3) = -1.51$ eV, and so on, approaching zero energy from below as $n \rightarrow \infty$. Combining equations 38.6 and 38.7 gives $E = -E_0 a_0/r$. Figure 38.5 shows E plotted versus r. We indicate in this figure that only certain values of E and r, corresponding to n being an integer in equations 38.6 and 38.7, are allowed.

In many textbooks, the Bohr radius is defined in terms of the mass of the electron, m_e, rather than the reduced mass, μ, of the electron in the hydrogen atom:

$$a_{0,m_e} = \frac{\hbar^2}{m_e ke^2}.$$

The Bohr radius has the value $a_{0,m_e} = 5.292 \cdot 10^{-11}$ m using this definition.

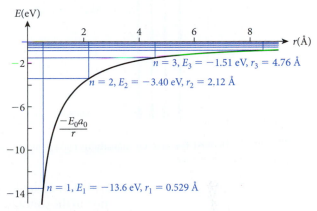

FIGURE 38.5 Energies and radii for allowed electron orbits in the Bohr model of hydrogen.

Spectral Lines in the Bohr Model

In the Bohr model, an electron cannot radiate away small amounts of energy and thus spiral into the nucleus, because of the condition for quantization of angular momentum. However, transitions between states are still allowed. Bohr postulated that an electron in a higher energy state with quantum number n_2 could "jump" to a lower energy state with quantum number n_1 and would emit a photon with energy equal to the difference in the energies between the two states. The photon energy is related to its frequency via Planck's relation $E = hf$: $E_{n_2} = E_{n_1} + hf$. For light, the frequency is related to the wavelength via $f = c/\lambda$, with $c =$ the speed of light, giving

$$E_{n_2} = E_{n_1} + \frac{hc}{\lambda}. \tag{38.8}$$

The formula for the energies (equation 38.7) then gives

$$-\frac{ke^2}{2a_0}\frac{1}{n_2^2} = -\frac{ke^2}{2a_0}\frac{1}{n_1^2} + \frac{hc}{\lambda} \Rightarrow$$

$$\frac{1}{\lambda} = \frac{ke^2}{2hca_0}\left(\frac{1}{n_1^2} - \frac{1}{n_2^2}\right).$$

This equation is structurally identical to equation 38.2. Furthermore, the product of the constants in this equation evaluates to

$$\frac{ke^2}{2hca_0} = \frac{\left(8.988 \cdot 10^9 \text{ N m}^2/\text{C}^2\right)\left(1.602 \cdot 10^{-19} \text{ C}\right)^2}{2\left(6.626 \cdot 10^{-34} \text{ J s}\right)\left(2.998 \cdot 10^8 \text{ m/s}\right)\left(5.295 \cdot 10^{-11} \text{ m}\right)} = 1.097 \cdot 10^7 \text{ m}^{-1}.$$

This agrees with the value of the Rydberg constant, which was determined from the experimental data, to four significant figures. That is, we can identify

$$R_\text{H} = \frac{ke^2}{2hca_0}.$$

Given $a_0 = \hbar^2/\mu k e^2$, the Rydberg constant can also be written

$$R_H = \frac{ke^2}{2hc(\hbar^2/\mu ke^2)} = \frac{\mu k^2 e^4}{4\pi c\hbar^3}.$$

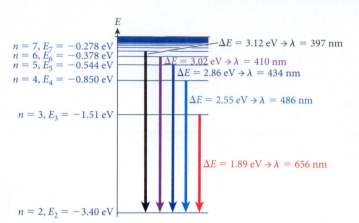

FIGURE 38.6 The first few electron transitions corresponding to the Balmer series in hydrogen.

In other words, the Bohr model is able to explain the structure of the hydrogen line spectrum and can be used to derive the experimentally found value of the Rydberg constant from the fundamental constants of the electron mass m (contained in the reduced mass μ), charge quantum e, Planck's constant h (equal to $2\pi\hbar$), the speed of light c, and Coulomb's constant k. This was properly celebrated as an astounding success and lent considerable credence to the seemingly *ad hoc* assumption of a quantized angular momentum of the electron in its orbit around the central nucleus. Transitions between energy levels can be depicted in an **energy-level diagram**, as is done in Figure 38.6 for the Balmer series.

However, at its heart the Bohr model assumes that the electron is a classical point particle. The model was refined several times, but after Louis de Broglie postulated matter waves in 1923 (see Chapter 36) and Schrödinger published his wave equation in 1926 (see Chapter 37), it became clear that the Bohr model had to be replaced by a proper quantum mechanical theory of the hydrogen atom. In addition to the conceptual problems of a classical point particle, the Bohr model was also flawed in that it postulated an orbital angular momentum of $1\,\hbar$ for the lowest energy state—the ground state—of hydrogen, in contradiction to experimental evidence that points to orbital angular momentum of zero for this state. Nevertheless, the Bohr model is a very instructive first attempt at solving the hydrogen problem, and its impressive success in explaining line spectra hints that the real solution to the hydrogen atom must somehow be close to what the Bohr model postulates.

38.3 Hydrogen Electron Wave Function

If we want to improve on the Bohr model of the hydrogen atom, we have to return to what we learned about quantum mechanics in Chapter 37 and solve the Schrödinger equation for the electron in a Coulomb potential. To find the bound states of the electron, we need to solve the time-independent Schrödinger equation and find the energy eigenvalues E_n. We hope our solution will resemble equation 38.7, $E_n = -13.6/n^2$ eV, because this prediction of the Bohr model was found to be in agreement with experimental data.

A word of caution before we start on this task: Part of the mathematics involved in solving this problem is tedious and perhaps as advanced as any we will present. This is the bad news. The good news is that this problem is very instructive and allows us to get a glimpse of many deep features of quantum mechanics and of ways to illuminate them. The math will not be worked out in all its details, and sometimes the outcome of a calculation will simply be stated. Do not become discouraged as we study this essential quantum system. It not only will help us understand the hydrogen atom, but gives us insight into the entire periodic table of the elements.

The potential in the Schrödinger equation is the Coulomb potential, $U(\vec{r}) = -ke^2/r$. Note that this potential depends only on the radial distance to the origin (where the nucleus is located), and not on the angular direction of the electron relative to the nucleus. The hydrogen atom is an object in the real world and thus exists in three-dimensional space. According to Chapter 37, the Schrödinger equation in three-dimensional space must be written as

$$-\frac{\hbar^2}{2\mu}\frac{\partial^2\psi}{\partial x^2} - \frac{\hbar^2}{2\mu}\frac{\partial^2\psi}{\partial y^2} - \frac{\hbar^2}{2\mu}\frac{\partial^2\psi}{\partial z^2} + U\psi = -\frac{\hbar^2}{2\mu}\nabla^2\psi + U\psi = E\psi.$$

The Laplacian operator ∇^2 in Cartesian coordinates is $\nabla^2 = \partial^2/\partial x^2 + \partial^2/\partial xy^2 + \partial^2/\partial z^2$. This operator appeared in Chapter 37 in the section on multidimensional infinite wells. It is a seemingly straightforward generalization of the second derivative d^2/dx^2, which appears in the one-dimensional Schrödinger equation.

Now the coordinate system must be chosen. The potential energy term U depends only on the radial coordinate r, so it is advantageous to use spherical coordinates r, θ, ϕ, as shown in Figure 38.7. Then the Schrödinger equation reads

$$-\frac{\hbar^2}{2\mu}\nabla^2\psi(r,\theta,\phi)-k\frac{e^2}{r}\psi(r,\theta,\phi)=E\psi(r,\theta,\phi),\qquad(38.9)$$

where the Coulomb potential is used for the potential U.

In spherical coordinates, however, there is a price to pay: The Laplacian operator looks much more complicated:

$$\nabla^2\psi=\frac{1}{r}\frac{\partial^2}{\partial r^2}\left(r\psi\right)+\frac{1}{r^2\sin\theta}\frac{\partial}{\partial\theta}\left(\sin\theta\frac{\partial}{\partial\theta}\psi\right)+\frac{1}{r^2\sin^2\theta}\frac{\partial^2}{\partial\phi^2}\psi.\qquad(38.10)$$

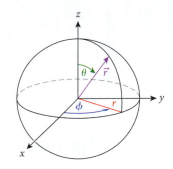

FIGURE 38.7 Definition of the spherical coordinates.

Is this worth the effort? The answer certainly would be yes if the wave function does not depend on the angular coordinates but only on the radial one. Therefore, we try this method first and look for spherically symmetric solutions.

Spherically Symmetric Solutions

Motivated by the Bohr model's success, we first investigate if there are spherically symmetric solutions. If these exist, they can only be a function of the radial coordinate r. So let us work with this assumption right now and see where it leads us. The Laplacian operator then simplifies greatly because the derivatives with respect to θ and ϕ vanish, leaving us

$$\nabla^2\psi(r)=\frac{1}{r}\frac{\partial^2}{\partial r^2}\left(r\psi(r)\right)=\frac{1}{r}\frac{d^2}{dr^2}\left(r\psi(r)\right).$$

Notice that we replaced the partial derivative by the conventional derivative, because the wave function in this case depends on only one variable, r. Thus, for the Schrödinger equation,

$$-\frac{\hbar^2}{2\mu}\frac{1}{r}\frac{d^2}{dr^2}\left(r\psi(r)\right)-k\frac{e^2}{r}\psi(r)=E\psi(r).\qquad(38.11)$$

The solution of this differential equation is tedious, but the result is very interesting. We find that, similar to the particle in a box or the harmonic oscillator, the energy can assume only certain values of E_n, given by

$$E_n=-\frac{\mu k^2e^4}{2\hbar^2}\frac{1}{n^2}\text{ with }n=1,2,3,....$$

This is exactly the form of the energy eigenvalues we obtained in the Bohr quantization procedure (equation 38.7) with $E_0=\mu k^2e^4/2\hbar^2=13.6$ eV, an energy unit sometimes also called 1 Rydberg.

The wave functions that correspond to the lowest values of the quantum number n are

$$\psi_1(r)=A_1e^{-r/a_0}$$

$$\psi_2(r)=A_2\left(1-\frac{r}{2a_0}\right)e^{-r/2a_0}\qquad(38.12)$$

$$\psi_3(r)=A_3\left(1-\frac{2r}{3a_0}+\frac{2r^2}{27a_0^2}\right)e^{-r/3a_0},$$

where the subscript on ψ represents the corresponding value of n. Here A_1, A_2, A_3 are normalization constants that can be determined from the condition that the integral of the absolute square of the wave function is 1,

$$\int\left|\psi_n(r)\right|^2d^3r=\int\limits_0^\infty 4\pi r^2\left|\psi_n(r)\right|^2dr=1.$$

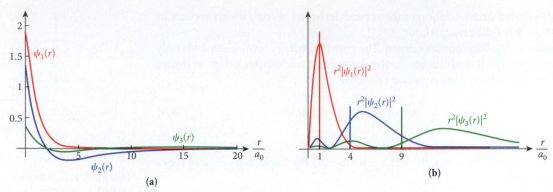

FIGURE 38.8 Spherically symmetric solutions to the hydrogen Schrödinger equation for the lowest three values of the radial quantum number. (a) Wave functions; (b) weighted probability densities. The vertical lines in (b) represent the Bohr model predictions.

Figure 38.8a shows these solutions and Figure 38.8b shows their probability densities, weighted with the factor r^2 to account for the effect that the volume element increases with rising radius. The solutions for the lower three values of the radial quantum number n are shown as a function of the radial coordinate divided by the Bohr radius a_0. Vertical lines indicate the locations of the semiclassical Bohr orbits that correspond to these values of n. For $n = 1$, the maximum of the probability density for the wave function corresponds to the Bohr orbit, whereas this is not the case for $n > 1$.

EXAMPLE 38.2 | Normalization of the Hydrogen Wave Function

PROBLEM

The hydrogen electron wave function corresponding to $n = 1$ has the form given in equation 38.12, $\psi_1(r) = A_1 e^{-r/a_0}$. Using the condition that the integral of the absolute square of the wave function is 1, determine A_1.

SOLUTION

The condition that the integral of the absolute square of the wave function is 1 gives us

$$\int \left|\psi_n(r)\right|^2 d^3r = \int_0^\infty 4\pi r^2 \left|\psi_n(r)\right|^2 dr = 1.$$

Putting in our particular wave function leads to

$$\int_0^\infty 4\pi r^2 \left|A_1 e^{-r/a_0}\right|^2 dr = 4\pi A_1^2 \int_0^\infty r^2 e^{-2r/a_0} dr.$$

The definite integral can be expressed as $\int_0^\infty x^2 e^{-cx} dx = 2/c^3$. Taking $c = 2/a_0$, we get

$$4\pi A_1^2 \int_0^\infty r^2 e^{-2r/a_0} dr = 4\pi A_1^2 \left[2\left(\frac{a_0}{2}\right)^3\right] = \pi a_0^3 A_1^2 = 1.$$

We can solve this equation for A_1:

$$A_1 = \frac{1}{\sqrt{\pi a_0^3}} = \frac{1}{\sqrt{\pi} a_0^{3/2}}.$$

Thus, the hydrogen electron wave function corresponding to $n = 1$ is

$$\psi_1(r) = \frac{1}{\sqrt{\pi} a_0^{3/2}} e^{-r/a_0}.$$

Angular Momentum

Chapter 37 introduced quantum operators for the momentum and kinetic energy, where the classical value of the momentum is replaced by a constant times the derivative in coordinate space, $-i\hbar d/dx$. In three-dimensional space, the momentum operator is a vector, for which each individual component is a partial derivative in the respective direction. In Cartesian coordinates, this is written as

$$\mathbf{p}\psi(x,y,z) = -i\hbar\left(\frac{\partial}{\partial x}, \frac{\partial}{\partial y}, \frac{\partial}{\partial z}\right)\psi(x,y,z) \equiv -i\hbar\vec{\nabla}\psi(x,y,z).$$

This derivative operator is called the *gradient,* where we use the common shorthand notation $\vec{\nabla}$. In spherical coordinates the gradient is given by

$$\vec{\nabla}\psi(r,\theta,\phi) = \left(\frac{\partial}{\partial r}, \frac{1}{r}\frac{\partial}{\partial\theta}, \frac{1}{r\sin\theta}\frac{\partial}{\partial\phi}\right)\psi(r,\theta,\phi).$$

Then the equation for the momentum operator becomes

$$\mathbf{p}\psi(x,y,z) = -i\hbar\vec{\nabla}\psi(r,\theta,\phi) = -i\hbar\left(\frac{\partial}{\partial r}, \frac{1}{r}\frac{\partial}{\partial\theta}, \frac{1}{r\sin\theta}\frac{\partial}{\partial\phi}\right)\psi(r,\theta,\phi).$$

Now the angular momentum operator can be written by simply replacing the momentum vector by the gradient operator we just introduced:

$$\mathbf{L}\psi = -i\hbar\vec{r}\times\vec{\nabla}\psi. \tag{38.13}$$

(Remember, classically the angular momentum is $\vec{L} = \vec{r}\times\vec{p}$.) According to our rules for quantum measurements, we can now calculate the expectation value of the angular momentum operator,

$$\langle L\rangle = \int \psi^*\mathbf{L}\psi d^3r = -i\hbar\int \psi^*\vec{r}\times\vec{\nabla}\psi d^3r. \tag{38.14}$$

It is particularly interesting to look at the outcome of this measurement of the angular momentum for the case of spherically symmetric wave functions, that is, wave functions that depend on only the radial coordinate r, and not on any angular coordinates θ or ϕ. Then we obtain

$$\langle L\rangle = \int \psi^*(r)\mathbf{L}\psi(r)d^3r = -i\hbar\int \psi^*(r)\vec{r}\times\vec{\nabla}\psi(r)d^3r = 0. \tag{38.15}$$

DERIVATION 38.1 / Angular Momentum Expectation Value

To show that equation 38.15 is true, we use the gradient operator in spherical coordinates. Applying it to our spherically symmetric wave function results in

$$\vec{\nabla}\psi(r) = \left(\frac{\partial}{\partial r}, \frac{1}{r}\frac{\partial}{\partial\theta}, \frac{1}{r\sin\theta}\frac{\partial}{\partial\phi}\right)\psi(r) = \hat{r}\frac{\partial}{\partial r}\psi(r),$$

because the partial derivatives in the θ and ϕ directions are both zero when the wave function does not depend on either of these angular variables. We have again used the notation of the unit vector in the radial direction as $\hat{r}$. In terms of the same unit vector, we can write the vector $\vec{r}$ as $\vec{r} = r\hat{r}$. Now our vector product in the integral (equation 38.15) simplifies to

$$-i\hbar\int \psi^*(r)\vec{r}\times\vec{\nabla}\psi(r)d^3r = -i\hbar\int \psi^*(r)r\hat{r}\times\hat{r}\frac{\partial}{\partial r}\psi(r)d^3r.$$

For any vector, the vector product with itself has the value zero. Thus, $\hat{r}\times\hat{r} = 0$ and the entire integral has the value zero.

This is a very general result that holds for any spherically symmetric wave function: A wave function that does not depend on the angular coordinates θ or ϕ must have an expectation value of zero for the angular momentum.

Our spherically symmetric solutions $\psi_n(r)$ for the hydrogen atom, obtained in the previous section, depend only on the radius, so all have angular momentum zero. It is not permissible to interpret the quantum number n as one of angular momentum, as we did in the Bohr model, despite the fact that the energy eigenvalues of our quantum mechanical spherically symmetric solutions reproduce exactly the discrete energy values of the Bohr model. This fact implies that *the Bohr model is fatally flawed*: We cannot start with a classical particle in orbit around a central nucleus, demand quantization of angular momentum, and then obtain consistent results. Despite its astounding success in explaining the hydrogen line spectra in great detail and to great precision, the Bohr model is wrong, and its success is accidental.

Full Solution

Does the Schrödinger equation for hydrogen have solutions that are not spherically symmetric and that have nonzero angular momentum? The answer is yes. Here we will only sketch how to get to these solutions. Typically, the mathematics needed to arrive at these results is lengthy and complicated. However, it is also instructive and something to look forward to in a more advanced course, where this topic will be covered in greater detail. In the following discussion, we will introduce many different functions. Keep in mind that these are not meant to be memorized. The goal here is for you to understand the general features of the solutions, not their exact form. Even the most seasoned professionals in the field do not know all the exact functions by heart and look them up in tables when they are needed.

Separation of Variables

We start with an assumption that we used successfully in finding solutions for the time-dependent Schrödinger equation in one spatial dimension, that of the separation of variables. To achieve this, we assume that the full wave function can be written as a product of three functions, each of which is a function of only one variable:

$$\psi(r,\theta,\phi) = f(r)g(\theta)h(\phi). \tag{38.16}$$

We insert this trial solution into the Schrödinger equation 38.9 to find

$$-\frac{\hbar^2}{2\mu}\nabla^2\Big(f(r)g(\theta)h(\phi)\Big) - k\frac{e^2}{r}\Big(f(r)g(\theta)h(\phi)\Big) = E\Big(f(r)g(\theta)h(\phi)\Big).$$

Now the action of the Laplacian operator (in spherical coordinates, equation 38.10) on this product of functions must be evaluated:

$$\nabla^2\Big(f(r)g(\theta)h(\phi)\Big) = \frac{1}{r}\frac{\partial^2}{\partial r^2}\Big(r\big(f(r)g(\theta)h(\phi)\big)\Big)$$

$$+ \frac{1}{r^2\sin\theta}\frac{\partial}{\partial\theta}\left(\sin\theta\frac{\partial}{\partial\theta}\big(f(r)g(\theta)h(\phi)\big)\right)$$

$$+ \frac{1}{r^2\sin^2\theta}\frac{\partial^2}{\partial\phi^2}\big(f(r)g(\theta)h(\phi)\big)$$

$$= g(\theta)h(\phi)\frac{1}{r}\frac{\partial^2}{\partial r^2}\big(rf(r)\big)$$

$$+ \frac{f(r)h(\phi)}{r^2\sin\theta}\frac{\partial}{\partial\theta}\left(\sin\theta\frac{\partial}{\partial\theta}g(\theta)\right)$$

$$+ \frac{f(r)g(\theta)}{r^2\sin^2\theta}\frac{\partial^2}{\partial\phi^2}h(\phi).$$

Inserting this result into the Schrödinger equation and multiplying both sides of the equation by the product $-2\mu r^2/\hbar^2 f(r)g(\theta)h(\phi)$, we arrive at

$$\frac{r}{f(r)}\frac{\partial^2}{\partial r^2}\big(rf(r)\big)+\frac{1}{g(\theta)\sin\theta}\frac{\partial}{\partial\theta}\left(\sin\theta\frac{\partial}{\partial\theta}g(\theta)\right)$$
$$+\frac{1}{h(\phi)\sin^2\theta}\frac{\partial^2}{\partial\phi^2}h(\phi)+\frac{2\mu r^2}{\hbar^2}\left(E+k\frac{e^2}{r}\right)=0.$$

Now we can see that some terms depend only on the radial variable r, and not on the angular variables θ, ϕ, while other terms depend on θ, ϕ, but not on r. We rearrange the preceding equation to get all r-dependent terms on the left-hand side and all others on the right-hand side:

$$\frac{r}{f(r)}\frac{\partial^2}{\partial r^2}\big(rf(r)\big)+\frac{2\mu r^2}{\hbar^2}\left(E+k\frac{e^2}{r}\right)=$$

$$-\frac{1}{g(\theta)\sin\theta}\frac{\partial}{\partial\theta}\left(\sin\theta\frac{\partial}{\partial\theta}g(\theta)\right)-\frac{1}{h(\phi)\sin^2\theta}\frac{\partial^2}{\partial\phi^2}h(\phi). \qquad (38.17)$$

This equation can be true only if each side is equal to the same constant. We make a choice that may look strange at first, but we call this constant $\ell(\ell+1)$.

Radial Part

The choice of the integration constant $\ell(\ell+1)$ results in the radial equation (left-hand side of equation 38.17):

$$\frac{r}{f(r)}\frac{d^2}{dr^2}\big(rf(r)\big)+\frac{2\mu r^2}{\hbar^2}\left(E+k\frac{e^2}{r}\right)=\ell(\ell+1).$$

(Again, we replaced the partial derivatives with conventional ones because the function f depends on only one variable.) If we multiply this equation by $-f(r)\hbar^2/2\mu r^2$, this equation has the form

$$-\frac{\hbar^2}{2\mu}\frac{1}{r}\frac{d^2}{dr^2}\big(rf(r)\big)-k\frac{e^2}{r}f(r)+\frac{\ell(\ell+1)\hbar^2}{2\mu r^2}f(r)=Ef(r). \qquad (38.18)$$

Compare this equation with the Schrödinger equation for the spherically symmetric solution (equation 38.11), and you see that the only difference is the term $\ell(\ell+1)\hbar^2/2\mu r^2$ that plays the role of an additional potential, in addition to the Coulomb potential. (For $\ell=0$ we recover the spherically symmetric solution.) What is this additional effective potential term?

An argument from the orbital motion of a classical particle in a central potential may help. To work out our classical argument, let's start with energy conservation, $\frac{1}{2}\mu v^2+U(r)=E$. (Remember, this is the classical equivalent of the Schrödinger equation.) Now we can split the velocity into radial and tangential parts and get $\frac{1}{2}\mu v^2=\frac{1}{2}\mu v_r^2+\frac{1}{2}\mu v_t^2=\frac{1}{2}\mu v_r^2+\frac{1}{2}\mu(\omega r)^2$. Further note that the angular momentum for the case of a point particle in circular motion is $L=\mu rv_t=\mu r^2\omega$. This definition also holds for all motion in a central potential, not just circular motion, and the angular momentum is a conserved quantity. Thus, we can write $\frac{1}{2}\mu(\omega r)^2=\frac{1}{2}L^2/\mu r^2$. Therefore, our classical equation for energy conservation in this case reads $\frac{1}{2}\mu v_r^2+U(r)+L^2/2\mu r^2=E$.

The term $\ell(\ell+1)\hbar^2/2\mu r^2$ now seems plausible as a term originating from conservation of angular momentum, if we are allowed to substitute the quantum mechanical $\ell(\ell+1)\hbar^2$ for the classical L^2. Later we will see that these solution functions are also the eigenfunctions for the quantum mechanical operator corresponding to L^2 with eigenvalue $\ell(\ell+1)\hbar^2$; so this substitution actually works out just as advertised. Now you can begin to understand why the seemingly arbitrary choice for the integration constant $\ell(\ell+1)$ was made. The term $\ell(\ell+1)\hbar^2/2\mu r^2$ represents the quantum mechanical angular momentum barrier in the rotating system, and we will soon see that ℓ is an integer. We say "barrier" because this term is positive and increases as r decreases, just like a potential energy barrier. Figure 38.9 shows the effective potential term from equation 38.18 for various values of ℓ.

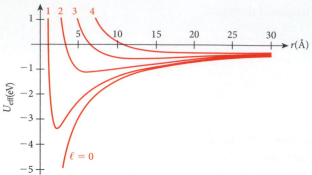

FIGURE 38.9 Effective potential for (from bottom to top) angular momentum quantum numbers $\ell = 0, 1, 2, 3,$ and 4.

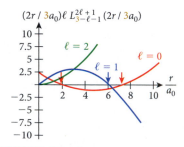

FIGURE 38.10 Polynomial terms in the radial wave function for $n = 3$.

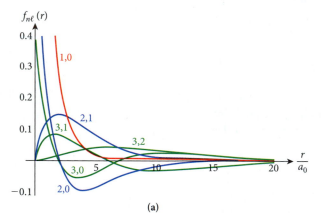

(a)

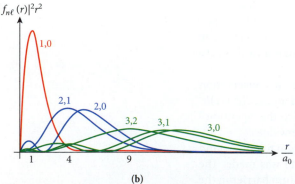

(b)

FIGURE 38.11 (a) Radial wave functions for the lowest values of the quantum numbers n and ℓ; (b) corresponding weighted probability densities.

What are the solutions to the differential equation for the radial function, equation 38.18? In general, they are products of exponential functions and polynomials. For very large r, the exponential function dominates the asymptotic behavior of the solutions. The solutions depend on two quantum numbers. First is the quantum number n, which was present in the spherically symmetric solutions studied earlier. Second is the quantum number ℓ that we just introduced in the separation of variables, which we motivated as the quantum number of angular momentum. (The next section will give further evidence for this assignment.) For the sake of completeness, we write out the solutions for the radial functions, but you are not expected to memorize them.

$$f_{n\ell}(r) = \sqrt{\frac{4(n-\ell-1)!}{a_0^3 n^4 (n+1)!}} e^{-r/na_0} \left(\frac{2r}{na_0}\right)^{\ell} L_{n-\ell-1}^{2\ell+1}\left(\frac{2r}{na_0}\right). \quad (38.19)$$

These solutions exist only for integers $n > \ell$ and $\ell \geq 0$. Therefore, for a given value of n the quantum number ℓ can have the values

$$0 \leq \ell \leq n-1.$$

The term $L_{n-\ell-1}^{2\ell+1}\left(2r/na_0\right)$ in equation 38.19 is an associated Laguerre polynomial. These polynomials have been tabulated in reference books, and you can look them up. Their exact form is not so important for us, except for the following general considerations. The product $\left(2r/na_0\right)^{\ell} L_{n-\ell-1}^{2\ell+1}\left(2r/na_0\right)$ is a polynomial of rank $n-1$ for any value of ℓ. This polynomial has $n-\ell-1$ positive roots and ℓ roots at $r = 0$. An example for $n = 3$ is shown in Figure 38.10, with the positive roots of the three possible polynomials indicated by the colored arrows.

Instead of examining equation 38.19 further in abstract terms, it is perhaps more instructive to write down the radial solutions for the lowest values of n and ℓ:

$$f_{10}(r) = 2a_0^{-3/2} e^{-r/a_0}$$

$$f_{20}(r) = \frac{1}{\sqrt{2}} a_0^{-3/2} \left(1 - \frac{r}{2a_0}\right) e^{-r/2a_0}$$

$$f_{21}(r) = \frac{1}{2\sqrt{6}} a_0^{-3/2} \left(\frac{r}{a_0}\right) e^{-r/2a_0}$$

$$f_{30}(r) = \frac{2}{3\sqrt{3}} a_0^{-3/2} \left(1 - \frac{2r}{3a_0} + \frac{2r^2}{27a_0^2}\right) e^{-r/3a_0}$$

$$f_{31}(r) = \frac{8}{27\sqrt{6}} a_0^{-3/2} \left(\frac{r}{a_0} - \frac{r^2}{6a_0^2}\right) e^{-r/3a_0}$$

$$f_{32}(r) = \frac{4}{81\sqrt{30}} a_0^{-3/2} \left(\frac{r^2}{a_0^2}\right) e^{-r/3a_0}.$$

Comparing these solutions to those of the spherically symmetric problem (equation 38.12) shows that for each value of n, the solutions with $\ell = 0$ correspond to the spherically symmetric solutions.

Because the lowest values of n correspond to the lowest values of the energy, the radial functions listed here are overwhelmingly the most important for the hydrogen atom.

If we plot the radial functions for the lowest allowed values of the quantum numbers n and ℓ, an interesting picture emerges. Figure 38.11a shows the six wave functions listed above. It is important to note that the solutions for a radial quantum number n are polynomials of rank $n - 1$, multiplied by an exponentially falling term that assures that the wave functions approach 0 for very large values of the radius r. All wave functions with $\ell > 0$ have a value of 0 at $r = 0$; only those with $\ell = 0$ assume a finite nonzero value at $r = 0$. (In the figure, it may appear that the $\ell = 0$ solutions diverge as they approach $r = 0$, but this is misleading. If the scale were expanded, we would see them turn over, because the exponential function has a value of 1 at $r = 0$.)

Figure 38.11b shows the probability of finding an electron in a state with quantum numbers n and ℓ at a distance r from the nucleus of the hydrogen atom. The wave function f_{10}—the only wave function for $n = 1$—is colored red, and the two possible wave functions for $n = 2$ (f_{20} and f_{21}) blue. The three possible wave functions for $n = 3$, which are f_{30}, f_{31}, and f_{32}, are shown in green. For the group of wave functions with a given value of n, all weighted probability densities plotted in part b peak at a similar distance from the origin. They form a **shell.** This appearance of shells is a universal phenomenon in atomic and in nuclear physics and is a unique quantum effect. We will return to the concept of shells repeatedly in our discussions of these subjects.

An interesting side note: The radial wave functions $f_{n,n-1}$ all peak at values of r that correspond to the radii predicted by the Bohr model for the corresponding quantum number n.

Angular Part

Let's return to the Schrödinger equation in spherical coordinates (equation 38.17): If the left-hand side is equal to $\ell(\ell + 1)$, so is the right-hand side.

$$\ell(\ell+1)=-\frac{1}{g(\theta)\sin\theta}\frac{\partial}{\partial\theta}\left(\sin\theta\frac{\partial}{\partial\theta}g(\theta)\right)-\frac{1}{h(\phi)\sin^2\theta}\frac{\partial^2}{\partial\phi^2}h(\phi). \tag{38.20}$$

This equation is still a function of the two variables θ and ϕ. We can rearrange the terms and obtain an equation for which all θ-dependent terms are on the left-hand side, and all ϕ-dependent terms are on the right-hand side. This is accomplished by multiplying both sides of equation 38.20 by $\sin^2\theta$ and then moving the first term on the right to the left:

$$\ell(\ell+1)\sin^2\theta+\frac{\sin\theta}{g(\theta)}\frac{\partial}{\partial\theta}\left(\sin\theta\frac{\partial}{\partial\theta}g(\theta)\right)=-\frac{1}{h(\phi)}\frac{\partial^2}{\partial\phi^2}h(\phi).$$

Again, we have separated the variables, and for this equation to be true, both sides need to be equal to the same constant. Let us call this constant m^2. (Remember, we use μ for the mass in this chapter, not m.)

The two resulting ordinary differential equations for the angular variables are

$$\frac{d^2}{d\phi^2}h(\phi)=-m^2h(\phi) \tag{38.21}$$

$$\ell(\ell+1)\sin^2\theta+\frac{\sin\theta}{g(\theta)}\frac{d}{d\theta}\left(\sin\theta\frac{d}{d\theta}g(\theta)\right)=m^2. \tag{38.22}$$

Note that the form of the differential equation 38.21 is exactly the same as that of a simple harmonic oscillator, which we studied extensively in Chapter 14. Perhaps you can then guess the solution to equation 38.21 right away: It is a linear combination of $\sin(m\phi)$ and $\cos(m\phi)$. Alternatively, following the commonly used complex notation convention, it is

$$h(\phi)=Ae^{im\phi}. \tag{38.23}$$

Here A is a normalization constant, which we can deal with later when we worry about the overall normalization of the wave function. What is more important, though, is that the same wave function must be obtained when 2π is added to the angle ϕ, because this corresponds to one complete 360° rotation in the xy-plane. This implies that

$$e^{im\phi}=e^{im(\phi+2\pi)}\Rightarrow e^{2\pi im}=1\Rightarrow m=0,\pm1,\pm2,.... $$

Thus, we have found that our integration constant m must be an integer.

The solution to equation 38.22 for θ is not so easy, and we simply state it here:

$$g(\theta)=BP_\ell^m(\cos\theta)=B(-1)^{|m|}\sin^{|m|/2}\theta\left(\frac{d}{d(\cos\theta)}\right)^{|m|}P_\ell(\cos\theta). \tag{38.24}$$

Here B is another normalization constant. Just as for the constant A above, we ignore it for now. The functions P_ℓ^m are called associated Legendre functions, and the functions P_ℓ are Legendre polynomials,

$$P_\ell(x) \equiv \frac{1}{2^\ell\,\ell!}\left(\frac{d}{dx}\right)^\ell (x^2-1)^\ell \text{ and } P_\ell^m(x) = (-1)^m \left(1-x^2\right)^{m/2}\left(\frac{d}{dx}\right)^m P_\ell(x)$$

for integer non-negative values of ℓ. The first few Legendre polynomials are

$$P_0(x) = 1$$

$$P_1(x) = \frac{1}{2}\frac{d}{dx}\left(x^2-1\right) = x$$

$$P_2(x) = \frac{1}{8}\frac{d^2}{dx^2}\left(x^2-1\right)^2 = \frac{1}{2}\left(3x^2-1\right)$$

$$P_3(x) = \frac{1}{48}\frac{d^3}{dx^3}\left(x^2-1\right)^3 = \frac{1}{2}\left(5x^3-3x\right)$$

$$P_4(x) = \frac{1}{384}\frac{d^4}{dx^4}\left(x^2-1\right)^4 = \frac{1}{8}\left(35x^4-30x^2+3\right)$$

$$P_5(x) = \frac{1}{3840}\frac{d^5}{dx^5}\left(x^2-1\right)^5 = \frac{1}{8}\left(63x^5-70x^3+15x\right).$$

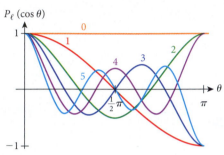

$P_\ell(\cos\theta)$

FIGURE 38.12 Legendre polynomials for $\ell = 1, 2, 3, 4,$ and 5.

and are shown in Figure 38.12.

You can see that a Legendre polynomial P_ℓ is a polynomial of rank ℓ. If you take the nth-derivative of a polynomial of rank ℓ, with $n > \ell$, then you obtain zero. This means that an upper limit exists for the integer $|m|$:

$$|m| \le \ell \Rightarrow -\ell \le m \le \ell.$$

Therefore, for any value of ℓ there are $2\ell + 1$ possible values of m. The associated Legendre functions $P_\ell^m(\cos\theta)$ with the lowest values of ℓ are

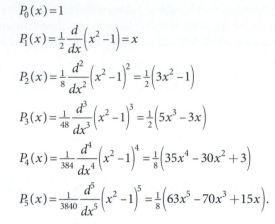

	$m=0$	$m=1$	$m=2$	$m=3$
$\ell=0$	1			
$\ell=1$	$\cos\theta$	$-\sin\theta$		
$\ell=2$	$\frac{1}{2}\left(3\cos^2\theta-1\right)$	$-3\cos\theta\sin\theta$	$3\sin^2\theta$	
$\ell=3$	$\frac{1}{2}\left(5\cos^3\theta-3\cos\theta\right)$	$-\frac{3}{2}\sin\theta\left(5\cos^2\theta-1\right)$	$15\cos\theta\sin^2\theta$	$-15\sin^3\theta.$

The expressions for negative values of m do not have to be written separately, because only the absolute value of the quantum number m appears in the defining equation 38.24.

The products of the functions $g(\theta)h(\phi)$, properly normalized, are called the **spherical harmonics** Y_ℓ^m,

$$Y_\ell^m(\theta,\phi) = \sqrt{\frac{(\ell-|m|)!}{(\ell+|m|)!}\frac{(2\ell+1)}{4\pi}}P_\ell^m(\cos\theta)e^{im\phi}. \tag{38.25}$$

The spherical harmonics describe the angular dependence of the electron wave functions for the hydrogen atom. They have a complex phase factor $e^{im\phi}$, but are otherwise real-valued. In Figure 38.13, we plot the absolute values of the spherical harmonics. The absolute values of these spherical harmonics are symmetric under rotation about the z-axis, but in general, for $m \neq 0$ this is not the case for the real part or for the imaginary part.

What else is special about the spherical harmonics? It can be shown (we will not do this here) that they are also eigenfunctions of the operator of the square of the angular

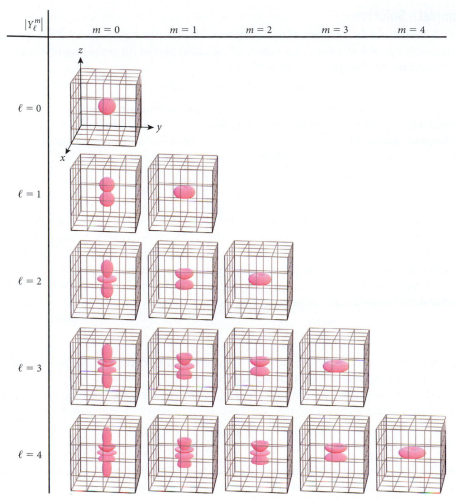

FIGURE 38.13 Absolute value of the spherical harmonics for the lowest values of the angular momentum quantum numbers. Each box shown extends from −1 to 1 in each Cartesian direction.

momentum, $\mathbf{L}^2$, with eigenvalues $\ell(\ell + 1)\hbar^2$ and of the operator of the projection of the angular momentum on the z-axis, $\mathbf{L}_z$, with eigenvalue $m\hbar$:

$$\mathbf{L}^2 Y_\ell^m(\theta,\phi) = \ell(\ell+1)\hbar^2 Y_\ell^m(\theta,\phi) \tag{38.26}$$

$$\mathbf{L}_z Y_\ell^m(\theta,\phi) = m\hbar Y_\ell^m(\theta,\phi). \tag{38.27}$$

Thus, the quantum number ℓ is a measure of the absolute value of the total orbital angular momentum—that is, the length of the angular momentum vector—and the quantum number m measures the length of the projection of the angular momentum vector along the z-axis.

You might now be able to guess what the operators $\mathbf{L}^2$ and $\mathbf{L}_z$ look like in spherical coordinates. They are

$$\mathbf{L}^2 = -\hbar^2 \frac{1}{\sin\theta} \frac{\partial}{\partial\theta}\left(\sin\theta \frac{\partial}{\partial\theta}\right) - \hbar^2 \frac{1}{\sin^2\theta} \frac{\partial^2}{\partial\phi^2}$$

$$\mathbf{L}_z = -i\hbar \frac{\partial}{\partial\phi}.$$

Comparing these two expressions to equations 38.20 and 38.21 shows why equations 38.26 and 38.27 must be true. The construction of the spherical harmonics as solutions to the angular part of a rotationally invariant Schrödinger equation means that they must be eigenfunctions to the operators of the square of the angular momentum and of the projection of the angular momentum on the symmetry axis.

Complete Solution

Let's collect the different parts of our complete solution. We have separated the variables and constructed the solutions as products of the radial part of the wave function (equation 38.19) and the angular part (equation 38.25):

$$\psi_{n\ell m}(r,\theta,\phi) = f_{n\ell}(r)Y_{\ell m}(\theta,\phi). \tag{38.28}$$

The quantum numbers of these solutions are: the **radial quantum number** $n = 1,2,3,...$, the **orbital angular momentum quantum number** $\ell = 0,1, ...,n-1$, and the so-called **magnetic quantum number** $m = -\ell,...,\ell$. The energy eigenvalue corresponding to a particular solution eigenfunction depends only on the radial quantum number,

$$E_n = -\frac{1}{n^2}E_0 = -\frac{1}{n^2}\frac{\mu k^2 e^4}{2\hbar^2}. \tag{38.29}$$

The solutions for a given value of the radial quantum number n all peak at a similar distance from the center, approximately given by the radius of the corresponding semiclassical Bohr orbit $r_n = n^2 a_0 = n^2\hbar^2/\mu k e^2$. (For the wave functions with maximum possible angular momentum quantum number $\ell = n-1$, this is an exact statement.) Thus, the wave functions for a given radial quantum number n form a shell. The shells are sometimes (especially in chemistry) labeled with capital letters according to their radial quantum number:

letter: K L M N O P Q ...

$n =$ 1   2   3   4   5   6   7   ....

Therefore, the lowest energy shell is the K-shell, and the shells with successively higher radial quantum number n follow alphabetically.

Within a given shell, all wave functions have the same energy. Traditionally, they are labeled with letters according to their orbital angular momentum quantum number:

letter: s p d e f g h ...

$\ell =$ 0   1   2   3   4   5   6   ....

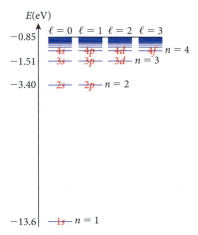

FIGURE 38.14 Wave function quantum numbers, energies, and label assignments.

Here the first two angular momentum states receive the letters s (monopole) and p (dipole), and from $\ell = 2$ (quadrupole) on the angular momentum states follow the alphabet successively, starting with the letter d. (Why this ordering? No real reason, except the historical ways in which the nomenclature was developed.) Thus, a $4d$ state implies that the radial quantum number is 4 and the orbital angular momentum quantum number is 2. The energies and level assignments are shown in Figure 38.14.

We can also plot the solution wave functions. We have the choice of displaying the wave function, its real part or imaginary part, or its absolute value. All wave functions with $m = 0$ are real, and for these we can easily plot the wave function—for example, in slices through space. Figure 38.15 shows the wave functions $\psi_{n\ell m}$ for the lowest values of the quantum numbers n and ℓ with $m = 0$ in a slice through coordinate space, in the xz-plane with $y = 0$. Blue colors imply positive values of the wave function, and red colors indicate negative values. Yellow represents values close to zero; a scale is given on the right side. In each case, a region between $\pm 30\,a_0$ is shown in each direction, x and z. As you can see, all s-states ($\ell = 0$ states) appear as concentric circles in this contour plot, because they are all spherically symmetric. The states with nonzero angular momentum show beautiful patterns of alternating regions of positive and negative values.

We can see regularities by looking at the patterns that emerge. As n increases, an outer shell becomes populated, while the structure of the inner shell wave function with the same angular momentum gets preserved and squeezed down, with a spherical node (yellow rings) separating it from the neighboring shells. All p-waves show twofold symmetries, as expected from dipolar shapes. In the same way, the d-waves exhibit fourfold quadrupole structure, and f-waves show a characteristic sixfold symmetry.

Finally, we list the wave functions for the innermost shells—those with the lowest values for the radial quantum numbers and thus lowest energies. In the $n = 1$ shell, there is only one possible wave function, and it is in an s-state:

$$\psi_{100}(r,\theta,\phi) = \frac{1}{\sqrt{\pi}a_0^{3/2}}e^{-r/a_0}. \tag{38.30}$$

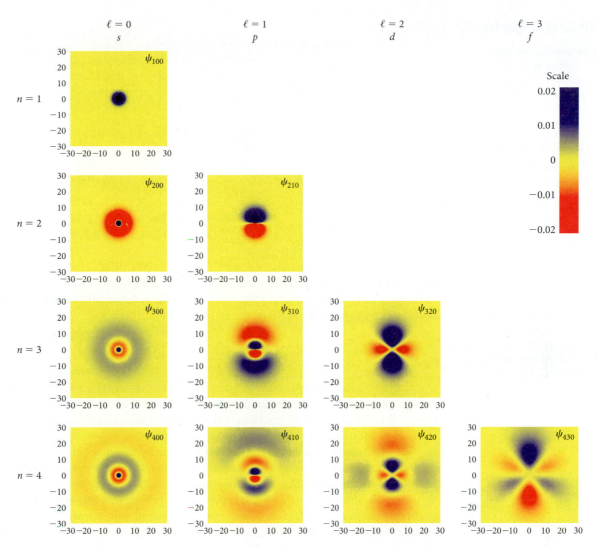

FIGURE 38.15 Hydrogen atom electron wave functions in the xz-plane for y = 0, $\psi_{n\ell m}$ (x, 0, z). The coordinates in the xz-plane are displayed in multiples of the classical Bohr radius a_0. The scale on the right indicates how the colors in the plots represent the value of $\psi_{n\ell m}$.

Note that the value of the energy eigenvalue that corresponds to this wave function is the lowest possible, $E_1 = -13.6$ eV. An electron that resides in this state cannot radiate away a photon to go to a lower energy state. Therefore, the state with this energy is the ground state of the electron in the hydrogen atom, and equation 38.30 is the ground state wave function. You can compare equation 38.30 with the results of Example 38.2.

In the $n = 2$ shell, one wave function in the s-state is possible, plus three p-waves, for a total of four wave functions, all of them corresponding to an energy eigenvalue of $E_2 = -3.40$ eV:

$$\psi_{200}(r, \theta, \phi) = \frac{1}{2\sqrt{2\pi}a_0^{3/2}} e^{-r/2a_0} \left(1 - \frac{r}{2a_0}\right)$$

$$\psi_{210}(r, \theta, \phi) = \frac{1}{2\sqrt{2\pi}a_0^{3/2}} e^{-r/2a_0} \left(\frac{r}{2a_0}\right) \cos\theta$$

$$\psi_{21\pm1}(r, \theta, \phi) = \mp \frac{1}{4\sqrt{\pi}a_0^{3/2}} e^{-r/2a_0} \left(\frac{r}{2a_0}\right) \sin\theta e^{\pm i\phi}.$$

In the $n = 3$ shell, there is one wave function in the s-subshell, three wave functions in the p-subshell, and five in the d-subshell, for a total of $1 + 3 + 5 = 9$ wave functions, all with energy $E_3 = -1.51$ eV:

$$\psi_{300}(r, \theta, \phi) = \frac{1}{3\sqrt{3\pi}a_0^{3/2}} e^{-r/3a_0} \left(1 - \frac{2r}{3a_0} + \frac{2r^2}{27a_0^2}\right)$$

Continued—

38.3 Self-Test Opportunity

Now that we have written down the explicit form of the solutions, sketch the result of plotting $\psi_{2,1,1}(r,\theta,\phi)$ or $\psi_{2,1,-1}(r,\theta,\phi)$ in the way of Figure 38.15.

38.2 In-Class Exercise

How many wave functions are possible in the $n = 5$ shell?

a) 9   d) 21

b) 14   e) 25

c) 16

38.4 Self-Test Opportunity

Verify that equation 38.30 is indeed a solution to the hydrogen problem by inserting it into the Schrödinger equation. For added difficulty, try the same exercise for one of the functions with $n = 2$.

$$\psi_{310}(r,\theta,\phi)=\frac{2\sqrt{2}}{27\sqrt{\pi}a_0^{3/2}}e^{-r/3a_0}\left(\frac{r}{a_0}-\frac{r^2}{6a_0^2}\right)\cos\theta$$

$$\psi_{31\pm1}(r,\theta,\phi)=\mp\frac{2}{27\sqrt{\pi}a_0^{3/2}}e^{-r/3a_0}\left(\frac{r}{a_0}-\frac{r^2}{6a_0^2}\right)\sin\theta e^{\pm i\phi}$$

$$\psi_{320}(r,\theta,\phi)=\frac{1}{81\sqrt{6\pi}a_0^{3/2}}e^{-r/3a_0}\left(\frac{r^2}{a_0^2}\right)\left(3\cos^2\theta-1\right)$$

$$\psi_{32\pm1}(r,\theta,\phi)=\mp\frac{1}{81\sqrt{\pi}a_0^{3/2}}e^{-r/3a_0}\left(\frac{r^2}{a_0^2}\right)\cos\theta\sin\theta e^{\pm i\phi}$$

$$\psi_{32\pm2}(r,\theta,\phi)=\frac{1}{162\sqrt{\pi}a_0^{3/2}}e^{-r/3a_0}\left(\frac{r^2}{a_0^2}\right)\sin^2\theta e^{\pm 2i\phi}.$$

In general, we can have $2\ell+1$ wave functions with different m for orbital angular momentum ℓ. In addition, since we can have $n-1$ different orbital angular momentum states for a given radial quantum number n, the total number $N(n)$ of possible wave functions in a given shell with radial quantum number n is

$$N(n)=\sum_{\ell=0}^{n-1}\left(2\ell+1\right)=n^2. \tag{38.31}$$

Does our procedure of separation of variables ensure that we have found *all* solutions? We will not prove this, but the answer is yes, because the set of solutions that we have found form a complete basis for all functions in this space. We will not elaborate on this statement; this will be left to an advanced course on quantum physics.

We add here a postscript to the discussion of the hydrogen atom. This system is surely the best-studied system in all of quantum physics, and one of the few for which an exact solution is possible. The differential equations that result, as well as their solutions, show an incredible amount of structure. It is perhaps surprising that this arguably simplest real quantum system exhibits such richness, and perhaps even more surprising that we can construct a mathematical description that captures this complexity—which is visible, for example, in the plots of Figure 38.15. The fact that the natural world arranges itself in this way and then can be captured by our mathematical constructions seems absolutely miraculous. We hope that the somewhat tedious mathematical derivations did not obscure the beauty that underlies this physical system and the theory that describes it.

38.4 Other Atoms

Now that we have a good understanding of the wave functions, energies, and possible transitions between levels for the hydrogen atom, we can ask if we can explain other atoms as well. What do we need to change in our formalism to explain other atoms? First, the charge Z of the nucleus changes. For hydrogen, we have $Z = 1$, but for higher values of Z our potential needs to change to

$$U(r)=k\frac{Ze^2}{r}.$$

To be electrically neutral, an atom with Z protons also must have Z electrons. These electrons will reside in the lowest energy state available to them. It is easy to see into which state the first electron can be put: It is the $1s$ state. We can generalize equation 38.29 from the treatment of the hydrogen atom to see that the energy of this first electron is

$$E_1=-\frac{\mu k^2 Z^2 e^4}{2\hbar^2}. \tag{38.32}$$

We arrive at this result by simply replacing e^2 by Ze^2 in the Schrödinger equation. It is a useful exercise to examine the mathematics for the hydrogen atom and to convince yourself that equation 38.32 indeed holds. (In equation 38.32 the value of μ is the reduced mass, calculated using equation 38.4 with M now the mass of the nucleus.)

Where do we place the second electron? Now the answer is not so straightforward, because any given electron will also interact with all other electrons of the same atom, which partially serves to shield the charge Z of the nucleus. This problem cannot be solved with the exact analytical methods developed earlier. It needs approximation methods that are beyond the scope of this book. However, we can still gain some qualitative understanding with the tools at our disposal.

The general idea of radial shells and subshells with fixed angular momentum quantum numbers, which we developed for the hydrogen atom, remains valid for other atoms. The partial screening of the Coulomb potential of the nucleus acts differently on different levels. The states with $\ell = 0$ have probability distributions that are peaked at the origin. For these states, we have a low probability that another electron lies closer to the origin, partially shielding the central potential. For states with larger values of the angular momentum quantum number, however, shielding plays a more important role. This leads to the distribution of energy levels sketched in Figure 38.16. At first glance, this figure looks similar to Figure 38.14, which showed the plot for the hydrogen atom. Now, however, the energy scale is different by a factor of Z^2. The locations of energy levels in the absence of shielding are shown by the dashed blue lines, while the proper locations of the levels are indicated by the dark blue solid lines. For a few levels we have also included small black arrows that show the shifts. What is the main difference between Figure 38.16 and Figure 38.14? In Figure 38.16, different angular momentum quantum numbers ℓ for the same values of the main quantum number n result in different energy values, whereas they all have the same value for the hydrogen atom. This is due to the other electrons partially shielding the nuclear charge in atoms with more than one electron. If we sort the levels by increasing energy, starting with the lowest one, we find the progression $1s$, $2s$, $2p$, $3s$, $3p$, $4s$, $3d$, $4p$, $5s$, $4d$, $5p$, $6s$, $5d$, $4f$, $6p$, $7s$, $6d$, and $5f$.

Electrons have a spin quantum number $\frac{1}{2}$. Thus, $2(\frac{1}{2}) + 1 = 2$ electrons can occupy each wave function with its set of fixed quantum numbers—one electron with "spin up" and the other with "spin down." This fact is due to the Pauli exclusion principle. With the Pauli exclusion principle and Figure 38.16 in hand, we can determine in which order the electrons are distributed over the energy levels, as we fill each level in Figure 38.16 with two electrons successively.

Appendix D contains the ground state electron configurations for all elements. We use the conventional shorthand notation of writing the number of electrons in a given level as a superscript right after the notation for the level. For example, fluorine ($Z=9$) is listed with the ground state electron configuration of $1s^2 2s^2 2p^5$. This means that two electrons occupy each of the $1s$ and $2s$ levels, and that the remaining 5 electrons are in the $2p$ state. The electron configuration of aluminum ($Z=13$) is listed as $[Ne]3s^2 3p$. This means that the occupation of the $1s$, $2s$, and $2p$ levels in aluminum is the same as in neon (that is, fully occupied), and that aluminum has two additional electrons in the $3s$ state, plus one additional electron in the $3p$ state.

When all electrons for a given atom are in the lowest possible energy states available to them, the atom is in its ground state. Note that as Z increases, electrons sometimes fill an s state in a higher n shell rather than a state with lower n and higher angular momentum. (For example, see $Z = 19$, 37, and 55.) This occurs because the angular momentum barrier has made the lower n state correspond to a higher electron energy.

Just as in the hydrogen atom, electrons in other atoms can be excited into higher energy states by photons. However, the resulting line spectra are generally much more complicated than that of the hydrogen atom. It is considerably easier to answer the question of how much energy it takes to remove the least-bound electron from an atom. The process of removing an electron from an atom is called **ionization,** so we want to know the single-electron **ionization energy.** To lift the least-bound electron from its state with energy lower than zero to an energy just above zero, the ionization energy is the same as the magnitude of the energy of the least-bound occupied level in an atom. Figure 38.17 presents this information for every known atom (except for astatine, $Z = 85$, for which the ionization energy has not been measured yet). Now the question is: Can we explain the regularities of the ionization energies in this figure?

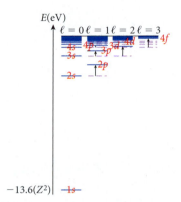

FIGURE 38.16 Energy level scheme in atoms with several electrons.

38.3 In-Class Exercise

The element 111 was discovered in 1994 and named Roentgenium in 2004. It exists for only a few seconds before it decays. The electron configuration of Roentgenium has not been determined yet. What would you predict it to be?

a) $[Xe]\ 4f^{14}5d^{10}6s^2 6p^6$

b) $[Rn]\ 4f^{14}5d^{10}6s^2 6p^6$

c) $[Rn]\ 5f^{14}6d^9 7s^2$

d) $[Xe]\ 5f^{14}6d^9 7s^2$

FIGURE 38.17 Single-electron ionization energies for all elements.

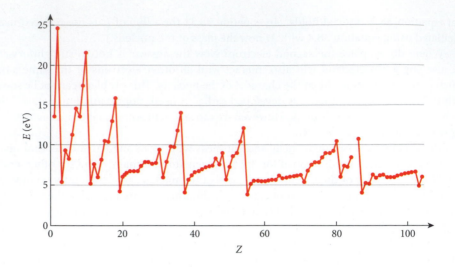

FIGURE 38.17 Single-electron ionization energies for all elements.

Let's start with hydrogen. The single electron of hydrogen resides in the 1s state in the hydrogen ground state. Its energy is –13.6 eV, so it takes +13.6 eV to liberate this single electron.

Helium: Helium has two electrons, and they can both be accommodated in the 1s state. With $Z = 2$, the nuclear Coulomb potential is twice as strong as that of the hydrogen atom. The presence of the other electron partially screens this potential, but the second electron in the helium atom is still bound more deeply than the single electron in hydrogen. Experimentally, we find a very large value of 24.6 eV for the ionization energy of helium, the largest value for any element. What does this mean for the chemical properties of helium? It is very difficult to remove an electron from helium and add it to any other element to form some kind of chemical bond. Conversely, it is also not possible to add another electron to the $n = 1$ shell in helium, because it is already completely filled with two electrons. An additional electron would have to be added to the $n = 2$ shell and thus be much more weakly bound to helium. Therefore, helium is chemically completely inert, and is called a **noble gas.**

Lithium: Because $Z = 3$, lithium atoms have three electrons as well. The first two reside in the 1s state and completely fill the $n = 1$ level, just as in helium. The third electron needs to go in the next higher state, the 2s state. From Figure 38.11b, we see that the radial wave function of the 2s state is localized farther away from the center of the atom. We then expect it to be a much weaker bond, and indeed, we find an ionization energy of only 5.39 eV for lithium. Consequently, lithium acts as an excellent electron donor in chemical reactions.

Beryllium: This element has four protons and four electrons. The fourth electron also fits into the 2s state, filling this angular momentum subshell completely. However, remember that for $n = 2$, we have angular momentum quantum numbers of 0 and 1. Thus, beryllium is not a closed-shell atom like helium. The ionization energy of beryllium is 9.32 eV, much higher than that of lithium, but also much lower than that of helium.

Boron through Neon: Now we populate the 2p subshell, which can hold $2(2\ell+1) = 2(2 \cdot 1+1) = 6$ electrons. As we add the first electron into the 2p subshell, it is alone in this subshell. Again examining Figure 38.11b for guidance, we see that the 2s and 2p states have similar average distance of the radial wave functions from the center. Therefore, the ionization energy in boron ($Z = 5$) is slightly lower than in beryllium, but not nearly as low as in lithium. Adding more electrons to the 2p subshell increases the ionization energy successively. Thus, it is increasingly difficult to remove an electron from the elements carbon, nitrogen, oxygen, and fluorine, so they are not good electron donors in chemical reactions. However, they become better electron receptors as we add more electrons to the 2p subshell. In particular, fluorine, with only one electron missing from an otherwise closed $n = 2$ shell, is chemically very aggressive. Neon ($Z = 10$) completes the 2p subshell as well as the $n = 2$ shell. Neon thus has chemical properties very similar to those of helium—it is also chemically inert and is a so-called noble gas. Because the radial wave functions in the $n = 2$ shell are on average separated farther from the center than those in the $n = 1$ shell, the electrons in this shell are expected to experience less attraction to the central nucleus and thus have lower ionization energies. Comparing the values for helium (24.6 eV) and neon (21.6 eV), this expectation is borne out.

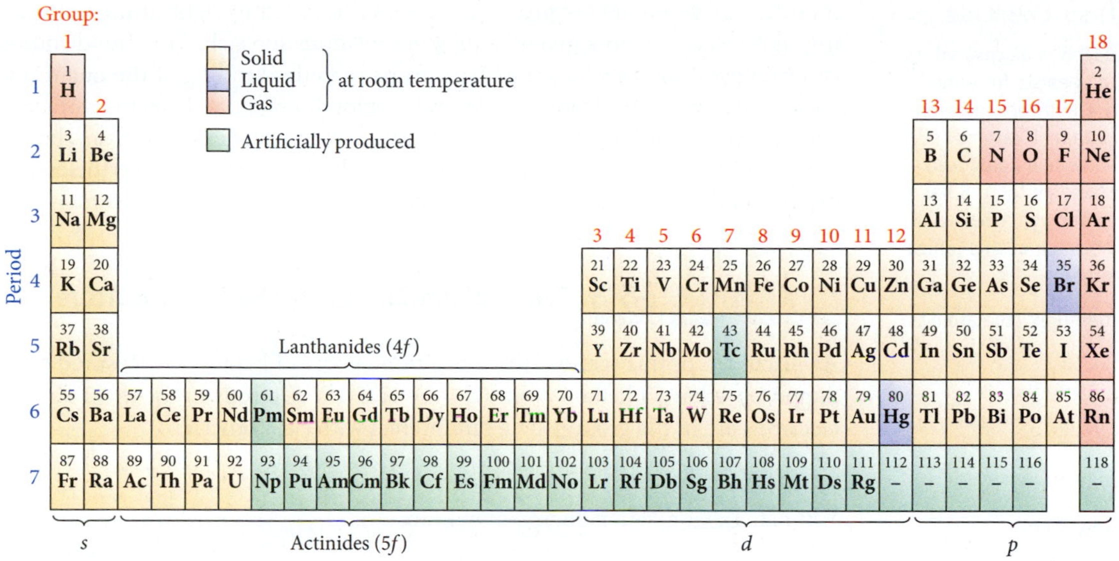

FIGURE 38.18 Periodic table of the elements.

Group:

Legend:
- Solid
- Liquid } at room temperature
- Gas
- Artificially produced

Period	1	2	3	4	5	6	7	8	9	10	11	12	13	14	15	16	17	18
1	1 H																	2 He
2	3 Li	4 Be											5 B	6 C	7 N	8 O	9 F	10 Ne
3	11 Na	12 Mg											13 Al	14 Si	15 P	16 S	17 Cl	18 Ar
4	19 K	20 Ca	21 Sc	22 Ti	23 V	24 Cr	25 Mn	26 Fe	27 Co	28 Ni	29 Cu	30 Zn	31 Ga	32 Ge	33 As	34 Se	35 Br	36 Kr
5	37 Rb	38 Sr	39 Y	40 Zr	41 Nb	42 Mo	43 Tc	44 Ru	45 Rh	46 Pd	47 Ag	48 Cd	49 In	50 Sn	51 Sb	52 Te	53 I	54 Xe
6	55 Cs	56 Ba	71 Lu	72 Hf	73 Ta	74 W	75 Re	76 Os	77 Ir	78 Pt	79 Au	80 Hg	81 Tl	82 Pb	83 Bi	84 Po	85 At	86 Rn
7	87 Fr	88 Ra	103 Lr	104 Rf	105 Db	106 Sg	107 Bh	108 Hs	109 Mt	110 Ds	111 Rg	112 –	113 –	114 –	115 –	116 –		118 –

Lanthanides (4f): 57 La, 58 Ce, 59 Pr, 60 Nd, 61 Pm, 62 Sm, 63 Eu, 64 Gd, 65 Tb, 66 Dy, 67 Ho, 68 Er, 69 Tm, 70 Yb

Actinides (5f): 89 Ac, 90 Th, 91 Pa, 92 U, 93 Np, 94 Pu, 95 Am, 96 Cm, 97 Bk, 98 Cf, 99 Es, 100 Fm, 101 Md, 102 No

s   Actinides (5f)   d   p

Some systematic trends in the progression of ionization energies are becoming evident. What looked like a wild and irregular zigzag curve in Figure 38.17 begins to exhibit structure. It is this structure that underlies the **periodic table of chemical elements** (Figure 38.18). In the periodic table, elements that fill the $n = 1$ shell are in the first row, and those that fill the $n = 2$ shell are in the second row. When an electron is added to a new shell, this corresponds to a new row in the periodic table. In each row, the electrons are added one by one, going from left to right. The elements with completely filled subshells, so that any additional electron would have to be placed into the next-higher major shell, end up in the rightmost column; those that start a new shell are in the leftmost column.

The first two rows in the periodic table are identical to the first two major shells, but in general, a given row in the periodic table is not identical to a major shell. Already in the third row, the 3d electrons are missing. Because their energy levels are higher than those of the 4s electrons, the 4s electrons are added before the 3d. Because the 4s electrons start a new row, the atoms scandium through zinc, in which electrons fill the 3d subshell, are members of the fourth row of the periodic table, not the third.

Elements in a given column have a similar electron configuration in their outermost shell. The completely filled inner shells have very little influence on the atom's interactions with other atoms or photons. The electrons in the outermost shells of the atoms, called the **valence electrons,** primarily determine the atom's chemical behavior. Therefore, elements in a given column are chemically similar in many ways.

Note that the lowest row in Figure 38.18 consists mostly of artificially produced atoms. Atoms with atomic number larger than 92 (uranium) are not stable and thus not found in nature. They can be created in the lab and exist for some time before they decay. As a general rule of thumb, the higher the atomic number is, the shorter the artificially produced atom exists before it decays. Elements 113, 115, 116, and 118 were discovered only in the first decade of the 21st century, and each exists for only a few milliseconds to seconds before decay. (Chapter 40 investigates the reasons for these short lifetimes of atoms with high atomic numbers.)

Using the arrangement of the elements in the periodic table, we can display the ionization energies for all elements up to $Z = 104$ in a three-dimensional representation (Figure 38.19). In general, the ionization energy falls as we move to higher number n in the radial shells. Also, this figure shows

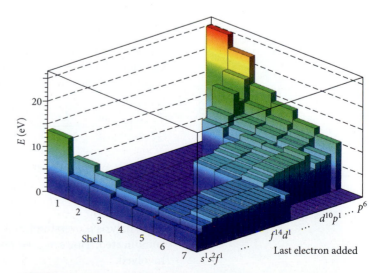

FIGURE 38.19 Ionization energies for the elements up to $Z = 104$ (Rutherfordium) in the periodic table.

38.4 In-Class Exercise

The element 118 was discovered only in 2006 and exists for only approximately 1 millisecond before it decays. Less than 10 atoms of this element have been produced. The ionization energy for 118 has not been measured yet. What would you predict the value of this ionization energy to be?

a) approximately 0 eV (the atom is unstable)

b) approximately 2 eV

c) approximately 5 eV

d) approximately 10 eV

e) approximately 20 eV

that the ionization energy grows as we move from left to right in the periodic table, adding more electrons to a given angular momentum subshell. This trend indicates how we can gain great insights into chemistry from an understanding of the quantum mechanical wave functions of electrons in a central Coulomb potential. It seems miraculous that this general understanding of atoms with many electrons can be gained from the solution of the comparatively simple hydrogen atom with only one electron. Nevertheless, judging from experimental data on the ionization energies, the general principle works.

EXAMPLE 38.3 | Ionization Energy of the Helium Atom

As discussed in this section, the ionization energy of helium is measured to be 24.6 eV. Suppose we have a helium atom ($Z = 2$) with two electrons in the $n = 1$ shell. These two electrons are in the ground state but interact with each other.

PROBLEM
Estimate the ionization energy of helium.

SOLUTION
The ground state energy of electrons, if they do not interact with each other, is given by equation 38.32:

$$E_1 = -\frac{\mu k^2 Z^2 e^4}{2\hbar^2}.$$

The reduced mass is now closer to the mass of the electron because the mass of the helium nucleus is approximately four times more than the mass of a proton. The ground state energy for one of the electrons in the helium atom is

$$E_1 = -\frac{\mu k^2 Z^2 e^4}{2\hbar^2} = -Z^2 \frac{\mu k^2 e^4}{2\hbar^2} = -Z^2 \left(13.6 \text{ eV}\right)$$

$$E_1 = -4\left(13.6 \text{ eV}\right) = -54.4 \text{ eV}.$$

We can estimate the energy of the interaction between the two electrons if we imagine that the two electrons repel each other because they are both negatively charged. Thus, the electrons will try to be as far apart as possible. The Bohr model assumes that the electrons are in an orbit with a fixed radius, so the two electrons will be on opposite sides of the orbit, which means they are separated by the diameter of the orbit. The radius of the orbit is given by equation 38.6 with e^2 replaced by Ze^2:

$$r = \frac{\hbar^2}{\mu k Z e^2} = \frac{a_0}{Z} = a_0/2.$$

The electric potential energy of the two electrons separated by a distance d is given in Chapter 23 as $U = ke^2/d$. This energy is positive, which means it decreases the energy required to ionize the helium atom. The electric potential energy is

$$U = \frac{ke^2}{d} = \frac{ke^2}{2(a_0/2)} = \frac{\left(8.99 \cdot 10^9 \text{ N m}^2/\text{C}^2\right)\left(1.602 \cdot 10^{-19}\right)^2}{\left(5.295 \cdot 10^{-11} \text{ m}\right)} = 4.36 \cdot 10^{-18} \text{ J} = 27.2 \text{ eV}.$$

The difference between the ground state energy and the electric potential energy between the two electrons is 27.2 eV, which is close to the measured ionization energy of 24.6 eV.

DISCUSSION
This analysis is oversimplified. The actual interaction between the two electrons is more complicated. However, we do gain some insight into the magnitude of the energies involved.

X-Ray Production

X-rays have been discussed several times in this book. Arguably the discovery of X-rays by Wilhelm Conrad Röntgen (1845–1923) in 1895 was one of the events that kicked off the modern physics era. X-rays are penetrating radiation consisting of high-energy photons. Typical photon energies used for diagnostic medical X-rays are between 25 keV and 140 keV, with mammograms using a typical maximum energy of 25 keV and dental X-rays using a typical maximum energy of 60 keV. X-rays for baggage screening at airports have energies up to 160 keV. All of these X-rays can be produced by X-ray tubes (Figure 38.20).

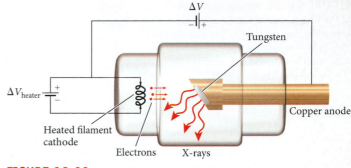

FIGURE 38.20 Schematic diagram of an X-ray tube.

How does an X-ray tube work? The physical principle is straightforward: A metal filament is heated and emits electrons from its surface. These electrons are then accelerated across an electrostatic potential difference ΔV. These electrons hit the surface of the anode, which is typically made of metal, conventionally tungsten. As the electrons hit the surface of the metal, they can produce X-rays in two different ways. One is bremsstrahlung (German word for "braking radiation"). As the electrons penetrate the surface of the metal, they experience a strong deceleration, which in turn causes the emission of photons. This process generates a continuous distribution of photons that is sketched via the dashed line in Figure 38.21. Note that this continuum distribution terminates at a maximum energy, which is given by the total kinetic energy gained by the electron from the process of acceleration through the electrostatic potential, $E_{max} = e\Delta V$ being converted to a single photon of energy $hf_{max} = \dfrac{hc}{\lambda_{min}}$. The low-energy X-rays cannot penetrate through the walls of the X-ray tube. Therefore the X-ray continuum is cut off at lower energies as well, as indicated in the figure.

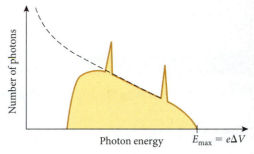

FIGURE 38.21 Sketch of an X-ray spectrum with a bremsstrahlung continuum plus sharp lines from electron transitions between atomic shells.

The other process of producing X-rays is due to the accelerated electrons colliding with electrons in the atoms of the anode and knocking them out of their atomic shells. If a deeply bound electron from the inner shells gets knocked out, then an electron from the outer shells can make a transition into those inner shells and emit a photon of a fixed energy. This process is exactly the same as the one that leads to the spectral lines of hydrogen, which was discussed extensively in Section 38.2. However, the most important difference is that atoms with $Z > 1$ have electrons that are bound more strongly, and correspondingly the photon energies are much larger than in the hydrogen line spectrum. The resulting discrete lines appear superimposed as spikes above the bremsstrahlung continuum, as sketched for two transitions in Figure 38.21. Since the photon energy is given as the difference between the energies of the initial and final states of the electron, these discrete peaks are characteristic of the anode material used.

These lines are named according to the shell (K shell for $n = 1$, L shell for $n = 2$, ...) that the electron transitions into, and an index (α, β, ...) corresponding to the shell above, from which the electron transitions. A K_α X-ray corresponds to the transition from the $n = 2$ to the $n = 1$ shell.

We can even make predictions for the energies of the X-rays from these transitions of the electrons between the atomic shells. In equation 38.32 we have stated that the energy of the innermost electron is $E_1 = -\mu k^2 Z^2 e^4 / 2\hbar^2 = -(13.6 \text{ eV})Z^2$. Our transition energies should then be a factor of Z^2 bigger than those in the hydrogen atom. We can also take into account that for the outer shells the inner electrons partially screen that nuclear charge, which should result in a factor of $(Z-Z_{screen})^2$ enhancement over the hydrogen case. The screening charge Z_{screen} depends only on the electron shell under consideration, but not on the charge of the nucleus, Z. Therefore we can predict from our simple considerations that the X-ray energy corresponding to a particular transition in the atom should depend on the square of the nuclear charge. This means that a plot of the square root of the X-ray energy resulting from a particular transition should show a linear dependence on the nuclear charge. This type of plot was first produced by Henry Moseley (1887–1915) in 1913, and contributed tremendously to convincing the scientific community of the validity of atomic models.

38.5 In-Class Exercise

As you observe the X-ray spectrum in Figure 38.21, you are told that one of the lines corresponds to a K-shell transition and the other to an L-shell transition. Which of the following is true?

a) The line with the higher photon energy corresponds to the K-shell transition.

b) The line with the higher photon energy corresponds to the L-shell transition.

c) Either line could correspond to either transition, depending on the electrostatic potential difference applied to the X-ray tube.

d) This question cannot be decided, unless we know what material was used for the anode.

FIGURE 38.22 Moseley-type plot of the square root of the X-ray energy versus the nuclear charge. The blue dashed line represents the experimental data and the red solid line depicts the theoretical prediction.

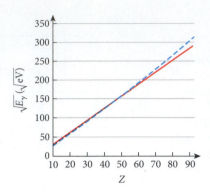

Figure 38.22 shows a plot of the type introduced by Moseley for the square root of the X-ray energy of the K_α transition for all atoms from neon to uranium. The blue dashed line represents the experimental data, as collected by the National Institute of Standards and Technology. The red line is based on the theory that the dependence of the X-ray energy on the charge is given by $(Z - Z_{screen})^2$, with Z_{screen} independent of the nucleus in question. (For this plot, $Z_{screen} = 1$, because the other electron still in the K shell will partially screen the nuclear charge so that it appears 1 less than its original value.) One can see that our theory is a good approximation to the experimental data, with small deviations visible only for the heaviest atoms.

38.5   Lasers

Originally, the word **laser** was an acronym for *Light Amplification by Stimulated Emission of Radiation*. Currently the word *laser* is so common that it is not written in all-capital letters, as would be the case for an acronym. Lasers, invented in 1960, are now used in all kinds of practical applications, such as Blu-ray, DVD and CD players and recorders, laser surgery, laser pointers, guidance systems, and precision measurement and survey systems. Lasers come in a large variety of sizes, power outputs, beam colors, and materials used. However, a few characteristics are common to practically all lasers. First, all lasers must have a medium (a gas, liquid, or solid) in which atoms can be excited to a higher energy state. Then it must be possible to create a population inversion, in which more atoms are in the excited state than in the ground state. Second, a laser has to have a resonator cavity, usually simply a pair of parallel mirrors. Third, a laser has to have a means to pump energy into the lasing medium.

We will explain the basic physics involved by using the example of the helium-neon (He–Ne) gas laser. The lowest energy levels of helium and neon are sketched in Figure 38.23. The two valence electrons of helium reside in the lowest possible shell, the 1s shell, as discussed in the previous section. The ten electrons of neon fill the 1s, 2s, and 2p subshells. The valence electrons of neon in its ground state reside in the 2p shell.

Figure 38.23 shows the ground states of the two atoms (blue for helium, green for neon) as horizontal lines at zero energy. The lowest possible excited state for a single electron in helium is the 2s state, which lies at an energy of 20.61 eV above the ground state. Photons carry angular momentum $\ell = 1\hbar$, and the 1s and 2s states both have angular momentum zero, so it is not possible for a photon to cause a transition from the 1s to the 2s state or from the 2s to the 1s state.

Figure 38.23b shows the lowest energy levels of a single electron in neon atoms. (The other nine electrons remain in their respective shells: two each in the 1s and 2s angular momentum states, and five in the 2p angular momentum state.) We have visually separated into two columns the states of highest interest for the present purpose (2p, 3p, and 5s; solid green lines) from those that are of lesser interest for the present process (3s, 4s, and 4p; dotted green lines). The energy difference between the 2p ground state and the 5s excited state in neon is 20.66 eV, very close to the 20.61 eV by which the 2s state lies above the 1s state in helium. The difference between these two energies is only 0.05 eV, which is very close to the average kinetic energy of gas molecules at room temperature (see Example 19.4). The energy of the 3p state in neon is 18.70 eV above the ground state. A transition between the 3p (angular momentum $\ell = 1\hbar$) to the 5s state (angular momentum 0) is thus possible by absorption of a photon of wavelength (see equation 38.8):

$$E_{5s} = E_{3p} + \frac{hc}{\lambda} \Rightarrow$$

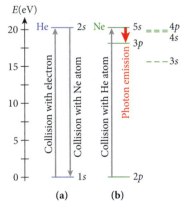

FIGURE 38.23 Energy levels relative to their respective ground states for (a) helium and (b) neon atoms.

$$\lambda = \frac{hc}{E_{5s} - E_{3p}} = \frac{1240 \text{ eV nm}}{(20.66 - 18.70) \text{ eV}} = \frac{1240 \text{ eV nm}}{1.96 \text{ eV}} = 633 \text{ nm}.$$

Conversely, the transition from the 5s to the 3p state proceeds via the emission of a photon of wavelength 633 nm. This process causes the emission of red light, which is characteristic for this type of laser.

You may ask why the 5s state cannot emit a photon and transition to the 4p state. In fact, this transition is also observed in practice and corresponds to a photon wavelength of 3390 nm. Another possible transition is between the 4s and 3p states, which causes the emission of a photon with wavelength of 1150 nm. In total, more than 10 laser transitions are known in this system, but for practical purposes the most useful one is the 5s-to-3p transition, because it is in the visible region.

Stimulated Emission and Population Inversion

Figure 38.24 illustrates the possible interaction processes of photons with atoms. The discussion of spectral lines in the first two sections of this chapter examined the connection between the transition between states of different energy in atoms and the emission and absorption of photons.

In Figure 38.24a, we depict the emission of a photon from an excited state of an atom and the absorption of a photon with the appropriate energy that leads to the formation of the excited state in the atom. Figure 38.24b shows what happens if a large number of atoms is present in a system, a few of which are in the excited state. The excited atoms decay by emitting a photon of fixed energy, but the direction of emission is random. Finally, Figure 38.24c shows the conditions for **stimulated emission:** A number of coherent photons, all with the same energy and moving in the same spatial direction, are sent into a system of atoms, for which the population of atoms in the state with higher energy is greater than the population with lower energy. Photons are bosons, so they prefer to occupy quantum states that are already occupied by other photons. Thus, the presence of the large number of coherent photons causes the preferred emission of photons into the same quantum state (same energy and same direction of motion) already occupied by the existing photons.

For stimulated emission to occur, the coherent photons sent into the system must have the same energy as the emitted photons, which is the energy difference between the two states in the atom. Then each photon of coherent light sent into the system of atoms can also be absorbed by an atom in the lower energy state, lifting it into the higher energy state. If more atoms are in the lower energy state than in the higher energy state, the net effect is an overall reduction in the number of photons. Thus, "light amplification through stimulated emission" can only be successful if the population of atoms in higher energy states is greater than the population with lower energy.

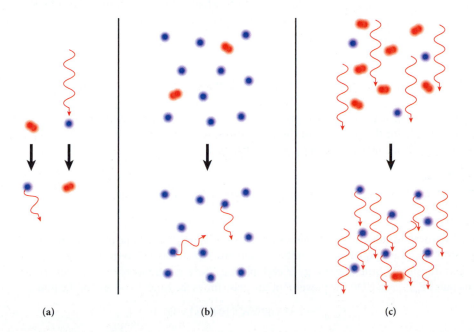

(a)   (b)   (c)

FIGURE 38.24 Interaction of photons (red sine waves) with atoms (blue dots for ground state, red dots for excited state). (a) Emission and absorption processes for an isolated atom. (b) Decay of excited atoms in a mixture of atoms in the ground state and excited state. (c) Stimulated emission of photons from a population-inverted mixture of atoms in the excited and ground states.

38.6 In-Class Exercise

What is the population ratio of the 5s state relative to the 3p state at a temperature of 5000 K?

a) $1.2 \cdot 10^{-33}$ d) 0.51

b) $4.3 \cdot 10^{-23}$ e) 2.4

c) 0.011

Chapter 19 showed that the probability for atoms (or molecules) in a gas to have an energy E at a given temperature T is proportional to the exponential factor $e^{-E/k_B T}$. Thus, the ratio of the population of the higher energy state to the population of the lower energy state is

$$\frac{n_{\text{higher}}}{n_{\text{lower}}} = e^{-\Delta E/k_B T},$$

where ΔE is the energy difference between the higher and the lower energy states. Let's put in some typical numbers. Previously, we found that the energy difference between the 5s and 3p states in neon is $\Delta E = 1.96$ eV. At room temperature ($T = 300$ K), this means that for every atom found in the 3p state, there are only $e^{-1.96/(300 \cdot 8.617 \cdot 10^{-5})} = 1.2 \cdot 10^{-33}$ atoms in the 5s state. (Here we used the expression of the Boltzmann constant in the convenient unit of $k_B = 8.617 \cdot 10^{-5}$ eV/K.)

Since a positive energy difference ΔE always implies that $e^{-\Delta E/k_B T} < 1$, the population of the state with higher energy is always less than that with lower energy, if a system of atoms is in thermal equilibrium.

Achieving the population inversion required for light amplification through stimulated emission requires systems for which **metastable states** exist—that is, states that have long lifetimes because a simple decay is forbidden. The level scheme of helium and neon in Figure 38.23 shows that the 2s state in helium is just such a metastable state, because it cannot decay via emission of a single photon. This state can be populated through collisions of the helium atoms with electrons, thus "pumping" the excited state. This part of the laser works similarly to a simple neon sign, where gas atoms are excited via an electron discharge. The helium atoms can then transfer this energy to the neon atoms in the 5s state, which has approximately the same energy relative to its ground state, through collisions between excited helium atoms with neon atoms in the ground state. If enough energy is pumped into the system, population inversion between the 5s and 3p states in neon can be achieved by populating the 5s state faster than it can decay into the 3p state. The utilization of a metastable state to achieve population inversion is a very common theme in all kinds of laser systems.

Figure 38.25 shows a demonstration model of a helium-neon laser. The discharge to pump the lasing medium takes place in the gas cell. It is enclosed between a pair of parallel mirrors, which bounce the coherent photons back and forth many times to maximize the stimulated emission. The mirror on the right side is not perfectly reflective, but only 99%; this allows the laser beam to exit from the cell.

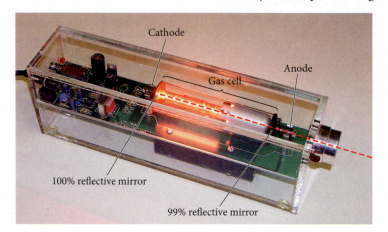

FIGURE 38.25 Demonstration model of a laser. The lasing medium, a mixture of helium and neon gas, is enclosed between two parallel mirrors. The laser beam itself is not visible but is indicated by the red dashed line.

Labels on figure: Cathode, Gas cell, Anode, 100% reflective mirror, 99% reflective mirror

---

EXAMPLE 38.4 Number of Photons from a Pulsed Ruby Laser

A ruby laser consists mostly of alumina (Al_2O_3) and a small amount of chromium, which is responsible for the red color of the ruby. This laser emits a pulse of light ($\lambda = 685$ nm) with a power of 3.00 kW over a period of 10.0 ns.

PROBLEM

How many chromium atoms undergo stimulated emission to produce this pulse?

SOLUTION

Each chromium atom that undergoes stimulated emission gives off one photon. The total energy of the pulse is equal to the number of photons, which is equal to the number of excited chromium atoms, times the energy of each photon. The total energy of the light contained in the pulse E_{total} is equal to the power P of the pulse times the time duration t of the pulse

$$E_{\text{total}} = Pt = (3.00 \text{ kW})(10.0 \text{ ns}) = 3.0 \cdot 10^{-5} \text{ J}.$$

The energy of each photon is

$$E = hf = h\frac{c}{\lambda} = \left(6.626\cdot10^{-34}\text{ J s}\right)\frac{3.00\cdot10^{8}\text{ m/s}}{685\cdot10^{-9}\text{ m}} = 2.90\cdot10^{-19}\text{ J}.$$

Thus, the number of chromium atoms N undergoing stimulated emission during the pulse is

$$N = \frac{E_{\text{total}}}{E} = \frac{3.00\cdot10^{-5}\text{ J}}{2.90\cdot10^{-19}\text{ J}} = 1.03\cdot10^{14}.$$

This number of chromium atoms compares with the total number of atoms in the ruby of around 10^{23}.

Research with Lasers

Lasers come in a huge variety. Typical laser pointers emit light with a power of up to 5 mW and typically use semiconductor devices called diodes. Carbon-dioxide lasers, another type of gas laser, emit laser beams with wavelengths of 9.6 and 10.6 μm at a power of up to 100 kW. These lasers are used in industrial applications for cutting and for welding. Other devices include chemical lasers, dye lasers, and solid-state lasers.

Current research on advanced lasers focuses on the generation of extremely short laser pulses of duration less than one picosecond (10^{-12} s). At this timing resolution, it begins to become possible to obtain information on the establishment of chemical bonds in real time.

Another field of current research interest is that of free-electron lasers, in which the lasing medium is a relativistic electron beam. The electron beam passes through an alternating periodic magnetic field, which causes transverse "wiggles" in the electron motion that generates coherent electromagnetic radiation. A laser beam generated in this way is tunable over a wide range of wavelengths, from millimeters to the visible range.

By using ions instead of atoms, it is possible to produce X-ray lasers, which are lasers with a much higher energy beam because each photon has a higher energy than in the case of optical lasers. This is also a very active area of research, in particular because of its potential medical and military applications.

The largest laser in the world was completed in 2009 and is housed at the National Ignition Facility (NIF; see Figures 38.26 and 38.27) at the Lawrence Livermore National Laboratory in California. It is a pulsed laser consisting of 192 separate beams, and produces

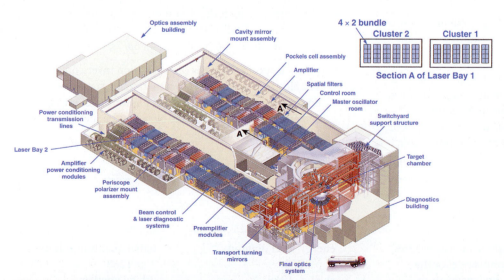

FIGURE 38.26 Cutaway drawing of the National Ignition Facility (NIF) laser system. The lasers are focused on the fusion pellets in the target chamber.

FIGURE 38.27 The Cluster 3 beam path in Laser Bay 2 of the NIF, which contains 6 bundles of 8 beams for a total of 48 beam lines. It was completed in October 2006, and all 6 bundles were commissioned as operational by December 2006. Laser beams travel over 1000 feet before they reach the target chamber.

short-lived laser pulses of total energy 1.8 MJ and duration 4 ns. Thus, the peak power of this laser is

$$P_{max} = E / \Delta t = (1.8 \cdot 10^6 \text{ J}) / (4 \cdot 10^{-9} \text{ s}) = 5 \cdot 10^{14} \text{ W} = 500 \text{ TW}.$$

This laser at the NIF is used to study nuclear fusion.

WHAT WE HAVE LEARNED | EXAM STUDY GUIDE

- The observed narrow lines in the spectra of atoms are due to the transitions of electrons from one state to another.

- The wavelengths of the photons emitted by an excited hydrogen atom are given by

$$\frac{1}{\lambda} = R_H \left(\frac{1}{n_1^2} - \frac{1}{n_2^2} \right) \text{ with } n_1 < n_2,$$

where $R_H = 1.097373 \cdot 10^7 \text{ m}^{-1}$ is the Rydberg constant.

- The Bohr model of the atom was an early success in explaining the observed line spectra. It is based on the quantization condition for the angular momentum

$$L = |\vec{r} \times \vec{p}| = r\mu v = n\hbar, \text{ with } n = 1, 2, 3, \ldots.$$

- The orbital radii in the Bohr model are given by

$$r = \frac{\hbar^2}{\mu k e^2} n^2 = a_0 n^2.$$

The Bohr radius is $a_0 = 5.295 \cdot 10^{-11}$ m = 0.05295 nm = 0.5295 Å.

- To go beyond the Bohr model and solve for the correct electron wave functions of the hydrogen atom, we need to solve the Schrödinger equation for the electron in the Coulomb potential of the nucleus,

$$-\frac{\hbar^2}{2\mu} \nabla^2 \psi(r,\theta,\phi) - k\frac{e^2}{r}\psi(r,\theta,\phi) = E\psi(r,\theta,\phi).$$

- The full solution of the Schrödinger equation for this problem can be obtained by separation of variables, $\psi(r, \theta, \phi) = f(r)g(\theta)h(\phi)$.

- The angular parts of the solution are spherical harmonics, which are simultaneously eigenfunctions of the squared angular momentum operator, $\mathbf{L}^2 Y_\ell^m(\theta,\phi) = \ell(\ell+1)\hbar^2 Y_\ell^m(\theta,\phi)$, and the operator of the z-projection of the angular momentum vector, $\mathbf{L}_z Y_\ell^m(\theta,\phi) = m\hbar Y_\ell^m(\theta,\phi)$.

- The complete solution for the hydrogen eigenfunction is the product of the angular part and the radial part, $\psi_{n\ell m}(r, \theta, \phi) = f_{n\ell}(r) Y_{\ell m}(\theta, \phi)$.

- The quantum numbers of these solutions are the radial quantum number $n = 1, 2, 3, \ldots$; the orbital angular momentum quantum number $\ell = 0, 1, \ldots, n-1$; and the magnetic quantum number $m = -\ell, \ldots, \ell$. The energy eigenvalue corresponding to a particular solution eigenfunction depends only on the radial quantum number and is the same as for the Bohr model,

$$E_n = -\frac{1}{n^2}E_0 = -\frac{1}{n^2}\frac{\mu k^2 e^4}{2\hbar^2}.$$

- The hydrogen ground state wave function is

$$\psi_{100}(r,\theta,\phi) = \frac{1}{\sqrt{\pi}a_0^{3/2}}e^{-r/a_0}.$$

- The total number $N(n)$ of possible wave functions in a given shell with radial quantum number n is

$$N(n) = \sum_{\ell=0}^{n-1}(2\ell+1) = n^2.$$

- The wave functions in other atoms can by understood in a manner similar to the hydrogen wave functions. The Pauli exclusion principle states that each possible state is occupied by at most two electrons, one with spin down, and one with spin up.

- The order in which the different angular momentum states are filled follows their energies and is 1s, 2s, 2p, 3s, 3p, 4s, 3d, 4p, 5s, 4d, 5p, 6s, 5d, 4f, 6p, 7s, 6d, and 5f.

- Lasers work on the basis of population inversion. The electrons are lifted into a metastable state, and undergo stimulated emission of photons with energy corresponding to transitions between the metastable state and the ground state.

KEY TERMS

magnetic quantum
  number, p. 1268
ionization, p. 1271

ionization energy, p. 1271
noble gas, p. 1272

periodic table of chemical
  elements, p. 1273
valence electrons, p. 1273

laser, p. 1276
stimulated emission, p. 1277
metastable states, p. 1278

NEW SYMBOLS AND EQUATIONS

$R_H = 1.097373 \cdot 10^7 \text{ m}^{-1}$, Rydberg constant

$a_0 = 5.295 \cdot 10^{-11} \text{ m} = 0.05295 \text{ nm} = 0.5295 \text{ Å}$, Bohr radius

$Y_{\ell m}(\theta, \phi)$, spherical harmonics

$\psi_{n\ell m}(r, \theta, \phi) = f_{n\ell}(r) Y_{\ell m}(\theta, \phi)$, complete wave function for an electron in the hydrogen atom

ANSWERS TO SELF-TEST OPPORTUNITIES

38.1 We need to take the limit of $n_2 \to \infty$ and find

$\dfrac{1}{\lambda} = R_H \left(\dfrac{1}{n_1^2} - \dfrac{1}{n_2^2} \right) \xrightarrow[n_2 \to \infty]{} \dfrac{R_H}{n_1^2}$. With $R_H = 1.097373 \cdot 10^7 \text{ m}^{-1}$

we find for the Lyman ($n_1 = 1$), Palmer (2), Paschen (3), and Brackett (4) series: $\lambda_{\min} = 91.1 \text{ nm}, 365 \text{ nm}, 820 \text{ nm}, 1458 \text{ nm}$.

38.2 Balmer $n_1 = 2$, $n_2 = n$: $\dfrac{1}{\lambda} = R_H \left(\dfrac{1}{4} - \dfrac{1}{n^2} \right) = R_H \left(\dfrac{n^2 - 4}{4n^2} \right)$.

Therefore, the wavelengths in this series are $\lambda = \dfrac{4}{R_H} \dfrac{n^2}{n^2 - 4} = (364.5 \text{ nm}) \dfrac{n^2}{n^2 - 4}$.

38.3 $\psi_{211}(x,0,z)$ has a shape like $\psi_{210}(x,0,z)$, but rotated 90° counterclockwise in the xz-plane.

38.4 Insert the wave function $\psi_{100}(r, \theta, \phi) = \dfrac{1}{\sqrt{\pi} a_0^{3/2}} e^{-r/a_0}$

into the Schrödinger equation

$-\dfrac{\hbar^2}{2\mu} \dfrac{1}{r} \dfrac{d^2}{dr^2} \left(r\psi(r) \right) - k \dfrac{e^2}{r} \psi(r) = E\psi(r)$. First, take the

derivative: $\dfrac{d}{dr} \left(r \dfrac{1}{\sqrt{\pi} a_0^{3/2}} e^{-r/a_0} \right) = \dfrac{1}{\sqrt{\pi} a_0^{3/2}} e^{-r/a_0} \left(1 - \dfrac{r}{a_0} \right)$;

then take the second derivative:

$\dfrac{d}{dr} \left[\dfrac{1}{\sqrt{\pi} a_0^{3/2}} e^{-r/a_0} \left(1 - \dfrac{r}{a_0} \right) \right] = -\dfrac{e^{-r/a_0}}{a_0 \sqrt{\pi} a_0^{3/2}} \left(2 - \dfrac{r}{a_0} \right)$. Insert back:

$\left(-\dfrac{\hbar^2}{2\mu} \dfrac{1}{r} \right) \left[-\dfrac{e^{-r/a_0}}{a_0 \sqrt{\pi} a_0^{3/2}} \left(2 - \dfrac{r}{a_0} \right) \right] - k \dfrac{e^2}{r} \left(\dfrac{1}{\sqrt{\pi} a_0^{3/2}} e^{-r/a_0} \right) = E \left(\dfrac{1}{\sqrt{\pi} a_0^{3/2}} e^{-r/a_0} \right)$. Now cancel common factors and find

$E = -\dfrac{\mu k^2 e^4}{2\hbar^2}$, which means that our function $\psi_{100}(r, \theta, \phi)$ is indeed a solution to the Schrödinger equation with the correct value for the energy.

PROBLEM-SOLVING PRACTICE

SOLVED PROBLEM 38.1 | "Paschen Series" for Doubly Ionized Lithium

We learned that the Lyman, Balmer, Paschen, Bracket, ... series in the hydrogen spectrum correspond to transitions in the hydrogen atom, in which the (single) electron of the hydrogen atom "jumps" between two discrete energy states. We can also observe the equivalent line spectra in ions of other atoms with just one electron. Doubly ionized lithium is such an ion.

PROBLEM
What is the wavelength of the second line of the Paschen series of doubly ionized lithium?

SOLUTION

THINK
Lithium has an atomic number of $Z = 3$. Doubly ionized lithium ions have lost two electrons from the neutral atom. This leaves one electron in a Coulomb potential, and we obtain hydrogen-like wave functions. The only difference in our treatment of the hydrogen

Continued—

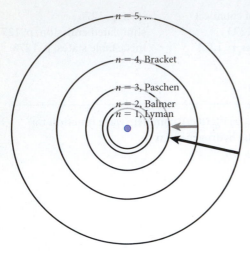

FIGURE 38.28 Second line of the Paschen series (in black).

atom is that we have $-kZe^2/r = -3ke^2/r$ for the Coulomb potential in this case. Thus, our systematics for the line spectra obtained for the hydrogen atom still hold, but we need to find the Rydberg constant appropriate to the present case.

SKETCH

Though the Bohr model of the hydrogen atom has fatal shortcomings, it is still useful to sketch a Bohr model (Figure 38.28) to obtain clarity on what transition we are trying to calculate.

RESEARCH

The wavelengths for hydrogen-like line spectra are given by equation 38.2:

$$\frac{1}{\lambda} = R_\text{H}\left(\frac{1}{n_1^2} - \frac{1}{n_2^2}\right) \quad \text{with } n_1 < n_2.$$

For the Rydberg constant R_H we had correctly obtained from the Bohr model

$$R_\text{H} = \frac{\mu k^2 e^4}{4\pi c\hbar^3},$$

where μ is the reduced mass of the electron (which for this problem is even closer to the exact electron mass than in the hydrogen atom, where it only deviated by 1 part in 2000 from the real value).

As already discussed, in this case we need to replace ke^2 by $3ke^2$ to obtain a Rydberg constant for the doubly ionized lithium ion. It is 9 times bigger $((3ke^2)^2 = 9(ke^2)^2 = 9k^2e^4)$ than the value $R_\text{H} = 1.097373 \cdot 10^7$ m^{-1} for the hydrogen atom.

Finally, from the sketch we can see that $n_1 = 3$ for the Paschen series, and that the second line in the Paschen series must have $n_2 = 5$.

SIMPLIFY

Inserting $9R_\text{H}$ instead of R_H and $n_1 = 3$, $n_2 = 5$ into our formula for the wavelengths of the transition photons we obtain

$$\frac{1}{\lambda} = 9R_\text{H}\left(\frac{1}{9} - \frac{1}{25}\right)$$

or

$$\lambda = \left(9R_\text{H}\left(\frac{1}{9} - \frac{1}{25}\right)\right)^{-1}.$$

CALCULATE

Now it is time to put in the numbers:

$$\lambda = \left(9(1.097373 \cdot 10^7 \text{ m}^{-1})\left(\frac{1}{9} - \frac{1}{25}\right)\right)^{-1} = 1.423855 \cdot 10^{-7} \text{ m}.$$

ROUND

Within the assumptions stated, our result is exact in the sense that we used only the measured value of the Rydberg constant and integer value for the transition numbers. However, the reduced mass of the electron is not exactly the same for the present problem as it is for the hydrogen atom, so we have a deviation on the order of 1 part in 2000, and it is prudent to round our result to 3 or at most 4 digits. Our final answer is thus

$$\lambda = 1.424 \cdot 10^{-7} \text{ m} = 142.4 \text{ nm}.$$

DOUBLE-CHECK

Figure 38.4 shows that the Paschen lines in hydrogen start at 820 nm. The second line in the Paschen spectrum of hydrogen thus must have a wavelength greater than 820 nm. After our previous discussion, it is clear that our result should be a factor of 9 smaller in wavelength than this value, which it is. So we have additional confidence that our result is realistic.

MULTIPLE-CHOICE QUESTIONS

38.1 The wavelength of the fourth line in the Lyman series is

a) 80.0 nm.   b) 85.0 nm.   c) 90.2 nm.   d) 94.9 nm.

38.2 In a hydrogen atom, the electron is in the $n = 5$ state. Which of the following sets could correspond to the ℓ, m states of the electron?

a) 5, −3   b) 4, −5   c) 3, −2   d) 4, −6

38.3 The muon has the same charge as an electron but a mass that is 207 times greater. The negatively charged muon can bind to a proton to form a new type of hydrogen atom. How does the binding energy $E_{B\mu}$ of the muon in the ground state of a muonic hydrogen atom compare with the binding energy E_{Be} of an electron in the ground state of a conventional hydrogen atom?

a) $|E_{B\mu}| \approx |E_{Be}|$ d) $|E_{B\mu}| \approx 200\,|E_{Be}|$
b) $|E_{B\mu}| \approx 100\,|E_{Be}|$ e) $|E_{B\mu}| \approx |E_{Be}|/200$
c) $|E_{B\mu}| \approx |E_{Be}|/100$

38.4 Which of the following can be used to explain why you can't walk through walls?

a) Coulomb repulsion
b) the strong nuclear force
c) gravity
d) the Pauli exclusion principle
e) none of the above

38.5 How many antinodes are there in a system with a quantum number $n = 6$?

a) 12   b) 6   c) 3   d) 5

38.6 Transition metals can be defined as elements where the d shell goes from empty to full. How many transition metals are there in each period?

a) 2   b) 6   c) 10   d) 14

38.7 What is the shortest wavelength photon that can be emitted by singly ionized helium (He^+)?

a) 0.00 nm   d) 91.0 nm
b) 23.0 nm   e) 365 nm
c) 46.0 nm

38.8 An electron made a transition between allowed states emitting a photon. What physical constants are needed to calculate the energy of photon from the measured wavelength? (select all that apply)

a) the Plank constant, h
b) the basic electric charge, e
c) the speed of light in vacuum, c
d) the Stefan-Boltzmann constant, σ

QUESTIONS

38.9 The common depiction of an atom, with electrons tracing elliptical orbits centered on the nucleus, is an icon of the Atomic Age. Given what you know of the physics of atoms, what's wrong with this picture?

38.10 Given that the hydrogen atom has an infinite number of energy levels, why can't a hydrogen atom in the ground state absorb all possible wavelengths of light?

38.11 Compare the difference in energy levels for increasing n for an infinite square well, a simple harmonic oscillator, and the hydrogen atom.

38.12 What would happen to the energy levels of a hydrogen atom if the Coulomb force doubled in strength? What would happen to the sizes of atoms?

38.13 Which model of the hydrogen atom—the Bohr model or the quantum mechanical model—predicts that the electron spends more time near the nucleus?

38.14 For $\ell < 4$, which values of ℓ and m correspond to wave functions that have their maximum probability in the xy-plane?

38.15 Hund's rule, a component of the *Aufbauprinzip* (construction principle), states that as one moves across the periodic table, with increasing atomic number, the available electron subshells are filled successively with one electron in each orbital, their spins all parallel; only when all orbitals in a subshell contain one electron are second electrons, with spins opposite to the first, placed in the orbitals. Explain why the ground state electron configurations of successive elements should follow this pattern.

38.16 The energy diagram of a four-level laser is presented in the figure, along with the time rates of the various transitions involved in the operation of such a laser. Explain why the nonradiative transitions have to be fast (higher rate of transitions per second) compared to the slower laser transition?

38.17 The hydrogen atom wave function ψ_{200} is zero when $r = 2a_0$. Does this mean that the electron in that state can never be observed at a distance of $2a_0$ from the nucleus or that the electron can never be observed passing through the spherical surface defined by $r = 2a_0$? Is there a difference between those two descriptions?

38.18 A 10-eV electron collides with (but is not captured by) a hydrogen atom in its ground state. Calculate the wavelengths of all photons, if any, that might be emitted.

PROBLEMS

A blue problem number indicates a worked-out solution is available in the Student Solutions Manual. One • and two •• indicate increasing level of problem difficulty.

Section 38.1

38.19 What is the shortest wavelength of light that a hydrogen atom will emit?

38.20 Determine the wavelength of the second line in the Paschen series.

38.21 The Pfund series results from emission/absorption of photons due to transitions of electrons in a hydrogen to/from the $n = 5$ energy level from/to higher energy levels. What are the shortest and longest wavelength photons emitted in the Pfund series? Are any of these photons visible?

38.22 An electron in the second excited state of a hydrogen atom jumps to the ground state. What are the possible photon colors and wavelengths emitted by the electron as it makes the jump?

Section 38.2

38.23 Calculate the energy of the fifth excited state of a hydrogen atom.

38.24 Hydrogen atoms are bombarded with 13.1-eV electrons. Determine the shortest wavelength line the atom will emit.

38.25 The Rydberg constant with a finite mass of the nucleus is given modified by $R_{modified} = R_H/(1+m/M)$, where m and M are masses of an electron and a nucleus, respectively. Calculate the modified value of the Rydberg constant for

a) a hydrogen atom, and

b) a positronium (in a positronium, the "nucleus" is a positron, which has the same mass as an electron).

38.26 A muon is a particle very similar to an electron. It has the same charge but its mass is $1.88 \cdot 10^{-28}$ kg.

a) Calculate the reduced mass for a hydrogen-like muonic atom consisting of a single proton and a muon.

b) Calculate the ionization energy for such an atom, assuming the muon starts off in its ground state.

•**38.27** An excited hydrogen atom emits a photon with an energy of 1.133 eV. What were the initial and final states of the hydrogen atom before and after emitting the photon?

•**38.28** An 8-eV photon is absorbed by an electron in the $n = 2$ state of a hydrogen atom. Calculate the final speed of the electron.

•**38.29** Assume that Bohr's quantized energy levels applied to planetary orbits. Derive an equation similar to equation 38.6 to find the allowed radius of the orbits and estimate the principal quantum number for the Earth's orbit.

•**38.30** Prove that the period of rotation of electron on the nth Bohr orbit is given by: $T = n^3/(2cR_H)$, with $n = 1,2,3,...$.

Section 38.3

38.31 What are the largest and smallest possible values for the angular momentum L of an electron in the $n = 5$ shell?

38.32 Electrons with the same value of quantum number n are said to occupy the same electron shell K, L, M, N, etc. Calculate the maximum allowed number of electrons for the

a) K shell,   b) L shell, and   c) M shell.

•**38.33** What is the angle between the total angular momentum vector and the z-axis for a hydrogen atom in the stationary state (3,2,1)?

•**38.34** A hydrogen atom is in its fifth excited state, with principal quantum number $n = 6$. The atom emits a photon with a wavelength of 410 nm. Determine the maximum possible orbital angular momentum of the electron after emission.

•**38.35** The radial wave function for hydrogen in the 1s state is given by $R_{1s} = A_1 e^{-r/a_0}$.

a) Calculate the normalization constant A_1.

b) Calculate the probability density at $r = a_0/2$.

c) The 1s wave function has a maximum at $r = 0$ but the 1s radial density peaks at $r = a_0$. Explain this difference.

•**38.36** For the wave function $\psi_{100}(r)$ shown in equation 38.30, find the value of r for which the function $P(r) = 4\pi r^2 |\psi_{100}(r)|^2$ is a maximum.

••**38.37** An electron in a hydrogen atom is in the 2s state. Calculate the probability of finding the electron within a Bohr radius ($a_0 = 0.05295$ nm) of the proton. The ground-state wave function for hydrogen is:

$$\psi_{2s}(r) = \frac{1}{4\sqrt{2\pi a_0^3}}\left(2 - \frac{r}{a_0}\right)e^{-r/2a_0}.$$

The integral is a bit tedious, so you may want consider using mathematical programs such as Mathcad, Mathematica, etc., or doing the integral online at http://integrals.wolfram.com/index.jsp.

Section 38.4

38.38 Calculate the energy needed to change a single ionized helium atom into a double ionized helium atom (that is, change it from He^+ into He^{2+}). Compare it to the energy needed to ionize the hydrogen atom. Assume that both atoms are in their fundamental state.

38.39 A He^+ ion consists of a nucleus (containing two protons and two neutrons) and a single electron. Obtain the Bohr radius for this system.

•**38.40** Obtain the wavelength of the three lowest energy lines in the Paschen series spectrum of the He^+ ion.

•**38.41** The binding energy of an extra electron when As atoms are doped in a Si crystal may be approximately calculated by considering the Bohr model of a hydrogen atom.

a) Show the ground energy of hydrogen-like atoms in terms of the dielectric constant and the ground state energy of a hydrogen atom.

b) Calculate the binding energy of the extra electron in a Si crystal. (The dielectric constant of Si is about 10, and the effective mass of extra electrons in a Si crystal is about 20% of that of free electrons.)

•**38.42** What is the wavelength of the first visible line in the spectrum of doubly ionized lithium? Begin by writing down the formula for the energy levels of the electron in doubly ionized lithium—then consider energy-level differences that give energies in the appropriate (visible) range. Express the answer in terms of "the transition from state n to state n' is the first visible, with wavelength X."

•**38.43** Following the steps outlined in our treatment of the hydrogen atom, apply the Bohr model of the atom to derive an expression for

a) the radius of the nth orbit,

b) the speed of the electron in the nth orbit, and

c) the energy levels in a hydrogen-like ionized atom of charge number Z that has lost all of its electrons except for one electron. Compare the results with the corresponding ones for the hydrogen atom.

•**38.44** Apply the results in the previous problem to determine the maximum and minimum wavelengths of the spectral lines in the Lyman, Balmer, and Paschen series for a singly ionized helium atom (He^{1+}).

Section 38.5

38.45 Consider an electron in the hydrogen atom. If you are able to excite its electron from the $n = 1$ shell to the $n = 2$ shell with a given laser, what kind of a laser (that is, compare wavelengths) will you need to excite that electron again from the $n = 2$ to the $n = 3$ shell? Explain.

38.46 A low-power laser has a power of 0.50 mW and a beam diameter of 3.0 mm.

a) Calculate the average light intensity of the laser beam, and

b) compare it to the intensity of a 100-W light bulb producing light viewed from 2.0 m.

38.47 A ruby laser consists mostly of alumina (Al_2O_3) and a small amount of chromium ions, responsible for its red color. One such laser of power 3.00 kW emits light pulse of duration 10.0 ns and of wavelength 685 nm.

a) What is the energy of the photons in the pulse?

b) Determine the number of chromium atoms undergoing stimulated emission to produce this pulse.

38.48 You hold in your hands both a green 543-nm, 5.00-mW laser and a red, 633-nm, 4.00-mW laser. Which one will produce more photons per second, and why?

Additional Problems

38.49 What is the shortest possible wavelength of the Lyman series in hydrogen?

38.50 How much energy is required to ionize hydrogen when the electron is in the nth level?

38.51 By what percentage is the electron mass changed in using the reduced mass for the hydrogen atom? What would the reduced mass be if the proton had the same mass as the electron?

38.52 Show that the number of different electron states possible for a given value of n is $2n^2$.

38.53 Section 38.2 established that an electron, if observed in the ground state of hydrogen, would be expected to have an observed speed of $0.0073c$. For what atomic charge Z would an innermost electron have a speed of approximately $0.500c$, when considered classically?

38.54 A collection of hydrogen atoms have all been placed into the $n = 4$ excited state. What wavelengths of photons will be emitted by the hydrogen atoms as they decay back to the ground state?

38.55 Consider a muonic hydrogen atom, in which an electron is replaced by a muon of mass 105.66 MeV/c^2 that orbits a proton. What are the first three energy levels of the muon in this type of atom?

38.56 What is the ionization energy of a hydrogen atom excited to the $n = 2$ state?

38.57 He^+ is a helium atom with one electron missing. Treating this like a hydrogen atom, what are the first three energy levels of this atom?

38.58 What is the energy of a transition capable of producing light of wavelength 10.6 μm? (This is the wavelength of light associated with a commonly available infrared laser.)

38.59 What is the energy of the orbiting electron in a hydrogen atom with a quantum number of 45?

•**38.60** A beam of electrons is incident upon a gas of hydrogen atoms. What minimum speed must the electrons have to cause the emission of light from the $n = 3$ to $n = 2$ transition of hydrogen?

•**38.61** Find the energy difference between the ground state of hydrogen and deuterium (hydrogen with an extra neutron in the nucleus).

•**38.62** Find the ratio of hydrogen atoms in the $n = 2$ state to the number of hydrogen atoms in the $n = 3$ state at room temperature.

•**38.63** An excited hydrogen atom, whose electron is in an $n = 4$ state, is motionless. When the electron falls into the ground state, does it set the atom in motion? If so, with what speed?

•**38.64** The radius of the $n = 1$ orbit in the hydrogen atom is $a_0 = 0.053$ nm.

a) Compute the radius of the $n = 6$ orbit. How many times larger is this compared to the $n = 1$ radius?

b) If an $n = 6$ electron relaxes to an $n = 1$ orbit (ground state), what is the frequency and wavelength of the emitted radiation? What kind of radiation was emitted (visible, infrared, etc.)?

c) How would your answer in (a) change if the atom was a singly ionized helium atom (He^+), instead?

•**38.65** An electron in a hydrogen atom is in the ground state (1s). Calculate the probability of finding the electron within a Bohr radius ($a_0 = 0.05295$ nm) of the proton. The ground state wave function for hydrogen is:

$$\psi_{1s}(r) = A_{1s}e^{-r/a_0} = e^{-r/a_0}/\sqrt{\pi a_0^3}.$$

39

Elementary Particle Physics

FIGURE 39.1 Two galaxies colliding as photographed by the Hubble Space Telescope.

WHAT WE WILL LEARN

- The quest to reduce systems into their components is a main theme in science and has its present culmination in particle physics. While it is equally valid to understand complexity and emergent structure, the present chapter focuses on reductionism in particle physics.

- Substructure is probed using scattering experiments. This chapter defines the concept of a scattering cross section.

- Elementary fermions have spin $\frac{1}{2}\hbar$ and consist of the six quarks (up, down, strange, charm, bottom, top), the electron, muon, and tau leptons, and the electron-, muon-, and tau-neutrinos. Each of these 12 fermions has an antiparticle. Quarks have non-integer charges of $-\frac{1}{3}e$ or $+\frac{2}{3}e$ and cannot be observed in isolation.

- Elementary bosons are the mediators of the interactions between the fermions. These bosons include the photon (electromagnetic), W- and Z-bosons (electroweak), gluon (strong), and graviton (gravitational). The graviton is yet to be observed experimentally. Gluons can also interact with other gluons.

- Feynman diagrams can represent the interaction of fundamental particles pictorially, and calculations can be based on an expansion in Feynman diagrams.

- The currently widely accepted model of particle interactions is called the standard model. Extensions of the standard model, which itself is incomplete, are grand unified theories, supersymmetric theories, and string theories.

- Elementary quarks and antiquarks can combine to form particles, which can be observed in isolation. A quark and an antiquark can form a meson (pion, kaon, etc.). Three quarks can form a baryon (proton, neutron, delta, lambda, etc.).

- Particle physics and astrophysics intersect in the study of Big Bang cosmology, its inflationary phase, and its electroweak and quark-gluon-plasma phase transitions, all of which happened in the first three minutes of the existence of the universe. We also discuss the origin of the cosmic microwave background radiation.

Cosmology is the study of the largest things we know about: the universe and its time evolution. Cosmology is principally a part of astronomy, and finds clues about the early universe from studying far-distant galaxies and clusters of galaxies. For example, the collision of two galaxies (Figure 39.1) indicates how galaxies might have grown and evolved billions of years ago. Strangely enough, some of the other major contributions to cosmology come from studying the smallest things we know about: elementary particles and their interactions. Understanding how matter and energy interact at the most basic level gives us some idea of what the earliest moments of the universe must have been like.

This chapter examines the most basic components of matter, the elementary particles that make up atoms and even the particles within atoms. What is known about these basic units is based on quantum mechanics, but we will skip over most of the mathematics involved and emphasize the basic concepts instead. Experimental evidence to back up these ideas is hard to come by, and usually requires analysis of debris formed by colliding particles at speeds very close to the speed of light in enormous particle accelerators. The combination of theory and experiment has led to a continual increase in the energy needed for further research in this field. Today's particle colliders are among the largest and most powerful instruments ever built.

However, the study of the tiniest particles and the highest energies involves some of the most speculative areas of physics. Particle physicists have developed a model of the most basic particles and their interactions, usually referred to as the "standard model." But even the standard model does not explain everything, and new ideas seem to arise every day about how to extend or improve the model. This chapter will look at some of these theories that may become the standard model of the future.

39.1 Reductionism

At least since the times of the ancient Greek philosophers, a central theme has guided our attempts to achieve a quantitative scientific understanding of the world around us. This central theme is **reductionism,** which assumes that a system can be understood in terms of its subsystems. Figure 39.2 illustrates reductionism in particle physics.

In astronomy, we understand that the universe contains hundreds of billions of galaxies, which are each composed of hundreds of billions of star systems. These star systems can consist of a central star (and sometimes two or more stars), which is orbited by planets, which in turn are orbited by their moons.

Biologists are trying to understand the entire ecosystem as composed of populations of species. Individual organisms of these species contain specialized organs and cells, which have nuclei containing chromosomes consisting of DNA. The DNA, which contains the genetic code that governs all traits passed on from one generation to the next, is understood to be segmented into thousands of genes. Genes are made of usefully readable content, called exons, interspersed with "junk," called introns. The exons are made of "words"— so-called codons—each of which contains three amino acid base pairs: either the pair of adenine (A) and thymine (T), or the pair of cytosine (C) and guanine (G).

Arguably, physicists have led the charge toward reductionism. Physicists now know that matter is composed of molecules, which in turn are made of atoms. The word *atom* derives from the Greek word *átomos* (ατομος: individual, indivisible). In particular, the Greek philosopher Democritus was the earliest prominent proponent of the atomistic theory of matter. The four elements of the Greek atomistic theory were Earth, Air, Fire, and Water, which were arranged in a two-by-two matrix of hot-cold and wet-dry opposites (see Figure 39.3), and which were thought to be immutable. Aristotle added to these four a fifth element, quintessence, which he thought to be the material of which the heavens were made.

The Greek atomistic theory was forgotten until the Renaissance. Then philosophers, physicists, and chemists created a new atomistic view of the world, and it proved to be incredibly successful and productive. By 1869 many different kinds of atoms had been identified as the elementary building blocks of the chemical elements. The Russian chemist Dimitri Mendeleev (1834–1907) took a giant step forward in that year by arranging the atoms of the elements in the periodic table. The current count of the different known elements is 117: the 111 elements from hydrogen through roentgenium, plus the yet unnamed elements 112, 113, 114, 115, 116, and 118. (The element with charge 117 remains undiscovered up to now, and perhaps does not exist.)

The idea of simplicity is intimately connected with reductionism. According to the idea known as Occam's razor, when an observation can be explained by several alternative theories, the simplest explanation is usually preferable. In addition, physical theories are often guided by the search for symmetry and abstract beauty, as well as the more practical demands of consistency and reproducibility. While physics is *the* basic science and is usually determined only by measurable facts, at the level discussed here it is also influenced by concepts from philosophy and even the arts.

The fact that so many kinds of atoms exist led many people to speculate at the end of the 19th century that atoms were not really the most basic fundamental building blocks of matter, and that there must be an underlying, more fundamental structure. This more fundamental structure turned out to be a hierarchy of several layers. The rest of this chapter is devoted to peeling back these layers one at a time.

Nuclear physics and elementary particle physics are devoted to this research and to explaining the phenomena that we encounter at the respective size scales. The reductionist

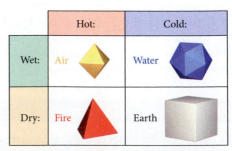

FIGURE 39.3 The four basic elements of ancient Greek philosophy, each shown with its regular polyhedron, as assigned by Plato.

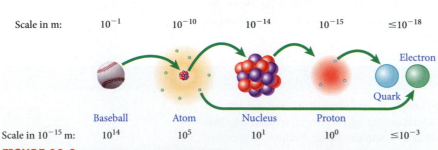

FIGURE 39.2 Reductionism in particle physics.

quest to discover the most fundamental level is far from complete. For example, nuclear and particle physicists are now using the Large Hadron Collider (LHC, Figure 39.4) at CERN (straddling the Swiss-French border, near Geneva), which became operational in 2008.

Cathedral-size caves located approximately 100 m underground house the largest particle detectors ever built. Figure 39.5 shows a construction photo of the partially finished ATLAS detector at CERN. Figure 39.6 shows a cutaway construction drawing of the ALICE detector at the same LHC complex at CERN. At the LHC, physicists hope to discover hints of the next more fundamental level in the hierarchy (if it exists!), and to perform research on the most fundamental structure currently known.

Complexity

The remainder of this chapter will focus on atomic nuclei, elementary particles, and smallest constituents, but reductionism is only one part of science research. The complementary research goal is just as important: Finding out how small constituents on one level of the size hierarchy work together to create the structure observed on the next-larger level. This is the science of **complexity**—pattern formation, self-organization of systems, and emergence of order out of chaos. This part of science is much less mature than the reductionist thread

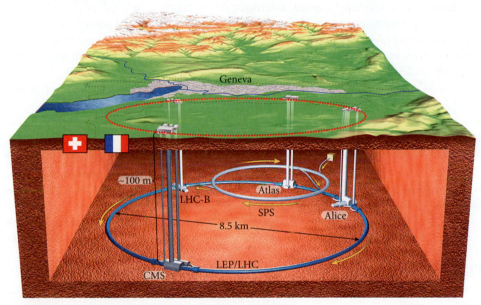

FIGURE 39.4 Schematic layout of the LHC accelerator and detector complex at CERN.

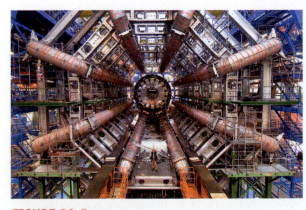

FIGURE 39.5 Partially finished construction of the ATLAS detector at CERN-LHC.

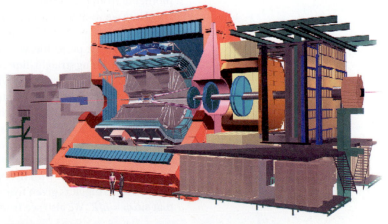

FIGURE 39.6 Construction drawing of the ALICE detector.

that science has followed for the last few centuries. It often relies on computer modeling, and is thus not usually included in introductory textbooks. Nevertheless, many stunning investigations of complex systems are currently ongoing.

Examples of complexity on the atomic scale include attempts to understand pattern formation in complex materials and the self-assembly of nanostructures. Physicists, biochemists, and microbiologists are forming interdisciplinary collaborations to try to understand the mechanism of protein folding to form three-dimensional structure from the encoding of our genetic sequence in DNA; the interaction of cells in large multicellular organisms; the precise mechanism of evolution and propagation of genetic mutations; and the origins of life and of intelligence (biological, as well as artificial). In biology, the establishment of order in animal populations (examples are anthills or bird-flight formations) is under intense study. The complexities in the spread of epidemics need to be understood. Almost everything in nature shows complex patterns, and scientists seek to understand which universal principles govern their formation. In meteorology, the goal is to come closer to being able to describe and predict the complexity of weather patterns. Human interaction has many complex dimensions that can be modeled, such as traffic flow in cities, the rise and fall of social trends and fashions, and the movement of market indicators in economies. Even the universe as a whole shows fractal patterns in the distribution of its galaxy clusters and superclusters.

For many or perhaps most of these fields of research, training in just one discipline of science is not sufficient. Instead, interdisciplinary teams composed of physical scientists, mathematicians, engineers, biologists, computer scientists, social scientists, and others need to work together to obtain a deeper understanding of these areas. In particular, as our ability to shape nature grows faster, the risk of unintended consequences from intervention in poorly understood systems increases and demands a deeper exploration.

But first things first! Let's take things apart, before we can put them back together. What is an elementary particle?

39.2   Probing Substructure

For a particle to be considered elementary, it cannot have a substructure of other particles. How is substructure discovered? Visual examination can reveal substructure for systems that are large enough, but as Chapter 38 showed, atomic radii are only on the order of 10^{-10} m and thus too small to see. The only means available to explore dimensions that are smaller than this scale is to shoot small particles at atomic targets and observe how they are deflected and scattered.

Classical Scattering

Before we proceed to use quantum mechanics, let's extend the classical theory of momentum and collisions, which we studied in Chapters 7 and 8. In Chapter 7 we considered elastic two-body scattering in two and three spatial dimensions and found that conservation of total energy and momentum did not provide enough constraints to predict the final momenta from the initial ones.

Here we want to examine a very simple case of scattering a very small low mass particle off a larger and much higher mass spherical target. In the limit that the target is much more massive than the projectile, we can consider the target to be stationary. Further, we consider only motion along straight-line trajectories, which implies the absence of a force other than the contact force at the instant when the particles collide.

We define the impact parameter b as the perpendicular distance from the incoming trajectory to the trajectory that would result in a direct head-on collision. What is the deflection angle θ of the new trajectory after the collision relative to the trajectory before the collision? A side view is shown in Figure 39.7a. In this figure, we draw in black the acute angle ϕ, which is the angle of the perpendicular to the target at the point at which the small particle makes contact with the target relative to the position vector of the low mass particle before scattering. It is given from trigonometry (see the blue right triangle in Figure 39.7a) that $\sin \phi = b/R$. We also see from Figure 39.7a that the deflection angle θ is given by

$$\theta = \pi - 2\phi = \pi - 2\sin^{-1}(b/R).$$

Therefore, given the impact parameter and the radius of the target, the deflection angle can be calculated. Alternatively, given the impact parameter and the deflection angle, the radius of the target can be calculated.

In practice, physicists construct macroscopic scattering targets from very thin foils of material. Because nuclei are so small, each particle hits at most one nucleus in a thin foil. The impact parameter cannot be predicted or measured. However, one can observe the angles at which particles scatter from the foil and make conclusions based on how many particles scatter at a particular angle. This situation is sketched in Figure 39.7b. Because a tiny nucleus will not get hit multiple times, the more realistic situation is shown in Figure 39.7c, which shows a thin foil composed of many very small scattering targets, many of which are not hit, and some of which are hit (but very likely only once). The angular distribution of these events is the same as the one shown in Figure 39.7b.

In addition to a side view of the scattering centers inside the foil, we can also imagine a head-on frontal view of the foil as it presents itself to the incoming projectile particles. This situation would look something like Figure 39.8, which shows some unit area A of the foil (gray square) with the scattering centers indicated by the small blue circles. Assume the projectile particles impinge perpendicular to the page. Each scattering center has an effective area perpendicular to the stream of projectile particles associated with it. This effective area is called the **cross section,** denoted with the letter σ. It has the dimension of an area, that is, $[\sigma] = \text{m}^2$. However, in nuclear and particle physics the unit m^2 is not very practical because the relevant cross sections are such small fractions of a square meter. Instead, the (rather ironically named!) units **barn** (b) and millibarn (mb) have been introduced as

$$1\ \text{b} = 10^{-28}\ \text{m}^2$$
$$1\ \text{mb} = 10^{-31}\ \text{m}^2.$$

Assume that the cross section for each scattering center in the target is the same, and that it is small enough that we do not have to worry about the cross sections of different scattering centers overlapping. Then the probability Π that any given projectile particle has a scattering event can be stated as the sum of the areas of all scattering centers divided by the area A over which they are spread. Since each scattering center has the same area σ, the sum of the areas is then $n_t\sigma$, where n_t is the number of scattering centers contained in the target area A. Thus

$$\Pi = \frac{n_t\sigma}{A}.$$

If a large number n_p of projectile particles fall on this area A, then the total number of reactions N_{pt} is the product of the probability that a projectile particle has a scattering event times the number n_p:

$$N_{pt} = n_p\Pi = \frac{n_p n_t\sigma}{A}.$$

This equation can be solved for the cross section σ:

$$\sigma = \frac{N_{pt}A}{n_p n_t}.$$

The same basic relationship is also useful for quantum mechanical problems. In general, a cross section is defined as the number of reactions of a particular kind per second divided by the number of projectile particles falling on the target per second and per unit area:

$$\sigma = \frac{\text{Number of reactions per scattering center/s}}{\text{Number of impinging particles/s/m}^2}.$$

A differential cross section $d\sigma/d\Omega$ can be defined as the number of reactions that lead to a scattering of a particle into some solid angle $d\Omega$ per second, divided by the same denominator as in the definition of σ above:

$$\frac{d\sigma}{d\Omega} = \frac{\text{Number of scatterings into solid angle } d\Omega \text{ per scattering center/s}}{\text{Number of impinging particles/s/m}^2}. \quad (39.1)$$

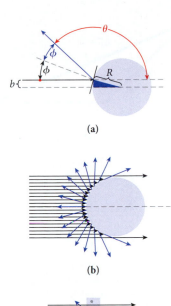

(a)

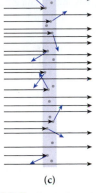

(b)

(c)

FIGURE 39.7 Classical scattering of a very small low mass particle (red dot) off a high mass and large one (blue circle). (a) Scattering at one particular fixed impact parameter; (b) superposition of many scattering events at different impact parameters; (c) superposition of many scattering events of projectile particles off small targets within a thin foil.

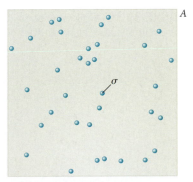

FIGURE 39.8 Frontal view of a target foil composed of many individual, small, randomly distributed scattering centers, as it presents itself to the impinging projectile particles.

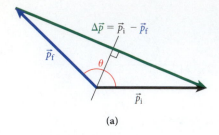

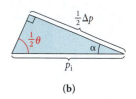

FIGURE 39.9 (a) Momentum transfer for the same deflection angle as in Figure 39.7a. (b) Right triangle formed by the initial momentum and one-half of the momentum transfer.

The unit of the differential cross section is $[d\sigma/d\Omega]$ = mb/sr. The abbreviation sr stands for the unit of solid angle, steradian. Remember that a sphere subtends a solid angle of 4π sr. Integrating $d\sigma/d\Omega$ over the unit sphere gives the total cross section

$$\sigma = \int\limits_{4\pi} \left(\frac{d\sigma}{d\Omega}\right) d\Omega.$$

Before we leave the classical considerations, let's examine the impulse (momentum transfer) that the target receives from the projectile. Since the collision is elastic and the target is assumed to be much heavier than the projectile, the lengths of the initial and final momentum vectors of the projectile are the same, $\left|\vec{p}_i\right| = \left|\vec{p}_f\right|$ and thus form an isosceles triangle.

Figure 39.9 shows the initial and final momentum vectors for the scattering event depicted in Figure 39.7a, together with the momentum transfer (green vector arrow). If the deflection angle is bisected, the line crosses the momentum transfer arrow at a right angle. This forms a right triangle with angles $\frac{1}{2}\theta, \frac{1}{2}\pi, \alpha$. Because the sum of the angles in a triangle is always π, it follows that $\frac{1}{2}\theta + \frac{1}{2}\pi + \alpha = \pi$, and thus $\alpha = \frac{1}{2}\pi - \frac{1}{2}\theta$. The absolute value of the momentum transfer $\Delta p = \left|\Delta\vec{p}\right| \equiv \left|\vec{p}_i - \vec{p}_f\right|$ can then be expressed as a function of the deflection angle θ as

$$\Delta p = 2p_i \cos(\alpha) = 2p_i \cos(\tfrac{1}{2}\pi - \tfrac{1}{2}\theta) = 2p_i \sin(\tfrac{1}{2}\theta). \tag{39.2}$$

For backward scattering ($\theta = \pi$), the momentum transfer is then at its maximum with value $2p_i$, and for forward scattering ($\theta = 0$, no collisions) the momentum transfer is zero. Keep in mind that we have assumed that the target is much heavier than the projectile. If this is not the case, then the above considerations are still true if, instead of using the laboratory frame as was done here, the center-of-mass frame is used to calculate the deflection angle and momentum transfer. In particular, if the projectile is as heavy or heavier than the target, then no backward scattering can be observed in the laboratory frame. (This observation was made in Chapter 7 for one-dimensional collisions of two carts with different masses on a frictionless track.)

Rutherford Scattering

As mentioned in Chapter 38, the experiments of Hans Geiger and Ernest Marsden, and their interpretation by Ernest Rutherford, gave definite proof that the atom consists of an outer layer of electrons surrounding a nucleus. What did Geiger and (then undergraduate student) Marsden actually do?

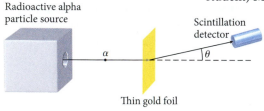

FIGURE 39.10 Basic setup of the Geiger-Marsden experiment to scatter alpha particles off gold foils.

Figure 39.10 shows the basic schematic setup of the Geiger-Marsden experiment. They used a radioactive polonium source that emitted alpha particles. (Alpha emission as one possible radioactive decay mechanism will be discussed in Chapter 40.) For now, all we need to know is that the alpha particle is the nucleus of a helium atom and has a charge of $+2e$. The kinetic energy of the alpha particle is typically on the order of 5 MeV. Geiger and Marsden scattered these alpha particles off thin gold foils and observed the resulting angular distribution. What can we expect from the scattering in this setup?

Let's first calculate the de Broglie wavelength of a 5-MeV alpha particle. The alpha particle has a mass of 3.73 GeV/c^2, so a 5-MeV kinetic energy corresponds to a momentum of 193 MeV/c. (Because the mass is very large compared to the kinetic energy in this case, we can use the nonrelativistic approximation $p = \sqrt{2Em}$.) Thus, the de Broglie wavelength of a 5-MeV alpha is

$$\lambda = \frac{h}{p} = \frac{4.136 \cdot 10^{-15} \text{ eV s}}{193 \text{ MeV}/c} = 6.4 \cdot 10^{-15} \text{ m}.$$

The discussion of diffractive scattering of waves in Chapter 34 shows that the angle of the diffraction maximum is approximately $\theta_d \sim \lambda/R$, where R is the radius of the object the waves diffract off—in this case the atom. At the time, the most prominent model of the atom was the "plum pudding" model, in which negatively charged electrons were thought to be randomly distributed over the volume of the atom, similar to raisins sticking in a plum pudding. The "dough" of the plum pudding was supposed to be the distribution of positive charge. Geiger and Marsden expected to see diffractive scattering with angles

$$\theta_d \sim 6 \cdot 10^{-15} \, \text{m} / 10^{-10} \, \text{m} = 6 \cdot 10^{-5} \, \text{rad} = 0.003°.$$

In other words, they expected extremely forward-peaked angular distributions. (The limitations imposed by quantum mechanics are discussed in more detail in the next subsection.)

However, to their surprise, Geiger and Marsden observed a significant amount of scattering at large angles, and even some scattering at backward angles. When Ernest Rutherford saw the data, he compared the alpha particles to artillery shells reflected backward by tissue paper. This observation led him to formulate the Rutherford model of the atom, with a very small positively charged nucleus in the center.

Rutherford considered the scattering problem for a charged particle with (positive) charge $Z_p e$ off a target nucleus with (also positive) charge $Z_t e$. The interaction between these two particles takes place via the Coulomb potential (see Chapter 23), and the potential energy is

$$U = k \frac{Z_p Z_t e^2}{r}$$

(where k is Coulomb's constant). He found that the differential scattering cross section can be written as

$$\frac{d\sigma}{d\Omega} = \left(\frac{k Z_p Z_t e^2}{4K} \right)^2 \frac{1}{\sin^4 \left(\frac{1}{2} \theta \right)}, \tag{39.3}$$

where K is the kinetic energy of the projectile.

Note that the Rutherford formula (equation 39.3) assumes that the projectile and target are both point charges. As Figure 39.11 shows, this formula explains extremely well the number of scintillation counts per minute (which are proportional to the differential cross section) as a function of the deflection angle.

The relationship between the scattering angle and the momentum transfer that was derived above (equation 39.2) then gives

$$\frac{d\sigma}{d\Omega} = \left(2k Z_p Z_t e^2 m_p \right)^2 \frac{1}{(\Delta p)^4}, \tag{39.4}$$

where we have used the nonrelativistic approximation $K = p^2/2m_p$, and where m_p is the mass of the projectile. (If we drop the assumption of an infinitely heavy target, then we have to use the reduced mass $\mu = m_p m_t/(m_p + m_t)$ instead of m_p.)

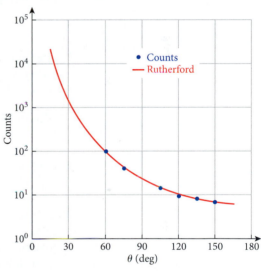

FIGURE 39.11 Comparison of the Rutherford scattering formula (red line) to the angular dependence of the scintillation counts per minute (blue dots), as reported by Geiger and Marsden in *Phil. Mag.* 25 (1913) 604.

EXAMPLE 39.1 | Backward Scattering of Alpha Particles

We can get a useful upper limit for the size of the atomic nucleus just from observing that 5.00-MeV alpha particles can get scattered in the backward direction.

PROBLEM
How close can a 5.00-MeV alpha particle get to a gold nucleus if the only interaction between them is the Coulomb force?

SOLUTION
We can use a straightforward application of conservation of total mechanical energy. Far from the nucleus, the alpha particle has kinetic energy of 5.00 MeV and zero potential energy. As it approaches the gold nucleus, more and more kinetic energy is converted into potential energy. At the point of closest approach, the alpha particle has zero kinetic

Continued—

energy, and all initial kinetic energy has been converted to potential energy. The alpha particle has a charge of $+2e$ and the gold nucleus has a charge of $+79e$, so the electrostatic potential energy between them is

$$U(r) = k\frac{(2e)(79e)}{r} = ke^2 \frac{2(79)}{r}.$$

This gives us the equation

$$ke^2 \frac{2(79)}{r_{min}} = 5.00 \text{ MeV},$$

which we need to solve for the minimum distance r_{min}. The product ke^2 has the dimension of energy times length. We can evaluate it as

$$ke^2 = (8.9876 \cdot 10^9 \text{ J m/C}^2)(1.602 \cdot 10^{-19} \text{ C})^2$$

$$= 2.307 \cdot 10^{-28} \text{ J m}$$

$$= 1.440 \cdot 10^{-9} \text{ eV m}$$

$$ke^2 = 1.44 \text{ MeV fm}.$$

Therefore, we find for the minimum distance:

$$r_{min} = \frac{(1.44 \text{ MeV fm}) \cdot 2 \cdot 79}{5.00 \text{ MeV}} = 45.5 \text{ fm}.$$

The upper bound for the radius of the gold nucleus that we obtain from the backscattering of a 5-MeV alpha particle is $4.55 \cdot 10^{-14}$ m, which is approximately a factor of 3000 smaller than the radius of a gold atom, which is $1.4 \cdot 10^{-10}$ m.

DISCUSSION

Later in this chapter we will learn that the radius of a gold nucleus is approximately 7 fm. Our upper bound is actually a factor of 7 bigger than the real radius of the gold nucleus, but it is astonishing that such a simple experiment and calculation can yield a result for the radius of the atomic nucleus of the right order of magnitude.

Why did we use a 5-MeV alpha particle, and not one with 10 MeV kinetic energy, in this example? One reason is that Geiger and Marsden used alpha particles of approximately 5 MeV kinetic energy. But more importantly, if we use an alpha particle kinetic energy that is too high, the Rutherford scattering approximation breaks down, as we will discuss in the following.

The Rutherford scattering formula (equation 39.3) predicts a $1/K^2$ dependence of the differential cross section on the kinetic energy. However, as the kinetic energy becomes higher, the alpha particle comes closer to the nucleus. At some point the two nuclei "touch" and the Rutherford formula breaks down. This result is shown very impressively in Figure 39.12, where the differential cross section for alpha particles deflected off lead at a scattering angle $\theta = 60°$ is shown as a function of the kinetic energy of the alpha particle projectile. Clearly, small kinetic energies show agreement with the Rutherford formula, but above 28 MeV, a clear deviation occurs.

Quantum Limitations

Let's now use what we've learned so far about quantum mechanics to estimate the order of magnitude involved of the various quantities involved in particle scattering. To resolve structure of size Δx, projectile particles are needed that have a momentum that is larger than the momentum uncertainty Δp_x associated with a coordinate space uncertainty Δx. The Heisenberg uncertainty relation requires $\Delta x \cdot \Delta p_x \geq \frac{1}{2}\hbar$ (see Chapter 36), so the minimum momentum needed for a particle to probe a spatial size of Δx is of the order

$$p_{min} \sim \frac{\hbar}{\Delta x}.$$

The minimum particle energy is then

$$E_{\min} = \sqrt{p_{\min}^2 c^2 + m^2 c^4} = \sqrt{\frac{\hbar^2 c^2}{\Delta x^2} + m^2 c^4}\,.$$

The product of the constants $\hbar$ and c appears frequently enough that it is useful to know it by heart,

$$\hbar c = (6.58212 \cdot 10^{-16}\ \text{eV s})(2.99792 \cdot 10^8\ \text{m/s}) = 197.327\ \text{MeV fm}.$$

In Example 39.1 we calculated the combination of the Coulomb constant k times the square of the elementary charge quantum e and found that this product also has the unit MeV · fm. We can take the ratio of these two products with the same units to find a dimensionless number called the **fine-structure constant** α, so named because it first appeared in a theory by Arnold Sommerfield to explain the fine structure in spectral lines. His theory was incorrect but the constant is significant, as we will see later, as the coupling constant for electromagnetic interactions:

$$\alpha = \frac{ke^2}{\hbar c} = \frac{e^2}{4\pi\epsilon_0 \hbar c} = \frac{1}{137.036}\,.$$

This number α is known to much greater precision than stated here, to about one part in a billion. However, for most purposes is it sufficient to remember that $\alpha = 1/137$.

Probing smaller dimensions means using projectile particles with higher momenta and energies. Figure 39.13 shows the minimum kinetic energies, $K = E - mc^2$, for electrons (mass 511 keV/c^2, red curve), photons (mass 0, blue line), and alpha particles (mass 3.73 GeV/c^2, green line) needed to probe structure of size Δx. Note that the curve for the electron kinetic energy merges with that of the photon for energies high enough that the mass of the electron becomes negligible as compared to its kinetic energy. (The same would happen for the alpha particle once the kinetic energy exceeds its rest mass of 3.73 GeV/c^2, but as we will see below, the alpha particle is itself a composite particle, so it does not make sense to use it for scattering experiments at these high energies, if one wants to probe the substructure of some other object.) Chapter 38 established that the atomic size is on the order of 10^{-10} m. Thus, Figure 39.13 shows that probing structure on the atomic scale requires photons with at least a few keV energy or electrons with at least a few tens of eV kinetic energy.

Alternatively, the same order-of-magnitude result can be obtained by using the postulate that the de Broglie wavelength, $\lambda = h/p$, of the probe needs to be smaller than the spatial resolution to be explored. Within a factor of π, the two answers are the same. (We are only interested in the order of magnitude, so a factor of 2 or π is not relevant.)

Quantum Wave Scattering

The Rutherford scattering formula (equation 39.3) was derived with the classical considerations employed above, but it turns out to be correct in the quantum mechanical case as well. For the scattering of quantum waves, consider a plane wave $e^{i\kappa z}$ incident on a target. The target then emits a spherical wave that depends on the polar angle θ, as depicted in Figure 39.14. (Just as for the classical scattering process, assume no dependence on the azimuthal angle, implying symmetry with respect to rotation about the z-axis.) Then the wave function for the stationary (time-independent) solution of the scattering problem can be written as

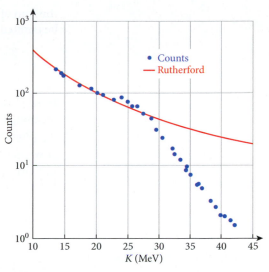

FIGURE 39.12 Dependence of the differential cross section at 60° on the kinetic energy of the alpha particle scattering off lead nuclei (blue circles), as reported by Eisberg and Porter, *Rev. Mod. Phys.* 33 (1961) 190. Red line: Expectation from Rutherford formula.

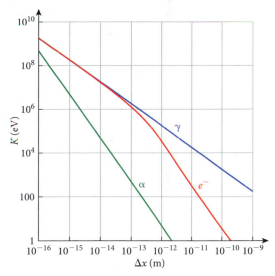

FIGURE 39.13 Minimum kinetic energy required to probe a given distance scale for electrons (red), photons (blue), and alpha particles (green).

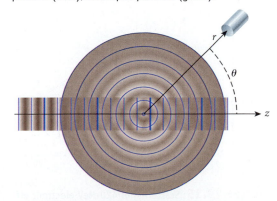

FIGURE 39.14 Scattering of a plane wave, moving from left to right along the z-direction, with a target emitting a spherical wave.

the sum of the incident plane wave plus the outgoing spherical wave, which is assumed to be emitted from a pointlike distribution without measurable spatial size,

$$\psi_{total}(\vec{r}) = \psi_i(\vec{r}) + \psi_f(\vec{r}) = N\left(e^{i\kappa z} + f(\theta)\frac{e^{i\kappa r}}{r}\right). \tag{39.5}$$

Here N is a normalization constant and $f(\theta)$ is the scattering amplitude. The physical meaning of the scattering amplitude can be extracted as follows. The number of impinging projectile particles per unit area per unit time used for the definition of the differential cross section (equation 39.1) is the particle density of the incident wave, $\left|\psi_i(\vec{r})\right|^2 = \left|Ne^{i\kappa z}\right|^2 = |N|^2$. The number of particles per unit time scattered into the area $dA = r^2 d\Omega$ is

$$\left|\psi_f(\vec{r})\right|^2 dA = \left|N\left(f(\theta)\frac{e^{i\kappa r}}{r}\right)\right|^2 dA$$

$$= |N|^2\left|f(\theta)\right|^2 \frac{1}{r^2}dA$$

$$= |N|^2\left|f(\theta)\right|^2 \frac{1}{r^2}r^2 d\Omega$$

$$= |N|^2\left|f(\theta)\right|^2 d\Omega.$$

Therefore, we find from our definition of the differential cross section (equation 39.1) that

$$\frac{d\sigma}{d\Omega} = \frac{\left|\psi_f(\vec{r})\right|^2 dA}{\left|\psi_i(\vec{r})\right|^2 d\Omega} = \frac{|N|^2\left|f(\theta)\right|^2 d\Omega}{|N|^2 d\Omega} = \left|f(\theta)\right|^2. \tag{39.6}$$

That is, the differential cross section is the absolute square of the scattering amplitude.

If we now relax the condition that the emission of the outgoing spherical wave results from a single point, then we can ask how the Rutherford scattering cross section has to be modified. The **form factor** $F^2(\Delta p)$ is the square of the Fourier transform of the charge distribution $\rho(\vec{r})$ of the scattering target,

$$F^2(\Delta p) = \left|\frac{1}{e}\int \rho(\vec{r})e^{i\Delta\vec{p}\cdot\vec{r}/\hbar}dV\right|^2, \tag{39.7}$$

where the volume integral extends over all three-dimensional coordinate space. A Fourier transform is a technique that allows us to express a function of one variable, such as $\vec{r}$, as a function of another, related variable, such as Δp.

The form factor determines how the actual charge (or mass) distribution influences the cross section. In other words, the form factor measures the deviation from the Rutherford pointlike scattering formula (equation 39.4) with the Δp^{-4} dependence on the momentum transfer,

$$\frac{d\sigma}{d\Omega} = \left(\frac{d\sigma}{d\Omega}\right)_{point}\cdot F^2(\Delta p). \tag{39.8}$$

In general terms, the scattering of a point particle shows the least-steep falloff as a function of increasing angle or momentum transfer. Any extended source results in a steeper fall-off of the differential cross section, due to the form factor. If the form factor could be measured at all values of the momentum transfer, then the Fourier transform (equation 39.7) could be inverted directly. However, usually this is not feasible with the necessary accuracy. Instead, the usual procedure is to use a charge density distribution with adjustable parameters—for example, the mean radius—and fit these parameters to provide the best match with the experimental scattering data. One example for this procedure is shown in Figure 39.15.

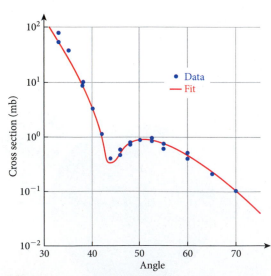

FIGURE 39.15 Scattering of 420-MeV electrons off oxygen. Data: blue circles; fit with form factor: red line, as reported by R. Hofstadter in *Annual Reviews of Nuclear Science* 7 (1957) 231.

39.3   Elementary Particles

A short time after Rutherford's discovery that the nucleus of an atom is smaller than the radius of an atom by a factor of 10,000, it became clear that the atomic nucleus consists of positively charged protons and uncharged neutrons. (Neutrons were discovered in 1932 by the British physicist James Chadwick (1891–1974).)

For a brief period in the 1930s, reductionists were very content with the picture presenting itself. Only three fundamental indivisible fermionic particles were known—the positively charged proton, the uncharged neutron, and the negatively charged electron. In addition, there was only one known boson, the photon. In 1930 the German physicist Wolfgang Pauli (1900–1958) postulated another fermion, the neutrino, to explain the missing energy in so-called beta decays. (Beta decays are explained later in this chapter.) In addition, in 1935 the Japanese physicist Hideki Yukawa (1907–1981) postulated yet one more boson, the pion, as the particle responsible for the so-called strong interaction (also addressed later in this chapter). Dirac had predicted antiparticles in 1928, and Anderson had found the positron as the antiparticle to the electron in 1932 (see Section 37.10), so it was generally accepted that each of the elementary particles also has its antiparticle. It seemed that physicists were very close to achieving a satisfactory simple, elegant picture of the layer of truly indivisible elementary particles, similar to the Greek ideal of only four or five elements.

However, small problems began to appear in this picture. In 1936 Anderson made another discovery of an elementary particle, the muon, by examining cosmic rays. At first this muon was thought to be the pion postulated by Yukawa, because it had a mass of approximately 100 MeV/c^2, as demanded by the Yukawa theory. The real pion was only discovered 11 years later by Cecil Powell and his collaborators. In 1947, George Rochester and Clifford Butler discovered the kaon in cosmic rays as well. Elementary particles were now grouped into baryons (neutron, proton, and others), mesons (pion, kaon, and others), and leptons (electron, neutrino, muon, and others). Baryons and mesons together are classified as hadrons, which are particles that experience the strong force.

In 1934, Ernest Lawrence (1901–1958) invented the cyclotron, a powerful way to accelerate particles to higher energies, and the era of accelerator-based physics began. Everbigger accelerators reaching ever-higher energies led to more discoveries of an ever-larger zoo of "elementary" particles. Eventually, scientists came to believe that this large number of particles meant that another level of substructure existed underneath the layer of what was previously considered elementary. In the years 1953 to 1957, Robert Hofstadter (1915–1990) and collaborators used the newly invented linear accelerators to perform a basic variation of Rutherford scattering with high-energy electrons on protons. Their work revealed the charge distributions of nuclei (see, for example, Figure 39.15), but also showed that the proton has a substructure. This means that the proton is composed of even smaller, more fundamental particles. In 1964, Murray Gell-Mann gave this group of particles that make up the proton and other hadrons the name **quarks** (named after the phrase "Three quarks for Muster Mark" from the novel *Finnegan's Wake* by James Joyce).

In the following discussion, we will summarize the ingredients of the so-called **standard model** of elementary particle physics. It has been in place in its basic form for three decades and has withstood a large number of experimental tests. First, we list the basic constituents, the set of particles currently considered fundamental/elementary/indivisible. These elementary particles are of size less than 10^{-16} m, which is the approximate experimental limit we can reach, and are probably pointlike—that is, of a size much smaller than the 10^{-16} m that we can resolve. Next, we discuss how these particles interact. After these steps are completed, we can investigate how the fundamental particles combine to form the next-higher level in the hierarchy of matter.

Elementary Fermions

The elementary fermions are particles of spin $\frac{1}{2}\hbar$, or simply "spin $\frac{1}{2}$." (When we say that a particle has a spin that is a number, such as the number $\frac{1}{2}$, we imply that this number is to be multiplied by $\hbar$.) Chapter 36 showed that a particle's spin is the maximum measurable value for the component of the particle's spin angular momentum in any direction. Fermions include six quarks having names (or *flavors*) of up, down, strange, charm, bottom, and top. These quarks all have non-integer charges of $-\frac{1}{3}e$ or $+\frac{2}{3}e$. The other fermions are the electron, muon, and tau leptons, and the electron, muon, and tau neutrinos. Each of these 12 fermions has an antiparticle.

The original quark flavors of up, down, and strange were introduced by Gell-Mann, and the other three flavors were discovered subsequently. The charm flavor was predicted theoretically in 1970 by Sheldon Glashow, John Iliopoulos, and Luciano Maiani, and was discovered experimentally in the form of the J/ψ (a meson composed of the charm quark and its antiquark) particle in 1974 by groups at Stanford Linear Accelerator Center led by Burton Richter and at Brookhaven National Laboratory led by Samuel Ting. The bottom flavor was discovered in 1977 at Fermilab by a collaboration led by Leon Lederman. The top quark, finally, was discovered simultaneously by two large collaborations, DØ and CDF, at Fermilab in 1995.

Quarks are grouped into three different generations, together with the corresponding leptons and neutrinos, as shown in Figure 39.16. (In Figure 39.16, the quarks are indicated by the first letter of their name.) Each particle has an antiparticle with the same mass and opposite charge. Table 39.1 lists the charges, masses, and antiparticles for all elementary fermions. The notation for the antiparticle is the same as for the particle, with a horizontal bar added above the symbol. The only exceptions are the antiparticles to the electron, muon, and tau lepton. For all of these, their charge is indicated by a minus sign in the exponent and their positively charged antiparticles are identified by their charge via a plus sign in the exponent.

Two fundamental conservation laws can be stated at this point: the conservation laws for total lepton number and for total quark number. These laws hold for any reaction or decay process. The **total lepton number** ℓ_{net} is simply the number of leptons minus the number of anti-leptons and is constant in time:

$$\ell_{net} = n_\ell - n_{\bar\ell} = \text{constant.} \tag{39.9}$$

An example of the conservation of total lepton number occurs in the decay of the neutron. A neutron decays into a proton, an electron, and an anti-electron-neutrino

$$n \rightarrow p + e^- + \bar\nu_e. \tag{39.10}$$

Generation:	1	2	3
Charge −1/3 quark:	d	s	b
Charge +2/3 quark:	u	c	t
Lepton:	e^-	μ^-	τ^-
Neutrino:	ν_e	ν_μ	ν_τ

FIGURE 39.16 Elementary fermions of the three generations.

Table 39.1 Elementary Fermions

Name	Particle Symbol	Particle Charge [e]	Particle Mass [MeV/c²]	Antiparticle* Name	Antiparticle* Symbol
Down quark	d	−1/3	5±2	Anti-down	$\bar{d}$
Up quark	u	+2/3	2.2±0.8	Anti-up	$\bar{u}$
Electron	e^-	−1	0.510999	Positron	e^+
Electron-neutrino	ν_e	0	< 0.0000005	Anti-electron-neutrino	$\bar{\nu}_e$
Strange quark	s	−1/3	100±30	Anti-strange	$\bar{s}$
Charm quark	c	+2/3	1250±100	Anti-charm	$\bar{c}$
Muon	μ^-	−1	105.66	Muon	μ^+
Muon-neutrino	ν_μ	0	< 0.19	Anti-muon-neutrino	$\bar{\nu}_\mu$
Bottom quark	b	−1/3	4200±100	Anti-bottom	$\bar{b}$
Top quark	t	+2/3	$(1.74\pm0.03)\cdot 10^5$	Anti-top	$\bar{t}$
Tau lepton	τ^-	−1	1776.99	Tau lepton	τ^+
Tau-neutrino	ν_τ	0	< 18.2	Anti-tau-neutrino	$\bar{\nu}_\tau$

*The antiparticle to a given particle has the same mass, but opposite charge.

The total lepton number before the neutron decays is zero. The total lepton number after the decay is $\ell_{net} = 1$ lepton (e^-) – 1 anti-lepton $\bar{\nu}_e = 0$, so that the total lepton number before the neutron decays is equal to the total lepton number after the decay. The process (equation 39.10) is the prototypical beta-decay process, which we will study in Chapter 40.

Until recently, physicists thought not only that ℓ_{net} is conserved, but that lepton number is conserved separately for each generation. This would mean that the net number of electron-leptons (number of electrons plus number of electron-neutrinos minus number of positrons minus number of anti-electron-neutrinos) is conserved, and separately the net number of muon-leptons is conserved, and separately the net number of tau-leptons is conserved. Indeed, these separate conservation laws are valid if the standard model of elementary particle physics holds. However, the process of neutrino oscillations, discussed later in this section, invalidates the standard model and the strict conservation laws for each individual lepton species, and only the ℓ_{net} conservation law in the form of equation 39.9 survives.

In the same way, the **net quark number** Q_{net} can be defined as the difference between the total number of quarks and the total number of antiquarks. It is also constant in time,

$$Q_{net} = n_q - n_{\bar{q}} = \text{constant.} \tag{39.11}$$

All physical processes have to obey the fundamental conservation laws of lepton number (equation 39.9) and quark number (equation 39.11), in addition to all of the previously introduced conservation laws of charge (Chapter 21), energy (Chapter 5), momentum (Chapter 7), and angular momentum (Chapter 10). The conservation law of net quark number is more often expressed as the law of **baryon number** conservation. The law of baryon number conservation states that the difference between the number of baryons and the number of anti-baryons is constant in time. This conservation law was found before quarks were discovered. Because baryons are composed of quarks, the baryon number conservation law is equivalent to the quark number conservation law stated in equation 39.11.

Elementary particles have a huge range of masses, as shown in Figure 39.17. From the observation of neutrino oscillations, we know that at least one species of neutrino must have a nonzero mass, but presently we can state only upper bounds for the neutrino masses, with the electron neutrino mass being smaller than 0.5 eV. The light quark masses (u and d) and the lepton mass (e^-) of the first generation are all of the order 1 MeV. The masses of the second-generation quarks (s and c) and leptons (μ^-) are two to three orders of magnitude larger, in the range 100 MeV to GeV. The masses of the third-generation quarks (b and t) and leptons (τ^-) are in the multi-GeV to 100-GeV range. Why these particles have the masses that they do is currently not known and is an open research topic. It is thought that a particle called the Higgs boson (discussed in the following section) is responsible for the generation of the mass of the elementary particles, and a mounting body of evidence suggests that this Higgs boson will be discovered at the Large Hadron Collider at CERN. A next-generation linear

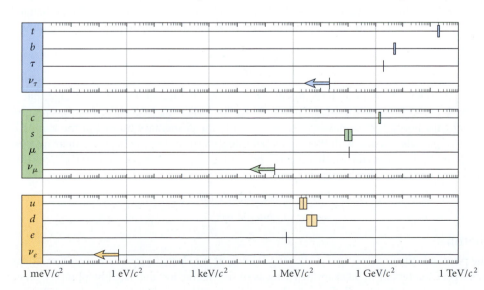

FIGURE 39.17 Currently best values for the masses of the elementary fermions. The widths of the boxes indicate the measurement uncertainty. The arrows indicate that the neutrino masses are upper limits.

collider in the works for the future, with the working name International Linear Collider, would enable physicists to study the Higgs particle in detail (if it exists at all!). They hope to then gain a deeper understanding into what sets the scale for the different particle masses.

Is it possible that more than these three generations of fundamental fermions exist? Yes, it is possible, but we have strong experimental evidence from the decay of W and Z bosons that there are indeed only exactly three generations.

Before we leave the topic of elementary fermions, it is important to point out several additional observations that are either poorly understood or point to problems in the standard model, possibly hinting at the need for a more fundamental framework.

First, we need to understand why isolated quarks have never been observed, despite a long history of high-precision searches. We will come back to the reason for this when we discuss the interaction of quarks with one another. While we can understand the reasons why it is impossible to isolate a single free quark, it is still legitimate to ask if it is philosophically consistent to speak of a fundamental indivisible particle, if we cannot isolate it.

Second, it is a puzzle why the combined electrical charge of two up quarks and one down quark (we will see below that they combine to make a proton) is exactly the same as the charge of the positron (= the opposite of the charge of an electron). Physicists have tested this equality to an incredible precision, better than one part in a trillion, but have never found any violation of this equality.

Finally, it is now assured through experimental observations that the three neutrino species ν_e, ν_μ, ν_τ can change from one species to the other and back again. This phenomenon of neutrino species changes, or **neutrino oscillations,** cannot be explained in the standard model of particle physics.

Fundamental Bosons and Interactions

What about interactions between the fundamental fermions? How do they exert forces on each other? The answer is that on the fundamental level, all forces are mediated through the exchange of elementary bosons. A full understanding of this concept requires a grasp of quantum field theory, which goes beyond the scope of this book. Some people find it helpful to think of this particle exchange as being similar to two people standing on skateboards and throwing a heavy ball back and forth, thus exerting forces on each other due to the exchange of momentum that goes with the transfer of the ball. However, this picture can really be considered only a crude illustration, because it neglects the wave character of the bosons exchanged. It also can account only for repulsion, and not attractive interactions.

In classical mechanics or in relativistic mechanics (see Chapter 35), a free particle cannot emit another particle and conserve energy and momentum in the process. However, the Heisenberg uncertainty relation (see Chapter 36) allows a violation of energy conservation of order ΔE over a time $\Delta t < \hbar/\Delta E$. Because this "borrowed" energy has to include the mass energy $m_b c^2$ of the exchange boson, we find $\Delta t < \hbar/\Delta E < \hbar/m_b c^2$. Also, because the exchanged boson can travel with at most the speed of light, we then find for the range $\Delta x \le c\Delta t$ and thus

$$\Delta x < \frac{\hbar c}{m_b c^2}.$$

Thus, an exchange boson of mass m_b can mediate an interaction only over a range that is inversely proportional to this mass. Infinite-range forces like the Coulomb force thus require an exchange boson that is massless. Table 39.2 lists the elementary bosons of the standard model of particle physics, with their spins (all integers, since they are bosons), their masses, and their charges. You can see that the W^- boson has a mass of more than 80 GeV/c^2. This implies an extremely short-range force of

$$\Delta x(W^-) < \frac{\hbar c}{m_{W^-} c^2} = \frac{197.33 \text{ MeV fm}}{80,403 \text{ MeV}} = 2.5 \cdot 10^{-3} \text{ fm.}$$

Photon

The photon is the massless spin-1 exchange boson of the electromagnetic interaction. It has 0 charge. Chapters 21 to 38 examined aspects of the interaction mediated by photons. Since the photon spin is 1, it is called a *vector boson,* as opposed to a *scalar boson* with spin 0.

Table 39.2	Elementary Bosons				
Name	Symbol	Spin [$\hbar$]	Mass [GeV/c^2]	Charge [e]	Interaction*
Photon	γ	1	0	0	Electromagnetic
W boson	W^+	1	80.403±0.029	+1	Weak
W boson	W^-	1	80.403±0.029	−1	Weak
Z boson	Z	1	91.1876±0.0021	0	Weak
Gluon	g	1	0	0	Strong
Higgs boson (?)	H^0	0	>150	0	(Mass)
Graviton (?)	G	2	0	0	Gravitation

*Interaction mediated by this particle.

W Boson

Quarks and leptons can also interact via the so-called **weak interaction.** This interaction can change one quark flavor into another by the emission of a lepton and a neutrino. The weak interaction is the *only* interaction that can change quark flavor. One example of such a reaction is

$$d \rightarrow u + e^- + \overline{\nu}_e. \tag{39.12}$$

The decay of the down quark is similar to the decay of a neutron (equation 39.10), as discussed previously in this chapter, but occurring on the fundamental level. The down quark has a charge of $-\frac{1}{3}e$ and thus needs to emit a particle with charge $-e$ to convert itself to an up quark with charge $+\frac{2}{3}e$. This is accomplished by emitting a W^- boson, which in turn decays into an electron and an anti-electron-neutrino. You can convince yourself easily that both sides of equation 39.12 have the same net charge, quark number (equation 39.11), and total lepton number (equation 39.9).

It is also important to note that the masses of the particles on the right side of equation 39.12 must sum up to a number no greater than the mass on the left side. This is required by the conservation of total energy. Table 39.1 shows that the mass of the down quark is 5 MeV/c^2. This is greater than the mass of the up quark at 2.2 MeV/c^2 plus the mass of the electron of 0.511 MeV/c^2 and the negligible mass of the anti-electron-neutrino. In the intermediate stage, the beta-decay process of equation 39.12 may require the emission of a W^- boson with mass of more 80 GeV/c^2, but in going from the initial state of this reaction to the final state, energy still must be conserved. The intermediate creation and subsequent decay of the W^- boson is said to be "off-shell" (in the sense of "off the energy shell," that is, not at constant energy), and thus the intermediate W^- boson is called a **virtual particle.**

Figure 39.18 shows a slightly different view of the quark masses than presented in Figure 39.17, where now the quarks are sorted not by generations but by charge. This enables us to generate a picture of all possible beta decays. Since all beta decays have to conserve energy, and since the quark mass increases from left to right, only those reactions are possible that have an arrow that points toward the left side, that is, from a higher-mass quark to a lower-mass one. In addition, beta decay involves the emission of a charged lepton, and thus only arrows between quarks of different charge are allowed, that is, no horizontal arrows are possible in the figure. The green arrows show all possible β^- decays. One of them was already written down in equation 39.12. The red arrows show all possible β^+ decays; these are decays that involve a virtual W^+ boson creation. Beta decays within the same generation of quarks are most common (solid lines in Figure 39.18). Beta decays that involve the jump into a neighboring generation of quarks are much more rare (dashed lines), and beta decays that involve a conversion from generation 3 to generation 1 (dotted lines) are extremely rare.

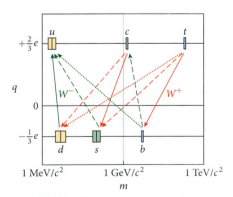

FIGURE 39.18 All possible beta decays of quarks. The red arrows represent beta+ decays, and the green arrows beta− decays. Energy conservation requires that all arrows point only toward the left, never to the right.

Z Boson

So far we have talked about only charged leptons and quarks and their electromagnetic and weak interactions. What about uncharged particles? The chapters on electromagnetism showed that uncharged particles do not participate in the electromagnetic interaction. (The neutron has a magnetic moment, but this is due to the fact that it is a composite particle.) Uncharged elementary particles can, however, participate in the weak interaction. The mediator of this interaction then necessarily needs to be a charge-0 boson, the Z boson. An exchange of an intermediate Z boson is the underlying mechanism of the scattering of a neutrino off an electron.

The W and Z bosons were both postulated in 1968 by a theory developed by the Americans Sheldon Glashow and Steven Weinberg and the Pakistani Abdus Salam. Their formulation provided a unified description of the electromagnetic interaction and the weak interaction, which is now called the electroweak interaction. The scattering of neutrinos off electrons was discovered by the Gargamelle collaboration in 1973 at CERN, which provided indirect evidence for the existence of the Z boson and convinced the physics community of the electroweak theory. The W and Z bosons were discovered in early 1983, right after the completion of the CERN SPS accelerator (which is now used as the smaller injector ring for the LHC; see Figure 39.4), by the UA1 and UA2 detector collaborations. Glashow, Weinberg, and Salam received the 1979 Nobel Prize in Physics for their theoretical work, and the 1984 Nobel Prize in Physics was awarded to the Italian Carlo Rubbia, the leader of the UA1 collaboration, and the Dutch accelerator physicist Simon van der Meer, whose work on stochastic beam cooling made possible the operation of the SPS and the discovery of the W and Z bosons.

Table 39.2 shows that the mass of the Z boson is 91.2 GeV/c^2, which is similar to the mass of a niobium atom. The Z boson can decay into an electron and a positron, which provides a clear experimental signature of the Z boson, allowing it to be detected and studied directly.

Gluon

The gluon is the mediator of the strong interaction between quarks. Before we come to the gluons, we have to add one more detail about quarks. So far, we have told you that quarks come in six flavors. However, they also have another property, which is whimsically called color. Quark colors come in three varieties: red (r), blue (b), and green (g). Antiquarks have anti-colors: anti-red ($\bar{r}$), anti-blue ($\bar{b}$), and anti-green ($\bar{g}$).

A gluon carries a combination of one anti-color and one color. From this information alone, you might conclude that a total of nine different gluons occur. However, a totally color-neutral gluon consisting of equal admixtures of $r\bar{r}, b\bar{b}, g\bar{g}$ is not possible, so there are eight possible gluons. For example, a blue quark can interact with a red quark by exchanging a $b\bar{r}$ gluon, turning the blue quark into a red one, and the red into a blue one. Since gluons carry the color charge, they can couple to one another. Thus, gluons can radiate other gluons, which no other exchange boson can do.

Higgs Boson

In 1964 the British theorist Peter Higgs postulated a boson that couples to all other elementary particles, including to itself, in such a way as to give them mass. This boson now is simply called the Higgs boson. It has not yet been found. This means that lower boundaries on its mass can be established. Currently this lower boundary is on the order of 150 GeV/c^2, mainly due to experimental results from Higgs particle searches at Fermilab's Tevatron accelerator.

Chapter 4 discussed the Higgs boson because of the relationship between force and mass. The search for the Higgs boson now enters a new phase with the operation of the LHC. The energy of the LHC should be high enough to create the Higgs boson; thus, the next few years promise to be incredibly exciting for particle physics.

Graviton

We have not yet talked about gravity in the context of quantum mechanics and elementary particles. As with the electromagnetic force, the gravitational force has infinite range. Thus, the exchange boson must have mass zero. In addition, because of the structure of the gravitational interaction as described by general relativity, the spin of this exchange particle must be 2. However, the standard model of particle physics does not address gravity at all, for two reasons, one experimental and one theoretical. Experimentally, gravity is so weak that it has not been possible to observe a graviton directly. However, a detector called LIGO may detect gravitons of cosmic origin, such as gravitons emitted by colliding galaxies (see Section 15.9). LIGO currently has installations in Louisiana and Washington State. Other similar detectors are under construction or already in operation in other parts of the world. However, much more important is that in its present formulation, the standard model (especially quantum mechanics) and gravity (as described by general relativity) are not compatible theories. This conflict has given rise to a whole new set of theories, including string theories, which we discuss further in Section 39.4.

Feynman Diagrams

If virtual particles can be created and destroyed as intermediate steps in reactions, possibly multiple times, then calculating anything that can be connected to experimentally observable consequences can become very complicated. Calculations might have to take into account infinitely many possible processes. Even the question of how to distinguish between processes that contribute a lot and those that contribute very little to the final result can be an insurmountable challenge.

The American physicist Richard Feynman (1918–1988) found an ingenious way to solve this problem by dividing complex interactions into a few elementary pieces, each of which can be represented by a **Feynman diagram,** which is just a simple drawing. He then formulated the rules by which the diagrams can be computed. In this way, he developed quantum electrodynamics, for which he shared the 1965 Nobel Prize in Physics with Sin-Itiro Tomonaga from Japan and the American Julian Schwinger, who made independent contributions to quantum electrodynamics. Even though quantum electrodynamics describes only the interaction of photons and charged particles, the Feynman diagram techniques have proven to be incredibly useful for all interactions of all particles.

We cannot pretend to give you a working knowledge of using Feynman diagrams, but they provide an excellent pictorial representation of the underlying physics of particle interactions. We introduce them here with this goal in mind. Start with freely moving particles propagating through the vacuum. They are simply represented by a line with an arrow at the end, indicating the direction of travel. This line is called a *propagator.* Conventionally, time moves from left to right and the vertical dimension of the diagram is a spatial coordinate (Figure 39.19). (Usually it is just assumed that time and space are the standard coordinates, so Feynman diagrams do not show the coordinate axes explicitly. After showing the spatial coordinate in this figure, we will drop it for the following figures.) A lepton or a quark is then simply represented by an arrow from left to right, a photon, W or Z boson by a wavy line, and a gluon by a coiled line. (There is no deeper meaning to the different line styles, just historical convention.)

Antiparticles move from right to left—that is to say, backward in time. This statement is not just a computational shortcut; it represents a deep underlying symmetry in the physics that governs the particle-antiparticle relationship: The fundamental theoretical construct of physics is invariant under simultaneous reversal of CPT (charge, parity, time). The parity of a particle is an intrinsic property characterized by behavior of the wave function of the particle under reflection through the origin of coordinates, when the x, y, and z coordinates are replaced by $-x$, $-y$, and $-z$. If the reflected wave function is identical to the original wave function, the parity is positive. If the reflected wave function is the negative of the original wave function, the parity is negative.

Inverting the charge and parity of an electron gives a positron. In order to then obtain a physically valid process, time must be reversed as well. The propagator lines are usually not horizontal, because the spatial coordinate is represented in the vertical direction in a Feynman diagram. However, the angle relative to the horizontal is not a measure for the momentum or velocity, because a Feynman graph is not intended as a plot from which some velocity can be extracted via $\Delta x / \Delta t$-type arguments. The lengths of the lines that represent the different particles also have no physical meaning and are arbitrary.

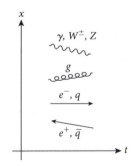

FIGURE 39.19 Feynman representations of a freely propagating photon, gluon, electron, positron, quark, anti-quark, W boson, and Z boson.

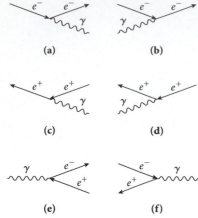

FIGURE 39.20 Elementary Feynman vertices for six different processes: (a) emission of a photon by an electron, (b) absorption of a photon on an electron, (c) emission of a photon by a positron, (d) absorption of a photon on a positron, (e) electron-positron pair creation from a photon, (f) electron-positron annihilation into a photon.

Next we want a diagram that describes the interaction of different particles. *Any given physical interaction has to have at least two vertices.* A *vertex* is defined as the intersection of three propagator lines. Figure 39.20 shows elementary vertices for six different processes. It is important to realize that none of these vertices shows a physical process that can happen in isolation. For each of these processes, it is impossible to conserve energy and momentum, provided all lines represent real particles. For example, let's look at photon emission from an electron, process (a) in Figure 39.20. In the rest frame of the electron before the emission, all the energy in the system is the rest mass of the electron times c^2, because the electron has no kinetic energy in this frame. After the emission, the photon receives some energy, and the electron experiences the recoil, thus also acquiring kinetic energy. Now the total energy in the system in the rest frame is the electron mass energy plus its kinetic energy plus the photon energy. Therefore, this process violates energy conservation. From this observation comes an important insight: Each vertex must have at least one virtual particle, which is off energy shell so that $E^2 \neq p^2c^2 + m^2c^4$. Thus, at least two vertices must be combined to create a Feynman diagram that can represent a physically possible process.

Now we are ready to formulate the rules that govern the construction of Feynman diagrams representing real processes:

- All lines that enter or exit a Feynman diagram must represent a real particle, for which the energy-momentum relationship $E^2 = p^2c^2 + m^2c^4$ holds.

- All lines that represent a virtual particle, for which $E^2 = p^2c^2 + m^2c^4$ does not hold, must start and end inside the diagram—that is, must have both ends connected to a vertex. This ensures that these virtual particles cannot be observed in experiments.

- At each vertex, energy, momentum, charge, quark number, and all three lepton numbers must be strictly conserved.

After these preliminary considerations, we can start to put together real Feynman diagrams. The first physical process we address is the scattering of an electron off a positron (or vice versa).

Figure 39.21 shows four Feynman diagrams, which all have in common that an electron and a positron come in and an electron and a positron go out. In all four cases, a virtual photon is exchanged. Thus, all four diagrams represent versions of the same physical process (called *amplitudes*) and need to be added. The cross section for electron-positron scattering can then be calculated by taking the absolute square of this amplitude sum. Note that diagram (a) and diagram (b) in Figure 39.21 differ only by the time ordering of the emission of the photon. In (a) the photon is first emitted by the electron and then absorbed by the positron, and in (b) it is the other way around. Thus, we see that all diagrams that differ only in the internal time ordering belong to the same physical process, because there is no way of knowing the internal time ordering of the process. Therefore both processes (a) and (b) can simply be written as (c), where the instantaneous exchange of the photon is implied. Process (d) can be interpreted as an electron-positron annihilation into a virtual photon, followed by a pair-creation event.

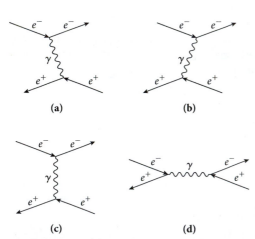

FIGURE 39.21 Four simplest Feynman diagrams for electron-positron scattering.

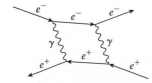

FIGURE 39.22 One possible two-photon exchange diagrams for electron-positron scattering.

Unfortunately, this is not where the process ends. The electron and positron can also exchange two photons, as is shown in Figure 39.22 for one of many possible diagrams, or they can exchange even more photons. For any diagram that involves closed loops of the kind shown in this figure, momentum and energy can travel around this loop without constraints. As a result, we have to integrate over all possible values. Clearly this summation of diagrams with loops can become arbitrarily complicated very fast.

Here is what saves us: For each virtual photon exchanged, the amplitude for the diagram acquires a multiplicative factor of $\alpha = ke^2/\hbar c = 1/137$. Thus, the contributions of multiphoton diagrams to the overall amplitude sum quickly become smaller, and the series converges.

What we have said so far still does not amount to an actual calculation, but at least it lets you understand what is involved in performing this task. In addition, it allows you to draw a quick picture of the lowest-order contributions to a physical process. For example, we can draw a Feynman diagram for the beta decay (equation 39.12) of the down quark (Figure 39.23). With the aid of this graph it is straightforward to check that the conservation laws of charge as well as quark and lepton numbers are observed in this process, that is, that the Feynman diagram represents a process that can actually happen physically.

The weak and the electromagnetic interactions can be treated beautifully with Feynman diagrams, because the series of amplitudes represented by different graphs converge quickly. We can also construct similar Feynman diagrams for the strong interaction. One example is shown in Figure 39.24, where a red up-quark exchanges a blue/anti-red gluon with a blue down-quark. (More on color will follow in Section 39.5.)

39.1 In-Class Exercise

Which of the following drawings is a valid Feynman diagram for the beta decay of the charm quark?

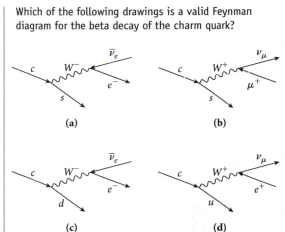

(a)   (b)

(c)   (d)

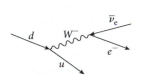

FIGURE 39.23 Feynman diagram for the beta decay of the down quark.

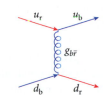

FIGURE 39.24 Simplest Feynman graph for a one-gluon exchange in the strong interaction.

However, the picture is much more complicated for the strong interaction, because the coupling constant is not α (which is small compared to 1 in the electromagnetic case), but is of the order unity. Thus, in general, series expansions in terms of the Feynman diagrams do not converge. Only for scattering at high momentum are these techniques applicable, as the expansion of the physical process in terms of Feynman diagrams converges quickly enough to be of practical utility. Other methods are needed to perform calculations for the strong interaction.

39.4 Extensions of the Standard Model

Our current understanding of the interactions in the world around us includes three different forces: the strong force, the electroweak force, and gravity. A general guiding principle in our reductionist quest is that different forces can be shown to be just different aspects of the same force on the next more fundamental level of the reductionist hierarchy. This approach has worked repeatedly to great satisfaction. Figure 39.25 shows a timeline of the unification of forces, a figure already shown in Chapter 21.

The chapters on electricity and magnetism described these two forces as just two aspects of the electromagnetic force, as described by the set of four Maxwell equations. In Section 39.3 we learned that the weak force can also be unified with electromagnetism to form the electroweak force. Consequently, there have been, and continue to be, new attempts to create theories that provide further unification of all known interactions into one.

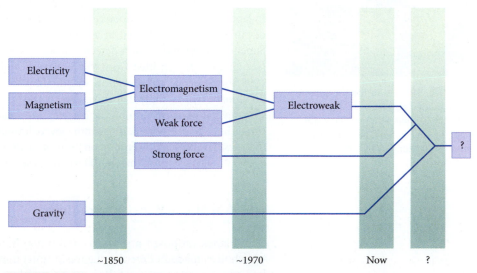

FIGURE 39.25 Time line of the unification of basic forces.

In addition, although the standard model of particle interactions has been very successful, many scientists are bothered by the seemingly arbitrary values of the parameters in the standard model, such as the quark and lepton masses and the coupling constants. Several candidates for a unified theory free of arbitrary parameters have been proposed. These theories are mostly based on arguments from a branch of mathematics called group theory. However, two essential features are worth discussing: One is the unification energy scale, and another is the decay of the proton.

Unification Energy Scale

The coupling constant of the electromagnetic interaction is the fine-structure constant α, which has the value of 1/137. The coupling constant of the weak interaction has a value of ~1/30, and the strong interaction has a coupling constant close to 1. The Glashow-Weinberg-Salam theory shows that at high energies, the coupling constants for the electromagnetic and weak force merge. Based on this idea, **grand unified theories** have been proposed that merge all three fundamental interactions at very high energies. Thus, if we were able to scatter particles at extremely high momentum transfer Δp, then we would be able to see all coupling constants merge (Figure 39.26). Unfortunately, the kind of momentum transfer required to test these theories is way beyond what can be achieved with present-day accelerators. However, possible hints on grand unified theories may be hidden in the cosmic ray spectrum, which reaches energies above 10^{20} eV.

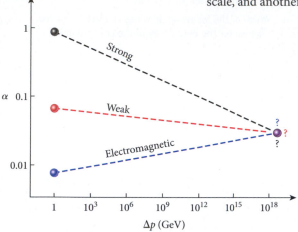

FIGURE 39.26 Proposed change of the coupling constants of fundamental interactions as a function of momentum transfer in a grand unified theory.

Proton Decay

In the preceding section on elementary particles, we introduced absolute conservation laws separately for the total number of quarks and for the total number of leptons. However, most grand unified theories predict that at the highest energies, quarks and leptons merge to be just different aspects of the same fundamental particle, and that the separation between quarks and leptons that we observe is only a consequence of the manifestation of the low-energy limit of these unified particles. If this is true, then a quark will be able to decay into a lepton, and the proton (composed of two up quarks and one down quark; see next section) will not be an absolutely stable particle; instead, it will decay exponentially with a very long (but *not* infinite) lifetime on the order of 10^{30} years.

Because the age of the universe is only on the order of 10^{10} years, a very large number of protons must be observed to have a chance to observe even one proton decay event. This is one of the basic purposes for which the Super-Kamiokande (Super-K) detector was built in Japan (Figure 39.27). Super-K is located 1000 m underground and consists of 50,000 tons of purified water contained in a cylindrical steel tank of 40 m diameter and 40 m height. Super-K has 11,200 photon detectors inside the tank walls. It can detect the Cerenkov light (see Chapter 16) emitted from the motion of high-energy particles through the water. In 1998 the Super-K collaboration was the first to observe neutrino oscillations, alluded to previously, which imply that at least one of the neutrino species has nonzero mass. However, after several years of running the detector and looking for possible signals, no proton decay events have been reported, thus setting severe limits for entire classes of possible grand unified theories. The present lower limit on the proton lifetime has been established experimentally as $>1.6 \cdot 10^{25}$ years, approximately 100 trillion times the current age of the universe.

FIGURE 39.27 Inside the Super-Kamiokande detector. Normally the detector is filled with purified water, but here the work of the maintenance crew lets us appreciate the scale of the detector.

EXAMPLE 39.2 | Planck Units

Max Planck proposed a set of natural units that do not depend on the anthropocentric choices made in determining the SI units that we currently use (see Chapter 1). He looked for combinations of the basic universal physics constants of nature that have the dimensions of length, time, mass, and so on.

PROBLEM

Using the universal constants of the speed of light in vacuum, c, Planck's constant divided by 2π, $\hbar$, Coulomb's constant $k = 1/4\pi\epsilon_0$, the universal gravitational constant G, and Boltzmann's constant k_B, what are the values of the Planck length, time, temperature, mass, and charge in SI units?

SOLUTION

The SI unit of the speed of light is $[c]$ = m/s, of Planck's constant is $[\hbar]$ = kg m^2/s, of Coulomb's constant is $[k]$ = kg m^3/C^2/s^2, and of Boltzmann's constant is $[k_B]$ = kg m^2/s^2/K, and the unit of the gravitational constant is $[G]$ = kg^{-1} m^3/s^2.

Our first observation is that the ampere, A, appears only in Coulomb's constant, and that the kelvin, K, appears only in Boltzmann's constant. Therefore, the kinematics-related quantities of length, time, mass, and energy cannot contain either of these two constants, because they are defined in terms of the MKS base units. If we multiply $\hbar\, G$, the kg cancels out in the unit, and we only obtain powers of the m and s units. Multiplication with the appropriate power of the speed of light enables us to cancel either one of these. Thus, we obtain for the Planck length and the Planck time:

$$x_p = \sqrt{\frac{\hbar G}{c^3}}$$

$$t_p = \frac{x_p}{c} = \sqrt{\frac{\hbar G}{c^5}}.$$

For the Planck mass, we must find a combination of $\hbar$ and G that has the kg in some power in it, times m/s to some power, which we can cancel out with the appropriate power of the speed of light. This can be done by taking the ratio $\hbar/G$, which has the units $[\hbar/G] =$ kg$^2 \cdot$ s/m. Then we obtain the Planck mass by taking

$$m_p = \sqrt{\frac{\hbar c}{G}}.$$

The Planck temperature has to have the Boltzmann constant in the denominator, because this is the only way we can obtain a K as the unit. The appropriate powers of the Planck length, time, and mass can then be used to cancel out the non-K parts of the unit of the Boltzmann constant. This results in

$$T_p = \frac{1}{k_B}\sqrt{\frac{\hbar c^5}{G}}.$$

A similar procedure finally leads to the Planck charge of

$$q_p = \sqrt{\frac{\hbar c}{k}}.$$

Numerically, these Planck units work out to the following numbers:

$$x_p = 1.61 \cdot 10^{-35}\ \text{m}$$

$$t_p = 5.39 \cdot 10^{-44}\ \text{s}$$

$$m_p = 2.18 \cdot 10^{-8}\ \text{kg}$$

$$T_p = 1.42 \cdot 10^{32}\ \text{K}$$

$$q_p = 1.88 \cdot 10^{-18}\ \text{C}.$$

The Planck scales for temperature, length, and time are very far away from our usual experiences and either astronomically high (temperature) or unimaginably small (length and time). The Planck charge, however, is relatively close to the elementary charge quantum e and differs only by a factor of $1/\sqrt{\alpha} = \sqrt{137} = 11.7$.

39.2 In-Class Exercise

The Planck mass is approximately the same as the mass of a

a) Z boson.

b) gold atom.

c) leg of a fruit fly.

d) baseball.

e) aircraft carrier.

39.2 Self-Test Opportunity

What is the value of the Planck energy in SI units?

Supersymmetry and String Theory

The incredible success of the Dirac theory motivated further extension of the standard model of particle physics. Simply based on symmetries, Dirac predicted that visible matter has undiscovered partners with negative energy states. This led to the discovery of antimatter and to the realization that every particle has a partner in the antiparticle world, with the same mass but opposite charge.

Supersymmetric (abbreviated as SUSY) theories propose to do the same for fermions and bosons. **Supersymmetry** postulates for each fermion a supersymmetric bosonic partner, and for each boson a supersymmetric fermionic partner. This idea was originally proposed in 1973 by the Italian-American Bruno Zumino and by the Austrian-German Julius Wess.

A severe constraint to SUSY theories is that experimental evidence for these supersymmetric partners has not yet been discovered. This fact excludes all versions of SUSY theories that predict masses below 100 GeV/c^2, the energy-mass region that we have thoroughly explored with existing particle accelerators, for any of the postulated new particles. However, some SUSY extensions of the standard model predict particle masses just above this limit. Thus, there is some hope that the rest-energy regime between 100 GeV and 1 TeV, which is accessible by the Large Hadron Collider at CERN, will reveal discoveries of magnitude comparable to the discovery of the positron.

In addition, many SUSY theories predict the existence of very heavy neutralinos. The neutralinos are thought to be mixtures of zinos (SUSY partner of the Z boson), photinos (SUSY partner of the photon), and higgsinos (SUSY partner of the Higgs boson). The lightest variant of the neutralino is absolutely stable in these theories and is a leading candidate for the possible composition of cold dark matter. It is called a WIMP (weakly interacting massive particle). Chapter 12 introduced the concept of dark matter as needed to explain effects like gravitational lensing and the fact that stars' orbital velocities around their galaxies are in general too large relative to the amount of visible matter.

The branch of particle physics theory that has unquestionably attracted more attention than any other during the last two decades is **string theory.** At its most basic level, string theory replaces the elementary (zero-dimensional) point particle as the most basic constituent of matter with one-dimensional strings of very small lengths, possibly on the Planck scale. The elementary particles that we observe in experiment are then simply strings vibrating at particular resonance frequencies. Strings can merge and split, giving rise to the particle interactions of the standard model of particle physics discussed in the preceding section (Figure 39.28).

While string theory has been around since the late 1960s, it really became dominant after the British theorist Michael Green and the American John Schwartz worked out a version of the theory in 1984 that started superstring theory as a path to unify all known forces, including gravity. In the popular press, string theory was called "The Theory of Everything." In the mid-1990s, through the work of Edward Witten and many others, it was discovered that string theory can be thought of as a certain limit of a perhaps more profound theory, which is called M-theory.

String theories predict additional spatial dimensions. The most popular versions postulate a 10- or 11-dimensional space-time. Why, then, can we observe only 3+1 dimensions? The answer given by string theories is that the extra dimensions are compacted to a size so small that we cannot observe them. This can happen at the Planck scale, but it is not excluded that the compactification length scale can be as large as a few micrometers. If this is correct, then we might expect to find distortion of gravity and deviation from the r^{-2} force law on this scale. Almost two decades of high-precision searches, however, have not come up with positive experimental evidence. String theories are very attractive mathematical frameworks in the eyes of particle theorists, but it has not yet been possible to establish any irrefutable connection to experimental observables. In 1997, however, Juan Maldacena from Argentina and the United States proved that string theory defined in a certain space is equivalent to a quantum field theory without gravity on the lower-dimensional boundary of that space. In the end, this may turn out to be the best connection to the standard model of particle physics and thus connect string theory to observable consequences.

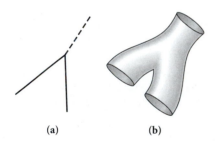

(a)   (b)

FIGURE 39.28 (a) Conventional Feynman diagram of the interaction of point particles; (b) representation of the same process as string fusion in string theory.

39.5 Composite Particles

Once we have a reasonable understanding of the basic fermionic building blocks of matter and their interactions via the exchange of fundamental bosons, then we can begin to try to understand how they form composite particles. All composite particles formed by quarks are called **hadrons.** Either they are quark-antiquark pairs, called **mesons,** or they are a group of three quarks, called **baryons.** Each quark has another trait that can be one of three values. As already alluded to in Section 39.3, this trait is arbitrarily called **color,** even though it has nothing to do with the common use of the word *color* as a visual property. Finally, all composite particles are color singlets.

What is a color singlet? Quarks carry something called a color charge, coming in three varieties: red (r), green (g), and blue (b). Antiquarks carry anti-colors, which are marked with horizontal bars above the letter. Of course, quarks cannot really be imagined as having a physical color, because what we see as color is only light from a very limited range of photon energies, energies that are millions of times smaller than the characteristic scales of the particle physics world. However, the algebra of additive color mixing works as a perfect analogy for the algebra of combining quarks into hadrons.

Figure 39.29 shows RGB-color selectors from a typical computer graphics program, where any color can be selected as a mixture of red, green, and blue by picking numbers between 0 and 255 for each of the three color-components. For example, selecting a value of 255 for red, 0 for green, and 0 for blue produces a pure red (upper left corner). Reversing all three sliders—0 for red, 255 for green, and 255 for blue—produces cyan. In a certain sense, it is the anti-color to red, $\bar{r}$, because if the RGB values of cyan and red are added, we obtain 255 for all three RGB values, which is the "color" white. Pure red, pure green, and pure blue can also be added to obtain white. Thus, white represents the color singlet that a combination of equal amounts of all colors or equal amounts of a color and its anti-color add up to. In particle physics, we also can only have color singlets as physically observable particles. Thus, we need either three quarks of different colors, or a quark of a certain color and an antiquark with the quark's anti-color, to obtain particles that we can observe. The first combination is called a baryon and is a fermion (either with spin $\frac{1}{2}\hbar$ or spin $\frac{3}{2}\hbar$), and the second is a meson, which is always a boson (either with spin 0 or with spin $1\hbar$). Three antiquarks with three different anti-colors are also a possibility to obtain a color singlet, and they form the antibaryons.

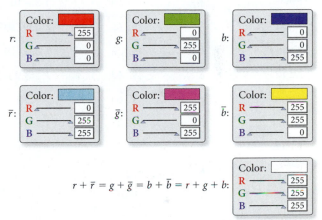

$$r + \bar{r} = g + \bar{g} = b + \bar{b} = r + g + b:$$

FIGURE 39.29 Additive color sliders in a computer graphics program as a model for quark colors (top row), antiquark anti-colors (middle row), and the color singlet (bottom panel).

Mesons

Table 39.3 gives a partial listing of the known mesons. They include all mesons with mass less than 1 GeV/c^2, and a few mesons with special properties above this mass. Hundreds of other mesons are known.

No meson is absolutely stable; they all decay. The lightest mesons are the pions π^+, π^-, π^0, all with masses of around 140 MeV/c^2. The neutral pion decays predominantly via the reaction

$$\pi^0 \rightarrow 2\gamma. \tag{39.13}$$

This decay proceeds relatively fast, on a time scale of the order of 10^{-16} s, via the electromagnetic interaction. The charged pions, however, cannot decay in this way because of the conservation law of charge. Instead, they have to decay via the weak interaction, predominantly via

$$\pi^+ \rightarrow \mu^+ + \nu_\mu$$
$$\pi^- \rightarrow \mu^- + \bar{\nu}_\mu. \tag{39.14}$$

Table 39.3	The Most Important Mesons, Including Their Masses, Lifetimes, Quantum Numbers, and Quark Composition							
Name	**Sym.**	$m[\text{MeV}/c^2]$	**Lifetime [s]**	**s**	**S**	**C**	**B**	**Quarks**
Pion	π^+	139.570	$2.60 \cdot 10^{-8}$	0	0	0	0	$u\bar{d}$
Pion	π^-	139.570	$2.60 \cdot 10^{-8}$	0	0	0	0	$\bar{u}d$
Pion	π^0	134.977	$8.4 \cdot 10^{-17}$	0	0	0	0	$\frac{1}{\sqrt{2}}(u\bar{u} - d\bar{d})$
Eta	η	547.5	$5.1 \cdot 10^{-19}$	0	0	0	0	$\frac{1}{\sqrt{6}}(u\bar{u} + d\bar{d} - 2s\bar{s})$
Eta Prime	η'	957.8	$3.2 \cdot 10^{-21}$	0	0	0	0	$\frac{1}{\sqrt{3}}(u\bar{u} + d\bar{d} + s\bar{s})$
Rho	ρ^+	775.5	$4.42 \cdot 10^{-24}$	1	0	0	0	$u\bar{d}$
Rho	ρ^-	775.5	$4.42 \cdot 10^{-24}$	1	0	0	0	$\bar{u}d$
Rho	ρ^0	775.5	$4.42 \cdot 10^{-24}$	1	0	0	0	$\frac{1}{\sqrt{2}}(u\bar{u} - d\bar{d})$
Omega	ω	782.6	$7.75 \cdot 10^{-23}$	1	0	0	0	$\frac{1}{\sqrt{2}}(u\bar{u} + d\bar{d})$
Phi	ϕ	1019.46	$1.55 \cdot 10^{-22}$	1	0	0	0	$s\bar{s}$
Kaon	K^+	493.68	$1.239 \cdot 10^{-8}$	0	1	0	0	$u\bar{s}$
Kaon	K^-	493.68	$1.239 \cdot 10^{-8}$	0	−1	0	0	$\bar{u}s$
Kaon	K^0	497.65	$K_S^0 : 8.96 \cdot 10^{-11}$	0	1	0	0	$d\bar{s}$
Kaon	$\bar{K}^0$	497.65	$K_L^0 : 5.11 \cdot 10^{-8}$	0	−1	0	0	$\bar{d}s$
K-star	K^{*+}	891.7	$1.30 \cdot 10^{-23}$	1	1	0	0	$u\bar{s}$
K-star	K^{*-}	891.7	$1.30 \cdot 10^{-23}$	1	−1	0	0	$\bar{u}s$
K-star	K^{*0}	896.0	$1.31 \cdot 10^{-23}$	1	1	0	0	$d\bar{s}$
K-star	$\bar{K}^{*0}$	896.0	$1.31 \cdot 10^{-23}$	1	−1	0	0	$\bar{d}s$
D meson	D^+	1869.3	$1.04 \cdot 10^{-12}$	0	0	1	0	$c\bar{d}$
D meson	D^-	1869.3	$1.04 \cdot 10^{-12}$	0	0	−1	0	$\bar{c}d$
D meson	D^0	1864.5	$4.10 \cdot 10^{-13}$	0	0	1	0	$c\bar{u}$
D meson	$\bar{D}^0$	1864.5	$4.10 \cdot 10^{-13}$	0	0	−1	0	$\bar{c}u$
D_s meson	D_s^+	1968.2	$5.0 \cdot 10^{-13}$	0	1	1	0	$c\bar{s}$
D_s meson	D_s^-	1968.2	$5.0 \cdot 10^{-13}$	0	−1	−1	0	$\bar{c}s$
J/Psi	J/ψ	3096.9	$7.05 \cdot 10^{-21}$	1	0	0	0	$c\bar{c}$
B-meson	B^+	5279.0	$1.64 \cdot 10^{-12}$	0	0	0	1	$u\bar{b}$
B-meson	B^-	5279.0	$1.64 \cdot 10^{-12}$	0	0	0	−1	$\bar{u}b$
B-meson	B^0	5279.4	$1.53 \cdot 10^{-12}$	0	0	0	1	$d\bar{b}$
B-meson	$\bar{B}^0$	5279.4	$1.53 \cdot 10^{-12}$	0	0	0	−1	$\bar{d}b$
B_s-meson	B_s^0	5367	$1.47 \cdot 10^{-12}$	0	−1	0	1	$s\bar{b}$
B_s-meson	$\bar{B}_s^0$	5367	$1.47 \cdot 10^{-12}$	0	1	0	−1	$s\bar{b}$
B_c-meson	B_c^+	6286	$5 \cdot 10^{-13}$	0	0	1	1	$c\bar{b}$
B_c-meson	B_c^-	6286	$5 \cdot 10^{-13}$	0	0	−1	−1	$\bar{c}b$
Upsilon	Y	9460.3	$1.22 \cdot 10^{-20}$	1	0	0	0	$b\bar{b}$

Data from W.-M. Yao et al. (Particle Data Group), *J. Phys.* G 33, 1 (2006), (http://pdg.lbl.gov).

In comparison with neutral pion decay, the decay of the charged pions takes practically forever, on the order of 10^{-8} s, which is approximately 100 million times longer that the neutral pion decay time.

Let's examine the lifetimes of the other mesons. Figure 39.30 shows three vastly different regions of lifetimes. The area highlighted in yellow shows the decays that proceed via the weak interaction; in white via the electromagnetic interaction; and in pink via the strong interaction. Because energy must be conserved in every decay, heavier mesons can decay into lighter mesons but not vice versa. This decay into other mesons happens very fast, if it proceeds via the strong interaction. There must be a good reason why the kaons and D and B mesons live such a comparatively long time. After our discussion of the quark model, you can now understand that it is because they contain strange, charm, and bottom quarks, respectively, which can only decay via the weak interaction into lower-mass quarks of other flavors.

The strangeness quantum number S counts the number of anti-strange quarks in a hadron, minus the number of strange quarks. (It is a purely historical artifact that anti-strange quarks are counted with positive numbers. For the quantum numbers of charm, C, and bottom, B, which count the net number of charm and bottom quarks, the sign is reversed.) Another useful quantum number is the isospin, t, and its projection, t_z. Even though no real angular momentum is involved in this case, the same algebra can be used for isospin as was used for conventional spin. In particular, a given isospin t has $(2t+1)$ states of different projection t_z, just as a given spin s has $(2s+1)$ states of different projection s_z. The charge number Z, the isospin, and the strangeness quantum number for the low-mass mesons (below the mass threshold where they start to contain quarks other than up, down, and strange) are related via

$$Z = t_z + \tfrac{1}{2}S. \tag{39.15}$$

Figure 39.31 shows the arrangement of the low-mass mesons of spin 0 (part a) and spin $1\hbar$ (part b) as a function of isospin and strangeness. They form characteristic nonets, which the quark model is able to explain beautifully. Figure 39.32 shows the same plot, but now in terms of the constituent quarks of the different mesons. In each case, the quark composition of the three neutral mesons with $S = t_z = 0$ is a linear combination of the three quark-antiquark pairs. The quark charges in Table 39.1 confirm that their charges always add up to the corresponding charge of the meson.

The up and down quarks have isospin $\tfrac{1}{2}$, with projections $t_z = \tfrac{1}{2}$ for the up quark and $t_z = -\tfrac{1}{2}$ for the down quark. The isospin for all other quarks and antiquarks (strange, charm,

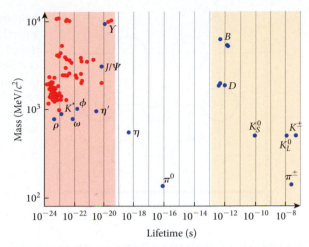

FIGURE 39.30 Plot of the lifetimes and masses for all known mesons. The blue dots represent the entries in Table 39.3, and the red dots are other mesons.

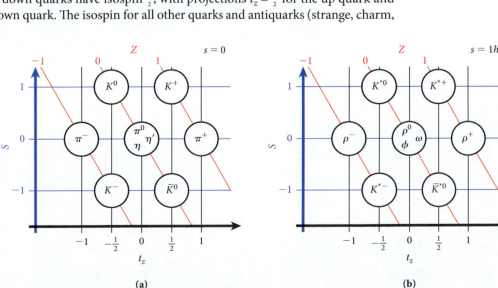

(a)   (b)

FIGURE 39.31 Meson nonets with (a) spin 0 and (b) spin $1\hbar$.

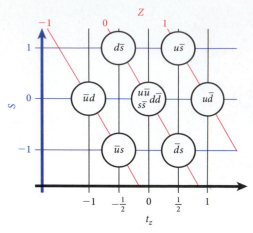

FIGURE 39.32 Quark composition of the two meson nonets.

bottom, top) is 0. Similar to the case for angular momentum and regular spin, if a particle with isospin $\frac{1}{2}$ is combined with a particle with isospin 0, the resulting isospin is $\frac{1}{2}$. If two particles with isospin $\frac{1}{2}$ are combined, the resulting isospin can be 0 (where the two isospin vectors are antiparallel) or 1 (where the two isospin vectors are parallel). If three particles with isospin $\frac{1}{2}$ are combined, the resulting isospin can be $\frac{3}{2}$ (all three isospin vectors parallel) or $\frac{1}{2}$ (one antiparallel to the other two). This "algebra of isospin" allows us to construct the observed isospin of all mesons from the quark isospins, as is done in Figure 39.32.

39.3 In-Class Exercise

Examining Figure 39.32 leads us to conclude that the isospin projections t_z of the anti-up quark ($\bar{u}$) and ($\bar{d}$) are

a) $t_z(\bar{u}) = -\frac{1}{2}, \; t_z(\bar{d}) = -\frac{1}{2}.$

b) $t_z(\bar{u}) = -\frac{1}{2}, \; t_z(\bar{d}) = \frac{1}{2}.$

c) $t_z(\bar{u}) = \frac{1}{2}, \; t_z(\bar{d}) = -\frac{1}{2}.$

d) $t_z(\bar{u}) = \frac{1}{2}, \; t_z(\bar{d}) = \frac{1}{2}.$

e) $t_z(\bar{u}) = 0, \; t_z(\bar{d}) = 0.$

Baryons

The most important baryons are listed in Table 39.4. As we have already stated, baryons are composed of three quarks each, with one quark of each color. The only stable baryon is the proton, the lightest baryon. As discussed previously, a neutron can decay into a proton via a beta decay,

$$n \rightarrow p + e^- + \bar{\nu}_e. \tag{39.16}$$

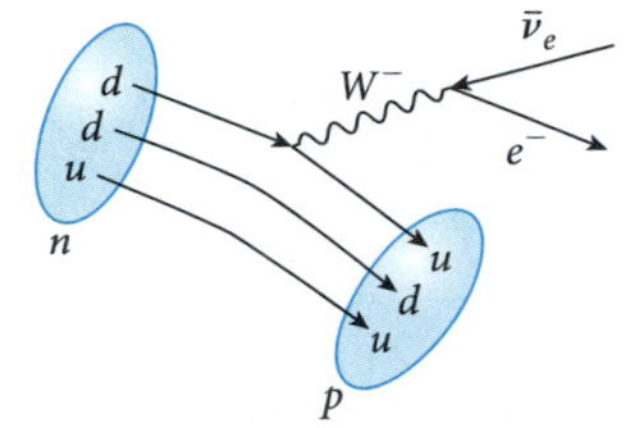

FIGURE 39.33 Feynman diagram for the beta decay of the neutron.

This is the same process for the beta decay of the down quark expressed in equation 39.12, because in the process in equation 39.16 one of the two down quarks of the neutron decays into an up quark, and the other up and down quarks remain unchanged Figure 39.33.

Table 39.4	The Most Important Baryons, Including Their Masses, Lifetimes, Quantum Numbers, and Quark Composition									
Name	Symb.	$m[\text{MeV}/c^2]$	Lifetime [s]	Z	t_z	s	S	C	B	Quarks
Proton	p	938.272	∞	1	$\frac{1}{2}$	$\frac{1}{2}$	0	0	0	uud
Neutron	n	939.565	885.7	0	$-\frac{1}{2}$	$\frac{1}{2}$	0	0	0	udd
Delta	Δ^{++}	1232	$5.6 \cdot 10^{-24}$	2	$\frac{3}{2}$	$\frac{3}{2}$	0	0	0	uuu
Delta	Δ^{+}	1232	$5.6 \cdot 10^{-24}$	1	$\frac{1}{2}$	$\frac{3}{2}$	0	0	0	uud
Delta	Δ^{-}	1232	$5.6 \cdot 10^{-24}$	-1	$-\frac{1}{2}$	$\frac{3}{2}$	0	0	0	udd
Delta	Δ^{--}	1232	$5.6 \cdot 10^{-24}$	-2	$-\frac{3}{2}$	$\frac{3}{2}$	0	0	0	ddd
Lambda	Λ^0	1115.683	$2.63 \cdot 10^{-10}$	0	0	$\frac{1}{2}$	-1	0	0	uds
Sigma	Σ^{+}	1189.37	$8.02 \cdot 10^{-11}$	1	1	$\frac{1}{2}$	-1	0	0	uus
Sigma	Σ^{0}	1192.64	$7.4 \cdot 10^{-20}$	0	0	$\frac{1}{2}$	-1	0	0	uds
Sigma	Σ^{-}	1197.45	$1.48 \cdot 10^{-10}$	-1	-1	$\frac{1}{2}$	-1	0	0	dds
Xi	Ξ^0	1314.8	$2.90 \cdot 10^{-10}$	0	$\frac{1}{2}$	$\frac{1}{2}$	-2	0	0	uss
Xi	Ξ^-	1321.3	$1.64 \cdot 10^{-10}$	-1	$-\frac{1}{2}$	$\frac{1}{2}$	-2	0	0	dss
Omega	Ω^-	1672.5	$8.21 \cdot 10^{-11}$	-1	0	$\frac{3}{2}$	-3	0	0	sss
Lambda-c	Λ^+_c	2286.5	$2.0 \cdot 10^{-13}$	1	0	$\frac{1}{2}$	0	1	0	udc
Lambda-b	Λ_b^0	5624	$1.23 \cdot 10^{-12}$	0	0	$\frac{1}{2}$	0	0	1	udb

Data from W.-M. Yao et al. (Particle Data Group), *J. Phys.* G 33, 1 (2006), (http://pdg.lbl.gov).

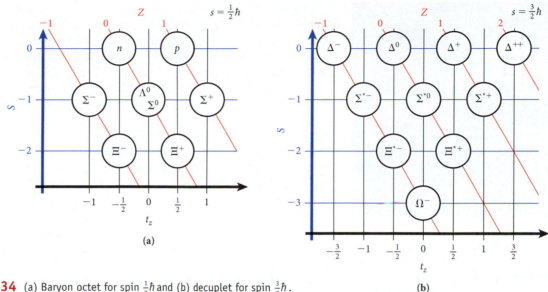

FIGURE 39.34 (a) Baryon octet for spin $\frac{1}{2}\hbar$ and (b) decuplet for spin $\frac{3}{2}\hbar$.

The dominant decays for all heavier baryons are via emission of one or more pions. This can be a process that involves the strong interaction only, in which case lifetimes of the order of 10^{-20} s or shorter are found. If, in addition, a quark needs to change flavor in the process, then again much longer lifetimes are seen, typical of the weak interaction.

In Figure 39.34, the lowest-mass baryons of spin $\frac{1}{2}\hbar$ and spin $\frac{3}{2}\hbar$ are arranged as an octet and a decuplet as functions of isospin and strangeness, similar to the diagram for mesons in Figure 39.31. This arrangement into nonets (for the mesons) and octet and decuplet (for the baryons) derives naturally from the underlying quark structure. In particular note the omega baryon. Because it has strangeness –3, it is composed of three strange quarks. Its existence was predicted by the quark model, and it was subsequently discovered in 1964. This was the first great success of the quark model.

39.3 Self-Test Opportunity

Construct the quark composition for the baryons of Figure 39.34 in the same way as done in Figure 39.32.

Lattice-QCD, Confinement, and Asymptotic Freedom

The sum of the masses of the constituent quarks inside a proton is less than 15 MeV/c^2, which is only 1.5% of the proton mass. Where does the rest of the proton mass come from? Answering this question is very difficult and requires mathematical tools that go beyond the scope of this book. However, we can sketch the answer.

The fundamental theory, including relativity and quantum physics, that describes the interaction of quarks and gluons with one another, with themselves, and with the vacuum, is called **quantum chromodynamics (QCD).** As an illustrative example for the kinds of processes QCD needs to take into account, Figure 39.35 shows a Feynman diagram of a strange quark propagating through the vacuum, emitting and absorbing gluons and virtual quark-antiquark pairs.

Physicists understand the underlying framework of QCD in principle, but it is very hard to obtain numerical results. To obtain any prediction that can be compared to experiment, physicists have built customized supercomputers to simulate QCD. Despite the exponential growth of computer power during the last few decades, it is fair to say that progress in computing observables such as hadron masses has been slow.

Solving the equations of QCD is difficult. One method that allows solutions to the equations of QCD is **lattice-QCD.** In lattice-QCD, quarks are assumed to reside only on the vertices of a crystal-like lattice and the gluons must travel in straight lines between the quarks. This technique builds in cutoffs that allow the calculations to be done. Of course, the real world is not a crystal lattice. However, the aim is to carry out the calculations for smaller and smaller lattice spacings, and larger and larger lattices to do a better job approximating physical reality.

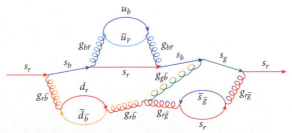

FIGURE 39.35 Example of a Feynman diagram of a red strange quark propagating through the QCD vacuum.

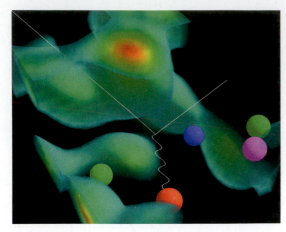

FIGURE 39.36 Artist's rendition of a proton. The red, green, and blue balls represent the location of the constituent quarks. The green-magenta ("anti-green") pair of balls on the right represents a virtual quark-antiquark pair. The white line represents a high-energy electron that scatters off one of the quarks through exchange of a photon (wavy line). All these objects are displayed on the background of a lattice-QCD calculation of the gluon field, as perfomed by the Adelaide, Australia, group.

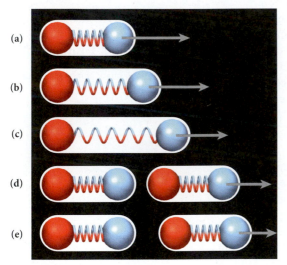

FIGURE 39.37 Schematic illustration of the fragmentaton of a color flux-tube.

Figure 39.36 shows the result of a lattice-QCD calculation for a snapshot of the spatial configuration of the gluon field inside a proton. Overlaid on this numerical result are drawings of three balls marking the constituent quarks of the proton, as well as two balls marking the position of a temporarily created virtual quark-antiquark pair.

Quarks, antiquarks, and gluons all carry color charge. QCD predicts that only color-neutral objects can exist in isolation. This means that a single quark or gluon cannot be observed by itself. Very sensitive searches for free quarks have been conducted, all of them with negative results.

Then what happens if we try to remove a quark from a baryon—for example, by shooting an electron with very high energy into a proton, trying to knock out a quark? The quark can acquire a very high momentum after the collision with the electron and move rapidly away from the other two quarks inside the proton. However, it remains tethered to them by a string of gluons called a *flux-tube*. QCD predicts that the energy contained in this flux-tube increases in proportion to its length. Once it is long enough and energetic enough, it can rupture and split into two or more color-neutral objects—mainly pions. This is illustrated schematically in Figure 39.37, which shows a meson consisting of a red quark and an anti-red antiquark. The antiquark receives an impulse and moves to the right, stretching the gluon flux-tube between them until it ruptures into another quark-antiquark pair. These two newly produced particles pair up with the original quark and antiquark into two new mesons, which can now move freely away from each other. If the impulse that the original quark receives becomes high enough, the color flux-tube can also rupture in multiple places, producing a quark-antiquark pair at each breakpoint. The result is a group of mesons, which are produced in the process of string fragmentation, and which are all traveling in roughly the same direction. Provided that the initial impulse is large enough, the jet formed by this group of newly produced particles can then be observed in particle physics detectors.

This means that all elementary particles carrying color charge—all gluons, quarks, and antiquarks—are absolutely confined. Confinement also implies that any attempt to remove a quark from its color singlet will only produce additional color-singlets, that is, hadrons.

The color force acts in such a way that the farther away objects with color charge move from each other, the more strongly they are attached to each other. Does this also mean the opposite, that is to say, that when color-charged particles inside a singlet are close to each other, can they then move freely? Indeed this is the case. At close distances or high relative momenta, color-charged objects experience so-called asymptotic freedom. This was predicted by David Gross, Frank Wilczek, and David Politzer in 1973 and subsequently confirmed experimentally. They shared the 2004 Nobel Prize in Physics for this achievement.

EXAMPLE 39.3 | **Feynman Diagrams of the Decay of Positive Pions and Positive Muons**

When high-energy cosmic rays consisting mainly of protons are incident on the Earth's atmosphere, reactions take place that produce positive pions. These pions decay with a meanlife of 26 ns into a positive muon and a muon neutrino

$$\pi^+ \rightarrow \mu^+ + \nu_\mu.$$

In turn, the positive muons decay with a meanlife of 2.2 μs into a positron, an electron neutrino, and an anti-muon-neutrino

$$\mu^+ \rightarrow e^+ + \nu_e + \bar{\nu}_\mu.$$

PROBLEM

Draw the Feynman diagrams representing these two decays.

SOLUTION

The positive pion is a meson composed of an up quark and an anti-down quark as shown in Figure 39.38.

This quark-antiquark pair $u/\bar{d}$ annihilates to produce a W^+. The W^+ then materializes as the lepton/anti-lepton pair ν_μ/μ^+.

The positive muon that is produced in the decay now emits a W^+ and transforms into an anti-muon-neutrino as shown in Figure 39.39.

The W^+ then materializes as the lepton/anti-lepton pair ν_e/e^+.

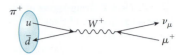

FIGURE 39.38 Feynman diagram of the decay of a positive pion.

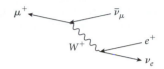

FIGURE 39.39 Feynman diagram for the decay of a positive muon.

DISCUSSION

The decay of positive pions is particularly interesting because the positive pions are produced at the top of the Earth's atmosphere with relatively high energies by reactions of cosmic ray protons with the nuclei of nitrogen and oxygen atoms. These pions decay into muons and the resulting muons travel at a large fraction of the speed of light. The muons should travel at most only 0.66 km before they decay into positrons. However, muons produced by cosmic rays are observed at the surface of the Earth. About 10 muons per second pass through your hand. One reason the muons reach the surface of the Earth is relativistic time dilation, as discussed in Chapter 35, Example 35.2. For a speed $\beta = v/c = 0.998$, the time dilation factor is

$$\gamma = \frac{1}{\sqrt{1-\beta^2}} = \frac{1}{\sqrt{1-(0.998)^2}} = 15.8.$$

Thus, the muons will travel a distance of 10 km before they decay, allowing the muons to reach the surface of the Earth. A second reason the muons reach the Earth's surface is that the muons can traverse the Earth's atmosphere without interacting with the nuclei in the atoms of the atmospheric gases, because the muon interacts only via the weak force and the electromagnetic force, and not via the strong force.

39.6 Big Bang Cosmology

The night sky presents an amazing display of light that has traveled for many years from distant stars (Figure 39.40). Since the dawn of civilization, humans have asked where all of this came from (and how, and why). Since the middle of the 20th century, scientists have been reasonably sure that they know the answer, and the picture is becoming clearer in ever more detail. Through the interplay of observational astronomy, cosmology, and nuclear and particle physics, the models have been refined.

The entire universe came into being through a singular event that we call the Big Bang. This event happened $(13.73\pm0.12)\cdot10^9$ years ago. Although the age of the universe cannot be fixed to an uncertainty smaller than 120 million years, physicists still are able to state how our universe developed in the first fractions of a second after the Big Bang.

The Planck time is the earliest time after the Big Bang for which it makes any sense to try to deduce a history of the universe (Figure 39.41). (See Example 39.2.) At a time of 10^{-43} s, the temperature was on the order of 10^{32} K. All forces were unified at that time, with gravity separating from the other forces a short time thereafter. The era of grand unification of the strong and electroweak forces is estimated to have lasted to about 10^{-34} s, when the strong force became separated from the electroweak force. At the end of this era, the universe underwent a rapid inflationary expansion. To understand

FIGURE 39.40 View of the universe from Earth in the visible spectrum, using the Aitoff projection in galactic coordinates. This projection shows the view over the entire night sky projected onto the ellipse shown here. The bright stars and clouds in the central horizontal band belong to the Milky Way.

FIGURE 39.41 Temperature of the universe as a function of time. The red circle with the cross marks the current universe, and the vertical gray lines mark the approximate boundaries between the different epochs. For each epoch, we sketch the basic constituents at that time. Radiation was present at all times and in equilibrium with matter until 10^{13} seconds, when the universe became transparent to radiation.

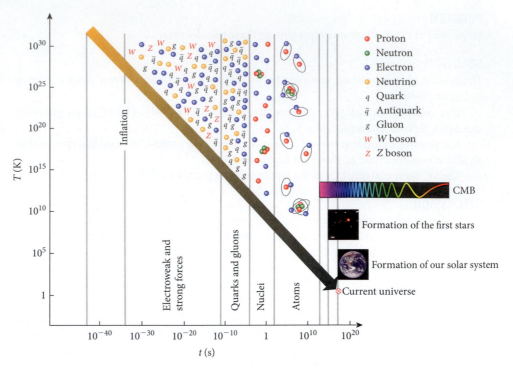

the necessity for this process to have been part of the history of our universe, we need to discuss the concept of inflation in cosmology a little further.

Inflation

Three important questions have puzzled astronomers for a long time:

1. Why is the cosmic microwave background radiation so incredibly isotropic, which is to say, smooth and very much the same in every direction?

2. Why is the universe made entirely of matter and not of antimatter?

3. Why is the universe so incredibly flat (a term we will explain shortly)?

All three questions deserve some elaboration to point out how truly startling these facts really are.

First, let's think about the smoothness of the cosmic microwave background. As we saw in Chapter 36, the cosmic microwave background (CMB) radiation shows a perfect black-body spectrum at a temperature of 2.725 K. CMB radiation was discovered in 1965 by Arno Penzias and Robert Woodrow Wilson, who received the Nobel Prize in Physics for this discovery in 1978. This radiation had been predicted in 1948 by George Gamov, Ralph Alpher, and Robert Herman as a consequence of the Big Bang model, which at that time was still very speculative and only one of several possible models of the time evolution of the universe. Only through the COBE and subsequently the WMAP satellite missions was it possible to measure just how isotropic this radiation is. This discovery earned the 2006 Nobel Prize in Physics for George Smoot and John C. Mather. Figure 39.42 shows the picture of the universe at microwave wavelengths. This image was produced by the team that operated the WMAP satellite. It shows that the fluctuations of the temperature of space as a function of direction are only of the order of $\left|\delta T/T\right| \sim 10^{-4}$ to 10^{-5}. For comparison, the WMAP team also produced a temperature map of the surface of Earth for June 1992, for which the fluctuations are on the order of $\left|\delta T/T\right| \sim 0.2$.

The smoothness of the CMB is so stunning because the radiation comes to us from all directions and has the same value for all points in the sky, even if they could not have been in causal contact during the history of the universe (that is, without inflation).

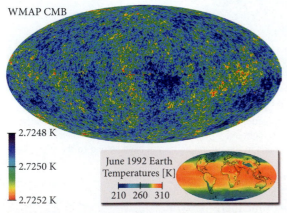

WMAP CMB

FIGURE 39.42 Temperature map of the Cosmic Microwave Background (CMB) radiation of the universe. Inset: temperature on the surface of Earth in June 1992.

Second, presently the universe is filled with matter and almost no antimatter. We know that initially matter and antimatter were produced in almost equal amounts and then almost completely annihilated into radiation, which now constitutes the cosmic microwave background. Estimates are that presently approximately 10^{10} photons exist for each hadron. This means that initially there were 10,000,000,001 matter particles for each 10,000,000,000 antimatter particles. Where does this very small asymmetry come from?

Third, after more than 13 billion years, the universe is still expanding, and we still cannot decide if it will eventually stop expanding or if it will go on forever. That means that the total kinetic energy of the universe is almost exactly equal to the total potential energy, but opposite in sign, so that the total energy of the universe must be unbelievably close to zero. We thus say that the universe is flat. This is analogous to the situation of shooting off a gun from the surface of Earth. If you shoot the gun so that the bullet has an initial speed that is large compared to the escape speed, then it will easily leave Earth, whereas if the initial speed is smaller than the escape speed, then the bullet will fall down again. Now imagine that after billions of years we still cannot decide if the bullet will eventually fall back down to Earth or will keep going! (As an aside, we note that very recent results seem to imply that the universe is dominated by dark energy, which is slowly causing the re-acceleration of the universe, presenting yet another mystery. We have already mentioned this result in Chapter 12. Thus it seems likely that, according to what we know now, that the universe will continue to expand forever.)

A model proposed by Alan Guth in 1980 and later expanded on by others answers all three questions in the same model framework, one that sounds like science fiction: A very short fraction of time after the Big Bang, the universe went through a very rapid period of inflation. Figure 39.43 depicts this inflation. The idea is that approximately 10^{-34} s after the Big Bang, the universe was at a temperature of around 10^{27} K and in a process of rapid cooling. It then became supercooled. This is analogous to freezing rain still being liquid while at a temperature at which water should be frozen. When the freezing rain hits the ground, the relatively small impact shock creates fluctuations that increase exponentially and cause the formation of very smooth ice. The supercooled state of the freezing raindrop is a false ground state, and small fluctuations cause it to make a transition to the true ground state, ice. At the onset of the period of inflation, the universe also entered a false ground state and became supercooled (albeit still at astronomically high temperature). The transition to the true ground state caused an exponential expansion in size, resulting in the smoothness of the distribution of matter and radiation still seen today in the cosmic microwave background radiation.

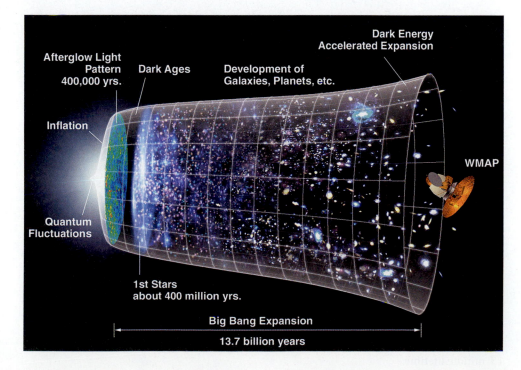

FIGURE 39.43 Illustration for the concept of inflation in the early universe.

After the reheating of the inflationary period, which may have lasted until 10^{-32} s, the universe entered a radiation-dominated era that lasted until approximately 10^{11} s. During this time the temperature T continued to fall gradually and inversely proportional to the square root of the time t,

$$T(t) = \frac{1.5 \cdot 10^{10} \text{ K s}^{1/2}}{\sqrt{t}}. \tag{39.17}$$

The electromagnetic and weak forces parted ways at approximately 10^{-11} s. At that point, the W and Z bosons became massive, through a process called spontaneous symmetry breaking, and subsequently decayed away rapidly. (Perhaps it is instructive to look at another example of spontaneous symmetry breaking, to illustrate the concept: In liquid water, the water molecules can have any arrangement and orientation. However, in the process of freezing, this symmetry has to be broken to force the molecules into the crystalline structure of ice.)

Quark-Gluon Plasma

After 10^{-11} s, and lasting until about 10^{-4} s, the universe was a mixture of quarks, gluons, and leptons, forming a plasma. Despite carrying color charges, the quarks and gluons were asymptotically free due to the high temperatures. At about 10^{-4} s the universe had cooled to a temperature of approximately $2.1 \cdot 10^{12}$ K (180 MeV). Lattice-QCD calculations indicate that at this temperature the quarks and gluons coalesced into color singlets.

Amazingly, the quark-gluon state of matter that dominated the early universe from 10^{-11} s to 10^{-4} s after the Big Bang can be recreated in accelerator-based experiments today. Experiments at the Relativistic Heavy Ion Collider in Brookhaven, Long Island (and starting in 2010 at the Large Hadron Collider), use colliding beams of gold nuclei with total kinetic energy of up to 20 TeV each at RHIC (and up to 550 TeV at the LHC) to probe the conditions during the early time evolution of the universe and to produce small volumes of the quark-gluon matter. However, this state of matter only lasts less than 10^{-23} s in the laboratory before it explodes into more than 5000 particles, mainly pions. A RHIC event is displayed in Figure 39.44.

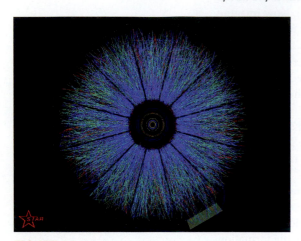

FIGURE 39.44 Event display from a gold + gold collision at the Relativistic Heavy Ion Collider. Each line represents the track that one of the over 5000 produced particles has left in the STAR detector. From these tracks scientists are trying to determine the properties of the quark-gluon state that was created transiently during the nuclear collision and learn about the history of our universe.

Nucleosynthesis

As the universe emerged from its quark-gluon phase at a time of 10^{-4} s after the Big Bang, it consisted overwhelmingly of photons, electrons, positrons, neutrinos, and anti-neutrinos. A small number of protons and neutrons—the lightest color singlets into which the quarks could condense—were also present. The protons and neutrons were constantly converted into each other via the reactions

$$n + \nu_e \rightleftarrows p + e^-$$
$$n + e^+ \rightleftarrows p + \bar{\nu}_e. \tag{39.18}$$

This was easily possible because the temperature was still on the order of 10^{11} or 10^{12} K, which corresponds to 10 to 100 MeV, much larger than the mass difference of the neutron and the proton, which is only 1.293 MeV/c^2. This means that the protons and neutrons were in equilibrium, with their relative numbers determined by the ratio of their Boltzmann factors:

$$\frac{n_n}{n_p} = e^{(m_n c^2 - m_p c^2)/k_B T}. \tag{39.19}$$

As the universe reached a temperature of approximately 10^{10} K (~0.86 MeV), the reactions of equation 39.18 proceeded too slowly to maintain equilibrium. At this point, the ratio of neutrons to protons froze out at the value $e^{-1.293/0.86} = 0.222$. This means that there were then about 22 neutrons for every 100 protons present. At this time, the universe was at an age of 1 second. However, the temperature was still too high at this time for nuclei to form. Instead, all neutrons remained free, and thus free to beta decay according to equation 39.16, with a 15-minute lifetime.

After 100 s, the universe had cooled down to 10^9 K, and due to beta decay, only about 16 neutrons were left for every 100 protons. This was the point where alpha particles (consisting of two neutrons and two protons each) could form, and essentially all available neutrons were bound in them. Small traces of deuterium, tritium, helium-3, and lithium (more on these in the next chapter) were also formed. Sixteen neutrons and 16 protons can form 8 alpha particles. Therefore, the total mass fraction of nucleons trapped inside alpha particles at that time was 16 + 16 out of 100 + 16: (16+16)/(100 + 16) ~ 27%. This rough estimate is consistent with the observed primordial mass fraction of 23% helium (alpha nucleus + electrons). This successful prediction is considered one of the greatest successes of the Big Bang theory.

The radiation dominance ended at approximately 10^{11} s (~ 3000 years), when the temperature had reached 10^5 K. As the temperature fell further below approximately 3000 K (~ 0.25 eV), electrons were captured by the protons and alpha particles to form neutral hydrogen and helium atoms. A photon needs to have at least 10.2 eV energy to excite a hydrogen atom electron from the 1s to the 2s state (see Chapter 38), so the photons then had many fewer electrons to scatter from, and thus the opacity of the universe dropped. At that time, photons were able to roam freely through the universe and the radiation decoupled from matter. The photons thus showed a perfect blackbody spectrum with the temperature of the universe at the time of decoupling. The universe was then about 10^{13} s (~ 300,000 years) old. These freely roaming photons are still visible today, and they form the cosmic microwave background (CMB).

Why are the CMB photons now showing a temperature of 2.725 K, approximately 1000 times lower than the original temperature? The answer is that the photons have been stretched due to the expansion of the universe. The scale factor of the universe is now approximately a factor $R = 1100$ times bigger than it was at the time the CMB decoupled. This means that the wavelengths of these photons are now $\lambda = R\lambda_0$. According to Wien's displacement law (see Chapter 36), the wavelength λ_m at which the blackbody spectrum has its maximum is related to the temperature as $\lambda_m T = $ const. Consequently, with the expansion of the wavelengths by a factor of 1100, the temperature has to fall by the same factor. (Figure 39.41 indicates this effect schematically by stretching the wavelength of the CMB photons and shifting the color toward the longer wavelengths.)

What followed in the universe was a period of darkness, of a few million years perhaps, before the first stars were formed through the gravitational interaction in the gas of hydrogen and helium atoms. Our own Solar System formed approximately 4.5 billion years ago, about 9 billion years after the Big Bang. In Figure 39.41, this point in time corresponds to the last gray vertical line, which in this logarithmic scale is only ever so slightly to the left of the present, which is marked by the red dot at a time of $4.3 \cdot 10^{17}$ s after the Big Bang. Chapter 40 will return to the nuclear physics that went into the formation of the heavy elements that make up the majority of our Earth. Their presence cannot be explained based on Big Bang cosmology alone and requires other events, which we will explore.

From this short narration of the Big Bang cosmology, it is clear how particle physics at the smallest imaginable scales is intimately connected to astronomy at the largest possible scales. This amazing connection between the smallest and largest has fascinated physicists for the last 60 years and continues to do so, with new results being discovered each year.

WHAT WE HAVE LEARNED | EXAM STUDY GUIDE

- The reductionist quest to reduce systems into their component parts is a main theme in science, and has its present culmination in particle physics. While an equally valid methodology exists to understand complexity and emergent structure, the present chapter focuses on reductionism in particle physics.

- Substructure is probed using scattering experiments. The scattering cross section is defined as

$$\sigma = \frac{\text{Number of reactions per scattering center/s}}{\text{Number of impinging particles/s/m}^2}.$$

- A cross section has the physical dimension of area and is measured in units of barn (b), or millibarn (mb): 1 b = 10^{-28} m^2, 1 mb = 10^{-31} m^2.

- The classical Rutherford cross section for scattering from a pointlike target by the Coulomb interaction is

$$\frac{d\sigma}{d\Omega} = \left(\frac{kZ_p Z_t e^2}{4K}\right)^2 \frac{1}{\sin^4\left(\frac{1}{2}\theta\right)}.$$

- For scattering of a plane wave off a point source, the scattering wave function is

$$\psi_{total}(\vec{r}) = \psi_i(\vec{r}) + \psi_f(\vec{r}) = N\left(e^{i\kappa z} + f(\theta)\frac{e^{i\kappa r}}{r}\right),$$

and the cross section is related to the scattering amplitude as

$$\frac{d\sigma}{d\Omega} = |f(\theta)|^2.$$

- The form factor is the Fourier transform of the density distribution,

$$F^2(\Delta p) = \left|\frac{1}{e}\int \rho(\vec{r})e^{i\Delta\vec{p}\cdot\vec{r}/\hbar}dV\right|^2$$

and measures the deviation of the scattering cross section from the Rutherford cross section of a pointlike target,

$$\frac{d\sigma}{d\Omega} = \left(\frac{d\sigma}{d\Omega}\right)_{point} \cdot F^2(\Delta p).$$

- Elementary fermions have spin $\frac{1}{2}\hbar$ and include the six quarks (up, down, strange, charm, bottom, top), the electron-, muon-, and tau-leptons, and the electron-, muon-, and tau-neutrinos. Each of these 12 fermions has an antiparticle. Quarks all have non-integer charges of $-\frac{1}{3}e$ or $+\frac{2}{3}e$ and cannot be observed in isolation.

- Elementary bosons are the mediators of the interactions between the fermions. They are the photon (electromagnetic), W and Z boson (electroweak), gluon (strong), and graviton (gravitational). The graviton is yet to be found experimentally. Gluons can also interact with other gluons.

- The interaction of fundamental particles can be represented pictorially by Feynman diagrams, and calculations can be performed based on expressing each step of the interaction in terms of Feynman diagrams.

- The standard model of particle physics has several possible extensions, none of which has received experimental confirmation up to now. At the Planck scale, a grand unified theory of all forces is postulated, which allows for decay of the proton on very long time scales. Supersymmetry is another extension of the standard model, which postulates the existence of a supersymmetric partner for every known fermion and boson. Other theories with fundamental appeal are string theories, which hold that the dimensionality of space-time is 11, with the other dimensions compactified on very small scales.

- Elementary quarks and antiquarks can combine to form color singlets, which are particles that can be observed in isolation. A quark and an antiquark can form a meson (pion, kaon, etc.). Three quarks can form baryons (proton, neutron, delta, lambda, etc.). The only stable baryon is the proton, and none of the mesons is stable. Lifetimes of the unstable particles vary from 10^{-23} s to 15 minutes.

- Because gluons carry color charge, they can interact with other gluons. This leads to confinement of quarks. The theory of the interaction of quarks and gluons, quantum chromodynamics, can only be solved analytically with perturbation theory in the realm of large momentum transfer and short distances. There the elementary particles show asymptotic freedom. For low-momentum transfers and large distances, the only solution method is lattice-QCD simulated on a computer.

- The history of the early universe at fractions of a second after the Big Bang is dominated by particle physics. The early inflationary phase has left traces in the isotropy of the cosmic microwave background radiation, which originated 300,000 years after the Big Bang at a temperature of 3000 K and cooled down to the presently observed 2.725 K due to the expansion of the universe. The quark-gluon plasma phase transition of the early universe can be probed in the laboratory with relativistic heavy ion collisions. The primordial fraction of 23% helium relative to hydrogen in the universe can be explained from the neutron-proton mass difference, which fixed the ratio of proton and neutron numbers to

$$n_n/n_p = e^{(m_n c^2 - m_p c^2)/k_B T}.$$

KEY TERMS

NEW SYMBOLS AND EQUATIONS

$\sigma = \dfrac{\text{Number of reactions per scattering center/s}}{\text{Number of impinging particles/s/m}^2}$, scattering

cross section

$\alpha = \dfrac{ke^2}{\hbar c} = \dfrac{e^2}{4\pi\epsilon_0 \hbar c} = \dfrac{1}{137.036} \approx \dfrac{1}{137}$, fine-structure constant

$\ell_{\text{net}} = n_\ell - n_{\bar{\ell}} = $ constant, total lepton number

$Q_{\text{net}} = n_q - n_{\bar{q}} = $ constant, net quark number

ANSWERS TO SELF-TEST OPPORTUNITIES

39.1 $c \rightarrow s + \mu^+ + \nu_\mu$ and $t \rightarrow b + \tau^+ + \nu_\tau$

39.2 The Planck energy is most straightforwardly obtained by multiplying the Planck mass with two powers of the speed of light:

$$E_P = m_P c^2 = \sqrt{\dfrac{\hbar c^5}{G}} = 1.96 \cdot 10^9 \, \text{J} = 1.22 \cdot 10^{19} \, \text{GeV}.$$

39.3 Most of these are already listed in Table 39.4, with the exception of the Σ^* and Ξ^*, which have the same constituent quark composition as the Σ and Ξ, respectively.

MULTIPLE-CHOICE QUESTIONS

39.1 The de Broglie wavelength, λ, of a 5-MeV alpha particle is 6.4 fm, as shown in this chapter, and the closest distance, r_{min}, to the gold nucleus this alpha particle can get is 45.5 fm (calculated in Example 39.1). Based on the fact that $\lambda \ll r_{\text{min}}$, one can conclude that, for this Rutherford scattering experiment, it is adequate to treat the alpha particle as a

a) particle.   b) wave.

39.2 Which of the following is a composite particle? (select all that apply)

a) electron   c) proton

b) neutrino   d) muon

39.3 Which of the following formed latest in the universe?

a) quarks   d) helium nuclei

b) protons and neutrons   e) gluons

c) hydrogen atoms

39.4 An exchange particle for the weak force is the

a) photon.   c) W boson.   e) gluon.

b) meson.   d) graviton.

39.5 Which of the following particles does not have integer spins?

a) photon   c) ω meson

b) π meson   d) ν_e lepton

39.6 At about what kinetic energy will the length scale probed by the α particle cross over from the classical formula to the relativistic formula?

a) 0.3 GeV   c) 3 GeV

b) 1 GeV   d) 10 GeV

39.7 Which of the following experiments proved the existence of the nucleus?

a) the photoelectric effect

b) the Millikan oil-drop experiment

c) the Rutherford scattering experiment

d) the Stern-Gerlach experiment

QUESTIONS

39.8 Which of the following reactions cannot occur, and why?

a) $p \rightarrow \pi^+ \pi^0$

b) $p\pi^0 \rightarrow ne^+$

c) $\Lambda^0(1116) \rightarrow pK^- \pi^+$

d) $\Lambda^0(1450) \rightarrow pK^- \pi^+$

39.9 Between neutron scattering and electromagnetic waves scattering (like X-rays or light), which of the two would be more appropriate for investigating the scattering cross section of the atom as a whole, and which would be more appropriate for investigating the nucleus of an atom? Which one will depend on Z, the atomic number?

39.10 Consider a hypothetical force mediated by the exchange of bosons that have the same mass as protons. Approximately what would be the maximum range of such a force? You may assume that the total energy of these particles is simply the rest-mass energy and that they travel close to the speed of light. If you do not make these assumptions and instead use the relativistic expression for total energy, what happens to your estimate of the maximum range of the force?

39.11 Looking at Table 39.3, do the constituent quarks uniquely define the type of meson?

39.12 A free neutron decays into a proton and an electron (and an anti-neutrino). A free proton has never been observed to decay into anything. Why then do we consider the neutron to be as "fundamental" (at the nuclear level) a particle as the proton? Why do we not consider a neutron to be a proton-electron composite?

39.13 In a positron annihilation experiment, positrons are directed toward a material such as a metal. What are we likely to observe in such an experiment, and how might it provide information about the momentum of electrons in the metal?

39.14 If the energy of the virtual photon mediating an electron-proton scattering, $e^- + p \to e^- + p$, is given by E, what is the range of this electromagnetic interaction in terms of E?

39.15 Describe the following Feynman diagrams and the physical processes they represent:

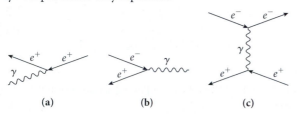

    (a)   (b)   (c)

39.16 The text describes and sketches the basic Feynman diagram for the fundamental process involved in the decay of the free neutron: One of the neutron's d-quarks converts to a u-quark, emitting a virtual W^- boson, which decays into an electron and an electron anti-neutrino (the only decay energetically possible). Similarly describe and sketch the basic (tree-level) Feynman diagram for the fundamental process involved in each of the following decays:

a) $\mu^- \to e^- + \nu_\mu + \bar{\nu}_e$ d) $K^+ \to \mu^+ + \nu_\mu$

b) $\tau^- \to \pi^- + \nu_\tau$ e) $\Lambda \to p + \pi^-$

c) $\Delta^{++} \to p + \pi^+$

39.17 Does the proposed decay $n \to p + \pi^-$ violate any conservation rules?

39.18 Consider the proposed decay $\pi^+ \to \mu^+ + \nu_\mu + \nu_e$. Can this decay occur?

39.19 Consider the proposed reaction $\pi^0 + n \to K^- + \Sigma^+$. Can this reaction occur?

39.20 How do we know for certain that the scattering process $e^+ + \nu_\mu \to e^+ + \nu_\mu$ proceeds through an intermediate Z boson, and cannot proceed through an intermediate charged W boson, while both options are possible for $e^+ + \nu_e \to e^+ + \nu_e$?

39.21 What baryons have a quark content of uds? What is the mass of these baryons?

39.22 In the following Feynman diagram for proton-neutron scattering, what is the virtual particle?

PROBLEMS

A blue problem number indicates a worked-out solution is available in the Student Solutions Manual. One • and two •• indicate increasing level of problem difficulty.

Section 39.2

39.23 A 4.50-MeV alpha particle is incident on a platinum nucleus ($Z = 78$). What is the minimum distance of approach, r_{min}?

39.24 A 6.50-MeV alpha particle is incident on a lead nucleus. Because of the Coulomb force between them, the alpha particle will approach the nucleus only to a minimum distance, r_{min}.

a) Determine r_{min}.

b) If the kinetic energy of the alpha particle is increased, will the particle's distance of approach increase, decrease, or remain the same? Explain.

39.25 A 6.50-MeV alpha particle scatters at a 60° angle off a lead nucleus. Determine the differential scattering cross section of the alpha particle.

39.26 Protons with kinetic energy of 2 MeV scatter off gold nuclei in a foil target. Each gold nucleus contains 79 protons. If both the incoming protons and the gold nuclei can be treated as point objects, what is the differential scattering cross section for the protons to scatter off the gold nuclei at an angle of 30° from their initial velocity?

•**39.27** The de Broglie wavelength, λ, of a 5.00-MeV alpha particle is 6.40 fm, as shown in this chapter, and the closest distance, r_{min}, to the gold nucleus this alpha particle can get is 45.5 fm (calculated in Example 39.1). How does the ratio r_{min}/λ vary with the kinetic energy of the alpha particle?

•**39.28** A Geiger-Marsden type experiment is done by bombarding a 1-μm thick gold foil with 8-MeV alpha rays. Calculate the fraction of particles scattered to an angle

a) between 5° and 6° and

b) between 30° and 31°.

(The atomic mass number of gold is 197 and its density is 19.3 g/cm³.)

•**39.29** The differential scattering cross section for particles to scatter by 55° off a target is $4.0 \cdot 10^{-18} \frac{m^2}{sr}$. A detector with an area of 1.0 cm² is placed 1.0 m away from the target in order to detect particles that have been scattered at 55°. If $3.0 \cdot 10^{17}$

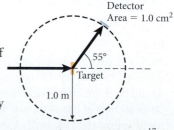

particles hit the 1-mm^2-area target every second, how many will strike the detector every second?

•**39.30** Some particle detectors measure the total number of particles integrated over part of a sphere of radius R from the target. Assuming symmetry about the axis of the incoming particle beam, use the Rutherford scattering formula to obtain the total number of particles detected as a function of the scattering angle θ.

••**39.31** Evaluate the form factor and the Coulomb scattering differential cross section $d\sigma/d\Omega$ for a beam of electrons scattering off a *uniform-density* charged sphere of total charge Ze and radius R. Describe the scattering pattern.

Section 39.3

39.32 A proton is made of 2 up quarks and a down quark (*uud*). Calculate its charge.

39.33 Use the fact that the observed magnetic moment of the proton is $1.4 \cdot 10^{-26}$ A m^2 to estimate the speed of the quarks. For this estimate, assume that the quarks move in circular orbits of radius 0.80 fm and that they all move at the same speed and direction. Ignore any relativistic effects.

39.34 Determine the approximate probing distance of a photon with an energy of 2.0 keV.

39.35 Draw a Feynman diagram for an electron-proton scattering, $e^- + p \rightarrow e^- + p$, mediated by photon exchange.

39.36 Based on the information in Table 39.2, what is the approximate upper bound on the range of a reaction mediated by the Higgs boson?

39.37 Draw possible Feynman diagrams for the following phenomena:

a) protons scattering off each other

b) neutron beta decays to a proton: $n \rightarrow p + e^- + \bar{\nu}_e$

•**39.38** A proton and neutron interact via the strong nuclear force. Their interaction is mediated by a particle called a meson, much like the interaction between charged particles is mediated by photons—the particles of the electromagnetic field.

a) Perform a rough estimate of the mass of the meson from the uncertainty principle and the known dimensions of a nucleus ($\sim 10^{-15}$ m). Assume the meson travels at relativistic speed.

b) Use a line of reasoning similar to the one in part (a) to prove that the theoretically expected rest mass of the photon is zero.

Section 39.5

39.39 How many fundamental fermions are there in a carbon dioxide molecule?

39.40 Suppose a neutral pion at rest decays into two identical photons.

a) What is the energy of each photon?

b) What is the frequency of each photon?

c) To what part of the electromagnetic spectrum does this correspond?

•**39.41** Draw a quark-level Feynman diagram for the decay of a neutral kaon into two charged pions, $K^0 \rightarrow \pi^+ + \pi^-$.

Section 39.6

39.42 During the radiation-dominated era of the universe, the temperature was falling gradually according to equation 39.17. Using Stefan's Law, find the time dependence of background-radiation intensity during the radiation-dominated era.

39.43 Use equation 39.17 to estimate the temperature and age of the universe at about the time protons and neutrons began to form.

39.44 Three hundred thousand years after the Big Bang, the average temperature of the universe was about 3000 K.

a) At what wavelength of radiation would the blackbody spectrum peak for this temperature?

b) To what portion of the electromagnetic spectrum does this correspond?

•**39.45** At about 10^{-6} s after the Big Bang, the universe had cooled to a temperature of approximately 10^{13} K.

a) Calculate the thermal energy.

b) Explain what would happen to most of the hadrons—protons and neutrons.

c) Explain also about the electrons and positrons in terms of temperature and time.

•**39.46** Three hundred thousand years after the Big Bang, the temperature of the universe was 3000 K. Because of expansion the temperature of the universe is now 2.75 K. Modeling the universe as an ideal gas and assuming that the expansion of the universe is adiabatic, calculate how much the volume of the universe has changed. If the process is irreversible, estimate the change in the entropy of the universe based on the change in volume.

•**39.47** The fundamental observation underlying the Big Bang theory of cosmology is Edwin Hubble's 1929 discovery that the arrangement of galaxies throughout space is expanding. Like the photons of the cosmic microwave background, the light from distant galaxies is stretched to longer wavelengths by the expansion of the universe. This is *not* a Doppler shift: Except for their local motions around each other, the galaxies are essentially at rest in space; it is the space itself that expands. The ratio of the wavelength of light λ_{rec} Earth receives from a galaxy to its wavelength λ_{emit} at emission is equal to the ratio of the scale factor (e.g., radius of curvature) a of the universe at reception to its value at emission. The redshift z of the light—which is what Hubble could measure—is defined by $1 + z = \lambda_{rec}/\lambda_{emit} = a_{rec}/a_{emit}$.

a) *Hubble's Law* states that the redshift z of light from a galaxy is proportional to the galaxy's distance from us (for reasonably nearby galaxies): $z \cong c^{-1}H\Delta s$, where c is the vacuum speed of light, H is the *Hubble constant,* and Δs is the distance of the galaxy. Derive this law from the first relationships stated in the problem, and determine the Hubble constant in terms of the scale-factor function $a(t)$.

b) If the present Hubble constant has the value $H_0 = 72$ (km/s)/Mpc, how far away is a galaxy, the light from which has redshift $z = 0.10$? (The megaparsec (Mpc) is a unit of length equal to $3.26 \cdot 10^6$ light-years. For comparison, the Great Nebula in Andromeda is approximately 0.60 Mpc from us.)

Additional Problems

39.48 What is the minimum energy of a photon capable of producing an electron-positron pair? What is the wavelength of this photon?

39.49 a) Calculate the kinetic energy of a neutron that has a de Broglie's wavelength of 0.15 nm. Compare this with the energy of an X-ray photon that has the same wavelength.

b) Comment on how this would be relevant for investigating biological samples with neutrons vs. X-rays.

39.50 A photon can interact with matter by producing a proton-antiproton pair. What is the minimum energy the photon must have?

39.51 Suppose you had been doing an experiment to probe structure on a scale such that you needed electrons of 100. eV kinetic energy. Then a neutron beam became available to you to do the experiment. What energy of neutrons would you need to get the same resolution?

39.52 What is the de Broglie wavelength of an α particle that has a kinetic energy of 100. MeV? According to Figure 39.13, how does this wavelength compare to the distance that can be probed by this α particle?

39.53 One of the elementary bosons for the electroweak interaction is the Z^0 boson having the mass of 91.1876 GeV/c^2. Find the order of magnitude of the range of the weak interaction.

39.54 What are the wavelengths of the two photons produced when a proton and antiproton at rest annihilate?

39.55 Estimate the cross section of a Λ particle decay (into $p\pi^0$, $pe^-\nu_e$) if the time it takes for this weak interaction to occur is $\sim 10^{-10}$ s.

39.56 Determine the classical differential cross section for Rutherford scattering of alpha particles of energy 5 MeV projected at uranium atoms at an angle of 35° from the initial direction. Assume point charges for both the target and the projectile atoms.

39.57 The Rutherford experiment successfully demonstrated the existence of the nucleus and put limits on the size of the nucleus from the scattering of alpha particles from gold

foils. Assume the alpha particle was fired with a speed of about 5% of the speed of light.

a) Derive the upper bound of the radius of nucleus in terms of speed of the alpha particle that is scattered in the backward direction.

b) Calculate the approximate radius of the gold nucleus with this information.

•**39.58** If a neutrino beam of 300 GeV energy is passed through a 1-cm-thick sheet of aluminum, how much of the beam is attenuated, if the cross section of interactions is given by $\sigma(E) = (10^{-38}\ \mathrm{cm}^2\ \mathrm{GeV}^{-1})E$?

•**39.59** An electron-positron pair, traveling toward each other with a speed of $0.99c$ with respect to their center of mass, collide and annihilate according to $e^- + e^+ \to \gamma + \gamma$. Assuming the observer is at rest with respect to the center of mass of the electron-positron pair, what is the wavelength of the photons?

•**39.60** A Geiger-Marsden experiment, where α particles are scattered off of a thin gold film, yields an intensity of particles of $I(90°) = 100$ counts/s at a scattering angle of $90° \pm 1°$. What will be the intensity of particles at a scattering angle of $60° \pm 1°$ if the scattering obeys the Rutherford formula?

•**39.61** Electron and positron beams are brought together into collision, and pairs of tau-leptons are produced. If the angular distribution in the laboratory of the tau-leptons varies as $(1 + \cos^2\theta)$, what fraction of the tau-lepton pairs would be captured in a detector that covers only the angles 60° to 120°?

•**39.62** Within three years after it begins operation, the proton beam at the Large Hadron Collider at CERN is expected to reach a luminosity of $10^{34}\ \mathrm{cm}^{-2}\mathrm{s}^{-1}$ (this means that in a 1-cm^2 area, 10^{34} protons encounter each other every second). The cross section for collisions, which could lead to direct evidence of the Higgs boson, is approximately 1 pb (picobarn). [These numbers were obtained from "Introduction to LHC physics," by G. Polesello, *Journal of Physics: Conference Series* **53** (2006), 107–116.] If the accelerator runs without interruption, approximately how many of these Higgs events can one expect in one year at the LHC?

••**39.63** Evaluate the form factor and the Coulomb-scattering differential cross section $d\sigma/d\Omega$ for a beam of electrons scattering off a *thin spherical shell* of total charge Ze and radius a. Could this scattering experiment distinguish between the thin-shell and solid-sphere charge distributions? Explain.

Nuclear Physics

40

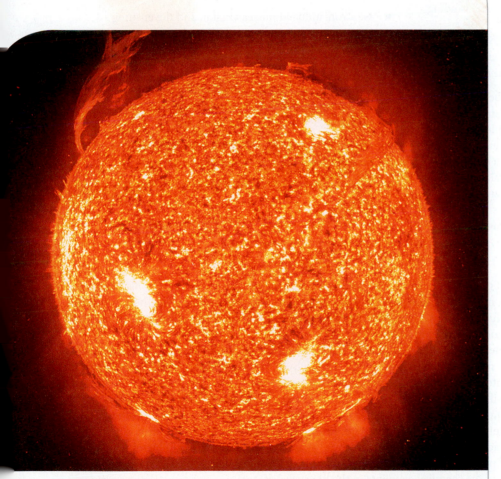

FIGURE 40.1 Our Sun, the giant nuclear fusion reactor that provides almost all of our energy resources, taken with the Extreme Ultraviolet Imaging Telescope on the *SOHO* spacecraft using UV light with wavelength 30.4 nm.

1325

WHAT WE WILL LEARN

- The atomic nucleus is composed of nearly equal numbers of protons and neutrons (nucleons), densely packed so that the nuclear volume increases approximately linearly with the mass number (which equals the number of nucleons).

- Nucleons interact primarily via the strong interaction, which has a short range, on the order of 10^{-15} m.

- Only 251 stable isotopes are known to exist. More than 2400 unstable isotopes have been discovered up to now, with several thousand additional isotopes yet to be discovered.

- Nuclear masses can be measured very precisely, typically with the precision of one part in 100 million. From knowing the masses, we can calculate the binding energy of the isotopes, as well as the Q-values of nuclear reactions. The Q-value determines whether or not a reaction is allowed by energy conservation, and is used to find how much energy the reaction releases.

- Several different models of the nucleus can be constructed. One model is based on the assumption that nuclear matter behaves like a liquid drop. This liquid-drop model explains the systematics of the nuclear binding energy. Another model is based on a quantum gas of particles trapped inside the nuclear walls. A third class of models takes into account the angular momentum quantum numbers and yields a shell model picture not unlike the electron shells in the atom.

- Unstable isotopes and their excited states can decay in three primary ways. Alpha decay is the emission of a helium-4 nucleus. Beta decay is a weak interaction process and results in either the emission of an anti-neutrino and an electron, the emission of a neutrino and a positron, or the capture of an electron by the nucleus along with the emission of a neutrino. Gamma decay is the emission of a high-energy photon.

- Energy can be released from nuclear reactions in two primary ways—fission and fusion. In fission, a massive nucleus splits into two medium-mass nuclei and a few neutrons. In fusion, two light nuclei merge into a heavier one.

- Neutron-induced nuclear fission and the ensuing chain reactions can be used for nuclear power plants, but also for nuclear weapons.

- Nuclear fusion powers most stars and is responsible for the creation of the light emitted by stars as well as the medium-mass elements up to iron and nickel.

- Elements heavier than iron are predominantly produced in supernova explosions.

- Nuclear physics has many applications in medicine, both in diagnostics and in the radiation treatment of cancers.

Almost all of the energy resources used on Earth trace back to nuclear reactions. The overwhelming majority of these resources are due to our Sun (Figure 40.1), which at its core is a nuclear fusion reactor. Given this dependence on nuclear reactions, it becomes important to try to understand them better.

Chapter 39 explained that most of the mass of an atom resides in a small central region called the nucleus. This chapter explores the physics of the nucleus, which offers unique insights into the behavior of matter at its most basic level.

We will examine several models that help explain the properties of the nucleus, particularly those that explain the stability of some nuclei and the decay modes of others. That most nuclei are unstable, or radioactive, has led to many important applications of nuclear physics, including nuclear power, nuclear weapons, and nuclear medicine. In addition, studies of nuclear properties have contributed to our understanding of astronomy and of the origin of the chemical elements in the life cycles of stars.

40.1 Nuclear Properties

Isotopes

The atomic nucleus consists of two kinds of **nucleons**: protons and neutrons. These nucleons are held together by the strong interaction. However, not all combinations of protons and neutrons are possible, because of the limitations imposed by the strong interaction, the Coulomb interaction, and quantum mechanics. The number of protons, Z, inside a nucleus determines the element that is formed. Z is the charge number, which is sometimes called the atomic number. For a given element, different numbers of neutrons, N, are possible. Nuclei of the same element with different numbers of neutrons are called **isotopes.** The total number of protons and neutrons combined is called the **mass number,** A

$$A = Z + N. \tag{40.1}$$

Conventionally, a given nucleus is denoted by the symbol that represents its atom in the periodic table, with a superscript for the mass number of the isotope on the left and a subscript for the charge number on the left, and sometimes a subscript for the number of neutrons on the right. For example, the nucleus formed by 8 protons and 9 neutrons is written as $^{17}_{8}O_{9}$, and the nucleus formed by 92 protons and 146 neutrons is written as $^{238}_{92}U_{146}$. This is the most complete notation for isotopes. However, most of the time the subscript on the right is omitted, because the neutron number is simply $N = A - Z$ and thus can be computed easily with the information provided by the sub- and superscripts on the left. Most practicing nuclear physicists also omit the subscript for the charge number; they simply assume we know that uranium has 92 protons and oxygen has 8 protons. Thus, you can see the above two isotopes written in many nuclear physics textbooks as ^{17}O (pronounced: "oxygen seventeen") and ^{238}U (pronounced: "uranium two-thirty-eight"). In this book, we will carry the subscript for the charge number, but not that for the neutron number. The above isotopes of oxygen and uranium are thus written as $^{17}_{8}O$ and $^{238}_{92}U$.

The only two isotopes that have historically received a special name are the hydrogen isotopes of deuterium ($^{2}_{1}H \equiv d$, an atom made up of a nucleus with a proton and a neutron, and a single electron) and tritium ($^{3}_{1}H \equiv t$, an atom made up of a nucleus with a proton and two neutrons, and a single electron).

What kinds of isotopes are found in nature? The answer to this question will occupy us several times during the course of this chapter. The overwhelming majority of isotopes decay in time. Only 251 stable isotopes are known, and they are shown in Figure 40.2, where each of them is marked by a purple square at the point that marks its neutron and proton number.

Right away, we can see that there is no stable isotope with $Z > 83$ (bismuth). Also, note that there are no stable isotopes for $Z = 43$ (technetium) or for $Z = 61$ (promethium). Further, it is apparent that only a few stable isotopes occur for each Z value, and they are located along a narrow valley in this plot, the "valley of stability." For small values of N, this valley follows the $N = Z$ line, which is shown as the diagonal gray line in the plot. However, for $N \sim 20$ the valley of stability begins to veer off this line toward the neutron-rich ($N > Z$) side. This effect is due to the Coulomb interaction among the protons, which also limits the size of the largest nucleus. The next section discusses this feature in more quantitative detail.

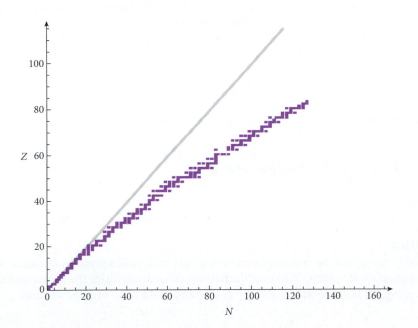

FIGURE 40.2 Stable isotopes, marked with purple squares, as a function of the neutron and proton number. The gray line marks the $N = Z$ line.

Nuclear Interactions

It is not convenient to describe the strong interaction between nucleons based on gluon exchange using the formalism of QCD, described in Chapter 39. Instead, it is more efficient to construct effective interactions based on the exchange of color-singlets between the nucleons (which are also color-singlets). Since the least massive color-singlets are mesons, and the least massive of these is the pion, description of the nucleon-nucleon interaction in terms of pion exchange has been very successful. Historically, these pion exchange potentials had already been formulated before the strong interaction was understood in terms of quarks and gluons, but now we begin to understand why they have been so successful.

The Japanese physicist Hideki Yukawa (1907–1981) invented the pion exchange potential theory in 1935 and won the 1949 Nobel Prize in Physics for this achievement. The Yukawa potential is conventionally written in the form

$$U(r) = -g^2 \frac{e^{-r/R_\pi}}{r},$$ (40.2)

where g is a real number and is the effective coupling constant in the same way that ke^2 is the coupling constant for the Coulomb potential of the electrostatic interaction. Here R_π is the range of the potential and is given by the pion mass as

$$R_\pi = \frac{\hbar}{m_\pi c} = \frac{\hbar c}{m_\pi c^2} = \frac{197.3 \text{ MeV fm}}{139.6 \text{ MeV}} = 1.413 \text{ fm}.$$

The derivative of the one-pion exchange potential with respect to r is greater than zero, so the corresponding force ($F(r) = -dU(r)/dr$) is attractive at all distances r.

The one-pion exchange potential is a very good approximation to the nucleon-nucleon potential at large separation. However, at shorter distances, two- and three-pion exchanges dominate the nucleon-nucleon interaction. Here the effective nucleon-nucleon potential is strongly repulsive at short distances, thus preventing nucleons from penetrating each other. Schematically, the nucleon-nucleon potential is shown in Figure 40.3. It varies depending on the spin and isospin projections (proton or neutron) of the two nucleons involved in the interaction, but has the general shape shown in the graph, with the minimum located at approximately 1 fm.

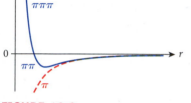

FIGURE 40.3 Schematic drawing of the nucleon-nucleon potential (blue). At the point labeled $\pi\pi$, two-pion exchange begins to dominate. At the point labeled $\pi\pi\pi$, three-pion exchange becomes important. The one-pion exchange Yukawa potential is shown in red.

Nuclear Radius and Nuclear Density

Because the strong nuclear interaction is very short-range, what is most important for nuclear physics is the nearest-neighbor interaction between nucleons. The repulsive core of the potential prevents the nucleons from penetrating each other, so it has become common to visualize the nucleus as a roughly spherically shaped and densely packed collection of nucleons. This means that the volume of the nucleus should be proportional to the mass number A. Since the volume of a sphere is proportional to the third power of the radius, we find that the third power of the nuclear radius is proportional to its mass number, or, alternatively,

$$R(A) = R_0 A^{1/3},$$ (40.3)

where the constant $R_0 = 1.12$ fm has been determined experimentally.

EXAMPLE **40.1** / **Nuclear Density**

PROBLEM

What is the nuclear matter density, that is, the mass density inside an atomic nucleus?

SOLUTION

Since the nucleus can be approximated by a sphere with radius given by equation 40.3, we can calculate its volume. We know the number of nucleons inside the sphere and their mass, so we can find the nuclear density as the ratio of the mass over the volume.

The volume of a nucleus of mass number A is

$$V = \frac{4\pi R(A)^3}{3} = \frac{4\pi R_0^3 A}{3} = (5.88 \text{ fm}^3)A.$$

This tells us that on average a nucleon occupies approximately 5.9 fm^3 of space inside a nucleus. We can then give the number density of nucleons inside the nucleus as

$$n = \frac{A}{V} = \frac{A}{(5.88 \text{ fm}^3)A} = 0.170 \text{ fm}^{-3}.$$

Thus, an atomic nucleus has 0.17 nucleons per cubic-femtometer. The mass of a nucleon is approximately $1.67 \cdot 10^{-27}$ kg. Multiplying this mass of a single nucleon with the number density of nucleons gives us the mass density of atomic nuclei:

$$\rho = m_{\text{nucleon}} n = m_{\text{nucleon}} \frac{A}{V} = \left(1.67 \cdot 10^{-27} \text{ kg}\right)\left(0.170 \, (10^{-15} \text{ m})^{-3}\right) = 2.84 \cdot 10^{17} \text{ kg/m}^3.$$

For comparison, the density of liquid water is 10^3 kg/m^3; thus nuclear matter density is approximately 280 trillion times higher than the density of liquid water.

Electron-scattering experiments of the kind described in Chapter 39 have established that the density is approximately constant in the interior of a heavy nucleus, and it falls off gradually at the surface. The dependence of the number density on the radial coordinate r can be described by the Fermi function

$$n(r) = \frac{n_0}{1 + e^{(r - R(A))/a}} \qquad (40.4)$$

where $R(A)$ is given by equation 40.3, and the constant a has the value of 0.54 fm (Figure 40.4). The distance over which the density falls from 90% of its central value to 10% of the central value is conventionally defined as the nuclear surface thickness t. By using equation 40.4, we can show that $t \approx 4.4a$ (see Solved Problem 40.2). Also, using the numerical value $a = 0.54$ fm, the nuclear surface thickness for large nuclei is found to be approximately $t = 2.4$ fm.

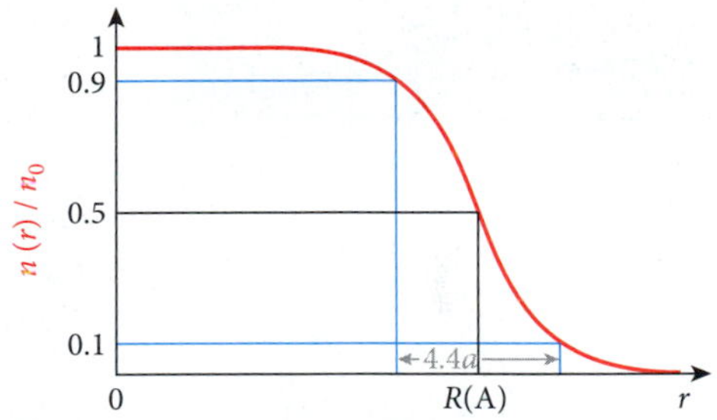

Nuclear Lifetimes

The stable isotopes shown in Figure 40.2 are not the only isotopes that can exist. A huge number of unstable isotopes are produced through natural nuclear decays and interactions, or in the laboratory. The mean lifetime of an unstable isotope, which we will also call its **nuclear lifetime,** is the average time it exists prior to decaying. A quantitative discussion of lifetimes is presented in Section 40.3. Lifetimes vary over an incredible range, from greater than the age of the universe to less than a microsecond. So far, approximately 2400 unstable isotopes are known to exist in addition to the 251 stable ones. Theoretical predictions for the number of isotopes that can possibly exist range up to approximately 6000, so there are many yet to be discovered.

Figure 40.5 shows the isotopes for which lifetimes have been measured. Each square represents an isotope, and the color of each square indicates the isotope's lifetime, according to the scale shown on the right of the figure. In general, lifetimes are longest for isotopes that are near the stable isotopes, with some lifetimes (shown in dark red) even exceeding the age of the universe (which is approximately $4.3 \cdot 10^{17}$ s). Farther away from the stable isotopes, lifetimes become rapidly shorter. Note in particular the very short lifetimes of all isotopes with neutron

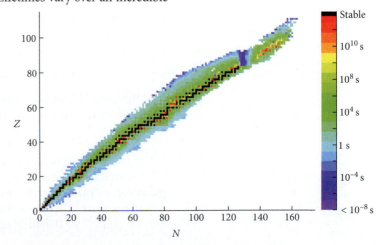

FIGURE 40.4 Nuclear density profile as a function of the radial coordinate.

FIGURE 40.5 Measured lifetimes of the known isotopes.

numbers around 130, and then the longer lifetimes for larger neutron numbers and proton numbers around 90. These longer lifetimes include the isotopes of the actinides, most notably thorium, uranium, and plutonium. Uranium has no stable isotopes, but the isotopes $^{235}_{92}\text{U}$ and $^{238}_{92}\text{U}$ have lifetimes of 700 million years and 4.5 billion years, respectively, and thus live long enough to still be found in large quantities on Earth. Part of our task in the following discussion is to understand the systematic trends of the observed lifetimes.

Nuclear and Atomic Mass

The masses of the proton and of the neutron are known to great precision. The values for the proton mass and the proton mass-energy equivalent are

$$m_p = 1.672\ 621\ 637(83) \cdot 10^{-27}\ \text{kg}$$

$$m_p c^2 = 1.503\ 277\ 359(75) \cdot 10^{-10}\ \text{J}$$

$$= 938.272\ 013(23)\ \text{MeV}$$

and for the neutron mass and neutron mass-energy equivalent

$$m_n = 1.674\ 927\ 211(84) \cdot 10^{-27}\ \text{kg}$$

$$m_n c^2 = 1.505\ 349\ 505(75) \cdot 10^{-10}\ \text{J}$$

$$= 939.565\ 346(23)\ \text{MeV}.$$

The numbers in brackets ( ) indicate that the last two digits are uncertain by the amount in brackets, which is a concise notation for the standard ± way of denoting the uncertainties. For example, the notation 1.672 621 637(83) is equivalent to writing 1.672 621 637 ± 0.000 000 083. The stated uncertainties in the proton and neutron masses mean that they are known to approximately one part in 10 million. Interestingly, what limits the precision for the proton and neutron mass is the accuracy to which Avogadro's constant (Chapter 13) is determined, or equivalently, to what precision the SI system standard kg measure is known (Chapter 1).

Mass measurements for nuclei also reach this precision, even if the isotopes in question live only a few seconds or even a fraction of a second. One way to achieve this measurement is to trap a single ion of a given atom in an electromagnetic trap inside a magnetic field (Figure 40.6a) and then measure its cyclotron frequency $\omega = qB/m \Rightarrow m = qB/\omega$ (see Chapter 27). Ion manipulation and storage are made possible by the electrode configurations shown in Figure 40.6b.

Because the charge of the ion is known precisely, because it is an integer multiple of the electron charge, and because frequencies can be measured to essentially arbitrary accuracy by simply counting cycles, the limit of this type of mass measurement depends on the accuracy with which the strength of the magnetic field B can be measured. The magnetic field cannot be measured to the necessary accuracy, so we use a known reference atom and measure its cyclotron frequency in the same trap, thus measuring the unknown mass relative to the known one.

The reference mass usually chosen is the isotope $^{12}_{6}\text{C}$. Because this isotope consists of 12 nucleons, the **atomic mass unit (u)** is defined as exactly $\frac{1}{12}$ of the mass of a $^{12}_{6}\text{C}$ atom (mass of the nucleus of $^{12}_{6}\text{C}$ plus the mass of the six electrons bound to the nucleus via the Coulomb interaction). The conversion between the u, kg, and MeV/c^2 is

$$1\ \text{u} = 1.660538782(83) \cdot 10^{-27}\ \text{kg} = 931.494028(23)\ \text{MeV}/c^2. \tag{40.5}$$

Note that some older references use "amu" instead of "u," and in chemistry the term dalton (Da) is often used instead of u. The atomic mass unit is 1 gram divided by Avogadro's number,

$$1\ \text{u} = 1\ \text{g}/N_A.$$

If masses are stated in terms of the atomic mass unit and the above definition of u is used, then the masses of other atoms can be measured relative to that of $^{12}_{6}\text{C}$. This enables measurement of high precision because it is not limited by the precision to which Avogadro's number is known. For example, the proton and neutron masses can be specified to 10 significant digits (1 in 10 billion accuracy!) as

$$m_p = 1.007\ 276\ 466\ 77(10)\ \text{u}$$

$$m_n = 1.008\ 664\ 915\ 97(43)\ \text{u}. \tag{40.6}$$

(a)

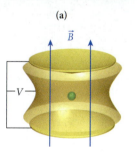

(b)

FIGURE 40.6 (a) Ion trap, which is used for mass measurements by measuring the ion's cyclotron frequency. A U.S. silver dollar is shown for scale comparison. (b) The configuration of the electrodes and magnetic field of the ion trap. The green sphere signifies the region where the ions are trapped.

Why use the mass of the neutral atom of $^{12}_6C$, including its six electrons, as a reference value, instead of the mass of only the nucleus of $^{12}_6C$? This is simply for convenience, because it is very hard to strip all the electrons off an atom and only measure the mass of the nucleus. For the same reason, masses of all isotopes are always listed as atomic masses and contain the same number of electrons as protons.

An ion trap like the one shown in Figure 40.6 can yield a precision of one part in 100 million for the mass measurement of an atom with a lifetime of only 1 second. This precision is equivalent to measuring the mass of a convoy of 10 large 18-wheel trucks, each of mass 20 tons, to the accuracy of the weight of a single dime in the pocket of one of the truck drivers!

The mass of a nucleus is not simply the sum of the masses of the protons and neutrons contained in it. Instead, the nucleus is a bound object, and it takes energy to pull it apart into its constituents. Chapter 35 showed that energy is stored in the form of mass, and that energy and mass are related through the famous Einstein formula $E = mc^2$. Thus, the **binding energy** $B(N, Z)$ of a nucleus that consists of N neutrons and Z protons can be written as the difference between the mass-energy of the collection of N neutrons plus Z hydrogen atoms (consisting of 1 proton and 1 electron each) and the mass-energy of the atom of mass $m(N, Z)$, consisting of N neutrons, Z protons, and Z electrons:

$$B(N,Z) = Zm(0,1)c^2 + Nm_n c^2 - m(N,Z)c^2. \qquad (40.7)$$

Here $m(0,1)$ is the mass of the hydrogen atom with 0 neutrons, 1 proton, and 1 electron,

$$m(0,1) = 1.007825032 \text{ u}.$$

Note that this value is slightly bigger than the proton mass given in equation 40.6; the difference is due to the electron mass.

While equation 40.7 gives an expression for the total binding energy of a nucleus, it is more instructive to examine the binding energy per nucleon,

$$\frac{B(N,Z)}{(N+Z)} = \frac{B(N,Z)}{A}. \qquad (40.8)$$

Figure 40.7 shows the binding energy per nucleon (blue dots) for all stable isotopes as a function of the mass number, A. There is a strong increase for small Z, with a spike at $Z = 2$, a data point representing the binding energy of the nucleus of the helium atom—that is, the alpha particle. The value of the binding energy per nucleon of the alpha particle is $B(^4_2He)/A = 7.074$ MeV. The curve reaches a maximum at iron ($Z = 26$) and nickel ($Z = 28$). The highest experimentally measured values of the binding energy per nucleon are $B(^{62}_{28}Ni)/A = 8.795$ MeV, $B(^{58}_{26}Fe)/A = 8.792$ MeV, and $B(^{56}_{26}Fe)/A = 8.790$ MeV (indicated by the yellow circles in Figure 40.7). For $Z > 28$ and $A > 60$, the binding energy per nucleon falls gradually to a value slightly below 8 MeV. For $A > 100$ the binding energy per nucleon falls very nearly linearly with A, with a slope of

$$\left. \frac{\Delta(B/A)}{\Delta A} \right|_{A>100} = -7.1 \cdot 10^{-3} \text{ MeV}.$$

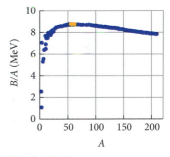

FIGURE 40.7 Binding energy per nucleon as a function of the mass number for all stable isotopes.

In Section 40.3, we will construct models for the nucleus and try to understand why the binding energies of nuclei exhibit the trends shown in Figure 40.7.

Another way to express how well a nucleus is bound is the **mass excess,** defined as the difference between the mass of a nucleus and the mass number times the atomic mass unit:
Mass excess = $m(N,Z) - A$ u.

The mass excess can be expressed in terms of energy units by converting the atomic mass units using equation 40.5. The binding energy is defined in terms of the mass of the neutron and the hydrogen atom, while the mass excess is defined in terms of the mass of $^{12}_6C$. The mass of $^{12}_6C$ is defined to be 12 u, so that its mass excess is zero. Thus, the mass excess and the binding energy are similar but are not the same. If the mass excess of a nucleus is very negative, it will have a large binding energy per nucleon, as defined by equations 40.7 and 40.8. For example, one of the most tightly bound nuclei, $^{56}_{26}Fe$, has a mass excess of -60.6 MeV/c^2 and a binding energy per nucleon of 8.79 MeV/c^2. The mass excesses for nuclei up to $Z = 40$ are shown in Figure 40.8. The valley of stability described in Section 40.1 is clearly visible in this plot.

FIGURE 40.8 Mass excesses for nuclei up to $Z = 40$.

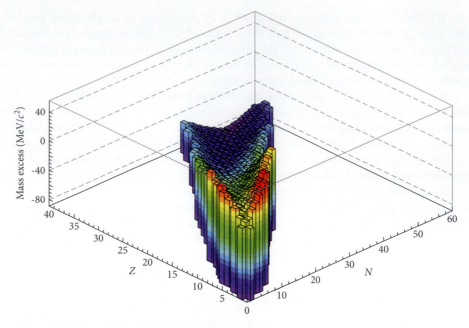

FIGURE 40.8 Mass excesses for nuclei up to $Z = 40$.

Nuclear Reactions and Q-Values

The calculation of the binding energy of a nuclear isotope in equation 40.7 is a special case of a larger class of problems. The binding energy is simply the energy that must be supplied to break up one nucleus consisting of N neutrons and Z protons into its individual nucleons. In general, we can ask about the net energy change due to any rearrangement of an arbitrary group of neutrons and protons from an initial distribution into a final distribution. This rearrangement is called a **nuclear reaction,** in analogy with chemical reactions in which atoms get redistributed among different molecules. In chemical reactions, the number of atoms of a given species on the right-hand side (final state) of a reaction equation is exactly the same as that on the left-hand side (initial state). In nuclear reactions a similar conservation law is observed: Because baryon number is a conserved quantity, the number of nucleons on the left-hand side and right-hand side is the same. In addition, the number of protons and the number of neutrons are also separately conserved, with reactions that involve the weak force constituting the only exception. (The weak force was discussed in Chapter 39, in the form of the beta decay of the neutron, or equivalently, of the down quark.)

In practically all nuclear reactions, the initial state consists of either one or two nuclei, but not more, because nuclear sizes and therefore nuclear cross sections (see Chapter 39) are so small that the probability of three or more nuclei running into each other simultaneously is negligible.

The energy difference between the initial and final states is conventionally called the **Q-value** of the reaction. If the masses of all isotopes involved are known, then the Q-value is easily computed as the difference in the sum of the masses of the initial state nuclei minus the sum of the masses of the final state nuclei. For example, for an initial state composed of a deuteron ($^2_1$H nucleus) and $^{12}_6$C, and a final state composed of an isolated proton and $^{13}_6$C, this reaction can be written as $^2_1$H + $^{12}_6$C $\rightarrow$ p + $^{13}_6$C, and the Q-value of this reaction is computed as $Q = m(1,1)c^2 + m(6,6)c^2 - (m(0,1)c^2 + m(7,6)c^2)$. In this calculation, we have used the mass of a hydrogen atom $m(0,1)$ and the mass of a deuterium atom $m(1,1)$.

Note: An often-used alternative notation for the same reaction is $^{12}_6$C($^2_1$H,p)$^{13}_6$C or also $^{12}_6$C(d,p)$^{13}_6$C. Such "d,p" reactions are very popular tools for exploring nuclear structure.

Why is the Q-value an interesting quantity? The answer is the same as in chemistry: The Q-value indicates whether the reaction is exothermic (Q > 0) or endothermic (Q < 0)—in other words, is energy derived from this reaction, or must energy be put in to make the reaction work. Many concepts and applications of nuclear physics depend on the Q-value, and we will repeatedly return to it in this chapter.

If we know the masses of the isotopes, then we can also ask how much energy it takes to separate some part of a particular isotope away from the remainder of that nucleus. In general, for a division of a nucleus with N neutrons and Z protons into two smaller nuclei with neutron numbers N_1 and N_2, and proton numbers Z_1 and Z_2, the Q-value of this process can be computed as

$$Q_{12} = m(N,Z)c^2 - m(N_1,Z_1)c^2 - m(N_2,Z_2)c^2 \qquad (40.9)$$

where $N_1 + N_2 = N$ and $Z_1 + Z_2 = Z$. The negative of the Q-value associated with this separation process, $-Q_{12}$, is called the **separation energy,** denoted by S; so $S = -Q_{12}$. If $S > 0$, then energy is required to separate the nucleus into parts 1 and 2, while if $Q_{12} > 0$, then energy is released when the separation takes place.

In these separation reactions the number of protons and the numbers of neutrons remain the same in the initial and final states, so equation 40.7 can be used and the separation energy can be expressed as the difference in binding energies:

$$S = B(N_1+N_2, Z_1+Z_2) - B(N_1,Z_1) - B(N_2,Z_2). \qquad (40.10)$$

In the special case that one of the two nuclei is an alpha particle, this separation energy is usually denoted by the symbol S_α. Other conventionally quoted separation energies are that for proton emission, S_p, and single- and double-neutron emission, S_n and S_{2n}.

EXAMPLE 40.2 Separation Energy

PROBLEM
The binding energy per nucleon of the tin isotopes $^{136}_{50}\text{Sn}$, $^{134}_{50}\text{Sn}$, $^{132}_{50}\text{Sn}$, $^{130}_{50}\text{Sn}$, $^{128}_{50}\text{Sn}$, and $^{126}_{50}\text{Sn}$ are measured as 8.1991 MeV, 8.2778 MeV, 8.3549 MeV, 8.3868 MeV, 8.4167 MeV, and 8.4435 MeV, respectively. What are the two-neutron separation energies of the first five of these isotopes?

SOLUTION
From the given values of the binding energy per nucleon, we can obtain the total binding energy of the isotopes by multiplication with their respective number of nucleons. We thus find

$$^{136}_{50}\text{Sn:}\quad B(86,50) = 136 \cdot 8.1991\ \text{MeV} = 1115.08\ \text{MeV}$$

$$^{134}_{50}\text{Sn:}\quad B(84,50) = 134 \cdot 8.2778\ \text{MeV} = 1109.23\ \text{MeV}$$

$$^{132}_{50}\text{Sn:}\quad B(82,50) = 132 \cdot 8.3549\ \text{MeV} = 1102.85\ \text{MeV}$$

$$^{130}_{50}\text{Sn:}\quad B(80,50) = 130 \cdot 8.3868\ \text{MeV} = 1090.28\ \text{MeV}$$

$$^{128}_{50}\text{Sn:}\quad B(78,50) = 128 \cdot 8.4167\ \text{MeV} = 1077.34\ \text{MeV}$$

$$^{126}_{50}\text{Sn:}\quad B(76,50) = 126 \cdot 8.4435\ \text{MeV} = 1063.88\ \text{MeV}.$$

Since two neutrons do not form a bound state, the binding energy of the two neutrons is zero. Thus, in general, we have for the two-neutron separation energy the simple formula

$$S_{2n}(N,Z) = B(N,Z) - B(N-2,Z). \qquad (40.11)$$

Using the values for the total binding energies that we have just computed, and inserting them into equation 40.11, we then find

$$S_{2n}(^{136}_{50}\text{Sn}) = (1115.08 - 1109.23)\ \text{MeV} = 5.85\ \text{MeV}$$

$$S_{2n}(^{134}_{50}\text{Sn}) = (1109.23 - 1102.85)\ \text{MeV} = 6.38\ \text{MeV}$$

$$S_{2n}(^{132}_{50}\text{Sn}) = (1102.85 - 1090.28)\ \text{MeV} = 12.57\ \text{MeV}$$

$$S_{2n}(^{130}_{50}\text{Sn}) = (1090.28 - 1077.34)\ \text{MeV} = 12.94\ \text{MeV}$$

$$S_{2n}(^{128}_{50}\text{Sn}) = (1077.34 - 1063.88)\ \text{MeV} = 13.46\ \text{MeV}.$$

This is a very interesting result. It shows that it suddenly becomes much harder to remove a pair of neutrons from a tin isotope as the neutron number reaches 82. Why is there a big jump in the value of the two-neutron separation energy at this neutron number? This question will be answered in Section 40.3's discussion of the nuclear shell model.

40.2   Nuclear Decay

As we've noted, not all nuclear isotopes found in nature are stable. An example is uranium, which can be found on Earth in three naturally occurring isotopes: $^{238}_{92}U$ (99.3% abundance), $^{235}_{92}U$ (0.7%), and traces of $^{234}_{92}U$. They all decay naturally over very long times, and thus are still present in appreciable quantities that have survived since the time Earth was formed approximately 4.5 billion years ago. In this section, we look at what processes make nuclei unstable and cause them to decay over time. We will discuss α, β, and γ decays, as well as other decays. These nuclear decays are collectively called **radioactivity.** Radioactivity was discovered in 1896 by Pierre (1859–1906) and Marie Curie (1867–1934) and by Henry Becquerel (1852–1908), for which the three shared the 1903 Nobel Prize in Chemistry.

Exponential Decay Law

Because the laws of quantum mechanics govern atomic nuclei, all decays can be viewed as transitions from one quantum state to another. Thus, they follow quantum mechanical probability rules like those for tunneling in Chapters 36 and 37. It is possible to calculate most of these decays by using quantum mechanics, but we do not need to do this here. All we need to know to understand radioactive decays is that the probability of observing a decay in a given set of atomic nuclei in a given time interval dt is proportional to the number of nuclei present. Letting the rate of change of the number of nuclei be dN/dt, this proportionality can be expressed as

$$dN = -\lambda N dt \Leftrightarrow \frac{dN}{dt} = -\lambda N,$$

where λ is the decay constant. (The minus sign indicates that nuclei are lost as a function of time.) The solution of this differential equation leads to the exponential decay law

$$N(t) = N_0 e^{-\lambda t}, \qquad (40.12)$$

where N_0 is the initial number of nuclei and $N(t)$ is the number of nuclei that remain as a function of time. Figure 40.9 shows plots of equation 40.12.

The **half-life,** $t_{1/2}$, is defined as the time it takes a quantity of nuclei of a given material to decay to half of its number,

$$N(t_{1/2}) = \frac{1}{2} N_0. \qquad (40.13)$$

After two half-lives, the population has decreased to one-quarter of its initial value, and after three half-lives, to one-eighth. The decay constant can be related to the half-life by inserting equation 40.13 into equation 40.12. This results in

$$\frac{1}{2} N_0 = N_0 e^{-\lambda t_{1/2}} \Rightarrow$$

$$\frac{1}{2} = e^{-\lambda t_{1/2}} \Rightarrow$$

$$\ln\frac{1}{2} = -\lambda t_{1/2}$$

$$t_{1/2} = \frac{\ln 2}{\lambda}. \qquad (40.14)$$

It is also common to talk about the **mean lifetime,** τ. This is defined as the average time it takes a nucleus to decay if the population of nuclei obeys the exponential decay law (equation 40.12). The mean lifetime is obtained by integration:

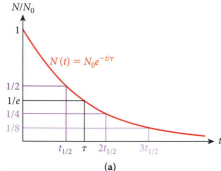

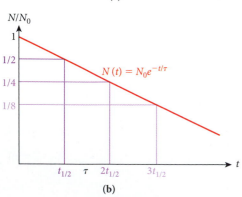

FIGURE 40.9 Exponential decay in time: (a) linear plot; (b) logarithmic plot.

$$\tau = \left\langle N(t) \right\rangle_t = \frac{\int_0^\infty t\, N(t)\, dt}{\int_0^\infty N(t)\, dt} = \frac{\int_0^\infty t\, N_0 e^{-\lambda t}\, dt}{\int_0^\infty N_0 e^{-\lambda t}\, dt} = \frac{N_0(-1/\lambda^2)e^{-\lambda t}(1+\lambda t)\Big|_0^\infty}{N_0(-1/\lambda)e^{-\lambda t}\Big|_0^\infty} = \frac{1}{\lambda}. \qquad (40.15)$$

Thus, the mean lifetime τ is simply the inverse of the decay constant λ. Therefore, as an alternative to equation 40.12, the exponential decay law can be written as

$$N(t) = N_0 e^{-t/\tau}.$$

After one mean lifetime, the population has been reduced by a factor of $1/e$:

$$N(\tau) = N_0 e^{-\tau/\tau} = N_0/e.$$

Finally, combining equations 40.15 and 40.14, the half-life $t_{1/2}$ and the mean lifetime τ are related via

$$t_{1/2} = \frac{\ln 2}{\lambda} = \tau \ln 2.$$

Thus, the half-life is not a half of the lifetime, but a factor $\ln 2 \approx 0.693$ of the lifetime. For example, Chapter 39 quoted the lifetime of the neutron as 885.7 s. This means that its half-life is $(885.7 \text{ s})\ln 2 = 613.9$ s.

The mean lifetime τ of isotopes is the physical quantity displayed in Figure 40.5. What kinds of nuclear decays leading to the lifetimes shown in Figure 40.5 are possible in nature?

The three main nuclear decays are the emission of an alpha particle, the emission of an electron or positron (or, equivalently, the capture of an electron), and the emission of a photon. These decays constitute the three components of what is commonly called radioactivity or radioactive decays. They can be very harmful to human health, completely harmless, or in some cases even very helpful in medical diagnostics and treatment. Their effect on health depends on the type of decay, the energy of the decay product, and the radiation dose, that is, the amount of radioactive material present and the amount of radiation emitted. Figure 40.10 shows how the different radioactive decays change the nucleus that decays. The following sections discuss each decay channel in more detail.

In all decays, the decaying nucleus is called the *parent* and the nucleus it decays into is the *daughter*. If the parent and daughter nucleus are different elements, the process is known as **transmutation.** Only those decays that obey the conservation laws, in particular those of energy, charge, and baryon number, are possible.

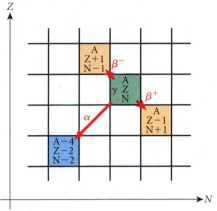

FIGURE 40.10 Nuclear decays in the chart of isotopes.

Alpha Decay

In an **alpha decay** (α-decay), the nucleus emits an α-particle, which is the nucleus of a helium atom $_2^4\text{He}$. This means that the mass number of the parent nucleus decreases by four and the charge number by two,

$$_Z^A\text{Nuc} \rightarrow {}_2^4\text{He} + {}_{Z-2}^{A-4}\text{Nuc}'. \tag{40.16}$$

(The notation Nuc' is meant to indicate a nucleus that is different in its composition from the initial nucleus before the decay.) In general, alpha decays are possible when the energy contained in the mass of the α-particle plus the mass of the daughter nucleus $m\left({}_{Z-2}^{A-4}\text{Nuc}'\right)$ is smaller than the mass of the nucleus that undergoes the alpha decay:

$$m\left({}_Z^A\text{Nuc}\right) > m_\alpha + m\left({}_{Z-2}^{A-4}\text{Nuc}'\right). \tag{40.17}$$

Since the binding energy per nucleon of the alpha particle (of mass m_α) is very large, 7.074 MeV, the mass of the alpha particle is comparatively low. In addition, as shown in Figure 40.7, the binding energy per nucleon falls very gradually as a function of the mass number for nuclei with large mass number A. This makes alpha decay possible for almost all unstable isotopes with $A > 150$.

Chapter 37 mentioned in passing that alpha decay is an example of tunneling—the transmission of the wave function of the alpha particle through a classically forbidden region. This is illustrated in Figure 40.11, where the total energy of the nuclear system is sketched as a function of their separation r between the center of the alpha particle and the center of the daughter nucleus. If the four nucleons that constitute the alpha particle are approximately in the center of the parent nucleus, then the total energy is approximately just the energy

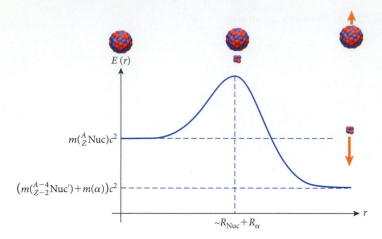

FIGURE 40.11 Sketch of the potential energy of the alpha and the daughter nucleus, showing the potential barrier (located at the vertical dashed line) in alpha decay.

contained in the mass of the parent nucleus, $m(^A_Z\text{Nuc})c^2$, as shown in Figure 40.11 at $r = 0$. When the alpha particle and the daughter nucleus are widely separated, $r \to \infty$, the interaction between the two becomes negligible, and the total energy is the mass-energy of the daughter nucleus plus that of the alpha particle, $\left(m\left(^{A-4}_{Z-2}\text{Nuc}'\right)+m(\alpha)\right)c^2$. For many heavy nuclei, this value of the total energy is *lower* than the value at the center. Therefore, the emission of the alpha particle from the parent nucleus is energetically favorable.

However, first the alpha particle has to get out of the parent nucleus. Moving the four nucleons of the alpha particle to one side or the other deforms the nucleus. This deformation adds excitation to the nucleus, and the potential energy increases relative to the value at $r = 0$. At a configuration where the alpha particle and daughter nucleus barely touch, indicated by the vertical dashed line in Figure 40.11, the potential energy has a maximum, because of the Coulomb repulsion between the alpha particle and the daughter nucleus. This increase in potential energy forms a potential barrier and prevents spontaneous alpha decay. However, the wave function of the alpha particle can tunnel through this potential barrier, leading to the emission of the alpha particle. The tunneling probability and thus the lifetime of the nucleus against alpha decay depends very strongly on the shape (mainly width, but also height) of the barrier.

When the alpha particle escapes from the nucleus, the energy difference between the mass-energy of the parent and daughter nuclei, which is the Q-value of the reaction, is converted into kinetic energy of the alpha particle and the heavy remnant,

$$K_\alpha + K_{\text{Nuc}'} = Q = \left(m\left(^A_Z\text{Nuc}\right) - m\left(^{A-4}_{Z-2}\text{Nuc}'\right) - m_\alpha\right)c^2. \tag{40.18}$$

Total momentum is conserved, so the momentum of the alpha particle and the momentum of the daughter nucleus must be equal in magnitude and opposite in direction in the rest frame of the parent nucleus, $|\vec{p}_\alpha| = |\vec{p}_{\text{Nuc}'}|$. For the low kinetic energies at work here, a non-relativistic approximation of $K = p^2/2m$ is sufficient. Momentum conservation then means that the kinetic energy of the daughter nucleus is related to the kinetic energy of the alpha particle via

$$K_{\text{Nuc}'}m\left(^{A-4}_{Z-2}\text{Nuc}'\right) = K_\alpha m_\alpha \Leftrightarrow K_{\text{Nuc}'} = K_\alpha \frac{m_\alpha}{m\left(^{A-4}_{Z-2}\text{Nuc}'\right)}.$$

Inserting this result into equation 40.18, the kinetic energy of the alpha particle in the rest frame of the parent nucleus is

$$K_\alpha = \frac{m\left(^{A-4}_{Z-2}\text{Nuc}'\right)}{m\left(^{A-4}_{Z-2}\text{Nuc}'\right) + m_\alpha}\left(m\left(^A_Z\text{Nuc}\right) - m\left(^{A-4}_{Z-2}\text{Nuc}'\right) - m_\alpha\right)c^2. \tag{40.19}$$

EXAMPLE 40.3 | Roentgenium Decay

By shooting a beam of $^{64}_{28}\text{Ni}$ nuclei on a $^{209}_{83}\text{Bi}$ target in December 1994, a group at the GSI national laboratory in Germany was able to produce a new element with 111 protons and 161 neutrons through the reaction $^{64}_{28}\text{Ni} + ^{209}_{83}\text{Bi} \to ^{272}_{111}\text{Rg} + n$. Here the Rg stands for the name Roentgenium, a name that this new element officially received in November 2006. The GSI group detected three events in which the new element was produced through the signature of successive alpha decays of the new element and its known daughter nuclei. One of the three events, with the measured alpha decay times, is shown in Figure 40.12.

PROBLEM

The masses of the nuclei in the decay chain are listed in the nuclear data tables as 272.1536, 268.138728, 264.1246, 260.1113, 256.098629, and 252.08656 in units of u. What alpha energies would you predict?

SOLUTION

Since the masses of the isotopes are given, we can calculate the Q-value for each decay by using equation 40.18. We can then use equation 40.19 to calculate the expected kinetic energy of the alpha particle. This is done in the table below for each of the decays.

The last column of the table compares our predictions with the experimental results that were reported by the GSI group. As you can see, these numbers are in reasonably good agreement with the calculated values.

DISCUSSION

In practice, the masses of the so-called superheavy elements with triple-digit charge numbers are mostly determined from the measurements of the kinetic energies of the alpha particles of the isotopes' decays. What we list in the table is only the result of one particular event in one particular experiment. The isotope masses are determined from best fits to all available data. For the heaviest elements, these data consist of only a few events, but for charge numbers below 100, millions of events have been recorded. The fact that the observed alpha kinetic energies of the daughter nuclei and their decay times agreed so well with previously measured data served to convince the GSI group that they had indeed seen the first events of the production of the new element Roentgenium.

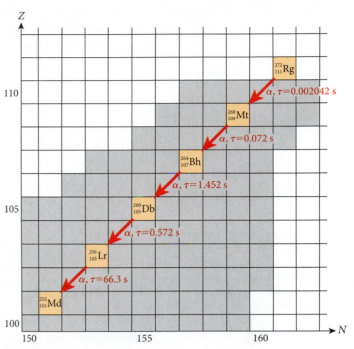

FIGURE 40.12 Roentgenium decay chain as observed on Dec. 17, 1994, at the GSI. The alpha decay chain proceeds through Meitnerium (Mt), Bohrium (Bh), Dubnium (Db), and Lawrencium (Lr) to Mendelevium (Md). The gray boxes indicate isotopes that have been observed up until now.

Name	A	Z	m [u]	$m + m$[u]	Q [MeV]	K_α [MeV]	K_α data [MeV]
Rg	272	111	272.1536				
Mt	268	109	268.138728	272.1413313	11.42826	11.3	10.82
Bh	264	107	264.1246	268.1272033	10.73523	10.6	10.221
Db	260	105	260.1113	264.1139033	9.96395	9.8	9.621
Lr	256	103	256.098629	260.1012323	9.37804	9.2	9.2
Md	252	101	252.08656	256.0891633	8.81729	8.7	8.463

Beta Decay

In a **beta decay** (β-decay), the nucleus emits an electron e^- or a positron e^+ or captures one of its own atomic electrons. Chapter 39 discussed the beta decay of quarks, particularly the β^--decay of the down quark into the up quark: $d \rightarrow u + e^- + \bar{\nu}_e$. Since the neutron is composed of two down quarks and an up quark, one way that the β^--decay of the down quark manifests itself is the β^--decay of the neutron, $n \rightarrow p + e^- + \bar{\nu}_e$, discussed earlier. The general formula of a nuclear β^--decay can be written as

$$_Z^A \text{Nuc} \rightarrow _{Z+1}^A \text{Nuc}' + e^- + \bar{\nu}_e. \qquad (40.20)$$

This means that in nuclear β^--decay, the mass number of the nucleus remains the same, but the charge number increases by 1.

Because nuclei are made of neutrons and protons, it might seem that beta decays should always be possible. However, the decay can happen only if it is allowed by energy

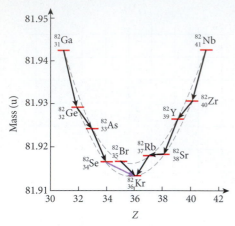

FIGURE 40.13 Nuclei of mass number 82. Their measured masses are displayed in units of the atomic mass unit u and are shown as a function of the charge number. The black arrows represent beta-decay processes. The purple arrow represents a double beta decay.

conservation—that is, the combined mass of the electron and the daughter nucleus must be smaller than the mass of the parent nucleus. It might be expected that this will always work out because of the mass difference of 1.293 MeV/c^2 (or 0.00139 u) between neutron and proton, which is quite a bit larger than the electron mass of 0.511 MeV/c^2. Although this argument is correct for free neutrons, it is not always true for neutrons bound inside a nucleus. The reason for this is that the nuclear interaction favors configurations with equal numbers of neutrons and protons. As an example for the effect that this part of the nuclear interaction has, we plot in Figure 40.13 the experimentally measured masses (horizontal red lines) of all known $A = 82$ isotopes as a function of the charge number Z. Clearly it is energetically possible for the bromine ($Z = 35$) isotope with mass number 82 to undergo a beta-minus decay $^{82}_{35}\text{Br} \rightarrow\ ^{82}_{36}\text{Kr} + e^- + \bar{\nu}_e$, but the beta-minus decay of $^{82}_{36}\text{Kr}$ into $^{82}_{37}\text{Rb}$ is energetically forbidden, because $^{82}_{37}\text{Rb}$ has a greater mass than $^{82}_{36}\text{Kr}$.

Furthermore, it is also possible that beta decays proceed in the opposite direction inside nuclei. In this β^+-decay, a proton is converted into a neutron via the emission of a positron and an electron neutrino, $p \rightarrow n + e^+ + \nu_e$. For a free proton, this process is energetically not possible because the neutron has a higher mass than the proton. However, inside the nucleus the reaction

$$^A_Z\text{Nuc} \rightarrow\ ^A_{Z-1}\text{Nuc}' + e^+ + \nu_e \qquad (40.21)$$

can proceed when the masses of initial and final isotopes are such that the Q-value of this reaction is positive. In addition, a proton inside the nucleus can be converted into a neutron in another way: electron capture, $e^- + p \rightarrow n + \nu_e$. In this β^+-process, the nucleus captures one of its own or biting electrons:

$$e^- +\ ^A_Z\text{Nuc} \rightarrow\ ^A_{Z-1}\text{Nuc}' + \nu_e. \qquad (40.22)$$

We use the notation "β^+-decay" for both processes, positron emission (equation 40.21) and electron capture (equation 40.22). Theoretically, a third way to convert a proton into a neutron can occur, via anti-neutrino capture, $\bar{\nu}_e + p \rightarrow n + e^+$. However, this process is negligible for our nuclear physics considerations because the cross section is extremely small and because an atom does not have a source of anti-neutrinos present. Consulting Figure 40.13 once again, you see that the nuclei on the right-hand side of the figure can all undergo β^+-processes.

Note that the Q-values for the reactions in equations 40.21 and 40.22 are not the same. The initial state of the reaction in equation 40.21 consists of Z protons, $N = A - Z$ neutrons, and Z electrons, all of which are accounted for in the mass of the initial atom, $m(N, Z)$. The final state consists of an atom with $Z - 1$ protons, $N + 1$ neutrons, and $Z - 1$ electrons, plus one additional electron, plus the newly created positron and neutrino. This atom has a mass $m(N + 1, Z - 1)$, and the electron and positron each have a mass of $m_e = 0.511$ MeV/c^2. Neglecting the binding energy of the electron and the mass of the neutrino, each of which are on the order of 1 eV or less, we then obtain for the Q-value of the positron-emitting β^+-reaction (equation 40.21)

$$Q(e^+) = m(N, Z)c^2 - m(N + 1, Z - 1)c^2 - 2(0.511\ \text{MeV}).$$

The initial state of the reaction in equation 40.22 consists of the same atom with Z protons, $N = A - Z$ neutrons, and Z electrons, and the final state consists of a neutrino and an atom with $Z - 1$ protons, $N + 1$ neutrons, and $Z - 1$ electrons. Thus, in this case the Q-value is simply

$$Q(ec) = m(N, Z)c^2 - m(N + 1, Z - 1)c^2.$$

(Remember, the nucleus captured one of its atom's own electrons!) Therefore, the Q-value of the electron capture (ec) reaction is always 1.022 MeV larger than that for the same β^+-process that involves positron emission (e^+). This implies that for some isotopes only the electron-capture process is possible, not positron emission.

For alpha decay, we have seen that this process leads to a characteristic energy (equation 40.19) for the emitted alpha particle, because energy and momentum need to be conserved in the decay. However, in beta decays the situation is more complicated. In β^--decays as well as in positron-emitting β^+-decays, the final state consists of three particles that can share the

40.2 In-Class Exercise

The isotopes $^{82}_{36}\text{Kr}$, $^{82}_{37}\text{Rb}$, $^{82}_{38}\text{Sr}$ have masses of 81.9134836 u, 81.9182086 u, and 81.91840164 u, respectively. For which of the β^+-decays of $^{82}_{37}\text{Rb}$, $^{82}_{38}\text{Sr}$ is positron emission possible?

a) for neither

b) only for the $^{82}_{37}\text{Rb}$ decay

c) only for the $^{82}_{38}\text{Sr}$ decay

d) for both

decay energy. The emitted neutrinos cannot be observed directly and can carry different amounts of energy. Thus, the observed electrons or positrons from these decays do not have well-defined kinetic energy values, but instead show a continuous distribution of energies.

Gamma Decay

A **gamma decay** (γ-decay) is the emission of a photon from a nucleus and is always the product of a de-excitation of an excited nuclear state. Gamma decays are qualitatively different from alpha or beta decays, because they are the only decay mode that does not cause transmutation. Nuclei can enter into excited states in ways similar to how atoms can get into excited states, by collisions with other objects. The kinetic energy from these processes can be converted into excitation energy, lifting nucleons into higher shells, or causing collective vibrations or rotations of the entire nucleus.

The process of gamma decay in nuclei is similar to the emission of photons in the de-excitation of atoms. However, the characteristic electron energies in the atom are on the order of eV, whereas the characteristic energies of nuclear excited states are on the order of MeV. Thus, the photons emitted in gamma decays typically are a million times more energetic than the photons emitted in atomic decays.

The initial and final states of the nucleus have well-defined energies, so the photon energy is also, in principle, well defined. However, the excited nuclear states have finite lifetimes, τ. Thus, the uncertainty in the energy of the emitted photon, denoted as the width Γ, has a lower limit given by the uncertainty relation: $\Gamma > \hbar/\tau$. Thus, if you measure the energy of gamma rays resulting from the decay between two specific states with finite lifetimes in a nucleus, the distribution will not be a spike at a single energy, but rather the distribution will take the form of a peak in energy with a characteristic width in energy.

Most excited isotopes can de-excite via gamma decay. However, the larger the difference between the angular momentum of the initial and that of the final nuclear state, the lower the probability for gamma decay becomes. In some isotopes the lowest excited state has a large difference in angular momentum from that of the ground state, and so a gamma decay is very improbable. These isotopes then form very long-lived "isomer" states.

Gamma rays emitted from nuclei are very important diagnostic tools for learning about nuclear structure. Since a very large number of states exist in any given isotope, a large number of gamma rays can be detected at different energies. It then requires ingenious detective work and large photon detector arrays (Figure 40.14) to puzzle together the information on nuclear structure contained in these spectra.

Gamma rays are not only emitted in single-particle transitions resulting from a single nucleon jumping from an excited state to another. They can also be emitted from collective vibrations of nuclei, as well as from rotating deformed nuclei. Superdeformed nuclei, which are cigar-shaped with an axes ratio of 2:1, were discovered only recently with the aid of gamma-ray spectroscopy. Figure 40.15 shows one example of such a photon spectrum. The most prominent feature of this gamma-ray spectrum of $^{152}_{66}$Dy is the sequence of gamma-ray peaks with very regular spacing of $\Delta E = 47.5$ keV. These peaks result when a rotating nucleus makes transitions between angular momentum states of even-numbered multiples of $\hbar$.

Other Decays

While α-, β-, and γ-decays constitute the overwhelming majority of radioactive decay modes, we should also mention a few others.

Light, very neutron-rich isotopes can decay by emission of a single neutron. Light and very proton-rich isotopes can decay via the emission of a proton. Neutron and proton emitters usually have extremely short lifetimes and are not quite bound nuclei. An example of a neutron emitter is $^{10}_{3}$Li. The isotope with one more neutron than $^{10}_{3}$Li, $^{11}_{3}$Li, has been investigated in many labs around the world

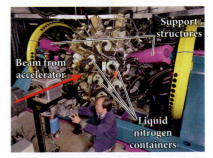

FIGURE 40.14 Gammasphere, the world's most sensitive detector for nuclear gamma rays. Gammasphere consists of 110 gamma-ray detectors cooled by liquid nitrogen surrounding the point where the nuclear interactions take place. In the photograph, only the support structure and the liquid nitrogen containers are visible.

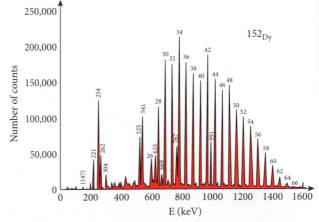

FIGURE 40.15 Gamma-ray spectrum of the superdeformed isotope dysprosium-152. The measured gamma rays are identified either by the energy (in keV, vertical numbers) or by the angular momentum (in units of $\hbar$, horizontal numbers).

40.3 In-Class Exercise

What types of nuclear decays are possible for the isotope $^{82}_{33}$As? Select all that apply. (*Hint:* The mass of $^{82}_{33}$As is 81.925 u, the mass of $^{82}_{33}$Ge is 81.930 u, the mass of $^{82}_{34}$Se is 81.917 u, the mass of $^{78}_{31}$Ga is 77.932 u, and the mass of $^{4}_{2}$He is 4.002u.)

a) alpha decay

b) beta decay

c) positron emission

d) electron capture

e) gamma decay

during the last ten years because it consists of three parts ($^{9}_{3}$Li + 2n) that form a bound state only if all three are together. There is no bound state of two neutrons, and the neutron emitter $^{10}_{3}$Li($^{9}_{3}$Li + 1n) also is not a bound nucleus. Thus, $^{11}_{3}$Li is a strange nucleus consisting of a core of $^{9}_{3}$Li and a very large "halo" made of two neutrons. The diameter of this halo has been measured to be almost as large as that of a lead nucleus, which consists of 208 nucleons. $^{11}_{3}$Li is therefore a stunning deviation from the nuclear size law expressed in equation 40.3.

A few isotopes are known to exhibit cluster decays. **Cluster decays** are the emission of nuclei heavier than helium. Almost all known cluster decays are in the form of carbon isotopes, $^{12}_{6}$C or $^{14}_{6}$C, and in some very rare cases oxygen, neon, or magnesium nuclei.

One very important decay mode of heavy nuclei, fission, will be discussed in Section 40.4.

Many isotopes can have two or more decay modes. Figure 40.16 presents an overview of the dominant decay modes for all known isotopes. Proton-rich isotopes decay via β^{+}-decays or proton emission, neutron-rich nuclei via β^{-}-decays or neutron emission. For heavier nuclei, α-emission becomes dominant, and the heaviest isotopes often decay predominantly via spontaneous fission.

An extremely rare decay mode of some isotopes is double beta decay, which was first observed in 1987 in an isotope of selenium, $^{82}_{34}$Se, by Michael Moe and colleagues at the University of California–Irvine. Referring back to Figure 40.13 once more, you can see that $^{82}_{34}$Se cannot undergo a simple β^{-}-decay to $^{82}_{35}$Br, because its mass is lower than that of $^{82}_{35}$Br. Thus, this reaction is forbidden by energy conservation. However, the mass of $^{82}_{36}$Kr is lower. So a nucleus of $^{82}_{34}$Se can be converted to a nucleus of $^{82}_{36}$Kr via the double beta-decay reaction

$$^{82}_{34}\text{Se} \rightarrow {}^{82}_{36}\text{Kr} + 2e^{-} + 2\bar{\nu}_{e}. \quad (40.23)$$

The double beta decay is the only reaction that prevents $^{82}_{34}$Se from being a stable isotope. This decay is depicted in Figure 40.17. Chapter 39 noted the beta-decay process involves the exchange of a W boson and thus leads to comparatively long lifetimes. A double beta decay requires two W boson exchanges and thus leads to extremely long lifetimes. The lifetime of $^{82}_{34}$Se is measured to be 10^{20} years, which is approximately 10 billion times the age of the universe! Only 12 isotopes are known to undergo double beta-decay processes (shown in yellow in Figure 40.16), and their average lifetimes are all of the same order as that of $^{82}_{34}$Se. Thus double beta-decay event rates are extremely small, making the detection of double beta decays very difficult experimentally.

Research into double beta decays is flourishing. A theoretical possibility of double β^{+}-decays exists, but no isotope showing this type of decay has yet been observed. In addition, there are ongoing searches for neutrino-less double beta decays. The reaction for this process would be the same as in equation 40.23, except that no anti-neutrino is emitted. This process can work only if the neutrino is its own antiparticle. If this decay mode would

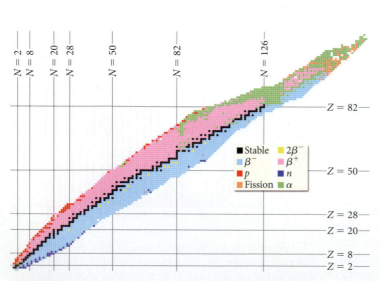

FIGURE 40.16 Dominant decay modes for the known isotopes.

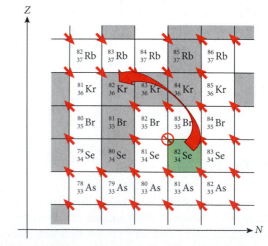

FIGURE 40.17 Part of the chart of nuclides around mass number 82. The gray shaded squares represent stable nuclei. The red arrows indicate the directions of the beta decays. The green shaded box represents the double beta-decay isotope 82-selenium.

be observed, it would violate the standard model of particle physics, introduced in Chapter 39, and it would point toward completely new physics that could be discovered.

Carbon Dating

Carbon has two stable isotopes: $^{12}_{6}C$ at 98.90% abundance and $^{13}_{6}C$ at 1.10% abundance. All other carbon isotopes, except $^{14}_{6}C$, decay with half-lives ranging from milliseconds to minutes. The exception, $^{14}_{6}C$, has a half-life of $t_{1/2}(^{14}_{6}C) = (5730 \pm 40)$ years. It decays via the β^--reaction $^{14}_{6}C \rightarrow {}^{14}_{7}N + e^- + \bar{\nu}_e$ into nitrogen. The Q-value of this reaction is 156.5 keV.

The $^{14}_{6}C$ isotope is produced in the upper atmosphere at a constant rate via the interaction of cosmic rays with atmospheric nitrogen. Because this isotope is produced at a constant rate and decays at a constant rate, the concentration of $^{14}_{6}C$ in the atmosphere, and thus the ratio of the number of $^{14}_{6}C$ atoms relative to the number of $^{12}_{6}C$ atoms, is constant in time, at a value of approximately $1.20 \cdot 10^{-12}$. The plants on Earth's surface consume this $^{14}_{6}C$, mainly in the form of CO_2 molecules. Thus, the ratio of $^{14}_{6}C/^{12}_{6}C$ is also constant in living plants, and consequently all the way up the food chain as well.

While plants and animals are living and consuming food, their carbon isotope ratio stays at a constant value. However, at the moment of death, the intake of $^{14}_{6}C$ ceases and thus the ratio $^{14}_{6}C/^{12}_{6}C$ decreases in time as $^{14}_{6}C$ decays. Measuring this ratio in a tissue sample can determine how long the plant or animal has been dead. This method can also be used on products made from plants or animals, such as textiles or wooden objects. The procedure of age determination from radioactive decays, called **radiocarbon dating,** or just carbon dating, revolutionized the field of archeology. It allows us to date samples up to an age of approximately 10 half-lifes of $^{14}_{6}C$, more than 50,000 years. The American physical chemist Willard Frank Libby (1908–1980) discovered the method of carbon dating in 1949 and was awarded the 1960 Nobel Prize in Chemistry for this achievement. In principle, any other long-lived isotope for which the initial concentration is known can be used as a tool for dating objects, but carbon dating is by far the most important.

SOLVED PROBLEM 40.1 | Carbon Dating

The Shroud of Turin, shown in Figure 40.18, is a large piece of linen cloth that some people claim is the burial shroud of Jesus of Nazareth. It is kept in the Cathedral of Saint John the Baptist in Turin, Italy, and displayed for viewing only on very rare occasions. However, some people expressed doubts concerning the authenticity of this object, and other people claim it to be a hoax of medieval origin. In 1988 the Vatican allowed radiocarbon dating of the shroud by three laboratories, in Zürich, Oxford, and Arizona, with the result that it was shown to be of medieval origin. The testing concluded with a 95% confidence level that the shroud was made between 1260 and 1390.

PROBLEM

If a textile sample contains $(1.08 \pm 0.01) \cdot 10^{-12}$ atoms of the $^{14}_{6}C$ isotope for each $^{12}_{6}C$ isotope, what is its age?

SOLUTION

THINK

We know that radioactive decays follow an exponential decay law governing the number of remaining $^{14}_{6}C$ isotopes as a function of time (equation 40.12). The number of $^{12}_{6}C$ isotopes

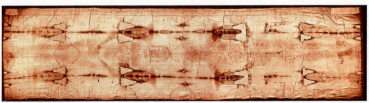

FIGURE 40.18 Example of carbon dating: the Shroud of Turin.

Continued—

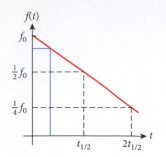

FIGURE 40.19 Carbon-14 fraction as a function time plotted on a semilogarithmic scale.

stays constant in time because this isotope is stable. Therefore, the ratio of the two isotopes, $f_{14/12}(t) = N(^{14}_6C,t)/N(^{12}_6C)$, follows the same exponential decay law. Because we know the half-life of $^{14}_6C$ and the initial and final fractions of $^{14}_6C$, we can solve the exponential decay law for the time and obtain our answer.

SKETCH

Figure 40.19 serves to remind us that the exponential decay law also holds for the fraction of $^{14}_6C$ relative to $^{12}_6C$. Given a certain number for the fraction, we can extract the age of the sample.

RESEARCH

We can use the exponential decay law (equation 40.12) for the number of $^{14}_6C$ atoms as a function of time,

$$N(t) = N_0 e^{-\lambda t}.$$

Since we are given the half-life of the isotope, we need to relate the decay constant λ to the half-life by using the relationship in equation 40.14,

$$t_{1/2} = \ln 2/\lambda.$$

SIMPLIFY

First we write the decay law in terms of the half-life,

$$N(^{14}_6C,t) = N_0(^{14}_6C)e^{-\lambda t} = N_0(^{14}_6C)e^{-t\ln 2/t_{1/2}}.$$

Dividing both sides by the number of $^{12}_6C$ isotopes gives us the fraction of $^{14}_6C$ relative to $^{12}_6C$:

$$f_{14/12}(t) = f_{14/12}(t=0)e^{-t\ln 2/t_{1/2}}.$$

Now we can solve this equation for the time and find

$$\frac{f_{14/12}(t)}{f_{14/12}(t=0)} = e^{-t\ln 2/t_{1/2}} \Rightarrow$$

$$\ln\left(\frac{f_{14/12}(t)}{f_{14/12}(t=0)}\right) = -t\frac{\ln 2}{t_{1/2}} \Rightarrow$$

$$t = -\frac{t_{1/2}}{\ln 2}\ln\left(\frac{f_{14/12}(t)}{f_{14/12}(t=0)}\right).$$

CALCULATE

Now we are in a position to insert numbers. As we stated above, the half-life of $^{14}_6C$ is $t_{1/2}(^{14}_6C) = (5730\pm40)$ years and $f_{14/12}(t=0) = 1.20\cdot10^{-12}$. The problem specifies $f_{14/12}(t) = (1.08\pm0.01)\cdot10^{-12}$. Thus, we insert the average value into the above equation to find the mean value and then the upper and lower quoted uncertainty to find the error bars for our solution:

$$t = -\frac{5730 \text{ years}}{\ln 2}\ln\left(\frac{1.08}{1.20}\right) = 870.978 \text{ years}.$$

Inserting a value of 1.09 in the same equation results in a time of 794.787 years (76 years less), and a value of 1.07 yields 947.877 years (77 years more).

ROUND

From the given uncertainty in our $^{14}_6C$ fraction, we have extracted the uncertainty in our final answer. Thus, we round our final answer to a value of 870 years and quote the uncertainty as 80 years:

$$t = (870\pm80) \text{ years}.$$

The textile sample in question would have to have been produced in the 12th century, and perhaps as early as the 11th and as late as the 13th century.

DOUBLE-CHECK

It is quite reasonable that we find a time that is small compared to the half-life of $^{14}_{6}C$, because after one half-life we would have had a remaining concentration of $0.60\cdot10^{-12}$ $(=\frac{1}{2}\cdot1.20\cdot10^{-12})$, and our problem stated that our sample had only lost 10% of its $^{14}_{6}C$ content. In fact, for small values of $t/t_{1/2}$ the exponential decay function can be expanded as $e^{-\ln2\cdot t/t_{1/2}}\approx1-\ln2\cdot t/t_{1/2}$. Since 90% of the initial $^{14}_{6}C$ concentration is left over at that time, this linear approximation results in

$$0.9=1-\ln2\cdot t/t_{1/2}\Rightarrow t=t_{1/2}\cdot0.1/\ln2=0.144t_{1/2}=830\text{ years.}$$

Note: The assumption that the atmospheric concentration of $^{14}_{6}C$ has always been constant is not quite true. Historic changes in the cosmic-ray flux and other events have resulted in temporal fluctuations. In addition, since the 1950s the atmospheric concentrations of $^{14}_{6}C$ and other isotopes have been changed by atmospheric tests of nuclear weapons. However, by correlating the results of carbon dating with other sources of age determination (ice cores, tree rings, etc.), calibration curves for the numerical results obtained with carbon dating have been developed.

Units of Radioactivity

How can the radioactivity of a sample be measured? There are two main quantities of interest. First is the intensity of the products emitted from a particular nuclear decay, that is, the number of decays per unit time. Second is the effect that a given type of radiation has on the human body. (Note that Figure 40.20 is a radioactivity warning sign.)

The SI unit of radioactivity is the **becquerel (Bq)**, named in honor of the French physicist Henry Becquerel (1852–1908), co-discoverer of radioactivity:

$$1\text{ Bq}=1\text{ nuclear decay/s.}\tag{40.24}$$

FIGURE 40.20 The trefoil symbol is internationally recognized to indicate radioactive material.

This is an extremely small unit of radioactivity. Another common, non-SI, but generally accepted unit of radioactivity is the **curie (Ci)**, named in honor of the French husband-and-wife team of Pierre Curie (1859–1906) and his Polish-born wife Marie Skłodowska-Curie ("Madame Curie," 1867–1934), who were the other two co-discoverers of radioactivity. Initially, the curie was defined as the activity of 1 gram of the isotope $^{226}_{88}Ra$, which decays via α-emission. Currently the curie is defined in terms of the becquerel as

$$1\text{ Ci}=3.7\cdot10^{10}\text{ Bq}$$
$$1\text{ Bq}=2.7\cdot10^{-11}\text{ Ci.}\tag{40.25}$$

Older textbooks still show the rutherford (Rd) unit, 1 Rd = 1 MBq.

As we stated previously, the number of decays alone is not enough to measure the effects of radioactivity. Much more important is the absorbed radiation dose. The SI unit for the absorbed dose is the gray (Gy), and it is defined in terms of other SI units as an absorbed energy of 1 joule per kilogram of absorbing material,

$$1\text{ Gy}=1\text{ J/kg.}\tag{40.26}$$

A commonly used non-SI unit is the rad (rd). Its conversion to the gray is simply

$$1\text{ rd}=0.01\text{ Gy}$$
$$1\text{ Gy}=100\text{ rd.}\tag{40.27}$$

For X-rays and gamma rays, it is also important to measure the amount of ionization (separation of electrons from their previously neutral atoms). This quantity is called exposure and is measured in Coulomb/kilogram. A non-SI unit for exposure is the röntgen (R), and it is defined when the material is in air as

$$1\text{ R}=2.58\cdot10^{-4}\text{ C/kg of air.}\tag{40.28}$$

The exposure and absorbed dose are closely related, and for biological tissue we find that 1 röntgen corresponds to 1 rad, in very good approximation.

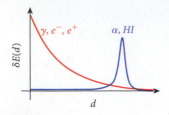

FIGURE 40.21 Profile of energy deposition as a function of penetration depth, d. Blue curve: profile for alpha particles and heavy ions. Red curve: profile for photons and leptons.

Damage of biological tissue depends not only on the deposited energy (absorbed dose), but also on the type of particle the nuclear decay emitted. Different types of radiation interact in very different ways with matter, in particular with biological tissue, and have very different depth profiles in which they deposit energy. When photons penetrate into matter, they lose energy exponentially as a function of depth into the material. Alpha particles and heavy nuclei, on the other hand, lose a very small fraction of their energy at the entry point, but deposit almost all of their energy at their maximum penetration depth, which is a function of the initial energy. Thus, for example, α-particles of a given initial energy all penetrate to approximately the same depth in biological tissue and then deposit a very significant fraction of their energy at that depth (Figure 40.21).

Because of the different energy deposition as a function of penetration depth, an α-particle is much more damaging to the cell walls of biological tissue than a photon of the same energy. Therefore, we introduce the radiation weight factor w_r. It is 1 for all photons and leptons, independent of their energy. Also, $w_r = 5$ for protons with energy greater than 2 MeV, and for neutrons with energy either greater than 20 MeV or less than 10 keV. In the energy intervals between 10 keV and 100 keV, as well as between 2 MeV and 20 MeV, the weight factor has the value of 10 for neutrons. Neutrons are at their most damaging in the energy interval between 100 keV and 2 MeV, where the weight factor is $w_r = 20$. The same weight factor of 20 is also assigned to α-particles and heavy nuclei. With this weight factor, we can then measure the dose equivalent, which is the product of absorbed dose times the weight factor.

The SI unit for the dose equivalent is the sievert (Sv), defined as

$$1 \text{ Sv} = w_r (1 \text{ Gy}). \tag{40.29}$$

Again, a non-SI unit is commonly used, the rem (röntgen equivalent, man). It is defined as

$$1 \text{ rem} = w_r (1 \text{ rd}). \tag{40.30}$$

Comparing equations 40.30, 40.29, and 40.27, we see that

$$1 \text{ Sv} = 100 \text{ rem}. \tag{40.31}$$

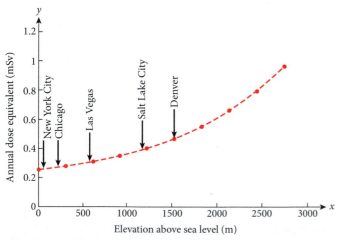

FIGURE 40.22 Annual dose equivalent from cosmic rays as a function of elevation of your hometown.

Radiation Exposure

What is the average dose of radiation that a typical U.S. resident receives each year? To answer this question, we need to consider many sources of radiation. Perhaps the easiest to quantify is the dose we receive from cosmic radiation, or cosmic rays. This dose strongly depends on the elevation above sea level of the city or town in which you live (Figure 40.22). At sea level you receive an annual dose equivalent of 0.26 mSv, whereas in Denver your annual dose equivalent is almost 0.5 mSv. If you build your dream house on a mountaintop, you may get up to 1 mSv per year.

Among other sources of natural radiation, the most important is radon gas. This gas is part of the air you breathe, and it deposits its radiation on the inside of your body, in your lungs. This radiation contributes 2 mSv per year to your dose equivalent. You also ingest radioactive isotopes (mainly $^{14}_{6}\text{C}$ and $^{40}_{19}\text{K}$) with your food, another 0.4 mSv annually. The Earth itself contributes radiation: 0.16 mSv at the Atlantic Coast, 0.3 mSv in the continental United States, except for the area around Denver, where the dose equivalent is 0.63 mSv per year.

Medical X-rays (dental imaging is 0.01 mSv at the lower end to upper-GI-tract imaging of 2.5 mSv at the upper end) and other procedures add an average of 0.5 mSv to your annual dose equivalent. Each hour of air travel contributes approximately 0.005 mSv to your dose equivalent; the passenger enjoying Gold Elite status (more than 50,000 miles/year flown) pays for this with an additional 0.5 to 1 mSv. Other minute sources of radiation (watching TV, working on a computer, airport screening, using a camping gas lantern or a smoke detector, etc.) contribute a combined 0.1 mSv per year. The total annual exposure dose equivalent for the average U.S. resident, the sum of all of the above contributions, is approximately 3.6 mSv and is broken down by source in Figure 40.23.

People who work with radiation sources such as X-ray machines or nuclear reactors are strictly monitored for radiation exposure. Their maximum annual dose equivalent is not allowed to exceed 50 mSv, which is approximately 15 times the dose they receive from the natural environmental sources listed here.

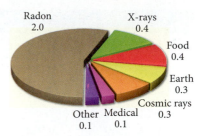

FIGURE 40.23 Average annual radiation exposure dose equivalent for a U.S. resident, in units of mSv.

EXAMPLE 40.4 | Chest X-Ray

Suppose you had a chest X-ray, during which you received a dose of 60. μSv, during your last physical.

PROBLEM 1
What was your dose in mrem?

SOLUTION
The conversion factor for sievert to rem is given by equation 40.31, so you can convert your dose as

$$\left(60. \cdot 10^{-6} \ \text{Sv}\right)\frac{100 \ \text{rem}}{1 \ \text{Sv}} = 6.0 \cdot 10^{-3} \ \text{rem} = 6.0 \ \text{mrem}.$$

PROBLEM 2
What was the absorbed dose in μGy and mrad?

SOLUTION
The radiation weight factor w_r for X-rays is 1. Therefore, the absorbed dose according to equation 40.29 was

$$\frac{60. \ \mu\text{Sv}}{1 \ \text{Sv/Gy}} = 60. \ \mu\text{Gy} \qquad \frac{6.0 \ \text{mrem}}{1 \ \text{rem/rad}} = 6.0 \ \text{mrad}.$$

PROBLEM 3
How much energy did you absorb, assuming that the X-rays illuminated 15 kg of your body?

SOLUTION
The Gy (equation 40.26) is defined in terms of the energy absorbed divided by the mass over which the energy is absorbed. Thus, the energy you absorbed was

$$\left(60. \ \mu\text{Gy}\right)\left(15 \ \text{kg}\right) = \left(60. \cdot 10^{-6} \ \text{J/kg}\right)\left(15 \ \text{kg}\right) = 9.0 \cdot 10^{-4} \ \text{J}.$$

PROBLEM 4
What fraction of your average annual dose of radiation did this chest X-ray represent?

SOLUTION
The average annual radiation dose is 3.6 mSv. Thus, the fraction of your annual dose was

$$\frac{60. \ \mu\text{Sv}}{3.6 \ \text{mSv}} = \frac{60. \cdot 10^{-6} \ \text{Sv}}{3.6 \cdot 10^{-3} \ \text{Sv}} = 1.7\%.$$

Liquid-Drop Model and Empirical Mass Formula

How can we understand the systematic trends of the binding energy per nucleon, equation 40.7, shown in Figure 40.7? One model of the nucleus that yields astonishingly good results is the **liquid-drop model.** In this model, the nucleus is treated as a spherical drop of quantum liquid composed of individual nucleons. Because the strong interaction between the nucleons inside the nucleus is attractive, each nucleon in the bulk receives a positive contribution to the binding energy. Conventionally, this contribution is called the volume term and is proportional to the number of nucleons A,

$$B_v(N,Z) = a_v A = a_v(N+Z). \tag{40.32}$$

Here a_v is a positive constant that is adjusted to the data.

Nucleons at the surface of the liquid drop are not surrounded by other nucleons. Thus, they have fewer nearest-neighbor interactions and therefore are less bound. Accordingly, the model must include a negative term that is proportional to the nuclear surface area. The surface area of a spherical drop is proportional to R^2, the square of its radius. According to equation 40.3, the nuclear radius is proportional to $A^{1/3}$. Therefore the surface area of the nucleus is proportional to $A^{2/3}$, and the contribution of the nuclear surface to the overall binding energy can be parameterized as

$$B_s(N,Z) = -a_s A^{2/3} = -a_s(N+Z)^{2/3} \tag{40.33}$$

with a positive fit constant a_s.

Next, we need to take into account that the protons are all positively charged and thus have a repulsive Coulomb interaction with one another. Since the Coulomb potential is proportional to the square of the charge number and inversely proportional to the radius, we can write for this term:

$$B_c(N,Z) = -a_c \frac{Z^2}{A^{1/3}} = -a_c \frac{Z^2}{(N+Z)^{1/3}} \tag{40.34}$$

where we have again made use of $R \propto A^{1/3}$ from equation 40.3. The constant a_c can be calculated if the protons are assumed to be evenly distributed throughout the nucleus. It has the value of $a_c = 0.71$ MeV.

The Coulomb interaction alone might make us conclude that it is most advantageous just to put neutrons into the nucleus. Because neutrons do not feel the Coulomb interaction, the term in equation 40.34 does not contribute to their binding energy, so a number of neutrons alone would have a higher binding energy than the same number of mixed neutrons and protons. However, we also have to take into consideration that protons and neutrons are fermions and thus need to respect the Pauli exclusion principle (see Chapter 36) separately for spin-up and spin-down protons, and for spin-up and spin-down neutrons. Just as in the case of the electrons filling shells in the atom at higher and higher angular momentum and energy, we also have to fill higher and higher energy levels the more protons and neutrons we add. Simply adding neutrons will become energetically unfavorable at some point. Thus, filling the nucleus approximately equally with protons and neutrons is encouraged by the action of the Pauli exclusion principle (Figure 40.24).

This effect can be parameterized via an asymmetry term, which is negative (representing less binding) when the number of protons is different from the number of neutrons. This motivates the following expression for the asymmetry term, where $A = Z + N$ in the second equality:

$$B_a(N,Z) = -a_a \frac{\left(Z - \frac{1}{2}A\right)^2}{A} = -\frac{a_a}{4} \frac{(Z-N)^2}{N+Z}. \tag{40.35}$$

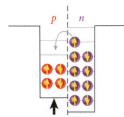

FIGURE 40.24 Illustration for the asymmetry term in the mass formula of the liquid-drop model.

Finally, the two-body interaction of nucleon pairs depends on their relative spin. This leads to higher binding energies for nuclei that have all of their protons paired up and all of their neutrons paired up. Thus, an even number of protons in a nucleus results in an additional positive contribution to the binding energy. In the same way, an even number of neutrons creates a higher binding energy as well. Empirically good results are found with this parameterization for the pairing term:

$$B_{\text{p}}(N,Z) = +a_{\text{p}}\frac{(-1)^Z + (-1)^N}{\sqrt{A}} = +a_{\text{p}}\frac{(-1)^Z + (-1)^N}{\sqrt{N+Z}}. \tag{40.36}$$

By combining equations 40.32 to 40.36, we can write an expression for the binding energy as a function of the mass number and charge number of a given nuclide as

$$B(N,Z) = B_{\text{v}}(N,Z) + B_{\text{s}}(N,Z) + B_{\text{c}}(N,Z) + B_{\text{a}}(N,Z) + B_{\text{p}}(N,Z)$$

$$= a_{\text{v}}A - a_{\text{s}}A^{2/3} - a_{\text{c}}\frac{Z^2}{A^{1/3}} - a_{\text{a}}\frac{\left(Z - \frac{1}{2}A\right)^2}{A} + a_{\text{p}}\frac{(-1)^Z + (-1)^N}{\sqrt{A}}.$$

Dividing this expression by the mass number, the binding energy per nucleon is

$$\frac{B(N,Z)}{A} = a_{\text{v}} - a_{\text{s}}A^{-1/3} - a_{\text{c}}\frac{Z^2}{A^{4/3}} - a_{\text{a}}\left(\frac{Z}{A} - \frac{1}{2}\right)^2 + a_{\text{p}}\frac{(-1)^Z + (-1)^N}{A^{3/2}}. \tag{40.37}$$

Now the constants in this expression can be used as fit parameters to obtain the best agreement with all binding energies for all nuclei. This procedure to explain the systematic trends in the binding energy per nucleon as a function of mass number was first invented by the German physicists Hans Bethe and Carl Friedrich von Weizsäcker in 1935, and the resulting **empirical mass formula** (equation 40.37) carries their name. Several fits have been published in the literature, all fairly successful. We follow Bertulani and Schechter (2002) and use the values

$$a_{\text{v}} = 15.85 \text{ MeV}, \; a_{\text{s}} = 18.34 \text{ MeV}, \; a_{\text{c}} = 0.71 \text{ MeV},$$
$$a_{\text{a}} = 92.86 \text{ MeV}, \; a_{\text{p}} = 11.46 \text{ MeV}. \tag{40.38}$$

Figure 40.25 shows the same experimental values of the binding energies as displayed in Figure 40.7, but only for the odd-mass nuclei. For these, the last term in equation 40.37 is zero, and then the effect of the other terms in the fit can be compared to the experimental data as a function of the mass number A. (The curve is more complicated for even-mass nuclei, because they can have either an even number of protons and an even number of neutrons, or an odd number of neutrons and an odd number of protons, which changes the sign of the term in equation 40.36 and thus does not produce a single curve like the one shown in Figure 40.7.) The yellow line shows the (constant) volume term. Adding the surface term to it leads to the orange line, which already shows very good agreement with the experimental data for low mass numbers. Adding the Coulomb term leads to the red line, which is very close to the experimental data, showing that the Coulomb interaction becomes very important for large nuclei. Finally, adding the asymmetry term leads to the green line, which gives a very good overall description of all binding energies of all odd-mass nuclei, at least at the resolution of the plot shown here.

The success of the Bethe-Weizsäcker formula (equation 40.37) for the binding energy is conventionally interpreted as strong support for the liquid-drop model of the nucleus. However, it is really only a consequence of the approximately spherical shape of nuclei, combined with the fact that the strong interaction is short-range and of the order of the nearest-neighbor spacing of the nucleons inside a nucleus.

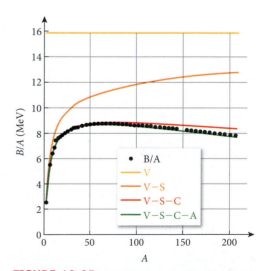

FIGURE 40.25 Binding energy per nucleon for odd-A nuclei. Black dots: experimental data. Lines are fits including different terms in the Bethe-Weizsäcker mass formula.

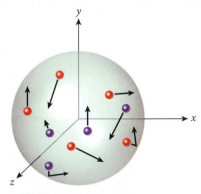

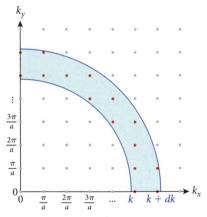

FIGURE 40.26 The Fermi gas model of the nucleus. Protons (purple) and neutrons (red) move independently and freely throughout the interior of the nucleus and are reflected off the nuclear surface that confines them.

FIGURE 40.27 Density of states in the Fermi gas model. Each dot represents a possible quantum state. The third momentum coordinate is not shown in this illustration.

Fermi Gas Model

Judging from the success of the liquid-drop model of the nucleus in reproducing the binding energies of nuclei, one might be tempted to think about nucleons as resting in a fixed and approximately spherical arrangement inside the nucleus. However, the Heisenberg uncertainty relation (see Chapter 36) imposes constraints on quantum particles/waves confined to such a small volume. Since we need to demand that $\Delta p_x \cdot \Delta x \geq \frac{1}{2}\hbar$, and since we have found out in the present chapter that we need to localize a nucleon inside a nucleus to within a few fm, this implies a high uncertainty in momentum, on the order of 100 MeV/c. (If you remember our handy rule that $\hbar c = 197.33$ MeV/fm, this is easy to see.) That means that nucleons inside the nucleus cannot be considered to be resting in some fixed configuration, but that they have fairly high momentum uncertainty and thus momentum inside the nucleus.

The **Fermi gas model** of the nucleus (invented by the Italian-American physicist Enrico Fermi, 1901–1953) considers this and assumes that nucleons can move freely like a gas inside the nucleus (Figure 40.26).

Chapter 37 showed that a particle with mass m and momentum k confined to a cubic box of width a in three dimensions, but that can move freely inside the box, has a total energy of

$$E = \frac{\hbar^2}{2m}k^2 = \frac{\hbar^2}{2m}\left(k_x^2 + k_y^2 + k_z^2\right) = \frac{\hbar^2 \pi^2}{2ma^2}\left(n_x^2 + n_y^2 + n_z^2\right). \tag{40.39}$$

The Pauli exclusion principle allows each quantum state to be occupied by exactly four nucleons (one spin-up and one spin-down proton, and one spin-up and one spin-down neutron, for each state).

We now want to use this information to calculate the density of states $dN(E)$, which will tell us how many quantum states reside in an interval dE around a given energy E, for the Fermi gas model.

Figure 40.27 illustrates the possible momentum states that can be occupied by the nucleons in a Fermi gas. Each possible quantum state occupies a volume of $(\pi/a)^3$. Then the number of states $dN(k)$ between the absolute value of the momentum k and $k+dk$ is given by the ratio of the volume of the spherical shell with radius k and thickness dk and the volume occupied by each quantum state:

$$dN(k) = \frac{\frac{1}{8}4\pi k^2 dk}{(\pi/a)^3} = \frac{a^3 k^2}{2\pi^2}dk. \tag{40.40}$$

(The factor $\frac{1}{8}$ appears in this equation because we examine only the octant of the three-dimensional sphere, in which all three momentum coordinates are positive. Negative values of the momentum coordinates do not add additional solutions, because the energy is proportional to k^2.)

Now we need to convert the density of momentum states $dN(k)$, which we just calculated, into a density of states $dN(E)$. Equation 40.39 shows that $E = \hbar^2 k^2/2m$, and therefore $k = \sqrt{2mE}/\hbar$, so the derivative of the energy with respect to the wave number can be calculated: $dE/dk = \hbar^2 k/m$. The differential dk can be expressed in terms of the differential dE via

$$dk = \frac{dE}{\hbar\sqrt{2E/m}}.$$

Combining this result with equation 40.40, the density of states as a function of energy is

$$dN(E) = \frac{a^3 m^{3/2}}{2^{1/2}\pi^2\hbar^3}E^{1/2}dE. \tag{40.41}$$

The maximum energy required to accommodate all A nucleons in a Fermi gas model is called the **Fermi energy**, E_F. To find the Fermi energy, we need to integrate the density of states (equation 40.41) up to the Fermi energy and set this number equal to $\frac{1}{4}A$. (Remember that each quantum state can be occupied by 4 nucleons.)

$$\int_0^{n(E_F)} dN(E) = \tfrac{1}{4}A \Rightarrow$$

$$\int_0^{E_F} \frac{a^3 m^{3/2}}{2^{1/2}\pi^2 \hbar^3} E^{1/2} dE = \frac{2^{1/2} a^3 m^{3/2}}{3\pi^2 \hbar^3} E_F^{3/2} = \tfrac{1}{4}A \Rightarrow$$

$$E_F = \frac{\hbar^2}{2m}\left(\frac{3A\pi^2}{2a^3}\right)^{2/3} = \frac{\hbar^2}{2m}\left(\frac{3\pi^2}{2}n\right)^{2/3}, \qquad (40.42)$$

where $n = A/a^3$ is the nucleon density inside the nucleus. Using the number $n = 0.17$ fm^{-3} for the nucleon density (Example 40.1), a value of 938.9 MeV/c^2 as the average of the proton and neutron mass, and our handy value of $\hbar c = 197.33$ MeV fm, we arrive at the numerical value of the Fermi energy:

$$E_F = \frac{(197.33 \text{ MeV fm}/c)^2}{2(938.9 \text{ MeV}/c^2)}\left(\frac{3\pi^2}{2}(0.17 \text{ fm}^{-3})\right)^{2/3} = 38 \text{ MeV}.$$

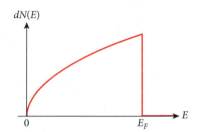

FIGURE 40.28 Density of states as a function of energy in the Fermi gas model. Here the energy is measured relative to the bottom of the potential well.

The density of states for the states occupied by nucleons inside the nuclear potential well is sketched in Figure 40.28 as a function of the energy above the bottom of the well. You can see that the largest probability for finding a nucleon inside a nucleus corresponds to an energy just below the Fermi energy. Note that the case displayed here is for temperature $T = 0$; for nonzero temperatures the vertical drop-off at the Fermi energy is replaced by a more gradual decline.

The Fermi momentum p_F is the momentum (magnitude) that corresponds to the Fermi energy,

$$p_F = \sqrt{2mE_F} = 270 \text{ MeV}/c$$

$$k_F = p_F/\hbar = 1.36 \text{ fm}^{-1}.$$

Dividing the Fermi momentum by the nucleon mass gives the maximum velocity with which nucleons move around inside the nucleus: $v_F = p_F/m = 0.29c$. That is, nucleons move around inside the nucleus with speeds of up to almost 30% of the speed of light!

Are these nucleon momentum values inside the nucleus real, or are they an artifact of a false interpretation of a quantum result in terms of classical physics? The answer is that they are real. This can be shown via particle production in heavy ion collisions, where the maximum energy of the produced particle can exceed the beam energy, a result that can be traced back directly to the presence of the Fermi motion of the nucleons inside the nucleus.

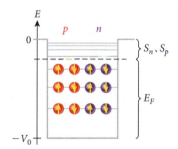

FIGURE 40.29 Approximation for the nuclear potential in the Fermi gas model.

The picture of the nucleus that emerges in this model can be illustrated as shown in Figure 40.29: Nucleons inside the nucleus are bound by a potential of depth V_0, which is the sum of the Fermi energy and the separation energy of the least-bound nucleon (S_n or S_p). Separation energies are typically on the order of 8 MeV. Thus, the depth of the nuclear well is approximately $V_0 = 8$ MeV + 38 MeV = 46 MeV. This result is valid for all nuclei with A larger than about 12, because from then on the nuclear saturation density is a constant $n = 0.17$ fm^{-3}.

This picture can be refined by introducing a separate Fermi energy for neutrons and protons. This gives rise to the asymmetry term (equation 40.35) in the Bethe-Weizsäcker mass formula (compare also Figure 40.24).

Shell Model

Chapter 38 showed that the electrons in atoms arrange themselves in shells because of the combined effects of the Coulomb interaction of the electrons with the positively charged nucleus, quantum mechanics, and the Pauli exclusion principle. One essential piece of experimental evidence that supports the existence of electron shells is the separation (ionization) energy of the outermost electron as a function of the charge number, which shows a pronounced maximum at a shell closure. In Example 40.2, we calculated the two-neutron separation energies for a series of tin isotopes and found a pronounced jump of the separation

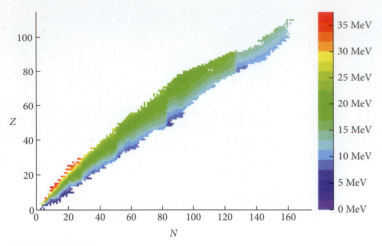

FIGURE 40.30 Experimentally measured two-neutron separation energies.

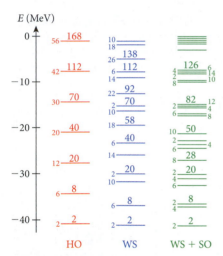

FIGURE 40.31 Energy levels for neutrons in a lead nucleus, calculated for three different potentials: harmonic oscillator (HO), Woods-Saxon (WS), and Woods-Saxon with strong spin-orbit coupling (WS+SO). The small numbers on the sides of the levels indicate the maximum occupation for the level, and the large numbers above the states indicate the cumulative sum of the occupation numbers.

energy at a neutron number of 82. Figure 40.30 shows a systematic compilation of two-neutron separation energies for all known isotopes. It is clear that it becomes harder to remove neutron pairs from a nucleus with a given charge number Z and with fewer neutrons N. This observation is a consequence of the asymmetry term (equation 40.35) discussed in the context of the liquid-drop model. However, the color coding in Figure 40.30 shows sudden jumps at certain neutron numbers, which are superimposed on this general trend of rising two-neutron separation energies as we move from right to left in a given isotope chain. It is clear from this figure that the neutron numbers 50, 82, and 126 are special, because they exhibit these sudden jumps. When two-proton separation energies are compiled, a similar picture emerges, with sudden jumps at the so-called magic numbers of 50 and 82 as we move along lines of constant neutron number. (There is no nucleus with charge number $Z = 126$.)

These and other data indicate that when certain numbers of protons or neutrons occur in a nucleus, a shell is filled. These numbers are called **magic numbers.** The magic numbers indicate that shell closures exist in nuclei just as they exist for the electron configurations in atoms. The magic numbers in nuclei are

$$2, 8, 20, 28, 50, 82, 126,$$

and they are the same for neutrons and protons, separately. For comparison, the magic numbers for the electron configurations are the charge numbers of the noble gases,

$$2, 10, 18, 36, 54, 86.$$

These sets of numbers are similar, but obviously not identical. Chapter 38 showed that these magic numbers indicate shell closures. The shells are formed by maximally occupying states with a given orbital angular momentum ℓ, where the maximum occupation of a given state is $2\ell + 1$. In the case of the electrons in an atom, the potential is provided by the Coulomb potential of the nucleus.

The potential of the nucleons inside the nucleus, on the other hand, is provided by the collective action of the other nucleons. Nevertheless, the quantum numbers of the nucleons inside the nuclear potential can also be calculated. Figure 40.31 shows the energy levels of neutrons inside the lead nucleus, calculated for three different potentials. The left side represents the harmonic oscillator potential, for which the solutions were calculated in Chapter 37. These energy levels are equally spaced. Interestingly, this simple harmonic oscillator potential already reproduces the three lowest magic numbers. A similar agreement is achieved by using the Woods-Saxon potential $V(r)$, which has the same dependence on the radial coordinate r as the nucleon density has (equation 40.4):

$$V(r) = -\frac{V_0}{1 + e^{(r-R(A))/a}}. \tag{40.43}$$

Again, the surface thickness constant $a = 0.54$ fm, and the radius is given by equation 40.3 as $R(A) = R_0 A^{1/3}$. The depth of the potential is $V_0 = 50$ MeV, close to what we obtained in the Fermi gas model. The energy levels in this potential cannot be calculated analytically, but only with the aid of a computer. It is still possible to assign angular momentum quantum numbers to these numerical solutions, though, and so the maximum occupation numbers for this Woods-Saxon potential can be calculated. The results of this calculation are shown in blue in Figure 40.31. For each level, we again show the maximum occupation number. The cumulative occupation numbers are shown only where there is a larger-than-average energy gap to the next level.

The decisive insight into the nuclear shell model was achieved in 1949 by the Germans Maria Goeppert Mayer, Otto Hahn, J. Hans, D. Jensen, and Hans Suess. They realized that an essential part of the nuclear interaction is the coupling between the spin and the orbital angular momentum. (We do not derive this interaction term here, but merely state that it exists; a course on advanced quantum physics will fill this gap.) This spin-orbit coupling is also present for electrons, but for nucleons it is much larger. If a spin-orbit interaction potential with the correct strength is added to the Woods-Saxon potential, then we obtain the energy levels shown in green on the right-hand side of Figure 40.31. Now we see that the large energy gaps between the levels occur at the magic numbers observed in experiment.

This simple shell model has been refined repeatedly during the last half century, and it remains the main workhorse of low-energy nuclear structure calculations. Because the numerical solution of the shell model for heavy nuclei involves inversion of extremely large matrices, this problem still cannot be solved exactly, even using the largest available computers. Research on the nuclear shell model thus is still a vibrant field, focusing on novel approximation techniques and steady refinement of nuclear interactions.

Other Models of the Nucleus

The three models presented so far are by no means the only ones. Nuclei are a very interesting testing ground for many theoretical models, in particular when it comes to understanding collisions of nuclei, in which nuclei can be heated and compressed. Models of nuclear collisions involve several subfields of physics we have introduced previously: quantum physics, fluid dynamics, thermodynamics, and electromagnetism. Nuclei can even undergo thermodynamic phase transitions between liquid-like and gas-like phases. The study of these phenomena, their technical applications, and their interdisciplinary relevance has been going on for the past two decades and promises to yield more interesting results in the coming decades.

As one example of a current nuclear model, we show an example from our own research. Figure 40.32 shows the time sequence of a computer model simulation of a collision of krypton-86 with niobium-93. The nuclei experience an off-center collision at an impact parameter of 5 fm. The time between two successive frames is 8.0 fm/c = $2.7 \cdot 10^{-23}$ s. Displayed is the nuclear density, which is color-coded according to the legend on the right-hand side of the figure. The compression, deformation, and sideways deflection of the nuclei are clearly visible. From the comparison of these and other model simulations to experimental data, we are able to draw conclusions on the dynamics and thermodynamics of nuclei, and on the properties of nuclear matter.

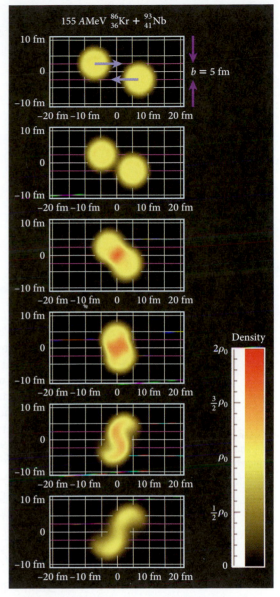

FIGURE 40.32 Time sequence of a nuclear collision, as calculated with a nuclear transport model.

40.4   Nuclear Energy: Fission and Fusion

Let's return once again to Figure 40.7, which shows the binding energy per nucleon as a function of the mass number for stable isotopes. The iron and nickel nuclei have the largest binding energy per nucleon and are thus most deeply bound. Any other arrangement of nucleons into nuclei, lighter or heavier, is less deeply bound. This suggests that very heavy nuclei can be split into two more deeply bound medium-mass nuclei, and obtain a positive Q-value (Figure 40.33). This positive Q-value yields useful energy, in a process called **nuclear fission.** In the same way, very light nuclei can be combined into heavier ones to give a positive Q-value with associated useful energy output, in the process of **nuclear fusion.** This is the entire physical basis for the energy production in our Sun and almost all other stars, for fission and fusion reactor power plants, as well as nuclear weapons.

Nuclear Fission

The nuclear fission process represents the splitting of an atomic nucleus into two smaller nuclei, often with the emission of one or more neutrons. The two smaller nuclei have mass

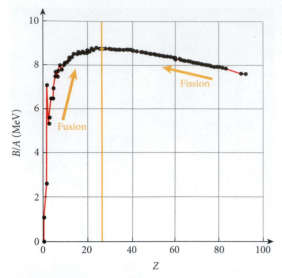

FIGURE 40.33 Energy gain through fission and fusion reactions.

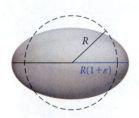

FIGURE 40.34 Nuclear deformation and definition of the deformation parameter.

40.4 In-Class Exercise

Which of the following isotopes has the largest fissionability?

a) $^{235}_{92}U$ c) $^{254}_{98}Cf$

b) $^{240}_{94}Pu$ d) $^{258}_{100}Fm$

numbers close to half that of the parent nucleus. As we will see, nuclei that undergo fission typically have mass numbers of 230 or greater, so typical isotopes resulting from nuclear fission are typically in the mass range of 100 to 150. Important examples of fission fragments are the isotopes of krypton through barium, and some of the lanthanides. These fission fragments tend to undergo subsequent radioactive decays, sometimes with very long half-lives. Many of them are easily ingested and accumulated in our bodies, so they pose great health risks. Examples are technetium-99 (more on this isotope later), cesium-137 (30-year half-life) and iodine-131 (8-day half-life), which is stored in the human thyroid.

A nucleus that is undergoing fission has to pass through a deformed configuration not unlike the one we discussed for α-decay, which is sketched in Figure 40.11. A quantitative understanding of the onset of spontaneous fission in heavy nuclei requires a criterion in terms of mass number and charge number. To develop this criterion, we first parameterize the deformation of the nucleus (Figure 40.34) in terms of an ellipsoid with semimajor axis length $R(1+\epsilon)$, where R is the radius of the same nucleus if it were spherical, and ϵ is the **deformation parameter**.

When a nucleus is deformed from its spherical configuration, how is the binding energy per nucleon (equation 40.37) affected? The volume term is not changed, because it depends only on the number of nucleons, which stays the same. Since the charge-to-mass ratio is not changed, the symmetry term should also remain the same, independent of deformation. The number of neutron pairs and proton pairs, and therefore the pairing term in equation 40.37, also remains constant. However, the surface is slightly increased when the sphere is deformed, so the surface term increases as a function of deformation. In addition, the protons are on average farther apart in the deformed nucleus, so the Coulomb term decreases. Niels Bohr (1885–1962) and John Archibald Wheeler (1911–2008) showed in 1939 that the change in binding energy as a function of deformation for the surface term is (to leading order) given by $\Delta B_s(\epsilon) = \frac{2}{5}\epsilon^2 B_s (N,Z) = -\frac{2}{5}\epsilon^2 a_s A^{2/3}$, and the change in the Coulomb term by $\Delta B_c(\epsilon) = -\frac{1}{5}\epsilon^2 B_c (N,Z) = \frac{1}{5}\epsilon^2 a_c A^{-1/3} Z^2$. Therefore the total change in the binding energy is

$$\Delta B(\epsilon) = \Delta B_c(\epsilon) + \Delta B_s(\epsilon)$$
$$= \frac{1}{5}\epsilon^2 a_c A^{-1/3} Z^2 - \frac{2}{5}\epsilon^2 a_s A^{2/3}$$
$$= \frac{1}{5}\epsilon^2 (a_c A^{-1/3} Z^2 - 2a_s A^{2/3}).$$

If $\Delta B(\epsilon) > 0$, energy can be gained by deforming a nucleus, and it can fission instantaneously. This transition point to spontaneous fission is reached when

$$\Delta B(\epsilon) = 0 \Rightarrow a_c A^{-1/3} Z^2 - 2a_s A^{2/3} = 0 \Rightarrow$$
$$\frac{Z^2}{A} = \frac{2a_s}{a_c} = \frac{2(18.34 \text{ MeV})}{0.71 \text{ MeV}} \approx 51.7,$$

using the values given in equation 40.38.

The quantity Z^2/A is called the fissionability parameter. Higher fissionability means higher likelihood for fission and thus a shorter lifetime for the isotope.

In Figure 40.16 you can see that spontaneous fission is the dominant decay mode for some of the heaviest isotopes. However, in almost all cases where spontaneous fission dominates, the half-life of the isotope is so short that it does not exist in nature. Much more important for the use of nuclear fission in nuclear reactors and nuclear weapons is the very long-lived isotope $^{235}_{92}U$, which has a half-life of 700 million years and which can be found in natural uranium deposits at 0.7% abundance. The most abundant uranium isotope is $^{238}_{92}U$, with a natural abundance of 99.3% and a half-life of 4.5 billion years. Both isotopes decay overwhelmingly via α-emission and have only a very small probability for spontaneous fission. However, $^{235}_{92}U$ fissions almost instantaneously after being hit by a neutron of fairly low energy. In this process, it releases two or three neutrons, which can then trigger further induced fission reactions in other $^{235}_{92}$U nuclei. If sufficient $^{235}_{92}$U is present, the resulting **chain reaction** will exponentially multiply the number of neutrons in the sample and cause a nuclear explosion. However, for this to happen, the uranium needs to be predominantly composed of $^{235}_{92}$U and not $^{238}_{92}$U (which does not show induced fission with low-energy neutrons). Thus, to be useful for nuclear weapons, the concentration of $^{235}_{92}$U needs to be *enriched*. We speak of *weapons-grade* uranium if it contains 80% or more $^{235}_{92}$U. Enrichment can be done by various means, but the dominant technique is to use gas centrifuges for isotope separation.

How much weapons-grade uranium is needed to make a nuclear weapon, that is, what is the **critical mass?** The optimum geometrical arrangement of the uranium is in the shape of a sphere. In the bulk of the sphere the neutrons trigger further induced fission reactions, whereas on the surface of the sphere, on average, half of the neutrons escape. The critical mass of a sphere of $^{235}_{92}$U is approximately 50 kg. Since uranium has a mass density of 19.2 times that of water, the radius of a sphere of uranium $^{235}_{92}$U with critical mass is only approximately 8.6 cm. In a nuclear weapon, a sufficient quantity of $^{235}_{92}$U is kept separated in several pieces and then detonated by pushing these separate pieces together via the use of conventional chemical explosives. The exact details of this process are closely guarded military secrets.

Quantities of $^{235}_{92}$U can also be used for nuclear power production in nuclear fission reactors. The uranium needed for this is much less enriched than for nuclear weapons. The art of reactor construction is to keep the chain reaction controlled and thus avoid runaway reactions. This is achieved by using moderator materials, which capture neutrons and thus make them unavailable for further induced fission reactions. The April 26, 1986, disaster of the reactor core meltdown at the Ukrainian city of Chernobyl was the result of such a runaway reaction in which the moderator system failed.

A second isotope that is very important for nuclear weapons is plutonium-239. $^{239}_{94}$Pu is produced in reactors from $^{238}_{92}$U via neutron capture and subsequent β^- decays:

$$^{238}_{92}\text{U} + n \rightarrow {}^{239}_{92}\text{U} + \gamma$$
$$^{239}_{92}\text{U} \rightarrow {}^{239}_{93}\text{Np} + e^- + \bar{\nu}_e$$
$$^{239}_{93}\text{Np} \rightarrow {}^{239}_{94}\text{Pu} + e^- + \bar{\nu}_e.$$

This production process of $^{239}_{94}$Pu is often referred to as **breeding.** $^{239}_{94}$Pu also undergoes neutron-induced fission and has a higher yield of neutrons than $^{235}_{92}$U. Thus its critical mass is only 10 kg, and it is much more valuable for nuclear weapons than $^{235}_{92}$U. However, breeding weapons-grade plutonium is not easy because $^{239}_{94}$Pu can easily capture another neutron and turn into $^{240}_{94}$Pu, which does not show spontaneous fission. Too high a fraction of $^{240}_{94}$Pu will lead to precritical detonation (fizzle) of a warhead, so the $^{239}_{94}$Pu needs to be sufficiently pure to be useful for weapons purposes.

Clearly, nuclear weapons pose a great threat, and increasing numbers of countries are acquiring this dangerous technology. After the United States and Russia in the 1950s, China, France, Great Britain, Israel, India, Pakistan, and perhaps others (North Korea, Iran, South Africa) have become nuclear powers. With every additional country that is able to produce nuclear weapons, the danger increases that these could fall into the wrong hands. Small nuclear warheads fit into suitcases, and terrorists have turned their attention to them. Even detonating a bomb of conventional explosives mixed with radioactive material, a so-called *dirty bomb,* could have devastating consequences.

Equally clearly, the peaceful use of nuclear power from fission reactions is a way to avoid the greenhouse gas emissions associated with conventional power plants that burn fossil fuels. However, the storage of radioactive waste with very long half-lives is still a problem that is not completely solved. Countries such as Germany, for example, have sworn off the use of fission reactors and are turning their attention to alternative carbon-neutral energy sources (wind, biofuels, hydro, photovoltaics, and others) to solve the greenhouse gas problem. However, with global energy consumption on a very steep rise, it remains to be seen if these alternatives to fission power can satisfy the energy needs of the world.

EXAMPLE 40.5 | Fission Energy Yield

PROBLEM

What is the energy yield per kilogram of uranium for the neutron-induced fission reaction of $^{235}_{92}$U to krypton-92 and three neutrons, if $^{235}_{92}$U is at 20% enrichment in the reactor?

SOLUTION

First we need to write down the reaction equation for this fission reaction. This will tell us what the second fission fragment is. The reactant contains a total number of 92 protons, the charge number of uranium. Since one of the fission fragments is krypton with 36 protons, the other fission fragment has $Z = 92 - 36 = 56$ protons. This means that it is a barium nucleus. Counting the total mass number for the reactant, we find $A = 236$. Subtracting the mass number of the three neutrons and the mass number 92 of the krypton isotope, we find for the mass number of the barium isotope $A = 236 - 3 - 92 = 141$. Thus, the induced fission reaction equation is in this case

$$n + {}^{235}_{92}\text{U} \rightarrow {}^{141}_{56}\text{Ba} + {}^{92}_{36}\text{Kr} + 3n.$$

The mass of uranium-235 is 235.0439299 u, the mass of barium-141 is 140.914411 u, the mass of krypton-92 is 91.92615621 u, and the mass of a neutron is 1.008664916 u. Therefore, the mass difference between the reactants and products is

$$235.0439299 - (140.914411 + 91.92615621 + 2 \cdot 1.008664916) = 0.186033$$

atomic mass units. Converting to MeV, the mass-energy difference between the initial and final states is 173.3 MeV. This is also the sum of the kinetic energies of the final three neutrons plus the two fission fragments minus the kinetic energy of the initial neutron.

This energy output can also be converted into joules:

$$\Delta E = (173.3 \text{ MeV})(1.602 \cdot 10^{-13} \text{ J/MeV}) = 2.776 \cdot 10^{-11} \text{ J}.$$

Uranium at a mixture of 80% U-238 and 20% U-235 has an average mass number of $0.8(238) + 0.2(235) = 237.4$. Thus, 237.4 grams of uranium at our given enrichment has $6.022 \cdot 10^{23}$ uranium atoms (Avogadro's number, see Chapter 13). Of these atoms, 20% are uranium-235 and can undergo neutron-induced fission. Therefore, 237.4 g of our uranium sample contains $0.2(6.022 \cdot 10^{23}) = 1.2044 \cdot 10^{23}$ uranium-235 atoms, and consequently 1 kg of our sample contains $1.2044 \cdot 10^{23}/0.2374 = 5.073 \cdot 10^{23}$ uranium-235 atoms.

All that is left to do is to multiply the number of atoms times the energy yield per atom. Then the total fission energy yield of our uranium mixture is

$$(2.776 \cdot 10^{-11} \text{ J})(5.073 \cdot 10^{23}) = 14.08 \cdot 10^{12} \text{ J}.$$

Keep in mind that not all of this energy is available for feeding into the electric grid. Some uranium-235 will always be left nonfissioned, and some of the neutrons released will hit uranium-238 and breed it into plutonium.

For comparison, the combustion of 1 kg of gasoline yields a maximum of $4.6 \cdot 10^7$ J. This means that the nuclear fission of enriched uranium yields a factor of more than 300,000 more energy than the combustion of gasoline for the same mass of fuel.

Nuclear Fusion

The nuclear fusion process merges light nuclei into heavier ones and thus also increases the binding energy per nucleon, as long as the end product of this fusion process is lighter than iron.

Stellar Fusion

Nuclear fusion is the basic process of energy production in stars. Two main reaction chains, discovered in 1938 by Hans Bethe, accomplish this. The first is the proton-proton chain, and the second is the CNO cycle. Both of these serve to fuse four hydrogen atomic nuclei into one helium nucleus.

The **proton-proton chain** starts with two protons that form a deuteron plus a positron. This is a weak interaction process and takes a long time, approximately 10 billion years on average (half-life). This explains why the Sun has already existed for approximately 5 billion years and did not explode instantaneously after its formation. The newly formed deuteron can then quickly capture another proton and form helium-3; the freshly created positron annihilates almost instantaneously with an electron:

$$p + p \rightarrow d + e^+ + \nu_e \qquad Q = 0.42 \text{ MeV}$$
$$d + p \rightarrow {}^3_2\text{He} + \gamma \qquad Q = 5.49 \text{ MeV}$$
$$e^+ + e^- \rightarrow 2\gamma \qquad Q = 1.02 \text{ MeV}.$$

The net result of these three reactions is the fusion of three protons into one helium-3 nucleus plus three photons and one neutrino. The combined Q-value of the three reactions is $Q_{net} = (0.42 + 5.49 + 1.02) \text{ MeV} = 6.93 \text{ MeV}$.

The next step involves three different reactions through which fusion can proceed toward helium-4. In the Sun, 86% of the time, two helium-3 nuclei have a fusion reaction to yield

$${}^3_2\text{He} + {}^3_2\text{He} \rightarrow {}^4_2\text{He} + p + p$$

with a Q-value of 12.86 MeV. Adding the Q-value of this reaction to the Q-values that were obtained from the formation of helium-3 gives

$$Q_{net} = (2 \cdot 6.93 + 12.86) \text{ MeV} = 26.7 \text{ MeV}.$$

This is exactly the mass difference between four protons and one helium-4, as expected from conservation of energy.

Alternatively, 14% of the time the helium-3 finds an existing helium-4 in the Sun and fuses to

$${}^3_2\text{He} + {}^4_2\text{He} \rightarrow {}^7_4\text{Be} + \gamma.$$

This beryllium-7 nucleus then captures an electron to form lithium-7, which in turn has a fusion reaction to form two helium-4 nuclei:

$${}^7_4\text{Be} + e^- \rightarrow {}^7_3\text{Li} + \nu_e$$
$${}^7_3\text{Li} + p \rightarrow 2\,{}^4_2\text{He}.$$

The production of 26.7 MeV of kinetic energy from the fusion of four protons into one helium-4 nucleus is the basic reason why the Sun shines. What happens to this energy? Part of it is carried away by neutrinos. Due to the low-interaction cross sections, these neutrinos can leave the Sun right away on straight paths. The photons, on the other hand, scatter off the electrons and ions in the Sun, become absorbed and re-emitted in random directions, and take, on average, approximately 50,000 years to reach the Sun's surface.

The **CNO cycle** is named for the chemical symbols of the elements carbon, nitrogen, and oxygen. This cycle is dominant in stars that are significantly more massive than our Sun. In the CNO cycle, carbon serves as a catalyst to fuse four protons into one helium-4 nucleus. This means that at the end of the cycle a carbon-12 nucleus is obtained, just like the one that we started with:

$${}^{12}_6\text{C} + p \rightarrow {}^{13}_7\text{N} + \gamma \qquad Q = 1.95 \text{ MeV}$$
$${}^{13}_7\text{N} \rightarrow {}^{13}_6\text{C} + e^+ + \nu_e \qquad Q = 1.20 \text{ MeV}$$
$$e^+ + e^- \rightarrow 2\gamma \qquad Q = 1.02 \text{ MeV}$$
$${}^{13}_6\text{C} + p \rightarrow {}^{14}_7\text{N} + \gamma \qquad Q = 7.54 \text{ MeV}$$
$${}^{14}_7\text{N} + p \rightarrow {}^{15}_8\text{O} + \gamma \qquad Q = 7.35 \text{ MeV}$$
$${}^{15}_8\text{O} \rightarrow {}^{15}_7\text{N} + e^+ + \nu_e \qquad Q = 1.73 \text{ MeV}$$
$$e^+ + e^- \rightarrow 2\gamma \qquad Q = 1.02 \text{ MeV}$$
$${}^{15}_7\text{N} + p \rightarrow {}^{12}_6\text{C} + {}^4_2\text{He} \qquad Q = 4.96 \text{ MeV}$$
$$\underline{\text{net} \quad 4p \rightarrow {}^4_2\text{He} + 7\gamma + 2\nu_e} \qquad Q_{net} = 26.8 \text{ MeV}$$

These fusion processes need to overcome the Coulomb repulsion between the positively charged atomic nuclei. Thus, fusion is restricted to the inner core of the Sun, which extends from the center to about 20% of the solar radius. The core contains about 10% of the mass of the Sun. In the core, the density is up to 150 tons per cubic meter and the temperature is approximately 13.6 million K. The high temperatures and densities in the Sun's core are sufficient for the proton-proton fusion chain to proceed. By contrast, the CNO cycle requires somewhat higher temperatures than the Sun provides, and so contributes relatively little to the Sun's total energy output. However, as the Sun ages, it will reach stages of stellar evolution when the densities and temperatures in the core will be high enough for the CNO cycle to proceed.

EXAMPLE 40.6 | Fusion in the Sun

At the beginning of this chapter, we stated that nuclear fusion reactions are responsible for almost all of the energy resources at our disposal on Earth. Let's see how the numbers work out.

PROBLEM

Knowing that the Sun radiates approximately 1370 W/m² on Earth, how much total energy do fusion reactions inside the solar core need to produce per second, and how many protons does it take per second to generate this much power?

SOLUTION

The Sun is an almost perfect sphere and radiates its power uniformly over the 4π solid angle. Thus, if we calculate the surface area of a sphere with the radius of the Earth's orbit around the Sun, we can calculate the total power output. The orbital radius is 1 astronomical unit, 149.6 million km. Therefore the area of the surface of the sphere we are looking for is

$$A = 4\pi r^2 = 4\pi(1.496 \cdot 10^{11} \text{ m})^2 = 2.812 \cdot 10^{23} \text{ m}^2.$$

Given that each square meter on Earth receives approximately 1370 W of power, the total power output of the Sun is

$$P_{\text{total}} = (1.370 \cdot 10^3 \text{ W/m}^2)(2.812 \cdot 10^{23} \text{ m}^2) = 3.85 \cdot 10^{26} \text{ W (385 yottaWatt)}.$$

As a reminder, 1 W = 1 J/s and 1 eV = $1.602178 \cdot 10^{-19}$ J. The energy released by the fusion of four protons into one helium-4 in joules can be calculated:

$$26.8 \text{ MeV} = (26.8 \cdot 10^6)(1.602178 \cdot 10^{-19} \text{ J}) = 4.29 \cdot 10^{-12} \text{ J}.$$

One-quarter of this energy is therefore what we obtain for each proton involved in the fusion cycle, $1.07 \cdot 10^{-12}$ J. Therefore, $(3.85 \cdot 10^{26} \text{ W})/(1.07 \cdot 10^{-12} \text{ J}) = 3.60 \cdot 10^{38}$ protons are needed each second to fuse into helium-4. With a proton mass of $1.6726 \cdot 10^{-27}$ kg, this means that a total mass of $(3.60 \cdot 10^{38})(1.6726 \cdot 10^{-27} \text{ kg}) = 6.02 \cdot 10^{11}$ kg (600 million metric tons) of protons is converted into helium each second!

40.5 In-Class Exercise

With this much hydrogen being converted into helium, how much longer can the Sun keep shining at the current burn rate? (*Hint:* The mass of the Sun is $1.99 \cdot 10^{30}$ kg, the core contains approximately 10% of that mass, and approximately half of the core mass currently is still hydrogen.)

a) 30,000 years

b) 2 million years

c) 100 million years

d) 5 billion years

e) $3.3 \cdot 10^{18}$ years

Terrestrial Fusion

Can the power of fusion be harnessed on Earth? The answer is yes and maybe. In 1952 the first hydrogen bomb was detonated by the United States. Hydrogen bombs use nuclear fusion reactions to achieve maximum energy output. The high temperatures and compressions needed to achieve fusion are obtained by using a nuclear bomb of the fission type as a trigger. So yes, it is possible to achieve nuclear fusion reactions and to obtain a very large energy release. However, the detonation of hydrogen bombs cannot be considered "controlled" fusion. Controlled fusion as a means of production of useful energy could be the solution to global energy problems. During the past four or five decades, very large sums of money have been dedicated to obtaining controlled fusion, so far with little success. Achieving the high pressures and temperatures needed for thermonuclear fusion has been an elusive goal.

Presently, two leading designs promise great progress. One is ITER, the international collaboration to build a thermonuclear fusion reactor with magnetic confinement technology. Figure 40.35 shows a drawing of the central fusion core of ITER. The toroidal vacuum chamber of ITER is lined with a shield that absorbs heat and neutrons produced in the fusion reactions. The magnet systems that are used to confine the heated plasma surround the vacuum chamber. There is also a divertor to handle neutral particles that are not contained by the magnetic field.

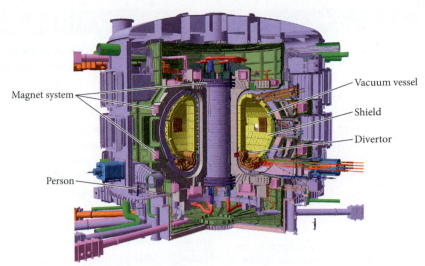

FIGURE 40.35 Drawing of the planned central fusion reactor of the ITER facility, now under construction in Cadarache, France.

In 2006 the European Union, the United States, China, Russia, Japan, India, and South Korea agreed to fund this project jointly and selected the construction site, Cadarache, in the south of France. The project costs will be approximately $15 billion, and it will take 10 years to complete. It is expected that ITER will provide 500 MW of power sustained for approximately 10 minutes. This is an improvement of several orders of magnitude over what has been achieved with magnetic confinement fusion reactors up to now, and a big step toward break-even—that is, extracting at least as much power as has been put in.

The other very promising approach to fusion is laser fusion. In Livermore, California, the U.S. Department of Energy has constructed the National Ignition Facility. Here the world's most powerful lasers are shot into small hollow cylinders (*hohlraum*) and the X-rays released from the interaction of the lasers with the hohlraum walls compress the nuclear fusion fuel in the interior, thus achieving fusion ignition. An overview of the laser assembly of the National Ignition Facility is shown at the end of Chapter 38. Figure 40.36a gives a schematic overview of the laser-induced fusion process, Figure 40.36b shows one of the hohlraum cylinders, and Figure 40.36c shows the target chamber, in which the hohlraum targets will be ignited and in which the fusion energy that is released will be recaptured.

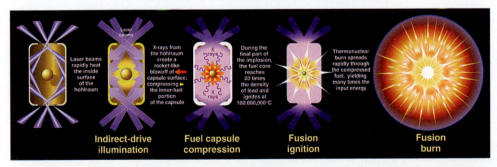

(a)

FIGURE 40.36 (a) Description of nuclear fusion in the National Ignition Facility. (b) A hohlraum cylinder is just a few millimeters wide (less than the size of a dime), with beam entrance holes at either end. It contains the fusion fuel capsule, which is the size of a small pea or pencil eraser. (c) A technician checks the NIF target chamber while in the Target Chamber Service System lift. Technicians must dress in full body suits to protect the chamber from any lint or microscopic particles. The target positioner, which holds the target during a shot, is on the right.

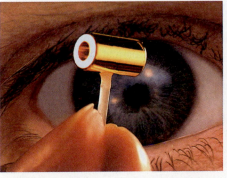

(b)

(c)

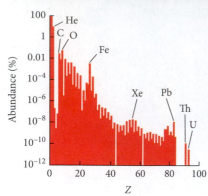

FIGURE 40.37 Abundance of the chemical elements in our Solar System by weight.

The interface of nuclear physics, particle physics, and astrophysics is one of the most fascinating areas of current physics research. Laboratory experiments, computer simulations, model studies, and calculations can be performed to answer basic questions about the universe. Several aspects of this interdisciplinary field were discussed in Chapter 39. However, one of the most interesting questions to be studied is the origin of the chemical elements. So far, we have discovered that the Big Bang (see Chapter 39) was able to create almost all of the hydrogen and most of the helium in the universe. Fusion reactions in stars, as we have just seen, generate helium as well. On the other hand, the elements that we observe around us are different. The Earth is predominantly composed of iron, silicon, oxygen, aluminum, sodium, and other elements. Elements as heavy as uranium can be found in sizable quantities. Our own bodies consist mainly of oxygen (at 65% of our weight), carbon (18%), hydrogen (3%), calcium (1.5%), phosphorus (1%), and traces of many other elements. Figure 40.37 shows the percentage abundance by weight of the elements in the Solar System. Hydrogen and helium dominate, but all stable elements can be found, and even thorium and uranium exist in appreciable quantities.

Where do the heavier elements come from? To answer this question, we have to examine the later stages in a star's life. For example, let's study a star of the mass of about 20 times the mass of the Sun. The combination of nuclear physics and stellar modeling shows that a star this massive uses up its hydrogen fuel much faster than the Sun does, and after only 10 million years its core is predominantly composed of helium. These helium-4 nuclei then fuse to carbon-12 through the so-called triple-alpha process,

$$^4_2\text{He} + {}^4_2\text{He} \leftrightarrow {}^8_4\text{Be}$$

$$^8_4\text{Be} + {}^4_2\text{He} \rightarrow {}^{12}_6\text{C} + \gamma.$$

This two-step process is necessary because ^8_4Be is not stable and has a half-life of only $6.7 \cdot 10^{-17}$ s and decays back into two helium-4 nuclei. Therefore, the density of helium-4 needs to be very high for this process to proceed. It then takes approximately 1 million years for much of the helium to be converted into carbon. As the core temperature rises, it eventually becomes sufficient to overcome the Coulomb repulsion of ever-heavier nuclei. This leads to the production of nuclei such as oxygen, neon, silicon, and then finally iron and nickel.

As we have seen, the stable iron and nickel isotopes have the highest binding energy per nucleon. The production of elements heavier than iron and nickel through fusion processes is thus not possible. This also means that the core of the star can no longer produce energy through nuclear fusion processes. This iron core of our 20-solar-mass star has a mass of 1 solar mass. Once the fusion energy output ceases, the thermal pressure due to the fusion photons vanishes, and the core starts collapsing from the gravitational pull. Capture of a significant fraction of the electrons on the protons, $p + e^- \rightarrow n + \nu_e$, accelerates the collapse. After only a few milliseconds, the center of the core has been compressed to the density of nuclear matter and a shock wave begins to propagate radially outward through the core, causing the star to explode and become a **supernova.**

Exactly how the core-collapse supernova explosion process works in detail is still under intensive investigation. However, the experimental observations of supernovas are clear. The total energy emitted in the form of photons and neutrinos is on the order of 10^{44} J to 10^{45} J, which corresponds to the energy release of more than a billion billion billion (10^{27}) hydrogen bombs ignited simultaneously. What's left behind is a neutron star that is very small (radius approximately 10 km) and incredibly dense (mass of approximately 1.5 solar masses at nuclear matter density). As the neutron star is formed, angular momentum is conserved, so that it rotates very quickly. As a neutron star spins, it can emit "lighthouse beams" of radio waves, light, and/or X-rays. Such a rotating neutron star is called a pulsar. Pulsars have been observed to rotate as fast as 40 times per second! The rotational speed of the Sun is $4.6 \cdot 10^{-7}$ rotations/s (one rotation every 25 days).

As a result of the supernova explosion, matter is thrown into interstellar space and forms clouds of dust and gas, which are some of the most picturesque objects in the universe. Figure 40.38 shows the famous Crab Nebula, which is the remnant of a supernova

FIGURE 40.38 The Crab Nebula, a supernova remnant.

that exploded 6500 light-years from Earth and was observed in the year 1054. It was so bright that it could be seen during bright daylight, even though the explosion happened at a distance almost half a billion times greater than the distance to the Sun.

In the process of a supernova explosion, the isotopes that exist outside the core are bombarded by an incredibly high flux of neutrons. They capture these neutrons very rapidly and also undergo fast β^--decays, thus adding to their neutron and proton numbers. The path of this so-called *r-process* (r = rapid) through the isotope chart is sketched in Figure 40.39. It takes only seconds to populate the r-process isotopes, which then decay back toward the valley of stability. This process is understood qualitatively, but not in quantitative detail. However, astronomers generally agree that the elemental abundance observed in the Solar System (see Figure 40.37) is the result of such a supernova explosion more than 5 billion years ago. Our Solar System formed from the ashes of that supernova. The vast majority of the atoms around us and inside our bodies are thus at least 5 billion years old. These atoms are simply recycled again.

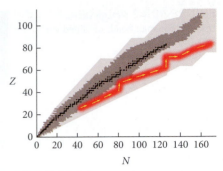

FIGURE 40.39 Path of the r-process (red with yellow arrows) through the isotope landscape. Light gray: postulated isotopes. Gray: known isotopes. Black: stable isotopes.

40.6 Nuclear Medicine

Nuclear medicine is a flourishing subfield of medicine, where nuclear physics finds direct applications in diagnostics and the treatment of patients. Great progress continues to be made by using more-precise tools for delivering radiation to the relevant areas of patients' bodies. There has also been steady improvement of detection equipment, resulting in the continual lowering of the minimum radiation doses needed for diagnostic tests. Progress in basic nuclear physics research thus translates directly into medical advances.

Radiation of various kinds is used for cancer treatment. The idea is straightforward: Concentrate the radiation on tumor cells that need to be destroyed, while at the same time trying to avoid irradiating healthy tissues. This treatment option is particularly attractive in cases where the tumor cannot be removed surgically or where the surgery area needs to be treated postoperatively. Brain cancers in particular have been successfully treated with radiation methods, a process that is being refined steadily.

The most common radiation treatment of cancers is based on gamma rays. This technology uses a very strong radioactive photon source, usually $^{60}_{27}$Co. This isotope β^--decays into an excited state of nickel via $^{60}_{27}\text{Co} \rightarrow {}^{60}_{28}\text{Ni} + e^- + \bar{\nu}_e$ with a Q-value of 318 keV. The half-life for this process is 5.27 years. Once populated, the excited state of $^{60}_{28}$Ni then decays promptly into the ground state via the emission of two high-energy gamma rays (Figure 40.40). Thus, this isotope delivers two high-energy gamma rays, but still has a very long half-life. Since $^{60}_{27}$Co is straightforwardly produced with 99% purity in reactors via neutron capture on the stable isotope $^{59}_{27}$Co, it is an ideal isotope for medical purposes, but also for other purposes, such as food irradiation.

The radioactive cobalt is contained behind thick shielding. Narrow channels in the shielding allow photons to escape in well-defined directions to be used as a beam for radiation. This is the basis for the **gamma-knife** setup (invented in 1968), mainly used to treat inoperable brain cancers (Figure 40.41). For each patient, a custom radiation collimator in the shape of a very thick helmet is designed, for which the many channels through the shielding all point to a certain point inside the patient's head where the cancer tumor is located.

The last few years have seen a very strong increase in research into cancer treatment with heavy ion beams, which, like α-particles, deposit most of their energy near their maximum penetration depth and so allow for greater depth control in the deposited radiation dose. Great advances in radiation treatment can be expected from the use of this method in the future.

In the field of imaging, the most important technology is based on nuclear magnetic resonance (NMR). The imaging technology based on NMR is called magnetic resonance imaging (MRI). How does NMR/MRI work? Particles such as protons have an intrinsic magnetic dipole moment, as discussed in Chapter 28. When protons are placed in a strong

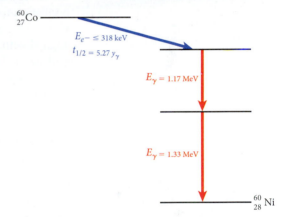

FIGURE 40.40 Decay scheme of cobalt-60.

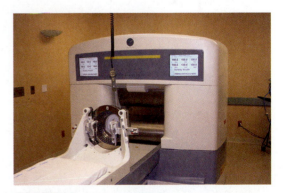

FIGURE 40.41 Gamma-knife cancer treatment facility at Scripps Memorial Hospital, La Jolla, CA.

FIGURE 40.42 Image slices through a human head, obtained via MRI techniques.

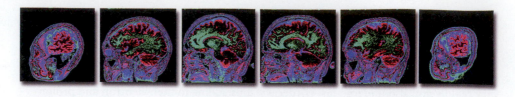

magnetic field, their spin and therefore their magnetic dipole moment can have only two directions: parallel or antiparallel to the external field. The difference in energy between the two states is given by the difference in magnetic potential energy, which is $2\mu B$, where μ is the component of the proton's magnetic moment along the direction of the external field. This potential energy difference was discussed in Chapter 27.

A time-varying electric field at the proper frequency can induce some of the protons to flip the direction of their magnetic dipole moments from parallel to antiparallel to the external field, and the protons thereby gain potential energy. Because the magnetic potential energy can have only two possible values, the energy required to flip the direction is a discrete value, depending on the magnitude of the external field. Chapter 36 showed that the energy delivered by the field is proportional to the frequency. Thus, only one given oscillation frequency will cause the dipole moment to flip. If the time-varying electric field is switched off, the protons in the higher energy state, with dipole moments that are not aligned with the field, will flip back to being parallel with the field, emitting photons of a well-defined energy that can be detected.

A magnetic resonance imaging device uses the physical principle of nuclear magnetic resonance just described. This technique can image the location of the protons in a human body by introducing a time-varying electric field, and then varying the magnetic field in a known, precise manner to produce a three-dimensional picture of the distribution of tissue containing hydrogen. The quality of this imaging depends on the strength of the external magnetic field. Figure 40.42 shows results of an MRI.

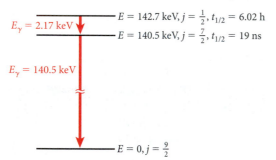

FIGURE 40.43 Lowest energy levels of technetium, their angular momenta in unit of Planck's constant, and their half-lives.

We close with an interesting example of the diagnostic use of nuclear radiation. The element technetium ($Z = 43$) has no stable isotopes. The longest-lived technetium isotope is $^{99}_{43}$Tc, with a half-life of 4.2 million years. For one medical diagnostic purpose, however, the isotope $^{99}_{43}$Tc is most valuable. Its lowest energy levels, their angular momentum values, and their half-lives are shown in Figure 40.43. The state with angular momentum $\frac{1}{2}\hbar$ is an isomeric state, because its angular momentum is more than $1\hbar$ different from either of the two lower-energy states. Thus, it is relatively long-lived and decays with a half-life of 6.02 h into the $j = \frac{7}{2}\hbar$ state via the emission of a photon with energy of 2.17 keV. This is followed by a very quick decay to the ground state, with a half-life of only 19 ns, via the emission of a 140.5-keV photon. Thus, the decay of $^{99}_{43}$Tc delivers high-energy photons over several hours after the initial preparation of the isotope. Since a 140.5-keV photon easily penetrates biological tissue, this isotope can be used for diagnostic purposes.

In a technetium scan, $^{99}_{43}$Tc is injected into the patient. After a few minutes to a few hours, a picture can be taken of the patient with a gamma-ray camera to monitor where in the body the technetium has traveled. Certain cancers result in an increased local concentration of $^{99}_{43}$Tc near the tumor, and appear as black areas in the gamma-ray picture. The patient shown in Figure 40.44 fortunately turned out to be cancer-free. (The black dot near the left elbow is the injection point.) $^{99}_{43}$Tc is thus a great example of the medical use of specific isotopes.

FIGURE 40.44 Technetium scan of a female patient. Dark areas show an increased concentration of the technetium-99 isotope in the body.

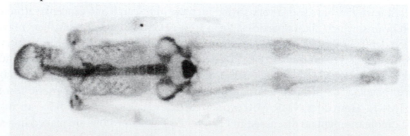

WHAT WE HAVE LEARNED | EXAM STUDY GUIDE

- The atomic nucleus consists of protons and neutrons, collectively called nucleons. Nuclei with different numbers of neutrons, but the same number of protons, are called isotopes of the same element. The mass number of an isotope is the sum of the number of protons and the number of neutrons, $A = Z + N$.

- The interaction between the nucleons is short-range and can effectively be described by potentials based on pion exchange.

- Nuclei are approximately spherical in shape, with the radius of the sphere depending on the mass number, $R(A) = R_0 A^{1/3}$, and $R_0 = 1.12$ fm.

- The number density of nuclear matter in the interior of a nucleus is $n = 0.17$ fm^{-3}, and the mass density is $\rho = m_{\text{nucleon}} n = 2.8 \cdot 10^{17}$ kg/m^3. The dependence of the number/mass density on the radial coordinate is given by the Fermi function $n(r) = n_0 / [1 + e^{(r - R(A))/a}]$, with $a = 0.54$ fm.

- There are 251 stable isotopes and more than 2400 known unstable isotopes, with lifetimes from fractions of seconds to many times the age of the universe.

- Nuclear mass is measured in multiples of the atomic mass unit, 1 u = $1.660538782 \cdot 10^{-27}$ kg = 931.494028 MeV/c^2, with the definition of 1 u as 1/12 of the mass of the $^{12}_{6}$C atom.

- The mass of a nucleus with Z protons and N neutrons is smaller than the sum of the masses of the individual nucleons, and the binding energy is defined as that mass difference times c^2, $B(N, Z) = Z m_p c^2 + N m_n c^2 - m(N, Z)c^2$.

- The mass excess of a nucleus is defined as the difference between the mass of a nucleus expressed in atomic mass units and the mass number: mass excess = $m(N, Z) - A$ u.

- The binding energy of different isotopes can be reproduced well by the Bethe-Weizsäcker formula for the liquid-drop model, as the sum of volume, surface, Coulomb, asymmetry, and pairing contributions,

$$B(N, Z) = a_v A - a_s A^{2/3} - a_c Z^2 A^{-1/3}$$
$$- a_a \left(Z - \tfrac{1}{2} A \right)^2 A^{-1} - a_p \left((-1)^Z + (-1)^N \right) A^{-1/2}.$$

- The Q-value of a given nuclear reaction is the difference between the sum of the mass-energies of the initial nuclei minus that of the final nuclei.

- The Fermi gas model approximates the nucleus as a quantum gas of nucleons that can move freely inside the nucleus but are confined by the nuclear surface. The density of states in the Fermi gas model is $dN(E) = m^{3/2} E^{1/2} dE / (2^{1/2} \pi^2 a^3 \hbar^3)$.

- The Fermi energy is

$$E_{\text{F}} = (\hbar^2 / 2m) \left(\tfrac{3}{2} \pi^2 n_0 \right)^{2/3} = 38 \text{ MeV}.$$

- The nuclear shell model predicts angular momentum shells inside the nucleus similar to the electron shells in the atom. It is able to reproduce the magic numbers of 2, 8, 20, 28, 50, 82, 126.

- Nuclear decays show exponential decay laws, $N(t) = N_0 e^{-\lambda t}$. The decay constant λ, half-life $t_{1/2}$, and mean lifetime τ are related via $t_{1/2} = \ln 2 / \lambda = \tau \ln 2$.

- In α-decay, a nucleus emits the nucleus of the helium-4:

$$^A_Z \text{Nuc} \rightarrow \, ^4_2 \text{He} + \, ^{A-4}_{Z-2} \text{Nuc}'.$$

In a β^--decay, an electron and an anti-neutrino are emitted:

$$^A_Z \text{Nuc} \rightarrow \, ^A_{Z+1} \text{Nuc}' + e^- + \bar{\nu}_e.$$

A β^+-decay can proceed via a positron emission:

$$^A_Z \text{Nuc} \rightarrow \, ^A_{Z-1} \text{Nuc}' + e^+ + \nu_e$$

or the capture of an electron:

$$e^- + \, ^A_Z \text{Nuc} \rightarrow \, ^A_{Z-1} \text{Nuc}' + \nu_e.$$

A γ-decay is the emission of a high-energy photon from an excited state of a nucleus, a process that does not transmute the nucleus.

- The SI unit for radioactivity is the becquerel, 1 Bq = 1 nuclear decay/s. The SI unit for the absorbed dose is the gray, 1 Gy = 1 J/kg. The SI unit for the dose equivalent is the sievert, 1 Sv = w_r(1 Gy), where the radiation weight factor w_r varies between 1 and 20, depending on the type and energy of the emitted particle.

- Nuclear fission is the splitting of a very heavy nucleus into two medium-mass nuclei, usually with the associated emission of one or a few neutrons. This process is the physical basis for nuclear power plants, as well as nuclear weapons.

- Nuclear fusion is the merging of two light nuclei into a heavier one. Nuclear fusion is the basic process that powers the stars. The two most common fusion chains are the proton-proton chain and the CNO chain.

- Nuclear astrophysics research is employed to explain, among other things, the abundance of the chemical elements in our Solar System.

- Nuclear medicine uses nuclear physics for medical diagnostics and cancer radiation treatment.

KEY TERMS

ANSWERS TO SELF-TEST OPPORTUNITIES

40.1

Particle	Mass (u)
d	2.014101778
$^{12}_{6}C$	12
p	1.007825032
$^{13}_{6}C$	13.00573861

$$Q = \left(2.014101778 \text{ u}\right)c^2 + \left(12 \text{ u}\right)c^2 -$$

$$\left(\left(1.007825032 \text{ u}\right)c^2 + \left(13.00573861 \text{ u}\right)c^2\right) =$$

$$\left(0.000538128 \text{ u}\right)c^2 = 0.501263 \text{ MeV, positive, exothermic.}$$

40.2 First let's look at only the odd-charge nuclei. Using the mass formula, we see that they have the same values for the volume term, surface term, and pairing term. The asymmetry term and the Coulomb term together combine to yield the parabolic shape. The even-charge nuclei have a similar parabola that is offset vertically relative to the odd-charge nuclei by the pairing term.

40.3

$$dE = \frac{\hbar^2}{m}k\,dk = \frac{\hbar^2}{m}\frac{\sqrt{2mE}}{\hbar}dk = \hbar\sqrt{2E/m}\,dk \Rightarrow \frac{dE}{\hbar\sqrt{2E/m}} = dk$$

$$dN(E) = \frac{a^3 k^2}{2\pi^2}\frac{dE}{\hbar\sqrt{2E/m}} = \frac{a^3\,2mE/\hbar^2}{2\pi^2}\frac{dE}{\hbar\sqrt{2E/m}} = \frac{a^3 m^{3/2}}{2^{1/2}\pi^2\hbar^3}E^{1/2}dE.$$

PROBLEM-SOLVING PRACTICE

Problem-Solving Guidelines

1. Remember that the nucleus is the prototypical quantum system. The Pauli exclusion principle and the uncertainty relation provide essential tools for working out relationships of momenta, energies, and sizes.

2. When it comes to binding energies, you should first consider the options provided by the liquid-drop formula.

3. If you need to solve a problem involving nuclear decays, first make sure which decay process is involved. You can do this by comparing the initial and final nuclei involved in the decay. Then you need to make sure you respect the conservation laws (lepton number, baryon number, momentum, energy, charge, etc.) in your decay process. Calculating the Q-value is a good idea, because it lets you find out if a decay is possible or not.

SOLVED PROBLEM **40.2**   **Nuclear Surface**

PROBLEM
The nuclear surface thickness t is defined as the distance over which the nuclear density falls from 90% to 10% of its central value. Derive a relationship between this surface thickness and the parameters of the Fermi function (equation 40.4). What are the distances over which the nuclear density falls from 80% to 20%, and from 70% to 30%, of its central value?

SOLUTION

THINK
Equation 40.4 gives the Fermi function for the nuclear density $n(r)$ as a function of the radial coordinate r. It contains the 3 parameters n_0, $R(A)$, and a,

$$n(r) = \frac{n_0}{1 + e^{(r-R(A))/a}}.$$

Section 40.1 established that the empirical value of the parameters are $a = 0.54$ fm, $n_0 = 0.17$ fm^{-3}, and $R(A) = R_0 A^{1/3}$ with $R_0 = 1.12$ fm. Since n_0 is a multiplicative constant in the Fermi function, it cannot have any influence on the thickness of the nuclear surface. However, it is not quite as easy to see if $R(A)$ or a determine the surface thickness.

SKETCH

Figure 40.45 shows the Fermi function for three different values of the nuclear radius R. It is clear from this sketch that the nuclear surface thickness t is independent of R. Thus our desired expression for the nuclear surface thickness can only be a function of the parameter a.

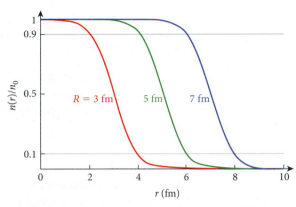

FIGURE 40.45 Fermi function for three different nuclear radii.

RESEARCH

Our starting point is the Fermi function, which gives us the density as a function the radial coordinate, $n(r)$,

$$\frac{n(r)}{n_0} = \frac{1}{1 + e^{(r-R)/a}}.$$

Here we have simply written R instead of $R(A)$ to simplify our notation, recognizing from our sketch that the particular value of the nuclear radius will have no influence on the nuclear surface thickness. We need to invert this function so that we obtain the radial coordinate as a function of the density, $r(n/n_0)$. Then the thickness is defined as

$$t = r(10\%) - r(90\%).$$

SIMPLIFY

We solve our expression above for r by taking the inverse of both sides and then subtracting 1 from both sides

$$\frac{n_0}{n} = 1 + e^{(r-R)/a} \Rightarrow$$

$$\frac{n_0}{n} - 1 = e^{(r-R)/a}$$

Now we can take the natural log and find

$$\ln\left(\frac{n_0}{n} - 1\right) = (r - R)/a.$$

Multiplication of both sides with a and addition of R then results in

$$r = R + a \ln\left(\frac{n_0}{n} - 1\right),$$

which is our desired expression of the radial coordinate as a function of the density, $r(n/n_0)$. By using this expression for $n = 0.1 n_0$ and $n = 0.9 n_0$ we can express the surface thickness as

$$t = r(0.1) - r(0.9) = r = \left[R + a \ln\left(\frac{1}{0.1} - 1\right)\right] - \left[R + a \ln\left(\frac{1}{0.9} - 1\right)\right]$$

$$t = a\left[\ln\left(\frac{1}{0.1} - 1\right) - \ln\left(\frac{1}{0.9} - 1\right)\right] = a\left[\ln 9 - \ln 0.\overline{1}\right] = 2a\ln 9$$

$$t = 4.39445a.$$

CALCULATE

We have obtained the desired relationship between the parameter a of the Fermi function and the nuclear surface thickness, $t \approx 4.4a$. Using $a = 0.54$ fm, we find that the surface

Continued—

thickness of nuclei is 2.4 fm. What remains to be calculated are the thicknesses of the regions over which the density falls off from 80% to 20%, and from 70% to 30%:

$$r(0.2) - r(0.8) = (0.54 \text{ fm})\left[\ln\left(\frac{1}{0.2} - 1\right) - \ln\left(\frac{1}{0.8} - 1\right)\right] = 1.4972 \text{ fm}$$

$$r(0.3) - r(0.7) = (0.54 \text{ fm})\left[\ln\left(\frac{1}{0.3} - 1\right) - \ln\left(\frac{1}{0.7} - 1\right)\right] = 0.91508 \text{ fm}.$$

ROUND

Since the constant $a = 0.54$ fm is given only to 2 significant figures, it does not make sense to specify our result to a greater precision than this, and our final answer is

$$r(0.2) - r(0.8) = 1.5 \text{ fm}$$
$$r(0.3) - r(0.7) = 0.92 \text{ fm}.$$

DOUBLE-CHECK

As you can see from the sketch, the Fermi function falls off almost linearly between $n/n_0 = 0.9$ and $n/n_0 = 0.1$. Since the density interval between $n/n_0 = 0.7$ and $n/n_0 = 0.3$ is half of that between 0.9 and 0.1, we expect that the distance over which this fall-off occurs is approximately half of the nuclear surface thickness. (Remember: The surface thickness is defined as the distance over which the density falls off from 0.9 to 0.1 of the central density!) We found that the thickness is $t = 2.4$ fm, so it is thus not unreasonable to find $r(0.3) - r(0.7) = 0.92$ fm.

MULTIPLE-CHOICE QUESTIONS

40.1 Radium-226 decays by emitting an alpha particle. What is the daughter nucleus?

a) Rd   b) Rn   c) Bi   d) Pb

40.2 Which of the following decay modes is due to a transition between states of the same nucleus?

a) alpha decay   c) gamma decay

b) beta decay   d) none of the above

40.3 In neutron stars, which are roughly 90% neutrons and supported almost entirely by nuclear forces, which of the following binding-energy terms becomes relatively dominant compared to ordinary nuclei?

a) the Coulomb term   d) all of the above

b) the asymmetry term   e) none of the above

c) the pairing term

40.4 When a target nucleus is bombarded by an appropriate beam of particles, it is possible to produce

a) a less massive nucleus, but not a more massive one.

b) a more massive nucleus, but not a less massive one.

c) a nucleus with smaller charge number, but not one with a greater charge number.

d) a nucleus with greater charge number, but not one with a smaller charge number.

e) a nucleus with either greater or smaller charge number.

40.5 The strong force (select all that apply)

a) is only attractive.   d) All of the above are true.

b) does not act on electrons.   e) None of the above are

c) only acts over a few fm.   true.

40.6 Cobalt has a stable isotope, ^{59}Co, and 22 radioactive isotopes. The most stable radioactive isotope is ^{60}Co. What is the dominant decay mode of this ^{60}Co isotope?

a) β^+ b) β^- c) p   d) n

40.7 The mass of an atom (atomic mass) is equal to

a) the sum of the masses of the protons.

b) the sum of the masses of protons and neutrons.

c) the sum of the masses of protons, neutrons and electrons.

d) the sum of the masses of protons, neutrons, and electrons minus the atom's binding energy.

QUESTIONS

40.8 What is more dangerous, a radioactive material with a short half-life or a long one?

40.9 Apart from fatigue, what is another reason the Federal Aviation Administration limits the number of hours intercontinental pilots can travel annually?

40.10 Why are there magic numbers in the nuclear shell model?

40.11 The binding energy of $^3_2$He is lower than that of $^3_1$H. Provide a plausible explanation, considering the Coulomb interaction between two protons in $^3_2$He.

40.12 Which of the following quantities is conserved during a nuclear reaction, and how?

a) charge

d) linear momentum

b) the number of nucleons, A

e) angular momentum

c) mass-energy

40.13 Some food is treated with gamma radiation to kill bacteria. Why is there not a concern that people who eat such food might be consuming food containing gamma radiation?

40.14 Refer to the subsection "Terrestrial Fusion" in Section 40.4 to see how achieving controlled fusion would be the solution to mankind's energy problems, and how difficult it is to do. Why is it so hard? The Sun does it all the time (see the previous subsection, "Stellar Fusion"). Do we need to understand better how the Sun works to build a useful nuclear fusion reactor?

40.15 Why are atomic nuclei more or less limited in size and neutron-proton ratios? That is, why are there no stable nuclei with 10 times as many neutrons as protons, and why are there no atomic nuclei the size of marbles?

40.16 A nuclear reaction of the kind $^3_2$He + $^{12}_6$C $\rightarrow$ X + α is called a *pick-up* nuclear reaction.

a) Why is it called a *pick-up* reaction, that is, what is picked up, what picked it up, and where did it come from?

b) What is the resulting nucleus X?

c) What is the Q-value of this reaction?

d) Is this reaction endothermic or exothermic?

40.17 Isospin, or *isotopic spin,* is a quantum variable describing the relationship between protons and neutrons in nuclear and particle physics. (Strictly, it describes the relationship between *u-* and *d*-quarks, as described in Chapter 39, but isospin was introduced before the advent of the quark model.) It has the same algebraic properties as quantum angular momentum: The proton and neutron form an iso-doublet of states, with total-isospin quantum number $\frac{1}{2}$; the proton is the $t_z = +\frac{1}{2}$ state, the neutron the $t_z = -\frac{1}{2}$ state, where z refers to a direction in an abstract isospin space.

a) What isospin states can be constructed from two nucleons, that is, two $t = \frac{1}{2}$ objects? To what nuclei do these states correspond?

b) What isospin states can be constructed from three nucleons? To what nuclei do these correspond?

40.18 Before you look it up, make a prediction of the spin (intrinsic spin, i.e., actual angular momentum) of the deuteron, $^2_1$H. Explain your reasoning. *Hint:* Nucleons are fermions.

40.19 ^{39}Ar is an isotope with a half-life of 269 yr. If it decays through beta-minus emission, what isotope will result?

40.20 A neutron star is essentially a gigantic nucleus with mass 1.35 times that of the Sun, or mass number of order 10^{57}. It consists of approximately 99% neutrons, the rest being protons and an equal number of electrons. Explain the physics that determines these features.

40.21 What is the nuclear configuration of the daughter nucleus associated with the alpha decay of Hf ($A = 157$, $Z = 72$)?

PROBLEMS

A blue problem number indicates a worked-out solution is available in the Student Solutions Manual. One • and two •• indicate increasing level of problem difficulty.

Section 40.1

40.22 Estimate the volume of the uranium-235 nucleus.

40.23 Using the table of isotopes in Appendix B, calculate the binding energies of the following nuclei.

a) ^{7}Li   b) ^{12}C   c) ^{56}Fe   d) ^{85}Rb

40.24 According to the standard notation, find the number of protons, nucleons, neutrons, and electrons of $^{134}_{54}$Xe.

•40.25 Using the Fermi function, determine the relative change in density $(dn(r)/dr)/n_0$ at the nuclear surface, $r = R(A)$.

•40.26 Calculate the binding energy for the following two uranium isotopes:

a) $^{238}_{92}$U, which consists of 92 protons, 92 electrons, and 146 neutrons, with a total mass of 238.0507826 u.

b) $^{235}_{92}$U, which consists of 92 protons, 92 electrons, and 143 neutrons, with a total mass of 235.0439299 u.

The atomic mass unit u = $1.66 \cdot 10^{-27}$ kg. Which isotope is more stable (or less unstable)?

Section 40.2

40.27 Write down equations to describe the β^--decay of the following atoms:

a) ^{60}Co   b) ^{3}H   c) ^{14}C

40.28 Write down equations to describe the alpha decay of the following atoms:

a) ^{212}Rn   b) ^{241}Am

40.29 How much energy is released in the beta decay of ^{14}C?

40.30 A certain radioactive isotope decays to one-eighth its original amount in 5.0 h.

a) What is its half-life?   b) What is its mean lifetime?

40.31 A certain radioactive isotope decays to one-eighth its original amount in 5.0 h. How long would it take for 10% of it to decay?

40.32 Determine the decay constant of radium-226, which has a half-life of 1600 yr.

40.33 A 1-gram sample of radioactively decaying thorium-228 decays via β^--emission, and 75 counts are recorded in one day in a detector that has 10% efficiency (that is, 10% of all events that occur are actually recorded by the detector). What is the lifetime of this isotope?

40.34 The half-life of a sample of 10^{11} atoms that decay by alpha emission is 10 min. How many alpha particles are emitted between the time interval 100 min and 200 min?

•40.35 The specific activity of a radioactive material is the number of disintegrations per second per gram of radioactive atoms.

a) Given the half-life of ^{14}C of 5730 yr, calculate the specific activity of ^{14}C. Express your result in disintegrations per second per gram, becquerel per gram, and curie per gram.

b) Calculate the initial activity of a 5-g piece of wood.

c) How many ^{14}C disintegrations have occurred in a 5-g piece of wood that was cut from a tree January 1, 1700?

•40.36 During a trip to a historic excavation site, an archeologist found a piece of charcoal. Analysis of the charcoal found the activity of ^{14}C in the sample to be 0.42 Bq. If the mass of the charcoal is 7.2 g, estimate the approximate age of the site.

•40.37 In 2008, crime scene investigators discover the bones of a person who appeared to have been the victim of a brutal crime, which had occurred a long time ago. They would like to know the year when the person was murdered. Using carbon dating, they determine that the rate of change of the $^{14}_{6}$C is 0.268 Bq per gram of carbon. The rate of change of $^{14}_{6}$C in the bones of a person who had just died is 0.270 Bq per gram of carbon. What year was the victim killed? The half-life of $^{14}_{6}$C is $5.73 \cdot 10^3$ yr.

•40.38 Physicists blow stuff up better than anyone else. The figure of merit for blowing stuff up is the fraction of initial rest mass converted into energy in the process. Looking up the necessary data, calculate this fraction for the following processes:

a) chemical combustion of hydrogen: $2H_2 + O_2 \rightarrow 2H_2O$

b) nuclear fission: $n + ^{235}_{92}U \rightarrow ^{89}_{36}Kr + ^{142}_{56}Ba + 5n$

c) thermonuclear fusion: $^{6}_{3}Li + ^{2}_{1}H \rightarrow ^{7}_{4}Be + n$

d) decay of free neutron: $n \rightarrow p + e^- + \bar{\nu}_e$

e) decay of muon: $\mu^- \rightarrow e^- + \nu_\mu + \bar{\nu}_e$

f) electron-positron annihilation: $e^- + e^+ \rightarrow 2\gamma$

••40.39 An unstable nucleus A decays to an unstable nucleus B, which in turn decays to a stable nucleus. If at $t = 0$ s there are N_{A0} and N_{B0} nuclei present, derive an expression for N_B, the number of B nuclei present, as a function of time.

••40.40 In a simple case of *chain* radioactive decay, a *parent* radioactive species of nuclei, A, decays with a decay constant λ_1 into a *daughter* radioactive species of nuclei, B, which then decays with a decay constant λ_2 to a *stable* element C.

a) Write the equations describing the number of nuclei in each of the three species as a function of time, and derive an expression for the number of daughter nuclei, N_2, as a function of time, and for the activity of the daughter nuclei, A_2, as a function of time.

b) Discuss the results in the case when $\lambda_2 > \lambda_1$ ($\lambda_2 \approx 10\lambda_1$) and when $\lambda_2 >> \lambda_1$ ($\lambda_2 \approx 100\lambda_1$)

Section 40.3

•40.41 Consider the Bethe-Weizsäcker formula for the case of odd A nuclei. Show that the formula can be written as a quadratic in Z—and thus, that for any given A, the binding energies of the isotopes having that A take a quadratic form, $B = a + bZ + cZ^2$. Find the most deeply bound isotope (the most stable one) having $A = 117$ using your result.

••40.42 The neutron drip line is defined to be the point at which the neutron separation energy for any isotope of an element is negative. That is, the neutron is unbound. Using the Bethe-Weizsäcker mass formula, find the neutron drip line for the element Sn. Find this value using S_n and S_{2n}. Plot both S_n and $S_{2n}/2$ as a function of neutron number.

Section 40.4

40.43 A nuclear fission power plant produces about 1.50 GW of electrical power. Assume that the plant has an overall efficiency of 35.0% and that each fission event produces 200 MeV of energy. Calculate the mass of ^{235}U consumed each day.

40.44 a) What is the energy released in the fusion reaction $^{2}_{1}H + ^{2}_{1}H \rightarrow ^{4}_{2}He + Q$?

b) The oceans have a total mass of water of $1.50 \cdot 10^{16}$ kg, and 0.0300% of this quantity is deuterium, $^{2}_{1}H$. If all the deuterium in the oceans were fused by controlled fusion into $^{4}_{2}He$, how many joules of energy would be released?

c) World power consumption is about 10^{13} W. If consumption were to stay constant and all problems arising from ocean water consumption (including those of political, meteorological, and ecological nature) could be avoided, how many years would the energy calculated in part (b) last?

40.45 The Sun radiates energy at the rate of $3.85 \cdot 10^{26}$ W.

a) At what rate, in kg/s, is the Sun's mass converted into energy?

b) Why is this result different from the rate calculated in Example 40.6, $6.02 \cdot 10^{11}$ kg protons being converted into helium each second?

c) Assuming that the current mass of the Sun is $1.99 \cdot 10^{30}$ kg and that it radiated at the same rate for its entire lifetime of $4.50 \cdot 10^{9}$ yr, what percentage of the Sun's mass was converted into energy during its entire lifetime?

40.46 Consider the following fusion reaction, which allows stars to produce progressively heavier elements: $_2^3\text{He} + _2^4\text{He} \rightarrow _4^7\text{Be} + \gamma$. The mass of $_2^3\text{He}$ is 3.016029 u, the mass of $_2^4\text{He}$ is 4.002602 u, and the mass of $_4^7\text{Be}$ is 7.0169298 u. The atomic mass unit is $u = 1.66 \cdot 10^{-27}$ kg. Assuming the Be atom is at rest after the reaction and neglecting any potential energy between the atoms and kinetic energy of the He nuclei, calculate the minimum possible energy and maximum possible wavelength of the photon γ that is released in this reaction.

•**40.47** Estimate the temperature that would be needed to make the fusion reaction $_2^3\text{He} + _2^3\text{He} \rightarrow _2^4\text{He} + p + p$ go.

•**40.48** Consider a hypothetical fission process where a $_{52}^{120}\text{Te}$ nucleus splits into two identical $_{26}^{60}\text{Fe}$ nuclei without producing any other particles or radiation. The mass of $_{52}^{120}\text{Te}$ is 119.904040 u, and the mass of $_{26}^{60}\text{Fe}$ is 59.934078 u. At the moment when the two Fe nuclei form, but before they start moving away due to Coulomb repulsion, how far apart are the two Fe nuclei?

•**40.49** The *mass excess* of a nucleus is defined as the difference between the atomic mass (in atomic mass units, u), and the mass number of the nucleus, A. Using the mass-energy conversion 1 u = 931.49 MeV/c^2, this mass excess is usually expressed in units of keV. The table below presents the mass excess for several nuclides (per Berkeley National Lab *NuBase* data base):

No	Nuclide	Mass Number A	Mass Excess Δm (keV/c^2)	Atomic mass (u)
1	$_0^1 n$	1	8071.3	1.00866491
2	$_{98}^{252}\text{Cf}$	252	76034	
3	$_{100}^{256}\text{Fm}$	256	85496	
4	$_{56}^{140}\text{Ba}$	140	−83271	
5	$_{54}^{140}\text{Xe}$	140	−72990	
6	$_{46}^{112}\text{Pd}$	112	−86336	
7	$_{42}^{109}\text{Mo}$	109	−67250	

a) Calculate the atomic mass (in atomic mass units) for each of the elements in the table. For reference, the atomic mass of the neutron is included.

b) Using your results in (a), determine the mass-energy difference between the initial and final state for the following possible fission reactions:

$_{98}^{252}\text{Cf} \rightarrow _{56}^{140}\text{Ba} + _{42}^{109}\text{Mo} + 3n$ and

$_{100}^{256}\text{Fm} \xrightarrow{\text{sf}} _{54}^{140}\text{Xe} + _{46}^{112}\text{Pd} + 4n$

c) Will these reactions occur spontaneously?

Section 40.5

40.50 Neutron stars are sometimes approximated to be nothing more than large atomic nuclei (but with many more neutrons). Assuming that a neutron star is as dense as an atomic nucleus, estimate the number of nucleons in the star for a 10-km-diameter star.

40.51 What is the average kinetic energy of protons at the center of a star where the temperature is 10^7 K? What is the average velocity of those protons?

•**40.52** Billions of years ago, our Solar System was created out of the remnants of exploding stars. Nuclear scientists believe that two isotopes of uranium, ^{235}U and ^{238}U, were created in equal amounts at the time of a stellar explosion. However, today 99.28% of uranium is in the form of ^{238}U and only 0.72% is in the form of ^{235}U. Assuming a simplified model in which all of the matter in the Solar System originated in a single exploding star, estimate the approximate time of this explosion.

Section 40.6

40.53 A drug containing $_{43}^{99}\text{Tc}$ ($t_{1/2} = 6.05$ h) with an activity of 1.50 μCi is to be injected into a patient at 9.30 a.m. You are to prepare the sample 2.50 h before the injection (at 7:00 a.m.). What activity should the drug have at the preparation time (7:00 a.m.)?

40.54 Consider a 42.58-MHz photon needed to produce NMR transition in free protons in a magnetic field of 1.000 T. What is the wavelength of the photon, its energy, and the region of the spectrum in which it lies? Could it be harmful to the human body?

40.55 The radon isotope ^{222}Rn, which has a half-life of 3.825 days, is used for medical purposes such as radiotherapy. How long does it take until ^{222}Rn decays to 10% of its initial quantity?

40.56 Radiation therapy is one of the modalities for cancer treatment. Based on the approximate mass of a tumor, oncologists can calculate the radiation dose necessary to treat their patients. Suppose a patient has a 50-g tumor and needs to receive 0.18 J of energy to kill the cancer cells. What rad (radiation absorbed dose) should the patient receive?

Additional Problems

40.57 The nucleus of sodium-22 ($_{11}^{22}\text{Na}$) has a mass of 21.994435 u. How much work would be needed to take this nucleus completely apart into its constituent pieces (protons and neutrons)?

40.58 A Geiger counter initially records 7210 counts per second. After 45 min it records 4585 counts/s. Ignore any uncertainty in the counts and find the half-life of the material.

40.59 How close can a 5-MeV alpha particle get to a uranium-238 nucleus, assuming the only interaction is Coulomb?

40.60 ^{239}Pu decays with a half-life of 24,100 yr via a 5.25-MeV alpha particle. If you have a 1.00 kg spherical sample of ^{239}Pu, find the initial activity in Bq.

40.61 The activity of a sample of ^{210}Bi (with a half-life of 5.01 days) was measured to be 1.000 μCi. What is the activity of this sample after 1 yr?

40.62 Assuming carbon makes up 14% of the mass of a human body, calculate the activity of a 75-kg person considering only the beta decays of carbon-14.

40.63 ^{8}Li is an isotope that has a lifetime of less than one second. Its mass is 8.022485 u. Calculate its binding energy in MeV.

40.64 What is the total energy released in the decay $n \rightarrow p + e^{-} + \bar{\nu}_e$?

40.65 A gallon of regular gasoline (density of 737 kg/m^3) contains about 131 MJ of chemical energy. How much energy is contained in the rest mass of this gallon?

40.66 10^{30} atoms of a radioactive sample remain after 10 half-lives. How many atoms remain after 20 half-lives?

40.67 Calculate the binding energy per nucleon of

a) $^{4}_{2}$He (4.002603 u).   c) $^{3}_{1}$H (3.016050 u).

b) $^{3}_{2}$He (3.016030 u).   d) $^{2}_{1}$H (2.014102 u).

40.68 The time constant for a radioactive nucleus is 4300 s. What is its half-life?

40.69 ^{214}Pb has a half-life of 26.8 min. How many minutes must elapse for 90% of a given sample of ^{214}Pb atoms to decay?

•**40.70** A 10-g fragment of charcoal is to be carbon dated. Measurements show a ^{14}C activity of 100 decays/min. Date the tree that this charcoal came from.

•**40.71** After a tree has been chopped down and burned to produce ash, the carbon isotopes in the ash are found to have a ratio for ^{14}C to ^{12}C of $1.300 \cdot 10^{-12}$. Experimental tests on the ^{14}C atoms reveal that ^{14}C is a beta emitter with a half-life of 5730 years. At an archeological excavation, a skeleton is found next to a campfire with some wood ash. If 50 g of carbon from the ash in the campfire emits betas at a rate of 20 per h, how old is the campfire?

•**40.72** If your mass is 70 kg and you have a lifetime of three score and ten (70 yr), how many proton decays would you expect to have in your body during your life (assume your body is entirely composed of water)? Use a half-life of 10^{30} yr.

•**40.73** You have developed a grand unified theory which predicts the following things about the decay of the proton: (1) protons never get any older, in the sense that their probability of decay per unit time never changes, and (2) half the protons in any given collection of protons will have decayed in $1.8 \cdot 10^{29}$ yr. You are given experimental facilities to test your theory: A tank containing 10,000 tons of water and sensors to record proton decays. You will be allowed access to this facility for two years. How many proton decays will occur in this period if your theory is correct?

•**40.74** The precession frequency of the protons in a laboratory NMR spectrometer is 15.35850 MHz. The magnetic moment of the proton is $1.410608 \cdot 10^{-26}$ J/T, while its spin angular momentum is $0.5272863 \cdot 10^{-34}$ J s. Calculate the magnitude of the magnetic field in which the protons are immersed.

•**40.75** Two species of radioactive nuclei, A and B, each with an initial population N_0, start decaying. After a time of 100 s it is observed that $N_A = 100 N_B$. If $\tau_A = 2\tau_B$, find the value of τ_B.

••**40.76** The most common isotope of uranium, $^{238}_{92}$U, produces radon $^{222}_{86}$Rn through the following sequence of decays:

$$^{238}_{92}\text{U} \rightarrow {}^{234}_{90}\text{Th} + \alpha, \quad {}^{234}_{90}\text{Th} \rightarrow {}^{234}_{91}\text{Pa} + \beta^{-} + \bar{\nu}_e,$$

$$^{234}_{91}\text{Pa} \rightarrow {}^{234}_{92}\text{U} + \beta^{-} + \bar{\nu}_e, \quad {}^{234}_{92}\text{U} \rightarrow {}^{230}_{90}\text{Th} + \alpha,$$

$$^{230}_{90}\text{Th} \rightarrow {}^{226}_{88}\text{Ra} + \alpha, \quad {}^{226}_{88}\text{Ra} \rightarrow {}^{222}_{86}\text{Rn} + \alpha.$$

A sample of $^{238}_{92}$U will build up equilibrium concentrations of its daughter nuclei down to $^{226}_{88}$Ra; the concentrations of each are such that each daughter is produced as fast as it decays. The $^{226}_{88}$Ra decays to $^{222}_{86}$Rn, which escapes as a gas. (The α particles also escape, as helium; this is a source of much of the helium found on Earth.) In high concentrations, the radon is a health hazard in buildings built on soil or foundations containing uranium ores, as it can be inhaled.

a) Look up the necessary data, and calculate the rate at which 1.00 kg of an equilibrium mixture of $^{238}_{92}$U and its first five daughters produces $^{222}_{86}$Rn (mass per unit time).

b) What activity (in curies per unit time) of radon does this represent?

Appendix A

Mathematics Primer

Notation:

The letters a, b, c, x, and y represent real numbers.

The letters i, j, m, and n represent integer numbers.

The Greek letters α, β, and γ represent angles, which are measured in radians.

1. Algebra

1.1 Basics

Factors:

$$ax + bx + cx = (a + b + c)x \tag{A.1}$$

$$(a+b)^2 = a^2 + 2ab + b^2 \tag{A.2}$$

$$(a-b)^2 = a^2 - 2ab + b^2 \tag{A.3}$$

$$(a+b)(a-b) = a^2 - b^2 \tag{A.4}$$

Quadratic equation:
An equation of the form

$$ax^2 + bx + c = 0 \tag{A.5}$$

for given values of a, b, and c has the two solutions:

$$x = \frac{-b + \sqrt{b^2 - 4ac}}{2a}$$

and $$\tag{A.6}$$

$$x = \frac{-b - \sqrt{b^2 - 4ac}}{2a}$$

The solutions of this quadratic equation are called *roots*. The roots are real numbers if $b^2 \geq 4ac$.

1.2 Exponents

If a is a number, a^n is the product of a with itself n times:

$$a^n = \underbrace{a \times a \times a \times \cdots \times a}_{n \text{ factors}} \tag{A.7}$$

The number n is called the *exponent*. However, an exponent does not have to be a positive number or an integer. Any real number x can be used as an exponent.

$$a^{-x} = \frac{1}{a^x} \tag{A.8}$$

$$a^0 = 1 \tag{A.9}$$

$$a^1 = a \tag{A.10}$$

Roots:

$$a^{1/2} = \sqrt{a} \tag{A.11}$$

$$a^{1/n} = \sqrt[n]{a} \tag{A.12}$$

Multiplication and division:

$$a^x a^y = a^{x+y} \tag{A.13}$$

$$\frac{a^x}{a^y} = a^{x-y} \tag{A.14}$$

$$\left(a^x\right)^y = a^{xy} \tag{A.15}$$

1.3 Logarithms

The logarithm is the inverse function of the exponential function of the previous section:

$$y = a^x \Leftrightarrow x = \log_a y \tag{A.16}$$

The notation $\log_a y$ indicates the logarithm of y with respect to the base a. Since the exponential and logarithm are inverse functions of each other, we can also write the identity:

$$x = \log_a(a^x) = a^{\log_a x} \quad \text{(for any base } a\text{)} \tag{A.17}$$

The two bases most commonly used are base 10, the common logarithm base, and base e, the natural logarithm base. The numerical value of e is

$$e = 2.718281828\ldots \tag{A.18}$$

Base 10:

$$y = 10^x \Leftrightarrow x = \log_{10} y \tag{A.19}$$

Base e:

$$y = e^x \Leftrightarrow x = \ln y \tag{A.20}$$

This book follows the convention of using ln to indicate the logarithm with respect to the base e.

The rules for calculating with logarithms follow from the rules of calculating with exponents:

$$\log(ab) = \log a + \log b \tag{A.21}$$

$$\log\left(\frac{a}{b}\right) = \log a - \log b \tag{A.22}$$

$$\log(a^x) = x \log a \tag{A.23}$$

$$\log 1 = 0 \tag{A.24}$$

Since these rules are valid for any base, the subscript indicating the base is omitted.

1.4 Linear Equations

The general form of a linear equation is

$$y = ax + b \qquad \text{(A.25)}$$

where a and b are constants. The graph of y versus x is a straight line; a is the slope of this line, and b is the y-intercept. See Figure A.1.

The slope of the line can be calculated by inserting two different values, x_1 and x_2, into the linear equation and calculating the resulting values, y_1 and y_2:

$$a = \frac{y_2 - y_1}{x_2 - x_1} = \frac{\Delta y}{\Delta x} \qquad \text{(A.26)}$$

If $a = 0$, then the line will be horizontal; if $a > 0$, then the line will rise as x increases as shown in the example of Figure A.1; if $a < 0$, then the line will fall as x increases.

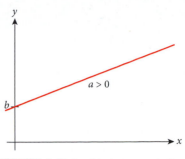

FIGURE A.1 Graphical representation of a linear equation.

2. Geometry

2.1 Geometrical Shapes in Two Dimensions

Figure A.2 lists the area, A, and perimeter length or circumference, C, of common two-dimensional geometrical objects.

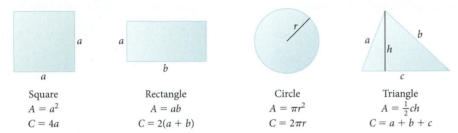

Square	Rectangle	Circle	Triangle
$A = a^2$	$A = ab$	$A = \pi r^2$	$A = \frac{1}{2}ch$
$C = 4a$	$C = 2(a+b)$	$C = 2\pi r$	$C = a + b + c$

FIGURE A.2 Area, A, and perimeter length, C, for square, rectangle, circle, and triangle.

2.2 Geometrical Shapes in Three Dimensions

Figure A.3 lists the volume, V, and surface area, A, of common three-dimensional geometrical objects.

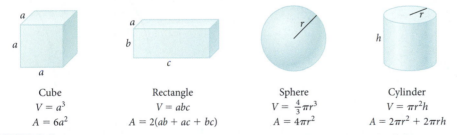

Cube	Rectangle	Sphere	Cylinder
$V = a^3$	$V = abc$	$V = \frac{4}{3}\pi r^3$	$V = \pi r^2 h$
$A = 6a^2$	$A = 2(ab + ac + bc)$	$A = 4\pi r^2$	$A = 2\pi r^2 + 2\pi rh$

FIGURE A.3 Volume, V, and surface area, A, for cube, rectangular box, sphere, and cylinder.

3. Trigonometry

It is important to note that for the following all angles need to be measured in radians.

3.1 Right Triangles

A right triangle is a triangle for which one of the three angles is a right angle, that is, an angle of exactly 90° ($\pi/2$ rad) (indicated by the small square in Figure A.4). The hypotenuse is the side opposite the 90° angle. Conventionally, one uses the letter c to mark the hypotenuse.

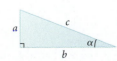

FIGURE A.4 Definition of the side lengths a, b, c, and angles for the right triangle.

Pythagorean Theorem:

$$a^2 + b^2 = c^2 \tag{A.27}$$

Definition of trigonometric functions (see Figure A.5):

$$\sin\alpha = \frac{a}{c} = \frac{\text{opposite side}}{\text{hypotenuse}} \tag{A.28}$$

$$\cos\alpha = \frac{b}{c} = \frac{\text{adjacent side}}{\text{hypotenuse}} \tag{A.29}$$

$$\tan\alpha = \frac{\sin\alpha}{\cos\alpha} = \frac{a}{b} \tag{A.30}$$

$$\cot\alpha = \frac{\cos\alpha}{\sin\alpha} = \frac{1}{\tan\alpha} = \frac{b}{a} \tag{A.31}$$

$$\csc\alpha = \frac{1}{\sin\alpha} = \frac{c}{a} \tag{A.32}$$

$$\sec\alpha = \frac{1}{\cos\alpha} = \frac{c}{b} \tag{A.33}$$

Inverse trigonometric functions (the notations $\sin^{-1}$, $\cos^{-1}$, etc., are used in this book):

$$\sin^{-1}\frac{a}{c} \equiv \arcsin\frac{a}{c} = \alpha \tag{A.34}$$

$$\cos^{-1}\frac{b}{c} \equiv \arccos\frac{b}{c} = \alpha \tag{A.35}$$

FIGURE A.5 The trigonometric functions sin, cos, tan, and cot.

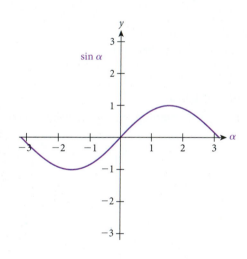

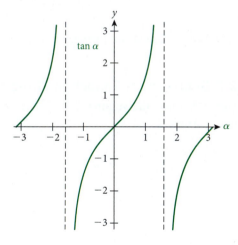

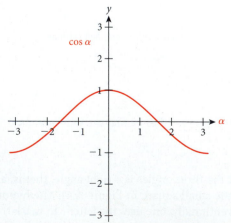

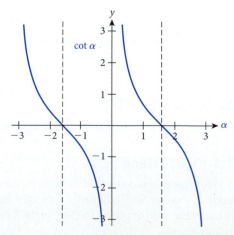

$$\tan^{-1}\frac{a}{b} \equiv \arctan\frac{a}{b} = \alpha \qquad\qquad \text{(A.36)}$$

$$\cot^{-1}\frac{b}{a} \equiv \operatorname{arccot}\frac{b}{a} = \alpha \qquad\qquad \text{(A.37)}$$

$$\csc^{-1}\frac{c}{a} \equiv \operatorname{arccsc}\frac{c}{a} = \alpha \qquad\qquad \text{(A.38)}$$

$$\sec^{-1}\frac{c}{b} \equiv \operatorname{arcsec}\frac{c}{b} = \alpha \qquad\qquad \text{(A.39)}$$

All trigonometric functions are periodic:

$$\sin\left(\alpha + 2\pi\right) = \sin\alpha \qquad\qquad \text{(A.40)}$$

$$\cos\left(\alpha + 2\pi\right) = \cos\alpha \qquad\qquad \text{(A.41)}$$

$$\tan\left(\alpha + \pi\right) = \tan\alpha \qquad\qquad \text{(A.42)}$$

$$\cot\left(\alpha + \pi\right) = \cot\alpha \qquad\qquad \text{(A.43)}$$

Other relations between the trigonometric functions:

$$\sin^2\alpha + \cos^2\alpha = 1 \qquad\qquad \text{(A.44)}$$

$$\sin(-\alpha) = -\sin\alpha \qquad\qquad \text{(A.45)}$$

$$\cos(-\alpha) = \cos\alpha \qquad\qquad \text{(A.46)}$$

$$\sin(\alpha \pm \pi/2) = \pm\cos\alpha \qquad\qquad \text{(A.47)}$$

$$\sin(\alpha \pm \pi) = -\sin\alpha \qquad\qquad \text{(A.48)}$$

$$\cos(\alpha \pm \pi/2) = \mp\sin\alpha \qquad\qquad \text{(A.49)}$$

$$\cos(\alpha \pm \pi) = -\cos\alpha \qquad\qquad \text{(A.50)}$$

Addition formulas:

$$\sin(\alpha \pm \beta) = \sin\alpha\cos\beta \pm \cos\alpha\sin\beta \qquad\qquad \text{(A.51)}$$

$$\cos(\alpha \pm \beta) = \cos\alpha\cos\beta \mp \sin\alpha\sin\beta \qquad\qquad \text{(A.52)}$$

Small-angle approximations:

$$\sin\alpha \approx \alpha - \tfrac{1}{6}\alpha^3 + \cdots \quad \left(\text{for } |\alpha| \ll 1\right) \qquad \text{(A.53)}$$

$$\cos\alpha \approx 1 - \tfrac{1}{2}\alpha^2 + \cdots \quad \left(\text{for } |\alpha| \ll 1\right) \qquad \text{(A.54)}$$

For small angles, for $|\alpha| \ll 1$, it is often acceptable to use the small-angle approximations $\cos\alpha = 1$ and $\sin\alpha = \tan\alpha = \alpha$.

3.2 General Triangles

The three angles of any triangle add up to π rads (see Figure A.6):

$$\alpha + \beta + \gamma = \pi \qquad\qquad \text{(A.55)}$$

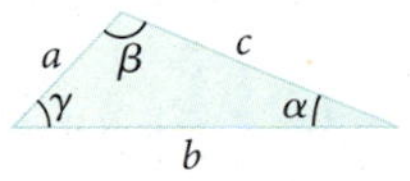

Law of cosines:

$$c^2 = a^2 + b^2 - 2ab\cos\gamma \qquad\qquad \text{(A.56)}$$

(This is a generalization of the Pythagorean Theorem for the case where the angle γ has a value that is not 90°, or $\pi/2$ rads.)

Law of sines:

$$\frac{\sin\alpha}{a} = \frac{\sin\beta}{b} = \frac{\sin\gamma}{c} \qquad\qquad \text{(A.57)}$$

FIGURE A.6 Definition of the sides and angles for a general triangle.

4. Calculus

4.1 Derivatives

Polynomials:

$$\frac{d}{dx}x^n = nx^{n-1} \tag{A.58}$$

Trigonometric functions:

$$\frac{d}{dx}\sin(ax) = a\cos(ax) \tag{A.59}$$

$$\frac{d}{dx}\cos(ax) = -a\sin(ax) \tag{A.60}$$

$$\frac{d}{dx}\tan(ax) = \frac{a}{\cos^2(ax)} \tag{A.61}$$

$$\frac{d}{dx}\cot(ax) = -\frac{a}{\sin^2(ax)} \tag{A.62}$$

Exponentials and logarithms:

$$\frac{d}{dx}e^{ax} = ae^{ax} \tag{A.63}$$

$$\frac{d}{dx}\ln(ax) = \frac{1}{x} \tag{A.64}$$

$$\frac{d}{dx}a^x = a^x \ln a \tag{A.65}$$

Product rule:

$$\frac{d}{dx}\big(f(x)g(x)\big) = \left(\frac{df(x)}{dx}\right)g(x) + f(x)\left(\frac{dg(x)}{dx}\right) \tag{A.66}$$

Chain rule:

$$\frac{dy}{dx} = \frac{dy}{du}\frac{du}{dx} \tag{A.67}$$

4.2 Integrals

All indefinite integrals have an additive integration constant, c.
Polynomials:

$$\int x^n dx = \frac{1}{n+1}x^{n+1} + c \quad \text{(for } n \neq -1\text{)} \tag{A.68}$$

$$\int x^{-1} dx = \ln|x| + c \tag{A.69}$$

$$\int \frac{1}{a^2 + x^2} dx = \frac{1}{a}\tan^{-1}\frac{x}{a} + c \tag{A.70}$$

$$\int \frac{1}{\sqrt{a^2 + x^2}} dx = \ln\left|x + \sqrt{a^2 + x^2}\right| + c \tag{A.71}$$

$$\int \frac{1}{\sqrt{a^2 - x^2}} dx = \sin^{-1}\frac{x}{|a|} + c \equiv \tan^{-1}\frac{x}{\sqrt{a^2 - x^2}} + c \tag{A.72}$$

$$\int \frac{1}{\left(a^2+x^2\right)^{3/2}}dx = \frac{1}{a^2}\frac{x}{\sqrt{a^2+x^2}}+c \qquad \text{(A.73)}$$

$$\int \frac{x}{\left(a^2+x^2\right)^{3/2}}dx = -\frac{1}{\sqrt{a^2+x^2}}+c \qquad \text{(A.74)}$$

Trigonometric functions:

$$\int \sin\left(ax\right)dx = -\frac{1}{a}\cos\left(ax\right)+c \qquad \text{(A.75)}$$

$$\int \cos\left(ax\right)dx = \frac{1}{a}\sin\left(ax\right)+c \qquad \text{(A.76)}$$

Exponentials:

$$\int e^{ax}dx = \frac{1}{a}e^{ax}+c \qquad \text{(A.77)}$$

5. Complex Numbers

We are all familiar with real numbers, which can be sorted along a number line in order of increasing value, from $-\infty$ to $+\infty$. These real numbers are embedded in a much larger set of numbers, called the *complex numbers*. Complex numbers are defined in terms of their real part and their imaginary part. The space of complex numbers is a plane, for which the real numbers form one axis, labeled $\Re(z)$ in Figure A.7. The imaginary part forms the other axis, labeled $\Im(z)$ in Figure A.7. (It is conventional to use the Old German script letters R and I to represent the real and imaginary parts of complex numbers.)

A complex number z is defined in terms of its real part, x, its imaginary part, y, and Euler's constant, i:

$$z = x+iy \qquad \text{(A.78)}$$

Euler's constant is defined as:

$$i^2 = -1 \qquad \text{(A.79)}$$

Both the real part, $x = \Re(z)$, and the imaginary part, $y = \Im(z)$, of a complex number are real numbers. Addition, subtraction, multiplication, and division of complex numbers are defined in analogy to those operations for real numbers, with $i^2 = -1$:

$$(a+ib)+(c+id) = (a+c)+i(b+d) \qquad \text{(A.80)}$$

$$(a+ib)-(c+id) = (a-c)+i(b-d) \qquad \text{(A.81)}$$

$$(a+ib)(c+id) = (ac-bd)+i(ad+bc) \qquad \text{(A.82)}$$

$$\frac{a+ib}{c+id} = \frac{(cd+bd)+i(bc-ad)}{c^2+d^2}. \qquad \text{(A.83)}$$

For each complex number z there exists a complex conjugate z^*, which has the same real part, but an imaginary part with the opposite sign:

$$z = x+iy \Leftrightarrow z^* = x-iy \qquad \text{(A.84)}$$

We can then express the real and imaginary parts of a complex number in terms of the number and its complex conjugate:

$$\Re(z) = \tfrac{1}{2}(z+z^*) \qquad \text{(A.85)}$$

$$\Im(z) = \tfrac{1}{2}i(z^*-z). \qquad \text{(A.86)}$$

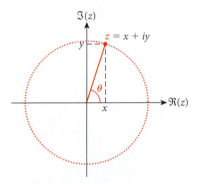

FIGURE A.7 The complex plane. The horizontal axis is formed by the real part of complex numbers and the vertical axis by the imaginary part.

Just like a two-dimensional vector, a complex number, $z = x + iy$, has the magnitude $|z|$ as well as angle θ with respect to the axis of positive real numbers, as indicated in Figure A.7:

$$|z|^2 = zz^*$$ (A.87)

$$\theta = \tan^{-1}\frac{\Im(z)}{\Re(z)} = \tan^{-1}\frac{i(z^*-z)}{(z^*+z)}$$ (A.88)

We can thus write the complex number $z = x + iy$ in terms of the magnitude and the "phase angle":

$$z = |z|(\cos\theta + i\sin\theta)$$ (A.89)

An interesting and most useful identity is *Euler's formula*:

$$e^{i\theta} = \cos\theta + i\sin\theta$$ (A.90)

With the aid of this identity, we can write, for any complex number, z,

$$z = |z|e^{i\theta}$$ (A.91)

We can thus take a complex number z to any power n:

$$z^n = |z|^n e^{in\theta}$$ (A.92)

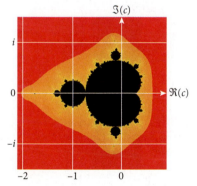

$\Im(c)$

$\Re(c)$

FIGURE A.8 Mandelbrot set in the complex plane.

EXAMPLE A.1 | Mandelbrot Set

We can put our knowledge of complex numbers and their multiplication to good use by examining the *Mandelbrot set*, defined as the set of all points c in the complex plane for which the series of iterations

$$z_{n+1} = z_n^2 + c, \quad \text{with } z_0 = c$$

does not escape to infinity; that is, for which $|z_n|$ remains finite for all iterations.

This iteration prescription is seemingly simple. For example, we can see that any number for which $|c| > 2$ cannot be part of the Mandelbrot set. However, if we plot the points of the Mandelbrot set in the complex plane, a strangely beautiful object emerges. In Figure A.8, the black points are part of the Mandelbrot set, and the remaining points are color-coded by how fast z_n escapes to infinity.

Appendix B

Isotope Masses, Binding Energies, and Half–Lives

Only isotopes with half–lives longer than 1 h are listed.

Z	N	Sym	m (amu)	B (MeV)	Spin	%	τ(s)	Z	N	Sym	m (amu)	B (MeV)	Spin	%	τ(s)
1	0	H	1.007825032	0.000	1/2+	99.985	stable	14	16	Si	29.9737702	8.521	0+	3.0872	stable
1	1	H	2.014101778	1.112	1+	0.0115	stable	14	17	Si	30.97536323	8.458	3/2+		9.44E+03
1	2	H	3.016049278	2.827	1/2+		3.89E+08	14	18	Si	31.97414808	8.482	0+		5.42E+09
2	1	He	3.016029319	2.573	1/2+	0.0001	stable	15	16	P	30.9737615	8.481	1/2+	100	stable
2	2	He	4.002603254	7.074	0+	100	stable	15	17	P	31.9739072	8.464	1+		1.23E+06
3	3	Li	6.0151223	5.332	1+	7.5	stable	15	18	P	32.9717253	8.514	1/2+		2.19E+06
3	4	Li	7.0160040	5.606	3/2-	92.41	stable	16	16	S	31.9720707	8.493	0+	94.93	stable
4	3	Be	7.0169292	5.371	3/2-		4.59E+06	16	17	S	32.97145876	8.498	3/2+	0.76	stable
4	5	Be	9.0121821	6.463	3/2-	100	stable	16	18	S	33.9678668	8.583	0+	4.29	stable
4	6	Be	10.0135337	6.498	0+		4.76E+13	16	19	S	34.9690322	8.538	3/2+		7.56E+06
5	5	B	10.01293699	6.475	3+	19.9	stable	16	20	S	35.96708076	8.575	0+	0.02	stable
5	6	B	11.00930541	6.928	3/2-	80.1	stable	16	22	S	37.9711634	8.449	0+		1.02E+04
6	6	C	12	7.680	0+	98.89	stable	17	18	Cl	34.9688527	8.520	3/2+	75.78	stable
6	7	C	13.00335484	7.470	1/2-	1.11	stable	17	19	Cl	35.9683069	8.522	2+		9.49E+12
6	8	C	14.0032420	7.520	0+		1.81E+11	17	20	Cl	36.9659026	8.570	3/2+	24.22	stable
7	7	N	14.003074	7.476	1+	99.632	stable	18	18	Ar	35.96754511	8.520	0+	0.3365	stable
7	8	N	15.0001089	7.699	1/2-	0.368	stable	18	19	Ar	36.966776	8.527	3/2+		3.02E+06
8	8	O	15.99491463	7.976	0+	99.757	stable	18	20	Ar	37.96273239	8.614	0+	0.0632	stable
8	9	O	16.999131	7.751	5/2+	0.038	stable	18	21	Ar	38.9643134	8.563	7/2-		8.48E+09
8	10	O	17.999163	7.767	0+	0.205	stable	18	22	Ar	39.9623831	8.595	0+	99.6	stable
9	9	F	18.0009377	7.632	1+		6.59E+03	18	23	Ar	40.9645008	8.534	7/2-		6.56E+03
9	10	F	18.99840322	7.779	1/2+	100	stable	18	24	Ar	41.963046	8.556	0+		1.04E+09
10	10	Ne	19.99244018	8.032	0+	90.48	stable	19	20	K	38.9637069	8.557	3/2+	93.258	stable
10	11	Ne	20.99384668	7.972	3/2+	0.27	stable	19	21	K	39.9639987	8.538	4-	0.0117	4.03E+16
10	12	Ne	21.9913855	8.080	0+	9.25	stable	19	22	K	40.9618254	8.576	3/2+	6.7302	stable
11	11	Na	21.9944368	7.916	3+		8.21E+07	19	23	K	41.962403	8.551	2-		4.45E+04
11	12	Na	22.9897697	8.111	3/2+	100	stable	19	24	K	42.960716	8.577	3/2+		8.03E+04
11	13	Na	23.9909633	8.063	4+		5.39E+04	20	20	Ca	39.96259098	8.551	0+	96.941	stable
12	12	Mg	23.9850419	8.261	0+	78.99	stable	20	21	Ca	40.9622783	8.547	7/2-		3.25E+12
12	13	Mg	24.9858370	8.223	5/2+	10	stable	20	22	Ca	41.9586183	8.617	0+	0.647	stable
12	14	Mg	25.9825930	8.334	0+	11.01	stable	20	23	Ca	42.95876663	8.601	7/2-	0.135	stable
12	16	Mg	27.9838767	8.272	0+		7.53E+04	20	24	Ca	43.9554811	8.658	0+	2.086	stable
13	13	Al	25.98689169	8.150	5+		2.33E+13	20	25	Ca	44.956186	8.631	7/2-		1.40E+07
13	14	Al	26.9815384	8.332	5/2+	100	stable	20	26	Ca	45.9536928	8.669	0+	0.004	stable
14	14	Si	27.97692653	8.448	0+	92.23	stable	20	27	Ca	46.9545465	8.639	7/2-		3.92E+05
14	15	Si	28.9764947	8.449	1/2+	4.6832	stable	20	28	Ca	47.9525335	8.666	0+	0.187	1.89E+26

Continued—

Z	N	Sym	m (amu)	B (MeV)	Spin	%	τ(s)	Z	N	Sym	m (amu)	B (MeV)	Spin	%	τ(s)
21	22	Sc	42.9611507	8.531	7/2−		1.40E+04	28	31	Ni	58.9343516	8.737	3/2−		2.40E+12
21	23	Sc	43.9594030	8.557	2+		1.41E+04	28	32	Ni	59.93078637	8.781	0+	26.223	stable
21	24	Sc	44.9559102	8.619	7/2−	100	stable	28	33	Ni	60.93105603	8.765	3/2−	1.1399	stable
21	25	Sc	45.9551703	8.622	4+		7.24E+06	28	34	Ni	61.92834512	8.795	0+	3.6345	stable
21	26	Sc	46.9524080	8.665	7/2−		2.89E+05	28	35	Ni	62.9296729	8.763	1/2−		3.19E+09
21	27	Sc	47.952231	8.656	6+		1.57E+05	28	36	Ni	63.92796596	8.777	0+	0.9256	stable
22	22	Ti	43.9596902	8.533	0+		1.89E+09	28	37	Ni	64.9300880	8.736	5/2−		9.06E+03
22	23	Ti	44.9581243	8.556	7/2−		1.11E+04	28	38	Ni	65.92913933	8.739	0+		1.97E+05
22	24	Ti	45.9526295	8.656	0+	8.25	stable	29	32	Cu	60.9334622	8.715	3/2−		1.20E+04
22	25	Ti	46.9517638	8.661	5/2−	7.44	stable	29	34	Cu	62.92959747	8.752	3/2−	69.17	stable
22	26	Ti	47.9479471	8.723	0+	73.72	stable	29	35	Cu	63.9297679	8.739	1+		4.57E+04
22	27	Ti	48.9478700	8.711	7/2−	5.41	stable	29	36	Cu	64.9277929	8.757	3/2−	30.83	stable
22	28	Ti	49.9447921	8.756	0+	5.18	stable	29	38	Cu	66.9277503	8.737	3/2−		2.23E+05
23	25	V	47.9522545	8.623	4+		1.38E+06	30	32	Zn	61.93432976	8.679	0+		3.31E+04
23	26	V	48.9485161	8.683	7/2−		2.85E+07	30	34	Zn	63.9291466	8.736	0+	48.63	stable
23	27	V	49.9471609	8.696	6+	0.25	4.42E+24	30	35	Zn	64.929245	8.724	5/2−		2.11E+07
23	28	V	50.9439617	8.742	7/2−	99.75	stable	30	36	Zn	65.92603342	8.760	0+	27.9	stable
24	24	Cr	47.95403032	8.572	0+		7.76E+04	30	37	Zn	66.92712730	8.734	5/2−	4.1	stable
24	26	Cr	49.94604462	8.701	0+	4.345	4.10E+25	30	38	Zn	67.92484949	8.756	0+	18.75	stable
24	27	Cr	50.9447718	8.712	7/2−		2.39E+06	30	40	Zn	69.9253193	8.730	0+	0.62	stable
24	28	Cr	51.9405119	8.776	0+	83.789	stable	30	42	Zn	71.926858	8.692	0+		1.68E+05
24	29	Cr	52.9406513	8.760	3/2−	9.501	stable	31	35	Ga	65.93158901	8.669	0+		3.42E+04
24	30	Cr	53.9388804	8.778	0+	2.365	stable	31	36	Ga	66.9282049	8.708	3/2−		2.82E+05
25	27	Mn	51.9455655	8.670	6+		4.83E+05	31	37	Ga	67.92798008	8.701	1+		4.06E+03
25	28	Mn	52.9412947	8.734	7/2−		1.18E+14	31	38	Ga	68.9255736	8.725	3/2−	60.108	stable
25	29	Mn	53.9403589	8.738	3+		2.70E+07	31	40	Ga	70.9247013	8.718	3/2−	39.892	stable
25	30	Mn	54.9380471	8.765	5/2−	100	stable	31	41	Ga	71.9263663	8.687	3−		5.08E+04
25	31	Mn	55.9389094	8.738	3+		9.28E+03	31	42	Ga	72.92517468	8.694	3/2−		1.75E+04
26	26	Fe	51.948114	8.610	0+		2.98E+04	32	34	Ge	65.93384345	8.626	0+		8.14E+03
26	28	Fe	53.9396127	8.736	0+	5.845	stable	32	36	Ge	67.92809424	8.688	0+		2.34E+07
26	29	Fe	54.9382980	8.747	3/2−		8.61E+07	32	37	Ge	68.927972	8.681	5/2−		1.41E+05
26	30	Fe	55.93493748	8.790	0+	91.754	stable	32	38	Ge	69.92424	8.722	0+	20.84	stable
26	31	Fe	56.93539397	8.770	1/2−	2.119	stable	32	39	Ge	70.9249540	8.703	1/2−		9.88E+05
26	32	Fe	57.93327556	8.792	0+	0.282	stable	32	40	Ge	71.92207582	8.732	0+	27.54	stable
26	33	Fe	58.9348880	8.755	3/2−		3.85E+06	32	41	Ge	72.92345895	8.705	9/2+	7.73	stable
26	34	Fe	59.934072	8.756	0+		4.73E+13	32	42	Ge	73.92117777	8.725	0+	36.28	stable
27	28	Co	54.942003	8.670	7/2−		6.31E+04	32	43	Ge	74.92285895	8.696	1/2−		4.97E+03
27	29	Co	55.9398439	8.695	4+		6.68E+06	32	44	Ge	75.92140256	8.705	0+	7.61	stable
27	30	Co	56.936296	8.742	7/2−		2.35E+07	32	45	Ge	76.92354859	8.671	7/2+		4.07E+04
27	31	Co	57.935757	8.739	2+		6.12E+06	32	46	Ge	77.922853	8.672	0+		5.29E+03
27	32	Co	58.93319505	8.768	7/2−	100	stable	33	38	As	70.92711243	8.664	5/2−		2.35E+05
27	33	Co	59.9338222	8.747	5+		1.66E+08	33	39	As	71.92675228	8.660	2−		9.33E+04
27	34	Co	60.9324758	8.756	7/2−		5.94E+03	33	40	As	72.92382484	8.690	3/2−		6.94E+06
28	28	Ni	55.94213202	8.643	0+		5.25E+05	33	41	As	73.92392869	8.680	2−		1.54E+06
28	29	Ni	56.939800	8.671	3/2−		1.28E+05	33	42	As	74.92159648	8.701	3/2−	100	stable
28	30	Ni	57.9353462	8.732	0+	68.077	stable	33	43	As	75.92239402	8.683	2−		9.31E+04

Z	N	Sym	m (amu)	B (MeV)	Spin	%	τ(s)	Z	N	Sym	m (amu)	B (MeV)	Spin	%	τ(s)
33	44	As	76.92064729	8.696	3/2−		1.40E+05	38	50	Sr	87.9056143	8.733	0+	82.58	stable
33	45	As	77.92182728	8.674	2−		5.44E+03	38	51	Sr	88.9074529	8.706	5/2+		4.37E+06
34	38	Se	71.92711235	8.645	0+		7.26E+05	38	52	Sr	89.907738	8.696	0+		9.08E+08
34	39	Se	72.92676535	8.641	9/2+		2.57E+04	38	53	Sr	90.9102031	8.664	5/2+		3.47E+04
34	40	Se	73.92247644	8.688	0+	0.89	stable	38	54	Sr	91.9110299	8.649	0+		9.76E+03
34	41	Se	74.92252337	8.679	5/2+		1.03E+07	39	46	Y	84.91643304	8.628	(1/2)−		9.65E+03
34	42	Se	75.9192141	8.711	0+	9.37	stable	39	47	Y	85.914886	8.638	4−		5.31E+04
34	43	Se	76.91991404	8.695	1/2−	7.63	stable	39	48	Y	86.9108778	8.675	1/2−		2.88E+05
34	44	Se	77.91730909	8.718	0+	23.77	stable	39	49	Y	87.9095034	8.683	4−		9.21E+06
34	45	Se	78.9184998	8.696	7/2+		2.05E+13	39	50	Y	88.9058483	8.714	1/2−	100	stable
34	46	Se	79.9165213	8.711	0+	49.61	stable	39	51	Y	89.90715189	8.693	2−		2.31E+05
34	48	Se	81.9166994	8.693	0+	8.73	2.62E+27	39	52	Y	90.907305	8.685	1/2−		5.06E+06
35	40	Br	74.92577621	8.628	3/2−		5.80E+03	39	53	Y	91.9089468	8.662	2−		1.27E+04
35	41	Br	75.924541	8.636	1−		5.83E+04	39	54	Y	92.909583	8.649	1/2−		3.66E+04
35	42	Br	76.92137908	8.667	3/2−		2.06E+05	40	46	Zr	85.91647359	8.612	0+		5.94E+04
35	44	Br	78.91833709	8.688	3/2−	50.69	stable	40	47	Zr	86.91481625	8.624	(9/2)+		6.05E+03
35	46	Br	80.9162906	8.696	3/2−	49.31	stable	40	48	Zr	87.9102269	8.666	0+		7.21E+06
35	47	Br	81.9168047	8.682	5−		1.27E+05	40	49	Zr	88.908889	8.673	9/2+		2.83E+05
35	48	Br	82.915180	8.693	3/2−		8.64E+03	40	50	Zr	89.9047037	8.710	0+	51.45	stable
36	40	Kr	75.9259483	8.609	0+		5.33E+04	40	51	Zr	90.90564577	8.693	5/2+	11.22	stable
36	41	Kr	76.92467	8.617	5/2+		4.46E+03	40	52	Zr	91.9050401	8.693	0+	17.15	stable
36	42	Kr	77.9203948	8.661	0+	0.35	6.31E+28	40	53	Zr	92.9064756	8.672	5/2+		4.83E+13
36	43	Kr	78.920083	8.657	1/2−		1.26E+05	40	54	Zr	93.90631519	8.667	0+	17.38	stable
36	44	Kr	79.9163790	8.693	0+	2.25	stable	40	55	Zr	94.9080426	8.644	5/2+		5.53E+06
36	45	Kr	80.9165923	8.683	7/2+		7.22E+12	40	56	Zr	95.9082757	8.635	0+	2.8	1.23E+27
36	46	Kr	81.9134836	8.711	0+	11.58	stable	40	57	Zr	96.9109507	8.604	1/2+		6.08E+04
36	47	Kr	82.9141361	8.696	9/2+	11.49	stable	41	48	Nb			(9/2+)		6.84E+03
36	48	Kr	83.911507	8.717	0+	57	stable	41	48	Nb	88.9134955	8.617	(1/2)−		4.25E+03
36	49	Kr	84.9125270	8.699	9/2+		3.40E+08	41	49	Nb	89.911265	8.633	8+		5.26E+04
36	50	Kr	85.91061073	8.712	0+	17.3	stable	41	50	Nb	90.9069905	8.671	9/2+		2.14E+10
36	51	Kr	86.9133543	8.675	5/2+		4.57E+03	41	51	Nb	91.9071924	8.662	7+		1.09E+15
36	52	Kr	87.914447	8.657	0+		1.02E+04	41	52	Nb	92.90637806	8.664	9/2+	100	stable
37	44	Rb	80.918996	8.645	3/2−		1.65E+04	41	53	Nb	93.9072839	8.649	6+		6.40E+11
37	46	Rb	82.915110	8.675	5/2−		7.45E+06	41	54	Nb	94.9068352	8.647	9/2+		3.02E+06
37	47	Rb	83.91438482	8.676	2−		2.83E+06	41	55	Nb	95.9081001	8.629	6+		8.41E+04
37	48	Rb	84.9117893	8.697	5/2−	72.17	stable	41	56	Nb	96.9080971	8.623	9/2+		4.32E+03
37	49	Rb	85.91116742	8.697	2−		1.61E+06	42	48	Mo	89.9139369	8.597	0+		2.04E+04
37	50	Rb	86.9091835	8.711	3/2−	27.83	1.50E+18	42	50	Mo	91.9068105	8.658	0+	14.84	stable
38	42	Sr	79.92452101	8.579	0+		6.38E+03	42	51	Mo	92.90681261	8.651	5/2+		1.26E+11
38	44	Sr	81.918402	8.636	0+		2.21E+06	42	52	Mo	93.9050876	8.662	0+	9.25	stable
38	45	Sr	82.9175567	8.638	7/2+		1.17E+05	42	53	Mo	94.9058415	8.649	5/2+	15.92	stable
38	46	Sr	83.91342528	8.677	0+	0.56	stable	42	54	Mo	95.90467890	8.654	0+	16.68	stable
38	47	Sr	84.9129328	8.676	9/2+		5.60E+06	42	55	Mo	96.90602147	8.635	5/2+	9.55	stable
38	48	Sr	85.9092602	8.708	0+	9.86	stable	42	56	Mo	97.9054078	8.635	0+	24.13	stable
38	49	Sr	86.9088793	8.705	9/2+	7	stable	42	57	Mo	98.90771187	8.608	1/2+		2.37E+05

Continued—

Z	N	Sym	m (amu)	B (MeV)	Spin	%	τ(s)	Z	N	Sym	m (amu)	B (MeV)	Spin	%	τ(s)
42	58	Mo	99.90747734	8.605	0+	9.63	3.78E+26	48	58	Cd	105.9064594	8.539	0+	1.25	stable
43	50	Tc	92.91024898	8.609	9/2+		9.90E+03	48	59	Cd	106.9066179	8.533	5/2+		2.34E+04
43	51	Tc	93.9096563	8.609	7+		1.76E+04	48	60	Cd	107.9041837	8.550	0+	0.89	stable
43	52	Tc	94.90765708	8.623	9/2+		7.20E+04	48	61	Cd	108.904982	8.539	5/2+		4.00E+07
43	53	Tc	95.907871	8.615	7+		3.70E+05	48	62	Cd	109.9030056	8.551	0+	12.49	stable
43	54	Tc	96.90636536	8.624	9/2+		1.33E+14	48	63	Cd	110.9041781	8.537	1/2+	12.8	stable
43	55	Tc	97.90721597	8.610	(6)+		1.32E+14	48	64	Cd	111.9027578	8.545	0+	24.13	stable
43	56	Tc	98.90625475	8.614	9/2+		6.65E+12	48	65	Cd	112.9044017	8.527	1/2+	12.22	2.93E+23
44	51	Ru	94.91041293	8.587	5/2+		5.91E+03	48	66	Cd	113.9033585	8.532	0+	28.73	stable
44	52	Ru	95.90759784	8.609	0+	5.54	stable	48	67	Cd	114.905431	8.511	1/2+		1.93E+05
44	53	Ru	96.9075547	8.604	5/2+		2.51E+05	48	68	Cd	115.9047558	8.512	0+	7.49	9.15E+26
44	54	Ru	97.90528713	8.620	0+	1.87	stable	48	69	Cd	116.9072186	8.489	1/2+		8.96E+03
44	55	Ru	98.9059393	8.609	5/2+	12.76	stable	49	60	In	108.9071505	8.513	9/2+		1.51E+04
44	56	Ru	99.90421948	8.619	0+	12.6	stable	49	61	In	109.9071653	8.509	7+		1.76E+04
44	57	Ru	100.9055821	8.601	5/2+	17.06	stable	49	62	In	110.90511	8.522	9/2+		2.42E+05
44	58	Ru	101.9043493	8.607	0+	31.55	stable	49	64	In	112.904061	8.523	9/2+	4.29	stable
44	59	Ru	102.9063238	8.584	3/2+		3.39E+06	49	66	In	114.9038785	8.517	9/2+	95.71	1.39E+22
44	60	Ru	103.9054301	8.587	0+	18.62	stable	50	60	Sn	109.9078428	8.496	0+		1.48E+04
44	61	Ru	104.9077503	8.562	3/2+		1.60E+04	50	62	Sn	111.9048208	8.514	0+	0.97	stable
44	62	Ru	105.9073269	8.561	0+		3.23E+07	50	63	Sn	112.9051734	8.507	1/2+		9.94E+06
45	54	Rh	98.9081321	8.580	1/2–		1.39E+06	50	64	Sn	113.9027818	8.523	0+	0.66	stable
45	55	Rh	99.90812155	8.575	1–		7.49E+04	50	65	Sn	114.9033424	8.514	1/2+	0.34	stable
45	56	Rh	100.9061636	8.588	1/2–		1.04E+08	50	66	Sn	115.9017441	8.523	0+	14.54	stable
45	57	Rh	101.9068432	8.577	2–		1.79E+07	50	67	Sn	116.9029517	8.510	1/2+	7.68	stable
45	58	Rh	102.9055043	8.584	1/2–	100	stable	50	68	Sn	117.9016063	8.517	0+	24.22	stable
45	60	Rh	104.9056938	8.573	7/2+		1.27E+05	50	69	Sn	118.9033076	8.499	1/2+	8.59	stable
46	54	Pd	99.90850589	8.564	0+		3.14E+05	50	70	Sn	119.9021966	8.505	0+	32.58	stable
46	55	Pd	100.9082892	8.561	(5/2+)		3.05E+04	50	71	Sn	120.9042369	8.485	3/2+		9.76E+04
46	56	Pd	101.9056077	8.580	0+	1.02	stable	50	72	Sn	121.9034401	8.488	0+	4.63	stable
46	57	Pd	102.9060873	8.571	5/2+		1.47E+06	50	73	Sn	122.9057208	8.467	11/2–		1.12E+07
46	58	Pd	103.9040358	8.585	0+	11.14	stable	50	74	Sn	123.9052739	8.467	0+	5.79	stable
46	59	Pd	104.9050840	8.571	5/2+	22.33	stable	50	75	Sn	124.907785	8.446	11/2–		8.33E+05
46	60	Pd	105.9034857	8.580	0+	27.33	stable	50	76	Sn	125.9076533	8.444	0+		3.15E+12
46	61	Pd	106.9051285	8.561	5/2+		2.05E+14	50	77	Sn	126.9103510	8.421	(11/2–)		7.56E+03
46	62	Pd	107.9038945	8.567	0+	26.46	stable	51	66	Sb	116.9048359	8.488	5/2+		1.01E+04
46	63	Pd	108.9059535	8.545	5/2+		4.93E+04	51	68	Sb	118.9039465	8.488	5/2+		1.37E+05
46	64	Pd	109.9051533	8.547	0+	11.72	stable	51	69	Sb	119.905072	8.476	8–		4.98E+05
46	66	Pd	111.9073141	8.521	0+		7.57E+04	51	70	Sb	120.9038180	8.482	5/2+	57.21	stable
47	56	Ag	102.9089727	8.538	7/2+		3.96E+03	51	71	Sb	121.9051754	8.468	2–		2.35E+05
47	57	Ag	103.9086282	8.536	5+		4.14E+03	51	72	Sb	122.9042157	8.472	7/2+	42.79	stable
47	58	Ag	104.9065287	8.550	1/2–		3.57E+06	51	73	Sb	123.9059375	8.456	3–		5.20E+06
47	60	Ag	106.905093	8.554	1/2–	51.839	stable	51	74	Sb	124.9052478	8.458	7/2+		8.70E+07
47	62	Ag	108.9047555	8.548	1/2–	48.161	stable	51	75	Sb	125.9072482	8.440	(8–)		1.08E+06
47	64	Ag	110.9052947	8.535	1/2–		6.44E+05	51	76	Sb	126.9069146	8.440	7/2+		3.33E+05
47	65	Ag	111.9070048	8.516	2(–)		1.13E+04	51	77	Sb	127.9091673	8.421	8–		3.24E+04
47	66	Ag	112.9065666	8.516	1/2–		1.93E+04	51	78	Sb	128.9091501	8.418	7/2+		1.58E+04

Z	N	Sym	m (amu)	B (MeV)	Spin	%	τ(s)	Z	N	Sym	m (amu)	B (MeV)	Spin	%	τ(s)
52	64	Te	115.9084203	8.456	0+		8.96E+03	55	74	Cs	128.9060634	8.416	1/2+		1.15E+05
52	65	Te	116.90864	8.451	1/2+		3.72E+03	55	76	Cs	130.9054639	8.415	5/2+		8.37E+05
52	66	Te	117.9058276	8.470	0+		5.18E+05	55	77	Cs	131.906430	8.406	2+		5.60E+05
52	67	Te	118.9064081	8.462	1/2+		5.77E+04	55	78	Cs	132.9054469	8.410	7/2+	100	stable
52	68	Te	119.9040202	8.477	0+	0.09	stable	55	79	Cs	133.9067134	8.399	4+		6.51E+07
52	69	Te	120.9049364	8.467	1/2+		1.45E+06	55	80	Cs	134.905972	8.401	7/2+		7.25E+13
52	70	Te	121.9030471	8.478	0+	2.55	stable	55	81	Cs	135.907307	8.390	5+		1.14E+06
52	71	Te	122.9042730	8.466	1/2+	0.89	1.89E+22	55	82	Cs	136.9070895	8.389	7/2+		9.48E+08
52	72	Te	123.9028180	8.473	0+	4.74	stable	56	70	Ba	125.9112502	8.380	0+		6.01E+03
52	73	Te	124.9044285	8.458	1/2+	7.07	stable	56	72	Ba	127.90831	8.396	0+		2.10E+05
52	74	Te	125.9033095	8.463	0+	18.84	stable	56	73	Ba	128.9086794	8.391	1/2+		8.03E+03
52	75	Te	126.905217	8.446	3/2+		3.37E+04	56	74	Ba	129.9063105	8.406	0+	0.106	stable
52	76	Te	127.9044631	8.449	0+	31.74	2.43E+32	56	75	Ba	130.9069308	8.399	1/2+		9.94E+05
52	77	Te	128.906596	8.430	3/2+		4.18E+03	56	76	Ba	131.9050562	8.409	0+	0.101	stable
52	78	Te	129.9062244	8.430	0+	34.08	8.51E+28	56	77	Ba	132.9060024	8.400	1/2+		3.32E+08
52	80	Te	131.9085238	8.408	0+		2.77E+05	56	78	Ba	133.9045033	8.408	0+	2.417	stable
53	67	I	119.9100482	8.424	2−		4.86E+03	56	79	Ba	134.9056827	8.397	3/2+	6.592	stable
53	68	I	120.9073668	8.442	5/2+		7.63E+03	56	80	Ba	135.9045701	8.403	0+	7.854	stable
53	70	I	122.9055979	8.449	5/2+		4.78E+04	56	81	Ba	136.905824	8.392	3/2+	11.232	stable
53	71	I	123.9062114	8.441	2−		3.61E+05	56	82	Ba	137.9052413	8.393	0+	71.698	stable
53	72	I	124.9046242	8.450	5/2+		5.13E+06	56	83	Ba	138.908836	8.367	7/2−		4.98E+03
53	73	I	125.9056242	8.440	2−		1.13E+06	56	84	Ba	139.91060	8.353	0+		1.10E+06
53	74	I	126.9044727	8.445	5/2+	100	stable	57	75	La	131.910110	8.368	2−		1.73E+04
53	76	I	128.9049877	8.436	7/2+		4.95E+14	57	76	La	132.908218	8.379	5/2+		1.41E+04
53	77	I	129.9066742	8.421	5+		4.45E+04	57	78	La	134.9069768	8.383	5/2+		7.02E+04
53	78	I	130.9061246	8.422	7/2+		6.93E+05	57	80	La	136.90647	8.382	7/2+		1.89E+12
53	79	I	131.9079945	8.406	4+		8.26E+03	57	81	La	137.9071068	8.375	5+	0.09	3.31E+18
53	80	I	132.9078065	8.405	7/2+		7.49E+04	57	82	La	138.9063482	8.378	7/2+	99.91	stable
53	82	I	134.91005	8.385	7/2+		2.37E+04	57	83	La	139.9094726	8.355	3−		1.45E+05
54	68	Xe	121.9085484	8.425	0+		7.24E+04	57	84	La	140.910958	8.343	(7/2+)		1.41E+04
54	69	Xe	122.908480	8.421	(1/2)+		7.49E+03	57	85	La	141.9140791	8.321	2−		5.46E+03
54	70	Xe	123.9058942	8.438	0+	0.09	stable	58	74	Ce	131.9114605	8.352	0+		1.26E+04
54	71	Xe	124.906398	8.431	(1/2)+		6.08E+04	58	75	Ce	132.911515	8.350	9/2−		1.76E+04
54	72	Xe	125.9042736	8.444	0+	0.09	stable	58	75	Ce	132.9115515	8.350	1/2+		5.83E+03
54	73	Xe	126.905184	8.434	1/2+		3.14E+06	58	76	Ce	133.9089248	8.366	0+		2.73E+05
54	74	Xe	127.9035313	8.443	0+	1.92	stable	58	77	Ce	134.9091514	8.362	1/2(+)		6.37E+04
54	75	Xe	128.9047794	8.431	1/2+	26.44	stable	58	78	Ce	135.907172	8.373	0+	0.185	stable
54	76	Xe	129.903508	8.438	0+	4.08	stable	58	79	Ce	136.9078056	8.367	3/2+		3.24E+04
54	77	Xe	130.9050824	8.424	3/2+	21.18	stable	58	80	Ce	137.9059913	8.377	0+	0.251	stable
54	78	Xe	131.9041535	8.428	0+	26.89	stable	58	81	Ce	138.9066466	8.370	3/2+		1.19E+07
54	79	Xe	132.905906	8.413	3/2+		4.53E+05	58	82	Ce	139.905434	8.376	0+	88.45	stable
54	80	Xe	133.9053945	8.414	0+	10.44	stable	58	83	Ce	140.908271	8.355	7/2−		2.81E+06
54	81	Xe	134.90721	8.398	3/2+		3.29E+04	58	84	Ce	141.909241	8.347	0+	11.114	1.58E+24
54	82	Xe	135.9072188	8.396	0+	8.87	2.93E+27	58	85	Ce	142.9123812	8.325	3/2−		1.19E+05
55	72	Cs	126.9074175	8.412	1/2+		2.25E+04	58	86	Ce	143.913643	8.315	0+		2.46E+07

Continued—

Z	N	Sym	m (amu)	B (MeV)	Spin	%	τ(s)	Z	N	Sym	m (amu)	B (MeV)	Spin	%	τ(s)
59	78	Pr	136.910687	8.341	5/2+		4.61E+03	63	88	Eu	150.919848	8.239	5/2+	47.81	stable
59	80	Pr	138.9089384	8.349	5/2+		1.59E+04	63	89	Eu	151.921744	8.227	3–		4.27E+08
59	82	Pr	140.9076477	8.354	5/2+	100	stable	63	90	Eu	152.921229	8.229	5/2+	52.19	stable
59	83	Pr	141.910041	8.336	2–		6.88E+04	63	91	Eu	153.922976	8.217	3–		2.71E+08
59	84	Pr	142.9108122	8.329	7/2+		1.17E+06	63	92	Eu	154.92289	8.217	5/2+		1.50E+08
59	86	Pr	144.9145069	8.302	7/2+		2.15E+04	63	93	Eu	155.9247522	8.205	0+		1.31E+06
60	78	Nd	137.91195	8.325	0+		1.81E+04	63	94	Eu	156.9254236	8.200	5/2+		5.46E+04
60	80	Nd	139.90931	8.338	0+		2.91E+05	64	82	Gd	145.9183106	8.250	0+		4.17E+06
60	81	Nd	140.9096099	8.336	3/2+		8.96E+03	64	83	Gd	146.919090	8.243	7/2–		1.37E+05
60	82	Nd	141.907719	8.346	0+	27.2	stable	64	84	Gd	147.918110	8.248	0+		2.35E+09
60	83	Nd	142.90981	8.330	7/2–	12.2	stable	64	85	Gd	148.919339	8.239	7/2–		8.02E+05
60	84	Nd	143.910083	8.327	0+	23.8	7.22E+22	64	86	Gd	149.9186589	8.243	0+		5.64E+13
60	85	Nd	144.91257	8.309	7/2–	8.3	stable	64	87	Gd	150.9203485	8.231	7/2–		1.07E+07
60	86	Nd	145.913116	8.304	0+	17.2	stable	64	88	Gd	151.919789	8.233	0+	0.2	3.41E+21
60	87	Nd	146.916096	8.284	5/2–		9.49E+05	64	89	Gd	152.9217495	8.220	3/2–		2.09E+07
60	88	Nd	147.916889	8.277	0+	5.7	stable	64	90	Gd	153.9208623	8.225	0+	2.18	stable
60	89	Nd	148.920145	8.255	5/2–		6.22E+03	64	91	Gd	154.922619	8.213	3/2–	14.8	stable
60	90	Nd	149.920887	8.250	0+	5.6	3.47E+26	64	92	Gd	155.922122	8.215	0+	20.47	stable
61	82	Pm	142.9109276	8.318	5/2+		2.29E+07	64	93	Gd	156.9239567	8.204	3/2–	15.65	stable
61	83	Pm	143.912586	8.305	5–		3.14E+07	64	94	Gd	157.924103	8.202	0+	24.84	stable
61	84	Pm	144.9127439	8.303	5/2+		5.58E+08	64	95	Gd	158.9263861	8.188	3/2–		6.65E+04
61	85	Pm	145.914696	8.289	3–		1.74E+08	64	96	Gd	159.9270541	8.183	0+	21.86	stable
61	86	Pm	146.9151339	8.284	7/2+		8.27E+07	65	82	Tb	146.9240446	8.207	(1/2+)		6.12E+03
61	87	Pm	147.9174746	8.268	1–		4.64E+05	65	83	Tb	147.9242717	8.204	2–		3.60E+03
61	88	Pm	148.91833	8.262	7/2+		1.91E+05	65	84	Tb	148.9232459	8.210	1/2+		1.48E+04
61	89	Pm	149.92098	8.244	(1–)		9.65E+03	65	85	Tb	149.9236597	8.206	(2)–		1.25E+04
61	90	Pm	150.921207	8.241	5/2+		1.02E+05	65	86	Tb	150.9230982	8.209	1/2(+)		6.34E+04
62	80	Sm	141.9151976	8.286	0+		4.35E+03	65	87	Tb	151.9240744	8.202	2–		6.30E+04
62	82	Sm	143.911998	8.304	0+	3.07	stable	65	88	Tb	152.9234346	8.205	5/2+		2.02E+05
62	83	Sm	144.913407	8.293	7/2–		2.94E+07	65	89	Tb	153.9246862	8.197	0(+)		7.74E+04
62	84	Sm	145.913038	8.294	0+		3.25E+15	65	90	Tb	154.9235052	8.203	3/2+		4.60E+05
62	85	Sm	146.914894	8.281	7/2–	14.99	3.34E+18	65	91	Tb	155.924744	8.195	3–		4.62E+05
62	86	Sm	147.914819	8.280	0+	11.24	2.21E+23	65	92	Tb	156.9240212	8.198	3/2+		2.24E+09
62	87	Sm	148.91718	8.263	7/2–	13.82	6.31E+22	65	93	Tb	157.9254103	8.189	3–		5.68E+09
62	88	Sm	149.9172730	8.262	0+	7.38	stable	65	94	Tb	158.9253431	8.189	3/2+	100	stable
62	89	Sm	150.919929	8.244	5/2–		2.84E+09	65	95	Tb	159.9271640	8.177	3–		6.25E+06
62	90	Sm	151.9197282	8.244	0+	26.75	stable	65	96	Tb	160.9275663	8.174	3/2+		5.94E+05
62	91	Sm	152.922097	8.229	3/2+		1.67E+05	66	86	Dy	151.9247140	8.193	0+		8.57E+03
62	92	Sm	153.9222053	8.227	0+	22.75	stable	66	87	Dy	152.9257647	8.186	7/2(–)		2.30E+04
62	94	Sm	155.9255279	8.205	0+		3.38E+04	66	88	Dy	153.9244220	8.193	0+		9.46E+13
63	82	Eu	144.9162652	8.269	5/2+		5.12E+05	66	89	Dy	154.9257538	8.184	3/2–		3.56E+04
63	83	Eu	145.91720	8.262	4–		3.97E+05	66	90	Dy	155.9242783	8.192	0+	0.06	stable
63	84	Eu	146.916742	8.264	5/2+		2.08E+06	66	91	Dy	156.9254661	8.185	3/2–		2.93E+04
63	85	Eu	147.91815	8.254	5–		4.71E+06	66	92	Dy	157.924405	8.190	0+	0.1	stable
63	86	Eu	148.917930	8.254	5/2+		8.04E+06	66	93	Dy	158.925736	8.182	3/2–		1.25E+07
63	87	Eu	149.9197018	8.241	5(–)		1.16E+09	66	94	Dy	159.925194	8.184	0+	2.34	stable

Z	N	Sym	m (amu)	B (MeV)	Spin	%	τ(s)	Z	N	Sym	m (amu)	B (MeV)	Spin	%	τ(s)
66	95	Dy	160.926930	8.173	5/2+	18.91	stable	70	106	Yb	175.942571	8.064	0+	12.76	stable
66	96	Dy	161.926795	8.173	0+	25.51	stable	70	107	Yb	176.9452571	8.050	9/2+		6.88E+03
66	97	Dy	162.928728	8.162	5/2+	24.9	stable	70	108	Yb	177.9466467	8.043	0+		4.43E+03
66	98	Dy	163.9291712	8.159	0+	28.18	stable	71	98	Lu	168.937649	8.086	7/2+		1.23E+05
66	99	Dy	164.93170	8.144	7/2+		8.40E+03	71	99	Lu	169.9384722	8.082	0+		1.74E+05
66	100	Dy	165.9328032	8.137	0+		2.94E+05	71	100	Lu	170.93791	8.085	7/2+		7.12E+05
67	94	Ho	160.9278548	8.163	7/2−		8.93E+03	71	101	Lu	171.9390822	8.078	4−		5.79E+05
67	96	Ho	162.9287303	8.157	7/2−		1.44E+11	71	102	Lu	172.938927	8.079	7/2+		4.32E+07
67	98	Ho	164.9303221	8.147	7/2−	100	stable	71	103	Lu	173.940334	8.071	(1)−		1.04E+08
67	99	Ho	165.9322842	8.135	0−		9.63E+04	71	104	Lu	174.94077	8.069	7/2+	97.41	stable
67	100	Ho	166.933127	8.130	7/2−		1.12E+04	71	105	Lu	175.9426824	8.059	7−	2.59	1.29E+18
68	90	Er	157.9298935	8.148	0+		8.24E+03	71	106	Lu	176.9437550	8.053	7/2+		5.82E+05
68	92	Er	159.92908	8.152	0+		1.03E+05	71	108	Lu	178.9473274	8.035	7/2(+)		1.65E+04
68	93	Er	160.93	8.146	3/2−		1.16E+04	72	98	Hf	169.939609	8.071	0+		5.76E+04
68	94	Er	161.928775	8.152	0+	0.14	stable	72	99	Hf	170.940492	8.066	7/2+		4.36E+04
68	95	Er	162.9300327	8.145	5/2−		4.50E+03	72	100	Hf	171.9394483	8.072	0+		5.90E+07
68	96	Er	163.929198	8.149	0+	1.61	stable	72	101	Hf	172.940513	8.066	1/2−		8.50E+04
68	97	Er	164.930726	8.140	5/2−		3.73E+04	72	102	Hf	173.940044	8.069	0+	0.16	6.31E+22
68	98	Er	165.9302900	8.142	0+	33.61	stable	72	103	Hf	174.9415024	8.061	5/2−		6.05E+06
68	99	Er	166.932046	8.132	7/2+	22.93	stable	72	104	Hf	175.941406	8.061	0+	5.26	stable
68	100	Er	167.9323702	8.130	0+	26.78	stable	72	105	Hf	176.9432207	8.052	7/2−	18.6	stable
68	101	Er	168.9345881	8.117	1/2−		8.12E+05	72	106	Hf	177.9436988	8.049	0+	27.28	stable
68	102	Er	169.935461	8.112	0+	14.93	stable	72	107	Hf	178.9458161	8.039	9/2+	13.62	stable
68	103	Er	170.938026	8.098	5/2−		2.71E+04	72	108	Hf	179.94655	8.035	0+	35.08	stable
68	104	Er	171.9393521	8.090	0+		1.77E+05	72	109	Hf	180.9490991	8.022	1/2−		3.66E+06
69	94	Tm	162.9326500	8.125	1/2+		6.52E+03	72	110	Hf	181.9505541	8.015	0+		2.84E+14
69	96	Tm	164.932433	8.126	1/2+		1.08E+05	72	111	Hf	182.9535304	8.000	(3/2−)		3.84E+03
69	97	Tm	165.9335541	8.119	2+		2.77E+04	72	112	Hf	183.9554465	7.991	0+		1.48E+04
69	98	Tm	166.9328516	8.123	1/2+		7.99E+05	73	100	Ta	172.94354	8.044	5/2−		1.13E+04
69	99	Tm	167.9341728	8.115	3+		8.04E+06	73	101	Ta	173.944256	8.040	3(+)		3.78E+03
69	100	Tm	168.934212	8.114	1/2+	100	stable	73	102	Ta	174.9437	8.044	7/2+		3.78E+04
69	101	Tm	169.9358014	8.106	1−		1.11E+07	73	103	Ta	175.944857	8.039	1−		2.91E+04
69	102	Tm	170.936426	8.102	1/2+		6.05E+07	73	104	Ta	176.9444724	8.041	7/2+		2.04E+05
69	103	Tm	171.9384	8.091	2−		2.29E+05	73	105	Ta	177.9457782	8.034	7−		8.50E+03
69	104	Tm	172.9396036	8.084	1/2+		2.97E+04	73	106	Ta	178.94593	8.034	7/2+		5.74E+07
70	94	Yb	163.9344894	8.109	0+		4.54E+03	73	107	Ta	179.9474648	8.026	1+	0.012	2.93E+04
70	96	Yb	165.9338796	8.112	0+		2.04E+05	73	108	Ta	180.9479958	8.023	7/2+	99.988	stable
70	98	Yb	167.9338969	8.112	0+	0.13	stable	73	109	Ta	181.9501518	8.013	3−		9.89E+06
70	99	Yb	168.9351871	8.104	7/2+		2.77E+06	73	110	Ta	182.9513726	8.007	7/2+		4.41E+05
70	100	Yb	169.934759	8.107	0+	3.04	stable	73	111	Ta	183.954008	7.994	(5−)		3.13E+04
70	101	Yb	170.936323	8.098	1/2−	14.28	stable	74	102	W	175.945634	8.030	0+		9.00E+03
70	102	Yb	171.9363777	8.097	0+	21.83	stable	74	103	W	176.946643	8.025	(1/2−)		8.10E+03
70	103	Yb	172.938208	8.087	5/2−	16.13	stable	74	104	W	177.9458762	8.029	0+		1.87E+06
70	104	Yb	173.9388621	8.084	0+	31.83	stable	74	106	W	179.9467045	8.025	0+	0.12	stable
70	105	Yb	174.941273	8.071	7/2−		3.62E+05	74	107	W	180.9481972	8.018	9/2+		1.05E+07

Continued—

Z	N	Sym	m (amu)	B (MeV)	Spin	%	τ(s)	Z	N	Sym	m (amu)	B (MeV)	Spin	%	τ(s)
74	108	W	181.9482042	8.018	0+	26.5	stable	78	109	Pt	186.960587	7.941	3/2−		8.46E+03
74	109	W	182.950223	8.008	1/2−	14.31	stable	78	110	Pt	187.9593954	7.948	0+		8.81E+05
74	110	W	183.9509312	8.005	0+	30.64	stable	78	111	Pt	188.9608337	7.941	3/2−		3.91E+04
74	111	W	184.9534193	7.993	3/2−		6.49E+06	78	112	Pt	189.9599317	7.947	0+	0.014	2.05E+19
74	112	W	185.9543641	7.989	0+	28.43	stable	78	113	Pt	190.9616767	7.939	3/2−		2.47E+05
74	113	W	186.9571605	7.975	3/2−		8.54E+04	78	114	Pt	191.961038	7.942	0+	0.782	stable
74	114	W	187.9584891	7.969	0+		6.00E+06	78	115	Pt	192.9629874	7.934	1/2−		1.58E+09
75	106	Re	180.9500679	8.004	5/2+		7.16E+04	78	116	Pt	193.9626803	7.936	0+	32.967	stable
75	107	Re	181.9512101	7.999	7+		2.31E+05	78	117	Pt	194.9647911	7.927	1/2−	33.832	stable
75	107	Re			2+		4.57E+04	78	118	Pt	195.9649515	7.927	0+	25.242	stable
75	108	Re	182.9508198	8.001	5/2+		6.05E+06	78	119	Pt	196.9673402	7.916	1/2−		7.16E+04
75	109	Re	183.9525208	7.993	3−		3.28E+06	78	120	Pt	197.9678928	7.914	0+	7.163	stable
75	110	Re	184.952955	7.991	5/2+	37.4	stable	78	122	Pt	199.9714407	7.899	0+		4.50E+04
75	111	Re	185.9549861	7.981	1−		3.21E+05	78	124	Pt	201.97574	7.881	0+		1.56E+05
75	112	Re	186.9557531	7.978	5/2+	62.6	1.37E+18	79	112	Au	190.9637042	7.925	3/2+		1.14E+04
75	113	Re	187.9581144	7.967	1−		6.13E+04	79	113	Au	191.964813	7.920	1−		1.78E+04
75	114	Re	188.959229	7.962	5/2+		8.75E+04	79	114	Au	192.9641497	7.924	3/2+		6.35E+04
76	105	Os	180.953244	7.983	1/2−		6.30E+03	79	115	Au	193.9653653	7.919	1−		1.37E+05
76	106	Os	181.9521102	7.990	0+		7.96E+04	79	116	Au	194.9650346	7.921	3/2+		1.61E+07
76	107	Os	182.9531261	7.985	9/2+		4.68E+04	79	117	Au	195.9665698	7.915	2−		5.33E+05
76	108	Os	183.9524891	7.989	0+	0.02	stable	79	118	Au	196.9665687	7.916	3/2+	100	stable
76	109	Os	184.9540423	7.981	1/2−		8.09E+06	79	119	Au	197.9682423	7.909	2−		2.33E+05
76	110	Os	185.9538382	7.983	0+	1.59	6.31E+22	79	120	Au	198.9687652	7.907	3/2+		2.71E+05
76	111	Os	186.9557505	7.974	1/2−	1.96	stable	80	112	Hg	191.9656343	7.912	0+		1.75E+04
76	112	Os	187.9558382	7.974	0+	13.24	stable	80	113	Hg	192.9666654	7.908	3/2−		1.37E+04
76	113	Os	188.9581475	7.963	3/2−	16.15	stable	80	114	Hg	193.9654394	7.915	0+		1.64E+10
76	114	Os	189.958447	7.962	0+	26.26	stable	80	115	Hg	194.9667201	7.909	1/2−		3.79E+04
76	115	Os	190.9609297	7.951	9/2+		1.33E+06	80	116	Hg	195.9658326	7.914	0+	0.15	stable
76	116	Os	191.9614807	7.948	0+	40.78	stable	80	117	Hg	196.9672129	7.909	1/2−		2.34E+05
76	117	Os	192.9641516	7.936	3/2−		1.08E+05	80	118	Hg	197.966769	7.912	0+	9.97	stable
76	118	Os	193.9651821	7.932	0+		1.89E+08	80	119	Hg	198.9682799	7.905	1/2−	16.87	stable
77	107	Ir	183.957476	7.959	5−		1.11E+04	80	120	Hg	199.968326	7.906	0+	23.1	stable
77	108	Ir	184.956698	7.964	5/2−		5.18E+04	80	121	Hg	200.9703023	7.898	3/2−	13.18	stable
77	109	Ir	185.9579461	7.958	5+		5.99E+04	80	122	Hg	201.970643	7.897	0+	29.86	stable
77	109	Ir			2−		7.20E+03	80	123	Hg	202.9728725	7.887	5/2−		4.03E+06
77	110	Ir	186.9573634	7.962	3/2+		3.78E+04	80	124	Hg	203.9734939	7.886	0+	6.87	stable
77	111	Ir	187.9588531	7.955	1−		1.49E+05	81	114	Tl	194.9697743	7.891	1/2+		4.18E+03
77	112	Ir	188.9587189	7.956	3/2+		1.14E+06	81	115	Tl	195.9704812	7.888	2−		6.62E+03
77	113	Ir	189.960546	7.948	(4)+		1.02E+06	81	116	Tl	196.9695745	7.893	1/2+		1.02E+04
77	114	Ir	190.960594	7.948	3/2+	37.3	stable	81	117	Tl	197.9704835	7.890	2−		1.91E+04
77	115	Ir	191.962605	7.939	4(+)		6.38E+06	81	118	Tl	198.969877	7.894	1/2+		2.67E+04
77	116	Ir	192.9629264	7.938	3/2+	62.7	stable	81	119	Tl	199.9709627	7.890	2−		9.42E+04
77	117	Ir	193.9650784	7.928	1−		6.89E+04	81	120	Tl	200.9708189	7.891	1/2+		2.62E+05
77	118	Ir	194.9659796	7.925	3/2+		9.00E+03	81	121	Tl	201.9721058	7.886	2−		1.06E+06
78	107	Pt	184.960619	7.940	9/2+		4.25E+03	81	122	Tl	202.9723442	7.886	1/2+	29.524	stable
78	108	Pt	185.9593508	7.947	0+		7.20E+03	81	123	Tl	203.9738635	7.880	2−		1.19E+08

Z	N	Sym	m (amu)	B (MeV)	Spin	%	τ(s)	Z	N	Sym	m (amu)	B (MeV)	Spin	%	τ(s)
81	124	Tl	204.9744275	7.878	1/2+	70.476	stable	88	138	Ra	226.0254098	7.662	0+		5.05E+10
82	116	Pb	197.972034	7.879	0+		8.64E+03	88	140	Ra	228.0310703	7.642	0+		1.81E+08
82	117	Pb	198.9729167	7.876	3/2−		5.40E+03	88	142	Ra	230.0370564	7.622	0+		5.58E+03
82	118	Pb	199.9718267	7.882	0+		7.74E+04	89	135	Ac	224.0217229	7.670	0−		1.04E+04
82	119	Pb	200.9728845	7.878	5/2−		3.36E+04	89	136	Ac	225.0232296	7.666	(3/2−)		8.64E+05
82	120	Pb	201.9721591	7.882	0+		1.66E+12	89	137	Ac	226.0260981	7.656	(1−)		1.06E+05
82	121	Pb	202.9733905	7.877	5/2−		1.87E+05	89	138	Ac	227.0277521	7.651	3/2−		6.87E+08
82	122	Pb	203.9730436	7.880	0+	1.4	4.42E+24	89	139	Ac	228.0310211	7.639	3(+)		2.21E+04
82	123	Pb	204.9744818	7.874	5/2−		4.83E+14	89	140	Ac	229.0330152	7.633	(3/2+)		3.78E+03
82	124	Pb	205.9744653	7.875	0+	24.1	stable	90	137	Th	227.0277041	7.647	3/2+		1.62E+06
82	125	Pb	206.9758969	7.870	1/2−	22.1	stable	90	138	Th	228.0287411	7.645	0+		6.03E+07
82	126	Pb	207.9766521	7.867	0+	52.4	stable	90	139	Th	229.0317624	7.635	5/2+		2.49E+11
82	127	Pb	208.9810901	7.849	9/2+		1.17E+04	90	140	Th	230.0331338	7.631	0+		2.38E+12
82	128	Pb	209.9841885	7.836	0+		7.03E+08	90	141	Th	231.0363043	7.620	5/2+		9.18E+04
82	130	Pb	211.9918975	7.804	0+		3.83E+04	90	142	Th	232.0380553	7.615	0+	100	4.43E+17
83	118	Bi	200.977009	7.855	9/2−		6.48E+03	90	144	Th	234.0436012	7.597	0+		2.08E+06
83	119	Bi	201.9777423	7.852	5+		6.19E+03	91	137	Pa	228.0310514	7.632	(3+)		7.92E+04
83	120	Bi	202.976876	7.858	9/2−		4.23E+04	91	138	Pa	229.0320968	7.630	(5/2+)		1.30E+05
83	121	Bi	203.9778127	7.854	6+		4.04E+04	91	139	Pa	230.0345408	7.622	(2−)		1.50E+06
83	122	Bi	204.9773894	7.857	9/2−		1.32E+06	91	140	Pa	231.035884	7.618	3/2−	100	1.03E+12
83	123	Bi	205.9784991	7.853	6+		5.39E+05	91	141	Pa	232.0385916	7.609	(2−)		1.13E+05
83	124	Bi	206.9784707	7.854	9/2−		9.95E+08	91	142	Pa	233.0402473	7.605	3/2−		2.33E+06
83	125	Bi	207.9797422	7.850	(5)+		1.16E+13	91	143	Pa	234.0433081	7.595	4+		2.41E+04
83	126	Bi	208.9803987	7.848	9/2−	100	stable	91	148	Pa	239.05726	7.550	(3/2)(−)		6.37E+03
83	127	Bi	209.9841204	7.833	1−		4.33E+05	92	138	U	230.0339398	7.621	0+		1.80E+06
83	129	Bi	211.9912857	7.803	1(−)		3.63E+03	92	139	U	231.0362937	7.613	(5/2)		3.63E+05
84	120	Po	203.9803181	7.839	0+		1.27E+04	92	140	U	232.0371562	7.612	0+		2.17E+09
84	121	Po	204.9812033	7.836	5/2−		5.98E+03	92	141	U	233.0396352	7.604	5/2+		5.01E+12
84	122	Po	205.9804811	7.841	0+		7.60E+05	92	142	U	234.0409521	7.601	0+	0.0055	7.74E+12
84	123	Po	206.9815932	7.837	5/2−		2.09E+04	92	143	U	235.0439299	7.591	7/2−	0.72	2.22E+16
84	124	Po	207.9812457	7.839	0+		9.14E+07	92	144	U	236.045568	7.586	0+		7.38E+14
84	125	Po	208.9824304	7.835	1/2−		3.22E+09	92	145	U	237.0487302	7.576	1/2+		5.83E+05
84	126	Po	209.9828737	7.834	0+		1.20E+07	92	146	U	238.0507882	7.570	0+	99.275	1.41E+17
85	122	At	206.9857835	7.814	9/2−		6.48E+03	92	148	U	240.056592	7.552	0+		5.08E+04
85	123	At	207.98659	7.812	6+		5.87E+03	93	141	Np	234.042895	7.590	(0+)		3.80E+05
85	124	At	208.9861731	7.815	9/2−		1.95E+04	93	142	Np	235.0440633	7.587	5/2+		3.42E+07
85	125	At	209.9871477	7.812	5+		2.92E+04	93	143	Np	236.0465696	7.579	(6−)		4.86E+12
85	126	At	210.9874963	7.811	9/2−		2.60E+04	93	143	Np			1(−)		8.10E+04
86	124	Rn	209.9896962	7.797	0+		8.71E+03	93	144	Np	237.0481734	7.575	5/2+		6.75E+13
86	125	Rn	210.9906005	7.794	1/2−		5.26E+04	93	145	Np	238.0509464	7.566	2+		1.83E+05
86	136	Rn	222.0175777	7.694	0+		3.30E+05	93	146	Np	239.052939	7.561	5/2+		2.04E+05
86	138	Rn	224.02409	7.671	0+		6.41E+03	93	147	Np	240.0561622	7.550	(5+)		3.72E+03
88	135	Ra	223.0185022	7.685	3/2+		9.88E+05	94	140	Pu	234.0433171	7.585	0+		3.17E+04
88	136	Ra	224.0202118	7.680	0+		3.16E+05	94	142	Pu	236.046058	7.578	0+		9.01E+07
88	137	Ra	225.0236116	7.668	1/2+		1.29E+06	94	143	Pu	237.0484097	7.571	7/2−		3.91E+06

Continued—

Z	N	Sym	m (amu)	B (MeV)	Spin	%	τ(s)	Z	N	Sym	m (amu)	B (MeV)	Spin	%	τ(s)
94	144	Pu	238.0495599	7.568	0+		2.77E+09	97	151	Bk	248.073086	7.491	(6+)		2.84E+08
94	145	Pu	239.0521634	7.560	1/2+		7.60E+11	97	151	Bk			1(−)		8.53E+04
94	146	Pu	240.0538135	7.556	0+		2.07E+11	97	152	Bk	249.0749867	7.486	7/2+		2.76E+07
94	147	Pu	241.0568515	7.546	5/2+		4.53E+08	97	153	Bk	250.0783165	7.476	2−		1.16E+04
94	148	Pu	242.0587426	7.541	0+		1.18E+13	98	148	Cf	246.0688053	7.499	0+		1.29E+05
94	149	Pu	243.0620031	7.531	7/2+		1.78E+04	98	149	Cf	247.0710006	7.493	(7/2+)		1.12E+04
94	150	Pu	244.0642039	7.525	0+		2.55E+15	98	150	Cf	248.0721849	7.491	0+		2.88E+07
94	151	Pu	245.0677472	7.514	(9/2−)		3.78E+04	98	151	Cf	249.0748535	7.483	9/2−		1.11E+10
94	152	Pu	246.0702046	7.507	0+		9.37E+05	98	152	Cf	250.0764061	7.480	0+		4.12E+08
94	153	Pu	247.07407	7.494	1/2+		1.96E+05	98	153	Cf	251.0795868	7.470	1/2+		2.83E+10
95	142	Am	237.049996	7.561	5/2(−)		4.39E+03	98	154	Cf	252.0816258	7.465	0+		8.34E+07
95	143	Am	238.0519843	7.556	1+		5.87E+03	98	155	Cf	253.0851331	7.455	(7/2+)		1.54E+06
95	144	Am	239.0530245	7.554	5/2−		4.28E+04	98	156	Cf	254.0873229	7.449	0+		5.23E+06
95	145	Am	240.0553002	7.547	(3−)		1.83E+05	98	157	Cf	255.091046	7.438	(9/2+)		5.04E+03
95	146	Am	241.0568291	7.543	5/2−		1.36E+10	99	150	Es	249.076411	7.474	7/2(+)		6.12E+03
95	147	Am	242.0595492	7.535	1−		5.77E+04	99	151	Es	250.078612	7.469	(6+)		3.10E+04
95	148	Am	243.0613811	7.530	5/2−		2.32E+11	99	151	Es			1(−)		7.99E+03
95	149	Am	244.0642848	7.521	(6−)		3.64E+04	99	152	Es	251.0799921	7.466	(3/2−)		1.19E+05
95	150	Am	245.0664521	7.515	(5/2)+		7.38E+03	99	153	Es	252.0829785	7.457	(5−)		4.07E+07
96	142	Cm	238.0530287	7.548	0+		8.64E+03	99	154	Es	253.0848247	7.453	7/2+		1.77E+06
96	143	Cm	239.054957	7.543	(7/2−)		1.04E+04	99	155	Es	254.088022	7.444	(7+)		2.38E+07
96	144	Cm	240.0555295	7.543	0+		2.33E+06	99	156	Es	255.0902731	7.438	(7/2+)		3.44E+06
96	145	Cm	241.057653	7.537	1/2+		2.83E+06	99	157	Es			(8+)		2.74E+04
96	146	Cm	242.0588358	7.534	0+		1.41E+07	100	151	Fm	251.081575	7.457	(9/2−)		1.91E+04
96	147	Cm	243.0613891	7.527	5/2+		9.18E+08	100	152	Fm	252.0824669	7.456	0+		9.14E+04
96	148	Cm	244.0627526	7.524	0+		5.71E+08	100	153	Fm	253.0851852	7.448	1/2+		2.59E+05
96	149	Cm	245.0654912	7.516	7/2+		2.68E+11	100	154	Fm	254.0868542	7.445	0+		1.17E+04
96	150	Cm	246.0672237	7.511	0+		1.49E+11	100	155	Fm	255.0899622	7.436	7/2+		7.23E+04
96	151	Cm	247.0703535	7.502	9/2−		4.92E+14	100	156	Fm	256.0917731	7.432	0+		9.46E+03
96	152	Cm	248.0723485	7.497	0+		1.07E+13	100	157	Fm	257.0951047	7.422	(9/2+)		8.68E+06
96	153	Cm	249.0759534	7.486	1/2+		3.85E+03	101	155	Md	256.094059	7.420	(0−,1−)		4.69E+03
96	154	Cm	250.078357	7.479	0+		3.06E+11	101	156	Md	257.0955414	7.418	(7/2−)		1.99E+04
97	146	Bk	243.0630076	7.517	(3/2−)		1.62E+04	101	157	Md	258.0984313	7.410	(8−)		4.45E+06
97	147	Bk	244.0651808	7.511	(1−)		1.57E+04	101	157	Md	256.09360		(1−)		3.60E+03
97	148	Bk	245.0663616	7.509	3/2−		4.27E+05	101	158	Md	259.100509	7.405	(7/2−)		5.76E+03
97	149	Bk	246.0686729	7.503	2(−)		1.56E+05	101	159	Md	260.103652	7.396			2.40E+06
97	150	Bk	247.0703071	7.499	(3/2−)		4.35E+10	103	159	Lr	262.109634	7.374			1.30E+04

Appendix C

Element Properties

Z	Charge number (number of protons in the nucleus = number of electrons)	
ρ	Mass density at standard temperature (20 °C = 293.15 K) and pressure (1 atmosphere)	
m	Standard atomic weight (average mass of an atom, abundance-weighted average of the isotope masses)	
$T_{melting}$	Temperature of melting point (transition point between solid and liquid phase) at 1 atm pressure	
$T_{boiling}$	Temperature of boiling point (transition point between liquid and gas phase) at 1 atm pressure	
L_m	Heat of melting/fusion	
L_v	Heat of vaporization	
E_1	Ionization energy (energy to remove least bound electron)	

Z	Sym	Name	Electron Configuration	ρ(g/cm³)	m(g/mol)	$T_{melting}$ (K)	$T_{boiling}$ (K)	L_m (kJ/mol)	L_v (kJ/mol)	E_1(eV)
1	H	Hydrogen$_{gas}$	$1s^1$	$8.988 \cdot 10^{-5}$	1.00794	14.01	20.28	0.117	0.904	13.5984
2	He	Helium$_{gas}$	$1s^2$	$1.786 \cdot 10^{-4}$	4.002602	—	4.22	—	0.0829	24.5874
3	Li	Lithium	$[He]2s^1$	0.534	6.941	453.69	1615	3.00	147.1	5.3917
4	Be	Beryllium	$[He]2s^2$	1.85	9.012182	1560	2742	7.895	297	9.3227
5	B	Boron	$[He]2s^2\,2p^1$	2.34	10.811	2349	4200	50.2	480	8.2980
6	C	Carbon$_{graphite}$	$[He]2s^2\,2p^2$	2.267	12.0107	3800	4300	117	710.9	11.2603
7	N	Nitrogen$_{gas}$	$[He]2s^2\,2p^3$	$1.251 \cdot 10^{-3}$	14.0067	63.1526	77.36	0.72	5.56	14.5341
8	O	Oxygen$_{gas}$	$[He]2s^2\,2p^4$	$1.429 \cdot 10^{-3}$	15.9994	54.36	90.20	0.444	6.82	13.6181
9	F	Fluorine$_{gas}$	$[He]2s^2\,2p^5$	$1.7 \cdot 10^{-3}$	18.998403	53.53	85.03	0.510	6.62	17.4228
10	Ne	Neon$_{gas}$	$[He]2s^2\,2p^6$	$9.002 \cdot 10^{-4}$	20.1797	24.56	27.07	0.335	1.71	21.5645
11	Na	Sodium	$[Ne]3s^1$	0.968	22.989770	370.87	1156	2.60	97.42	5.1391
12	Mg	Magnesium	$[Ne]3s^2$	1.738	24.3050	923	1363	8.48	128	7.6462
13	Al	Aluminum	$[Ne]3s^2\,3p^1$	2.70	26.981538	933.47	2792	10.71	294.0	5.9858
14	Si	Silicon	$[Ne]3s^2\,3p^2$	2.3290	28.0855	1687	3538	50.21	359	8.1517
15	P	Phosphorus$_{white}$	$[Ne]3s^2\,3p^3$	1.823	30.973761	317.3	550	0.66	12.4	10.4867
16	S	Sulfur	$[Ne]3s^2\,3p^4$	1.92–2.07	32.065	388.36	717.8	1.727	45	10.3600
17	Cl	Chlorine	$[Ne]3s^2\,3p^5$	$3.2 \cdot 10^{-3}$	35.453	171.6	239.11	6.406	20.41	12.9676
18	Ar	Argon	$[Ne]3s^2\,3p^6$	$1.784 \cdot 10^{-3}$	39.948	83.80	87.30	1.18	6.43	15.7596
19	K	Potassium	$[Ar]4s^1$	0.89	39.0983	336.53	1032	2.4	79.1	4.3407

Z	Sym	Name	Electron Configuration	ρ(g/cm^3)	m(g/mol)	$T_{melting}$ (K)	$T_{boiling}$ (K)	L_m (kJ/mol)	L_v (kJ/mol)	E_1(eV)
20	Ca	Calcium	[Ar]$4s^2$	1.55	40.078	1115	1757	8.54	154.7	6.1132
21	Sc	Scandium	[Ar]$3d^1\,4s^2$	2.985	44.955910	1814	3109	14.1	332.7	6.5615
22	Ti	Titanium	[Ar]$3d^2\,4s^2$	4.506	47.867	1941	3560	14.15	425	6.8281
23	V	Vanadium	[Ar]$3d^3\,4s^2$	6.0	50.9415	2183	3680	21.5	459	6.7462
24	Cr	Chromium	[Ar]$3d^5\,4s^1$	7.19	51.9961	2180	2944	21.0	339.5	6.7665
25	Mn	Manganese	[Ar]$3d^5\,4s^2$	7.21	54.938049	1519	2334	12.91	221	7.4340
26	Fe	Iron	[Ar]$3d^6\,4s^2$	7.874	55.845	1811	3134	13.81	340	7.9024
27	Co	Cobalt	[Ar]$3d^7\,4s^2$	8.90	58.933200	1768	3200	16.06	377	7.8810
28	Ni	Nickel	[Ar]$3d^8\,4s^2$	8.908	58.6934	1728	3186	17.48	377.5	7.6398
29	Cu	Copper	[Ar]$3d^{10}\,4s^1$	8.94	63.546	1357.77	2835	13.26	300.4	7.7264
30	Zn	Zinc	[Ar]$3d^{10}\,4s^2$	7.14	65.409	692.68	1180	7.32	123.6	9.3942
31	Ga	Gallium	[Ar]$3d^{10}\,4s^2\,4p^1$	5.91	69.723	302.9146	2477	5.59	254	5.9993
32	Ge	Germanium	[Ar]$3d^{10}\,4s^2\,4p^2$	5.323	72.64	1211.40	3106	36.94	334	7.8994
33	As	Arsenic	[Ar]$3d^{10}\,4s^2\,4p^3$	5.727	74.92160	1090	887	24.44	34.76	9.7886
34	Se	Selenium	[Ar]$3d^{10}\,4s^2\,4p^4$	4.28–4.81	78.96	494	958	6.69	95.48	9.7524
35	Br	Bromine$_{liquid}$	[Ar]$3d^{10}\,4s^2\,4p^5$	3.1028	79.904	265.8	332.0	10.571	29.96	11.8138
36	Kr	Krypton$_{gas}$	[Ar]$3d^{10}\,4s^2\,4p^6$	$3.749\cdot10^{-3}$	83.798	115.79	119.93	1.64	9.08	13.9996
37	Rb	Rubidium	[Kr]$5s^1$	1.532	85.4678	312.46	961	2.19	75.77	4.1771
38	Sr	Strontium	[Kr]$5s^2$	2.64	87.62	1050	1655	7.43	136.9	5.6949
39	Y	Yttrium	[Kr]$4d^1\,5s^2$	4.472	88.90585	1799	3609	11.42	365	6.2173
40	Zr	Zirconium	[Kr]$4d^2\,5s^2$	6.52	91.224	2128	4682	14	573	6.6339
41	Nb	Niobium	[Kr]$4d^4\,5s^1$	8.57	92.90638	2750	5017	30	689.9	6.7589
42	Mo	Molybdenum	[Kr]$4d^5\,5s^1$	10.28	95.94	2896	4912	37.48	617	7.0924
43	Tc	Technetium	[Kr]$4d^5\,5s^2$	11	(98)	2430	4538	33.29	585.2	7.28
44	Ru	Ruthenium	[Kr]$4d^7\,5s^1$	12.45	101.07	2607	4423	38.59	591.6	7.3605
45	Rh	Rhodium	[Kr]$4d^8\,5s^1$	12.41	102.90550	2237	3968	26.59	494	7.4589
46	Pd	Palladium	[Kr]$4d^{10}$	12.023	106.42	1828.05	3236	16.74	362	8.3369
47	Ag	Silver	[Kr]$4d^{10}\,5s^1$	10.49	107.8682	1234.93	2435	11.28	250.58	7.5762
48	Cd	Cadmium	[Kr]$4d^{10}\,5s^2$	8.65	112.411	594.22	1040	6.21	99.87	8.9938
49	In	Indium	[Kr]$4d^{10}\,5s^2\,5p^1$	7.31	114.818	429.7485	2345	3.281	231.8	5.7864
50	Sn	Tin$_{white}$	[Kr]$4d^{10}\,5s^2\,5p^2$	7.365	118.710	505.08	2875	7.03	296.1	7.3439
51	Sb	Antimony	[Kr]$4d^{10}\,5s^2\,5p^3$	6.697	121.760	903.78	1860	19.79	193.43	8.6084
52	Te	Tellurium	[Kr]$4d^{10}\,5s^2\,5p^4$	6.24	127.60	722.66	1261	17.49	114.1	9.0096
53	I	Iodine	[Kr]$4d^{10}\,5s^2\,5p^5$	4.933	126.90447	386.85	457.4	15.52	41.57	10.4513
54	Xe	Xenon$_{gas}$	[Kr]$4d^{10}\,5s^2\,5p^6$	$5.894\cdot10^{-3}$	131.293	161.4	165.03	2.27	12.64	12.1298
55	Cs	Cesium	[Xe]$6s^1$	1.93	132.90545	301.59	944	2.09	63.9	3.8939
56	Ba	Barium	[Xe]$6s^2$	3.51	137.327	1000	2170	7.12	140.3	5.2117
57	La	Lanthanum	[Xe]$5d^1\,6s^2$	6.162	138.9055	1193	3737	6.20	402.1	5.5769
58	Ce	Cerium	[Xe]$4f^1\,5d^1\,6s^2$	6.770	140.116	1068	3716	5.46	398	5.5387
59	Pr	Praseodymium	[Xe]$4f^3\,6s^2$	6.77	140.90765	1208	3793	6.89	331	5.473
60	Nd	Neodymium	[Xe]$4f^4\,6s^2$	7.01	144.24	1297	3347	7.14	289	5.5250
61	Pm	Promethium	[Xe]$4f^5\,6s^2$	7.26	(145)	1315	3273	7.13	289	5.582

Z	Sym	Name	Electron Configuration	ρ(g/cm^3)	m(g/mol)	$T_{melting}$ (K)	$T_{boiling}$ (K)	L_m (kJ/mol)	L_v (kJ/mol)	E_1(eV)
62	Sm	Samarium	[Xe]$4f^6\,6s^2$	7.52	150.36	1345	2067	8.62	165	5.6437
63	Eu	Europium	[Xe]$4f^7\,6s^2$	5.264	151.964	1099	1802	9.21	176	5.6704
64	Gd	Gadolinium	[Xe]$4f^7\,5d^1\,6s^2$	7.90	157.25	1585	3546	10.05	301.3	6.1498
65	Tb	Terbium	[Xe]$4f^9\,6s^2$	8.23	158.92534	1629	3503	10.15	293	5.8638
66	Dy	Dysprosium	[Xe]$4f^{10}\,6s^2$	8.540	162.500	1680	2840	11.06	280	5.9389
67	Ho	Holmium	[Xe]$4f^{11}\,6s^2$	8.79	164.93032	1734	2993	17.0	265	6.0215
68	Er	Erbium	[Xe]$4f^{12}\,6s^2$	9.066	167.259	1802	3141	19.90	280	6.1077
69	Tm	Thulium	[Xe]$4f^{13}\,6s^2$	9.32	168.93421	1818	2223	16.84	247	6.1843
70	Yb	Ytterbium	[Xe]$4f^{14}\,6s^2$	6.90	173.04	1097	1469	7.66	159	6.2542
71	Lu	Lutetium	[Xe]$4f^{14}\,5d^1\,6s^2$	9.841	174.967	1925	3675	22	414	5.4259
72	Hf	Hafnium	[Xe]$4f^{14}\,5d^2\,6s^2$	13.31	178.49	2506	4876	27.2	571	6.8251
73	Ta	Tantalum	[Xe]$4f^{14}\,5d^3\,6s^2$	16.69	180.9479	3290	5731	36.57	732.8	7.5496
74	W	Tungsten	[Xe]$4f^{14}\,5d^4\,6s^2$	19.25	183.84	3695	5828	52.31	806.7	7.8640
75	Re	Rhenium	[Xe]$4f^{14}\,5d^5\,6s^2$	21.02	186.207	3459	5869	60.3	704	7.8335
76	Os	Osmium	[Xe]$4f^{14}\,5d^6\,6s^2$	22.61	190.23	3306	5285	57.85	738	8.4382
77	Ir	Iridium	[Xe]$4f^{14}\,5d^7\,6s^2$	22.56	192.217	2739	4701	41.12	563	8.9670
78	Pt	Platinum	[Xe]$4f^{14}\,5d^9\,6s^1$	21.45	195.078	2041.4	4098	22.17	469	8.9588
79	Au	Gold	[Xe]$4f^{14}\,5d^{10}\,6s^1$	19.3	196.96655	1337.33	3129	12.55	324	9.2255
80	Hg	Mercury$_{liquid}$	[Xe]$4f^{14}\,5d^{10}\,6s^2$	13.534	200.59	234.32	629.88	2.29	59.11	10.4375
81	Tl	Thallium	[Xe]$4f^{14}\,5d^{10}\,6s^2\,6p^1$	11.85	204.3833	577	1746	4.14	165	6.1082
82	Pb	Lead	[Xe]$4f^{14}\,5d^{10}\,6s^2\,6p^2$	11.34	207.2	600.61	2022	4.77	179.5	7.4167
83	Bi	Bismuth	[Xe]$4f^{14}\,5d^{10}\,6s^2\,6p^3$	9.78	208.98038	544.7	1837	11.30	151	7.2855
84	Po	Polonium	[Xe]$4f^{14}\,5d^{10}\,6s^2\,6p^4$	9.320	(209)	527	1235	13	102.91	8.414
85	At	Astatine	[Xe]$4f^{14}\,5d^{10}\,6s^2\,6p^5$	?	(210)	?	?	?	?	?
86	Rn	Radon	[Xe]$4f^{14}\,5d^{10}\,6s^2\,6p^6$	$9.73 \cdot 10^{-3}$	(222)	202	211.3	3.247	18.10	10.7485
87	Fr	Francium	[Rn]$7s^1$	1.87	(223)	~300	~950	~2	~65	4.0727
88	Ra	Radium	[Rn]$7s^2$	5.5	(226)	973	2010	8.5	113	5.2784
89	Ac	Actinium	[Rn]$6d^1\,7s^2$	10	(227)	1323	3471	14	400	5.17
90	Th	Thorium	[Rn]$6d^2\,7s^2$	11.7	232.0381	2115	5061	13.81	514	6.3067
91	Pa	Protactinum	[Rn]$5f^2\,6d^1\,7s^2$	15.37	231.03588	1841	~4300	12.34	481	5.89
92	U	Uranium	[Rn]$5f^3\,6d^1\,7s^2$	19.1	238.02891	1405.3	4404	9.14	417.1	6.1941
93	Np	Neptunium	[Rn]$5f^4\,6d^1\,7s^2$	20.45	(237)	910	4273	3.20	336	6.2657
94	Pu	Plutonium	[Rn]$5f^6\,7s^2$	19.816	(244)	912.5	3505	2.82	333.5	6.0260
95	Am	Americium	[Rn]$5f^7\,7s^2$	12	(243)	1449	2880	14.39	238.5	5.9738
96	Cm	Curium	[Rn]$5f^7\,6d^1\,7s^2$	13.51	(247)	1613	3383	~15	?	5.9914
97	Bk	Berkelium	[Rn]$5f^9\,7s^2$	~14	(247)	1259	?	?	?	6.1979
98	Cf	Californium	[Rn]$5f^{10}\,7s^2$	15.1	(251)	1173	1743	?	?	6.2817
99	Es	Einsteinium	[Rn]$5f^{11}\,7s^2$	8.84	(252)	1133	?	?	?	6.42
100	Fm	Fermium	[Rn]$5f^{12}\,7s^2$	?	(257)	1800	?	?	?	6.50
101	Md	Mendelevium	[Rn]$5f^{13}\,7s^2$	?	(258)	1100	?	?	?	6.58
102	No	Nobelium	[Rn]$5f^{14}\,7s^2$	?	(259)	?	?	?	?	6.65

Continued—

Z	Sym	Name	Electron Configuration	$\rho(\text{g/cm}^3)$	$m(\text{g/mol})$	T_{melting} (K)	T_{boiling} (K)	L_{m} (kJ/mol)	L_{v} (kJ/mol)	$E_1(\text{eV})$
103	Lr	Lawrencium	$[\text{Rn}]5f^{14}\,7s^2\,7p^1$	?	(262)	?	?	?	?	4.9
104	Rf	Rutherfordium	$[\text{Rn}]5f^{14}\,6d^2\,7s^2$	?	(261)	?	?	?	?	6
105	Db	Dubnium	$[\text{Rn}]5f^{14}\,6d^3\,7s^2$	?	(262)	?	?	?	?	?
106	Sg	Seaborgium	$[\text{Rn}]5f^{14}\,6d^4\,7s^2$	?	(266)	?	?	?	?	?
107	Bh	Bohrium	$[\text{Rn}]5f^{14}\,6d^5\,7s^2$	?	(264)	?	?	?	?	?
108	Hs	Hassium	$[\text{Rn}]5f^{14}\,6d^6\,7s^2$	?	(277)	?	?	?	?	?
109	Mt	Meitnerium	$[\text{Rn}]5f^{14}\,6d^7\,7s^2$	?	(276)	?	?	?	?	?
110	Ds	Darmstadtium	$*[\text{Rn}]5f^{14}\,6d^9\,7s^1$	?	(281)	?	?	?	?	?
111	Rg	Roentgenium	$*[\text{Rn}]5f^{14}\,6d^9\,7s^2$	?	(280)	?	?	?	?	?
112			$*[\text{Rn}]5f^{14}\,6d^{10}\,7s^2$	?	(285)	?	?	?	?	?
113			$*[\text{Rn}]5f^{14}\,6d^{10}\,7s^2\,7p^1$	?	(284)	?	?	?	?	?
114			$*[\text{Rn}]5f^{14}\,6d^{10}\,7s^2\,7p^2$	?	(289)	?	?	?	?	?
115			$*[\text{Rn}]5f^{14}\,6d^{10}\,7s^2\,7p^3$	?	(288)	?	?	?	?	?
116			$*[\text{Rn}]5f^{14}\,6d^{10}\,7s^2\,7p^4$	?	(293)	?	?	?	?	?
118			$*[\text{Rn}]5f^{14}\,6d^{10}\,7s^2\,7p^6$	?	(294)	?	?	?	?	?

*Predicted

(longest-lived isotope)

Answers to Selected Questions and Problems

Chapter 21: Electrostatics

Multiple Choice

21.1 b. **21.3** b. **21.5** b. **21.7** a. **21.9** c.

Problems

21.27 96470 C. **21.29** $3 \cdot 10^{17}$ electrons. **21.31** 32 C. **21.33** (a) $5.00 \cdot 10^{16}$ conduction electrons/cm^3. (b) There are $5.88 \cdot 10^{-7}$ conduction electrons in the doped silicon sample for every conduction electron in the copper sample. **21.35** $1 \cdot 10^{-5}$ C; The force is attractive. **21.37** $-2.9 \cdot 10^{-9}$ N.

21.39 100 N. **21.41** $q = 2.02 \cdot 10^{-5}$C. **21.43** 3.1 N. **21.45** (a) 0. (b) $\pm a/\sqrt{2}$.

21.47 $\vec{F}_{net,2} = (-1 \cdot 10^8 \text{ N})\hat{x} + (7 \cdot 10^7 \text{ N})\hat{y}$; $\left|\vec{F}_{net,2}\right| = 1 \cdot 10^8$ N. **21.49** (a) No.

(b) -0.6 N. **21.51** $-3.7 \cdot 10^{-29}$C $= -2.3 \cdot 10^{-10}$ e. **21.55** $6 \cdot 10^{12}$ C.

21.57 $n = 1$: $F_1 = 8.24 \cdot 10^{-8}$ N; $F_{g,1} = 3.63 \cdot 10^{-47}$ N
$n = 2$: $F_2 = 5.15 \cdot 10^{-9}$ N; $F_{g,2} = 2.27 \cdot 10^{-48}$ N
$n = 3$: $F_3 = 1.02 \cdot 10^{-9}$ N; $F_{g,3} = 4.49 \cdot 10^{-49}$ N
$n = 4$: $F_4 = 3.22 \cdot 10^{-10}$ N; $F_{g,4} = 1.42 \cdot 10^{-49}$ N.

21.59 $4.41 \cdot 10^{-40}$.

Additional Problems

21.61 -5 N$\hat{y}$. **21.63** (a) 70 N. (b) $4 \cdot 10^{28}$ m/s. **21.65** 114 N. **21.67** -65 μC. **21.69** $2 \cdot 10^{-7}$ e; $9 \cdot 10^{-19}$ kg. **21.71** $2 \cdot 10^{-7}$C $= 0.2$ μe. **21.73** $m_2 = 50.4$ g. **21.75** $q = 1.1$ pC. **21.77** -24 cm. **21.79** $\left|Q_0\right| = 3$ nC. **21.81** 2.8 m.

Chapter 22: Electric Fields and Gauss's Law

Multiple Choice

22.1 e. **22.3** a. **22.5** d. **22.7** c. **22.9** a.

Problems

22.23 $5.75 \cdot 10^4$ N/C. **22.25** 192.53° counterclockwise from the positive x-axis. **22.27** 0.56 m and 4.4 m. **22.29** $E = -kp/x^3$; The field strength falls off more rapidly perpendicular to the dipole axis.

22.31 $(3.7 \text{ m/s})\hat{x} + (2.4 \text{ m/s})\hat{y}$. **22.33** $\vec{E} = (-Q/\pi\epsilon_0\pi R^2)\hat{j}$.

22.35 $\vec{E}(d) = \left(\dfrac{kQ}{d\sqrt{d^2 + L^2}}\right)\hat{x} - \left(\dfrac{kQ}{dL} - \dfrac{kQ}{L\sqrt{d^2 + L^2}}\right)\hat{y}$.

22.37 $4.1 \cdot 10^3$ N/C. **22.39** $3.5 \cdot 10^{-15}$ N m. **22.41** 0.189 m.
22.43 (a) $v = \sqrt{2h(g - QE/M)}$. (b) If the value $g - QE/M$ is less than zero the value is non-real and the body does not move.
22.45 (a) 0.0141 N/C. (b) 135,000 m/s^2. (c) toward the wire.

22.47 $-1 \cdot 10^{-8}$ C. **22.49** 60 N/C into the face of the cube.
22.51 Since the radius of the balloon never reaches R, the charge enclosed is constant and the electric field does not change.
22.53 (a) 50 N/C. (b) 0 N/C. (c) 0.4 N/C. (d) $5 \cdot 10^{-12}$ C/m^2.
22.55 $-6.8 \cdot 10^5$ C. **22.57** $1 \cdot 10^5$ N/C directed from the positive plate to the negative plate. **22.59** $Q = (4/5)\pi AR^5$. **22.61** (a) $4.5 \cdot 10^8$ N/C.
(b) $2.0 \cdot 10^8$ N/C. (c) 0, since it is in the conducting shell. (d) $-1.6 \cdot 10^7$ N/C.
22.63 (a) $1 \cdot 10^5$ N/C. (b) $5 \cdot 10^5$ N/C.
22.65 (a) $\vec{E} = (Qr/4\pi a^3\epsilon_0)\hat{r}$. (b) $\vec{E} = (Q/4\pi\epsilon_0 r^2)\hat{r}$.
(c)

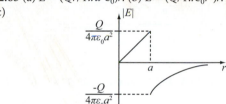

The discontinuity at $r = a$ is due to the surface charge density of the gold. The charge on the gold layer causes a sudden spike in the total charge resulting in a discontinuity in the electric fields.
22.67 $E_x \approx 2 \cdot 10^3$ N/C, $E_y \approx 1 \cdot 10^3$ N/C.

Additional Problems

22.69 0. **22.71** $1.45 \cdot 10^2$ N/C directed away from the y-axis.
22.73 (a) -5 μC. (b) 0. **22.75** $2.6 \cdot 10^{13}$ m/s^2. **22.77** $4.31 \cdot 10^{-5}$ C/m.
22.79 $3.1 \cdot 10^{16}$ electrons. **22.81** 274.76 N/C. **22.83** $7.13 \cdot 10^4$ N/C.
22.85 (a)

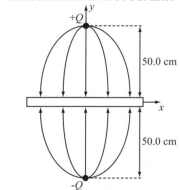

(b) $-8.99 \cdot 10^{-3}$ N. (c) $E_{total} = \dfrac{1}{2\pi\epsilon_0}\left|\dfrac{Qa}{\left(a^2 + \rho^2\right)^{3/2}}\right|$.

(d) $\sigma(\rho) = E\epsilon_0 = \dfrac{1}{2\pi}\left|\dfrac{Qa}{\left(a^2 + \rho^2\right)^{3/2}}\right|$.

(e) The total charge induced is equal to the charge in magnitude.

Chapter 23: Electric Potential

Multiple Choice

23.1 a. **23.3** c. **23.5** a. **23.7** a. **23.9** a & c.

Problems

23.21 0.3 m. **23.23** $3.84 \cdot 10^{-17}$ J. **23.25** $3.10 \cdot 10^5$ m/s. **23.27** (a) 0.0293 m. (b) 7.9 cm. (c) 247 km/s. **23.29** (a) 18 kV. (b) –72 kV. **23.31** $7 \cdot 10^{12}$ electrons. **23.33** 11.2 MV. **23.35** 847 V. **23.37** (a) $V(R) = 7$ kV. (b) $V(0) = 10$ kV. **23.39** 480 V. **23.41** (a) $x_{max} = 1$. (b) $E_0 (2e^{-1} - 1)$. **23.43** $E_x = 70.2$ V/m. **23.45** 11.2 m/s². **23.47** (a) $10x$ V/m². (b) $4 \cdot 10^9$ m/s². (c) $3 \cdot 10^5$ m/s.

23.49 $\vec{E}(\vec{r}) = \dfrac{kq}{r^3}(x\hat{x} + y\hat{y} + z\hat{z})$.

23.51 (a) $\vec{E}(\vec{r}) = \dfrac{2V_0\vec{r}}{a^2} e^{-r^2/a^2}$. (b) $\rho(\vec{r}) = \dfrac{2\epsilon_0 V_0}{a^2}\left[1 - 2\left(\dfrac{r^2}{a^2}\right)\right]e^{-r^2/a^2}$. (c) 0.

(d)
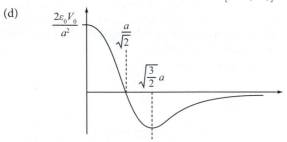

23.53 $2.31 \cdot 10^{-13}$ J = 1.44 MeV. **23.55** 140000 eV. **23.57** $v_1 = 0.1$ m/s, $v_2 = 0.07$ m/s.

Additional Problems

23.59 field: 0; potential: 12 V. **23.61** 0.247 J. **23.63** $2.3 \cdot 10^5$ V. **23.65** $5 \cdot 10^5$ V; $1 \cdot 10^5$ C. **23.67** (a) 4:1. (b) 100/3 μC. **23.69** 2.0 J. **23.71** first sphere: $3 \cdot 10^3$ N/C; second sphere: $6 \cdot 10^3$ N/C. **23.73** (a) 46.8 V. (b) 7.29 cm. **23.75** (a) $9.4 \cdot 10^4$ V. (b) $0.84 \cdot 10^{-6}$ C.

23.77 (a) $V(x) = \dfrac{1}{4\pi\epsilon_0}\left(\dfrac{q_1}{x - x_1} + \dfrac{q_2}{x - x_2}\right)$ for $x > x_1, x_2$,

$V(x) = \dfrac{1}{4\pi\epsilon_0}\left(\dfrac{q_1}{x_1 - x} + \dfrac{q_2}{x - x_2}\right)$ for $x_1 < x < x_2$, and

$V(x) = \dfrac{1}{4\pi\epsilon_0}\left(\dfrac{q_1}{x_1 - x} + \dfrac{q_2}{x_2 - x}\right)$ for $x < x_1, x_2$. (b) $x = 11$ m, $x = 0.25$ m.

(c) $E = \dfrac{1}{4\pi\epsilon_0}\left[\dfrac{q_1}{(x - x_1)^2} + \dfrac{q_2}{(x - x_2)^2}\right]$ for $x > x_1, x_2$,

$E = \dfrac{1}{4\pi\epsilon_0}\left[-\dfrac{q_1}{(x_1 - x)^2} + \dfrac{q_2}{(x - x_2)^2}\right]$ for $x_1 < x < x_2$, and

$E = \dfrac{1}{4\pi\epsilon_0}\left[-\dfrac{q_1}{(x_1 - x)^2} - \dfrac{q_2}{(x_2 - x)^2}\right]$ for $x < x_1, x_2$.

23.79 (a) $V = \dfrac{q}{4\pi\epsilon_0}\left(\dfrac{1}{R}\right)$. (b) $V = \dfrac{q}{4\pi\epsilon_0}\left(\dfrac{1}{R}\right)$. (c) The electric field is dependent on direction, so this is not possible.

Chapter 24: Capacitors

Multiple Choice

24.1 b. **24.3** c. **24.5** c. **24.7** d. **24.9.** a.

Problems

24.25 $1 \cdot 10^2$ km². **24.27** $9 \cdot 10^9$ m. **24.29** $7.09 \cdot 10^{-4}$ F. **24.31** (a) 64.0 F. (b) 4.5 V^{-7}. **24.33** 1 pF. **24.35** 4.3 nF. **24.37** (a) 2.7 nF. (b) 3.101 nF. **24.39** $C_1/1275$. **24.41** 0.04 J. **24.43** $1.0 \cdot 10^{-8}$ FV²/m³. **24.45** 1 F; $7 \cdot 10^{13}$ J. **24.47** (a) 70 nJ. (b) 40 nJ. (c) 9 nJ. (d) 20 nJ; The energy lost is the difference between the initial and final energies of the system. **24.49** $3.89 \cdot 10^4$. **24.51** $2.2 \cdot 10^{-5}$ C/m **24.53** 70 pF.

24.55 $q = \dfrac{\pi\epsilon_0 L(\kappa + 1)V}{2\ln(r_2/r_1)}$; $\dfrac{\kappa + 1}{2}$. **24.57** (a) 12550 V/M. (b) 6 nC. (c) 4.9 nC.

24.59 $9 \cdot 10^{-2}\, q_1(0.1\text{ m} - h) = 7(q_2 h)$

$q_1 + q_2 = 9 \cdot 10^{-8}$

$4 \cdot 10^{-5} = \dfrac{q_1^2}{1 \cdot 10^{-7}h} + \dfrac{q_2^2}{2 \cdot 10^{-9}(0.1\text{ m} - h)} + 5 \cdot 10^{-3}h^2$.

Additional Problems

24.61 $5.5 \cdot 10^{-8}$ C. **24.63** 210%. **24.65** 1.5 km. **24.67** $V_A = 0.3$ V. **24.69** $4 \cdot 10^{-13}$ F. **24.71** 0.79 nF. **24.73** (a) 5.4. (b) 0.13 μC. (c) $5.4 \cdot 10^5$ N/C. (d) $5.4 \cdot 10^5$ N/C. **24.75** $1 \cdot 10^{-8}$ N in the direction of the motion of the dielectric material. **24.77** (a) $C_0 = 35$ pF, $U_0 = 1.0 \cdot 10^{-7}$ J. (b) $C' = 92$ pF, $U = 3.8 \cdot 10^{-8}$ J. (c) No. **24.79** (a) 0.74 nC. (b) 36 nJ. (c) $1.2 \cdot 10^5$ V/m. **24.81** 45.0 μC. **24.83** (b)

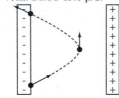

(c) $V = mv_i^2 \sin^2\theta/(2q)$. (d) $v_i = 2.0 \cdot 10^5$.

Chapter 25: Current and Resistance

Multiple Choice

25.1 d. **25.3** c. **25.5** c. **25.7** c. **25.9** c. **25.11** c.

Problems

25.27 $6.02 \cdot 10^{28}$ electrons/m³; $3 \cdot 10^{-8}$ m/s. **25.29** (a) $5.85 \cdot 10^{22}$ cm^{-3}. (b) 133 A/m². (c) $1.42 \cdot 10^{-8}$ m/s. **25.31** 2.24 mm. **25.33** between 9 and 10 gauge wire. **25.35** 3. **25.39** 0.02 Ω. **25.41** (a) –84%. (b) 540%. (c) At room temperature, $V_d = 0.4$ mm/s. At temperature $T = -196$ °C, $V_d = 3$ mm/s. The speed at 77 K is 7.5 times faster than the speed at 293 K. **25.43** 2.571 Ω. **25.45** (a) 1.5 Ω. (b) 1.1 Ω. (c) When two bulbs are put in series, it is expected that they glow dimmer than only one bulb. This would mean the one bulb would be hotter and thus have a larger resistance. **25.47** 60 Ω. **25.49** (a) 10 Ω. (b) 1 A. (c) 1 V. **25.51** (a) 1 V. (b) 0.5 A. **25.53** 86% brighter. **25.55** 7.56 Ω. **25.57** (a) 8 W. (b) 40 V.

25.59 $\dfrac{V_{emf}^2}{(R_i + R_2)^2} R_L$; 16 W; 18 W; 17 W.

25.61 (a) $R_C = 64.6$ mΩ. (b) $i = 9.99$ mA. (c) $b = 1.31$ mm. (d) 0.772 s. (e) 4.99 mW. (f) Electric power is lost via heat.

Additional Problems

25.63 10 A. **25.65** 1 m. **25.67** (a) 0.045 W. (b) 0.41; The power is 9 times greater in parallel than it is in series. **25.69** 25 W. **25.71** 9.00 W. **25.73** (a) $3.2 \cdot 10^{-5}$ A. (b) 640 kW. (c) $6.3 \cdot 10^{14}$ Ω. **25.75** (a) 60 V. (b) 30 A. (c) 1 kW. **25.77** 2.7 s. **25.79** 24:1. **25.81** (a) EJ. (b) ρJ^2; σE^2.

Chapter 26: Direct Current Circuits

Multiple Choice

26.1 a. **26.3** b. **26.5** b. **26.7** c. **26.9** d, e & f.

Problems

26.23 $V_1 = \left(\dfrac{R_1}{R_1 + R_2}\right)\Delta V$, $V_2 = \left(\dfrac{R_2}{R_1 + R_2}\right)\Delta V$; The resistors in series construct a voltage divider. The voltage ΔV is divided between the two resistors with potential drop proportional to their respective resistances. **26.25** $R = 40$ Ω, $V_{emf} = 120$ V.

26.27 (a)

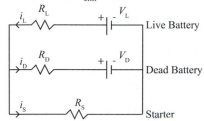

(b) starter: 150. A; live battery: 150. A; dead battery: 0.496 A.
26.29 $i_1 = 0.20$ A, $i_2 = 0.20$ A, $i_3 = 0.40$ A, $P_A = 1.2$ W, $P_B = 2.4$ W.

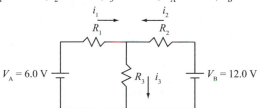

26.31 $i_1 = 0.3$ A clockwise, $i_2 = 0.4$ A counterclockwise, $i_3 = 0.2$ A counterclockwise, $i_4 = 0.6$ A clockwise, $i_5 = 0.2$ A clockwise, $i_6 = 0.4$ A clockwise, $P(V_{emf,1}) = 1$ W, $P(V_{emf,2}) = 5$ W.
26.33 $R_L = \left(1 + \sqrt{3}\right)R$.
26.35 $R_{Shunt} = \dfrac{R_{Ammeter}}{N-1}$; $R_{Shunt} = 10.1$ mΩ; 1/100 through the ammeter and 99/100 through the shunt.
26.37 $V_{ab} = 2.985$ V, $V_{bc} = 3.015$ V. Increasing R_{volt} will reduce the error since the voltmeter will draw less current. **26.39** (a) 6.00 A each.
(b) 0.012 A. **26.41** 9 s. **26.43** $2.4 \cdot 10^{-4}$ C; 40 μs, **26.45** 4 ms. **26.47** 89 s.
26.49 22 kV, 0.020 μF. **26.51** (a) 2 V. (b) 5 μJ. (c) 1 μJ. **26.53** $\left(\sqrt{3} - 1\right)C/2$.

Additional Problems

26.55 (a) The shunt resistor carries most of the load so that the ammeter is not damaged.

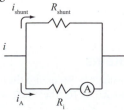

(b) 7.5 mΩ. **26.57** $C = 80$ nF, $R = 10$ kΩ. **26.59** (a) 200 mA. (b) 20 mA. **26.61** 3.8 s. **26.63** $P_1 = 7$ mW, $P_2 = 4$ W, $P_3 = 7$ W. **26.65** (a) $1 \cdot 10^{-4}$ s. (b) 70 μC. **26.67** (a) 0.7 τ. (b) 1:4. (c) 10 μs. (d) 7 μs. **26.69** $i_2 = 469$ mA. **26.71** $2C$.

Chapter 27: Magnetism

Multiple Choice

27.1 a. **27.3** e. **27.5** a. **27.7** a,c,d,e.

Problems

27.21 $9.4 \cdot 10^{-5}$ T. **27.23** $7.616 \cdot 10^{-4}$ N at 23.20° above the negative x-axis. **27.25** $3.4 \cdot 10^{-4}$ T.
27.27 $\dfrac{d\vec{p}}{dt} = -q\left(\vec{E} + \vec{v} \times \vec{B}\right)$, $\dfrac{dK}{dt} = -q\vec{E} \bullet \vec{v}$.
27.29 $6.8\hat{z}$ T. **27.31** The proton will create a spiral with a velocity of $3.0 \cdot 10^5$ m/s along the z-axis, with the circular motion having a speed of $2.2 \cdot 10^5$ m/s and a radius of 4.7 mm. **27.33** 1:4; The electric field must have magnitude $E = vB$ and point in the positive y-direction in order for the particles to move in a straight line. **27.35** (a) 2 T. (b) 5 cm. (c) $2 \cdot 10^6$ m/s. **27.37** 0.4 kg. **27.39** 15.0 N; This is the same as the force on a wire of the same length with the same current and magnetic field.
27.41 $0.92 \cdot 10^3$ rad/s². **27.43** $\dfrac{iB_0 l^2}{2a}\hat{x}$. **27.45** 0.14 T. **27.47** 1.8 rad/s².
27.51 (a) $\vec{F}_{ab} = -0.32\hat{y}$ N. (b) 0.62 N directed 15° from the x-axis toward the negative z-axis. (c) $F_{net} = 0$. (d) $\tau = 0.025$ N m and rotates along the y-axis in counterclockwise fashion. (e) The coil rotates in a counterclockwise fashion as seen from above. **27.53** (a) holes. (b) $2.3 \cdot 10^{24}$ e/m³.

Additional Problems

27.55 $7.4 \cdot 10^{-2}$ N. **27.57** $2.33 \cdot 10^6$ A. **27.59** $2.00 \cdot 10^5$ m/s in the positive x-direction. **27.61** (a) $1.89 \cdot 10^{-6}$ T. (b) none. (c) 5 μs. **27.63** 0.031 m, $3.8 \cdot 10^{-7}$ s. **27.65** $2 \cdot 10^5$ m/s. **27.67** $-4 \cdot 10^{-3} \hat{z}$ T. **27.69** 6.97°.
27.71 (b) 20 mm. (c) $T = 1 \cdot 10^{-7}$ s, $f = 8 \cdot 10^6$ Hz. (d) 70 mm.

Chapter 28: Magnetic Fields of Moving Charges

Multiple Choice

28.1 b. **28.3** c. **28.5** d. **28.7** c. **28.9** a.

Problems

28.27 parallel to the z-axis through (0,4,0), carrying a current in the opposite direction of the first wire. **28.29** 7.5° below the x-axis. **28.31** 45° in the x-y plane at a height of b. **28.33** $i_1 = 2.0 \cdot 10^2$ A. **28.35** (a) $-2 \cdot 10^{-2}$ N$\hat{y}$. (b) $\vec{F} = 2 \cdot 10^{-2}$ N $\hat{x}$. (c) $\vec{F} = 0$ N.
28.37 $9.42 \cdot 10^{-5}$ T at 45° from the negative x-direction toward the positive y-axis. **28.39** $B_a = 0$; $B_b = 1.08 \cdot 10^{-6}$ T; $B_c = 2.70 \cdot 10^{-6}$ T; $B_d = 1.69 \cdot 10^{-6}$ T. **28.41** Yes. **28.43** $9.4 \cdot 10^{-5}$ T.
28.45 If $r \leq R$ $B_{r<R} = \dfrac{\mu_0 J_0}{r}[R^2 - R(R+r)e^{-r/R}]$;
If $r \geq R$ then
$B_{r>R} = \dfrac{\mu_0 J_0 R^2}{r}[1^2 - 2e^{-1}]$.

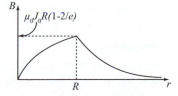

28.47 4:3. **28.49** (a) $1.3 \cdot 10^{-5}$ T. (b) $1.0 \cdot 10^{-2}$ T.
28.51 $1.9 \cdot 10^{-19}$ kg m/s. **28.53** 0.20 K. **28.55** 55. **28.57** $1.12 \cdot 10^{-5}$ Am^2.
28.59 (a) $L = I\overline{\omega} = (2/5)mR^2\overline{\omega}$. (b) $\mu = q\overline{\omega}R^2/5$. (c) $\gamma_e = -e/(2m)$.

Additional Problems

28.61 4 mT into the page. **28.63** $4.1 \cdot 10^9$ A. **28.65** 18,000 turns.
28.67 30 mT. **28.69** $q = 0.5$ C. **28.71** $1 \cdot 10^{-7}\hat{x}$ N m.
28.73 (a) $-7.3 \cdot 10^{10}$ m/s^2. (b) $-1.2 \cdot 10^{11}$ m/s^2. **28.75** (a) $1.06 \cdot 10^{-4}$ N m.
(b) 1.72 rad/s. **28.77** 4 A counterclockwise (as viewed from the bar magnet).

28.79 $B = \mu_0 J_0 \left[\dfrac{r}{2} - \dfrac{r^2}{3R} \right]$.

Chapter 29: Electromagnetic Induction

Multiple Choice

29.1 d. **29.3** a. **29.5** a. **29.7** d.

Problems

29.23 $2 \cdot 10^{-5}$ V. **29.25** 0.
29.27 $\Delta V_{\text{ind}} = \dfrac{\mu_0 \pi a^2 V_0 \omega}{2bR_1}\cos\omega t; \quad i = -\dfrac{\mu_0 \pi a^2 V_0 \omega}{2bR_1 R_2}\cos\omega t.$
29.29 (a) 7.1 mA. (b) clockwise. **29.31** 0.6 V. **29.33** $v_{\text{term}} = \dfrac{mgR}{\omega^2 B^2}$.
29.35 $6 \cdot 10^{-7}$ V. **29.37** 20 Hz. **29.39** (a) 0.4 A. (b) 0.3 A, 100 W.
29.41

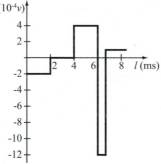

29.43 (a) 1 μs. (b) 0; 9 μA; 10 μA. **29.45** 11 V. **29.47** (a) 6.0 A. (b) 3.0 A.
(c) 3.0 A. (d) −18 V. (e) −18 V. (f) 0. (g) 0. **29.49** $1.0 \cdot 10^3$ m^3.
29.51 (a) $6.4 \cdot 10^{26}$ J/m^3. (b) $7.1 \cdot 10^9$ kg/m^3. **29.53** $4 \cdot 10^{-5}$ °C. **29.55** 1:1.

Additional Problems

29.57 $8 \cdot 10^{-3}$ V. **29.59** It does not change. **29.61** $u_B = 9.95 \cdot 10^{-4}$ J/m^3,
$u_E = 9.95 \cdot 10^{-8}$ J/m^3; $\dfrac{u_B}{u_E} = 1.00 \cdot 10^5$. **29.63** 87 mH. **29.65** 10 V; The
induced current is counterclockwise. **29.67** 2.0 m^2. **29.69** 0.3 H.
29.71 (a) $i_1 = -4$ mA; $i_2 = -2$ mA. (b) 2 mW. (c) 0.3 mN. **29.73** 0.
29.75 (a) 0.3 mN. (b) 30 μW. (c) 30 μW.

Chapter 30: Electromagnetic Oscillations and Currents

Multiple Choice

30.1 d. **30.3** b. **30.5** d. **30.7** d.

Problems

30.23 0.942 ms. **30.25** 7 mF. **30.27** 1 ms, 3 ms, 6 ms. **30.29** $t = (L/R)\ln(2)$.
30.31 300 Hz. **30.33** 500 rad/s. **30.35** 300 Ω; 40 mA. **30.37** (a) 356 Hz.

(b) 0.400 A. **30.39** 0.5 rads; 100 Ω. **30.41** 500 V; Any voltage is
permissible across the inductor. **30.43** (a) $I_{\text{max}} = 34$ A. (b) $\theta = 0.816$ rad.
(c) $C' = 760$ μF, $I'_{\text{max}} = 50$. A, $\varphi' = 0$ rad. **30.45** (a) $C = 1$ nF; Use a
capacitor of capacitance 1 nF. (b) 100 kHz. **30.47** $Q = (1/R)\sqrt{L/C}$.
30.49 (a) 18.4 kHz. (b) 2.25 W. **30.51** (a) 0.392 pF. (b) 11.9 Ω.
30.53 (a) 1 V. (b) 0. **30.55** 2 W. **30.57** (a) 14 V. (b) 9.0 V.

Additional Problems

30.59 $3 \cdot 10^{-12}$ F. **30.61** (a) 0.771 A. (b) 100 V. **30.63** 1 Ω. **30.65** (a) 0.1 A.
(b) $7 \cdot 10^{-4}$ s. **30.67** (a) 10 Ω. (b) 8 Ω. (c) $6 \cdot 10^{-5}$ H. (d) 30 kHz.
30.69 (a) $1 \cdot 10^{-3}$ J/m^3. (b) 10 A. **30.71** $P = \frac{1}{2}I_R^2 R(1 - \cos(2\omega t))$.
30.73 400 Hz. **30.75** 2000 Hz.

Chapter 31: Electromagnetic Waves

Multiple Choice

31.1 c. **31.3** b. **31.5** a. **31.7** c.

Problems

31.21 $2.5 \cdot 10^{-6}$ T. **31.23** 10.0 μA.
31.25 $i_d = \epsilon_0 R \left(\dfrac{A}{L}\right)\dfrac{di}{dt} = \epsilon_0 \rho \dfrac{di}{dt}.$
31.27 2.0 ft. **31.29** (a) 0.03 s. (b) 0.24 s; When the signal travels by
the cable is not noticeable. However, via satellite, Alice will receive a
response from her fiancé after 0.5 s, which is quite noticeable.
31.31 $4 \cdot 10^{14}$ Hz to $8 \cdot 10^{14}$ Hz. **31.33** 3.2 mH. **31.35** (a) 1.3 W/m^2.
(b) 1.2 W/m^2. (c) 0.12 W/m^2. **31.37** $2 \cdot 10^6$ V/m. **31.39** (a) $4.43 \cdot 10^{-8}$ J/m^3.
(b) $3.33 \cdot 10^{-7}$ T. **31.41** (a) $1 \cdot 10^7$ W/m^2; The intensity is much larger
than the intensity of sunlight on Earth (1400 W/m^2). (b) $7 \cdot 10^4$ V/m.
(c) $1 \cdot 10^7$ W/m^2. (d) $S(x,t) = 3 \cdot 10^7$ W/m^2 $\sin^2(10^7 x\,\text{m}^{-1} - 4 \cdot 10^{15}t$ Hz$)$.
(e) $2 \cdot 10^{-4}$ T. **31.43** (a) $E = 713$ V/m, $B = 2.38$ μT. (b) 3.3 μPa, 2.5 μN.
31.45 $3.274 \cdot 10^{-6}$ N. **31.47** 15 mW. **31.49** (a) 3 nN. (b) 2 kW/m^2,
5 μN/m^2. (c) 200. **31.51** 3 mW. **31.53** $1.1 \cdot 10^4$ W/m^2, $2.9 \cdot 10^3$ V/m,
$9.7 \cdot 10^{-6}$ T.

Additional Problems

31.55 500 kW h. **31.57** $4 \cdot 10^5$ V/m. **31.59** 136 W. **31.61** 100 V/m.
31.63 6.3 cm. **31.65** (a) $7.07 \cdot 10^{-8}$ T. (b) 38.0 W. **31.67** (a) 800 V/m.
(b) $8 \cdot 10^{-12}$ J. **31.69** $1 \cdot 10^7$ m/s^2. **31.71** (a) The radio waves will, in
general, be polarized. (b) $6 \cdot 10^{-2}$ V/m. **31.73** 9 h; $4 \cdot 10^{23}$.

Chapter 32: Geometric Optics

Multiple Choice

32.1 c. **32.3** a. **32.5** b.

Problems

32.21 −1.0 m. **32.23** 6 m. **32.25** −13 cm. **32.27** 0.389. **32.29** −3.0 m.
32.33 $\theta_{c,\text{air}} = 42°$, $\theta_{c,\text{water}} = 63°$, $\theta_{c,\text{oil}} = 90.°$. **32.35** 52.4°. **32.37** 16.3°.
32.39 1.36 %.

Additional Problems

32.43 (a) The image distance is 50 cm into the mirror. (b) The image
has the same height, $h = 2.0$ m. (c) The image is upright. (d) The

image is virtual. **32.45** If the object is placed at 15 cm, the image will be real and at a distance of 30 cm from the mirror. If the object is placed at 5.0 cm, the image will be virtual and –10 cm from the mirror. **32.47** 126°. **32.49** 31.4 %. **32.51** 40° with respect to the normal. **32.53** 2 m. **32.55** (a) $\vec{B}(\vec{x},t) \rightarrow -\vec{B}(\vec{x},-t)$. (b) They do not transmit light unidirectionally. **32.57** 1.40 rad/s.

Chapter 33: Lenses and Optical Instruments

Multiple Choice

33.1 b. **33.3** c. **33.5** a. **33.7** b. **33.9** b. **33.11** b.

Problems

33.29 3.0. **33.31** 0.200 m. **33.33** 1.5; inverted.
33.35 (a) 48 cm.
(b)

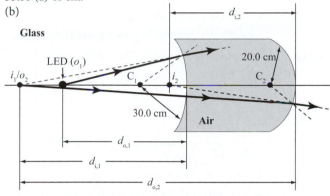

33.37 5. **33.39** 6.00 cm. **33.41** –8.5 mm, inverted, virtual.
33.43 (a) 2f. (b) –2.
(c)

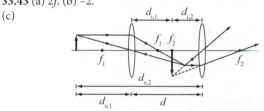

(d) virtual, inverted, larger.
33.45 18.2 cm. **33.47** (a) 2.50 cm. (b) 2.3 cm. **33.49** (a) 40. cm.
(b) –125 cm; The negative indicates the image is on the same side as the object. (c) 59 cm. (d) 1.7 diopter. (e) converging.
33.51 39 cm; 50. cm. **33.53** 28.75 cm; 9.86 cm; The ratio of angular magnifications is $\dfrac{m_{\text{near}}}{m_{\text{norm}}} = 0.34$.
33.55 $4.0 \cdot 10^2$ mm; $3.0 \cdot 10^2$ mm. **33.57** 4 times that of the original lens. **33.59** 2.4 cm. **33.61** 4 cm. **33.63** (a) 20.0 mm. (b) 9.7 mm.
33.65 20.0, inverted. **33.67** 20, inverted. **33.69** 1.3 m.

Additional Problems

33.73 –4 diopter. **33.75** 13 cm. **33.77** 18 cm from the lens, on the same side of the lens as the object. **33.79** 1.4 diopter. **33.81** 8.84 cm.
33.83 0.243 mm. **33.85** 190 cm to the right of the object; 5.0 cm.
33.87 nearsighted; 0.12 m. **33.89** 3.31 cm; magnification –0.0698.
33.91 38.0 cm. **33.93** 8 cm. **33.95** 1.0 diopter.

Chapter 34: Wave Optics

Multiple Choice

34.1 c. **34.3** d. **34.5** a. **34.7** b.

Problems

34.19 (a) 240 nm. (b) $2.0 \cdot 10^8$ m/s. **34.21** 87.5 nm. **34.23** 1.00 m.
34.25 1.05 mm. **34.27** 420 nm. **34.29** 139 nm. **34.31** 1.50. **34.33** 700 nm.
34.35 $17.0 \cdot 10^2$. **34.37** 1200 nm. **34.39** 600 nm. **34.41** $2.7 \cdot 10^{-6}$ degrees.
34.43 Hubble Space Telescope: $1.3 \cdot 10^{-5}$ degrees; Keck Telescope: $3.1 \cdot 10^{-6}$ degrees; Arecibo radio telescope: 0.048 degrees; The Arecibo radio telescope is worse than the other telescopes in terms of angular resolution. The Keck Telescope is better than the Hubble Space Telescope due to its larger diameter. **34.45** (a) $7.7 \cdot 10^{-3}$ degrees. (b) 11.1 km. **34.47** 31. **34.49** (a) $a = 10\lambda$. (b) $d = 100\lambda$.
(c) $a/d = 1:10$. (d) Without λ, there is insufficient information to find a or d. **34.51** 449 nm. **34.53** 400 nm, 600 nm.

Additional Problems

34.55 $1.25 \cdot 10^4$ lines/cm. **34.57** 0.6 m. **34.59** 70 nm. **34.61** 30.0 mm.
34.63 $1.0 \cdot 10^2$ nm. **34.65** 667. **34.67** 2.1 km. **34.69** 120 nm. **34.71** 2.1 μm.
34.73 (3) 400.

Chapter 35: Relativity

Multiple Choice

35.1 a. **35.3** c. **35.5** a. **35.7** c.

Problems

35.21 0.984 ft/ns. **35.23** $(1/v)\sqrt{v^2 - u^2}$. **35.25** (a) one meter. (b) 0.87 m.
35.27 0.94c. **35.29** (a) $4.44 \cdot 10^{-8}$ s. (b) $5.55 \cdot 10^{-8}$ s. **35.31** (a) 0.814 c.
(b) 20.3 ly. **35.33** 80°.

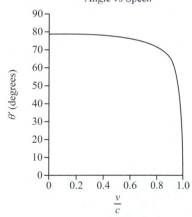

35.35 0.22c. **35.37** lower by 310 Hz. **35.39** (a) B. (b) $3.73 \cdot 10^{-5}$s.
35.41 –100 m/s. **35.43** (a) 4.92 ly. (b) 7.6 y. **35.45** (a) 66 m. (b) 0.86c.
(c) 2.1 μs. (d) 530 m. **35.47** 0.707c. **35.49** 4.02 PeV. **35.51** 1200 MeV.
35.53 (a) 5 keV. (b) 0.1c. (c) energies: 5 keV and 500 keV. The momenta are the same: 50 keV/c . **35.55** (a) $c/\sqrt{2}$. (b) $\sqrt{17}m$. (c) $\sqrt{18}$.
35.59 (a) 2.954 km. (b) $2.485 \cdot 10^{-54}$ m; If a proton had a classical space-time description it would be a bizarre, unphysical object, known as naked singularity. **35.61** 0.073 AU.

Additional Problems

35.63 relative momenta: $2.44 \cdot 10^{-22}$ kg m/s, $1.82 \cdot 10^{-27}$ kg m/s; The former differs from its classical value by 34%. The latter is the same as its classical value. **35.65** 5.03 keV; 32.5 MeV. **35.67** 0.67 g. **35.69** $0.9c$. **35.71** 0.25 ns, too small to detect. **35.73** 32.9 years; 105 years. **35.75** $0.9998c$. **35.77** (a) $2.2 \cdot 10^8$ m/s. (b) The photons would appear to be travelling at the speed of light. **35.79** 18 hours.

35.81 (a) $dv = g\left(1 - \dfrac{v^2}{c^2}\right) d\tau$. (b) $v(t) = \dfrac{gt}{\left[1 + \left(gt/c\right)^2\right]^{1/2}}$.

(c) In the time $gt \ll c$, i.e., the Newtonian limit, $v(t) \cong gt$. In the Einsteinian (relativistic) limit, $\lim\limits_{t \to +\infty} v(t) = c$. (d) $x(t) = \dfrac{c^2}{g}\sqrt{1 + \left(\dfrac{gt}{c}\right)^2}$.

(e) The trajectory is part of a branch of a hyperbola on a Minkowski diagram. (f) 354 days, 0.401 light years.

Chapter 36: Quantum Physics

Multiple Choice

36.1 b. **36.3** a, c. **36.5** c. **36.7** c.

Problems

36.19 (a) $5 \cdot 10^{-7}$ m. (b) $10 \cdot 10^{-6}$ m. **36.21** $3.5 \cdot 10^{-19}$ m; The energy of the gamma ray is 3700 times greater than the rest mass of a proton. **36.23** (a) 9.42 μm. (b) 1 kW. (c) Your wavelength is not in the visible spectrum. **36.25** (a) $E = 0.105$ J. (b) $5.24 \cdot 10^{18}$ photons per second. (c) $6.01 \cdot 10^{-7}$ m³. **36.27** 0.34 eV. **36.29** (a) 2.30 eV. (b) potassium or sodium. **36.31** (a) 0.622 eV. (b) No. (c) 500 nm. **36.33** $2.42 \cdot 10^{-12}$ m. **36.35** 64 eV. **36.37** (a) $1 \cdot 10^{-4}$ (b) 9 MeV. **36.39** (a) 600 nm. (b) 0.9 nm. **36.41** (a) 47.52 pm. (b) 11 nm. **36.43** (a) 1300 m/s. (b) This speed is much less than the speed of light, so the non-relativistic approximation is sufficient. (c) 5.0 μeV.

36.45 (a) $\omega(k) = \dfrac{hk^2}{4\pi m}$. (b) $v_P = \dfrac{p}{2m}$, $v_g = \dfrac{p}{m}$, phase velocity.

36.47 (a) $6.63 \cdot 10^{-35}$ m/s. (b) $5.27 \cdot 10^{-36}$ m/s. **36.49** $7 \cdot 10^{-55}$ kg. **36.51** 1 m/s.

36.53 $\langle n \rangle = \dfrac{\exp\left(\dfrac{-E}{k_B T}\right)}{1 - \exp\left(\dfrac{-E}{k_B T}\right)}$.

36.55 $C = N k_B \left(\dfrac{E}{k_B T}\right)^2 \dfrac{\exp\left(\dfrac{E}{k_B T}\right)}{\left(\exp\left(\dfrac{E}{k_B T}\right) + 1\right)^2}$. For $k_B T \gg 1$, $C \approx \dfrac{N k_B}{4}\left(\dfrac{E}{k_B T}\right)^2$.

For $0 < k_B T \ll 1$, $C \approx N k_B \left(\dfrac{E}{k_B T}\right)^2 \exp\left(-\dfrac{E}{k_B T}\right)$.

Additional Problems

36.57 0.0233. **36.59** $7.6 \cdot 10^{-41}$ m. **36.61** $1 \cdot 10^{19}$ s⁻¹. **36.63** 60 pm. **36.65** The uncertainty in the electron's position is at least $\Delta x = 0.6$ m. **36.67** 61. **36.69** 400 nm. **36.71** $3 \cdot 10^{43}$. **36.73** $p = 1.89 \cdot 10^{-19}$ N m, $E = 5.68 \cdot 10^{-11}$ J.

Chapter 37: Quantum Mechanics

Multiple Choice

37.1 c. **37.3** d. **37.5** b. **37.7** d. **37.9** a.

Problems

37.23 9.0 fm; Since protons and neutrons have a diameter of about 1 fm, they would be useful targets to demonstrate the wave nature of 10-MeV neutrons. **37.25** 0.094 eV, 0.38 eV. **37.27** 4.

37.29

$$\psi(x) = \begin{cases} 0 & \text{for } x < -a/2 \text{ and } x > a/2 \\ \sqrt{\dfrac{2}{a}}\sin\left(\dfrac{n\pi x}{a}\right) & \text{for } -a/2 \le x \le a/2 \text{ with even } n \\ \sqrt{\dfrac{2}{a}}\cos\left(\dfrac{n\pi x}{a}\right) & \text{for } -a/2 \le x \le a/2 \text{ with odd } n \end{cases}$$

37.31 No. **37.33** 5.9. **37.35** 6.99 eV. **37.37** 50 eV.

37.39 (a) $\psi(x) = \begin{cases} Ae^{\gamma x} & \text{for } x \le -a/2 \\ C\cos(\kappa x) + D\sin(\kappa x) & \text{for } -a/2 \le x \le a/2 \\ Fe^{-\gamma x} & \text{for } x \ge a/2 \end{cases}$

$E = \dfrac{\hbar^2 \kappa^2}{2m}$.

(b) $\eta = \dfrac{1}{\gamma} = \sqrt{\dfrac{\hbar^2}{2m(U_0 - E)}} = \dfrac{\hbar}{\sqrt{2m(U_0 - E)}}$. (c) $2.93 \cdot 10^{-9}$ m.

37.41 720 nm. **37.43** 101 eV; 85.5 kN/m.

37.45 (a) $A_2 = \dfrac{1}{\sqrt[4]{\pi}\sqrt{b}}$. (b) 0.157. **37.47** (a) $A = \sqrt{\dfrac{2}{L}}$. (b) 0.402.

37.49 0.25. **37.51** $\Psi(\vec{r},t) = Ae^{i(\vec{p}\cdot\vec{r})/\hbar}e^{-ip^2 t/2m\hbar} = A^2$.

37.53 $|\Psi(x,t)|^2 = \dfrac{1}{a}\sin^2\left(\dfrac{\pi x}{a}\right)\left[1 + 4\cos^2\left(\dfrac{\pi x}{a}\right) + 4\cos\left(\dfrac{\pi x}{a}\right)\cos\left(\dfrac{3\hbar\pi^2 t}{2ma^2}\right)\right]$.

37.55 (a) $3.22 \cdot 10^{-22}$ s. (b) $1.75 \cdot 10^{-25}$ s.

Additional Problems

37.57 28.2 eV.

37.59 (a) $\psi = \left(\sqrt{\dfrac{2}{L}}\right)^3 \sin\dfrac{\pi x}{L}\sin\dfrac{\pi y}{L}\sin\dfrac{\pi z}{L}$; $0 < x, y, z < L$. (b) 14.

37.61 $8 \cdot 10^{-4}$ eV. **37.63** 4.0 pm. **37.65** (a) 1.2 eV. (b) 0.19 eV. **37.67** 2.9 MeV. **37.69** 1000 nm. **37.71** 38 eV. **37.73** (2,2) and (1,4).

Chapter 38: Atomic Physics

Multiple Choice

38.1 d. **38.3** d. **38.5** b. **38.7** b.

Problems

38.19 91.2 nm. **38.21** 2279 nm, 7458 nm; No. **38.23** −0.378 eV. **38.25** (a) $1.0968 \cdot 10^7$ m⁻¹. (b) $5.4869 \cdot 10^6$ m⁻¹. **38.27** initial: 6; final: 3.

38.29 $r = \dfrac{\hbar^2}{GMm^2}n^2$; $2.523 \cdot 10^{74}$.

38.31 $4.716 \cdot 10^{-34}$ J s and 0 J s. **38.33** 65.9°.

38.35 (a) $A_1 = \dfrac{2}{a_0^{3/2}}$. (b) e^{-1}/a_0.

38.37 0.03432. **38.39** $a_0/2$.

38.41 (a) $\dfrac{E_0}{\kappa^2}$. (b) 0.0272 eV.

38.43 (a) $r_n = \dfrac{a_0 n^2}{Z}$. (b) $v_n = \dfrac{Z}{n}\sqrt{\dfrac{ke^2}{\mu a_0}}$. (c) $E_0' = \dfrac{-Z^2}{n^2} E_0$.

38.45 656 nm; A laser with a wavelength of about 5.5 times bigger is needed. **38.47** (a) 30 μJ. (b) $1\cdot10^{14}$ atoms.

Additional Problems

38.49 91 nm. **38.51** 0.05445%, $4.554\cdot10^{-31}$ kg. **38.53** 4694.
38.55 −2528 eV, −632 eV, −281 eV. **38.57** −54.438 eV, −13.609 eV, −6.049 eV. **38.59** 6.72 meV. **38.61** $3.71\cdot10^{-3}$ eV.
38.63 Yes. $v = 4.077$ m/s. **38.65** 0.323.

Chapter 39: Elementary Particle Physics

Multiple Choice

39.1 a. **39.3** c. **39.5** d. **39.7** c.

Problems

39.23 49.9 fm. **39.25** $1\cdot10^{-27}$ m²/sr. **39.27** r_{min}/λ is proportional to $1/\sqrt{K}$. **39.29** 120 particles per second.

39.31 $F^2(\Delta p) = \dfrac{4\pi\rho\hbar^3}{Ze(\Delta p)^3}\left[\sin\left(\dfrac{\Delta p R}{\hbar}\right) - \left(\dfrac{\Delta p R}{\hbar}\right)\cos\left(\dfrac{\Delta p R}{\hbar}\right)\right]$

$\dfrac{d\sigma}{d\Omega} = \dfrac{(2kZ_p Z_t e^2 m_p)^2}{(\Delta p)^4}\left[\dfrac{3\hbar^3}{\Delta p^3 R^3}\sin\left(\dfrac{\Delta p R}{\hbar}\right) - \left(\dfrac{\Delta p R}{\hbar}\right)\cos\left(\dfrac{\Delta p R}{\hbar}\right)\right]^2$.

39.33 $2.2\cdot10^8$ m/s.
39.35

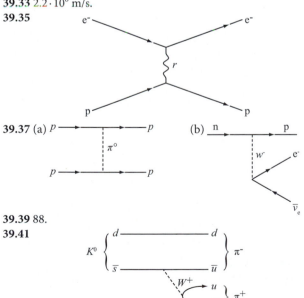

39.37 (a) $p \rightarrow p$, π^0, $p \rightarrow p$ (b) $n \rightarrow p$, w^-, e^-, $\bar{v}_e$

39.39 88.
39.41

39.43 $2.1\cdot10^{12}$ K; $5.1\cdot10^{-5}$ s. **39.45** (a) $9\cdot10^5$ eV.

39.47 (a) $H = \left(\dfrac{1}{a}\dfrac{da}{dt}\right)_{rec}$. (b) 420 Mpc.

Additional Problems

39.49 (a) 0.03636 eV. **39.51** 54.4 meV. **39.53** 10^{-18} m. **39.55** $9\cdot10^{-4}$ m².
39.57 (a) $16\,ke^2/mv^2$. (b) 49 fm. **39.59** 340 fm. **39.61** 13/32.

39.63 $F(\Delta p) = \dfrac{\sin[(\Delta p)a/\hbar]}{(\Delta p)a/\hbar}$; $\dfrac{d\sigma}{d\Omega} = \left(\dfrac{Ze^2 m_e}{4\pi\epsilon_0}\right)^2 \dfrac{1}{(\Delta p)^4}\left(\dfrac{\sin[(\Delta p)a/\hbar]}{(\Delta p)a/\hbar}\right)^2$;

Yes.

Chapter 40: Nuclear Physics

Multiple Choice

40.1 b. **40.3** b. **40.5** d. **40.7** d.

Problems

40.23 (a) 39 MeV. (b) 92 MeV. (c) 489 MeV. (d) 731 MeV.
40.25 −0.46 fm⁻¹. **40.27** (a) $^{60}_{27}\text{Co} \rightarrow \,^{60}_{28}\text{Ni}^+ + \,^{0}_{-1}e^- + \bar{v}_e$.
(b) $^3_1\text{H} \rightarrow \,^3_2\text{He}^+ + \,^{0}_{-1}e^- + \bar{v}_e$. (c) $^{14}_6\text{C} \rightarrow \,^{14}_7\text{N}^+ + \,^{0}_{-1}e^- + \bar{v}_e$.
40.29 0.351 MeV. **40.31** 0.25 h. **40.33** $1\cdot10^{16}$ years.
40.35 (a)$1.65\cdot10^{11}$ dist/(g s) = $1.65\cdot10^{11}$ Bq/g = 4.45 Ci/g.
(b) $8\cdot10^{11}$ Bq. (c) $8\cdot10^{21}$ dist. **40.37** 1946 or 1947.

40.39 $N_B(t) = \dfrac{\lambda_A}{\lambda_B - \lambda_A}N_{A0}e^{-\lambda_A t} + \left(N_{B0} - \dfrac{\lambda_A}{\lambda_B - \lambda_A}N_{A0}\right)e^{-\lambda_B t}$.

40.41 $^{117}_{50}\text{Sn}$. **40.43** 4.51 kg. **40.45** (a) $4.2778\cdot10^9$ kg/s. (c) 0.030%.
40.47 13.8 GK.
40.49

No	Nuclide	Mass number, A	Mass excess, Δm (keV/c²)	Atomic mass (u)
1	$^1_0 n$	1	8071.3	1.0087
2	$^{252}_{98}\text{Cf}$	252	76034	252.08
3	$^{256}_{100}\text{Fm}$	256	85496	256.09
4	$^{140}_{56}\text{Ba}$	140	−83271	139.91
5	$^{140}_{54}\text{Xe}$	140	−72990	139.92
6	$^{112}_{46}\text{Pd}$	112	−86336	111.91
7	$^{109}_{42}\text{Mo}$	109	−67250	108.93

(b) ΔE_{Cf} = 202341 keV and ΔE_{Fm} = 212537 keV. (c) Yes.
40.51 $2\cdot10^{-16}$ J; $5\cdot10^5$ m/s. **40.53** 2.00 μCi. **40.55** 12.71 days.

Additional Problems

40.57 168.512 MeV. **40.59** 50 fm. **40.61** $1.17\cdot10^{-22}$ μCi.
40.63 41.279 MeV. **40.65** $2.51\cdot10^{17}$ J. **40.67** (a) 7.074 MeV.
(b) 2.573 MeV. (c) 2.827 MeV. (d) 1.112 MeV. **40.69** 90 min.
40.71 $6\cdot10^4$ years. **40.73** $3\cdot10^8$ decays **40.75** 10 s.

Credits

Photographs

About the Authors

Photo courtesy of Okemos Studio of Photography.

Chapter 21

Figure 21.1a-c: © W. Bauer and G. D. Westfall; **21.2:** © R. Morley/PhotoLink/Getty Images RF; **21.7a:** © Kim Steele/Getty Images RF; **21.7b:** © Geostock/Getty Images RF; **21.8-21.11, 21.18, 21.19:** © W. Bauer and G. D. Westfall.

Chapter 22

Figure 22.1: © Royalty-Free/Corbis; **22.2:** National Weather Service; **22.17:** Brookhaven National Laboratory; **22.19:** © The McGraw-Hill Companies, Inc./Mark Dierker, photographer; **22.21:** © Royalty-Free/Corbis; **22.31a-b:** © W. Bauer and G. D. Westfall; **22.32:** © Gerd Kortemeyer; **22.41:** © Royalty-Free/Corbis.

Chapter 23

Figure 23.1 (left)-(right): Springer Netherlands/ *Annals of Biomedical Engineering,* Volume 30, Number 9, October 2002, 1172–80, "Induced current impedance technique for monitoring brain cryosurgery in a two-dimensional model of the head," Zlochiver S, Radai MM, Rosenfeld M, Abboud S., Fig. 6. With kind permission from Springer Science and Business Media; **23.5a-b:** © W. Bauer and G. D. Westfall; **23.7a:** © Regis Duvignau/Reuters; **23.7b:** © Candy Lab Studios/ Mendola Artists; **23.8:** © 2009 Tesla Motors, Inc. All rights reserved. 'Tesla Motors' and Tesla Roadster' are trademarks of Tesla Motors, Inc. (version 1-5-0296-228); **23.9a-b:** © W. Bauer and G. D. Westfall.

Chapter 24

Figure 24.1-24.2: © W. Bauer and G. D. Westfall; **24.21:** © Lawrence Livermore National Lab.

Chapter 25

Figure 25.1: © ImageSource/agefotostock RF; **25.2a-f:** © W. Bauer and G. D. Westfall; **25.3a:** © IBM; **25.3c:** © Brand X Pictures/PunchStock RF; **25.3d:** © Don Farrall/Getty Images RF; **25.3e:** © W. Bauer and G. D. Westfall; **25.3f:** © Craig Bickford/International Light Technologies; **25.3g:** © 1000Bulbs.com; **25.3h:** © Josh Slaymaker/Grant Heilman Photography, Inc.; **25.3i:** © Thomas Allen/Getty Images RF; **25.3j:** NASA; **25.3k:** © Lawrence Livermore National Lab; **25.3l:** NASA artist Werner Heil; **25.4:** © W. Bauer and G. D. Westfall; **25.8a:** Photo courtesy Mike Smith, www.mds975.co.uk; **25.8b:** © Josh Slaymaker/ Grant Heilman Photography, Inc.; **25.15:** Printed with permission from Mayfield Clinic; **25.21a-b:** Courtesy of ABB; **25.22a-b:** © W. Bauer and G. D. Westfall; **25.24:** © Julie Jacobson/AP Photo; **25.25:** http://en.wikipedia.org/wiki/File:Path_65_ P0002014.jpg.

Chapter 26

Figure 26.1: © Royalty-Free/Corbis.

Chapter 27

Figure 27.1: NASA/TRACE; **27.4b:** © Steve Cole/Getty Images RF; **27.7:** NASA/courtesy of nasaimages.org; **27.8:** NASA/Hubble/Z. Levay and J. Clarke; **27.11:** © W. Bauer and G. D. Westfall; **27.12:** © The McGraw-Hill Companies, Inc./ Mark Dierker, photographer; **27.13:** National Geophysical Data Center; **27.15:** © W. Bauer and G. D. Westfall; **27.16:** Lawrence Berkeley Nat'l Lab; **27.17a:** Brookhaven National Laboratory; **27.17b:** STAR collaboration/RHIC/Brookhaven National Laboratory; **27.18:** Brookhaven National Laboratory; **27.19a:** © Michigan State University; **27.22:** © W. Bauer and G. D. Westfall; **27.23a:** © Ren Long, Xinhua/AP Photo; **27.24b, 27.28:** © The McGraw-Hill Companies, Inc./Mark Dierker, photographer.

Chapter 28

Figure 28.1: © Photolibrary/agefotostock RF; **28.3b, 28.5:** © The McGraw-Hill Companies, Inc./Mark Dierker, photographer; **28.9:** U.S. Navy Photograph by Mr. John F. Williams; **28.18a:** © W. Bauer and G. D. Westfall; **28.22 (hand):** © The McGraw-Hill Companies, Inc./Mark Dierker, photographer; **28.25:** © High Field Magnet Laboratory, Radboud University Nijmegen, The Netherlands; **28.28a-e:** © W. Bauer and G. D. Westfall; **28.30:** © Royalty-Free/Corbis; **28.31a- d:** © The McGraw-Hill Companies, Inc./Mark Dierker, photographer.

Chapter 29

Figure 29.1: © PhotoLink/Getty Images RF; **29.11:** © W. Bauer and G. D. Westfall; **29.14a:** NASA; **29.14b:** NASA/courtesy of nasaimages.org; **29.16:** © W. Bauer and G. D. Westfall; **29.19 & (inset):** © Tom Watson; **29.29:** © Daisuke Morita/Getty Images RF; **p. 955:** © Royalty-Free/Corbis.

Chapter 30

Figure 30.1a: © Ryan McVay/Getty Images RF; **30.1b:** © Royalty-Free/Corbis; **30.18:** © W. Bauer and G. D. Westfall; **30.21:** © The McGraw-Hill Companies, Inc./Mark Dierker, photographer; **30.31a:** © Edmond Van Hoorick/Getty Images RF; **30.31b, 30.32:** © W. Bauer and G. D. Westfall.

Chapter 31

Figure 31.1: E. Kolmhofer, H. Raab; Johannes- Kepler-Observatory, Linz, Austria (http://www. sternwarte.at); **31.9a:** © Kim Steele/Getty Images RF; **31.9b:** NASA/courtesy of nasaimages.org; **31.9c:** © W. Bauer and G. D. Westfall; **31.10b:** © C. Borland/PhotoLink/Getty Images RF; **31.10c:** © W. Bauer and G. D. Westfall; **31.10d:** © Don Tremain/ Getty Images RF; **31.10e:** © PhotoLink/Getty Images RF; **31.10f:** © Geostock/Getty Images RF; **31.10h-i:** © Russell Illig/Getty Images RF; **31.10j:** © Royalty-Free/Corbis; **31.10k:** © David R. Frazier Photography/Alamy RF; **31.10l:** © W. Bauer and G. D. Westfall; **31.10m:** © PhotoDisc/Getty Images RF; **31.10n:** © W. Bauer and G. D. Westfall; **31.10o:** © ImageState/Alamy RF; **31.14a:** © John Keating/ Photo Researchers, Inc.; **31.14b:** © General Motors Corp. Used with permission, GM Media Archives; **31.23:** © Photographer's Choice/Getty Images RF.

Chapter 32

Figure 32.1, 32.7, 32.9a-b, 32.10 (top, center, bottom), 32.18a-b, 32.25a-b, 32.31a: © W. Bauer and G. D. Westfall; **32.33:** © P. K. Chen; **32.35a-b, 32.37a-b, 32.40-32.41:** © W. Bauer and G. D. Westfall; **32.43a:** Courtesy IT Concepts GmbH; **32.43b:** © David M. Martin, M.D./Photo Researchers, Inc.; **32.45a, 32.51:** © W. Bauer and G. D. Westfall.

Chapter 33

Figure 33.1a: NASA and The Hubble Heritage Team (STScI/AURA); **33.1b:** Prof. Gordon

T. Taylor, Stony Brook University/NSF Polar Programs; **33.6a-c, 33.10a-c:** © W. Bauer and G. D. Westfall; **33.14:** © Robert George Young/Getty Images; **33.20:** © Scott Bodell/Getty Images RF; **33.26 (left)-(right), 33.28a-c, 33.29, 33.31:** © W. Bauer and G. D. Westfall; **33.32:** © Comstock/PunchStock RF; **33.33:** © Meade; **33.35:** © Victor Krabbendam/Southern Astrophysical Research Telescope; **33.37a:** STS-82 Crew/STScI/NASA; **33.37b:** NASA, ESA, STScI, J. Hester and P. Scowen (Arizona State University); **33.38a-b:** NASA, STScI; **33.39:** NASA; **33.40a:** NASA/CXC/D.Berry; **33.40b:** NASA/CXC/D.Berry & A.Hobart; **33.40c:** NASA/CXC/MIT/F.K.Baganoff et al.

Chapter 34

Figure 34.1: © MilaZinkova; **34.12a, 34.17, 34.19a-b, 34.23, 34.28a, 34.31, 34.32a-c:** © W. Bauer and G. D. Westfall; **34.33:** NASA, 1990; **34.35, 34.38, 34.41, 34.42a-b:** © W. Bauer and G. D. Westfall; **34.51:** © Dr. Bernard Santarsiero, University of Illinois at Chicago; **34.52:** Argonne National Laboratory, managed and operated by UChicago Argonne, LLC, for the U.S. Department of Energy under Contract No. DE-AC02-06CH11357.

Chapter 35

Figure 35.1: NASA, ESA, and STScI; **35.8:** U.S. Air Force photo by Master Sgt Michael A. Kaplan, NASCAR Race; **35.17:** © W. Bauer and G. D. Westfall; **35.19:** F. W. Dyson, A. S. Eddington, and C. Davidson, "A Determination of the Deflection of Light by the Sun's Gravitational Field, from Observations Made at the Total Eclipse of May 29, 1919," *Philosophical Transactions of the Royal Society of London.* Series A, Containing Papers of a Mathematical or Physical Character (1920): 291–333, on 332; **35.21b:** NASA, Andrew Fruchter and the ERO Team [Sylvia Baggett (STScI), Richard Hook (ST-ECF), Zoltan Levay (STScI)] (STScI); **35.22:** NASA/JPL/Caltech; **35.23:** Image courtesy of Lockheed Martin Corporation.

Chapter 36

Figure 36.1: © Kai Pfaffenbach/Reuters/Corbis; **36.2:** © Royalty-Free/Corbis; **36.10a:** © Lencho Guerra; **36.16a:** © W. Bauer and G. D. Westfall; **36.17:** Reprinted with permission from P.G. Merli, G.F. Missiroli, G. Pozzi, *American Journal of Physics,* "On the Statistical Aspect of Electron Interference Phenomena," Vol. 44, Issue 3, pp. 306–307, March 1976. © 1976, American Association of Physics Teachers; **36.26:** © Mike Matthews, JILA.

Chapter 37

Figure 37.1a: Photo courtesy of Erik Lucero and Max Hofheinz; **37.9a-b, 37.17b:** Image reproduced by permission of IBM Research, Almaden Research Center. Unauthorized use not permitted; **37.27:** From R. Blatt "Quantum Information Processing: Dream and Realization," *Entangled World,* pp. 235–270, Wiley-VCH, Weinheim 2006. Image courtesy R. Blatt, University of Innsbruck.

Chapter 38

Figure 38.1: © WMKO; **38.2-38.3, 38.25:** © W. Bauer and G. D. Westfall; **38.26-38.27:** Image courtesy of University of California, Lawrence Livermore National Laboratory, and the Department of Energy.

Chapter 39

Figure 39.1: NASA/ESA/STScI/AURA; **39.2a:** © Royalty-Free/Corbis; **39.4-39.6:** © CERN; **39.27:** © Kamioka Observatory, ICRR (Institute for Cosmic Ray Research), The University of Tokyo; **39.36:** Image courtesy of Derek Leinweber, CSSM, University of Adelaide; **39.40:** © Axel Mellinger, University of Potsdam, Germany; **39.41b:** NASA, ESA, and The Hubble Heritage Team (STScI/AURA); Hubble Space Telescope ACS; STScI-PRC05-20; **39.41c:** © Royalty-Free/Corbis; **39.42-39.43:** NASA/WMAP Science Team; **39.44:** STAR collaboration/RHIC/Brookhaven National Laboratory.

Chapter 40

Figure 40.1: © Brand X Pictures/PunchStock RF; **40.6a:** © MSU National Superconducting Cyclotron Laboratory; **40.14:** Lawrence Berkeley National Lab; **40.18:** http://en.wikipedia.org/wiki/File:Shroudofturin.jpg; **40.35:** © ITER; **40.36a-c:** Image courtesy of University of California, Lawrence Livermore National Laboratory, and the U.S. Department of Energy; **40.38:** NASA/ESA/JPL/Arizona State Univ.; **40.41:** Image courtesy of Steve Goetsch; **40.42, 40.44:** © W. Bauer and G. D. Westfall.

Line Art and Text

Chapter 25

Table 25.2: Data from American Wire Gauge convention.

Chapter 27

Figure 27.9: Image data from NOAA's National Geophysical Data Center (NGDC), Boulder, Colorado, based on the International Goemagnetic Reference Field (IGRF), Epoch 2000 updated to December 31, 2004. The IGRF is developed by the International Association of Geomagnetism and Aeronomy (IAGA) Division V.

Chapter 36

Figure 36.6: COBE data from NASA / COBE Science Team.

Chapter 39

Figure 39.11: Data as reported by Geiger and Marsden in Phil. Mag. 25 (1913), p. 604; **Figure 39.12:** Data as reported by Eisberg and Porter, Rev. Mod. Phys. 33 (1961), p. 190; **Figure 39.15:** Data as reported by R. Hofstadter in Annual Reviews of Nuclear Science 7 (1957), p. 231; **Table 39.3:** Data from W.-M. Yao, et al. (Particle Data Group), J. Phys. G 33, 1 (2006), http://pdg.lbl.gov; **Figure 39.30:** Data from W.-M. Yao et al. (Particle Data Group), J. Phys. G 33, 1 (2006), http://pdg.lbl.gov); **Table 39.4:** Data from W.-M. Yao et al. (Particle Data Group), J. Phys. G 33, 1 (2006), http://pdg.lbl.gov).

Chapter 40

Figure 40.12: Data from the GSI National Laboratory in Germany, December 17, 1994; **Figure 40.15:** Figure courtesy of Argonne National Laboratory, adapted from T. Lauritzen et al., Phys. Rev. Lett. 88, 042501 (2002); **Figure 40.31:** Calculations by B. Alex Brown, Michigan State University; **Problem 40.49 Table:** Data per Berkeley National Lab *NuBase* data base.

Appendix B

Data source: David R. Lide (ed.), Norman E. Holden in *CRC Handbook of Chemistry and Physics* 85th Edition, online version. CRC Press. Boca Raton, Florida (2005). Section 11, Table of the Isotopes.

Appendix C

Data sources: http://physics.nist.gov/PhysRefData/PerTable/periodic-table.pdf

http://www.wikipedia.org/ and Generalic, Eni. "EniG. Periodic Table of the Elements." 31 Mar. 2008. KTF-Split. <http://www.periodni.com/>.

Inside Front Cover

Fundamental Constants from National Institute of Standards and Technology, http://physics.nist.gov/constants.

Other Useful Constants from National Institute of Standards and Technology, http://physics.nist.gov/constants.

Index